GATE PREP SERIES

GATE

Graduate Aptitude Test in Engineering

2024

Previous Years'

Solved Papers

(Section-Wise & Year-Wise)

(2009-2023)

Engineering Sciences

GATE PREP SERIES

GATE

Graduate Aptitude Test in Engineering

2024

Previous Years'

Solved Papers

(Section-Wise & Year-Wise)

(2009-2023)

Engineering Sciences

G. K. PUBLICATIONS (P) LTD.

Title : GATE : Engineering Sciences - Previous Years' Solved Papers 2009-2023 (Section-Wise)

Language : English

Editor's Name : Vinit Garg

Typeset & Published by :

Career Launcher Infrastructure (P) Ltd.

A-45, Mohan Cooperative Industrial Area, Near Mohan Estate Metro Station, New Delhi - 110044

Marketed by :

G.K. Publications (P) Ltd.

Plot No. 9A, Sector-27A, Mathura Road, Faridabad, Haryana-121003

ISBN : **978-93-56811-93-5**

Printer's Details : Printed in India, New Delhi.

For product information :

Visit ***www.gkpublications.com*** or email to ***gkp@gkpublications.com***

GATE Papers
Engineering Sciences

Contents

GENERAL APTITUDE

ENGINEERING MATHEMATICS

FLUID MECHANICS

MATERIAL SCIENCE

SOLID MECHANICS

THERMODYNAMICS

POLYMER SCIENCE & ENGINEERING

FOOD TECHNOLOGY

ATMOSPHERIC AND OCEANIC SCIENCES

GATE

Graduate Aptitude Test in Engineering

IIT Institutes

GATE 2024 will be conducted by
Indian Institute of Science, IISc Bangalore

GATE 2023 conducted by
Indian Institute of Technology, Kanpur

GATE 2022 conducted by
Indian Institute of Technology, Kharagpur

Indian Institute of Technology, Mumbai

Indian Institute of Technology, Delhi

Indian Institute of Technology, Chennai

Indian Institute of Technology, Guwahati

Indian Institute of Technology, Roorkee

Preface

The Graduate Aptitude Test in Engineering (GATE) is an online exam conducted by the IITs for admissions to PG courses in IITs, IISc Bangalore, NITs and many state run universities as well as private universities. Also there are more than 37 PSUs that use GATE score for recruitments. A large number of corporates are also using GATE score as a tool to screen students for placements.

GK Publications is well known as the "publisher of choice" to students preparing for GATE and other technical test prep examinations in the country. We published first set of books in 1994 when GATE exam, both objective and conventional, was conducted in the paper and pencil environment, and used as a check point for entry to post graduate courses in IITs and IISCs. At that time, students had little access to technology and relied mainly on instructor led learning followed by practice with books available for these examinations.

A lot has changed since then!

Today, GATE is conducted in an online only mode with Multiple Choice Questions, Multiple Select Questions and Numerical Based Questions. The score is valid for three years and is used not only for post graduate courses but is also used by major PSUs for recruitment. Today's students have easy access to technology and the concept of a monologue within the classroom has changed to dialogue where students come prepared with concepts and then discuss topics. They learn a lot of things on the go with their mobile devices and practice for mock tests online.

We, as a leading publisher of GATE books, have also embraced change. Today, our books are no more guides only but come with a fully supported mobile app and a web portal. The mobile App provides access to video lectures, short tests and regular updates about the exam. The web portal in additional to what is available on the App provides full length mock tests to mimic the actual exam and help you gauge your level of preparedness. The combination of practice content in print, video lectures, and short and full length tests on mobile and web makes this product a complete courseware for GATE preparation.

We also know that improvement is a never ending process and hence we welcome your suggestions and feedback or spelling and technical errors if any. Please write to us at gkp@gkpublications.com

We hope that our small effort will help you prepare well for the examination.

We wish you all the best!

GK Publications Pvt. Ltd.

About GATE

The Graduate Aptitude Test in Engineering (GATE) conducted by IISc and IITs has emerged as one of the bench mark tests for engineering and science aptitude in facilitating admissions for higher education (M.Tech./Ph.D.) in IITs, IISc and various other Institutes/Universities/Laboratories in India. With the standard and high quality of the GATE examination in 29 disciplines of engineering and science, Humanities and Social Sciences subjects, it identifies the candidate's understanding of a subject and aptitude and eligibility for higher studies. During the last few years, GATE score is also being used as one of the criteria for recruitment in Government Organizations such as Cabinet Secretariat, and National/State Public Sector Undertakings in India. Because of the importance of the GATE examination, the number of candidates taking up GATE exams has increased tremendously. GATE exams are conducted by the IITs and IISc as a computer based test having multiple choice questions and numerical answer type questions. The questions are mostly fundamental, concept based and thought provoking. From 2017 onwards GATE Exam is being held in Bangladesh, Ethiopia, Nepal, Singapore, Sri Lanka and United Arab Emirates. An Institute with various nationalities in its campus widens the horizons of an academic environment. A foreign student brings with him/her a great diversity, culture and wisdom to share. Many GATE qualified candidates are paid scholarships/assistantship, especially funded by Ministry of Human Resources Development, Government of India and by other Ministries. IISc, Bangalore is the Organizing Institute for GATE 2024.

Why GATE?

Admission to Post Graduate and Doctoral Programmes

Admission to postgraduate programmes with MHRD and some other government scholarships/ assistantships in engineering colleges/institutes is open to those who qualify through GATE. GATE qualified candidates with Bachelor's degree in Engineering/Technology/Architecture or Master's degree in any branch of Science/Mathematics/Statistics/Computer Applications are eligible for admission to Master/Doctoral programmes in Engineering/Technology/Architecture as well as for Doctoral programmes in relevant branches of Science with MHRD or other government scholarships/ assistantships. Candidates with Master's degree in Engineering/Technology/Architecture may seek admission to relevant Ph.D programmes with scholarship/assistantship without appearing in the GATE examination.

Financial Assistance

A valid GATE score is essential for obtaining financial assistance during Master's programs and direct Doctoral programs in Engineering/Technology/Architecture, and Doctoral programs in relevant branches of Science in Institutes supported by the MHRD or other Government agencies. As per the directives of the MHRD, the following procedure is to be adopted for admission to the post-graduate programs (Master's and Doctoral) with MHRD scholarship/assistantship. Depending upon the norms adopted by a specific institute or department of the Institute, a candidate may be admitted directly into a course based on his/her performance in GATE only **or** based on his/her performance in GATE **and** an admission test/interview conducted by the department to which he/she has applied **and/or** the candidate's academic record. If the candidate is to be selected through test/interview for post-graduate programs, a minimum of 70% weightage will be given to the performance in GATE and the remaining 30% weightage will be given to the candidate's performance in test/interview and/or academic record, as per MHRD guidelines. The admitting institutes could however prescribe a minimum passing percentage of marks in the test/interview. Some colleges/institutes specify GATE qualification as the mandatory requirement even for admission without MHRD scholarship/assistantship.

To avail of the financial assistance (scholarship), the candidate must first secure admission to a program in these Institutes, by a procedure that could vary from institute to institute. Qualification in GATE is also a minimum requirement to apply for various fellowships awarded by many Government organizations. Candidates are advised to seek complete details of admission procedures and availability of MHRD scholarship/assistantship from the concerned admitting institution. The criteria for postgraduate admission with scholarship/assistantship could be different for different institutions. The management of the post-graduate scholarship/assistantship is also the responsibility of the admitting institution. Similarly, reservation of seats under different categories is as per the policies and norms prevailing at the admitting institution and Government of India rules. *GATE offices will usually not entertain any enquiry about admission, reservation of seats and / or award of scholarship / assistantship.*

PSU Recruitments

As many as 37 PSUs are using GATE score for recruitment. It is likely that more number of PSUs may start doing so by next year. Below is the list of PSUs:

MDL, BPCL, GAIL, NLC LTD, CEL, Indian Oil, HPCL, NBPC, NECC, BHEL, WBSEDCL, NTPC, ONGC, Oil India, Power Grid, Cabinet Secretariat, Govt. of India, BAARC, NFL, IPR, PSPCL, PSTCL, DRDO, OPGC Ltd., THDC India Ltd., BBNL, RITES, IRCON, GHECL, NHAI, KRIBHCO, Mumbai Railway Vikas Corporation Ltd. (MRVC Ltd.), National Textile Corporation, Coal India Ltd., BNPM, AAI, NALCO, EdCIL India.

Important :

1. Admissions in IITs/IISc or other Institutes for M.Tech./Ph.D. through GATE scores shall be advertised separately by the Institutes and GATE does not take the responsibility of admissions.
2. Cabinet Secretariat has decided to recruit officers for the post of Senior Field Officer (Tele) (From GATE papers of EC, CS, PH), Senior Research Officer (Crypto) (From GATE papers of EC, CS, MA), Senior Research Officer (S&T) (From GATE papers EC, CS, CY, PH, AE, BT) in the Telecommunication Cadre, Cryptographic Cadre and Science & Technology Unit respectively of Cabinet Secretariat. The details of the scheme of recruitment shall be published in National Newspaper/Employment News by the concerned authority.
3. Some PSUs in India have expressed their interest to utilize GATE scores for their recruitment purpose. The Organizations who intend to utilize GATE scores shall make separate advertisement for this purpose in Newspapers etc.

Who Can Appear for GATE?

Eligibility for GATE

Before starting the application process, the candidate must ensure that he/she meets the eligibility criteria of GATE 2024 given in Table.

Eligibility Criteria for GATE 2024

Degree/Program	Qualifying Degree/Examination	Description of Eligible Candidates	Expected Year of Completion
B.E. / B.Tech. / B.Pharm.	Bachelor's degree in Engineering / Technology (4 years after 10+2 or 3 years after B.Sc. / Diploma in Engineering / Technology)	Currently in the 3rd year or higher grade or already completed	2025
B. Arch.	Bachelor's degree of Architecture (5- year course) / Naval Architecture (4- year course) / Planning (4- year course)	Currently in the 3rd year or higher grade or already completed	2026 (for 5-year program), 2025 (for 4-year program)
B.Sc. (Research) / B.S.	Bachelor's degree in Science (Post-Diploma/4 years after 10+2)	Currently in the 3rd year or higher grade or already completed	2025
Pharm. D. (after 10+2)	6 years degree program, consisting of internship or residency training, during third year onwards	Currently in the 3rd/4th/5th/6th year or already completed	2027
M.B.B.S.	Degree holders of M.B.B.S. and those who are in the 5th/6th/7th semester or higher semester of such programme.	5th, 6th, 7th or higher semester or already completed	2025
M. Sc. / M.A. / MCA or equivalent	Master's degree in any branch of Arts/Science/Mathematics/Statistics/ Computer Applications or equivalent	Currently in the first year or higher or already Completed	2025
Int. M.E./ M.Tech. (Post-B.Sc.)	Post-B.Sc Integrated Master's degree programs in Engineering/ Technology (4-year program)	Currently in the 1st / 2nd/3rd/4th year or already completed	Any Year
Int. M.E./ M.Tech. or Dual Degree (after Diploma or 10+2)	Integrated Master's degree program or Dual Degree program in Engineering/Technology (5-year program)	Currently in the 3rd /4th/5th year or alreadycompleted	2026
B.Sc. / B.A. / B.Com.	Bachelor degree in any branch of Science / Arts / Commerce (3 years program)	Currently in the 3rd year or already completed	2024
Int. M.Sc. / Int. B.S. / M.S.	Integrated M.Sc. or 5-year integrated B.S.-M.S. program	Currently in the 3rd year or higher or already completed	2025
Professional Society Examinations (equivalent to B.E. / B.Tech. / B.Arch.)	B.E./B.Tech./B.Arch. equivalent examinations of Professional Societies, recognized by MoE/UPSC/AICTE (e.g. AMIE by Institution of Engineers-India, AMICE by the Institute of Civil Engineers-India and so on)	Completed Section A or equivalent of such professional courses	Enrolled upto 31st May 2013
B.Sc (Agriculture, Horticulture, forestry)	4-year Program	Currently in the 3rd/4th year or already completed	2025

In case a candidate has passed one of the qualifying examinations as mentioned above in 2021 or earlier, the candidate has to submit the degree certificate / provisional certificate / course completion certificate / professional certificate / membership certificate issued by the society or institute. In case, the candidate is expected to complete one of the qualifying criteria in 2024 or later as mentioned above, he/she has to submit a certificate from Principal or a copy of marks card for section A of AMIE.

Certificate From Principal

Candidates who have to submit a certificate from their college Principal have to obtain one from his/her institution beforehand and upload the same during the online submission of the application form.

Candidates With Backlogs

Candidates, who have appeared in the final semester/year exam in 2023, but with a backlog (arrears/failed subjects) in any of the papers in their qualifying degree should upload a copy of any one of the mark sheets of the final year,

OR

obtain a declaration from their Principal along with the signature and seal beforehand and upload the same during the online submission of the application form.

GATE Structure

Structure of GATE

GATE 2024 will be conducted on 29 subjects (papers). Table below shows the list of papers and paper codes for GATE 2024. A candidate is allowed to appear in ANY ONE or UP TO TWO papers of the GATE examination. However, note that the combination of TWO papers in which a candidate can appear MUST be from the pre-defined list as given in Table. Also note that for a paper running in multiple sessions, a candidate will be mapped to appear for the examination in one of the sessions ONLY.

List of GATE Papers and Corresponding Codes

GATE Paper	Code	GATE Paper	Code
Aerospace Engineering	AE	Instrumentation Engineering	IN
Agricultural Engineering	AG	Mathematics	MA
Architecture and Planning (New Pattern)	AR	Mechanical Engineering	ME
Biomedical Engineering	BM	Mining Engineering	MN
Biotechnology	BT	Metallurgical Engineering	MT
Civil Engineering	CE	Naval Architecture and Marine Engineering	NM
Chemical Engineering	CH	Petroleum Engineering	PE
Computer Science and Information Technology	CS	Physics	PH
Chemistry	CY	Production and Industrial Engineering	PI
Electronics and Communication Engineering	EC	Statistics	ST
Electrical Engineering	EE	Textile Engineering and Fibre Science	TF
Environmental Science & Engineering	ES	Engineering Sciences	XE
Ecology and Evolution	EY	Humanities & Social Sciences	XH
Geomatics Engineering	GE	Life Sciences	XL
Geology and Geophysics	GG		

*XE Paper Sections	Code	**XH Paper Sections	Code	***XL Paper Sections	Code
Engineering Mathematics (Compulsory) (15 marks)	A	Reasoning and Comprehension (Compulsory) (25 marks)	B1	Chemistry (Compulsory) (25 marks)	P
Any TWO optional Sections (2x35 = 70 marks)		**Any ONE optional Section (60 marks)**		**Any TWO optional Sections (2x30 = 60 marks)**	
Fluid Mechanics	B	Economics	C1	Biochemistry	Q
Materials Science	C	English	C2	Botany	R
Solid Mechanics	D	Linguistics	C3	Microbiology	S
Thermodynamics	E	Philosophy	C4	Zoology	T
Polymer Science and Engineering	F	Psychology	C5	Food Technology	U
Food Technology	G	Sociology	C6		
Atmospheric and Oceanic Sciences	H				

*XE (Engineering Sciences), **XH (Humanities & Social Sciences), ***XL (Life Sciences), papers are of general nature and will be comprised of Sections listed in the above table

Note: *Each subject/paper is of total 100 marks. General Aptitude (GA) section of 15 marks is common for all papers. Hence remaining 85 marks are for the respective subject/paper code.*

Combination of Two Papers Allowed to Appear in GATE 2024 (subject to availability of infrastructure and schedule)

Code of the First (Primary) Paper	Codes of Papers Allowed as the Second Paper	Code of the First (Primary) Paper	Codes of Papers Allowed as the Second Paper
AE	CE, ME, XE	GG	GE
AG	CE	IN	BM, EC, EE, ME
AR	CE, GE	MA	CS, PH, ST
BM	BT, IN	ME	AE, IN, NM, PI, XE
BT	BM, XL	MT	XE
CE	AE, AG, AR, ES, GE, NM, XE	NM	CE, ME
CH	ES, PE, XE	PE	CH
CS	EC, GE, MA, PH, ST	PH	CS, EC, EE, MA, XE
CY	XE, XL	PI	ME, XE
EC	CS, EE, IN, PH	ST	CS, MA, XH
EE	EC, IN, PH	XE	AE, CE, CH, CY, ME, MT, PH, PI
ES	CE, CH, GE	XH	ST
EY	XL	XL	BT, CY, EY
GE	AR, CE, CS, ES, GG	-	-

General Aptitude Questions

All the papers will have a few questions that test the General Aptitude (Language and Analytical Skills), apart from the core subject of the paper.

Duration and Examination Type

All the papers of the GATE 2024 examination will be for 3 hours duration and they consist of 65 questions for a total of 100 marks. Since the examination is an ONLINE computer based test (CBT), at the end of the stipulated time (3-hours), the computer screen will automatically close the examination inhibiting any further action.

Candidates will be permitted to occupy their allotted seats 40 minutes before the scheduled start of the examination. Candidates can login and start reading the instructions 20 minutes before the start of examination. The late login time (if any) recorded by the computer system MUST NOT be beyond 30 minutes from the actual starting time of the examination. Under NO circumstances, will a candidate be permitted to login after 30 minutes from the actual examination starting time. Candidates will NOT be permitted to leave the examination hall before the end of the examination.

Pattern of Question Papers

GATE 2024 may contain questions of THREE different types in all the papers:

(i) **Multiple Choice Questions (MCQ)** carrying 1 or 2 marks each, in all the papers and sections. These questions are objective in nature, and each will have choice of four answers, out of which ONLY ONE choice is correct.

Negative Marking for Wrong Answers: For a wrong answer chosen in a MCQ, there will be negative marking. For 1-mark MCQ, 1/3 mark will be deducted for a wrong answer. Likewise, for 2-mark MCQ, 2/3 mark will be deducted for a wrong answer.

(ii) **Multiple Select Questions (MSQ)** carrying 1 or 2 marks each in all the papers and sections. These questions are objective in nature, and each will have choice of four answers, out of which ONE or MORE than ONE choice(s) is / are correct.

Note: There is NO negative marking for a wrong answer in MSQ questions. However, there is NO partial credit for choosing partially correct combinations of choices or any single wrong choice.

(iii) **Numerical Answer Type (NAT)** Questions carrying 1 or 2 marks each in most of the papers and sections. For these questions, the answer is a signed real number, which needs to be entered by the candidate using the virtual numeric keypad on the monitor (keyboard of the computer will be disabled). No choices will be shown for these types of questions. The answer can be a number such as 10 or –10 (an integer only). The answer may be in decimals as well, for example, 10.1 (one decimal) or 10.01 (two decimals) or –10.001 (three decimals). These questions will be mentioned with, up to which decimal places, the candidates need to present the answer. Also, for some NAT type problems an appropriate range will be considered while evaluating these questions so that the candidate is not unduly penalized due to the usual round-off errors. Candidates are advised to do the rounding off at the end of the calculation (not in between steps). Wherever required and possible, it is better to give NAT answer up to a maximum of three decimal places.

Note: *There is NO negative marking for a wrong answer in NAT questions.*
Also, there is NO partial credit in NAT questions.

Pattern of Examination

Paper Code	General Aptitude (GA) Marks	Subject Marks	Total Marks	Total Time (Minutes)*
AE, AG, BM, BT, CE, CH, CS, CY, EC, EE, ES, EY, IN, MA, ME, MN, MT, NM, PE, PH, PI, ST, TF	15	85	100	180
AR [Part A + Part B1 or B2] (B1: Architecture or B2: Planning)	15	60 + 25	100	180
GE [Part A + Part B1 or B2] (B1: Surveying and Mapping or B2: Image Processing and Analysis)	15	55 + 30	100	180
GG [Part A + Part B] (Section 1:Geology or Section 2: Geophysics)	15	25 + 60	100	180
XE (Section A + Any TWO Sections)	15	15 + (2 x 35)	100	180
XH (Section B1 + Any ONE Section)	15	25 + 60	100	180
XL (Section P + Any TWO Sections)	15	25 + (2 x 30)	100	180

*PwD candidates with benchmark disability are eligible for the compensatory time of 20 minutes per hour. Thus, they will get one hour extra for a three hours examination

Marking Scheme - Distribution of Marks and Questions

General Aptitude (GA) Questions

In all papers, GA questions carry a total of 15 marks. The GA section includes 5 questions carrying 1-mark each (sub-total 5 marks) and 5 questions carrying 2-marks each (sub-total 10 marks).

Question Papers other than AR, GE, GG, XE, XH and XL

These papers would contain 25 questions carrying 1-mark each (sub-total 25 marks) and 30 questions carrying 2-marks each (sub-total 60 marks) consisting of some MCQ type questions, while the remaining may be MSQ and / or NAT questions.

AR (Architecture and Planning) Paper

Apart from the General Aptitude (GA) section, the question paper consists of two parts: Part A (60 marks) and Part B (25 marks). Part A is compulsory for all the candidates. Part B contains two parts: Part B1 (Architecture) and Part B2 (Planning). Candidates will have to attempt questions in Part A and questions in either Part B1 or Part B2 of Part B. The choice of Part B1 OR Part B2 can be made during examination.

Part A consists of 39 questions carrying a total of 60 marks: 18 questions carrying 1-mark each (sub- total 18 marks) and 21 questions carrying 2-marks each (sub-total 42 marks). Some are MCQ type questions, while the remaining may be MSQ and / or NAT questions. Either section of Part B [Part B1 (Architecture) and Part B2 (Planning)] consists of 16 questions carrying a total of 25 marks: 7 questions carrying 1-mark each (sub-total 7 marks) and 9 questions carrying 2-marks each (sub-total 18 marks). Some are MCQ type questions, while the remaining may be MSQ and / or NAT questions.

GE (Geomatics Engineering) Paper

Apart from the General Aptitude (GA) section, the question paper consists of two parts: Part A (55 marks) and Part B (30 marks). Part A - Engineering Mathematics and Basic Geomatics is compulsory for all the candidates. Part B contains two sections: Section I - Surveying and Mapping and Section II - Image Processing and Analysis. Candidates will have to attempt questions in Part A and questions in either Section I or Section II of Part B. The choice of Section I OR Section II of Part B can be made during examination.

Part A consists of 36 questions carrying a total of 55 marks: 17 questions carrying 1-mark each (sub- total 17 marks) and 19 questions carrying 2-marks each (sub-total 38 marks). Some are MCQ type questions, while the remaining may be MSQ and / or NAT questions. Either section of Part B [Section I (Surveying and Mapping) or Section II (Image Processing and Analysis))] consists of 19 questions carrying a total of 30 marks: 8 questions carrying 1-mark each (sub-total 8 marks) and 11 questions carrying 2-marks each (sub-total 22 marks). Some are MCQ type questions, while the remaining may be MSQ and / or NAT questions.

GG (Geology and Geophysics) Paper

Apart from the General Aptitude (GA) section, the GG question paper consists of two parts: Part A and Part B. Part A is compulsory for all the candidates. Part B contains two sections: Section 1 (Geology) and Section 2 (Geophysics). Candidates will have to attempt questions in Part A and questions in either Section 1 or Section 2 of Part B. The choice of Section 1 OR Section 2 of Part B has to be made **at the time of filling online application form**. At the examination hall, candidate cannot request for change of section.

Part A consists of 16 questions carrying a total of 25 marks: 7 questions carrying 1-mark each (sub-total 7 marks) and 9 questions carrying 2-marks each (sub-total 18 marks). Some are MCQ type questions, while the remaining may be MSQ and / or NAT questions. Either section of Part B (Section 1: Geology and Section 2: Geophysics) consists of 39 questions carrying a total of 60 marks: 18 questions carrying 1-mark each (sub-total 18 marks) and 21 questions carrying 2-marks each (sub-total 42 marks). Some are MCQ type questions, while the remaining may be MSQ and / or NAT questions.

XE Paper (Engineering Sciences)

A candidate appearing in the XE paper has to answer the following:

- **GA - General Aptitude** carrying a total of 15 marks.
- **Section A - Engineering Mathematics (Compulsory):** This section contains 11 questions carrying a total of 15 marks: 7 questions carrying 1-mark each (sub-total 7 marks), and 4 questions carrying 2-marks each (sub-total 8 marks). Some are MCQ type questions, while the remaining may be MSQ and / or NAT questions.
- **Any two of XE Sections B to H:** The choice of two sections from B to H can be made during the examination after viewing the questions. Only TWO optional sections can be answered at a time. A candidate wishing to change midway of the examination to another optional section must first choose to deselect one of the previously chosen optional sections (B to H). Each of the optional sections of the XE paper (Sections B through H) contains 22 questions carrying a total of 35 marks: 9 questions carrying 1-mark each (sub-total 9 marks) and 13 questions carrying 2-marks each (sub total 26 marks). Some are MCQ type questions, while the remaining may be MSQ and / or NAT questions.

XH Paper (Humanities and Social Sciences)

A candidate appearing in the XH paper has to answer the following:

- **GA - General Aptitude** carrying a total of 15 marks
- **Section B1 - Reasoning and Comprehension (Compulsory):** This section contains 15 questions carrying a total of 25 marks: 7 questions carrying 1-mark each (sub-total 7 marks) and 9 questions carrying 2-marks each (sub-total 18 marks). Some are MCQ type questions, while the remaining may be MSQ and / or NAT questions.
- **Any ONE of XH Sections C1 to C6:** The ONE choice of section from C1 to C6 has to be made **at the time of filling online application form**. Candidate cannot request for change of section at the examination hall. Each of the optional sections of the XH paper (Sections C1 through C6) contains 39 questions carrying a total of 60 marks:18 questions carrying 1-mark each (sub-total 18 marks) and 21 questions carrying 2-marks each (sub-total 42 marks). Some are MCQ type questions, while the remaining may be MSQ and / or NAT questions.

XL Paper (Life Sciences)

A candidate appearing in the XL paper has to answer the following:

- **GA – General Aptitude** carrying a total of 15 marks.
- **Section P–Chemistry (Compulsory):** This section contains 17 questions carrying a total of 25 marks: 9 questions carrying 1-mark each (sub-total 9 marks) and 8 questions carrying 2-marks each (sub-total 16 marks). Some questions will be of MSQ and/or numerical answer type while remaining questions will be MCQ type.
- **Any two of XL Sections Q to U:** The choice of two sections from Q to U can be made during the examination after viewing the questions. Only TWO optional sections can be answered at a time. A candidate wishing to change midway of the examination to another optional section must first choose to deselect one of the previously chosen optional sections (Q to U). Each of the optional sections of the XL paper (Sections Q through U) contains 19 questions carrying a total of 30 marks: 8 questions carrying 1-mark each (sub-total 8 marks) and 11 questions carrying 2-marks each (sub-total 22 marks). Some questions will be of MSQ and/or numerical answer type while remaining questions will be MCQ type.

GATE Score

After the evaluation of the answers, the actual (raw) marks obtained by a candidate will be considered for computing the GATE Score. For multi-session papers (subjects), raw marks obtained by the candidates in different sessions will be converted to Normalized marks for that particular subject. Thus, raw marks (for single session papers) or normalized marks (for multi-session papers) will be used for computing the GATE Score, based on the qualifying marks.

Calculation of Normalized Marks for Multi-Session Papers

In GATE 2024 examination, some papers may be conducted in multi-sessions. Hence, for these papers, a suitable normalization is applied to take into account any variation in the difficulty levels of the question papers across different sessions. The normalization is done based on the fundamental assumption that "in all multi-session GATE papers, the distribution of abilities of candidates is the same across all the sessions". This assumption is justified since the number of candidates appearing in multi-session papers in GATE 2024 is large and the procedure for allocation of session to candidates is random. Further, it is also ensured that for the same multi-session paper, the number of candidates allotted in each session is of the same order of magnitude.

Based on the above, and considering various normalization methods, the committee arrived at the following formula for calculating the normalized marks for the multi-session papers.

Normalization mark of jth candidate in the ith session $\widehat{M_{ij}}$ is given by

$$\widehat{M_{ij}} = \frac{\bar{M}_t^g - M_q^g}{\bar{M}_{ti} - M_{iq}}(M_{ij} - M_{iq}) + M_q^g$$

where

M_{ij} : is the actual marks obtained by the jth candidate in i^{th} session

$\bar{M}_t^g$: is the average marks of the top 0.1% of the candidates considering all sessions

M_q^g : is the sum of mean and standard deviation marks of the candidates in the paper considering all sessions

$\bar{M}_{ti}$: is the average marks of the top 0.1% of the candidates in the ith session

M_{iq} : is the sum of the mean marks and standard deviation of the ith session

Calculation of GATE Score for All Papers

For all papers for which there is only one session, actual marks obtained by the candidates will be used for calculating the GATE 2024 Score. For papers in multi-sessions, normalized marks will be calculated corresponding to the raw marks obtained by a candidate and the GATE 2024 Score will be calculated based on the normalized marks.

The GATE 2024 score will be computed using the formula given below.

$$\text{GATE Score} = S_q + (S_t - S_q)\frac{(M - M_q)}{(\bar{M}_t - M_q)}$$

where

- M : marks obtained by the candidate (actual marks for single session papers and normalized marks for multi-session papers)
- M_q : is the qualifying marks for general category candidate in the paper
- $\bar{M}_t$: is the mean of marks of top 0.1% or top 10 (whichever is larger) of the candidates who appeared in the paper (in case of multi-session papers including all sessions)
- S_q : 350, is the score assigned to M_q
- S_t : 900, is the score assigned to $\bar{M}_t$

In the GATE 2024 the qualifying marks (M_q) for general category student in each subject will be 25 marks (out of 100) or $\mu + \sigma$, whichever is larger. Here μ is the mean and σ is the standard deviation of marks of all the candidates who appeared in the paper.

After the declaration of results, GATE Scorecards can be downloaded by the GATE qualified candidates ONLY.

The GATE 2024 Committee has the authority to decide the qualifying mark/score for each GATE paper. In case of any claim or dispute with respect to GATE 2024 examination or score, the Courts and Tribunals in Mumbai alone will have the exclusive jurisdiction to entertain and settle them.

GATE Syllabus

GENERAL APTITUDE

Verbal Aptitude

Basic English Grammar: Tenses, articles, adjectives, prepositions, conjunctions, verb-noun agreement, and other parts of speech.

Basic Vocabulary: Words, idioms, and phrases in context, Reading and comprehension, Narrative sequencing.

Quantitative Aptitude

Data Interpretation: Data graphs (bar graphs, pie charts, and other graphs representing data), 2- and 3-dimensional plots, maps, and tables.

Numerical Computation and Estimation: Ratios, percentages, powers, exponents and logarithms, permutations and combinations, and series, Mensuration and geometry, Elementary statistics and probability.

Analytical Aptitude

Logic: Deduction and induction, Analogy, Numerical relations and reasoning.

Spatial Aptitude

Transformation of Shapes: Translation, rotation, scaling, mirroring, assembling, and grouping Paper folding, cutting, and patterns in 2 and 3 dimensions.

XE - A – ENGINEERING MATHEMATICS

Section 1: Linear Algebra

Algebra of real matrices: Determinant, inverse and rank of a matrix; System of linear equations (conditions for unique solution, no solution and infinite number of solutions); Eigenvalues and eigenvectors of matrices; Properties of eigenvalues and eigenvectors of symmetric matrices, diagonalization of matrices; Cayley-Hamilton Theorem.

Section 2: Calculus

Functions of single variable: Limit, indeterminate forms and L'Hospital's rule; Continuity and differentiability; Mean value theorems; Maxima and minima; Taylor's theorem; Fundamental theorem and mean value theorem of integral calculus; Evaluation of definite and improper integrals; Applications of definite integrals to evaluate areas and volumes (rotation of a curve about an axis).

Functions of two variables: Limit, continuity and partial derivatives; Directional derivative; Total derivative; Maxima, minima and saddle points; Method of Lagrange multipliers; Double integrals and their applications.

Sequences and series: Convergence of sequences and series; Tests of convergence of series with non-negative terms (ratio, root and integral tests); Power series; Taylor's series; Fourier Series of functions of period 2π.

Section 3: Vector Calculus

Gradient, divergence and curl; Line integrals and Green's theorem.

Section 4: Complex Variables

Complex numbers, Argand plane and polar representation of complex numbers; De Moivre's theorem; Analytic functions; Cauchy-Riemann equations.

Section 5: Ordinary Differential Equations

First order equations (linear and nonlinear); Second order linear differential equations with constant coefficients; Cauchy-Euler equation; Second order linear differential equations with variable coefficients; Wronskian; Method of variation of parameters; Eigenvalue problem for second order equations with constant coefficients; Power series solutions for ordinary points.

Section 6: Partial Differential Equations

Classification of second order linear partial differential equations; Method of separation of variables: One dimensional heat equation and two dimensional Laplace equation.

Section 7: Probability and Statistics

Axioms of probability; Conditional probability; Bayes' Theorem; Mean, variance and standard deviation of random variables; Binomial, Poisson and Normal distributions; Correlation and linear regression.

Section 8: Numerical Methods

Solution of systems of linear equations using LU decomposition, Gauss elimination method; Lagrange and Newton's interpolations; Solution of polynomial and transcendental equations by Newton-Raphson method; Numerical integration by trapezoidal rule and Simpson's rule; Numerical solutions of first order differential equations by explicit Euler's method.

XE - B – FLUID MECHANICS

Section 1: Flow and Fluid Properties

Fluid Properties: Density, viscosity, surface tension, relationship between stress and strain-rate for Newtonian fluids.

Classification of Flows: Viscous versus inviscid flows, incompressible versus compressible flows, internal versus external flows, steady versus unsteady flows, laminar versus turbulent flows, 1-D, 2-D and 3-D flows, Newtonian versus non-Newtonian fluid flow.

Hydrostatics: Buoyancy, manometry, forces on submerged bodies and its stability.

Section 2: Kinematics of Fluid Motion

Eulerian and Lagrangian descriptions of fluid motion.

Concept of local, convective and material derivatives. Streamline, streakline, pathline and timeline.

Section 3: Integral Analysis for a Control Volume

Reynolds Transport Theorem (RTT) for conservation of mass, linear and angular momentum.

Section 4: Differential Analysis

Differential equations of mass and momentum for incompressible flows.

Inviscid flows - Euler equations and viscous flows - Navier-Stokes equations.

Concept of fluid rotation, vorticity, stream function and circulation.

Exact solutions of Navier-Stokes equations for Couette flow and Poiseuille flow, thin film flow.

Section 5: Dimensional Analysis

Concept of geometric, kinematic and dynamic similarity.

Buckingham Pi theorem and its applications.

Non-dimensional parameters and their physical significance - Reynolds number, Froude number and Mach number.

Section 6: Internal Flows

Fully developed pipe flow.

Empirical relations for laminar and turbulent flows: friction factor, Darcy-Weisbach relation and Moody's chart.

Major and minor losses.

Section 7: Bernoulli's Equation and its Applications, Potential Flows

Bernoulli's equation: Assumptions and applications.

Flow measurements - Venturi meter, Pitot-static tube and orifice meter.

Elementary potential flows: Velocity potential function.

Uniform flow, source, sink and vortex, and their superposition for flow past simple geometries.

Section 8: External Flows

Prandtl boundary layer equations: Concept and assumptions.

Boundary layer characteristics: Boundary layer thickness, displacement thickness and momentum thickness.

Qualitative idea of boundary layer separation, streamlined and bluff bodies, and drag and lift forces.

XE - C – MATERIALS SCIENCE

Section 1: Classification and Structure of Materials

Classification of materials: metals, ceramics, polymers and composites.

Nature of bonding in materials:metallic,ionic, covalent and mixed bonding; structure of materials:fundamentals of crystallography, symmetry operations, crystal systems, Bravais lattices, unit cells, primitive cells, crystallographic planes and directions; structures of metals, ceramics, polymers, amorphous materials and glasses.

Defects in crystalline materials: 0-D, 1-D and 2-D defects; vacancies, interstitials, solid solutions in metals and ceramics, Frenkel and Schottky defects; dislocations; grain boundaries, twins, stacking faults; surfaces and interfaces.

Section 2: Thermodynamics, Kinetics and Phase Transformations

Extensive and intensive thermodynamic properties, laws of thermodynamics, phase equilibria, phase rule, phase diagrams (unary and binary), basic electrochemistry.

Reaction kinetics, fundamentals of diffusion, Fick's laws, their solutions and applications.

Solidification of pure metals and alloys, nucleation andgrowth, diffusional solid-state phase transformations (precipitation and eutectoid), martensitic transformation.

Section 3: Properties and Applications of Materials

Mechanical properties of metals, ceramics, polymers and composites at room temperature; stress-strain response (elastic, anelastic and plastic deformation).

Electronic properties: free electron theory, Fermi energy, density of states, elements of band theory, semiconductors, Hall effect, dielectric behaviour, piezo- and ferro-electric behaviour.

Magnetic properties:Origin of magnetism in materials, para-, dia-, ferro- and ferri-magnetism.

Thermal properties: Specific heat, heat conduction, thermal diffusivity, thermal expansion, and thermoelectricity.

Optical properties: Refractive index, absorption and transmission of electromagnetic radiation.

Examples of materials exhibiting the above properties, and their typical/common applications.

Section 4: Characterization and Measurements of Properties

X-ray diffraction;spectroscopic techniques such as UV-Vis, IR and Raman; optical microscopy, electron microscopy, composition analysisin electron microscopes.

Tensile test, hardness measurement.

Electrical conductivity, carrier mobility and concentrations.

Thermal analysis techniques: thermogravimetry and calorimetry.

Section 5: Processing of Materials

Heat treatment of ferrous and aluminium alloys; preparation of ceramic powders, sintering; thin film deposition: evaporation and sputtering techniques, and chemical vapour deposition, thin film growth phenomena.

Section 6: Degradation of Materials

Corrosion and its prevention; embrittlement of metals; polymer degradation.

XE - D – SOLID MECHANICS

Section 1: Mechanics of Rigid Bodies

Equivalent force systems; free-body diagrams; equilibrium equations; analysis of determinate trusses and frames; friction; principle of minimum potential energy; particle kinematics and dynamics; dynamics of rigid bodies under planar motion; law of conservation of energy; law of conservation of momentum.

Section 2: Mechanics of Deformable Bodies

Stresses and strains; transformation of stresses and strains, principal stresses and strains; Mohr's circle for plane stress and plane strain; generalized Hooke's Law; elastic constants; thermal stresses; theories of failure.

Axial force, shear force and bending moment diagrams; axial, shear and bending stresses; combined stresses; deflection (for symmetric bending); torsion in circular shafts; thin walled pressure vessels; energy methods (Castigliano's Theorems); Euler buckling.

Section 3: Vibrations

Free vibration of undamped single degree of freedom systems.

XE - E – THERMODYNAMICS

Section 1: Basic Concepts

Continuum and macroscopic approach; thermodynamic systems (closed and open); thermodynamic properties and equilibrium; state of a system, state postulate for simple compressible substances, state diagrams, paths and processes on state diagrams; concepts of heat and work, different modes of work; zeroth law of thermodynamics; concept of temperature.

Section 2: First Law of Thermodynamics

Concept of energy and various forms of energy; internal energy, enthalpy; specific heats; first law applied to elementary processes, closed systems and control volumes, steady and unsteady flow analysis.

Section 3: Second Law of Thermodynamics

Limitations of the first law of thermodynamics, concepts of heat engines and heat pumps/refrigerators, Kelvin-Planck and Clausius statements and their equivalence; reversible and irreversible processes; Carnot cycle and Carnot principles/theorems; thermodynamic temperature scale; Clausius inequality and concept of entropy; microscopic interpretation of entropy, the principle of increase of entropy, T-s diagrams; second law analysis of control volume; availability and irreversibility; third law of thermodynamics.

Section 4: Properties of Pure Substances

Thermodynamic properties of pure substances in solid, liquid and vapor phases; P-v-T behaviour of simple compressible substances, phase rule, thermodynamic property tables and charts, ideal and real gases, ideal gas equation of state and van der Waals equation of state; law of corresponding states, compressibility factor and generalized compressibility chart.

Section 5: Thermodynamic Relations

T-ds relations, Helmholtz and Gibbs functions, Gibbs relations, Maxwell relations, Joule-Thomson coefficient, coefficient of volume expansion, adiabatic and isothermal compressibilities, Clapeyron and Clapeyron-Clausius equations.

Section 6: Thermodynamic Cycles

Carnot vapor cycle, ideal Rankine cycle, Rankine reheat cycle, air-standard Otto cycle, air-standard Diesel cycle, air-standard Brayton cycle, vapor-compression refrigeration cycle.

Section 7: Ideal Gas Mixtures

Dalton's and Amagat's laws, properties of ideal gas mixtures, air-water vapor mixtures and simple thermodynamic processes involving them; specific and relative humidities, dew point and wet bulb temperature, adiabatic saturation temperature, psychrometric chart.

XE - F – POLYMER SCIENCE AND ENGINEERING

Section 1: Chemistry of High Polymers

Monomers, functionality, degree of polymerizations, classification of polymers, glass transition, melting transition, criteria for rubberiness, polymerization methods: addition and condensation; their kinetics, metallocene polymers and other newer methods of polymerization, copolymerization, monomer reactivity ratios and its significance, kinetics, different copolymers, random, alternating, azeotropic copolymerization, block and graft copolymers, techniques for polymerization-bulk, solution, suspension, emulsion. Concept of intermolecular order (morphology) – amorphous, crystalline, orientation states. Factor affecting crystallinity. Crystalline transition. Effect of morphology on polymer properties.

Section 2: Polymer Characterization

Solubility and swelling, Concept of molecular weight distribution and its significance, concept of average molecular weight, determination of number average, weight average, viscosity average and Z-average molecular weights, polymer crystallinity, analysis of polymers using IR, XRD, thermal (DSC, DMTA, TGA), microscopic (optical and electronic) techniques, Molecular wt. distribution: Broad and Narrow, GPC, mooney viscosity.

Section 3: Synthesis, Manufacturing and Properties

Commodity and general purpose thermoplastics: PE, PP, PS, PVC, Polyesters, Acrylic, PU polymers. Engineering Plastics: Nylon, PC, PBT, PSU, PPO, ABS, Fluoropolymers Thermosetting polymers: Polyurethane, PF, MF, UF, Epoxy, Unsaturated polyester, Alkyds. Natural and synthetic rubbers: Recovery of NR hydrocarbon from latex; SBR, Nitrile, CR, CSM, EPDM, IIR, BR, Silicone, TPE, Speciality plastics: PEK, PEEK, PPS, PSU, PES etc. Biopolymers such as PLA, PHA/PHB.

Section 4: Polymer Blends and Composites

Difference between blends and composites, their significance, choice of polymers for blending, blend miscibility-miscible and immiscible blends, thermodynamics, phase morphology, polymer alloys, polymer eutectics, plastic-plastic, rubber-plastic and rubber-rubber blends, FRP, particulate, long and short fibre reinforced composites. Polymer reinforcement, reinforcing fibres – natural and synthetic, base polymer for reinforcement (unsaturated polyester), ingredients / recipes for reinforced polymer composite.

Section 5: Polymer Technology

Polymer compounding-need and significance, different compounding ingredients for rubber and plastics (Antioxidants, Light stabilizers, UV stabilizers, Lubricants, Processing aids, Impact modifiers, Flame retardant, antistatic agents. PVC stabilizers and Plasticizers) and their function, use of carbon black, polymer mixing equipments, cross-linking and vulcanization, vulcanization kinetics.

Section 6: Polymer Rheology

Flow of Newtonian and non-Newtonian fluids, different flow equations, dependence of shear modulus on temperature, molecular/segmental deformations at different zones and transitions. Measurements of rheological parameters by capillary rotating, parallel plate, cone-plate rheometer. Visco-elasticity-creep and stress relaxations, mechanical models, control of rheological characteristics through compounding, rubber curing in parallel plate viscometer, ODR and MDR.

Section 7: Polymer Processing

Compression molding, transfer molding, injection molding, blow molding, reaction injection molding, filament winding, SMC, BMC, DMC, extrusion, pultrusion, calendaring, rotational molding, thermoforming, powder coating, rubber processing in two-roll mill, internal mixer, Twin screw extruder.

Section 8: Polymer Testing

Mechanical-static and dynamic tensile, flexural, compressive, abrasion, endurance, fatigue, hardness, tear, resilience, impact, toughness. Conductivity-thermal and electrical, dielectric constant, dissipation factor, power factor, electric resistance, surface resistivity, volume resistivity, swelling, ageing resistance, environmental stress cracking resistance, limiting oxygen index. Heat deflection temperature –Vicat softening temperature, Brittleness temperature, Glass transition temperature, Co-efficient of thermal expansion, Shrinkage, Flammability, dielectric constant, dissipation factor, power factor, Optical Properties - Refractive Index, Luminous Transmittance and Haze, Melt flow index.

Section 9: Polymer Recycling and Waste management

Polymer waste, and its impact on environment, Sources, Identification and Separation techniques, recycling classification, recycling of thermoplastics, thermosets and rubbers, applications of recycled materials. Life cycle assessment of polymer products (case studies like PET bottles, packaging bags)

XE - G – FOOD TECHNOLOGY

Section 1: Food Chemistry and Nutrition

Carbohydrates: structure and functional properties of mono-, oligo-, & poly- saccharides including starch, cellulose, pectic substances and dietary fibre, gelatinization and retrogradation of starch. Proteins: classification and structure of proteins in food, biochemical changes in post mortem and tenderization of muscles. Lipids: classification and structure of lipids, rancidity, polymerization and polymorphism. Pigments: carotenoids, chlorophylls, anthocyanins, tannins and myoglobin. Food flavours: terpenes, esters, aldehydes, ketones and quinines. Enzymes: specificity, simple and inhibition kinetics, coenzymes, enzymatic and non- enzymatic browning. Nutrition: balanced diet, essential amino acids and essential fatty acids, protein efficiency ratio, water soluble and fat soluble vitamins, role of minerals in nutrition, co-factors, anti-nutrients, nutraceuticals, nutrient deficiency diseases. Chemical and biochemical changes: changes occur in foods during differentprocessing.

Section 2: Food Microbiology

Characteristics of microorganisms: morphology of bacteria, yeast, mold and actinomycetes, spores and vegetative cells, gram-staining. Microbial growth: growth and death kinetics, serial dilution technique. Food spoilage: spoilage microorganisms in different food products including milk, fish, meat, egg, cereals and their products. Toxins from microbes: pathogens and non-pathogens including Staphylococcus, Salmonella, Shigella, Escherichia, Bacillus, Clostridium, and Aspergillus genera. Fermented foods and beverages: curd, yoghurt, cheese, pickles, soya-sauce, sauerkraut, idli, dosa, vinegar, alcoholic beverages and sausage.

Section 3: Food Products Technology

Processing principles: thermal processing, chilling, freezing, dehydration, addition of preservatives and food additives, irradiation, fermentation, hurdle technology, intermediate moisture foods. Food pack aging and storage: packaging materials, aseptic packaging, controlled and modified atmosphere storage. Cereal processing and products: milling of rice, wheat, and maize, parboiling of paddy, bread, biscuits, extruded products and ready to eat breakfast cereals. Oil processing: expelling, solvent extraction, refining and hydrogenation. Fruits and vegetables p processing: extraction, clarification, concentration and packaging of fruit juice, jam, jelly, marmalade, squash, candies, tomato sauce, ketchup, and puree, potato chips, pickles. Plantation crops processing and products: tea, coffee, cocoa, spice, extraction of essential oils and oleoresins from spices. Milk and milk products processing: pasteurization and sterilization, cream, butter, ghee, ice- cream, cheese and milk powder. Processing of animal products: drying, canning, and freezing of fish and meat; production of egg powder. Waste utilization: pectin from fruit wastes, uses of by-products from rice milling. Food standards and quality maintenance: FPO, PFA, A-Mark, ISI, HACCP, food plant sanitation and cleaning in place (CIP).

Section 4: Food Engineering

Mass and energy balance; Momentum transfer: Flow rate and pressure drop relationships for Newtonian fluids flowing through pipe, Reynolds number. Heat transfer: heat transfer by conduction, convection, radiation, heat exchangers. Mass transfer: molecular diffusion and Flick's law, conduction and convective mass transfer, permeability through single and multilayer films. Mechanical operations: size reduction of solids, high pressure homogenization, filtration, centrifugation, settling, sieving, mixing & agitation of liquid. Thermal operations: thermal sterilization, evaporation of liquid foods, hot air drying of solids, spray and freeze-drying, freezing and crystallization. Mass transfer operations: psychometric, humidification and dehumidification operations.

XE - H – ATMOSPHERIC & OCEAN SCIENCE

Section A: Atmospheric Science

Vertical Structure and Composition of the Atmosphere; Blackbody Radiation and Radiation Balance; Modes of Heat Transfer in the Atmosphere; Greenhouse Effect; Cloud Types; Laws of Thermodynamics; Gas

Laws; Hydrostatic Equation; Clausius Clapeyron Equation; Adiabatic Processes, Humidity in the Atmosphere, Atmospheric Stability; Weather and Climate.

Navier-Stokes and Continuity Equations; Compressible and Incompressible Fluids; Pressure Gradient, Centripetal, Centrifugal and Coriolis Forces; Geostrophic, Gradient and Cyclostrophic Balances; Circulations and Vorticity, General Circulation of the Atmosphere. Broad Features of Indian Monsoons, Monsoon Depressions; Tropical Convergence Zones; Tropical Cyclones.

Section B: Ocean Sciences

Vertical Profiles of Temperature and Salinity; Stability and Double Diffusion; Equation of State, Equations for Conservation of Mass, Momentum, Heat and Salt; Inertial Currents; Geostrophic Motion; Air-Sea Surface Fluxes; Wind-driven Circulation, Ekman and Sverdrup Transports; Storm Surges, Tides, Tsunamis and Wind Waves; Eddies and Gyres; Eastern and Western Boundary Currents, Equatorial Currents, Indian Ocean Current Systems; Thermohaline Circulation.

Chemical Properties of Seawater, Major and Minor Elements, Ocean Acidification, Biochemical Cycling of Nutrients, Trace Metals and Organic Matter. Biological Pump; Primary and Secondary Biological Productivity; Air-sea Exchange of Biogenic Dissolved Gases; Marine Ecology.

GENERAL APTITUDE

Syllabus

Verbal Ability

English grammar, sentence completion, verbal analogies, word groups, instructions, critical reasoning and verbal deduction.

Numerical Ability

Numerical computation, numerical estimation, numerical reasoning and data interpretation.

Subject-wise Analysis

Year	1 mark Questions	2 marks Questions	Total Marks
2009	–	–	–
2010	5	5	10
2011	5	5	10
2012	5	5	10
2013	5	5	10
2014	5	5	10
2015	5	5	10
2016	5	5	10
2017	5	5	10
2018	5	5	10

2010

Q.(1 – 5) carry one mark each.

1. *The question below consists of a pair of related words followed by four paris of words. Select the pair that best expresses the relation in the original pair.*

Unemployed : Worker

(*a*) fallow : land (*b*) unaware : sleeper

(*c*) wit : jester (*d*) renovated : house

2. *Choose the most appropriate word from the options given below to complete the following sentence:*

His rather casual remarks on politics ______ his lack of seriousness about the subject.

(*a*) masked (*b*) belied

(*c*) betrayed (*d*) suppressed

3. *Which of the following options is the closest in meaning to the word below:*

Circuitous

(*a*) cyclic (*b*) indirect

(*c*) confusing (*d*) crooked

4. 25 persons are in a room. 15 of them play hockey, 17 of them play football and 10 of them play both hockey and football. Then the number of persons playing neither hockey nor football is:

(*a*) 2 (*b*) 17

(*c*) 13 (*d*) 3

5. *Choose the most appropriate word from the options given below to complete the following sentence:*

If we manage to ______ our natural resources, we would leave a better planet for our children.

(*a*) uphold (*b*) restrain

(*c*) cherish (*d*) conserve

Q.(6 – 10) carry two marks each.

6. 5 skilled workers can build a wall in 20 days; 8 semi-skilled workers can build a wall in 25 days; 10 unskilled workers can build a wall in 30 days. If a term has 2 skilled, 6 semi-skilled and 5 unskilled workers, how long will it take to build the wall?

(*a*) 20 days (*b*) 18 days

(*c*) 16 days (*d*) 15 days

7. Given digits 2, 2, 3, 3, 3, 4, 4, 4, 4 how many distinct 4 digit numbers greater than 3000 can be formed?

(*a*) 50 (*b*) 51

(*c*) 52 (*d*) 54

8. If 137 + 276 = 435 how much is 731 + 672?

(*a*) 534

(*b*) 1403

(*c*) 1623

(*d*) 1513

9. Hari (H), Gita (G), Irfan (I) and Saira (S) are siblings (i.e., brothers and sisters). All were born on I[st] January. The age difference between any two successive siblings (that is born one after another) is less than 3 years. Given the following facts :

I. Hari's age + Gita's age > Irfan's age + Saira's age

II. The age difference between Gita and Saira is 1 year. However, Gita is not the oldest and Saira is not the youngest

III. There are no twins.

In what order were they born (oldest first)?

(*a*) HSIG

(*b*) SGHI

(*c*) IGSH

(*d*) IHSG

10. Modern warfare has changed from large scale clashes of armies to suppression of civilian populations. Chemical agents that do their work silently appear to be suited to such warfare; and regretfully, there exist people in military establishments who think that chemical agents are useful tools for their cause.

Which of the following statements best sums up the meaning of the above passage:

(*a*) Modern warfare has resulted in civil strife

(*b*) Chemical agents are useful in modern warfare

(*c*) Use of chemical agents in warfare would be undersirable

(*d*) People in military establishments like to use chemical agents in war

ANSWERS

1. (a)	**2.** (c)	**3.** (b)	**4.** (d)	**5.** (d)	**6.** (d)	**7.** (b)
8. (a)	**9.** (b)	**10.** (d)				

EXPLANATIONS

1. (a) A worker who is not working is called unemployed. Similarly the land that is not used is called fallow.

2. (c) The blank will take in a word that means revealed or showed. Hence, (c) is the correct answer.

3. (b) The word 'circuitous' means indirect, which makes option (b) correct.

4. (d) Hockey (15)

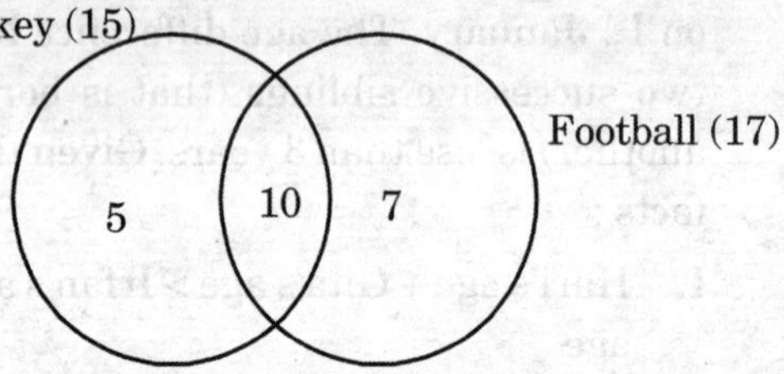

Total persons in a room are = 25

$\therefore$ No of persons playing neither hockey nor footbal is $25 - (5 + 10 + 7) = 3$

5. (d) 'Conserve' is the most appropriate word to be used with 'natural resources' as there is a need to preserve natural resources.

6. (d) Let skilled workers, semi-skilled workers and unskilled workers be represented by A, B and C respectvely,

$\therefore 5A \times 20 = 8B \times 25$

$\Rightarrow A = 2B$

and $8B \times 25 = 10C \times 30$

$\Rightarrow 2B = 3C$

$\therefore A = 2B = 3C$

If the team has $2A + 6B + 5C$

workers i,e. $2A + 3A + \frac{5}{3}A = \frac{20}{3}A$

and take d days to fuild the wall

So, $5A \times 20 = \frac{20A}{30} \times d$

d = 15 days

7. (b) Given digits are 2, 2, 3, 3, 3, 4, 4, 4, 4,

Total possible number greater than 3000 are 51.

3223, 3224, 3232, 3234, 3233, 3242, 3243, 3244, 3322, 3422, 3423, 3424, 3432, 3433, 3434, 3442, 3443, 3444, 4223, 4224, 4323, 4233, 4234, 4242, 4243, 4244, 4322, 4322, 4323, 4324, 4332, 4333, 4334, 4342, 4432, 4433, 4434, 4442, 4443, 4444.

8. (a) $137 + 276 = 435$ ...(1)

On reversing the digits of above statement we will get a new one i.e.

$731 + 672 = 534$

9. (b) SGHI is the correct order (oldest first)

Suppose S's age is 10 years then G's age will be 9 years.

Now, the age difference between any two successive siblings is less than 3 years.

So

S G H I

10 9 8 6

$H + G = 17, I + S = 16$

$H + G > I + S.$

or

S G H I

10 9 7 5

$H + G = 16, I + S = 15$

$H + G > I + S.$

10. (d) The last line of the paragraph states that there are people in military establishments who think that chemical agents are useful for warfare, which makes option D correct. Civil Strife is a term used to describe unrest caused by a group of people, which is not true in the light of the given passage. So, Option A is not the correct option. Option B is incorrect because there is no evidence that justifies the usefulness of chemical agents for warfare. Option C is incorrect because it is mentioned in the second line that chemical agents appear to be suited for such warfare, which directly means that they are desirable.

2011

Q.(1 – 5) carry one mark each.

1. Choose the most appropriate word from the options given below to complete the following sentence:

Under ethical guidelines recently adopted by the Indian Medical Association, human genes are to be manipulated only to correct diseases for which ____ treatments are unsatisfactory.

(*a*) similar (*b*) most

(*c*) uncommon (*d*) available

2. Choose the word from the options given below that is most nearly opposite in meaning to the given word:

Frequency

(*a*) periodicity (*b*) rarity

(*c*) gradualness (*d*) persistency

3. Choose the most appropriate word from the options given below to complete the following sentence :

It was her view that the country's problems had been_____ by foreign technocrats, so that to invite them to come back would be counter-productive.

(*a*) identified (*b*) ascertained

(*c*) exacerbated (*d*) analysed

4. There are two candidates P and Q in an election. During the campaign, 40% of the voters promised to vote for P, and rest for Q. However, on the day of election 15% of the voters went back on their promise to vote for P and instead voted for Q. 25% of the voters went back on their promise to vote for Q and instead voted for P. Suppose, P lost by 2 votes, then what was the total number of voters?

(*a*) 100 (*b*) 110

(*c*) 90 (*d*) 95

5. The question below consists of a pair of related words followed by four pairs of words. Select the pair that best expresses the relation in the original pair :

Gladiator : Arena

(*a*) dance : stage

(*b*) commuter : train

(*c*) teacher : classroom

(*d*) lawyer : courtroom

Q.(6-10) carry two marks each.

6. The sum of n terms of the series 4 + 44 + 444 + ... is

(*a*) $(4/81)\,[10^{n+1} - 9n - 1]$

(*b*) $(4/81)\,[10^{n-1} - 9n - 1]$

(*c*) $(4/81)\,[10^{n+1} - 9n - 10]$

(*d*) $(4/81)\,[10^{n} - 9n - 10]$

7. Given that $f(y) = \frac{|y|}{y}$, and q is any non-zero real number, the value of $|f(q) - f(-q)|$ is

(*a*) 0 (*b*) –1

(*c*) 1 (*d*) 2

8. Three friends, R, S and T shared toffee from a bowl, R took 1/3rd of the toffees, but returned four to the bowl. S took 1/4th of what was left but returned three toffees to the bowl. T took half of the remainder but returned two back into the bowl. If the bowl had 17 toffees left, how many toffees were originally there in the bowl?

(*a*) 38 (*b*) 31

(*c*) 48 (*d*) 41

9. The fuel consumed by a motorcycle during a journey while traveling at various speeds is indicated in the graph below.

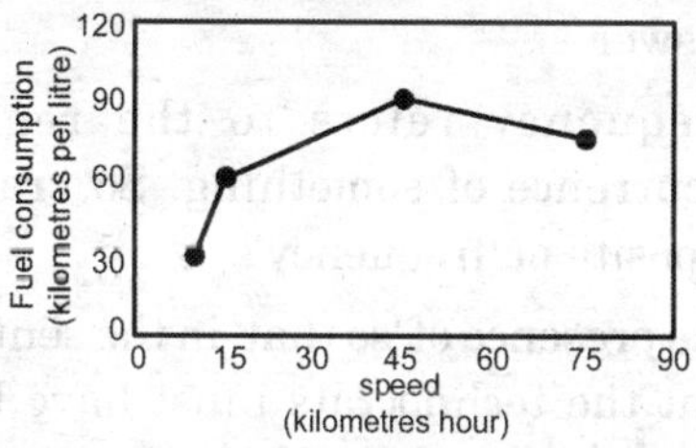

The distances covered during four laps of the journey are listed in the table below

Lap	Distance (kilometers)	Average speed (kilometers per hour)
P	15	15
Q	75	45
R	40	75
S	10	10

From the given data, we can conclude that the fuel consumed per kilometre was least during the lap

(*a*) P (*b*) Q

(*c*) R (*d*) S

10. The horse has played a little known but very important role in the field of medicine. Horses were injected with toxins of diseases until their blood built up immunities. Then a serum was made from their blood. Serums to fight with diphtheria and tetanus were developed this way.

It can be inferred from the passage, that horses were

(*a*) given immunity to diseases

(*b*) generally quite immune to diseases

(*c*) given medicines to fight toxins

(*d*) given diphtheria and tetanus serums

ANSWERS

1. (d)	**2.** (b)	**3.** (c)	**4.** (a)	**5.** (d)	**6.** (c)	**7.** (d)
8. (c)	**9.** (b)	**10.** (a)				

EXPLANATIONS

1. (d) The given blank needs a word, which indicates that the existing treatments are unsatisfactory. Therefore, option D 'available' is the best answer.

2. (b) Frequency refers to the regularity of occurrence of something. So, 'rarity' is the opposite of 'frequency'.

3. (c) The presence of 'so that' in the sentence shows that the technocrats must have bad effects, that is, they must have 'exacerbated' the country's problems. Hence, C is the correct answer.

4. (a) Let the total number of voters be 100

No. of voters promised to vote for P = 40

No. of voters promised to vote for Q = 60

Now, according to the question,

15% of the voters who promised to vote for P, gave their vote to Q and 25% of those whose who promised to vote for q, gave it to P.

$\therefore$ 15% of 40 = 6

25% of 60 = 15

So P has 40 – 6 + 15 = 49 votes

Q has 60 – 15 + 6 = 51 votes

Clearly, P lost by 2 votes, which satisfies the given condition.

$\therefore$ The total number of voters was 100.

5. (d) A gladiator fights or performs in an arena. Similarly a layer fights or performs in a courtroom.

6. (c) 4 + 44 + 444 + ... n terms

4 (1 + 11 + 111 + ... n terms)

$\frac{4}{9}$ (9 + 99 + 999 + ... n terms)

$\frac{4}{9}$ (10 – 1 + 100 – 1 + 1000 – 1 + ... n terms)

$$\frac{4}{9}\left[\left(10+10^2+10^3+\ldots\ldots\text{ n terms}\right)-n\right]$$

$$=\frac{4}{9}\left(\frac{10(10^n-1)}{10-1}-n\right)$$

$$=\frac{4}{81}\left(10^{n+1}-10-9n\right)$$

7. (d) $f(y)=\frac{|y|}{y}$

$$f(q)=\frac{|q|}{q}=1$$

$$f(-q) = \frac{|-q|}{-q} = \frac{q}{-q} = -1$$

$$\therefore \quad |f(q) - f(-q)| = |1 - (-1)| = |2| = 2$$

8. (c) Let original number of toffees in the bowl = x

R has number of toffees = $\frac{1}{3}x - 4$

$\therefore$ Remaining toffees in bowl = $\frac{2}{3}x + 4$

S took $\frac{1}{4}^{th}$ of the remaining but returned three toffees.

$\therefore$ S has = $\frac{1}{4}\left(\frac{2}{3}x + 4\right) - 3$ toffees

Now remaining toffees in the bowl

$$= \frac{3}{4}\left(\frac{2}{3}x + 4\right) + 3$$

$$= \frac{1}{2}x + 3 + 3 = \frac{1}{2}x + 6$$

Now T took half of the remainder but returned two.

$\therefore$ T has $\frac{1}{4}x + 3 - 2$

Now remaining toffees in the bowl = $\frac{1}{4}x + 3 + 2$

$$\therefore \frac{1}{4}x + 5 = 17$$

$$\Rightarrow x = 48$$

9. (b) For Lap P, distance covered is 15 km and average speed is 15 km/hr.

$\therefore$ full consumption is 60 km/litre.

So, fuel consumed per km = $\frac{1}{60}$ litre

Similarly lap Q, fuel consumed per km

$= \frac{1}{90}$ litre

Similarly lap R, fuel consumed per km

$= \frac{1}{75}$ litre

Similarly lap S, fuel consumed per km

$= \frac{1}{30}$ litre

Clearly, it was least for Lap Q.

10. (a) The given paragraph mentions- "Horses were injected with toxins of diseases until their blood built up immunities". This clearly means that immunity was given to horses, which makes option A correct.

2012

Q. (1 – 5) carry one mark each.

1. One of the parts (a, b, c, d) in the sentence given below contains an ERROR. Which one of the following is **INCORRECT**?

I requested that he should be given the driving test today instead of tomorrow.

(*a*) requested that (*b*) should be given

(*c*) the driving test (*d*) instead of tomorrow

2. Which one of the following options is the closest in meaning to the word given below?

Latitude

(*a*) Eligibility (*b*) Freedom

(*c*) Coercion (*d*) Meticulousness

3. Choose the most appropriate word from the options given below to complete the following sentence:

Given the seriousness of the situation that he had to face, his ___ was impressive.

(*a*) beggary (*b*) nomenclature

(*c*) jealousy (*d*) nonchalance

4. Choose the most appropriate alternative from the options given below to complete the following sentence:

If the tired soldier wanted to lie down, he ___ the mattress out on the balcony.

(*a*) should take (*b*) shall take

(*c*) should have taken (*d*) will have taken

5. If $(1.001)^{1259} = 3.52$ and $(1.001)^{2062} = 7.85$, then $(1.001)^{3321} =$

(*a*) 2.23 (*b*) 4.33

(*c*) 11.37 (*d*) 27.64

Q. (6 – 10) carry two marks each.

6. A and B are friends. They decide to meet between 1 PM and 2 PM on a given day. There is a condition that whoever arrives first will not wait for the other for more than 15 minutes. The probability that they will meet on that day is

(*a*) 1/4

(*b*) 1/16

(*c*) 7/16

(*d*) 9/16

7. The data given in the following table summarizes the monthly budget of an average household.

Category	Amount (₹)
Food	4000
Clothing	1200
Rent	2000
Savings	1500
Other expenses	1800

The approximate percentage of the monthly budget **NOT** spent on savings is

(*a*) 10% (*b*) 14%

(*c*) 81% (*d*) 86%

8. There are eight bags of rice looking alike, seven of which have equal weight and one is slightly heavier. The weighing balance is of unlimited capacity. Using this balance, the minimum number of weighings required to identify the heavier bag is

(*a*) 2 (*b*) 3

(*c*) 4 (*d*) 8

9. Raju has 14 currency notes in his pocket consisting of only ₹ 20 notes and ₹ 10 notes. The total money value of the notes is ₹ 230. The number of ₹ 10 notes that Raju has is

(*a*) 5 (*b*) 6

(*c*) 9 (*d*) 10

10. One of the legacies of the Roman legions was discipline. In the legions, military law prevailed and discipline was brutal. Discipline on the battlefield kept units obedient, intact and fighting, even when the odds and conditions were against them.

Which one of the following statements best sums up the meaning of the above passage?

(*a*) Thorough regimentation was the main reason for the efficiency of the Roman legions even in adverse circumstances.

(*b*) The legions were treated inhumanly as if the men were animals.

(*c*) Discipline was the armies' inheritance from their seniors.

(*d*) The harsh discipline to which the legions were subjected to led to the odds and conditions being against them.

ANSWERS

1. (b)	**2.** (b)	**3.** (d)	**4.** (c)	**5.** (d)	**6.** (c)	**7.** (d)
8. (a)	**9.** (a)	**10.** (a)				

EXPLANATIONS

1. (b) The error is in part B of the sentence. It should be 'he should give'.

2. (b) 'Latitude' is a word that provides a sense of beyond the boundary. Hence, 'freedom' is the best answer.

3. (d) The blank would take in a word that expresses the ability to face a serious situation. Hence, 'nonchalance', which means bold, is the correct answer.

4. (c) Option (c) provides the correct verb form that should appear in the blank.

5. (d) $(1.001)^{1259} = 3.52$

$(1.001)^{2062} = 7.85$

$(1.001)^{1259} \times (1.001)^{2062} = 3.52 \times 7.85$

$(1.001)^{3321} = 27.64$

7. (d) Budget spent on everything except saving is:-

$4000 + 1200 + 2000 + 1800 = 9000$

Total monthly budget = 10500

$\Rightarrow 10500 \times \dfrac{x}{100} - 9000$

$\Rightarrow x = 85.7 \approx 86\%$

8. (a) Divide the 8 bags into the sets of 2, 3 and 3 bags respectively. First put 3 bags on each side of weighing balance.

Case I. If they come equal.

Then put 1 bag each on weighing balance from the set of 2 bags.

Case II. If they are not equal

Then which ever side is heavier select 2 bag and put 1 bag each on each side of weighing balance.

In two turns.

9. (a) $20x + 10(14 - x) = 230$

$20x + 140 - 10x = 230$

$10x = 90$

$x = 9$

Answer is $(14 - x) = (14 - 9) = 5$

10. (a) Option (d) is vague. Option (c) doesn't sum up the above passage. It merely states an insignificant fact stated in the passage. Option (b) wrongly compares men to animals. Such a development is vague. The entire paragraph suggests what is mentioned in option (a). Regimentation refers to the strict discipline of military groups.

2013

Q. 1 – 5 carry one mark each.

1. If $3 \le X \le 5$ and $8 \le Y \le 11$ then which of the following options is TRUE?

(a) $\frac{3}{5} \le \frac{X}{Y} \le \frac{8}{5}$ (b) $\frac{3}{11} \le \frac{X}{Y} \le \frac{5}{8}$

(c) $\frac{3}{11} \le \frac{X}{Y} \le \frac{8}{5}$ (d) $\frac{3}{5} \le \frac{X}{Y} \le \frac{8}{11}$

2. The Headmaster ______ to speak to you.
Which of the following options is incorrect to complete the above sentence?

(a) is wanting (b) wants

(c) want (d) was wanting

3. Mahatama Gandhi was known for his humility as

(a) he played an important role in humiliating exit of British from India.

(b) he worked for humanitarian causes.

(c) he displayed modesty in his interactions.

(d) he was a fine human being.

4. All engineering students (I) should learn mechanics, (II) mathematics and (III) how to do computation. (IV)

Which of the above underlined parts of the sentence is not appropriate?

(a) I (b) II

(c) III (d) IV

5. Select the pair that best expresses a relationship similar to that expressed in the pair :

water : pipe::

(a) cart : road (b) electricity : wire

(c) sea : beach (d) music : instrument

Q. 6 – 10 carry two marks each.

6. Velocity of an object fired directly in upward direction is given by $V = 80 - 32t$, where t (time) is in seconds. When will the velocity be between 32 m/sec and 64 m/sec ?

(a) (1, 3/2) (b) (1/2, 1)

(c) (1/2, 3/2) (d) (1, 3)

7. In a factory, two machines M1 and M2 manufacture 60% and 40% of the autocomponents respectively. Out of the total production, 2% of M1 and 3% of M2 are found to be defective. If a randomly drawn autocomponent from the combined lot is found defective, what is the probability that it was manufactured by M2 ?

(a) 0.35 (b) 0.45

(c) 0.5 (d) 0.4

8. Following table gives data on tourists from different countries visiting India in the year 2011.

Country	Number of Tourists
USA	2000
England	3500
Germany	1200
Italy	1100
Japan	2400
Australia	2300
France	1000

Which two countries contributed to the one third of the total number of tourists who visited India in 2011 ?

(a) USA and Japan

(b) USA and Australia

(c) England and France

(d) Japan and Australia

9. If $|-2X + 9| = 3$ then the possible value of $|-X| - X^2$ would be :

(a) 30 (b) –30

(c) –42 (d) 42

10. All professors are researchers
Some scientists are professors
Which of the given conclusions is logically valid and is inferred from the above arguments :

(a) All scientists are researchers

(b) All professors are scientists

(c) Some researchers are scientists

(d) No conclusion follows

ANSWERS

1. (b)	2. (b)	3. (c)	4. (d)	5. (b)	6. (c)	7. (c)
8. (c)	9. (b)	10. (c)				

EXPLANATIONS

1. (b) $3 \le x \le 5$

$8 \le y \le 11$

$\Rightarrow \frac{1}{8} \ge \frac{1}{y} \ge \frac{1}{11}$

$\Rightarrow \frac{1}{11} \le \frac{1}{y} \le \frac{1}{8}$

$\therefore \boxed{\frac{3}{11} \le \frac{x}{y} \le \frac{5}{8}}$

2. (b) The sentence is in present tense and the blank should maintain the tense structure. Additionally 'principal' is in singular form the verb should also be in the singular form. Hence, (b) is the correct answer.

3. (c) Mahatma Gandhi was known for his humility as he displayed modesty in his interactions.

4. (d) In part IV, the 'how to do' is redundant.

5. (b) Water flows through a pipe and similarly electricity flows through a wire.

6. (c) V = 80 – 32t

V 32m/sec $\Rightarrow t = \frac{8}{2}$

V = 64 m/sec $\rightarrow t = \frac{1}{2}$

$\therefore \boxed{t \in \left(\frac{1}{2}, \frac{3}{2}\right)}$

7. (c) Let P(M) denote the probability that the component is manufactured by Machine Mi & P (def) denote the probability that the component is defective.

$$P(M_2 | def) = \frac{P(M_2 \cap def)}{P(def)}$$

$$= \frac{0.4 \times 0.03}{(0.4 \times 0.3) + (0.6 \times .02)}$$

$$= 0.5$$

8. (c) Total No. of tourists = 13500.

$\frac{1}{3} \times 13500 = 4500$

Option (c) England (3500) + France(1000) = 4500

9. (b) $|-2x + 9| = 3$

$\Rightarrow +(2x + 9) = 3$

$\Rightarrow x = 3$

$-(-2x + 9) = 3$

$\Rightarrow x = 6$

$x = 3$

$\Rightarrow |-x| - x^2$

$\Rightarrow |-3| - (3)^2.$

$\Rightarrow 3 - 9 = -6$

$x = 6$

$\Rightarrow |-6| - 6^2$

$\Rightarrow 6 - 36 = -30$

Option (b)

10. (c) From clear logical deduction, (c) is the correct choice.

2014

Q. (1 – 5) Carry One Mark Each.

1. A student is required to demonstrate a high level of <u>comprehension</u> of the subject, especially in the social sciences.

   The word closest in meaning to <u>comprehension</u> is

   (*a*) understanding (*b*) meaning

   (*c*) concentration (*d*) stability

2. Choose the most appropriate word from the options given below to complete the following sentence.

   One of his biggest ______ was his ability to forgive.

   (*a*) vice (*b*) virtues

   (*c*) choices (*d*) strength

3. Rajan was not happy that Sajan decided to do the project on his own. On observing his unhappiness, Sajan explained to Rajan that he preferred to work independently.

   Which one of the statements below is logically valid and can be inferred from the above sentences?

   (*a*) Rajan has decided to work only in a group.

   (*b*) Rajan and Sajan were formed into a group against their wishes.

   (*c*) Sajan had decided to give in to Rajan's request to work with him.

   (*d*) Rajan had believed that Sajan and he would be working together.

4. If $y = 5x^2 + 3$, then the tangent at $x = 0$, $y = 3$

   (*a*) passes through $x = 0$, $y = 0$

   (*b*) has a slope of +1

   (*c*) is parallel to the *x*-axis

   (*d*) has a slope of –1

5. A foundry has a fixed daily cost of ₹ 50,000 whenever it operates and a variable cost of ₹ 800 Q, where Q is the daily production in tonnes. What is the cost of production in ₹ per tonne for a daily production of 100 tonnes ?

Q. (6 – 10) Carry Two Marks Each.

6. Find the odd one in the following group: ALRVX, EPVZB, ITZDF, OYEIK

   (*a*) ALRVX (*b*) EPVZB

   (*c*) ITZDF (*d*) OYEIK

7. Anuj, Bhola, Chandan, Dilip, Eswar and Faisal live on different floors in a six-storied building (the ground floor is numbered 1, the floor above it 2, and so on). Anuj lives on an even-numbered floor. Bhola does not live on an odd numbered floor. Chandan does not live on any of the floors below Faisal's floor. Dilip does not live on floor number 2. Eswar does not live on a floor immediately above or immediately below Bhola. Faisal lives three floors above Dilip. Which of the following floor-person combinations is correct?

	Anuj	Bhola	Chandan	Dilip	Eswar	Faisal
(A)	6	2	5	1	3	4
(B)	2	6	5	1	3	4
(C)	4	2	6	3	1	5
(D)	2	4	6	1	3	5

8. The smallest angle of a triangle is equal to two thirds of the smallest angle of a quadrilateral. The ratio between the angles of the quadrilateral is 3 : 4 : 5 : 6. The largest angle of the triangle is twice its smallest angle. What is the sum, in degrees, of the second largest angle of the triangle and the largest angle of the quadrilateral?

9. One percent of the people of country X are taller than 6 ft. Two percent of the people of country Y are taller than 6 ft. There are thrice as many people in country X as in country Y. Taking both countries together, what is the percentage of people taller than 6 ft?

   (*a*) 3.0 (*b*) 2.5

   (*c*) 1.5 (*d*) 1.25

10. The monthly rainfall chart based on 50 years of rainfall in Agra is shown in the following figure. Which of the following are true? (*k* percentile is the value such that *k* percent of the data fall below that value)

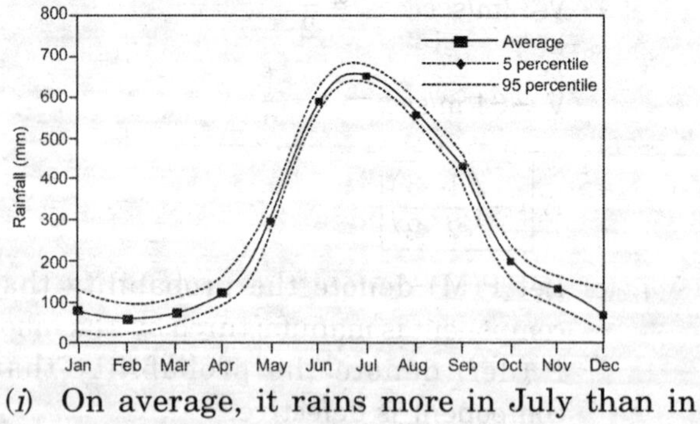

   (*i*) On average, it rains more in July than in December

   (*ii*) Every year, the amount of rainfall in August is more than that in January

   (*iii*) July rainfall can be estimated with better confidence than February rainfall

   (*iv*) In August, there is at least 500 mm of rainfall

   (*a*) (*i*) and (*ii*) (*b*) (*i*) and (*iii*)

   (*c*) (*ii*) and (*iii*) (*d*) (*iii*) and (*iv*)

ANSWERS

1. (a)	**2.** (b)	**3.** (d)	**4.** (c)	**5.** (1300 to 1300)	**6.** (d)	**7.** (b)
8. (180 to 180)	**9.** (d)	**10.** (b)				

EXPLANATIONS

1. (a) To comprehend is to understand making A the correct answer.

2. (b) The blank requires a plural form making B the correct choice. D is incorrect as it is in the singular form.

3. (d) By clear deduction, D follows.

4. (c) $y = 5x^2 + 3$

$\frac{dy}{dx} = 10x$, at $x = 0$

$\frac{dy}{dx} = 0$

therefore are lope at x = 0 is zero, therefore the tangent passes through x = 0, y = 3 & is parallel to the x-axis.

5. (1300 to 1300)

The total production cost will be:-

50,000 + 800 Q where Q = 100

$\Rightarrow$ 5000 + 80,000 = 130,000

So, person production cost $= \frac{1,30,000}{100} = 1300$

6. (d) $A \xrightarrow{+4} E \xrightarrow{+4} I \xrightarrow{+6} O$

$L \longrightarrow P \longrightarrow T \xrightarrow{+5} Y$

$R \longrightarrow V \longrightarrow Z \xrightarrow{+5} E$

$V \longrightarrow Z \longrightarrow D \xrightarrow{+5} I$

$X \longrightarrow B \longrightarrow F \xrightarrow{+5} K$

(D) YOKE

7. (b) Anuj, Bhola, Chandan and Dilip satisfy their respective conditions, but, Eshwar does not live on a floor immediately above or immediately below Bhola. Therefore, only in option B, Eshwar is following this condition.

8. (180 to 180)

Let the angles of Quadrilaterals be, 3x, 4x, 5x, 6x.

$\therefore\ 3x + 4x + 5x + 6x = 18x = 360$

$\Rightarrow\ x = 20$

$\therefore$ angles are 60, 80, 100, 120.

smallest angle of a triangle $= \frac{2}{3} \times 60 = 40$

largest angle $= 2 \times 40 = 80$.

third angle = 60.

$\therefore$ required sum = 120 + 60 = 180°.

9. (d) Let the number of people in country y be p. Then country x has '3p' people.

Also, is country y, $\frac{2p}{100}$, people are above 6ft and in country x, $\frac{3p}{100}$, people are above 6ft.

Total people in both countries = p + 3p = 4p

and people above 6ft $= \frac{2p}{100} + \frac{3p}{100} = \frac{5p}{100}$

$(4p) \times \frac{q}{100} = \frac{6p}{100}$

$9 = \frac{5p}{100} \times \frac{100}{40} = \frac{5}{4} = 1.25\%$

10. (b) (i) On average, it rains rnose in July than in December.

(ii) July rainfall confidence is better than Febraury rainfall confidence.

2015

Q. (1 – 5) Carry One Mark Each.

1. Choose the most appropriate word from the options given below to complete the following sentence.

The principal presented the chief guest with a ________ as token of appreciation.

(*a*) momento (*b*) memento

(*c*) momentum (*d*) moment

2. Choose the appropriate word/phrase, out of the four options given below, to complete the following sentence:

Frogs ________.

(*a*) croak (*b*) roar

(*c*) hiss (*d*) patter

3. Choose the word most similar in meaning to the given word:

Educe

(*a*) Exert (*b*) Educate

(*c*) Extract (*d*) Extend

4. Operators □ , ◇ and → are defined by:

$a \square b = \frac{a-b}{a+b}$; $a \diamond b = \frac{a+b}{a-b}$; $a \rightarrow b = ab$.

Find the value of $(66 \square 6) \rightarrow (66 \diamond 6)$.

(*a*) – 2 (*b*) – 1

(*c*) 1 (*d*) 2

5. If $\log_x (5/7) = -1/3$, then the value of x is

(*a*) 343/125 (*b*) 125/343

(*c*) – 25/49 (*d*) – 49/25

Q. (6 – 10) Carry Two Marks Each.

6. The following question presents a sentence, part of which is underlined. Beneath the sentence you find four ways of phrasing the underlined part. Following the requirements of the standard written English, select the answer that produces the most effective sentence.

Tuberculosis, together with its effects, <u>ranks one of the leading causes of death</u> in India.

(*a*) ranks as one of the leading causes of death

(*b*) rank as one of the leading causes of death

(*c*) has the rank of one of the leading causes of death

(*d*) are one of the leading causes of death

7. Read the following paragraph and choose the correct statement.

Climate change has reduced human security and threatened human well being. An ignored reality of human progress is that human security largely depends upon environmental security. But on the contrary, human progress seems contradictory to environmental security. To keep up both at the required level is a challenge to be addressed by one and all. One of the ways to curb the climate change may be suitable scientific mnnovations, while the other may be the Gandhian perspective on small scale progress with focus on sustainability.

(*a*) Human progress and security are positively associated with environmental security.

(*b*) Human progress is contradictory to environmental security.

(*c*) Human security is contradictory to environmental security.

(*d*) Human progress depends upon environmental security.

8. Fill in the missing value

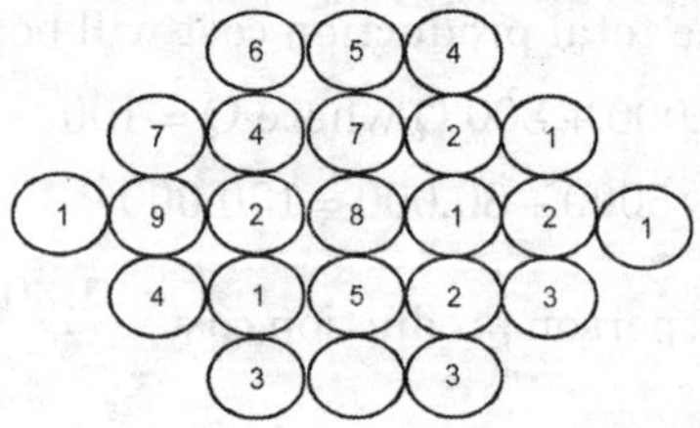

9. A cube of side 3 units is formed using a set of smaller cubes of side 1 unit. Find the proportion of the number of faces of the smaller cubes visible to those which are NOT visible.

(*a*) 1 : 4 (*b*) 1 : 3

(*c*) 1 : 2 (*d*) 2 : 3

10. Humpty Dumpty sits on a wall every day while having lunch. The wall sometimes breaks. A person sitting on the wall falls if the wall breaks. Which one of the statements below is logically valid and can be inferred from the above sentences?

(*a*) Humpty Dumpty always falls while having lunch

(*b*) Humpty Dumpty does not fall sometimes while having lunch

(*c*) Humpty Dumpty never falls during dinner

(*d*) When Humpty Dumpty does not sit on the wall, the wall does not break

ANSWERS

1. (b)	2. (a)	3. (c)	4. (c)	5. (a)	6. (a)	7. (b)
8. (3)	9. (c)	10. (b)				

EXPLANATIONS

1. (b) 'Memento' means an object kept as a reminder of a person or event.

2. (a) The sound that frogs make is known as 'croaking'.

3. (c) 'Educe' means to infer.

4. (c) $(66 \,\square\, 6) \longrightarrow (66 \diamond 6)$

$$\Rightarrow \left(\frac{66-6}{66+6}\right)\left(\frac{66+6}{66-6}\right) = 1$$

5. (a) $\log_x\left(\frac{5}{7}\right) = -\frac{1}{3}$

$$\log\frac{5}{7} = (\log x)\left(-\frac{1}{3}\right)$$

$$\log\frac{5}{7} = \log x^{\frac{-1}{3}}$$

$$\Rightarrow \left(\frac{5}{7}\right)^{-3} = x$$

$$\Rightarrow \boxed{\frac{343}{125} = x}$$

6. (a) Only A is written correctly. In B 'rank' cause subject-verb disagreement, in C there is tense error.

7. (b) Only B can follow from the given paragraph.

8. (3) Middle value first row $= \frac{6+4}{2} = 5$

Middle value of 2nd row $= \frac{7+4+2+1}{2} = 7$

Similarly, for last row $= \frac{3+3}{2} = 3$ is the missing value

9. (c) Small cubes are of 1 unit, therefore,

total cubes = 9 × 3 = 27

total faces = 27 × 6 = 162

total faces visible = 9 × 6 = 54

faces that are not visible = 162 – 54 = 108

$\therefore$ Required Ratio $= \frac{54}{108} = \frac{1}{2}$

10. (b) Only B follows logically from the given premises.

2016

Q. 1 – 5) : Carry One Mark Each.

1. The chairman requested the aggrieved shareholders to _______________ him.

   (*a*) bare with (*b*) bore with

   (*c*) bear with (*d*) bare

2. Identify the correct spelling out of the given options:

   (*a*) Managable (*b*) Manageable

   (*c*) Mangaeble (*d*) Managible

3. Pick the odd one out in the following:

   13, 23, 33, 43, 53

   (*a*) 23 (*b*) 33

   (*c*) 43 (*d*) 53

4. R2D2 is a robot. R2D2 can repair aeroplanes. No other robot can repair aeroplanes.

   Which of the following can be logically inferred from the above statements?

   (*a*) R2D2 is a robot which can only repair aeroplanes.

   (*b*) R2D2 is the only robot which can repair aeroplanes.

   (*c*) R2D2 is a robot which can repair only aeroplanes.

   (*d*) Only R2D2 is a robot.

5. If $|9y - 6| = 3$, then $y^2 - 4y/3$ is _____.

   (*a*) 0 (*b*) + 1/3

   (*c*) – 1/3 (*d*) undefined

(Q. 6 – 10) : Carry Two Marks Each.

6. The following graph represents the installed capacity for cement production (in tonnes) and the actual production (in tonnes) of nine cement plants of a cement company. Capacity utilization of a plant is defined as ratio of actual production of cement to installed capacity. A plant with installed capacity of at least 200 tonnes is called a large plant and a plant with lesser capacity is called a small plant. The difference between total production of large plants and small plants, in tonnes is _________.

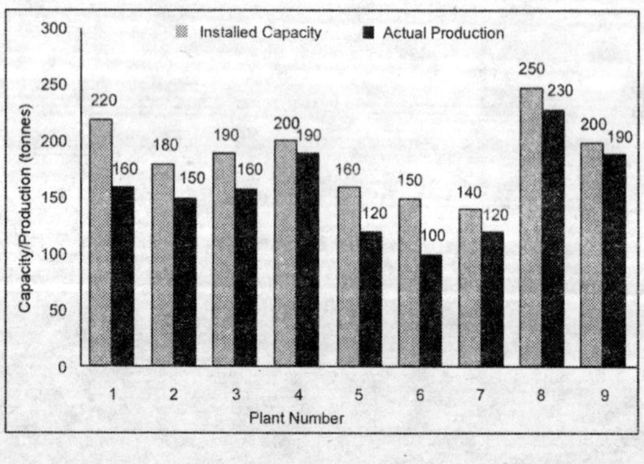

7. A poll of students appearing for masters in engineering indicated that 60 % of the students believed that mechanical engineering is a profession unsuitable for women. A research study on women with masters or higher degrees in mechanical engineering found that 99 % of such women were successful in their professions.

   Which of the following can be logically inferred from the above paragraph?

   (*a*) Many students have misconceptions regarding various engineering disciplines.

   (*b*) Men with advanced degrees in mechanical engineering believe women are well suited to be mechanical engineers.

   (*c*) Mechanical engineering is a profession well suited for women with masters or higher degrees in mechanical engineering.

   (*d*) The number of women pursuing higher degrees in mechanical engineering is small.

8. Sourya committee had proposed the establishment of Sourya Institutes of Technology (SITs) in line with Indian Institutes of Technology (IITs) to cater to the technological and industrial needs of a developing country.

   Which of the following can be logically inferred from the above sentence?

   Based on the proposal,

   (i) In the initial years, SIT students will get degrees from IIT.

   (ii) SITs will have a distinct national objective.

   (iii) SIT like institutions can only be established in consultation with IIT.

   (iv) SITs will serve technological needs of a developing country.

   (*a*) (iii) and (iv) only

   (*b*) (i) and (iv) only

   (*c*) (ii) and (iv) only

   (*d*) (ii) and (iii) only

9. Shaquille O' Neal is a 60% career free throw shooter, meaning that he successfully makes 60 free throws out of 100 attempts on average. What is the probability that he will successfully make <u>exactly</u> 6 free throws in 10 attempts?

   (*a*) 0.2508 (*b*) 0.2816

   (*c*) 0.2934 (*d*) 0.6000

10. The numeral in the units position of $211^{870} + 146^{127} \times 3^{424}$ is _____.

ANSWERS

1. (c) **2.** (b) **3.** (b) **4.** (b) **5.** (c) **6.** (120 to 120) **7.** (c)

8. (c) **9.** (a) **10.** (7 to 7)

EXPLANATIONS

1. (c) "Bear with" means "to have patience with."

2. (b) The other spellings are wrong.

3. (b) All are prime numbers except 33.33 is divisible by 3 & 11.

4. (b) Since no other robot except R2D2 robot can repair aeroplanes. Therefore, option (b) is correct.

5. (c) $|9y - 6| = 3$

Possible values of y are $1 \& \frac{1}{3}$.

For y = 1 : $y^2 - \frac{4y}{3}$

$\Rightarrow 1 - \frac{4}{3} = -\frac{1}{3}$

For $y = \frac{1}{3}$: $y^2 - \frac{4y}{3}$

$\frac{1}{9} - \frac{4}{3} \times \frac{1}{3} = \frac{-1}{3}$

6. (120 to 120)

Large plant production = 160 + 190 + 230 + 190 = 770 ...(1)

Small plant production = 650 ...(ii)

Difference = (i) – (ii) = 770 – 550 = 120

7. (c) From the last line, option C can be inferred.

8. (c) SITs will have a national character like IITs. Statement iv can be inferred from what is stated in the given premise.

9. (a) Here we have,

n = number of trials,

p = success probability = 0.6,

q = failure probability = 0.4 and

k = number of shots hit = 6

p (6 hits in 10 attempts) = $C(10, 6)p^6q^4$

$= \frac{10!}{6!4!}(0.6)^6(0.4)^4$

$= 210 \times .046656 \times .0256$

$= .2508$

10. (7 to 7)

$211^{870} + 146^{127} \times 3^{424}$ unit digits

$= 1^{870} + 6^{127} + 3^{424}$ units digit

$= 1 + 6 \times 1 = 7$

2017

Q. 1 – Q. 5 carry one mark each.

1. The event would have been successful if you _______ able to come.

(a) are (b) had been

(c) have been (d) would have been

2. There was no doubt that their work was <u>thorough</u>.

Which of the words below is closest in meaning to the underlined word above?

(a) pretty (b) complete

(c) sloppy (d) haphazard

3. Four cards lie on a table. Each card has a number printed on one side and a colour on the other. The faces visible on the cards are 2, 3 red, and blue.

Proposition: If a card has an even value on one side, then its opposite face is red.

The cards which MUST be turned over to verify the above proposition are

(a) 2, red (b) 2, 3 red

(c) 2, blue (d) 2, red, blue

4. What is the value of x when

$$81 \times \left(\frac{16}{25}\right)^{x+2} \div \left(\frac{3}{5}\right)^{2x+4} = 144?$$

(a) 1 (b) –1

(c) –2 (d) Cannot be determined

5. Two dice are thrown simultaneously. The probability that the product of the numbers appearing on the top faces of the dice is a perfect square is

(a) 1/9 (b) 2/9

(c) 1/3 (d) 4/9

Q. 6 – Q. 10 carry two marks each

6. Bhaichung was observing the pattern of people entering and leaving a car service centre. There was a single window where customers were being served. He saw that people inevitably came out of the centre in the order that they went in. However, the time they spent inside seemed to vary a lot:

some people came out in a matter of minutes while for others it took much longer.

From this, what can one could?

(a) The centre operates on a first-come-first-served basis, but with variable service times, depending on specific customer needs.

(b) Customers were served in an arbitrary order, since they took varying amount of time for service completion in the centre.

(c) Since some people came out within a few minutes of entering the centre, the system is likely to operate on a last-come-first-served basis.

(d) Entering the centre early ensured that one would have shorter service times and most people attempted to do this.

7. A map shows the elevations of Darjeeling, Gangtok, Kalimpong, Pelling, and Siliguri. Kalimpong is at a lower elevation than Gangtok. Pelling is at a lower elevation than Gangtok. Pelling is at a higher elevation than Siliguri. Darjeeling is at a higher elevation than Gangtok. Which of the following statements can be inferred from the paragraph above?

i. Pelling is at a higher elevation than Kalimpong

ii. Kalimpong is at a lower elevation than Darjeeling

iii. Kalimpong is at a higher elevation than Siliguri

iv. Siliguri is at a lower elevation than Gangtok

(a) Only ii

(b) Only ii and iii

(c) Only ii and iv

(d) Only iii and iv

8. P, Q, R, S, T and U are seated around a circular table. R is seated two places to the right of Q. P is seated three places to the left of R. S is seated opposite U. If P and U now switch seats, which of the following must necessarily be true?

(a) P is immediately to the right of R

(b) T is immediately to the left of P

(c) T is immediately to the left of P or P is immediately to the right of Q

(d) U is immediately to the right of R or P is immediately to the left of T

9. Budhan covers a distance of 19 km is 2 hours cycling one fourth of the time and walking the rest. The next day he cycles (at the same speed as before) for half the time and walks the rest (at the same speed as before) and covers 26 km in 2 hours. The speed in km/h at which Budhan walks is

(a) 1

(b) 4

(c) 5

(d) 6

10. The points in the graph below represent the halts of a lift for durations of 1 minute, over a period of 1 hour.

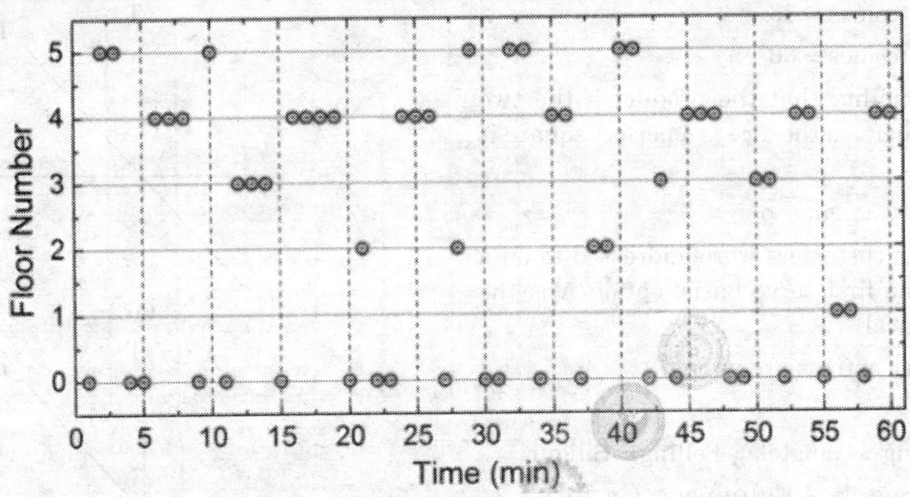

Which of the following statements are correct?

i. The elevator never moves directly from any non-ground floor to another non-ground floor over the one hour period.

ii. The elevator stays on the fourth floor for the longest duration over the one hour period

(*a*) Only i
(*b*) Only ii
(*c*) Both i and ii
(*d*) Neither i nor ii

ANSWERS

1. (b)	**2.** (b)	**3.** (c)	**4.** (b)	**5.** (b)	**6.** (a)	**7.** (c)
8. (c)	**9.** (d)	**10.** (d)				

EXPLANATIONS

1. (b) The past perfect tense should be used in the sentence.

2. (b) "Thorough" means complete.

3. (*) The faces visible on the cards are 2, 3, red and blue

If the card has an even value on one side, then its opposite face is red.

∴ The cards which must be turned over to verify the above proposition are 2, blue.

4. (b) $81 \times \left(\frac{16}{25}\right)^{x+2} \div \left(\frac{3}{5}\right)^{2x+4} = 144$

$$81 \times \frac{2^{4x+8}}{5^{2x+4}} \times \frac{5^{2x+4}}{3^{2x+4}} = 144$$

$$3^4 \times \frac{2^{4x+8}}{3^{2x+4}} = 2^4 \times 3^2$$

$$\frac{2^{4x+8}}{3^{2x+4}} = \frac{2^4}{3^2}$$

$$\frac{2^{4x+4}}{3^{2x+2}} = 1$$

$$\left(\frac{4}{3}\right)^{2x+2} = \left(\frac{4}{3}\right)^0$$

$$\Rightarrow 2x + 2 = 0$$

$$x = -1$$

5. (b) **11** 21 31 41 51 61
12 **22** 32 42 52 62
13 23 **33** 43 53 63
14 24 34 **44** 54 64
15 25 35 45 **55** 65
16 26 36 46 56 **66**

No. of favourable outcomes = 8

Total outcomes = 36

Total outcomes = 36

$\therefore$ Probability that the product if the two numbers on the dice is a perfect square is

$$\frac{8}{36} = \frac{2}{9}$$

6. (a) Since the customers were addressed to on a first come first serve basis, option A is the most logical.

7. (c) The other options are wrong.

Alternate

Darjeeling > Gangtok > Pelling > Siliguri and Gangtok > Kalimpong (in terms of elevation)

$\Rightarrow$ The elevation of Kalimpong can be greater than that of feeling or between feeling and Siliguri or lower than that of Siliguri

Only (ii) and (iv) statements can be inferred.

8. (c)

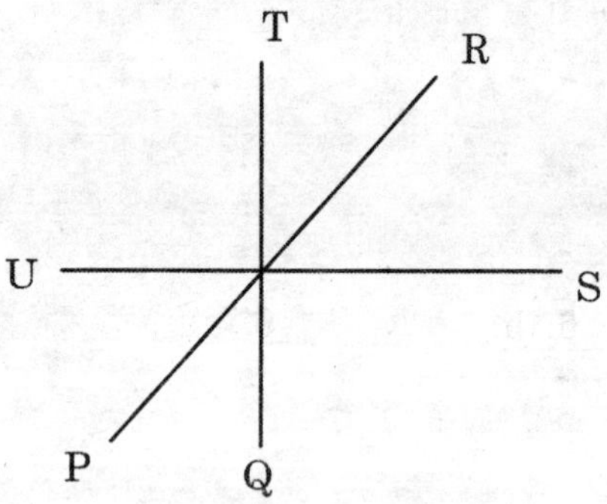

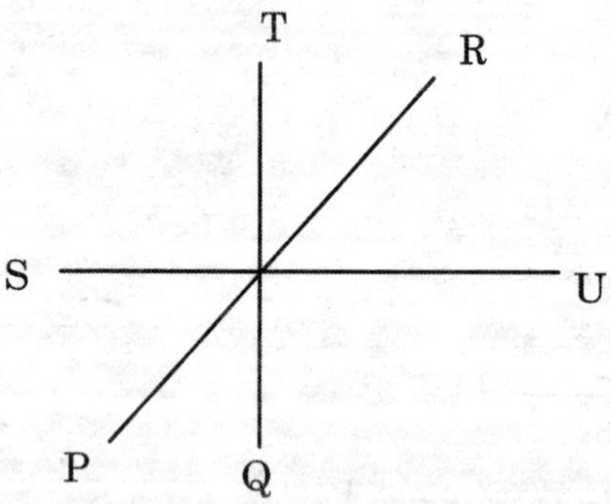

It P and U nor with seats,

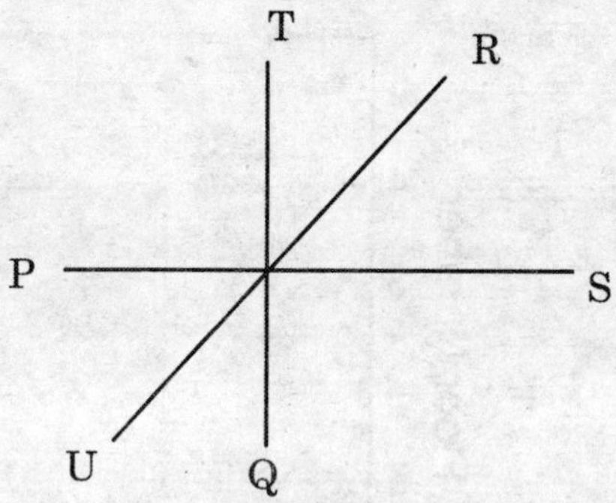

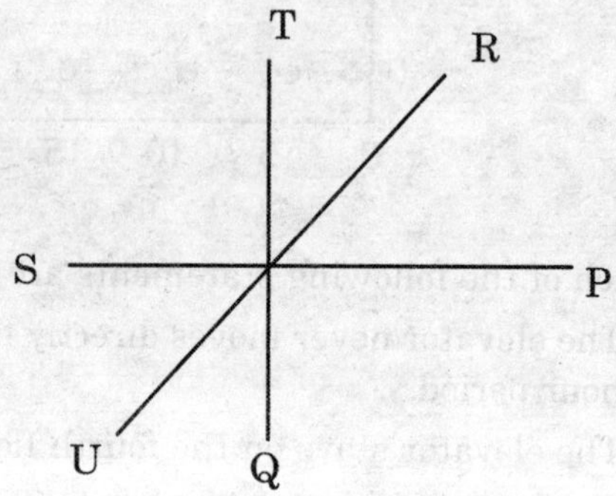

Clearly option C must necessarily be true.

9. (d) Let the speed at which Budhan cycles and walls be s_c and s_w respectively.

$$\therefore\ 19 = s_c \times \frac{1}{2} + \frac{3}{2} \times s_w$$

$$s_c + 3s_w = 38 \quad ...(i)$$

$$s_c \times 1 + s_w \times 1 = 26$$

$$s_c + s_w = 26 \quad ...(ii)$$

On solving both eq. (i) and (ii)

$s_c = 20$ km/hr, $s_w = 6$ km/hr.

10. (d) Statement (i) is not correct as between 25 and 30 min it moves directly from 2nd floor to 5th floor once.

Statement (ii) is also not correct as it stays on the ground floor for the longest duration.

$\therefore$ Neither (i) nor (ii) is correct.

2018

Q. 1 – Q. 5 carry one mark each.

1. Going by the ______ that many hands make light work, the school ______ involved all the students in the task."

(*a*) principle, principal (*b*) principal, principle

(*c*) principle, principle (*d*) principal, principal

2. "Her ____ should not be confused with miserliness; she is ever willing to assist those in need."

The word that best fills the blank in the above sentence is

(*a*) cleanliness (*b*) punctuality

(*c*) frugality (*d*) greatness

3. Seven machines take 7 minutes to make 7 identical toys. At the same rate, how many minutes would it take for 100 machines to make 100 toys?

(*a*) 1 (*b*) 7

(*c*) 100 (*d*) 700

4. A rectangle becomes a square when its length and breadth are reduced by 10 m and 5 m, respectively, during this process, the rectangle loses 650 m^2 of area. What is the area of the original rectangle in square meters?

(*a*) 1125 (*b*) 2250

(*c*) 2924 (*d*) 4500

5. A number consists of two digits. The sum of the digits is 9. If 45 is subtracted from the number, its digits are interchanged. What is the number?

(*a*) 63 (*b*) 72

(*c*) 81 (*d*) 90

Q. 6 – Q. 10 carry two marks each

6. For integers a, b and c, what would be the minimum and maximum values respectively of a + b + c if $\log|a| + \log|b| + \log|c| = 0$

(*a*) –3 and 3 (*b*) –1 and 1

(*c*) –1 and 3 (*d*) 1 and 3

7. Given that a and b are integers and $a + a^2b^3$ is odd, which one of the following statements is correct?

(*a*) a and b are both odd

(*b*) a and b are both even

(*c*) a is even and b is odd

(*d*) a is odd and b is even

8. From the time the front of train enters a platform, it takes 25 seconds for the back of the train to leave the platform, while travelling at a constant speed of 54 km/h. At the same speed, it takes 14 seconds to pass a man running at 9 km/h in the same direction as the train. What is the length of the train and that of the platform in meters, respectively?

(*a*) 210 and 140

(*b*) 162.5 and 187.5

(*c*) 245 and 130

(*d*) 175 and 200

9. Which of the following functions describe the graph shown in the below figure.

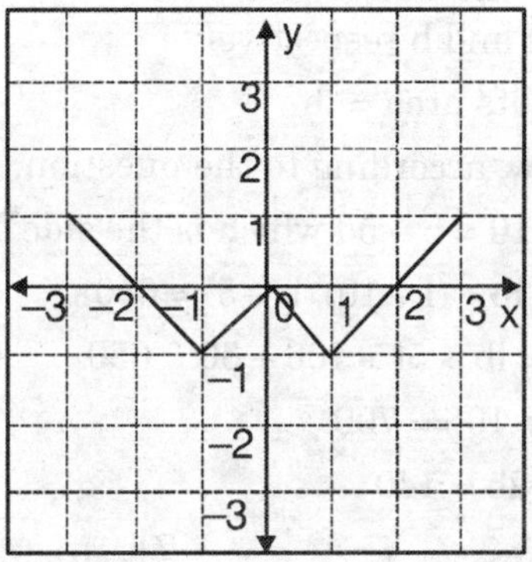

(*a*) $y = ||x| + 1| - 2$

(*b*) $y = ||x| - 1| - 1$

(*c*) $y = ||x| + 1| - 1$

(*d*) $y = ||x - 1| - 1|$

10. Consider the following three statements:

(i) Some roses are red

(ii) All red flowers fade quickly

(iii) Some roses fade quickly

Which of the following statements can be logically inferred from the above statements?

(*a*) If (i) is true and (ii) is false, then (iii) is false.

(*b*) If (i) is true and (ii) is false, then (iii) is true.

(*c*) If (i) and (ii) are true, then (iii) is true.

(*d*) If (i) and (ii) are false, then (iii) is false.

ANSWERS

1. (a)	2. (c)	3. (b)	4. (b)	5. (b)	6. (a)	7. (d)
8. (d)	9. (b)	10. (c)				

EXPLANATIONS

1. (a) "Principle" means "ethics."

2. (c) "Frugality" refers to "thriftiness."

3. (b) $\frac{M_1 \times H_1}{\omega_1} = \frac{M_2 \times H_2}{\omega_2}$

$\frac{7 \times 7}{7} = \frac{m \times 100}{100} = 7$

4. (b) Let the length and breadth of the rectangle be l and b respectively.

So, its area = lb

Now, according to the question,

l – 10 = b – 5d which is the side of the square

$\therefore$ lb – (l – 10) (b – 5) = 650

lb = lb + 5l + 106 – 50 = 650

5l + 106 = 700

l + 2b = 140

b + 5 + 2b = 140 $\left(\begin{array}{l} l-10=b-5 \\ \therefore l=b+5 \end{array}\right)$

3b = 135

b = 42 m

l = 50 m

So, area of the original rectangle

$= 50 \times 45 = 2250 m^2$

5. (b) Let the number of xy

x + y = 9

10 x + y – 45 = 10y + x

9x = 45 + 9y

x = 5 + y

x – y = 5 ...(i)

and x + y = 9 ...(ii)

on solving eq (i) and (ii)

x = 7

y = 2

$\therefore$ Number is 72.

6. (a) If log | a | + log | b | + log | c | = 0

$\Rightarrow$ log | a | = log | b | = log | c | =0

$\Rightarrow$ a = b = c ± 1

$\therefore$ Minimum value of a + b + c = –1 –1 –1 = –3

Maximum value of a + b + c = 1 + 1 + 1 = 3.

7. (d) $a + a^2 b^3$ = odd.

When a is odd and 6 is even then only at $a^2 b^3$ will give the odd value.

Fore e.g.

If a = 3, b = 2

$a + a^2 b^3 = 3 + 9 \times 8 = 75$ which is odd.

8. (d) Let the lengths of the train and platform be l_t and l_p respectively.

$\therefore l_t + l_p = 54 \times \frac{5}{18} \times 25$

$l_t + l_p = 375$ m

$l_t = (54 - 9) \times \frac{5}{18} \times 14$

$45 \times \frac{5}{18} \times 14$

$l_t = 175$ m

$l_p = 200$ m

9. (b) Clearly the function

| | x | –1 | –1 describes the given graph.

At x = 0,	y = 0
x = 1,	y = –1
x = 2,	y = 0
x = –,	y = –1

10. (c) A comes close but if ii is false, we cannot get any information about iii.

XE-A

ENGINEERING MATHEMATICS

Syllabus

Section 1: Linear Algebra

Algebra of matrices; Inverse and rank of a matrix; System of linear equations; Symmetric, skew-symmetric and orthogonal matrices; Determinants; Eigenvalues and eigenvectors; Diagonalisation of matrices; Cayley-Hamilton Theorem.

Section 2: Calculus

Functions of single variable: Limit, continuity and differentiability; Mean value theorems; Indeterminate forms and L'Hospital's rule; Maxima and minima; Taylor's theorem; Fundamental theorem and mean value-theorems of integral calculus; Evaluation of definite and improper integrals; Applications of definite integrals to evaluate areas and volumes.

Functions of two variables: Limit, continuity and partial derivatives; Directional derivative; Total derivative; Tangent plane and normal line; Maxima, minima and saddle points; Method of Lagrange multipliers; Double and triple integrals, and their applications.

Sequence and series: Convergence of sequence and series; Tests for convergence; Power series; Taylor's series; Fourier Series; Half range sine and cosine series.

Section 3: Vector Calculus

Gradient, divergence and curl; Line and surface integrals; Green's theorem, Stokes theorem and Gauss divergence theorem (without proofs).

Section 4: Complex variables

Analytic functions; Cauchy-Riemann equations; Line integral, Cauchy's integral theorem and integral formula (without proof); Taylor's series and Laurent series; Residue theorem (without proof) and its applications.

Section 5: Ordinary Differential Equations

First order equations (linear and nonlinear); Higher order linear differential equations with constant coefficients; Second order linear differential equations with variable coefficients; Method of variation of parameters; Cauchy-Euler equation; Power series solutions; Legendre polynomials, Bessel functions of the first kind and their properties.

Section 6: Partial Differential Equations

Classification of second order linear partial differential equations; Method of separation of variables; Laplace equation; Solutions of one dimensional heat and wave equations.

Section 7: Probability and Statistics

Axioms of probability; Conditional probability; Bayes' Theorem; Discrete and continuous random variables: Binomial, Poisson and normal distributions; Correlation and linear regression.

Section 8: Numerical Methods

Solution of systems of linear equations using LU decomposition, Gauss elimination and Gauss-Seidel methods; Lagrange and Newton's interpolations, Solution of polynomial and transcendental equations by Newton-Raphson method; Numerical integration by trapezoidal rule, Simpson's rule and Gaussian quadrature rule; Numerical solutions of first order differential equations by Euler's method and 4th order Runge-Kutta method.

Subject-wise Analysis

XE -A : Engineering Mathematics

Year	1 mark Questions	2 marks Questions	Total Marks
2009	4	8	20
2010	7	4	15
2011	7	4	15
2012	7	4	15
2013	7	4	15
2014	7	4	15
2015	7	4	15
2016	7	4	15
2017	7	4	15
2018	7	4	15

2009

Q. 1 – 4 carry one mark each

1. Let A and B be two similar square matrices of order two. If 1 and – 2 are the eigenvalues of A, the Trace of B is

(*a*) – 2 (*b*) – 1
(*c*) 1 (*d*) 2

2. The root of $ax + b = 0$ (a, b constants), can be found by the Newton-Raphson method with a minimum of

(*a*) 1 iteration
(*b*) 2 iterations
(*c*) 3 iterations
(*d*) an undeterminable number of iterations

3. The solution $u(x, t)$ of the one-dimensional heat equation,

$$\frac{\partial u}{\partial t} = c^2 \frac{\partial^2 u}{\partial x^2},\ x \in \mathbb{R}$$

with a Gaussian initial condition,

(*a*) travels with finite constant wave-speed
(*b*) travels with finite variable wave-speed
(*c*) spreads in both directions, with the magnitude of the peak increasing with time
(*d*) spreads in both directions, with the magnitude of the peak decreasing with time

4. Let C be the boundary of the square given by $0 \le x \le 1, 0 \le y \le 1$.

Then $\oint_C (xdy - ydx)$ equals

(*a*) – 2
(*b*) 0
(*a*) 1
(*d*) 2

Q. 5 –12 carry two marks each

5. Let the eigenvalues of a square matrix A of order two be 1 and 2. The corresponding eigen vectors are $\begin{pmatrix} 0.6 \\ 0.8 \end{pmatrix}$ and $\begin{pmatrix} 0.8 \\ -0.6 \end{pmatrix}$, respectively.

Then, the element A(2,2) is

(*a*) – 0.48
(*b*) 0.48
(*c*) 1.36
(*d*) 1.64

6. Let $y_1(x)$ and $y_2(x)$ be two linearly independent solutions of $\frac{d^2y}{dx^2} + \frac{6}{x}\frac{dy}{dx} + q(x)y = 0,\ \lambda \in (1,3)$ where $q(x)$ is a continuous function in (1,3). If the Wronskian of $y_1(x)$ and $y_2(x)$, at $x = 1$, denoted by $w(y_1, y_2)$ (1), is 1, then $w(y_1, y_2)$ (2) is

(*a*) $\frac{1}{2^6}$
(*b*) $\frac{1}{2^3}$
(*c*) $\frac{1}{2}$
(*d*) 2

7. Simpson's 1/3 rule applied to $\int_{-1}^{1} (3x^2 + 5)dx$, with subinterval $h = 1$, will give

(*a*) the exact result
(*b*) error between 0.01% to 0.1%
(*c*) error between 0.1% to 1.0%
(*d*) error > 1.0%

8. The probability that a six-sided dice is thrown n times without giving a '6' even once, is

(*a*) $\left(\frac{5}{6}\right)^n$ (*b*) $\frac{5}{6}\left(\frac{1}{6}\right)^n$
(*c*) $\frac{(n-1)!}{n!}\left(\frac{1}{6}\right)^n$ (*d*) $1 - \frac{(n-1)!}{n!}\left(\frac{5}{6}\right)^n$

9. If a complex function $f(z) = u(x, y) + i\ v(x, y)$ is analytic, then

(*a*) $i\frac{\partial u}{\partial x} + \frac{\partial v}{\partial x} = \frac{\partial u}{\partial y} + \frac{\partial v}{\partial y}$
(*b*) $i\frac{\partial u}{\partial x} + \frac{\partial v}{\partial x} = -\frac{\partial u}{\partial y} - i\frac{\partial v}{\partial y}$
(*c*) $\frac{\partial u}{\partial x} + i\frac{\partial v}{\partial x} = -i\frac{\partial u}{\partial y} + \frac{\partial v}{\partial y}$
(*d*) $\frac{\partial u}{\partial x} + i\frac{\partial v}{\partial x} = i\frac{\partial u}{\partial y} + \frac{\partial v}{\partial y}$

10. Let $\vec{u} = \omega y\hat{i} + \omega x\hat{j}$ and $\vec{v} = \omega z\hat{j} - \omega y\hat{k}$ be two given vectors, where ω is a constant. Then $\text{div}(\vec{u} \times \vec{v})$ equals

(*a*) 0
(*b*) $2\omega^2 y$
(*c*) $4\omega^2 y$
(*d*) $-4\omega^2 y$

11. The infinite series

$$\sum_{m=1}^{\infty} \frac{(-1)^m}{(1+x^2)^m} \text{ is}$$

(*a*) Divergent for all x

(*b*) Convergent only for $x \geq 1$

(*c*) Convergent for all x

(*d*) Divergent only for $-1 \leq x \leq 1$

12. Let $f(x)$ be continuous and satisfy $m \leq f(x) \leq M$ in $1 \leq x \leq 10$. Then

$$\mu = \frac{\int_1^{10} (f(x)x^2)\,dx}{\int_1^{10} (x^2)\,dx} \text{ satisfies}$$

(*a*) $\mu \leq 333m$ (*b*) $333\,\mu \geq M$

(*c*) $m \leq \mu \leq M$ (*d*) $m \leq \mu \leq \frac{M}{333}$

ANSWERS

1. (b)	**2.** (a)	**3.** (a)	**4.** (b)	**5.** (c)	**6.** (*)	**7.** (d)
8. (a)	**9.** (c)	**10.** (d)	**11.** (c)	**12.** (b)		

EXPLANATIONS

1. (b) If there are two similiar matrices A & B, then,

Trace (A) = Trace (B) ...(i)

Trace of any matrix = Sum of its diagonal elements = sum of eigen values.

$$\text{Trace}(A) = \lambda_{1A} + \lambda_{2A} = 1 + (-2) = -1$$

From (i), we can say that

Trace (B) = Trace (A) = −1

2. (a) $f(x) = 4x+3$

$f'(x) = 4$

$$x_{n+1} = x_n - \frac{f(x_n)}{f'(x_n)} = x_n - \left[\frac{4x_n + 3}{4}\right]$$

$$= \cancel{x_n} - \cancel{x_n} - \frac{3}{4}$$

Hence cnly one iteration is required.

3. (a) Gaussian pulse wave travels with finite constant wave speed.

4. (b)

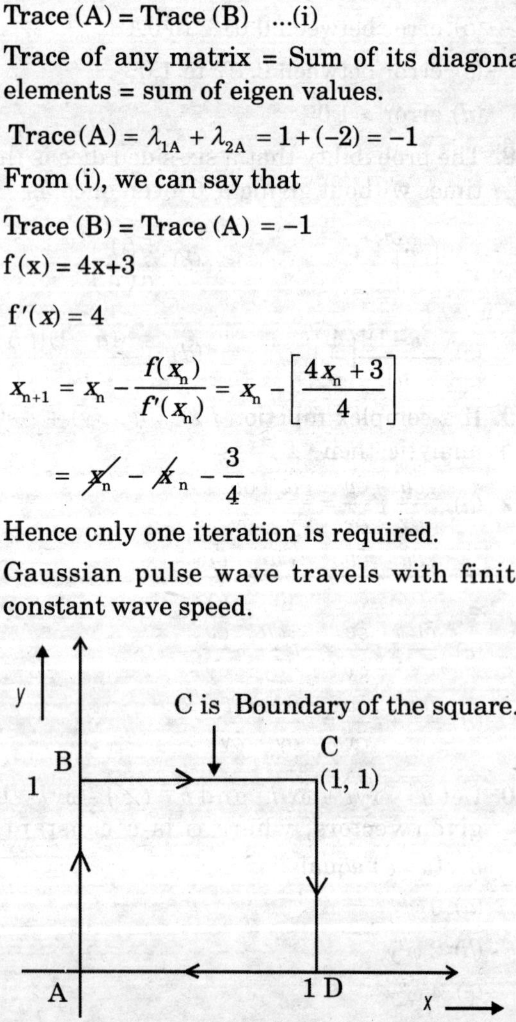

Let D be the square enclosed by the path C.

By Green's theorem, we get

$$\oint_c (x\,dy - y\,dx) = \iint \left(\frac{\partial y}{\partial x} - \frac{\partial x}{\partial y}\right) \cdot dA$$

$$= \iint_D (0-0)\,dx\,dy$$

$$= 0$$

5. (c) $\lambda_1 = 1$

$\lambda_2 = 2$

$$A = \begin{bmatrix} a & b \\ c & d \end{bmatrix}$$

Equation developed due to 1st eigen vector

$$\begin{bmatrix} a-1 & b \\ c & d-1 \end{bmatrix}\begin{bmatrix} 0.6 \\ 0.8 \end{bmatrix} = \begin{bmatrix} 0 \\ 0 \end{bmatrix}$$

Equation developed due to 2nd eight vector

$$\begin{bmatrix} a-2 & b \\ c & d-2 \end{bmatrix}\begin{bmatrix} 0.8 \\ -0.6 \end{bmatrix} = \begin{bmatrix} 0 \\ 0 \end{bmatrix}$$

$$(a-1) \times 0.6 + 0.8b = 0 \quad \text{(i)}$$

$$(a-2) \times 0.8 - 0.6b = 0 \quad \text{(ii)}$$

$$0.6c + (d-1)0.8 = 0$$

$$0.8c - (d-2)0.6 = 0$$

$$0.6c + 0.8d = 0.8 \quad \text{(iii)}$$

$$0.8c - 0.6d = -1.2 \quad \text{(iv)}$$

By solving (iii) & (iv), we get

$$d = 1.36$$

7. (d)

x	-1	0	1
f(x)	8	5	8

$f(x) = 3x^2 + 5$

$$I_s = \int_{-1}^{1}(3x^2+5)dx = \frac{h}{3}[y_0 + 4y_1 + y_2]$$

$$= \frac{1}{3}[8+4\times5+8] = \frac{32}{3} = 10.67$$

$$I = \int_{-1}^{1}(3x^2+5)dx = \left[\frac{3x^3}{3}+5x\right]_{-1}^{1} = 12$$

$$\%\,\text{error} = \left[\frac{12-10.67}{12}\right]\times 100\% = 11.08\%$$

8. (a) The probability of not seeing 6 in a single roll is $\left(5/6\right)$ & if the rolls are independent the probability of not having 6 in 'n' successive rolls will be $\left(5/6\right)^n$.

9. (c) Analytic function,

$f(x+iy) = u(x, y) + iv(x, y)$

$$\frac{\partial u}{\partial x} = \frac{\partial v}{\partial y}$$

$$\frac{\partial u}{\partial y} = \frac{-\partial v}{\partial x}$$

10. (d) $\overline{U} = -wy\hat{i} + wx\hat{j}$ and $\overline{V} = wz\hat{j} - wy\hat{k}$

$$\text{Now div}\left(\overline{U}\times\overline{V}\right) = \begin{vmatrix} \frac{\partial}{\partial x} & \frac{\partial}{\partial y} & \frac{\partial}{\partial z} \\ -wy & wx & 0 \\ 0 & wz & -wy \end{vmatrix}$$

$$= \frac{\partial}{\partial x}\left(-w^2xy\right) - \frac{\partial}{\partial y}\left(w^2y^2\right) + \frac{\partial}{\partial z}\left(-w^2yz\right)$$

$$= -w^2y - w^2 2y - w^2y$$

$$= -4w^2y$$

11. (c) $$\sum_{m=1}^{\infty}\frac{(-1)^m x^2}{(1+x^2)^m} = \frac{-x^2}{1+x^2} + \frac{x^2}{(1+x^2)^2} - \frac{x^2}{(1+x^2)^3} + \ldots$$

$$= x^2\left[-\frac{1}{(1+x^2)} + \frac{1}{(1+x^2)^2} - \frac{1}{(1+x^2)^3} + \frac{1}{(1+x^2)^4} - \frac{1}{(1+x^2)^5} + \ldots\right]$$

$$= x^2\left[\frac{-1 - x^2 + 1}{(1+x^2)^2} + \frac{-1 - x^2 + 1}{(1+x^2)^4}\ldots\right]$$

$$= x^2\left[\frac{-x^2}{(1+x^2)^2} + \frac{-x^2}{(1+x^2)^4} + \ldots\right]$$

$$= -x^4\left[\frac{1}{(1+x^2)^2} + \frac{1}{(1+x^2)^4} + \frac{1}{(1+x^2)^6} + \ldots\right]$$

As we can observe from above equation that the series is convergent in nature. It is also evident that series contains even terms in power of $(1 + x^2)$ therefore it will be convergent for all x.

2010

Q.(1 – 7) carry one mark each.

1. If $P = \begin{pmatrix} 1 & 1 \\ 1 & 1 \end{pmatrix}$, then $P^8 - 2P^7 + 2P^6 - 4P^5 + 3P^4 - 6P^3 + 2P^2$ equals

(*a*) P (*b*) 2 P

(*c*) 3 P (*d*) 4 P

2. Which one of the following matrices has the same eigen values as that of $\begin{pmatrix} 1 & 2 \\ 4 & 3 \end{pmatrix}$?

(*a*) $\begin{pmatrix} 3 & 4 \\ 1 & 2 \end{pmatrix}$ (*b*) $\begin{pmatrix} 1 & 4 \\ 2 & 3 \end{pmatrix}$

(*c*) $\begin{pmatrix} 4 & 2 \\ 1 & 3 \end{pmatrix}$ (*d*) $\begin{pmatrix} 2 & 4 \\ 1 & 3 \end{pmatrix}$

3. The integral $\int_{-1}^{1} x^{-4/3}\, dx$ is

(*a*) an improper integral converging to –6.

(*b*) an improper integral converging to 0.

(*c*) not an improper integral but has value – 6.

(*d*) a divergent improper integral.

4. The residue of the function

$$f(z) = \frac{\sin^4 z}{\left(z + \pi/4\right)^3} \text{ at } z = -\pi/4 \text{ is}$$

(*a*) 2 (*b*) 1

(*c*) –1 (*d*) –2

5. The variance of the number of heads resulting from ten independent tosses of a fair coin is

(*a*) $5/4$ (*b*) $5/2$

(*c*) $3/4$ (*d*) $3/2$

6. If the quadrature rule $\int_0^3 f(x)\, dx = \alpha f(1) + \beta f(3)$ is exact for all polynomials of degree 2 or less, then

(*a*) $\alpha = 3/4, \beta = 3/4$ (*b*) $\alpha = 3/4, \beta = 9/4$

(*c*) $\alpha = 9/4, \beta = 3/4$ (*d*) $\alpha = 9/4, \beta = 9/4$

7. Given that $\frac{dy}{dx} = 1 + y^2, y(0) = 0$, which one of the following is nearest to $y(0.4)$ computed by Euler's method with step size of 0.2?

(*a*) 0.408 (*b*) 0.404

(*c*) 0.208 (*d*) 0.204

Q.(8 – 11) carry one mark each.

8. Let

$$f(x) = \begin{cases} \dfrac{\sin x}{x} & \text{if } x \neq 0 \\ 1 & \text{if } x = 0. \end{cases}$$

Then

(*a*) f is not continuous at $x = 0$.

(*b*) f is continuous at $x = 0$ but not differentiable at $x = 0$.

(*c*) f is differentiable at $x = 0$ and $f'(0) = 0$.

(*d*) f is differentiable at $x = 0$ and $f'(0) = 1$.

9. Let $u(x, y) = \tan \{xy (x + y)\}$, Then

(*a*) $x\frac{\partial u}{\partial x} + y\frac{\partial u}{\partial y} = \frac{1}{3} xy (x + y) \sec^2 \{xy (x + y)\}$.

(*b*) $x\frac{\partial u}{\partial x} - y\frac{\partial u}{\partial y} = \frac{1}{3} xy (x + y) \sec^2 \{xy (x + y)\}$.

(*c*) $x\frac{\partial u}{\partial x} + y\frac{\partial u}{\partial y} = 3\, xy (x + y) \sec^2 \{xy (x + y)\}$.

(*d*) $x\frac{\partial u}{\partial x} - y\frac{\partial u}{\partial y} = 3\, xy (x + y) \sec^2 \{xy (x + y)\}$.

10. Which one of the following is a particular solution of the ordinary differential equation

$$x\frac{d^2 y}{dx^2} - \frac{dy}{dx} = 2x^2 f(x)\,?$$

(*a*) $x^2 \int x f(x)\, dx + \int x^3 f(x)\, dx$

(*b*) $x^2 \int x f(x)\, dx + \int x^2 f(x)\, dx$

(*c*) $x^2 \int x f(x)\, dx - \int x^3 f(x)\, dx$

(*d*) $x^2 \int x f(x)\, dx - \int x^2 f(x)\, dx$

11. Which one of the following is a possible solution to the partial differential equation $\frac{\partial^2 u}{\partial t^2} - \frac{\partial^2 u}{\partial x^2} = 0$ with boundary conditions

$u(0, t) = 0, \quad \frac{\partial u(\pi, t)}{\partial x} = 0$, for $t \geq 0$,

$u(x, 0) = 0, \quad \frac{\partial u(x, 0)}{\partial t} = \pi$, for $0 \leq x \leq \pi$?

(*a*) $u(x, t) = \sum_{n=0}^{\infty} a_n \sin\left(\left(n + \frac{1}{2}\right) t\right) \sin\left(\left(n + \frac{1}{2}\right) x\right)$

(*b*) $u(x, t) = \sum_{n=0}^{\infty} a_n \cos\left(\left(n + \frac{1}{2}\right) t\right) \sin\left(\left(n + \frac{1}{2}\right) x\right)$

(*c*) $u(x, t) = \sum_{n=0}^{\infty} a_n \sin\left(\left(n + \frac{1}{2}\right) t\right) \cos\left(\left(n + \frac{1}{2}\right) x\right)$

(*d*) $u(x, t) = \sum_{n=0}^{\infty} a_n \cos\left(\left(n + \frac{1}{2}\right) t\right) \cos\left(\left(n + \frac{1}{2}\right) x\right)$

ANSWERS

1. (d)	**2.** (b)	**3.** (c)	**4.** (a)	**5.** (c)	**6.** (c)	**7.** (a)
8. (b)	**9.** (c)	**10.** (b)	**11.** (b)			

EXPLANATIONS

1. (d) If $P = \begin{bmatrix} 1 & 1 \\ 1 & 1 \end{bmatrix}$

find $P^8 - 2P^7 + 2P^6 - 4P^5 + 3P^4 - 6P^3 + 2P^2$

Find eigen values first

$$\begin{vmatrix} (1-\lambda) & 1 \\ 1 & (1-\lambda) \end{vmatrix} = 0$$

$(1-\lambda)^2 - 1 = 0$

$(1-\lambda)^2 = 1$

$1-\lambda = \pm 1$

$1-\lambda = 1 \Rightarrow \lambda = 0$

$1-\lambda = -1 \Rightarrow \lambda = 2$

$P^8 - 2P^7 + 2P^6 - 4P^5 + 3P^4 - 6P^3 + 2P^2$

$= 2^8 - 2^8 + 2^7 - 2^7 + 3 \times 2^4 - 2^4 + 8 = 8$

& $4P = 4 \times 2$ (By putting eigen value = 2)

$= 8$

2. (b) Eigen values of matrix $\begin{bmatrix} 1 & 2 \\ 4 & 3 \end{bmatrix}$

Sum of main Diagonal elements = Sum of eigen values

$\therefore \begin{bmatrix} 1 & 2 \\ 4 & 3 \end{bmatrix}$ and $\begin{bmatrix} 1 & 4 \\ 2 & 3 \end{bmatrix}$ will have same eigen values

3. (c) $I = \int_{-1}^{1} x^{-4/3} dx = \left[\frac{x^{\frac{-4}{3}+1}}{\frac{-4}{3}+1} \right]_{-1}^{1}$

$$= \left[\frac{x^{\frac{-1}{3}}}{\frac{-1}{3}} \right]_{-1}^{1} = \left[-3x^{\frac{-1}{3}} \right]_{-1}^{1} = -6$$

4. (a) Residence of f(z) at $z = -\frac{\pi}{4}$

$$= \lim_{z \to \frac{-\pi}{4}} \frac{d}{dz^2} \cdot \sin^4 z$$

$$= \lim_{z \to \frac{-\pi}{4}} \frac{d}{dz}\left(4\sin^3 z \cdot \cos z\right)$$

$$= \lim_{z \to \frac{-\pi}{4}} \left[4 \times 3\sin^2 z \cdot \cos^2 z - 4\sin^4 z \right]$$

$= 2$

6. (c) Let $f(x) = x^2$ be a polynomial of degree 'z'.

Now $\int_0^3 f(x) = \alpha f(1) + \beta f(3)$

$\therefore \int_0^3 x^2 = \alpha + \beta 9$

$\therefore \left[\frac{x^3}{3} \right]_0^3 = \alpha + \beta 9$

$\therefore 9 = \alpha + \beta 9$

Now option (c) statisfies the above equation.

$\therefore$ Option (c) is correct.

7. (a) $\frac{dy}{dx} = 1 + y^2;\ y(0) = 0$

Here $x_0 = 0, y_0 = 0$

$\therefore x_1 = x_0 + h = 0 + 0.2 = 0.2$

$\therefore y_1 = y_0 + hf(x_0, y_0) = y_0 + h(1 + y_0^2)$

$\therefore y_1 - 0 + 0.2(1+0) = 0.2$

$(x_1, y_1) = (0.2, 0.2)$

Now $x_2 - x_1 + h = 0.4$

$\therefore y_2 = y_1 + hf(x_1, y_1) = 0.2 + 0.2 \times (1 + 0.2^2)$

$= 0.2 + 0.2 \times 1.04$

$= 0.2 + .208 = .408$

9. (c) $u(x, y) = \tan\{xy(x+y)\}$

$$\frac{\partial u}{\partial x} = \sec^2\{xy(x+y)\} \cdot \left[2xy + y^2\right]$$

$$\frac{\partial u}{\partial y} = \sec^2\{xy(x+y)\} \cdot \left[x^2 + 2xy\right]$$

$$x\frac{\partial u}{\partial x} + y\frac{\partial u}{\partial y} = 3(x^2y + xy^2)\sec^2\left[xy(x+y)\right]$$

$= 3xy(x+y)\sec^2\{xy(x+y)\}$

2011

Q.(1-7) carry one mark each.

1. A vector field is called *solenoidal* if its divergence is zero. Consider the vector fields $\vec{P}$ and $\vec{Q}$ given by $\vec{P}(x, y, z) = (2x^2 + 8xy^2 z)\hat{i} + (3x^3y - 3xy)\hat{j} - (4y^2 z^2 + 2x^3 z)\hat{k}$ and $\vec{Q}(x, y, z) = xyz^2\ \vec{P}(x, y, z)$. Then

(*a*) $\vec{P}$ and $\vec{Q}$ are both solenoidal

(*b*) both $\vec{P}$ and $\vec{Q}$ are not solenoidal

(*c*) $\vec{P}$ is solenoidal but not $\vec{Q}$

(*d*) $\vec{Q}$ is solenoidal but not $\vec{P}$

2. The eigenvalues of a 3×3 matrix P are 2, 2 and –1. Then P^{-1} is equal to

(*a*) $\frac{1}{4}(3P - P^2)$ (*b*) $\frac{1}{2}(P^2 - 2P)$

(*c*) $\frac{-1}{2}(P^2 + 3P)$ (*d*) $\frac{-1}{4}(P^2 + 2P)$

3. The integral $\int_{-5\pi/2}^{5\pi/2} f(x)\,dx$, where

$f(x) = e^{\pi x^2}\sin^3 x + 4\cos x$, equals

(*a*) 4 (*b*) 8

(*c*) $\frac{5\pi}{2}$ (*d*) $\frac{-5\pi}{2}$

4. The integral $\oint_C \frac{z^3 e^z}{(z-1)^3}\,dz$ along the curve C : $|z| = \frac{\pi}{2}$, oriented counter-clockwise, equals

(*a*) 0 (*b*) $2\pi i$

(*c*) $13e\pi i$ (*d*) $20e\pi i$

5. Consider the function $f(x, y, z) = x^3 e^y \sin z$ and the point $P = \left(1, 0, \frac{\pi}{2}\right)$. The value of f DOES NOT change due to a small displacement of P along the direction of

(*a*) $\left[1, 0, \frac{\pi}{2}\right]$ (*b*) $[1, -1, 1]$

(*c*) $[1, -3, 0]$ (*d*) $[2, 0, -1]$

6. A solution of the differential equation $\frac{d^2y}{dx^2} - 5\frac{dy}{dx} + 6y = 36x$ is

(*a*) $e^{2x} + e^{3x} + 6x + 5$ (*b*) $e^{-3x} + 6x + 5$

(*c*) $e^{2x} + e^{-3x} + 6x + \frac{5}{6}$ (*d*) $e^{2x} + e^{3x} + x + \frac{5}{6}$

7. For any positive numbers a and b, the matrix

$$P = \begin{bmatrix}1\\ a\\ b\end{bmatrix}[4\ 5\ 6] \text{ is}$$

(*a*) orthogonal (*b*) diagonalizable

(*c*) nonsingular (*d*) of rank 2

Q.(8-11) carry two mark each.

8. Suppose x_n is the n-th iterated value while finding the positive square root of 7 by the Newton Raphson method with a positive initial guess x_0 $(\neq\sqrt{7})$. If $e_n = \sqrt{7} - x_n$, for $n \geq 1$, then

(*a*) $e_{n+1} = \frac{e_n}{2x_n^2}$ (*b*) $e_{n+1} = \frac{-\sqrt{7}\,e_n^2}{2x_n}$

(*c*) $e_{n+1} = \frac{e_n^2}{\sqrt{7}}$ (*d*) $e_{n+1} = \frac{-e_n^2}{2x_n}$

9. The solution of the initial boundary value problem $\frac{\partial u}{\partial t} = \frac{\partial^2 u}{\partial x^2}$, $0 < x < \pi, t > 0$ with boundary and initial conditions $\frac{\partial u}{\partial x}(0, t) = 0 = u(\pi, t)$, $t > 0$ and $u(x, 0) = f(x)$, $0 < x < \pi$, is

(*a*) $u(x, t) = \sum_{n=0}^{\infty} A_n \exp\left(-\left(\frac{2n+1}{2}\right)^2 t\right)\cos\left(\frac{2n+1}{2}x\right)dx$

with $A_n = \frac{2}{\pi}\int_0^{\pi} f(x)\cos\left(\frac{2n+1}{\pi}x\right)dx$

(*b*) $u(x, t) = \sum_{n=0}^{\infty} A_n \exp(-n^2 t)\cos(nx)$,

with $A_n = \frac{2}{\pi}\int_0^{\pi} f(x)\cos(nx)\,dx$

(*c*) $u(x, t) \sum_{n=1}^{\infty} A_n \exp\left(-\left(\frac{2n+1}{2}\right)^2 t\right)\sin\left(\frac{2n+1}{2}x\right)$,

with $A_n = \frac{2}{\pi}\int_0^{\pi} f(x)\sin\left(\frac{2n+1}{\pi}x\right)dx$

(*d*) $u(x, t) \sum_{n=1}^{\infty} A_n \exp(-n^2 t)\sin(nx)$,

with $A_n = \frac{2}{\pi}\int_0^{\pi} f(x)\sin(nx)\,dx$

10. The function $f(x)$ defined by

$$f(x) = \begin{cases} 3 - x^2, & x \leq 1, \\ 3 - x, & 1 < x \leq 2, \\ x - 1, & x > 2, \end{cases}$$

has

(*a*) a local maxima at $x = 3$ and a local minima at $x = 0$

(*b*) a local maxima at $x = 0$ and no local minima

(*c*) a local maxima at $x = 0$ and a local minima at $x = 2$

(*d*) no local maxima and a local minima at $x = 1$

11. In a biased die experiment, the random variable x of the outcome has the (cumulative) distribution function $F(x)$ as shown below.

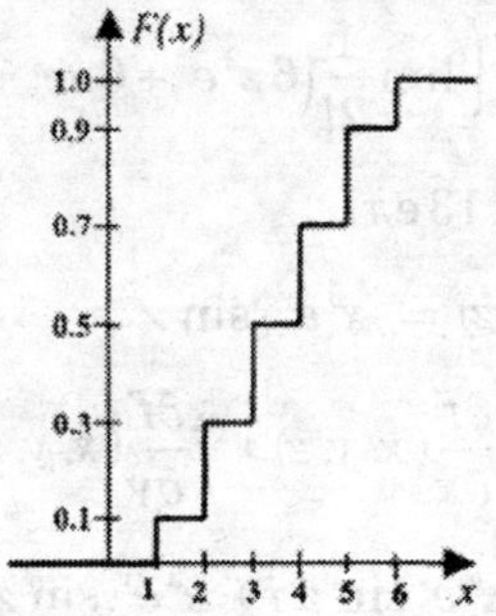

The variance of x is

(*a*) 1.5 (*b*) 2.25

(*c*) 3.5 (*d*) 4.25

ANSWERS

1. (d)	**2.** (a)	**3.** (b)	**4.** (c)	**5.** (c)	**6.** (a)	**7.** (b)
8. (d)	**9.** (d)	**10.** (a)	**11.** (c)			

EXPLANATIONS

1. (b) $\nabla\cdot(\overline{P}) = \left(i\frac{\partial}{\partial x} + j\frac{\partial}{\partial y} + k\frac{\partial}{\partial z}\right)$

$\cdot\left[\left(2x^2+8xy^2\right)i+\left(3x^3y-3xy\right)\hat{j}-(4y^2z^2+2x^3z)\hat{k}\right]$

$= 4x+8y^2+3x^3-3x-8y^2z-2x^3 \neq 0$

$\nabla\cdot(\overline{Q}) = 6x^2yz^2 + 16xy^3z^3 + 6x^4yz^2 - 6x^2yz^2$

$- 4xy^3 \times 4z^3 - 6x^4yz^2 = 0$

$\overrightarrow{Q}$ is solenoidal but not $\overline{P}$.

2. (a) $(P-2)^2(P+1) = 0$

$(P^2+4-4P)(P+1) = 0$

$P^3 + P^2 + \cancel{4P} + 4 - 4P^2 - \cancel{4P} = 0$

$P^3 + P^2 - 4P^2 + 4 = 0$

$P^3 - 3P^2 + 4 = 0$

Multiply above equation by P^{-1}

$P^2 - 3P + 4P^{-1} = 0$

$4P^{-1} = 3P - P^2$

$P^{-1} = \frac{\left(3P-P^2\right)}{4}$

3. (b) Evaluate $\int_{-\frac{5\pi}{2}}^{\frac{5\pi}{2}} f(x)dx$

where $f(x) = e^{\pi x^2}\cdot\sin^3 x + 4\cos x$

$e^{\pi x^2}\cdot\sin^3 x$ is an odd function because $f(x) \neq f(-x)$

$4\int_{-5\pi/2}^{5\pi/2} \cos x = 4\sin x\Big]_{-5\pi/2}^{5\pi/2} = 8$

$\cos x$ is an even function because

$f(x) = f(-x)$

$\therefore I = 2\int_0^{5\pi/2} f(x)dx$

$= 2\int_0^{5\pi/2} 4\cos x\,dx = 2\times 4\times \sin x\Big]_0^{5\pi/2}$

$= 8$

4. (c) $I = 2\pi i\left(\lim_{z\to 1}\frac{1}{2!}\frac{d^2}{dz^2}\left[z^3e^z\right]\right)$

$Z = 1$ lies inside $|Z| = \frac{\Pi}{2}$

$$\therefore \quad I = \left(\lim_{z \to 1} \frac{1}{2!} \left(6z^2 e^z + 6ze^z + e^z \cdot z^3 \right) \right) \times 2\pi i$$

$$I = 13 e\pi i$$

5. (c) $f(x, y, z) = x^3 e^y \cdot \sin z$

$$\vec{A} = \hat{i}\frac{\partial f}{\partial x}(x, y, z) + \hat{j}\frac{\partial f}{\partial y}(x, y, z) + \hat{k}\frac{\partial f}{\partial z}(x, y, z)$$

$$= 3x^2 e^y \sin z\, \hat{i} + x^3 e^y . \sin z\, \hat{j} + x^3 e^y \cos z\, \hat{k}$$

$$\vec{A}\left(1, 0, \pi/2\right) = 3\hat{i} + \hat{j}$$

Let $\vec{B}$ in the direction of $[1, -3, 0] = \hat{i} - 3\hat{j}$

$$\vec{A} \cdot \vec{B} = \left(3\hat{i} + \hat{j}\right) \cdot \left(\hat{j} - 3\hat{j}\right)$$

$$= 0$$

6. (a) $\dfrac{d^2 y}{dx^2} - 5\dfrac{dy}{dx} + 6y = 36x$

$$(D^2 - 50 + 6) = 36x$$

C.F.

$$(D^2 - 5D + 6) = 0$$

$$(D - 3)(D - 2) = 0$$

$$C.F = C_1 e^{3x} + C_2 e^{2x}$$

$$\text{P.I} = \frac{36x}{D^2 - 5D + 6}$$

$$\text{P.I} = \frac{1}{6}\left(\frac{36x}{\frac{D^2}{6} - \frac{5D}{6} + 1} \right)$$

$$= \left[1 - \left(\frac{5D}{6} - \frac{D}{6} \right)^2 \right]^{-1} 6x$$

$$= \left(1 + \frac{5D}{6} - \frac{D^2}{6} \right) 6x$$

($\because$ Higher powers of expansion series are neglected)

$$= 6x + 5$$

$$y = \text{C.F.} + \text{P.I.} = c_1 e^{3x} + c_2 e^{2x} + 6x + 5$$

7. (b) $P = \begin{bmatrix} 1 \\ a \\ b \end{bmatrix} \begin{bmatrix} 4 & 5 & 6 \end{bmatrix} = \begin{bmatrix} 4 & 5 & 6 \\ 4a & 5a & 6a \\ 4b & 5b & 6b \end{bmatrix}$

Since determininant of matrix P can't be zero, therefore matrix P is non singular.

8. (a) $e_n = \sqrt{7} - x_n$

Let ξ be an exact root of the equation f(x) = 0 & x_n be the approximate root at the nth iteration. Then $f(\xi) = 0$. Let e_n be the error occurs at the nth iteration.

Then $e_n = x_n - \xi$

We know

$$e_{n+1} = e_n - \frac{f(e_n + \xi)}{f'(x_0)}$$

$$e_{n+1} = e_n \left(1 - \frac{f'(\xi)}{f'(x_0)} \right)$$

Let $C = 1 - \dfrac{f'(\xi)}{f'C_0\, e_{n+1 = Ce_n}}$, which is free from e_n

So $e_{n+1} = \text{C}\ e_n$

9. (d) Tutorial.math.lamar.edu/Classes/DE/Solving Heat equation.aspx

11. (*)

x	1	2	3	4	5	6
F(x)	0.1	0.3	0.5	0.7	0.9	1.0

$$\mu = 0.1 + 0.6 + 1.5 + 2.8 + 4.5 + 6 = 15.5$$

2012

Q. (1 – 7) carry one mark each.

1. For the matrix $M = \begin{bmatrix} 1 & 4 & 5 \\ 0 & 2 & 6 \\ 0 & 0 & 3 \end{bmatrix}$, consider the following statements:

P : 3 is an eigenvalue of M.

Q : $\begin{bmatrix} 4 \\ 1 \\ 0 \end{bmatrix}$ is an eigenvector of M.

R : $\begin{bmatrix} 4 \\ 2 \\ 0 \end{bmatrix}$ is an eigenvector of M.

Which of the above statements are **TRUE**?

(*a*) P and Q, but not R

(*b*) Q and R, but not P

(*c*) P and R, but not Q

(*d*) P, Q, and R

2. Taylor series of $f(x) = \dfrac{1}{1-x}$ about $x = 0$ is given as $TS|_f\ 1 + x + x^2 + x^3 + ...$ This series can be used to evaluate $f(x)$ for

(*a*) $|x| \neq 1$ (*b*) $x < -1$

(*c*) $|x| < 1$ (*d*) $-1 \leq x < 1$

3. Let $f(u, v) = u \ln(v)$ and $F(x, y) = f(u(x, y), v(x, y))$, where $u = x/y$ and $v = x - y$. Then $\partial F/\partial y$ is

(*a*) $-\dfrac{x}{y^2} \ln(x-y) + \dfrac{x}{y(x-y)}$

(*b*) $-\dfrac{x}{y^2} \ln v - \dfrac{u}{v}$

(*c*) $\dfrac{x}{y^2} \ln v - \dfrac{u}{v}$

(*d*) $\dfrac{x}{y^2} \ln v + \dfrac{x}{yv}$

4. Consider two functions $f(z) = |z|$ and $g(z) = \bar{z}$ (conjugate of z). Using Cauchy-Riemann conditions, choose the correct answer

(*a*) Both f and g are analytic

(*b*) f is analytic but g is not analytic

(*c*) g is analytic but f is not analytic

(*d*) Neither f nor g is analytic

5. For $f = x^4 - 5xy^2$, the direction of maximum increase of $f(x, y)$ at the point (2, 2) is along

(*a*) $3\hat{i} + 10\hat{j}$ (*b*) $-12\hat{i} - 40\hat{j}$

(*c*) $3\hat{i} - 10\hat{j}$ (*d*) $-12\hat{i} + 40\hat{j}$

6. Suppose 50% of the population of a village like oranges, 70% of the population like apples, and 40% like both. If a person is picked at random who likes at least one of these fruits, what is the probability that the person likes oranges?

(*a*) 1/8 (*b*) 5/12

(*c*) 1/2 (*d*) 5/8

7. For the solution of $\nabla^2 u = 0$, the domain and boundary conditions are shown below.

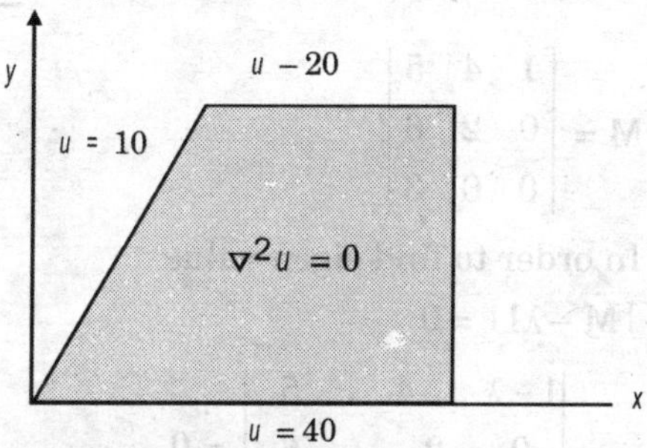

Which of the following statements is **TRUE**?

(*a*) The solution cannot be obtained using separation of variables because the governing equation is non-separable.

(*b*) The solution cannot be obtained using separation of variables because all the boundary values are non-zero.

(*c*) The solution cannot be obtained using separation of variables because not all the boundaries are along constant coordinate lines.

(*d*) The solution can be obtained by separation of variables.

Q. (8 – 11) carry two marks each.

8. If $f(x) = x \sin(x)$ and $g(x) = |x| \sin(x)$, then

(*a*) $g(x) = |f(x)|$

(*b*) $g(x)$ is an even function

(*c*) The x-coordinates corresponding to the various local maxima are identical for both $f(x)$ and $g(x)$

(*d*) $g(x)$ is differentiable at $x = 0$

9. The general solution of

$$\frac{d^4y}{dx^4} - 2\frac{d^3y}{dx^3} + 2\frac{d^2y}{dx^2} - 2\frac{dy}{dx} + y = 0 \text{ is}$$

(a) $c_1e^x + c_2xe^x + c_3 \cosh(x) + c_4 \sinh(x)$

(b) $c_1e^x + c_2e^{-x} + c_3e^{ix} + c_4e^{-ix}$

(c) $c_1e^x + c_2xe^x + c_3 \cos(x) + c_4 \sin(x)$

(d) $c_1e^x + c_2xe^{-x} + c_3e^{ix} + c_4e^{-ix}$

10. Evaluation of $\iint_S \left(e^x\hat{i}+3y\hat{j}-ze^x\hat{k}\right).\hat{n}\,dA$ over a surface S : $x^2 + y^2 + z^2 = 1$, using Gauss divergence theorem, gives

(a) 0 (b) 4π

(c) $4\pi/3$ (d) 12π

11. The exact solution of the integral $\int_0^4 \left(x^2-4\right)dx$ is denoted by I_E. The same integral evaluated numerically by the trapezoidal rule and the Simpson's 1/3 rule are denoted by I_T and I_S, respectively. The subinterval used in the numerical methods is $h = 2$. Then

(a) $I_E = I_S > I_T$ (b) $I_E = I_S < I_T$

(c) $I_E < I_S < I_T$ (d) $I_E > I_S > I_T$

ANSWERS

1. (a)	**2.** (c)	**3.** (b)	**4.** (d)	**5.** (c)	**6.** (d)	**7.** (c)
8. (d)	**9.** (c)	**10.** (b)	**11.** (b)			

EXPLANATIONS

1. (a) $M = \begin{bmatrix}1 & 4 & 5\\0 & 2 & 6\\0 & 0 & 3\end{bmatrix}$

In order to find eigen value

$|M - \lambda I| = 0$

$$\therefore \begin{vmatrix}1-\lambda & 4 & 5\\0 & 2-\lambda & 6\\0 & 0 & 3-\lambda\end{vmatrix} = 0$$

$$\therefore (1-\lambda)[(2-\lambda)(3-\lambda)]-4[0]+5[0]=0$$

$\therefore \lambda = 1, \lambda = 2, \lambda = 3$

$\therefore$ Statement-P is true.

Now for $\lambda = 1$,

$$\begin{bmatrix}0 & 4 & 5\\0 & 2 & 6\\0 & 0 & 2\end{bmatrix}\begin{bmatrix}x\\y\\z\end{bmatrix}=\begin{bmatrix}0\\0\\0\end{bmatrix}$$

$$4y + 5z = 0 \quad ...(i)$$

$$2x + 6y = 0 \quad ...(ii)$$

$$2z = 0$$

$$\Rightarrow z = 0 \quad ...(iii)$$

from equation (i) and (iii) we get $\boxed{y=0}$

Hence eigen vector is $\begin{bmatrix}0\\0\\0\end{bmatrix}$

Now for $\lambda = 2$

$$\begin{bmatrix}-1 & 4 & 5\\0 & 0 & 6\\0 & 0 & 1\end{bmatrix}\begin{bmatrix}x\\y\\z\end{bmatrix}=\begin{bmatrix}0\\0\\0\end{bmatrix}$$

$$\therefore -x + 4y + 5z = 0 \quad ...(i)$$

$$6z = 0$$

$$\Rightarrow z = 0 \quad ...(ii)$$

Hence 'Q' satisfies above equation

$\therefore$ Q is eigen vector of 'M'.

Now for $\lambda = 3$

$$\begin{bmatrix}-2 & 4 & 5\\0 & -1 & 6\\0 & 0 & 0\end{bmatrix}\begin{bmatrix}x\\y\\z\end{bmatrix}=\begin{bmatrix}0\\0\\0\end{bmatrix}$$

$$-2x + 4y + 5z = 0 \quad ...(i)$$

$$-y + 6z = 0 \quad ...(ii)$$

$$z = 0 \quad ...(iii)$$

Hence P and Q are true but not R

$\therefore$ Option (a) is correct.

2. (c) $x^2 + 4xl = 108$

$v^{2/3} + 4v^{1/3}l = 108$

$v = x^3$

$(v)^{1/3} = x$

$v^{2/3} = x^2$

3. (b) $F = \frac{x}{y}\ln(x-y)$

$$\frac{\partial F}{\partial y} = \frac{-x}{y^2}\ln(x-y) - \frac{x}{y}\frac{1}{y(x-y)}$$

$$= \frac{-x}{y^2}\ln V - \frac{U}{V}$$

5. (c) $f = x^4 - 5xy^2$

(2, 2)

$$f_x = 4x^3 - 5y^2 \text{ at } (2, 2) = 12$$

$$f_y = -10xy \text{ at } (2, 2) = -40$$

$$\frac{\nabla f}{|\nabla f|} = \frac{(12, -40)}{\sqrt{144+1600}} = \frac{12i - 40j}{\sqrt{1744}}$$

$$= \frac{4}{\sqrt{1744}}\left(3\hat{i} - 10\hat{j}\right)$$

6. (d) 10% like only oranges

30% like only apples

40% like both

80% of the population like atleast one.

20% of the population like none.

So when we pick 80% population who like atleast one fruit, out of these 80%, 50% of the total population like oranges.

So, probability that the person like oranges will be $= 5/8$

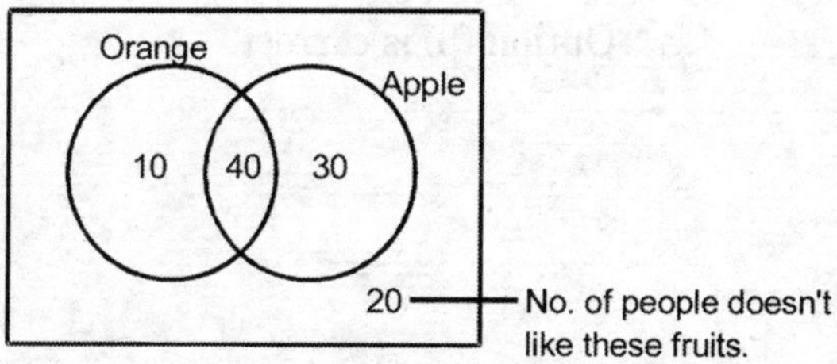

8. (d) $f(x) = x\sin(x)$ and $g(x) = |x| \sin x$

Now from option (*a*)

$$g(x) = |x| \sin x$$

and $\quad |f(x)| = |x \sin(x)| \neq |x| \sin x$

$\therefore$ Option (*a*) is wrong.

Now from option (*b*)

$$g(x) = |x| \sin(x)$$

Now $\quad g\left(\frac{\pi}{2}\right) = \left|\frac{\pi}{2}\right|\sin\left(\frac{\pi}{2}\right) = \frac{\pi}{2}$

and $\quad g\left(-\frac{\pi}{2}\right) = \left|-\frac{\pi}{2}\right|\sin\left(-\frac{\pi}{2}\right)$

$$= -\frac{\pi}{2}(1) = -\frac{\pi}{2}$$

$\therefore \quad g(\pi) \neq g\left(-\frac{\pi}{2}\right)$

$\therefore$ $g(x)$ is not an even function.

$\therefore$ Option (*b*) is wrong.

$$g(x) = |x| \sin x$$

$\therefore \quad g(x) = \begin{cases} x\sin x, & x \geq 0 \\ -x\sin x, & x < 0 \end{cases}$

$$f'(0) = \lim_{h\to 0}\frac{f(h) - f(0)}{h}$$

$$= \lim_{h\to 0}\frac{f(h)}{h}$$

Now for $h \to 0$,

$$f'(0) = \lim_{h\to 0}+\frac{h\sinh}{h} = 0$$

and for $\quad h < 0$

$$f'(0) = \lim_{h\to 0}\frac{-h\sinh}{h} = 0$$

Hence $g(x)$ is continuous and differentiable at $x = 0$

$\therefore$ Option (*d*) is correct.

9. (c) $\frac{d^4y}{dx^4} - 2\frac{d^3y}{dx^3} + 2\frac{d^2y}{dx^2} - 2\frac{dy}{dx} + y = 0$

$$\left(D^4 - 2D^3 + 2D^2 - 2D + 1\right)y = 0$$

Let D = + 1

(D – 1) is a factor

Alternate

9. (c) D - 1) D⁴ - 2 D³ + 2D² - 2D + 1 (D³ - D² + D - 1

$$\begin{array}{r} D^4 - D^3 \\ \hline -D^3 + 2D^2 \\ -D^3 + D^2 \\ \hline D^2 - 2D \\ D^2 - D \\ \hline -D + 1 \\ -D + 1 \\ \hline 0 \\ \hline \end{array}$$

$$(D-1)(D^3 - D^2 + D - 1)$$

$$(D-1)(D^2(D-1) + (D-1)) = 0$$

$$(D-1)^2(D^2+1) = 0$$

$$D = 1, 1, \pm i$$

$\therefore \quad y(x) = c_1e^x + xc_2e^x + c_3\cos x + c_4\sin x$

11. (b) $\quad I_E = \int_0^4 \left(x^2 - 4\right)dx$

where I_E denotes exact solution of integral

$\therefore \quad I_E = \left[\frac{x^3}{3} - \frac{4x}{0}\right]_0^4 = \left[\frac{x^3}{3} - 4x\right]_0^4$

$$= \frac{64}{3} - 16 = \frac{64-48}{3}$$

$$= \frac{16}{3}$$

Simpson $\frac{1}{3}$ formula

$$I_S = \int_a^b f(x)dx$$

$$= \frac{h}{3}\left[(y_0 + y_n) + 4(y_1 + y_3 + ... + y_{n-1}) + 2(y_2 + y_4 + ... + y_{n-2})\right]$$

$\therefore$ Now $\int_0^4 (x^2 - 4)\,dx$

$$\therefore \quad h = \frac{b-a}{n}$$

$$\therefore \quad 2 = \frac{4-0}{n}$$

$$\therefore \quad 2n = 4$$

$$\therefore \quad \boxed{n = 2}$$

Now $x_0 = 0$

$x_1 = 0 + 2 = 2$

and $x_2 = x_1 + 2 = 2 + 2 = 4$

$\Rightarrow y_0 = -4, y_1 = 0$

and $y_2 = 12$

$$\therefore \int_0^4 (x^2 - 4)\,dx = \frac{2}{3}\left[(-4 + 12) + 4(0)\right]$$

$$= \frac{2}{3} \times 8$$

$$\therefore \quad I_S = \frac{16}{3}$$

Now by trapezoidal rule

For $n = 2$, $\Delta x = \frac{4-0}{2} = 2$

x	0	2	4
$x^2 - 4$	-4	0	12

$$\therefore \int_0^4 (x^2 - 4)\,dx = \frac{2}{2}(-4 + 0 + 12) = 8$$

$\therefore$ $I_T = 8$

$\Rightarrow$ $I_E = I_S < I_T$

$\therefore$ Option (b) is correct.

2013

Q. 1 – 7 carry one mark each.

1. The value of the integral $\int_0^1 \frac{dt}{\sqrt{(-\log_e t)}}$ is

(a) $\frac{\sqrt{\pi}}{2}$ (b) $\sqrt{\pi}$

(c) $-\sqrt{\pi}$ (d) $-\frac{\sqrt{\pi}}{2}$

2. Which one of the following partial differential equations **CAN NOT** be reduced to two ordinary differential equations by the method of separation of variables ?

(a) $\frac{\partial u}{\partial t}-\frac{\partial^2 u}{\partial x^2}=0$ (b) $\frac{\partial^2 u}{\partial t^2}-\frac{\partial^2 u}{\partial x^2}=0$

(c) $\frac{\partial^2 u}{\partial t^2}+\frac{\partial^2 u}{\partial t\,\partial x}+\frac{\partial u}{\partial x}=0$ (d) $\frac{\partial^2 u}{\partial t^2}+\frac{\partial^2 u}{\partial t\,\partial x}+\frac{\partial^2 u}{\partial x^2}=0$

3. The Fourier series of the periodic function

$f(x) = |x|, -1 < x < 1, f(x+2) = f(x), x \in \mathbb{R}$ is given by

$$\frac{1}{2}-\sum_{n=1}^{\infty}\frac{4\cos(2n-1)\pi x}{(2n-1)^2\pi^2}$$

Using the above, the sum of the infinite series $1+\frac{1}{3^2}+\frac{1}{5^2}+\ldots$ is

(a) $\frac{\pi^2}{4}$ (b) $\frac{3\pi^2}{8}$

(c) $\frac{\pi^2}{8}$ (d) $\frac{\pi^2}{2}$

4. Consider the function $f(z) = z^2\,\bar{z} \in \mathbb{C}$. At $z = 0$, the function f

(a) does not satisfy the Cauchy-Riemann equations

(b) satisfies the Cauchy-Riemann equations but is not differentiable

(c) is differentiable but not analytic

(d) is analytic

5. The integral $\oint_C \frac{(z+4)}{(z+1)(z-2)^3}\,dz$ along the contour C : $|z-(1+i)| = 2$, oriented anticlockwise, is equal to

(a) 0 (b) $\frac{2\pi i}{9}$

(c) $-\frac{2\pi i}{9}$ (d) $\frac{4\pi i}{9}$

6. The integral $\int_0^1\int_{x^2}^{x}\left(\frac{x}{y}\right)e^{-x^2/y}\,dy\,dx$ equals

(a) $\frac{e-2}{e}$ (b) $\frac{e-1}{2e}$

(c) $\frac{e-1}{2}$ (d) $\frac{e-2}{2e}$

7. If the mean and variance of a binomial distribution are 6 and 2 respectively, then the probability of two failures is

(a) $4\left(\frac{2}{3}\right)^7$ (b) $4\left(\frac{2^2}{3^7}\right)$

(c) $17\left(\frac{2}{3}\right)^7$ (d) $17\left(\frac{2^2}{3^7}\right)$

Q. 8 – 11 carry two marks each.

8. For the matrix $M = \begin{pmatrix}1 & 0 & -1\\ 0 & 1 & -1\\ 1 & 1 & -2\end{pmatrix}$, consider the following statements :

(P) The characteristic equation of M is $\lambda^3-\lambda=0$.

(Q) M^{-1} does not exist.

(R) The matrix M is diagonalizable.

Which of the above statements are true?

(a) P, Q and R

(b) P and R but not Q

(c) P and Q but not R

(d) Q and R but not P

9. The work done by the force $\vec{F}=(x+x^2)\hat{i}+(x^2+y^3)\hat{j}$ in moving a particle once along the triangle with vertices (0,0), (1,0) and (0,1) in the anti-clockwise direction is

(a) 0 (b) 1/6

(c) 1/3 (d) 5/3

10. The general solution of the differential equation

$$x^3\frac{d^3y}{dx^3}+x^2\frac{d^2y}{dx^2}+x\frac{dy}{dx}-y=0,\quad x>0 \text{ is}$$

(a) $C_1e^x+e^{x/2}\left\{C_2\cos\left(\frac{\sqrt{3}}{2}x\right)+C_3\sin\left(\frac{\sqrt{3}}{2}x\right)\right\}$

(b) $C_1x+x^{-1/2}\left\{C_2\cos\left(\frac{\sqrt{3}}{2}\log_e x\right)+C_3\sin\left(\frac{\sqrt{3}}{2}\log_e x\right)\right\}$

(c) $C_1e^x+e^{-x/2}\left\{C_2\cos\left(\frac{\sqrt{3}}{2}x\right)+C_3\sin\left(\frac{\sqrt{3}}{2}x\right)\right\}$

(d) $C_1x+x^{1/2}\left\{C_2\cos\left(\frac{\sqrt{3}}{2}\log_e x\right)+C_3\sin\left(\frac{\sqrt{3}}{2}\log_e x\right)\right\}$

11. Using Euler's method to solve the differential equation

$$\frac{dy}{dx} = 2\cos\left(\frac{4\pi x}{3}\right) - y, y(0) = 1$$

with step-size $h = 0.25$, the value of $y(0.5)$ is

(a) 1.3125 (b) 1.1875

(c) 1.125 (d) 1.0625

ANSWERS

1. (b)	**2.** (d)	**3.** (c)	**4.** (c)	**5.** (b)	**6.** (d)	**7.** (a)
8. (a)	**9.** (c)	**10.** (d)	**11.** (b)			

EXPLANATIONS

1. (b) We have, $\int_0^1 \frac{dt}{\sqrt{(-\log_e t)}}$

Let $-\log_e t = z$

$t = e^{-z}$

$dt = -e^{-z}\, dz$

$$= -\int_0^1 \frac{e^{-z}dz}{\sqrt{z}} = -\int_0 z^{-1/2}e^{-z}dz$$

By using Gamma function

$$= -\overline{|1/2} \qquad \therefore \overline{|1/2} = \sqrt{\pi}$$

$$= -\sqrt{\pi}$$

4. (c) We have, $f(z) = z^2\bar{z}, z \in \ell$ at $z = 0$

$$\frac{\partial f}{\partial \bar{z}} = z^2 = 0 \quad \Leftrightarrow z = 0$$

$\Rightarrow$ differentiable at $z = 0$ only

$\Rightarrow$ not Analytic

Hence, the funtion f is differentiable but not Analytic

5. (b) $\oint_c \frac{(z+4)}{(z+1)(z-2)^3}dz$ along the contour

$c : |z-(1+i)| = 2$

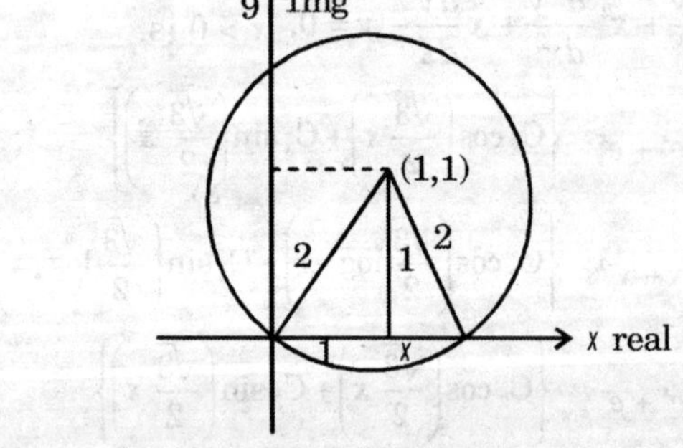

Center = 1 + i

radius = 2

z = 2

$$f(z) = \left(\frac{z+4}{z+1}\right)$$

$$\oint_c \frac{f(z)}{(z-2)^3}dz = 2\Pi i \frac{d^2}{dz^2} f(z)\Big|_{+z=2}$$

$$= 2\Pi i \frac{d}{dz} \frac{(z+1).1-(z+4).1}{(z+1)^2}$$

$$= 2\Pi i \frac{d}{dz}\left(\frac{-3}{(z+1)^2}\right) = 2\Pi i \frac{3}{(z+1)^3}$$

$$= 2\Pi i \frac{3}{(2+1)^3} = \frac{2\Pi i}{9}$$

6. (d) The integral $\int_0^1 \int_{x^2}^{x} \left(\frac{x}{y}\right) e^{-x^2/y} dydx$ equals

We have, $\int_0^1 \int_{x^2}^{x} \left(\frac{x}{y}\right) e^{-x^2/y} dydx$

Give that : $y = x^2, y = x; x = 0, x = 1$

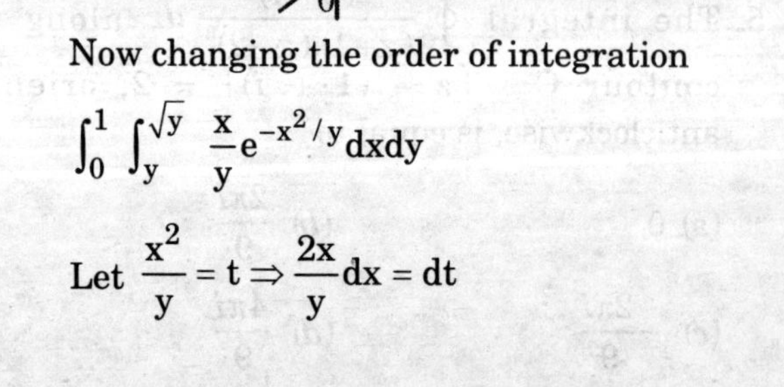

Now changing the order of integration

$$\int_0^1 \int_y^{\sqrt{y}} \frac{x}{y} e^{-x^2/y} dxdy$$

Let $\frac{x^2}{y} = t \Rightarrow \frac{2x}{y}dx = dt$

$$= \int_0^1 \frac{1}{2}\int_y^{1-t} e\, dtdy$$

$$= \frac{1}{2}\int_0^1 \left[-e^{-t}\right]_y^1 dy$$

$$= \frac{1}{2}\left[\frac{-y}{e} - e^{-y}\right]_0^1$$

$$= \frac{1}{2}\left(\frac{-2}{e}+1\right) = \frac{e-2}{2e}$$

7. (a) We have

mean np = 6

variance npq = 2

$\Rightarrow q = 1/3, \quad p = 1-q = 2/3$

n = 9

Now, We can find probability of 2 failure, by using Binomial mass function:

$$P(\times = \chi) = n_{c_\chi} P^{\chi} q^{n-\chi}$$

Therefore probability of 2 failures

$\chi = 2$ thus, We Get

$$P(x=2) = {}^9C_2 (2/3)^2 (1/3)^7$$

$$= \frac{9!}{2!7!} \times \frac{2^2}{9} \times \frac{1}{3^7} = 4 \times \frac{2^2}{3^7}$$

$$p(x=2) = 4\left(\frac{2^2}{3^7}\right)$$

Hence, Probability of 2 failure is $4\left(\frac{2^2}{3^7}\right)$

8. (d) We have $M = \begin{bmatrix} 1 & 0 & -1 \\ 0 & 1 & -1 \\ 1 & 1 & -2 \end{bmatrix}$

(P) The Characteristic equation of M

$$|m = \lambda I| = o$$

$$\Rightarrow \lambda^3 - (0)\lambda^2 + (-1)\lambda - 0 = 0$$

$$\Rightarrow \lambda^3 - \lambda = 0$$

$$|M| = \begin{vmatrix} 1 & 0 & -1 \\ 0 & 1 & -1 \\ 1 & 1 & -2 \end{vmatrix}$$

$$= 1(-2+1) - 0 - 1(0-1)$$

$$= -1 + 1 = 0$$

$$\Rightarrow |M| = 0$$

So M^{-1} dose not exist

(R) The characteristic equation is $\lambda^3 - \lambda = 0$

$$\Rightarrow \lambda(\lambda^2 - 1) = 0$$

$$\Rightarrow \lambda = 0, \lambda = \pm 1$$

Here eigen values of M are distinct there Matrix M is diagonalizable.

Hence (P),(Q) and (R) is True.

9. (c) We have $\vec{F} = \left(x + x^2\right)i + \left(x^2 + y^3\right)i$

Along the tringle with vertical (0,0),(1,0),

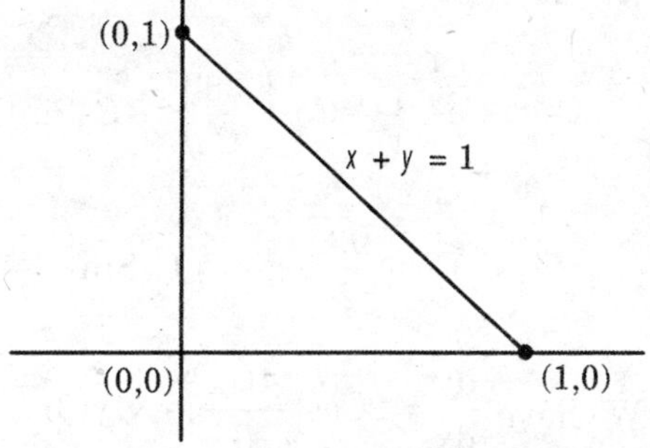

$$\vec{F}.d\vec{r} = \left(x + x^2\right)dx + \left(x^2 + y^3\right)dy$$

Work done is $\oint_c \vec{F}.d\vec{r}$

$$\Rightarrow \oint_e \left(x + x^2\right)dx + \left(x^2 + y^3\right)dy$$

$$\int_c \left[\left(x + x^2\right)\frac{dx}{dt} + \left(x^2 + y^3\right)\frac{dy}{dt}\right]dt$$

$$x + y = 1 : y = t, x = 1 - t$$

$$= \int_0^1 \left[\left[(1-t) + (1-t)^2\right](-1) + [(1-t)^2 + t^3](1)\right]dt$$

$$= \int_0^1 \left(t^3 + t - 1\right)dt = \left[\frac{t^4}{4} + \frac{t^2}{2} - t\right]_0^1$$

$$= \frac{1}{4} + \frac{1}{2} - 1 = -\frac{1}{4}$$

in the antilock wise direction work done is $\frac{1}{4}$

10. (d) We have, $x^3\frac{d^3y}{dx^3} + x^2\frac{d^2y}{dx^2} + x\frac{dy}{dx} - y = 0, x > 0$

$\left(x^3D^3 + x^2D^2 + xD - 1\right)y = 0$, Where $D = \frac{d}{dx}$

Let $\log_x = z \Rightarrow x = e^z$

$$xD = D_1, x^2D^2 = D_1(D_1 - 1)$$

$$x^3D^3 = D_1(D_1 - 1)(D_1 - 2),$$

Where $D_1 = \frac{d}{dz}$

$\left(D_1^3 - 2D_1^2 + 2D_1 - 1\right)y = 0$

The Auxiliary Equation is

$D_1^3 - 2D_1^2 + 2D_1 - 1 = 0$

$(D_1 - 1)\left(D_1^2 - D_1 + 1\right) = 0$

$D_1 = 1, D_1 = \frac{1}{2} \pm \frac{\sqrt{3}}{2} i$

Hence, general solution of differential equation is

$$y = C_1 e^z + e^{\frac{1}{2}z}\left(C_2 \text{Cos}\left(\frac{\sqrt{3}}{2}\right)z + C_3 \text{Sin}\left(\frac{\sqrt{3}}{2}\right)z\right)$$

$$y = C_1 x + x^{\frac{1}{2}}\left(C_2 \text{Cos}\left(\frac{\sqrt{3}}{2}\log x\right) + C_3 \text{Sin}\left(\frac{\sqrt{3}}{2}\log x\right)\right)$$

11. (b) We have $\frac{dy}{dx} = 2\text{Cos}\left(\frac{4\Pi x}{3}\right) - y, y(0) = 1$

Step size h = 0.25

Where $\frac{dy}{dx} = f(x_n, y_n)$

Using Euler's method

$y^*_{n+1} = y_n + hf\left(x_{n,}y_n\right)$

$y_{n+1} = y_n + \frac{h}{2}\left(f\left(x_n, y_n\right) + f(x_{n+1}, y^*_{n+1}\right)$

When n = 0

$y^*_1 = y_0 + hf\left(x_0, y_0\right)$

$= 1 + 0.25 \times 1$

$= 1.25$

$y_1 = y_0 + \frac{h}{2}\left(f\left(x_0, y_0\right) + f\left(x_1, y^*_1\right)\right)$

$= 1 + \frac{0.25}{2}(1 - 0.25) \qquad \therefore x_1 = x_{0+h}$

$= 1 + 0.125 \times (0.75)$

$= 1 + 0.0938 = 1.0938$

When n = 1

$y^*_2 = y_1 + hf\left(x_1, y_1\right)$

$= 1.0938 + 0.25 \times (-0.0938)$

$= 1.0938 - 0.0235$

$= 1.0704$

$y_2 = y1 + \frac{h}{2}\left(f\left(x_1, y_1\right) + f\left(x_2, y^*_2\right)\right)$

$= 1.0938 + \frac{0.25}{2}(-0.0938 - 2.0704)$

$= 1.0938 - 0.2705 \qquad \therefore x_2 = x_0 + 2h$

$= 0.8233$

When n = 2

$y^*_3 = y_2 + hf(x_2, y_2)$

$= 0.8233 + 0.25 \times (-1.8233)$

$= 0.8233 - 0.4558$

$= 0.3674$

$y_3 = y_2 + \frac{h}{2}\left(f\left(x_2, y_2\right) + f\left(x_3, y^*_3\right)\right)$

$= 0.8233 + 0.125 \times (-1.8233 - 2.3674)$

$= 0.8233 - 0.5238$

$= 0.2995$

2014

Q. (1 – 7) Carry One Mark Each.

1. If 1, 0, and –1 are the eigenvalues of a 3 × 3 matrix A, then the trace of A^2 $5A$ is equal to ________.

2. Which of the following is a solution of the differential equation
$x^2 y'' + xy' + y = 4 \sin(\ln x), x > 0$?
(a) $y = 2x \sin(\ln x)$ (b) $y = -2x \sin(\ln x)$
(c) $y = -2 \ln x \cos(\ln x)$ (d) $y = 2 \ln x \cos(\ln x)$

3. At $z = 0$, the complex function $f(z) = z|z|^2$
(a) satisfies the Cauchy-Riemann equations and is differentiable
(b) satisfies the Cauchy-Riemann equations but is not differentiable
(c) does not satisfy the Cauchy-Riemann equations but is differentiable
(d) does not satisfy the Cauchy-Riemann equations and is not differentiable

4. Ten chocolates are distributed randomly among three children standing in a row. The probability that the first child receives exactly three chocolates is
(a) $\frac{5 \times 2^{11}}{3^9}$ (b) $\frac{5 \times 2^{10}}{3^9}$
(c) $\frac{1}{3^9}$ (d) $\frac{1}{3}$

5. Let the function $f:[0, 5] \to R$ be defined by

$$f(x) = \begin{cases} 2x + 5, & 0 \leq x < 1 \\ 2x^2 + 5, & 1 \leq x < 2 \\ \frac{2}{3}x^3 + \frac{23}{3}, & 2 \leq x \leq 5 \end{cases}$$

The number of points where f is not differentiable in (0,5), is ________.

6. An integrating factor of the differential equation $(3x^2 y^3 e^y + y^3 + y^2)\, dx + (x^3 y^3 e^y - xy)\, dy = 0$ is
(a) $\frac{1}{y}$ (b) $\frac{1}{y^2}$
(c) $\frac{1}{y^3}$ (d) $\ln y$

7. If a cubic polynomial passes through the points (0, 1), (1, 0), (2, 1) and (3, 10), then it also passes through the point
(a) (–2, – 11) (b) (–1, – 2)
(c) (–1, – 4) (d) (–2, – 23)

Q. (8 – 11) Carry Two Mark Each.

8. Let the function $f: [0, \infty] \to$ R be such that $f'(x) = \frac{8}{x^2 + 3x + 4}$ for $x > 0$ and $f(0) = 1$. Then $f(1)$ lies in the interval
(a) [0, 1] (b) [2, 3]
(c) [4, 5] (d) [6, 7]

9. The perimeter of a rectangle having the largest area that can be inscribed in the ellipse $\frac{x^2}{8} + \frac{y^2}{32}$ =, is ________

10. If the work done in moving a particle once around a circle $x^2 + y^2 = 4$ under the force field $\vec{F}(x, y) = (2x - ay)\hat{i} + (2y + ax)\hat{j}$ is 16π, then $|a|$ is equal to ____.

11. Let r and s be real numbers. If

$$A = \begin{pmatrix} 1 & 2 & 0 \\ 2 & 0 & 3 \\ r & s & 0 \end{pmatrix} \text{ and } b = \begin{pmatrix} 1 \\ 1 \\ s-1 \end{pmatrix},$$

then the system of linear equations $AX = b$ has
(a) no solutions for $s \neq 2r$.
(b) infinitely many solutions for $s = 2r \neq 2$
(c) a unique solution for $s = 2r = 2$
(d) infinitely many solutions for $s = 2r = 2$

ANSWERS

1. (2)	2. (c)	3. (a)	4. (b)	5. (1)	6. (c)	7. (b)
8. (b)	9. (24)	10. (2)	11. (d)			

EXPLANATIONS

1. (2) Here, 1, 0 and –1 are eigenvalues of A. Then

$$A=\begin{bmatrix}1&0&0\\0&0&0\\0&0&-1\end{bmatrix}$$

$$\Rightarrow A^2 = AA = \begin{bmatrix}1&0&0\\0&0&0\\0&0&1\end{bmatrix}$$

$$\therefore A^2+5A = \begin{bmatrix}1&0&0\\0&0&0\\0&0&1\end{bmatrix}+5\begin{bmatrix}1&0&0\\0&0&0\\0&0&-1\end{bmatrix}$$

$$\begin{bmatrix}1&0&0\\0&0&0\\0&0&1\end{bmatrix}+\begin{bmatrix}5&0&0\\0&0&0\\0&0&-5\end{bmatrix}\begin{bmatrix}6&0&0\\0&0&0\\0&0&-4\end{bmatrix}$$

$\therefore$ Trace $(A^2 + 5A) = 6 + 0 - 4 = 2$

2. (c) Let $y = -2 \ln x \cos(\ln x)$

$$\Rightarrow y' = -2\,[\ln x\,(-\sin(\ln x)].\ \frac{1}{x}+\frac{1}{x}\cos(\ln x)\,)$$

$$=\frac{2}{x}\ln x\,\sin(\ln x)-\frac{2}{x}\cos(\ln x)$$

$$\Rightarrow y'' = 2\left[\frac{\ln x}{x}\cos(\ln x).\frac{1}{x}+\left(\frac{x.\frac{1}{x}-\ln x}{x^2}\right)\sin(\ln x)\right]$$

$$-2\left[\frac{x(-\sin(\ln x)).\frac{1}{x}-\cos(\ln x)}{x^2}\right]$$

$$=\frac{2}{x^2}\ln x\cos(\ln x)+\frac{2}{x^2}(1-\ln x)\sin(\ln x)$$

$$+\frac{2}{x^2}\sin(\ln x)+\frac{2}{x^2}\cos(\ln x)$$

Now, L.H.S

$$x^2y''+xy'+y$$

$= 2 \ln x \cos(\ln x) + 2(1 - \ln x)\sin(\ln x)$

$\quad + 2\sin(\ln x) + 2\cos(\ln x) + 2\ln x\sin(\ln x)$

$\quad - 2\cos(\ln x) - 2(\ln x)\cos(\ln x)$

$= 4\sin(\ln x) =$ R.H.S

$\therefore$ option (c) is correct.

3. (a) Let $z = x + iy$

$$\Rightarrow |z| = \sqrt{x^2+y^2}$$

$\therefore f(z) = z\,|z|^2$

$= x^3 + ix^2y + xy^2 + iy^3$

$= (x^3 + xy^{2)} + i\,(x^2y + y^3)$

$= u + iv$

$\Rightarrow x = x^3 + xy^3,\ x = x^2y + y^3$

$$\therefore\ \frac{\partial u}{\partial x}=3x^2+y^2,\ \frac{\partial u}{\partial y}=2xy$$

$$\frac{\partial v}{\partial x}=2xy,\ \frac{\partial v}{\partial y}=x^2+3y^2$$

$$\Rightarrow\left(\frac{\partial u}{\partial x}\right)_{(0,0)}=0,\left(\frac{\partial u}{\partial y}\right)_{(0,0)}=0$$

$$\left(\frac{\partial v}{\partial x}\right)_{(0,0)}=0,\left(\frac{\partial v}{\partial y}\right)_{(0,0)}=0$$

$\therefore$ f(z) at z = 0 satisfies the C-R equations

Also, u, x, u_x, x_y, u_x, u_y are continuously differentiable function.

Hence f(z) at z = 0 is differentiable function.

$\therefore$ option (a) is correct.

4. (b) Probability the each three boy gets chocolate is $\frac{1}{3}$

$$\therefore\ p=\frac{1}{3},\ q=1-\frac{1}{3}=\frac{2}{3}$$

$\therefore$ Required probability

$$= {}^{10}C_3\left(\frac{1}{3}\right)^3\left(\frac{2}{3}\right)^7=5\times\frac{2^{10}}{3^9}$$

$\therefore$ option (b) is correct.

5. (1)

$$f(x)=\begin{cases}2x+5, & 0\le x<1\\ 2x^2+5, & 1\le x<2\\ \frac{2}{3}x^3+\frac{23}{3}, & 2\le x\le 5\end{cases}$$

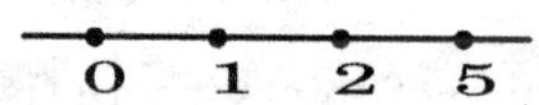

Now L.H.D at $x = 1$,

$$\lim_{h\to0}\frac{f(a-h)-f(a)}{-h}=\lim_{h\to0}\frac{f(1-h)-f(1)}{-h}$$

$$=\lim_{h\to0}\frac{2(1-h)+5-7}{-h}=\lim_{h\to0}\frac{2-2h-2}{-h}$$

$$=\lim_{h\to0}\frac{-2h}{-h}=2$$

and R.H.D. at $x = 1$

$$\lim_{h\to0}\frac{f(a+h)-f(a)}{h}=\lim_{h\to0}\frac{f(1+h)-f(1)}{h}$$

$$=\lim_{h\to0}\frac{2(1+h)^2+5-7}{h}=\lim_{h\to0}\frac{2(1+2h+h^2)-2}{h}$$

$$= \lim_{h\to 0}\frac{2+4h+2h^2-2}{h} = \lim_{h\to 0}\frac{h(4+2h)}{h}$$

$= 4$

$\therefore$ L.H.D $\neq$ R.H.D at $x = 1$

Hence given function is not differentiable at $x = 1$.

Now, L.H.D at $x = 2$

$$\lim_{h\to 0}\frac{f(2-h)-f(2)}{-h}$$

$$\lim_{h\to 0}\frac{2(2-h)^2+5-\left(\frac{2}{3}(2)^3+\frac{23}{3}\right)}{-h}$$

$$\lim_{h\to 0}\frac{2\left[4-4h+h^2\right]+5-\left[\frac{16+23}{3}\right]}{-h}$$

$$= \lim_{h\to 0}\frac{2\left[4-4h+h^2\right]+5-13}{-h}$$

$$= \lim_{h\to 0}\frac{8-8h+2h^2-8}{-h}$$

$$= \lim_{h\to 0}\frac{h[-8+2h]}{-h}$$

$= 8$

and R.H.D at $x = 2$

$$\lim_{h\to 0}\frac{f(2+h)-f(2)}{h}$$

$$= \lim_{h\to 0}\frac{\left[\frac{2}{3}(2+h)^3+\frac{23}{3}\right]-13}{h}$$

$$= \lim_{h\to 0}\frac{\left[\frac{2}{3}\left(8+h^3+3\times 4h+3(2)h^2\right)+\frac{23}{3}\right]-13}{h}$$

$$= \lim_{h\to 0}\frac{\left[\frac{16}{3}+\frac{2}{3}\left(h^3+12h+6h^2\right)+\frac{23}{3}-13\right]}{h}$$

$$= \lim_{h\to 0}\frac{\left[\frac{2}{3}h\left(h^2+12+6h\right)\right]}{h}$$

$\therefore$ L.H.D = R.H.D

at $x = 2$

Hence at $x = 2$,

$f(x)$ is differentiative

and similarly at $x = 5$

$f(x)$ is differentiable

$\therefore$ The one point i.e. $x = 1$ where 'f' is not differentiable.

6. (c) The given differential equation is $(3x^2y^3e^y + y^3 + y^2)\,dx + (x^3y^3e^y - xy)\,dy = 0$

$$\Rightarrow \left(3x^2e^y+1+\frac{1}{y}\right)dx+\left(x^3e^y-\frac{x}{y^2}\right)dy = 0$$

$$\Rightarrow M = 3x^2e^y+1+\frac{1}{y},$$

$$\text{and } N = x^3e^y-\frac{x}{y^2}$$

$$\therefore \frac{\partial M}{\partial y} = 3x^2e^y-\frac{1}{y^2},$$

$$\text{and } \frac{\partial N}{\partial x} = 3x^2e^y-\frac{1}{y^2}$$

$$\Rightarrow \frac{\partial M}{\partial x} = \frac{\partial N}{\partial x}$$

$\therefore \frac{1}{y^3}$ is the integrating factor of given differential equation.

7. (b) Let $y = a + bx + cx^2 + dx^3$ passes the points (0, 1), (1, 0), (2, 1) and (3, 10). Then

$$\begin{cases} 1 = a & (1) \\ 0 = a+b+c+d & (2) \\ 1 = a+2b+4c+8d & (3) \\ 10 = a+3b+9c+27d & (4) \end{cases}$$

$$\Rightarrow \begin{cases} b+c+d = -1 & (5) \\ b+2c+4d = 0 & (6) \\ 3b+9c+27d = 9 & (7) \end{cases}$$

$$\Rightarrow \begin{cases} c+3d = 1 & (8) \\ 3b+9c+27d = 9 & (9) \end{cases}$$

$$\Rightarrow \begin{cases} 9c+27d = 9 & (10) \\ 3b+9c+27d = 9 & (11) \end{cases}$$

$\Rightarrow$ $3b = 0$

$\Rightarrow$ $b = 0$

From (2), we have

$c + d = -1$ (12)

from (8) and (12), we have

$2d = 2 \Rightarrow d = 1$

$\therefore C = -2$

$\therefore$ The cubic polynomial is

$y = 1 - 2x^2 + x^3$.

Which satisfies by point $(-2, -1)$.

$\therefore$ option (b) is correct.

9. (24) Note: Consider the following rectangle:

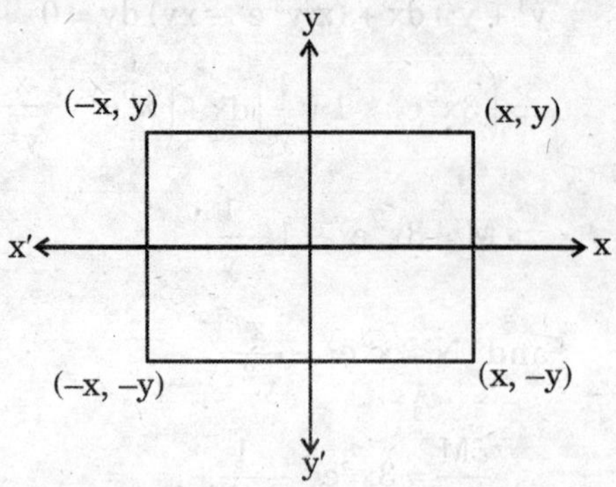

$\therefore$ Area A = 4xy and let the ellipse be

$$\frac{x^2}{a^2}+\frac{y^2}{b^2}=1 \quad ...(1)$$

Now $\frac{dA}{dx}=4x\frac{dy}{dx}+4y$

Now, by differentiating (1) w.r.t. to 'x', we get,

$$\frac{2x}{a^2}+\frac{2y}{b^2}\frac{dy}{dx}=0$$

$$\Rightarrow \frac{dy}{dx}=-\frac{b^2x}{a^2y}$$

$$\therefore \frac{dA}{dx}=4y-\frac{4b^2x^2}{a^2y}=0$$

$$\Rightarrow y^2=\frac{b^2x^2}{a^2}$$

$\therefore$ from (1), we have

$$y^2=b^2-\frac{b^2x^2}{a^2}$$

$$\Rightarrow y^2=b^2-y^2, 2y^2=b^2$$

and $\frac{y^2}{b^2}=\frac{1}{2}$

$$\Rightarrow \frac{x^2}{a^2}=\frac{1}{2}$$

Hence area is maximum when

$$x=\frac{a\sqrt{2}}{2}, y=\frac{b\sqrt{2}}{2}.$$

Here $a=\sqrt{8}$ and $b=\sqrt{32}$.

$\Rightarrow x = 2, y = 4$

$\therefore$ required perimeter

$= 4(x + y) = 4(2 + 4)$

$= 24.$

2015

Q. (1 – 7) Carry One Mark Each.

1. Considering the matrix $\begin{pmatrix} 0 & -1 & 2 \\ 1 & 0 & 3 \\ -2 & -3 & 0 \end{pmatrix}$

which one of the following statements is INCORRECT?

(*a*) One of its eigenvalues is zero.
(*b*) It has two purely imaginary eigenvalues.
(*c*) It has a non-zero real eigenvalue.
(*d*) The sum of its eigenvalues is zero.

2. The value of x where the function $f(x) = \sin(x) + \cos(x)$, defined over the domain $0 \le x \le 2\pi$, attains a minimum is

3. The radius of convergence of the following power series is $\sum_{n=0}^{\infty} \frac{(x-3)^n}{3^n n!}$

(*a*) zero
(*b*) 1
(*c*) 3
(*d*) ∞

4. For a complex number k and a complex variable z, the complex function $(z + k)^2$ is analytic

(*a*) for all k.
(*b*) only for all k whose imaginary component is zero.
(*c*) only for all k whose real component is zero.
(*d*) only when k is zero.

5. The divergence of a vector field $\vec{v}(x, y, z) = 2^x\vec{i} + 2^y\vec{j} + 2^z\vec{k}$ at a point (1, 1, 1) is

6. The type of the differential equation

$$(1-x)\frac{d^3y}{dx^3} - \sqrt{1 + \left(\frac{dy}{dx}\right)^2} + 5y = \cos(x) \text{ is}$$

(*a*) linear and first order.
(*b*) non-linear and first order.
(*c*) linear and third order.
(*d*) non-linear and third order.

7. A box has ten light bulbs out of which two are defective. Two light bulbs are drawn from this box one after the other without replacement. The probability that both light bulbs drawn are NOT DEFECTIVE is

(*a*) $\frac{8}{45}$
(*b*) $\frac{28}{45}$
(*c*) $\frac{16}{25}$
(*d*) $\frac{4}{5}$

Q. (8 – 11) Carry Two Marks Each.

8. The value of the following limit is

$$\lim_{x \to \infty} \left(\frac{x - \sin(x)}{x + \sin(x)} + \frac{\ln x}{x} \right)$$

(*a*) zero (*b*) 1
(*c*) 2 (*d*) ∞

9. The value of the Fourier coefficient, A_0, in the series

$$\left\{A_0 + \sum_{n=1}^{\infty}(a_n \cos(nx) + b_n \sin(nx))\right\}$$

of a function $f(x) = x^2$ with a period 2π defined over an interval $0 \le x \le 2\pi$ is

(*a*) $\frac{4\pi^2}{3}$ (*b*) $\frac{2\pi^2}{3}$

(*c*) $\frac{\pi^2}{3}$ (*d*) $\frac{\pi^2}{6}$

10. The general solution, $y(x)$, for the differential equation $x\frac{d^2y}{dx^2} - \frac{dy}{dx} - 1 = 0$ is (c_1 and c_2 are real constants)

(*a*) $c_1\frac{x^2}{2} + 2x + c_2$ (*b*) $c_1\frac{x^2}{2} - x + c_2$

(*c*) $c_1\frac{x^2}{2} + x + c_2$ (*d*) $c_1\frac{x^2}{2} - 2x + c_2$

11. The minimum number of iterations required to evaluate $\sqrt{28}$, correct up to three decimal places, by the Newton-Raphson method with an initial guess of 5 is

ANSWERS

1. (c)	**2.** (3.92)	**3.** (d)	**4.** (a)	**5.** (4.15)	**6.** (d)	**7.** (b)
8. (b)	**9.** (a)	**10.** (b)	**11.** (2)			

EXPLANATIONS

1. (c) $A = \begin{bmatrix} 0 & -1 & 2 \\ 1 & 0 & 3 \\ -2 & -3 & 0 \end{bmatrix}$

$A - \lambda I = 0$

$$\begin{bmatrix} 0 & -1 & 2 \\ 1 & 0 & 3 \\ -2 & -3 & 0 \end{bmatrix} - \begin{bmatrix} \lambda & 0 & 0 \\ 0 & \lambda & 0 \\ 0 & 0 & \lambda \end{bmatrix} = 0$$

$$\therefore \begin{bmatrix} -\lambda & -1 & 2 \\ 1 & -\lambda & 3 \\ -2 & -3 & -\lambda \end{bmatrix} = 0$$

$\therefore -\lambda(\lambda^2 + 9) + 1(-\lambda + 6) + 2(-3 - 2\lambda) = 0$

$\therefore -\lambda^3 - 9\lambda - \lambda + \not{6} - \not{6} - 4\lambda = 0$

$\therefore -\lambda^3 - 14\lambda = 0$

$\therefore -\lambda^{\not{3}} = 14\not{\lambda}$

$-\lambda(\lambda^2 + 14) = 0$

$\therefore\ \lambda = 0$ or $\lambda^2 = 74$

$\therefore\ \lambda = \pm\sqrt{14}\, i$

$\therefore\ \lambda_1 = 0;\ \lambda_2 = \pm\sqrt{14}\, i$ and $\lambda_3 = -\sqrt{14}\, i$

Comment :

(i) One of the eigen values is zero i.e. ($\lambda = 0$) (True)

(ii) It has two purely imaginary eigen values (True)

(iii) Sum of $\lambda_1 + \lambda_2 + \lambda_3 = 0$

2. (3.92) $f(x) = \sin x + \cos x$

$f'(x) = \cos x - \sin x$

put $f'(x) = 0$

$\therefore\ x = \frac{\pi}{4}, \frac{5\pi}{4}$ $\quad (\because x \in [0, 2\pi])$

Now $f''(x) = -\sin x - \cos x$

Now, putting $x = \pi/4$ & $x = 5\pi/4$ in above equation

we get $\quad (\because f''(x) < 0)$

$f''(\pi/4) = -\sin\frac{x}{y} - \sqrt{2}$,

which is maximum

$$\left[\because f''(x) > 0 - \cos\frac{\pi}{4}\right]$$

$$= -\frac{1}{\sqrt{2}} - \frac{1}{\sqrt{2}} = \frac{-2}{\sqrt{2}}$$

$f''(5\pi/4) = -\sin 5\frac{\pi}{4} - \cos 5\frac{\pi}{4}\sqrt{2}$,

which is minimum

$x = 3.925$

3. (d) $\sum_{n=0}^{\infty} \frac{(x-3)^n}{3^n.n!}$

Let us assume, $4_n = \frac{(x-3)^n}{3^n.n!}$

For convergence,

$$4_{n+1} = \frac{(x-3)^{n+1}}{3^{n+1}.(n+1)!}$$

$$\frac{4_n}{4_{n+1}} = \frac{(x-3)^n}{3^n.n!} \times \frac{3^{n+1}.(n+1)!}{(n-3)^{n+1}}$$

$$\frac{4_n}{4_{n+1}} = \frac{3}{(n-3)}.(n+1)$$

$$\underset{n\to\infty}{\text{Lt}} \frac{4_n}{4_{n+1}} = \left(\frac{3}{n-3}\right)(n+1) = \infty$$

4. (a) For all value of k the complex function $(z + k)^2$ is analytical, according to its properties.

5. (4.15) $\vec{V} = 2^x\hat{i} + 2^y\hat{j} + 2^z\hat{k}$

$$\text{div}(\nabla.\vec{V}) = \left(\frac{\partial}{\partial x}\hat{i} + \frac{\partial}{\partial y}\hat{j} + \frac{\partial}{\partial z}\hat{k}\right).\left(2^x\hat{i} + 2^y\hat{j} + 2^z\hat{k}\right)$$

$$\text{div}(\nabla.\vec{V}) = 2^x.\ln 2 + 2^y.\ln 2 + 2^z.\ln 2$$

$$= \ln 2\,(2^x + 2^y + 2^z)$$

We get $(x, y, z) = (1, 1, 1)$

$$\text{div}(\nabla.\vec{V}) = \ln 2\,(2 + 2 + 2)$$

$$= 6 \ln 2$$

$$= 4.158$$

6. (d) $(1-x)\frac{d^3y}{dx^3} - \sqrt{1 + \left(\frac{dy}{dx}\right)^2} + 5y = \cos(x)$

The given equation having

order = 3

degree = 2

Non-linear equation.

7. (b) p (Both bulbs drawn are not defective without replacement)

$$= \frac{8}{10} \times \frac{7}{9} = \frac{28}{45}$$

9. (a) Fourier coefficient's is given by

$$A_0 = \frac{1}{2\pi}\int_0^{2\pi} f(x).dx$$

$$= \frac{1}{2\pi}\int_0^{2\pi} x^2.dx = \frac{1}{2\pi}\left[\frac{x^3}{3}\right]_0^{2\pi}$$

$$\boxed{A_0 = \frac{8\pi^2}{2\times 3}}$$

$$A_0 = \frac{4\pi^2}{3}$$

10. (b) $x\frac{d^2y}{dx^2} - \frac{dy}{dx} - 1 = 0$ ....(1)

Put $x = e^z$

$z = \log x$

Multiply (1) with x

$$x^2\frac{d^2y}{dx^2} - x\frac{dy}{dx} = x$$

$[D(D-1) - D]\, y = x$

Now complementery function :

$D^2 - D - D = 0$

$D^2 - 2D = 0$

$\therefore\ D(D-2) = 0$

$\therefore\ D = 0, 2$

Hence $y_1(x) = c_1e^{oz} + c_2e^{2z} = c_1 + c_2x^2$

$(\because e^z = x)$

P.T.

$$y_2(x) = \frac{x}{D^2 - 2D} = \frac{2\times x}{2D} = \frac{x^2}{2}$$

$$\therefore\ y = y_1(x) + y_2(x) = c_1\frac{x^2}{2} - x + c_2$$

11. (2) $x_{n+1} = \frac{1}{b}\left\{(b-1)x_n + \frac{a}{x_nb-1}\right\}$

$b = 2,\ a = 28$

$x_0 = 5$

$$x_1 = \frac{1}{2}\left\{x_0 + \frac{28}{x_0}\right\} = \frac{1}{2}\left\{5 + \frac{28}{5}\right\} = 5.3$$

$$x_2 = \frac{1}{2}\left\{x_1 + \frac{28}{x_1}\right\} = \frac{1}{2}\left\{5.3 + \frac{28}{5.3}\right\} = 5.2915$$

$$x_3 = \frac{1}{2}\left\{x_2 + \frac{28}{x_2}\right\}$$

$$= \frac{1}{2}\left\{5.2915 + \frac{28}{5.2915}\right\} = 5.2915$$

x_3 is same as x_2, it means iteration gets stabilized after x_2.

Hence, correct answer is (2).

2016

Q. 1 – 7) : Carry One Mark Each

1. A company records heights of all employees. Let X and Y denote the errors in the average height of male and female employees respectively. Assume that $X \sim N(0,4)$ and $Y \sim N(0,9)$ and they are independent. Then the distribution of $Z = (X + Y)/2$ is

(*a*) N(0, 6.5) (*b*) N(0, 3.25)

(*c*) N(0, 2) (*d*) N(0,1)

2. The volume of the solid obtained by revolving the curve $y^2 = x, 0 \le x \le 1$ around y-axis is

(*a*) π (*b*) 2

(*c*) $\frac{\pi}{2}$ (*d*) $\frac{\pi}{5}$

3. Let y(x) be the solution of the initial value problem $\frac{dy}{dx} + 2xy = x;\ y(0) = 0$. Find the value of $\lim_{x\to\infty}$ y(x).

4. Which of the following is a quasi-linear partial differential equation?

(*a*) $\frac{\partial^2 u}{\partial t^2} + u^2 = 0$

(*b*) $\left(\frac{\partial u}{\partial t}\right)^2 + \frac{\partial u}{\partial x} = 0$

(*c*) $\left(\frac{\partial u}{\partial t}\right)^2 - \left(\frac{\partial u}{\partial x}\right)^2 = 0$

(*d*) $\left(\frac{\partial u}{\partial t}\right)^4 - \left(\frac{\partial u}{\partial x}\right)^3 = 0$

5. Let P(x) and Q(x) be the polynomials of degree 5, generated by Lagrange and Newton interpolation methods respectively, both passing through given six distinct points on the xy-plane. Which of the following is correct?

(*a*) $P(x) \equiv Q(x)$

(*b*) P(x) – Q(x) is a polynomial of degree 1

(*c*) P(x) – Q(x) is a polynomial of degree 2

(*d*) P(x) – Q(x) is a polynomial of degree 3

6. The Laurent series of $f(z) = 1/(z^3 - z^4)$ with center at z = 0 in the region $|z| > 1$ is

(*a*) $\sum_{n=0}^{\infty} z^{n-3}$ (*b*) $-\sum_{n=0}^{\infty} \frac{1}{z^{n+4}}$

(*c*) $\sum_{n=0}^{\infty} z^{n}$ (*d*) $\sum_{n=0}^{\infty} \frac{1}{z^{n}}$

7. The value of the surface integral $\iint_\Gamma \vec{F} \cdot n\, dS$ over the sphere Γ given by $x_2 + y_2 + z_2 = 1$, where $\vec{F} = 4x\,\hat{i} - z\,\hat{k}$, and n denotes the outward unit normal, is

(*a*) π (*b*) 2π

(*c*) 3π (*d*) 4π

Q. 8 – 11) : Carry Two Marks Each.

8. A diagnostic test for a certain disease is 90% accurate. That is, the probability of a person having (respectively, not having) the disease tested positive (respectively, negative) is 0.9. Fifty percent of the population has the disease. What is the probability that a randomly chosen person has the disease given that the person tested negative?

9. Let $M = \begin{pmatrix} 1 & 1 \\ 0 & 1 \end{pmatrix}$. Which of the following is correct?

(*a*) Rank of M is 1 and M is not diagonalizable

(*b*) Rank of M is 2 and M is diagonalizable

(*c*) 1 is the only eigenvalue and M is not diagonalizable

(*d*) 1 is the only eigenvalue and M is diagonalizable

10. Let $f(x) = 2x^3 - 3x^2 + 69, -5 \le x \le 5$. Find the point at which f attains the global maximum.

11. Calculate $\int_{C_1} \vec{F} \cdot d\vec{r} - \int_{C_2} \vec{F} \cdot d\vec{r}$,

where $C_1 : \vec{r}(t) = (t, t^2)$

and $C_2 : \vec{r}(t) = (t, \sqrt{t}), t$ varying from 0 to 1 and $\vec{F} = xy\,\hat{j}$.

ANSWERS

1. (b)	**2.** (d)	**3.** (0.5)	**4.** (a)	**5.** (a)	**6.** (b)	**7.** (d)
8. (0.010)	**9.** (c)	**10.** (5)	**11.** (0.150)			

EXPLANATIONS

1. (b) We have

X ~ N (0, 4) and y ~ N (0, 9)

$\Rightarrow$Distribution of $z = \dfrac{(x+y)}{2}$

~N(0, 3.25).

2. (d) We have

$$V = \int_0^1 -\pi(\sqrt{x})^2\,dx = \int_0^1 \pi x\,dx$$

$$= \pi\left[\frac{x^2}{2}\right]_0^1 = -\pi\left(\frac{1^2}{2}-0\right) = \frac{-\pi}{2}.$$

3. (0.5) We have

$$\frac{dy}{dx} + 2xy = x \text{ with } y(0) = 0$$

$$\therefore \quad \text{If} = e^{\int ex\,dx} = e^{x^2}$$

$$\therefore \text{ Solution is } y.e^{x^2} = \int x.e^{x^2}\,dx + c$$

$$\Rightarrow y.e^{x^2} = \frac{1}{2}e^{x^2} + c$$

$$\Rightarrow y = \frac{1}{2} + ce^{-x^2}$$

But y(0) = 0

$$\Rightarrow 0 = \frac{1}{2} + c$$

$$\Rightarrow c = -\frac{1}{2}$$

$$\therefore \; y = \frac{1}{2} - \frac{1}{2}e^{-x^2} = \frac{1}{2}\left(1-e^{-x^2}\right)$$

$$\therefore \lim_{x\to\infty} y(x) = \lim_{x\to\infty}\frac{1}{2}\left(1-e^{-x^2}\right) = \frac{1}{2}.$$

4. (a) A partial differential equation of the form

$$a(x, y, u(x, y))\frac{\partial u(x,y)}{\partial x} + b(x, y, u(x,y))\frac{\partial v(x,y)}{\partial y}$$

$$= c(x. y, u(x, y))$$

is known as quasi linear differential equation

This equation

$$\frac{\partial u}{\partial + 2} + x^2 = 0$$

is quasi linear partial differential equation.

5. (a) If P(x) and Q(x) are the polynomials of degree 5, generated by hagrange and Newton interpolation methods respectively, both passing through given six district points on the xy-plane, then

$$P(x) \cong Q(x)$$

$\therefore$option (a) is correct.

6. (b) We have

$$f(z) = \frac{1}{z^3 - z^4},\; z = 0, |z| > l$$

$$= \frac{1}{z^3(1-2)} = \frac{1}{z^3}\left(-\sum_{n=1}^{\infty}\frac{1}{z^n}\right), |z| > l$$

$$= \sum_{n=1}^{\infty}\frac{1}{z^{n+3}} = -\sum_{n=1}^{\infty}\frac{1}{z^{n+4}}$$

$\therefore$ option (*b*) is correct.

7. (d) given that $\vec{F} = ux\hat{i} + z\hat{k}$

$$\text{div}\vec{F} = \left(\frac{\partial}{\partial x}\hat{i} + \frac{\partial}{\partial y}\hat{j} + \frac{\partial}{\partial z}\hat{k}\right).(ux\hat{i} - z\hat{k})$$

$$\frac{\partial}{\partial x}(ux) + \frac{\partial}{\partial z}(-z) = 4 - 1 = 3$$

$$\Rightarrow \text{div}\vec{F} = 3$$

$$\iint_T \vec{F}.\,\hat{n}\,ds$$

$$\Rightarrow \iint_T \vec{F}.\,\hat{n}\,ds = \iint_T \text{div}\vec{F}\,dv$$

$$= \iiint 3\,dv = 3\iiint dv$$

= 3 × volume of the sphere ($X^2 + y^2 + z^2 = 1$)

$$= 3 \times \frac{4}{3}\pi(1)^3 = 4\pi.$$

2017

1. If $\int_0^{\frac{\pi}{\alpha}}\int_0^{\frac{\pi}{\alpha}} \frac{\sin y}{y}\, dy\, dx = \frac{1}{2}$

for same $\alpha \geq 1$, then the value of α is _____

2. Three fair dice are rolled simultaneously. The probability of getting a sum of 5 is

(*a*) $\frac{1}{108}$

(*b*) $\frac{1}{72}$

(*c*) $\frac{1}{54}$

(*d*) $\frac{1}{36}$

3. Suppose α, β, γ and δ are constant such that $p(x) = \delta + \gamma(x + 1) + \beta x(x + 1) + \alpha x(x + 1)(x - 1)$ is the interpolating polynomial for the data (–1, –3) (0, 1), (1, –1) and (2, –3). Then the value of $\gamma - \beta$ is _________.

4. Consider the ordinary differential equation

$y'' + \alpha y' + \beta y = 0,$

where α and β are constants. If $y(x) = xe^x$ is a solution of the above equation, then the value of $\beta - \alpha$ is _________.

5. Consider the system of linear equations

$2x_2 + x_3 = 0,$

$-2x_1 - x_3 = 0,$

$-x_1 + x_2 = 1.$

The above system has

(*a*) a unique solution

(*b*) infinite number of solutions

(*c*) no solution

(*d*) only two distinct solutions

6. Let C be a simple smooth closed curve enclosing the region R in the xy-plane. Let C be oriented counterclockwise. If the value of the integral

$\oint_c \left(y + e^{x^2}\right) dx + (3x + \cos y)\, dy$

is 16, then the area of R is _________.

7. Consider the ordinary differential equation

$x^2 y'' + xy' - y = x, x > 0.$

In terms of arbitrary constants c_1 and c_2, the general solution of the above equation is

(*a*) $y(x) = c_1x + c_2x^{-1} + x^3$

(*b*) $y(x) = c_1x^2 + c_2x^{-1} + \frac{1}{2}x$

(*c*) $y(x) = c_1x + c_2x^{-1} + \frac{1}{2}x \ln x$

(*d*) $y(x) = c_1x + c_2 + x^{-1}$

8. Let $f: \mathbb{R} \to \mathbb{R}$ and $g: \mathbb{R} \to \mathbb{R}$ be defined by

$$f(x) = \begin{cases} x(\sin x)\cos\frac{1}{x}, & x \neq 0 \\ 0, & x = 0 \end{cases}$$ and

$$g(x) = \begin{cases} x\cos\frac{1}{x}, & x \neq 0 \\ 0, & x = 0 \end{cases}$$

where $\mathbb{R}$ denotes the set of real numbers. Then, at x = 0

(*a*) f is differentiable but g is NOT differentiable

(*b*) f is NOT differentiable but g is differentiable

(*c*) both f and g are differentiable

(*d*) neither f nor g is differentiable

9. If $u(x, t) = g(t) \sin x$ is the solution of the wave equation $u_{tt} = u_{xx}$, $t > 0$, $0 < x < \pi$, with the initial conditions

$u(x, 0) = 2 \sin x,$

$u_t(x, 0) = 0$

$0 \leq x \leq \pi,$

and the boundary conditions

$u(0, t) = u(\pi, t) = 0, t \geq 0,$

then the value of $g\left(\frac{\pi}{3}\right)$ is _________.

10. Let $I = \int_0^1 \frac{1}{1+t}\, dt + \frac{\pi i}{2}\int_0^1 \frac{e^{\frac{i\pi t}{2}}}{1+e^{\frac{i\pi t}{2}}} dt - i\int_0^1 \frac{1}{1+it}\, dt,$

where t is a real variable and $i = \sqrt{-1}$. The value of I is _____.

11. Let $a_k = 2^{-k}k^4 \sin k$ and $b_k = 2^{-k^2} k \sin^2 k$ for k = 1, 2, ... Then

(*a*) $\sum_{k=1}^{\infty} a_k$ converges but $\sum_{k=1}^{\infty} b_k$ does NOT converge

(b) $\sum_{k=1}^{\infty} a_k$ does NOT converge but $\sum_{k=1}^{\infty} b_k$ converge

(*c*) both $\sum_{k=1}^{\infty} a_k$ and $\sum_{k=1}^{\infty} b_k$ converge

(*d*) neither $\sum_{k=1}^{\infty} a_k$ nor $\sum_{k=1}^{\infty} b_k$ converge

ANSWERS

1. (2.98 to 3.02) **2.** (d) **3.** (6.98 to 7.02) **4.** (2.98 to 3.02) **5.** (c) **6.** (7.98 to 8.02)

7. (c) **8.** (a) **9.** (.98 to 1.02) **10.** (– .02 to .02) **11.** (c)

EXPLANATIONS

1. (2.98 to 3.02)

Question is based on Fubini Series.

$$\int_0^{\frac{\pi}{\alpha}}\int_x^{\frac{\pi}{\alpha}}\frac{\sin y}{y}dydx=\frac{1}{2}$$

$$\int_0^{\frac{\pi}{\alpha}}\frac{\sin y}{y}\left(\int_0^y dx\right)dy=\frac{1}{2}$$

$$\int_0^{\pi/\alpha}\frac{\sin y}{y}ydy=\frac{1}{2}$$

$$[-\cos y]_0^{\pi/\alpha}=1/2$$

$$-\cos\left(\frac{\pi}{\alpha}\right)+1=1/2+\cos(\pi/\alpha)=+1/2$$

$$\frac{\pi}{\alpha}=\frac{\Pi}{3}$$

$$\alpha=3$$

2. (d) Number of events of getting a total of 5 = 6

i.e. (1, 1, 3), (1, 3, 1), (3, 1, 1), (2, 2, 1), (2, 1, 2) & (1, 2, 2)

Total number of possible outcomes will be $6^3 = 216$.

Therefore, probability of getting a total 1 of 5

$$P(E)=\frac{\text{No. of favorable outcomes}}{\text{Total number of Possible outcome}}$$

$$=\frac{6}{216}=\frac{1}{36}$$

3. (6.98 to 7.02)

$$P(x)=\delta+r(x+1)+\beta x(x+1)+\alpha x(x+1)(x-1)$$

$$[(x,P(x))]=[(-1,-3),(0,1),(1,-1),(2,-3)]$$

$$-3=\delta+0(\because x=-1, P(x)=-3)$$
$$\delta=-3$$

$$1=\delta+\gamma\times1(\because x=0, P(x)=1)$$
$$1=-3+\gamma$$
$$\gamma=4$$

$$-1=\delta+2\gamma+\beta$$

$$-1=-3+8+2\beta$$

$$\beta=-3$$

$$\gamma-\beta=4+3=7$$

4. (2.98 to 3.02)

Ordinary differential equation

$$y''+\alpha y'+\beta y=0$$

$$\left(D^2+\alpha D+\beta\right)y=0$$

$$\text{Root}_{1,2}=\frac{-\alpha\pm\sqrt{\alpha^2-4\beta}}{2}$$

Roots are repetitive in nature

$$\text{C.F.}=C_1e^{r_1x}+C_2xe^{r_1x}$$

$$\alpha^2-4\beta=0$$

$$\frac{-\alpha}{2}=1$$

$$\alpha=-2,\ \beta=1$$

$$\therefore \beta-\alpha=1-(-2)=3$$ ...Ans

5. (c) $0x_1+2x_2+x_3=0$

$$-2x_1+0x_2-x_3=0$$

$$-x_1+x_2+0x_3=1$$

$$\begin{bmatrix}0&2&1\\-2&0&-1\\-1&1&0\end{bmatrix}\begin{bmatrix}x_1\\x_2\\x_3\end{bmatrix}=\begin{bmatrix}0\\0\\1\end{bmatrix}$$

$$A=\begin{bmatrix}0&2&1\\-2&0&-1\\-1&1&0\end{bmatrix}$$

$$B = \begin{bmatrix} 0 \\ 0 \\ 1 \end{bmatrix}$$

$$[A/B] = \begin{bmatrix} 0 & 2 & 1 & 0 \\ -2 & 0 & -1 & 0 \\ -1 & 1 & 0 & 1 \end{bmatrix}$$

$$[A] = \begin{bmatrix} 0 & 2 & 1 \\ -2 & 0 & -1 \\ -1 & 1 & 0 \end{bmatrix}$$

Rank of $[A/B] \neq$ Rank of $[A]$

$\therefore$ No Solution exist

6. (7.98 to 8.02)

$$I = \oint \left(y + e^{x^2} \right) dx + (3x + \cos y)\, dy = 16$$

Find area of R.

7. (*c*) $x^2 y'' + xy' - y = x, x > 0$

$$\left(x^2 D^2 + xD - 1 \right) y = x$$

Put $x = e^z$

$$\left(D(D-1) + D - 1 \right) y = x$$

$$\left(D^2 - \not{D} + \not{D} - 1 \right) y = x$$

$$(D+1)(D-1)\, y = x$$

C.F:

$$y_1(x) = C_1 e^z + C_2 e^{-z}$$

$$y_1(x) = C_1 x + \frac{C_2}{x}$$

P.I.

$$\left(D^2 - 1 \right) y = e^z$$

$$y_2(x) = \frac{e^z}{D^2 - 1} = \frac{ze^z}{2D} \text{Pu} + D = 1$$

$$= \frac{z \cdot e^z}{2} = \frac{\ln(x) \cdot x}{2}$$

$$y(x) = y_1(x) + y_2(x)$$

$$= c_1 x + \frac{C_2}{x} + \frac{x \ln x}{2}$$

10. (– .02 to .02)

$$I_1 = \ln(1+t)\Big]_0^1 = \ln 2 - \ln 1 = \ln 2$$

$$I_2 = i \int_0^1 \frac{dt}{1 + it} = -\ln(1 + it)\Big]_0^1$$

Let $1 + it = y = -\ln(1 + i) + \ln 1$

$= -\ln(1 + i)$

$i\, dt = dy$

when $t = 0$, $y = 1$

$t = 1$, $y = 1 + i$

$$I_3 = \ln\left(1 + e^{\frac{i\pi t}{2}} \right) = \ln\left(1 + e^{\frac{i\pi}{2}} \right) - \ln 2$$

$$= \ln(1 + i) - \ln 2$$

$$I = I_1 + I_2 + I_3 = 0$$

2018

Q. 1 – Q. 7 carry one mark each

1. The largest interval in which the initial value problem

$$e^x \frac{d^2y}{dx^2} + \frac{1}{(x-5)}\frac{dy}{dx} + (\sqrt{x})y = \ln(x),$$

$$y(1) = 0 \text{ and } \frac{dy}{dx}(1) = 1,$$

has a unique solution is

(a) $(-\infty, \infty)$

(b) $(-5, 5)$

(c) $(0, \infty)$

(d) $(0, 5)$

2. The sum of the roots of the indicial equation at $x = 0$ of the differential equation

$$x^3 \frac{d^2y}{dx^2} + (x\sin x)\frac{dy}{dx} - (\tan x)y = 0,\ x > 0, \text{ is}$$

(a) 0

(b) 1

(c) 2

(d) –2

3. Let f be a three times continuously differentiable real valued function on (0, 5) such that its third derivative $f'''(x) = \frac{1}{100}$ for all $x \in (0,5)$. If P(x) is a polynomial of degree ≤ 2 such that P(1) = f(1), P(2) = f(2) and P(3) = f(3) then $|f(4) - P(4)|$ equals ______

4. For real numbers α_1 and α_2, if the formula $\int_{-1}^{1} f(x)\,dx = \alpha_1 f\left(\frac{-1}{2}\right) + \alpha_2 f\left(\frac{1}{2}\right)$ is exact for all polynomials of degree ≤ 1 then $2\alpha_1 + 3\alpha_2$ equals ______

5. Raju has four fair coins and one fair dice. At first Raju tosses a coin. If the coin shows head then he rolls the dice and the number that dice shows is taken as his score. If the coin shows tail then he tosses three more coins and the total number of tails shown (including the first one) is taken as his score. If Raju tells that his score is 2 then the probability that he rolled the dice is (up to two decimal places) ______

6. Let *f* be a continuously differentiable real valued function defined by

$$f(x) = \begin{cases} bx + a & \text{if } x < 1, \\ 5x^2 & \text{if } x \geq 1. \end{cases}$$

Then the value of a^2b is ______

7. A rectangular box without top cover having a square base is to be made from a sheet of 108 square meters. Then the largest possible volume of the box in cubic meters is ______

Q. 8 – 11 carry two marks each.

8. Let $A = \begin{pmatrix} 5 & -3 \\ 6 & -4 \end{pmatrix}$. Then the trace of A^{1000} equals

(a) $2^{1000} - 1$ (b) $2^{1000} + 1$

(c) 1 (d) 2^{1000}

9. Let $\mathbb{C}$ denote the set of complex numbers and $i^2 = -1$. Let γ be the simple positively oriented circle $|z| = 1$ and $S = \{z \in \mathbb{C} | 0 < |z| < 2\}$. If $f: S \to \mathbb{C}$ is analytic in S and is given by

$$f(z) = \frac{1}{8z^2} + \frac{7}{2z} + \sum_{n=0}^{\infty} a_n z^n,\ z \in S$$

then the value of the contour integral

$$\frac{1}{\pi i}\int_{\gamma}\left(\frac{e^z}{\cos z} + f(z)\right)dz \text{ is}$$

(a) 0

(b) $\frac{1}{8}$

(c) 7

(d) $\frac{7}{2}$

10. Let $\mathbb{R}^3$ denote the three dimensional Euclidean space and $F(x, y, z) = y\hat{i} + x\hat{j} + z\hat{k}$ for all $(x, y, z) \in \mathbb{R}^3$. If C is the curve described by the parametric equation $r(t) = \cos t\,\hat{i} + \sin t\,\hat{j} + 2t^2\,\hat{k}$, $0 \leq + \leq 1$, then the value of the line integral $\int_C F.dr$ is ______

11. Let u(x, t) satisfy the initial and boundary value problem

$$\frac{\partial u}{\partial t} = 2\frac{\partial^2 u}{\partial x^2},\ 0 < x < \pi,\ t > 0,$$

$u(0, t) = 0\ u(\pi, t), t > 0$

$u(x, 0) = \sin x + 2 \sin 4x, 0 < x < \pi.$

Then the value of $u\left(\frac{\pi}{2}, \ln(5)\right)$ is ______

ANSWERS

1. (d)	**2.** (a)	**3.** (.01 to .01)	**4.** (5 to 5)	**5.** (.29 to .32)	**6.** (250 to 250)	**7.** (108 to 108)
8. (b)	**9.** (c)	**10.** (3 to 3)	**11.** (.04 to .04)			

EXPLANATIONS

1. (d) We have

$$e^x \frac{d^2y}{dn^2} + \frac{1}{(x-5)}\frac{dy}{dx} + \sqrt{x}y = \ln x$$

with $y(1) = 0, \frac{dy}{dx}(1) = 1$

$$\Rightarrow \frac{d^2y}{dn^2} + \frac{1}{e^x(x-5)}\frac{dy}{dx} + \frac{\sqrt{x}}{e^x}y = \frac{\ln x}{e^x}$$

On comparing

$$\frac{d^2y}{dx^2} + p(x)\frac{dy}{dx} + q(x)y = g(x)$$

we get $p(x) = \frac{1}{e^x(x-5)}, q(x) = \frac{\sqrt{x}}{e^x},$

$$g(x) = \frac{\ln x}{e^x}.$$

$\therefore$ p(x) is continuous for $x \neq 5$

q(x) is continuous for $x \geq 0$

and g(x) is continuous for $x \neq 0$

so possible interval are (0, 5) and (5, ∞).

But interval must contains 1.

So required interval is (0, 5)

2. (a) The given differential equation is

$$x^3\frac{d^2y}{dx^2} + (x\sin x)\frac{dy}{dx} - (\tan x)y = 0,\ x > 0$$

$$\Rightarrow x^2\frac{d^2y}{dx^2} + \sin x\frac{dy}{dx} - \frac{\tan x}{x}y = 0$$

$$\Rightarrow p(x) = \frac{\sin x}{x}, q(x) = -\frac{\tan x}{x}$$

$\therefore$ Indicial equation is

$r^2 + (p(0) - 1)r + q(0) = 0$

$\Rightarrow r^2 - r = 0$

$\Rightarrow r(r-1) = 0$

$\Rightarrow r = 0, 1$

$\therefore$ sum of roots = 0 + 1 = 1.

3. (.01 to .01)

We have

$$f(x) = \frac{x^3}{600}$$

Which is three times continuously differentiable real valued function on (0, 5) such that

$$f''(x) = \frac{1}{100}.$$

net $p(x) = a + bx + cx^2$.

p(1) = f(1)

$$\Rightarrow a + b + c = \frac{1}{600} \quad ...(1)$$

p(2) = f(2)

$$\Rightarrow a + 2b + 4c = \frac{1}{75} \quad ...(2)$$

and p(3) = f(3)

$$\Rightarrow a + 3b + 9c = \frac{9}{200} \quad ...(3)$$

on solving (1), (2) and (3) we get

$$a = \frac{1}{100}, b = \frac{11}{600}, c = \frac{1}{100}$$

$$\therefore\ p(x) = \frac{1}{100} - \frac{11}{600}x + \frac{1}{100}x^2.$$

$\therefore |f(4) - p(4)|$

$$= \left|\frac{64}{600} - \left(\frac{1}{100} - \frac{44}{600} + \frac{16}{100}\right)\right|$$

$$= \left|\frac{64}{600} - \left(\frac{17}{100} - \frac{44}{600}\right)\right|$$

$$= \left|\frac{64}{600} + \frac{58}{600}\right| = \frac{6}{600} = 0.01$$

4. (5 to 5)

Let f(x) = 1 then

$$\int_{-1}^{1} f(x)\,dx = \alpha_1 f\left(-\frac{1}{2}\right) + \alpha_2 f\left(\frac{1}{2}\right)$$

$$\int_{-1}^{1} 1\,dx = \alpha_1 1 + \alpha_2(1)$$

$$\therefore\ [x]_{-1}^{1} = \alpha_1 + \alpha_2$$

$$\Rightarrow \alpha_1 + \alpha_2 = 2$$

Let f(x) = x, then

$$\int_{-1}^{1} x\,dx = -\frac{1}{2}\alpha_1 + \frac{1}{2}\alpha_2$$

$$\Rightarrow \left[\frac{x^2}{2}\right]_{-1}^{1} = \frac{1}{2}(\alpha_2 - \alpha_1)$$

$$\therefore \frac{1}{2} - \frac{1}{2} = \frac{1}{2}(\alpha_2 - \alpha_1)$$

$$\Rightarrow \alpha_2 - \alpha_1 = 0 \quad ...(2)$$

from (1) and (2)

$$2\,\alpha_1 = 2$$

$$\Rightarrow \alpha_1 = 1$$

$$\therefore \alpha_2 = 1$$

$$\therefore 2\alpha_1 + 3\alpha_2 = 2 \times 1 + 3 \times 1 = 5$$

5. (.29 to .32)

Required probability

$$= \frac{0.1 \times \frac{1}{2}}{0. \times \frac{1}{2} + 0.9 \times \frac{1}{2}} = \frac{0.1 \times 0.5}{0.1 \times 0.5 + 0.9 \times 0.5}$$

$$= \frac{0.05}{0.05 + 0.45} = \frac{0.05}{0.50} = 0.1$$

6. (250 to 250)

$$\text{L. H.} = \lim_{x \to 1^{-1}} f(x)$$

$$= \lim_{x \to 0} f(1-h)$$

$$= \lim_{h \to 0} b(1-h) + a$$

$$= \lim_{h \to 0} b + a - bh \; = b + a$$

$$\text{R. H.} = \lim_{x \to 1^+} f(x) = \lim_{h \to 0} f(1+h)$$

$$= \lim_{h \to 0} 5(1+h)^2 = 5$$

But given function is continuous.

$$\therefore a + b = 5$$

and clearly it is differentiable if

$$a = -5 \text{ and } b = 10$$

So $a^2 b = (-5)^2 \times 10 = 250$.

7. (108 to 108) Volume of box = $x^2 \times h$

Now total suface area of box open at top

$$\therefore \quad 108 = x^2 + 4xh$$

$$\therefore \quad h = \frac{108 - x^2}{4x} \quad \text{eq. (i)}$$

Now, volume of box = x^2h

$$= x^2\left(\frac{108 - x^2}{4x}\right)$$

$$= x\,27 - \frac{x^3}{4}$$

$$\text{Now} \quad \frac{dy}{dx} = 27 - \frac{1}{4} \times 3x^2$$

$$\text{Now} \quad \frac{dy}{dx} = 0$$

$$\therefore \quad x = 6$$

Putting the value of 'x' in eq. (i) we get

$$h = \frac{108 - 36}{24}$$

$$h = \frac{72}{24}$$

$$\therefore \quad h = 3$$

$\therefore$ Largest possible volume = x^2h

$$= 36 \times 3$$

$$= 108 \text{ m}^3$$

8. (b) We have

$$A = \begin{bmatrix} 5 & -3 \\ 6 & -4 \end{bmatrix}$$

Let $|A - \lambda\, I| = 0$

$$\Rightarrow \begin{vmatrix} 5-\lambda & -3 \\ 6 & -4-\lambda-9 \end{vmatrix} = 0$$

$$\Rightarrow -(5-\lambda)(4+\lambda) + 18 = 0$$

$$\Rightarrow -(20 - 4\lambda + 5\lambda - \lambda^2) + 18 = 0$$

$$\Rightarrow \lambda^2 - \lambda - 20 + 18 = 0$$

$$\Rightarrow \lambda^2 - \lambda - 2 = 0$$

$$\Rightarrow \lambda^2 - 2\lambda + \lambda - 2 = 0$$

$$\Rightarrow \lambda(\lambda - 2) + 1(\lambda - 2) = 0$$

$$\Rightarrow (\lambda - 2)(\lambda + 1) = 0$$

$$\Rightarrow \lambda = -1, 2$$

$\therefore$ eigen values of

$$A^{1000} = (-1)^{1000}, 2^{1000} = 1, 2^{1000}$$

$\therefore$ trace (A) = $1 + 2^{1000}$

9. (c) We have

$$I = \frac{1}{\pi i}\int_\gamma \left(\frac{e^z}{\cos z} + f(z)\right) dz$$

$$= \frac{1}{\pi i}\int_\gamma \frac{e^z}{\cos z}\,dz + \frac{1}{\pi i}\int f(z)\,dz$$

where r : $|z| = 1$,

But $\frac{e^z}{\cos z}$ is analytic in $|z| = 1$.

$$\Rightarrow \int \frac{e^z}{\cos z} d_z = 0$$

$$\therefore\ I = \frac{1}{\pi i}\int_{\gamma} f(z)\,dz$$

$$= \frac{1}{\pi i} 2\pi i \,(\text{Residue of } f(z))$$

$$= \frac{1}{\pi i} 2\pi i\left(\frac{7}{2}\right) = 7$$

10. (3 to 3)

We have, $F(x, y, z) = -y\hat{i} + x\hat{j} + z\hat{k}$

given that

$r(t) = \cos t\hat{i} + \sin t\hat{j} + 2t^2\hat{k} \quad 0 \le t \le 1$

compare with $r(t) = x(t)\hat{i} + y(t)\hat{j} + z(t)\hat{k}$

$x(t) = \cos t,\ y(t) = \sin t.\ z(t) = 2t^2$.

$$\int_c \vec{F}.\,d\vec{r} = \int_c \left(\vec{F}.\frac{d\vec{r}}{dt}\right) dt$$

$$= \int_0^1 \left(-\sin t\hat{i} + \cos t\hat{j} + 2t^2\hat{k}\right).$$

$$\int_0^1 \left(-\sin \hat{i} + \cos t\hat{j} + 4t\hat{k}\right) dt$$

$$= \int_0^1 \left(\sin^2 t + \cos^2 + 8t^3\right) dt$$

$$= \int_0^1 \left(1 + 8t^3\right) dt = \left(t + 2t^4\right)_0^1$$

$= 1 + 2 = 3$

Hence $\int_c \vec{F}\,d\vec{r} = 3$

11. (.04 to .04)

The given heat equation is

$$\frac{\partial u}{\partial t} = 2\frac{\partial^2 u}{\partial x^2},\ 0 < x < \pi,\ t > 0$$

$u(0, t) = 0 = u(\pi, t),\ t > 0$

$u(x, 0) = \sin x + 2 \sin 4x,\ 0 < x < \pi$

The solution is

$$u(x,t) = \sum_{x=1}^{\infty} Bn \sin\left(\frac{n\pi x}{L}\right) e^{-K\left(\frac{n\pi}{L}\right)^2 t}$$

where $B_n = \frac{2}{L}\int_0^L f(x) \sin\left(\frac{n\pi x}{L}\right) dx$

Here $L = \pi,\ K = 2,\ f(x) = \sin x + 2 \sin 4x$

$$\therefore\ B_n = \frac{2}{\pi}\int_0^{\pi} (\sin x + 2\sin 4x) \sin(nx)\, dx$$

$$= \frac{2}{\pi}\left[\int_0^{\pi} \sin x \sin(x) dx + 2\int_0^{\pi} \sin 4x \sin(x)\, dx\right]$$

$$= \frac{2}{\pi} \times \frac{\pi}{2} = 1$$

$B_2 = 0,\ B_3 = 0$

$$B_4 = \frac{2}{\pi}\left[\int_0^{\pi} \sin x \sin 4x\, dx + 2\int_0^{\pi} \sin 4x \sin 4x\, dx\right]$$

$$= \frac{2}{\pi} \times 2 \times \frac{\pi}{2} = 2$$

$B_5 = B_6 =$ ------- 0

$$\therefore\ u(x,t) = \sin x\, e^{-2t} + 2\sin 4x\, e^{-32t}$$

$$\therefore\ u\left(\frac{\pi}{2}, \ln 5\right) =$$

$$\sin\left(\frac{\pi}{2}\right) e^{-2\ln 5} + 2\sin\left(4.\frac{\pi}{2}\right) e^{-32\ln 5} = 5^{-2}$$

$$= \frac{1}{25} = 0.04$$

XE-B

FLUID MECHANICS

Syllabus

Section 1: Flow and Fluid Properties

Viscosity, relationship between stress and strain-rate for Newtonian fluids, incompressible and compressible flows, differences between laminar and turbulent flows. Hydrostatics:

Buoyancy, manometry, forces on submerged bodies.

Section 2: Kinematics

Eulerian and Lagrangian description of fluids motion, concept of local and convective accelerations, steady and unsteady flows.

Section 3: Integral analysis

Control volume analysis for mass, momentum and energy.

Section 4: Differential Analysis

Differential equations of mass and momentum for incompressible flows: inviscid - Euler equation and viscous flows - Navier-Stokes equations, concept of fluid rotation, vorticity, stream function, Exact solutions of Navier-Stokes equation for Couette Flow and Poiseuille flow.

Section 5: Inviscid flows

Bernoulli's equation - assumptions and applications, potential function, Elementary plane flows - uniform flow, source, sink and doublet and their superposition for potential flow past simple geometries.

Section 6: Dimensional analysis

Concept of geometric, kinematic and dynamic similarity, some common non-dimensional parameters and their physical significance: Reynolds number, Froude number and Mach number.

Section 7: Internal flows

Fully developed pipe flow, empirical relations for laminar and turbulent flows: friction factor and Darcy-Weisbach relation.

Section 8: Prandtl boundary layer equations

Concept and assumptions, qualitative idea of boundary layer and separation, streamlined and bluff bodies, drag and lift forces. Flow measurements: Basic ideas of flow measurement using venturimeter, pitot-static tube and orifice plate.

Subject-wise Analysis

XE -B : FLUID MECHANICS

Year	1 mark Questions	2 marks Questions	Total Marks
2009	8	16	40
2010	9	13	35
2011	9	13	35
2012	9	13	35
2013	9	13	35
2014	9	13	35
2015	9	13	35
2016	9	13	35
2017	9	13	35
2018	9	13	35

2009

Useful Data

Acceleration due to gravity, $g = 10 \text{ m/s}^2$

Density of water, $\rho_w = 1000 \text{ kg/m}^3$

Density of air, $\rho_a = 1.2 \text{ kg/m}^3$

Density of mercury (Hg) $\rho_{Hg} = 13600 \text{ kg/m}^3$

Dynamic viscosity of water, $\mu_w = 10^{-3} \text{ kg/(m.s)}$

Dynamic viscosity of air, $\mu_a = 1.8 \times 10^{-5} \text{ kg/(m.s)}$

Q. 1 – 8 carry one mark each

1. Under what conditions is the equation $\Delta \cdot \rho \vec{V} = 0$ valid?

P : Steady incompressible flow

Q : Unsteady incompressible flow

R : Steady compressible flow

S : Unsteady compressible flow

(*a*) P, Q, R (*b*) Q, R, S

(*c*) P, R, S (*d*) P, Q, S

2. Stream function CANNOT be defined for

(*a*) two dimensional incompressible flow

(*b*) two dimensional compressible flow

(*c*) three dimensional incompressible flow

(*d*) axisymmetric incompressible flow

3. Which one of the following is an irrotational flow?

(*a*) Free vortex flow (*b*) Forced vortex flow

(*c*) Couette flow (*d*) Wake flow

4. Under strong wind conditions, electrical cables can be subjected to wind-induced oscillations. Which one of the following non-dimensional numbers is relevant to this problem?

(*a*) Froude number (*b*) Weber number

(*c*) Faraday number (*d*) Strouhal number

5. Dimples are made on golf balls for which of the following reasons?

P : to make the ball travel a longer distance

Q : to make the flow over the ball turbulent

R : to make the flow over the ball laminar

S : to create a separated boundary layer flow over the ball

(*a*) P, Q (*b*) Q, S

(*c*) R, S (*d*) P, R

6. In a 2-D boundary layer flow, x and y are the streamwise and wll-normal coordinates, respectively. If u denotes the velocity along the x direction, which one of the following represents the condition at the point of flow separation?

(*a*) $\frac{\partial u}{\partial x} = 0$ (*b*) $\frac{\partial u}{\partial y} = 0$

(*c*) $\frac{\partial^2 u}{\partial x^2} = 0$ (*d*) $\frac{\partial^2 u}{\partial y^2} = 0$

7. Which one among the following boundary layer flows is the LEAST susceptible to flow separation?

(*a*) Turbulent boundary layer in a favourable pressure gradient

(*b*) Laminar boundary layer in a favourable pressure gradient

(*c*) Turbulent boundary layer in an adverse pressure gradient

(*d*) Laminar boundary layer in an adverse pressure gradient

8. Air from the blower of a hair dryer flows between two identical elliptical cylinders suspended freely, for two cases shown in the figure. The cylinders would move

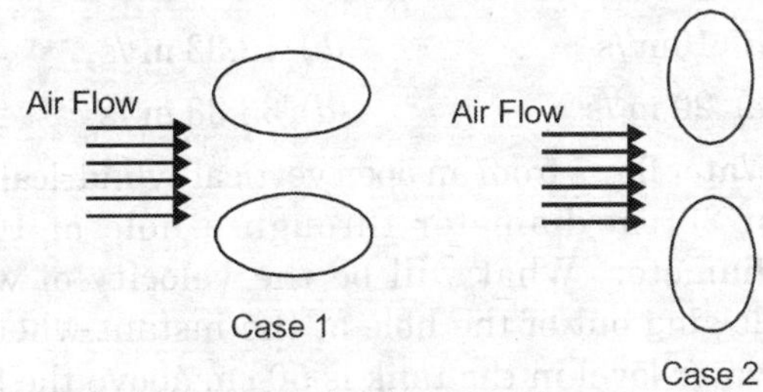

(*a*) away from each other for Case 1 and towards each other for Case 2

(*b*) towards each other for Case 1 and away from each other for Case 2

(*c*) away from each other for Case 1 and away from each other for Case 2

(*d*) towards each other for Case 1 and towards each other for Case 2

Q. 9 – 24 carry two marks each

9. A 40 cm cubical block slides on oil (viscosity = 0.80 Pa.s), over a large plane horizontal surface. If the oil film between the block and the surface has a uniform thickness of 0.4 mm, what will be the force required to drag the block at 4 m/s? Ignore the end effects and treat the flow as two dimensional.

(*a*) 1280 N (*b*) 1640 N

(*c*) 1920 N (*d*) 2560 N

10. For a floating body, G, B, and M represent the centre of gravity, centre of buoyancy, and the metacentre, respectively. The body will be stable if

(*a*) G is located above B

(*b*) B is located above M

(*c*) M is located above B

(*d*) M is located above G

11. A nozzle has inlet and outlet diameters of 10 cm and 5 cm, respectively. If it discharges air at the rate of 0.1 m³/s into the atmosphere, the gauge pressure (static) at the nozzle inlet will be

(*a*) 1.26 kPa (*b*) 1.46 kPa

(*c*) 3.52 kPa (*d*) 3.92 kPa

12. Consider incompressible flow through a two-dimensional open channel. At a certain section A-A, the velocity profile is parabolic. Neglecting air resistance at the free surface, find the volume flow rate per unit width of the channel.

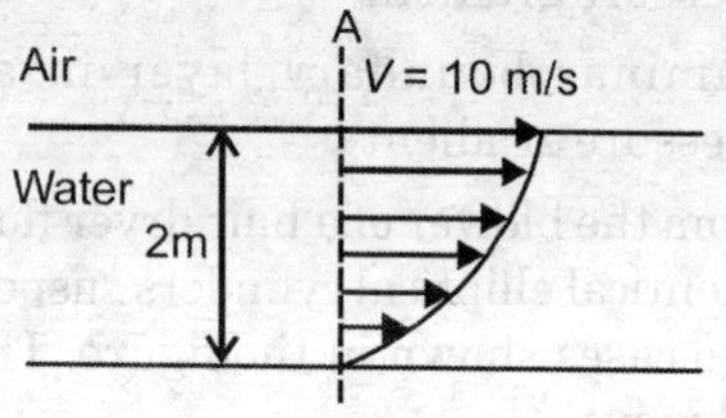

(*a*) 10m³/s (*b*) 13.33 m³/s

(*c*) 20 m³/s (*d*) 33.33 m³/s

13. Water flows from an open vertical cylindrical tank of 20 cm diameter through a hole of 10 cm diameter. What will be the velocity of water flowing out of the hole at the instant when the water level in the tank is 50 cm above the hole? Ignore unsteady effects.

(*a*) 3.16 m/s

(*b*) 3.26 m/s

(*c*) 3.36 m/s

(*d*) 3.46 m/s

14. In the manometer shown in the figure, the pressure p_A of the gas inside bulb A is approximately,

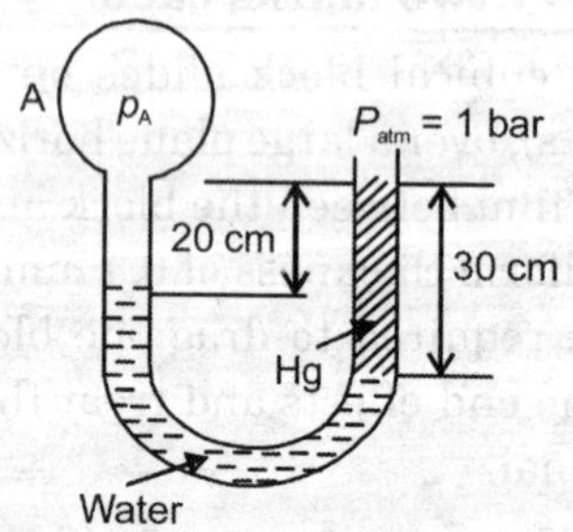

(*a*) 0.8 bar (*b*) 1.2 bar

(*c*) 1.4 bar (*d*) 1.6 bar

15. Consider a fully developed laminar flow in a circular pipe. If the diameter of the pipe is halved while the flow rate and length of the pipe are kept constant, the head loss increases by a factor of

(*a*) 4 (*b*) 8

(*c*) 16 (*d*) 32

16. A 1 : 20 model of a submarine is to be tested in a towing tank containing sea water. If the submarine velocity is 6 m/s, at what velocity should the model be towed for dynamic similarity?

(*a*) 60 m/s (*b*) 120 m/s

(*c*) 180 m/s (*d*) 240 m/s

17. An oil droplet (density = 800 kg/m³) is rising in still water at a constant velocity of 1mm/s. Its radius is approximately

(*a*) 21 micron

(*b*) 24 micron

(*c*) 34 micron

(*d*) 47 micron

18. Determine the correctness or otherwise of the following **Assertion [a]** and the **Reason [r]:**

Assertion [a] : The coefficient of discharge of orifice flow meter is less than that of venturi meter.

Reason [r] : Orifice flow meter is a differential pressure device.

(*a*) Both [a] and [r] are true and [r] is the correct reason for [a]

(*b*) Both [a] and [r] are true but [r] is not the correct reason for [a]

(*c*) Both [a] and [r] are false

(*d*) [a] is true but [r] is false

Common Data Questions

Common Data for Questions 19 and 20:

A long cylindrical object submerged in still water is moving at a constant speed of 5 m/s perpendicular to its axis, as shown in the figure. Neglect viscous effects and assume free stream pressure to be 100 kPa.

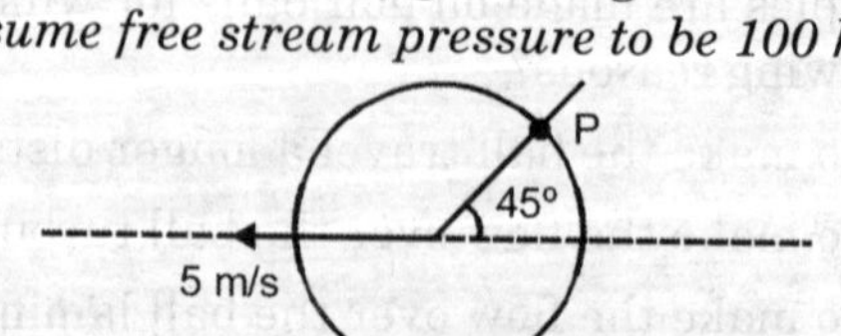

19. The fluid velocity at point P with respect to the cylinder will be approximately

(*a*) 3.5 m/s (*b*) 5 m/s

(*c*) 7 m/s (*d*) 10 m/s

20. The absolute pressure at point P will be approximately

(*a*) 137 kPa (*b*) 112 kPa

(*c*) 87 kPa (*d*) 62 kPa

Common Data for Questions 21 and 22:

The velocity field for a two dimensional flow is given by: $\vec{V}(x,y,t) = \frac{x}{t}\hat{i} - \frac{y}{t}\hat{j}$

21. The total acceleration is

(*a*) $\frac{x}{t^2}\hat{i} - \frac{y}{t^2}\hat{j}$ (*b*) $-\frac{x}{t^2}\hat{i} + \frac{y}{t^2}\hat{j}$

(*c*) $\frac{2x}{t^2}\hat{i}$ (*d*) $\frac{2y}{t^2}\hat{j}$

22. The given velocity field is

(*a*) incompressible and rotational

(*b*) compressible and rotational

(*c*) incompressible and irrotational

(*d*) compressible and irrotational

Linked Answer Questions

Statement for Linked Answer Questions 23 - 24:

An incompressible fluid is passed through a T-junction supported on wheels, as shown in the figure. The area at outlet A is twice that of outlet B. While the incoming mass flow rate is fixed, the distribution of flow at the two outlets can be varied by a suitable mechanism built in the system. Assume that the flexible tube offers no resistance to motion, and frictional effects in the pipes and wheels can be neglected. Now, consider the following two cases:

Case 1: The flow rates at sections A and B are equal.

Case 2 : The velocities at sections A and B are equal.

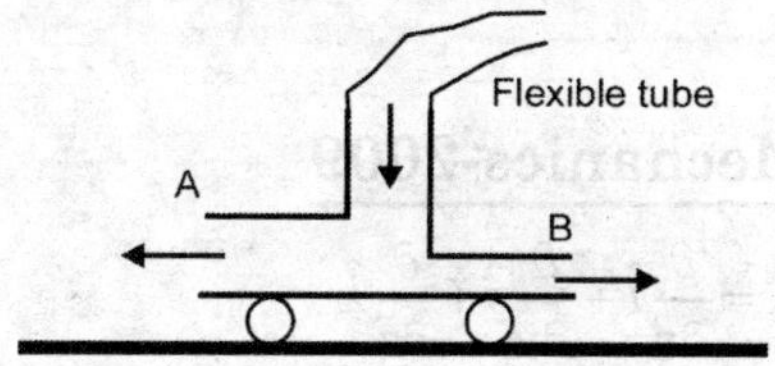

23. Which of the following statements are true?

P : In Case 1, the velocity at section A is twice the velocity at section B

Q : In Case 1, the velocity at section A is half the veloci ty at section B

R : In Case 2, the flow rate at section A is twice that at section B

S : In Case 2, the flow rate at section A is half that at section B

(*a*) P, R (*b*) P, S

(*c*) Q, R (*d*) Q, S

24. Which of the following statements are true?

P : In Case 1, the system moves to the left

Q : In Case 1, the system moves to the left

R : In Case 2, the system moves to the left

S : In Case 2, the system moves to the left

(*a*) P, R (*b*) P, S

(*c*) Q, R (*d*) Q, S

ANSWERS

1. (c)	**2.** (c)	**3.** (a)	**4.** (d)	**5.** (c)	**6.** (b)	**7.** (c)
8. (b)	**9.** (d)	**10.** (d)	**11.** (b)	**12.** (c)	**13.** (a)	**14.** (c)
15. (d)	**16.** (b)	**17.** (b)	**18.** (a)	**19.** (b)	**20.** (a)	**21.** (c)
22. (c)	**23.** (c)	**24.** (d)				

EXPLANATIONS

Fluid Mechanics-2009

1. (c) $\Delta = \frac{\partial}{\partial x}\hat{i} + \frac{\partial}{\partial y}\hat{j} + \frac{\partial}{\partial z}\hat{k}$

$\vec{V} = \rho\left(4x\hat{i} + v\hat{j} + w\hat{k}\right)$

$\Delta \cdot \vec{V} = 0$

$\rho\left(\frac{\partial u}{\partial i} + \frac{\partial v}{\partial y} + \frac{\partial w}{\partial z}\right) = 0$

if $\rho = c$, then above equation is valid for steady and incompressible flow.

if $\rho \neq c$, then above equation is valid for steady and compressible flow.

2. (c) Stream function is scalar function of space and time.

$\psi = f(x \cdot y \cdot t)$

It is valid for two dimensional flow.

3. (a) Free vertex flow is an irrotational flow, because in the free vortex flow, there is no any external torque are required.

In free vortex flow,

$\tau = 0$

4. (d) Under strong wind conditions, electrical cables can be subjected to wind-induced oscillation the non-dimensional number are strouhal number.

5. (c) Dimples are made on golf balls are due to,

(i) to make the flow over the ball laminar.

(ii) to create a separated boundary layer flow over the ball.

6. (b) Condition for flow separation

i) $\frac{\partial p}{\partial x} > 0$

ii) $\frac{\partial u}{\partial y} = 0$

iii) $\tau = \mu \cdot \frac{\partial u}{\partial y} = 0$

7. (c) Boundary layer separation occur in $\frac{\partial p}{\partial x} > 0$ - adverse pressure gradient where shear stress is zero, in turbulence condition.

8. (b)

9. (d) Given that,

For Cubical block

Side of block (a) = 40 cm = 0.4 m

For two dimensional flow

Area (A) = $2 \times a^2 = 2 \times 0.4^2$

Force required (F) = $\frac{\mu u A}{n}$

$$F = \frac{0.8 \times 4 \times 2 \times 0.4^2}{0.4 \times 10^{-3}}$$

$$\boxed{F = 2560 \text{ N}}$$

10. (d) Here,

G – Centre of gravity

B – Centre of Buoyancy

M – Metacentre

The floating body will be in stable equilibrium, if metacentre (M) lies above the C.G. (G)

Melacentric heigh (GM) = BM – BG

GM > 0 For stable condition

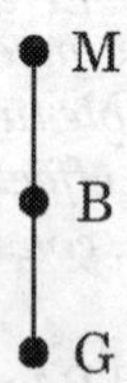

11. (b) Given that

Inlet diameter (d_1) = 10 cm = 0.1 m.

$$a_1 = \frac{\pi}{4}d_1^2 = \frac{\pi}{4} \times 0.1^2 = 7.85 \times 10^{-3} \text{ m}^2$$

Outlet diameter (d_2) = 5 cm = 0.05 m

$$a_2 = \frac{\pi}{4}d_2^2 = \frac{\pi}{4} \times 0.05^2 = 1.96 \times 10^{-3} \text{ m}^2$$

discharge (Q) = 0.1 m³/sec.

$$Q = \frac{c_d a_1 a_2 \sqrt{2gh}}{\sqrt{a_1^2 - a_2^2}}$$

$$0.1 = \frac{0.98 \times (9.85 \times 10^{-3})(1.96 \times 10^{-3})\sqrt{2gh}}{\sqrt{(7.85 \times 10^{-3})^2 - (1.96 \times 10^{-3})^2}}$$

2gh = 2544

gh = 1272

Garge Pressure (P) = ρ gh

$$\boxed{Pg = 1.2 \times 1272 = 1.50 \text{ kPa}}$$

12. (c)

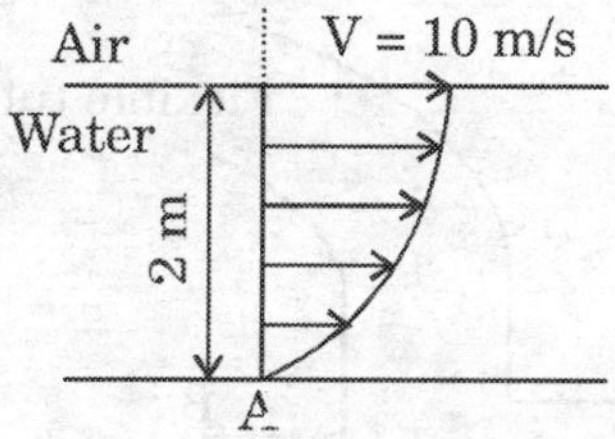

Volume flow rate per unit width of channel (Q)

$$Q = A \cdot V$$
$$= 2 \cdot 10$$
$$\boxed{Q = 20 \text{ m}^3/\text{sec.}}$$

13. (a) Given that

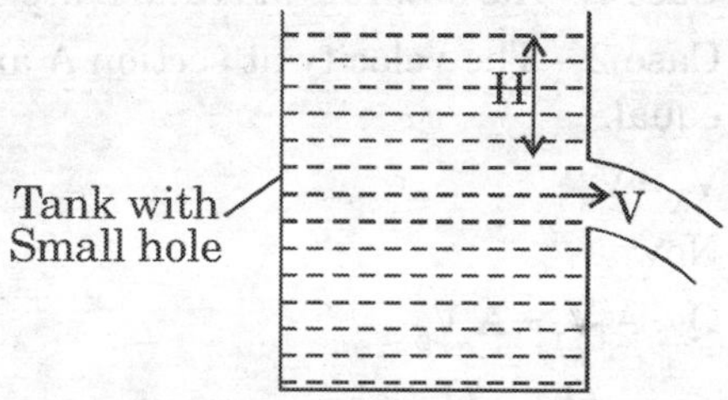

H = 50 cm = 0.5 m

Velocity of flow (V) = $C_V\sqrt{2gH}$

$$V = 0.98\sqrt{2 \times 9.81 \times 0.5}$$
$$\boxed{V = 3.16 \text{ m/sec.}}$$

14. (c)

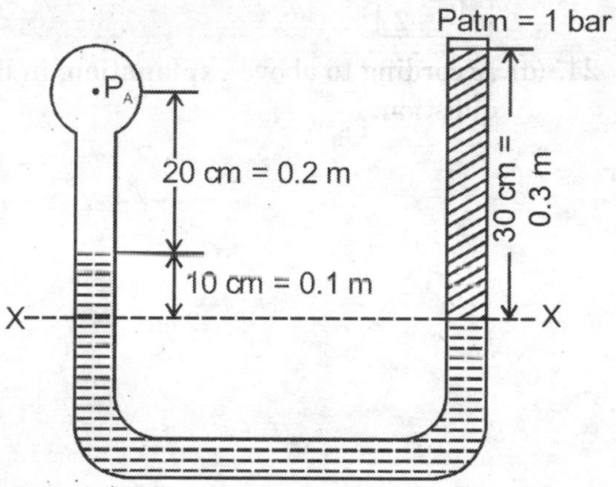

Pressure at a point A

$$P_A = \rho_{Hg} \cdot g \cdot h_{Hg} - \rho_w \cdot g \cdot h_w + P_{atm.}$$
$$P_A = (13600 \times 9.81 \times 0.3)$$
$$-(1000 \times 9.81 \times 0.1) + 1 \times 10^5$$
$$\boxed{P_A = 1.39 \text{ bar}}$$

15. (d) Head loss in fully developed, laminar flow

$$h_f = \frac{4fLV^2}{2gd} = \frac{32fLQ^2}{\pi^2 gd^5}$$

According to given data

$$d' = \frac{d}{2}$$

C = Constant

Q = Constant

then,

$$h_f' = \frac{32fLQ^2}{\pi^2 g \cdot \left(\frac{d}{2}\right)^5} = 32\left(\frac{32fLQ^2}{\pi^2 gd^5}\right)$$
$$\boxed{h_f' = 32 \cdot h_f}$$

16. (b) Given that, we need to find

$$V_m = ?$$
$$\frac{V_m}{V_s} = 20$$
$$V_m = 20 \times 6 = 120 \text{ m/s}$$

18. (a) The discharge of orifice meter is less than that of venturimeter.

$$(Q)_{\text{venturimeter}} > (Q)_{\text{orifice meter}}$$
$$(C_d)_{\text{venturimeter}} > (C_d)_{\text{orifice meter}}$$

Orifice meter are connected to the differential manometer.

19. (b)

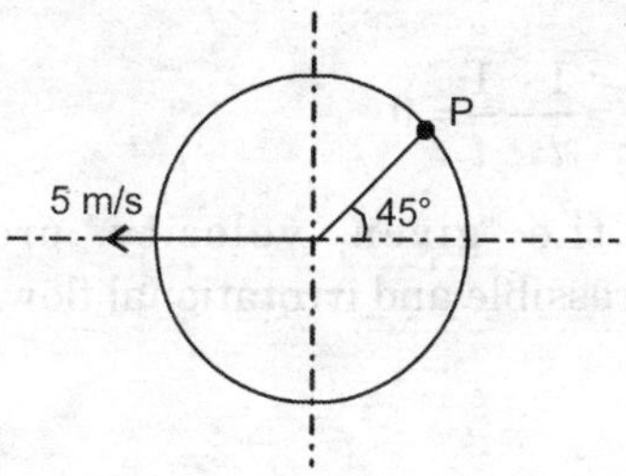

V= 5 m/s

Free stream pressure = 100 kPa

Velocity of the fluid at point P will be same at free stream velocity

$$V_p = 5 \text{ m/s}$$

20. (a) The absolute pressure at a point P

$$= 100 \cos 45° + 100 \sin 45°$$
$$P_a = 137 \text{ kPa}$$

21. (c) $\vec{V}(x, y, t) = \frac{x}{t}\hat{i} - \frac{y}{t}\hat{j}$

$$u = \frac{x}{t}, \quad v = -\frac{y}{t}$$
$$a_x = u \cdot \frac{\partial u}{\partial x} + v \cdot \frac{\partial u}{\partial y} + \frac{\partial u}{\partial t}$$
$$a_y = u \cdot \frac{\partial v}{\partial x} + v \cdot \frac{\partial v}{\partial y} + \frac{\partial v}{\partial t}$$
$$\frac{\partial u}{\partial x} = \frac{1}{t}, \quad \frac{\partial v}{\partial y} = -\frac{1}{t}$$

$$\frac{\partial u}{\partial t} = \frac{x}{t^2}, \frac{\partial v}{\partial t} = \frac{-y}{t^2}$$

$$a_x = \left(\frac{x}{t}\right)\left(\frac{1}{t}\right) + 0 + \frac{x}{t^2}$$

$$a_x = \frac{2x}{t^2}$$

$$a_y = 0 + \left(\frac{y}{t}\right)\left(\frac{-1}{t}\right) - \frac{y}{t^2}$$

$$a_y = \frac{y}{t^2} - \frac{y}{t^2} = 0$$

Total acceleration (a) = $a_x\hat{i} + a_y\hat{j}$

$$\vec{a} = \frac{2x}{t^2}\hat{i} + 0$$

$$\boxed{\vec{a} = \frac{2x}{t^2}\hat{i}}$$

22. (c) From continuity equation

$$\frac{\partial u}{\partial x} + \frac{\partial v}{\partial y} = 0$$

$$\frac{\partial u}{\partial x} + \frac{\partial v}{\partial y} = \frac{1}{t} - \frac{1}{t} = 0$$

hence the given velocity profile is incompressible and irrotational flow.

23. (c)

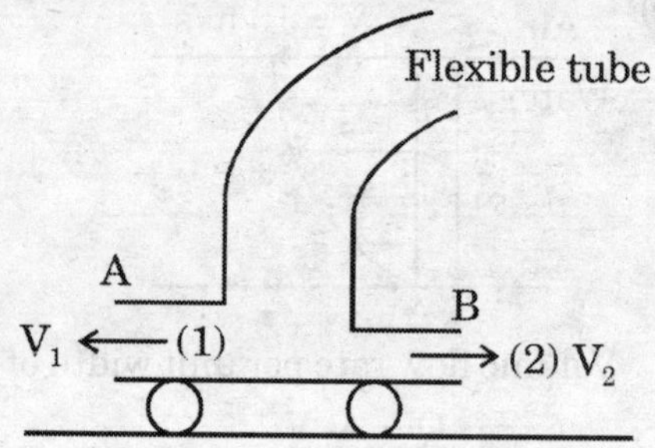

For incompressible flow T – Junction

Area of A = 2 × Area of B

$$\dot{m}_A = \dot{m}_B$$

Case 1 – The flow rate at A and B are constant.

Case 2 – The velocity at section A and B are equal.

$V_A = V_B$

Now

$Q = A_1V_1 = A_2V_2$

$$V_1 = \frac{Q}{A_1} = \frac{Q}{2A_2} \left[A_1 = 2A_2\right]$$

$$V_2 = \frac{Q}{A_2}$$

$$\boxed{V_1 = \frac{V_2}{2}}$$

24. (d) According to above explanation, in link-data question.

2010

Useful data

Acceleration due to gravity = 10 m/s^2

Density of water = 1000 kg/m^3

Density of mercury = 13570 kg/m^3

Q.(1 – 9) carry one mark each.

1. A cylindrical container is filled with a liquid up to half of its height. The container is mounted on the centre of a turn-table and is held fixed using a spindle. The turn-table is now rotated about its central axis with a certain angular velocity. After some time interval, the fluid attains rigid body rotation. Which of the following profiles best represents the constant pressure surfaces in the container?

(*a*)

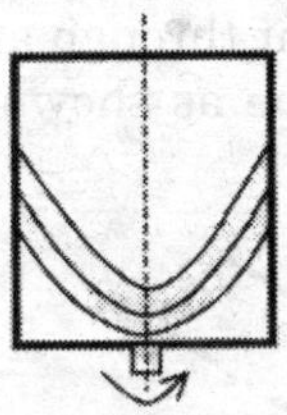

(*b*)

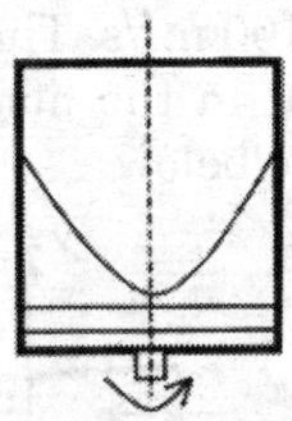

(*c*)

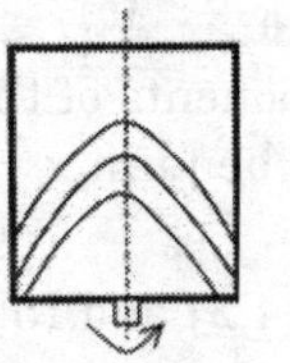

(*d*)

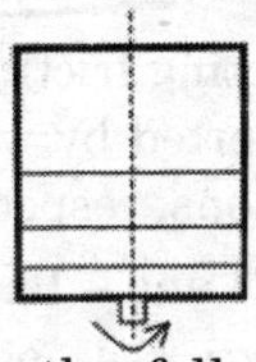

2. Match the items given in the following two columns using appropriate combinations :

Column 1	**Column 2**
(P) Ratio of inertial force to viscous force	1. Reynolds number (Re)
(Q) Ratio of momentum diffusivity to thermal diffusivity	2. Froude number (Fr)
(R) Ratio of inertial force to compressibility force	3. Prandtl number (Pr)
(S) Ratio of inertial force to gravity force	4. Mach number (Ma)

(*a*) P-1; R-2; Q-3; S-4 (*b*) P-1; Q-2; R-3; S-4

(*c*) P-1; R-2; S-3; Q-4 (*d*) P-1; S-2; Q-3; R-4

3. In the context of boundary layers, which one of the following statements is **FALSE** ?

(*a*) It is a frictional layer, close to the body

(*b*) It is a region where the fluid flow is irrotational

(*c*) It is a region across which the pressure gradient is negligible

(*d*) It is a diffusion layer of vorticity

4. Consider an ideal fluid flow past a circular cylinder shown in the figure below. The peripheral velocity at a point P on the surface of the cylinder is

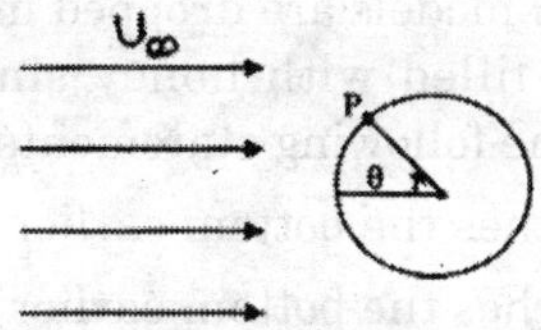

(*a*) 0 (*b*) U_∞

(*c*) $U_\infty \sin\theta$ (*d*) $2\,U_\infty \sin\theta$

5. The Rheological diagram depicting the relation between shear stress and strain rate for different types of fluids is shown in the figure below.

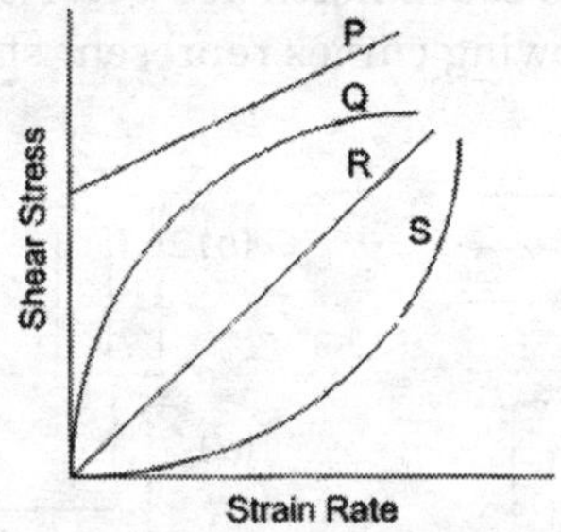

The most suitable relation for flow of tooth paste being squeezed out of the tube is given by the curve

(*a*) P (*b*) Q

(*c*) R (*d*) S

6. The diverging limb of a venturimeter is kept longer than the converging limb to

(*a*) ensure that the flow remains laminar

(*b*) avoid separation

(*c*) ensure that the flow remains turbulent

(*d*) avoid formation of boundary layer

7. The length scale of a model is kept as 1 : 64. The prototype fluid is water. Viscous and gravity forces are equally dominant in the prototype. The required kinematic viscosity (m^2/s) of the fluid used in the model is

(*a*) 0.100 E-07 (*b*) 0.195 E-08

(*c*) 0.156 E-07 (*d*) 0.125 E-07

8. Let φ and Ψ represent, respectively, the velocity potential and stream function of a flow field of an incompressible fluid. Which of the following statements are **TRUE?**

P : φ exists for irrotational flows only

Q : Ψ exists for both irrotational and rotational flows

R : φ exists for rotational flows only

S : φ exists for both rotational and irrotational flows

(*a*) P, R (*b*) Q, S

(*c*) Q, R (*d*) P, Q

9. Two models M_1 and M_2 have equal volumes and are made of steel. Model M_1 is an aerofoil (a streamlined body) and model M_2 a sphere (a bluff body). Both models are dropped in two identical short jars filled with honey simultaneously. Which of the following statements is **TRUE?**

(*a*) M_1 reaches the bottom earlier than M_2

(*b*) M_2 reaches the bottom earlier than M_1

(*c*) Both models reach the bottom at the same time

(*d*) Both models float on the surface

Q.(10 – 22) carry two marks each.

10. The wind is blowing east-west at time $t < T$, and switches to south-north at $t = T$. At $t > T$, which of the following curves represent streaklines?

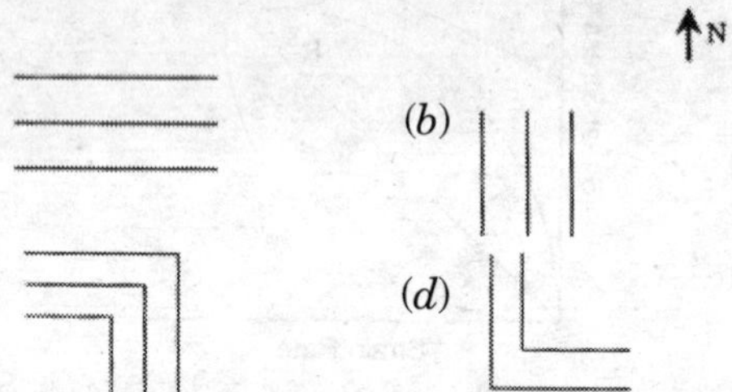

11. A cricket ball comprises of a seam running along its central section, which is essentially stitching of two hemi-spheres. The seam creates additional roughness. The bowler releases the ball with seam orientation as shown in the figure below.

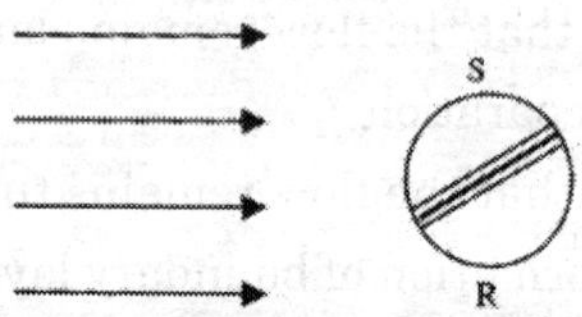

This would result in an out swinger with side forces on the cricket ball. These sides forces on the ball are attributed to

(*a*) flow having a laminar boundary layer separation on both sides.

(*b*) flow having turbulent boundary layer separation on both sides.

(*c*) flow having a laminar boundary layer separation on the side "S" and a turbulent boundary layer separation on the side "R".

(*d*) flow having a laminar boundary layer separation on the side "R" and a turbulent boundary layer separation on the side "S".

12. Ancients have designed water clocks based upon the head of the water in a circular section container with a hole at the bottom as shown in the figure below. The radius (r) varies as a function of head (H) to maintain a constant rate of decline of H.

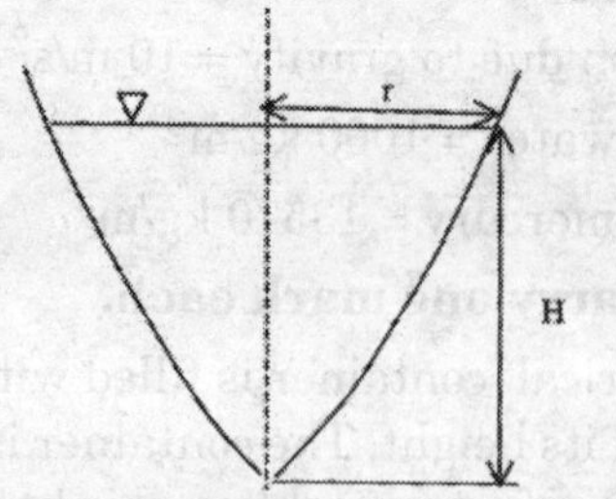

The relation between H and r is

(*a*) H is proportional to r

(*b*) H is proportional to r^2

(*c*) H is proportional to r^3

(*d*) H is proportional to r^4

13. A 20 cm diameter pipe carries a water discharge of $\pi/100$ m^3/s. The pipe is bent through an angle of 30° in the horizontal plane as shown in the figure below.

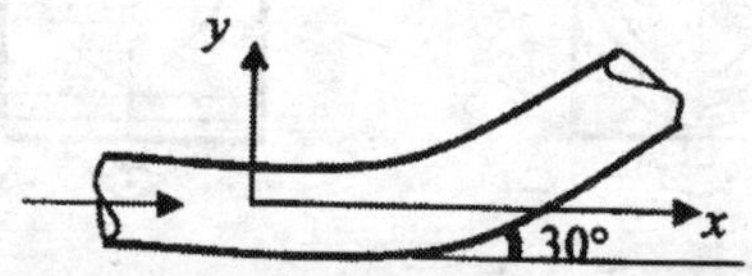

Neglecting friction, the components of the force (N) exerted by water on the bend in x - and y-directions, respectively, are

(*a*) 4.21 and – 15.71 (*b*) – 4.21 and 15.71

(*c*) 15.71 and – 27.2 (*d*) 4.21 and 15.71

14. A differential U-tube manometer with mercury as the manometric fluid is used to measure the pressure difference between two sections P and Q in a horizontal pipe carrying water at steady state as shown in the figure below. If the difference in mercury levels in the two limbs of the manometer is 0.75 m, the difference in pressure (kPa) between sections P and Q is

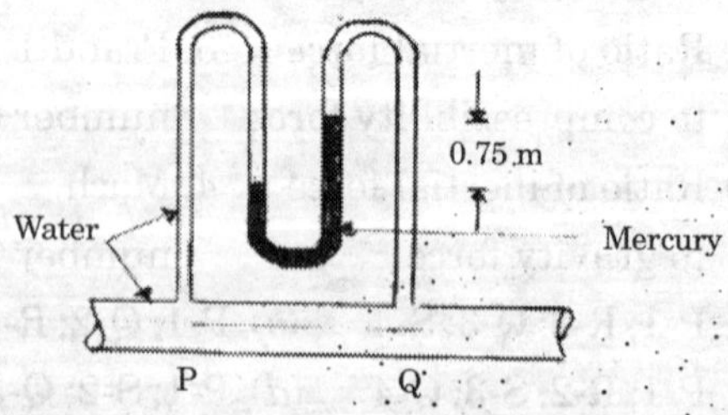

(*a*) 49.275 (*b*) 94.275

(*c*) 9.4275 (*d*) 492.75

15. Two walls are holding back water as shown in the figures below. The resisting moments per unit length of the walls at points P and Q are M_P and M_Q. Denoting the specific weight of water as γ, the difference in the moments ($M_Q - M_P$) is

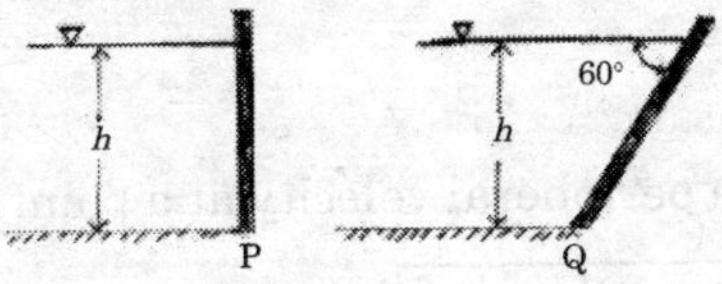

(a) $\frac{\sqrt{3}\gamma h^3}{2}$ (b) $\frac{2\gamma h^3}{\sqrt{3}}$

(c) $\frac{\gamma h^3}{18}$ (d) $\frac{\gamma h^3}{2}$

16. A 20 cm cubical box slides on oil (mass density = 800 kg/m^3), over a large plane surface with a steady state velocity of 0.4 m/s. The plane surface is inclined at an angle of 30° with the horizontal plane. The oil film between the block and the plane surface is 0.4 mm thick. The weight of the cubical box is 64 N. The kinematic viscosity of the oil is

(a) 0.8 Pa.s (b) 0.001 m^2/s

(c) 1.6 Pa.s (d) 0.002 m^2/s

Common Data Questions

Common Data for Q. (17 - 18) :

A 60% efficient pump is installed in a pipe of diameter 20 cm to lift water from a sump to an overhead tank at a discharge rate of π/100 m^3/s. Free surface level in the overhead tank is 20 m higher than the free surface level in the sump. The all-inclusive head losses (not including the lift) in the suction and delivery sides of the pump are 2 times and 28 times the velocity head, respectively.

17. The power (W) supplied to the pump is

(a) 10476.2 (b) 6285.7

(c) 6757.1 (d) 11261.9

18. The suction side of the pump is located L m above the free surface level in the sump. The minimum permissible pressure in the pipeline on the suction side of the pump is 8 m of water below atmospheric pressure. The maximum permissible value of L is

(a) 20.00 (b) 8.00

(c) 7.85 (d) 5.00

Common Data for Q. (19 - 20) :

The velocity field of a two-dimensional fluid flow is as follows :

$$u = U_0\frac{x}{L},\ v = -U_0\frac{y}{L}$$

where, U_0 and L are, respectively, the characteristic velocity and length.

19. If L = 0.2 m and the resultant of total accelerations in x- and y- directions at (x = L, y = L) is 10 m/s^2, the magnitude of U_0 (m/s) is

(a) 1.414 (b) 2.38

(c) 1.19 (d) 11.90

20. The above fluid flow can be described as

(a) rotational and compressible

(b) irrotational and compressible

(c) rotational and incompressible

(d) irrotational and incompressible

Linked Answer Question

Statement for Linked Answer Q. (21 - 22):

The boundary layer formation over a flat plate is shown in the figure below. The variation of horizontal velocity (u) with y at any x along the plate in the boundary layer is approximated as: u = P sin (Qy) + R

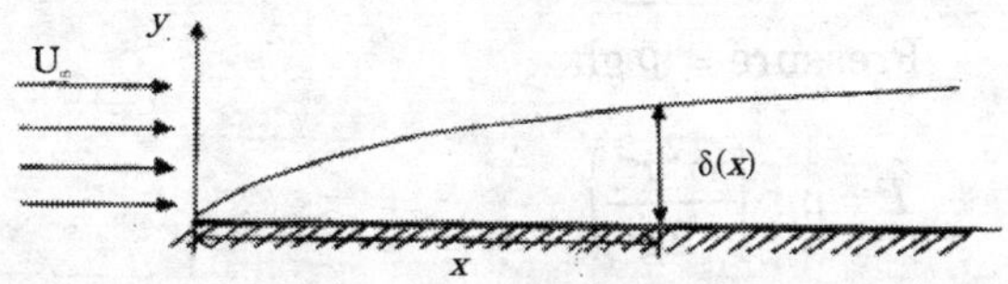

21. The most acceptable boundary conditions are

(a) at y = 0, u = 0; at y = δ, u = U_∞;

at y = 0, $\frac{du}{dy} = 0$

(b) at y = 0, u = U_∞; at y = δ, u = U_∞;

at y = 0, $\frac{du}{dy} = 0$

(c) at y = 0, u = 0; at y = δ, u = U_∞;

at y = δ, $\frac{du}{dy} = 0$

(d) at y = 0, u = U_∞; at y = δ, u = U_∞;

at y = δ, $\frac{du}{dy} = 0$

22. Expressions for P, Q and R are

(a) P = 0; Q = 0; R = 0

(b) P = U_∞; Q = 0; R = 0

(c) P = 0; Q = $\frac{\pi}{2\delta}$; R = U∞

(d) P = U_∞; Q = $\frac{\pi}{2\delta}$; R = 0

ANSWERS

1. (b)	**2.** (d)	**3.** (b)	**4.** (d)	**5.** (a)	**6.** (b)	**7.** (c)
8. (d)	**9.** (a)	**10.** (c)	**11.** (c)	**12.** (b)	**13.** (c)	**14.** (b)
15. (b)	**16.** (b)	**17.** (*)	**18.** (*)	**19.** (c)	**20.** (d)	**21.** (a)
22. (a)						

EXPLANATIONS

1. (b) Rising of liquid (h)

$$h = \frac{\omega^2 r^2}{2g}$$

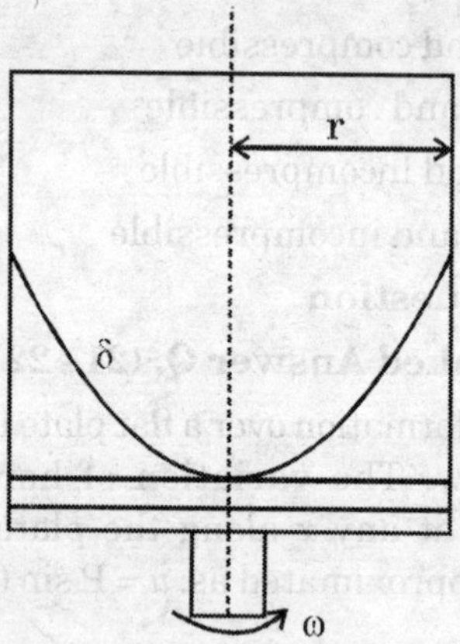

Pressure = ρgh

$$P = \rho g \cdot \left(\frac{\omega^2 r^2}{2g}\right)$$

$$\boxed{P = \frac{\rho \omega^2 r^2}{2}}$$

P = Constant [ρ = c, ω = C, r = C]

2. (d) Reynold's Number (Re) = $\dfrac{\text{Inertial Force } (F_i)}{\text{Viscous Force } (F_V)}$

Froude Number (Fr) = $\dfrac{\text{Inertia Force } (F_i)}{\text{Gravity Force } (F_g)}$

Prandtl Number (Pr)

$= \dfrac{\text{Momentum diffusivity}}{\text{Thermal diffuvisity}}$

Mach Number (Ma)

$= \dfrac{\text{Inertia Force (Fi)}}{\text{Compressibility Force (Fc)}}$

P – 1, Q – 3, R – 4, S – 2

3. (b) In the context of bounary layer viscous force must be there and fluid is rotational nature.

4. (d) Free stream velocity

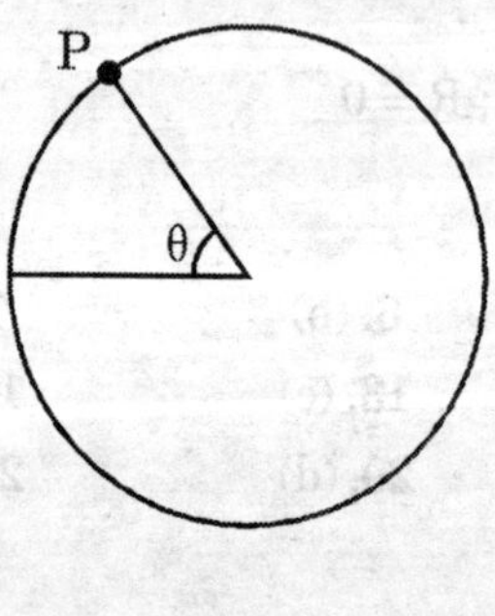

The peripheral velocity at a point (P)

$$\boxed{V_P == 2U \propto \sin\theta}$$

5. (a)

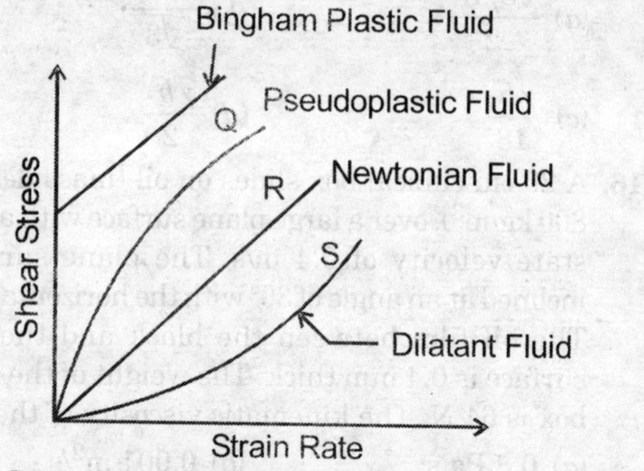

Line (P), represent the Bingham plastic fluid

Ex – Tooth paste

– All facial cream.

6. (b) The diverging limb of venturimeter is 3 to 4 times of Converging limb to avoid the flow separation, and minimize the eddy formation, so that losses is minimize and it give the more reading.

7. (c) Scale ration – 1 : 64

Kinematic Viscosity of water at normal temperature and pressure = 10^{-6} m²/s

Now,

$$\frac{\nu_m}{\nu_p} = \frac{1}{64}$$

Kinematic viscosity of model $(\nu_m) = \dfrac{\nu_p}{64}$

$$\nu_m = \frac{10^{-6}}{64}$$

$$\nu_m = 1.5625 \times 10^{-8} \text{ m}^2/\text{sec.}$$

$$\nu_m = 0.156E-07$$

8. (d) ϕ – velocity potential function.

ψ – stream function.

– velocity potential function (ϕ), exist for only irrotational flow.

– Stream function (ψ), exist for both rotational and irrotational flow.

9. (a) M_1 – Streamline body – A body is said to be streamline body, in which that the surface area of body coincides with streamline.

M_2 – Bluff body – A body is said to be Bluff body, in which that the surface area of body do not coincide with the streamline.

10. (c)

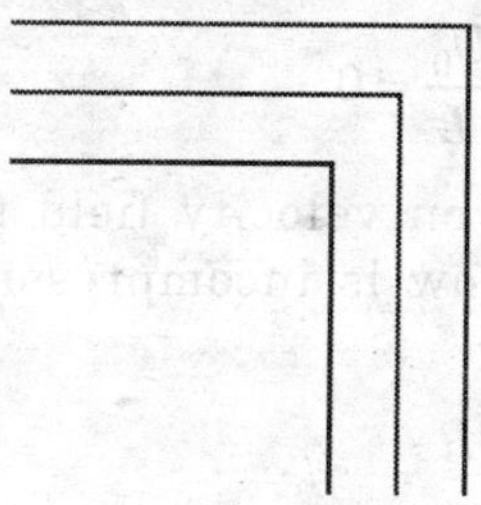

Curve (c), represent the streakline at the given condition.

Streakline → A line trace by the fluid particle passing through the fixed point, is called streak line.

11. (c)

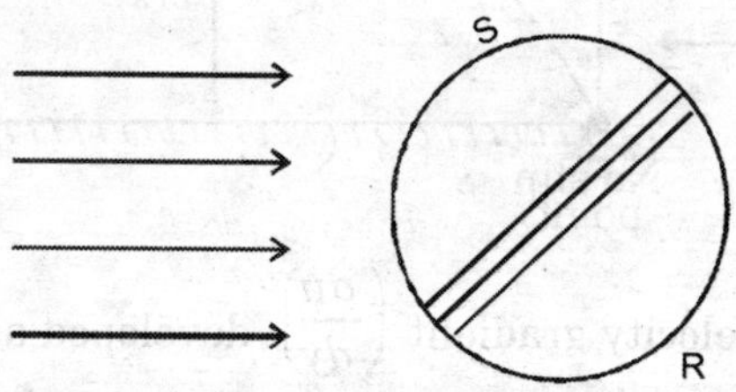

Flow having a laminar boundary layer on the side 's' and a turbulent boundary layer separation on the side "R".

12. (b)

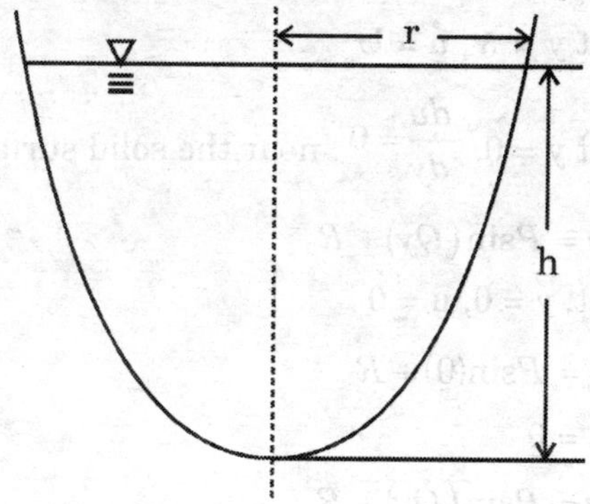

$$H = \frac{\omega^2 r^2}{2g}$$

ω – angular velocity

r – radius

$$\boxed{\begin{array}{c} H \propto \omega^2 \\ H \propto r^2 \end{array}}$$

13. (c) Discharge (Q) = $\frac{\pi}{100}$ m³/sec.

angle (θ) = 30º

diameter (d) = 20 cm = 0.2 m.

Density of water (ρ) = 1000 kg/m³

Q = AV

$$V = \frac{Q}{A} = \frac{4Q}{\pi d^2} = \frac{4 \times \frac{\pi}{400}}{\pi \times 0.2} = 1 \text{ m/s.}$$

Force in x-direction $(F_x) = \rho Q(V_1 - V_2 \cos\theta)$

$$F_x = 1000 \times \frac{\pi}{100} \times \left(-\frac{V_1^2}{2}\right)$$

$$\boxed{F_x = 15.71 \text{ N}}$$

Force in y-direction $(F_y) = -\delta Q V \cos\theta$

$$F_y = -1000 \times \frac{\pi}{100} \times 1 \times \cos 30°$$

$$\boxed{F_y = -27.2 \text{ N}}$$

14. (b)

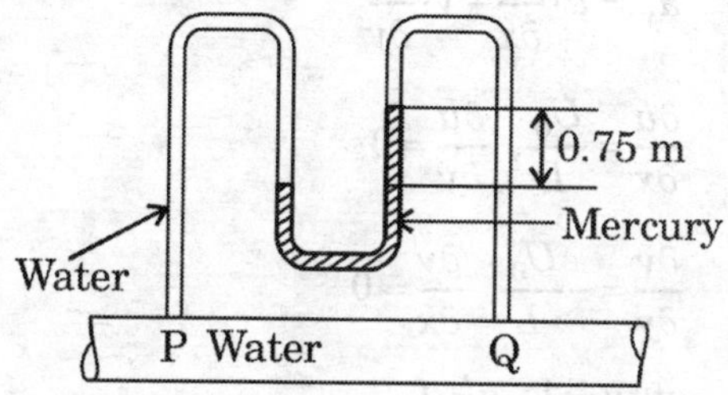

Pressure at point $(P_P) = \rho_\omega gh$

$$= 1000 \times 9.81 \times 0.75$$

Pressure at point $(P_Q) = \rho_{Hg} gh$

$$= 13600 \times 9.81 \times 0.75$$

$$P_P - P_Q = (13600 - 1000) \times 9.81 \times 0.75$$

$$\boxed{P_P - P_Q = 92.7045 \text{ kPa}}$$

16. (b)

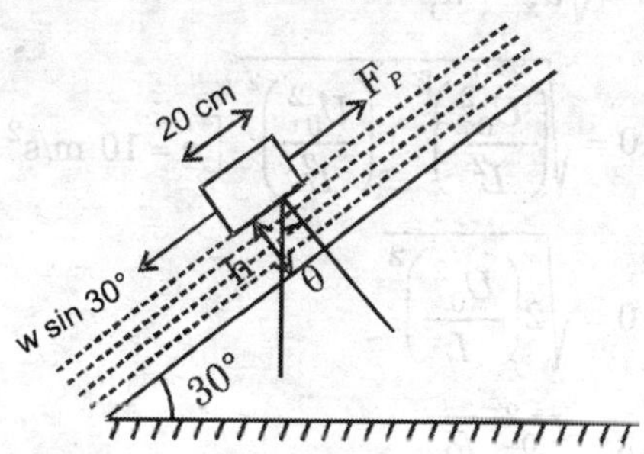

Given that

a = 20 cm = 0.2 m

Area (A) = 0.2² = 0.04 m²

$h = 0.4 \times 10^{-3}$ m

$w = 64$ N

$w\sin\theta = F_D$

$$64 \sin 30° = \frac{\mu u A}{n}$$

$$64 \sin 30° = \frac{\mu \times 0.4 \times 0.04}{0.4 \times 10^{-4}}$$

$\mu = 0.8$ Pa · s

Kinematic Viscosity $(\nu) = \frac{\mu}{\delta}$

$$\nu = \frac{0.8}{800}$$

$$\boxed{\nu = 0.001 \text{ m}^2/\text{sec.}}$$

19. (c) The velocity field of a two-dimensional flow

$$u = v_0\left(\frac{x}{L}\right)$$

$$v = -u_0\left(\frac{y}{L}\right)$$

$$a_x = u\cdot\frac{\partial u}{\partial x} + v\cdot\frac{\partial u}{\partial y}$$

$$a_y = u\cdot\frac{\partial v}{\partial x} + v\cdot\frac{\partial v}{\partial y}$$

$$\frac{\partial u}{\partial x} = \frac{U_0}{L},\ \frac{\partial u}{\partial y} = 0$$

$$\frac{\partial v}{\partial y} = -\frac{U_0}{L},\ \frac{\partial v}{\partial x} = 0$$

at $x = L,\ y = L$

$$u = u_0,\ v = -U_0$$

$$a_x = U_0\cdot\left(\frac{U_0}{L}\right) = \frac{U_0^2}{L}$$

$$a_y = 0 + U_0\left(\frac{U_0}{L}\right) = \frac{U_0^2}{L}$$

$$a = \sqrt{a_x^2 + a_y^2}$$

$$10 = \sqrt{\left(\frac{U_0^2}{L^2}\right)^2 + \left(\frac{U_0^2}{L^2}\right)^2}\quad \left[a = 10\ \text{m/s}^2\ \text{given}\right]$$

$$10 = \sqrt{2\left(\frac{U_0^2}{L}\right)^2}$$

$$10 = \frac{U_0^2}{L}\sqrt{2}$$

$$\left(\frac{10\times 0.2}{\sqrt{2}}\right)^{\frac{1}{2}} = U_0\quad [L = 0.2\ \text{m given}]$$

$$\boxed{U_0 = 1.189\ \text{m/s}}$$

20. (d) From continuity equation

$\frac{\partial u}{\partial x} + \frac{\partial v}{\partial y} = 0$, for incompressible streay and irrotational flow.

$$\frac{\partial u}{\partial x} = \frac{U_0}{L}$$

$$\frac{\partial v}{dy} = -\frac{U_0}{L}$$

$$\frac{\partial u}{\partial x} + \frac{\partial v}{\partial y} = \frac{U_0}{L} - \frac{U_0}{L} = 0$$

Hence, the given velocity field for two-dimensional flow is incompressible and irrotational flow.

21. (a) $u = P\sin(Qy) + R$

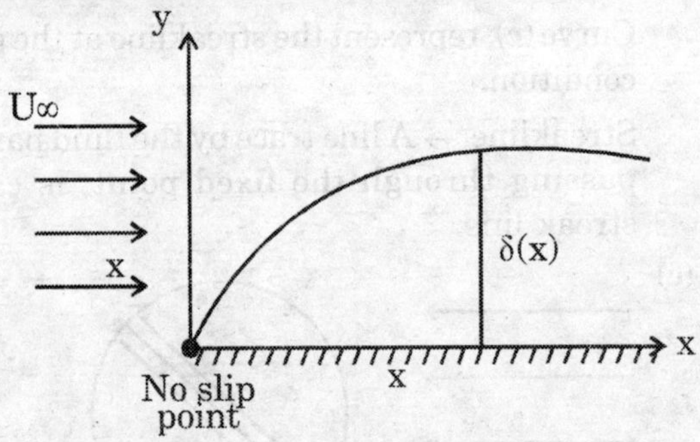

Velocity gradient $\left(\frac{du}{dy}\right)$ developed along the y-axis.

At No slip point, u = 0, y = 0

At y = 0, u = 0

At y = δ, u = U_g

At y = 0, $\frac{du}{dy} = 0$, near the solid surface.

22. (a) $u = P\sin(Qy) + R$

At, y = 0, u = 0

$$0 = P\sin(0) + R$$

$$R = 0$$

$$u = P\sin(Qy) + R$$

$$\frac{du}{dy} = P\cos(Qy)\cdot Q = 0$$

at y = 0, $\frac{du}{dy} = 0$

0 = PQ· 1

Q = 0

At y = δ, u = U_q

$U_g = P\sin(Q.\delta)$

$U_g = P\sin(0)$

P = 0

$$\boxed{P = 0,\ Q = 0,\ R = 0}$$

2011

Q. (1-9) carry one mark each.

1. For a boundary layer on a flat plate forces and _______ forces are of the same order of magnitude.

(*a*) body, inertia (*b*) viscous, body

(*c*) inertia, viscous (*d*) viscous, pressure

2. The temperature field in a fluid flow is given by $(60 - 0.2xy)$ °C. The velocity field is $\overline{V} = 2xy\hat{i} + ty\hat{j}$ m/s. The rate of change of the temperature measured by a thermometer moving along with the flow at (2, –4) m at t = 4s is

(*a*) – 12.8 °C/s (*b*) – 10.6°C/s

(*c*) – 6.4°C/s (*d*) – 4.8°C/s

3. Two tanks, A and B, with the same height are filled with water till the top. The volume of tank A is 10 times the volume of tank B. What can you say about the pressures p_A and p_B at the bottom of the tanks A and B respectively?

(*a*) $p_A = 10\ p_B$

(*b*) $p_B = 10\ p_A$

(*c*) $p_A = p_B$

(*d*) Additional data is required to compare the two pressures.

4. A velocity field in a plane flow is given by $\overline{V} = 2xy\hat{i} + 3y\hat{j}$ m/s. The vorticity at the point (2,4) m is

(*a*) $-4\hat{k}$ rad/s (*b*) $-3\hat{j}$ rad/s

(*c*) $-2\hat{k}$ rad/s (*d*) $-3\hat{i}$ rad/s

5. Separation is said to occur at a wall when _______ at the wall becomes zero.

(*a*) internal energy (*b*) pressure

(*c*) shear stress (*d*) density

6. A certain fluid is influenced by density (ρ), angular velocity (ω), dynamic viscosity (μ), and a characteristic length (L). A relevant non-dimensional parameter will be

(*a*) $\rho\omega\mu/L^2$ (*b*) $\rho\omega L^2/\mu$

(*c*) $\rho\omega\mu L^2$ (*d*) $\rho\omega\mu L$

7. The drags due to potential flow past a cylinder of diameter D cm and a slender airfoil of chord length D cm are compared. Assuming unit depth for both the bodies, which one of the following would be TRUE?

(*a*) The drag on the cylinder is greater

(*b*) The drag on the airfoil is greater

(*c*) Both the drags are equal

(*d*) Additional data is needed to compare the drags

8. For a fully developed flow between two parallel flat plates, the velocity gradient at a point is found to be 1000 s^{-1}. If the density of the fluid is 880 kg/m^3 and the kinematic viscosity of the fluid is 7.4×10^{-7} m^2/s, the shear stress at the same point is approximately

(*a*) 0 Pa (*b*) 1.30 Pa

(*c*) 0.32 Pa (*d*) 0.65 Pa

9. An open channel flow is to be simulated in the laboratory. For this purpose, a 1:25 scale model is constructed. If the flow velcoity in the prototype is 5 m/s, for dynamic similarity the model should have a flow velocity of

(*a*) 5 m/s (*b*) 1 m/s

(*c*) 125 m/s (*d*) 25 m/s

Q.(10-22) carry two marks each.

10. A pitot-static probe is inserted in an air flow. A monometer connected to this probe having Hg as the manometric fluid shows a difference of 30 mm. Assume a probe factor of 1. Assuming ρ_{air} = 1.23 kg/m^3, ρ_{Hg} = 13600 kg/m^3 and g = 10 m/s^2, the speed of the air flow is approximately

(*a*) 66.5 m/s (*b*) 81.5 m/s

(*c*) 76.5 m/s (*d*) 92.5 m/s

11. Consider an L-shaped gate with water level above the hinge as shown. At approximately what height D of the water level will the gate open? Neglect the mass of the gate. Assume g = 10 m/s^2.

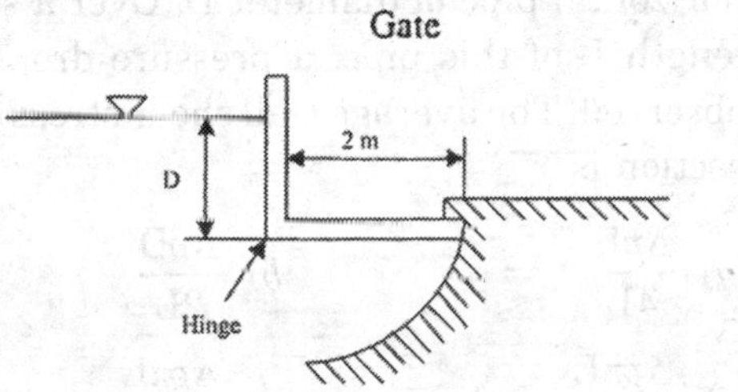

(*a*) 3.46 m (*b*) 4.36 m

(*c*) 6.43 m (*d*) 5.36 m

12. When a large tank containing water is placed on a weighing scale, a reading of 10000 N is obtained. The tank is fitted with an outlet pipe and a valve as shown. When the valve is opened, a jet of water with a velocity of 10 m/s issues out in the vertically upward direction. The diameter of the outlet pipe is 10 cm. Determine approximately the reading on the weighing scale at the instant the valve is opened and the water jet issues out. Density of water is 1000 kg/m^3.

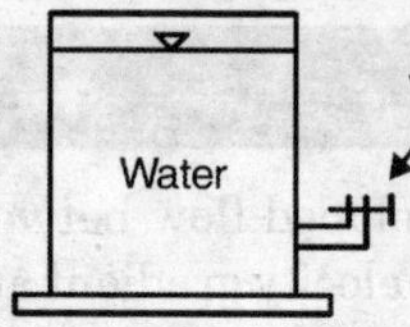

(a) 9215 N (b) 10000 N
(c) 10785 N (d) 12500 N

13. A fluid with a volumetric flow rate of 5 m^3/s enters the nozzle shown below. The cross-sectional area varies with x as $A(x) = 1/(1 + x^2)$. Assuming that the flow is parallel and uniform at each cross-section, the acceleration at any point in the nozzle is given by

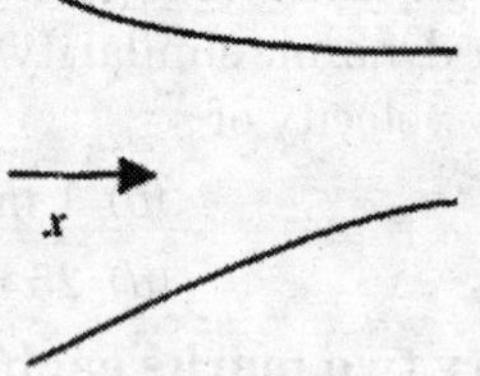

(a) $50(x + x^3)$ (b) $50(1 + x^2)$
(c) 0 (d) $50(x^2 + x^3)$

14. Consider fully developed flow of water in a pipe of diameter 2 cm. The average velocity of the flow is 2 m/s. The viscosity of the water is 10^{-3} kg/m-s and the density is 1000 kg/m^3. The friction factor can be calculated using $f = 64/Re$ for laminar flows and f = 0.3164/Re0.25 for turbulent flows. The pressure drop over a length of 0.5 m is

(a) 0.08 Pa (b) 325 Pa
(c) 1115 Pa (d) 9875 Pa

15. Consider a steady, fully developed flow in a horizontal pipe of diameter D. Over a section of length L of this pipe, a pressure drop of Δp is observed. The average wall shear stress over this section is

(a) $\frac{\Delta p D}{4L}$ b) $\frac{\Delta p D}{2L}$

(c) $\frac{\Delta p \pi L}{2D}$ (d) $\frac{\Delta p \pi L}{4D}$

16. In an inviscid incompressible flow, the velocity field is given by $\vec{V} = x\hat{i} + y\hat{j}$ m/s and the body force per unit mass is given by $\vec{g} = -10\hat{k}\, m/s^2$. The pressure at the point (0, 0, 0) is 101 Pa. Assuming that the density of the fluid is 1 kg/m^3, the pressure at the point (1, 1, 1) for this flow is

(a) 100 Pa (b) 105 Pa
(c) 95 Pa (d) 90 Pa

COMMON DATA QUESTIONS

Common Data for Questions 17 and 18 :

A two-dimensional rectangular water jet of velocity 10 m/s and area 5 cm^2 impinges normal to a flat plate and splits symmetrically into two half jets, each of area 2.5 cm^2 as shown. Assume steady flow and neglect viscous effects and the weight of the plate and the water. Density of water is 1000 kg/m^3.

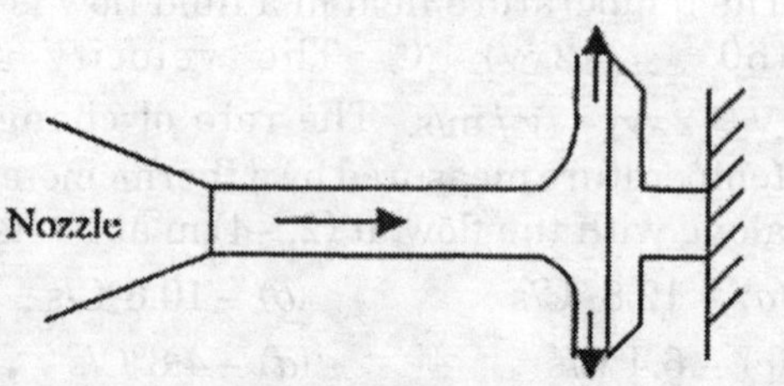

17. After splitting, the velocity of the upward half-jet along the plate is

(a) 5 m/s (b) 7.5 m/s
(c) 2.5 m/s (d) 10 m/s

18. The magnitude of the reaction force at the wall is

(a) 20 N (b) 25 N
(c) 35 N (d) 50 N

Common Data for Questions 19 and 20 :

A flow has a velocity field given by $\vec{V} = 2x\hat{i} - 2y\hat{j}$

19. The velocity potential $\phi(x, y)$ for the flow is

(a) $2x - 2y$ + const.
(b) $2xy$ + const.
(c) $x^2 + y^2$ + const.
(d) $x^2 - y^2$ + const.

20. The streamlines for the velocity field look like

(a) 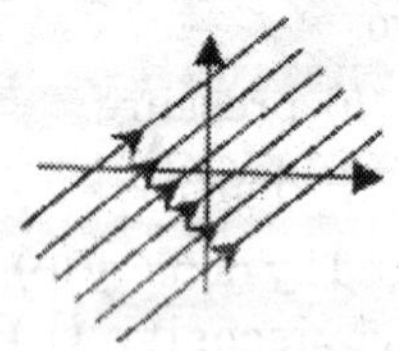 (b) 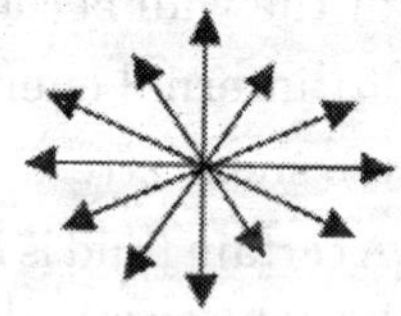

(c) 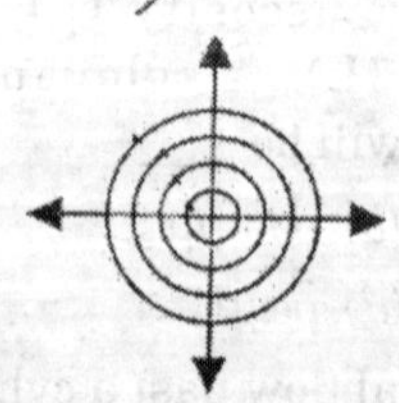 (d) 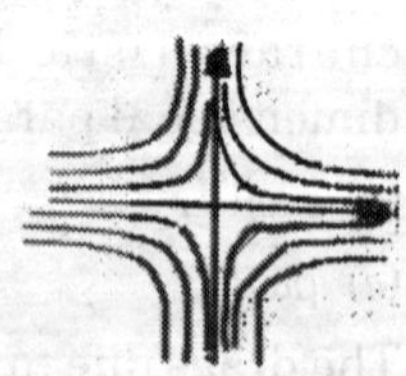

LINKED ANSWER QUESTIONS

Statement for Linked Answer Questions 21 & 22 :

Two flat parallel plates are separated by a small gap h filled with an incompressible fluid of viscosity μ. Assume that the length and width of the plates to be much larger than the gap h. The top plate moves horizontally while the bottom plate is held stationary. The magnitude of the difference between the shear stress at the top and bottom walls is found to be $\Delta\tau$.

21. The velocity of the top plate is

(a) $\frac{h\Delta\tau}{(2\mu)}$ (b) $\frac{h\Delta\tau}{\mu}$

(c) $\frac{2h\Delta\tau}{\mu}$ (d) $\frac{3h\Delta\tau}{(2\mu)}$

22. If a finite width slender object is introduced parallel to the plates in the middle of the gap, the time at which it would have rotated clockwise by 90° would be

(a) $\frac{2\pi\mu}{(\Delta\tau)}$ (b) $\frac{\pi\mu}{(\Delta\tau)}$

(c) $\frac{2\pi\mu}{(3\Delta\tau)}$ (d) $\frac{\pi\mu}{(4\Delta\tau)}$

ANSWERS

1. (c)	**2.** (c)	**3.** (c)	**4.** (a)	**5.** (c)	**6.** (b)	**7.** (c)
8. (d)	**9.** (c)	**10.** (b)	**11.** (a)	**12.** (c)	**13.** (a)	**14.** (c)
15. (a)	**16.** (d)	**17.** (d)	**18.** (d)	**19.** (d)	**20.** (d)	**21.** (a)
22. (a)						

EXPLANATIONS

1. (c) For a boundary layer on a flat plate the inertia force and viscous force are dominating in nature.

3. (c) Consider two tank A and B of same height.

Given that

Volume of tank A = 10 times of volume B

We know that

$P = \rho g h$

$P_A = \rho_w g h_A$

$P_B = \rho_w g h_B$

$P_A = P_B \quad [h_A = h_B \rightarrow Given]$

Pressure is independent of volume, it depend on the depth from the free surface.

4. (a) $\vec{V} = (2xy\hat{i} + 3y\hat{j})$ m/s

at point (2, 4)

u = 2xy, v = 3y

$\frac{\partial u}{\partial y} = 2x, \ \frac{\partial v}{\partial x} = 0$

$\left(\frac{\partial u}{\partial y}\right)_{(2,\,4)} = 2 \times 2 = 4$

$w_z = \frac{1}{2}\left(\frac{\partial v}{\partial x} - \frac{\partial u}{\partial y}\right)$

$w_z = \frac{1}{2}(0 - 4) = -2\hat{k}$

Vorticity $(\xi) = 2 \cdot w_z = 2(-2) = -4\hat{k}$ rad/sec.

5. (c) Separation of boundary layer occur in

i) $\frac{\partial p}{\partial x} > 0$

ii) $\frac{\partial u}{\partial y} = 0$

From Newton's law of viscosity,

$\tau = \mu \cdot \frac{\partial u}{\partial y}$

$\frac{\partial u}{\partial y} = 0$

$\boxed{\text{Shear stress } (\tau) = 0}$

6. (b) Given that

ρ – density

ω – angular velocity

μ – dynamic viscosity

L – characteristics length

Non-dimensional parameter (Re) = $\frac{\rho v L}{\mu}$

$V = \omega r = \omega L$

$Re = \frac{\rho \omega L \cdot L}{\mu}$

$\boxed{Re = \frac{\rho \omega L^2}{\mu}}$

7. (c) Drag force depend on the density, Area, and velocity of flow

$\boxed{F_D = \frac{1}{2}\rho A U^2 C_D}$

So that the drag force on body will be same, due to same parameter.

8. (d) Given that

$\frac{du}{dy} = 1000 \text{ s}^{-1}, \ \rho = 880 \text{ kg/m}^3$

$\nu = 7.4 \times 10^{-7} \text{ m}^2/\text{s}$

$\tau = \mu \frac{du}{dy} = \nu \times \rho \cdot \frac{du}{dy} \qquad \left[\nu = \frac{\mu}{\rho}\right]$

$\boxed{\tau = 7.4 \times 10^{-7} \times 880 \times 1000 = 0.651 \text{ Pa}}$

9. (c) Given that

Scale ratio = 1 : 25

V_m = 5 m/sec

$\frac{V_m}{V_p} = \frac{1}{25}$

$\frac{5}{V_p} = \frac{1}{25}$

$\boxed{V_p = 125 \text{ m/s}}$

12. (c) Weighing scale reading (F_1) = 10000 N

When the valve is opened,

V = 10 m/s

d = 10 cm = 0.1 m.

$\rho_m = 1000 \text{ kg/m}^3$

Force exerted by jet $(F_2) = \rho_w A V^2$

$F_2 = 1000 \times \frac{\pi}{4} \times (0.1)^2 \times 10^2$

$F_2 = 485$ N

Force weighing scale, when valve is opened

$= F_2 + F_1 = 10000 + 785$

$= 10785$ N

14. (c) Given that

$d = 2$ cm $= 0.02$ m

$v = 2$ m/s

$\mu = 10^{-3}$ kg/m-s

$\rho = 1000$ kg/m³

$$f = \frac{64}{Re}$$

$$Re = \frac{\rho v D}{\mu} = \frac{1000 \times 2 \times 0.02}{10^{-3}} = 40000$$

$$\text{Coefficient of friction } (f') = \frac{f'}{4} = 4 \times 10^{-4}$$

$$h_f = \frac{fLV^2}{2gd}$$

$$h_f = \frac{1.6 \times 10^{-3} \times 0.5 \times 2^2}{2 \times 9.81 \times 0.02}$$

$$h_f = 8.154 \times 10^{-3}$$

$$h_f = \frac{P_1 - P_2}{\rho g}$$

Pressure drop $= \rho g h_t$

$= 1000 \times 9.81 \times 8.154 \times 10^{-3}$

$$\boxed{P_1 - P_2 = 80 \text{ Pa}}$$

15. (a) Diameter of pipe = D

Length of pipe = L

Pressure drop = ΔP

$$\text{Force (F)} = \Delta P \times \frac{\pi}{4} D^2$$

$$\text{Shear stress } (\tau) = \frac{F}{A}$$

$$= \frac{\left(\Delta P \times \frac{\pi}{4} D^2\right)}{\pi D L}$$

Shear stress (τ) over the section,

$$\boxed{\tau = \frac{\Delta P \cdot D}{4L}}$$

17. (d) Given that

Velocity of jet = 10 m/s²

Area of jet = 5 cm²

After striking split into two jet having area – 2.5 cm² each of jet.

$\rho_\omega = 1000$ kg/m³

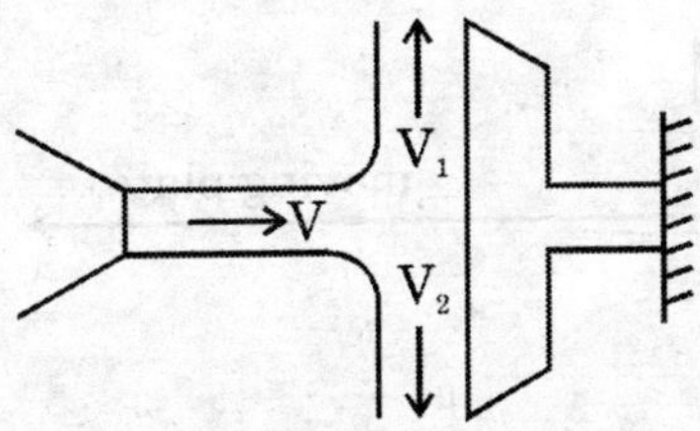

If plate is sationary, and jet strike normal to the plate, water striking after the plate which is move with same velocity as a jet.

$$\boxed{V = V_1 = V_2 = 10 \text{ m/s}}$$

18. (d) Force exerted (F) = $\rho A V^2$

$$F = 1000 \times 5 \times 10^{-4} \times 10^2$$

$$\boxed{F = 50 \text{ N}}$$

19. (d) Velocity field $(\vec{V}) = 2x\hat{i} - 2y\hat{j}$

Velocity potential function $\phi(x, y)$

$u = 2x$

$v = -2y$

$$u = -\frac{\partial\phi}{\partial x} = 2x$$

$$v = -\frac{\partial\phi}{\partial y} = -2y$$

$$\frac{\partial\phi}{\partial y} = 2y$$

$$\phi = \frac{xy^2}{x} + C$$

$$\phi_y = y^2 + C$$

$$\phi_x = -x^2 + C$$

$$\phi = f(x, y)$$

$$\boxed{\phi = x^2 - y^2 + C}$$

20. (d) $u = 2x$

$v = -2y$

For stream line

$$\frac{dx}{u} = \frac{dy}{v}$$

$$\frac{dx}{2x} = -\frac{dy}{2y}$$

Both velocity are perpendicular to each other

$$\log x = -\log y + C$$

$\log x + \log y = C$

$\boxed{xy = C}$

21. (a)

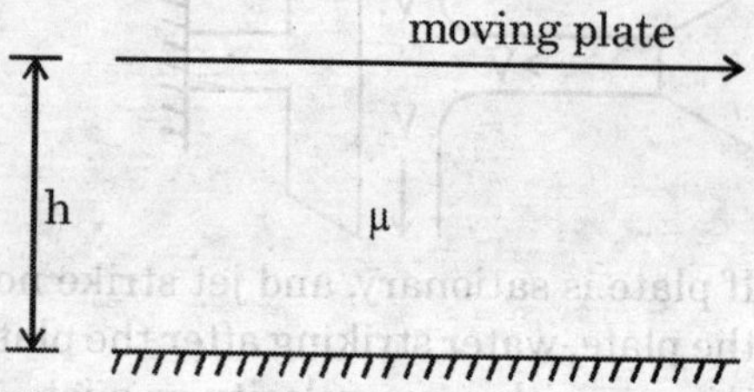

Shear stress at bottom of the plate,

$$\tau = \mu \cdot \frac{u}{h} = 0$$

Shear stress on moving plate is

$$\tau = \frac{\mu u}{h}$$ [Let u is velocity of upper plate.]

Change of shear stress $(\Delta\tau) = \dfrac{\tau_{top} - \tau_{bottom}}{2}$

$$\Delta\tau = \frac{\frac{\mu u}{h}}{2} - 0 = \frac{\mu u}{2h}$$

Velocity of the top plate (u) $= \dfrac{\Delta\tau \times h}{2\mu}$

22. (a) Time required to rotated

$$t = \frac{\text{Distance}}{\text{Velocity}}$$

$$= \frac{\pi h \times 2\mu}{\Delta\tau \times h}$$

[One to Rotation, Rotational distance = πh]

$\theta = \pi h$

$$= \frac{2\mu\pi}{\Delta\tau}$$

2012

Q. (1 – 9) carry one mark each.

1. In a two-dimensional flow field, the velocities in the x- and y- directions are u and v, respectively. The shear stress for a Newtonian fluid having dynamic viscosity μ is given by

(a) $\mu\left(\frac{\partial n}{\partial x}-\frac{\partial u}{\partial y}\right)$ (b) $2\mu\frac{\partial v}{\partial y}$

(c) $2\mu\frac{\partial u}{\partial x}$ (d) $\mu\left(\frac{\partial v}{\partial x}+\frac{\partial u}{\partial y}\right)$

2. In a potential flow, the superposition of the stream functions of a uniform flow and a line source gives rise to a dividing streamline representing

(a) Rankine's half-body

(b) infinite circular cylinder

(c) infinite rotating circular cylinder

(d) infinite elliptical cylinder

3. Given that V, L and g are the charact-eristic velocity, characteristic length and acceleration due to gravity, respectively, the expression $\frac{V}{\sqrt{Lg}}$ represents

(a) Weber number (b) Euler number

(c) Cavitation number (d) Froude number

4. Match the devices in **Column I** with the characteristics in **Column II**.

Column I	Column II
P. orifice meter	1. high head loss and low cost
Q. venturi meter	2. high head loss and high cost
	3. low head loss and high cost
	4. low head loss and low cost

	P	Q
(a)	2	4
(b)	1	2
(c)	3	1
(d)	1	3

5. Identify the visualization method that shows a PATHLINE in an unsteady flow, assuming that the camera covers the required field of view.

(a) A dye is continuously injected and a snap shot is taken

(b) A dye is continuously injected and a long-exposure picture is taken

(c) A blob (or drop) of dye is injected and a snap shot is taken

(d) A blob (or drop) of dye is injected and a long-exposure picture is taken

6. In the case of a fully developed flow through a pipe, the shear stress at the centerline is

(a) a function of the axial distance

(b) a function of the centerline velocity

(c) zero

(d) infinite

7. The velocity in a one-dimensional unsteady flow is given by $(x^2 - t)$, where x is the position and t is the time. The total acceleration at any x and t is

(a) $-1 + xt + x^3$ (b) $-1 + xt + 2x^3$

(c) $-1 - xt - x^3$ (d) $-1 - 2xt + 2x^3$

8. If Ψ is the stream function, the Laplace's equation $\nabla^2\Psi = 0$ is true when the flow is

(a) incompressible

(b) incompressible and irrotational

(c) irrotational

(d) compressible

9. A fully developed laminar flow is taking place through a pipe. If the flow velocity is doubled maintaining the flow laminar, the pressure loss would be

(a) halved (b) unaltered

(c) doubled (d) trebled

Q. (10 – 22) carry two marks each.

10. In the following equations, u and v are the velocities in the x- and y- directions, respectively and t is time. The flow field that CANNOT be termed as incompressible is

(a) $u = x^3 + xy^2$, $v = y^3 + yx^2$

(b) $u = 10xt$, $n = -10yt$

(c) $u = (y/\delta)^{1/7}$, $v = 0$ (δ = constant)

(d) $u = 2y$, $v = 2x$

11. A U-tube mercury (Hg) manometer as shown below is employed to measure the pressure of an oil-filled vessel. The densities of Hg and oil are 13600 kg/m³ and 800 kg/m³, respectively. The gravitational acceleration may be taken as 10 m/s². The gauge pressure (in Pa) at point A when h_1 = 0.5 m and h_2 = 0.9 m, is approximately

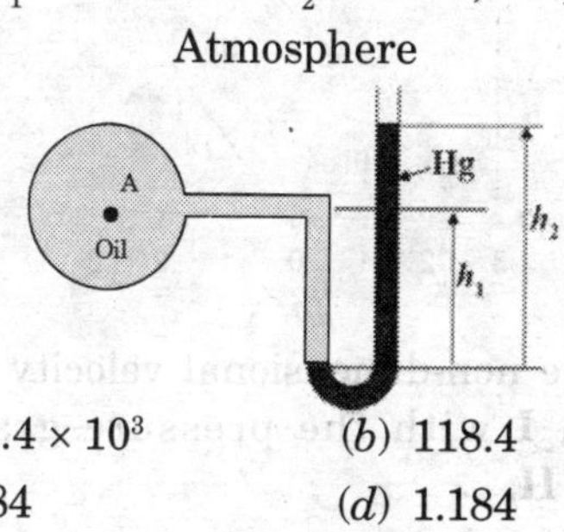

(a) 118.4×10^3 (b) 118.4

(c) 11.84 (d) 1.184

12. Water is supplied to a tank at the rate of 0.02 m^3/s, as shown in the figure below. The crosssectional area of the tank is 1 m^2 and the inner diameter of the outlet pipe is 60 mm. At a time when the water level in the tank is increasing at the rate of 5 mm/s, the average velocity (in m/s) of water in the outlet pipe is approximately

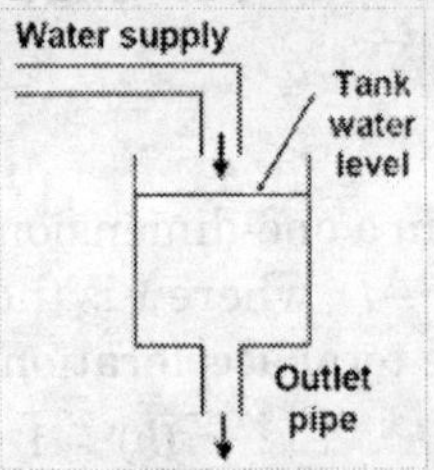

(*a*) 0.005 (*b*) 0.06

(*c*) 5.3 (*d*) 20

13. The water level in a gas-pressurized tank with a large cross-sectional area is maintained constant as shown in the figure below. The water level in the tank is 4.2 m above the pipe centerline as indicated in the figure. The gas pressure is 130 kPa. The atmospheric pressure, gravitational acceleration and density of water may be taken as 100 kPa, 10 m/s^2 and 1000 kg/m^3, respectively. Neglecting losses, the maximum velocity (in m/s) of water at any location in the horizontal portion of the delivery pipe for the pressure NOT to drop below atmospheric pressure, is

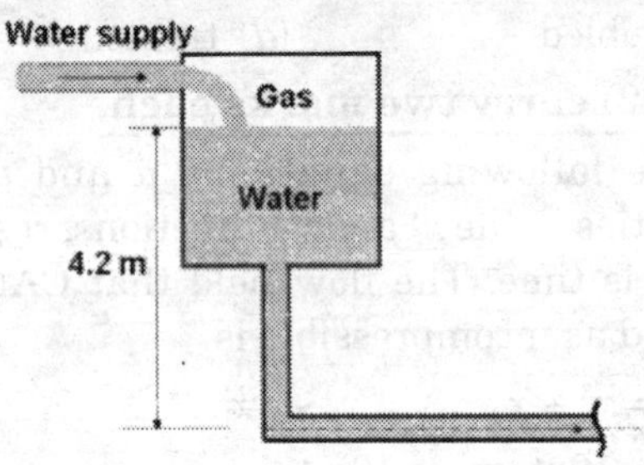

(*a*) 1.3 (*b*) 4.2

(*c*) 10 (*d*) 12

14. The figure given below shows typical non-dimensional velocity profiles for fully developed laminar flow between two infinitely long parallel plates separated by a distance *a* along *y*-direction. The upper plate is moving with a constant velocity U in the *x*-direction and the lower plate is stationary.

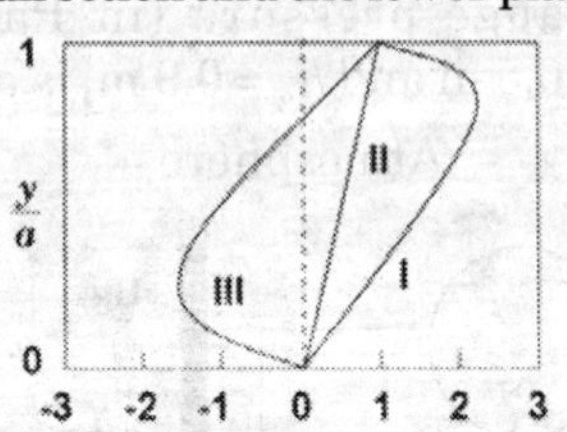

Match the non-dimensional velocity profiles in **Column I** with the pressure gradients in **Column II**.

Column I	Column II
P. profile I	1. $\frac{\partial p}{\partial x} > 0$
Q. profile II	2. $\frac{\partial p}{\partial x} < 0$
R. profile III	3. $\frac{\partial p}{\partial x} = 0$

	P	Q	R
(*a*)	2	3	1
(*b*)	3	2	1
(*c*)	3	1	2
(*d*)	1	2	3

15. Air flows over a spherical storage vessel of diameter 4 m at a speed of 1 m/s. To find the drag force on the vessel, a test run is to be carried out in water using a sphere of diameter 100 mm. The density and dynamic viscosity of air are 1.2 kg/m^3 and 1.8×10^{-5} Pa.s, respectively. The density and dynamic viscosity of water are1000 kg/m^3 and 10^{-3} Pa.s, respectively. The drag force on the model is 4 N under dynamically similar conditions. The drag force (in N) on the prototype is approximately

(*a*) 0.25 (*b*) 0.93

(*c*) 1.08 (*d*) 4

16. The velocity of an air stream is 20 m/s. The densities of mercury and air are 13600 kg/m^3 and 1.2 kg/m^3, respectively. The gravitational acceleration may be taken as 10 m/s^2. When a Pitot-static tube is placed in the stream, assuming the flow to be incompressible and frictionless, the difference between the stagnation and static pressure in the flow field (in mm Hg) would approximately be

(*a*) 1760 (*b*) 1.76

(*c*) 0.57 (*d*) 0.57×10^{-5}

Common Data Questions

Common Data Q. (17 – 18) :

A vessel containing water (density 1000 kg/m^3) and oil (density 800 kg/m^3), pressurized by gas, is shown in the figure below. Assume that the gravitational acceleration is 10 m/s^2.

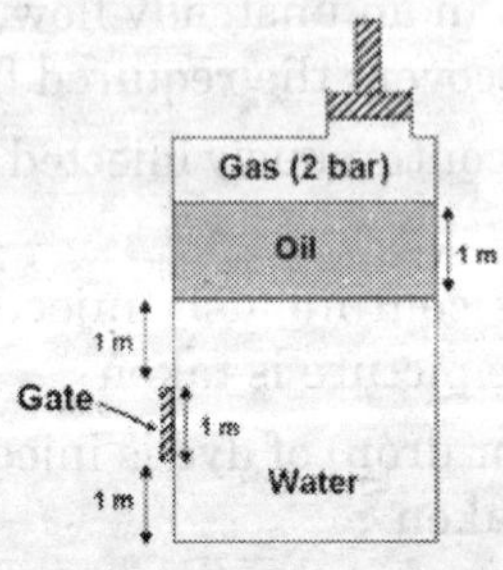

17. The pressure (in bar)exerted on the bottom wall inside the vessel is approximately

(*a*) 0.238 (*b*) 2.38

(*c*) 23.8 (*d*) 238

18. The gate is 1 m wide perpendicular to the plane of the paper. The force (in N) exerted on the gate is approximately

(*a*) 2.23×10^3 (*b*) 2.23×10^4

(*c*) 2.23×10^5 (*d*) 2.23×10^6

Common Data Q. (19 – 20) :

A boat is propelled in still water at a velocity of 5 m/s by taking water at the rate of 1 m³/s from the aft side and discharging it through the stern using a pump, as shown in the figure below. The velocity of the discharge jet relative to the boat is 9 m/s. The effect of pressure at the intake and discharge can be neglected. The density of water may be taken as 1000 kg/m³.

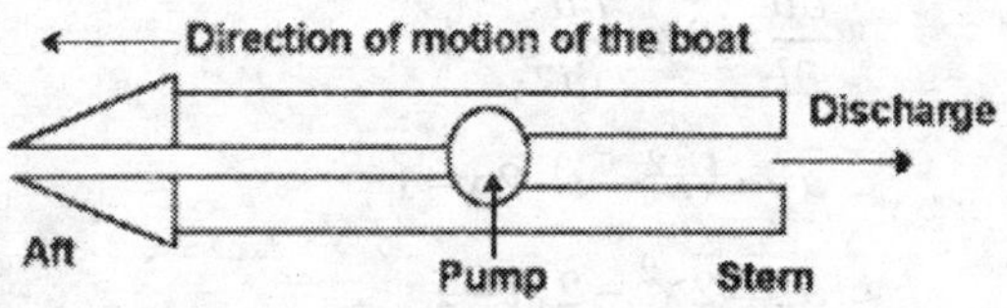

19. The power (in kW) required to propel the boat is

(*a*) 10 (*b*) 20

(*c*) 50 (*d*) 90

20. The total kinetic energy imparted to the water per second (in kW)by the pump is

(*a*) 10 (*b*) 25

(*c*) 28 (*d*) 81

Linked Answer Questions

Statement Q. (21 – 22) :

The hydrodynamic boundary layer over a flat plate is shown in the figure below. The velocity in the x-direction is approximated as $u = a + by + cy^2$, where a, b and c are constants. U is the free stream velocity and δ is the boundary-layer thickness at any point x on the plate.

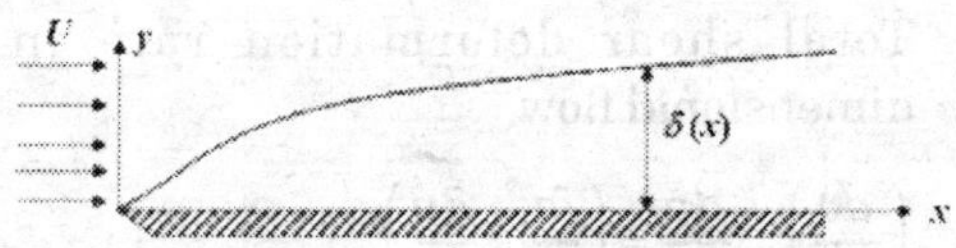

21. The dimensionless velocity profile is

(*a*) $\frac{u}{U} = 2\left(\frac{y}{\delta}\right) - \left(\frac{y}{\delta}\right)^2$

(*b*) $\frac{u}{U} = 2\left(\frac{y}{\delta}\right) + \left(\frac{y}{\delta}\right)^2$

(*c*) $\frac{u}{U} = 1.5\left(\frac{y}{\delta}\right) - 0.5\left(\frac{y}{\delta}\right)^2$

(*d*) $\frac{u}{U} = 1.5\left(\frac{y}{\delta}\right) + 0.5\left(\frac{y}{\delta}\right)^2$

22. The displacement thickness (in mm) when $\delta = 6$ mm, is

(*a*) 2.25 (*b*) 2

(*c*) –2 (*d*) –2.25

ANSWERS

1. (d)	**2.** (a)	**3.** (d)	**4.** (d)	**5.** (d)	**6.** (c)	**7.** (d)
8. (b)	**9.** (c)	**10.** (a)	**11.** (a)	**12.** (c)	**13.** (d)	**14.** (a)
15. (b)	**16.** (b)	**17.** (b)	**18.** (c)	**19.** (b)	**20.** (c)	**21.** (a)
22. (b)						

EXPLANATIONS

Fluid Mechanics – 2012

1. (d) μ – Dynamic viscosity

$\frac{du}{dy}$ – Velocity gradient/Shear deformation rate

Total shear deformation rate in two-dimensional flow

$$\left(\frac{d\theta}{dt}\right) = \frac{du}{dy} = \left(\frac{\partial v}{\partial x} + \frac{\partial u}{\partial y}\right)$$

Shear stress $(\tau) = \mu \cdot \frac{du}{dy}$

$$\boxed{\tau = \mu\left(\frac{\partial v}{\partial x} + \frac{\partial u}{\partial y}\right)}$$

2. (a) In a potential flow, the superposition of the stream function of a uniform flow and a line source gives rise to a dividing streamline represented by Rankine's half-body.

3. (d) Froude number (Fr)

$$Fr = \sqrt{\frac{F_i}{F_g}}$$

$$Fi = \rho L^2 V^2$$

$$Fg = \rho L^3 g$$

$$\boxed{\sqrt{\frac{Fi}{Fg}} = \sqrt{\frac{\rho L^2 V^2}{\rho L^3 g}} = \frac{V}{\sqrt{Lg}}}$$

4. (d) Orifice meter → It is a device which are used to measuring the rate of flowing fluid, and it is cheaper device as compared to venturimeter. It has high head loss and low lost.

Venturimeter → Measuring the rate of flowing fluid. It measured the large discharge through the given section. It has low head loss and high cost.

5. (d) Path line – A line trace by the single fluid particle, move over a period of time. It is described by Lagrangian concept.

6. (c) Fully developed flow through the pipe.

Shear stress $(\tau) = -\frac{1}{2}\frac{\partial p}{\partial x} \cdot R$

at the centre line, R = 0

then $\tau_c = 0$

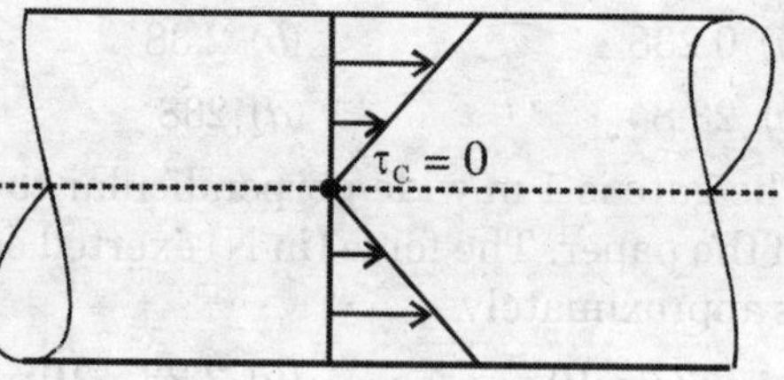

7. (d) Given that

For one dimensional flow $(u) = (x^2 - t)$

Where,

x – position, t – time

$$a_x = u \cdot \frac{\partial u}{\partial x} + v \cdot \frac{\partial u}{\partial y} + w \cdot \frac{\partial u}{\partial z} + \frac{\partial u}{\partial t}$$

$$\frac{\partial u}{\partial y} = 0, \quad \frac{\partial u}{\partial z} = 0, \quad v = 0, \quad w = 0$$

$$\frac{\partial u}{\partial l} = 2x, \quad \frac{\partial u}{\partial t} = -1$$

$$a_x = (x^2 - t) \cdot 2x - 1$$

$$a_x = 2x^2 - 2xt - 1$$

$$\boxed{a_x = -1 - 2xt + 2x^3}$$

8. (b) Ψ – Stream function

if $\Psi = f(x, y)$

$$u = -\frac{\partial \Psi}{\partial y}$$

$$v = \frac{\partial \Psi}{\partial x}$$

For irrotational flow

$$\frac{1}{2}\left[\frac{\partial v}{\partial x} - \frac{\partial u}{\partial y}\right] = 0$$

If $\nabla^2 \Psi = 0$, then flow is irrotational as well as incompressible flow.

9. (c) In fully developed Laminar flow

Pressure drop $(P_1 - P_2) = \frac{32\mu VL}{D^2}$

If V′ = 2V

Pressure drop $(P_1' - P_2') = \frac{32\mu 2VL}{D^2}$

$$\boxed{\text{Pressure drop } (dp') = 2 \times dp}$$

10. (a) For incompressible flow

$$\frac{\partial u}{\partial x} + \frac{\partial v}{\partial y} = 0$$

$$u = x^3 + xy^2 \qquad v = y^3 + x^2 y$$

$$\frac{\partial u}{\partial x} = 3x^2 + y^2 \qquad \frac{\partial v}{\partial y} = 3y^2 + x^2$$

$$\frac{\partial u}{\partial x} + \frac{\partial v}{\partial x} = 3x^2 + y^2 + 3y^2 + x^2$$

$$\frac{\partial u}{\partial x} + \frac{\partial v}{\partial x} = 4\left(x^2 + y^2\right)$$

Hence the given option (a) not satisfy the continuity equation for incompressible flow.

11. (a) Given that

$$\rho_{Hg} = 13600 \text{ kg/m}^3$$

$$\rho_{oil} = 800 \text{ kg/m}^3$$

$h_1 = 0.5$ m, $h_2 = 0.9$ m

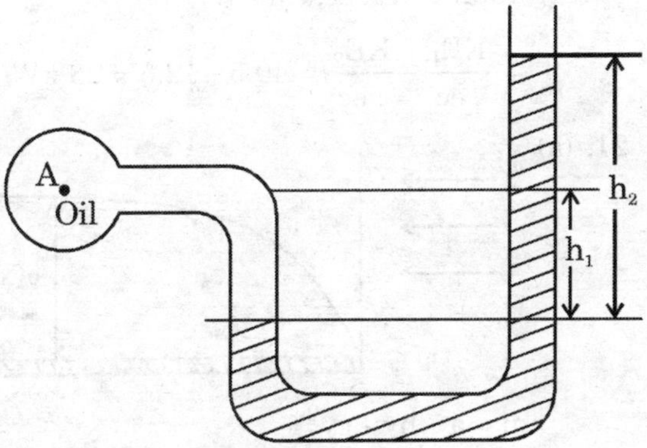

From Column balancing principle,

$$P_A + \rho_0 g h_1 = \rho_{Hg} g h_{Hg}$$

$$P_A + (800 \times 9.81 \times 0.5) = (13600 \times 9.81 \times 0.9)$$

$$\boxed{P_A = 116.15 \times 10^3 \text{ Pa}}$$

12. (c) $Q = 0.02 \text{ m}^3/\text{sec.}$

$Q = A_1 V_1$

$$V_1 = \frac{Q}{A_1} = 0.02 \text{ m/sec.}$$

Velocity is increases at the rate of 5 mm/s.

V = (0.02 + 0.005) m/s.

V = 0.025 m/s.

Actual discharge (Q) = A_1V

$= 1 \times 0.025 = 0.025 \text{ m}^2/\text{sec.}$

$Q = A_2 V_2$

$$V_2 = \frac{0.025}{\frac{\pi}{4}\left(10^{-3} \times 60\right)^2} = 8.84 \text{ m/sec.}$$

13. (d) Given that

$z_1 = 4.2$ m

$P_1 = 130$ kPa

$P_2 = 100$ kPa

$z_2 = 0$

$\rho_W = 1000 \text{ kg/m}^3$

Apply the Bernoulli's equation at section (1) & (2)

$$\frac{P_1}{\rho g} + \frac{V_2^2}{2g} + z_1 = \frac{P_2}{\rho g} + \frac{V_2^2}{2g} + z_2$$

$$\Rightarrow \frac{130 \times 10^3}{10^3 \times 9.81} + 0 + 4.2 = \frac{100 \times 10^3}{10^3 \times 9.81} + \frac{V_2^2}{2g} + 0$$

$$\Rightarrow \frac{10^3(130 - 100)}{10^3 \times 9.81} + 4.2 = \frac{V_2^2}{2g}$$

$$\Rightarrow \left(\frac{30}{9.81} + 4.2\right) \times 2 \times 9.81 = V_2^2$$

$$\Rightarrow V_2^2 = 142.404$$

$$\boxed{V_2 = 11.93 \text{ m/s.}}$$

14. (a) P. Profile 1 – $\frac{\partial p}{\partial x} < 0$

Q. Profile II – $\frac{\partial p}{\partial x} = 0$

R. Profile III – $\frac{\partial p}{\partial x} > 0$

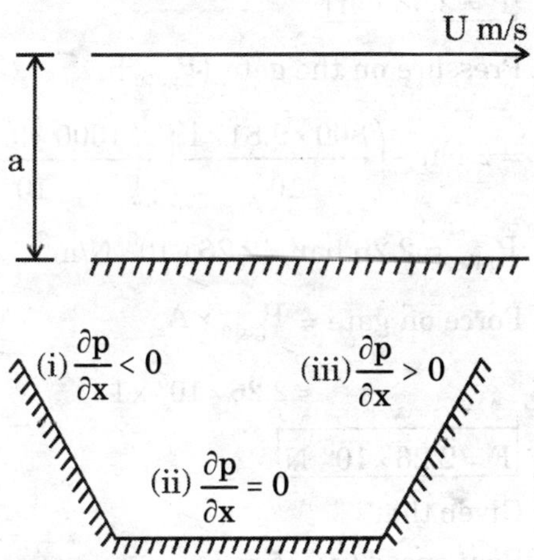

$$\boxed{P-2,\ Q-3,\ R-1}$$

16. (b) Given that

$\rho_{Hg} = 13600 \text{ kg/m}^3$

$\rho_A = 1.20 \text{ kg/m}^3$

V = 20 m/s.

Now, According to stagnation principle in pitot tube

$$\frac{P_1}{\rho g}+\frac{V_1^2}{2g}=\frac{P_2}{\rho g}+\frac{V_2^2}{2g}$$

$$\frac{P_1-P_2}{\rho g}=\frac{V_2^2}{2g} \quad [V_1=0]$$

$$P_1-P_2=\frac{\rho v^2}{2}=\frac{1.2\times 20^2}{2}$$

$$P_1-P_2=240\,Pa$$

Now, in Hg column

$$P_1-P_2=\rho gh$$

$$h_{Hg}=\frac{240}{13600\times 9.81}$$

$$\boxed{h_{Hg}=1.77 \text{ mm of hg}}$$

17. (b) Given that

$\rho_w = 1000$ kg/m^3

$\rho_{oil} = 800$ kg/m^3

Pressure exerted on the bottom wall (P)

$P = P_{gas} + P_{oil} + P_{water}$

$P_{gas} = 2$ bar

$P_{oil} = 800\times 9.81\times 1 = 0.07848$ bar

$P_w = 1000\times 9.81\times 3 = 0.2943$ bar

$P = (2+0.07848+0.2943)$ bar

$\boxed{P = 2.38 \text{ bar}}$

18. (c) Pressure on the gate (P_{gate})

$$= 2\text{ bar} + \left(\frac{800\times 9.81\times 1}{10^5}\right)+\left(\frac{1000\times 9.81\times 2}{10^2}\right)$$

$P_{gate} = 2.26$ bar $= 2.26\times 10^5$ N/m^3

Force on gate $= P_{gate}\times A$

$= 2.26\times 10^5\times 1$

$\boxed{F = 2.26\times 10^5 \text{ N}}$

19. (b) Given that

Boat speed (u) = 5 m/s

Q = 1 m³/sec.

Velocity of jet (V) = 9 m/s.

Density of water $(\rho_w) = 1000$ kg/m^3

Power required to propel the boat in (kW)

$P = \rho Q(v-u)\times u$

$P = 1000\times 1(9.5)\times 5$W

$\boxed{P = 20 \text{ kW}}$

20. (c) Total kinetic energy imported to the water (in kw) by the pump.

For jet,

$$\frac{K\cdot E_1}{sec}=\frac{1}{2}\rho A(V)^3$$

$$\frac{K\cdot E_1}{sec}=40.5\text{ kW}$$

For boat,

$$\frac{K\cdot E_2}{sec}=\frac{1}{2}\times 1000\times 1\times 5^2=12.5\text{ kW}$$

Total K· E/sec. (in kW)

$$=\frac{KE_1}{sec}-\frac{KE_2}{sec}=40.5-12.5=28\text{ kW}$$

21. (a)

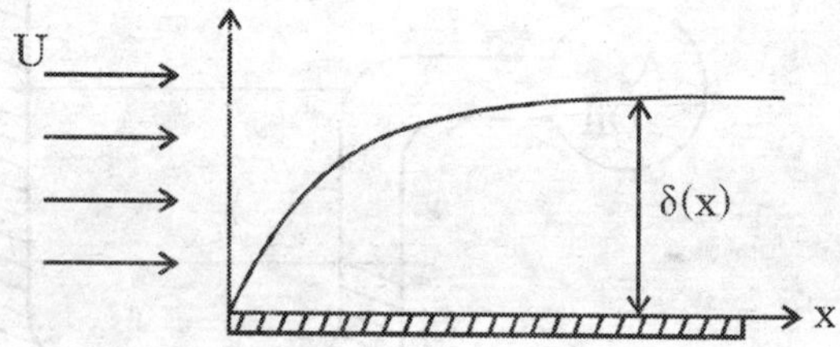

$u = a + by + cy^2$

δ = Boundary byer thickness

U – Free stream velocity

$$\boxed{\frac{u}{U}=2\left(\frac{y}{\delta}\right)-\left(\frac{y}{\delta}\right)^2}$$

Dimensionless velocity profile.

22. (b) $\frac{u}{U}=2\left(\frac{y}{\delta}\right)-\frac{y^2}{\delta^2}$

Displacement thickness $(\delta)=\int_0^\delta\left(1-\frac{u}{U}\right)\cdot dy$

$$\delta=\int_0^\delta\left[1-\frac{2y}{\delta}+\frac{y^2}{\delta^2}\right]\cdot dy$$

$$\delta=\left[y-\frac{2y^2}{2\delta}+\frac{y^3}{3\delta^2}\right]$$

at δ = 6 mm & y = 1

$\boxed{\delta = 2}$

2013

Q. 1 – 9 carry one mark each.

1. The gauge pressure inside a soap bubble of radius R, with σ denoting the surface tension between the soap solution and air, is.

(a) $\frac{\sigma}{2\pi R}$ (b) $\frac{4\sigma}{R}$

(c) $\frac{2\sigma}{R}$ (d) $\frac{\sigma}{4\pi R}$

2. Let M, B and G represent respectively the metacentre, centre of buoyancy and the centre of mass of a floating buoy. Which of the following statements is correct?

(a) M is above G; Buoy unstable

(b) B is above G; Buoy stable

(c) M is above G; Buoy stable

(d) B is above G; Buoy unstable

3. A reservoir connected to a pipe line is being filled with water, as shown in the Figure. At any time *t*, the free surface level in the reservoir is *h*. Find the time in seconds for the reservoir to get filled up to a height of 1 m, if the initial level is 0.2 m. ______

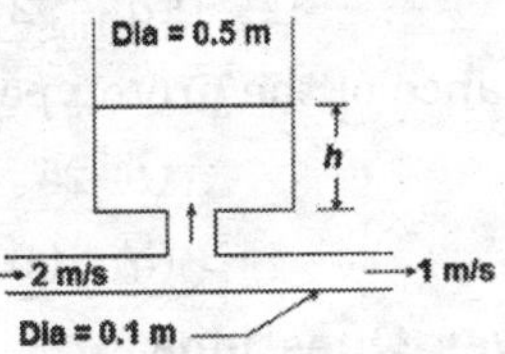

4. Bernoulli's equation is valid for the following type of flow :

(a) Compressible, steady, inviscid

(b) Incompressible, steady, viscous

(c) Compressible, unsteady, viscous

(d) Incompressible, steady, inviscid

5. If A is the area of a circle of radius *r* enclosing a plane forced vortex flow, with origin at the centre of the vortex and if ω is the angular velocity, ζ is the vorticity, $\overline{V}$ is the velocity vector, then the circulation around the contour of the circle is given by

(a) $2\omega A$ (b) $2\xi A$

(c) $2\overline{V}A$ (d) 0

6. Flow past a circular cylinder can be produced by superposition of the following elementary potential flows :

(a) Uniform flow, doublet

(b) Uniform flow, vortex

(c) Source, vortex

(d) Sink, vortex

7. Let δ, δ_1 and δ_2 denote respectively the boundary-layer thickness, displacement thickness and the momentum thickness for laminar boundary layer flow of an incompressible fluid over a flat plate. The correct relation among these quantities is

(a) $\delta < \delta_1 < \delta_2$ (b) $\delta > \delta_1 > \delta_2$

(c) $\delta > \delta_1 < \delta_2$ (d) $\delta < \delta_1 > \delta_2$

8. In the hydrodynamic entry region of a circular duct, the pressure forces balance the sum of

(a) viscous and buoyancy forces

(b) inertia and buoyancy forces

(c) inertia and surface tension forces

(d) inertia and viscous forces

9. Bodies with various cross-sectional shapes subjected to cross-flow of air are shown in the following figures. The characteristic dimension of all the shapes is the same. The cross-sectional shape with the largest coefficient of drag (i.e. sum of the pressure and skin friction drags), at any moderately large Reynolds number, is

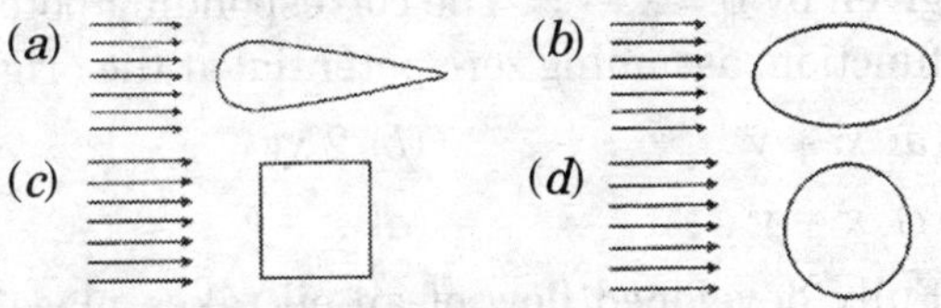

Q. 10 – 22 carry two marks each.

10. A U-tube of a very small bore, with its limbs in a vertical plane and filled with a liquid of density ρ, up to a height of *h*, is rotated about a vertical axis, with an angular velocity of ω, as shown in the Figure. The radius of each limb from the axis of rotation is R. Let p_a be the atmospheric pressure and *g*, the gravitational acceleration. The angular velocity at which the pressure at the point O becomes half of the atmospheric pressure is given by

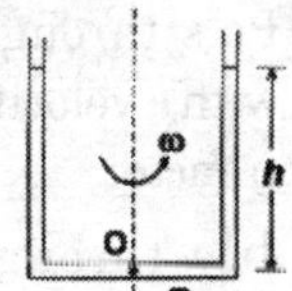

(a) $\sqrt{\frac{p_a + 2\rho gh}{\rho R^2}}$ (b) $\sqrt{\frac{2(p_a + 2\rho gh)}{\rho R^2}}$

(c) $\sqrt{\frac{p_a + 2\rho gh}{2\rho R^2}}$ (d) $\sqrt{\frac{p_a + \rho gh}{2\rho R^2}}$

11. An incompressible fluid at a pressure of 150 kPa (absolute) flows steadily through a two dimensional channel with a velocity of 5 m/s as shown in the Figure. The channel has a 90° bend. The fluid leaves the channel with a pressure of

100 kPa (absolute) and linearly varying velocity profile. υ_{max} is four times υ_{min}. The density of the fluid is 914.3 kg/m³. The velocity υ_{min}, in m/s, is

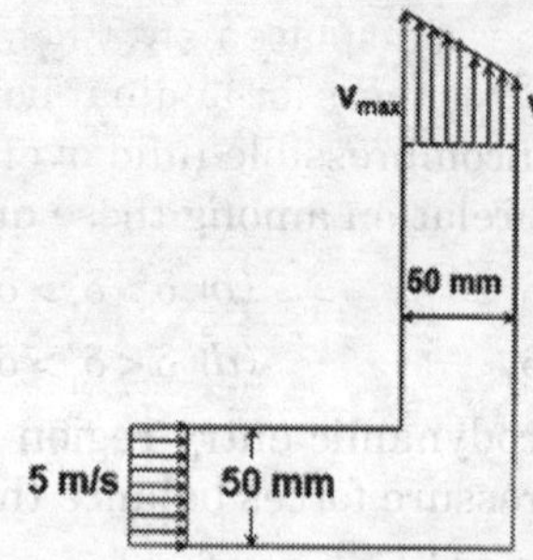

(a) 25 (b) 2.5

(c) 2.0 (d) 0.2

12. The velocity vector corresponding to a flow field is given, with usual notation, by $\bar{V} = 3x\hat{i} + 4xy\hat{j}$. The magnitude of rotation at the point (2,2) in rad/s is

(a) 0.75 (b) 1.33

(c) 2 (d) 4

13. The stream function for a potential flow field is given by $\psi = x^2 - y^2$. The corresponding potential function, assuming zero potential at the origin, is

(a) $x^2 + y^2$ (b) $2xy$

(c) $x^2 - y^2$ (d) $x - y$

14. Fully developed flow of an oil takes place in a pipe of inner diameter 50 mm. The pressure drop per metre length of the pipe is 2 kPa. Determine the shear stress, in Pa, at the pipe wall.

15. The Darcy friction factor f for a smooth pipe is given by $f = 64/\text{Re}$ for laminar flow and by $f = 0.3/\text{Re}^{0.25}$ for turbulent flow, where Re is the Reynolds number based on the diameter. For fully developed flow of a fluid of density 1000 kg/m³ and dynamic viscosity 0.001 Pa.s through a smooth pipe of diameter 10 mm with a velocity of 1 m/s, determine the Darcy friction factor. ______

16. Air flows steadily through a channel. The stagnation and static pressures at a point in the flow are measured by a Pitot tube and a wall pressure tap, respectively. The pressure difference is found to be 20 mm Hg. The densities of air, water and mercury, in kg/m³, are 1.18, 1000 and 13600, respectively. The gravitational acceleration is 9.81 m/s². Determine the air speed in m/s. ______

Common Data Questions

Common Data for Questions 17 and 18 : The velocity field within a laminar boundary layer is given by the expression :

$$\bar{V} = \frac{Bu_\infty y}{x^{3/2}}\hat{i} + \frac{Bu_\infty y^2}{4x^{5/2}}\hat{j}$$

where B = 100 $m^{1/2}$ and the free stream velocity u_∞ = 0.1 m/s.

17. Calculate the x-direction component of the acceleration in m/s² at the point x = 0.5 m and y = 50 mm. ______

18. Find the slope of the streamline passing through the point x = 0.5 m and y = 50 mm.

Common Data for Questions 19 and 20 :
The wave and eddy resistance of a sea-going vessel, 96 m in length, driven at a velocity of 12 m/s, is to be determined. For this purpose, a 1/16 th scale model is employed in fresh water and the coefficient of resistance C_{we} of the model is found to be 1.47×10^{-4}. The quantity C_{we} is defined as $F_{we}/(\rho V^2 L^2/2)$, where F_{we} is the wave and eddy resistance, ρ is the density, V is the velocity and is the characteristic length. The density of sea water is 1026 kg/m³.

19. The velocity in m/s, at which the model is towed, is

(a) 0.75 (b) 1.33

(c) 3 (d) 192

20. The resistance of the prototype, in kN, is

(a) 6 (b) 25

(c) 26.9 (d) 100.1

Linked Answer Questions

Statement for Linked Answer Questions 21 & 22 :
Water enters a symmetric forked pipe and discharges into atmosphere through the two branches as shown in the Figure. The cross-sectional area of section-1 is 0.2 m² and the velocity across section-1 is 3 m/s. The density of water may be taken as 1000 kg/m³. The viscous effects and elevation changes may be neglected.

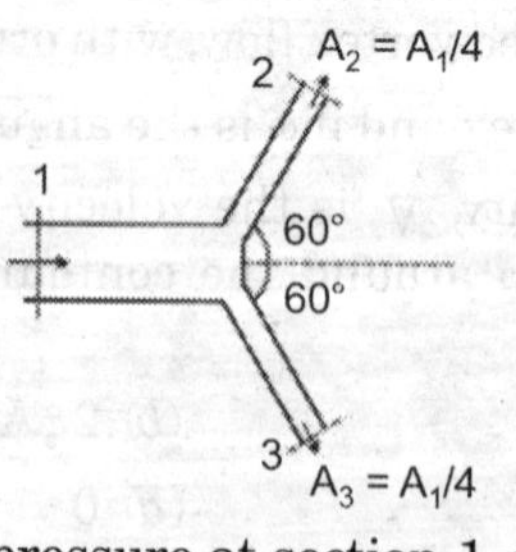

21. The gauge pressure at section-1, in kPa, is

(a) 0.6 (b) 13.5

(c) 135 (d) 600

22. The magnitude of the force, in kN, required to hold the pipe in place, is

(a) 2.7 (b) 5.4

(c) 19 (d) 27

ANSWERS

1. (b) 2. (c) 3. 20 4. (d) 5. (a) 6. (a) 7. (b)
8. (d) 9. (c) 10. (a) 11. (c) 12. (d) 13. (b) 14. 25
15. 0.0064 16. 65.89 to 67 17. 2 to 3 18. 0.09 19. (c) 20. (d) 21. (b)
22. (a)

EXPLANATIONS

1. (b) P_i – Inside pressure

P_0 – Outside pressure

P_i is little more than P_0.

$\Delta P = P_i - P_0$ – Gauge Pressure

D – Diameter of Soap bubbles

Force balancing in soap bubbles:

Force due to pressure = Force due to surface tension,

$$\Delta P \times \frac{\pi}{4} D^2 = 2\sigma AD$$

$$\Delta P = \frac{8\sigma}{D}$$

$$\boxed{\Delta P = \frac{4\sigma}{R}}$$

2. (c) Here,

M – Metacentre

B – Centre of Buoyancy

G – Centre of gravity

For floating body,

→ M above G for stable.

→ M below G for unstable.

→ M and G, both are coincides for neutral equilibrium.

3. Given that

D = 0.5 m

H = 1 m

d = 0.1 m

$$\text{Time required (T)} = \frac{2A\sqrt{H_1}}{Cd \times a \times \sqrt{2g}}$$

$$T = \frac{2 \times \frac{\pi}{4} \times 0.5^2 \times \sqrt{1}}{0.60 \times \frac{\pi}{4} \times 0.1^2 \times \sqrt{2 \times 9.81}}$$

$$\boxed{T = 18.81 \text{ sec.}}$$

4. (d) Bernoulli's equation work on the conservation of energy.

Following are the assumptions:

i) Incompressible.

ii) Irrotational

iii) Steady flow.

iv) Inviscid.

5. (a) Vorticity $(\xi) = 2 \cdot \omega$

$$\text{Vorticity } (\xi) = \frac{\text{Circulation}}{\text{Area}}$$

Circulation (Γ) = Vorticity × Area

$$\boxed{\text{Circulation } (\Gamma) = 2\omega A}$$

6. (a) For superimposed flow

i) A doublet in a uniform flow.

ii) Source and sink pair.

iii) A plane source in uniform flow.

iv) A source and sink pair in uniform flow.

7. (b) δ – Boundary layer thickness.

δ_1 – Displacement thickness.

δ_2 – momentum thickness.

So that,

$$\boxed{\delta > \delta_1 > \delta_2}$$

8. (d) In the hydrodynamic entry region of a circular duct the, the pressure force is the sum of

→ inertia force

→ viscous force

9. (c) For uniform flow

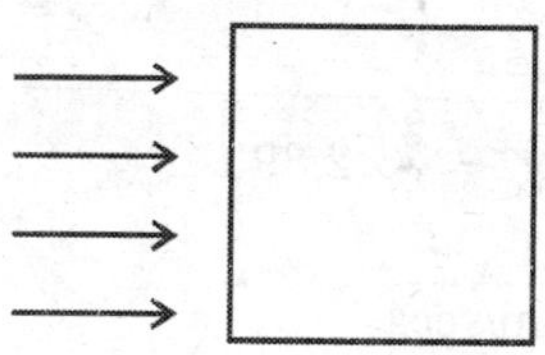

10. (a)

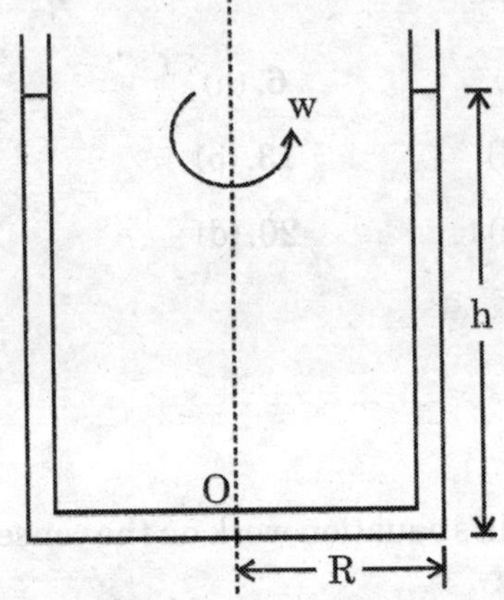

$$h = \frac{\omega^2 R^2}{2g}$$

$P_0 = \rho gh$

$\frac{Pa}{2} = \rho gh$ [Given]

$Pa = 2\rho gh$

$$P_0 = \frac{\omega^2 R^2}{2g} \times \rho g = \frac{\rho\omega^2 R^2}{2}$$

$$\frac{Pa}{2} = \frac{\rho\omega^2 R^2}{2}$$

Now, total pressure in the limb

$Pa + 2\rho gh = \rho\omega^2 R^2$

$$\omega = \sqrt{\frac{(Pa + 2\rho gh)}{\rho R^2}}$$

11. (c) Apply Bernoulli's equation,

$$\frac{P_1}{\rho g} + \frac{V_1^2}{2g} = \frac{P_2}{\rho g} + \frac{V_2^2}{2g}$$

$$\frac{(150 - 100) \times 1000}{\rho g} = \frac{V_2^2 - 5^2}{2g}$$

$$\frac{2 \times 50 \times 1000}{914.3} = (4V_{max})^2 - 25$$

$109.37 = 16V_{max}^2 - 25$

$V_{max} = 2.893$

$$\frac{V_{max}}{V_{min}} = \frac{3}{2}$$

$$V_{min} = \frac{2}{3} V_{max} = \frac{2}{3} \times 2.893$$

$\boxed{V_{min} = 1.93 \text{ m/sec.}}$

12. (d) Given that

$\vec{V} = 3x\hat{i} + 4xy\hat{j}$ at (2, 2)

$u = 3x,\ v = 4xy$

$$\frac{\partial v}{\partial x} = 4y,\ \frac{\partial u}{\partial y} = 0$$

$$w = \frac{1}{2}\left[\frac{\partial v}{\partial x} - \frac{\partial u}{\partial y}\right]$$

$$w = \frac{1}{2}[4y - 0]$$

$w = 2y = 4$ rad/s.

13. (b) Given that

Stream function $(\Psi) = x^2 - y^2$

$$u = \frac{\partial \Psi}{\partial y} = -\frac{\partial}{\partial y}(x^2 - y^2) = 2y$$

$$v = \frac{\partial \Psi}{\partial x} = \frac{\partial}{\partial x}(x^2 - y^2) = 2x$$

We know that,

$$u = -\frac{\partial \phi}{\partial x} \qquad v = -\frac{\partial \phi}{\partial y}$$

$$2y = -\frac{\partial \phi}{\partial x}$$

$\phi = -2xy + K$

at origin (0, 0)

$0 = 0 + K$

$K = 0$

$\boxed{\phi = -2xy}$

14. Given that

$D = 50$ mm $= 0.05$ m

$R = 0.025$ m

Pressure drop per meter length = 2 kPa

$$\text{Shear stress } (\tau) = -\frac{1}{2}\frac{(p_1 - p_2)}{L} \cdot R$$

$$\tau = \frac{1}{2} \cdot 2 \times 1000 \times 0.025$$

$\boxed{\tau = 25 \text{Pa}}$

15. Given that

$$f = \frac{64}{Re}$$

$\rho = 1000$ km/m^2

$\mu = 0.001$ Pa.s

$d = 10$ mm $= 10^{-2}$ m.

$v = 1$ m/s

$$Re = \frac{\rho v D}{\mu}$$

$$Re = \frac{1000 \times 1 \times 10^{-2}}{0.001}$$

$Re = 10000$

$$f = \frac{64}{Re} = \frac{64}{10000}$$

$$\boxed{f = 0.0064}$$

16. Given that

$x = 20$ mm

$$h = x\left(\frac{G_{Hg}}{G_{Air}} - 1\right)$$

$$h = 20 \times 10^{-3} \times \left(\frac{13600}{1.18} - 1\right)$$

$h = 230.48$ m

$$V = C_v\sqrt{2gh}$$

$$V = 0.98\sqrt{2 \times 9.81 \times 230.48}$$

$V = 65.89$ m/s

17. $$\vec{V} = \frac{BUgy}{x^{3/2}}\hat{i} + \frac{Bugy^2}{x^{5/2}}\hat{j}$$

$B = 100\ m^{1/2}$, $Ug = 0.1$ m/sec.

$$u = \frac{BUgy}{x^{3/2}} = \frac{100 \times 0.1y}{x^{3/2}} = \frac{10y}{x^{3/2}}$$

$$v = \frac{BUgy^2}{x^{5/2}} = \frac{100 \times 0.1y^2}{x^{5/2}} = \frac{10y^2}{x^{5/2}}$$

$$a_x = u \cdot \frac{\partial u}{\partial x} + v \cdot \frac{\partial u}{\partial y} \text{ at } (x = 0.5 \text{ m}, y = 0.05 \text{ m})$$

$$u = \frac{10 \times 0.05}{0.5^{3/2}} = 1.414$$

$v = 0.141$

$$\frac{\partial u}{\partial x} = 10y\left(-\frac{3}{2}\right) \cdot x^{-3/2-1} = -\frac{3}{2}\left(10yx^{-5/2}\right)$$

$$\left(\frac{\partial u}{\partial x}\right) = -\frac{3}{2} \times 10 \times 0.05 \times (0.5)^{-5/2} = -4.24$$

$$\left(\frac{\partial u}{\partial y}\right) = \frac{10}{x^{3/2}} = \frac{10}{0.5^{3/2}} = 28.28$$

$$a_x = (1.414 \times -4.24) + (0.141 \times 28.28)$$

$$\boxed{a_x = -2.007 \text{ m/s}^2}$$

18. Slope of stream line passing through

$x = 0.5$ m, $y = 50$ mm $= 0.05$ m

$$\frac{du}{dx} = \frac{v}{u} = \frac{0.141}{1.414} = 0.09$$

19. (c) According to given data

Model and prototype relation,

$$C_{we} = \frac{F_{we}}{(\rho V^2 L^2 / 2)}$$

Now,

$$\left(\frac{\rho V^2 L^2}{2}\right)_m = \left(\frac{\rho V^2 L^2}{2}\right)_p$$

$$\frac{V_m^{\ 2}}{V_p^{\ 2}} = \text{Scale ratio}$$

$$\frac{V_m^{\ 2}}{12^2} = \frac{1}{16}$$

$$V_m^{\ 2} = \frac{144}{16} = 9$$

$$\boxed{V_m = 3 \text{ m/sec.}}$$

20. (d) $$(F_{we})_{model} = \left(\frac{1}{2}\rho V^2 L^2 C_{we}\right)_{model}$$

$$(F_{we})_{model} = \frac{1}{2} \times 1026 \times 3^2 \times 96^2 \times 1.47 \times 10^{-4}$$

$$(F_{we})_{model} = 6.254 \text{ kN}$$

$$\frac{(F_{we})_{model}}{(F_{we})_{prototype}} = \frac{1}{16}$$

$$(F_{we})_{prototype} = 16 \times 6.254 = 100.1 \text{ kN.}$$

21. (b) Given that

$A_1 = 0.2\ m^2$

$V_1 = 3$ m/s

$\rho w = 1000\ kg/m^3$

Gauge pressure $(P_1 - P_2)$

$$= \frac{V_1^{\ 2} \times 1000}{2} \times 3^2 = \frac{3^2 \times 1000 \times 3}{2}$$

$$\boxed{\text{Gauge Pressure} = 13.5 \text{ KPa}}$$

22. (a) Magnitude of force = Pressure × Area

$$\boxed{F = 13.5 \times 0.2 = 2.7 \text{ KN}}$$

2014

Q. (1 – 9) Carry One Mark Each.

1. A dam with a curved shape is shown in the figure. The cross sectional area of the dam (shaded portion) is 100 m² and its centroid is at $\bar{x}$ = 10m. The vertical component of the hydrostatic force, F_z, is acting at a distance x_p. The value of x_p is _________ m.

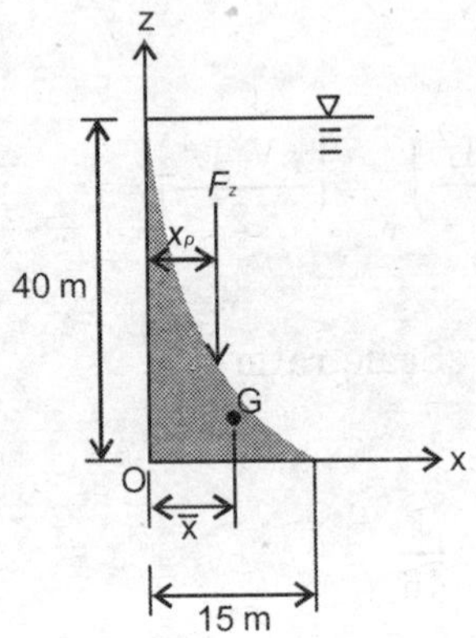

2. For an unsteady incompressible fluid flow, the velocity field is

$$\vec{V} = (3x^2 + 3)t\ \hat{i} - 6xyt\ \hat{j},$$

where *x*, *y* are in meters and *t* is in seconds. Acceleration in m/s² at the point *x* = 10m and *y* = 0, as measured by a stationary observer is

(*a*) 303 (*b*) 162
(*c*) 43 (*d*) 13

3. For an incompressible flow, the existence of components of acceleration for different types of flow is described in the table below.

Type of Flow

P. Steady and uniform
Q. Steady and non-uniform
R. Unsteady and uniform
S. Unsteady and non-uniform

Components of Acceleration

1. Local exists, convective does not exist
2. Both exist
3. Both do not exist
4. Local does not exist, convective exists

Which one of the following options connecting the left column with the right column is correct?

(*a*) P – 1; Q – 4; R – 3; S – 2
(*b*) P – 4; Q – 1; R – 2; S – 3
(*c*) P – 3; Q – 2; R – 1; S – 4
(*d*) P – 3; Q – 4; R – 1; S – 2

4. Velocity in a two-dimensional flow field is specified as: $u = x^2 y$; $v = -y^2 x$. The magnitude of the rate of angular deformation at a location (*x* = 2 m and *y* = 1 m) is ____ s^{-1}.

5. For a plane irrotational flow, equipotential lines and streamlines are

(*a*) parallel to each other
(*b*) at an angle of 90° to each other
(*c*) at an angle of 45° to each other
(*d*) at an angle of 60° to each other

6. Flow around a Rankine half-body is represented by the superposition of

(*a*) source and vortex flows
(*b*) source and uniform flows
(*c*) vortex and uniform flows
(*d*) source, vortex and uniform flows

7. It is required to carry out model studies on a boat having a characteristic length of 3.6 m and travelling at a speed of 3 m/s. Assume the acceleration due to gravity as 10 m/s² and neglect the effects due to viscous and surface tension forces. The value of appropriate non-dimensional number is ______.

8. Which one of the following velocity profiles typically represents a fully developed incompressible, turbulent flow in a pipe?

(*a*)  (*b*) 

(*c*) 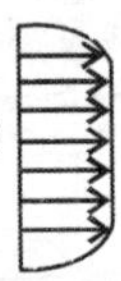 (*d*) 

9. Consider an incompressible, laminar flow past a circular cylinder of diameter *d*. The flow is uniform at the far upstream. Which one of the following figures typically represents the wake velocity profile just downstream of the cylinder?

(*a*) 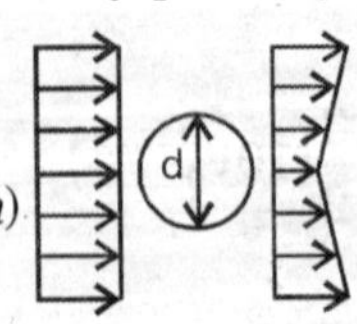

(*b*) 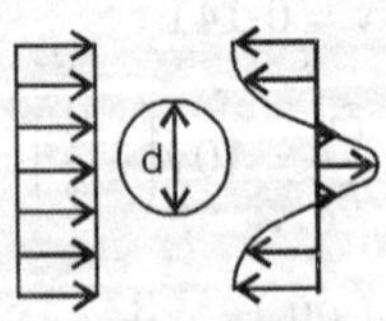

(*c*) 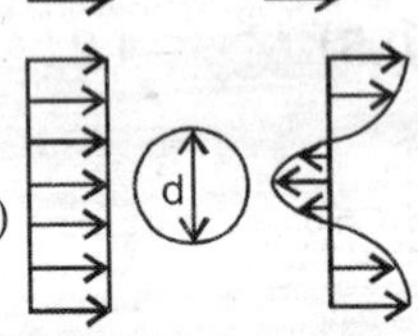

(*d*) 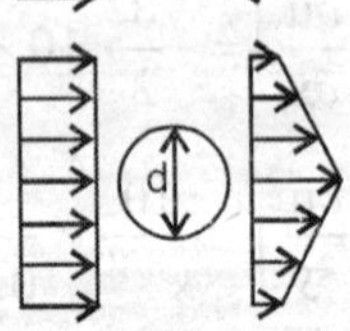

Q. (10 – 22) Carry Two Marks Each.

10. A container of square cross-section is partially filled with a liquid of density ρ_1. The cylinder is intended to float in another liquid of density ρ_2 as

shown in the figure. The distance between metacentre and centre of buoyancy is $\frac{I}{V_{sub}}$, where I and V_{sub} are area moment of inertia of the cross-section and submerged volume, respectively. Neglect the weight of the container.

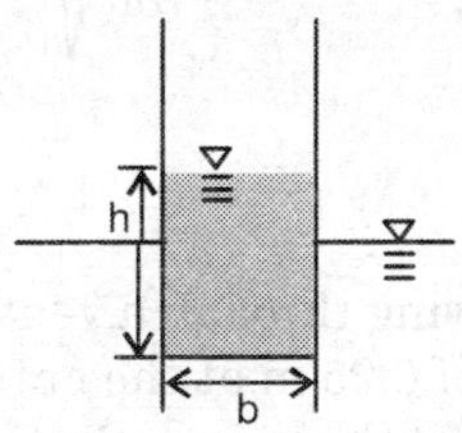

Which one of the following is the correct condition for stability?

(a) $\frac{\rho_2}{6\rho_1}\frac{b}{h}-\frac{h}{b}\left(1-\frac{\rho_1}{\rho_2}\right)>0$

(b) $\frac{\rho_2}{6\rho_1}\frac{b}{h}-\frac{h}{b}\left(1+\frac{\rho_1}{\rho_2}\right)>0$

(c) $\frac{\rho_2}{6\rho_1}\frac{b}{h}+\frac{h}{b}\left(1-\frac{\rho_1}{\rho_2}\right)>0$

(d) $\frac{\rho_2}{6\rho_1}\frac{b}{h}+\frac{h}{b}\left(1+\frac{\rho_1}{\rho_2}\right)>0$

11. In a steady state two-dimensional potential flow field due to a point source, the acceleration of a particle at a distance r from the point source is

(a) proportional to r^{-1} (b) proportional to r

(c) a constant (d) proportional to r^{-3}

12. Velocity in a two-dimensional flow at time t and location (x, y) is described as : $\vec{V}=3y^2\,\hat{i}+(x-1)\,\hat{j}$. The equation for the path line of a particle passing through the point (1,0) at $t = 0$ is

(a) $x^4-4y^3=0$ (b) $(x-1)^3-2y^4=0$

(c) $(x-1)^4-64y^3=0$ (d) $(x+1)^4-16y^3=0$

13. The gravity driven flow over a hump of height h in a canal is shown in the figure. The height of the free surface from the canal bed at upstream of the hump is H. The free surface height reduces to H_1 above the hump.

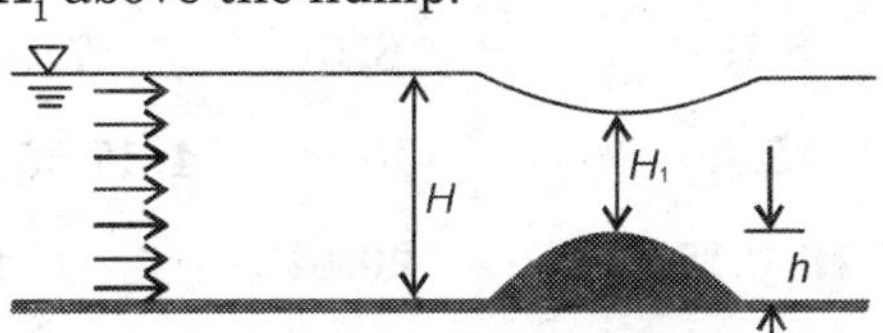

Assuming the canal bed to be horizontal, the discharge per unit width is given by

(a) $\sqrt{\dfrac{2g(H-H_1-h)}{\dfrac{1}{H_1^2}-\dfrac{1}{H^2}}}$ (b) $\sqrt{\dfrac{2gh}{\dfrac{1}{(H_1+h)^2}-\dfrac{1}{H^2}}}$

(c) $\sqrt{\dfrac{2g(H-H_1)}{\dfrac{1}{(H_1+h)^2}-\dfrac{1}{H^2}}}$ (d) $\dfrac{2g(H-H_1)}{\dfrac{1}{H_1^2}-\dfrac{1}{H^2}}$

14. Steady state incompressible flow through a pipe network is shown in the figure. Inlets marked as (1), (2) and (3) and exit marked as (4), are shown with their respective diameters. The exit flow rate at (4) is 0.1 m³/s. A 20% increase in flow rate through (3) results in a 10% increase in flow rate through (4). The original velocity through inlet (3) is _____ m/s.

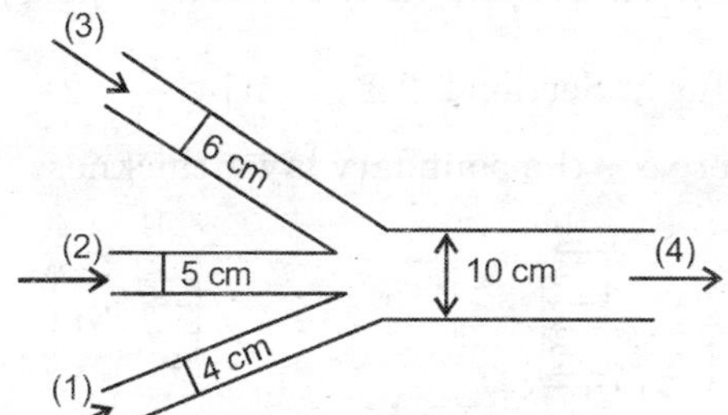

15. A reducing elbow is used to deflect water upward by 30° as shown in the figure. The mass flow rate at the inlet is 14 kg/s. Water is entering at a gauge pressure of 200 kPa and exits to the atmosphere. The cross-sectional area is 113 cm² at the inlet and 7 cm² at the exit. Density of water and acceleration due to gravity are 1000 kg/m³ and 10 m/s², respectively. Magnitude of x- component of the water force on the elbow is ____ N.

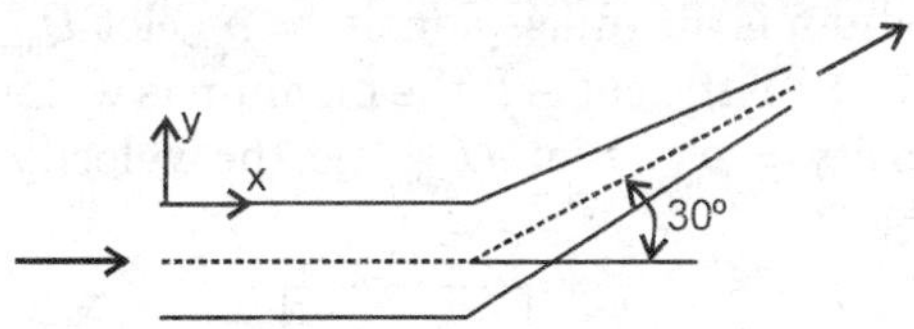

16. A source with a strength of k_1 and a vortex with a strength of k_2 are located at the origin. The resultant velocity at a radial distance r from the origin due to the superposition of the source and vortex is expressed as

(a) $\frac{k_1+k_2}{r}$ (b) $\frac{\sqrt{k_1^2+k_2^2}}{r}$

(c) $\frac{\sqrt{k_1^2-k_2^2}}{r}$ (d) $\frac{k_1-k_2}{r}$

17. Velocity potential for an incompressible fluid flow is given as: $\phi = 2(x^2+2y-y^2)$. Assume the value of stream function at the origin to be zero. The value of stream function at $[(x, y) \equiv (2, 2)]$ is ____.

18. The model of a conduit is scaled to 1/100 of the actual size. Seawater is used in the prototype and fresh water is used in the model. Velocity in the prototype is 0.5 m/s. Density and dynamic viscosity of the seawater are 1025 kg/m³ and 1.07×10^{-3} kg/m-s, respectively. Density and dynamic viscosity of fresh water are 1000 kg/m³ and 1×10^{-3} kg/m-s, respectively. Assume the viscous forces to be dominant. The velocity to be maintained in the model to ensure dynamic similarity is____ m/s.

19. A fluid is flowing through a pipe of circular cross-section. Reynolds number of the flow is 1600. The head loss over a 45 m length of the pipe is 0.6 m. The average flow velocity of the fluid is 1 m/s and the acceleration due to gravity is 10 m/s². The diameter of the pipe is _____ m.

20. Consider a laminar flow over a flat plate of width *w*. At Section 1-1, the velocity profile is uniform as shown in the figure. The *x*-direction velocity profile at Section 2-2 is given by $\frac{u}{U} = 2\frac{y}{\delta} - \left(\frac{y}{\delta}\right)^2$, where δ is the boundary layer thickness.

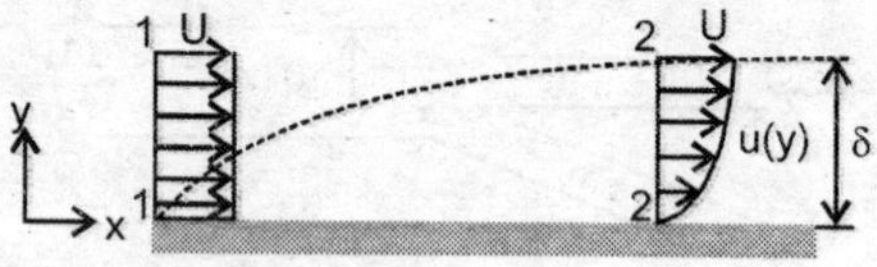

The volume flow rate through Section 2-2 is given by

(*a*) $\frac{1}{2}Uw\delta$ (*b*) $\frac{1}{3}Uw\delta$

(*c*) $Uw\delta$ (*d*) $\frac{2}{3}Uw\delta$

21. A cube of weight *W* and side a falls at a constant speed in a medium as shown in the figure. If the medium is air (mass density = ρ_{air}) let U_{air} be the velocity of the cube. If the medium is water (mass density = ρ_{water}) let U_{water} be the velocity of the cube.

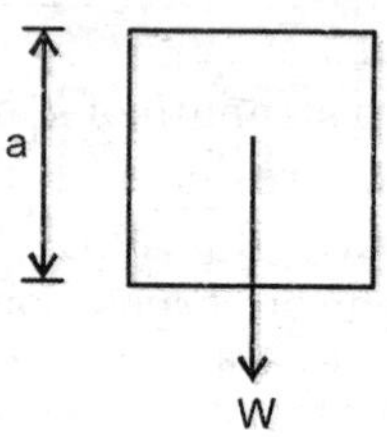

Neglecting the buoyancy force and assuming drag coefficient to be same for both cases, the ratio of velocities, $\left(\frac{U_{air}}{U_{water}}\right)$ is given by

(*a*) $\frac{\rho_{air}}{\rho_{water}}$ (*b*) $\sqrt{\frac{\rho_{air}}{\rho_{water}}}$

(*c*) $\sqrt{\frac{\rho_{water}}{\rho_{air}}}$ (*d*) 1

22. Water is flowing through a venturimeter having a diameter of 0.25 m at the entrance (Station 1) and 0.125 m at the throat (Station 2) as shown in the figure. A mercury manometer measures the piezometric head difference between Stations 1 and 2 as 1.3505 m. The loss of head between these two stations, is 1/7 times the velocity head at the Station 2. Assume the acceleration due to gravity to be 10 m/s². The velocity of water at the throat is _____ m/s.

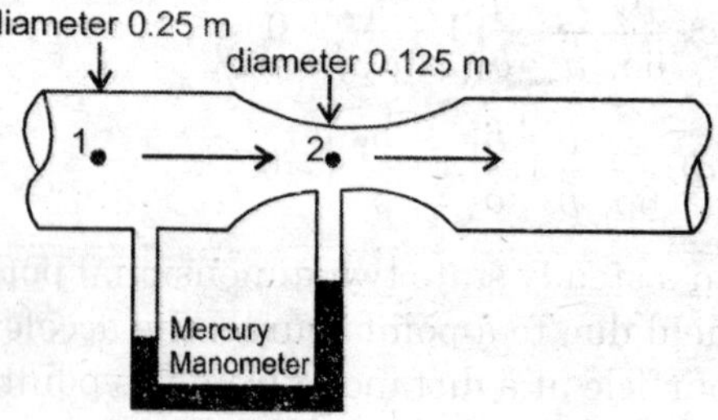

ANSWERS

1. 21.25	**2.** (a)	**3.** (d)	**4.** 3	**5.** (b)	**6.** (b)	**7.** .48 to .52
8. (c)	**9.** (c)	**10.** (a)	**11.** (d)	**12.** (c)	**13.** (a)	**14.** 17.55 to 17.75
15. 1973.44	**16.** (b)	**17.** 8	**18.** 47.89	**19.** 0.150	**20.** (d)	**21.** (c)
22. 18.47						

EXPLANATIONS

Fluid Mechanics – 2014

1. Given that

$A = 100\ m^2$

$\bar{x} = 10\ m$

$$I_G = \frac{bd^3}{12} = \frac{40 \times 15^3}{12} = 11250\ m^4$$

$$x_p = \frac{IG}{A\bar{x}} + \bar{x}$$

$$x_p = \frac{11250}{100 \times 10} + 10$$

$$\boxed{x_p = 21.25\ m}$$

2. (a) Given that

$$\vec{V} = (3x^2 + 3)t\hat{i} - 6xyt\hat{j}$$

$$u = (3x^2 + 3)t$$

$$v = -6xyt$$

$$a_x = \frac{\partial u}{\partial t} = 3x^2 + 3$$

$$a_y = \frac{\partial v}{\partial t} = -6xy$$

at, $x = 10$, $y = 0$

$$a_x = 3 \times 10^2 + 3 = 303$$

$$a_y = 0$$

$$a = \sqrt{{a_x}^2 + {a_y}^2} = \sqrt{(303)^2 + 0^2}$$

$$\boxed{a = 303}$$

3. (d) Total Acceleration = (Convective + Local)Acceleratio

→ For steady and uniform flow, both do not exist.

→ For steady and Non-uniform flow, only Convective acceleration exist.

→ Unsteady and uniform flow, only local acceleration exist.

→ Unsteady and Non-uniform flow, Both exist.

4. Velocity in two-dimensional flow,

$u = x^2y$, $v = -y^2x$

at ($x = 2$, $y = 1$)

$$\text{angular deformation} = \left(\frac{\partial v}{\partial x} + \frac{\partial u}{\partial y}\right)$$

$$\frac{\partial u}{\partial x} = -y^2 = -1, \quad \frac{\partial u}{\partial y} = x^2 = 4$$

angular deformation = $(-1 + 4) = 3$

5. (b) Slope of equipotential line $\left(\frac{du}{dx}\right)_{\phi=C} = -\frac{u}{v}$

Slope of stream function line $\left(\frac{dy}{dt}\right)_{\Psi=C} = \frac{v}{u}$

Hence, both equipotential line and stream function line are orthogonal to each other.

6. (b) Flow around a Rankine half-body is represented by the superposition of

(i) Source flow (ii) Uniform flow

Given that,

$L = 3.6\ m$

$v = 3\ m/s$

$a = 10\ m/s^2$

Froude Number (Fr) is a non-dimensional number.

$$Fr = \frac{V}{\sqrt{gL}} = \frac{3}{\sqrt{10 \times 3.6}}$$

$$\boxed{Fr = 0.50}$$

8. (c) The velocity profile for fully developed turbulent flow is Blunter and follow the logarithmic law.

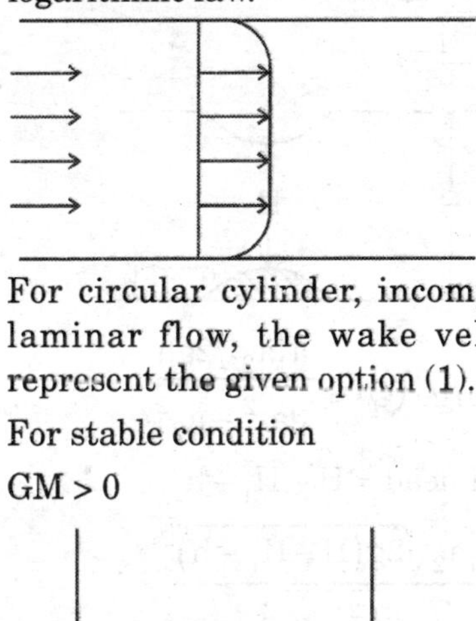

9. (c) For circular cylinder, incompressible and laminar flow, the wake velocity profile represent the given option (1).

10. (a) For stable condition

$GM > 0$

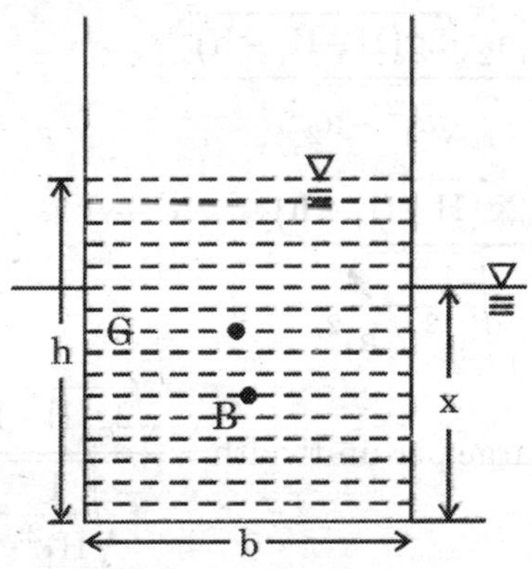

Consider x part is submerged in liquid of density ρ_2

$$GM = \frac{I}{\forall} - BG$$

$$\frac{I}{\forall} - BG > 0$$

$$\frac{I}{\forall} = \frac{b^4/12}{b^2 \cdot x} = \frac{b^2}{12x}$$

$$BG = \frac{h-x}{2} = \frac{1}{2}(h-x) = \frac{h}{2}\left(1 - \frac{\delta_1}{\delta_2}\right)$$

Weight of block = Weight of liquid displaced

$$\rho_1 g b^2 h = \rho_2 g b^2 x$$

$$x = \frac{\rho_1 h}{\delta_2}$$

$$\frac{I}{\forall} - BG > 0$$

$$\frac{\rho_2 h^2}{12\rho_1 h} - \frac{h}{2}\left(1 - \frac{\rho_1}{\rho_2}\right) > 0$$

$$\frac{\rho_2 h}{6\rho_1 h} - \frac{h}{b}\left(1 - \frac{\rho_1}{\rho_2}\right) > 0$$

11. (d) In steady state two-dimensional potential flow field due to a point source, the acceleration of a particle at a distance r from the point of source.

$$a \propto \frac{1}{r^3}$$

$$\boxed{a \propto r^{-3}}$$

13. (a)

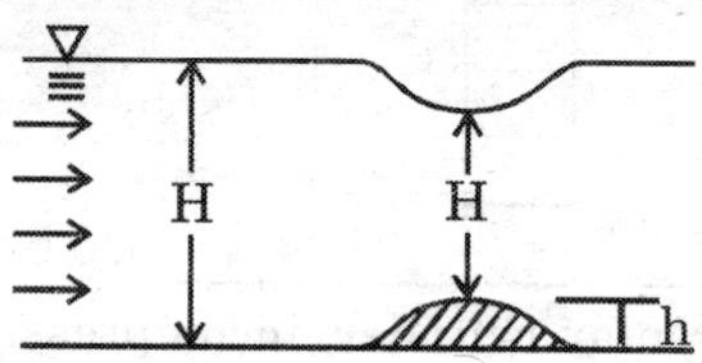

$$\text{Discharge (Q)} = \frac{a_1 a_2 \sqrt{2gh}}{\sqrt{a_1^2 - a_2^2}}$$

Actual head = $H - H_1 - h$

$$Q = \frac{a_1 a_2 \sqrt{2g(H - H_1 - h)}}{\sqrt{a_1^2 - a_2^2}}$$

$$Q = \frac{\sqrt{2g(H - H_1 - h)}}{\sqrt{\frac{1}{a_2^2} - \frac{1}{a_1^2}}}$$

$$\text{Discharge per unitwidth} = \frac{\sqrt{2g(H - H_1 - h)}}{\sqrt{\frac{1}{H_1^2} - \frac{1}{H^2}}}$$

15. Given that

$\theta = 30°$

$\overset{o}{m} = 14$ kg/sec.

$$\overset{o}{m} = \rho Q$$

$$Q = \frac{\overset{o}{m}}{1000} = \frac{14}{1000} = 0.014 \text{ m}^3/\text{sec.}$$

$$A_1 = 113 \text{ cm}^2 = 113 \times 10^{-4} \text{ m}^2$$

$$A_2 = 7 \text{ cm}^2 = 7 \times 10^{-4} \text{ m}^2$$

$$P_1 = 200 \text{ KPa} = 200 \times 10^3 \text{ N/m}^2$$

$$P_2 = 101.32 \text{ KPa} = 101.32 \times 10^3 \text{ N/m}^2$$

$$Q = A_1 V_1 = A_2 V_2$$

$$V_1 = 1.239 \text{ m/s.}$$

$$V_2 = 20 \text{ m/s.}$$

$$Fx = \delta Q(V_1 - V_2 \cos\theta) + p_1 A_1 - p_2 A_2 \cos\theta$$

$Fx = 1000 \times 0.014(1.239 - 20\cos 30°) + (200 \times 10^3 \times 113 \times 10^{-4}) - (101.32 \times 10^3 \times 7 \times 10^{-4} \cos 30°)$

$Fx = 1973.44$ N.

16. (b) Superimposed flow → The flow patterns due to uniform flow, a source flow, a sink flow and a free vortex flow can be supeimposed flow in any linear combination to get a resultant flow, which closely resembles the flow around bodies.

$$\text{Resultant velocity (V)} = \sqrt{u_r^2 + u_0^2}$$

$$V = \frac{\sqrt{K_1^2 + K_2^2}}{r}$$

17. Given that

Velocity potential function $(\phi) = 2(x^2 + 2y - y^2)$

$\Psi = 2$ at (2, 2)

$$u = -\frac{\partial \phi}{\partial x} = -2(2x) = -4x$$

$$v = -\frac{\partial \phi}{\partial y} = 2(2 - 2y) = 4 - 4y$$

We know that,

$$u = -\frac{\partial \Psi}{\partial y}, \ v = \frac{\partial \Psi}{\partial x}$$

$$\frac{\partial \Psi}{\partial x} = 4 - 4y$$

$$-\frac{\partial \Psi}{\partial y} = -4x$$

$$\partial \Psi = 4x \cdot \partial y$$

$$\Psi = 4xy + K$$

K is funtion of x.

$$\frac{\partial \Psi}{\partial x} = 4y + \frac{\partial K}{\partial x}$$

$$4 - 4y = 4y + \frac{\partial K}{\partial x}$$

$$\frac{\partial K}{\partial x} = 4 - 8y$$

$$K = 4x - 8xy$$

$$\Psi = 4xy + 4x - 8xy$$

At (2, 2)

$$\Psi = 16 + 8 - 32$$

$$\boxed{\Psi = -8}$$

18. Given that

Model scale $= \frac{1}{100}$

$V_p = 0.5$ m/sec.

$V_m = ?$

$$\frac{V_m}{V_p} = \frac{\delta p}{\delta m} \times \frac{\mu_m}{\mu_p} \times \frac{D_p}{D_m}$$

$$\frac{V_m}{0.5} = \left(\frac{1025}{1000}\right) \times \left(\frac{10^{-3}}{1.07 \times 10^{-3}}\right) \times \frac{1}{100}$$

$V_m = 47.89$ m/sec.

19. Given that

Re = 1600

L = 45 m

$h_f = 0.6$ m

V = 1 m/s

g = 10 m/s²

$$h_f = \frac{4fLV^2}{2gd}$$

$$f = \frac{16}{Re} = \frac{16}{1600} = 0.01$$

$$0.6 = \frac{4 \times 0.01 \times 45 \times 1^2}{2 \times 10 \times d}$$

$$\boxed{d = 0.150 \text{ m}}$$

20. (d) $\frac{u}{U} - 2\left(\frac{y}{\delta}\right) - \left(\frac{y}{\delta}\right)^2$

u – Average velocity of flow

U – Maximum velocity of flow

Flow over parallel plates.

$$\frac{U}{u} = \frac{3}{2}$$

$$u = \frac{2}{3}U$$

Volume flow rate (Q) = Au

$$\boxed{Q = \frac{2}{3}U\rho w}$$

21. (c)

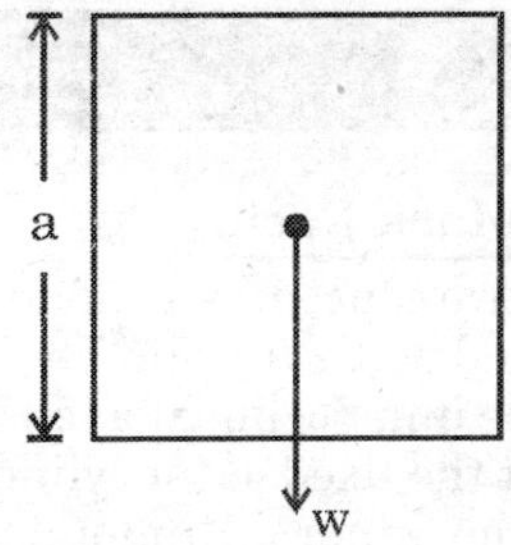

Velocity in air = U_a

Velocity in water = U_w

Density of air = ρ_a

Density of water = ρ_w

Drag force $(F_D) = \frac{1}{2}\rho AU^2C_D$

Drag force in air = Drag force in water

$$\frac{1}{2}\rho_a \cdot A \cdot U_a{}^2 \cdot C_{Da}$$

$$= \frac{1}{2}\rho_w \cdot A \cdot U_w{}^2 \cdot C_{Dw} \left[C_{Da} = C_{Dw}\right]$$

$$\frac{u_a}{u_w} = \sqrt{\frac{\rho_w}{\rho_a}}$$

22. Given that

For venturimeter,

$d_1 = 0.25$ m

$$a_1 = \frac{\pi}{4}d_1{}^2 = \frac{\pi}{4} \times 0.25^2 = 0.049 \text{ m}^2$$

$d_2 = 0.125$ m.

$$a_2 = \frac{\pi}{4}d_2{}^2 = \frac{\pi}{4} \times (0.125)^2 = 0.012 \text{ m}^2$$

x = 1.3505 m.

$$h = x\left(\frac{G_{Hg}}{G_w} - 1\right)$$

$$h = 1.3505\left(\frac{13.6}{1} - 1\right)$$

h = 14.0163 m.

$C_d = 0.98$

$$\text{Discharge} = \frac{C_d a_1 a_2 \sqrt{2gh}}{\sqrt{a_1{}^2 - a_2{}^2}}$$

$$Q = \frac{0.98 \times 0.049 \times 0.012 \times \sqrt{2 \times 9.81 \times 17.0163}}{\sqrt{0.049^2 - 0.012^2}}$$

Q = 0.2216 m³/sec.

$Q = a_2 v_2$

$0.2216 = 0.012 \times V_2$

$$\boxed{V_2 = 18.47 \text{ m/sec.}}$$

2015

Q. (1 – 9) Carry One Mark Each.

1. The gap δ between two concentric cylinders, each of height h, is filled with an oil. The torque required to rotate the inner cylinder at an angular velocity of ω against the fixed outer cylinder is T. The diameter of the inner cylinder is d and $\delta << d$. The dynamic viscosity of the oil is given by

(a) $\frac{4\pi\delta T}{d^3\omega h}$ (b) $\frac{4\delta T}{\pi d^3\omega h}$

(c) $\frac{4\pi\delta T}{d^2\omega h^2}$ (d) $\frac{4\delta T}{\pi d\omega h^3}$

2. Water is retained against a sluice gate in the form of a circular segment as shown in the figure. If ℓ and g are the density of water and gravitational acceleration respectively, the upward force exerted by the gate on the water per unit depth perpendicular to the plane of the figure is

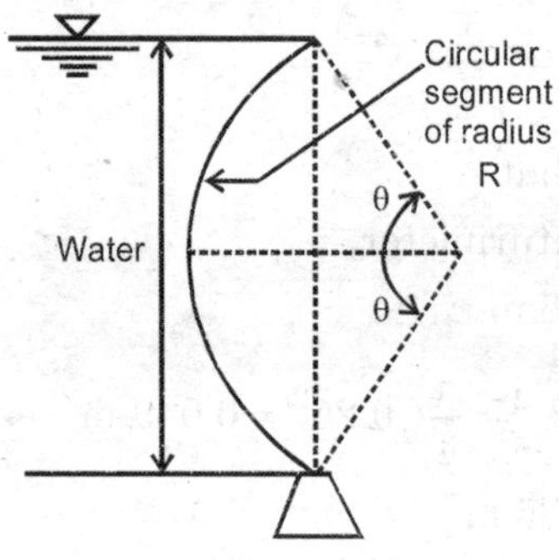

(a) $\ell R^2\left(\theta - \frac{1}{2}\sin 2\theta\right)g$

(b) $\ell R^2\left(\cos^2\theta - \frac{1}{2}\sin\theta\right)g$

(c) $\ell R^2\left(\cos\theta - \frac{1}{2}\sin\theta\right)g$

(d) $\ell R^2\left(\cos^2\theta - \frac{1}{2}\sin^2\theta\right)g$

3. A two-dimensional velocity field is given by $\vec{V} = 10(y^3 - x^2 y)\hat{i} + 2Cxy^2\hat{j}$, where $\hat{i}$ and $\hat{j}$ are the unit vectors in the directions of the rectangular Cartesian coordinates x and y, respectively. If the flow is incompressible, the constant C should be

(a) – 10 (b) 0

(c) 5 (d) 10

4. Let $\vec{V}$ and T denote the velocity vector and temperature in a flow field. The rate of change of temperature experienced by a fluid particle as it is passes through the point (x_1, y_1, z_1) at a time t_1 is 2.5 °C/s. The rate of change of temperature at a time t_1 at the point (x_1, y_1, z_1) is 4.8 °C/s. The quantity $\vec{V}\cdot\nabla T$ at (x_1, y_1, z_1) and t_1 in °C/s is______

5. In a simple Couette flow apparatus, the gap h between the parallel plates is filled with a liquid of density l and dynamic viscosity μ, and one plate is dragged at a velocity of U parallel to itself, while the other plate is fixed. The magnitude of vorticity at any point in the flow is

(a) $\frac{\mu}{lh^2}$ (b) 0

(c) $\frac{1}{h^2}\sqrt{\frac{\mu U h}{l}}$ (d) $\frac{U}{h}$

6. An open glass capillary tube of 2 mm bore is lowered into a cistern containing mercury (density = 13600 kg/m³) as shown in the figure. Given that the contact angle between mercury and glass = 140°, surface tension coefficient = 0.484 N/m and gravitational acceleration = 9.81 m/s², the depression of mercury in the capillary tube below the free surface in the cistern, in mm, is

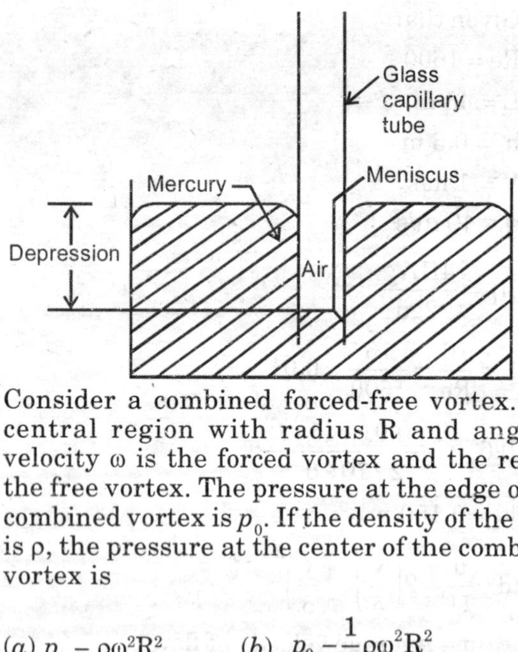

7. Consider a combined forced-free vortex. The central region with radius R and angular velocity ω is the forced vortex and the rest is the free vortex. The pressure at the edge of the combined vortex is p_0. If the density of the fluid is ρ, the pressure at the center of the combined vortex is

(a) $p_0 - \rho\omega^2 R^2$ (b) $p_0 - \frac{1}{2}\rho\omega^2 R^2$

(c) $p_0 + \frac{1}{2}\rho\omega^2 R^2$ (d) $p_0 + \rho\omega^2 R^2$

8. Which one of the following is true at the point of separation of a boundary layer:

(a) Transition occurs from laminar to turbulent flow

(b) The flow relaminarizes from turbulent regime

(c) The shear stress vanishes

(d) The relation between stress and rate of strain ceases to be linear

9. A certain fluid flow phenomenon is described by Reynolds, Weber and Ohnesorge numbers. The Weber and Ohnesorge numbers are defined as $We = \frac{\rho U^2 L}{\sigma}$, and $Oh = \frac{\mu}{\sqrt{\rho \sigma L}}$ respectively, where σ is the surface tension, ρ is the density, μ is the dynamic viscosity, U is the velocity and L is the characteristic dimension. If Re denotes the Reynolds number, which of the following relations is true?

(*a*) $We = Oh\, Re^2$ (*b*) $We = Oh^2 Re^2$

(*c*) $We = Oh^2 Re$ (*d*) $We = Oh\, Re$

Q. (10 – 22) Carry Two Marks Each.

10. A rectangular boat 6 m wide and 15 m long (dimension perpendicular to the plane of the figure) has a draught of 2 m. The side view of the boat is as shown in the figure. The centre of gravity G of the boat is at the free surface level. The metacentric height of the boat in m is

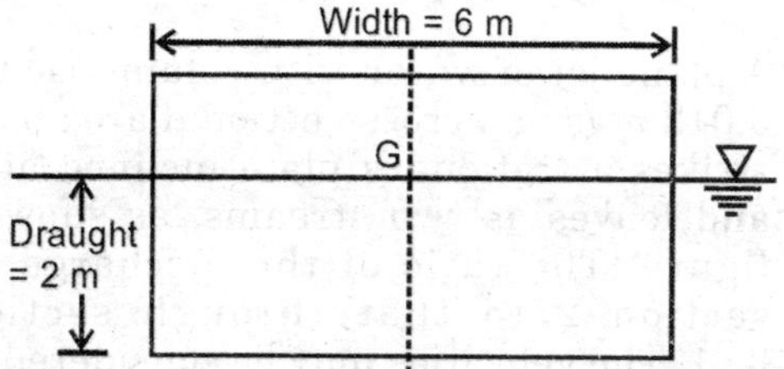

(*a*) – 1.0 (*b*) 0.5

(*c*) 1.5 (*d*) 2.0

11. Water drains out into atmosphere from a small orifice located at the bottom of a large open tank. If the initial height of the water column is H, the time taken to empty the tank is proportional to

(*a*) $H^{1/2}$ (*b*) H

(*c*) $H^{3/2}$ (*d*) H^2

12. Consider a two-dimensional velocity field given by $\vec{V} = \pi y\hat{i} - \pi x\hat{j}$, where $\hat{i}$ and $\hat{j}$ are the unit vectors in the directions of the rectangular Cartesian coordinates x and y, respectively. A fluid particle is located initially at the point (– 1, 1). Its position after unit time is

(*a*) (–2, – 2) (*b*) (1, – 1)

(*c*) (1, 1) (*d*) (3, – 1)

13. The figure shows a reducing area conduit carrying water. The pressure p and velocity V are uniform across sections 1 and 2. The density of water is 1000 kg/m³. If the total loss of head due to friction is just equal to the loss of potential head between the inlet and the outlet, then V_2 in m/s will be___

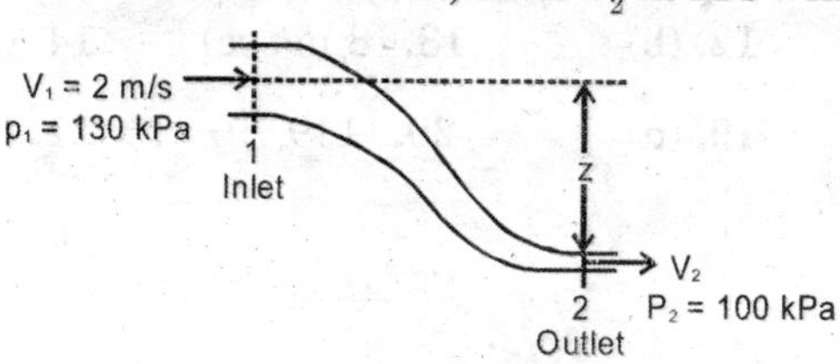

14. A fluid enters a control volume through an inlet port (denoted with subscript 'i') and leaves through two outlet ports (denoted with subscripts 'o, 1' and 'o, 2') as shown in the figure. The velocities may be assumed to be uniform across the ports. The rate of change of mass in the control volume in kg/s, at the instant shown in the figure, is

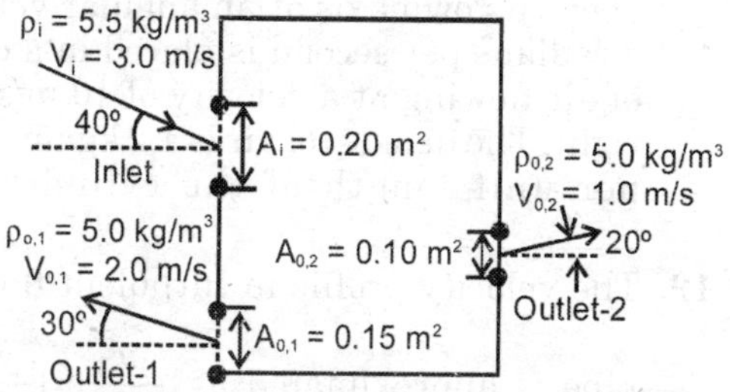

15. The total discharge of water through a lawn sprinkler shown in the figure is one liter per minute. The velocity of the jet at each end, relative to the arm, is $2\pi/30$ m/s. The density of water is 1000 kg/m³ and the length of each arm is 0.1 m. If the frictional torque of the pivot is $\pi/36$ mN.m, the rotational speed, in revolutions per minute, of the sprinkler is ________.

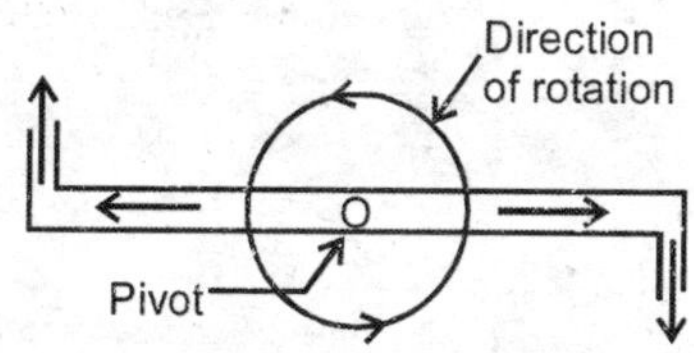

16. Consider a two-dimensional potential flow field with the radial and tangential velocity components, $v_r = \frac{m}{2\pi r}$ and $v_\theta = \frac{k}{2\pi r}$, respectively, where m and k are constants. The stream function is such that it increases along the direction of traverse of a line in the flow field if the flow is from left to right across that line. The stream function ψ for this flow field, with $\psi = 0$ at $r = a$ and $\theta = 0$, is

(*a*) $\frac{m\pi\theta}{2} + \frac{k\pi}{2}\ln\frac{r}{a}$ (*b*) $\frac{m\pi\theta}{2} - \frac{k\pi}{2}\ln\frac{r}{a}$

(*c*) $\frac{m\theta}{2\pi} + \frac{k}{2\pi}\ln\frac{r}{a}$ (*d*) $\frac{m\theta}{2\pi} - \frac{k}{2\pi}\ln\frac{r}{a}$

17. A steady, two-dimensional, inviscid and incompressible flow field is described in rectangular Cartesian coordinates as $u = ax$ and $v = -ay$, where u and v are the components of the velocity vector in the x and y directions, respectively. Grvity acts along the negative y-direction. The pressure distribution, with the reference pressure taken as zero at the origin, with usual notation, is given by

(*a*) $-\frac{1}{2}\rho a^2(x^2 + xy + y^2) - \rho g y$

(b) $-\frac{1}{2}\rho a^2(x^2 - xy + y^2) - \rho gy$

(c) $-\frac{1}{2}\rho a^2(x^2 + y^2) - \rho gy$

(d) $-\frac{1}{2}\rho a^2(x^2 - y^2) - \rho gy$

18. A cylinder of radius 0.1 m rotating clockwise about its own axis at an angular velocity of 100/π radians per second is placed in a cross-stream of air flowing at a velocity of 10 m/s from left to right. The density of air is 1.2 kg/m³. The lift force per unit length of the cylinder in N/m is ________.

19. The velocity profile in turbulent flow through a pipe is approximated as $\frac{u}{u_{max}} = \left(\frac{y}{R}\right)^{1/7}$, where u_{max} is the maximum velocity, R is the radius and y is the distance measured normal to the pipe wall towards the centerline. If u_{av} denotes the average velocity, the ratio $\frac{u_{av}}{u_{max}}$ is

(a) $\frac{2}{15}$

(b) $\frac{1}{5}$

(c) $\frac{1}{3}$

(d) $\frac{49}{60}$

20. A steel sphere (density = 7900 kg/m³) of diameter 0.1 m is dropped from rest in water (density = 100 kg/m³). The gravitational acceleration is 9.81 m/s². Assuming that the drag coefficient is constant and equal to 1.33, the terminal velocity attained by the sphere in m/s is ________.

21. An inclined venturimeter connected to an inverted manometer is shown in the figure. The cross-sectional areas at the inlet and the throat are 2×10^{-3} m² and 2×10^{-4} m², respectively. The densities of water and oil are 1000 kg/m³ and 800 kg/m³, respectively. The gravitational acceleration is 9.81 m/s². If the discharge of water through the venturimeter is 5×10^{-4} m³/s, neglecting viscous effects and assuming uniform velocities across the inlet and the throat, the manometer reading h, in m, will be ________

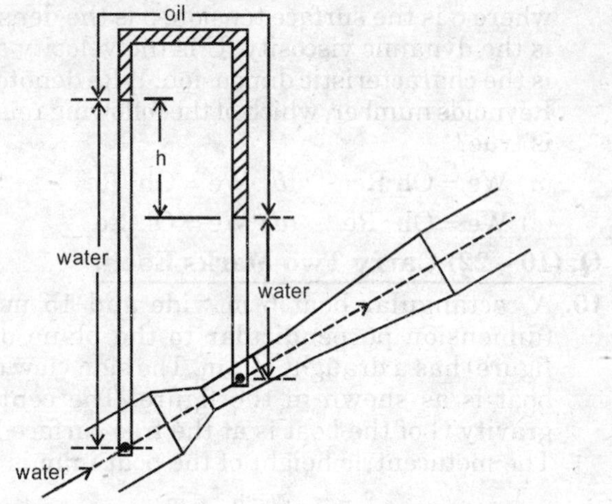

22. A plane jet of water with volumetric flow rate 0.012 m³/s and cross-sectional area 6×10^{-4} m² strikes a stationary plate inclined at angle θ and leaves as two streams, as shown in the figure. The ratio of the discharge through section 2 to that through section 3 is 3 : 1. The velocities may be considered uniform across the cross-sections and the effects of friction may be neglected. The density of water is 1000 kg/m³. Ignoring the effects of gravity, the magnitude of the normal force exerted on the plate, in N, is

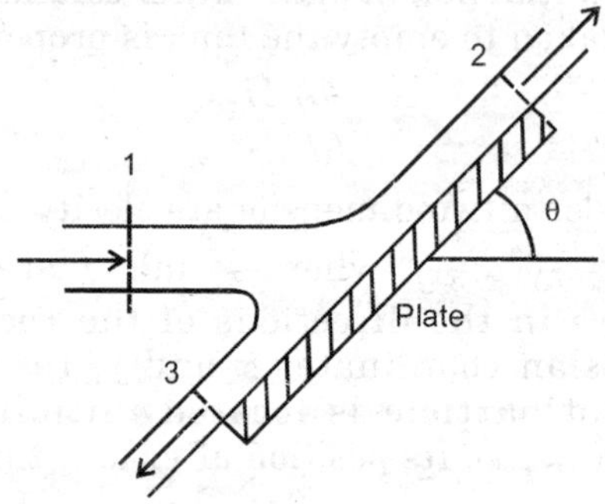

ANSWERS

1. (b)	**2.** (a)	**3.** (c)	**4.** (–2.3°C/s)	**5.** (d)	**6.** (h = –5.5 mm)	**7.** (b)
8. (c)	**9.** (b)	**10.** (b)	**11.** (a)	**12.** (b)	**13.** (8 m/sec)	**14.** (0.760)
15. (17.58)	**16.** (d)	**17.** (c)	**18.** (23.96)	**19.** (d)	**20.** (149.79)	**21.** (0.157)
22. (80)						

EXPLANATIONS

1. (b) Given that

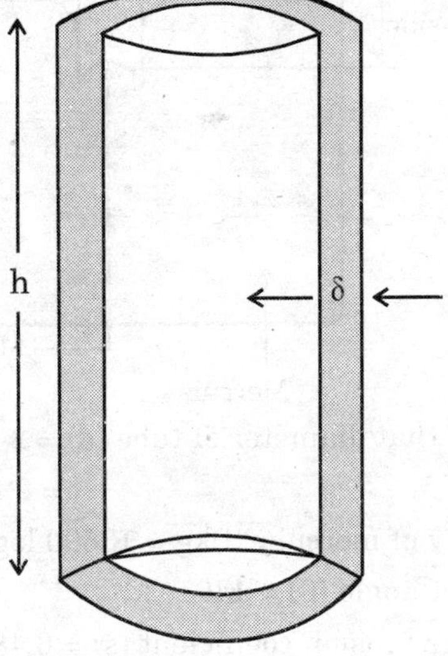

Gap between the cylinders = d
Height of the cylinders = h
Diameter of the inner cylinder = d
we know that

$$\text{shear stress}(\tau) = \mu \frac{du}{dy}$$

$$\Rightarrow \quad \tau = \mu \frac{u}{y} = \mu \frac{(\omega r)}{y} \qquad [u = \omega.r]$$

$$\Rightarrow \quad \tau = \frac{\mu.\omega\left(\frac{d}{2}\right)}{\delta} = \frac{\mu\omega d}{2\delta}$$

$$\Rightarrow \frac{f}{A} = \frac{\mu\omega d}{2\delta} \qquad \left[\because \quad \tau = \frac{F}{A} N/m^2\right]$$

$$\Rightarrow \quad f = \frac{\mu A\omega d}{2\delta} = \frac{\mu.(\Pi dh).\omega d}{2\delta}$$

$$\therefore \quad f = \frac{\mu\Pi d^2\omega h}{2\delta} \qquad ....(1)$$

$$\text{Now, Torque (T)} = f \times \frac{d}{2}$$

$$T = \frac{\mu\Pi d^2\omega h}{2\delta}.\frac{d}{2} \qquad [\text{from eq}^n \ (1)]$$

$$\therefore \quad \boxed{\mu = \frac{4T\delta}{\Pi d^3\omega h}}$$

Hence the option (b) is correct.

2. (a)

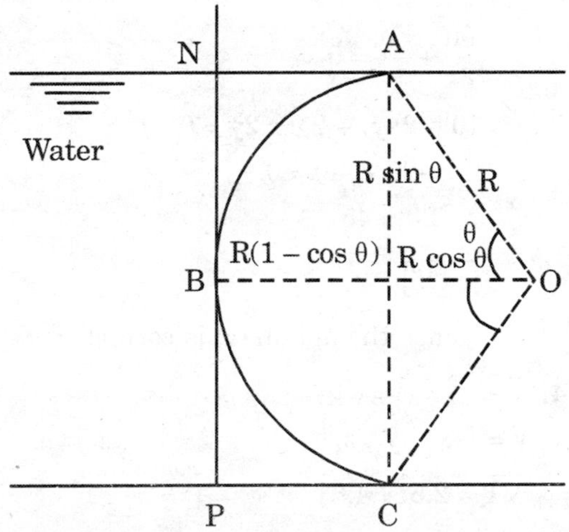

Given that,
density of water = f kg/m^3
Acceleration due to gravity = g m/s^2
Vertical force acting on NAB

$$(f_v)_{NAB} = (\text{vol. NAB}) \times w$$

$$= [\text{vol . (NADB)} - \text{vol . (ABD)}] \times w$$

$$= \left[R(1-\cos\theta)R\sin\theta - \begin{pmatrix}\frac{\theta}{2}\times R.R - \frac{1}{2}\times R\cos\theta \\ \times R\sin\theta\end{pmatrix}\right] \times w$$

$$= \begin{bmatrix}R^2(1-\cos\theta)\sin\theta - \frac{\theta}{2}.R^2 + \frac{1}{2} \\ R^2\cos\theta.\sin\theta\end{bmatrix} \times w \qquad ....(1)$$

Now, Vertical force acting on BCD

$$(f_v)_{BCD} = \text{vol.}(ADCBNA) \times w$$

$$= [\text{vol.}(ADBN) + \text{vol.}(BCD)] \times w$$

$$= \left[R^2(1-\cos\theta)\sin\theta + \frac{\theta}{2}R^2 - \frac{1}{2}R^2\cos\theta.\sin\theta\right] \times w \qquad ....(2)$$

Net upward force exerted by the gate on the water is

$$f_{Net} = (f_v)_{BCD} - (f_v)_{NAB}$$

$$= R^2(\theta - \cos\theta \sin\theta)\, eg \qquad [\because \ w = eg]$$

$$\boxed{f_{Net} = R^2\left(\theta - \frac{\sin 2\theta}{2}\right) eg}$$

Hence the option (a) is correct

3. (c) Given that

$$\overline{V} = 10\left(y^3 + x^2 y\right)\hat{i} + 2cxy^2\hat{j} \qquad \left[\overline{V} = \overrightarrow{u}\hat{i} + \overrightarrow{V}\hat{j}\right]$$

from the continiuty eq^n, if the flow is incompressible, then

$$\frac{\partial u}{\partial x}+\frac{\partial u}{\partial y}=0$$

10 (–2xy) + 2.c.x.2y = 0

xy (–20 + 4c) = 0

$\therefore$ 20 = 4c

$\boxed{c = 5}$

Hence the option (c) is correct.

4. $\vec{v} = 4\hat{i} + v\hat{j} + w\hat{k}$

v = f (x_1, y_1, z_1)

$\overrightarrow{\nabla T} = 2.5\hat{i} + 4.8\hat{j}$

$\boxed{\vec{v}.\overrightarrow{\nabla T} = 2.5u + 4.8u}$

u = f (x_1, y_1, z_1)

v = f (x_1, y_1, z_1)

Hence, the correct answer is –2.3°c/s.

5. (d) Given that

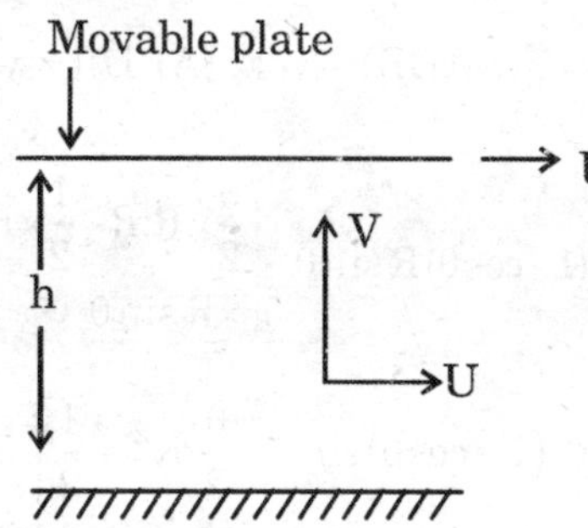

density of liquid = f

velocity of Movable plate =

we know that

Vorticity (z) = 2w ....(1)

$$\text{where } w = \frac{1}{2}\left[\frac{dy}{dx}-\frac{du}{dy}\right] = \frac{1}{2}\left[0-\frac{du}{dy}\right]$$

$$[\because \text{ flow across the plate is zero } -\frac{dv}{dx}=0]$$

$$=\frac{1}{2}\left[\frac{-u}{h}\right]$$

$$\text{Now, Vorticity (z)} = 2.\frac{1}{2}\left(\frac{-u}{h}\right) = \frac{-u}{h}$$

[from eq^n (1)]

$$\text{Magnitude of Vorticity (z)} = \left|\frac{-u}{h}\right| = \frac{u}{h}$$

Hence the option (d) is correct.

6.

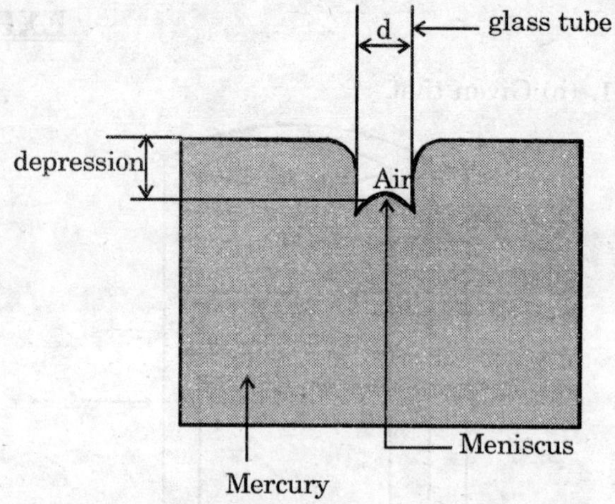

Given that diameter of tube (d) = 2 mm

$= 2 \times 10^{-3}$ m

density of mercury (f)kg = 13600 log/m^3

contact angle (θ) = 140°

Surface tension coefficient (s) = 0.484 N/m

we know that

$$\text{Depression of Mercury}(h) = \frac{4\sigma\cos\theta}{\rho g d}$$

$$= \frac{4\times 0.484\times\cos 140}{13600\times 9.81\times 2\times 10^{-3}}$$

$\boxed{h = -5.5 \text{ mm}}$

7. (b)

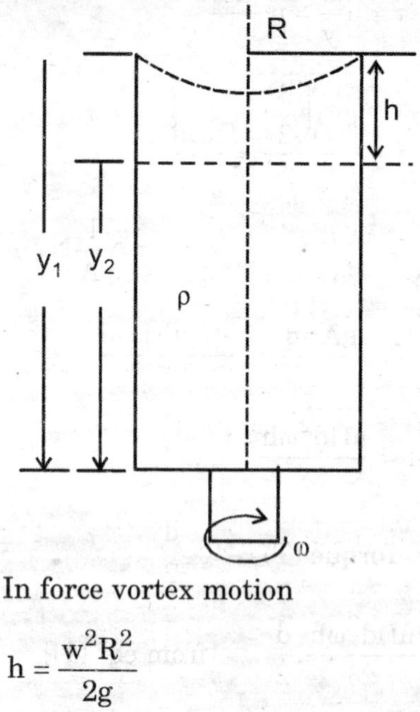

In force vortex motion

$$h = \frac{w^2R^2}{2g}$$

$$p = \rho g h$$

$$p = \rho g\frac{w^2R^2}{2g}$$

$$p = \frac{\rho w^2 R^2}{2}$$

Pressure at the centre of the combined vortex

$$= p_0 - p = p_0 - \frac{\rho w^2 R^2}{2}$$

8. (c) We know that

Shear stress $(\tau) = \mu \dfrac{\partial u}{\partial y}$ ....(1)

$\because$ At the point of separation the velocity gradient becomes zero

i.e.

$$\frac{\partial u}{\partial y} = 0$$

therefore,

shear stress $(\tau) = \mu - 0 = 0$ [from eq^n (1)]

$\boxed{\tau = 0}$

i.e. shear stress vanishes

Hence the option (c) is correct.

9. (b) Given that

weber Number $(w_e) = \dfrac{\ell u^2 L}{\sigma}$

Ohnesorge Number $(Oh) = \dfrac{\mu}{\sqrt{\rho \sigma L}}$

$$Oh^2 R_e^2 = \left(\frac{\mu}{\sqrt{\rho\sigma L}}\right)^2 \left(\frac{\ell u L}{\mu}\right)^2 \quad \left[\because Re = \frac{\ell u L}{\mu}\right]$$

$$= \frac{\mu^2}{\rho \sigma L} \times \frac{e^2 u^2 L^2}{\mu^2}$$

$$Oh^2 R_e^2 - \frac{eu^2 L}{\sigma} = w_c$$

$\therefore$ $\boxed{w_e - Oh^2 R_e^2}$

Hence the option (b) is correct.

10. (b)

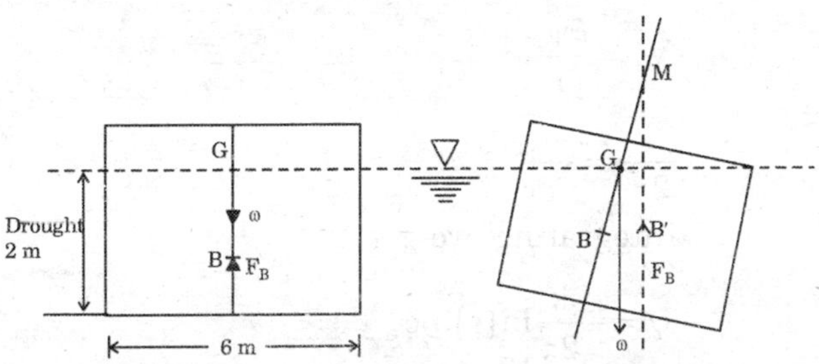

Given that

width of the boat = 6 m

length of the boat = 15 m

draught = 2m

we know that

Metacentric Height (GM) = BM – BG

$$GM = \frac{I}{\forall} - BG$$

$\left[\because BM = \dfrac{\delta}{\forall} \text{ where } \forall \to \text{displaced vol. } \delta = \dfrac{bh^3}{12}\right]$

$$= \frac{\frac{bh^3}{12}}{2 \times 6 \times 15} - \frac{2}{2} = \frac{15 \times 6^3}{12 \times 6 \times 2 \times 15} - 1$$

$\boxed{GM = 0.5 \text{ m}}$

Hence the option (b) is correct.

11. (a) we know that

Time taken to empty the tank through an orifice at the bottom of the tank is

$$t = \frac{2A(H)^{1/2}}{C_d\, a\sqrt{2g}}$$

$\boxed{t \propto H^{1/2}}$ $\left[\begin{array}{l} H \to \text{Height of the water column} \\ A \to \text{Area of tank} \\ a \to \text{Area of orifice} \\ C_d \to \text{Coefficient of discharge} \end{array}\right]$

Hence the option (a) is correct.

12. Option (b) is correct.

13.

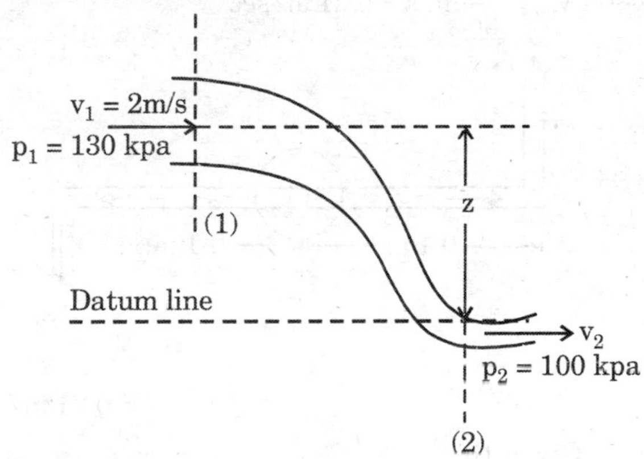

Given that

inlet velocity (v_1) = 2 m/s

inlet pressure (p_1) = 130 kPa

exit pressure (p_2) = 100 kPa

Apply Bernaulli equation at points (1) & (2)

$$\frac{p_1}{\rho g} + \frac{v_1^2}{2g} + z_1 = \frac{p_2}{\rho g} + \frac{v_2^2}{2g} + z_2 + h_f$$

$$\frac{p_1 - p_2}{\rho g} + \frac{v_1^2}{2g} + \cancel{z} = \frac{v_2^2}{2g} + 0 + \cancel{h_f} \quad [\text{Given } z = h_f]$$

$$\frac{(130-100) \times 10^3}{1000 \times 9.81} + \frac{2^2}{2 \times 9.81} = \frac{V^2}{2 \times 9.81}$$

$\boxed{V_2 = 8 \text{ m/s}}$

Hence the exit velocity of water V_2 = 8 m/sec

14.

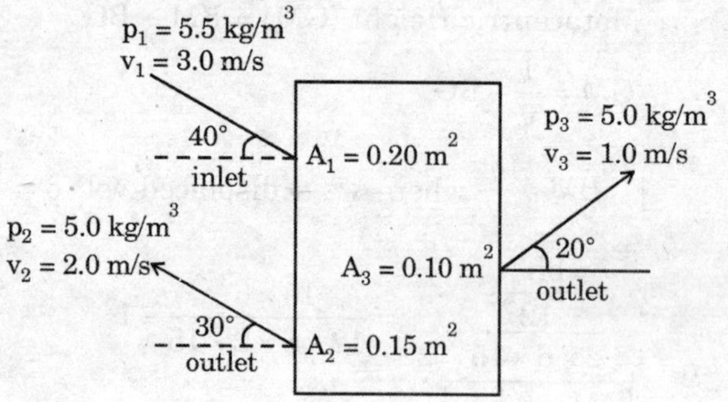

we know that

$$(\dot{m})_{inlet} = (\dot{m})_{outlet} + \left(\frac{dm}{dt}\right)_{Accumulated}$$

$$\ell_1 A_1 V_1 \cos 40° = (\ell_2 A_2 V_2 \cos 30° + \ell_3 A_3 V_3 \cos 20°) + \left(\frac{dm}{dt}\right)_{Accumulated}$$

$$(5.5 \times 0.20 \times 3 \times \cos 40) = (5 \times 0.15 \times 2 \times \cos 30° + 5 \times 0.10 \times 1 \times \cos 20) + \left(\frac{dm}{dt}\right)_{Accumulated}$$

$$\boxed{\left(\frac{dm}{dt}\right)_{Accumulated} = 0.760}$$

15. $V_{r_1} = \frac{2\pi}{30} m/s = 0.21\ m/sec$

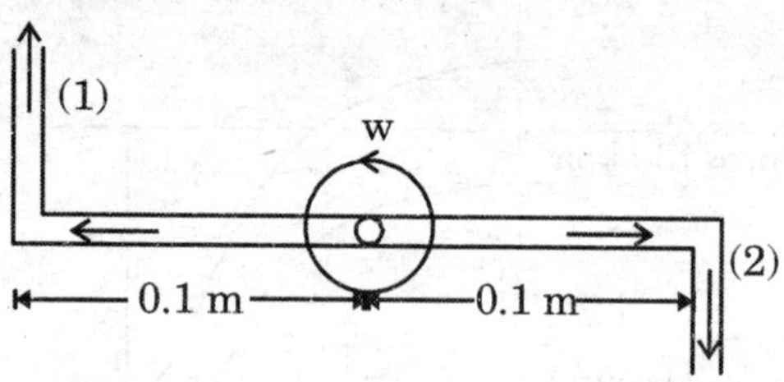

$$V_{r_2} = \frac{2\pi}{30} m/s$$
$$= 0.21\ m/sec$$

Given that

$Q = 1$ ltr/min

$Q = 10^{-3}/60$ m^3/sec

$Q = 1.66 \times 10^{-5}$ m^3/sec

$\rho_w = 1000$ kg/m^3

$T = \frac{\pi}{36}$ N.m.

Tangential velocity at 1 $(V_{t_1}) = V_{r_1} - w.r_1$

$= 0.21 - (w.\ 0.1)$

Similarly,

$$V_{t_2} = V_{r_2} + wr_2 = 0.21 - 0.1w$$

$$T = \rho Q_1 V_{t_1} r_1 + \rho Q_2 V_{t_2} r_2$$

$$T = 2\rho Q[(0.21 - 0.1w)0.1] \quad \begin{bmatrix} V_{t_1} = V_{t_2} \\ r_1 = r_2 \\ Q_1 = Q_2 = Q \end{bmatrix}$$

$$\frac{\pi \times 10^{-3}}{36v} = 2 \times 1000 \times 1.66 \times 10^{-5} \times 0.1(0.21 - 0.1w)$$

$0.21 - 0.1w = 26.27 \times 10^{-3}$

$0.21 - 0.1\ w = 0.026$

$0.1\ w = 0.184$

$w = 1.84$ rad/sec

$w = \frac{2\pi N}{60}$

$\boxed{N = 17.58 \text{ rpm}}$

16. (d) $V_r = \frac{m}{2\pi r}$, $V_\theta = \frac{k}{2\pi r}$

Boundry condition,

$\psi = 0$, at $r = a$

$\psi = 0$, at $\theta = 0$

Now

$$V_r = \frac{1}{r} \times \frac{r\psi}{r\theta}$$

$$\frac{m}{2\pi r} = \frac{1}{r} \cdot \frac{r\psi}{r\theta}$$

$$\partial\psi = \frac{m}{2\pi} . \partial\phi$$

Integrating both sides.

$$\Psi_1 = \frac{m\theta}{2\pi} + c$$

$0 = 0 + c$

$\boxed{c = 0}$, $\Psi_1 = \frac{m\theta}{2\pi}$

$$V_\theta = \frac{\partial\psi}{\partial r}$$

$$\frac{k}{2\pi r} = \frac{-\partial\psi}{\partial r}$$

Integrating, we get

$$\psi = -\frac{k}{2\pi}\ln(r) + c$$

$\psi = 0$, $r = a$

$c = -\ln(a)$

$$\psi = \frac{-k}{2\pi}[\ln(r) - \ln(a)]$$

$$\psi_2 = \frac{-k}{2\pi}\ln\left(\frac{r}{a}\right)$$

$$\left[\begin{array}{c}\psi = \psi_1 + \psi_2 \\ \boxed{\psi = \frac{m\theta}{2\pi} - \frac{k}{2\pi}\ln\left(\frac{r}{a}\right)}\end{array}\right]$$

17. (c) $p = \frac{-\rho a^2 x^2}{2} - \frac{\rho a^2 y^2}{2} - \rho g y$

Hence, option (c) is correct answer.

18. r = 0.1 m, D = 0.2 m

$$w = \frac{100}{\pi} \text{ rad/sec}$$

$$w = \frac{2\pi N}{60}$$

$$\frac{100}{\pi} = \frac{2\pi}{60} \times N$$

$$N = \frac{100 \times 60}{2\pi^2} \text{rpm}$$

v = 10 m/sec

$\rho_a = 1.2 \text{ kg/m}^3$

Lift force $(F_L) = \delta L V \Gamma$

Circulation $(\Gamma) = 2\pi r \times 4\theta$

$$4\theta = \frac{\pi DN}{60} = \frac{\not{\pi} \times 0.2}{60} \times \frac{100 \times 60}{2\pi^{\not{2}}}$$

$4\theta = 3.18$ m/sec

$\Gamma = 2 \times 3.14 \times 0.1 \times 3.18$

$\Gamma = 1.997 \text{ m}^2/\text{sec}$

$$\frac{F_L}{L} = 1.2 \times 10 \times 1.997$$

$$\boxed{\frac{F_L}{L} = 23.96 \text{ N/m}}$$

19. (d)

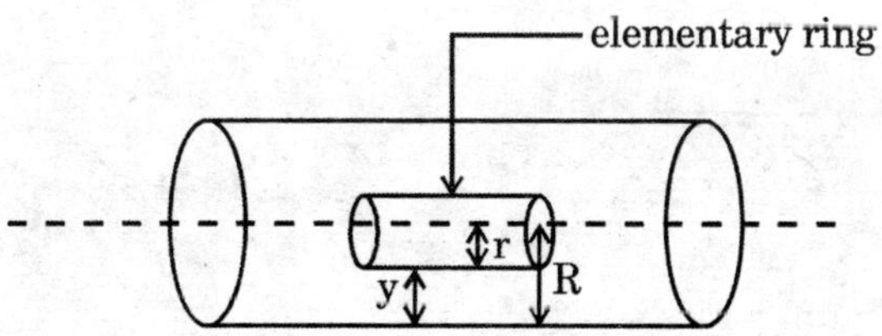

Given that

$$\frac{U}{U_{max}} = \left(\frac{y}{R}\right)^{1/7} \quad \text{....(1)}$$

Put y = R – r

$$\text{weget } \frac{U}{U_{max}} = \left(1 - \frac{r}{R}\right)^{1/7} \quad \text{....(2)}$$

Let the discharge 'dQ' through the elementry ring is

dQ = u(2∴pr) dr

∴ Total discharge

$$Q = \int_0^R u(2\pi r)\,dr$$

$$= \int_0^R U_{max}\left(1 - \frac{r}{R}\right)^{\frac{1}{7}}(2\pi r)\,dr \quad \text{[from eq}^n\text{ (2)]}$$

$$Q = U_{max}\frac{49}{60}\pi R^2 \quad \text{....(3)}$$

Now Average velocity $(v) = \frac{Q}{A} = \frac{Q}{\pi R^2}$

$$\therefore\ V = \frac{U_{max}\left(\frac{49}{60}\right)\cdot\not{\pi R^2}}{\not{\pi R^2}} = U_{max}\cdot\left(\frac{49}{60}\right)$$

$$\therefore\ \boxed{\frac{V}{U_{max}} = \frac{49}{60}}$$

Hence the option (d) is correct.

20. Given that

$\delta = 0.1$ m

$\rho_s = 7900 \text{ kg/m}^3$

$\rho_w = 1000 \text{ kg/m}^3$

$C_D = 1.33$

$w = F_D + F_B$

$$w = \rho_s \times g \times \frac{4}{3}\pi r^3$$

$$= 7900 \times 9.81 \times \frac{4}{3} \times 3.14 \times \left(\frac{0.1}{2}\right)^3$$

w = 40.56 N

Buoyant Force $(F_B) = \rho_w \times g \times V$

$$= 1000 \times 9.81 \times \frac{4}{3} \times 3.14\left(\frac{0.1}{2}\right)^3$$

$F_B = 5.13$ N

$$F_D = \frac{1}{2}\rho_a A U^2 C_D$$

$$= \frac{1}{2} \times 1.21 \times \frac{\pi}{4}\left(\frac{0.1}{2}\right)^2 \times U^2 \times 1.33$$

$F_D = 1.579 \times 10^{-3}\, U^2$

$w = F_D + F_B$

$F_D = w - F_B$

$1.579 \times 10^{-3}\, U^2 = 40.56 - 5.13$

$$\boxed{U = 149.79 \text{ m/sec}}$$

21. Given that

Area at the inlet of Venturimeter (A_1)

$= 2 \times 10^{-3} \text{ m}^2$

Area at the throat of Venturimeter (A2)

$= 2 \times 10^{-4}\ m^2$

density of water $(\ell)_w = 1000\ kg/m^3$

density of oil $(\ell)_{oil} = 800$ kg/m3

discharge (Q) = $5 \times 10^{-4}\ m^3/sec$

we know

$$Q = \frac{A_1 A_2 \sqrt{2g\,x}}{\sqrt{A_1^2 - A_2^2}}$$

$$\Rightarrow 5 \times 10^{-4} = \frac{2 \times 10^{-3} \times 2 \times 10^{-4} \times \sqrt{2 \times 9.81 \times x}}{\sqrt{(0.002)^2 - (0.0002^2)}}$$

$\boxed{x = 0.314\ m}$

Now, $x = h.\left(1 - \frac{Sm}{S}\right)$

$$0.314 = h\left(1 - \frac{800}{1000}\right)$$

$\boxed{h = 0.157\ m}$

Hence the Manometer reading h = 0.157 m

22. $Q = 0.012\ m^3/sec$

$A = 6 \times 10^{-4}\ m^2$

$\rho_w = 1000\ kg/m^3$

$\sin \delta\theta = 1/3$

$F_n = \rho A V^2 \sin\theta$

$Q = AV$

$V = Q/A$

Normal force

$$(F_n) = \rho A \left(\frac{Q}{A}\right)^2 \sin\theta = \frac{\rho Q^2}{A} \sin\theta$$

$$= \frac{1000 \times 0.012^2}{6 \times 10^{-4}} \times \frac{1}{3}$$

$\boxed{F_n = 80\ N}$

2016

Q. 1 – 9) : Carry one Mark Each.

1. In the parallel-plate configuration shown, steady-flow of an incompressible Newtonian fluid is established by moving the top plate with a constant speed, U_0 = 1m/s. If the force required on the top plate to support this motion is 0.5 N per unit area (in m²) of the plate then the viscosity of the fluid between the plates is ________ N-s/m²

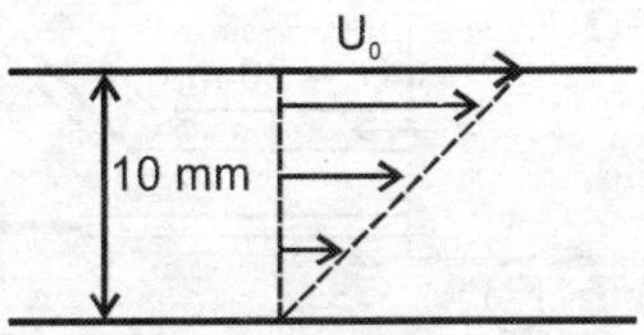

2. For a newly designed vehicle by some students, volume of fuel consumed per unit distance travelled (q_f in m³/m) depends upon the viscosity (μ) and density (ρ) of the fuel and, speed (U) and size (L) of the vehicle as

$$q_f = C\frac{\rho U^2 L}{\mu^3}$$

where C is a constant. The dimensions of the constant C are

(*a*) $M^0L^0T^0$ (*b*) $M^2L^{-1}T^{-1}$

(c) $M^2L^{-5}T^{-1}$ (*d*) $M^{-2}L^1T^1$

3. A semi-circular gate of radius 1m is placed at the bottom of a water reservoir as shown in figure below. The hydrostatic force per unit width of the cylindrical gate in y-direction is____kN. The gravitational acceleration, g = 9.8 m/s² and density of water = 1000 kg/m³.

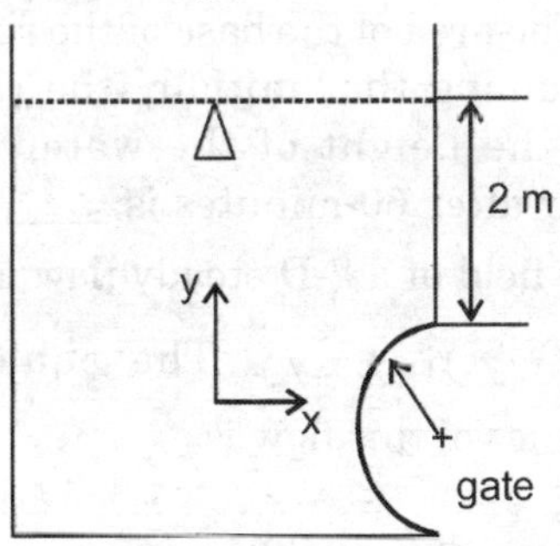

4. Velocity vector in m/s for a 2-D flow is given in Cartesian coordinate (x, y) as $\vec{V} = \left(\frac{x^2}{4}\hat{i} - \frac{xy}{2}\hat{j}\right)$.

Symbols bear usual meaning. At a point in the flow field, the x-and y-components of the acceleration vector are given as 1 m/s² and – 0.5 m/s², respectively. The velocity magnitude at that point is________m/s.

5. If Φ(x, y) is velocity potential and ψ(x, y) is stream function for a 2-D, steady, incompressible and irrotational flow, which one of the followings is incorrect?

(*a*) $\left(\frac{dy}{dx}\right)_{\Phi=\text{const}} = -\frac{1}{\left(\frac{ay}{dx}\right)_{\psi=\text{const}}}$

(*b*) $\frac{\partial^2\Psi}{\partial x^2} + \frac{\partial^2\Psi}{\partial y^2} = 0$

(*c*) $\left(\frac{dy}{dx}\right)_{\Phi=\text{const}} = \frac{1}{\left(\frac{dy}{dx}\right)_{\Phi=\text{const}}}$

(*d*) $\frac{\partial^2\Phi}{\partial x^2} + \frac{\partial^2\Phi}{\partial y^2} = 0$

6. The flow field shown over a bluff body has considerably curved streamlines. A student measures pressures at points A, B, C, and D and denotes them as PA, PB, PC, and PD respectively. State which one of the following statements is true. The arrow indicates the freestream flow direction.

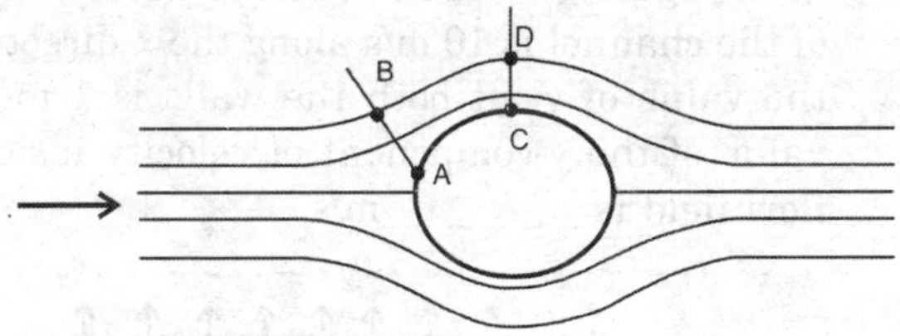

(*a*) $P_A = P_B$ and $P_C > P_D$

(*b*) $P_A > P_B$ and $P_C > P_D$

(*c*) $P_A = P_B$ and $P_C < P_D$

(*d*) $P_A > P_B$ and $P_C < P_D$

7. A 2-D incompressible flow is defined by its velocity components in m/s as $u = -\frac{cy}{x^2+y^2}$ and $v = \frac{cx}{x^2+y^2}$. If the value of the constnat c is equal to 0.1 m²/s, the numerical value of vorticity at the point x = 1 m and y = 2 m is________s^{-1}.

8. Two flow configurations are shown below for flow of incompressible, viscous flow. The inlet velocity for the diverging nozzle (Fig (i)) and free-stream velocity for flow past the bluff body (Fig(ii)) is constant. Points A and B are separation points and flow is laminar. The relation regarding velocity gradients at point A and B is (y is the direction normal to the surface at the point of separation)

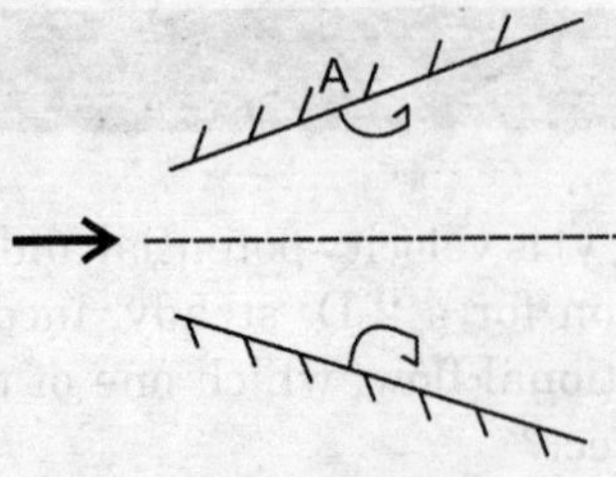

Fig (i)

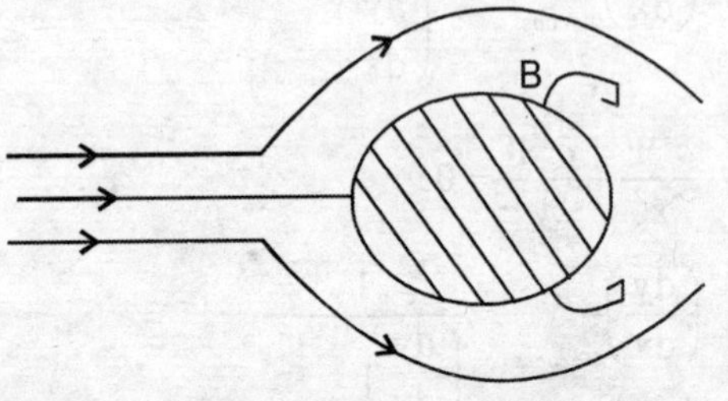

Fig (ii)

(a) $\left.\frac{\partial u}{\partial y}\right)_A = \left.\frac{\partial u}{\partial y}\right)_B$ (b) $\left.\frac{\partial u}{\partial y}\right)_A > \left.\frac{\partial u}{\partial y}\right)_B$

(c) $\left.\frac{\partial u}{\partial y}\right)_A < \left.\frac{\partial u}{\partial y}\right)_B$ (d) $\left.\frac{\partial^2 u}{\partial y^2}\right)_A = \left.\frac{\partial^2 u}{\partial y^2}\right)_B$

9. Consider a fully developed, steady, incompressible, 2-D, viscous channel flow with uniform suction and blowing velocity v_0, as shown in the figure given below. The centerline velocity of the channel is 10 m/s along the x-direction. If the value of v_0 at both the walls is 1 m/s, the value of the y-component of velocity inside the flow field is________m/s.

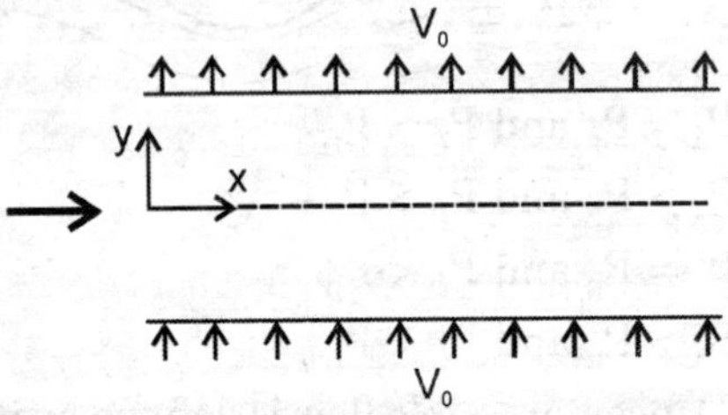

(Q. 10 – 22) : Carry Two Marks Each.

10. Exhaust from a kitchen goes into the atmosphere through a tapered chimney as shown. The area of cross-section of chimney at location-1 is twice of that at location-2. The flow can be assumed to be inviscid with constant exhaust density of 1 kg/m³ and acceleration due to gravity, g = 9.8 m/s². If the steady, uniform exhaust velocity at location-1 is U = 1m/s, the pressure drop across the chimney is ________ Pa.

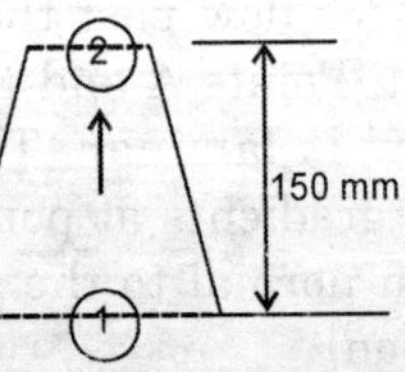

11. A jet of diameter 20 mm and velocity 6 m/s coming out of a water-tank standing on a frictionless cart hits a vane and gets deflected at an angle 45° as shown in the figure below. The density of water is 1000 kg/m³. Neglect all minor and viscous losses. If the cart remains stationary, the magnitude of tension in the supporting string connected to the wall is ________ N.

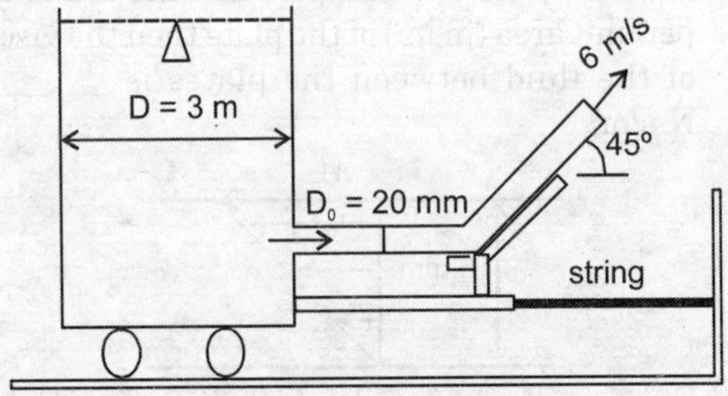

12. A block is floating at the oil-water interface as shown. The density of oil is two-thirds of that of water. Given that the density of the block is 800 kg/m³ and that of water is 1000 kg/m³, the fraction of the total height of block in oil is ________ .

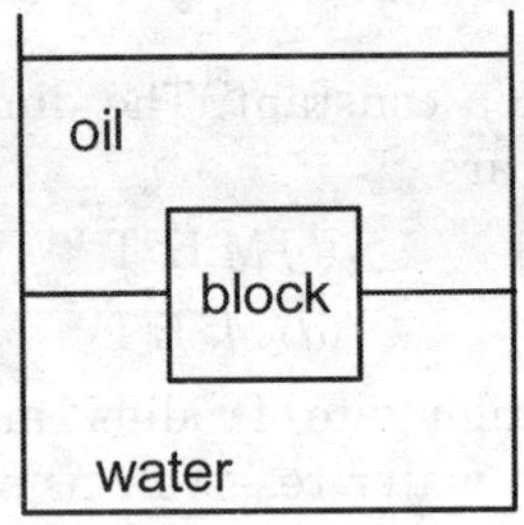

13. A horizontal pipe is feeding water into a reservoir from the top with a time-dependent volumetric flow-rate, Q (m³/h) = 1 + 0. 1 × t where t is time in hours. The area of the base of the reservoir is 0.5 m². Assuming that initially the reservoir was empty, the height of the water level in the reservoir after 60 minutes is __________m.

14. Velocity field of a 2-D steady flow is provided as $\vec{V} = c(x^2 - y^2)\hat{i} - 2cxy\,\hat{j}$. The equation of the streamlines of this flow is

(a) $x^2y - \frac{y^2}{3}$ = Constant

(b) $xy^2 - \frac{y^2}{3}$ = Constant

(c) $xy - \frac{y}{3}$ = Constant

(d) $x^2y - \frac{y^2}{3}$ = Constant

15. Velocity potential and stream function in polar coordinates (r, θ) for a potential flow over a cylinder with radius R is given as $\varnothing = U_\infty\left(r+\frac{R^2}{r}\right)\cos\theta$ and $\psi = U_\infty\left(r-\frac{R^2}{r}\right)\sin\theta$, respectively. Here, U_∞ denotes uniform freestream velocity, and θ is measured counter clockwise as shown in the figure. How does the velocity magnitude, q, over the surface of the cylinder will vary?

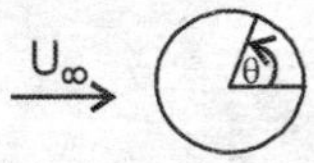

(*a*) $q = 2U_\infty \cos\theta$ (*b*) $q = 2U_\infty \sin 2\theta$

(*c*) $q = U_\infty \cos 2\theta$ (*d*) $q = 2U_\infty \sin\theta$

16. Consider a laminar flow over a flat plate of length L = 1m. The boundary layer thickness at the end of the plate is δ_w for water, and δ_a for air for the same freestream velocity. If the kinematic viscosities of water and air are 1×10^{-6} m²/s and 1.6×10^{-5} m²/s, respectively, the numerical value of the ratio, $\frac{\delta_w}{\delta_a}$ is________.

17. Prototype of a dam spillway (a structure used for controlled release of water from the dam) has characteristic length of 20m and characteristic velocity of 2 m/s. A small model is constructed by keeping Froude number same for dynamic similarity between the prototype and the model. What is the minimum length-scale ratio between prototype and the model such that the minimum Reynolds' number for the model is 100? The density of water is 1000 kg/m³ and viscosity is 10^{-3} Pa^{-s}.

(*a*) 1.8×10^{-4} (*b*) 1×10^{-4}

(*c*) 1.8×10^{-2} (*d*) 9.1×10^{-4}

18. An orifice meter, having orifice diameter of $d = \frac{20}{\sqrt{\pi}}$ mm is placed in a water pipeline having flow rate, $Q_{act} = 3 \times 10^{-4}$ m³/s. The ratio of orifice diameter to pipe diameter is 0.6. The contraction coefficient is also 0.6. The density of water is 1000 kg/m³. If the pressure drop across the orifice plate is 43.5kPa, the discharge co-efficient of the orifice meter at this flow Reynolds number is ___.

19. Consider the following figures shown below. The objects are marked as A1, A2, B1, B2 and C1, C2 and the flow directions over these objects are shown by the respective arrow placed to the left of the object. Freestream velocities are same for all the cases. Amongst these objects, A1, A2, B1 and C1 are having smooth surfaces while B2 and C2 are having rough surfaces. Reynolds number is such that flow over rough surfaces become turbulent and flow over smooth surfaces can be considered laminar. All the airfoils can be considered as thin slender airfoil. Among the statements (i) to (vi) made about the drag of these objects which is/are correct?

(i) Drag of object A1 is less than drag of object A2.

(ii) Drag of object A1 and A2 are same.

(iii) Drag of object B1 is more than drag of object B2.

(iv) Drag of object B2 is more than drag of object B1.

(v) Drag of object C1 is more than drag of object C2.

(vi) Drag of object C2 is more than drag of object C1.

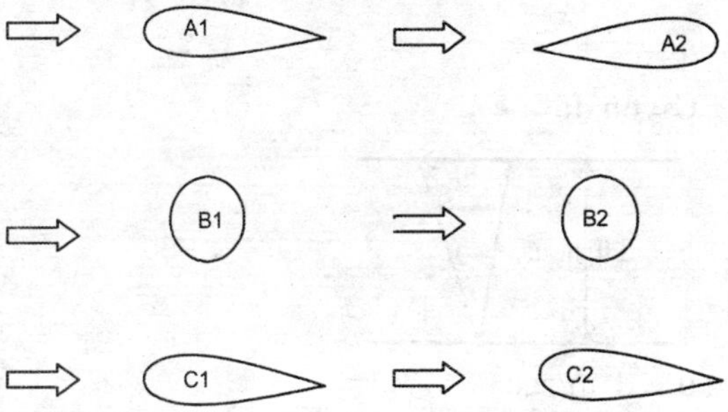

(*a*) (i), (iii) & (vi) (*b*) (ii), (iii) & (vi)

(*c*) (i), (iii) & (v) (*d*) (i), (iv) & (vi)

20. Consider 2-D, steady, incompressible, fully developed flow of viscous, Newtonian fluid through two stationary parallel plates, in Cartesian co-ordinate (x, y, z) system. Assume plates are very long in x-direction, wide in z-direction (also there is no variation of velocity in z direction) and distance between them is 2h. The velocity in such a channel is given as $U = U_{max}\left(1-\frac{y^2}{h^2}\right)$. The origin y = 0 is located at the center between the plates. If h = 48 mm and U_{max} = 100 mm/s, difference between values of stream functions passing through y = 0 and y = h/2 is______mm²/s.

21. A pump is used to deliver water to an overhead tank at a flow rate of $Q = 4 \times 10^{-3}$ m³/s. The pump adds 1.6 kW to water. If the density of water is 1000 kg/m³ and acceleration due to gravity is 10 m/s², the pump head added to the flow is_______m.

22. Water is discharged at atmospheric pressure from a large reservoir through a long pipe of diameter d and length L. The height of the free surface of the reservoir from the discharge point is h meters. The Darcy's friction factor of the pipe is 0.002. Neglect the velocity inside the reservoir as the reservoir is very large. Given, L = 20m, d = 40mm, density of water = 1000kg/m³ and flow rate is $Q = 4\pi + 10^{-3}$ m³/s. Assume gravitational acceleration, g = 10 m/s². The value of h is____________m.

ANSWERS

1. 0.0050	**2.** (b)	**3.** 15.40	**4.** 1.41	**5.** (c)	**6.** (d)	**7.** 0.00001
8. (a)	**9.** 0.9	**10.** 2.97	**11.** 5.652	**12.** 0.60	**13.** 2.10	**14.** (d)
15. (d)	**16.** 0.25	**17.** (a)	**18.** 0.31	**19.** (a)	**20.** 2200	**21.** 40
22. 10.19						

EXPLANATIONS

1. Given data

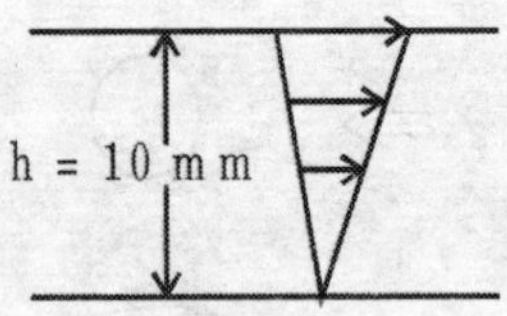

u = 1 m/sec

h = 10 mm

$\tau = 0.5\ \text{N/m}^2$

$$\text{Shear stress}(\tau) = \frac{\mu u}{h}$$

$$0.5 = \frac{\mu \times 1}{10 \times 18^3}$$

$$\boxed{\mu = 0.0050\ \text{Ns/m}^2}$$

2. (b) Given equation

$$q_f = \frac{c.\rho u^2 L}{\mu^3}$$

$$q_f - \frac{m_3}{m}$$

$$\rho - \frac{kg}{m^3}$$

v – m/sec

L – m

μ – kg/m.s

So that

$$\frac{m^3}{m} = c.\frac{\frac{kg}{m^3}\left(\frac{m}{s}\right)^2.m}{(kg/m.s)^3}$$

$c = kg^2/m.s$

$c = [M^2 L^{-1} T^{-1}]$

3. Given that

$\rho_w = 1000\ \text{kg/m}^3$

R = 1 m

$g = 9.81\ \text{m/s}^2$

Hydrostatic force

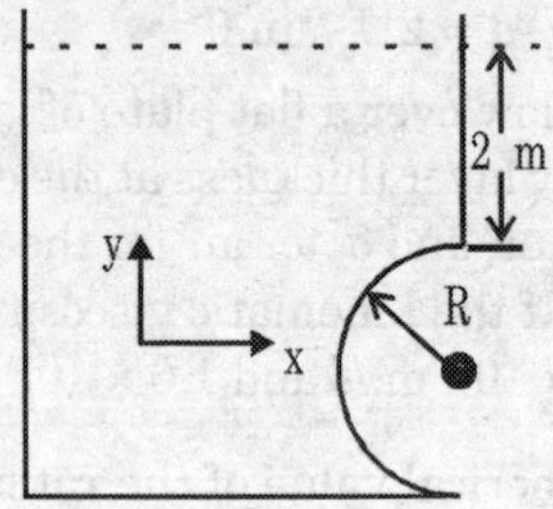

$$F_y = \rho g A \bar{h}$$

$$= 1000 \times 9.81 \times \frac{\pi}{4} \times 1^2 \times 2$$

$$\boxed{F_y = 15.40\ \text{kN}}$$

4. velocity profile $(\vec{v}) = \left(\frac{x^2}{4}\hat{i} - \frac{xy}{2}\hat{j}\right)$

$ax = 1\ \text{m/s}^2$ and $ay = -0.5\ \text{m/s}^2$

$$u = \frac{x^2}{4},\ v = \frac{-xy}{2}$$

$$\frac{\partial u}{\partial x} = \frac{x}{2},\ \frac{\partial v}{\partial y} = \frac{-x}{2}$$

$$a_x = u.\frac{\partial u}{\partial x} + v.\frac{\partial u}{\partial y}$$

$$1 = \left(\frac{x^2}{4}\right)\left(\frac{x}{2}\right) - \left(\frac{xy}{2}\right).0$$

$$8 = x^3,\ x = 2$$

$$ay = u.\frac{\partial u}{\partial x} + v.\frac{\partial u}{\partial y}$$

$$-0.5 = 1.\left(\frac{-y}{2}\right) + y(1)$$

y = –1

(2, –1)

u = 1 m/s

v = 1 m/s

$$v = \sqrt{u^2 - v^2}$$

$$\boxed{v = \sqrt{1^2 + 1^2} = 1.41\ \text{m/s}}$$

5. (c) $\phi = f(x, y), \psi = f(x, y)$

$$\left(\frac{dy}{dx}\right)_{\phi=c} = \frac{-u}{v}$$

$$\left(\frac{dy}{dx}\right)_{\psi=c} = \frac{u}{v}$$

$$\left(\frac{dy}{dx}\right)_{\phi=c} = \frac{-1}{\left(\frac{dy}{dx}\right)_{\psi=c}}$$

$$\frac{\partial^2\phi}{\partial x^2} + \frac{\partial^2\phi}{\partial y^2} = 0$$

Laplace equation for steady, Incompressible and irrotation flow.

$$\frac{\partial^2\psi}{\partial x^2} + \frac{\partial^2\psi}{\partial y^2} = 0$$

Hence option (C) is incorrect from above relation.

6. (d)

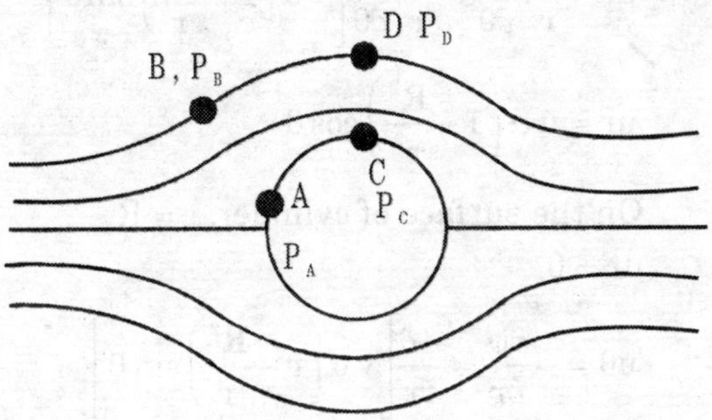

Bluff body:- A body is said to be bluff body in which that the surface area donot coincides with the stream line. The stream line is an imaginary which give the direction of velocity.

So, that

$$\boxed{P_A > P_B,\ P_D > P_C}$$

7. Given that for Two-dimensinal flow.

$$u = -\frac{cy}{x^2+y^2},\ v = \frac{cx}{x^2+y^2}$$

x = 1m, y = 2m, c = 0.1 m²/sec

Vorticity(ξ) = $2w_z$

$$w_z = \frac{1}{2}\left[\frac{\partial u}{\partial x} - \frac{\partial u}{\partial y}\right]$$

$$w_z = \frac{0.00001}{2}$$

$$\boxed{\xi = 0.00001}$$

8. (a)

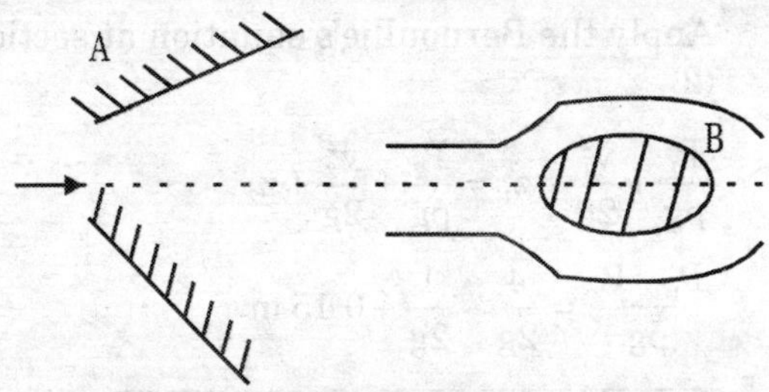

For flow seperation

(i) $\frac{\partial u}{\partial y} = 0$

(ii) $\frac{\partial p}{\partial x} > 0$

so that here,

$$\boxed{\left(\frac{\partial u}{\partial y}\right)_A = \left(\frac{\partial u}{\partial y}\right)_B}$$

9.

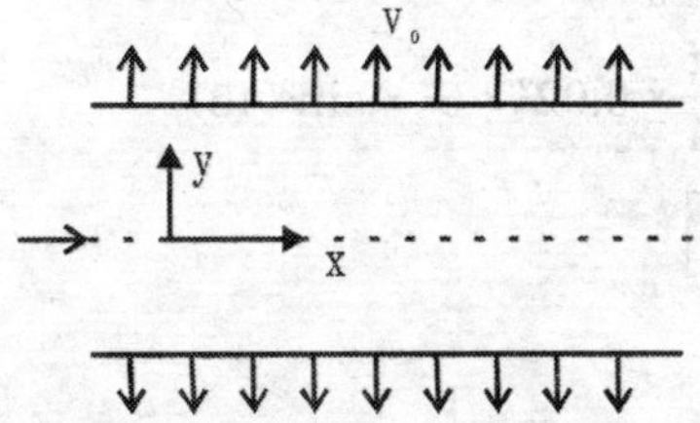

Given that

Centre line velocity(v) = 10 m/s

v_o = 1 m/s

Velocity of y-component = $1 - \frac{v_o}{v}$

$$= 1 - \frac{1}{10} = 0.9\ \text{m/sec}$$

10. Given that

$A_1 = 2A_2$

d = 1.2 kg/m³

g = 9.81 m/s²

v_1 = 1 m/sec

$z_1 = 0$

z_2 = 150 = 0.15 m

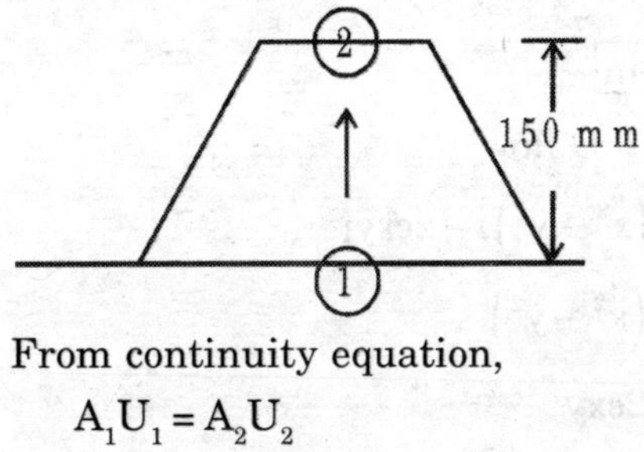

From continuity equation,

$A_1U_1 = A_2U_2$

$2A_2 \times 1 = A_2U_2$

U_2 = 2 m/s

Apply the Bernoullie's equation at section (1) & (2)

$$\frac{P_1}{\rho g}+\frac{v_1^2}{2g}+z_1=\frac{P_2}{\rho g}+\frac{v_2^2}{2g}+z_2$$

$$\frac{P_1-P_2}{\rho g}=\frac{4}{2g}-\frac{1}{2g}+0.15\text{ m}$$

$$\boxed{\text{Pressure drop}\left(P_1-P_2\right)=2.97\text{ Pa}}$$

11. Given that

$d = 20$ mm $= 0.02$ m

$v = 6$ m/sec

$Q = 45°$

$\delta w = 1000$ kg/m^3

Force exerted is equal to the tension in the supporting string, because the cart is stationary.

$Fx = \rho Av^2 \sin^2\theta$

$$Fx = 1000 \times \frac{\pi}{4} \times 0.02^2 \times 6^2 \times \sin^2(45)$$

$$\boxed{F_x = 5.652\text{ N}}$$

12. Given that

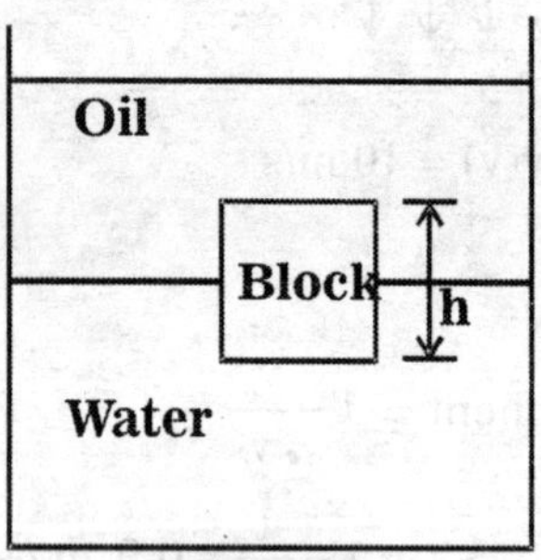

Here, h = 1m

$\rho_w = 1000$ kg/m^3

$\rho_B = 800$ kg/m^3

$$\rho_o = \frac{2}{3} \times 1000 = 666.66\text{ kg/m}^3$$

Consider unit volume of Block, and x unit submerged in oil.

$$\rho gAh = \rho wgAhw + \rho_o gAho$$

$800 \times 1 = 1000 \times (1 - x) + 666.66 \times x$

$$\boxed{x = 0.60\text{ m}}$$

14. (d) For 2 – D flow

$$\vec{v} = c\left(x^2-y^2\right)\hat{i} - 2cxy\hat{j}$$

$$u = c\left(x^2-y^2\right)$$

$$v = -2cxy$$

Now,

$$\frac{dx}{u}=\frac{dy}{v} \quad \text{[Streamline equation]}$$

$$\frac{dx}{c\left(x^2-y^2\right)}=\frac{dy}{-2cxy}$$

$$-2xy.dx = \left(x^2-y^2\right).dy$$

Integrating both sides.

$$-\int 2xy.dx = \int\left(x^2-y^2\right).dy$$

$$-x^2y + c = x^2y - \frac{y^3}{3} + c$$

$$\boxed{2x^2y - \frac{y^3}{3} = c}$$

15. (d) Given that

$$\phi = v\alpha\left(r+\frac{R^2}{r}\right)\cos\theta$$

$$\psi = v\alpha\left(r-\frac{R^2}{r}\right)\sin\theta$$

$$ur = \frac{1}{r}.\frac{\partial\psi}{\partial\theta} = \frac{1}{r}.\frac{\partial}{\partial\theta}\left[v\alpha\left(r-\frac{R^2}{r}\right)\sin\theta\right]$$

$$ur = u\alpha\left(1-\frac{R^2}{r^2}\right)\cos\theta$$

On the surface of cylinder, $r = R$

$ur = 0$

$$u\theta = -\frac{\partial\psi}{\partial r} = \frac{-\partial}{rr}\left[v\alpha\left(r-\frac{R^2}{r}\right)\sin\theta\right]$$

$$u\theta = -u\alpha\left(1+\frac{R^2}{r^2}\right)\sin\theta$$

at $r = R$

$u\theta = -2u\,\alpha\sin\theta$

$$q = \sqrt{4\theta^2 + 4\theta^2}$$

$$\boxed{q = 2u\ \alpha\sin\theta}$$

16. Given that

$L = 1$m

$\nu_w = 10^{-6}$ m^2 / sec

$\nu_a = 1.6\times10^{-5}$ m^2 / sec

$$\delta = \frac{5x}{\sqrt{Re_x}}$$

$$\frac{\rho w}{\rho a} = \sqrt{\frac{10^{-6}}{1.6\times10^{-5}}} = \frac{1}{4}$$

$$\boxed{\frac{\rho w}{\rho a} = 0.25}$$

18. Given that

$$d_0 = \frac{20}{\sqrt{\pi}} mm = 0.01125\,m$$

$$a_0 = 9.99 \times 10^{-5}\ m^2$$

$$\frac{a_0}{a_1} = 0.6$$

$$a_1 = 1.665 \times 10^{-4}\ m^2$$

$$h = \frac{P_1 - P_2}{\rho g} = \frac{43.5 \times 1000}{1000 \times 9.81}$$

$$= 4.434\ m$$

$$Q = 3 \times 10^{-4}\ m^3/s$$

$$Q = \frac{C_d a_1 a_0 \sqrt{2gh}}{\sqrt{a_1^2 - a_0^2}}$$

$$3\times10^{-4} = C_d.\frac{(1.665\times10^{-4})(9.99\times10^{-5})\sqrt{2\times9.81\times4.434}}{\sqrt{(1.665\times10^{-4})^2-(9.99\times10^{-5})^2}}$$

$$\boxed{C_d = 0.31}$$

19. (a) Drag force developed along the direction of motion.

$$\boxed{F_D = \frac{1}{2}\rho A\, v^2 C_D}$$

In the given figure

$$(F_D)_{A1} < (F_D)_{A2}$$

$$(F_D)_{B1} < (F_D)_{B2}$$

$$(F_D)_{C1} < (F_D)_{C2}$$

21. Given that

$$Q = 4 \times 10^{-3}\ m^3/s$$

$$P = 1.6\ kw = 1600\ w$$

$$\rho w = 1000\ kg/m^3$$

$$g = 10\ m/s^2$$

$$P = \delta g Q h$$

$$h = \frac{P}{\rho g Q} = \frac{1600}{1000 \times 9.81 \times 4 \times 10^{-3}}$$

$$\boxed{h = 40\ m}$$

22. Given that

$$Q = 4\pi \times 10^{-3}\ m^3/sec$$

$$f = 0.002$$

$$L = 20\ m$$

$$d = 40\ mm = 0.04\ m$$

$$h_f = \frac{fLv^2}{2gd}$$

$$h_f = \frac{2 \times 16 L Q^2 f}{2g\,\pi^2 d^5}$$

$$h_f = \frac{32 \times 20 \times (4\pi \times 10^{-3})^2 \times 0.002}{2 \times 9.81 \times \pi^2 \times (0.04)^5}$$

$$\boxed{h_f = 10.19\ m}$$

2017

1. In a given flow field, the velocity vector in Cartesian coordinate system is given as:

$$\overrightarrow{V}=\left(x^2+y^2+x^2\right)\hat{i}+\left(xy+yz+y^2\right)\hat{j}+\left(xz-z^2\right)\hat{k}$$

What is the volume dilation rate of the fluid at a point where x = 1, y = 2 and z = 3?

(*a*) 6 (*b*) 5

(*c*) 10 (*d*) 0

2. A steady, incompressible, two-dimensional velocity field in Cartesian coordinate system is represented by the following expression.

$$\overrightarrow{V}=(0.7+0.4x)\hat{i}+(1.2-0.4y)\hat{j}$$

The coordinate of the point (x, y) in the flow field having "zero" velocity is.

(*a*) (1.75, –3) (*b*) (–1.75, 3)

(*c*) (1.75, 3) (*d*) (–1.75, –3)

3. During an experiment, the position of a fluid particle is monitored by an instrument over a time period of 10 s. The trace of the particle given by the following figure represents a

Fluid particle at t = 4s

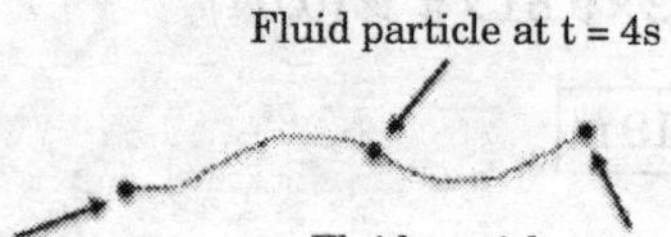

Fluid particle at t = 0 Fluid particle at t = 10 s

(*a*) streamline (*b*) streakline

(*c*) pathline (*d*) timeline

4. In a Cartesian two-dimensional coordinate system, u and v represent the velocities in x and y directions, respectively. For a certain flow, the velocity field is represented by the following expression:

$$\overrightarrow{V}=(ax+by)\hat{i}+(cx+dy)\hat{j}$$

(*a*) a + d = 0

(*b*) a + c = 0

(*c*) b + d = 0

(*d*) b + c = 0

5. Which one of the following figures represents potential flow past a circular cylinder with clockwise rotation of the cylinder?

(*a*) 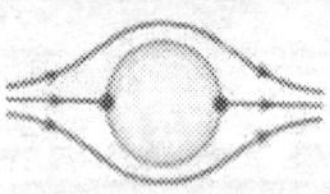 (*b*) 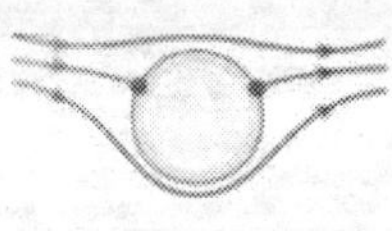

(*c*) 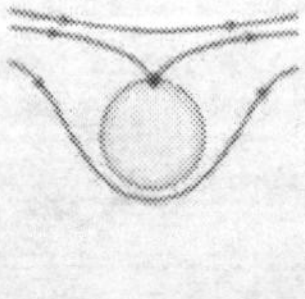 (*d*) 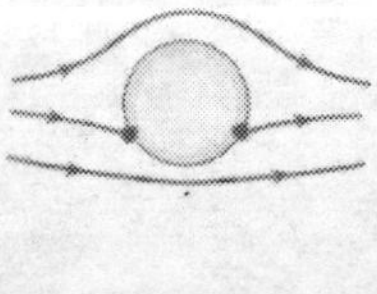

6. The stream function (Ψ) of a velocity field at any location (x, y) is given as, $\Psi = xy^2 - 2x^2y^2$. What is the rate of rotation of a fluid element located at (x = 2, y = 2)?

(*a*) 8 (*b*) 10

(*c*) 12 (*d*) 14

7. The nature of velocity profile within laminar viscous sublayer in a turbulent pipe flow is

(*a*) linear (*b*) parabolic

(*c*) logarithmic (*d*) exponential

8. In a 5 m deep vertical cylindrical tank, water is filled up to a level of 3 m from the bottom and the remaining space is filled with oil of specific gravity 0.88. Assume density of water as 1000 kg/m³ and acceleration due to gravity to be 10 m/s². The gauge pressure (in kN/m², rounded off to the first decimal place) at a depth of 2.5 m from the top of the tank will be ______.

9. In a two-dimensional potential flow, a point source is located at the origin (x = 0, y = 0) as shown in the figure. The strength of the point source is 2 cm²/s. A uniform flow with velocity 1 cm/s is approaching towards the point source at an angle of 30° from the horizontal axis. What is the distance (cm) of the stagnation point in the flow field from the point source?

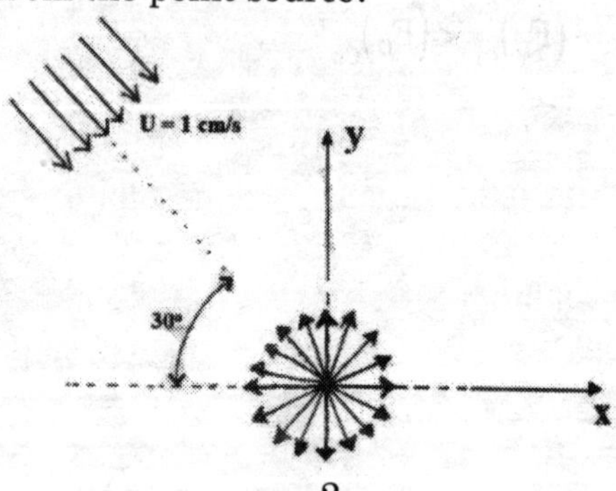

(*a*) $\dfrac{1}{\pi}$ (*b*) $\dfrac{2}{\pi}$

(*c*) $\dfrac{1}{2\pi}$ (*d*) $\dfrac{\sqrt{3}}{2\pi}$

10. Two infinite parallel horizontal plates are separated by a small gap (d = 20 mm) as shown in figure. The bottom plate is fixed and the gap between the plates is filled with oil having density of 890 kg/m³ and kinematic viscosity of 0.00033 m²/s. A shear flow is induced by moving the upper plate with a velocity of 5 m/s. Assume, linear velocity profile between the plates and the oil to be a Newtonian fluid. The shear stress (N/m²) at the upper plate is ______.

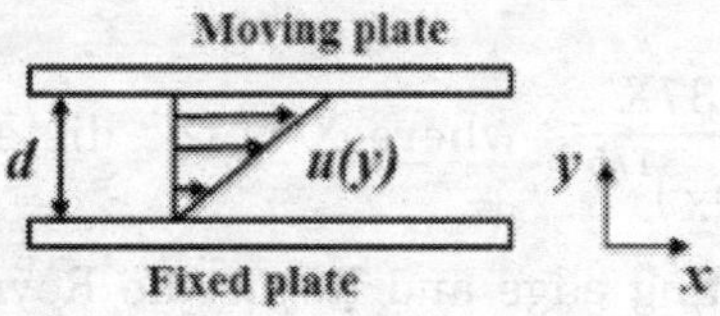

11. A spherical balloon of diameter 15 m is supposed to lift a load of 3000 N. The lifting of load is achieved by heating the air inside the balloon. Assume, air to be an ideal gas and atmospheric pressure either outside or inside the balloon. The value of acceleration due to gravity is 9.81 m/s^2 and the values of temperature and density of atmospheric air are 15°C and 1.2 kg/m^3, respectively. In order to lift the specified load, the air inside the balloon should be heated to a temperature (°C) of ______.

12. The velocity field in Cartesian coordinate system for a two-dimensional steady flow is given as:

$$\vec{V} = \left(\frac{V_0}{L}\right)(x\hat{i} - y\hat{j})$$

where V_0 and L are constants. Which one of the following expressions represents the acceleration field $(\vec{a})$ for this flow?

(*a*) a = 0 (*b*) $\vec{a} = \left(\frac{V_0}{L}\right)(x\hat{i} + y\hat{j})$

(*c*) $\vec{a} = \left(\frac{V_0^2}{L^2}\right)(x\hat{i} - y\hat{j})$ (*d*) $\vec{a} = \left(\frac{V_0^2}{L^2}\right)(x\hat{i} + y\hat{j})$

13. A cylindrical tank of 0.8 m diameter is completely filled with water and its top surface is open to atmosphere as shown in the figure. Water is being discharged to the atmosphere from a circular hole of 15 mm diameter located at the bottom of the tank. The value of acceleration due to gravity is 9.81 m/s^2. How much time (in seconds) would be required for water level to drop from a height of 1 m to 0.5 m?

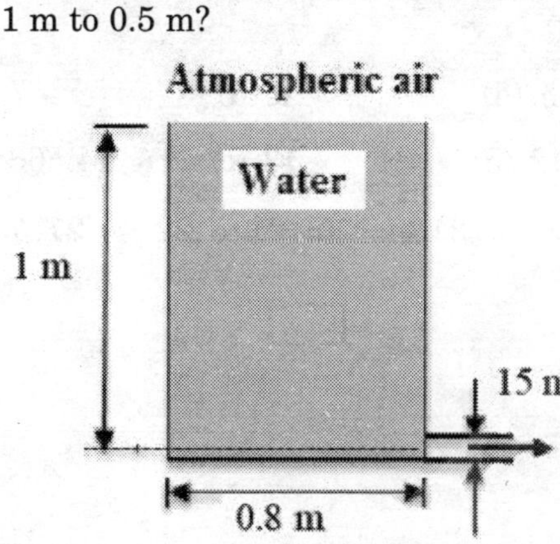

(*a*) 188 (*b*) 266

(*c*) 376 (*d*) 642

14. Consider steady laminar flow of an incompressible Newtonian fluid between two infinite parallel plates, separated by a distance of 1 m, as shown in the figure. The bottom plate is stationary but the top one is moving positive x-direction with a velocity of 3 m/s. The fluid pressure gradient in the flow direction is: $\frac{\partial P}{\partial x} = -18\text{N/m}^3$. If the viscosity of the fluid is 1 kg m^{-1} s^{-1} then the distance of the point of maximum velocity (in meters, rounded off to the second decimal place) from the bottom plate would be ______.

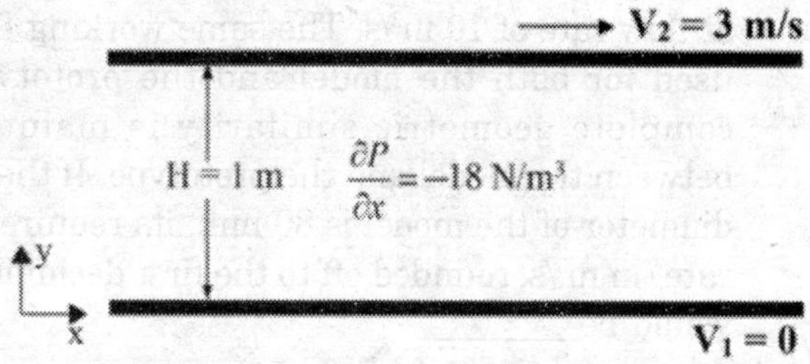

15. An inviscid incompressible fluid of density 1000 kg/m^3 is flowing in a horizontal pipe of tapered cross-section with a flow rate of 4000 cm^3/s. The area of cross-section at two different locations 'A' and 'B' are 10 cm^2 and 20 cm^2, respectively. The velocity of the fluid at the location 'A' is 4 m/s and pressure is 5 N/m^2. The pressure (N/m^2) at location 'B' would be ______.

16. A viscous, incompressible and Newtonian fluid following through the main branch of a circular pipe bifurcates into two daughter branches whose radii are 4 cm and 2 cm, respectively. The flow in both the daughter branches are laminar and fully developed. If the pressure gradients in both the daughter branches are same, then fraction of total volumetric flow rate (rounded off to the second decimal place) coming out from the branch with 4 cm diameter is ______

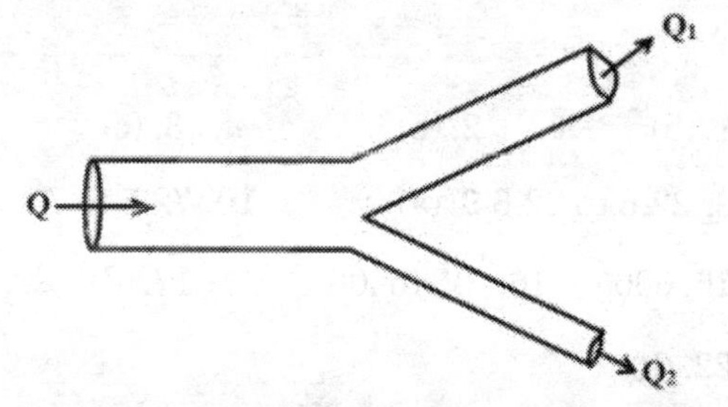

17. The volumetric flow rate (Q) of a triangular notch is a function of the upstream liquid surface elevation (H) measured from the bottom of the notch, acceleration due to gravity (g), notch angle (ϕ) and the approach velocity (V). Which one of the following is the correct expression for Q?

(a) $Q = H^{1/2} f\left(\frac{V}{\sqrt{H}}, \phi\sqrt{g}\right)$

(b) $Q = Hf\left(\frac{V}{\sqrt{H}}, \phi\sqrt{g}\right)$

(c) $Q = H^{3/2} f\left(\frac{V}{\sqrt{H}}, \phi\sqrt{g}\right)$

(d) $Q = H^{5/2} f\left(\frac{V}{\sqrt{H}}, \phi\sqrt{g}\right)$

18. Model tests are to be carried out to study the flow through a large prototype valve of 0.6 m diameter at flow rate of 10 m³/s. The same working fluid is used for both the model and the prototype. A complete geometric similarity is maintained between the model and the prototype. If the valve diameter of the model is 80 mm, its required flow rate (in m³/s, rounded off to the first decimal place would be ______.

19. Water is flowing at a rate of 0.5 m³/s in a horizontal pipeline of inside diameter 0.5 m. The density and kinematic viscosity of water is 1000 kg/m³ and 10^{-6} m²/s, respectively. Assume Darcy-Weisbach friction factor value to be 0.0093 and acceleration due to gravity as 9.81 m/s². To maintain constant flow rate, the required power per unit length of the pipeline (in W/m, rounded off to the first decimal place) would be ______.

20. Air flows over a smooth flat plate at a velocity of 4.39 m/s. The density of air is 1.031 kg/m³ and the kinematic viscosity is 1.34×10^{-5} m²/s. The plate length is 12.2 m in the direction of the flow. The boundary layer thickness (δ) is given as $\frac{0.37X}{(Re_X)^{1/5}}$, where X is the distance from the leading edge and Re_x is the Reynolds number. The boundary layer thickness (in meters, rounded off to the second decimal place) at 12.2 m from the leading edge will be ______.

21. A venturimeter of diameter 0.2 m at the entrance and 0.1 m at the throat is inclined upwards. The vertical elevation difference between the entrance and the throat is 0.5 m. The density of water is 1000 kg/m³ and the coefficient of velocity is 0.97. The differential U-tube manometer connected to the entrance and throat shows a pressure difference of 30 kN/m². Assume acceleration due to gravity as 9.81 m/s². The velocity of the water (in m/s, rounded off to the first decimal place) at the throat would be ______.

22. A spherical bubble of radius r is rising upward with a constant velocity U, in quiescent water of dynamic viscosity . The density of air and water are denoted by ρ_a and ρ_w, respectively, and g is acceleration due to gravity. The bubble motion is such that, the Reynolds number, Re <<1. The density of air can be neglected in comparison to the water density ($\rho_a << \rho_w$). Which one of the following expressions is TRUE for the density of water?

(a) $\rho_w = \frac{2}{9}\frac{U}{r^2 g}$ (b) $\rho_w = \frac{9}{2}\frac{U}{r^2 g}$

(c) $\rho_w = \frac{9}{4}\frac{U}{r^2 g}$ (d) $\rho_w = \frac{4}{9}\frac{U}{r^2 g}$

ANSWERS

1. (b) **2.** (b) **3.** (c) **4.** (a) **5.** (d) **6.** (d) **7.** (a)

8. 22.6 to 22.6 **9.** (a) **10.** 72.5 to 75 **11.** 62 to 66 **12.** (d) **13.** (c) **14.** .66 to .67

15. 6005 **16.** .05 to .06 **17.** (d) **18.** 1.3 to 1.4 **19.** 28.9 to 31.4 **20.** .21 to .22 **21.** 7 to 7.2

22. (b)

EXPLANATIONS

1. (b) $\vec{v} = (x^2 + y^2 + z^2)\hat{i} + (xy + yz + y^2)\hat{j} + (x^2 - z^2)\hat{k}$

at, $x = 1$, $y = 2$ and $z = 3$

Volume dilation rate $= \frac{du}{dx} + \frac{dv}{dy} + \frac{dw}{z}$

$= \frac{d}{dx}(x^2 + y^2 + z^2) + \frac{d}{dy}(xy + yz + y^2) + \frac{d}{dz}(x^2 - z^2)$

$= 2x + (x + z + 2y) + (x - 2z)$

At (1, 2, 3)

$= 2 + 1 + 3 + 4 + 1 - 6 = 5$

2. (b) For velocity field.

$\vec{v} = (0.7 + 0.4x)\hat{i} + (1.2 - 0.4y)\hat{j}$

$u = 0.7 + 0.4x$

$u = 1.2 - 0.4y$

$u = 0, \quad v = 0$ (Given)

$0.4 + 0.4x = 0$

$x = -1.75$

$1.2 - 0.4y = 0$

$y = 3$

Coordinate of the point $(x, y) - (-1.75, 3)$

3. (c)

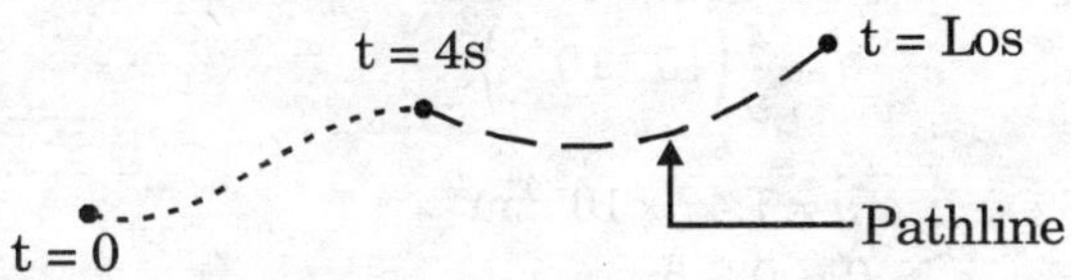

Pathine - A line traced by the single fluid particle over a period of time.

4. (a) $\vec{v} = (ax + by)\hat{i} + (Cx + dy)\hat{j}$

From Continuty equation, $\frac{du}{dx} + \frac{dv}{y} = 0$, For two dimensional and incompreesible flow.

$\frac{d}{dx}(ax + by) + \frac{d}{dy}(cx + dy) = 0$

$\boxed{a + d = 0}$

5. (d) Potential flow past a circular cylinder with clockwise rotation of the cylinder

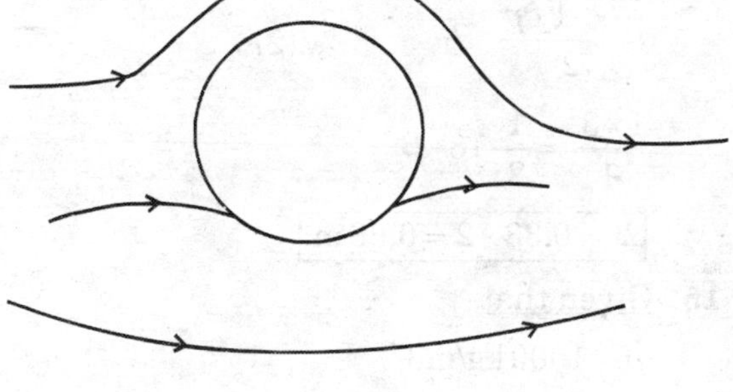

6. (d) Stream function

$\psi = xy^2 - 2x^2y^2$ at $(x = 2, y = 2)$

$u = \frac{-d\psi}{dy} = -(2xy - 4x^2y)$

$\frac{du}{dx} = -(2x - 4x^2)$

$\frac{du}{dx} = -(4 - 16) = 12$

$v = \frac{d\psi}{dx} = y^2 - 4xy^2$

$\frac{dv}{dx} = 0 - 4y^2 = -16$

$\omega_z = \frac{1}{2}\left(\frac{dv}{dx} - \frac{du}{dy}\right)$

$\omega_z = \frac{1}{2}(-16 - 12) = -14$

7. (a)

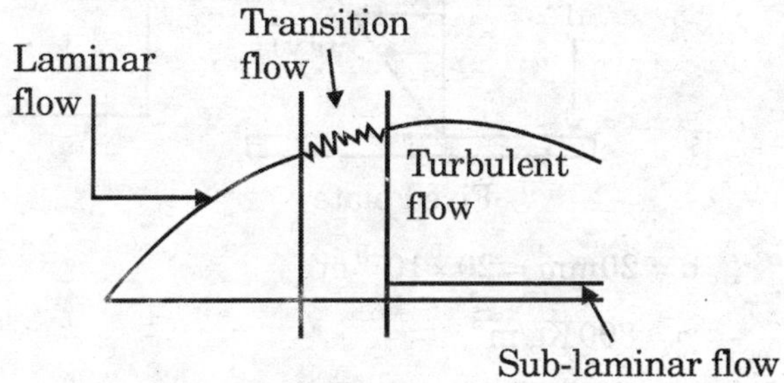

laminar Flow – parabolic natare

Turbulent flow – logarithmic nature.

→ The sub-laminar flow occur in the turbulent region, near to siold boundary and its is linear nature.

8. (a)

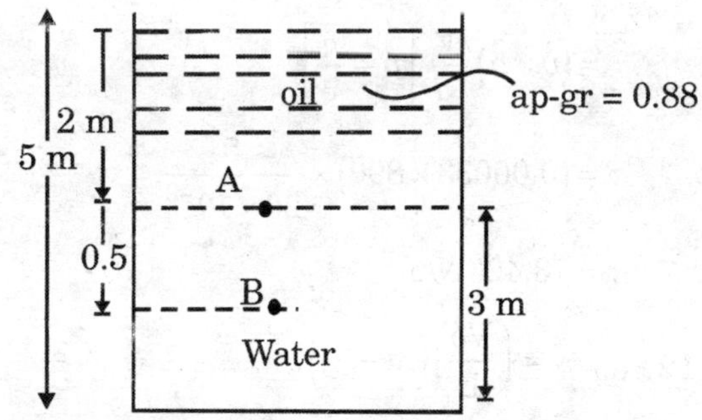

Goil = 0.88

Density of oil

$\rho_{soil} = 880\ kg/m^3$

$\rho_w = 1000\ kg/m^3$

$g = 10\ m/s^2$

$P_A = P_{oil} = \rho_o \times g \times h_{oil}$

$P_A = 880 \times 10 \times 2 = 17.6 \text{KN} / m^2$

$P_B = \rho_w \times g \times h$

$= 1000 \times 9.81 \times 0.5 \text{(up to 0.5)}$

$P_B = 4.95 \text{KN} / m^2$

Total Gange pressure at the 2.5m,

$P = P_A + P_B$

$P = 17.6 + 4.95$

$P = 22.5 \text{ KN/m}^2$

9. (a) Given that

$q = 2 \text{ cm}^2/\text{s}$

$u = 1 \text{ cm/s}$

$\psi = \frac{q.\alpha}{2\pi}$

Position from stagnation point $= \frac{2 \times 1}{2\pi} = \frac{1}{\pi}$

10.

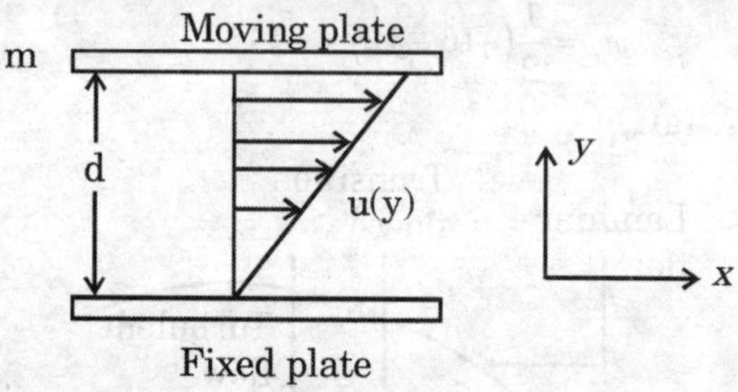

$d = 20\text{mm} = 20 \times 10^{-3} \text{m}$

$\rho = 890 \text{Kg/m}^3$

$\upsilon = 0.00033 \frac{m^2}{s}$

$V = 5 \text{ m/s}$

$\text{Shear Stress}(\tau) = \frac{\mu V}{d}$

$\tau = (\upsilon \times \rho) \cdot \frac{V}{d} \left[\upsilon = \frac{\mu}{\Im} \right]$

$\tau = (0.00033 \times 890) \times \frac{5}{20 \times 10^{-3}}$

$\tau = 73.425 \text{N/m}^2$

12. (c) $\vec{v} = \left(\frac{Vo}{L}\right)(x\hat{i} - y\hat{j})$

V_0 and L are constant

$u = \frac{V_0 x}{L}$

$\vartheta = -\frac{v_0 y}{L}$

$a_x = u \cdot \frac{\partial u}{\partial x} + u \cdot \frac{\partial u}{\partial y}$

$a_y = u \cdot \frac{\partial \vartheta}{\partial x} + u \cdot \frac{\partial \vartheta}{\partial y}$

$\frac{\partial u}{\partial x} = \frac{V_0}{L}, \frac{\partial u}{\partial y} = 0$

$\frac{\partial u}{\partial x} = 0, \frac{\partial v}{\partial y} = -\frac{V_0}{L}$

$\vec{a} = ax\hat{i} + ay\hat{j}$

$\vec{a} = \left(\frac{V_0 x}{L} \cdot \frac{V_0}{L}\right)\hat{i} + \left(\frac{V_0 y}{L} \cdot \frac{V_0}{L}\right)\hat{j}$

$\vec{a} = \left(\frac{V_0}{L}\right)^2 (x\hat{i} - y\hat{j})$

13. (c) Given that

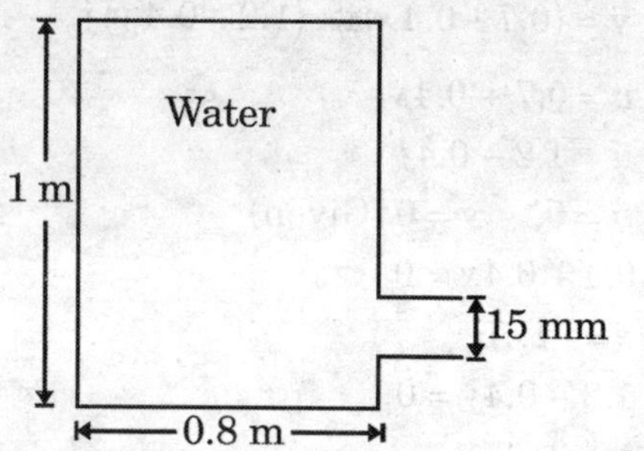

$d = 15 \times 10^{-3} \text{m}.$

$a = \frac{\pi}{u}(15 \times 10^{-3})^2$

$a = 1.47 \times 10^{-4} \text{m}^2$

$0 = 0.8 \text{ m}$

$T = \frac{2A\sqrt{H_1}}{Cd \cdot a \cdot \sqrt{2g}} = \frac{2 \times 0.5024\sqrt{1}}{0.6 \times 1.77 \times 10^{-4} \times \sqrt{2 \times 9.81}}$

$T = 376.6 \text{ sec.}$

14.

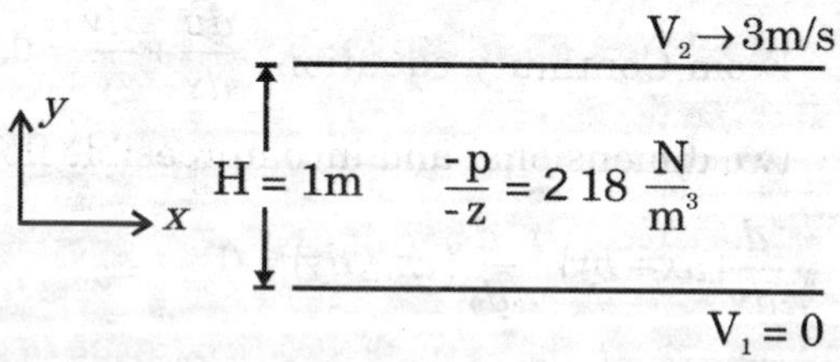

$\mu = 1 \text{kg} / \text{m} \cdot \text{s}$

$\tau = \frac{\mu v}{h}$ ...(1)

$\tau = \frac{-1}{2} \frac{\partial p}{\partial x} \cdot B$ ...(2)

$\frac{1 \times 3.}{1} = \frac{1}{2} \cdot 18 \cdot B$

$\boxed{B = 0.33 \times 2 = 0.66 \text{ m.}}$

15. Given that

$\rho = 1000 \text{kg/m3}$

$Q = 4000\,cm^3/s = 4\times10^{-3}\,m^3/s$

area at A = 10 cm² = $10^{-3}\,m^2$

area at B = 20 cm² = $2\times10^{3}\,m^2$

V_A = um/s

P_A = 5 N/m²

$Q = V_A . A_A = V_B . A_B$

$$V_B = \frac{4\times10^{-3}}{2\times15^3} = 2\,m/sec.$$

$$\frac{P_A}{\rho g} + \frac{V_A^2}{2g} = \frac{P_B}{\rho g} + \frac{V_B^2}{2g}$$

$$\frac{P_B}{\rho g} = \frac{5}{1000\times9.81} + \frac{4^2}{2\times9.81}$$

$$= \frac{2^2}{29.81},$$

$$P_B = 6005\frac{N}{m^2}$$

16. Given that

d_1 = 4 cm

d_2 = 4 cm

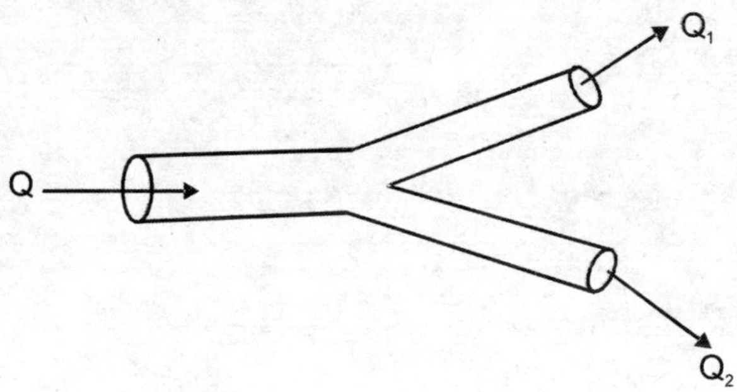

$Q = Q_1 + Q_2$

$Q = A_1V_1 + A_2V_2$

fraction of total volumetric flow.

$$\frac{2}{4\times10} = 0.05$$

17. (d) For Triangular Notch.

$$\text{Discharge}(Q) = \frac{8}{15}\text{Cd}\tan\frac{\theta}{2}\sqrt{2g}\text{H}^{5/2}$$

$$\boxed{Q = \text{H}^{5/2} f\left(\frac{V}{\sqrt{\text{H}}}\cdot\sqrt{g}\right)}$$

18. Given that

D_p = 0.6 m.

Q_p = 10m³/s

$D_m = 80\,mm = 80\times10^{-3}\,m$

Q_m = ?

$$\text{Scale ratio} = \frac{D_m}{D_p} = \frac{80\times10^{-3}}{0.6} = \frac{2}{15}$$

$$\frac{Q_m}{Q_p} = \frac{2}{15}$$

$$Q_m = \frac{2}{15}\times10 = 1.33\,m^3/s$$

18. Given that

Q = 0.5 m³/s

d = 0.5 m

$\rho = 1000\,Kg/m^3$

$\nu = 10^{-6}\,m^2/s$

$f' = 0.0093$

$$f = \frac{f'}{4} = 2.325\times10^{-3}$$

Q = AV

$$V = \frac{4Q}{\pi d^2}$$

$$V = \frac{4\times0.5}{3.14\times0.52}$$

V = 2.55 m/s

$$h_f = \frac{4fLV^2}{2gd}$$

$$h_f = \frac{4\times2.325\times10^{-3}\times2.55^2}{2\times9.81\times0.5}$$

$$h_f = 6.16\times10^{-3}$$

$$P = \rho g Q h_f$$

$$= 1000\times9.81\times0.5\times6.16\times10^{-3}$$

$$\boxed{P = 30.248\,w/m.}$$

20. Given that

v = 4.39 m/s

$\rho = 1.031\,kg/m^3$

$\upsilon = 1.34\times10^{-5}\,m^2/s$

L = 12.2 m.

$$Re = \frac{\rho vD}{\mu}$$

$$Re = \frac{VL}{\mu} = \frac{4.39\times12.2}{1.34\times10^{-5}}$$

Re = 3996865.67

$$\rho = \frac{0.37\,x}{(\mathrm{Re}_x)\frac{1}{5}}$$

$$\delta = \frac{0.37 \times 12.2}{(3996865.67)^{1/5}}$$

Boundary layer thickness (δ) = 0.216 m.

21. Given that

$P_1 - P_2 = 30$ KN/m²

$$h = \frac{P_1 - P_2}{\rho g} = 0.5$$

$$h = \frac{30 \times 1000}{1000 \times 9.81} - 0.5 = 2.56\,\text{m}.$$

$$V = Cv\sqrt{2gh} = 0.97\sqrt{2 \times 9.81 \times 2.56}$$

Velocity = 7.1 m/s

22. (b) According to given data.

$$\delta w = \left(\frac{3}{\sqrt{2}r}\right)^2 \left(\frac{\mu v}{g}\right)$$

$$\boxed{\delta_w = \frac{9}{2r^2}\left(\frac{\mu v}{g}\right)}$$

2018

Q. 1 – Q. 9 carry one mark each.

1. Rheological diagram of different types of fluids is shown in figure. Column I represents the nature of the fluid and column II represents the curve showing the variation of shear stress against shear strain rate.

Column I	Column II
(i) Newtonian	M
(ii) Shear thinning	N
(iii) Shear thickening	O
(iv) Bingham plastic	P

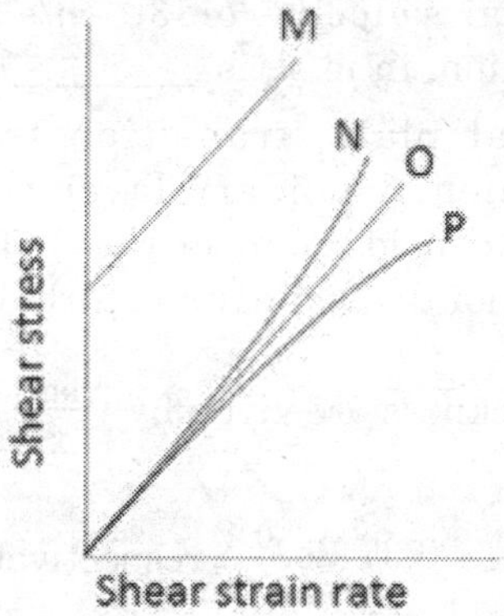

The most appropriate match between columns I and II is,

(*a*) (i) – O; (ii) – N ; (iii) – P; (iv) – M

(*b*) (i) – O; (ii) – P ; (iii) – N; (iv) – M

(*c*) (i) – P; (ii) – O ; (iii) – M; (iv) – N

(*d*) (i) – P; (ii) – O ; (iii) – N; (iv) – M

2. In a two-dimensional, incompressible and irrotational flow, stream function ($\Psi = \Psi(x, y)$) and velocity potential ($\phi = \phi(x, y)$) exist. The velocities in x and y directions are non-zero.

The product of $\left.\frac{dy}{dx}\right|_{\phi = \text{constant}}$ and $\left.\frac{dy}{dx}\right|_{\Psi = \text{constant}}$, is

(*a*) –1

(*b*) 0

(*c*) 1

(*d*) ∞

3. The inviscid flow past a rotating circular cylinder can be generated by the superposition of

(*a*) uniform flow, source and vortex

(*b*) uniform flow, doublet

(*c*) uniform flow, sink and vortex

(*d*) uniform flow, doublet and vortex Q.

4. The velocity field and the surface normal vector are given by, $\vec{V} = u\hat{i} + v\hat{j} + w\hat{k}$ and $\vec{n} = n_1\hat{i} + n_2\hat{j} + n_3\hat{k}$, respectively.

If Euler equations are to be solved, the boundary condition that must be satisfied at the wall is,

(*a*) $\vec{V}.\vec{n} = 0$ (*b*) $\vec{V} = 0$

(*c*) $\nabla . \vec{V} = 0$ (*d*) $\vec{V} \times \vec{n} = 0$

5. The influence of Froude number is most significant in

(*a*) capillary flows (*b*) creeping flows

(*c*) free surface flows (*d*) compressible flows

6. If the stream function ($\Psi(x, y)$) for a two-dimensional incompressible flow field is given as $2y(x^2 - y^2)$. the corresponding velocity field is

(*a*) $\vec{V} = 2\left(x^2 - 3y^2\right)\hat{i} + 4xy\,\hat{j}$

(*b*) $\vec{V} = 2\left(x^2 - 3y^2\right)\hat{i} - 4xy\,\hat{j}$

(*c*) $\vec{V} = 2\left(x^2 y\right)\hat{i} - 4xy\,\hat{j}$

(*d*) $\vec{V} = 2\left(x^2 y\right)\hat{i} + 4xy\,\hat{j}$

7. Water is flowing in two different tubes of diameters D and 2D, with the same velocity. The ratio of laminar friction factors for the larger diameter tube to the smaller diameter tube is

(*a*) 0.5 (*b*) 1.0

(*c*) 2.0 (*d*) 4.0

8. If the velocity field is $\vec{V} = xy^2\,\hat{i} + 4xy\,\hat{j}$ m/s, vorticity of the fluid element in the field at (x = 1, y = 2) in s^{-1} is ________.

9. A pitot-static tube is used to measure air velocity in a duct by neglecting losses. The density of air is 1.2 kg/m³. If the difference between the total and static pressures is 1 kPa, the velocity of air at the measuring location, in m/s, is _______.

Q. 10 – Q. 22 carry two marks each.

10. A parallelepiped of (2 m × 2 m) square cross-section and 10 m in length, is partially floating in water upto a depth of 1.2 m, with its longest side being horizontal. The specific gravity of the block is

(*a*) 0.8 (*b*) 0.6

(*c*) 0.5 (*d*) 0.4

11. The velocity field in a two-dimensional, unsteady flow is given by $\vec{V} = (x,y,t) = 2xy^2\,\hat{i} + 3xy\,\hat{j}$ m/s. The magnitude of acceleration of a fluid particle located at x = 1 m, y = 1 m at the time t = 1 s, in m/s², is

(a) 16.0 (b) 18.1
(c) 24.1 (d) 34.1

12. In a two-dimensional, incompressible and irrotational flow, fluid velocity (v) in the y direction is given by v = 2x – 5y. The velocity (u) in the x-direction is

(a) u = 2x – 5y (b) u = 2x + 5y
(c) u = 5x + 2y (d) u = 5x – 2y

13. A two-dimensional laminar viscous liquid film of constant thickness (h) steadily flows down an incline as shown in figure. Acceleration due to gravity is g. If the velocity profile in the liquid film is given as, u = ky(2h – y); v = 0, the value of constant k is

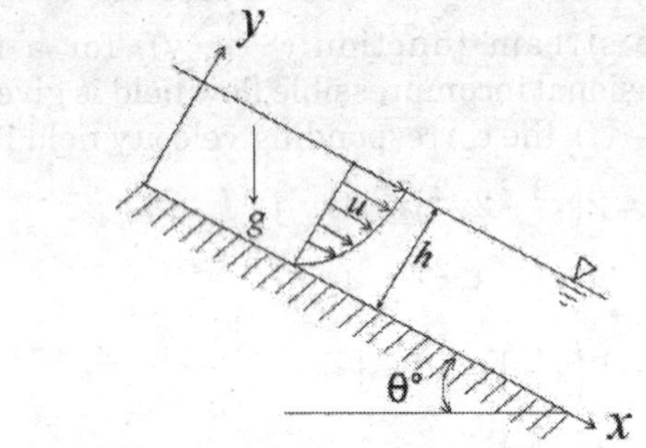

(a) $\frac{\rho g \sin\theta}{2}$ (b) $\frac{\rho g \cos\theta}{2}$
(c) ρg sin θ (d) ρg cos θ

14. A water jet of 100 mm diameter issuing out of a nozzle at a speed of 50 m/s strikes a vane and flows along it as shown in figure. The vane is attached to a cart which is moving at a constant speed of 20 m/s on a frictionless track. The jet is deflected at an angle of 30°. Take the density of water as 1000 kg/m³. Neglecting the friction between the vane and the fluid, the magnitude of the force exerted by water on the cart in the x-direction, in N, is ______.

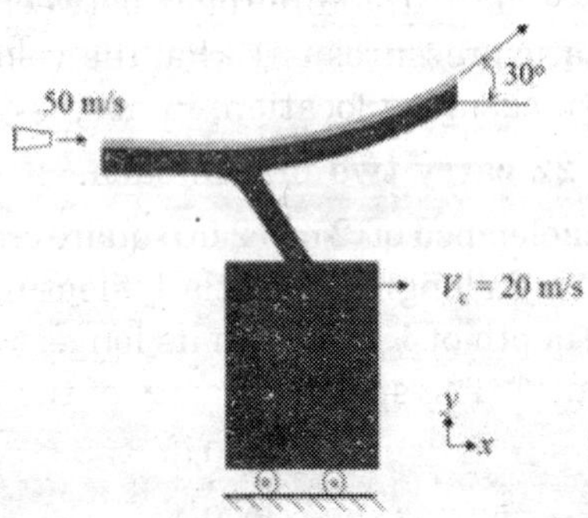

15. Capillary waves are generated in the sea. The speed of propagation (C) of these waves is known to be a function of density (ρ), wave length (λ), and surface tension (σ). Assume, ρ and λ to be constant. If the surface tension is doubled, in the functional form of the relevant non-dimensional group, the percentage increase in propagation speed (C) is ______ .

16. Consider a fully developed, two-dimensional and steady flow of a viscous fluid between two fixed parallel plates separated by a distance of 30 mm. The dynamic viscosity of the fluid is 0.01 kg/m-s and the pressure drop per unit length is 300 Pa/m. The fluid velocity at a distance of 10 mm from the bottom plate, in m/s, is ______.

17. A 2.6 gram smooth table-tennis (ping-pong) ball has a diameter of 38 mm. Density (ρ) of air is 1.2 kg/m³. Neglect the effect of gravity. Take coefficient of drag as 0.5. If the ball is struck with an initial velocity of 30 m/s, the initial deceleration, in m/s², is ______.

18. On a flat plate, transition from laminar to turbulent boundary layer occurred at a critical Reynolds number (Re_{cr}). The empirical relations for the laminar and turbulent boundary layer thickness are given by $\frac{\delta_{lam}}{x} = 5.48\,Re_x^{-0.5}$ and $\frac{\delta_{turb}}{x} = 0.37\,Re_x^{-0.2}$, respectively. The ratio of laminar to turbulent boundary layer thickness, at the location of transition, is 0.3. The value of Re_{cr} is ______.

19. In a capillary tube of radius R = 0.25 mm, a fully developed laminar velocity profile is defined as, $u = \frac{R^2}{4}\left(-\frac{dp}{dx}\right)\left(1-\frac{r^2}{R^2}\right)$. In this expression, $-\frac{dp}{dx} = 1$ MPa/m, is the dynamic viscosity of the fluid, and r is the radial position from the centerline of the tube. If the flow rate through the tube is 1000 mm³/s, the viscosity of the fluid, in Pa-s, is ______.

20. The skin friction coefficient for a turbulent pipe flow is defined as, $C_f = \frac{\tau_w}{1/2\rho V^2}$, where τ_w is the wall shear stress and V is the average flow velocity. The value of C_f is empirically given by the relation: $C_f = 0.065\left(2/Re\right)^{0.25}$, where Re is the Reynolds number. If the average flow velocity is 10 m/s, diameter of the pipe is 250 mm, kinematic viscosity of the fluid is 0.25 × 10⁻⁶ m²/s, and density of the fluid is 700 kg/m³, the skin friction drag induced by the flow over 1 m length of the pipe, in N, is ________.

21. A (150 mm × 150 mm) square pillar is located in a river with water flowing at a velocity of 2 m/s, as shown in figure. The height of the pillar in water is 8 m. Take density of water as 1000 kg/m³ and kinematic viscosity as 1×10^{-6} m/s². The coefficient of drag of the pillar is 2.0. The drag force exerted by water on the pillar in N is ________.

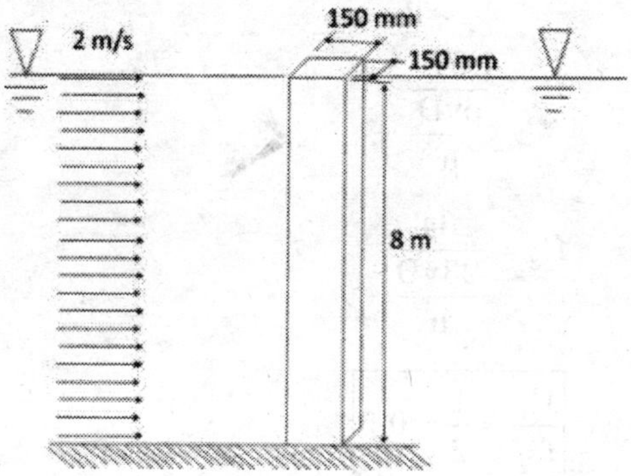

22. An orifice plate is used to measure flow rate of air (density = 1.23 kg/m³) in a duct of 250 mm diameter as shown in figure. The volume flow rate is 1 m³/s. Flow at sections 1 and 3 is uniform and section 2 is located at vena contracts. The diameter ratio, D_t/D_1, is 0.66. The flow area at vena contracts, $A_2 = 0.65A_t$, where A_t is area of the orifice. The pressure difference between locations 2 and 3 in N/m² is ________.

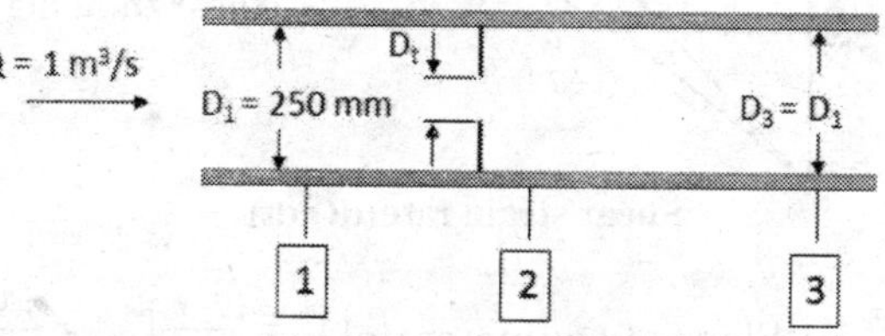

ANSWERS

1. (b)	**2.** (a)	**3.** (d)	**4.** (a)	**5.** (c)	**6.** (b)	**7.** (a)
8. 4	**9.** 40.82	**10.** (b)	**11.** (c)	**12.** (c)	**13.** (a)	**14.** 1466.25
15. 417	**16.** 2.25	**17.** 117.71	**18.** 440000 to 442000	**19.** 0.001	**20.** 37.64	**21.** 4800
22. 1200 to 1350						

EXPLANATIONS

1. (b)

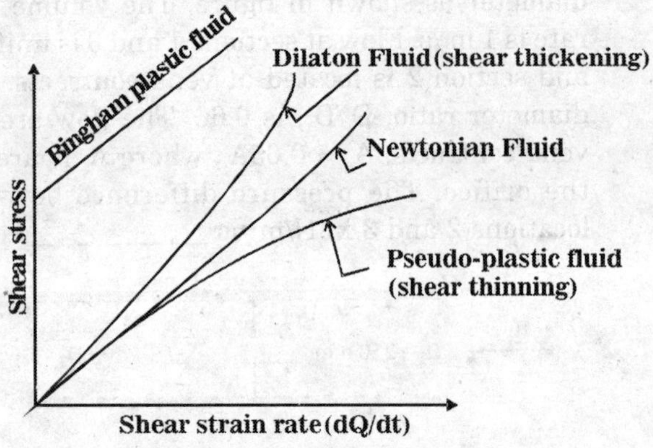

2. (a) Slope of equipotential line $\left(\frac{dy}{dx}\right)_{\phi=c} = \frac{-u}{v}$

Slope of stream function line $\left(\frac{dy}{dx}\right)_{\phi=c} = \frac{u}{v}$

$$\left(\frac{dy}{dx}\right)_{\phi=c} \times \left(\frac{dy}{dx}\right)_{\phi=c} = \frac{-u}{v} \times \frac{u}{v} = -1$$

3. (d) Inviscid flow past a rotating circular cylinder can be generated by the superposition of
   (i) Uniform flow.
   (ii) Doublet flow
   (iii) Vortex flow

4. (a) $\vec{v} = u\hat{i} + v\hat{j} + w\hat{k}$

$\vec{n} = n_1\hat{i} + n_2\hat{j} + n_3\hat{k}$

According Euler equation.

$\boxed{\vec{v}.\vec{n} = 0}$

For incompressible and steady flow

5. (c) Froude number (fr) is defined as the square root of inertia to gravity force.

$\boxed{Fr = \sqrt{\frac{F_i}{F_g}}}$

Froude number is applied in open channel flow where free surface is there.

6. (b) $\psi = f(x, y)$

$\psi = 2y(x^2 - y^2)$

$\psi = 2(x^2y - y^3)$

$$\psi u = \frac{-\partial\psi}{\partial y} = -\frac{\partial}{\partial y}\left[2\left(x^2y - y^3\right)\right]$$

$$u = -2\left(x^2 - 3y^2\right)$$

$$v = \frac{\partial\psi}{\partial x} = \frac{\partial}{\partial x}\left[2\left(x^2y - y^3\right)\right] = 2(2xy) = 4xy$$

$$\vec{v} = -2\left(x^2 - 3y^2\right)\hat{i} + 4xy\hat{j}$$

7. (a) laminar friction factor $(f') = \frac{64}{R_e}$

$$R_{e1} = \frac{\rho v D}{\mu}$$

$$R_{e2} = \frac{\rho v(2D)}{\mu}$$

$$f'_1 = \frac{64}{\frac{\rho v D}{\mu}}$$

$$f'_2 = \frac{64}{\frac{23vD}{\mu}}$$

$$\boxed{\frac{f'_1}{f'_2} = \frac{1}{2} = 0.5}$$

8. $\vec{v} = xy^2\hat{i} + 4xy\hat{j}$

(1, 2)

$u = xy^2$, $v = 4xy$

$$\omega_z = \frac{1}{2}\left[\frac{\partial v}{\partial x} - \frac{\partial u}{\partial y}\right]$$

$$\omega_z = \frac{1}{2}[4y - 2xy]$$

$$\omega_z = \frac{1}{2}[8 - 4] = 2$$

Vorticity(ξ) = $2.\omega_z$

$\boxed{\xi = 2 \times 2 = 4}$

9. $\rho a = 1.2$ kg/m3

$p_1 - p_2 = 1$KPa

$$h = \frac{p_1 - p_2}{\rho g} = \frac{1 \times 1000}{1.2 \times 9.81}$$

$= 84.95$ m

$$v = \sqrt{2gh}$$

$$v = \sqrt{2 \times 9.81 \times 84.95}$$

$\boxed{v = 40.82 \text{ m/sec}}$

10. (b) Specific gravity(G) = $\frac{2 \times 10 \times 1.2}{2 \times 10 \times 2}$

G = 0.6

11. (c) $\vec{v} = 2xy^2\hat{z} + 3xy + \hat{j}$

(1, 1), t = 1s

$a = \sqrt{a_x^2 + ay^2}$

$a_x = u.\frac{\partial u}{\partial x} + v\frac{\partial u}{\partial y}$

$a_y = u.\frac{\partial v}{\partial x} + v\frac{\partial v}{\partial y} + \frac{\partial v}{\partial t}$

$u = 2xy^2 = 2$

$v = 3xy\, t = 3$

$\frac{\partial u}{\partial x} = 2y^2 = 2$

$\frac{\partial u}{\partial y} = 4xy = 4$

$\frac{\partial v}{\partial x} = 3yt = 3$

$\frac{\partial v}{\partial y} = 3xt = 3$

$\frac{\partial v}{\partial t} = 3xy = 3$

ax = 16

ay = 18

$a = \sqrt{16^2 + 18^2}$

$a = 24.1\ m/s^2$

12. (c) $v = 2x - 5y$

velocity(u) in x-direction,

According to continuity equation,

$\frac{\partial u}{\partial x} + \frac{\partial u}{\partial y} = 0$

$\boxed{u = 5x + 2y}$

13. (a) $u = ky(2h - y)$

$u = 2kyh - ky^2$

$\frac{\partial u}{\partial h} = 2ky$

$f_D = mg \sin\theta$

$T.A = mg \sin\theta$

$u.2ky = \frac{mg}{v} y / \sin\theta$

$2\ uk = \rho g \sin\theta$

$\boxed{k = \frac{\rho g \sin\theta}{2u}}$

14. $d = 100\ mm = 0.1\ m$

$v = 50\ m/sec$

$u = 20\ m/sec$

$f_x = \rho A (v - u)^2 \sin^2\theta$

$f_x = 1000 \times \frac{\pi}{4} \times 0.1^2 \times (50 - 20)^2 \sin^2 30$

$\boxed{f_x = 1466.25\ N}$

15. Speed of propagation (C) = $\sqrt{\frac{g}{\lambda} + \frac{\sigma}{\rho}\lambda}$

$C_1 = \sqrt{\frac{g}{\lambda} + \frac{\sigma\lambda}{\rho}}$

$C_2 = \sqrt{\frac{g}{\lambda} + \frac{2\sigma\lambda}{\rho}}$

Percentage increase in propagation

$= \frac{\sqrt{2} - 1}{1} \times 100$

$= 417$

16. $\frac{P_1 - P_2}{L} = \frac{12\mu u}{B^2}$

[figure]

$u = \frac{(P_1 - P_2) \cdot B^2}{12\mu}$

$u = \frac{(300) \times 0.03^2}{12 \times 0.01}$

$\boxed{u = 2.25\ m/sec}$

17. Given that

$m = 2.6\ gm = 2.6 \times 10^{-3}\ kg$

$\rho_a = 1.2\ kg/m^3$

$C_D = 0.5,\ u = 30\ m/sec$

$d = 38\ mm = 0.038\ m$

$F_D = \frac{1}{2}\rho A u^2 C_D$

$F_D = \frac{1}{2} \times 1.2 \times \frac{\pi}{4} \times 0.038^2 \times (30)^2 \times 0.5$

$F_D = 0.306N$

$F_D = m.g.$

$0.306 = 2.6 \times 10^{-3} \times g$

$g = 117.71\ m/s^2$

19. Given that

$R = 0.25\ mm = 0.25 \times 10^{-3}\ m$

$-\frac{dp}{dx} = 1\ MPa/m = 10^6\ MPa/m$

$Q = 1000\ mm^3/sec = 10^{-6}\ m^3/sec$

Laminar flow through pipe,

$Q = \frac{-1}{8\mu} . \frac{\partial p}{\partial x} . R^4$

$$10^{-6} = \frac{1}{8\mu}.10^{6}.\left(0.25\times10^{-3}\right)^{4}$$

$$\boxed{\mu = 0.001\,\text{pa}\cdot\text{s}}$$

20. $C_f = \dfrac{\tau_w}{\frac{1}{2}\rho v^2}$

$$\tau_w = \frac{1}{2}C_f\,\rho v^2$$

$$C_f = 0.065\left(\frac{2}{Re}\right)^{0.25}$$

$$R_e = \frac{vD}{\nu} = \frac{10\times0.25}{0.25\times10^{-6}} = 10^{7}$$

$$C_f = 1.37\times10^{-3}$$

$$\tau_w = \frac{1}{2}\times1.37\times10^{-3}\times700\times10^{2}$$

$$\tau_w = 47.95\,\text{N/m}^2$$

$$f = \tau_w\times\pi dL = 47.95\times3.14\times0.25\times1$$

$$\boxed{f = 37.64\,\text{N}}$$

XE - C

MATERIAL SCIENCE

Syllabus

Section 1: Processing of Materials:

Powder synthesis, sintering, chemical methods, crystal growth techniques, zone refining, preparation of nanoparticles and thin films

Section 2: Characterisation Techniques:

X-ray diffraction, spectroscopic techniques like UV-vis, IR, Raman. Optical and Electron microscopy

Section 3: Structure and Imperfections:

Crystal symmetry, point groups, space groups, indices of planes, close packing in solids, bonding in materials, coordination and radius ratio concepts, point defects, dislocations, grain boundaries, surface energy and equilibrium shapes of crystals

Section 4: Thermodynamics and Kinetics:

Phase rule, phase diagrams, solid solutions, invariant reactions, lever rule, basic heat treatment of metals, solidification and phase transformations, Fick's laws of diffusion, mechanisms of diffusion, temperature dependence of diffusivity

Section 5: Properties of Materials:

Mechanical Properties: Stress-strain response of metallic, ceramic and polymer materials, yield strength, tensile strength and modulus of elasticity, toughness, plastic deformation, fatigue, creep and fracture Electronic Properties:Free electron theory, Fermi energy, density of states, elements of band theory, semiconductors, Hall effect, dielectric behaviour, piezo, ferro, pyroelectricmaterials Magnetic Properties: Origin of magnetism in metallic and ceramic materials, paramagnetism, diamagnetism, ferro and ferrimagnetism Thermal Properties: Specific heat, thermal conductivity and thermal expansion, thermoelectricity Optical Properties: Refractive index, absorption and transmission of electromagnetic radiation in solids, electrooptic and magnetoopticmaterials, spontaneous and stimulated emission, gas and solid state lasers

Section 6: Material types

Concept of amorphous, single crystals and polycrystalline materials, crystallinity and its effect on physical properties, metal, ceramic, polymers, classification of polymers, polymerization, structure and properties, additives for polymer products, processing and applications, effect of environment on materials, composites

Section 7: Environmental Degradation

Corrosion, oxidation and prevention

Section 8: Elements of Quantum Mechanics and Mathematics

Basics of quantum mechanics, quantum mechanical treatment of electrical, optical and thermal properties of materials, analytical solid geometry, differentiation and integration, differential equations, vectors and tensors, matrices, Fourier series, complex analysis, probability and statistics

Subject-wise Analysis

XE -C : MATERIALS SCIENCE

Year	1 mark Questions	2 marks Questions	Total Marks
2009	8	16	40
2010	9	13	35
2011	9	13	35
2012	9	13	35
2013	9	13	35
2014	9	13	35
2015	9	13	35
2016	9	13	35
2017	9	13	35
2018	9	13	35

2009

Useful Data

Avogadro's Number	: 6.023×10^{23} mol^{-1}
Boltzmann's constant	: 1.38×10^{-23} JK^{-1}
Electron Charge	: 1.6×10^{-19} C
Gas Constant	: 8.314 J mol^{-1} K^{-1}
Electron rest mass	: 9.1×10^{-31} kg
Permittivity of vacuum (ε_0)	: 8.854×10^{-12} F m^{-1}
Planck's constant (h)	: 6.62×10^{-34} J s^{-1}
Bohr Magneton (μ_B)	: 9.27×10^{-24} A–m^2

1 eV = 1.6×10^{-19} J

1 cal = 4.2 J

Atomic weight (in kg mol^{-1}) of:

Hydrogen	0.001
Carbon	0.012
Nitrogen	0.014

Q. 1 – 8 carry one mark each

1. Equal size spherical balls when packed together will yield maximum theoretical packing of

(*a*) ~ 52% (*b*) ~ 68%

(*c*) ~ 74% (*d*) ~ 86%

2. Steel containing 0.8% carbon cooled under equilibrium conditions from molten state to room temperature is soft, because it consists of lamellae of

(*a*) Ferrite and cementite

(*b*) Ferrite and austenite

(*c*) Ferrite and bainite

(*d*) Ferrite and martensite

3. Line broadening in X-ray diffraction pattern occurs on account of

(*a*) Coarse crystallite size

(*b*) Residual stresses

(*c*) Multiplicity of phases

(*d*) Coring of crystallites

4. Inter-granular corrosion of austenitic stainless steel is promoted by

(*a*) Fine grained microstructure

(*b*) Coarse grained microstructure

(*c*) Soaking steel at 700°C in air

(*d*) Quenching from 1000°C

5. Ferrites are preferred materials for use in high frequency applications (GHz range) as opposed to other ferromagnetic materials because ferrites also have

(*a*) High permeability

(*b*) High electrical resistivity

(*c*) High saturation magnetisation

(*d*) Low coercivity

6. During indirect intra-band transition, electrons undergo

(*a*) change in energy and momentum

(*b*) change in momentum but no change in energy

(*c*) change neither in energy nor in momentum

(*d*) change in energy but no change in momentum

7. A material has a band gap of 2.4 eV. Which of the following wavelengths of light will it absorb?

(*a*) 700 nm (*b*) 550 nm

(*c*) 650 nm (*d*) 400 nm

8. Thermal conductivity of a material at a temperature greater than Debye temperature

(*a*) is independent of temperature

(*b*) decreases inversely with temperature

(*c*) increases linearly with temperature

(*d*) increases exponentially with temperature

Q. 9 – 24 carry two marks each

9. Match the following classes of materials given in Column-I with the electron spin alignments in atoms shown in Column-II.

Column-I	**Column-II**
P. Ferromagnetic	1. ↑↓↑↓
Q. Anti-ferromagnetic	2. →↗↘↙↙←
R. Ferrimagnetic	3. ↑↑↑↑
S. Paramagnetic	4. ↓↓↓↓↓↓
	5. ↑↓↑↓

(*a*) P-3, Q-1, R-4, S-5

(*b*) P-4, Q-2, R-5, S-3

(*c*) P-3, Q-1, R-5, S-2

(*d*) P-3, Q-2, R-4, S-1

10. Match the following experimental techniques given in Column-I with applications given in Column-II.

Column I

P. Differential Scanning Calorimetry

Q. Atomic Absorption Spectroscopy

R. Scanning Electron Microscopy

S. Transmission Electron Microscopy

Column II

1. Dislocation studies

2. Surface Topography

3. Electrical Conductivity

4. Trace Element Analysis
5. Phase Transformation

(a) P-5, Q-4, R-2, S-1
(b) P-5, Q-1, R-3, S-2
(c) P-2, Q-5, R-3, S-1
(d) P-1, Q-5, R-4, S-2

11. Match the following materials given in Column-I with their applications given in Column-II.

Column I	Column II
P. Nylon	1. Electrical switch housing
Q. Urea formaldehyde	2. Conducting polymers
R. Pofyaniline	3. Heating Element
S. Alumina	4. Gears for toys
	5. Polishing material

(a) P-2, Q-4, R-3, S-5
(b) P-4, Q-1, R-2, S-5
(c) P-3, Q-4, R-2, S-1
(d) P-4, Q-5, R-3, S-2

12. Match the following materials given in Column-I with their applications given in Column-II.

Column I	Column II
P. Silicon carbide fibre	1. Fibre glass boat
Q. Polyester fibre	2. Heating element
R. Thoria doped tungsten	3. Magnetic material
S. Nichrome	4. Electric bulb filament
	5. Armour material

(a) P-5, Q-1, R-3, S-2
(b) P-1, Q-5, R-4, S-2
(c) P-5, Q-3, R-2, S-1
(d) P-5, Q-1, R-4, S-2

13. Correlate the material properties given in Column-I with the units given in Column-II.

Column I	Column II
P. Magnetic moment	1. $MN\ m^{-3/2}$
Q. Thermal conductivity	2. Hm^{-1}
R. Fracture toughness	3. $A - m^2$
S. Electron mobility	4. $m^2\ v^{-1}\ s^{-1}$
	5. $J\ s^{-1}\ m^{-1}\ K^{-1}$

(a) P-2, Q-5, R-1, S-4
(b) P-4, Q-5, R-1, S-3
(c) P-3,Q-5, R-1, S-4
(d) P-3, Q-2, R-4, S-1

14. If the spacing between two consecutive (110) planes in a BCC material is 0.203 nm, the lattice parameter and radius of the atom of the said material will be

(a) 0.242 nm and 0.110 nm
(b) 0.242 nm and 0.120 nm
(c) 0.287 nm and 0.134 nm
(d) 0.287 nm and 0.124 nm

15. A continuous and aligned carbon fibre reinforced composite is made up of 30 vol% carbon fibre having a modulus of elasticity of 300 GPa dispersed in a polymer matrix which on hardening has a modulus of elasticity of 4 GPa. What will be the modulus of elasticity of the composite in longitudinal and transverse directions of the carbon fibres respectively?

(a) 92.8 GPa and 5.7 GPa
(b) 211.0 GPa and 9.3 GPa
(c) 304.0 GPa and 7.5 GPa
(d) 92.8 GPa and 6.7 GPa

16. A potential of 10 volts is applied across a parallel plate capacitor which has a plate area of 10^{-4} m^2 and a plate separation of 2×10^{-3} m. If dielectric constant of the material placed between parallel plates is 10. The capacitance and the magnitude of the charge stored between the plates will be

(a) 4.425×10^{-13} F and 4.425×10^{-12} C
(b) 8.850×10^{-13} F and 8.850×10^{-12} C
(c) 4.425×10^{-12} F and 4.425×10^{-11} C
(d) 8.850×10^{-12} F and 8.850×10^{-11} C

17. Conductivity of Silicon at 300 K is 3.16×10^{-4} ohm^{-1} m^{-1} and that of Germanium is 2.12 ohm^{-1} m^{-1} at 300 K. At what temperature would the conductivity of intrinsic Silicon be the same as the conductivity of intrinsic Germanium at 300 K? (Given : E of Silicon at 300 K = 1.12 eV, E_g of Germanium at 300 K = 0.72 eV)

(a) ~ 506K (b) ~ 606 K
(c) ~ 726 K (d) ~ 816 K

18. Molecular weight distribution of a polystyrene polymer and the number fraction of polymer chains in the molecular weight range are given below

Range of Molecular weight (kg/mol)	Number fraction of polymer chain
5 – 10	0.05
10 – 15	0.15
15 – 20	0.20
20 – 25	0.30
25 – 30	0.20
30 – 35	0.08
35 – 40	0.02

The number average molecular weight and the number average degree of polymerization will be

(*a*) 15.750 kg/mol and 151

(*b*) 21.350 kg/mol and 203

(*c*) 15.750 kg/mol and 302

(*d*) 21.350 kg/mol and 205

Common Data Questions

Common Data for Questions 19 and 20:

Nickel has FCC structure and its lattice parameter is 0.353 nm. Weight of one mole of Nickel is 0.05871 kg.

19. The Ni- Ni nearest neighbour distance (in nm) is

(*a*) 0.173 (*b*) 0.223

(*c*) 0.250 (*d*) 0.273

20. Theoretical density of Nickel (in kg m^{-3}) is closer to

(*a*) 8700 (*b*) 8900

(*c*) 9100 (*d*) 9300

Common Data for Questions 21 and 22 :

The diffusivity of lithium in Silicon is 10^{-9} m^2 s^{-1} at 1400 K and 10^{-10} m^2 s^{-1} at 1000 K.

21. The value of activation energy (J mol^{-1}) of lithium diffusion in silicon is

(*a*) 66086 (*b*) 66986

(*c*) 67086 (*d*) 67986

22. The value of jump frequency factor of lithium in silicon in m^2 s^{-1} is

(*a*) 2.15×10^{-7} (*b*) 3.15×10^{-7}

(c) 3.15×10^{-8} (*d*) 2.15×10^{-8}

Linked Answer Questions

Statement for linked Answer Questions 23 - 24 :

Aluminum has a density of 2710 kg m^{-1} and weight of one mole of aluminum is 0.02698 kg. The collision time, τ, for electron scattering in Aluminum is 2×10^{-14} s at 300 K.

23. The number of free electrons per m^3 of Aluminum at 300 K is

(*a*) 6.05×10^{28} (*b*) 7.05×10^{28}

(*d*) 6.05×10^{27} (*d*) 7.05×10^{27}

24. The conductivity of aluminum (ohm^{-1} m^{-1}) at 300 K is

(*a*) 3.40×10^6 (*b*) 4.40×10^6

(c) 3.40×10^7 (*d*) 4.40×10^7

ANSWERS

1. (c)	**2.** (a)	**3.** (a)	**4.** (c)	**5.** (c)	**6.** (a)	**7.** (b)
8. (c)	**9.** (c)	**10.** (a)	**11.** (b)	**12.** (d)	**13.** (c)	**14.** (d)
15. (a)	**16.** (c)	**17.** (a)	**18.** (d)	**19.** (c)	**20.** (b)	**21.** (c)
22. (b)	**23.** (a)	**24.** (a)				

EXPLANATIONS

1. (c) For equal spheres in the three dimensions the densest packing uses approximately 74% of the volume.

2. (a) At about 900°C, this steel has austenite microstructure . When steel is slowly cooled, the transformation would start on reaching the point 'j'. At this point, the alloy enters a two-phase field of ferrite and austenite. On reaching the point, ferrite starts nucleating around the grain boundaries of austenite as shown in Fig. 1(a). By slowly cooling to point 'k', the ferrite grains grow in size and diffusion of carbon takes place from ferrite regions into the austenite regions as shown in Fig. 1 (b), since ferrite cannot retain carbon above 0.002% at room temperature.

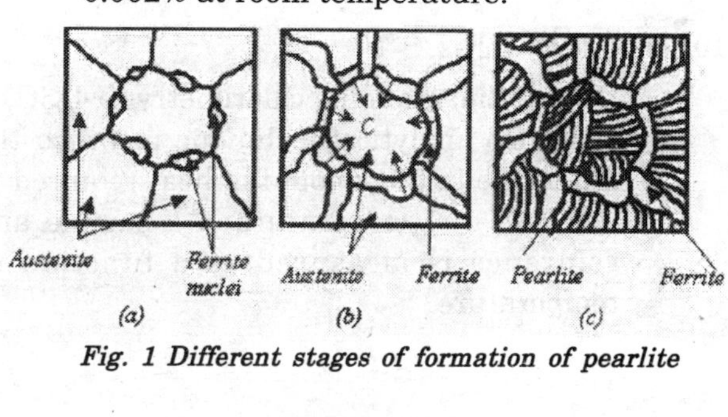

Fig. 1 Different stages of formation of pearlite

At this point it is seen that a netork of ferrite crystals surrounds each austenite grain. On slow cooling to point 'l' the remaining austenite gets transformed into 'pearlite' as shown in Fig. 1(c). Pearlite is a lamellar mixture of ferrite and cementite.

3. (a) Line broadening in X-ray diffraction pattern occurs on account of Coarse crystallite size.

4. (c) Inter-granular corrosion of austenitic stainless steel is promoted by soaking steel at 700°C in air.

5. (c) Ferrites are preferred materials for use in high frequency applications (GHz range) as opposed to other ferromagnetic materials because ferrites also have High saturation magnetization.

6. (a) Both energy and momentum change.

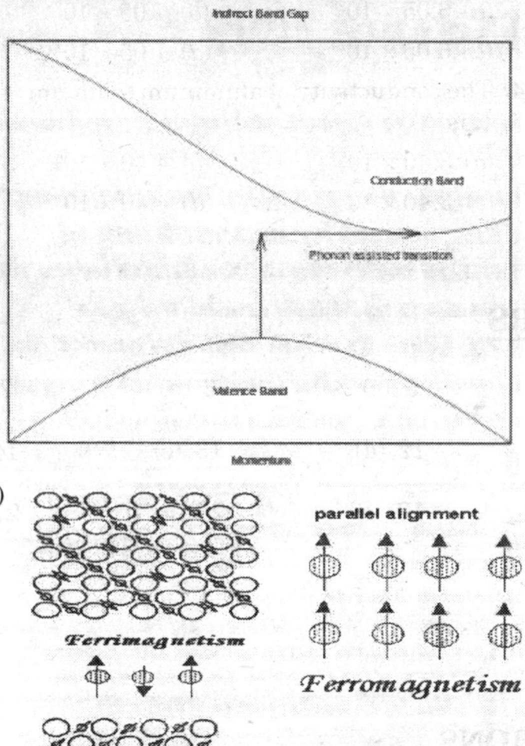

9. (c)

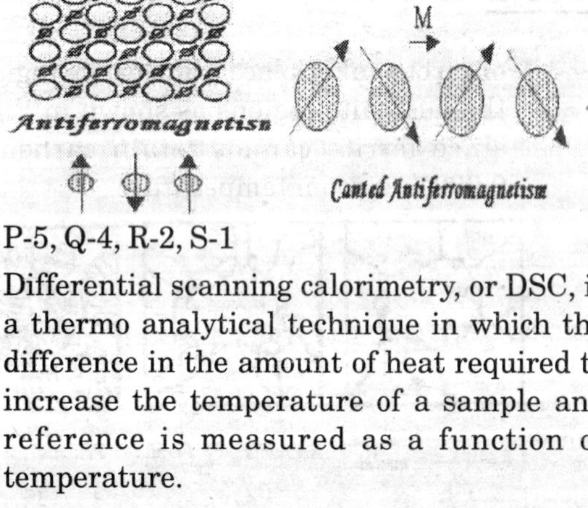

10. (a) P-5, Q-4, R-2, S-1

Differential scanning calorimetry, or DSC, is a thermo analytical technique in which the difference in the amount of heat required to increase the temperature of a sample and reference is measured as a function of temperature.

Atomic absorption spectroscopy (AAS) is a spectro analytical procedure for the quantitative determination of chemical elements using the absorption of optical radiation (light) by free atoms in the gaseous state.

Scanning Electron Microscopy (SEM): Accelerated electrons in an SEM carry significant amounts of kinetic energy, and this energy is dissipated as a variety of signals produced by electron-sample interactions when the incident electrons are decelerated in the solid sample. These signals include secondary electrons (that produce SEM images), backscattered electrons (BSE), diffracted backscattered electrons (EBSD that are used to determine crystal structures and orientations of minerals), photons (characteristic X-rays that are used for elemental analysis and continuum X-rays), visible light (cathodoluminescence-CL), and heat.

Transmission electron microscopy (TEM, also sometimes **conventional transmission electron microscopy** *or* **CTEM**) is a microscopy technique in which a beam of electrons is transmitted through a specimen to form an image. The specimen is most often an ultrathin section less than 100 nm thick or a suspension on a grid. An image is formed from the interaction of the electrons with the sample as the beam is transmitted through the specimen.

11. (b) P-4, Q-1, R-2, S-5

Conductive polymers or, more precisely, intrinsically **conducting polymers** (ICPs) are organicopolymers that conduct electricity. Such compounds may have metallic conductivity or can be semiconductors. The biggest advantage of conductive polymers is their processability, mainly by dispersion.

Plastic Gears – Lightweight and Rust-proof. Materials used in plastic gears are, in general, engineering plastics such as polyacetal (POM) and MC Nylon which is essentially polyamid resin.

Alumina is used for polishing temperature.

12. (d) P-5, Q-1, R-4, S-2

P. Silicon carbide fibre–

5. Armour material

Q. Polyester fibre–

1. Fibre glass boat

R. Thoria doped tungsten–
4. Electric bulb filament
S. Nichrome–
2. Heating element

14. (d) Let distance between two consecutive plane be x. Lattice paramater is 'a'.

$$a = x \times \sqrt{h^2 + k^2 + \ell^2}$$

$$a = x \times \sqrt{1^2 + 1^2 + 0^2}$$

$$a = 0.203 \times \sqrt{2}$$

$$a = 0.287 \text{ nm}$$

Relation between lattice parameter & radius of the atom is given by

$$\sqrt{3}a = 4R$$

$$R = \frac{\sqrt{3}a}{4} = \frac{\sqrt{3} \times 0.287}{4}$$

$$R = 0.124 \text{ nm}$$

15. (a) For longitudinal direction,

$$E_C = V_m E_m + V_f E_f$$

$$E_C = 0.7 \times 4 + 0.3 \times 300 = 92.8 \text{ GPa}$$

For transverse direction,

$$\frac{1}{E_C} = \frac{V_m}{E_m} + \frac{V_f}{E_f}$$

$$E_C = 5.7 \text{ GPa}$$

16. (c) $C = \dfrac{K \epsilon_0 A}{d}$

$\epsilon_0 = 8.854 \times 10^{-12}$ F/m

A = Area of the plate, m^2

d = Separation between the plates, m

K = dielectric constant

$$C = \frac{10 \times 8.85 \times 10^{-12} \times 10^{-4}}{2 \times 10^{-3}}$$

$$C = 4.427 \times 10^{-12} \text{ F}$$

18. (d) Number average molecular weight = 7.5 × 0.5 + 12.5 × 0.15 + 17.5 × 0.2 + 22.5 × 0.3 + 27.5 × 0.2 + 32.5 × 0.08 + 37.5 × 0.02

= 21.35 kg/mol

Average degree of polymerization

$$= \frac{21350}{104} = 205$$

19. (c) $\sqrt{2}a = 2d$

$$d = \frac{\sqrt{2}a}{2}$$

$$= \frac{\sqrt{2} \times 0.353}{2} = 0.249 \text{ nm}$$

20. (b) $\rho = \dfrac{Z \times M}{a^3 + N_A}$

$$\rho = \frac{4 \times 59}{\left(0.353 \times 10^{-9}\right)^3 \times 6.023 \times 10^{23}}$$

$= 8900$ kg/m^3

Z = number of atoms per unit cell

Z = 1 – SCC

= 2 – BCC

= 4 – FCC

ρ = Density (kg/m^2)

a = lattice parameter

21. (c) $D = D_0 e^{-\left(\frac{Q}{1400R}\right)}$ ...(1)

$10^{-9} = D_0 e^{-\left(\frac{Q}{1000R}\right)}$ ...(2)

$$10 = e^{\frac{Q}{3500R}}$$

$$2.3025 = \frac{Q}{3500R} \Rightarrow \frac{Q}{R} = 8059.05$$

& $D_0 = 3.162 \times 10^{-7}$ m^2/sec.

$$Q = R \times 8059.05$$

$$= 8.314 \times 8059.05 \ \frac{J}{mol}$$

= 6708.6 J/mol

22. (b) $3.162 \times 10^{-7} m^2 / sec$ is already solved in question no. 21.

2010

Useful Data

Avogadro's number	: 6.023×10^{23} mol^{-1}
Boltzmann's constant (k_B)	: 1.38×10^{-23} J K^{-1}
Electron change (e)	: 1.602×10^{-19} C
Gas Constant	: 8.314 J mol^{-1}K^{-1}
Electron rest mass	: 9.1×10^{-31} kg
Permittivity of vacuum (ε_0)	: 8.854×10^{-12} F m^{-1}
Planck's constant (h)	: 6.626×10^{-34} J s^{-1}
Bohr magneton (μ_B)	: 9.27×10^{-24} Am^2
Free space permeability (μ_0)	: $4\pi \times 10^{-7}$ H m^{-1}

1 J = 6.24×10^{18} eV

1 eV = 1.602×10^{-19} J

1 cal = 4.2 J

Q. (1 – 9) carry one mark each.

1. The number of lattice points in an ideal Perovskite unit cell is

(a) 1 (b) 2

(c) 4 (d) 5

2. A Frenkel defect is

(a) a pair of cation and anion vacancy

(b) a pair of cation interstitial and cation vacancy

(c) a cation vacancy

(d) an anion vacancy

3. The angle between the line vector of a screw dislocation and the Burgers vector is

(a) 0 degree (b) 45 degrees

(c) 60 degrees (d) 90 degrees

4. The addition of a network modifier to silica

(a) produces vacancies

(b) enhances the network structure

(c) disrupts the network structure

(d) increases the viscosity

5. The best semiconductor material for LED in the visible range is

(a) Si (b) Ge

(c) GaAs (d) $GaAs_{0.6}P_{0.4}$

6. A plain carbon steel sample is water-quenched from 900°C to room temperature. Its microstructure will consist of

(a) pearlite (b) bainite

(c) martensite (d) ferrite and pearlite

7. Graphite at zero Kelvin is a

(a) good conductor (b) insulator

(c) semiconductor (d) semi-metal

8. A high molecular weight polyethylene has an average molecular weight of 560,000 g/mol. Its average degree of polymerization is

(a) 15,000 (b) 18,660

(c) 19,310 (d) 20,000

9. In which region of the spectra crystal lattice absorption is very significant

(a) ultraviolet (b) visible

(c) microwave (d) infrared

Q. (10 – 22) carry two marks each.

10. Match the properties in **Column I** with appropriate units in **Column II**

Column I	**Column II**
P. viscosity	1. m^{2}s^{-1}
Q. diffusivity	2. Kg mm^{-2}
R. charge mobility	3. Nm^{-2} s
S. fracture toughness	4. m^2 V^{-1} s^{-1}
	5. MPa $\sqrt{m}$

(a) P-3, Q-4, R-1, S-2

(b) P-4, Q-1, R-2, S-5

(c) P-5, Q-4, R-1, S-2

(d) P-3, Q-1, R-4, S-5

11. Match the terms in **Column I** with the details of phase transformations in **Column II** (→ indicates cooling)

Column I	**Column II**
P. eutectic	1. $L + \alpha \rightarrow \beta$
Q. monotectic	2. $\gamma \rightarrow \alpha + \beta$
R. eutectoid	3. $L \rightarrow \alpha + \beta$
S. peritectic	4. $\alpha + \beta \rightarrow \gamma$
	5. $L1 \rightarrow \alpha + L2$

(a) P-1, Q-5, R-4, S-3

(b) P-3, Q-4, R-2, S-1

(c) P-3, Q-5, R-2, S-1

(d) P-5, Q-2, R-4, S-1

12. Match the following materials in **Column I** with appropriate preparation technique given in **Column II**

Column I	**Column II**
P. single crystals of laser materials	1. sol-gel
Q. highly dense fine grained ceramics	2. melt spinning
R. nanocrystalline oxide powders	3. Bridgman-Stockbarger
S. metallic glasses	4. hot pressing
	5. Czochralski

(a) P-5, Q-4, R-1, S-2 (b) P-3, Q-5, R-2, S-1

(c) P-2, Q-1, R-4, S-5 (d) P-5, Q-2, R-1, S-4

13. Match the statements given in **Column I** with the most suitable material given in **Column II**

Column I	Column II
P. biocompatible ceramic material	1. zinc
Q. magnetic material with very high B-H product	2. titanium
R. nonstick coating on aluminum	3. $Nd_2Fe_{14}B$
S. sacrificial coating on steel	4. $Ca_{10}(PO_4)_6(OH)_2$
	5. $BaFe_{12}O_{19}$
	6. polytetr afluo-roethylene
	7. polyethylene terephthalate

(*a*) P-4, Q-3, R-7, S-2 (*b*) P-2, Q-5, R-6, S-1
(*c*) P-4, Q-3, R-6, S-1 (*d*) P-6, Q-5, R-7, S-6

14. A 99% pure copper wire has resistance of 0.1 Ω at 0.1 K, and 20 Ω at 300 K. For 100% pure, perfect copper wire of the same size, the estimated resistance at 0.1 K and 300 K is

(*a*) ~ zero and 19.9 Ω (*b*) ~ zero and 20.1 Ω
(*c*) 0.1 and 19.9 Ω (*d*) 0.1 and 20.1 Ω

15. A 12.0 mm diameter aluminum alloy test bar is subjected to a load of 110 kN. If the diameter of the bar is 10.5 mm at this load, the true strain will be

(*a*) 0.134 (*b*) 0.306
(*c*) 0.267 (*d*) 0.767

16. If the effective magnetic moment of Fe^{3+} is equal to 5 μ_B, the magnetic moment in μ_B per formula of γ - Fe_2O_3 (which is inverse spinel with cation defects on the octahedral site) is

(*a*) zero (*b*) 2.5
(*c*) 5 (*d*) 10

Common Data Questions

Common Data for Q. (17 - 18):

A unidirectional carbon fiber epoxy matrix composite contains 60 vol% carbon fibers. The density of carbon fiber is 1790 kg/m^3 and that of the epoxy matrix is 1200 kg/m^3. The tensile moduli of the carbon fiber and the epoxy matrix are 340 GPa and 4.50 GPa respectively.

17. The density of the composite in the units of kg/m^3 is

(*a*) 1495 (*b*) 1554
(*c*) 1672 (*d*) 1790

18. The tensile modulus of elasticity of the composite under iso-strain condition is

(*a*) 5.5 GPa (*b*) 11.0 GPa
(*c*) 102.9 GPa (*d*) 205.8 GPa

Common Data for Q. (19 - 20):

For a type II superconductor (at 4 K), the lower critical field (B_{Cl}) and thermodynamic critical field (B_C) are respectively 0.001 Tesla and 0.10 Tesla.

19. The upper critical field (B_{C2}) in Tesla is

(*a*) 0.10 (*b*) 0.33
(*c*) 1.00 (*d*) 10.00

20. The maximum energy that can be stored per unit volume (Jm^{-3}) in the superconductor is

(*a*) 3.979×10^3 (*b*) 50.00×10^3
(*c*) 7.96×10^3 (*d*) 1.326×10^3

Linked Answer Questions:

Statement for linked Answer Q. (21 - 22):

For Cu metal, the conduction electron density, $n = 8.45 \times 10^{28}$ m^{-3}.

21. The energy of the electrons at the Fermi level (E_F) is

(*a*) 3.50 eV (*b*) 7.028 eV
(*c*) 8.45 eV (*d*) 49.0 eV

22. The density of states (DOS), for 1 cm^3 of Cu, at Fermi level per meV is

(*a*) 1.20×10^{19} (*b*) 1.80×10^{19}
(*c*) 1.20×10^{22} (*d*) 1.81×10^{22}

ANSWERS

1. (d)	**2.** (b)	**3.** (a)	**4.** (b)	**5.** (d)	**6.** (c)	**7.** (b)
8. (d)	**9.** (a)	**10.** (d)	**11.** (d)	**12.** (a)	**13.** (a)	**14.** (d)
15. (a)	**16.** (b)	**17.** (b)	**18.** (d)	**19.** (a)	**20.** (a)	**21.** (b)
22. (a)						

EXPLANATIONS

1. (d) It's got $\frac{1}{8}$ of a calcium atom at each of the 8 corner positions, 1 whole titanium atom at the body-center position, and $\frac{1}{2}$ of an oxygen atom at each of the 6 face positions.

So, it's not quite FCC and not quite BCC. The structure is known as "the perovskite structure".

2. (b) Frenkel defect is the defect in which a pair of cation interstitial and cation vacancy exists.

3. (a) b is the angle between the line direction of the dislocation and the Burgers vector. b is 0 degree for screw dislocation and b is 90 degree for edge dislocation.

4. (b) Network Modifiers affect the properties of a pure SiO_2 glass as follows:

A. Thermal Expansion Increases

B. Hardness Lowers

C. Chemical Durability Lowers

D. Density Increases

E. Surface Tension

F. Refractive Index Increases

G. Enhances network structure

5. (d) The best semiconductor material for LED in the visible range is $GaAs_{0.6}P_{0.4}$

Types of Light Emitting Diode

→ Gallium Arsenide (GaAs)–infra-red

→ Gallium Arsenide Phosphide (GaAsP)–red to infra-red, orange

→ Aluminium Gallium Arsenide Phosphide (AlGaAsP) – high-brightness red, orange-red, orange, and yellow

→ Gallium Phosphide (GaP)–red, yellow and green

→ Aluminium Gallium Phosphide (AlGaP)–green

→ Gallium Nitride (GaN)–green, emerald green

→ Gallium Indium Nitride (GaInN)–near ultraviolet, bluish-green and blue

→ Silicon Carbide (SiC)–blue as a substrate

→ Zinc Selenide (ZnSe)–blue

→ Aluminium Gallium Nitride (AlGaN) – ultraviolet

6. (c) A plain carbon steel sample is water- quenched from 900°C to room temperature. As microstructure will consist of Martensite. Structure will become hard.

7. (b) Graphite at zero Kelvin is an Insulator. As temperature increases, conductivity increases.

8. (d) Average degree of polymerization = (average molecular weight)/(Molecular weight of Polyethylene)

Average degree of polymerization =560000/28 = 20000

9. (a) Crystal lattice absorption happens at the scale of the crystal lattice, so the wavelength of the light should be of the same scale. Infrared, microwave and visible light have wavelengths that are too large to be absorbed by the lattice. Only UV has a small enough wavelength such that the length scales of the crystal lattice will become quite significant.

10. (d) 1. Viscosity has an SI unit of Ns/m^2

2. Diffusivity has an SI unit of m^2/s ($length^2$ / time)

3. Charge mobility has an SI unit of $m^2V^{-1}s^{-1}$

4. Fracture toughness has an SI unit of $MP\sqrt{m}$

11. (d) P-5, Q-2, R-4, S-1

Eutectoid Reaction

The *eutectoid* (eutectic-like) reaction is similar to the eutectic reaction but occurs from one solid phase to two *new* solid phases. It also shows as V on top of a horizontal line in the phase diagram. There are associated eutectoid temperature (or temperature), eutectoid phase, eutectoid and proeutectoid microstructures.

Solid Phase 1 → Solid Phase 2 + Solid Phase 3

Peritectic Reaction

The *peritectic* reaction also involves three solid in equilibrium, the transition is from a solid + liquid phase to a *different* solid phase when cooling. The inverse reaction occurs when heating.

Solid Phase 1 + liquid → Solid Phase 2

Eutectic Reaction

$$\text{Liquid} \xrightarrow[\text{cooling}]{\text{eutectic temperature}} \alpha \text{ solid solution} + \beta \text{ solid solution}$$

Monotectic Reaction

Liquid phase TRANSFORMS into solid phase and another liquid.

$$L_1 \xrightarrow{\text{cooling}} \alpha + \beta$$

12. (a) P-5, Q- 4, R- 1, S- 2

The Bridgman–Stockbarger technique is named after Harvard physicist Percy Williams Bridgman (1882-1961) and MIT physicist Donald C. Stockbarger (1895–1952). The technique includes two similar but distinct methods primarily used for growing boules (single crystal ingots), but which can be used for solidifying polycrystalline ingots as well.

The *Czochralski* process is a method of crystal growth used to obtain single crystals of semiconductors (e.g. silicon, germanium and gallium arsenide), metals (e.g. palladium, platinum, silver, gold), salts and synthetic gemstones.

Hot pressing is a high-pressure, low-strain-rate powder metallurgy process for forming of a powder or powder compact at a temperature high enough to induce sintering and creep processes. This is achieved by the simultaneous application of heat and pressure.

Melt spinning is a technique used for rapid cooling of liquids. A wheel is cooled internally, usually by water or liquid nitrogen, and rotated. A thin stream of liquid is then dripped onto the wheel and cooled, causing rapid solidification.

13. (a) P-4, Q- 3, R- 7, S-2

Biocompatible ceramic material is $Ca_{10}(PO_4)_6(0H)_2$

Sacrificial coating is done with the help of Titanium.

Magnetic materia] with very high B-H product is $Nd_2Fe_{14}B$

15. (a) True strain $= \ln\left(\frac{A_0}{A}\right)$

A_0 = Cross sectional area of the specimen

A = Cross sectional area of deformed specimen

$$\text{True strain, } \varepsilon_t = \ln\left(\frac{A_0}{A}\right)$$

$$\text{True strain, } \varepsilon_t = \ln\left(\frac{12}{10.5}\right)$$

True Strain = .134

17. (b) $\rho_c = \rho_F.V_F + \rho_M.V_M$

$$\rho_c = 1790 \times 0.6 + 1200 \times 0.4$$

$$\rho_c = 1554 \frac{kg}{m^3}$$

18. (d) Under isostrain conditions,

$$E_C V_C = E_F V_F + E_M V_M$$

or

$$\frac{\sigma_c V_c}{\varepsilon_c} = \frac{\sigma_F V_F}{\varepsilon_F} + \frac{\sigma_M V_M}{\varepsilon_M}$$

E_c = 340 × 0.6 + 4.5 × 0.4 = 205.8 GPa

21. (b) $E_F = \left(\frac{3\pi_2 n}{v}\right)^{\frac{2}{3}} \cdot \frac{h^2}{2m}$

m = 63.546 a.m.u

h = 6.626 × 10^{-34} J/s

n = 8.45 × 10^{28} m^{-3}

$$V = \sqrt{\frac{2E_F}{m}}$$

By substituting all values, we get

E_F = 7.028 eV

22. (a) Density of states (DOS) is Z(E).

$$Z(E) = \frac{2E_F.m}{4\pi^2}\left(\frac{1}{h^2}\right)^{\frac{3}{2}}$$

$$= \frac{2 \times 7.028 \times 63.546}{1.6 \times 10^{19} \times 4\pi^2 \times 10^6}\left(\frac{1}{6.626 \times 10^{-34}}\right)^{\frac{3}{2}}$$

per meter

2011

Useful Data

Boltzmann's constant : 1.38×10^{-23} J K^{-1}

Electron charge : 1.602×10^{-19} C

Gas constant : 8.314 $Jmol^{-1}$ K^{-1}

1 J = 6.242×10^{18} eV

Atomic weight (kg mol^{-1})

Hydrogen	0.0010
Carbon	0.0120
Chlorine	0.0355

Q.(1-9) carry one mark each.

1. Which one of the following pairs of crystal structures can have the same packing fraction of 0.74?

(*a*) FCC and BCC (*b*) HCP and BCC

(*c*) FCC and HCP (*d*) BCC and BCT

2. Which one of the following is NOT CORRECT?

(*a*) An edge dislocation can cross slip

(*b*) An edge dislocation can glide

(*c*) A screw dislocation can cross slip

(*d*) An edge dislocation can climb

3. Which one of the following is NOT CORRECT?

(*a*) Working of lead at 25°C is hot working

(*b*) Working of tungsten at 1000°C is hot working

(*c*) Working of lead at –100°C is cold working

(*d*) Working of tungsten at 25°C is cold working

4. Which one of the following is NOT a ceramic?

(*a*) SiC (*b*) MgO

(*c*) TiB_2 (*d*) TiAl

5. If the average of polymerisation of a polyvinyl chloride (PVC) polymer is 2000, then its average molecular weight (in g mol^{-1}) is

(*a*) 125000 (*b*) 119000

(*c*) 56000 (*d*) 2000

6. Which one of the following materials has the lowest coefficient of thermal expansion?

(*a*) Superalloy (*b*) Super Invar

(*c*) Spinel (*d*) α-brass

7. The colour of a metal is determined by the wavelength distribution of the radiation that is

(*a*) diffracted (*b*) transmitted

(*c*) reflected (*d*) refracted

8. Nickel ferrite is

(*a*) antiferromagnetic (*b*) ferromagnetic

(*c*) diamagnetic (*d*) ferrimagnetic

9. The oxide scale responsible for the excellent corrosion resistance of stainless steels is

(*a*) Cr_2O_3 (*b*) NiO

(*c*) Fe_2O_3 (*d*) Al_2O_3

Q.(10-22) carry two marks each.

10. A plain carbon steel was annealed just above the eutectoid temperature. Microstructural analysis revealed that the proeutectoid ferrite content was 30 wt%. The eutectoid reaction in the iron-iron carbide phase diagram is given below :

$$\gamma(0.76\text{ wt\% C}) \underset{\text{heating}}{\overset{\text{cooling}}{\rightleftharpoons}} \alpha\,(0.022\text{wt\%C}) + Fe_3C(6.7\text{ wt\% C})$$

The carbon content of the steel (in wt%) is

(*a*) 0.24 (*b*) 0.34

(*c*) 0.44 (*d*) 0.54

11. Match the materials in Column-I with the descriptions in Column-II.

Column-I	Column-II
P. Zirconia	1. Ultra-hard material
Q. Cubic boron nitride	2. High temperature superconductor
R. Hafnium carbide	3. Transformation toughening
S. Yttrium aluminium garnet	4. Ultra-high temperature material
	5. Host material for laser
	6. Micro-crack toughening

	P	Q	R	S
(*a*)	3	4	1	2
(*b*)	6	1	4	2
(*c*)	3	1	4	5
(*d*)	4	6	1	5

12. Match the materials in Column-I with the descriptions in Column-II.

Column-I	Column-II
P. Polyacrylonitrile	1. Hard and brittle material
Q. Nylon-6,6	2. Very high temperature resistant polymer
R. Polytetrafluoroethylene (PTFE)	3. H-bonding
S. Ebonite	4. Acrylic fibre
	5. Rubber
	6. Polyester fibre

	P	Q	R	S
(a)	6	3	2	1
(b)	2	6	4	5
(c)	4	2	6	5
(d)	4	3	2	1

13. Match the techniques is column-I with the descriptions in Column-II.

Column-I	Column-II
P. Differential scanning calorimetry	1. Residual stress measurement
Q. Atomic force microscopy	2. Surface morphology of a material
R. Scanning electron microscopy	3. Incident beam passes through a thin sample
S. X-ray diffraction	4. Thermal expansion measurement
	5. Resolution less than 1 nm is possible
	6. Measurement of enthalpy change

	P	Q	R	S
(a)	6	5	2	1
(b)	4	5	2	1
(c)	4	1	3	2
(d)	6	1	5	3

14. Match the properties in Column-I with the appropriate units in Column-II.

Column-I	Column-II
P. Thermal conductivity	1. Hm^{-1}
Q. Dielectric strength	2. $Wb\ m^{-2}$
R. Magnetic permeability	3. $V\ m^{-1}\ K^{-1}$
S. Capacitance	4. $V\ m^{-1}$
	5. $C\ V^{-1}$
	6. $J\ mol^{-1}\ K^{-1}$

	P	Q	R	S
(a)	6	4	2	5
(b)	3	5	1	4
(c)	3	4	1	5
(d)	6	5	1	4

15. It takes 4 h for carburising a steel at 900°C. If the same carburising is to be accomplished in 2 h, what should be the temperature? The activation energy of diffusion of carbon in the steel is 151 kJ mol^{-1}.

(a) 850°C (b) 955°C

(c) 1015°C (d) 1228°C

16. A steel specimen (12 mm diameter and 60 mm length) undergoes elastic deformation under tension. The deformed specimen experiences a longitudinal strain of 0.001. If the Poisson's ratio is 0.3, the diameter of the deformed specimen (in mm) is

(a) 12.0120 (b) 12.0036

(c) 11.9964 (d) 11.9880

Common Data Questions

Common Data for Questions 17 and 18 :

The first peak in the powder X-ray diffraction pattern of an FCC metal appears at a Bragg angle of 19.2°. The wavelength of Cu-K_α radiation used is 0.154 nm.

17. The lattice parameter of the metal (in nm) is

(a) 0.4505 (b) 0.4055

(c) 0.3505 (d) 0.3055

18. The full width at half maximum (FWHM) of the first peak is 0.35°. Ignoring micro-strain and instrumental broadening, the crystallite size of the sample (in nm) is

(a) 20 (b) 24

(c) 200 (d) 240

Common Data for Questions 19 and 20 :

For an intrinsic semiconductor, the mobilities of free electrons and holes are 0.14 $m^2V^{-1}s^{-1}$ and 0.038 $m^2 V^{-1}s^{-1}$, respectively. Its bandgap is 1.107 eV and electrical conductivity at 300 K is $3.99 \times 10^{-4}\ \Omega^{-1}\ m^{-1}$.

19. The free electron concentration (in m^{-3}) at 300 K is

(a) 13.99×10^{15} (b) 27.98×10^{15}

(c) 13.99×10^{17} (d) 27.98×10^{17}

20. What is the temperature at which the conductivity of the semiconductor is 0.399 $\Omega^{-1}m^{-1}$?

(a) 343 K (b) 443 K

(c) 493 K (d) 543 K

LINKED ANSWER QUESTIONS

Statement for Linked Answer Questions 21 & 22 :

A continuous and aligned glass fibre reinforced composite has a modulus of elasticity of 150 GPa in the longitudinal direction. The matrix is a polyester resin with a modulus of 4.5 GPa. The glass fibre has a modulus of 340 GPa.

21. The volume fraction of the glass fibres is

(a) 0.398 (b) 0.434

(c) 0.497 (d) 0.566

22. If the cross-sectional area of the composite is 300 mm^2, and a stress of 100 MPa is applied in the longitudinal direction, what will be the total load (in kN) carried by the glass fibres?

(a) 0.5 (b) 5

(c) 20.5 (d) 29.5

ANSWERS

1. (c)	**2.** (a)	**3.** (b)	**4.** (d)	**5.** (a)	**6.** (b)	**7.** (c)
8. (d)	**9.** (a)	**10.** (d)	**11.** (c)	**12.** (d)	**13.** (a)	**14.** (c)
15. (b)	**16.** (c)	**17.** (b)	**18.** (b)	**19.** (a)	**20.** (b)	**21.** (b)
22. (d)						

EXPLANATIONS

1. (c) Atomic Packing factor (APF)

$$APF = \frac{\text{Volume of the atom per unit cell}}{\text{Total volume of unit cell}}$$

Space lattice	Examples	APF
SC	Po	0.523
BCC	Li, Na, v, Cr, Mo, W, Fe	0.68
FCC	Ni, Pt, Cu, Ag, Au, Al, Pb	0.75
HCP	Zn, Mg, Co	0.75

2. (a) Edge-Dislocation : If in a crystal, above a particular plane the bond length of the atom is more than the equilibrium distance and below that plane bond length is smaller than the equilibrium distance. This plane is called slip plane, around which distortion is maximum is called edge dislocation.

3. (b) The working of tungsten at 1000°C is hot working is incorrect.

Because the hot working process occur above the recrystalization temperature.

4. (d) Ceramic is a brittle materials. They contain both metallic and non-metallic elements.

Ex.- Glass, Porcelain, Sic, Mgo, TiB_2

5. (a) Given data:

Average degree of polymerisation of PVC = 2000

Average molecular weight of PVC = 62.50

So, that 2000 × 62.50 = 125000

6. (b) Super Invar a magnetic, austenitic, solid solution alloy containing iron, nickel, and cobalt designed to provide minimum thermal expansion at room temperature. This alloy also exhibits austenite stability to service temperature at least –67°F and thermal expansion properties are less.

7. (c) The colour of a metal is determined by the wavelength distribution of the radiation that is reflected.

8. (d) Ferromagnetic materials contain magnetic moments. Nickel fertile is ferromagnetic material.

9. (a) The oxide scale responsible for corrosion resistance of stainless steel, because corrosion is a natural process, which converts a refined metal to a more chemically stable form such as oxides, sulfide etc. Cr_2O_3

11. (c) Zirconia – Transformation toughening.

Cubic bonn nitride – Ultra hard material

Hafnium Carbide – Ultra-hgh temperature material.

Yttrium aluminum garnet – Host material of laser.

12. (d) Polyaerylonitrile – Acrylic fibre.

Nylon – 6, 6 – H-bonding

Polytelrafluoroethylene (PTFE) – Very high temperature resistance polymer.

Ebonite – Hard and brittle material.

13. (a)
1. Differential scanning calorimetry – Measurement of enthalpy change
2. Atomic force microscopic – Resolution less than 1 mm is possible.
3. Scanning electron microscopy – Surface morphology of a material
4. X–ray diffraction – Residual stress measurement.

14. (c) Thermal conductivity – W m^{-1} k^{-1}

Dielectric strength – Vm^{-1}

Magnetic permeability – H m^{-1}

Capacitance – Cv^{-1}

16. (c) $z = \frac{-\text{Lateral strain}}{\text{Congitudinal strain}}$

$$= \frac{\frac{-\delta d}{d}}{\frac{\delta l}{l}}$$

$$+.3 \times .001 = \frac{-\delta d}{d}$$

$$.0003 = \frac{-\delta d}{d}$$

$\delta d = -3.6 \times 10^{-3}$ mm

$d_2 = d_1 + \delta d$

$d_2 = 11.9964$ mm

17. (b) Let FCC has (h, k, l) = (1, 1, 1)

$$\sin^2\theta = \frac{\lambda^2}{4a^2}\left(h^2 + k^2 + l^2\right)$$

$$\sin^2 19.2 = \frac{.154^2}{4a^2}(3)$$

$$\boxed{a = .4055 \text{ nm}}$$

18. (b) $B_c = \dfrac{k\lambda}{L\cos(\theta_B)}$

λ – Wavelength

L – Average crystalline size

k = 1

$$.35 \times \frac{\pi}{180} = \frac{1 \times .154}{L\cos(19.2)}$$

$$\boxed{L = 26 \text{ nm}}$$

19. (a) For intrinsic semiconductor

$$n = p = \frac{\sigma}{|e|\left(\mu_e + \mu_p\right)}$$

$$= \frac{3.99 \times 10^{-4}}{1.6 \times 10^{-19} \times .178}$$

$= 13.99 \times 10^{15}$

20. (b) $\sigma = \sigma_0 e^{-\left(\frac{Eg}{2kT}\right)}$

It means

$\sigma \alpha e$

$$\frac{\sigma_1}{\alpha_2} = e^{-\left(\frac{1}{T_1} - \frac{1}{T_2}\right)}$$

By solving above equation, we get

$T_2 = 443$ K

21. (b) $E_c = E_f V_f + EfV_m$

$150 = 340\, V_f + 4.5\, V_m$

$1 = 2.2666\, V_f + 4.5\, V_m$

let $V_c = V_f + V_m$

$1 = V_f + V_m$

$1 = 2.266\, V_f + .03(1 - V_f)$

$V_f = .4338$

22. (d) $P_c = 30$ kN $\quad \left(\because P_c = \sigma_c A_c\right)$

$$\frac{P_f}{P_c} = \frac{E_f V_f}{E_c V_c}$$

$$P_f = \frac{30 \times 340 \times .4338}{150}$$

$P_f = 29.5$ kN

2012

Useful data

Boltzmann's constant	: 1.38×10^{-23} J K^{-1}
Charge on an electron	: 1.602×10^{-19} C
Gas Constant	: 8.314 J mol^{-1} K^{-1}
Electron rest mass	: 9.1×10^{-31} kg
Permittivity of vacuum (ε_0)	: 8.854×10^{-12} F m^{-1}
Bohr Magneton	: 9.274×10^{-24} A/m^2

Q. (1 – 9) carry one mark each.

1. Which of the following is **NOT** a Bravais lattice?
   (*a*) Simple tetragonal
   (*b*) Body centred tetragonal
   (*c*) Base centred orthorhombic
   (*d*) Face centred tetragonal

2. A Schottky defect in an ionic crystal is a stochiometric defect of
   (*a*) Cation vacancy
   (*b*) Anion vacancy
   (*c*) Cation and anion vacancy
   (*d*) Cation and anion interstitial

3. Which of the following techniques is **NOT** used to grow single crystals of semiconductors?
   (*a*) Calendering (*b*) Czochralski
   (*c*) Float zone (*d*) Bridgman

4. Which of the following signals is produced due to the elastic scattering of electrons by a material?
   (*a*) Secondary electron
   (*b*) Backscattered electron
   (*c*) Auger electron
   (*d*) Photoelectron

5. The best magnetostrictive material is
   (*a*) $Nd_2Fe_{14}B$ (*b*) Fe_3O_4
   (*c*) Cu_2MnAl (*d*) $ZnFe_2O_4$

6. Of the following materials, which is the most suitable for an LED emitting at around 380 nm?
   (*a*) Direct bandgap material with a small bandgap
   (*b*) Indirect bandgap material with a large bandgap
   (*c*) Direct bandgap material with a large bandgap
   (*d*) Indirect bandgap material with a small bandgap

7. Which material has the lowest specific heat capacity at room temperature?
   (*a*) Water (*b*) Mercury
   (*c*) Copper (*d*) Silver

8. Microstrain can be measured by X-ray diffraction using peak
   (*a*) Area and intensity
   (*b*) Position and area
   (*c*) Broadening and intensity
   (*d*) Position and broadening

9. The Pilling-Bedworth ratio is defined as the ratio of
   (*a*) Volume of oxide to volume of metal
   (*b*) Weight of oxide to weight of metal
   (*c*) Density of oxide to density of metal
   (*d*) Surface area of oxide to surface area of metal

Q. (10 – 22) carry two marks each.

10. Match the properties in **Column I** with the appropriate units in **Column II**

Column I	Column II
P. Thermal diffusivity	1. Hm^{-1}
Q. Fracture toughness	2. m^{2}s^{-1}
R. Surface energy	3. Fm^{-1}
S. Magnetic permeability	4. Nm$^{-3/2}$
	5. Jm^{-2}

	P	Q	R	S
(*a*)	2	5	4	1
(*b*)	2	4	5	1
(*c*)	3	5	4	3
(*d*)	5	4	2	3

11. Match the characterization techniques in **Column I** with the options in **Column II**

Column I
P. Scanning tunneling microscopy
Q. Scanning electron microscopy
R. Transmission electron microscopy
S. Atomic force microscopy

Column II
1. No vacuum required
2. Backscattered electrons
3. Photoelectrons
4. Atomically sharp tip
5. Sub-Angstrom resolution

	P	Q	R	S
(*a*)	4	2	5	1
(*b*)	1	3	4	5
(*c*)	2	4	1	5
(*d*)	5	1	2	4

12. Match the materials in **Column I** with the applications in **Column II**

Column I

P. Titanium diboride

Q. Molybdenum disilicide

R. Hydroxyapatite

S. Nanocrystalline titanium oxide

Column II

1. Photocatalyst
2. Furnace heating element
3. Ultra high temperature material
4. Tough ceramic
5. Artificial bone implant

	P	Q	R	S
(a)	3	4	5	1
(b)	5	3	2	1
(c)	4	3	1	5
(d)	3	2	5	1

13. Match the properties in **Column I** with the options in **Column II**

Column I	Column II
P. Toughness	1. Resistance to plastic deformation
Q. Resilience	2. Time dependent perma-nent deformation under constant load
R. Creep	3. Total elongation at failure
S. Hardness	4. Area under the stress-strain curve
	5. Area under the elastic part of the stress-strain curve

	P	Q	R	S
(a)	5	1	3	2
(b)	4	3	2	1
(c)	4	5	2	1
(d)	5	4	3	2

14. Determine the mole fraction of vinyl chloride in a copolymer of vinyl chloride (CH_2CHCl) and vinyl acetate ($CH_2–CH–OCOCH_3$) having molecular weight of 10520g/mol and degree of polymerization of 160.

(a) 0.14 (b) 0.30

(c) 0.70 (d) 0.86

15. The electron concentration in an n-type semiconductor is $5 \times 10^{18}/m^3$. If the drift velocity of electrons is 100 m/s in an electric field of 500 V/m, calculate the conductivity of the semiconductor.

(a) 0.16×10^{-1} S/m (b) 1.60×10^{-1} S/m

(c) 2.50×10^{-1} S/m (d) 30.05×10^{-1} S/m

16. Calculate the saturation magnetization (M_{sat}) for bcc iron of lattice parameter 2.866 Å.

(a) 0.79×10^6 A/m (b) 1.5×10^6 A/m

(c) 3.15×10^6 A/m (d) 4.73×10^6 A/m

Common Data Questions

Common Data Q. (17 – 18) :

A plain 0.45 wt.% carbon steel is cooled slowly from 900 °C to just below the eutectoid temperature (723 °C) so that the following reaction occurs :

γ(0.8 wt.%C) $\leftrightarrow$ α(0.02 wt.%C) + Fe_3C (6.67 wt.%C)

17. During cooling from 900°C to 723°C, the proeutectoid α forms from γ. Find the volume % of proeutectoid α just below 723°C for the steel.

(a) 44.9% (b) 66.1%

(c) 55.1% (d) 34.9%

18. Find the volume % of pearlite for the steel just below 723°C for 0.45 wt.% carbon steel.

(a) 44.9% (b) 55.1%

(c) 40.9% (d) 59.1%

Common Data Q. (19 – 20) :

A 20 kN tensile load is applied axially to a steel bar of crosssectional area 8 cm² and 1m length. The Young's modulus of steel (E_{steel}) is 200 GPa, and of aluminium (E_{Al}) is 70 GPa. The Poisson's ratio can be taken as 0.3.

19. When the same load is applied to an aluminium bar, it is found to give same elastic strain as the steel. Calculate the cross-sectional area of the aluminium bar.

(a) 11.43 cm² (b) 14.93 cm²

(c) 18.26 cm² (d) 22.86 cm²

20. Calculate the final area of the steel bar after the deformation under the applied load of 20 kN.

(a) 7.9 cm² (b) 9.7 cm²

(c) 7.0 cm² (d) 8.1 cm²

Linked Answer Questions

Statement Q. (21 – 22) :

Chromium has the bcc structure with atomic diameter of 2.494 Å.

21. Calculate the lattice parameter of chromium assuming tight atomic bonding.

(a) 1.442 Å (b) 2.880 Å

(c) 4.323 Å (d) 5.764 Å

22. Find the first diffraction peak position (2θ) for Cu Kα radiation with a wavelength of 1.54 Å

(a) 21.76° (b) 33.05°

(c) 44.43° (d) 66.10°

ANSWERS

1. (d)	**2.** (c)	**3.** (a)	**4.** (b)	**5.** (a)	**6.** (c)	**7.** (b)
8. (d)	**9.** (a)	**10.** (b)	**11.** (a)	**12.** (d)	**13.** (c)	**14.** (d)
15. (b)	**16.** (c)	**17.** (a)	**18.** (b)	**19.** (d)	**20.** (Marks to All)	**21.** (b)
22. (c)						

EXPLANATIONS

1. (d) Bravais lattice are crystallographhy. Bravais lattice, any of 14 possible three-dimensional configuration of points used to describe the orderly arrangement of atoms, in a crystal. Simple tetragonal, Body centered tetragonal and base centred orthorhombic are bravais lattic.

2. (c) **Schottky defect:** When a action-anion pair is absent from an ionic crystal, the defect is called schottky defect.

3. (a) Calendaring techniques are not used to grow single crystals of semiconductor. Because a calendaring technique is a finishing process used to finish or smooth a sheet of material such as paper, textiles or plastics.

4. (b) Backscattered electron signals are produced due to the elastic scattering of electrons by a material.

5. (a) magnetostrictive material consist of tiny ferromagnetism. These ferromagnets usually iron, nickel or cobalt have small magnetic moment. $Nd_2Fe_{14}B$ is a best magnetostrictive material.

6. (c) Direct band gap material with a large bandgap is the most suitable for LED emitting at around 380 nm.

7. (b) mercury material has the lowest specific heat capacity at room temperature. This fact added to its liquid state at most for use in temperature.

8. (d) Microstrain can be measured by x-ray diffraction using peak position and broadening.

9. (a) Pilling-Bedworth is defined as ratio of volume of oxide to volume of metal.

$$\text{Pilling-Bedworth} = \frac{\text{Volume of oxide}}{\text{Volume of metal}}$$

10. (b) Thermal diffusivity – m^2s^{-1}

Fracture toughness – $Nm^{\frac{-3}{2}}$

Surface energy – Jm^{-2}

Magnetic permeability – Hm^{-1}

11. (a) Scanning tunneling microscopy – Atomically sharp tip.

Scanning electron microscopy – Backscahered electrons.

Transmission electron – Sub-Angstrom resolution

Atomic force microscopy – No vaccum required

12. (d) Titanium diboride – Ultra high temperature material.

Molybdenum disilicide – Furnace heating element.

Hydroxyapatite – Artificial bone implant.

Nanocrystalline titanium oxide – Photocatalyst

13. (c) Toughness – Area under the stress-strain curve

Resilience – Area under the elastic part of the stress-strain curve.

Creep – Time dependent permanent deformation under constant load.

Hardness – Resistance to plastic deformation.

14. (d) Given that:

Molecular write = 10520 g/mol

Degree of polymerisation = 160

$$\text{Average molecular weight} = \frac{10520}{160} = 65.75$$

Mole fraction of Vinyl Chloride

$$= \frac{\text{Average molecule write}}{\text{Molecule write of } CH_2CHCl} = \frac{65.75}{75} = .86$$

15. (b) $\sigma = n|e|\mu_e$

$$= 5\times10^{18}\times1.6\times10^{-19}\times\frac{100}{500}$$

$= 1.6 \times 10^{-1}$ S/m

16. (c) The maximum volume magnetization (M_{sat}) is the total magnetic moment per unit time

$$M_{sat} = \left[8.48\times10^{28}\,\frac{\text{atom}}{\text{m}^3}\right]\times\left[9.27\times10^{-24}\,\text{Am}^2\right].\left[4.\frac{\text{Bohr}}{\text{Atom}}\right]$$

$= 3.15 \times 10^{16}$ A/m

19. $$\in = \frac{\delta l}{l} = \frac{pA}{E}$$

$$\in_s = \frac{20\times10^3\times8\times10^{-4}}{200\times10^9} = 8\times10^{-11}\,\frac{\text{m}}{\text{m}}$$

$\in_s = \in_{Al}$ & $p_{Al} = p_s$

$$\in_{Al} = \frac{p_{Al}A_{Al}}{E_{Al}}$$

$$A_{Al} = \frac{\in_{Al}\cdot\in_{Al}}{p_{Al}} = \frac{8\times10^{-11}\times70\times10^9}{20\times10^3}$$

$A_{Al} = 28$ cm^2

21. (b) For chromium (BCC structure).

$\sqrt{3}a = 4R$

$$a = \frac{4R}{\sqrt{3}}$$

$$= \frac{2D}{\sqrt{3}} = \frac{2\times2.494}{\sqrt{3}}$$

$= 2.88$ A

22. (c) $n\lambda = 2d\sin\theta$

$$\theta = \sin^{-1}\left(\frac{n\lambda}{2d}\right)$$

$\theta = 44.43°$

2013

Useful Data :

Avogadro's Number	: 6.023×10^{23} mol^{-1}
Boltzmann's constant, k	: 1.38×10^{-23} J.K^{-1}
Electron Charge, e	: 1.6×10^{-19} C
Electron rest mass, m_o	: 9.1×10^{-31} kg
Gas Constant, R	: 8.314 J.mol^{-1}K^{-1}
Free Space Permittivity, ε_o	: 8.854×10^{-12} F.m^{-1}
Free Space magnetic permeability, μ_o	: $4\pi \times 10^{-7}$ H.m^{-1}
Speed of light, c	: 3×10^{8} m.s^{-1}
Planck's constant, h	: 6.63×10^{-34} J.s^{-1}
Bohr Magneton, μ_b	: 9.27×10^{-24} A m^2

1 eV = 1.6×10^{-19} J

1 calorie = 4.2 J

Q. 1 – 9 carry one mark each.

1. As temperature increases, diffusivity of an atom in a solid material,

(*a*) increases

(*b*) decreases

(*c*) remains constant

(*d*) depends on the specific material

2. Which of the following is NOT correct?

(*a*) Dislocations are thermodynamically unstable defects.

(*b*) Dislocations can move inside a crystal under the action of an applied stress.

(*c*) Screw dislocations can change the slip plane without climb

(*d*) Burger's vector of an edge dislocation is parallel to the dislocation line.

3. At a constant atmospheric pressure, the number of phases, P which coexist in a chosen system at equilibrium, is related to the number of components, C in the system and the degree of freedom, F by

(*a*) P + F =C – 2 (*b*) P + F = C + 2

(*c*) P + F =C + 1 (*d*) P + F = C – 1

4. Which one of the following metals is commonly alloyed with iron to improve its corrosion resistance?

(*a*) Co (*b*) Cr

(*c*) Ti (*d*) Nb

5. The number of slip systems in a metal with FCC crystal structure is

(*a*) 4 (*b*) 6

(*c*) 8 (*d*) 12

6. Upon recrystallization of a cold worked metal,

(*a*) strength increases and ductility decreases

(*b*) strength decreases but ductility increases

(*c*) both strength and ductility increase

(*d*) both strength and ductility decrease

7. In carbon fiber reinforced resin composites, for a given fiber volume content, Young's modulus depends on the orientation of the fiber with respect to the applied load. Which orientation of the fibers will give the maximum value of Young's modulus?

(*a*) transverse

(*b*) longitudinal

(*c*) random

(*d*) both transverse and longitudinal

8. Vulcanization is related to

(*a*) strengthening of rubber

(*b*) extrusion

(*c*) injection moulding

(*d*) addition polymerisation

9. Which one of the following oxides crystallizes into fluorite structure?

(*a*) UO_2 (*b*) MgO

(*c*) $BaTiO_3$ (*d*) $MgAl_2O_4$

Q. 10 – 22 carry two marks each.

10. Match the conventional ceramic materials listed in **Column I** with their respective common applications in **Column II**

Column I

P. Lead Zirconate Titanate (PZT)

Q. Zinc Oxide (ZnO)

R. Silicon Carbide (SiC)

S. Zirconia (ZrO_2)

Column II

1. cutting tool
2. thermal barrier coating
3. actuator
4. varistor
5. super conductor

(*a*) P-1, Q-2, R-3, S-5

(*b*) P-3, Q-2, R-1, S-5

(*c*) P-2, Q-1, R-5, S-3

(*d*) P-3, Q-4, R-1, S-2

11. Match the terminologies given in **Column I** with their relations listed in **Column II**

Column I	**Column II**
P. domain wall	1. superconductors
Q. Fick's law	2. mechanical properties
R. Matthiessen's rule	3. ferromagnetic materials
S. Hall-Petch relation	4. resistivity of impure metals
T. Meissner effect	5. diffusion

(*a*) P-1, Q-3, R-5, S-2, T-4
(*b*) P-3, Q-5, R-2, S-4, T-1
(*c*) P-3, Q- 5, R-4, S-2, T-1
(*d*) P-3, Q-4, R-3, S-2, T-4

12. Match the microscopes listed in **Column I** with their principle of operation listed in **Column II**

Column I

P. Scanning Electron Microscope (SEM)
Q. Transmission Electron Microscope (TEM)
R. Scanning Tunnelling Microscope (STM)
S. Atomic Force Microscope (AFM)

Column II

1. van der Waals forces between atoms
2. electrons to jump across a potential barrier
3. diffraction of electrons
4. detection of secondary electrons
5. photo emission of electrons

(*a*) P-2, Q-5, R-3, S-1
(*b*) P-3, Q-4, R-5, S-2
(*c*) P-4, Q-3, R-2, S-1
(*d*) P-4, Q-3, R-5, S-2

13. X-rays of unknown wavelength are diffracted by an FCC metal with a lattice parameter of 0.352 nm. The measured '2 θ' angle for the {200} peak is 61.08°. Calculate the wavelength of the X-ray used, in nm. ________

14. A metal with HCP crystal structure has lattice constants a = 0.30 nm and c = 0.56 nm. Determine the volume of the unit cell of this metal, in nm^3.

____

15. The band gap of a semiconducting material used to make an LED is 1.43 eV. What will be the minimum wavelength of the radiation emitted by this LED, in μm?

16. For automatic control of household electric water heater a relay switch is activated by thermal expansion of a brass rod of length 50 cm as shown in the schematic below.

The distance between the rod and the lever, *x*, is adjusted by moving the base of the rod. As the water gets heated the rod expands and as soon as the rod touches the lever, the circuit is broken disconnecting the heater from the power supply. Find the distance, *x*, in mm, to be set at water temperature of 20°C such that the circuit is broken at 70°C. The coefficient of linear thermal expansion of brass is 20×10^{-6} °C^{-1} ________

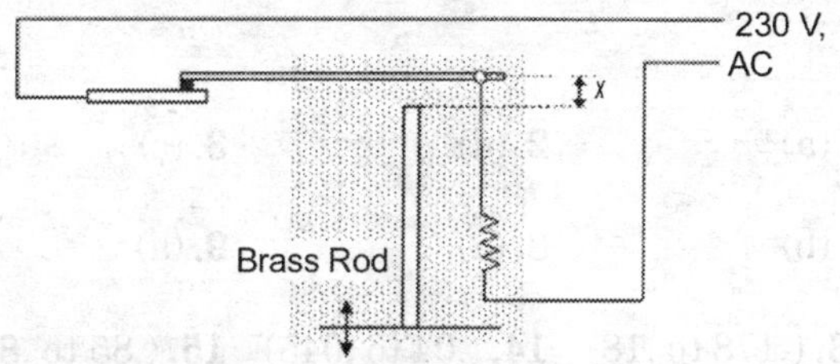

Common Data Questions

Common Data for Questions 17 and 18 :

From tensile test of a particular alloy the following values were obtained. The material exhibits linear work hardening as shown in the figure given below.

	At Yield	At Fracture
Stress, GPa	0.7	0.8
Strain, %	1	4

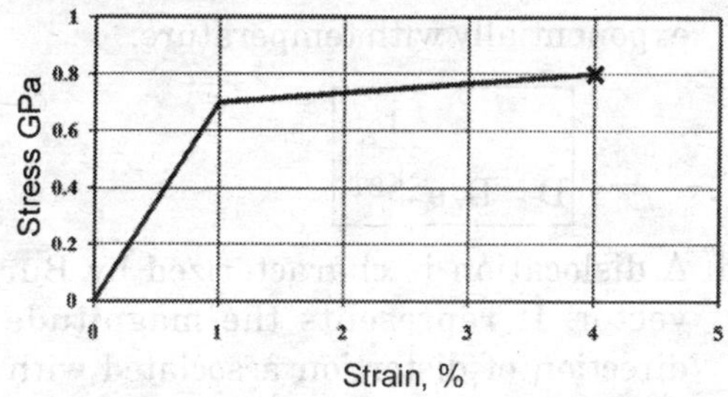

17. If the cylindrical specimen had a dimension of diameter 10 mm and length 50 mm, find the length of the specimen at the yield point, in mm. ________

18. Find the toughness of the material, in M J m^{-3}. ____

Common Data for Questions 19 and 20 :

An isomorphous alloy system contains 47 wt% of A and 53 wt % of B and is at 1300° C. Referring to the figure given below, answer the following :

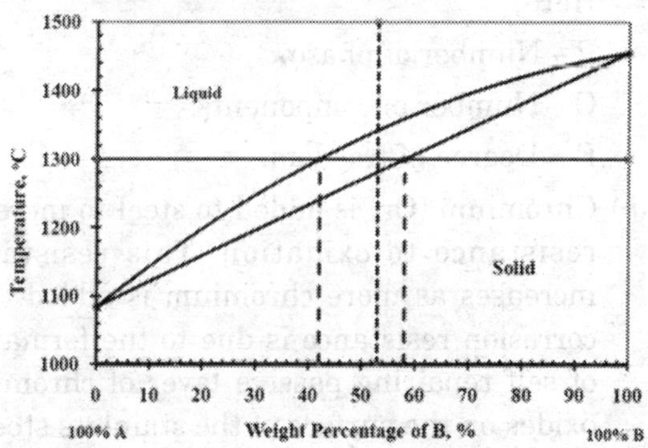

19. What is the weight percentage of A in solid phase at this temperature? ________

20. What weight percentage of this alloy is liquid ? _____

Linked Answer Questions

Statement for Linked Answer Questions 21 & 22 :

A stress of 10 MPa is applied to an elastomer to generate a strain of 50%. The strain is held constant at this value. After 40 days at 20°C, the stress decreases to 5 MPa.

21. What is the relaxation time constant for this material ? _____

22. What will be the stress after 60 days at 20°C? _____

ANSWERS

1. (a)	2. (d)	3. (c)	4. (b)	5. (d)	6. (b)
7. (b)	8. (a)	9. (a)	10. (d)	11. (c)	12. (c)
13. (.178 to .18)	14. (.04 to .045)	15. (.85 to .89)	16. (.4 to .6)	17. (50 to 51)	18. (22 to 28)
19. (41 to 43)	20. (30 to 33)	21. (*)	22. (*)		

EXPLANATIONS

1. (a) As the temperature increases, diffusivity of an atom in a solid material increases. The diffusion activation barrier energy is generally compared with the thermal energy supplied to the system. The diffusion increases exponentially with temperature.

$$\boxed{D = D_o e^{-\frac{E_A}{k_{BT}}}}$$

2. (d) A dislocation is characterized by Burger's vector. It represents the magnitude and direction of distortion associated with that particular dislocation. Two limiting cases of dislocations edge and screw are characterized by burger's vector perpendicular to the dislocation line respectively.

3. (c) F = C – P + 1

$\boxed{F + P = C + 1}$ At constant atmospheric pressure

Here,

P – Number of phase

C – Number of components.

F – Degree of freedom.

4. (b) Chromium (Cr) is added to steel to increase resistance to oxidation. This resistance increases as more chromium is added. The corrosion resistance is due to the formation of self repairing passive layer of chromium oxides on the surface of the stainless steel.

5. (d) The number of slip system is FCC structure is equal to 12.

6. (b) Cold work material done below the recrystalization temperature.

→ strength decreases.

→ Ductility increases.

7. (b) In carbon fiber reinforced resin composites, for a given fiber volume content, Young's modulus depend on the orientation of fibre with respect to applied load. The longitudinal orientation of the fibres will give maximum value of young's modulus.

8. (a) Vulcanization is a process used to harden and strengthen rubber.

9. (a) UO_2 oxides crystallized into fluorite structure.

10. (d) Lead Zirconate Titanate (PZT) – Actuator.

Zinc oxides (Zno) – Varistor

Silicon carbides (Sic) – Cutting tool.

Sirconia (ZrO_2) – Thermal barrier coating.

11. (c) Domain wall – Ferromagnetic materials

Fick's law – Diffusion

Matthiessen's rule – resistivity of impure metals.

Hall-petch relation – Mechanical properties.

Meissner effect – Superconductors.

12. (c) Scanning electron microscope (SEM) – Detection of secondary electron

Transmission Electron Microscope (TEM) – Diffraction of electrons.

Scanning Tunnelling Microscope (STM) – Electrons to jump across a potential barrier.

Atomic Force Microscope (AFM) – Van der waals forces between atoms.

2014

Useful constants

Avogadro's Number : 6.023×10^{23} mol^{-1}

Boltzmann's constant, k : 1.38×10^{23} J.K^{-1}

Electron Charge, e : 1.6×10^{-19} C

Electron rest mass, m_o : 9.1×10^{-31} kg

Universal gas constnat, R : 8.314 J.mol^{-1}. K^{-1}

Speed of light, c : 3×10^8 m.s^{-1}

Planck's constant, h : 6.63×10^{-34} J.s

1 eV = 1.6×10^{-19} J

Q. (1– 9) Carry One Mark Each.

1. Neoprene is rendered non-inflammable because
   (*a*) it has a highly cross-linked structure
   (*b*) it has a highly linear chain structure
   (*c*) of the presence of chlorine atom in the structure
   (*d*) of the absence of chlorine atom in the structure

2. Nylon-6 is manufactured from
   (*a*) caprolactum
   (*b*) adipic acid and hexamethylene diamine
   (*c*) maleic anhydride and hexamethylene diamine
   (*d*) sebasic acid and hexamethylene diamine

3. At room temperature, the typical barrier potential for silicon p-n junction in Volt (V) is
   (*a*) 0.7×10^{-23} (*b*) 0.07
   (*c*) 0.70 (*d*) 7.0

4. Quantitative measurement of the roughness of a polysilicon wafer can be performed with
   (*a*) scanning tunneling microscopy
   (*b*) scanning electron microscopy
   (*c*) transmission electron microscopy
   (*d*) atomic force microscopy

5. The temperature of the antiferro-magnetic-to-paramagnetic transition is called
   (*a*) Curie temperature
   (*b*) Curie-Weiss temperature
   (*c*) Neel temperature
   (*d*) Debye temperature

6. At low injection level, a forward biased p-n junction would have
   (*a*) no charge carriers
   (*b*) minority carrier concentration much more than majority carrier concentration
   (*c*) minority carrier concentration equal to majority carrier concentration
   (*d*) minority carrier concentration much less than majority carrier concen-tration

7. Which of the following mechanical properties of a material depend on the mobile dislocation density in it.
   (P) Young's modulus
   (Q) yield strength
   (R) ductility
   (S) fracture toughness
   (*a*) P, Q, R (*b*) Q, R, S
   (*c*) P, R, S (*d*) S, P, Q

8. The equilibrium concentration of vacancies in a pure metal
   (*a*) increases exponentially with temperature
   (*b*) decreases exponentially with temperature
   (*c*) varies linearly with temperature
   (*d*) is independent of temperature

9. The materials belonging to which one of the following crystal classes would be both piezoelectric and ferroelectric
   (*a*) 222 (*b*) 4mm
   (*c*) $\bar{1}$ (*d*) 2/m

Q. (10 – 22) Carry Two Marks Each.

10. Polymerized isotactic polybutadiene has a molecular weight of 3×10^5 g/mol. The degree of polymerization is _____.

11. A bar of Ti with Young's modulus of 110 GPa and yield strength of 880 MPa is tested in tension. It is noticed that the alloy does not exhibit any strain hardening and fails at a total strain of 0.108. The mechanical energy that is necessary to break the material in MJ/m^3 is __________.

12. A copper cup weighing 140 g contains 80 g of water at 4 °C. Specific heats of water and copper are 4.18 and 0.385 J/g °C, respectively. If 100 g of water that is at 90 °C is added to the cup, the final temperature of water in °C is ________.

13. Match the reaction in **Column I** with its name in **Column II.**

L – liquid, α, β, γ –different solid solution phases

Column–I	Column–II
P. $L \xrightarrow{\text{cooling}} \alpha + \beta$	1. peritectic
Q. $L + \beta \xrightarrow{\text{cooling}} \gamma$	2. eutectic
R. $\alpha \xrightarrow{\text{cooling}} \beta + \gamma$	3. monotectic
	4. eutectoid

Codes

	P	Q	R
(*a*)	1	4	3
(*b*)	2	1	4
(*c*)	2	3	1
(*d*)	4	2	3

14. The Young's modulus of a unidirectional SiC fiber reinforced Ti matrix composite is 185 GPa. If the Young's moduli of Ti and SiC are 110 and 360 GPa respectively, the volume fraction of fibers in the composite is ________.

15. Match the composite in **Column I** with the most suitable application in **Column II**.

Column I

P. Glass fibre reinforced plastic

Q. SiC particle reinforced A1 alloy

R. Carbon-carbon composite

S. Metal fibre reinforced rubber

Column II

1. Missile cone heads
2. Commercial automobile chasis
3. Airplane wheel tyres
4. Car piston rings
5. High performance skate boards

Codes

	P	Q	R	S
(*a*)	4	5	1	2
(*b*)	3	5	2	4
(*c*)	5	4	1	3
(*d*)	4	2	3	1

16. Which among the following rules need to be satisfied for obtaining an isomorphous phase diagram in a binary alloy system?

P. The atomic size difference should be less than 15%

Q. Both the end components should have the same crystal structure

R. The valency of the end components should be the same

S. The end components should have dissimilar electronegativities

(*a*) P, Q, R (*b*) Q, R, S

(*c*) R, S, P (*d*) S, P, Q

17. The energy in eV and the wavelength in m, respectively, of the photon emitted when an electron in a hydrogen atom falls from n = 4 to n = 2 state is

(*a*) 3.0, 0.413 (*b*) 2.55, 0.365

(*c*) 2.75, 0.451 (*d*) 2.55, 0.487

18. The weight in kg of gallium (Ga) to be mixed with arsenic (As) for obtaining 1.0 kg of gallium arsenide (GaAs) is ___.

(M_{Ga} = 69.72 g/mol; M_{As} = 74.92 g/mol)

19. Match the material in **Column I** with the property in **Column II**

Column I	**Column II**
P. $Pb(Zr, Ti)O_3$	1. Shape memory alloy
Q. $Ni_{50}Ti_{50}$	2. Piezoelectric ceramic
R. GaAs	3. High temperature superconductor
S. $YBa_2Cu_3O_7$	4. Optoelectronic semiconductor

(*a*) P-1, Q-2, R-3, S-4 (*b*) P-2, Q-3, R-4, S-1

(*c*) P-4, Q-1, R-2, S-3 (*d*) P-2, Q-1, R-4, S-3

20. Relevant portion of a binary phase diagram of elements A and B is shown below. The mass fraction of liquid phase at 1000 °C for an alloy with 15 wt.% B is ________.

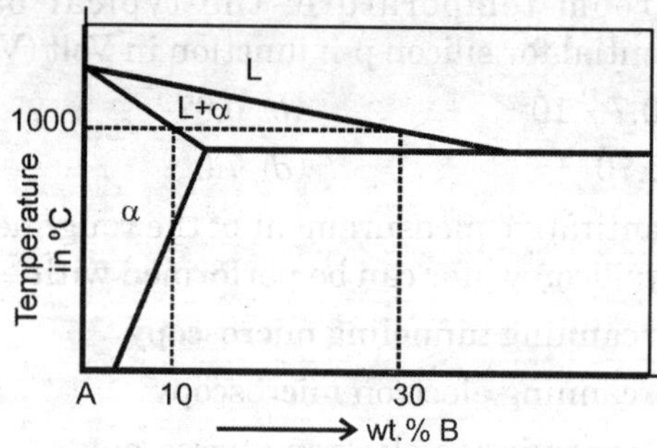

21. The expected diffraction angle (in degrees) for the first order reflection from the (113) set of planes for face centered cubic Pt (lattice parameter = 0.392 nm) using monochromatic radiation of wavelength 0.1542 nm is ________.

22. The diffusion coefficients of Mg in Al at 500 and 550 °C are 1.9×10^{-13} and 5.8×10^{-13} m^2/s respectively. The activation energy for diffusion of Mg in Al in kJ/mol is ________.

ANSWERS

1. (c)	**2.** (a)	**3.** (c)	**4.** (d)	**5.** (c)	**6.** (d)
7. (b)	**8.** (a)	**9.** (b)	**10.** (5555 to 5556)	**11.** (91 to 92)	**12.** (48 to 49)
13. (b)	**14.** (.29 to .31)	**15.** (c)	**16.** (a)	**17.** (d)	**18.** (.47 to .49)
19. (d)	**20.** (.24 to .26)	**21.** (40 to 41)	**22.** (115 to 120)		

EXPLANATIONS

1. (c) Neoprene is rendered non-inflammable, because of the presence of chlorine atom is the structure.

2. (a) Nylon-6 is only made from one kind of monomer, a monomer called caprolactan. Nylon-6 is made from two monomers adipolyl chloride and hexamethylene.

3. (c) Normally at room temperature, in a P-N junction the value of barrier potential is 0.3-0.7 V. The valve is 0.7 V for silicon and 0.3 V for germanium.

4. (d) Atomic force microscopy (AFM) or scanning force microscopy (SFM) is a very high. resolution type of scanning probe microscopy with demonstrated resolution on the order of fractions of a nanometer. AFM is 1000 times better than the optical diffraction limit. So that AFM is used for quantitative measurement of the roughness of a polysilicon water.

5. (c) The temperature of the antiferromagnetic to paramagnetic transition is called Neel temperature.

6. (d) At low junction level, a forwarded biased P–N junction would have, minority carrier concentration much less than the majority carrier concentration.

7. (b) Following are the mechanical property of a material depend on the mobile dislocation density.

→ Yield strength

→ Durtility

→ Fracture toughness

8. (a) The equilibrium concentration of vacancies in a pure metal, increase exponentially with temperature.

9. (b) The material belong to the crystal classes would be both piezoelectric and ferroelectric at 4 mm.

10. (5555 to 5556)

Given that Molecular weight of polymerized isotactic polybutadiene = 3 x 10^5 g/mol.

Molecular weight of monomer = 54

Degree of polymerization

$$= \frac{\text{Molecular weight}}{\text{Molecular weight of monomer}}$$

$$\text{Degree of polymerization} = \frac{3 \times 10^5}{54} = 5555.55$$

13. (b) $L \xrightarrow{\text{Cooling}} \alpha + \beta$ — Eutectic

$L + \beta \xrightarrow{\text{Cooling}} \gamma$ — Peritectic

$\alpha \xrightarrow{\text{Cooling}} \beta + \gamma$ — Eutectoid

14. (.29 to .31)

Young's Modulus of Composites (E_C)

$$E_C = E_P V_F + E_M V_M$$

$$V_M + V_F = 1$$

$$V_M = 1 - V_F$$

$$185 = 110 \cdot V_F + 360 \cdot V_M$$

$$\Rightarrow 185 = 1.10 V_F + 360(1 - V_F)$$

$$\Rightarrow 185 = 360 - 250 V_F$$

$$V_F = 0.7$$

$$\boxed{V_M = 0.3}$$

15. (c) Giass fibre reinforced plastic — High performance skate board.

Sic particle reinforced Ai alloy — Car piston ring

Carbon-Carbon composites — Missile cone head

Metal fibre reinforced rubber — Aeroplane wheel tyres

16. (a) Following rules need to be satisfied for obtaining an isomorphous phase diagram in a binary alloy system.

→ The atomic size difference should be less than 15%.

→ Both the end components should have the same crystal structure.

→ The valency of the end components should be the same.

17. (d) $\frac{1}{\lambda} = R\left(\frac{1}{n_f^2} - \frac{1}{n_i^2}\right)$

$$\frac{1}{\lambda} = 1.097 \times 10^7 \times \left(\frac{1}{2^2} - \frac{1}{4^2}\right)$$

$$\lambda = .487 \text{ mm}$$

$$E = 13.6 \text{ eV}\left(\frac{1}{2^2} - \frac{1}{4^2}\right)$$

$$E = 2.55 \text{ eV}$$

18. (.47 to .49)

$$\text{Weight of Ga (in kg)} = \frac{69.72}{69.72 + 74.92} = 48.2 \text{ kg}$$

19. (d) $Pb(Zr, Ti)O_3$ — Piezoelectric ceramic

$Ni_{50}Ti_{50}$ — Shape memory alloy

Ga As — Optoelectronic semiconductor

$YBa_2Cu_3O_7$ — High temperature superconductor

2015

Q. (1–10) Carry One Mark Each.

1. Arrange the following elements in order of increasing melting point:

P. gallium

Q. tungsten

R. aluminium

S. gold

(*a*) P < R < Q < S (*b*) S < P < R < Q

(*c*) P < R < S < Q (*d*) R < S < Q < P

2. When the atoms in a solid are separated by their equilibrium distance,

(*a*) the potential energy of the solid is lowest

(*b*) the force of attraction between the atoms is maximum

(*c*) the force of repulsion between the atoms is zero

(*d*) the potential energy of the solid is zero

3. To which of the following category of materials does Teflon (PTEE) belong?

(*a*) Thermosets (*b*) Thermoplastics

(*c*) Elastomers (*d*) Block copolymers

4. Which of the following statements is TRUE about the glass transition temperature (T_g)?

(*a*) T_g appears below the melting temperature in a perfectly crystalline material

(*b*) Upon heating through T_g, heat capacity remains constant but the thermal expansion coefficient changes

(*c*) Upon heating through T_g, heat capacity changes but the thermal expansion coefficient remains the same

(*d*) Upon heating through T_g, both the heat capacity and thermal expansion coefficient change

5. The slope of a graph of $\log_e$ (conductivity) versus 1/T (where T is the temperature) for an intrinsic semiconductor with energy gap E_g, is

(*a*) $E_g/2k$ (*b*) $-E_g/2k$

(*c*) E_g/k (*d*) $-E_g/k$

6. Which is NOT a ceramic forming process?

(*a*) extrusion (*b*) slip casting

(*c*) forging (*d*) tape casting

7. Which of the following is NOT a soft magnetic material?

(*a*) Iron-silicon steel

(*b*) Nickel zinc ferrite

(*c*) Nickel iron alloy

(*d*) Alnico

8. The eutectic reaction is [Note: S – solid; L – liquid]

(*a*) $S_1 \rightleftharpoons S_2 + S_3$

(*b*) $L \rightleftharpoons S_1 + S_2$

(*c*) $L_1 + S_1 \rightleftharpoons L_2 + S_2$

(*d*) $L_1 + S_1 \rightleftharpoons S_2 + S_3$

9. Vacancies play an important role in

(*a*) deformation twinning

(*b*) self diffusion

(*c*) strain hardening

(*d*) cross-slip

Q. (10 – 22) Carry Two Marks Each.

10. Match the techniques listed in Column I with the characteristics of the materials measured in Column II.

Column I	**Column II**
P. DSC	1. Density of states
Q. XRD	2. Glass transition temperature
R. STM	3. Cathodoluminescence
S. SEM	4. Crystal structure
	5. Thermal expansion coefficient

(*a*) P-2, Q-3, R-4, S-1

(*b*) P-5, Q-4, R-5, S-1

(*c*) P-2, Q-4, R-1, S-3

(*d*) P-3, Q-5, R-4, S-2

11. The mass of an electron would increase _____ times its original mass if it travels at 96% of the speed of light.

12. With increasing temperature from 15°C in winter to 45°C in summer, the length of an iron rail track increases by 0.05 cm. Calculate the original length of the iron rail track in cm. (linear thermal expansion coefficient of iron is $11.0 \times 10^{-6}\ K^{-1}$)

13. What is the thickness (in μm) of a germanium crystal layer that would be required for absorbing 80% of the incident radiation whose wavelength is 1.3 μm? The absorption coefficient (α) of germanium at 1.3 μm is $3.3 \times 10^5\ m^{-1}$.

14. A 1 kg sacrifical anode of Mg (atomic weight: 24.31 amu) is attached to the base of a ship. If the anode lasts for 60 days, what is the average corrosion current (in Amperes) during that period?

15. A capacitor has a 0.075 cm thick $BaTiO_3$ dielectric with a dielectric constant of 2000 and an electrode area of 0.2 cm^2. What is the capacitance of this capacitor in nF?

16. A hot pressed ceramic composite material consists of 30 volume % SiC whiskers in an Al_2O_3 matrix. The measured bulk density of this composite is 3.65 g cm^{-3}. If the theoretical density of SiC is 3.22 g cm^{-3} and that of Al_2O_3 is 3.95 g cm^{-3}, estimate the porosity (%) of the composite, assuming that the linear rule of mixtures is valid in this case.

17. Match the technical ceramics listed in Column I with their common applications listed in Column II

Column I	Column II
P. Y-doped ZrO_2	1. Lasers
Q. UO_2	2. Turbine engine
R. Si_3N_4	3. Integrated circuit substrate
S. AlN	4. Oxygen sensor
T. Cr doped Al_2O_3	5. Nuclear fuel
	6. Thermistor

(*a*) P-6, Q-4, R-5, S-1, T-3

(*b*) P-4, Q-5, R-2, S-3, T-1

(*c*) P-3, Q-1, R-2, S-6, T-5

(*d*) P-1, Q-4, R-5, S-2, T-1

18. Creep in metals is defined as

(*a*) the maximum energy a solid can absorb elastically

(*b*) the maximum energy a solid can absorb by plastic deformation

(c) the stress at which plastic deformation starts

(*d*) slow plastic deformation due to diffusion of atoms usually at high temperature (T > half the melting point)

19. Calculate the planar density of the (100) plane in an fcc crystal given that R is the atomic radius of the element.

(*a*) 0.25 R^2 (*b*) 0.25/R^2

(*c*) 1/R^2 (*d*) 4/R^2

20. The diffusion coefficient of copper atoms in aluminium is found to be 1.28×10^{-22} m^2s^{-1} at T = 400K and 5.75×10^{-19} m^2s^{-1} at T = 500K. Find the temperature (in Kelvin) at which the value of the diffusion coefficient is 10^{-16} m^2s^{-1}.

21. Calculate the density of copper in kg m^{-3} given that copper has an fcc lattice with a lattice parameter of 0.365 nm. Copper has an atomic weight of 63.54 amu.

22. What would be the maximum number of electron-hole pairs that can be generated using a silicon detector irradiated by x-ray of energy 1.54 keV. The band gap of silicon is 1.1 eV.

ANSWERS

1. (c)	**2.** (a)	**3.** (b)	**4.** (d)	**5.** (b)	**6.** (c)
7. (d)	**8.** (b)	**9.** (b)	**10.** (c)	**11.** (3.5 to 3.6)	
12. (151 to 152)	**13.** (4.8 to 4.9)	**14.** (1.50 to 1.55)	**15.** (0.47 to 0.475)	**16.** (2.02 to 2.3)	**17.** (b)
18. (3.5 to 3.6)	**19.** (b)	**20.** (588 to 592)	**21.** (8675 to 8680)	**22.** (1400)	

EXPLANATIONS

1. (c) Increasing order of melting point, gallium < aluminium < gold < tungsten

2. (a) The atoms in a solid are separated by their equilibrium distance, due to the potential energy of the solid is lowest.

3. (b) Following material are Teflon (PTFE)
   → Thermosets
   → Elastomers
   → Block copolymers
   But Thermoplastics is not a Teflon (PTFE).

4. (d) For the glass transition temperature (Tg), upon heating through Tg, both the heat capacity and thermal expansion coefficient change.

5. (b) The slope of a graph of $\log_e$(Conductivity) Vs $\frac{1}{T}$ for an intrinsic semiconductor with energy gap (Eg) is $= -\frac{Eg}{2K}$

6. (c) Following are the ceramic forming process
   → Extrusion
   → Slip casting
   → Tape casting
   Forging is not a ceramic forming process, because, used for increase the physical property of material.

7. (d) Following are the soft magnetic materials.
   → Iron-silicon steel
   → Nickel Zinc ferrite
   → Nickel iron alloy
   Alcino is not a soft magnetic material.

8. (b) In the eutectic reaction, one liquid phase is converteal into the two solid phase.
   At eutectic point
   $$L \rightarrow S_1 + S_2$$
   L – Liquid
   S – Solid

9. (b) The self-diffusion is generally controlled by the valency mechanism in FCC metal as the vacancy to adsorbate molecular size play a major role in determining diffusion properties.

10. (c) DSC – Glass transition temperature
    XRD – Crystal structure
    STM – Thermal expansion coefficient
    SEM – Density of states

12. (151 to 152)
    $$l = \frac{\delta l}{\Delta T.\alpha}$$
    $$= \frac{.05}{30 \times 11 \times 10^{-6}}$$
    $$= 151.51 \text{ cm}$$

15. (0.47 to 0.475)
    $$C = \frac{k\epsilon_0 A}{d}$$
    $\epsilon_0 = 8.854 \times 10^{-12}$ F/m
    k = dielectric constant
    $$C = \frac{2000 \times 8.854 \times 10^{-12} \times .2 \times 10^{-4}}{.074 \times 10^{-2}}$$
    $$C = .47 \text{ F}$$

17. (b) Y-doped ZrO_2 – Oxygen sensor
    UO_2 – Nuclear fuel
    Si_3N_4 – Turbine engine
    AlN – Integrated Circuit Substrate
    Cr doped Al_2O_3 – Lasers

18. (d) Creep is defined as slow plastic deformation due to diffusion of atoms usually at high temperature.
    → Creep is slow and progressive deformation of a material with time dependent properties.

19. (b) Planar density $= \frac{\text{No. of atoms}}{\text{Plane area}}$

2016

(Q. 1 – 9) : Carry One Mark Each.

1. Energy Dispersive Spectroscopy (EDS) in a typical scanning electron microscope enables elemental identification by collecting and examining which of the following:

(*a*) Secondary electrons from the sample

(*b*) Back scattered electrons from the sample

(*c*) Characteristic X-rays from the sample

(*d*) Diffraction pattern from the sample

2. Which of the following rotational symmetry is forbidden in a perfectly periodic 3-dimensional lattice?

(*a*) 1-fold (*b*) 3-fold

(*c*) 5-fold (*d*) 6-fold

3. Which of the following thermodynamic properties shows a discontinuity during a second-order phase transition?

(*a*) Volume (*b*) Enthalpy

(*c*) Entropy (*d*) Heat capacity

4. Cross slip is easily promoted in metals having

(*a*) a low stacking fault energy

(*b*) a low grain boundary energy

(*c*) a high stacking fault energy

(*d*) a high grain boundary energy

5. For a typical metal at room temperature and atmospheric pressure, the Fermi energy is defined as the energy level for which the probability of occupancy is:

(*a*) 0 (*b*) 0.25

(*c*) 0.5 (*d*) 1

6. Number of elements in a tensor of rank 4 is _____.

7. Which one of the following effects is the working principle of a thermocouple?

(*a*) Thomson

(*b*) Seebeck

(*c*) Peltier

(*d*) Meissner

8. At equilibrium, the maximum number of phases that can coexist in a ternary system at constant pressure is _________.

9. Defect-free single crystal alumina (sapphire) is

(*a*) opaque and white

(*b*) transparent

(*c*) translucent

(*d*) opaque and black

(Q. 10 – 22) Carry Two Marks Each.

10. Match the following processes and the products obtained:

P: Mechanical attrition	1: Thin films
Q: Physical vapour deposition	2: Plastics
R: Injection moulding	3: Nanoparticles
S: Sintering	4: Rails
	5: Carbide tools

(*a*) P-1, Q-2, R-3, S-5

(*b*) P-3, Q-1, R-2, S-5

(*c*) P-4, Q-1, R-3, S-2

(*d*) P-3, Q-4, R-1, S-2

11. In a diffraction experiment, monochromatic X-rays of wavelength 1.54 Å are used to examine a material with a BCC structure. If the lattice parameter is 4.1 Å, the angular position θ of the first diffraction peak is__________degrees.

12. The yield strength of a ferritic steel increases from 120 MPa to 150 MPa when the grain size is decreased from 256 μm to 64 μm. When the grain size is further reduced to 16 μm, the expected yield strength is______________MPa.

13. A direct bandgap semiconductor has a bandgap of 1.8 eV. The threshold value of the wavelength BELOW which this material will absorb radiation is______________Å. (Given: Planck's constant, $h = 6.626 \times 10^{-34}$ J s, the charge of an electron, $e = 1.6 \times 10^{-19}$ C, and speed of light, $c = 3 \times 10^{8}$ m s^{-1})

14. A half cell consisting of pure Ni immersed in an aqueous solution containing Ni^{2+} ions of unknown concentration, is galvanically coupled with another half cell consisting of pure Cd immersed in a 1 M aqueous solution of Cd^{2+} ions. The temperature is 25 °C and pressure is 1 atm. The standard electrode reduction potentials of Ni and Cd are –0.250 V and – 0.403 V, respectively. The voltage of the cell is found to be zero. The concentration of Ni^{2+} in the solution is_____$\times 10^{-6}$ M.

(Given: Universal gas constant, R = 8.31 J mol^{-1} K^{-1}, Faraday's constant, F = 96500 C mol^{-1})

15. Match the type of magnetism given in Group 1 with the material given in Group 2:

Group 1	**Group 2**
P: Ferromagnetic	1: Nickel oxide
Q: Ferrimagnetic	2: Sodium
R: Antiferromagnetic	3: Magnetite
S: Paramagnetic	4: Cobalt

(*a*) P-4, Q-3, R-1, S-2

(*b*) P-4, Q-1, R-3, S-2

(*c*) P-1, Q-2, R-4, S-3

(*d*) P-3, Q-2, R-1, S-4

16. Gallium is to be diffused into pure silicon wafer such that its concentration at a depth of 10^{-3} cm will be one half the surface concentration. Given that the diffusion coefficient (D) of gallium in silicon at 1355 °C is 6×10^{-11} cm^2 s^{-1}, the time the silicon wafer should be heated in contact with gallium vapour at 1355 °C is______s. (Given: erf(0.5) ≅ 0.5)

17. A batch of spherical titania nanoparticles, uniform in size, has a specific surface area of 125 m^2 g^{-1}. If the density of titania is 4.23 g cm^{-3}, the diameter of the particles is__________nm.

18. Given the probability distribution function

$$f(x) = \begin{cases} 0.25x \text{ for } 1 \le x \le 3 \\ 0 \text{ otherwise} \end{cases}$$

The probability that the random variable x takes a value between 1 and $\sqrt{5}$ is______.

19. In the vulcanization of 50 g of natural rubber, 10 g of sulfur is added. Assuming the mer to S ratio is 1:1, the maximum percentage of cross-linked sites that could be connected is__________%. (Given: atomic weight of S is 32 amu and molecular weight of a mer of natural rubber is 68 amu)

20. Match the heat treatment process of steels given in Group 1 with the microstructural feature given in Group 2:

Group 1	**Group 2**
P: Quenching	1: Bainite
Q: Normalizing	2: Martensite
R: Tempering	3: Pearlite
S: Austempering	4: Iron carbide precipitates
	5: Intermetallic precipitates

(*a*) P-2, Q-3, R-4, S-1

(*b*) P-3, Q-4, R-5, S-1

(*c*) P-4, Q-1, R-5, S-3

(*d*) P-2, Q-5, R-4, S-3

21. In the photoelectric effect, electrons are ejected

(*a*) at all wavelengths, as long as the intensity of the incident radiation is above a threshold value.

(*b*) at all wavelengths, as long as the intensity of the incident radiation is below a threshold value.

(*c*) at all intensities, as long as the wavelength of the incident radiation is below a threshold value.

(*d*) at all intensities, as long as the wavelength of the incident radiation is above a threshold value.

22. The angle between [110] and [111] directions in the cubic system is_____degrees.

ANSWERS

1. (c)	**2.** (c)	**3.** (d)	**4.** (c)	**5.** (c)	**6.** (81 to 81)
7. (b)	**8.** (4 to 4)	**9.** (b)	**10.** (b)	**11.** (15.35 – 15.45)	**12.** (210 to 210)
13. (6850 to 6950)	**14.** (6.50 to 6.75)	**15.** (a)	**16.** (16666 to 16668)	**17.** (11.30 to 11.40)	**18.** (0.5 to 0.5)
19. (42 to 43)	**20.** (a)	**21.** (c)	**22.** (35 to 36)		

EXPLANATIONS

1. (c) Energy Dispressive spectroscopy (EDS) in a typical scanning electron microscope enables elemental identification by collecting and examining by characteristic X-rays, from the sample.

2. (c) The 5-fold rotational symmetry is forbidden in a perfectly periodic 3-dimensional lattice.

3. (d) The thermodynamic properties heat capacity show discontinuity during a second-order phase transition.

4. (c) Cross slip is easily promoted in metals having a high stacking fault energy.

5. (c) For a typical metal at room temperature and atmospheric pressure, the Ferms energy is defined as the energy level for 0.5 of probability of occupancy.

6. (81 to 81)

Number of elements, in a tensor rank 4 is

No. of elements = $3^4 = 81$

7. (b) Thermocouple works on the principle of Seebeck effect.

8. (4 to 4)

At equilibrium, the maximum number of phases that can co-exist in a ternary system at constant pressure is equal to 4.

9. (b) Defect-free single crystal alumina (Sapphire) is transparent.

10. (b) Mechanical attrition – Nanoparticles

Physical Vapour deposition – Thin film

Injection moulding – Plastics

Sintering – Carbide Tools

15. (a) Ferromagnetic – Cobalt

Ferri magnetic – Magnetite

Antiferromagnetic – Nickel Oxide

Paramagnetic – Solidum

20. (a) Quenching – Martensite

Normalizing – Pearlite

Tempering – Iron Carbide precipitates

Austemping – Bainite

2017

1. Which of the following is a Frenkel defect?

   (*a*) One Cl vacancy and one Na vacancy in NaCl

   (*b*) One Zn vacancy and one Zn interstitial in ZnO

   (*c*) K at the Na site in NaCl

   (*d*) None of the above

2. Which processing technique is best suited for manufacturing decorative PVC floor tiles?

   (*a*) Blow molding

   (*b*) Filament winding

   (*c*) Rotational molding

   (*d*) Calendering

3. During deformation of a semi-crystalline polymer, with spherulitic morphology, stressed in tension, what happens to the amorphous and the crystalline regions at the later stages?

   (*a*) Amorphous regions remain intact and only crystallites experience bending and stretching of chains

   (*b*) Only amorphous regions elongate in the stress direction and crystallites remain intact

   (*c*) Amorphous regions elongate in the stress direction and crystallites experience bending and stretching of chains

   (*d*) None of the above

4. Which of the following statement(s) is/ are true regarding the structure-property correlation in polymers?

   (i) Polymers that are less coiled are more crystalline than those that are more coiled

   (ii) Branched polymers are more crystalline than the linear ones

   (iii) Polymers with inter-chain interactions have higher glass transition temper-ature than those without inter-chain interactions

   (iv) Polymers with inter-chain interactions are more crystalline than those without inter-chain interactions

   (*a*) (i) and (ii) (*b*) (i) and (iii)

   (*c*) (ii) and (iv) (*d*) (ii) and (iii)

5. The contrast obtained in scanning electron microscope using back scattered electrons depends on

   (*a*) Atomic number of the specimen material

   (*b*) Accelerating voltage of the microscope

   (*c*) Working distance in the microscope

   (*d*) Type of the electron emitter in the microscope

6. Ceramic materials fail at stresses much lower than their theoretical strength due to

   (*a*) Presence of dislocations

   (*b*) High elastic modulus

   (*c*) Presence of voids

   (*d*) Anisotropy in crystal structure

7. The miller indices of the first three Bragg peaks in the X-ray diffraction pattern obtained from a polycrystalline iron sample at room temperature are

   (*a*) (111), (200), (220)

   (*b*) (100), (110), (111)

   (*c*) (100), (110), (200)

   (*d*) (110), (200), (220)

8. The number of close packed planes in the lattice of an FCC metal is

   (*a*) 2 (*b*) 4

   (*c*) 6 (*d*) 12

9. Which of the following treatment(s) can increase the electrical conductivity of silicon

   (i) Heating

   (ii) Doping with arsenic

   (iii) Doping with aluminium

   (iv) Exposure to light

   (*a*) Only (i)

   (*b*) Only (i) and (ii)

   (*c*) Only (i), (ii) and (iv)

   (*d*) All (i), (ii), (iii) and (iv)

10. The unit cell volume of polyethylene (PE) is 0.0933 mm^3. Assuming two ethylene repeat units are contained within each unit cell, the density of a totally crystalline PE will be ......g/cm^3 (Take the atomic weights for carbon and hydrogen as 12.01 g/mol and 1.008 g/mol, respectively and the Avogadro's number as 6.023×10^{23} repeat units/mol)

11. A continuous, aligned carbon fibre (CF) reinforced polymer composite with 30 vol% of CF and rest resin was designed for a specific application. The modulus of elasticity of CF is 170 GPa and that of the resin is 3.0 GPa. The modulus of elasticity for this composite in the direction of fibre alignment is ....GPa.

12. Match the composites in Column I with the most suitable application in Column II

Column I	Column II
(P) Exfoliated silicates filled butyl rubber	(1) Automobile pistons
(Q) Fibre reinforced aluminium alloy	(2) Contact lenses
(R) Silicon carbide whiskers reinforced alumina	(3) Ski boards
(S) Carbon particles reinforced plastic composites	(4) Tennis balls
	(5) Cutting tool inserts for machining

(*a*) P-4; Q-1; R-5; S-3 (*b*) P-2; Q-3; R-4; S-5
(*c*) P-3; Q-5; R-5; S-3 (*d*) P-2; Q-1; R-3; S-5

13. Match the processes in Column I with products in Column II

Column I	Column II
(P) Slip casting	(1) Metal powders
(Q) Zone refining	(2) Thin films
(R) Sputtering	(3) Ceramic parts
(S) Atomization	(4) Single crystal
	(5) Metal sheets

(*a*) P-3; Q-4; R-2; S-1 (*b*) P-2; Q-1; R-2; S-1
(*c*) P-3; Q-4; R-5; S-1 (*d*) P-3; Q-4; R-1; S-5

14. The value of diffusivity (D) for the diffusion of carbon (C) in γ-iron at 727°C is ..... x10^{-13} m^2/s (Given D_0 = 2x 10^{-5} m^2/s; activation energy, Q = 142 kJ/ mol; R = 8.314 J/mol. K)

15. Refer to the figure below:

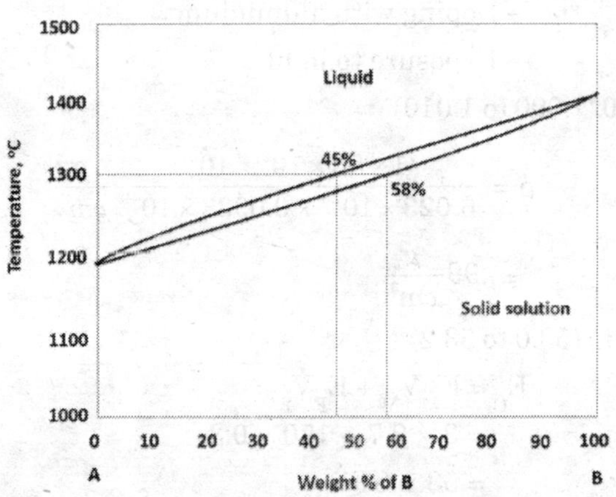

If the alloy contains 47% wt. % of A and 53 wt. % of B at 1300°C, the wt.% of liquid present in the alloy at this temperature will be ....

16. Which of the following statement(s) is/are true

(i) All piezoelectric materials are necessarily ferroelectric

(ii) All ferroelectric materials are necessarily piezoelectric

(iii) All pyroelectric materials are necessarily piezoelectric

(iv) All pyroelectric materials are necessarily ferroelectric

(*a*) (i) and (ii) (*b*) (ii) and (iii)
(*c*) (i) and (iv) (*d*) (ii) and (iv)

17. If the energy of formation of vacancies in pure copper is 0.9 eV, the fraction of vacancies in pure copper at 27°C will be ....x10^{-16} (Boltzmann's constant is 8.62x10^{-5} eV/K)

18. A ceramic material with a critical flaw size of 30 m has fracture stress of 300 MPa. For the same material the fracture stress for a critical flow size of 90 m will be ....MPa.

19. An inorganic material that is transparent under solar light appears coloured when doped with transition metal ions. The possible reason(s) for the color is/are

(i) The electronic energy levels of the host material changes significantly by doping

(ii) The doped element selectively absorbs certain wavelength of light other than the perceived colour

(iii) The doped element emits radiation of specific wave length

Which of the above stacment(s) is/are true?

(*a*) Only (i) (*b*) Both (i) and (ii)
(*c*) Both (i) and (iii) (*d*) Both (ii) and (iii)

20. Copper is an FCC metal with lattice parameter of 3.62Å. Hall effect measurement shows electron mobility to be 3.2x 10^{-3} $m^2V^{-1}s^{-1}$. Electrical resistivity of copper is 1.7 x10^{-8} Ωm. The average number of free electrons per atom in copper is .... (Charge of an electron: 1.6 x10^{-19} C)

21. In a an ionic solid the cation and the anion have ionic radii as 0.8Å and 1.6Å respectively. The maximum coordination number of the cation in the structure will be

(*a*) 3 (*b*) 4
(*c*) 6 (*d*) 8

22. Which of the following statement(s) is/are true regarding susceptibility of a material

(i) Magnetic susceptibility is positive for a diamagnetic material

(ii) Magnetic susceptibility is negative for a diamagnetic material

(iii) Magnetic susceptibility is negative for an antiferromagnetic material

(iv) Magnetic susceptibility is positive for a paramagnetic material

(*a*) (ii) and (iv) (*b*) (i) and (iii)

(*c*) (ii) and (iii) (*d*) (i) and (iv)

ANSWERS

1. (b)	**2.** (d)	**3.** (c)	**4.** (b)	**5.** (a)	**6.** (c)
7. (d)	**8.** (b)	**9.** (d)	**10.** (.990 to 1.010)	**11.** (53.0 to 53.2)	**12.** (a)
13. (a)	**14.** (7.5 to 7.7)	**15.** (37.5 to 38.5)	**16.** (b)	**17.** (7.50 to 7.80)	**18.** (172 to 174)
19. (d)	**21.** (1.35 to 1.37)	**22.** (c)			

EXPLANATIONS

1. (b) In ionic crystals where cations, smaller in size and onions are present, if cations are displaced from their regular location to interstitial location then the defect is called Frenkel defect. So that one Zn vacancy and one Zn interstitial is Zno.

2. (d) Calendering technique used for manufacturing foils, film sheet out of plastics such as pE, pET and rigid pvc, as well as rubber, metal and paper. A calender comprises several heated rolled made of polished steel or chilled cast iron. Calendering is also used for embossing, smoothing, compressing or satinizing as well as for consolidating polymer fibres into honwoven fabrics.

3. (c) During deformation of a semi-crystalline polymer, with spherulitic morphology, stressed intension the Amorphous regions elongate in the stress direction and crystalline experiences bending and stretching of chains.

4. (b) Following are the property, regarding the structure polymer correlation.

→ Polymer that are less coiled are more crystalline than those that are more coiled.

→ Polymer microscope usign back scattered electrons depend on, atomic number of the specimen material.

5. (a) Weaker than atomic number contrast requires good BSE detector.

BSE images show atomic number contrast.

6. (c) Ceramic materials fail at stress much lower than their theoretical strength due to presence of voids. because the ceramics tend to be weak in tension but strong in compression, and the ceramic having single phase. Due to this ceramic are brittle material.

7. (d) Shows (110), (200), (220)

8. (b) Face centred cubic (FCC) packing is a three dimensional packing of sphere, obtained from two-dimensional hexagonal close packing. This is one of the most effecient closed packing.

co-ordination Number of FCC = 12

Effective number of atom in FCC = 4

Atomic packing factor of FCC = 0.74

9. (d) Following are the treatments can increase the electrical conductivity of silicon.

→ Heating

→ Doping with arsenic

→ Doping with aluminium

→ Exposure to light

10. (.990 to 1.010)

$$\rho = \frac{1 \times 56 \times 10^9 \times 10^3}{6.023 \times 10^{23} \times 0.0933 \times 10^{-9}} \frac{g}{cm^3}$$

$$= .99 \frac{g}{cm^3}$$

11. (53.0 to 53.2)

$$E_C = E_M V_M + E_F V_F$$
$$= 3 \times 0.7 + 170 \times 0.3$$
$$= 53.1 \text{ GPa}$$

12. (a) Exforliated silicates filled butyl rubber – Tennis Ball.

Fiber reinforced aluminium alloy – Automobile piston.

Silicon carbide whisker reinforced alumina – cutting toward inserts for matching.

Carbon particles reinforced plastic composites – Ski boards

13. (a) Slip Casting – Ceramic parts

Zone refining – Single crystal

Sputtering – Thin films

Atomization – Metal powders

14. (7.5 to 7.7)

Given that

T = 727°C = 1000 K

$D_o = 2 \times 10^{-5}$ m²/s

Q = 142 kJ/mol

R = 8.314 J/mol.k

$$\text{Diffusivity (D)} = D_o e^{\frac{-Q}{RT}}$$

$$D = 2 \times 10^{-5} e^{\frac{-142 \times 1000}{1000 \times 8.314}}$$

$$\boxed{D = 7.64 \times 10^{-13}\ m^2/s}$$

16. (b) In the given statement (ii) and (iii) are correct.

(ii) All ferroelectric materials are necessarily piezoelectric.

(iii) All pyroelectric materials are necessarily piezoelectric.

17. (7.50 to 7.80)

Given that

E = 0.9 eV

T = 27°C = 300 K

$K = 8.62 \times 10^{-5}$ eV/k

$$\text{Energy formation} = e^{\frac{-E}{kT}}$$

$$= e^{\frac{-0.9}{8.62 \times 10^{-5} \times 300}}$$

$= 7.67 \times 10^{-16}$

18. (172 to 174)

Given that

$\sigma_1 = 300$ MPa

$a_1 = 30$ µm

$a_2 = 90$ µm

$\sigma_2 = ?$

$$k = \sigma\sqrt{\pi a_2}$$

$$\sigma_1\sqrt{\pi a_1} = \sqrt{2}\sqrt{\pi a_2}$$

$$\sigma_2 = 300\sqrt{\frac{\pi \times 30}{\pi \times 90}} = 173.2 \text{ MPa}$$

19. (d) An inorganic materials that is transparent under solar light apears coloured, when doped with transitions ions metal, due to,

→ The doped element selectively absorbs certain wavelength of light other than the perceived colour.

→ The doped element emits radiation specific wave length.

21. (1.35 to 1.37)

Given that

$r_{cation} = 0.8$ A°

$r_{anion} = 1.6$ A°

$$\text{radius ratio} = \frac{r_{cation}}{r_{anion}} = \frac{0.8}{1.6} = 0.5$$

If radius ratio lies between 0.414 – 0.732 having co-ordination number is 6.

22. (c) For susceptibility of a material following statement are correct.

→ Magnetic susceptibility is negative for a diamagnetic material.

→ Magnetic susceptibility is positive for a paramagnetic material.

2018

Q. 1 – Q. 9 carry one mark each.

1. The stress ratio for a completely reversed cyclic loading during a fatigue test is

(*a*) 0 (*b*) 1

(*c*) –1 (*d*) –1/2

2. Minimum symmetry that a cubic crystal must possess is

(*a*) four 3-fold rotation axes.

(*b*) three 4-fold rotation axes.

(*c*) three orthogonal mirror planes.

(*d*) centre of symmetry.

3. If a material is repelled in an external magnetic field then it is

(*a*) Ferromagnetic (*b*) Diamagnetic

(*c*) Paramagnetic (*d*) Antiferromagnetic

4. An electron makes a transition from the valence band to the conduction band in an indirect band gap semiconductor. Which one of the following is true?

(*a*) Energy of the electron decreases.

(*b*) A photon is emitted in the process.

(*c*) A photon is annihilated in the process.

(*d*) A photon is created in the process.

5. Which one of the following is the characteristic of a screw dislocation?

(*a*) Dislocation line and Burgers vector are parallel.

(*b*) Direction of motion of dislocation is parallel to the Burgers vector.

(*c*) Atomic displacement due to the movement of the dislocation is in the direction of the motion of the dislocation line.

(*d*) It has a unique slip plane.

6. The number of vibrational degrees of freedom for a non-linear triatomic molecule are

(*a*) 9

(*b*) 6

(*c*) 4

(*d*) 3

7. An atom is restricted to move in one dimension by making unit jumps either to the left or right, as shown in the figure. Assuming that a jump to the left or right is equally probable, the probability of the atom returning back to the starting point after four jumps is

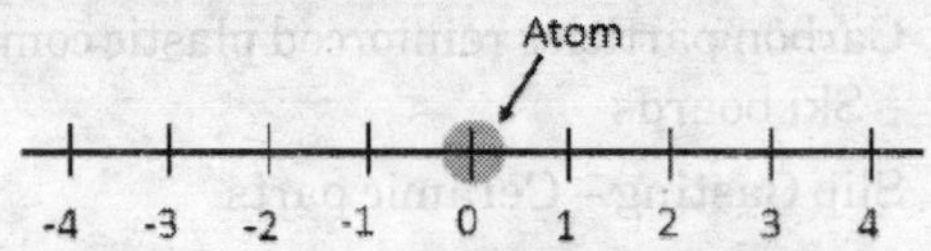

(*a*) 0.250 (*b*) 0.333

(*c*) 0.375 (*d*) 0.500

8. For a two-dimensional solid, the variation of lattice specific heat as a function of temperature T (in K, at low temperatures) is given as: $C_p = bT^n$, where b is a constant. The value of n is ________.

9. If the cation (C) to anion (A) radius ratio, r_C/r_A is 0.6, then the coordination number (i.e., number of A ions surrounding a C ion) is likely to be ________.

Q. 10 – Q. 22 carry two marks each.

10. Match the invariant reactions in Column I with the names in Column II (L is liquid phase, and α, β, γ are solid phases). All reactions proceed to the right on cooling.

Column I	Column II
(P) $L \rightleftarrows \alpha + \beta$	(1) Monotectic
(Q) $L + \alpha \rightleftarrows \beta$	(2) Peritectoid
(R) $\gamma \rightleftarrows \alpha + \beta$	(3) Peritectic
(S) $\alpha + \beta \rightleftarrows \gamma$	(4) Eutectoid
	(5) Eutectic

(*a*) P-5, Q-1, R-4, S-3 (*b*) P-5, Q-3, R-4, S-2

(*c*) P-5, Q-1, R-2, S-4 (*d*) P-2, Q-1, R-4, S-5

11. Consider the following anodic (oxidation) reaction in an acidic solution:

$$Mg \longrightarrow Mg^{+2} + 2e^-$$

If 48250 Coulomb charge is produced during this anodic reaction then the amount of Mg (in g) dissolved into the solution is

(Given: Faraday Constant =96500 C/mole of electrons, Atomic weight of Mg = 24)

(*a*) 6 (*b*) 12

(*c*) 24 (*d*) 48

12. An intrinsic semiconductor has conduction electron concentration, $n = 10^{12}\ cm^{-13}$. The mobility of both electrons and holes are identical $= 4 \times 10^4\ cm^2\ V^{-1}\ s^{-1}$. If a voltage of 100 V is applied on two parallel end faces of the cube (edge length 1 cm) through Ohmic contacts, the current through the cube would be (in mA)

(Given: charge of electron = 1.6×10^{-19} C)

(*a*) 640 (*b*) 1280

(*c*) 6400 (*d*) 12800

13. An infinite plate with a through-thickness crack of length 2 mm is subjected to a tensile stress (as shown in the figure). Assuming the plate to be linear elastic, the fracture stress is ________ MPa (round off to the nearest whole number)

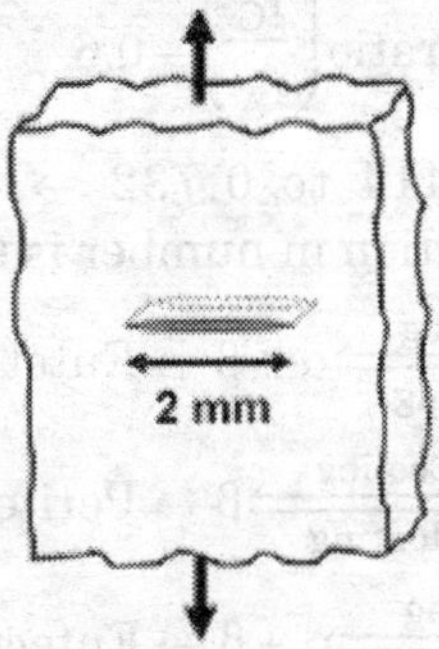

(Given: Fracture toughness, $K_{IC} = 25 MPa\sqrt{m}$)

14. A unidirectionally aligned carbon fibre reinforced epoxy composite is loaded as shown in the figure. The volume fraction of the fibre is 0.6. The Young's modulus of the composite is ________ GPa.

(Given: Young's Modulus of the fibre and the matrix are 200 GPa and 10 GPa, respectively)

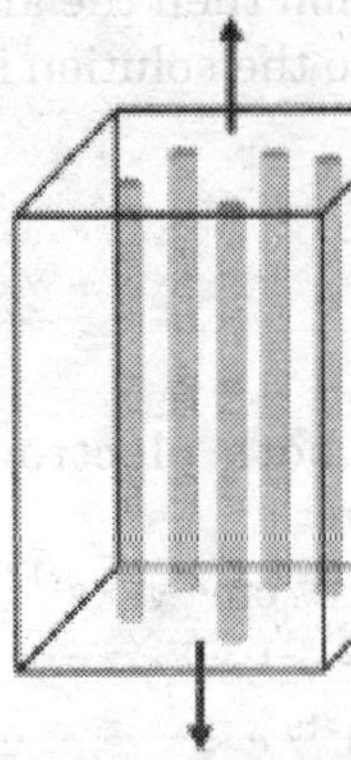

15. A sintered sample was weighed in air and water using an analytical balance. The mass of the sample in air is 2.67 g and its apparent mass in water is 1.67 g. The density of the sample is ________ g cm^{-3} (give answer up to 2 decimal places)

(Given: Density of water = 1.00 g cm^{-3})

16. The atoms in a gas laser have two energy levels such that a transition from the higher to the lower level releases a photon of wavelength 500 nm. If 7×10^{20} atoms are pumped into the upper level with 4×10^{20} atoms in the lower level, the amount of energy released in a single pulse is ________ Joules (give answer up to 2 decimal places)

(Given: Planck's constant, h = 6.6×10^{-34} J s; speed of light, c = 3×10^8 m s^{-1})

17. The speed of an electron is measured to be 300 m s^{-1} with an uncertainty of 0.01%. The fundamental accuracy with which the position of the electron can be determined simultaneously with the speed in the same experiment is ________ mm (give answer up to 2 decimal places)

(Given: Planck's constant, h = 6.6×10^{-34} J s; mass of electron = 9.1×10^{-31} kg)

18. When 3 identical non-interacting spin ½ particles are put in an infinite potential well, the ground state energy of the system is 18 meV. If instead, seven particles are put inside the potential well, the new ground state energy is ________ meV.

19. If the value of the integral (I) is 4, the value of the constant b is ________ (give answer up to 2 decimal places).

$$I = \int_{-\infty}^{\infty} e^{\frac{-x^2}{b}} dx$$

20. X-ray diffraction pattern is obtained from FCC polycrystalline aluminium (lattice parameter = 0.405 nm) using Cr-Kα radiation of wavelength 0.229 nm. The maximum number of peaks that can be observed in the pattern is ________.

21. The planar atomic density in the (110) plane of a BCC iron crystal is ________ nm^{-2} (give answer up to 2 decimal places)

(Given: lattice parameter of iron is 0.287 nm)

22. Mild steel is carburized at 1300 K for 1 hour to obtain a certain case depth. Keeping the time as 1 hour, the case depth can be doubled by increasing the temperature to ________ K (round off to the nearest whole number)

(Given: Activation energy Q = 148 kJ mol^{-1}, Gas constant, R = 8.314 J mol^{-1} K^{-1})

ANSWERS

1. (c)	**2.** (a)	**3.** (b)	**4.** (c)	**5.** (a)	**6.** (d)
7. (c)	**8.** (2 to 2)	**9.** (6 to 6)	**10.** (b)	**11.** (a)	**12.** (b)
13. (430 to 500)	**14.** (124 to 124)	**15.** (2.6 to 2.7)	**16.** (59 to 60)	**17.** (1.85 to 2)	**18.** (132 to 132)
19. (5.01 – 5.17)	**20.** (5 to 5)	**21.** (15.5 to 18.5)	**22.** (1420 to 1480)		

EXPLANATIONS

1. (c)

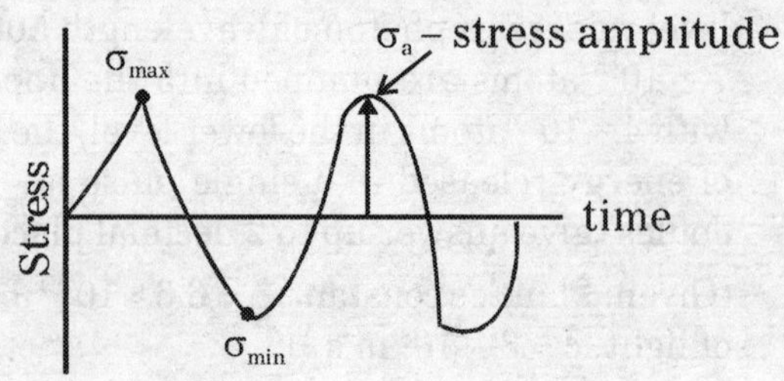

Stress ratio (R) = $\dfrac{\sigma_{min}}{\sigma_{max}}$

$|\sigma_{max}| = |\sigma_{min}|$

Stress ratio for completely reverse cyclic loading is

$\boxed{R = -1}$

2. (a) A cubic unit cell is described as having minimum symmetry of four 3-fold rotation axes.

3. (b) Diamagnetic substances create on induced magnetic field in a direction opposite to an externally applied magnetic field. They are repelled by the applied magnetic field.

4. (c) An electrons make a transition from the valence band to the conduction band in an indirect band gap semiconductor, A phonon is annihilated in the process.

5. (a) Two limiting cases of dislocations, edge and screw, are characterized by Burger's vector perpendicular to the dislocation line and Burger's vector parallel to the dislocation line respectively ordinary dislocation is of mixed character of edge and screw type.

6. (d) A non-linear triatomic molecule has total nine possible degree of free-dom

→ Three translational

→ Three rotational

→ Three vibrational

7. (c)

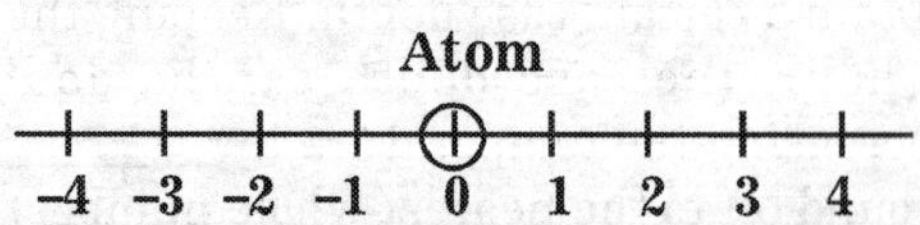

Probability of the atom returning back to the starting point after jump.

$$\boxed{p_r = \frac{9}{24} = 0.375}$$

8. (2 to 2)

For two-dimensional solid, the variation of lattice specific heat as a function of temperature T

$\boxed{C_p = bT^n}$

For two-dimensional

$\boxed{n = 2}$

9. (6 to 6)

Co-ordination number is the number of the nearest neighbours of an atom.

If radius ratio $\left(\dfrac{r_C}{r_A}\right) - 0.6$

r → 0.414 to 0.732 → octahedral and co-ordination m number is 6.

10. (b) $L \underset{\text{Heating}}{\overset{\text{Cooling}}{\rightleftharpoons}} \alpha + \beta \rightarrow$ Eutetic

$L + \alpha \underset{\text{Heating}}{\overset{\text{Cooling}}{\rightleftharpoons}} \beta \rightarrow$ Peritectic

$\gamma \underset{\text{Heating}}{\overset{\text{Cooling}}{\rightleftharpoons}} \alpha + \beta \rightarrow$ Eutectoid

$\alpha + \beta \underset{\text{Heating}}{\overset{\text{Cooling}}{\rightleftharpoons}} \gamma \rightarrow$ Peritectoid

11. (a) $Mg \rightarrow Mg^{+2} + 2e^-$

Amount of Mg dissolved into the solution, is

$= \dfrac{96500}{48250} = 2 = 2 \times 2 = 4$

Anodic reaction then the amount of Mg(in g) dissolved into the solution is,

$= \dfrac{24}{4} = 6$

12. (b) Given that

$n = 10^{12}\ cm^{-3}$

Mobility of both electron and holes are identical then

$m = 2 \times 4 \times 10^4\ cm^2\ v^{-1}.s^{-1}$

$= 8 \times 10^4\ cm^2\ v^{-1}.s^{-1}$

$e^- = 1.6 \times 10^{-19}\ c$

$l = 1\ cm$

$v = 100\ v$

$I = nelmv$

$= 10^{12} \times 8 \times 10^4 \times 1 \times 1.6 \times 10^{-19} \times 100$

$I = 1.28$

$I = 1280\ mA$

14. (124 to 124)

$E_C = E_M V_M + E_F V_F$

$= 0.4 \times 10 + 0.6 \times 200$

$= 124$ GPa

15. (2.6 to 2.7)

$$\rho = \frac{M_{air}}{M_{air} - M_w} = \frac{2.67}{.1} = 2.67 g/cm^3$$

XE-D

SOLID MECHANICS

Syllabus

Equivalent force systems; free-body diagrams; equilibrium equations; analysis of determinate trusses and frames; friction; particle kinematics and dynamics; dynamics of rigid bodies under planar motion; law of conservation of energy; law of conservation of momentum.

Stresses and strains; principal stresses and strains; Mohr's circle for plane stress and plane strain; generalized Hooke's Law; elastic constants; thermal stresses; theories of failure.

Axial, shear and bending moment diagrams; axial, shear and bending stresses; combined stresses; deflection (for symmetric bending); torsion in circular shafts; thin walled pressure vessels; energy methods (Castigliano's Theorems); Euler buckling.

Free vibration of single degree of freedom systems.

Subject-wise Analysis

XE - D : Solid Mechanics

Year	1 mark Questions	2 marks Questions	Total Marks
2009	8	16	40
2010	9	13	35
2011	9	13	35
2012	9	13	35
2013	9	13	35
2014	9	13	35
2015	9	13	35
2016	9	13	35
2017	9	13	35
2018	9	13	35

2009

Q. 1 – 8 carry one mark each

1. A small spherical ball fails at a normal load of 10 kN under the arrangement as shown below. The vertical force F required to crush the ball is

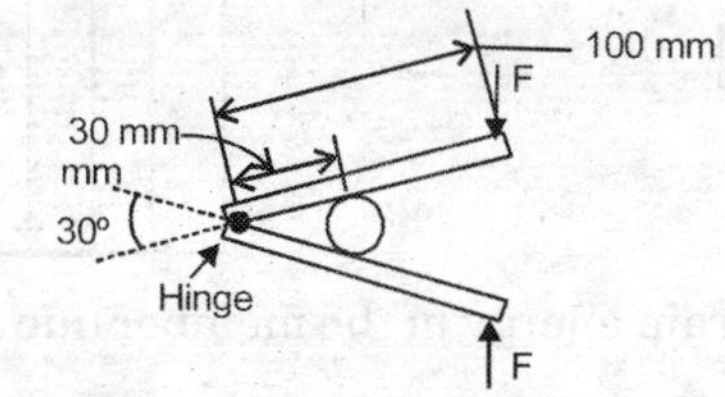

(*a*) 11.6 kN (*b*) 6.0 kN
(*c*) 3.5 kN (*d*) 3.1 kN

2. A projectile is fired from point P at an angle of 45° with horizontal as shown below. If g is the acceleration due to gravity, then the speed required to reach a point Q lying on the horizontal surface at a distance of R from point P is

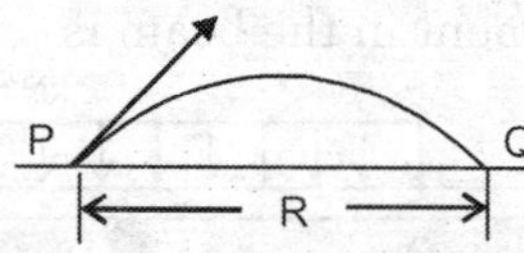

(*a*) $\sqrt{Rg/2}$ (*b*) $\sqrt{Rg}$
(*c*) $\sqrt{2Rg}$ (*d*) $\sqrt{3Rg}$

3. The state of stress at a point in a loaded body is given as σ_x = +40 MPa, σ_y = +60 MPa, τ_{xy} = +10 MPa. The sum of the principal stresses at that point is

(*a*) + 20 MPa (*b*) +50 MPa
(*c*) +100 MPa (*d*) +110 MPa

4. A composite system of two metal bars, as shown below is made of two dissimilar materials having areas of cross section A_1 and A_2, Young's moduli E_1 and E_2 and coefficients of thermal expansion α_1 and α_2. If the temperature of the system is raised by ΔT, then the resultant axial force required to be applied to the rigid end plates to maintain the same length L is

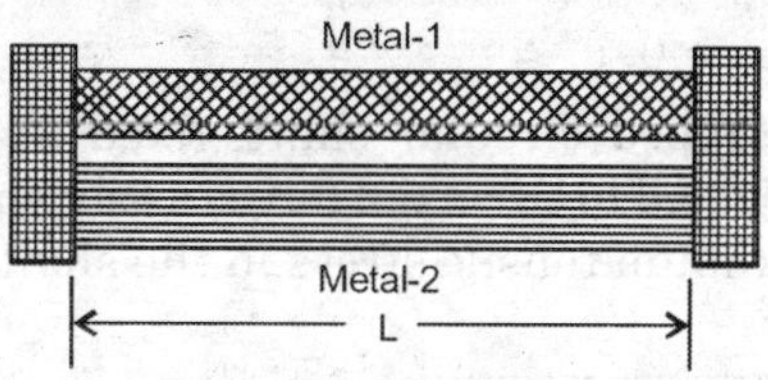

(*a*) $(E_1 \alpha_1 A_1 + E_2 \alpha_2 A_2) \Delta T$
(*b*) $(1/E_1 \alpha_1 A_1 + 1/E_2 \alpha_2 A_2)^{-1} \Delta T$
(*c*) $(E_1 + E_2)(\alpha_1 + \alpha_2)(A_1 + A_2) \Delta T$
(*d*) $(E_1 \alpha_1 A_1/E_2 \alpha_2 A_2) \Delta T$

5. The state of stress at a point is as shown below. Both the normal and shear stresses on a plane, inclined at an angle of 45° with horizontal are zero. If $\sigma_x = \sigma_y$ = 200 MPa,

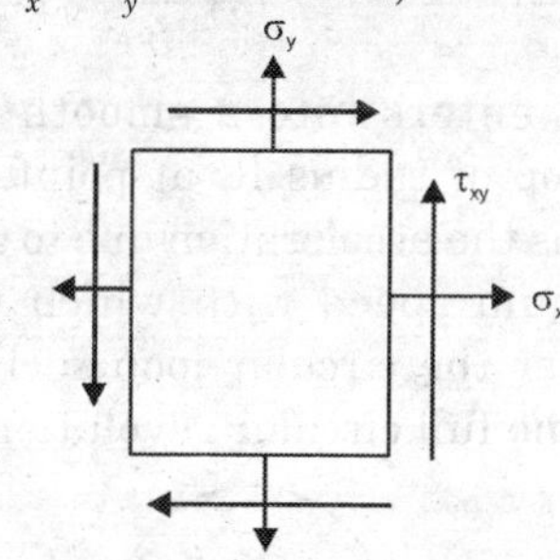

the shear stress τ_{xy} is

(*a*) 50 MPa (*b*) 70 MPa
(*c*) 100 MPa (*d*) 200 MPa

6. A simply supported beam of span L and flexural rigidity EI carries a uniformly distributed load w/unit length. The deflection at the mid span of the beam is

(*a*) $wL^4/(48EI)$ (*b*) $5wL^4/(96EI)$
(*c*) $5wL^4/(384EI)$ (*d*) $3wL^4/(16EI)$

7. During plastic impact of two bodies, which of the following statement is correct?

(*a*) Both energy and momentum are conserved
(*b*) Energy is not conserved; momentum is conserved
(*c*) Energy is conserved; momentum is not conserved
(*d*) Neither energy nor momentum is conserved

8. A disc of radius 1 m is rolling on the ground without slip as shown below. At a certain instant the center of the disc is moving with a velocity of 10 m/s and an acceleration of a = +10 m/s². The magnitude of acceleration of point P on the disc instantaneously touching the ground is

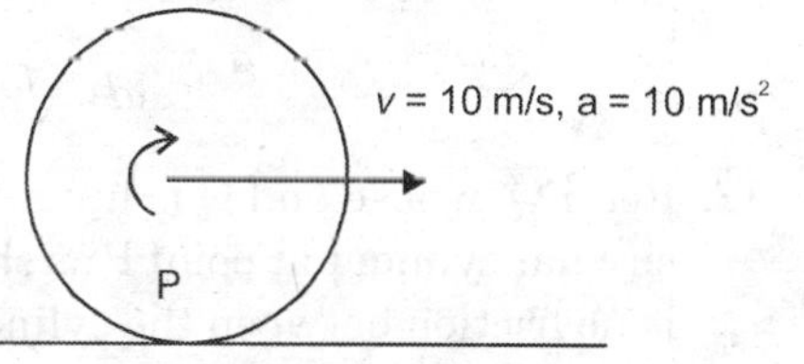

(*a*) 0.0 m/s² (*b*) 10.0 m/s²
(*c*) 20.0 m/s² (*d*) 100.0 m/s²

Q. 9 – 24 carry two marks each

9. A block of length *a* and height b rests on a rough inclined plane (coefficient of friction μ) as shown below. The length α of the inclined plane is slowly increased. The condition that the block will topple due to its own weight before it begins to slide is

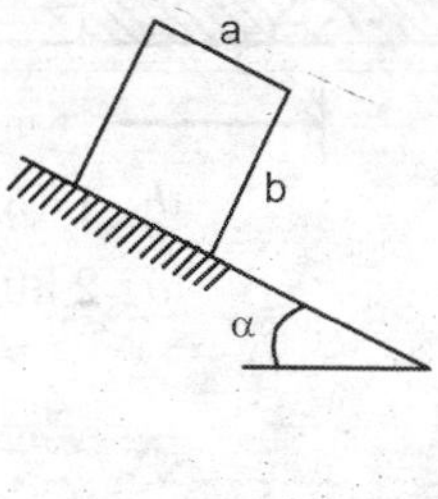

(a) $\frac{a}{b} < \mu$ (b) $\frac{a}{b} > \mu$

(c) $\frac{a}{b} > \sqrt{1-\mu}$ (d) $\frac{a}{b} < \sqrt{1-\mu}$

10. A particle enters into a smooth frictionless circular loop of radius R, at point P as shown below. If g is the acceleration due to gravity, then the minimum speed with which the particle should enter the circular loop such that it can complete one full circular revolution is

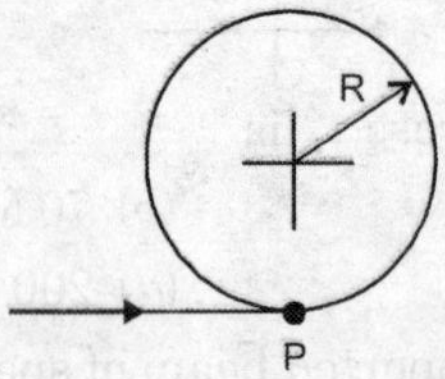

(a) $\sqrt{5Rg}$ (b) $\sqrt{3Rg}$

(c) $\sqrt{2Rg}$ (d) ∞

11. A circular cylinder of radius r and mass m, starts from the top of an inclined plane and rolls down without any slip. After its center moves to a point having a vertical height h as shown below, the velocity of the center of mass, with g as acceleration due to gravity is

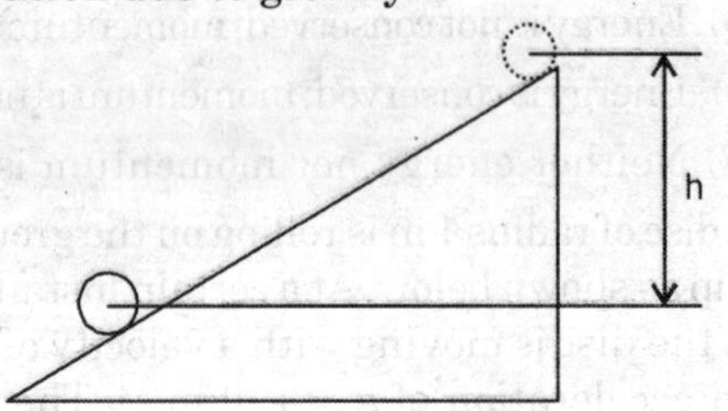

(a) $\sqrt{\frac{gh}{3}}$ (b) $\sqrt{\frac{2gh}{3}}$

(c) $\sqrt{\frac{4gh}{3}}$ (d) $\sqrt{2gh}$

12. Rod PQ, whose end Q is hinged, touches a semi-circular cylinder at point P as shown below. There is no friction between the cylindrical surface and the rod. If the cylinder moves with a constant velocity of 10 m/s in the horizontal direction as shown, then the angular velocity ω of the rod PQ at that instant is

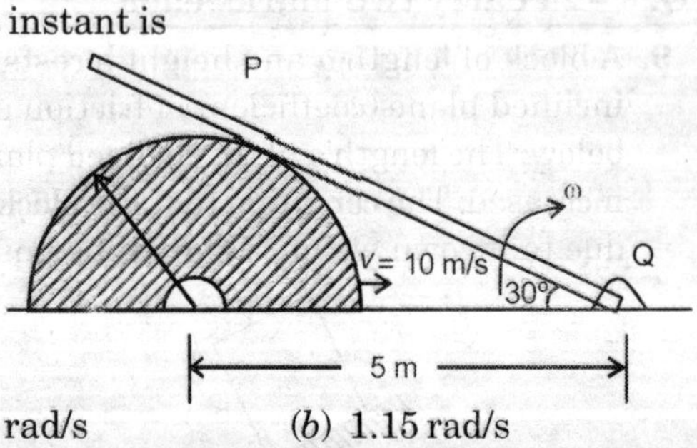

(a) 0.5 rad/s (b) 1.15 rad/s

(c) 2.0 rad/s (d) 2.30 rad/s

13. An L-shaped elastic member with flexural rigidity EI is loaded as shown below :

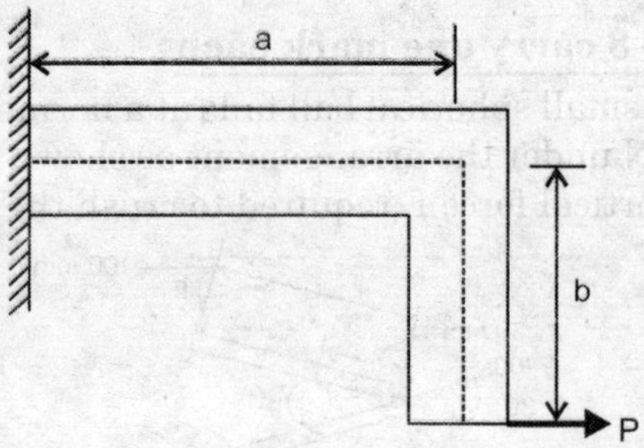

Total strain energy in the member due to bending is:

(a) $P^2 b^2 (b/3 + a) / (2EI)$

(b) $P^2 b^2 (a/3 + b) / (2EI)$

(c) $P^2 a^2 (b/3 + a) / (3EI)$

(d) $P^2 a^2 (a/3 + b) / (3EI)$

14. A simply supported beam with an overhanging end is loaded as shown below. The maximum bending moment in the beam is

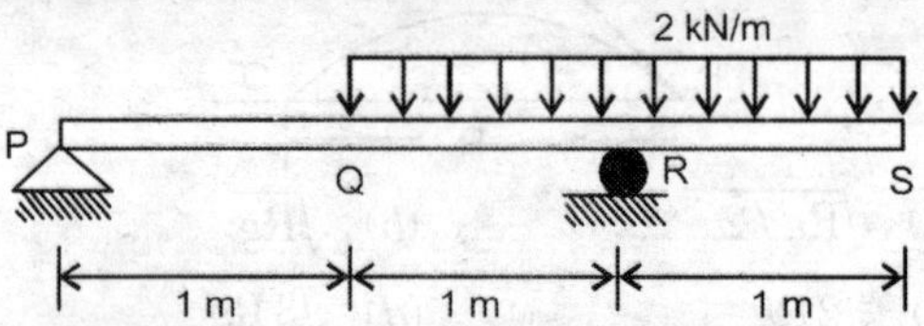

(a) 2 kNm (b) 1 kNm

(c) 0.75 kNm (d) 0.25 kNm

15. A body P while moving rectilinearly with velocity v_o collides directly with another body Q, which is at rest, as shown below. Assuming both the bodies have the same mass and the collision is elastic, the velocities of the bodies after the collision, measured positive towards right, are

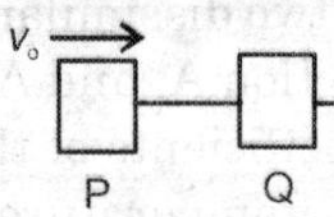

(a) $v_p = -v_o/2,\ v_Q = v_o/2$

(b) $v_p = v_o/2, v_Q = v_o/2$

(c) $v_p = 0,\ v_Q = v_o/2$

(d) $v_p = 0,\ v_Q = v_o$

16. A stepped circular shaft, fixed at one end, is subjected to two axial forces as shown below. The maximum tensile stress in the shaft is

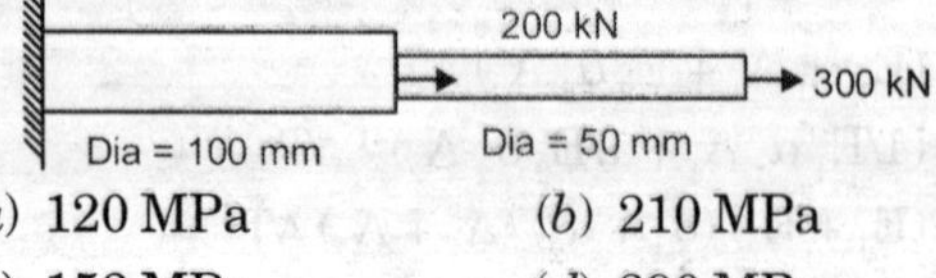

(a) 120 MPa (b) 210 MPa

(c) 153 MPa (d) 390 MPa

17. A thin string of negligible mass with one end fixed to the roof is wound around a circular disc of radius 2 m and mass 10 kg, as shown below. The disc rolls vertically down under the action of its own weight. Considering acceleration due to gravity as 10 m/s², the tension in the string is

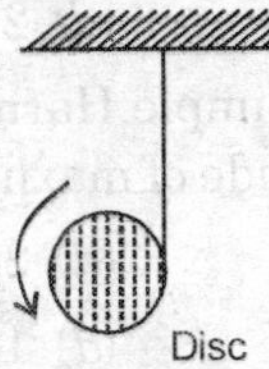

(*a*) 0N (*b*) 25.0 N

(*c*) 33.3 N (*d*) 50 N

18. A spring-mass system executes SHM in the vertical direction given by the equation

$$\frac{d^2y}{dt^2} + \omega^2 y = 0$$

The correct relation (including sense) between acceleration a, and velocity v, of the body is graphically represented by

(*a*)

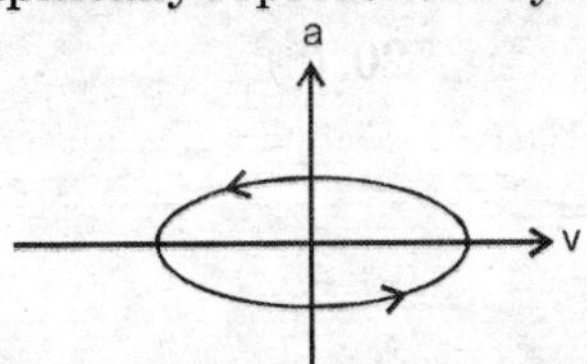

(*b*)

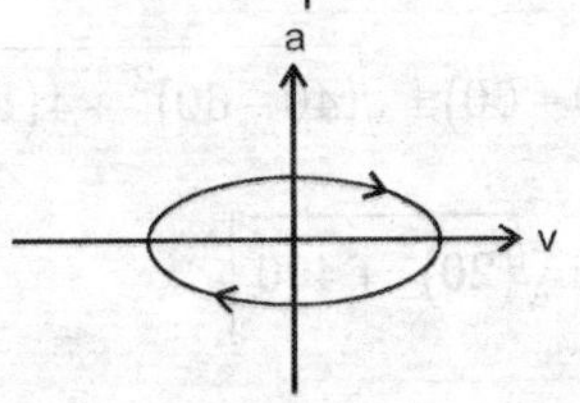

(*c*)

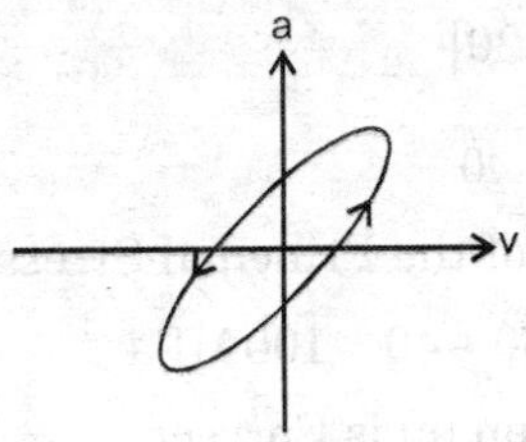

(*d*)

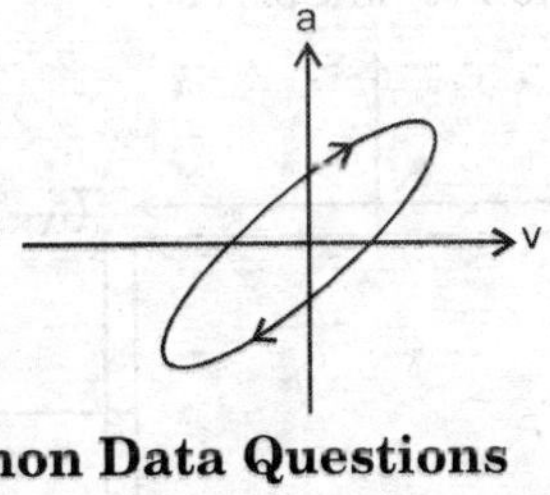

Common Data Questions

Common Data for Questions 19 and 20:

A 10 mm thick steel rectangular plate of size 100 mm × 200 mm is subjected to biaxial stresses of σ_x = 150 MPa, σ_y = 200 MPa, as shown below. The Young's modulus and Poisson's ratio are 200 GPa and 0.3 respectively.

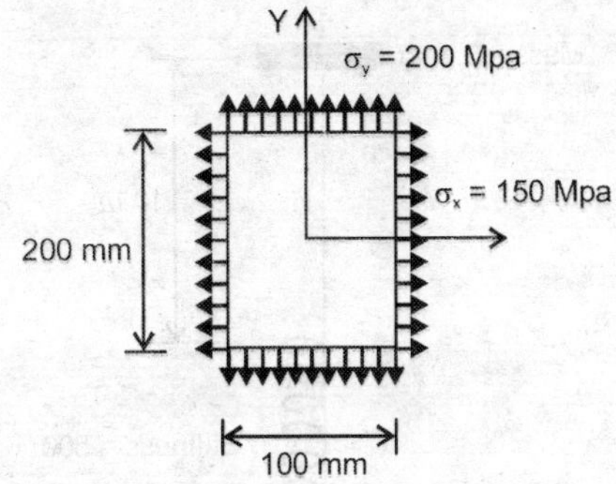

19. The change in the thickness of the plate is

(*a*) 2.39 μm (*b*) 5.25 μm

(*c*) 7.12 μm (*d*) 9.16 μm

20. The change in the surface area of the plate is

(*a*) 9.72 mm² (*b*) 13.61 mm²

(*c*) 17.52 mm² (*d*) 24.50 mm²

Common Data for Questions 21 and 22:

A solid circular steel shaft of 50 mm diameter, fixed at one end, is subjected to torques as shown below. The shearing modulus of the material is 80 GPa.

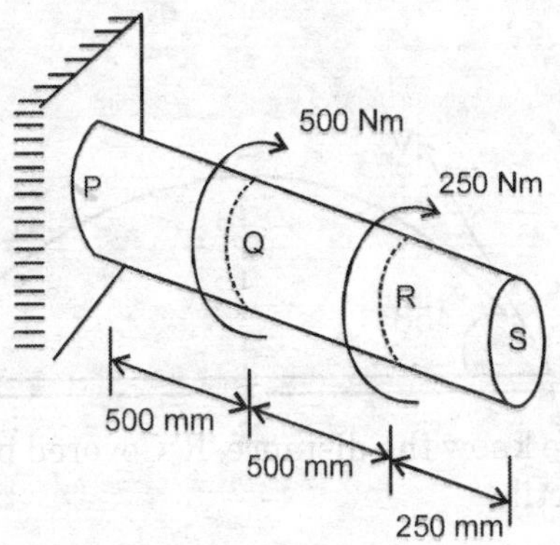

21. The maximum shear stress due to torsion in the length PQ is

(*a*) 15.75 MPa (*b*) 21.22 MPa

(*c*) 30.56 MPa (*d*) 51.21 MPa

22. The rotation of the free end S due to the torsion is

(*a*) 0.25° (*b*) 0.58°

(*c*) 1.22° (*d*) 1.25°

Linked Answer Questions

Statement for Linked Answer Questions 23 & 24:

A body of mass 0.1 kg is dropped from a height of 10 m above a spring of stiffness 500 N/m as shown below. The spring is initially in uncompressed natural state. The impact is without any energy loss and the body gets attached to the spring. The acceleration due to gravity is 10 m/s².

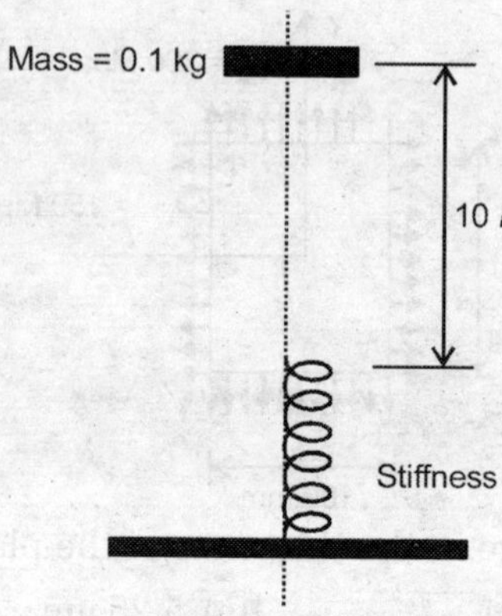

23. The maximum compression of the spring is

(a) 2 mm (b) 20.2 mm

(c) 202.0 mm (d) 2020 mm

24. In the ensuling Simple Harmonic Motion of the body, the magnitude of maximum acceleration is

(a) 100 m/s² (d) 200 m/s²

(c) 500 m/s² (d) 1000 m/s²

ANSWERS

1. (*)	**2.** (b)	**3.** (c)	**4.** (*)	**5.** (c)	**6.** (c)	**7.** (b)
8. (b)	**9.** (b)	**10.** (a)	**11.** (d)	**12.** (d)	**13.** (a)	**14.** (*)
15. (d)	**16.** (c)	**17.** (c)	**18.** (a)	**19.** (*)	**20.** (*)	**21.** (c)
22. (b)	**23.** (202.0 to 202.2)		**24.** (d)			

EXPLANATIONS

2. (b)

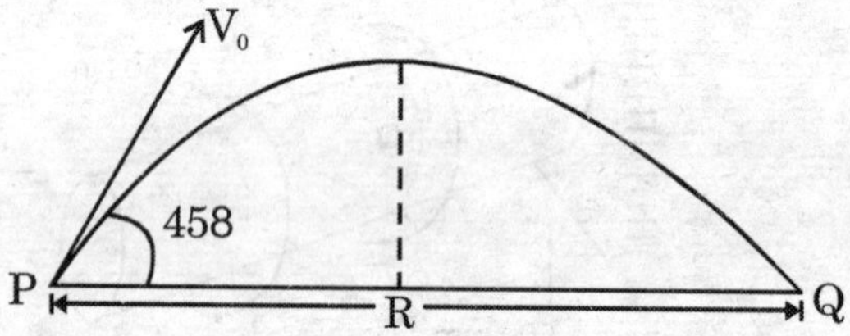

We know the distance 'R' Covered by the Projectile

$$R = \frac{V_0^2 \sin 20}{g} = \frac{V_0^2 \sin(2 \times 45)}{g}$$

$$= \frac{V_0^2 \sin 90}{g}$$

$$R = \frac{V_0^2}{g}$$

$$\therefore \boxed{V_0 = \sqrt{Rg}}$$

Hence the option (b) is Correct.

3. (c) Given that

$\sigma_n = 40\,\text{MPa}$

$\sigma_y = 60\,\text{MPa}$

$\tau_{xy} = 10\,\text{MPa}$

$\therefore$ Principal stresses

$$\sigma_{1,2} = \frac{1}{2}\left[(\sigma_x - \sigma_y)^2 \pm \sqrt{(\sigma_x - \sigma_y)^2 + 4\tau^2_{xy}}\right]$$

$$\sigma_{1,2} = \frac{1}{2}\left[(40+60) \pm \sqrt{(40-60)^2 + 4(10)^2}\right]$$

$$= \frac{1}{2}\left[100 \pm \sqrt{(20)^2 + 400}\right]$$

$$\sigma_{1,2} = \frac{1}{2}[100 \pm 20]$$

$\therefore \sigma_1 = 60$ & $\sigma_2 = 40$

therefore sum of the Principal Stresses

$\sigma = \sigma_1 + \sigma_2 = 60 + 40 = 100\,\text{MPa}$

Hence the option (c) is Correct.

5. (c)

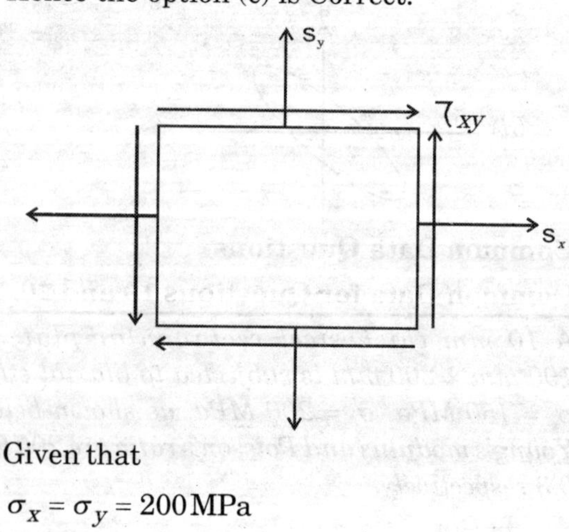

Given that

$\sigma_x = \sigma_y = 200\,\text{MPa}$

At $Q = 45°$,

Shear Stress $\tau = \frac{\sigma}{2}\sin(2 \times 45)° = \frac{\sigma}{2}\sin 90°$

$\therefore \tau = \tau_{Max} = \frac{\sigma}{2} = \frac{200}{2} = 100\,MPa$

$\therefore \boxed{\tau_{Max} = 100\,MPa}$

$\left[\tau_{Max} = \frac{\sigma}{2} \cdot \text{if } \sigma_x = \sigma_y = \text{Same in Nature}\right]$

Hence the option (*c*) is Correct.

6. (c)

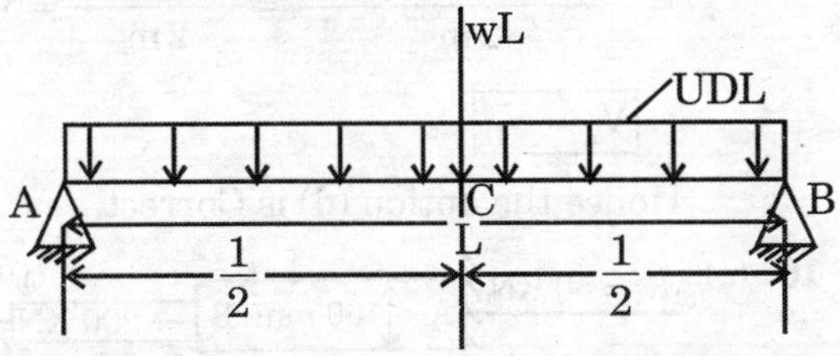

Total deflection at the mid span A the beam

$$\boxed{y_c = \frac{5\,wl^4}{384\,EI}}$$

Hence the option (c) is correct.

7. (b) An inelastic collision or plastic collision is one in which part of the kinetic energy is changed to some other from of energy in the collision, but momentum is conserved in inelastic collision.

Hence the option (*b*) is Correct.

8. (b)

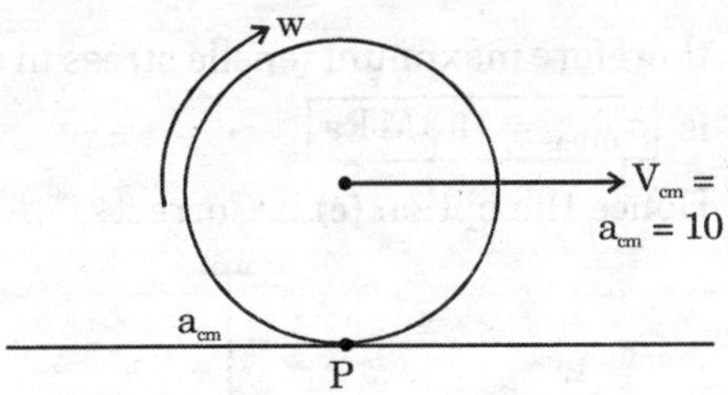

Acceleration of point 'P' would be

$$a_0 = (a_{max})_P = a_{cm}$$

$$\boxed{a_P = 10\,m/s^2}$$

Hence the option (b) is Correct.

10. (a)

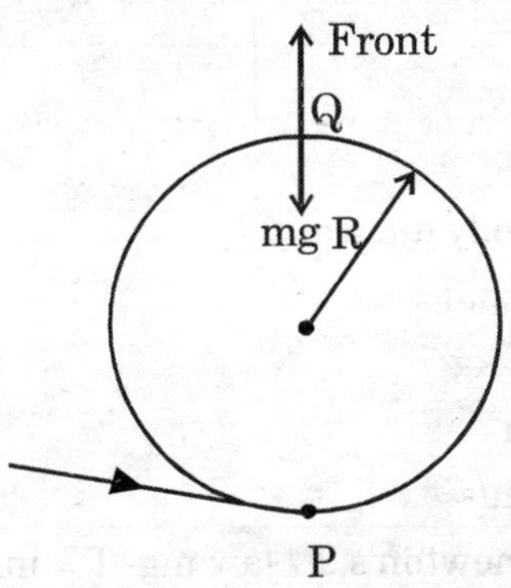

Let the particle is at point 'Q' then net normal force is equal to the centrifugal force

$\because f_{centri} = \frac{mv^2}{r}$

$\therefore \cancel{N}^\circ + \cancel{m}g = \frac{\cancel{m}v^2}{R}$ $[V_Q \to \text{top velocity at Q}]$

$\boxed{V_Q = \sqrt{gR}}$

Now total energy at the top of the loop is

$$E_Q = (k\cdot E)_Q + (P\cdot E)_Q$$

$$= \frac{1}{2}mv_Q{}^2 + mgh_Q$$

$$= \frac{1}{2}m\left(\sqrt{gR}\right)^2 + mg(2R)$$

$$E_Q = \frac{5}{2}mg\,R \quad ...(1)$$

Similarly, total energy at the sottom point P

$$E_P = (k\cdot E)_P + (P\cdot E)_P$$

$$= \frac{1}{2}mv_P{}^2 + mgh_Q = \frac{1}{2}mV_p{}^2 + 0$$

$$E_P = \frac{1}{2}mv_P^2 \quad ...(2)$$

By equating (1) & (2) weget

$$\frac{5}{2}\cancel{m}gR = \frac{1}{\cancel{2}}\cancel{m}v_P^2$$

$\therefore \boxed{V_P = \sqrt{5gR}}$

Hence option (a) is Correct.

11. (d)

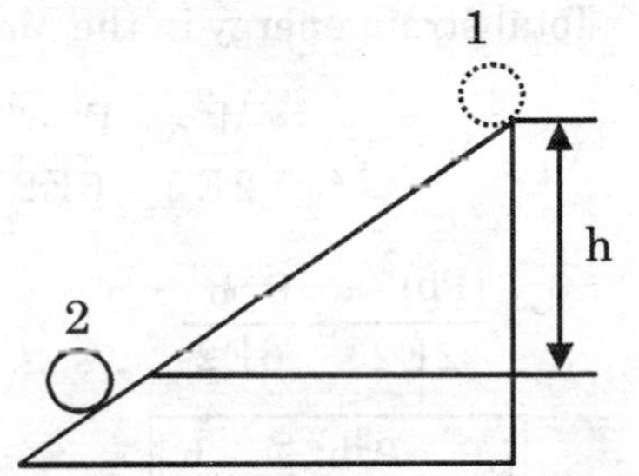

Applying Energy Conservation Methed

$$(P\cdot E)_1 + (K\cdot E)_1 = (P\cdot E)_2 + (K\cdot E)_2$$

$$mgh + 0 = 0 + \frac{1}{2}mv^2$$

$\therefore \boxed{V = \sqrt{2gh}}$

Hence the option (d) is Corrcet.

12. (d)

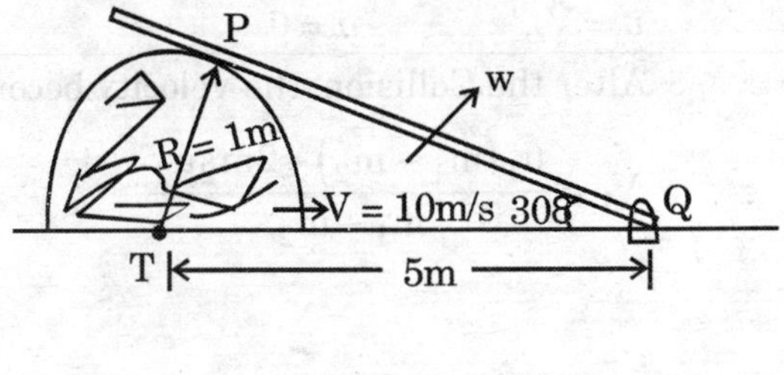

Length of $PQ = 5\cos 30°$

$$PQ = \frac{5\sqrt{3}}{2}$$

Angular velocity of $PQ = w_{PQ} = \frac{V}{PQ} = \frac{10}{\left(\frac{5\sqrt{3}}{2}\right)}$

$$\therefore \boxed{W_{PQ} = 2.30\,\text{rad/sec}}$$

Hence the option (*d*) is Correct.

13. (a)

I

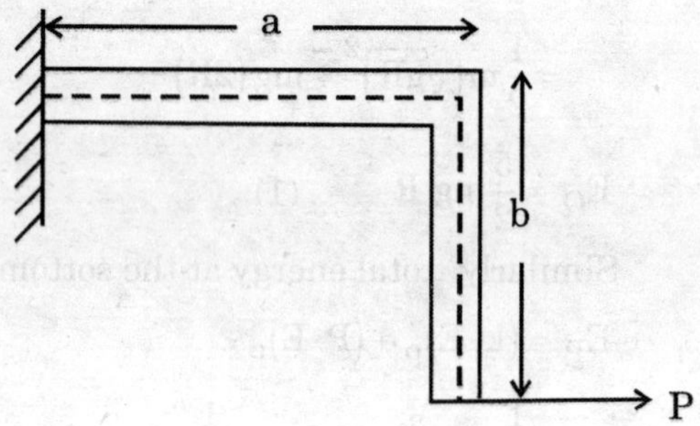

II

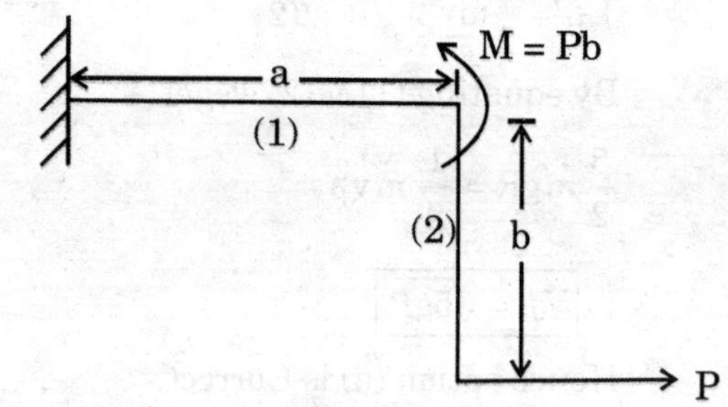

Total strain energy in the Member

$$V_T = V_1 + V_2 = \frac{M^2 a}{2E2} + \frac{P^2 b^3}{6E2}$$

$$= \frac{(Pb)^2 a}{2E2} + \frac{P^2 b^3}{6E2}$$

$$\boxed{U_T = \frac{P^2 b^2}{2E2}\left[a + \frac{b}{3}\right]}$$

Hence the option (a) is Correct.

15. (d)

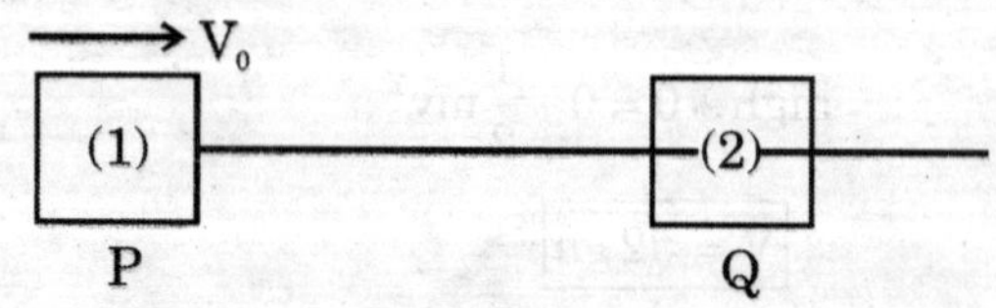

Let the velouities of bodies before Collision

$u_1 = V_0$, $u_2 = 0$

After the Collision, the velocity becomes

$$V_1 = \frac{u_1(m_1 - m_2) + 2m_2u_2}{m_1 + m_2}$$

$$= \frac{V_0 \times 0 + 0}{2m} \quad [\because m_1 = m_2 = m]$$

$$\boxed{V_1 = 0}$$

$$V_2 = \frac{u_2(m_1 - m_2) + 2m_1u_1}{m_1 + m_2}$$

$$= \frac{u_2(0) + 2 \times m \times V_0}{2m} = \frac{0 + 0mv_0}{2m} = V_0$$

$$\therefore \boxed{V_2 = V_0}$$

Hence the option (d) is Correct.

16. (c)

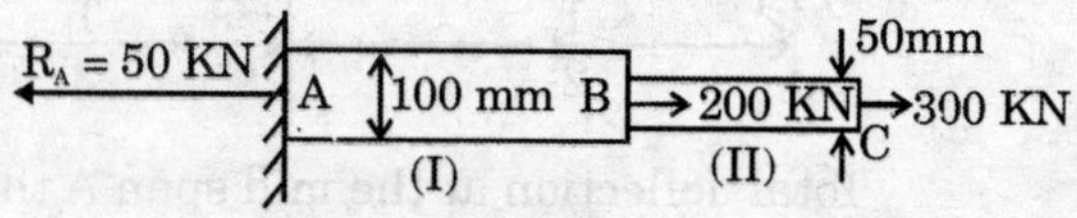

$P_1 = 500$ kN, $P_2 = 300$ kN

Stress developed in the bar ...(1)

$$\sigma_1 = \frac{P_1}{A_1} = \frac{500 \times 10^3}{\frac{\pi}{4}(100)^2} = 63.66\text{MPa}$$

Stress developed in the bar ...(2)

$$\sigma_2 = \frac{P_2}{A_2} = \frac{300 \times 10^3}{\frac{\pi}{4}(50)^2} = 152.78\text{Mpa}$$

therefore maximum tensile stress in the shaft is $\boxed{\sigma_{max} = 153\,\text{MPa}}$

Hence the option (c) is Correct.

17. (c)

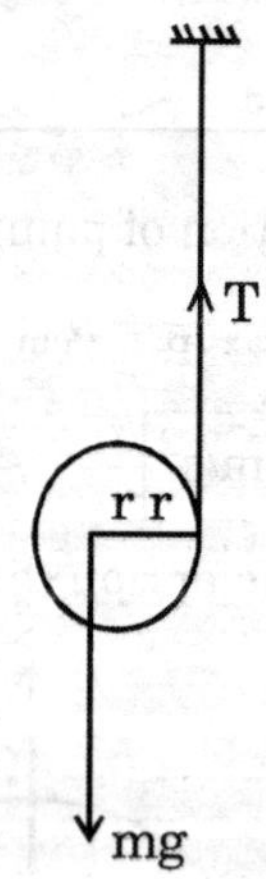

Free Body diagram

Given that

m = 10 kg

r = 2 m

g = 10m/s²

Apply newton's 2nd law mg–T = ma

$10 \times 10 - T = 10 \times a$

$100 - T = 100$ ...(1)

We know, Torque (T) $= I\alpha$

$$T \times 2 = \left(\frac{mR^2}{2}\right) \cdot \frac{a}{2}$$

$$T \times 2 = \frac{10 \times 4}{2} \times \frac{a}{2}$$

$$a = \frac{I}{5} \quad ...(2)$$

Put eq^n (2) in (1), weget

$$\boxed{T = 33.33\,N}$$

Hence the option (c) is Correct.

18. (a) Given eq^n of SHM

$$\frac{d^2y}{dt^2} + w^2 y = 0$$

the graphical representation between acceleration 'a' & velocity 'V' will be an ellipse.

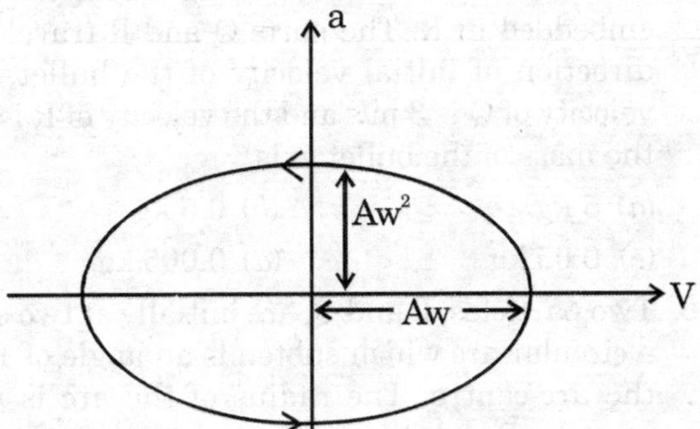

Hence the option (a) is Correct.

21. (c)

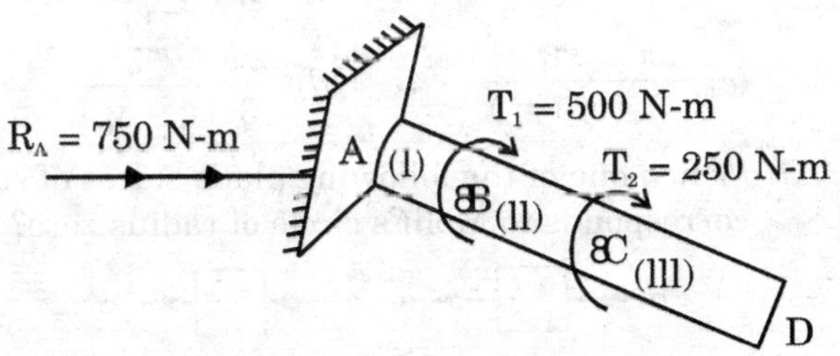

$T_1 = 750$ N-m, $T_2 = 250$ N-m $T_3 = 0$

Maximum shear stress developed in the bar

$$\tau_{max} = \frac{T_1}{Z} = \frac{750 \times 10^3}{\frac{\pi}{16}(50)^3}$$

$$\boxed{\tau_{max} = 30.56\,MPa}$$

Hence the option (c) is Correct.

22. (b) Total Angular twist at the end of the bar is

$$Q_D = Q_1 + Q_2 + Q_3$$

$$= \frac{T_1 L_1}{G_1 J_1} + \frac{T_2 L_2}{G_2 J_2} + \frac{T_3 L_3}{G_3 J_3}$$

$$= \frac{T_1 L_1}{G\left(\frac{\Pi}{34} d^4\right)} + \frac{T_2 L_2}{G_2\left(\frac{\Pi}{32} d^4\right)} + 0$$

$$= \frac{750 \times 500 \times 10^{-3}}{80 \times 10^9 \times 6.13 \times 10^{-7}} + \frac{250 \times 500 \times 10^{-3}}{80 \times 10^9 \times 6.13 \times 10^{-7}}$$

$Q_D = 0.010$ rad

$$Q_D = 0.010 \times \frac{180}{\pi}$$

$$\boxed{Q_D = 0.584°}$$

Hence the option (B) is Correct.

23. (*)

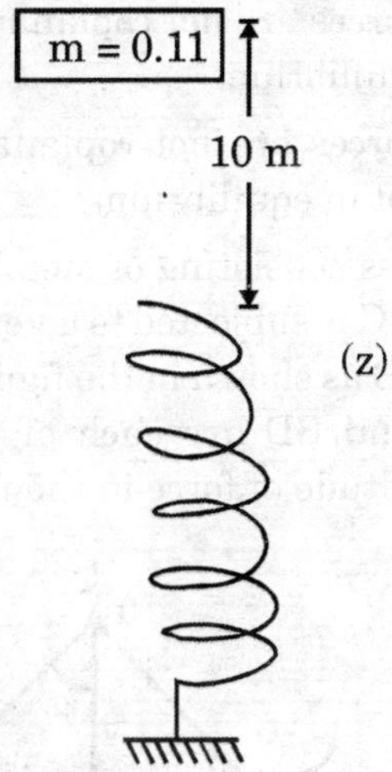

Applying energy Conservation

$$mg(h + x) = \frac{1}{2} kx^2$$

$$\frac{2 \times 0.1 \times 10(10 + x)}{K} = x^2$$

$10 + x = 250\,x^2$

$250\,x^2 - x - 10 = 0$

Solve the above eq^n

$$\boxed{x = 202.0\,mm}$$

24. (d) Applying Newtion's 2nd law

$f = ma$

$kx = ma$

$$\therefore a = \frac{kx}{m} = \frac{500 \times 202 \times 10^{-3}}{0.1}$$

$$\boxed{a = 1010\,m/s^2}$$

Hence the option (d) is Correct.

2010

Q.(1 – 9) carry one mark each.

1. Three forces acting on a particle are given as

$F_1 = (5i + 6j)$N, $F_2 = (-i + 4k)$ N

and $F_3 = (i + 6j + 16k)$ N.

where i, j, k are the unit vectors along Cartesian coordinate axes. Which one of the following statements is true?

(*a*) Forces are coplanar and the particle is in equilibrium

(*b*) Forces are coplanar but the particle is not in equilibrium

(*c*) Forces are not coplanar but the particle is in equilibrium

(*d*) Forces are not coplanar and the particle is not in equilibrium

2. A truss consisting of members AD, DC, AB, BD and BC is subjected to a vertical force of 120 N at joint B as shown in the figure. The members AD, DC and BD are each of 1 meter length. The magnitude of force in the member BD is

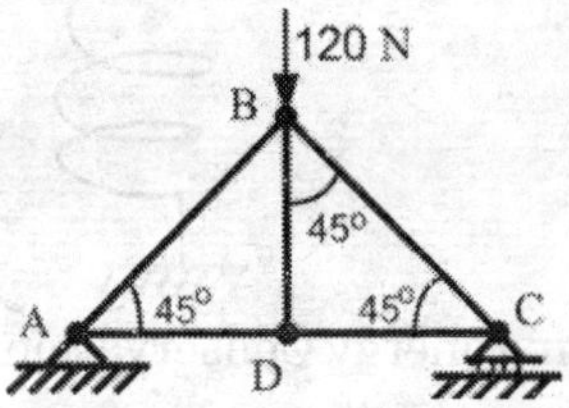

(*a*) 0 (*b*) $20\sqrt{2}$ N

(*c*) 40 N (*d*) 120 N

3. Two rigid bodies A and B are each weighing 30 N. Body A is kept on a floor and body B is kept on body A as shown in the figure. The coefficient of friction between two bodies, and between body A and the floor is 0.1. If a horizontal force of 2 N is applied on body A, the friction force at the interface of body A and body B will be

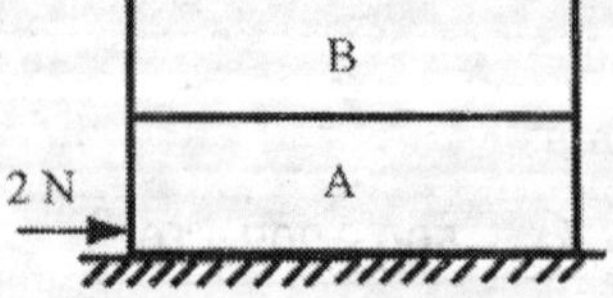

(*a*) 0 (*b*) 1 N

(*c*) 2 N (*d*) 3 N

4. A rigid link PQ is rotating about a revolute joint at P with a uniform angular velocity ω. A slider R is sliding on the link with a relative velocity v. Which one of the following figures represents the correct direction of the Coriolis acceleration a_c?

(*a*)

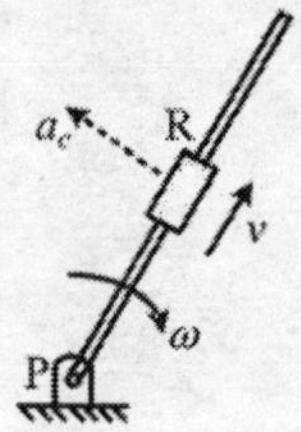

(*b*)

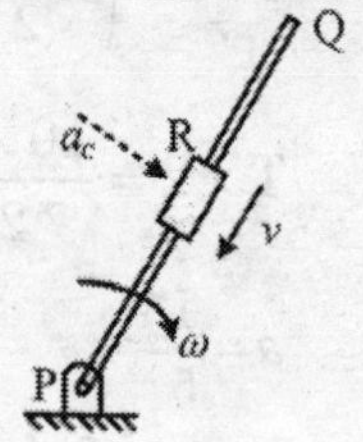

(*c*)

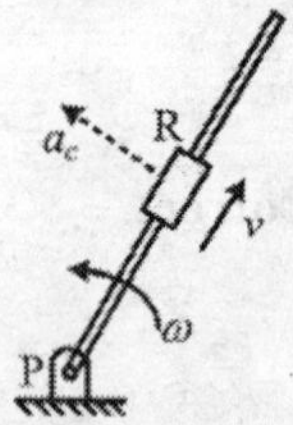

(*d*)

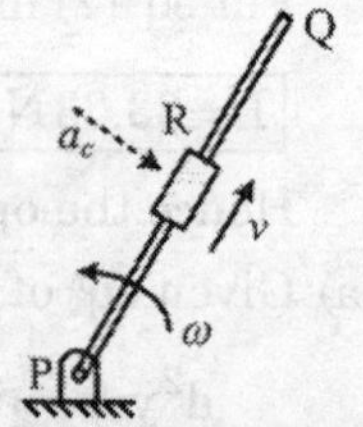

5. A bullet of mass m having a horizontal velocity of 500 m/s hits a stationary block of mass 6.15 kg. The block breaks into two parts viz. Q (mass of 3 kg) and R (mass of 3.15 kg), with the bullet embedded in R. The parts Q and R travel in the direction of initial velocity of the bullet. If the velocity of Q is 3 m/s and the velocity of R is 5 m/s, the mass of the bullet m is

(*a*) 5 kg (*b*) 0.5 kg

(*c*) 0.05 kg (*d*) 0.005 kg

6. Two particles, P and Q, are initially at two ends of a circular arc which subtends an angle of 120° at the arc-centre. The radius of the arc is r. The particles P and Q are moving along the arc towards each other with constant tangential velocities of v_P and v_Q respectively. The distance travelled by the particle P when it meets the particle Q is

(*a*) $\frac{2\pi}{3}\frac{r(v_P + v_Q)}{v_P}$ (*b*) $\frac{2\pi}{3}\frac{r(v_P + v_Q)}{v_Q}$

(*c*) $\frac{2\pi}{3}\frac{rv_P}{(v_P + v_Q)}$ (*d*) $\frac{2\pi}{3}\frac{rv_Q}{(v_P + v_Q)}$

7. Which one of the following plane states of stress corresponds to Mohr's circle of radius zero?

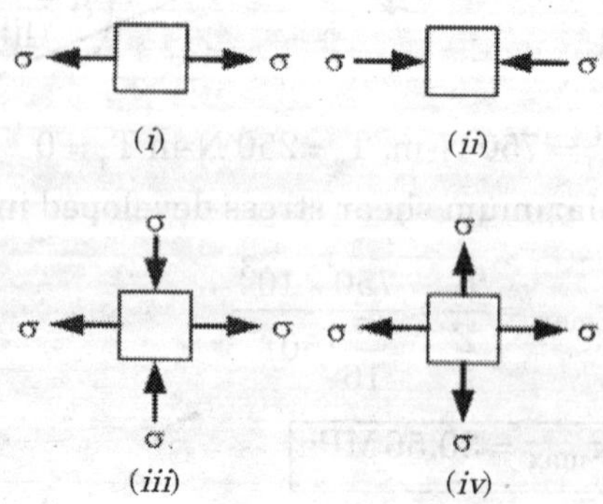

(*a*) (*i*) (*b*) (*ii*)

(*c*) (*iii*) (*d*) (*iv*)

8. Maximum shear stress theory for material yielding is known as

(a) Tresca's criterion

(b) von-Mises criterion

(c) Saint-Venant's theory

(d) Rankine's theory

9. In a cantilever beam of length 2 m, the shear force in newton (N) along the length is given by $V(x) = 5x^2$, where x is the distance in meter measured from the fixed end. The magnitude of the load intensity at the mid-span of the beam is

(a) 0 (b) 1 N/m

(c) 5 N/m (d) 10 N/m

Q.(10 – 22) carry two marks each.

10. The blocks P and Q are connected by a string, which passes over a pulley as shown in the figure. The block P is sliding on an inclined surface. Ignoring the masses of the string and the pulley, the tension in the string is (use gravitational acceleration $g = 9.81$ m/s^2 and neglect all friction)

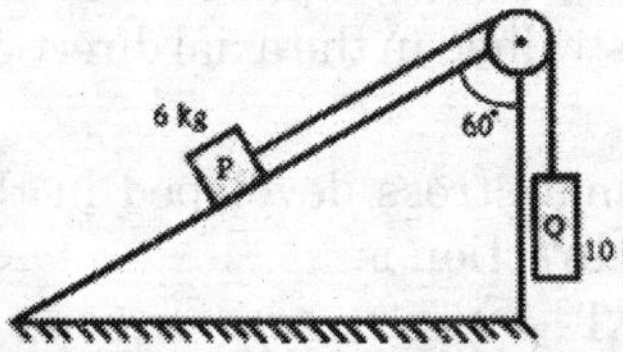

(a) 55.2 N (b) 62.5 N

(c) 74.3 N (d) 86.2 N

11. A hollow circular shaft of inside diameter 10 mm and outside diameter 20 mm is subjected to a pure symmetric-bending moment of 200 N-m. The magnitude of bending stress at a point in the plane of loading, which is at a distance of 5 mm from the neutral axis, is

(a) 0 (b) 68.8 MPa

(c) 135.8 MPa (d) 271.6 MPa

12. A stepped circular shaft made of steel is rigidly fixed at two supports A and C as shown in the figure. A torque of 680 N-m is applied on the shaft at point B. The diameter of portion AB is twice that of portion BC. The magnitudes of torque reactions at supports A and C respectively are

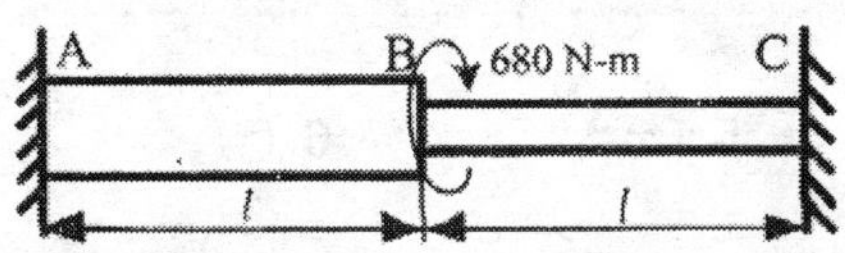

(a) 640 N-m, 40 N-m

(b) 40 N-m, 640 N-m

(c) 340 N-m, 340 N-m

(d) 544 N-m, 136 N-m

13. A thin-walled cylinder with open ends is subjected to uniform internal pressure p alone. The wall thickness is t, internal radius is r and the Young's modulus is E. The increase in radius of the cylinder due to the internal pressure is

(a) zero (b) $\dfrac{pt^2}{2Er}$

(c) $\dfrac{pr^2}{Et}$ (d) $\dfrac{pr^2}{Et} + r$

14. A cylindrical steel bar of uniform cross-sectional area is subjected to an axial tensile force P and a torque T. Assuming linear elastic deformation of the bar, the internal strain energy stored in the bar is $(20P^2 + 8T^2) \times 10^{-6}$ N-m. The axial extension of the bar for P = 10 N and T = 16 N-m is

(a) 256 μm (b) 400 μm

(c) 2000 μm (d) 2048 μm

15. The buckling load of a slender column clamped at both the ends is 4000 N. The column is subjected to an axial compression. During the course of service, one of the ends gets detached from the clamp and becomes free end. The absolute percentage change in the buckling load due to the change in the end condition is

(a) 50.00 (b) 75.00

(c) 83.25 (d) 93.75

16. A spring-mass system shown in the figure is vibrating with very small amplitude. The natural frequency of the system is

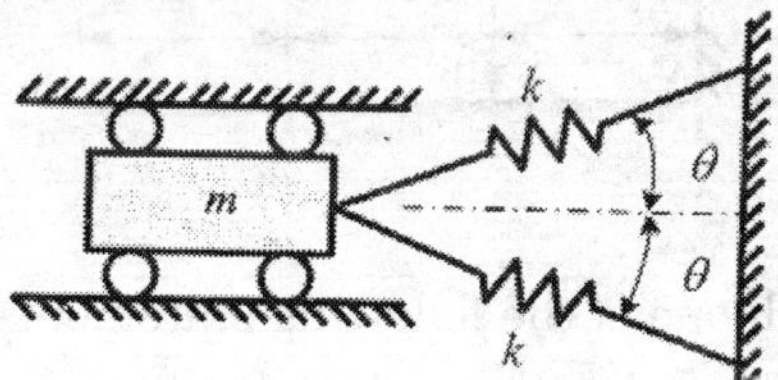

(a) $\sqrt{\dfrac{k}{m}}$ (b) $\sqrt{\dfrac{2k}{m}}$

(c) $\sqrt{\dfrac{2k\cos\theta}{m}}$ (d) $\sqrt{\dfrac{2k\cos^2\theta}{m}}$

Common Data Questions

Common Data for Q. (17 - 18)

Two particles P and Q are connected by a rigid link of negligible mass. The length of the link PQ is $\sqrt{2}\,r$. The inner radius of the ring is r and its centre is at O as shown in the figure. The particles are allowed to slide freely with negligible friction on the inner surface of a vertical circular ring. The angle θ, between OQ and horizontal X-axis, is measured from X-axis in the clockwise sense. Gravitational acceleration is g.

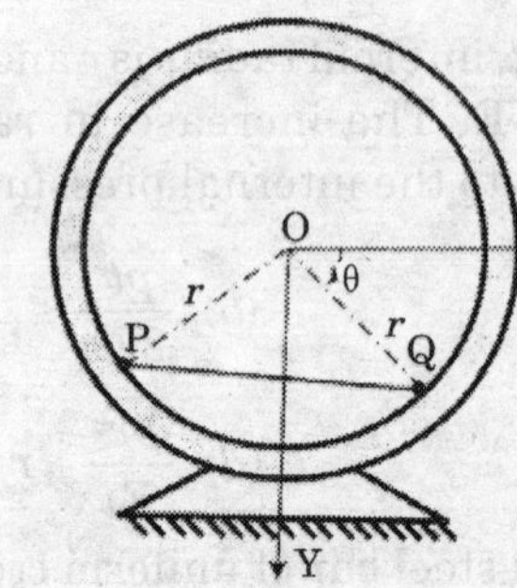

17. Mass of the particles P and Q are m and $2m$, respectively. The link PQ is released from $\theta = 0°$. When the link occupies the horizontal position, the magnitude of velocity of particle P is

(a) $0.865\sqrt{gr}$ (b) $1.865\sqrt{gr}$

(c) $0.086\sqrt{gr}$ (d) $2.865\sqrt{gr}$

18. If both the particles P and Q are of the mass m and the link PQ is released from $\theta = 0°$, the maximum possible value of θ during the oscillation of the link is

(a) 45° (b) 90°

(c) 135° (d) 180°

Common Data for Q.(19 - 20)

A cantilever beam of length $3l$ is subjected to two forces each of magnitude P as shown in the figure. The flexural rigidity of the beam is EI. Assume linear elastic material and small deflections.

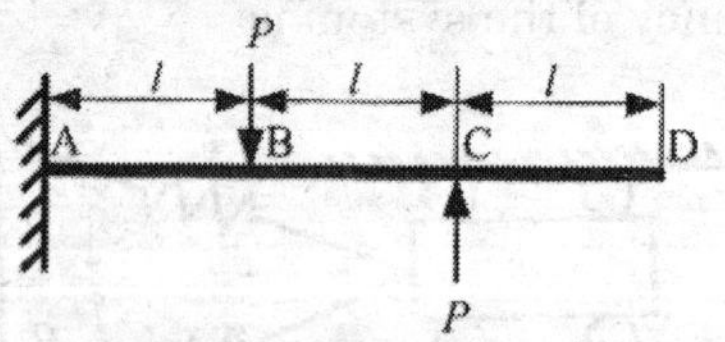

19. Which one of the following statements is true?

(a) The magnitude of the bending moment in portion AB is zero

(b) The magnitude of the bending moment in portion AB is Pl

(c) The magnitude of the bending moment in porition AB is $2Pl$

(d) The magnitude of the bending moment in portion AB varies linearly from 0 to Pl

20. The deflection due to bending at point B is

(a) $\frac{Pl^3}{3EI}\downarrow$ (downward)

(b) $\frac{Pl^3}{2EI}\uparrow$ (upward)

(c) $\frac{Pl^3}{6EI}\downarrow$ (downward)

(d) $\frac{Pl^3}{6EI}\uparrow$ (upward)

Linked Answer Questions

Statement for Linked Answer Q.(21 - 22)

A steel bar of rectangular cross-section is heated uniformly and the rise in the temperature is ΔT. The Young's modulus is E, the Poisson's ratio is ν and the coefficient of thermal expansion is α. The bar is completely restrained in the axial direction and lateral directions.

21. The thermal stress developed in the bar along the axial direction is

(a) $-\frac{E\alpha\Delta T}{1+2\nu}$ (b) $-E\alpha\Delta T$

(c) $-\frac{E\alpha\Delta T}{1-2\nu}$ (d) $-\frac{E\alpha\Delta T\nu}{1-2\nu}$

22. Assume that the bar is allowed to deform freely in the lateral directions, while keeping the axial direction restrained. The percentage change in the magnitude of axial thermal stress for $\nu = 0.25$ is

(a) 0 (b) 25

(c) 50 (d) 100

ANSWERS

1. (b)	2. (a)	3. (d)	4. (c)	5. (*)	6. (*)	7. (d)
8. (a)	9. (*)	10. (*)	11. (c)	12. (c)	13. (b)	14. (b)
15. (b)	16. (b)	17. (*)	18. (*)	19. (b)	20. (a)	21. (c)
22. (c)						

EXPLANATIONS

1. (b) Given that

$F_1 = (5i + 6j)$ N

$F_2 = (-i + 4k)$N

$F_3 = (i + 6j + 16k)$ N

Case 1 : The forces will be coplanar if

Scalar-Triple Product

$$\vec{a}\cdot(\vec{b}\times\vec{c}) = 0 \Rightarrow \begin{vmatrix} a_1 & a_2 & a_3 \\ b_1 & b_2 & b_3 \\ c_1 & c_2 & c_3 \end{vmatrix} = 0$$

i.e $F_1 \cdot (F_2 \times F_3) = 0$

$$1 \Rightarrow \begin{vmatrix} 5 & 6 & 0 \\ -1 & 0 & 4 \\ 1 & 6 & 16 \end{vmatrix}$$

$\Rightarrow 5(0-24) - 6(-16-4) + 0()$

$= -120 + 120 = 0$

Thus the given Vectors are coplanar

Case II : Vectors to be Equillibrium

$\vec{a}\times\vec{b} = \vec{b}\times\vec{c} = \vec{c}\times\vec{a}$

$$\vec{a}\times\vec{b}(\text{or})\vec{F}_1\times\vec{F}_2 = \begin{vmatrix} i & j & k \\ 5 & 6 & 0 \\ -1 & 0 & 4 \end{vmatrix}$$

$= i(24-0) - j(20-0) + k(0+6)$

$= 24i - 20j + 6k$

$$\vec{F}_2\times\vec{F}_3 = \begin{vmatrix} i & j & k \\ -1 & 0 & 4 \\ 1 & 6 & 16 \end{vmatrix}$$

$= i(0-24) - j(-16-4) + k(-6-0)$

$= -24i + 20j + 6k$

Here $\vec{F}_1\times\vec{F}_2 \neq \vec{F}_2\times F_3$

Thus the vectors are not equilibrium

Hence the option (b) is correct.

2. (a)

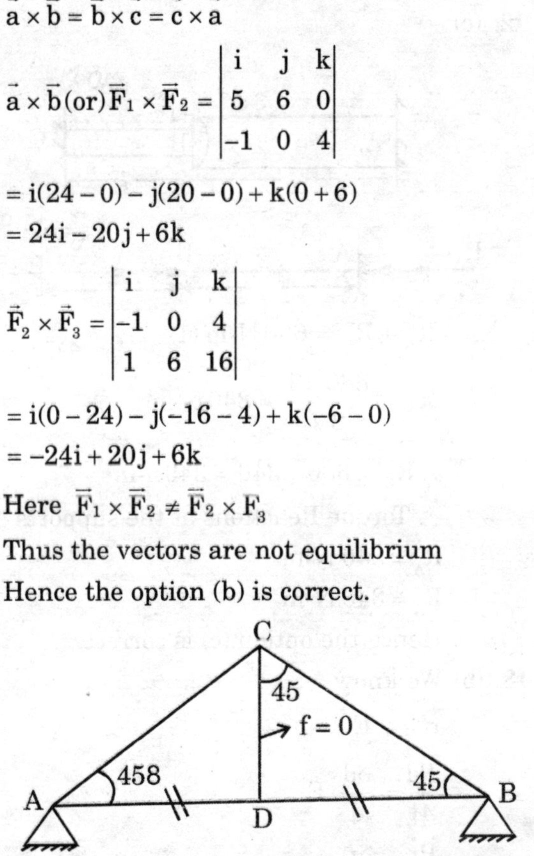

The members AD & BD are Collinear, therefore the force in the member CD will be zero. Since we know that, if at a Joint, three members or three forces or both acts simultaneously and if any two of them are collinear than the force on the third member will have zero forces.

Therefore $\boxed{f_{co} = 0}$

Hence the option (a) is correct.

3. (d)

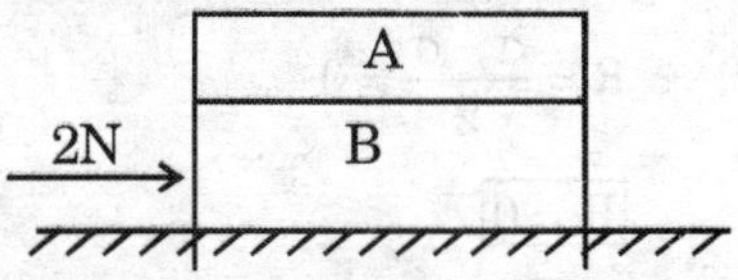

FBD for Block B:

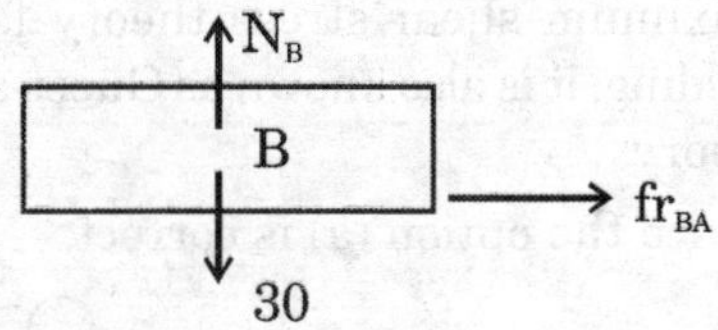

$fr_{BA} = \mu N_3 = 0.1 \times 30$

$\boxed{fr_{BA} = 3N}$

FBD for Block: A

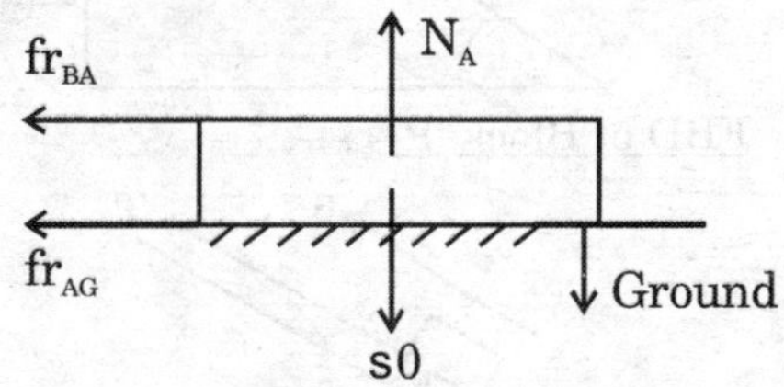

$fr_{BA} = \mu N_6 = 0.1 \times 30$

$fr_{BA} = 3N$

Therefore the force at the interface of A and body B will

$\boxed{(fr)_{BA} = 3N}$

Hence the option (d) is correct.

4. (c)

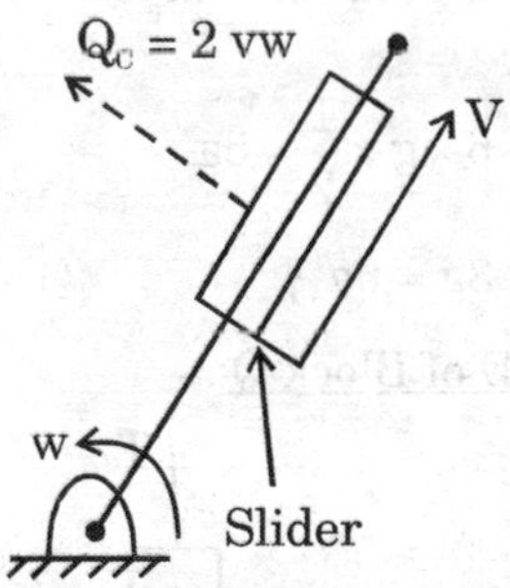

Corrioles Acceleration acts in the perpendicular direction of slider velocity and takes same direction of angular velocity.

Hence the option (c) is correct.

7. (d)

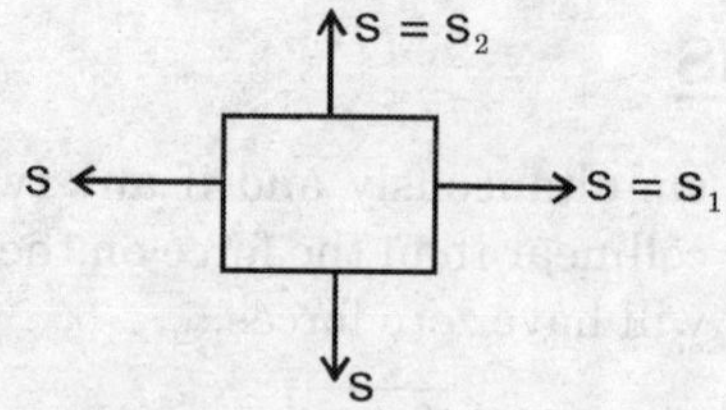

Radius of Mohr's Circle (R) = $\dfrac{\sigma_1 - \sigma_2}{2}$

$$R = \frac{\sigma - \sigma}{2} = 0$$

$$\boxed{R = 0}$$

Hence the option (d) is correct.

8. (a) Maximum shear stress theory for material yielding, it is also known as Guest-and Tresca's theory.

Hence the option (a) is correct.

10. (*)

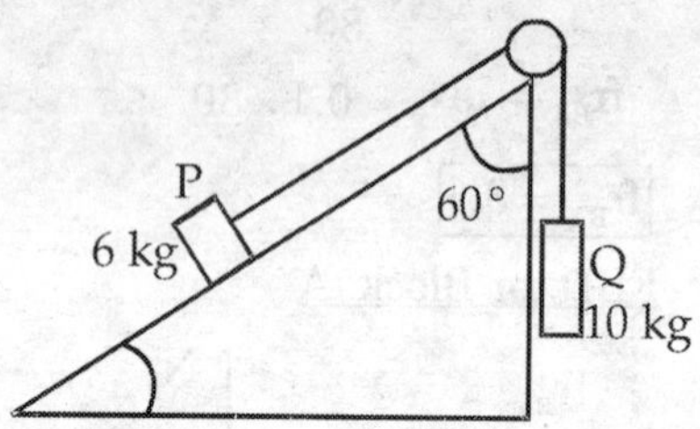

FBD of Block 'P'

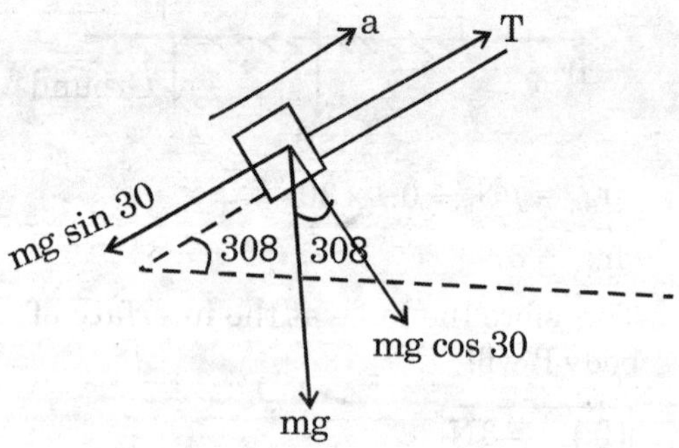

Applying Newton's 2nd law

$$T - mg\sin 30° = m_p a$$

$$T - 6 \times g \times \frac{1}{2} = 6a$$

$$T - 3g = 6a \qquad (1)$$

FBD of Block Q:

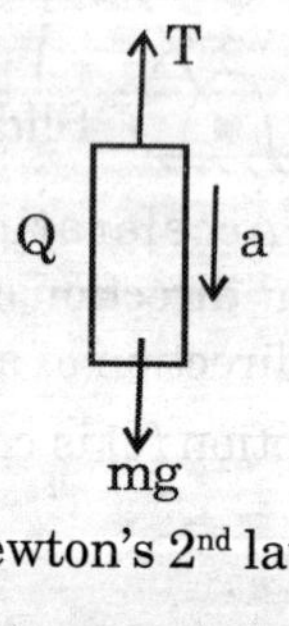

Applying Newton's 2nd law

$$mg - T = M_Q a$$

$$10g - T = 100a \qquad (2)$$

Solving eq^n (1) & (2) we get

$$\boxed{a = 4.3\,m/s^2}$$

11. (c) Given that

Inner diameter of shaft (D_i) = 10 mm

Outer diameter of shaft (D_o) = 20 mm

Bending moment (M) = 200 N-m

∴ Bending stress at a distance of r = 5 mm from the neutral Axis will be

$$\sigma_b = (\sigma_b)_{Max}\left[\frac{R}{r}\right]$$

$$= \frac{16M}{\pi D^3(1-k^4)}\left[\frac{5}{5}\right]$$

$$= \frac{16 \times 250 \times 10^3}{\pi (20)^3 (1 - 0.5^4)} \times 1 \qquad \left[K = \frac{D_i}{D_o} = \frac{10}{20} = 0.5\right]$$

$$\boxed{\sigma_b = 135.81\,MPa}$$

Hence the option (c) is correct.

12. (c)

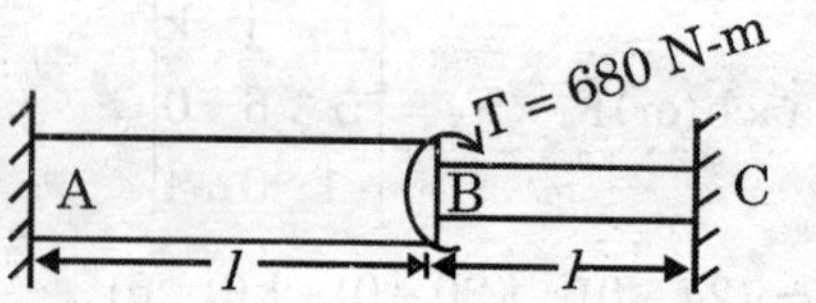

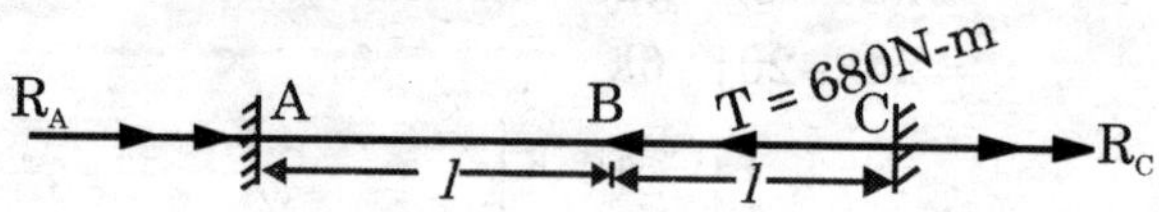

$$R_A + R_C = 680\,N\text{-}m \; (1)$$

$$R_A = \frac{680 \times l}{2\,l} = 340\,N\text{-}m$$

$$\therefore R_C = 680 - 340 = 340\,n\text{-}m$$

∴ Torque Reactions at the supports are

R_A = 340 N-m

R_A = 340 N-m

Hence the option (c) is correct.

13. (b) We know

$$\sigma = \in \cdot E$$

$$\frac{Pd}{4t} = \frac{\delta d}{d} \cdot E$$

$$\frac{Pr}{2t} = \frac{\delta r}{r} \cdot E$$

$$\boxed{\delta r = \frac{Pr^2}{2tE}}$$

Therefore increase in radius of the cylinder will be

$$\boxed{\delta r = \frac{Pr^2}{2tE}}$$

Hence option (b) is correct.

14. (b)

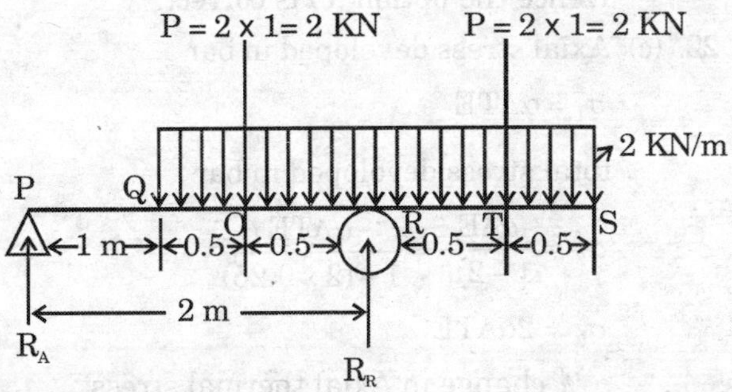

$R_A + R_R = 2 + 2 = 4$

$R_A + R_R = 41$

Taking moment about 'P'

$-2 \times 2.5 + R_R \times 2 - 2 \times 1.5 = 0$

$-5 + 2R_R - 3 = 0$

$\therefore \boxed{R_R = 4}$

$R \boxed{R_A = 0}$ $[\because \text{from } 1]$

Now Bending Moment at 'S'

$(B \cdot M)_S = 0$

$(B \cdot M)_R = -2 \times 0.5 = -1 kN$

$(B \cdot M)_U = -2 \times 1 + 4 \times 0.5 = 0$

$(B \cdot M)_Q = -2 \times 1.5 + 4 \times 1 - 2 \times 0.5 = 0$

$(B \cdot M)_P = -2 \times 2.5 + 4 \times 2 - 2 \times 1.5 = 0$

therefore the maximum bending moment in the beam is $\boxed{(B \cdot M)_{Max} = 1 kN\text{-}m}$

Hence the option (b) is correct.

15. (b) Buckling load for column clamped

$$P_1 = \frac{\pi^2 EI}{L_C^2}$$

$$P_1 = \frac{\pi^2 EI}{l^2} \quad (1) \quad [\because L_c = l]$$

Buckling load for column fixed-free

$$P_2 = \frac{\pi^2 EI}{L_C^2} = \frac{\pi^2 EI}{(2l)^2} \quad [lc = 2l]$$

$$P_2 = \frac{\pi^2 EI}{4l^2} \quad (2)$$

divide eq^n (1)/(2)

$$\frac{P_1}{P_2} = \frac{\pi^2 EI}{l^2} \times \frac{4l^2}{\pi^2 EI}$$

$$Q \frac{4000}{P_2} = 4 \quad [\because P_1 = 4000 \text{ N given}]$$

$\because P_2 = 1000 N$

$$\therefore \% \text{change in buckling load} = \frac{P_1 - P_2}{P_1} \times 100$$

$$= \frac{4000 - 1000}{4000} \times 100 = 75\%$$

Hence the option (b) is correct.

16. (b)

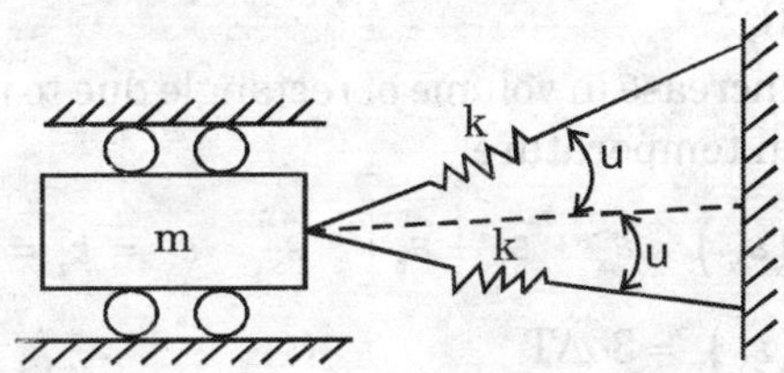

Natural frequency of the system

$$W_n = \sqrt{\frac{K_{eq}}{m}}$$

Here $K_{eq} = K_1 + K_2 = K + K = 2K$

$$\because \boxed{W_n = \sqrt{\frac{2K}{m}}}$$

Hence the option (b) is correct.

19. (b)

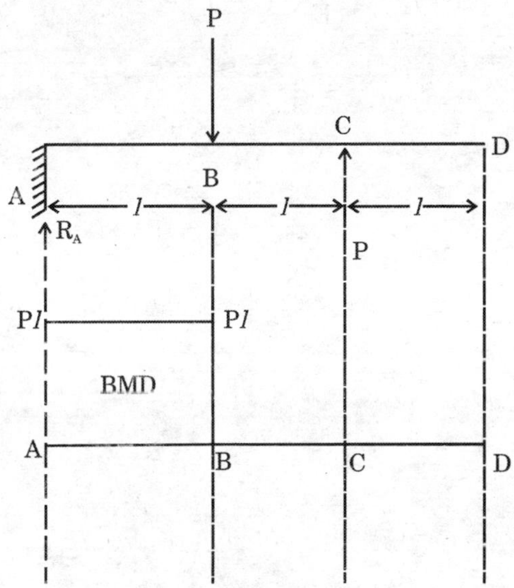

$R_A + P - P = 0 \Rightarrow \boxed{R_A = 0}$

Bending Moment at $D = (B \cdot M)_D = 0$

$(B \cdot M)_C = 0$

$(B \cdot M)_B = Pl$

$(B \cdot M)_A = P(2l) - Pl = Pl$

Hence the option (b) is correct

20. (a) The net deflection at point 'B' due to bending will be

$$\boxed{y_R = \frac{Pl^3}{3ER}\downarrow} \text{ downward}$$

Hence the option (a) is correct.

21. (c)

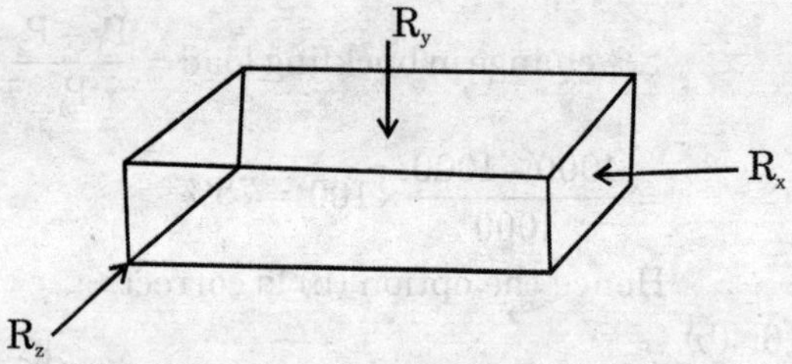

Increase in volume of rectangle due to increase in temperature

$$(\varepsilon_V) = \varepsilon_x + \varepsilon_y + \varepsilon_z \left[\because \varepsilon_x = \varepsilon_y = \varepsilon_z = \alpha T\right]$$

$$(\varepsilon_V)_1 = 3\alpha\Delta T \qquad (1)$$

decrease in Volume of Rectangle due to Reaction

$$(\varepsilon_V)_2 = \frac{1-2\mu}{E}(\sigma_x + \sigma_y + \sigma_z)$$

$$= \frac{(1-2\mu)}{E}[-\sigma-\sigma-\sigma]$$

$$(\varepsilon_N)_2 = \frac{(1-2\mu)}{E}(-3\sigma) \quad (2)$$

By equating the eq^n (1) & eq (2)

$$3\alpha T = \frac{1-2\mu}{E}\cdot(-3\sigma)$$

$$\therefore \boxed{\sigma = \frac{-\alpha\Delta TE}{1-2\mu}}$$

Hence the option (c) is correct.

22. (c) Axial stress developed in bar

$$\sigma_1 = \alpha\Delta TE$$

total stress developed in bar

$$\sigma_2 = \frac{\alpha\Delta E}{1-2\mu} = \frac{\alpha\Delta TE}{1-(2\times 0.25)}$$

$$\sigma_2 = 2\alpha\Delta TE$$

$\therefore$ % change in Axial thermal stress

$$= \frac{\sigma_2-\sigma_1}{\sigma_2}\times 100 = \frac{2\alpha\Delta TE - \alpha\Delta TE}{2\alpha\Delta TE}\times 100$$

$$= \frac{\cancel{\alpha\Delta TE}(2-1)}{2\cancel{\alpha\Delta TE}}\times 100 = \frac{1}{2}\times 100 = 50\%$$

Hence the option (c) is correct.

2011

Q.(1-9) carry one mark each.

1. A 10 kg rectangular steel block is placed over a smooth cast iron surface. The coefficient of friction between steel and cast iron is 0.4. Acceleration due to gravity is 10 m/s^2. If a force of 20 N is applied to the block parallel to the surface, the frictional resistance (in N) offered by the surface to the block is

(*a*) 0 (*b*) 4
(*c*) 20 (*d*) 40

2. The stress induced in an unconstrained rod that is heated uniformly depends

(*a*) only on the coefficient of expansion
(*b*) only on the temperature change
(*c*) neither on the coefficient of expansion nor on the temperature change
(*d*) both on the coefficient of expansion and on the temperature change

3. The Euler buckling load for a long slender column with fixed-fixed end conditions is 10000 N. If the end conditions are changed to fixed-free, Euler bucking load (in N) would be

(*a*) 625 (*b*) 1250
(*c*) 2500 (*d*) 5000

4. For the beam shown, if the maximum deflection occurs at a distance x from support P, which one of the following is TRUE?

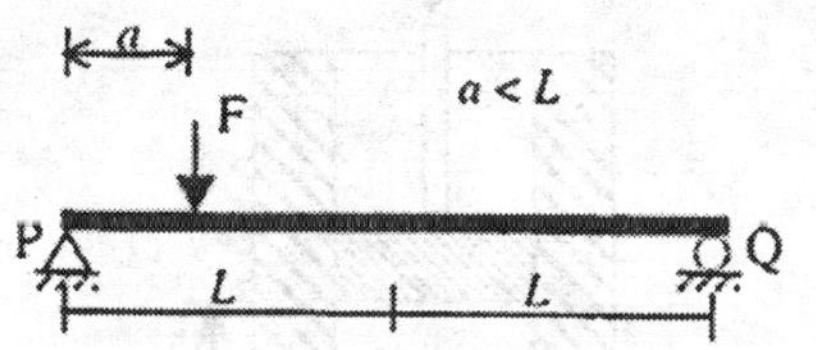

(*a*) $0 < x < a$ (*b*) $x = a$
(*c*) $a < x < L$ (*d*) $x = L$

5. For a biaxial state of stress defined by σ_{xx} = 40 MPa, σ_{yy} = – 20 MPa, τ_{xy} = 30 MPa, the Mohr's circle is centred at (in MPa)

(*a*) (0, 10)
(*b*) (0, 30)
(*c*) (10, 0)
(*d*) (60, 0)

6. A mass attached to a string is swung in a circular path with a uniform angular velocity about the horizontal plane, as shown. Seen directly from above, which path would the mass most closely follow if the string breaks suddenly at T?

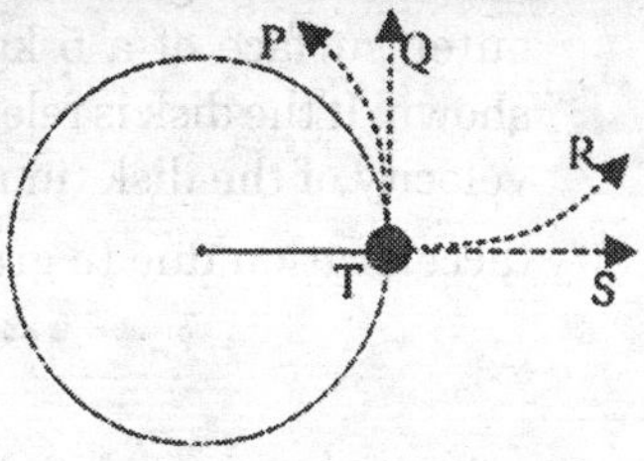

(*a*) P
(*b*) Q
(*c*) R
(*d*) S

7. For the beam shown, which of the following are discontinuous at the mid-span?

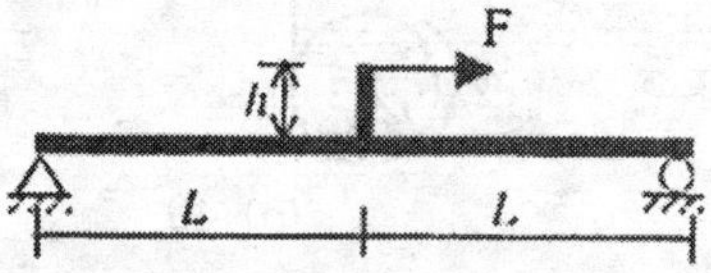

(*a*) bending moment only
(*b*) axial force and bending moment only
(*c*) bending moment and shear force only
(*d*) axial force, bending moment and shear force

8. A cylindrical shaft is subjected to pure torsion as shown. At point P, the only stress components expected to be non-zero are

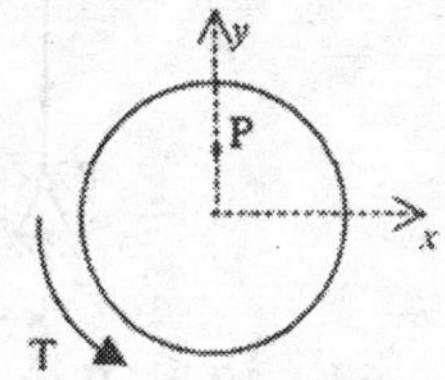

(*a*) τ_{yz}, τ_{zy}
(*b*) τ_{xy}, τ_{yx}, τ_{yz}, τ_{zy}
(*c*) τ_{xy}, τ_{yx}
(*d*) τ_{xz}, τ_{zx}

9. A non-magnetic light rigid frame with two magnets attached to its legs rests on a mild steel plate as shown. A spring, compressed by an amount δ from its unstretched length, connects the steel plate to the center of the frame as shown. Which of the figures below represents the CORRECT free-body diagram?

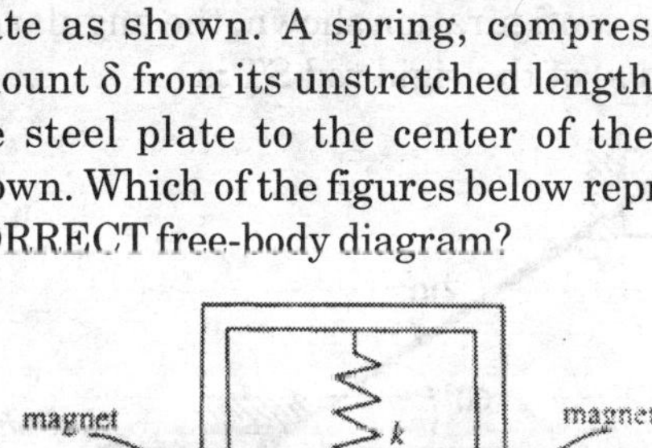

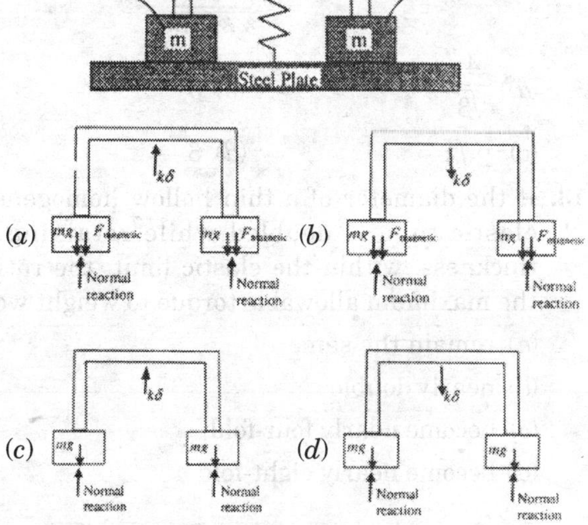

Q.(10-22) carry two marks each.

10. A cord of negligible mass is wrapped around the outer surface of a 5 kg disk of radius 0.5 m as shown. If the disk is released from rest, the angular velocity of the disk (in rad/s) after 3 seconds is (acceleration due to gravity, $g = 10$ m/s^2)

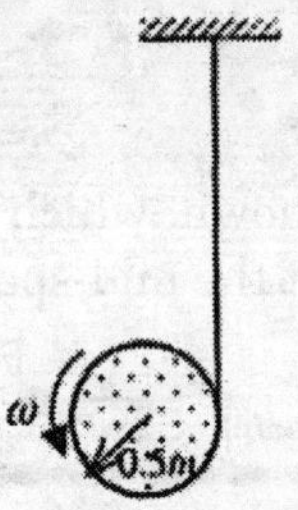

(*a*) 30 (*b*) 40
(*c*) 60 (*d*) 120

11. The force in the member PR for the truss shown is

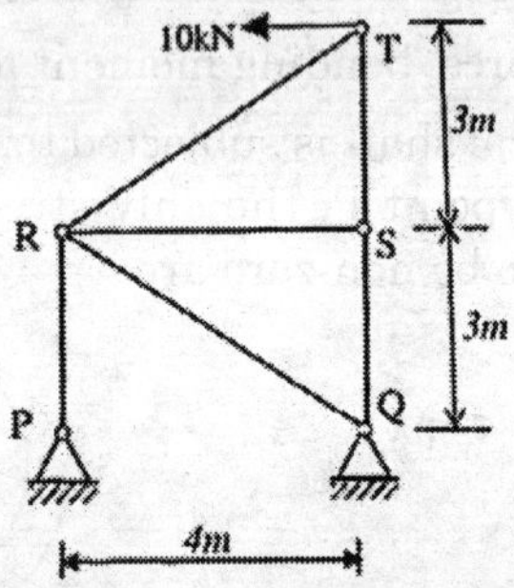

(*a*) 12 kN Tensile (*b*) 12 kN Compressive
(*c*) 15 kN Tensile (*d*) 15 kN Compressive

12. For the configuration shown, the angular velocity (in rad/s) of the rigid rod ST is

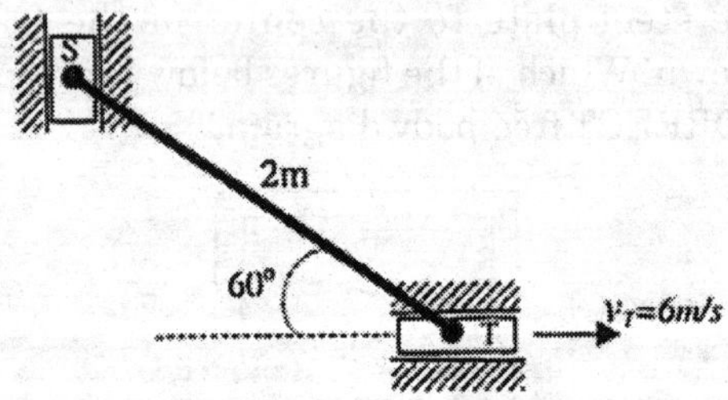

(*a*) $\dfrac{4}{\sqrt{3}}$ (*b*) 3
(*c*) $2\sqrt{3}$ (*d*) 6

13. If the diameter of a thin hollow homogeneous elastic tube is doubled while retaining the thickness, within the elastic limit, the ratio of the maximum allowable torque to weight would

(*a*) remain the same
(*b*) nearly double
(*c*) become nearly four-fold
(*d*) become nearly eight-fold

14. A simple mass-spring system shown has a natural frequency of 5 Hz. The static deflection of the spring due to the mass is (accelerations due to gravity is 9.81 m/s^2 and mass of the spring is negligible)

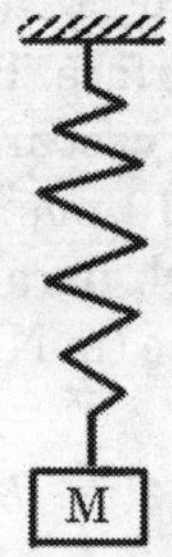

(*a*) 0.390 m (*b*) 0.039 m
(*c*) 0.019 m (*d*) 0.010 m

15. An unspecified pure bending moment is used to bend an aluminium rod of radius 2.5 mm elastically into a circular ring of radius 2 m. If the same bending moment is used to bend elastically a copper rod of radius 2 mm, the radius of the resulting ring (in m) is
(elastic modulus of aluminium is 70 GPa and elastic modulus of copper is 120 GPa)

(*a*) 0.702 (*b*) 1.404
(*c*) 1.755 (*d*) 2.808

16. A cylindrical chamber is filled with an elastic material of modulus E, and Poisson's ratio ν, as shown. A piston of diameter $2r$, is pushed down on the elastic material by a force F. Neglecting friction, and assuming that the chamber and the piston are rigid, the piston moves a distance of

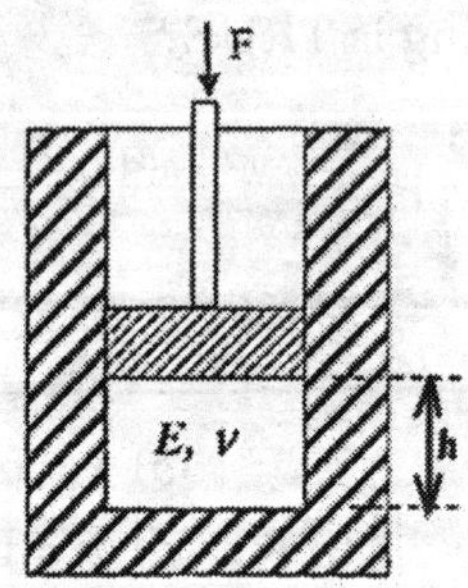

(*a*) $\dfrac{Fh}{\pi r^2 E}\left[1-\dfrac{2\nu^2}{1-\nu}\right]$ (*b*) $\dfrac{Fh}{\pi r^2 E}\left[1+\dfrac{2\nu^2}{1-\nu}\right]$

(*c*) $\dfrac{Fh}{\pi r^2 E}\left[1-\dfrac{2\nu^2}{1+\nu}\right]$ (*d*) $\dfrac{Fh}{\pi r^2 E}\left[1+\dfrac{2\nu^2}{1+\nu}\right]$

COMMON DATA QUESTIONS

Common Data for Questions 17 and 18 :

A ductile material has tensile yield strength of 200 MPa.

17. The stress at yield (in MPa) under equi-biaxial compression for this material according to Tresca criterion is

(*a*) 200 (*b*) 100
(*c*) 80 (*d*) 50

18. The stress at yield (in MPa) under pure shear for the same material according to von Mises criterion is

(*a*) $\frac{100}{\sqrt{3}}$ (*b*) $\frac{200}{\sqrt{3}}$

(*c*) $100\sqrt{2}$ (*d*) $100\sqrt{3}$

Common Data for Questions 19 and 20 :

A hub (outer radius is 35.00 mm and inner radius is 32.98 mm) is press fitted to a shaft (outer radius is 33.00 mm and inner radius is 31.00 mm) over a length of 40.00 mm. The hub and the shaft are made of steel (E = 200 GPa, ν = 0.3) and the coefficient of friction is 0.5. The pressure that arises at the interface due to the press fit is given by $p = \frac{Et\delta}{2r^2}$, where r is the nominal radius of the interface, δ is the radial interference and t = 2 mm is the nominal thickness. Local stress concentrations can be neglected in the calculations.

19. The hoop stress (in MPa) developed in the shaft is

(*a*) – 60.6 (*b*) – 30.3

(*c*) 15.7 (*d*) 60.6

20. The maximum torque (in Nm) that can be transmitted by this interference fit is

(*a*) 1005.3 (*b*) 754.0

(*c*) 601.0 (*d*) 502.7

LINKED ANSWER QUESTIONS

Statement for Linked Answer Questions 21 & 22 :

A 2 kg mass, initially at rest, is acted upon by a time-varying force whose profile is as shown.

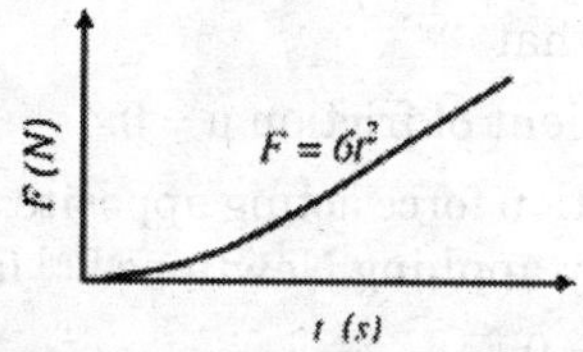

21. The velocity (in m/s) of the mass is

(*a*) t^3 (*b*) $2t^3$

(*c*) $6t$ (*d*) $12t$

22. The distance travelled by the mass in 2 seconds is

(*a*) 24 (*b*) 12

(*c*) 8 (*d*) 4

ANSWERS

1. (c)	**2.** (c)	**3.** (a)	**4.** (*c*)	**5.** (*c*)	**6.** (b)	**7.** (b)
8. (*d*)	**9.** (*a*)	**10.** (*b*)	**11.** (*d*)	**12.** (c)	**13.** (*b*)	**14.** (d)
15. (b)	**16.** (*a*)	**17.** (a)	**18.** (b)	**19.** (*a*)	**20.** (*d*)	**21.** (a)
22. (d)						

EXPLANATIONS

1. (c)

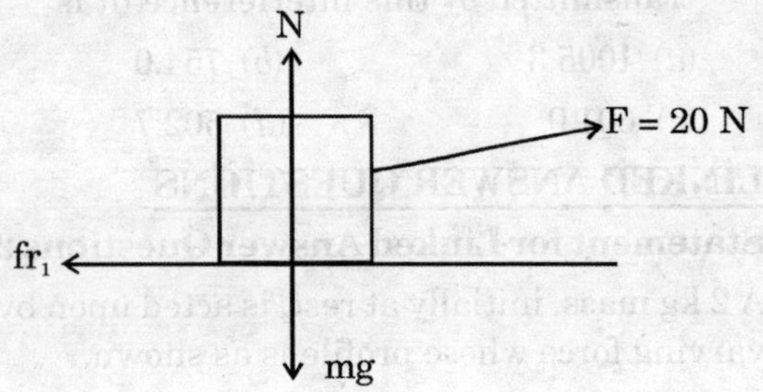

Given that

Coefficient of friction $\mu = 0.4$

$\because$ Friction force acting apposite to the applied force by appliing Newtions 2nd law

$F - fr = 0$

$\boxed{f = fr = 20\,N}$

Hence the option (c) is Correct.

2. (c) As we know for free expansion process, the stressed induced in the bar will be zero.

Therefore stress induced in the bar due to free expansion is independent of both thermal expansion coefficient and temperature change.

Hence the option (c) is Correct.

3. (a) Bucking load for fixed-fixed column

$$P_1 = \frac{\pi^2 \in I}{L_c^2} = \frac{\pi^2 \in I}{\left(\frac{1}{2}\right)^2} = \left[\because L_c = \frac{l}{2}\right]$$

$$P_1 = \frac{4\pi^2 \in I}{l^2} \quad ...(1)$$

Bucking load for fixed-free column

$$P_2 = \frac{\pi^2 \in I}{L_c^2} = \frac{\pi^2 \in I}{(2l)^2} [\because Lc = 2l]$$

$$P_2 = \frac{\pi^2 \in I}{4l^2} \quad ...(2)$$

$$(1)/(2)\ \frac{P_1}{P_2} = \frac{4\pi^2 \in I}{l^2} \times \frac{4l^2}{\pi^2 \in I}$$

$$\frac{1000}{P_2} = 16 \quad [\because P_1 = 10000\,N\,\text{given}]$$

$$\boxed{P_2 = 625\,N}$$

Hence the option (a) is Correct.

6. (b)

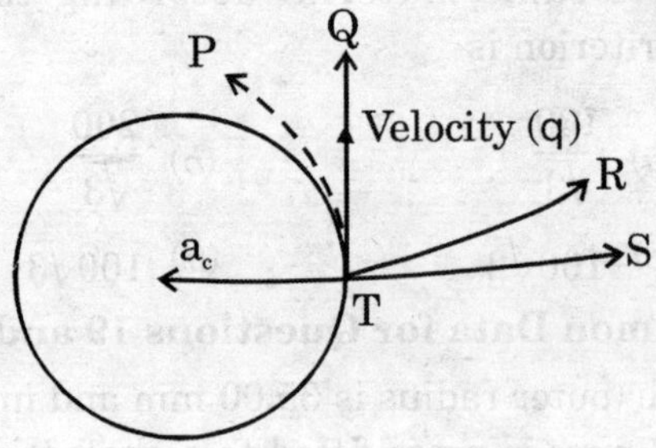

When the string breaks at point T the path followed by the mass will be in the direction of velocity (v). i.e. in the direction of 'Q'.

Hence the option (b) is Correct.

7. (b)

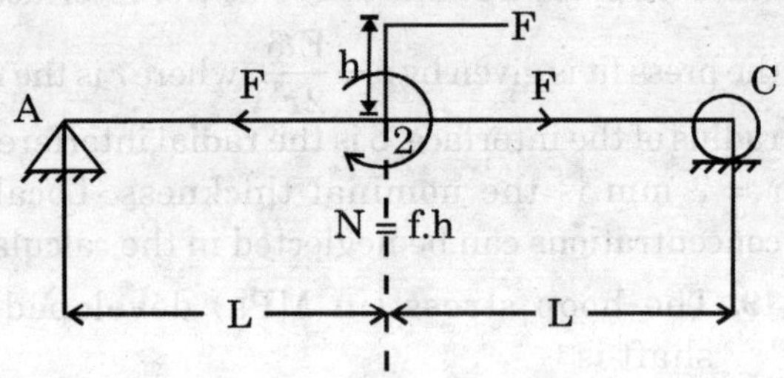

At the mid-space the Moment and Axial force are discotinuous

$(B.M)_B = M_B = f.h$

Axial force $(f_x) = F$

Hence the option (b) is Correct.

10. (*)

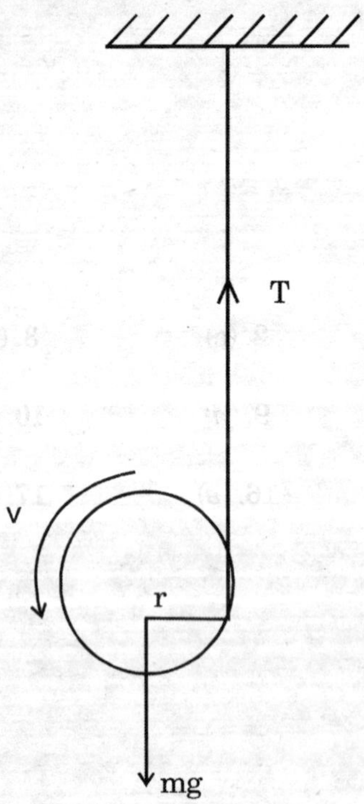

Given that Radius of disc (r) = 0.5 m

Mass of disc (m) = 5 kg

Acceleration due to gravity (g) = 10 m/s²

Now Applying Newton's 2nd law

$mg - T = mc$

$5 \times 10 - T = 5a$

$50 - T = 5a \quad (1)$

for solid disc

$$\left[\therefore x = \frac{MR^2}{2}\right]$$

Now Torque$(\tau) = I \propto$

$$T \times r = \frac{MR^2}{2}\left(\frac{a}{r}\right)$$

$$T \times 0.5 = \frac{5 \times (0.5)^2}{2} \cdot \frac{a}{0.5}$$

$$T = \frac{5a}{2}$$

from (1) $50 - \frac{5a}{2} = 5a \Rightarrow a = 6.66\, m/s^2$

$\Rightarrow a = 6.66\, m/s^2$

Velocity$(V) = a \times t = 6.66 \times 20\, m/s$

$\therefore$ Angular velocity of disc $(w) = \frac{V}{r}$

$$W = \frac{20}{0.5} = 40\, rad/sec$$

$w = 40\, rad/sec$

12. (c)

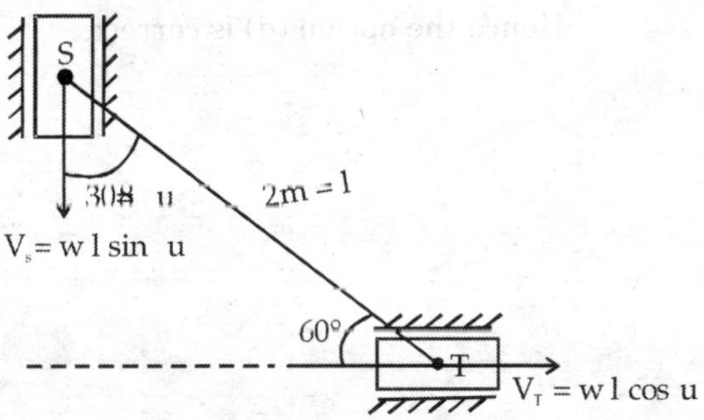

Given $V_T = 6\, m/s$

$wl \cos 30 = 6\, m/s$

$$w \times 2 \times \frac{\sqrt{3}}{2} = 6$$

$$w = \frac{6}{\sqrt{3}} = \frac{2 \times 3\sqrt{2}}{\sqrt{3}}$$

$\therefore \boxed{w = 2\sqrt{3}\, rad/sec}$

Hence option (c) is correct.

14. (d)

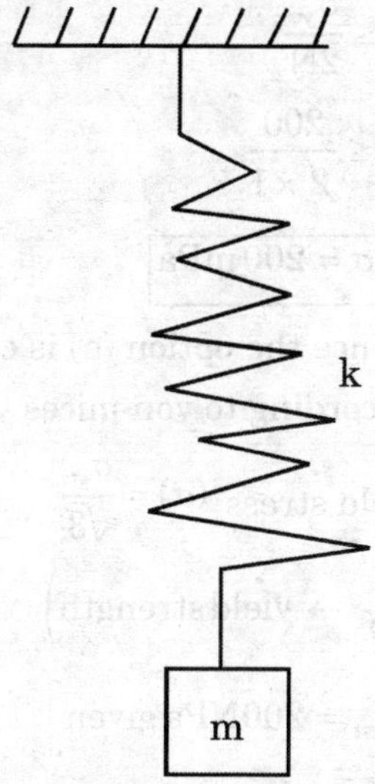

Given that

Natural frequency $(W_n) = 54$

We know $W_n = \frac{1}{2\pi}\sqrt{\frac{g}{\Delta}}$

$$5 = \frac{1}{2\pi}\sqrt{\frac{g}{\Delta}}$$

$$\Rightarrow (5 \times 2\pi)^2 = \frac{g}{\Delta}$$

$\boxed{D = 0.010\, m}$

Hence the static definition of the spring is

$\boxed{D = 0.010\, m}$

Hence the option (d) is correct.

15. (b) We know

Radius of curvature $(R) = \frac{EP}{M}$

Given that Bending Moment is constant

$\therefore R \propto E_2$

$$\therefore \frac{(R_1)}{(R_2)} = \frac{(E_1 I_1)\, \text{Alumi.}}{(E_2 I_2)\, \text{copper}}$$

$$\frac{2 \times 10^3}{R_2} = \frac{70 \times \frac{\pi}{64}(2.5)^4}{120 \times \frac{\pi}{64}(2)^4}$$

$R_2 = 1404.4\, mm$

$\boxed{R_2 = 1.404\, m}$

Hence the option (b) is correct.

17. (a) According to Tresca's theory

$$(\tau_{max})_{induced} \le (\tau_{max})_{Permissible}$$

$$\frac{\sigma}{2} \le \frac{\sigma_{yt}}{2N}$$

$$\frac{\sigma}{\not{2}} \le \frac{200}{\not{2} \times 1}$$

$$\therefore \boxed{\sigma = 200\,\text{mPa}}$$

Hence the option (a) is correct.

18. (b) According to von-mices yield criterion

yield stress $(\sigma) = \dfrac{\sigma_{yt}}{\sqrt{3}}$

$\left(\sigma_{yt} \to \text{yield strength}\right)$

$\left[\sigma_{yt} = 200\text{NPa given}\right]$

$$\boxed{\sigma = \frac{200}{\sqrt{3}}}$$

Hence the option (b) is correct.

21. (a)

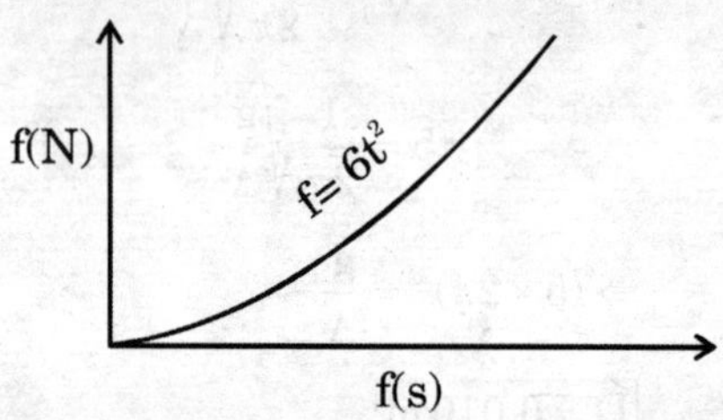

Given

Mass of body (m) = 2kg

force on the body (f) = $6t^2$

Applying Newton's 2nd law

$F = ma$

$$6t^2 = m\frac{dv}{dt}$$

$$3\not{6}t^2 = \not{2} \times \frac{dv}{dt}$$

$$\int dv = \int 3t^2\,dt$$

$$v = \frac{\not{3} \cdot t^3}{\not{3}}$$

$$\boxed{V = t^3 \text{ w/c}}$$

Hence option (a) is correct.

22. (d) We know $V = \dfrac{dx}{dt}$

$$\int dx = \int v\,dt$$

$$\int dx = \int t^3 dt$$

$$x = \frac{t^4}{4}$$

At t = 2 sec

distance travelled by body $(x) = \dfrac{(2)^4}{4}$

$$x = \frac{\not{16}}{\not{4}} = 4\,\text{m}$$

$$\therefore \boxed{x = 4\,\text{m}}$$

Hence the option (d) is correct.

2012

Q. (1 – 9) carry one mark each.

1. The axial force diagram for the weightless beam subjected to the inclined force P = 5 kN is

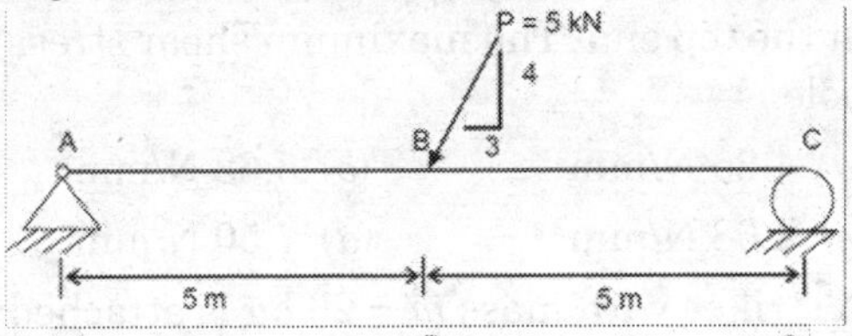

(a)
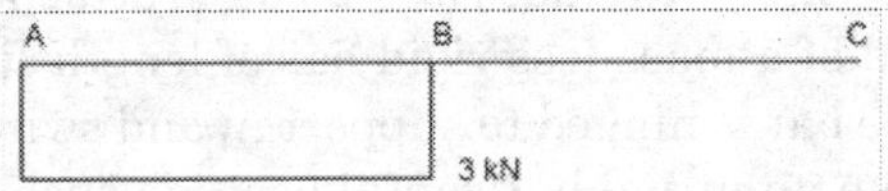

(b)
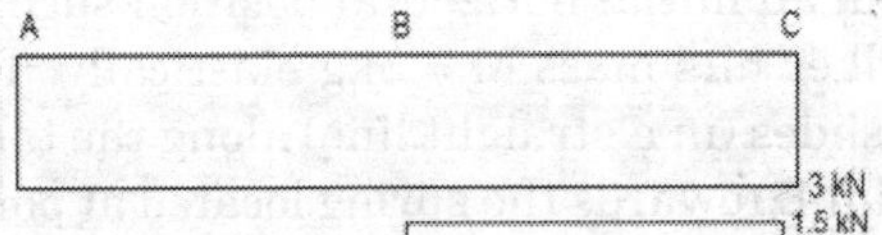

(c)
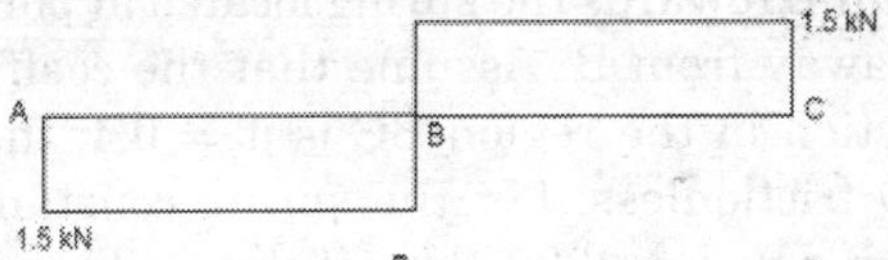

(d)
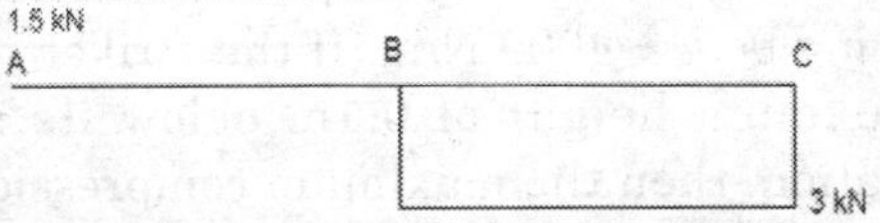

2. A block of weight W, connected to two springs with spring constants k_1 and k_2, rests initially on a horizontal frictional surface. The coefficient of static friction between the block and the surface is μ. Both springs are initially undeformed. The magnitude of force F, applied to the second spring, is now gradually increased. The block will start to slide when F becomes

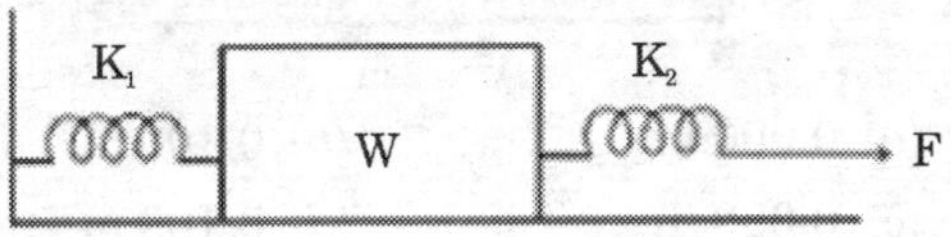

(a) μW
(b) $k_1\mu W/(k_1 + k_2)$
(c) $k_2\mu W/(k_1 + k_2)$
(d) $k_2\mu W/k_1$

3. Three connected railway coaches A, B and C of masses m_A, m_B and m_C, respectively are being pulled by a locomotive with force F over a horizontal track. The coaches may be assumed to move on frictionless wheels with negligible air resistance. The tension in the connector between coaches A and B is

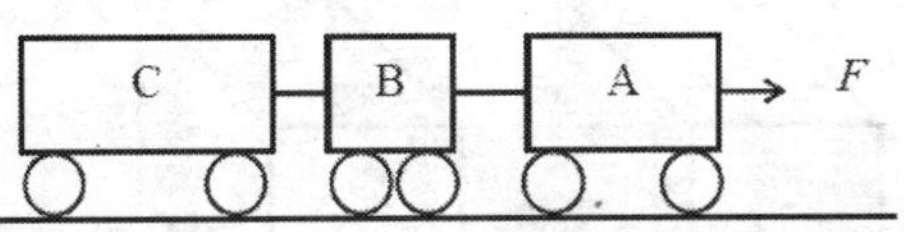

(a) F
(b) $FMA/(m_A + m_B + m_C)$
(c) $Fm_B/(m_A + m_B + m_C)$
(d) $F(\mu_B + \mu_C)/(m_A + m_B + m_C)$

4. For the beam-column configurations shown in figure, the minimum Euler buckling load is obtained for the case (Young's modulus and second moment of cross-sectional area are as indicated)

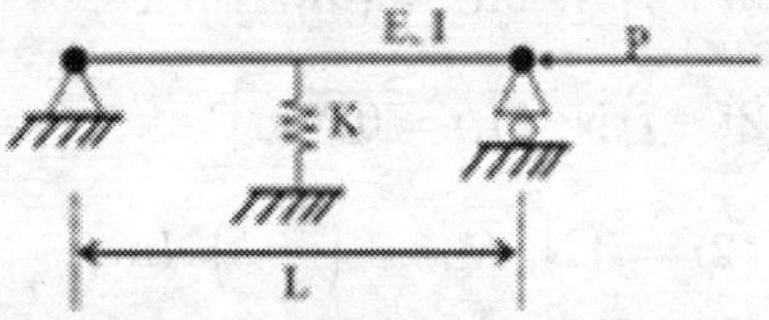

(i)

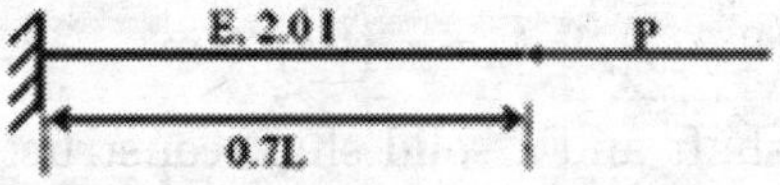

(ii)

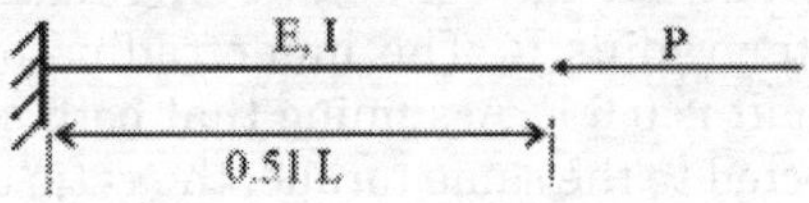

(ii)

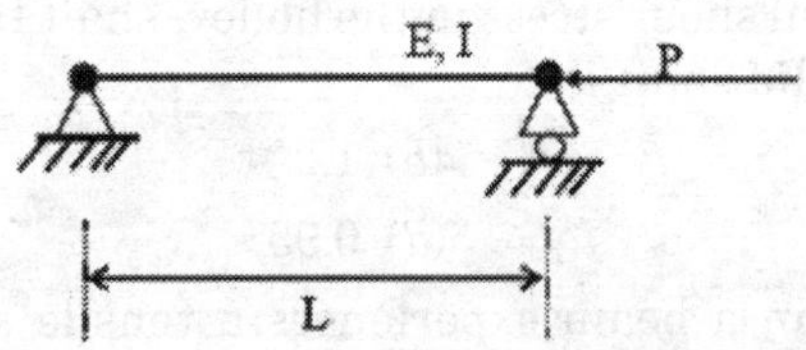

(iv)

(a) (i)
(b) (ii)
(c) (iii)
(d) (iv)

5. A disk of mass $m = 0.25$ g and radius $r = 10$ mm is at rest relative to a mass-less horizontal turntable spinning about a vertical axis at an angular speed of ω = 2 rad/s. The turntable is assumed to be mounted on frictionless bearings. Another identical, initially non-rotating disk is dropped onto the spinning disk. Friction causes both disks (and the turntable) to eventually rotate at the same angular speed. The eventual angular speed of the disks is

(a) 0.15 rad/s
(b) 1 rad/s
(c) 2 rad/s
(d) 4 rad/s

6. A rocket in the atmosphere is accelerating upwards with acceleration a m/s². The natural frequency of a spring-mass system (with mass m kg and spring constant k N/m), suspended vertically inside the rocket, is (take g m/s² to be the acceleration due to gravity)

(a) $\sqrt{\frac{\bar{k}}{m}}$; $(\bar{k} < k)$
(b) $\sqrt{\frac{k}{m}}$
(c) $\sqrt{\frac{kg}{ma}}$
(d) $\sqrt{\frac{k}{m}\left(1-\frac{a}{g}\right)}$

7. An irregular planar body in space is acted upon by a force $\vec{F} = \vec{F}\left(2\vec{i}+\vec{j}\right)$N at position $\vec{r}_1 = \left(\hat{i}+2\vec{j}\right)m$ and a moment $\vec{M} = 3\,\vec{k} = \left(-2\vec{i}\right)m$. The corresponding equivalent force $\vec{F}_O$ and moment $\vec{M}_O$ at the origin are

(a) $\vec{F}_O = \left(2\vec{i}+\vec{j}\right)$N; $\vec{M}_O = \left(3\vec{k}\right)$Nm

(b) $\vec{F}_O = \left(2\vec{i}+\vec{j}\right)$N; $\vec{M}_O = 0$ Nm

(c) $\vec{F}_O = \left(-2\vec{i}-\vec{j}\right)$N; $\vec{M}_O = \left(-6\vec{k}\right)$Nm

(d) $\vec{F}_O = \left(2\vec{i}+\vec{j}\right)$N; $\vec{M}_O = \left(6\vec{k}\right)$ Nm

8. A hollow shaft and a solid shaft constructed of the same material have the same length and the same outer radius R. The inner radius of the hollow shaft is 0.6 R. Assuming that both shafts are subjected to the same torque, the ratio of the maximum shear stress in the hollow shaft to that in the solid shaft is

(a) 1.1 (b) 1.2

(c) 1.15 (d) 0.95

9. A point in a beam experiences a tensile stress (due to bending) of 50 N/mm² and a shear stress of 20 N/mm². The principal stresses are

(a) 17 N/mm² tension, 67 N/mm² compression

(b) 0, 0

(c) 57 N/mm² tension, 7 N/mm² compression

(d) 52 N/mm² compression, 15 N/mm² tension

Q. (10 – 22) carry two marks each.

10. Consider a simply supported beam loaded either by a uniformly distributed transverse load or by a concentrated transverse load applied at the center such that the maximum bending stress in both cases is the same. The ratio of the strain energy for the two cases is

(a) 4/5 (b) 5/8

(c) 8/5 (d) 1

11. A small railway bridge is constructed from identical steel truss members, each of length l, crosssectional area A and Young's modulus E . A train stops on the bridge. The loads applied by the train on the truss on one side of the bridge may be assumed to act at pins A, B and C, as shown. The displacement of the support C due to this loading is

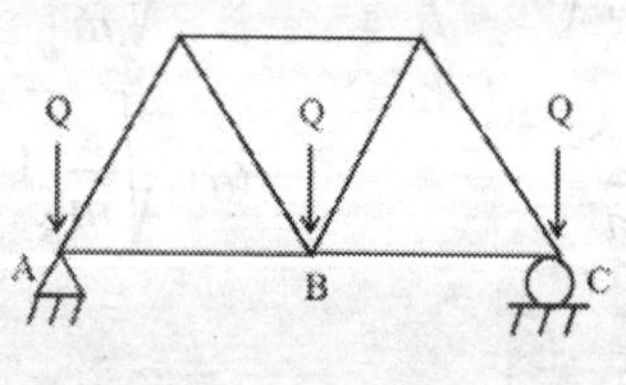

(a) 0 (b) $Ql/(AE)$

(c) $\sqrt{3}\,Ql/(4AE)$ (d) $Ql/\left(\sqrt{3}AE\right)$

12. A vertical pole, cantilevered at the bottom, has a solid circular cross-section of diameter d = 49.21 mm. It is loaded by a horizontal force P = 6675 N at the top end. The maximum shear stress in the pole is

(a) 4.25 N/mm² (b) 5.68 N/mm²

(c) 4.68 N/mm² (d) 7.50 N/mm²

13. A striker with mass m = 20 kg is attached to the end of a mass-less rigid bar of length R = 0.3m. The bar is hinged to support A, and swings down from an initial horizontal position such that the striker hits mass M = 5kg elastically. The mass M slides (in a straight line) along the table, from point B towards the spring located at point C, 0.2 m away from B. Assume that the coefficient of friction in the region BC is μ_d = 0.4; the region CD frictionless. Let the spring constant of the spring be k = 4000 N/m. If the striker rises to a maximum height of 0.1m below its starting location, then the maximum compression of the spring is (let acceleration due to gravity g = 10 m/s², the dimensions of the striker and mass are small)

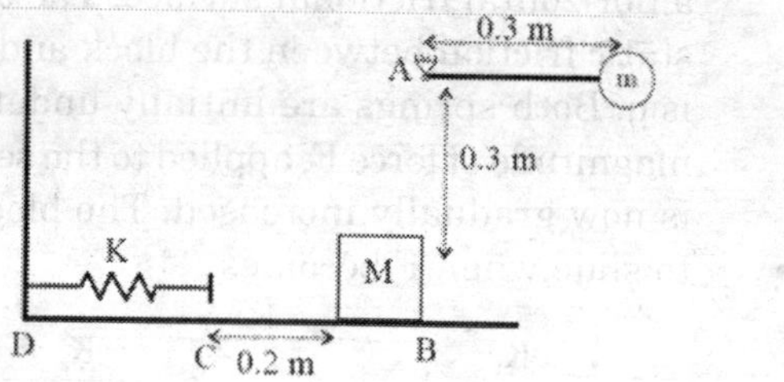

(a) 0.000 m (b) 0.100 m

(c) 0.089 m (d) 0.109 m

14. A cube, made of aluminum, of dimension 0.1 m × 0.1 m × 0.1 m, rests against a rigid wall (with normal in the y-direction), as shown in the figure. Another parallel rigid wall is located at a clearance of 0.2 mm from the block. Assuming all contacts to be frictionless, if the block is heated by ΔT = 150°C, the normal stress σ_{yy} induced in the block is (for aluminum E = 70 GPa; ν = 0.3; α = 20 × 10⁻⁶/°C)

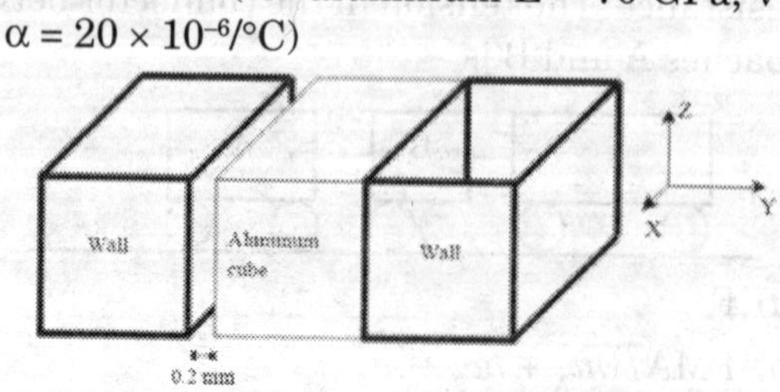

(a) σ_{yy} = –7 MPa (b) σ_{yy} = 7 MPa

(c) σ_{yy} = –70 MPa (d) σ_{yy} = 0

15. A thin walled spherical pressure vessel made of a linear elastic isotropic material has inner radius r and thickness t before pressurization. When subjected to internal pressure p, elements of the pressure vessel wall experience a state of stress described by a single point $(\sigma, \tau) = (pr/(2t), 0)$ in Mohr's circle. The reduction of the wall thickness due to pressurization

(*a*) increases with t

(*b*) remains independent of t

(*c*) decreases with t

(*d*) depends on the elastic properties.

16. For small oscillations, the natural frequency of the system in terms of K, a, b and M is (assuming ideal joints and mass-less rigid rod ABC)

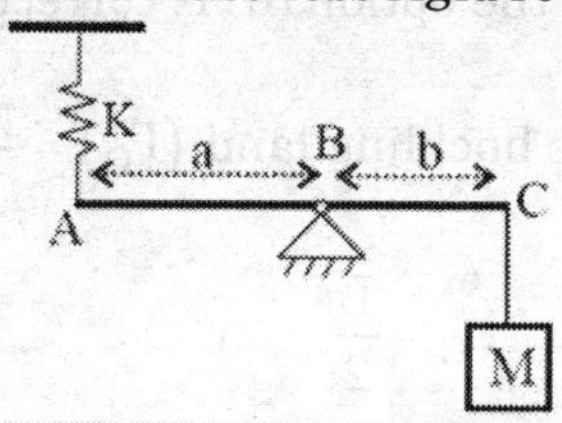

(*a*) $\sqrt{\frac{K}{M}}$ (*b*) $\sqrt{\frac{Ka}{Mb}}$

(*c*) $\sqrt{\frac{Kb^2}{Ma^2}}$ (*d*) $\sqrt{\frac{Ka^2}{Mb^2}}$

Common Data Q. (17 – 18) :

A steel cylindrical pressure vessel has an inner radius of 1.8 m and a wall thickness of 20 mm.

17. For an internal pressure of 800 kPa the maximum shear stress for the cylindrical part of the vessel is

(*a*) 16 MPa (*b*) 18 MPa

(*c*) 20 MPa (*d*) 0

18. At which of the following internal pressures will the cylindrical vessel yield as per the Tresca criterion if the yield strength of the material in tension is 320 MPa

(*a*) 3.55 MPa (*b*) 7.1 MPa

(*c*) 1.775 MPa (*d*) 4.0 MPa

Common Data Q. (19 – 20) :

A steel beam, of rectangular cross-section 25 mm wide and 75 mm deep, is pinned to supports at points A and B, where the support B is on rollers. The Young's modulus of steel may be assumed as 2.0 × 105 N/mm². The ends of the beam are loaded with 5 kN loads.

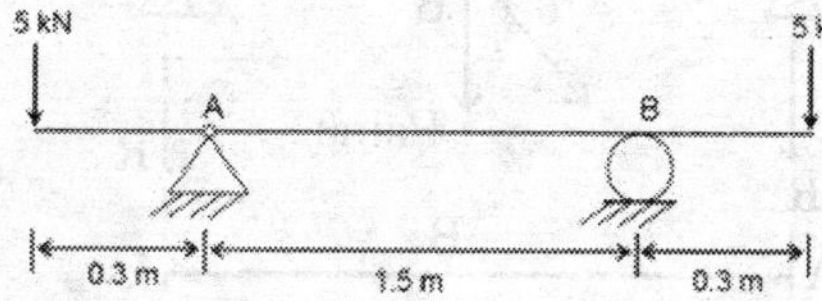

19. The maximum bending stress in the beam is

(*a*) 0 (*b*) 64 MN/m²

(*c*) 128 MN/m² (*d*) 32 MN/m²

20. The vertical deflection at the ends is

(*a*) 4.92 mm (*b*) 4.58 mm

(*c*) 9.31 mm (*d*) 9.84 mm

Linked Answer Q. (21– 22) :

A cantilevered beam of unknown material (which is homogeneous, linearly elastic and isotropic) and an unknown cross-section (which is uniform and symmetric) is given in the figure. The stiffness of the end spring is k = 2000 N/m and end load P = 1000 N; length of the beam L = 1m.

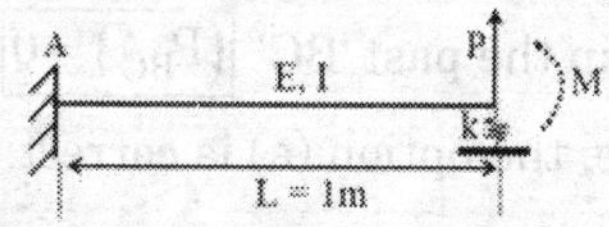

21. If the deflection at the free-end (under load P, with end moment M = 0) is measured as $\delta = 5$ mm, the flexural rigidity EI for the beam is (in Nm²)

(*a*) 66,666 (*b*) 66,000

(*c*) 67,300 (*d*) 64,000

22. The value of the additional end moment M (in N.m) required to obtain an upward deflection of 1 mm at the free end, is (moment is positive in counterclockwise direction)

(*a*) –533.33 (*b*) 533.33

(*c*) –528 (*d*) 528

ANSWERS

1. (a)	**2.** (a)	**3.** (d)	**4.** (c)	**5.** (b)	**6.** (b)	**7.** (b)
8. (c)	**9.** (c)	**10.** (c)	**11.** (d)	**12.** (c)	**13.** (*)	**14.** (c)
15. (*)	**16.** (d)	**17.** (b)	**18.** (b)	**19.** (b)	**20.** (*)	**21.** (b)
22. (a)						

EXPLANATIONS

1. (a)

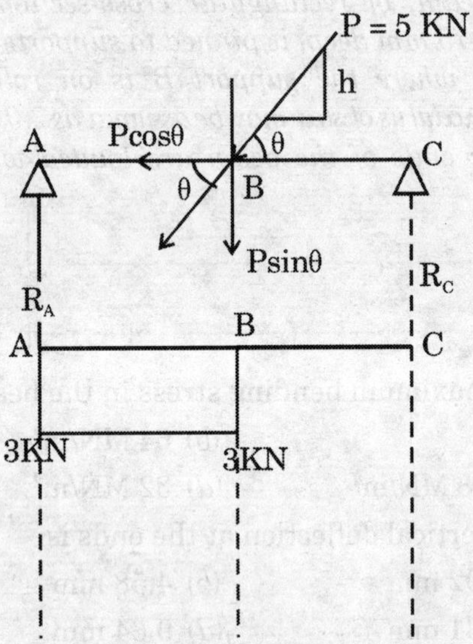

$\tan\theta = \frac{4}{3} \Rightarrow \boxed{\theta = 53.13°}$

$\therefore$ Axial load in the beam = $P\cos\theta$

$(P_{AB}) = 5 \times \cos 53.13$

$\boxed{P_{AB} = 3\ KN}$

Load in the past 'BC' $\boxed{(P_{BC}) = 0}$

Hence, the option (a) is correct.

2. (a)

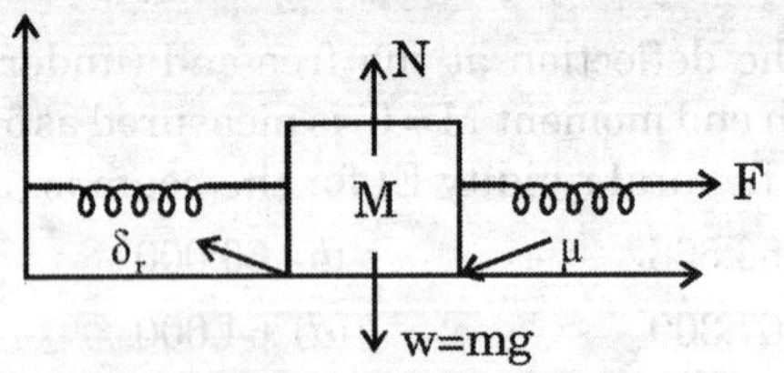

Applying Newton's 2nd law

$F - fr = 0 \qquad [\because a = 0]$

$F = fr = \mu N$

$\boxed{F = \mu\omega}$

Hence the option (a) is correct.

3. (d)

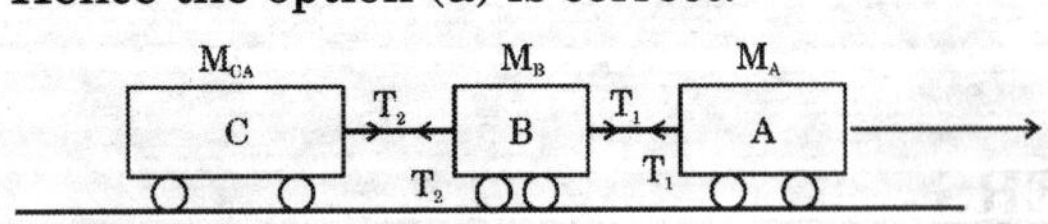

Applying Newton's 2nd law

$F = (m_C + m_B + m_A)^a$ ..(1)

FBD of Block B

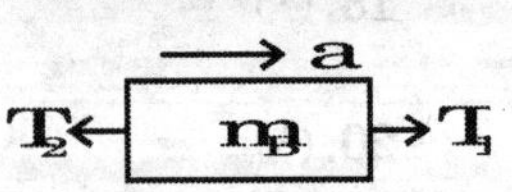

$T_1 - T_2 = m_B a$ ..(2)

FBD of Block C

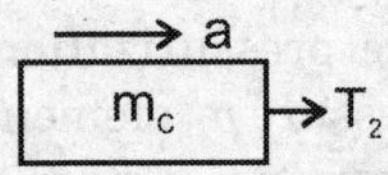

$T_2 = m_C a$ ...(3)

Solve equation (2) & (3), we get

$T_1 = (m_B + m_C)a$

$\therefore a = \frac{T_1}{m_B + m_C}$

Substitute the value of 'a' in equation (1)

$F = (m_C + m_B + m_A) \cdot \frac{T_1}{m_B + m_C}$

Hence the option (d) is correct.

4. (c) Euler's buckling land $(P_C) = \frac{\pi^2 E \cdot I}{L_C^2}$

Case 1 : $L_C = \frac{L}{2}$,

$$(P_C)_1 = \frac{\pi^2 EI}{\left(\frac{L}{2}\right)^2} = \frac{4\pi^2 EI}{L^2}$$

Case 2: $L_c = 2\ L$,

$$(P_C)_2 = \frac{\pi^2 EI}{(2 \times 0.7L)^2} = 1.02\frac{\pi^2 EL}{L^2}$$

Case 3: $L_c = 2\ L$,

$$(P_C)_3 = \frac{\pi^2 EI}{(2 \times 0.5L)^2} = 0.9615\frac{\pi^2 EI}{L^2}$$

Case 4: $L_c = \frac{L}{2}$,

$$(P_C)_4 = \frac{\pi^2 EI}{\left(\frac{L}{2}\right)^2} = 4 \cdot \frac{\pi^2 EI}{L^2}$$

The minimum Buckling load is

$P_{C_3} = 0.9615 P_C$.

Hence the option (c) is correct.

6. (b) Given that

Acceleration of the rocket a = m/s²

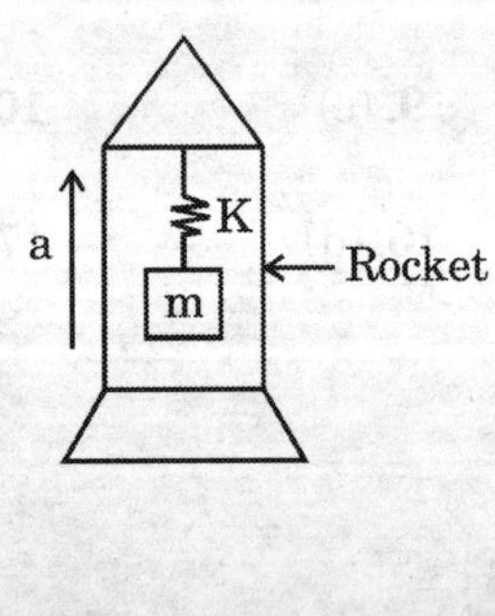

FBD of spring mass system

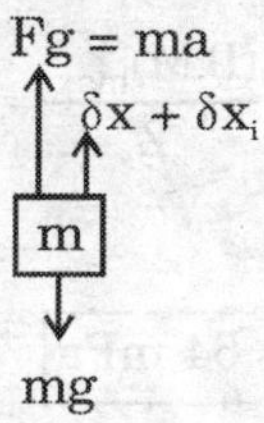

Here $mg = Sx_i$

Therefore mg will be neglected

Net force $F_2 + Sx = 0$

$$ma + Sx = 0$$

$$m\ddot{x} + Sx = 0 \quad [\ddot{x} = a]$$

$$\ddot{x} + \left(\frac{S}{m}\right)x = 0$$

Natural frequency $(\omega_n) = \sqrt{\frac{S}{m}}$

[S = K = stiffness of spring]

$$\boxed{\omega_n = \sqrt{\frac{K}{m}}}$$

Hence, the option (b) is correct.

8. (c) Given Radius of solid shaft = R

Inner Radius of Hollow shaft $(r_1) = 0.6R$

We know,

Shear stress of solid shaft $\tau_{solid} = \frac{16T}{\pi d^3}$

$$\tau_{solid} = \frac{16T}{\pi d^3} \quad ...(1)$$

Shear stress of Hollow shaft

$$\tau_{Hollow} = \frac{16T}{\pi d^3(1-K^4)}$$

$$\tau_{Hollow} = \frac{16T}{\pi d^3\left[1-\left(\frac{di}{dx}\right)^4\right]}$$

$$\tau_{Hollow} = \frac{16T}{\pi d^3\left[1-(0.6)^4\right]} \quad ...(2)$$

Divide equation (2) by (1), we get

$$\frac{\tau_{Hollow}}{\tau_{Solid}} = \frac{1}{1-(0.6)^4} = 1.148$$

Hence, the option (c) is correct.

14. (c) Given that

Thermal expansion coefficient

$(d) = 20 \times 10^{-6} / °C$

Young's Modulus of the body (E) = 70 GPa

We know, total deformation of the body δ_T

$$\delta_T = \delta_{axial} + \delta_{thermal}$$

$$0.2 \text{ mm} = \frac{-\sigma_y L}{E} + \alpha TL$$

$$0.2 \times 10^{-3} = \frac{-\sigma_y \cdot (0.1)}{70 \times 10^9} + 20 \times 10^{-6} \times 150 \times 0.1$$

$$-7 \times 10^6 = -\sigma y(0.1)$$

$$\sigma_y = -70 \times 10^6 \text{ N/m}^2$$

$\boxed{\sigma_y = 70 \text{ mPA}}$ compressive

Hence the option (c) is correct.

16. (d)

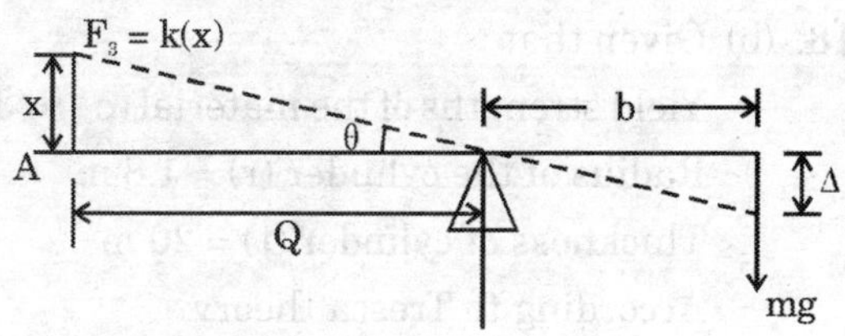

Taking moment about 'B'

$$F_s \times a = mg \times b$$

$$F_s = mg\left(\frac{b}{a}\right)$$

$$Kx = mg\left(\frac{b}{a}\right)$$

$$x = \frac{mg}{K}\left(\frac{b}{a}\right) \quad ...(1)$$

From equilateral triangles ΔABD & ΔCBE

$$\frac{AB}{BC} = \frac{x}{\Delta}$$

$$\frac{a}{b} = \frac{x}{\Delta} \Rightarrow x = \frac{a}{b}\Delta$$

Put the value of 'x' in (1), we get

$$\Delta = \frac{mg}{K}\left(\frac{b}{a}\right)^2 \quad ...(2)$$

Now, Natural frequency $(\omega_n) = \sqrt{\frac{g}{\Delta}}$

$$\omega_n = \sqrt{\frac{g}{\frac{mg}{K}\left(\frac{b}{a}\right)^2}} \Rightarrow \sqrt{\frac{Ka^2}{mb^2}}$$

$$\boxed{\therefore \omega_n = \sqrt{\frac{Ka^2}{Mb^2}}}$$

Hence, the option (d) is correct.

17. (b) Given that

Radius of cylinder (r) = 1.8 m

Wall thickness (t) = 20 mm

Internal Pressure = P) = 200 kPa

∴ Shear stress for the cylinder is

$$\tau_{max} = \frac{Pd}{8t} = \frac{P(3.6)\times 10^{-3}}{4\times 20}$$

$$= \frac{800\times 10^3 \times 366\times 10^3}{8\times 20}$$

$$\boxed{\tau_{max} = 18 \text{ mPA}}$$

Hence the option (b) is correct.

18. (b) Given that

Yield strengths of the material $(\sigma_{yt}) = 320$ mPA

Radius of the cylinder (r) = 1.8 m

Thickness of cylinder (t) = 20 m

According to Tresca theory

$$(\tau_{max})_{induced} \le (\tau_{max})_{permissible}$$

$$\frac{Pd}{4t} \le \frac{\sigma_{yt}}{2N}$$

$$\frac{P\times 3.6\times 10^3}{4\times 20} \le \frac{320}{2\times 1}$$

$$\therefore \boxed{P = 7.1 \text{ mPA}}$$

Therefore internal pressure of the cylinder

$$\boxed{P = 7.1 \text{ mPA}}$$

Hence, the option (b) is correct.

19. (b)

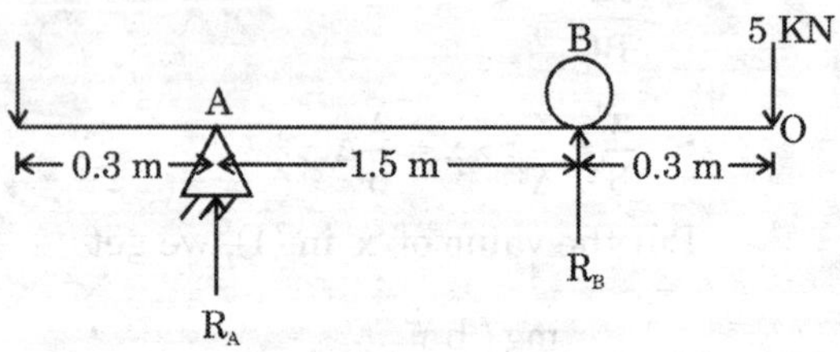

$R_A + R_B = 5 + 5 = 10$ kN

$R_A + R_B = 10$ kN

Taking moment about 'B'

$$5\times(0.3+1.5) - R_A \times (1.5) - 5\times 0.3 = 0$$

$$9 - 1.5 = R_A \times 1.5$$

$$\boxed{R_A = 5 \text{ kN}}$$

$$\boxed{R_B = 5 \text{ kN}}$$

$(BH)_D = 0$

$(BH)_B = -5 \times 0.3 = -1.5$ KN

$(BH)_A = -5 \times (1.5 + 0.3) + 5 \times 1.5 = -1.5$ kN

$(BH)_C = 0$

The maximum Bending stress is

$$(\sigma_b)_{max} = \frac{(B.M)_{max}}{Z} = \frac{(-1.5\times 10^3)}{\frac{25\times 10^{-3}\times (75)^2\times 10^{-6}}{6}}$$

$$\boxed{(\sigma_b)_{max} = 64 \text{ mPa}} \qquad \left[X = \frac{bd^2}{6}\right]$$

Hence, the option (b) is correct.

21. (b)

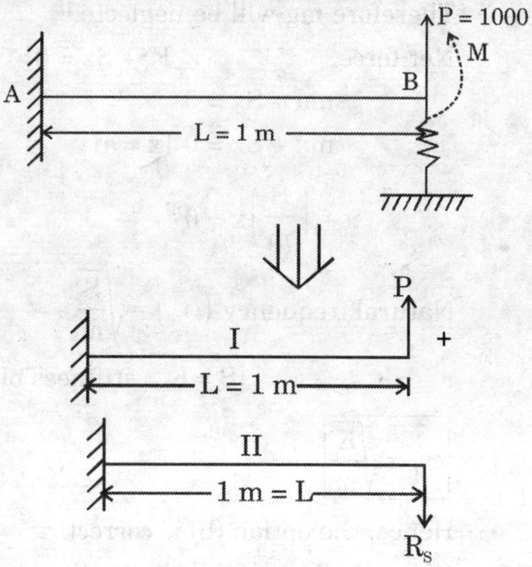

Total deflection at the end 'B'.

$$\delta_T = \delta_I + \delta_B$$

$$5\times 10^{-3} = \frac{PL^3}{3EI} + \frac{-R_\delta L^3}{3EI} \quad [\because R_S = K\delta]$$

$$5\times 10^{-3} = \frac{1000\times 1^3}{3EI} - \frac{(2000\times 5\times 10^{-3})\times 1^3}{EI}$$

$$EI = \frac{1000-(10)}{3\times 5\times 10^{-3}}$$

$$\boxed{EI = 66{,}000}$$

Hence, the option (b) is correct.

22. (a) Net upward deflection

$$\delta_{Net} = \delta_I + \delta_{II} + \delta_{III}$$

$$1 \text{ mm} = \frac{PL^2}{3EI} + \frac{-R_I L^3}{3EI} + \frac{ML^2}{2EI}$$

$$1\times 10^{-3} = \frac{PL^3 - 2RsL^3 + 3ML^2}{6EI}$$

$$\Rightarrow 1\times 10^{-3}\times 6EI = (2\times 1000\times 1^3) - (2\times 2000\times 10^{-3}\times 1^3) + 3MKI^2$$

After solving the above equation, we get

$$\boxed{M = -533.33 \text{ N}\cdot\text{m}}$$

Hence, the option (a) is correct.

2013

Q. 1 – 9 carry one mark each.

1. At a point in a body subjected to plane stress, the state of stress is as shown in the Figure. One of the principal stresses is 180 MPa. Find the unknown shear stress τ (in MPa). ________

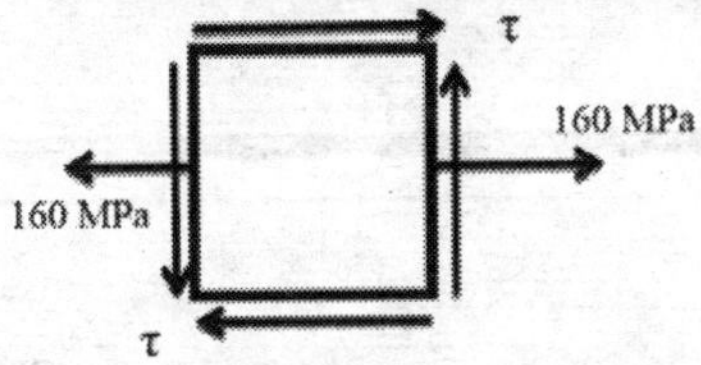

2. A point in a body is subjected to a hydrostatic pressure of 100 MPa. Find the maximum shear stress at this point in MPa. ________

3. A circular shaft of diameter 10 mm and length 3m is subjected to a torque of $T = \pi$ N-m at a location 2m away from the fixed end as shown in the Figure. Find out the angle of twist (in radians) at the free end. Shear modulus of the material of the shaft is 10 GPa. ________

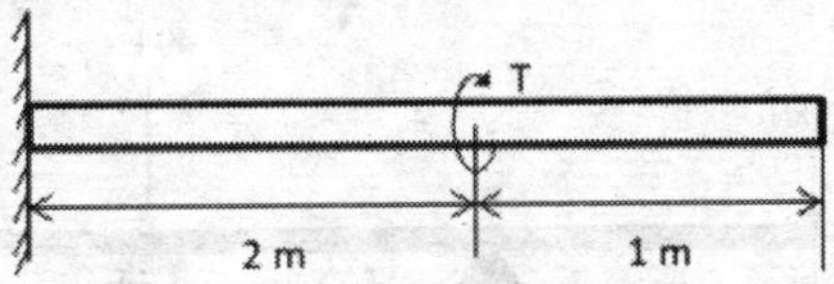

4. A rigid massless rod ABC is hinged at A and carries a point mass M (in kg) at C. Point B is connected to a linear spring with spring constant k (in N/m) as shown in the figure. The length AB and AC area and L, respectively. Neglecting the effect of gravity, the natural frequency of this spring-mass system in rad/s is

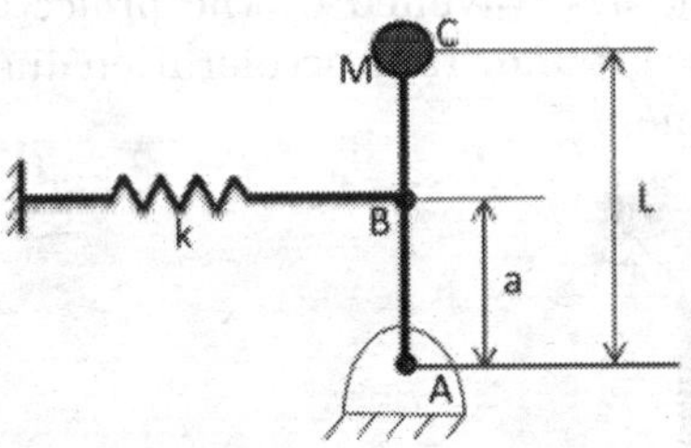

(a) $\sqrt{\dfrac{kL^2}{Ma^2}}$ (b) $\sqrt{\dfrac{ka^2}{ML^2}}$

(c) $\sqrt{\dfrac{kL}{Ma}}$ (d) $\sqrt{\dfrac{ka}{ML}}$

5. A two bar truss is shown in the Figure. The cross-sectional area and Young's modulus of bar 1 are 0.02 m² and 200GPa, respectively. The cross-sectional area and Young's modulus of bar 2 are 0.01 m² and 80GPa, respectively. The force F applied on the truss is 2 N. Find out the stress developed in bar 2 in Pa. ________

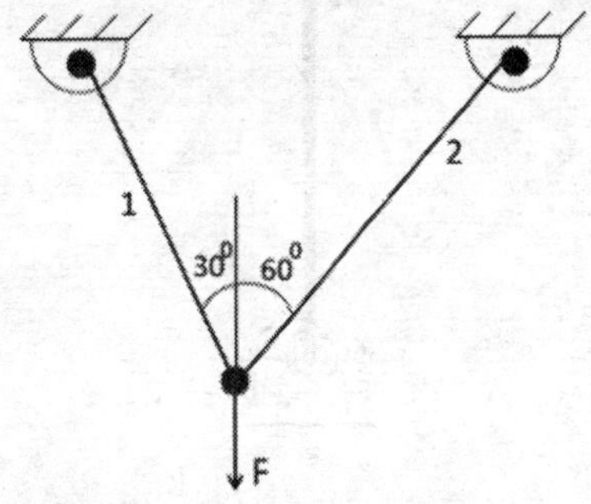

6. A spring balance reads 10 kg in a lift when the lift is stationary. When the lift starts moving with a constant acceleration, the new reading is 12.3 kg. If the upward acceleration is considered positive, what is the acceleration of the lift? Acceleration due to gravity may be taken as 10 m/s² downwards.

7. A force F = 2 N is applied on a block of mass M = 0.5 kg as shown in the figure. The block is constrained to move along the horizontal direction in a guideway. Find out the distance (in meters) travelled by the block in 2 s starting from rest. Neglect any friction between the block and the guideway. ________

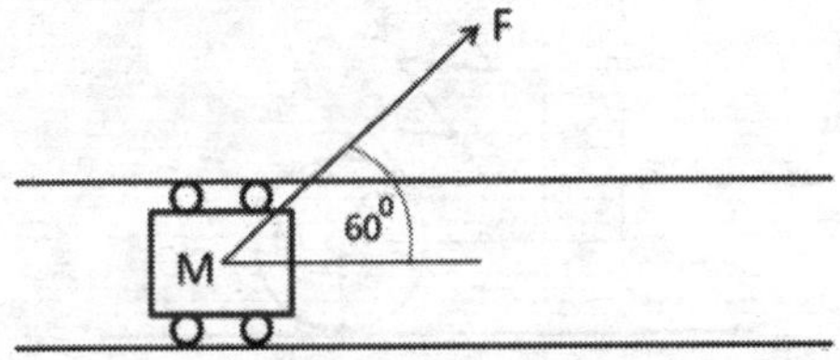

8. A man of mass 50 kg is walking on a long wooden board of mass 200 kg (as shown in the Figure). The wooden board is initially at rest on a frictionless ice surface. If the man walks with a velocity of V =1 m/s in the positive x direction relative to the wooden board, find the velocity of the board in m/s. Velocity is positive in the positive x direction. ________

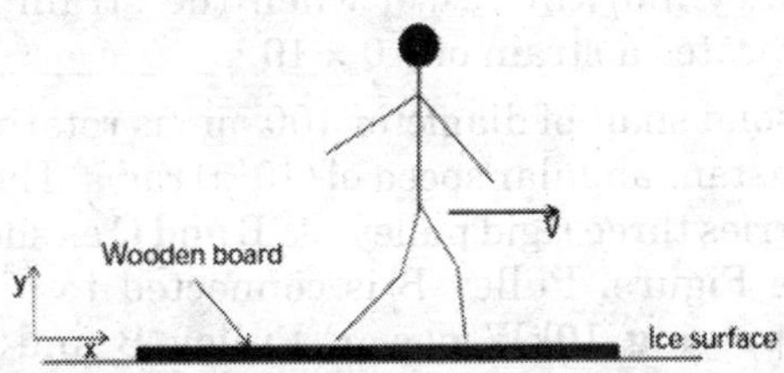

9. A rigid bar AB is hinged at B through a torsional spring with spring constant k_t. For small rotations of the bar AB about B, the critical load P_{cr} is given by

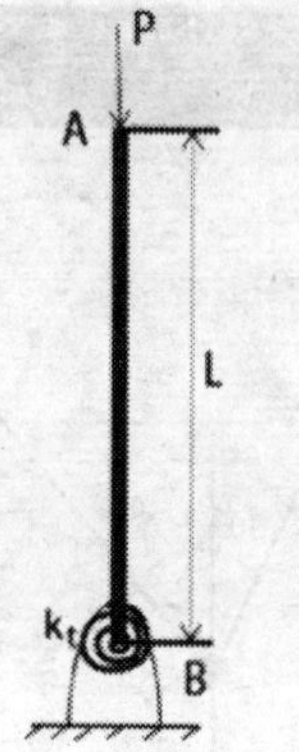

(a) $\frac{k_t}{2L}$ (b) $\frac{k_t}{L}$

(c) $\frac{2k_t}{L}$ (d) $\frac{4k_t}{L}$

Q. 10 - 22 carry two marks each.

10. A disk of mass M = 14kg and radius 1 m is attached to a spring which has a stiffness k = 75 N/m and an unstretched length of 1m. If the disk is released from rest in the position shown in the Figure and the disk rolls without slipping, find its angular velocity (in rad/s) at the instant the center of mass is displaced by 3 m. ________

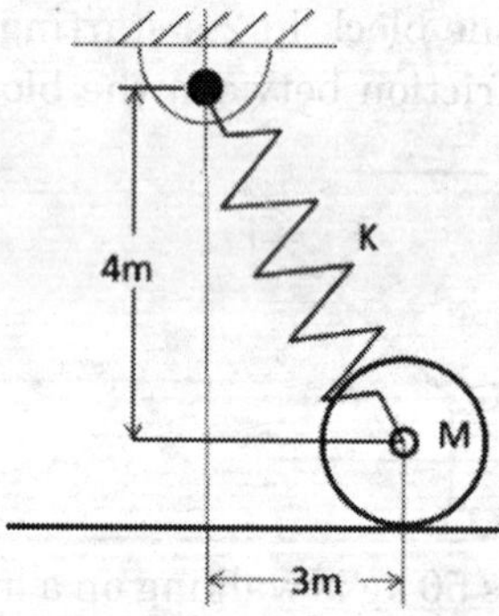

11. A strain gauge is mounted on the outer surface of a thin cylindrical pressure vessel in the circumferential direction. The mean diameter and thickness of the cylinder are 4.0 m and 20 mm, respectively. Young's modulus and Poisson's ratio of the material of the cylinder are 200 GPa and 0.25, respectively. Find the pressure in MPa inside the cylindrical vessel when the strain gauge indicates a strain of 7.0×10^{-4}. ________

12. A solid shaft of diameter 100 mm is rotating at a constant angular speed of $(10/\pi)$ rad/s. The shaft carries three rigid pulleys A, B and C as shown in the Figure. Pulley B is connected to a motor supplying 10kW power. Pulley B and C are connected to two pumps consuming 5kW each. Find the maximum shear stress (in MPa) in the shaft due to torsion alone. ________

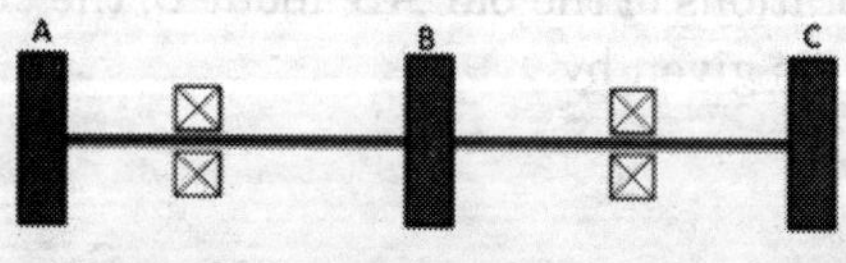

13. A beam is fixed at the left end and supported by a spring at the other end. The length of the beam is L and its flexural rigidity is EI. The spring constant of the spring is $k = \frac{3EI}{L^3}$. A vertical downward load P is applied at the right end. The deflection of the point under the load P is

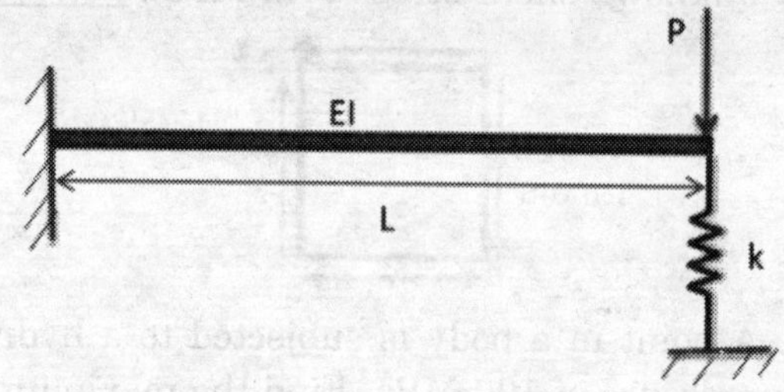

(a) $\frac{PL^3}{9EI}$ (b) $\frac{PL^3}{6EI}$

(c) $\frac{2PL^3}{9EI}$ (d) $\frac{5PL^3}{9EI}$

14. Find the maximum bending moment (magnitude wise) in kN-m for the beam shown in the Figure.______

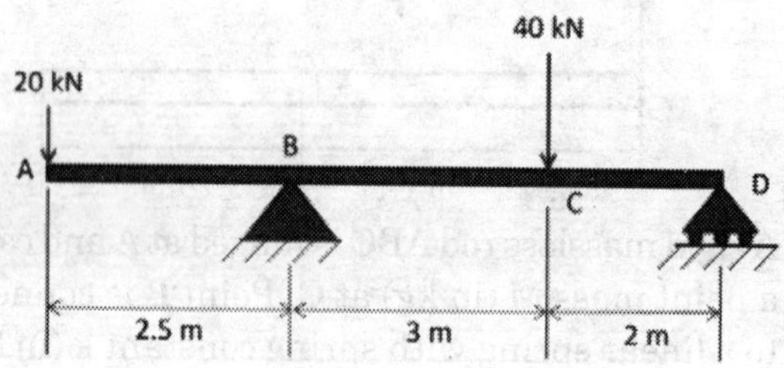

15. A projectile is fired with a velocity $V = 3\sqrt{2}$ m/s from a point at height H = 0.8 m at an angle of 45° with respect to the horizontal direction as shown in the Figure. Find the horizontal distance S in meters travelled by the projectile when it hits the ground. Take acceleration due to gravity as 10 m/s². ___

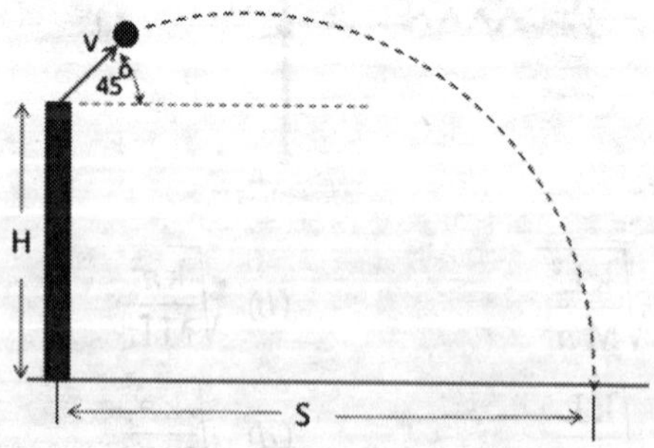

16. A particle P is moving on a circular path of radius r = 1m. The angular location θ of the particle is measured as shown in the Figure. The motion of the particle is described by θ = 2 sin(*t*). Find the magnitude of the total acceleration (in m/s²) of the particle at time t = π/3 seconds. ________

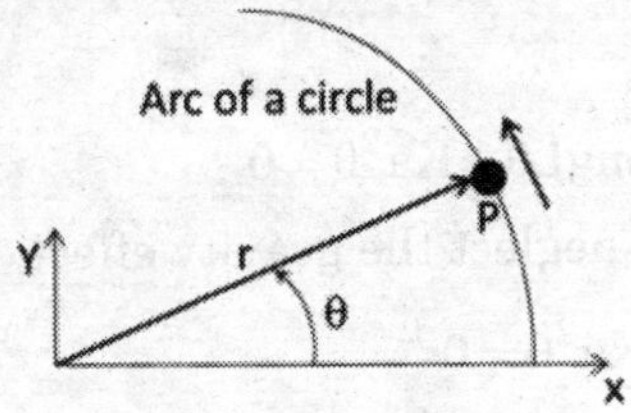

COMMON DATA QUESTIONS

Common Data for Questions 17 and 18 :

A frame ABC is shown in the Figure. Members AB and BC both have a length of L, and Young's modulus E. Members AB and BC both have a square cross-section of side *a*. A load P is applied at point C as shown in the figure.

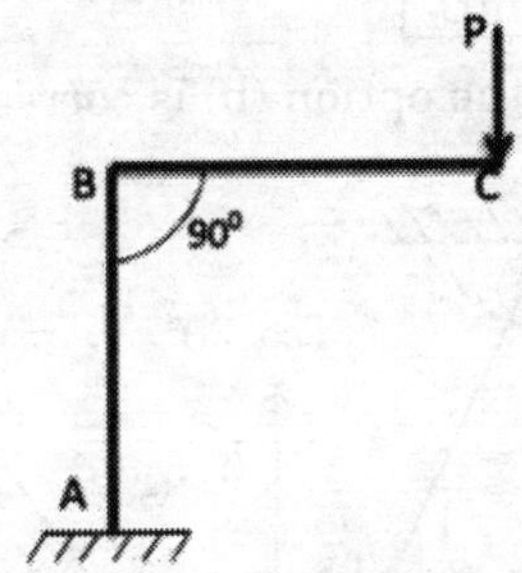

17. Neglecting the axial compression of member AB, the deflection of point C in the direction of the load is

(*a*) $\frac{2PL^3}{Ea^4}$ (*b*) $\frac{4PL^3}{Ea^4}$

(*c*) $\frac{8PL^3}{Ea^4}$ (*d*) $\frac{16PL^3}{Ea^4}$

18. The maximum bending stress in the frame is

(*a*) $\frac{3PL}{a^3}$ (*b*) $\frac{6PL}{a^3}$

(*c*) $\frac{9PL}{a^3}$ (*d*) $\frac{12PL}{a^3}$

Common Data for Questions 19 and 20 :

At a point in an object subjected to plane stress conditions, the state of stress is as shown in the Figure.

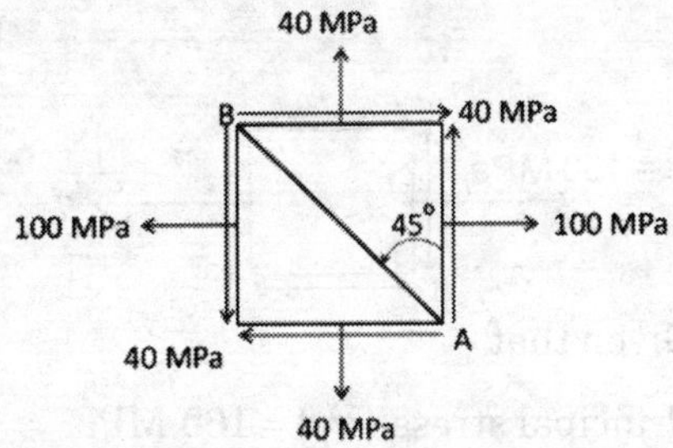

19. One of the principal stresses(in MPa) is

(*a*) 40 (*b*) 80

(*c*) 120 (*d*) 140

20. The normal stress on the plane AB (in MPa) is

(*a*) 30 (*b*) 70

(*c*) 100 (*d*) 110

LINKED ANSWER QUESTIONS

Statement for linked answer questions 21 & 22 :

Two rods are joined together and the entire assembly is supported between two rigid walls, as shown in the Figure. The cross-sectional area and Young's modulus for both the rods are 0.01 m² and 10 GPa, respectively. The coefficients of thermal expansion for the two rods are $a_1 = 4 \times 10^{-6}$ /°C and $a_2 = 10^{-6}$/°C, respectively. The entire assembly is heated by 100° C. Neglect the effect of Poisson's ratio.

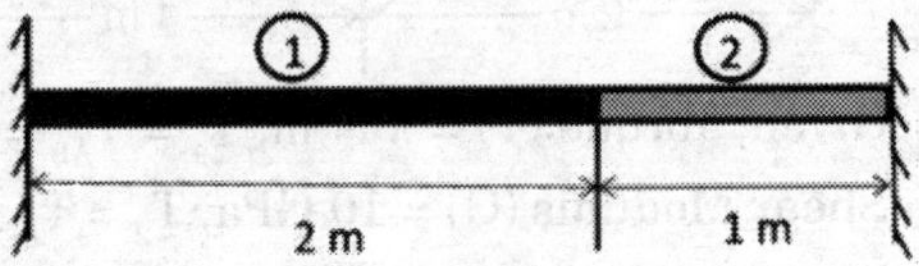

21. The stress in rod 1 (in MPa) is

(*a*) – 4.0 (*b*) – 3.0

(*c*) – 2.5 (*d*) – 1.0

22. Considering the displacement to the right as positive, the displacement (in mm) of the interface between the two rods is

(*a*) – 0.2 (*b*) – 0.1

(*c*) 0.1 (*d*) 0.2

ANSWERS

1. (*)	**2.** (–0.01 to 0.04)	**3.** (0.63 to 0.65)	**4.** (b)	**5.** (99 to 101)	**6.** (2.2 to 2.4)
7. (3.9 to 4.1)	**8.** (–0.3130 – 0.19)	**9.** (d)	**10.** (4.9 to 5.1)	**11.** (1.5 to 1.7)	**12.** (*)
13. (b)	**14.** (49.9 to 50.1)	**15.** (2.3 to 2.5)	**16.** (1.9 to 2.1)	**17.** (d)	**18.** (b)
19. (c)	**20.** (b)	**21.** (b)	**22.** (d)		

EXPLANATIONS

1. (*)

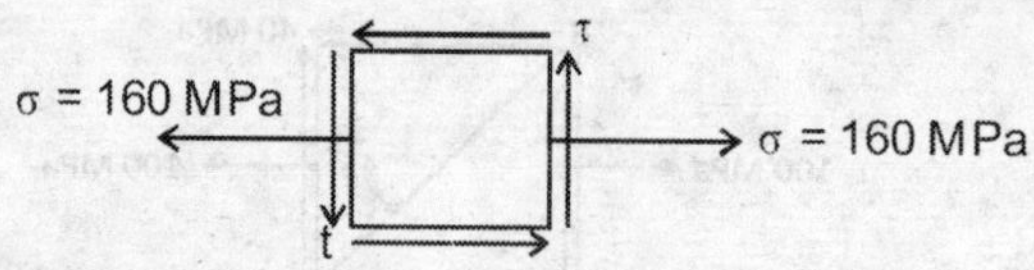

Given that

Principal stress $(\sigma_1) = 160$ MPa

Shear stress $(\tau) = \frac{\sigma_1}{2} = \frac{\sigma}{2} = \frac{160}{2} = 80$ MPa

$\boxed{\tau = 80 \text{ MPa}}$

2. (0) Given that

Hydrostatic pressure (P) = 100 MPa

Under hydrostatic state of stress every plane is a principal plane and in principle plane shear stress is zero.

Therefore, Maximum shear stress is a print is zero.

$\boxed{\therefore \tau_{max} = 0 \text{ MPa}}$

3. (0.64)

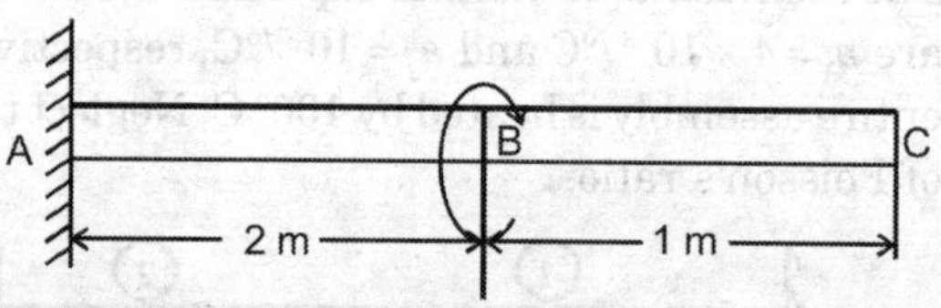

Given, Torque (T) = π N.m, $T_1 = T_{AB} = T$

Shear Modulus (G) = 10 GPa, $T_2 = T_{BC} = 0$

We know,

Angle of twist $(\theta)_C = \theta_{AB} + \theta_{BC}$

$$= \frac{T_1 L}{GJ} + \frac{T_2 L}{GJ}$$

$$= \frac{TX2}{G \times J} + 0 \qquad [T_2 = 0]$$

$$\theta_C = \frac{\pi \times 2}{10 \times 10^9 \times \frac{\pi}{32}(0.010)^4} \qquad \left[J = \frac{\pi}{32} d^4\right]$$

$\boxed{\therefore \theta_C = 0.64 \text{ rad}}$

4. (b)

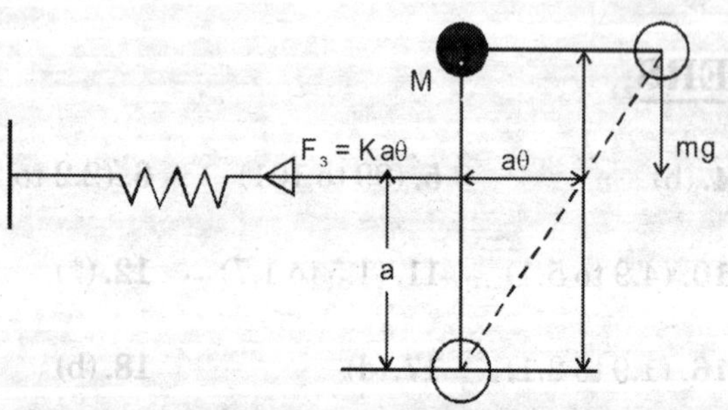

Taking the moment about '0'

$$I\theta = mgL\theta + Ka^2\theta = 0$$

If we neglect the gravity effect, we get

$$I\theta + Ka^2\theta = 0$$

$$\theta + \left(\frac{Ka^2}{I}\right)\theta = 0$$

Natural frequency $(\omega_n) = \sqrt{\frac{Ka^2}{I}}$

$$\boxed{\omega_n = \sqrt{\frac{Ka^2}{mL^2}}} \qquad [\because I = mL^2]$$

Hence, the option (b) is correct.

5. (100)

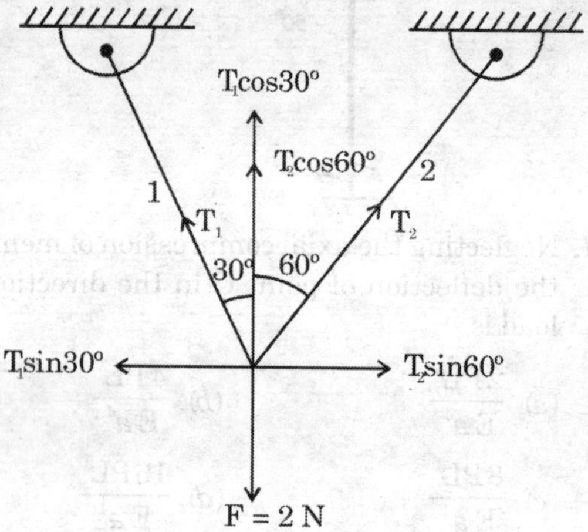

$$\sum F_V = 0$$

$$T_1 \cos 30° + T_2 \cos 60° - 2 = 0 \qquad ...(1)$$

$$\sum F_H = 0$$

$$T_1 \sin 30° - T_2 \sin 60° = 0$$

$$\frac{T_1}{T_2} = \frac{\sin 60°}{\sin 30°} = \sqrt{3}$$

$$T_1 = \sqrt{3} T_2 \qquad ..(2)$$

Now from (1)

$$T_1 \times \frac{\sqrt{3}}{2} + T_2 \times \frac{1}{2} = 2$$

$$\sqrt{3} T_1 + T_2 = 4 \qquad ..(3)$$

Solve the equation (2) & (3), we get

$$T_1 = \sqrt{3} \text{N} \;\&\; T_2 = 1 \text{ N}$$

∴ Stress developed in bar 2 is

$$\sigma_2 = \frac{T_2}{A_2} = \frac{1}{0.01} = 100 \text{ N/m}^2$$

$$\boxed{\sigma_2 = 100 \text{ N/m}^2 = 100 \text{ Pa}}$$

6. (2.3)

We know, when lift is stationary, the net load on spring is

$N = mg$

$N = 10 \times 10 \left[\because g = 10\ m/s^2\ \text{given}\right]$

$\boxed{N = 100\ N}$ ...(1)

Now when lift starts moving upward, then net load on spring will be

$N = Mg + Ma$

$N = (12.3 \times 10) + (10 \times a)\ [\because N = 10\ N \text{ from (i)}]$

$100 = 123 + 10a \Rightarrow -2.3 = a$

$\boxed{a = 2.3\ m/s^2}$

7. (4)

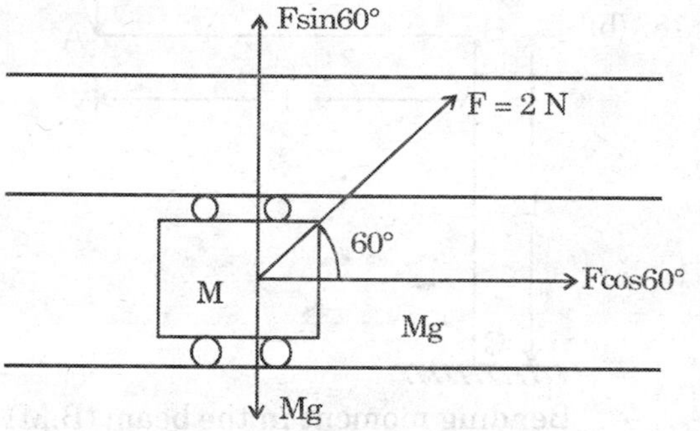

Given Mass of block (H) = 0.5 kg

Applied force (F) = 2 N

Applying Newton's 2nd law

$F \cos 60° = ma$

$2 \times \frac{1}{2} = 0.5 \times a$

$\boxed{a = 2\ m/s^2}$

The distance travelled by the block

$S = ut + \frac{1}{2}at^2 = 0 + \frac{1}{2} \times 2 \times (2)^2$

[u = 0 – initial velocity, t = 2 sec given]

$\boxed{S = 4\ m}$

13. (b)

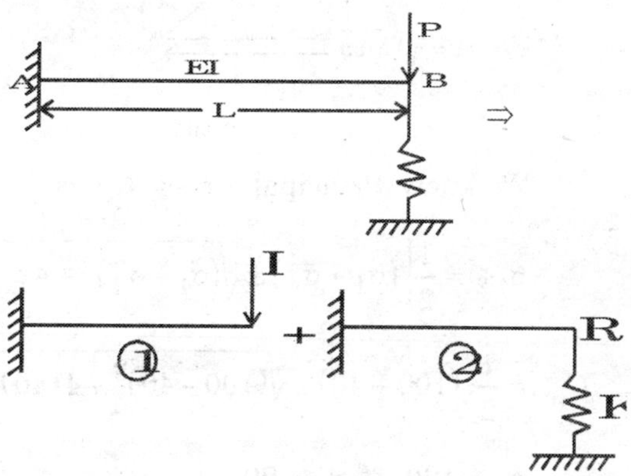

$Y_B = Y_1 + Y_2$

$$-\frac{R_B}{K} = -\frac{PL^3}{3EI} + \frac{R_1 L^2}{3EI}$$

$$\Rightarrow R_S\left[\frac{1}{K} + \frac{L^3}{3EI}\right] = \frac{PL^3}{3EI}$$

$$\Rightarrow R_S\left[\frac{3EI + KL^3}{K(3EI)}\right] = \frac{PL^3}{3EI}$$

$$\therefore R_S = \frac{KPL^3}{3EI + KL^3}$$

$$= \frac{\frac{3EI}{L^3} \times P \times L^3}{3EI + \frac{3EI}{L^3} \times L^3} \qquad \left[\because K = \frac{3EI}{L^3}\ \text{given}\right]$$

$$R_S = \frac{P}{2} \qquad ...(1)$$

Net deflection at 'B'

$$Y_B = \frac{-PL^3}{3EI} + \frac{R_S L^3}{3E} = \frac{-PL^3}{3EI} + \frac{\left(\frac{P}{2}\right) \cdot L^3}{3EI}$$

[From (1)]

$$= \frac{-PL^3}{3EI} + \frac{PL^3}{6EI} = \frac{-PL^3}{6EI}$$

$\therefore \boxed{Y_B = \frac{PL^3}{6EI}}$ (Downward/Hogging)

Hence, the option (b) is correct.

14. (50)

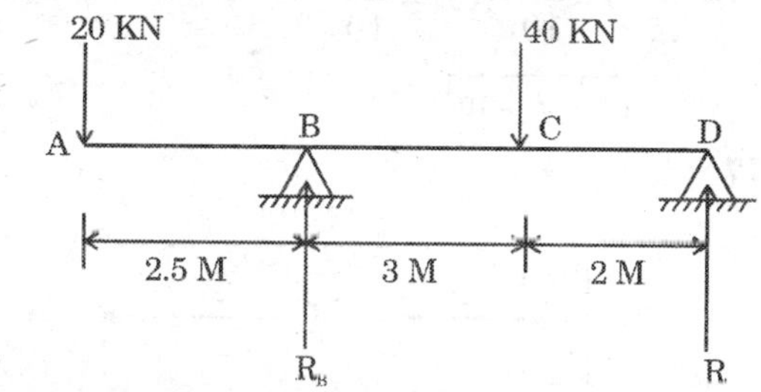

Net vertical force $\sum Fr = 0$

$R_B + R_D = 60$ ...(1)

Taking moment about 'D'

$20 \times 7.5 - R_B \times 5 + 40 \times 2 = 0$

$150 + 8 = R_B \times 5$

$230 = R_B \times 5$

$\therefore \boxed{R_B = 46\ N}$

From (1)

$R_B + R_D = 60$

$46 + R_D = 60 \Rightarrow \boxed{R_D = 14\ N}$

Now, Bending moment at A $(BM)_A = 0$

$(B.M)_B = 20 \times 2.5 = 50$ N-m

$(B.M)_C = 20 \times 5.5 - R_B \times 3$

$= 110 - 46 \times 3$

$= -28$ N-m

$(B.M)_D = 0$

Therefore, the maximum bending moment developed in the beam $(B.M)_{max} = 50$ N-m.

15. (2.4)

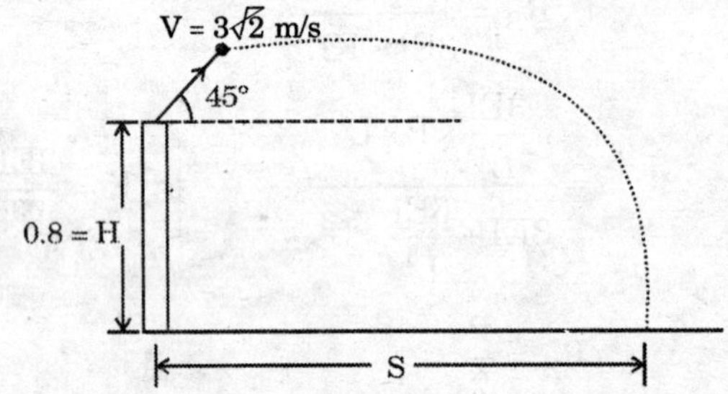

We know, the distance covered by the projectile motion 'S' is

$$S = \frac{V\cos\theta}{g}\left[V\sin\theta + \sqrt{(V\sin\theta)^2 + 2gH}\right]$$

$$= \frac{3\sqrt{2}\cos 45}{10}$$

$$\left[3\sqrt{2}\sin 45° + \sqrt{\left(3\sqrt{2}\sin 45°\right)^2 + (2\times 10\times 0.8)}\right]$$

$$= \frac{3}{10}\left[3 + \sqrt{9+16}\right]$$

$$S = \frac{3}{10} - (8) = \frac{24}{10}$$

$$\boxed{S = 2.4 \text{ m}}$$

17. (d)

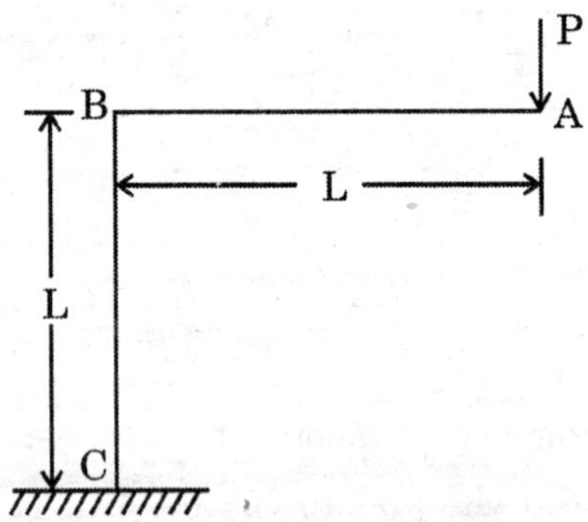

Total strain energy in the beam

$$U_r = U_1 + U_2 = \int_0^L \frac{M_x^2 dx}{2EI} + \int_0^L \frac{M^2 dx}{2EI}$$

$$= \int_0^L \frac{(Px)^2 dx}{2EI} + \int_0^L \frac{(PL)^2 dx}{2EI}$$

$$= \frac{P^2L^2}{6EI} + \frac{P^2L^3}{2EI}$$

$$= \frac{4P^2L^3}{6E \times \frac{b^4}{12}} \qquad \left[\because I = \frac{b^4}{12}\right]$$

$$U = \frac{8P^2L^3}{Eb^4}$$

Now deflection at point 'C'

$$Y_C = \frac{dU}{dp} = \frac{8 \times 2PL^3}{Eb^4} = \frac{16PL^3}{Eb^4}$$

$$\boxed{\therefore y_C = \frac{16PL^3}{Eb^4}}$$

Hence, the option (d) is correct.

18. (b)

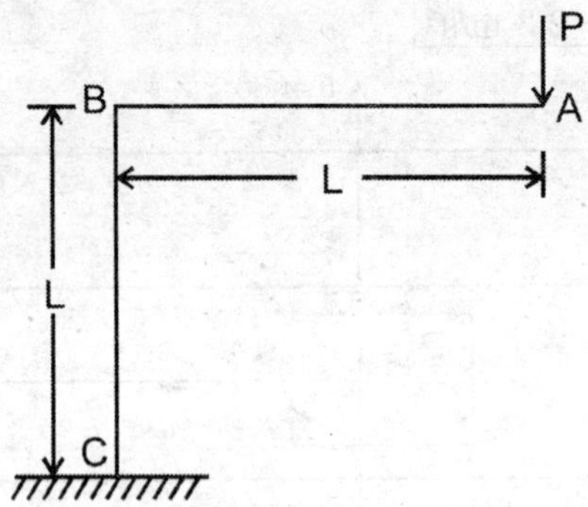

Bending moment in the beam (B.M) = PL

Bending stress $(\sigma_b) = \frac{(B.M.)}{2} = \frac{M}{Z}$

$$\sigma_b = \frac{PL}{\frac{a \cdot a^2}{6}} = \frac{6PL}{a^3}$$

$$\therefore \boxed{\sigma_b = \frac{6PL}{a^3}}$$

Hence, the option (b) is correct.

19. (c)

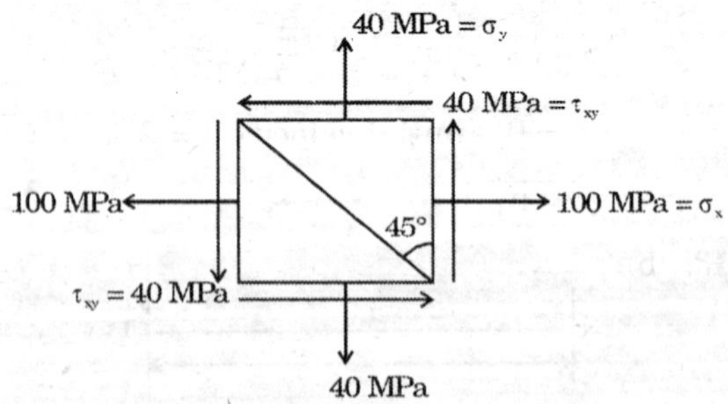

We know, Principal stress $\sigma_{1,2}$ is

$$\sigma_{1,2} = \frac{1}{2}\left[(\sigma_1 + \sigma_4) \pm \sqrt{(\sigma_1 - \sigma_4)^2 + 4\tau_{xy}^2}\right]$$

$$= \frac{1}{2}\left[(100 + 40) \pm \sqrt{(100 - 40)^2 + 4(40)^2}\right]$$

$$\sigma_1 = 120 \ \& \ \sigma_2 = 20$$

| Maximum principal stress $(\sigma) = 120$ MPa

Hence, option (c) is correct.

20. (b)

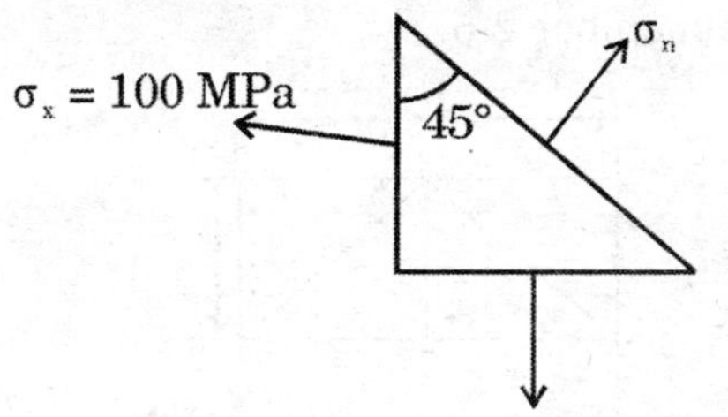

Where $\theta = 45°$

Normal stress $\left(\sigma_n^*\right) = \dfrac{\sigma_1 + \sigma_2}{2}$

$$\sigma_n^* = \frac{120+20}{2} = \frac{140}{2}$$

$$\boxed{\sigma_n^* = 70 \text{ MPa}}$$

Hence, option (b) is correct.

21. (b)

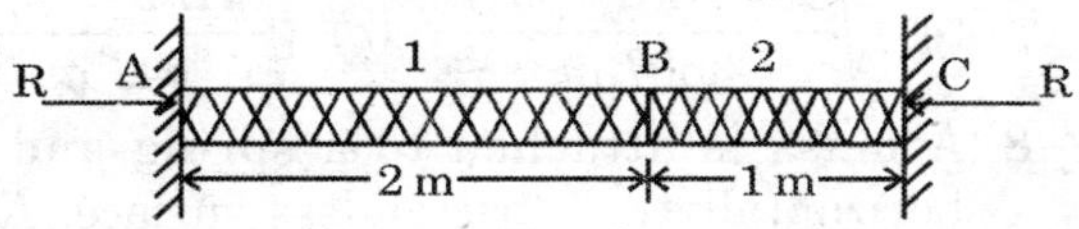

We know

Sum of Axial deformation = Sum of thermal deformation

$$\frac{\sigma_1 L_1}{E_1} + \frac{\sigma_2 L_2}{E_2} = \alpha_1 T L_1 + \alpha_2 T L_2$$

$$\frac{\sigma_1 \times 2}{10 \times 10^9} + \frac{\sigma_2 \times 1}{10 \times 10^9} = \left(4 \times 10^{-6} \times 150 \times 2\right) + \left(10^{-6} \times 100 \times 1\right)$$

$$2\sigma_1 + \sigma_2 = 9 \quad \ldots(1)$$

$$\frac{\sigma_1}{\sigma_2} = \frac{A_2}{A_1} = \frac{0.01}{0.01} = 1$$

$$\sigma_1 = \sigma_2 \quad \ldots(2)$$

From (1),

$$2\sigma_1 + \sigma_1 = 9$$

$\boxed{\sigma_1 = 3 \text{ MPa}}$ (Compressive)

Hence, the optin (b) is correct.

22. (d)

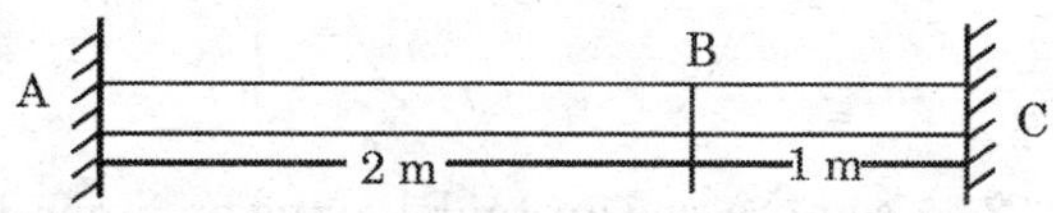

Deformation or displacement at the interface between the two rods

$$\delta_B = +\alpha_1 T L_1 - \frac{\sigma_1 L}{E}$$

$$= \left(+4 \times 10^{-6} \times 150 \times 2\right) - \frac{3 \times 2 \times 10^6}{10^9 \times 10}$$

$$= \left(+4 \times 10^{-6} \times 100 \times 2\right) - \left(6 \times 10^{-4}\right)$$

$$= -8 \times 10^{-4} - 6 \times 10^{-4}$$

$$= -2 \times 10^{-4} \text{ m}$$

$$= -2 \times 10^{-1} \text{ mm}$$

$$\boxed{\delta_B = 0.2 \text{ mm}}$$

Hence, the option (d) is correct.

2014

Q. (1 – 9) Carry One Mark Each.

1. A steel wire of diameter 5 mm is bent around a cylindrical drum of radius 0.5 m. The steel wire has modulus of elasticity of 200 GPa. Find the bending moment in the wire in N-m.

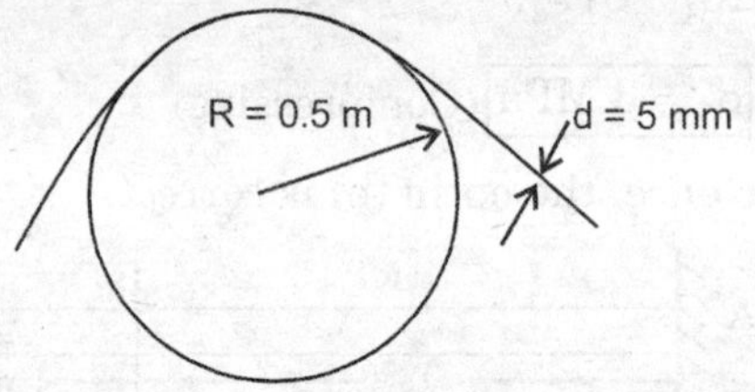

2. A compressed air tank having an inner diameter of 480 mm and a wall thickness of 8 mm is formed by welding two steel hemispheres. If the allowable shear stress in the steel is 40 MPa, find the maximum permissible pressure (in MPa) inside the tank.

3. The Euler's buckling load of a column fixed at both the ends is P. If one of the ends is made free, the buckling load shall change to

(*a*) P/16 (*b*) P/8

(*c*) P/4 (*d*) P/2

4. A point in a body is subjected to a bi-axial state of stress, equal in magnitude but opposite in nature. On a plane inclined at an angle 45° with respect to x-axis (passing through the point), the

(*a*) shear and normal stresses are zero

(*b*) normal stress is maximum and shear stress is zero

(*c*) shear stress is maximum and normal stress is zero

(*d*) shear stress is maximum and normal stress is non-zero

5. A weightless beam subjected to two point loads is shown in the figure below.

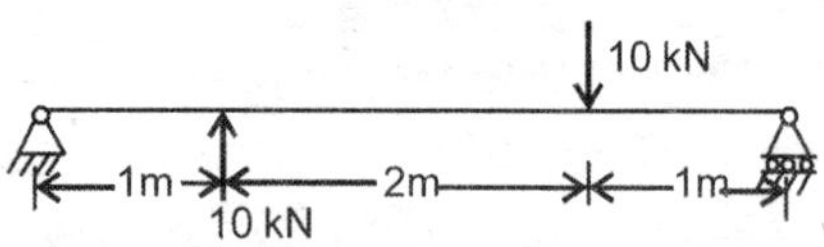

The shear force diagram of the beam is

(*a*)

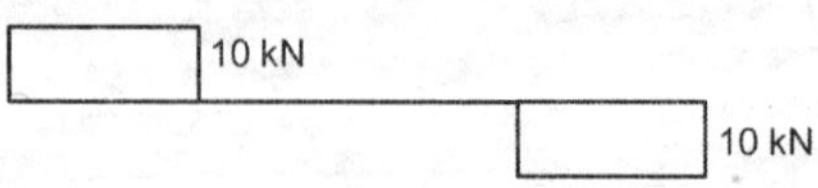

(*b*)

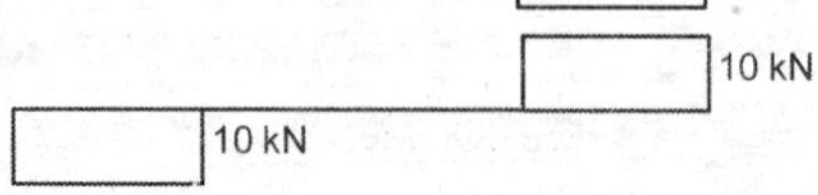

(*c*)

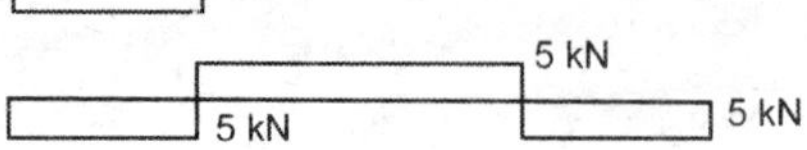

(*d*)

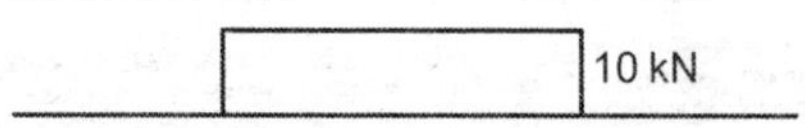

6. For the pin jointed truss, find the axial force (in kN) in the member 2-5.

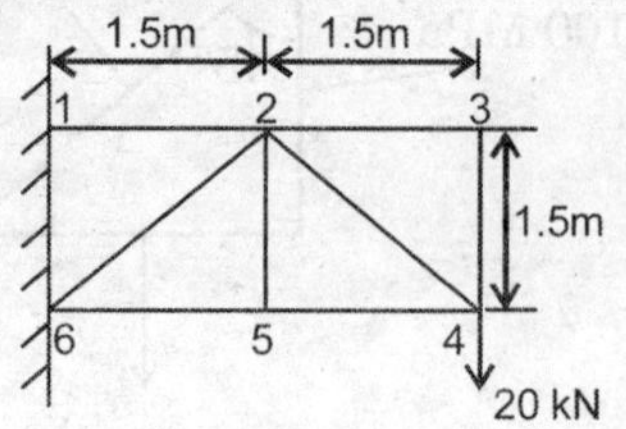

7. The supporting structure of a water tank is made of reinforced concrete (RC) with a tubular cross section of inner diameter d_i, outer diameter d_o, height *l*, and Young's modulus *E*. The mass of the tank is *m*. If mass of the supporting structure is neglected, then the natural frequency of the water tank in transverse direction is

(*a*) $\sqrt{\dfrac{3\pi E(d_o^4 - d_i^4)}{64 l^3 m}}$ (*b*) $\sqrt{\dfrac{\pi E(d_o^4 - d_i^4)}{8 l^3 m}}$

(*c*) $\sqrt{\dfrac{384\pi E(d_o^4 - d_i^4)}{360 l^3 m}}$ (*d*) $\sqrt{\dfrac{\pi E(d_o^4 - d_i^4)}{64 l^3 m}}$

8. A mass is attached to a spring and placed horizontally in a frictionless surface. A simple pendulum has been pivoted to the mass. The degree of freedom of this system is

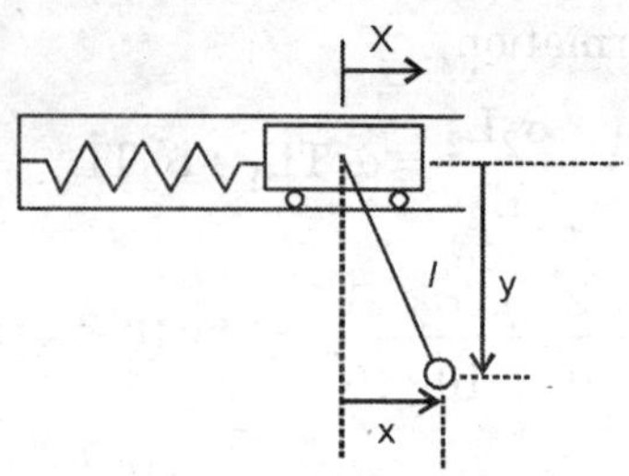

(*a*) 1 (*b*) 2

(*c*) 3 (*d*) 4

9. Consider the following two statements

Statement 1 : A body of weight W falls from a height *h* and strikes the ground. If the body starts from rest, the velocity with which it strikes the ground is $\sqrt{2gh}$, where *g* is the acceleration due to gravity.

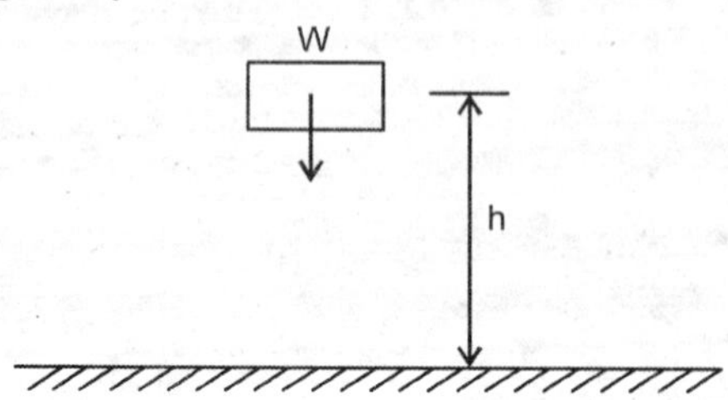

Statement 2: If the same body (initially at rest) slides without friction along an inclined plane PQ (angle of inclination α) starting from an elevation h above point Q, then its velocity at point Q is $\sqrt{2gh}$

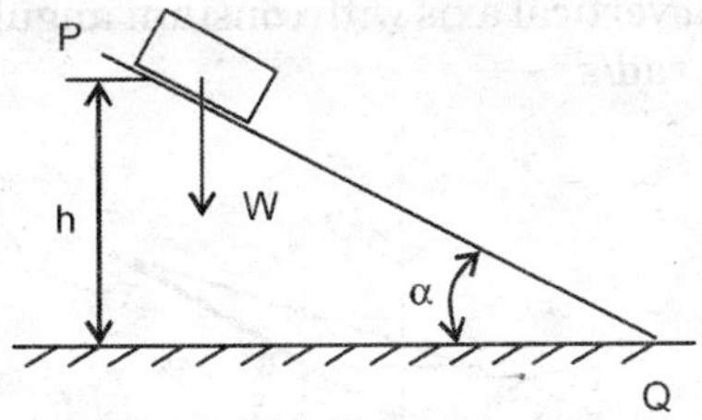

The correct option is

(*a*) Both statements 1 and 2 are true

(*b*) Statement 1 is true and 2 is false

(*c*) Statement 1 is false and 2 is true

(*d*) Both statements 1 and 2 are false

Q. (10 – 22) Carry Two Marks Each.

10. A composite bar of length 'L' is made of a centrally placed steel plate (50 mm wide × 10 mm thick) with two copper plates (each 30 mm wide × 5 mm thick) connected rigidly on each side. If the temperature of the composite bar is raised by 50°C, find the stress developed in each copper plate in MPa. (For Steel: $E_s = 2 \times 10^5$ MPa and $a_s = 12 \times 10^{-6}$/ C; For Copper: $E_c = 1 \times 10^5$ MPa and $a_c = 17 \times 10^{-6}$ / C)

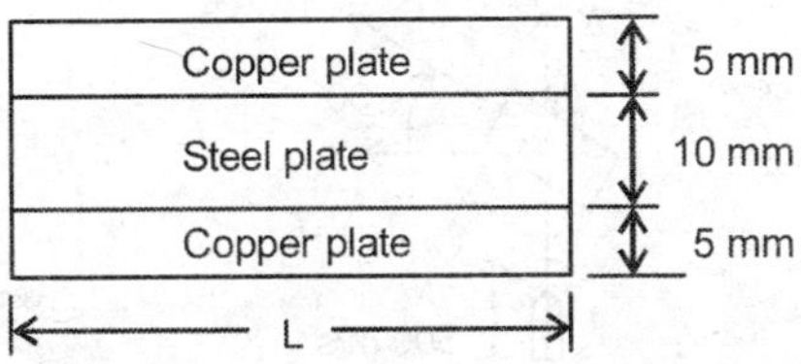

11. The vertical deflection at the free end of the cantilever beam as shown in figure is

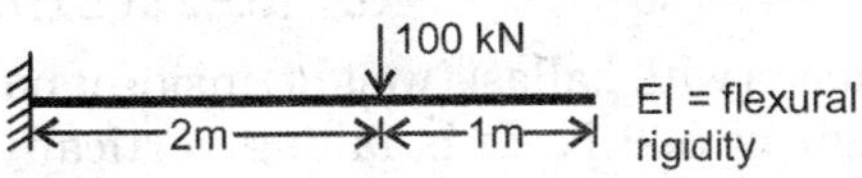

(*a*) 1400/EI (*b*) 1400/3 EI

(*c*) 200/EI (*d*) 100/EI

12. A hollow shaft and a solid shaft have the same length and the same outer radius R. The inner radius of the hollow shaft is 0.6 R. Assuming that both the shafts are made of same material and are subjected to the same torque, find the ratio of shear stress in hollow shaft to that in solid shaft.

13. A beam with overhangs carries one point load acting downwards and the other upward. The clockwise moment P*b* is applied at each support. The bending moment at the midpoint of the beam is

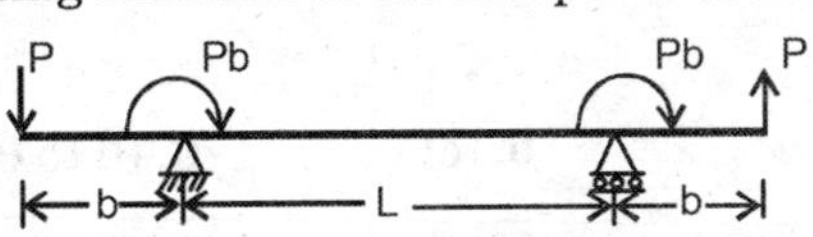

(*a*) 0 (*b*) PL/2

(*c*) PL (*d*) P*b*L

14. A cantilever beam of length *L* supports a concentrated load *P* at the free end. The cross section of the beam is rectangular with constant width b and varying depth *h*. The depth *h* of this idealized cantilever beam varies in such a way that the maximum normal stress at every cross section remains equal to the allowable bending stress. Considering only the bending stresses, the depth h_x of the fully stressed beam at any distance *x* from the free end shall vary

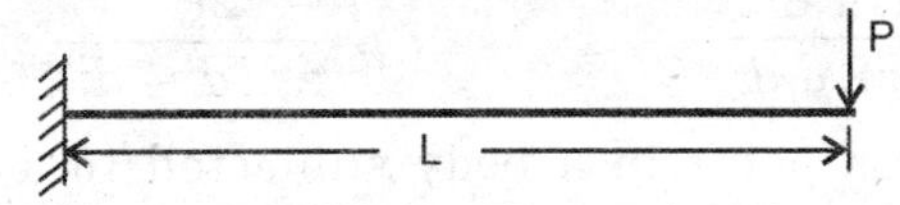

(*a*) with square of *x*

(*b*) with square root of *x*

(*c*) linearly with *x*

(*d*) with cube of *x*

15. A cantilever beam is subjected to following three different loading conditions:

(*a*) a concentrated load P at its free end,

(*b*) a couple M_o at its free end and

(*c*) both loads acting simultaneously

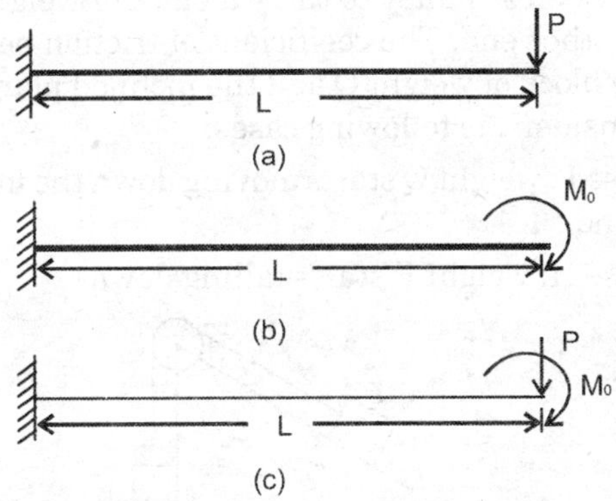

The flexural rigidity of the beam may be assumed as EI. The strain energy due to bending when both loads act simultaneously

(*a*) can be determined by applying the principle of superposition and the strain energy is $\dfrac{P^2L^3}{6EI} + \dfrac{M_o^2L}{2EI}$

(*b*) can be determined by applying the principle of superposition and the strain energy is $\dfrac{P^2L^2}{6EI} + \dfrac{M_oL^3}{2EI}$

(*c*) cannot be determined by applying the principle of superposition and the strain energy is $\dfrac{P^2L^3}{6EI} + \dfrac{M_o^2L}{2EI} + \dfrac{PM_oL^2}{2EI}$

(*d*) cannot be determined by applying the principle of superposition and the strain energy is $\dfrac{P^2L^3}{6EI} + \dfrac{M_oL^3}{2EI} + \dfrac{PM_oL^2}{2EI}$

16. A tapered rod has diameter d_1 at one end which reduces uniformly to a diameter d_2 over the length (L). If the modulus of elasticity of the material is E, the change in the length of the rod due to the application of axial force (P) is

(*a*) $\dfrac{4PL}{\pi E d_1 d_2}$ (*b*) $\dfrac{4PL}{\pi E(d_1^2 - d_1^2)}$

(*c*) $\dfrac{PL}{\pi E\, d_1 d_2}$ (*d*) $\dfrac{2PL}{\pi E(d_1^2 - d_2^2)}$

17. For a point in a body subjected to a plane stress condition (σ_x = 100 MPa, σ_y = 50 MPa and $\tau_{xy} = \tau_{yx}$ = 25 MPa), the maximum principal stress in MPa is __.

18. An isotropic body is subjected to a state of stress given by: σ_x = 10 MPa and $\tau_{xy} = \tau_{yx}$ = – 20 MPa. Assuming G = 0.4E, the volumetric strain is

(*a*) 5/E (*b*) 7.5/E

(*c*) 10/E (*d*) 15/E

19. A block of weight Q rests on an inclined plane and it is attached to a string which runs over a frictionless pulley to carry a block of weight P at its other end. The coefficient of friction between the block of weight Q and the inclined plane is . Consider the following cases:

Case I : weight Q starts moving down the inclined plane

Case II: weight P starts falling down

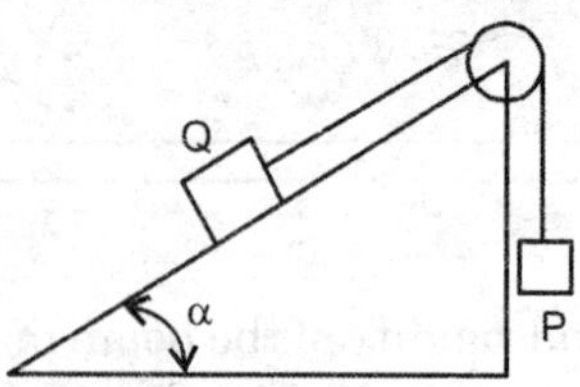

The limiting values of ratio P/Q for Case I and Case II respectively are

(*a*) (sin α – μ cos α) and (sin α + μ cos α)

(*b*) (μsin α – cos α) and (μsin α + cos α)

(*c*) (sin α + μ cos α) and (sin α – μ cos α)

(*d*) (μsin α + cos α) and μ(sin α – cos α)

20. To unload an item from a truck a crane boom is raised with a constant angular velocity of 1 rad/s relative to the cab and then the cab is rotated about a vertical axis with constant angular velocity of 0.5 rad/s.

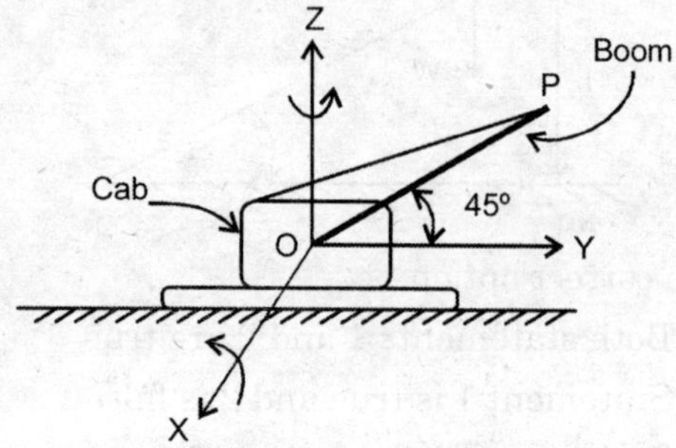

If the length of the boom (OP) is 10 m, the velocity of the tip (P) of the boom in m/s is

(*a*) $\dfrac{\sqrt{2}}{5}(-2\hat{i} - \hat{j} + 2\hat{k})$ (*b*) $\dfrac{\sqrt{2}}{5}(-\hat{i} - 2\hat{j} + \hat{k})$

(*c*) $\dfrac{5}{\sqrt{2}}(-2\hat{i} - \hat{j} + 2\hat{k})$ (*d*) $\dfrac{5}{\sqrt{2}}(-\hat{i} - 2\hat{j} + 2\hat{k})$

21. A block of mass 5 kg moves up on a smooth inclined plane with a velocity of 10 m/s in the direction shown. A bullet of mass 60 g travelling at 500 m/s strikes the block centrally and gets embedded in it. The velocity of the block and embedded bullet in m/s immediately after the impact is

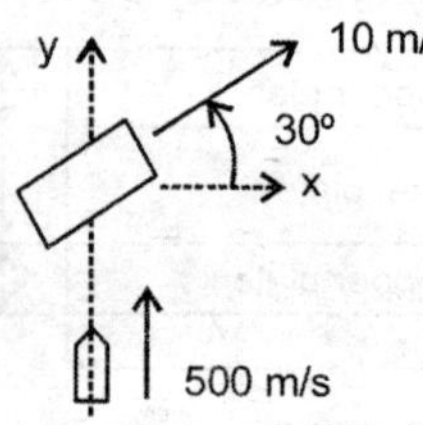

(*a*) 12.54 at 30° (*b*) 13.84 at 51.78°

(*c*) 13.84 at 30° (*d*) 15.62 at 51.78°

22. A balloon with ballast (weight) inside it has a gross weight of 500 N. It is falling vertically with a constant acceleration of 2 m/s². If air resistance is negligible, find the weight of ballast (in N) that must be thrown out in order to give the balloon an upward acceleration of 2 m/s². (Acceleration due to gravity, g = 9.81 m/s²)

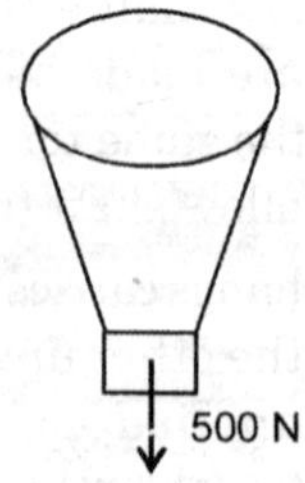

ANSWERS

1. (12.20 to 12.22)	**2.** (5.3 to 5.4)	**3.** (a)	**4.** (c)	**5.** (c)	**6.** (0 to 0)
7. (a)	**8.** (b)	**9.** (a)	**10.** (19.1 to 19.3)	**11.** (b)	**12.** (1.14 to 1.16)
13. (a)	**14.** (b)	**15.** (a)	**16.** (a)	**17.** (110.3 to 110.5)	
18. (a)	**19.** (a)	**20.** (d)	**21.** (b)	**22.** (169.3 to 169.4)	

EXPLANATIONS

1. (12.20)

Given steel wire diameter (d) = 5 mm

Radius of cylindrical drum (r) = 0.5 m

Modulus of rigidity (G) = 200 GPa

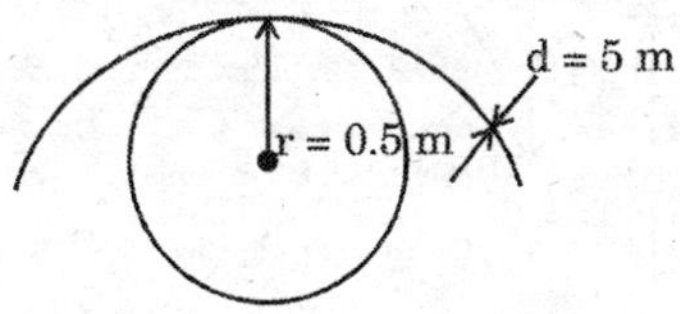

We know that

Bending Moment (M) = EI

[R_c → Radius of curvature]

$$M = \frac{EI}{R_o + \frac{d}{2}} = \frac{200 \times \frac{\pi}{64} \times (d)^4}{r + \frac{d}{3}}$$

$$M = \frac{200 \times \frac{\pi}{64} \times (0.005)^4 \times 10^9}{0.5 + \frac{5 \times 10^{-3}}{2}}$$

$\boxed{M = 12.20 \text{ N-m}}$

Therefore bending moment in the wire is 12.20 N-m.

$\boxed{M = 12.20 \text{ N-m}}$

2. (5.33)

Given diameter of tank (d) = 480 mm

Thickness of tank wall (t) = 8 mm

Allowable shear stress (ι_{Allow}) = 40 MPs

We know for safe condition

$\tau_{induced} \le \tau_{allowable}$

$$\frac{Pd}{80} \le 40 \qquad \left[\because \tau_{sphere} = \frac{Pd}{8t}\right]$$

$$\frac{P \times 480}{8 \times 8} \le 40$$

$\boxed{P \le 5.33 \text{ MPa}}$

Hence, the maximum permissible pressure inside the tank $\boxed{P = 5.33 \text{ MPa}}$.

3. (a) We know, Buckling load (P) = $\frac{\Pi^2 EI}{L_c^2}$ ...(1)

If the column is fixed at the ends then

$$L_c = \frac{\ell}{2}$$

$$\therefore P = \frac{\pi^2 EI}{\left(\frac{\ell}{2}\right)^2} = \frac{4\pi^2 EI}{\ell^2} \qquad ..(2)$$

If the column is free at one end. Then,

$L_c = 2\ell$

$$\therefore P' = \frac{\pi^2 EI}{(2\ell)^2} = \frac{\pi^2 EI}{4\ell^2} \qquad ...(3)$$

$$\therefore \frac{P'}{P} = \frac{\pi^2 EI}{4\ell^2} \times \frac{\ell^2}{4p^2 EI} = \frac{1}{16}$$

$$\therefore \boxed{P' = \frac{P}{16}}$$

Hence, the option (a) is correct.

4. (c)

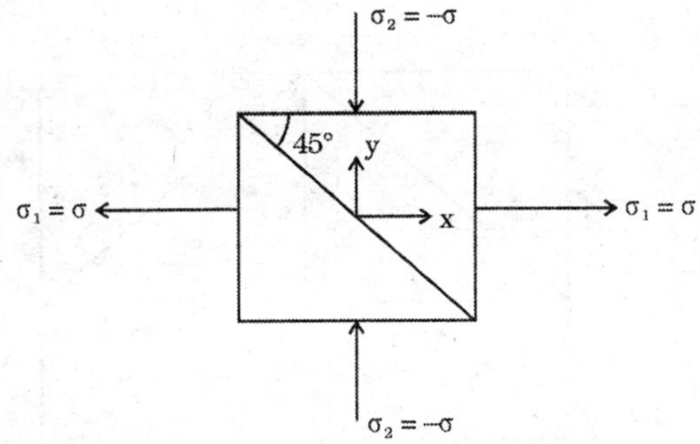

When θ = 45°

Shear stress $\tau_{max} = \frac{\sigma}{2}$ (Maximum)

$$\text{Normal stress } \sigma_n^* = \frac{\sigma_1 + \sigma_2}{2}$$

$$= \frac{\sigma_1 - \sigma_2}{2} = \frac{\sigma - \sigma}{2} = 0$$

$\boxed{\sigma_n^* = 0}$

Hence, the option (c) is correct.

5. (c)

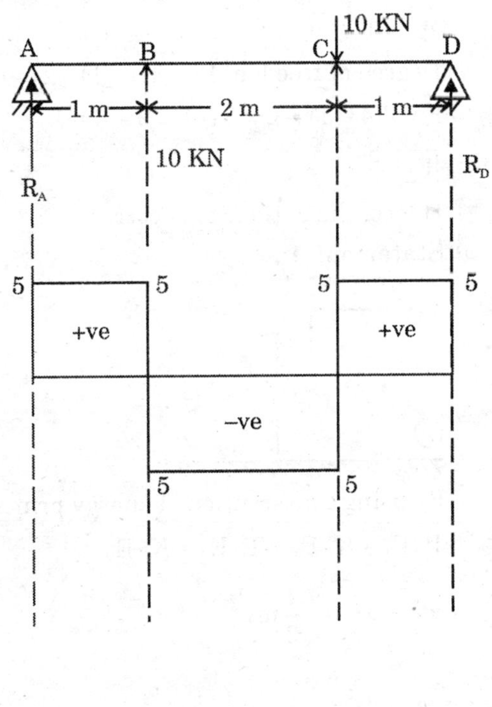

$R_A + R_D = 10 - 10 = 0$ ...(1)

Taking moment about 'A'

$R_D \times 4 - 10 \times 3 + 10 \times 1 = 0$

$\boxed{R_D = 5 \text{ kN}}$

From equation (1)

$R_A + R_D = 0$

$R_A = -R_D = -5 \text{ KN}$

$\boxed{R_A = -5 \text{ KN}}$

Hence, the option (c) is correct.

6. (0)

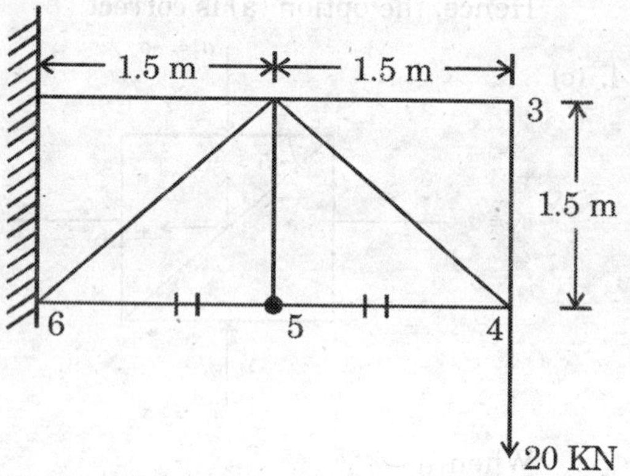

The axial force in the member 2-5 will be zero.

$\boxed{P_{2-5} = 0}$

Note : Since we know that if, three members or three forces or both meet at a joint and two of them are collinear, then third will have zero force.

8. (b) Here No. of links $(\ell) = 4$

Number of joints (J) = 3

Number of Higher pair (h) = 1

We know

Degree of freedom $F = 3(\ell - 1) - 2j - h$

$F = 3(4 - 1) - 2 \times 3 - 1 = 9 - 6 - 1$

$\boxed{F = 2}$

Hence, the option (b) is correct.

9. (a) **Statement 1 :**

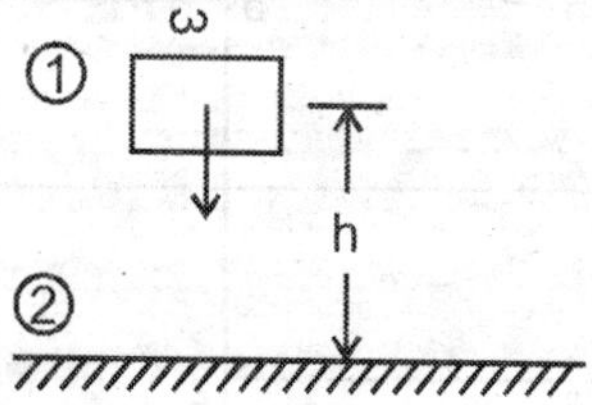

By using conservation of energy principle

$P \cdot E_1 + K \cdot E_1 = P \cdot E_2 + K \cdot E_2$

$wh + 0 = 0 + \frac{1}{2}mv^2$

$mgh = \frac{1}{2}mv^2$

$\therefore \boxed{v = \sqrt{2gh}}$

Statement 2 :

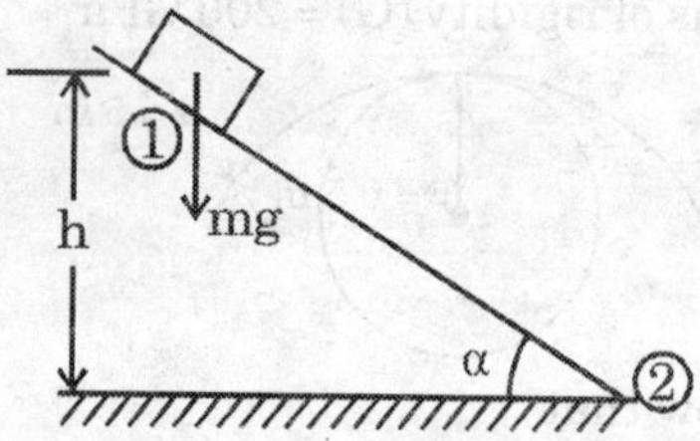

By using conservation of energy method

$P \cdot E_1 + K \cdot E_1 = P \cdot E_2 + K \cdot E_2$

$mgh + 0 = 0 + \frac{1}{2}mv^2$

$mgh = \frac{1}{2}mv^2$

$\therefore \boxed{V = \sqrt{2gh}}$

Therefore, both the statements are correct.

Hence, the option (a) is correct.

11. (b)

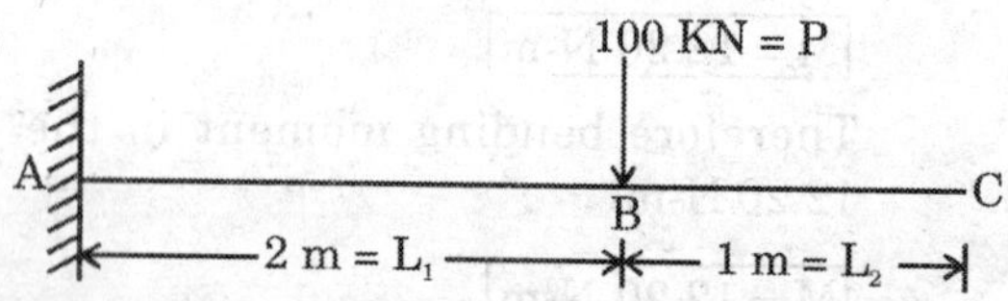

The deflection at the free end 'C' will be

$y_C = y_B + \theta_B \cdot L_2$

$= \frac{PL_1^{\,3}}{3EI} + \frac{PL_1^{\,2}}{2EI}$ ...(1)

$= \frac{100 \times (2)^3}{3EI} + \frac{100(2)^2}{2EI}$

$\boxed{y_C = \frac{1400}{3EI}}$

Hence, the option (b) is correct.

12. (1.14)

Given that

Outer Radius of Solid shaft $(r_0) = R$

Outer Radius of Hollow shaft $R_o = R$

Inner Radius of Hollow shaft $R_I = 0.6\,R$

We know

Shear stress for solid shaft $(\tau)_{solid} = \frac{T}{2}$

$\tau_{solid} = \frac{T}{\frac{\pi}{16}d^3} = \frac{16T}{\pi d^3}$

$$\tau_{solid} = \frac{16T}{\pi(2R_0)^3} = \frac{16T}{\pi R^3} = \frac{2T}{\pi R^3} \quad ...(1)$$

Now, shear stress for Hollow shaft

$$(\tau)_{Hollow} = \frac{T}{Z(1-K^4)}$$

$$\tau_{Hollow} = \frac{16T}{\pi D^3\left[1-\left(\frac{d_i}{d_o}\right)^4\right]} \quad \left[\because K = \frac{d_i}{d_o}\right]$$

$$= \frac{16T}{\pi(2R_o)^3\left[1-(0.6)^4\right]}$$

$$= \frac{16T}{\pi 8R^3 \times 0.8714} = \frac{2T}{\pi R^3(0.1704)} \quad ...(2)$$

By dividing equation (2) by (1), we have

$$\frac{\tau_{Hollow\ shaft}}{\tau_{solid\ shaft}} = \frac{2T}{\pi R^3 \times 0.8704} \times \frac{\pi R^3}{2T}$$

$$\boxed{\frac{\tau_{Hollow}}{\tau_{Solid}} = 1.14}$$

13. (a)

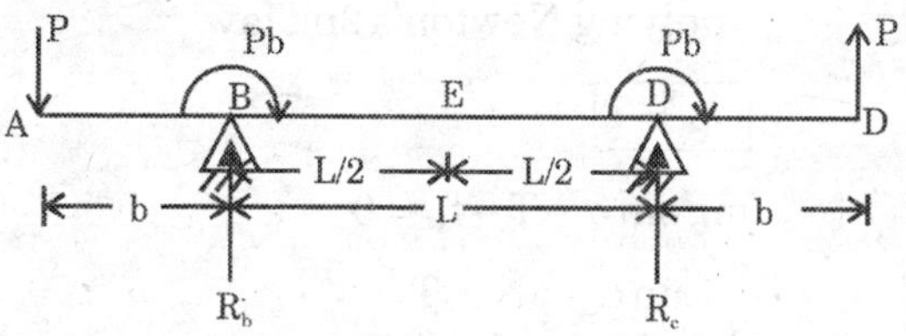

$$R_B + R_C + P - P = 0$$

$$\therefore R_B + R_C = 0 \quad ...(1)$$

Taking Moment about 'A'

$$P(L+b) + R_C \times L - Pb - Pb + Pb = 0$$

$$PL + Pb + R_C \times L - Pb = 0$$

$$PL + R_C \times L = 0$$

$$R_C \times L = -PL$$

$$\therefore \boxed{R_C = -P}$$

From equation (1),

$$R_B + R_C = 0$$

$$R_B - P = 0$$

$$\boxed{R_B = P}$$

The Bending Moment at the mid-point of the beam

$$M_E = P\left(\frac{L}{2}+b\right) + R_C \times \frac{L}{2} - Pb$$

$$= \frac{PL}{2} + Pb + (-P)\cdot\frac{L}{2} - Pb$$

$$\boxed{M_E = 0}$$

Therefore, Bending Moment at the midpoint of the beam will be zero.

14. (b)

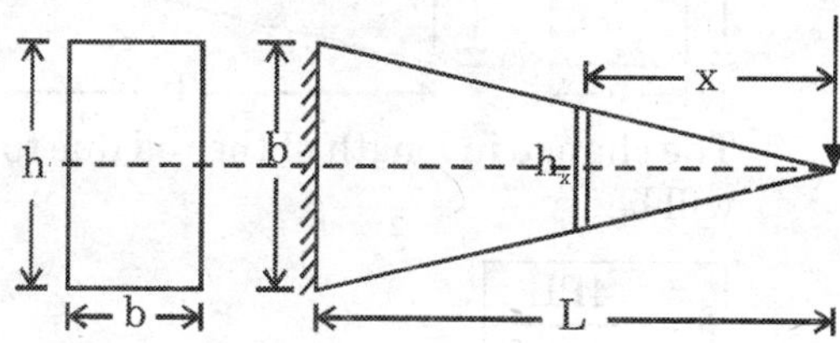

We know, for any distance 'x'

$$\sigma_{Allowable} = \frac{M_x}{Z_x} = \frac{Px}{\frac{bh_x^2}{6}} = \frac{6Px}{bh_x^2}$$

$$\therefore h_x^2 = \frac{6Px}{b\sigma_{Allow}}$$

$$h_x = \sqrt{\frac{6Px}{b\sigma_{Allow}}}$$

$$\therefore \boxed{h_x \propto \sqrt{x}}$$

Hence, the option (b) is correct.

15. (a)

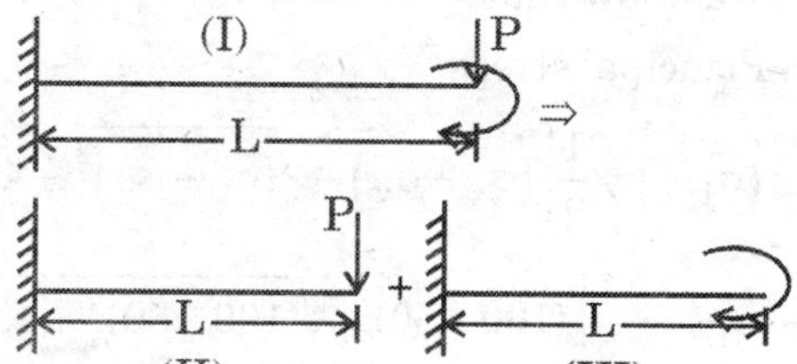

We know,

Strain energy for part II

$$U_{II} = \int_0^L \frac{M_x^2 dx}{2EI} = \int_0^L \frac{(Px)^2 dx}{2EI} = \int_0^L \frac{P^2 \cdot x^2 dx}{2EI}$$

$$U_{II} = \frac{P^2}{2EI}\cdot\frac{L^3}{3} = \frac{P^2L^3}{6EI}$$

Now, strain energy for part III

$$U_{III} = \frac{M_0^2 L}{2EI}$$

By using principal of superposition and strain energy

$$U_{Net} = U_{II} + U_{III}$$

$$\boxed{U_I = \frac{P^2L^3}{6EI} + \frac{M_0^2 L}{2EI}}$$

Hence, the option (a) is correct.

16. (a)

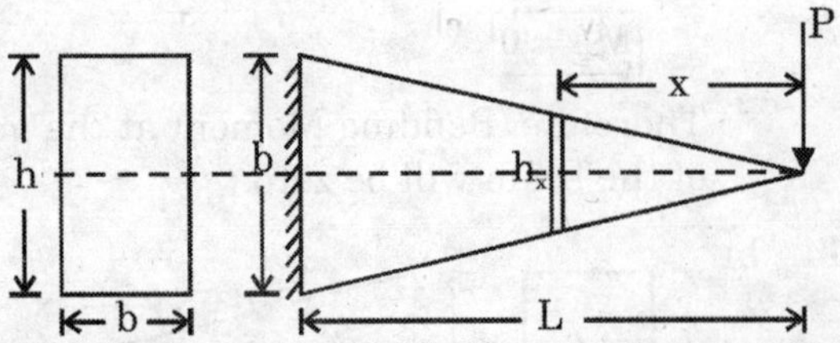

The change in length of the rod due to load 'P' will be

$$\boxed{\delta = \frac{4PL}{\Pi d_1 d_2 E}}$$

Hence, the option (a) is correct.

17. (110.25)

Given

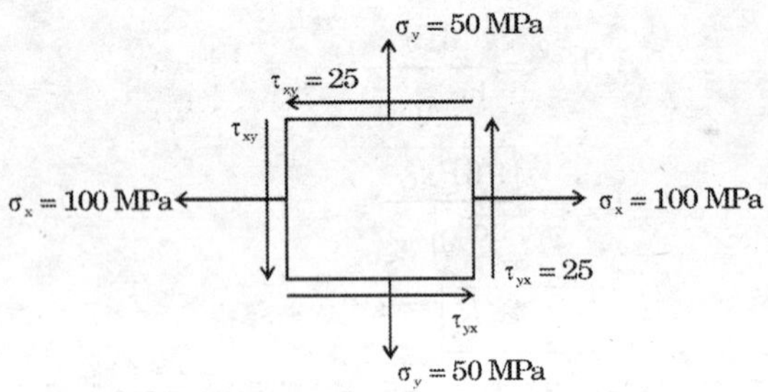

We know that

Principal stress

$$(\sigma_{1,x}) = \frac{1}{2}\left[(\sigma_x + \sigma_y) + \sqrt{(\sigma_x - \sigma_y)^2 + 4\tau_{xy}^2}\right]$$

$$\sigma_{1,0} = \frac{1}{2}\left[(100+50) + \sqrt{(100-50)^2 + 4(25)^2}\right]$$

$$= \frac{1}{2}\left[150 + \sqrt{(50)^2 + (4\times 625)}\right]$$

$$\sigma_{1,x} = 110.25 \text{ MPa}$$

Therefore the maximum principal stress

$$\boxed{\sigma_1 = 110.25}$$

18. (a) Given $\sigma_x = 10$ MPa

$\tau_{xy} = \tau_{yx} = -20$ MPa

$G = 0.4E$

We know,

$E = 2G(1 + m)$

$E = 2(0.4 \times E)(1+\mu) \quad [\because G = 0.4E \text{ given}]$

$$\boxed{\mu = 0.25}$$

$\because$ Volumetric strain

$$(\varepsilon_V) = \frac{1-2\mu}{E}(\sigma_x + \sigma_y + \sigma_z)$$

$$\varepsilon_V = \frac{1-2\mu}{E}(\sigma_x + 0 + 0) \quad [\because \sigma_y + \sigma_z = 0]$$

$$= \frac{1-(2\times 0.25)}{E}(10) = \frac{(1-0.5)\times 10}{E}$$

$$\boxed{\varepsilon_V = \frac{5}{E}}$$

Hence, the option (a) is correct.

19. (a)

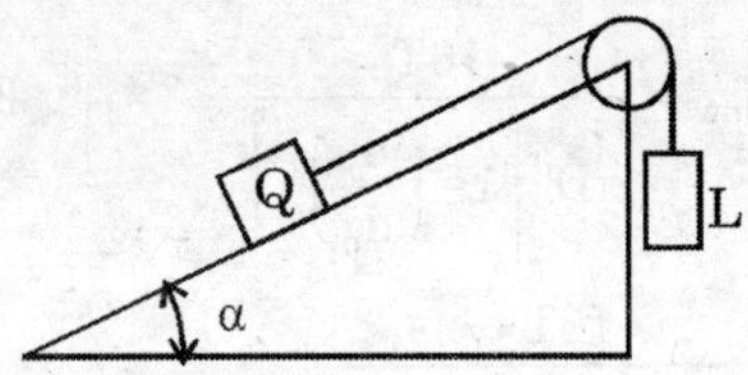

Case-I : When weight 'Q' moves down the plane

FBD:

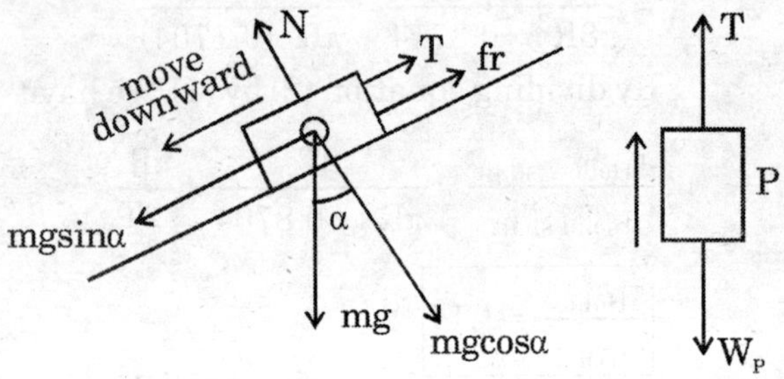

Applying Newton's 2nd law

$$\boxed{\omega_P = T} \quad \ldots(1)$$

$mg\sin\alpha - T - f_r = 0$

$\omega_\theta \sin\alpha - \mu N = T$

$\omega_\theta \sin\alpha - \mu(\omega_\theta \cos\alpha) = T$

$\omega_\theta(\sin\alpha - \mu\cos\alpha) = T$

$\omega_\theta(\sin\alpha - \mu\cos\alpha) = \omega_p \quad [\because \text{From (1) } T = \omega_P]$

$$\therefore \boxed{\frac{\omega_P}{\omega_Q} = \sin\alpha - \mu\cos\alpha}$$

Case II :

When weight P starts falling down

$T - fr - \omega_\theta \sin\alpha = 0$

$T = fr + \omega_\theta \sin\alpha = \mu N + \omega_\theta \sin\alpha$

$$\boxed{\omega_P = T} \quad \ldots(2)$$

$\omega_P = \mu(\omega_\theta \cos\alpha) + \omega_\theta \sin\alpha \quad [\therefore \omega_P = T]$

$\omega_P = \omega_Q(\mu\cos\alpha + \sin\alpha)$

$$\therefore \boxed{\frac{\omega_P}{\omega_Q} = \sin\alpha + \mu\cos\alpha}$$

Hence, the option (a) is correct.

21. (b)

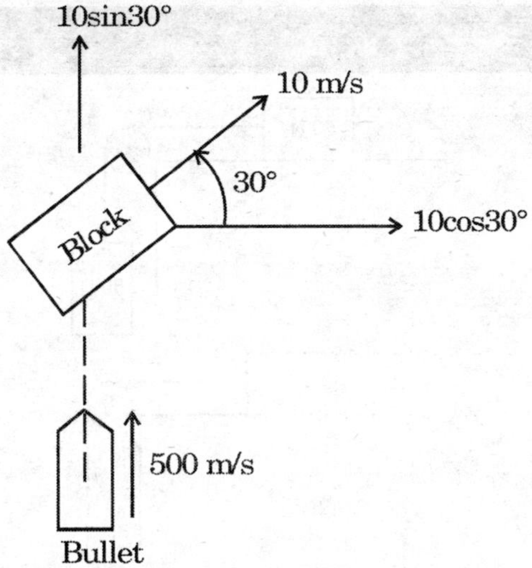

Given,

Mass of block $(m_1) = 5$ kg

Velocity of block $V_1 = 10\sin 51.78$ m/s.

Mass of the Bullet $(m_2) = 60$ g $= 0.060$ kg

Velocity of the Bullet $(V_2) = 500$ m/s.

By applying law of conservation of linear momentum,

$$m_1v_1 + m_2v_2 = (m_1 + m_2)V$$

$$5 \times 10\sin 51.78 + (0.060 \times 500) = (5 + 0.060) \times V$$

$$[\because \theta = 51.78]$$

$$\Rightarrow V = \frac{(5 \times 10\sin 51.78) + (0.060 + 500)}{(5 + 0.060)}$$

$\boxed{V = 13.69 \text{ m/s}}$ at $\theta = 51.78$

Hence, the option (b) is correct.

2015

Q. (1 – 9) Carry One Mark Each.

1. Find the force (in kN) in the member BH of the truss shown.

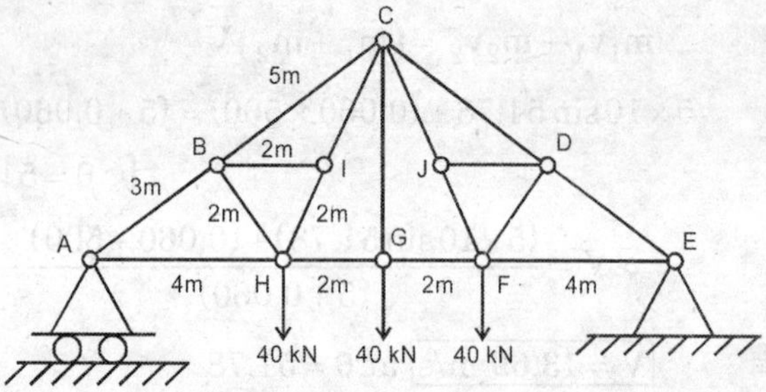

2. Consider the forces of magnitude F acting on the sides of the regular hexagon having side length a. At point B, the equivalent force and couple are, respectively,

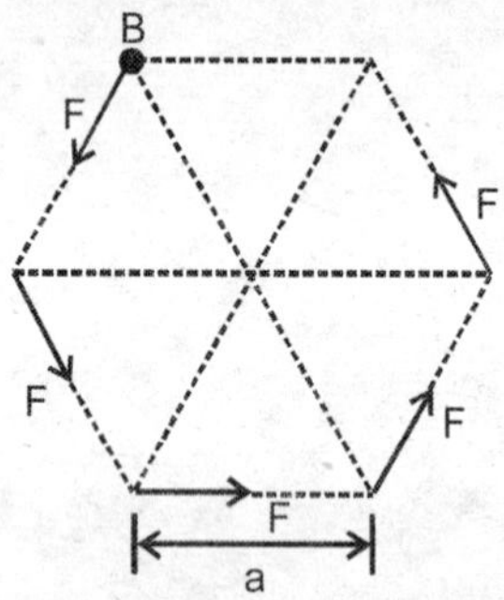

(*a*) F(←) and $3\sqrt{3}Fa$ (clockwise)

(*b*) F(→) and $\sqrt{3}Fa$ (clockwise)

(*c*) F(←) and $\sqrt{3}Fa$ (counter clockwise)

(*d*) F(→) and $3\sqrt{3}Fa$ (counter clockwise)

3. Bar-1 has a diameter d, length L, and elastic modulus E and subjected to tensile load P, resulting in an elongation of Δ_1. Bar-2 has diameter, 2d, length 2L, and elastic modulus 2E and subjected to tensile load 2P, resulting in an elongation of Δ_2. Find the ratio Δ_1/Δ_2.

4. In a plane stress problem, the principal stresses at a point are 30MPa and –15MPa. At the same point, on an element whose sides make an angle of 45° with respect to the principal axes, the normal stresses (in MPa) are

(*a*) 15/2 and 15/2 (*b*) 30/2 and 30/2

(*c*) 15/2 and –15/2 (*d*) 30/2 and –30/2

5. Two systems shown below start from rest. For the system shown on the left, two 50N blocks are connected by a cord. For the system shown on the right, the 50N block is pulled by a 50N downward force. Neglect friction. Which of the following is true?

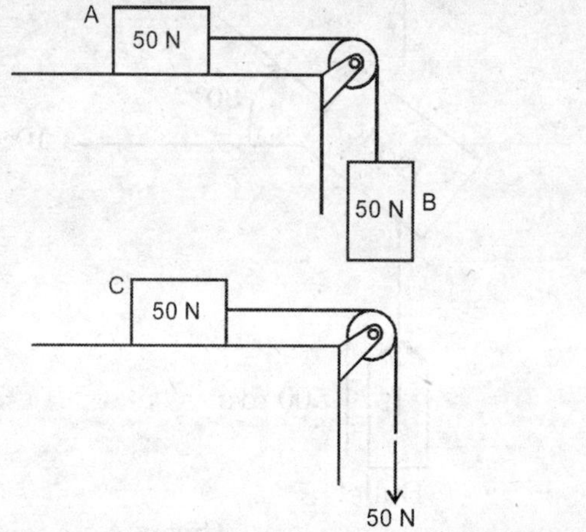

(*a*) Blocks A and C have the same acceleration.

(*b*) Block C will have a larger acceleration than block A.

(*c*) Block A will have a larger acceleration than block C.

(*d*) Block A will not move.

6. Two massless rigid bars, each of length a = 0.5m, are connected by a rotational spring having stiffness k = 1000 N.m/rad. Find the buckling load P (in kN).

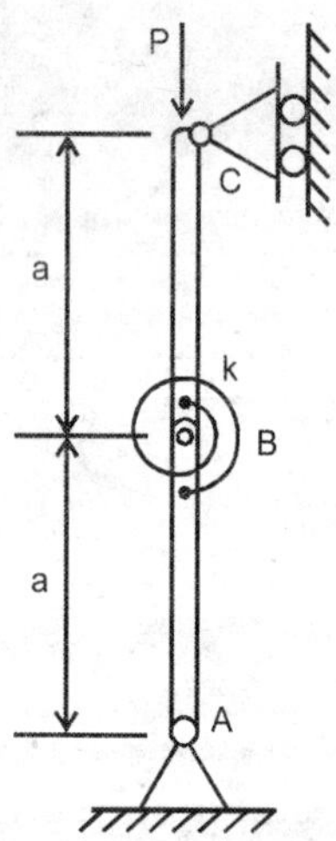

7. A simply supported beam having a rectangular cross-section of depth d is subjected to a vertical concentrated load P at the mid-span. The maximum shear stress in a section occurs at

(*a*) d/2 from the top of the cross-section

(*b*) d/3 from the top of the cross-section

(*c*) 2d/3 from the top of the cross-section

(*d*) Top of the cross-section

8. A steel block of size 100 × 50 × 25 mm³ is subjected to a uniform pressure on all faces. The dimension of the 100 mm edge reduces by 25μm (note 1 μm = 10^{-6}m). Find the applied pressure (in GPa). Use E = 240 GPa and ν = 0.3.

9. Which one of the following statements is true?

(*a*) In a tensile test on a rod made of ductile material, failure occurs along a plane making 45° with respect to the axis of the rod

(*b*) In a tensile test on a rod made of brittle material, failure occurs along a plane making 45° with respect to the axis of the rod

(*c*) In a torsion test on a rod made of ductile material, failure occurs along a plane making 45° with respect to the axis of the rod

(*d*) In a torsion test on a rod made of brittle material, failure occurs along a plane making 0° with respect to the axis of the rod

10. (10 – 22) Carry Two Marks Each.

10. A block is travelling with a constant speed v_o on a smooth surface when the surface suddenly becomes rough with a coefficient of friction μ, which causes the block to stop after a distance *d*. When the block travels twice as fast, i.e. at a speed $2v_o$, it travels a distance D on the rough surface before stopping. Find the ratio D/*d*.

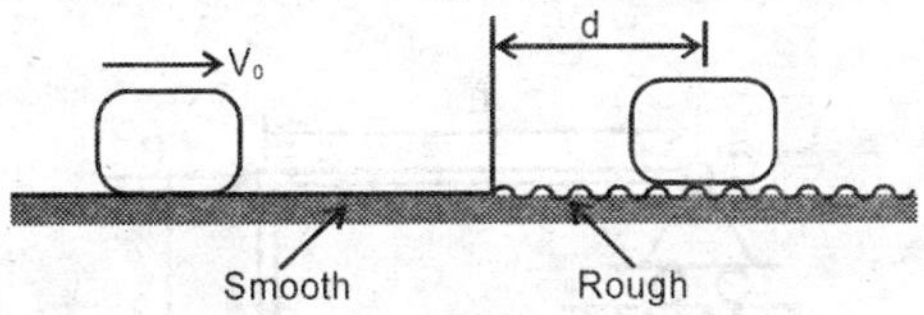

11. The beam shown below is loaded with a concentrated clockwise moment of 80kN-m at point B. The bending moment diagram (in kN-m) is

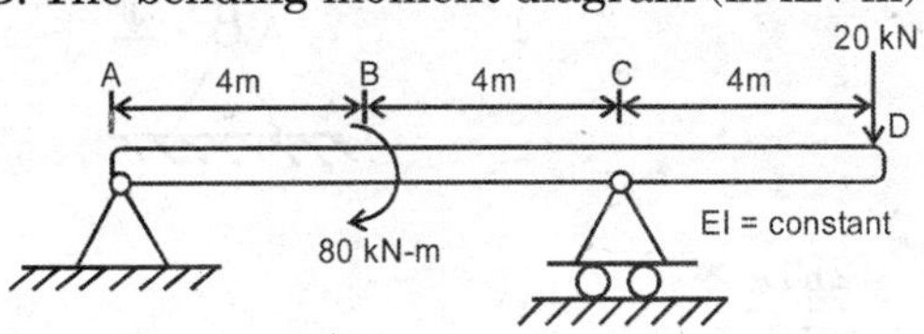

(*a*)
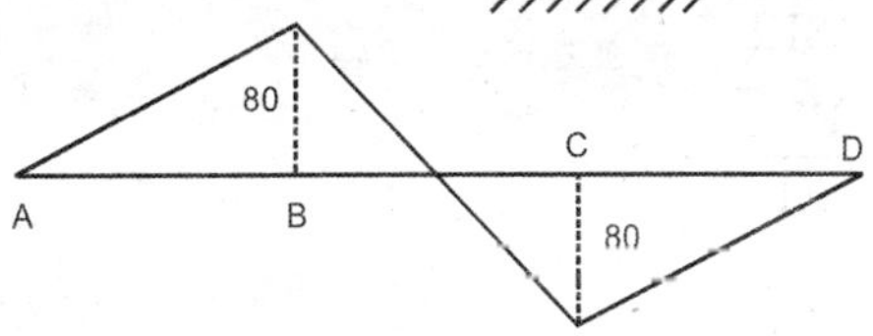

(*b*)
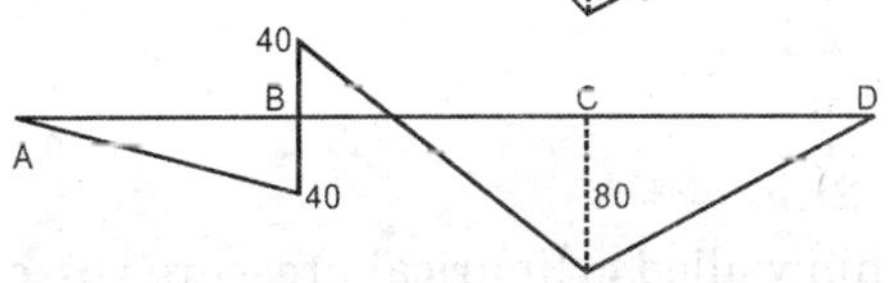

(*c*)
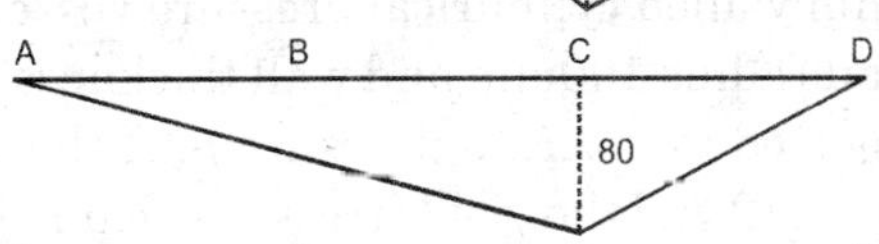

(*d*)
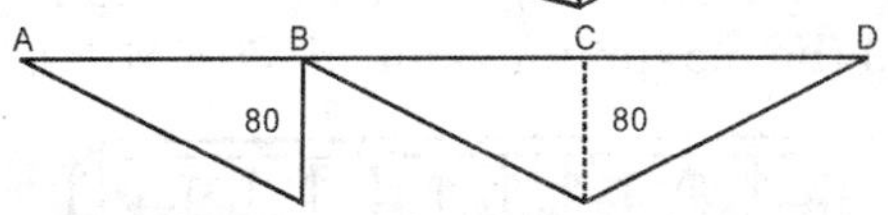

12. The beam shown has an internal hinge at B. A vertical load P = 25kN is applied at B. Use L = 2m. Magnitude of the reactions (i.e. forces and moments) at A and C are

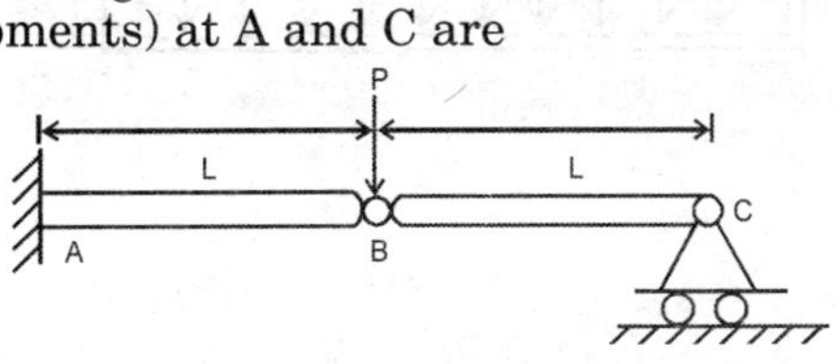

(*a*) Vertical reaction force at C is 12.5kN, vertical reaction force at A is 12.5kN, moment reaction at A is 0kN-m.

(*b*) Vertical reaction force at C is 0kN, vertical reaction force at A is 25kN, moment reaction at A is 50kN-m.

(c) Vertical reaction force at C is 25kN, vertical reaction force at A is 0kN, moment reaction at A is 50 kN-m.

(*d*) Vertical reaction froce at C is 0kN, vertical reaction force at A is 25kN, moment reaction at A is 25kN-m.

13. Blocks P and Q are released from rest in the positions shown. Neglect friction between all surfaces, i.e., both blocks can translate freely. Then the direction of the acceleration of block Q (i.e. a_Q) is

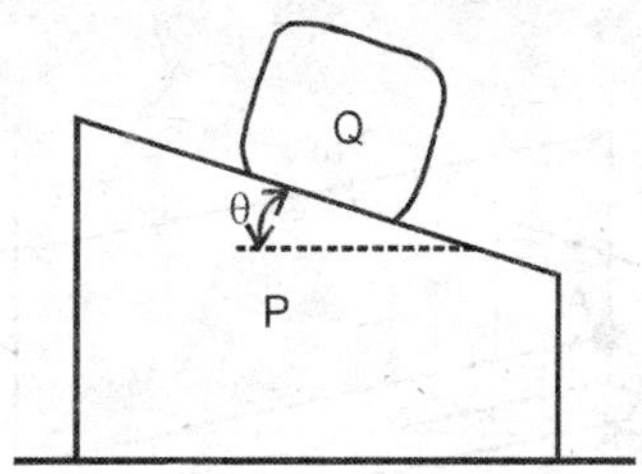

(*a*)
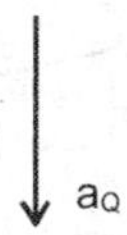

(*b*)
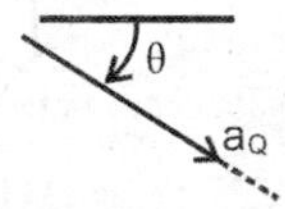

(c)
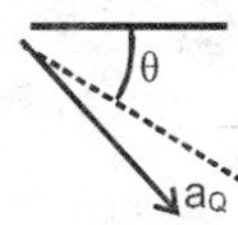

(*d*)
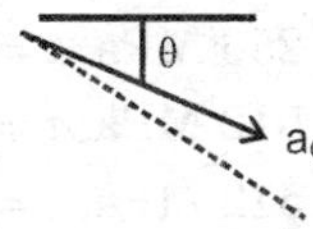

14. The acceleration, *a*, of a particle as a function of its position, *x*, is given by the relation $a = 0.1 + \sin\frac{x}{b}$, where *a* and *x* are expressed in m/s² and meters, respectively. Consider b = 1m. When $x = 0$, velocity is v = 1m/s. Find v (in m/s) when $x = \pi$ meters.

15. The 30 kg block B shown below is suspended by a 2m cord attached to the 60kg cart A. Friction is negligible. If the system is released from rest in the position shown, find the ratio of the velocity magnitudes $|v_A|/|v_B|$ when the cord is vertical.

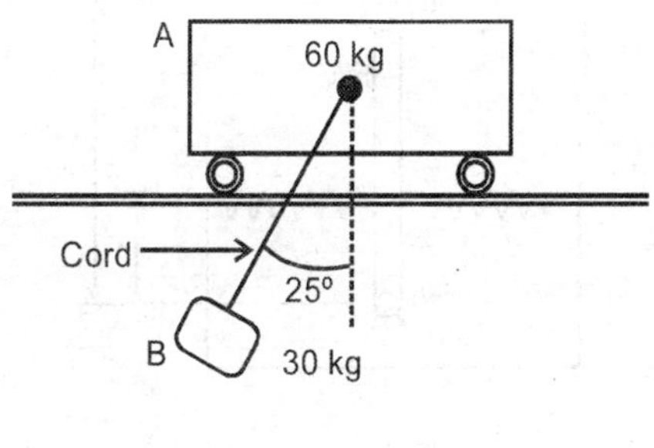

16. The plane frame shown has an internal hinge at C. Find the magnitude of axial force (in kN) in member BC.

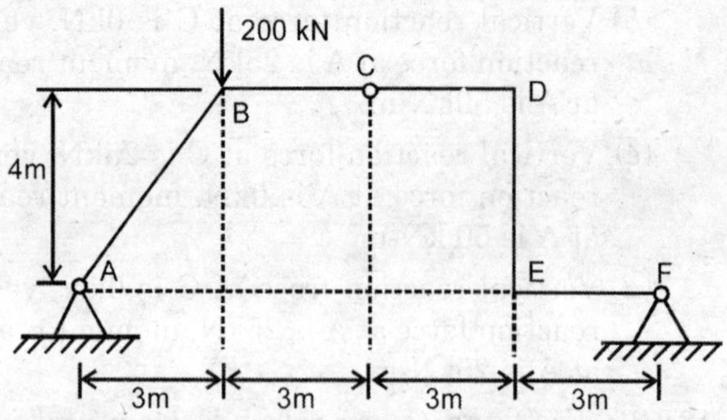

17. Two 50mm diameter solid steel rods are rigidly connected together at right angles and loaded as shown. Use P = 1000π kN. At point A, located at the top of the cross-section at the fixed end, the magnitude of bending stress (σ) and shear stress (τ) are

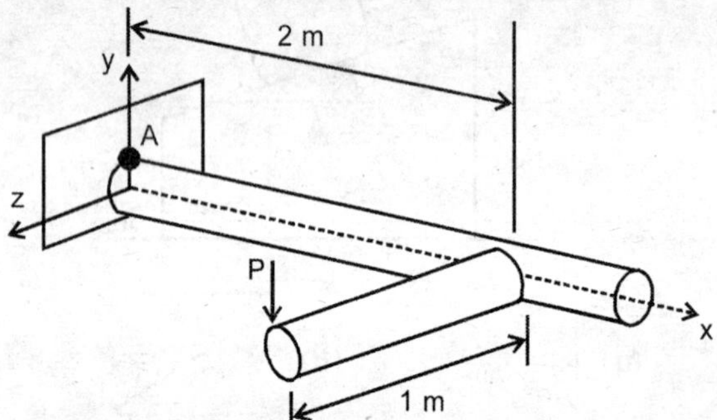

(*a*) σ = 256 MPa, τ = 512 MPa

(*b*) σ = 512 MPa, τ = 256 MPa

(*c*) σ = 512 MPa, τ = 128 MPa

(*d*) σ = 128 MPa, τ = 512 MPa

18. At a temperature of 40°C, a rod tightly fits between two rigid walls such that the compressive stress in the rod is 60 MPa. Given E = 200GPa and α = 20 × 10^{-6}/°C, find the temperature at which the rod will just lose contact with the walls.

19. A massless rigid rod AB of length *h* is pinned at end A and carries mass *m* at end B. The rod is also supported by two linear springs of stiffness *k* at a height *d* from the end A. Use *m* = 4kg, *h* = 0.5m, *d* = 0.2m, *k* = 600 N/m and *g* = 10 m/s². For small oscillations about the position shown, find the frequency of free vibration (in rad/s).

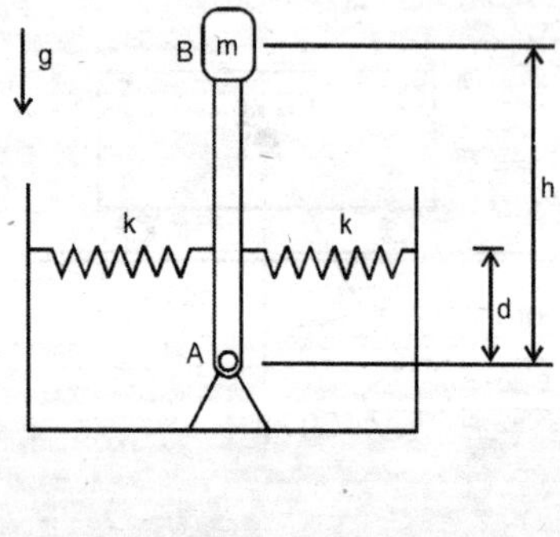

20. Find the maximum force P (in kN) that can be applied to the planar structure ABC so as to prevent buckling in any of the members. Consider buckling only in the plane of the structure. Joint B is a pin connection. Use E = 200 GPa for both members. The diameter of member AB is 10 mm and the diameter of member BC is 15 mm.

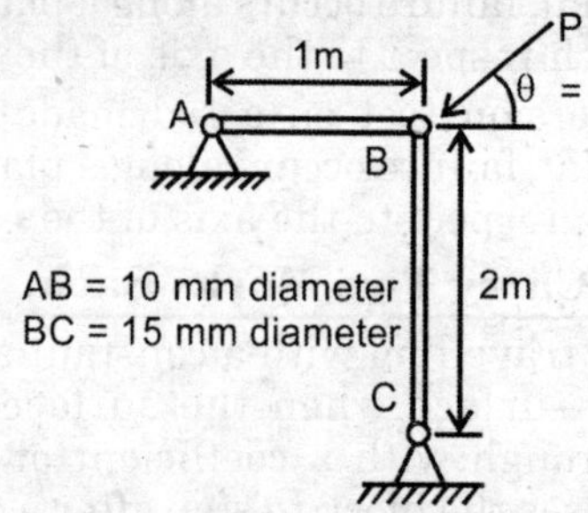

21. The plane frame shown is analyzed by neglecting axial and shear deformations. The horizontal displacement of joint B is

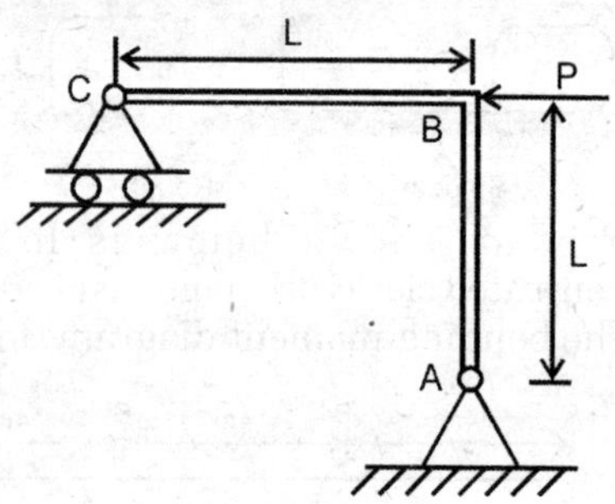

(*a*) $\frac{2PL^3}{3EI}$

(*b*) $\frac{PL^3}{EI}$

(*c*) $\frac{3PL^3}{2EI}$

(*d*) $\frac{PL^3}{2EI}$

22. A thin walled cylindrical pressure vessel having mean radius 100mm and wall thickness 5mm, is subjected to internal pressure *p*. If the factor of safety is 2 and the yield stress in shear is 100MPa, find the maximum value of *p* (in MPa).

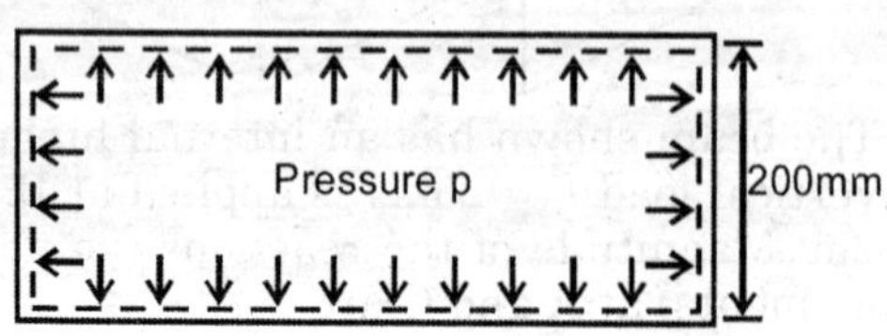

ANSWERS

1. (0)	**2.** (d)	**3.** (1.9 to 2.1)	**4.** (a)	**5.** (b)	**6.** (4)
7. (a)	**8.** (0.14 to 0.16)	**9.** (a)	**10.** (4)	**11.** (d)	**12.** (b)
13. (c)	**14.** (2.3 to 2.4)	**15.** (0.5)	**16.** (74 to 76)	**17.** (c)	**18.** (25°)
19. (5.2 to 5.4)	**20.** (1.3 to 1.45)	**21.** (a)	**22.** (5)		

EXPLANATIONS

1. (0)

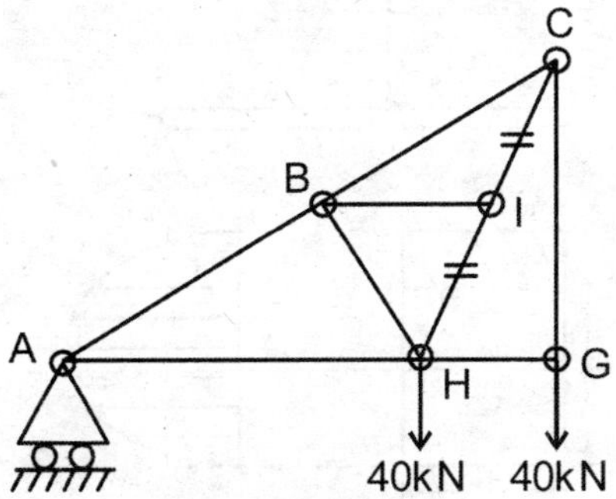

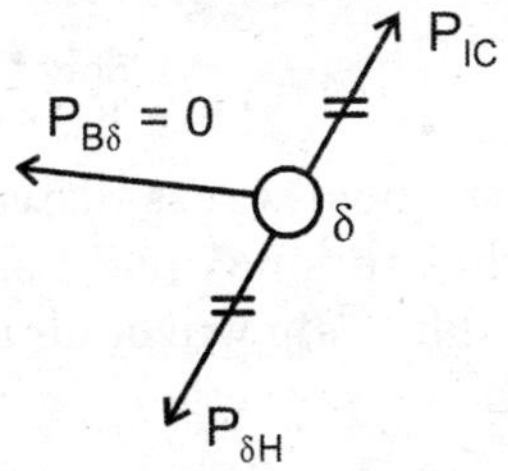

Take the joint 'δ'

Since these are three members (or) three forces meet at joint 'δ' and $P_{\delta C}$ & P_{IH} are collinear therefor P_{BI} will be zero.

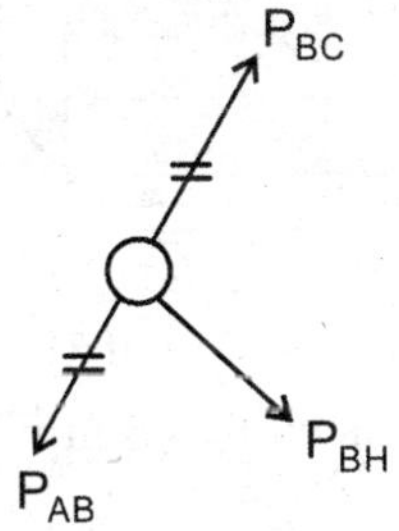

Similarly Joint 'B'

Here P_{BC} & P_{AB} are collinear therefore $P_{BH} = 0$

Thus the force on the member BH is zero.

2. (d) All 'F' Forces cancel each other except force F acting horizontally because regular hexagon has internal angle of 120°.

∴ Resultant force = F (→)

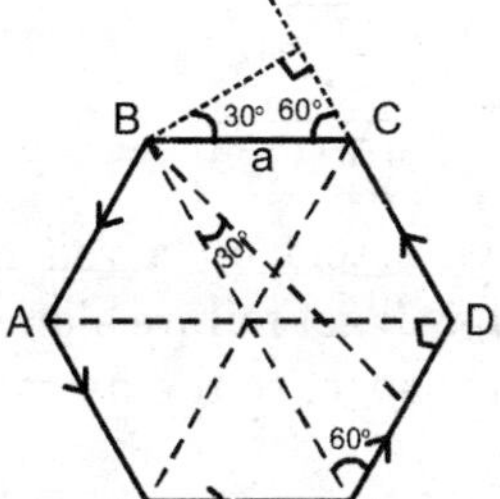

$$\Sigma M@B = Fxasin60° + 2asin60° \times F + 2asin60° \times F + Fasin60°$$

$$\Sigma M@B = 3\sqrt{3}\ Fa\ (\circlearrowleft)\ \text{counterclockwise}$$

3. (2) Given

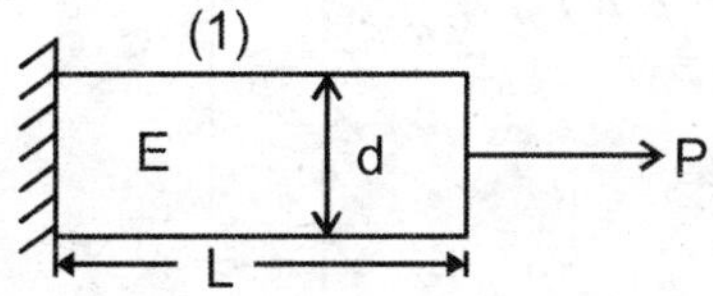

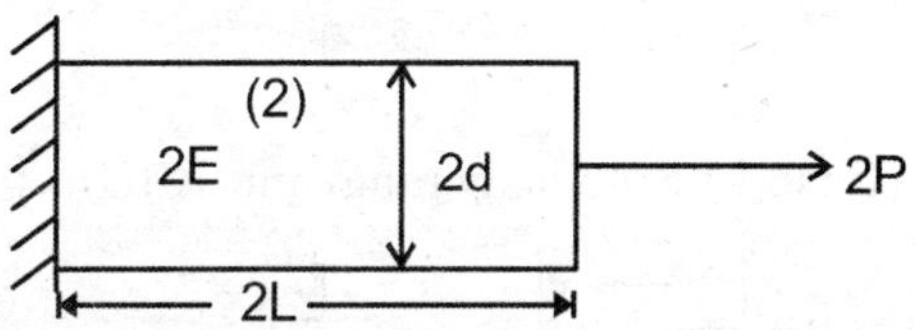

Elongation in bar-1 $(\Delta_1) = \dfrac{P_1L_1}{A_1E_1} = \dfrac{PL}{\frac{\pi}{4}d^2.E}$

$$\Delta_1 = \frac{4PL}{\pi d^2 E}$$

Elongation in bar-2

$$(\Delta_2) = \frac{P_2L_2}{A_2E_2} = \frac{(2P).(2L)}{\frac{\pi}{4}(2d)^2 \times 2E}$$

$$\Delta_2 = \frac{2\cancel{4}PL}{\frac{\pi}{\cancel{4}} \times \cancel{4}d^2 \times \cancel{2}E} = \frac{2PL}{\pi d^2 E}$$

$$\therefore \boxed{\frac{\Delta_1}{\Delta_2} = 2}$$

4. (a) Given that

Principal stresses are $\sigma_1 = 30$ Mpa

$\sigma_2 = -15$ Mpa

Normal stress (σ_n)

$$= \frac{\sigma_1 + \sigma_2}{2} = \frac{30 - 15}{2} = \frac{15}{2}\text{Mpa}$$

(at θ = 45°)

Normal stress (σ_n)

$$= \frac{\sigma_1 + \sigma_2}{2} = \frac{30 - 15}{2} = \frac{15}{2}\text{Mpa}$$

(at θ = 135°)

Hence the correct option is (a).

5. (b)

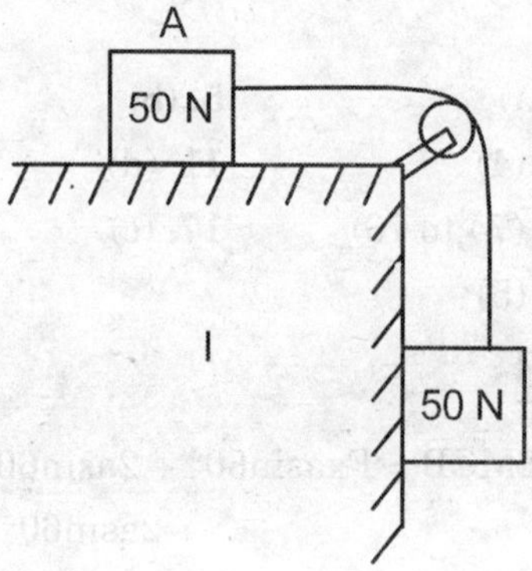

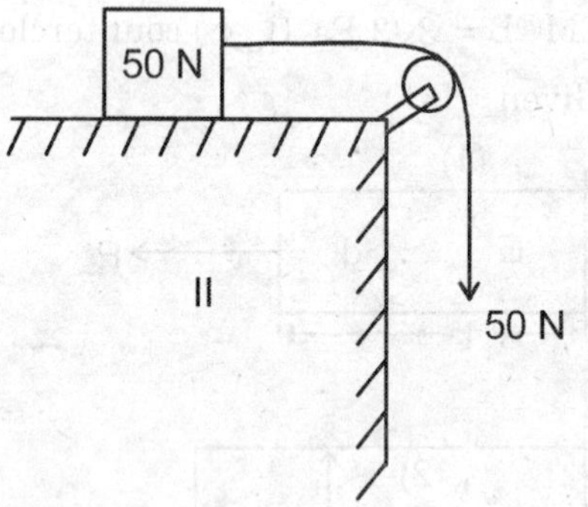

Case-I : Free body diagram (FBD)

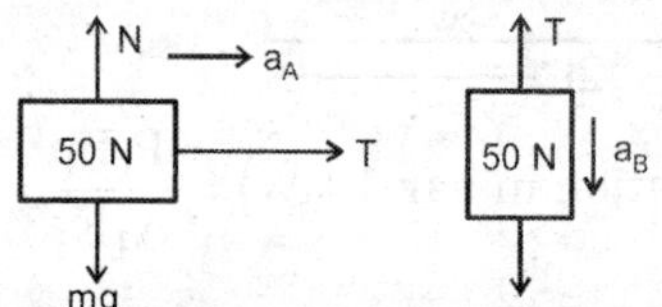

$mg = 50N \Rightarrow m = \frac{50}{10} = 5$ kg

Applying 2nd law of Newton.

($\because$ g = 10 m/s² Assume)

$50 - T = ma_B$ ....(1)

$T = ma_A$ ....(2)

Solving eq (1) & (2) we get

$\boxed{a_A = 5\ m/s^2}$

Case-II : FBD

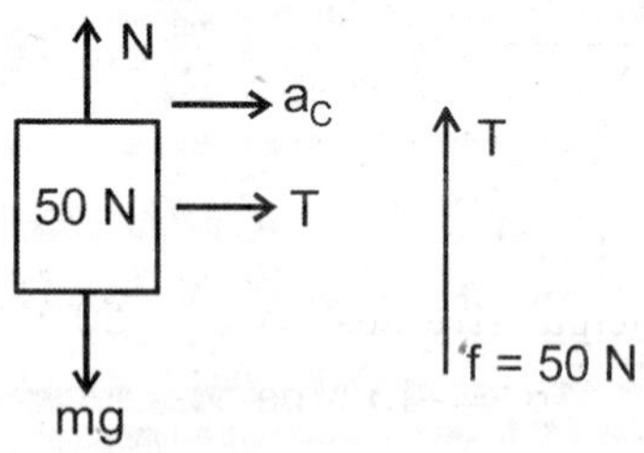

Applying Newton's 2nd law

$F - T = ma_C$ ....(1)

$50 - T = 0 \Rightarrow T = 50$ ....(2)

Solving (1) & (2) we get

$\boxed{a_c = 10 m/s^2}$

therefore $\boxed{a_c > a_A}$

Hence the option (b) is correct.

6. (4) $P = \frac{k}{a^2}$

$= \frac{1000N - m/rad}{.5^2}$

$= \frac{1000}{.25} = 4000N = 4kN$

7. (a)

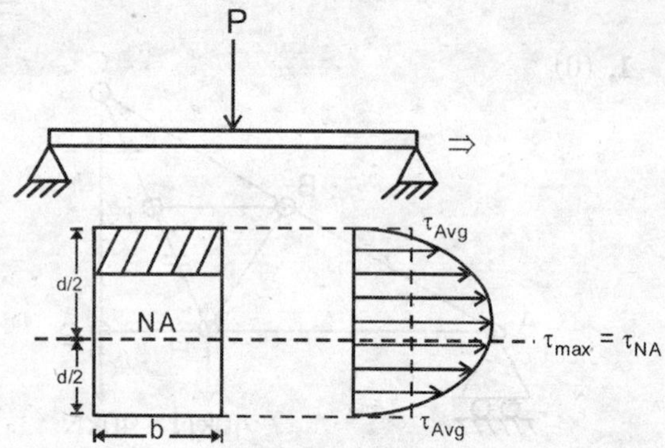

Shear stress distribution diagram

the above shear stress distribution diagram shows that the maximum shear stress in a rectangular beam will occur at "d/2" from the top of the cross-section.

Hence the option (a) is correct.

8. (0.15) Given that

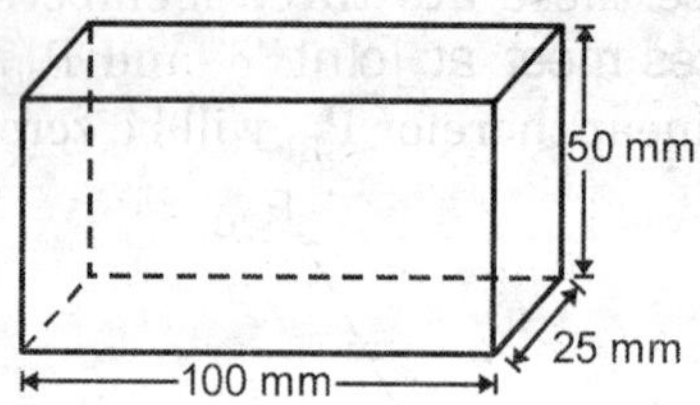

dimensions of block = $100 \times 50 \times 35$ mm³

since the dimension of 100 mm

edge is reduced by 25 mm

$\therefore \quad \frac{\delta\ell}{\ell} = \frac{25 \times 10^{-3}}{100}$

we know

$E_x = \frac{\delta\ell}{\ell} = \frac{(1-2\mu)}{E}(\sigma) \quad [\sigma_x = \sigma_y = \sigma_z = \sigma]$

$\frac{25 \times 10^{-3}}{100} = \frac{(1-2\times 0.3)}{240 \times 10^3}(\sigma)$

$\therefore \quad \boxed{\sigma = 150 \text{ Mpa}}$

or, $\boxed{\sigma = 0.150 \text{ Gpa}}$

9. (a) In a tensile test on a rod mode to ductile material, will get failure along a plane making 45° with respect to the axis of the rod.

10. (4) As mentioned in the problem, the body moves with constant speed which means constant acceleration.

$$\frac{D}{d} = \left(\frac{2V_0}{V_0}\right)^2 = \frac{1}{4}$$

11. (d)

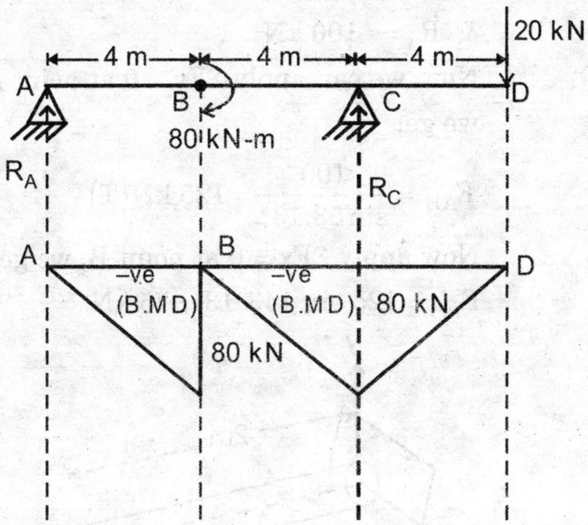

$R_A + R_C = 20$ ....(1)

Taking moment about 'A',

$-20 \times 12 + R_C \times 8 - 80 = 0$

$\boxed{R_C = 40\text{kN}}$

from (1)

$R_A + R_C = 20$ kN

$R_A = 20 - 40 = -20$ kN

$\boxed{R_A = -20\text{kN}}$

Now Bending moment at point "D" $(B.M.)_D = 0$

$(B.M)_C = -20 \times 4 = -80$ kN

$(B.M)_B = -20 \times 8 + R_C \times 4$

$= -160 + 160 = 0$

$(B.M)_B = 0 - 80$ kN

$= -80$ kN

$(B.M)_A = 0$

Hence the option (d) is correct.

12. (b)

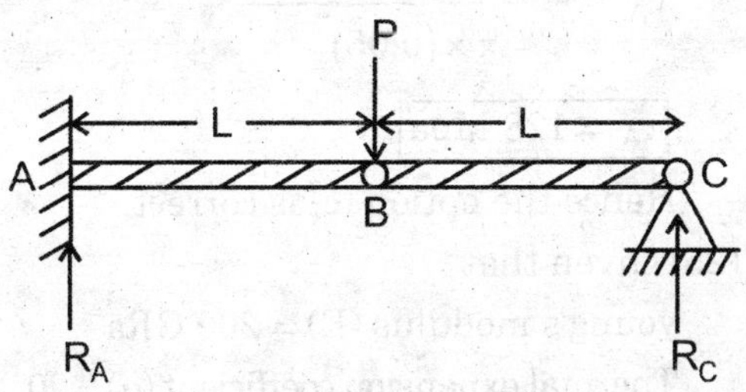

Given that

Load (P) = 25 kN

length (L) = 2 m

$R_A + R_C = P$ ....(1)

Bending moment at point B will be zero because it is hinged

$R_C \times L = 0 \Rightarrow \boxed{R_C = 0}$

from (1),

$R_A + R_C = P$

$R_A + 0 = P = 25$ kN

$\therefore \boxed{R_A = 25\text{ kN}}$

Now, Bending moment at 'A'

$(B.M)_A = PL = 25 \times 2 = 50$ kN

Hence the option (b) is correct.

13. (c)

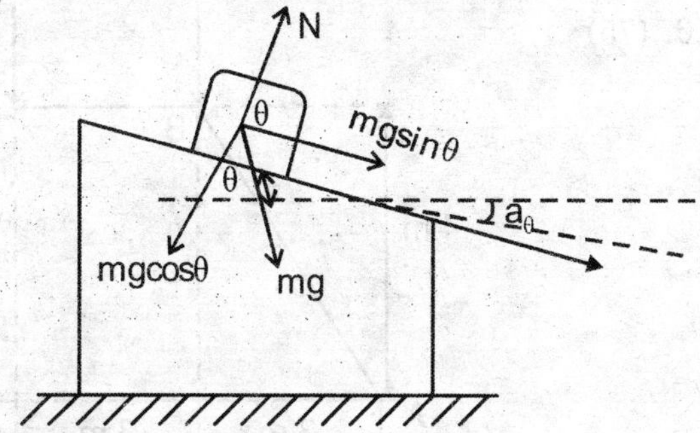

Hence the option (c) is correct.

14. (2.37) $a = 0.1 + \sin\frac{x}{b}$ Given b = 1 m

when x = 0, v = 1 m/s

$$a = v\frac{dv}{dx}$$

$$v\frac{dv}{dx} = 0.1 + \sin\left(\frac{x}{b}\right)$$

$vd_v = 0.1\,dx + \sin(x/b).dx$

$$\frac{v^2}{2} = 0.1x - \frac{\cos\left(x/b\right)}{1/b} + c_1$$

$$x = 0,\ v = 1,\ \frac{1}{2} = -1 + c_1$$

$c_1 = 3/2$, Find v when x = π meters

$$\frac{v^2}{2} = 0.1x - \cos x + \frac{3}{2} \quad (\because\ b = 1\text{m})$$

$$= 0.1 \times \pi + 1 + \frac{3}{2}$$

v = 2.37 m/s

15. (0.5) Conservation of linear momentum :

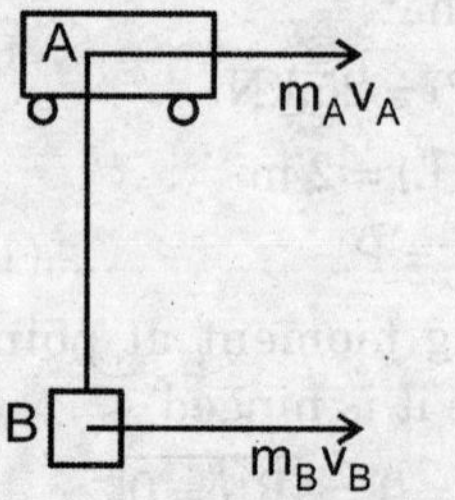

Since block and cart are initially at rest,

L0 = 0

Thus, as B passes under A,

L = mAvA + mBvB = 0

$\xrightarrow{+} m_Av_A + m_Bv_B = 0$

$$v_A = -\frac{m_B}{m_A}v_B \Rightarrow \frac{v_A}{v_B} = \frac{30}{60} = 0.5$$

16. (75)

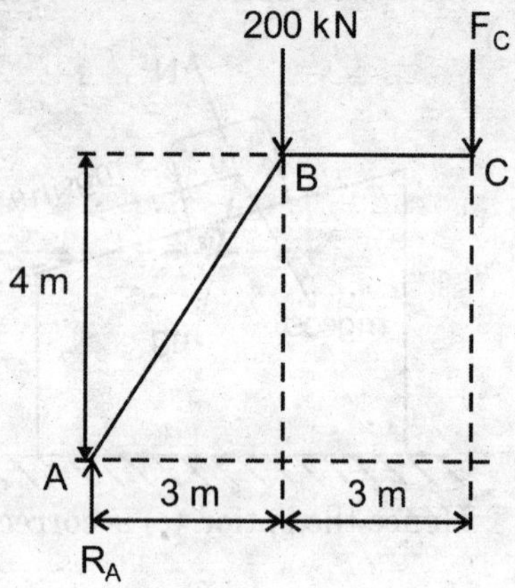

Point A

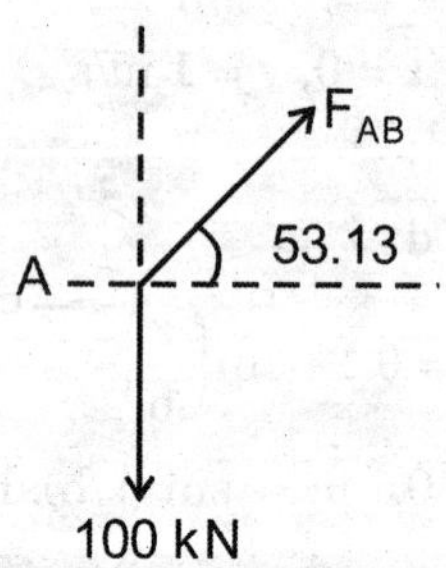

Point B

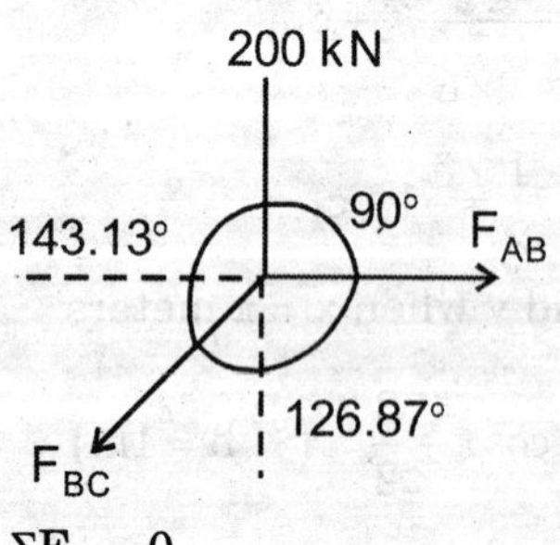

$\Sigma F_x = 0$

$\Sigma F_y = 0$

$\Sigma F @ B = 0$

By applying above conditions, we get

$R_A - F_C - 200 = 0$

& $-R_A \times \cancel{3} - F_C \times \cancel{3} = 0$

$R_A = -F_C$

$2F_C = -200$

$\boxed{F_C = 100 \text{ kN}}$

& $R_A = -100$ kN

Now, we can apply ΣFy = 0 at point A.

we get

$$F_{AB} = \frac{100}{\sin 53.13°} = 125 \text{ kN(T)}$$

Now apply ΣFx = 0 at point B, we get

$F_{BC} = 125 \sin 143°13 = 75$ kN

17. (c)

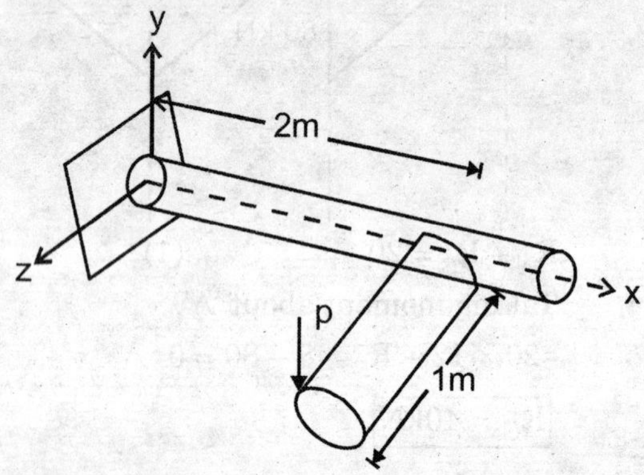

At point 'A'

$$\text{Bending stress } (\sigma_b) = \frac{M}{Z}$$

$$\sigma_b = \frac{(P \times 2)}{\frac{\pi}{32}(d)^3} = \frac{32 \times 100\cancel{\pi} \times 2 \times 10^3}{\cancel{\pi}(0.05)^3}$$

$= 5120000100$

$\boxed{\sigma_b = 512 \text{ Mpa}}$

$$\text{Shear stress } (\tau)_A = \frac{T}{z} = \frac{P \times 1}{\frac{\pi}{16}d^3}$$

$$\tau_A = \frac{1000 \times 10^3 \times \pi \times 16}{\pi \times (0.05)^3}$$

$\boxed{\tau_A = 128 \text{ Mpa}}$

Hence the option (c) is correct.

18. (25) Given that

young's modulus (E) = 200 GRa

Thermal expansion coefficient (α) = 20 × 10^{-6}/°C

Compressive stress (σ_c) = 60 Mpa

Temperature = 40°C

we know, $\sigma_{Total} = \sigma_{Thermal} + \sigma_{axial}$

when the rod will be detached from the wall

$\sigma_{Axial} = 0$

$\therefore \ \sigma_{Total} = \sigma_{Thermal}$

$60 \times 10^6 = \alpha(\Delta T)\, E$

$60 \times 10^6 = 20 \times 10^{-6} \times (\Delta T) \times 200 \times 10^9$

DT = 15°C

Ti – Tf = 15°C

40 – Tf = 15°C

Tf = 40 – 15 = 25°C

$\boxed{T_f = 25°C}$

Hence the final temperature at which the rod will just lose contact with the wall is

$\boxed{T_f = 25°C}$

19. (5.246)

Given that

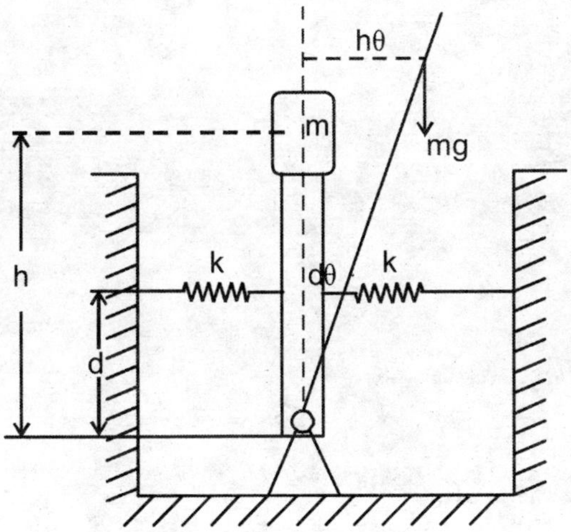

Mass of Block (m) = 4 kg

h = 0.5 m

d = 0.2 m

spring (k) = 600 N/m

stiffness g = 10 m/s²

Now, Taking Moment about 'o' weget

$$\delta\ddot{\theta} + mg(n\theta) + k(d\theta) - d + k(d\theta).d = 0$$

$$\delta\ddot{\theta} + \left(mgh + 2kd^2\right)\theta = 0$$

$\therefore$ frequency of vibration

$$(\omega_n) = \sqrt{\frac{mgh + 2kd^2}{\delta}}$$

$$(\omega_n) = \sqrt{\frac{(4 \times 10 \times 0.5) + 2 \times 600 \times (0.2)^2}{mh^2}}$$

$\boxed{\omega_n = 5.246 \text{ rad/sec}}$

20. (1.37)

Find euler's buckling load for column by

$$P_{AB} = \frac{\pi^2 EI}{\ell^2} = \frac{\pi^2 \times 200 \times 10^9 \times \pi \times (.01)^2}{1^2 \times 64}$$

$= 968.94$ N

Similarly, for

$$P_{BC} = \frac{\pi^2 EI}{\ell^2} = \frac{\pi^2 \times 200 \times 10^9 \times \pi \times (.015)^2}{1^2 \times 64}$$

By applying conditions of equilibrium,

$\Sigma F_x = 0 \Rightarrow P = P_{AB}/\sin 135° = 1.37$ kN –(i)

$\Sigma F_y = 0 \Rightarrow P = P_{BC}/\sin 135° = 3.47$ kN –(ii)

Choose minimum out of (i) & (ii)

$\therefore$ P = 1.37 kN

21. (a)

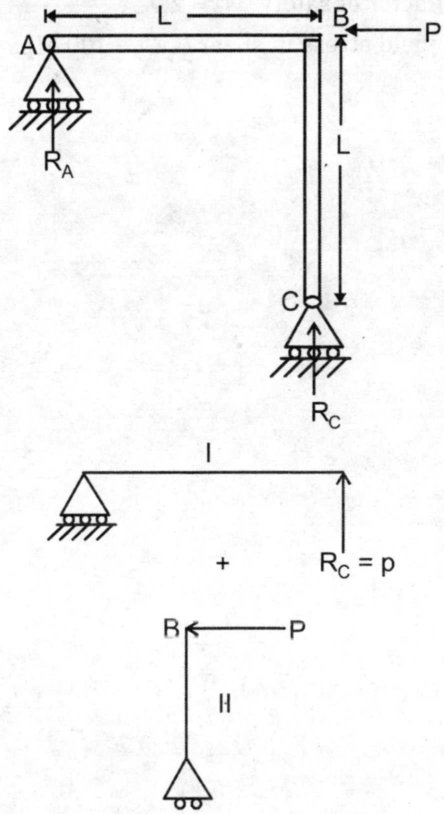

$R_A + R_C = 0$ ....(1)

At equilibrium condition

$Rc \times L = P \times L$

$\boxed{R_C = P}$ $\boxed{R_a = -P}$

Now deflection in the beam 2

$$y_1 = \frac{PL^3}{3E_2}$$

deflection of beam II

$$y_2 = \frac{PL^3}{3E_2}$$

Net deformation or deflection at point (B) is

$y_{Net} = y_1 + y_2$

$$\boxed{y = \frac{2PL^3}{3E\delta}}$$

Hence the option (a) is correct.

22. (5) Given that

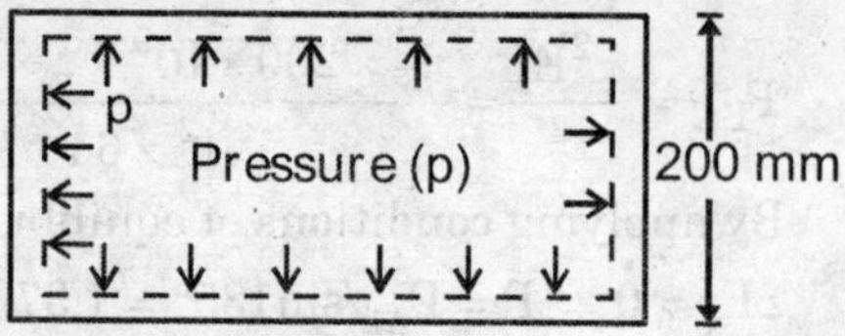

Mean Radius of cylinder

(r) = 100 mm

wall thickness (t) = 5 mm

factor of safety (N) = 2

yield stress in shear (sy_s) = 100 Mpa

we know,

$\sigma_{induced} \le \sigma_{permissible}$

$$\sigma_{Longitudinal} \le \frac{sy_s}{N}$$

$$\frac{Pd}{4t} \le \frac{sy_s}{N}$$

$$\frac{P \times 200}{4 \times 5} \le \frac{100}{2}$$

$$\boxed{P \le J\ Mpa}$$

therefore the maximum pressure inside the cylinder $\boxed{P_{ma\alpha} = 5\ Mpa}$

2016

(Q. 1 – 9) : Carry One Mark Each.

1. A single degree of freedom vibrating system has mass of 5 kg, stiffness of 500 N/m and damping coefficient of 100 N–s/m. To make the system critically damped

(*a*) only the mass is to be increased by 1.2 times.

(*b*) only the stiffness is to be reduced to half.

(*c*) only the damping coefficient is to be doubled.

(*d*) no change in any of the system parameters is required.

2. A "L" shaped robotic arm AB is connected to a motor at end A and a magnetic gripper at B as shown in the figure. If the arm is rotating with an angular velocity of 2 rad/s and an angular acceleration of 3 rad/s^2, the magnitude of the acceleration (in m/s^2) of the end B is

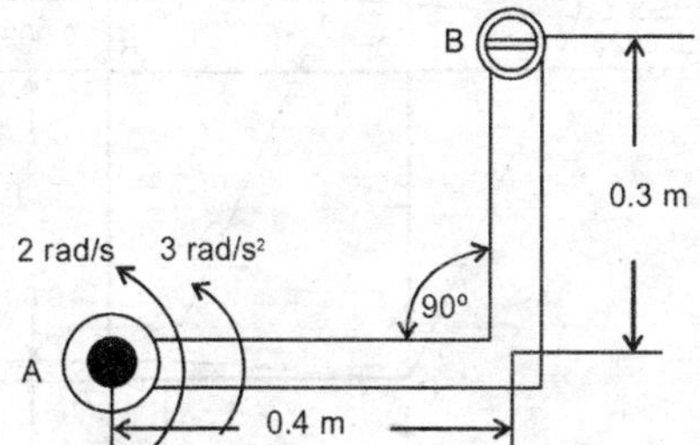

(*a*) 2.0 (*b*) 2.5

(*c*) 5.0 (*d*) 6.0

3. A block of weight 100 N is in static equilibrium on an inclined plane which makes an angle 15° with the horizontal. The coefficient of friction between the inclined plane and the block is 0.3. The magnitude of friction force (in N) acting on the block is ________

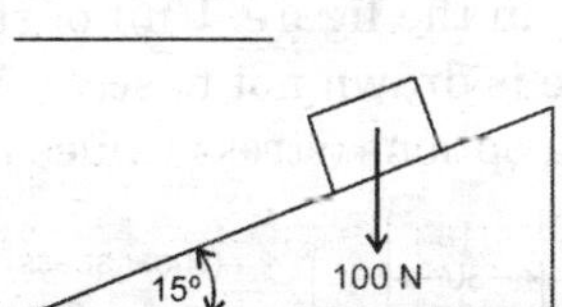

4. The lower end A of the rigid bar AB is moving horizontally on the floor towards right with a constant velocity of 5 m/s and the point B is sliding down the wall. The magnitude of the velocity of point B at the instant θ = 30° is

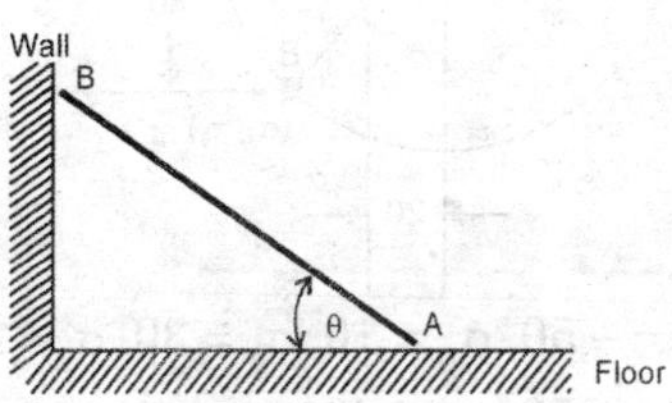

(*a*) zero (*b*) 4.34 m/s

(*c*) 7.25 m/s (*d*) 8.66 m/s

5. The state of plane stress at a point in a body is shown in the figure. The allowable shear stress of the material of the body is 200 MPa. According to the maximum shear stress theory of failure the maximum permissible value of σ (in MPa) is________

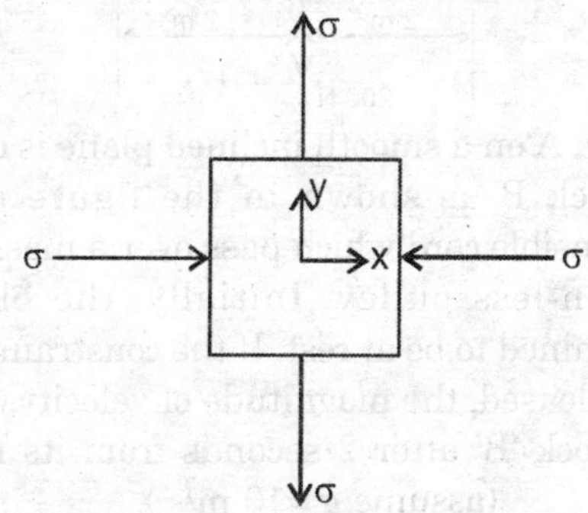

6. For a slender steel column of circular cross-section the critical buckling load is P_{cr}. If the diameter of the column is doubled (keeping other material and geometrical parameters same), then the critical buckling load of the column is

(*a*) $P_{cr}/16$ (*b*) 8 P_{cr}

(*c*) 2 P_{cr} (*d*) 16 P_{cr}

7. A closed thin cylindrical pressure vessel having an internal diameter of 1000 mm and a thickness of 10 mm is subjected to an internal pressure of 4 MPa. The maximum shear stress (in MPa) induced in the cylinder is ________ (neglect the radial stress).

8. A solid circular shaft subjected to pure torsion develops a maximum torsional shear stress of 120MPa. Keeping the torsional moment same, if the diameter of the shaft is doubled then the maximum shear stress (in MPa) induced in the shaft is ________.

9. On a single straight track, a vehicle of mass 500 kg moving with a velocity of 25 m/s strikes another vehicle of mass 250 kg moving with a velocity 10 m/s in the same direction. After the impact, if both the vehicles stick together, the common velocity (in m/s) with which both the vehicles will move together is ________.

(Q. 10 – 22) : Carry Two Marks Each.

10. A system with three forces and a concentrated moment at A is shown in the figure. The system is replaced by an equivalent force system with a single force and a single couple at point 'O'. The magnitude (in N-m) of the equivalent couple at 'O' is ________.

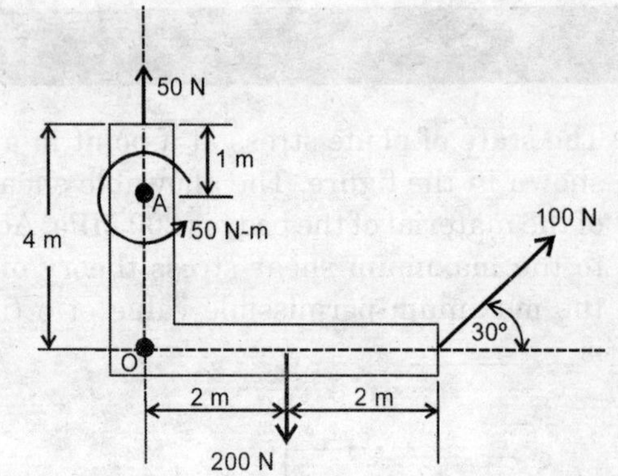

11. A block A on a smooth inclined plane is connected to block B as shown in the figure using an inextensible cord which pass over a mass-less and friction-less pulley. Initially, the block B is constrained to be at rest. If the constraint on block B is released, the magnitude of velocity (in m/s) of the block 'B' after 2 seconds from its release is ________ (assume g = 10 m/s²).

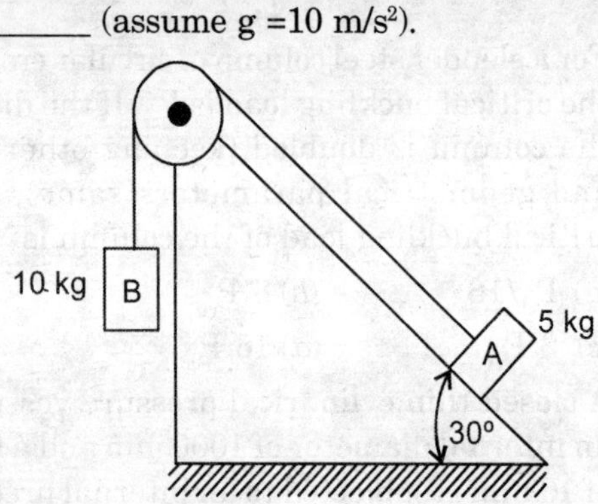

12. The vibrating system shown in the figure carries a mass of 10 kg at the free end, where the static deflection is 1 mm. This system is to be replaced by an equivalent vibrating spring mass system having equivalent mass of 2 kg (assume g = 10 m/s²). The natural frequency (in rad/s) and the stiffness (in kN/m) of the equivalent system respectively are.

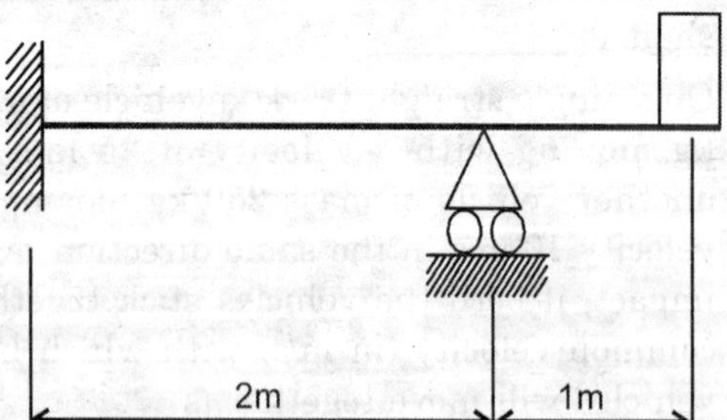

(*a*) 10 and 20 (*b*) 20 and 100
(*c*) 100 and 20 (*d*) 1000 and 20

13. A beam having flexural rigidity EI and length L is subjected to a concentrated end moment M_0 as shown in the figure. For EI = 4 × 10³ N-m², L = 1m and M_0 = 8 kN-m, the strain energy stored (in kN-m) in the beam and the rotation (in rad) at the free end respectively are

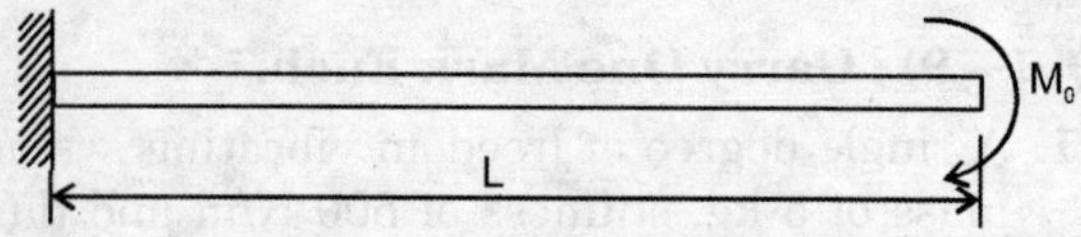

(*a*) 8.00 and 0.02 (*b*) 8.00 and 2.00
(*c*) 8.00 and 0.04 (*d*) 0.80 and 2.00

14. At a point 'O' on a metal sheet a square OABC of a unit side length is drawn. The square undergoes a small uniform elastic deformation and deforms to OA*B*C* (dashed lines) as shown in the figure. All dimensions are in mm and the figure is not to scale. The normal strains ε_x, ε_y and shear strain γ_{xy} developed in the square respectively are

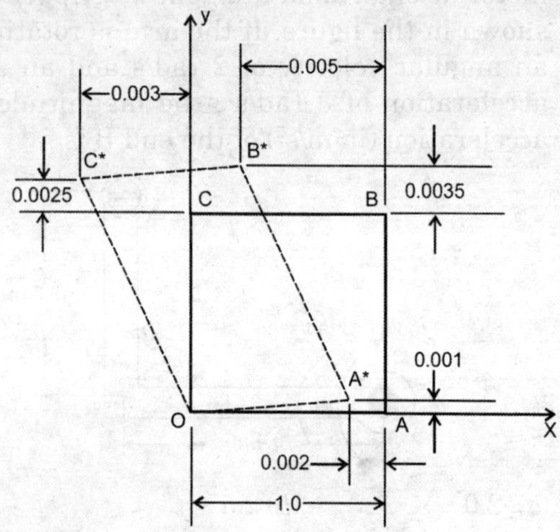

(*a*) – 0.0020, 0.0025 and 0.0020
(*b*) 0.0020, – 0.0025 and – 0.0020
(*c*) 0.0025, – 0.0020 and 0.0020
(*d*) – 0.0020, 0.0025 and – 0.0020

15. Mohr's circle for the state of plane stress at a point is shown in the figure. Unit of stress is MPa and the circle is drawn not to scale. Which one of the following options (stress values in MPa) is true?

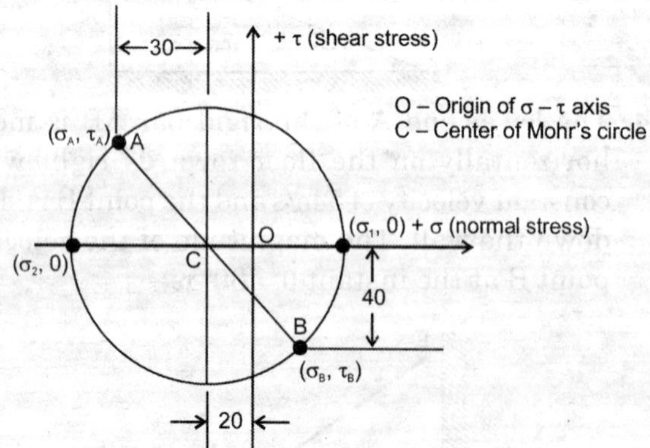

(*a*) $\sigma_A = -50, \sigma_B = 10, \sigma_1 = 30, \sigma_2 = -70$
(*b*) $\sigma_A = -50, \sigma_B = 20, \sigma_1 = 30, \sigma_2 = -50$
(*c*) $\sigma_A = -30, \sigma_B = 30, \sigma_1 = 30, \sigma_2 = -10$
(*d*) $\sigma_A = -20, \sigma_B = 10, \sigma_1 = 50, \sigma_2 = -30$

16. As shown in the figure, links AB and CD support the rigid member BD. Links AB and CD are made of aluminum alloy (E = 100GPa) and each has a cross-sectional area of $100 mm^2$. All the members are pin connected and all the dimensions are in mm. Neglecting the weights of the members, the elongation (in mm) of the link AB is__________.

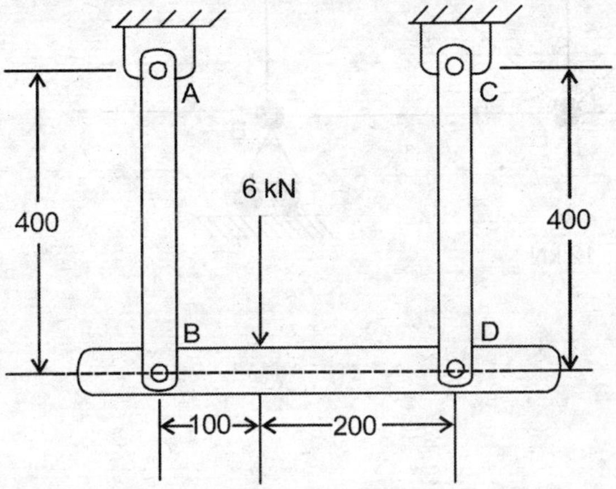

17. Figure shows an elastic beam of constant flexural rigidity EI and length L. The transverse deflection v(x) for the beam is represented by the equation $v(x) = M_0 (x^3 - x^2L)/(4EI\,L)$, where M_0 is the applied couple. If L = 100 mm and M_0 = 100N-mm, then the magnitude of the shear force (in N) at the middle of the beam (at x = L/2) is__________.

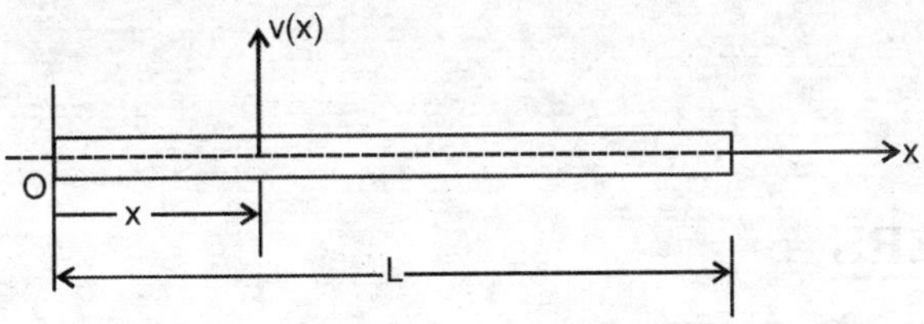

18. Which one of the following represents the correct bending moment diagram of the beam PQR loaded as shown in the figure?

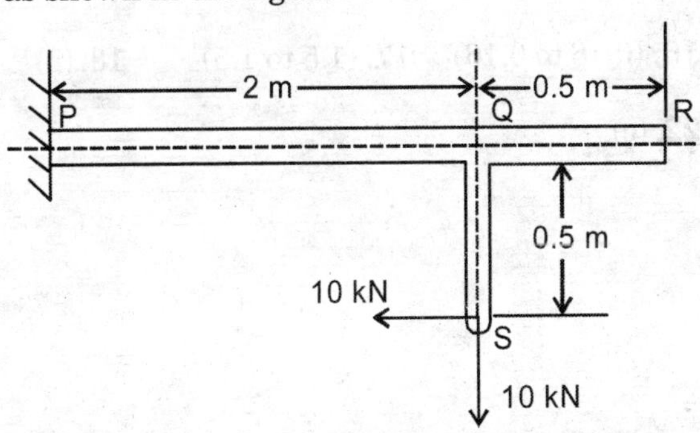

(*a*)

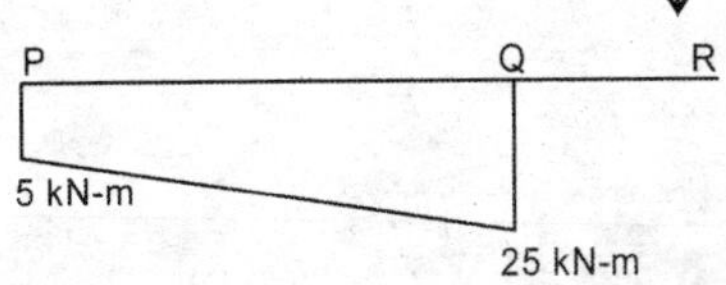

(*b*)

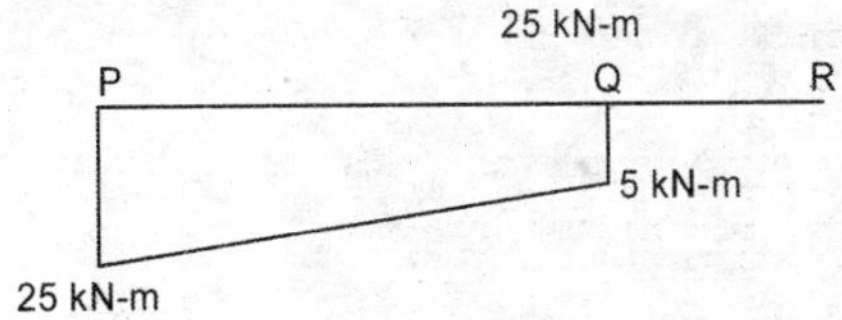

(*c*)

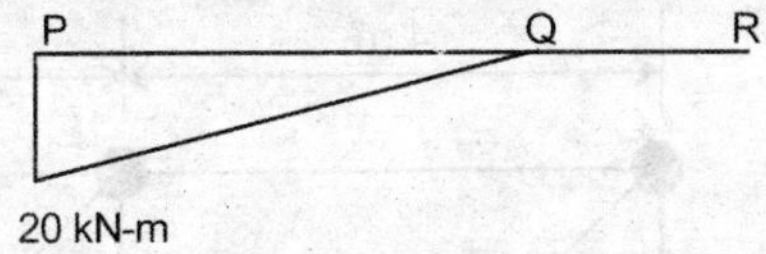

(*d*)

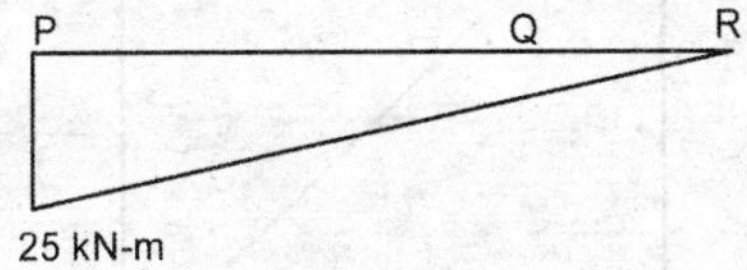

19. A point in a body is subjected to plane state of stress in XY plane. If σ_x = 140MPa, σ_y = 60 MPa and the major principal stress is 150 MPa, the magnitude of the in-plane shear stress τ_{xy} (in MPa) is

(*a*) 75 (*b*) 30

(*c*) 40 (*d*) 70

20. A 40 mm diameter rotor shaft of a helicopter transmits a torque T = 0.16 π kN-m and a tensile force P = 24 π kN. The maximum tensile stress (in MPa) induced in the shaft is __________. Use the value of π = 3.1416.

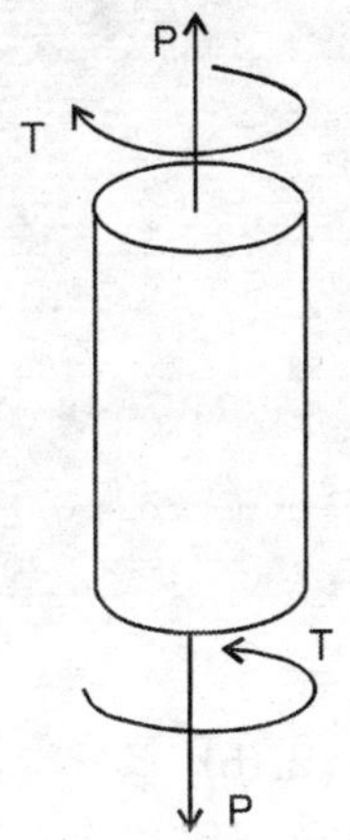

21. A wooden block of length 400 mm, width 50 mm and depth 100 mm is subjected to uniaxial load as shown in the figure. An inclined plane ABCD is shown which makes an angle θ with the XZ plane and the line CD is parallel to the Z-axis. The normal stress on the plane ABCD is a σ_{n1} when θ = 30° and the normal stress on the plane ABCD is σ_{n2} when θ = 120°. The value of $\frac{\sigma_{n2}}{\sigma_{n1}}$ is ________.

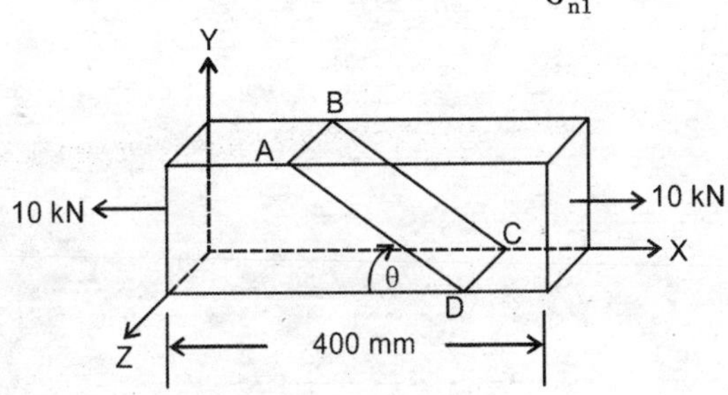

22. For the truss shown in the figure, which one of the following statements is true?

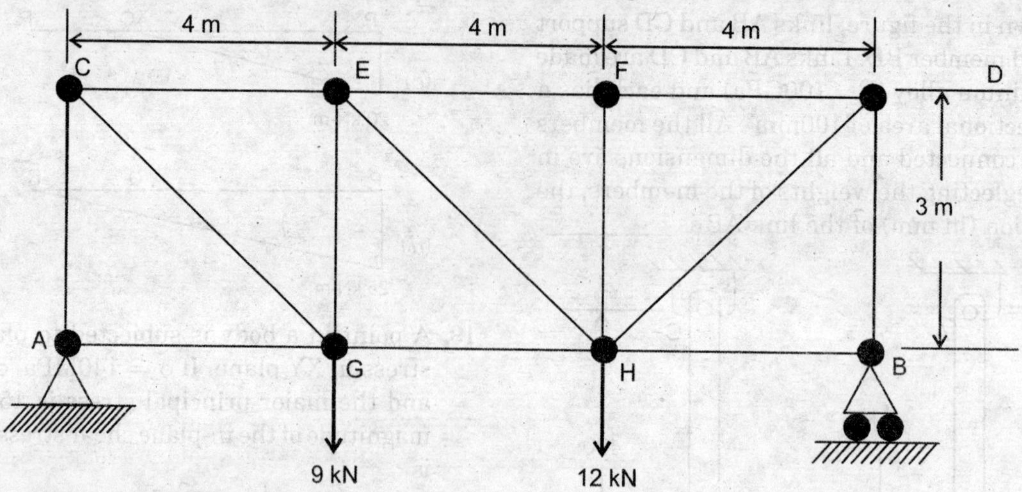

(*a*) AG is the only zero force member

(*b*) AG and BH are the only two zero force members

(*c*) AG, BH and HF are zero force members

(*d*) AG, BH, HF and GC are zero force members

ANSWERS

1. (d)	**2.** (b)	**3.** (25.5 to 26.0)	**4.** (d)	**5.** (200 to 200)	**6.** (d)
7. (100 to 100)	**8.** (15 to 15)	**9.** (20 to 20)	**10.** (150 to 150)	**11.** (10 to 10)	**12.** (c)
13. (b)	**14.** (d)	**15.** (a)	**16.** (0.16 to 0.16)	**17.** (1.5 to 1.5)	**18.** (b)
19. (b)	**20.** (80 to 80)	**21.** (3 to 3)	**22.** (c)		

EXPLANATIONS

1. (d) Given that

Mass(m) = 5 kg

Stiffness of spring (s) = 500 N/m

Damping Coefficient (c) = 100 N.s/m

We know that

$$\text{Damping factor}(z) = \frac{c}{2\sqrt{5m}} \qquad ...(i)$$

But for critical damping

$z = 1, \;\&\; c = C_c$

$\therefore$ from equation (i), we can write

$$1 = \frac{C_c}{2\sqrt{500 \times 5}}$$

$$1 = \frac{C_c}{2 \times 5 \times 10}$$

$\therefore \boxed{C_c = 100 \text{ N.s/m}}$ [Hence $C_c = C]_1$

Therefore, to make the system critically damped these should be no change in the system parameter.

Hence, the option (d) is correct.

2. (b)

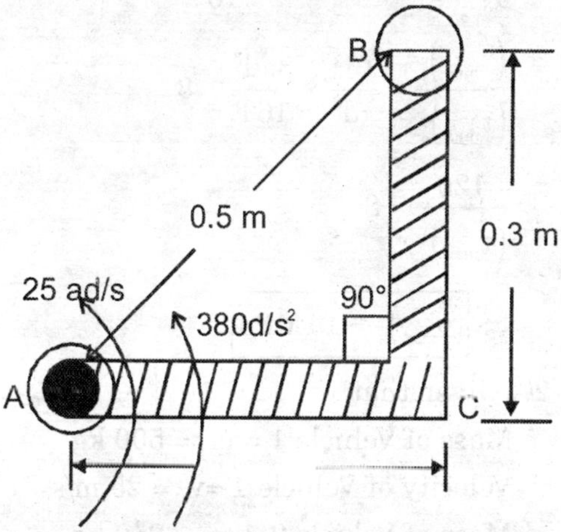

Given angular velocity of arm (w) = 2 rad/sec

Angular acceleration of arm (d) = 3 rad/sec²

$$\because AB = \sqrt{(AC)^2 + (BC)^2} = \sqrt{(0.4)^2 + (0.3)^2}$$

$= 0.5$ m

Now, Tangential Acceleration of arm

$$(a^t_{BA}) = \alpha AB \cdot (BA)$$

$= 3 \times 0.5 = 1.5$ m/s²

Radial Acceleration of arm

$$(a^r_{BA}) = \frac{V^2_{AB}}{BA} \qquad ...(i)$$

$\because V_{AB} = w_{AS} \times A_B = 2 \times 0.5 = 1$ m/s

from (i), we get

$$a^r_{BA} = \frac{1^2}{0.5} = 2 \text{ m/s}^2$$

$\therefore$ Resultant Accelaration of arm (a_{net})

$$= \sqrt{\left(a^t_{BA}\right)^2 + \left(a^r_{BA}\right)^2}$$

$$a_{net} = \sqrt{(1.5)^2 + (2)^2} = 2.5 \text{ m/s}^2$$

$\boxed{a_{net} = 2.5 \text{ m/s}^2}$

Hence the option (b) is correct

3. (25.88)

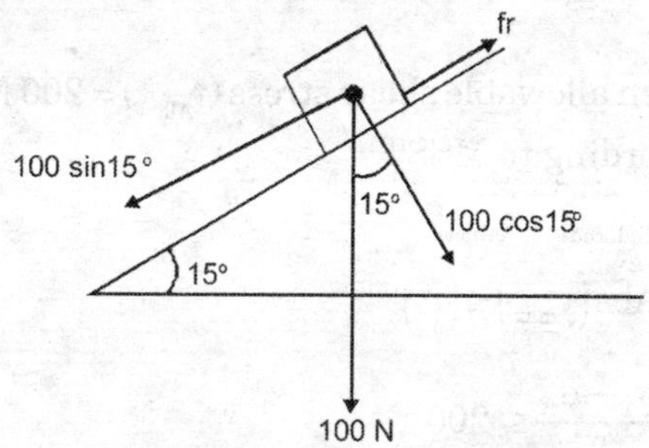

Now applying Newton's 2nd law

$f = ma$

$fr - 100 \sin 15 = 0$ [$fr \rightarrow$ friction force]

$fr = 100 \sin 15 = 25.88$ N

$\boxed{fr = 25.88 \text{ N}}$

Therefore Magnitude of friction force (fr) = 25.88 N

4. (d)

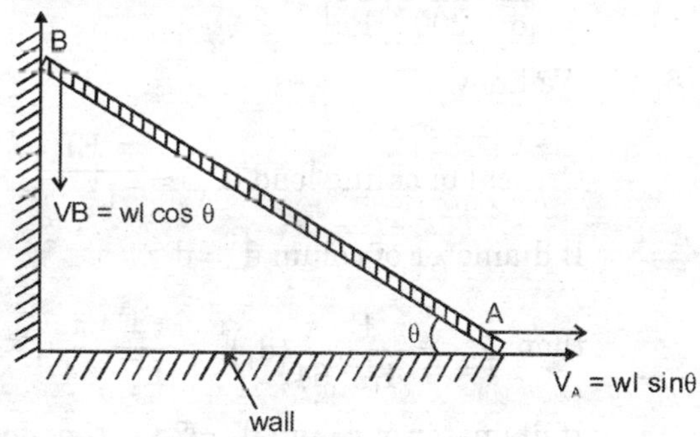

Given θ = 30°

Velocity of lower end A (V_A) = 5 m/s

Now, $V_A = wl \sin q$

$5 = w \times l \times \sin 30°$

$$5 = wl \cdot \frac{1}{2}$$

$\boxed{wl = 10}$ ...(i)

Velocity of end B(V_B) = wl cos30°

$$\therefore V_B = 10 \times \frac{\sqrt{3}}{2} \quad [wl = 10 \text{ from (i)}]$$

$$\boxed{V_B = 8.66\ m/s}$$

Hence the option (d) is correct.

5. (200)

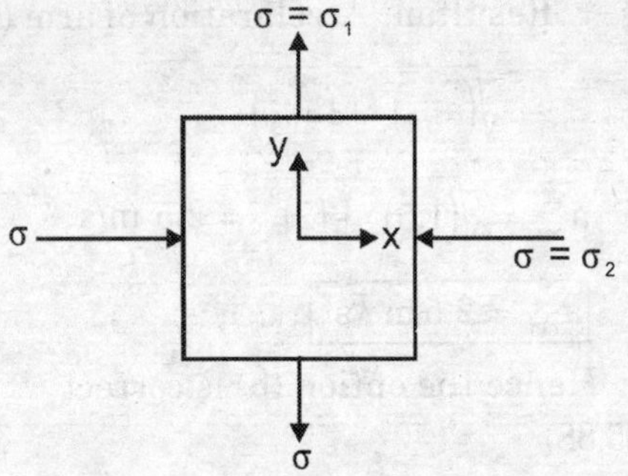

Given allowable shear stress (τ_{Allow}) = 200 MPa

According to MSST.

$$\tau_{Induced} \le \tau_{Allow}$$

$$Abs(\tau_{max}) \le (\tau_s)_{Allow}$$

$$\frac{\sigma_1 - \sigma_2}{2} \le 200$$

$$\frac{\sigma - (-\sigma)}{2} \le 200$$

$$\frac{7\sigma}{2} \le 200$$

$$\boxed{\sigma = 200\ MPa}$$

Therefore Maa^n permissible value of

$$\boxed{\sigma = 200\ MPa}$$

6. (d) We know

Critical buckling load (P_G) = $\dfrac{\pi^2 EI}{L_c^2}$

If diameter of colum $d_1 = d$

$$\text{then } P_{cr_1} = \frac{\pi^2 E}{L_c^2}.\frac{\pi}{64}(d_1)^4 = \frac{\pi^2 E}{L_c^2}.\frac{\pi}{64}d^4 \quad ...(i)$$

If diameter of colum $d_2 = 2d$ (i.e. double)

$$\text{then } P_{cr_2} = \frac{\pi^2 E}{L_c^2}.\frac{\pi}{64}(d_2)^4 = \frac{\pi^2 E}{L_c^2}.\frac{\pi}{64}(2d)^4 \quad ...(i)$$

$$= 16.\frac{\pi^2 E}{L_c^2}.\frac{\pi}{64}d^4$$

$$\boxed{(P_{cr})_2 = 16(P_{cr})_1}$$

Hence the option (d) is correct

7. (100) Given that

Internal diameter of cylinder (d) = 1000 mm

Thickness of cylinder (t) = 10 mm

Internal pressure (p) = 4 MPa

$\therefore$ Maximum shear stress(τmax) = $\dfrac{pd}{4t}$

$$\tau_{max} = \frac{4 \times 1000}{4 \times 10}$$

$$\boxed{\tau_{max} = 100\ MPa}$$

8. (15) Given that

Maximum torsional shear stress = $(t_{max})_1$

= 120 MPa

Torsional Moment = Constant

i.e. T = C

$$\text{We know } (\tau_{max})_1 = \frac{T}{z_1} = \frac{T}{\frac{\pi}{16}d^3} \Rightarrow \frac{16T}{\pi d^3} \quad ...(i)$$

If $d_2 = 2d$

$$\text{then } (\tau_{max})_2 = \frac{T}{z_2} = \frac{T}{\frac{\pi}{16}(2d)^3} = \frac{16T}{8\pi d^3} \quad ...(ii)$$

$$\frac{(\tau_{max})_1}{(\tau_{max})_2} = \frac{161}{\pi d^3} \times \frac{8\pi d^3}{16T} = 8$$

$$\frac{120}{(\tau_{max})_2} = 8$$

$$\Rightarrow \boxed{(\tau_{max})_2 = 15\ MPa}$$

9. (20) Given that

Mass of Vehicle 1 = m_1 = 500 kg

Velocity of Vehicle 1 = v_1 = 25 m/s

Mass of Vehicle 2 = m_2 = 250 kg

Velocity of Vehicle 2 = v_2 = 10 m/s

By applying law of conservation of linear momentum

$$m_1v_1 + m_2v_2 = (m_1 + m_2)V$$

$$(500 \times 25) + (250 \times 10) = (500 + 250) \times v$$

$$v = \frac{15000}{750}$$

$$\boxed{v = 20\ m/s}$$

Hence the common velocity of vehicles stick together is $\boxed{v = 20\ m/s}$

10. (150)

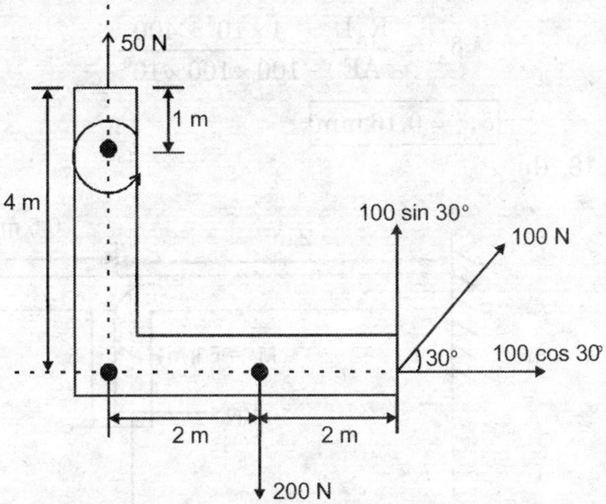

Taking moment about point'O'

$$M_o = 100\sin 30° \times 4 - (200 \times 2) + 50$$

$$= (50 \times 4) - 400 + 50 = -150 \text{ N-m}$$

Therefore the magnitude of net moment at 'O' is

$$\boxed{M_o = 150 \text{ N.m}}$$

11. (10)

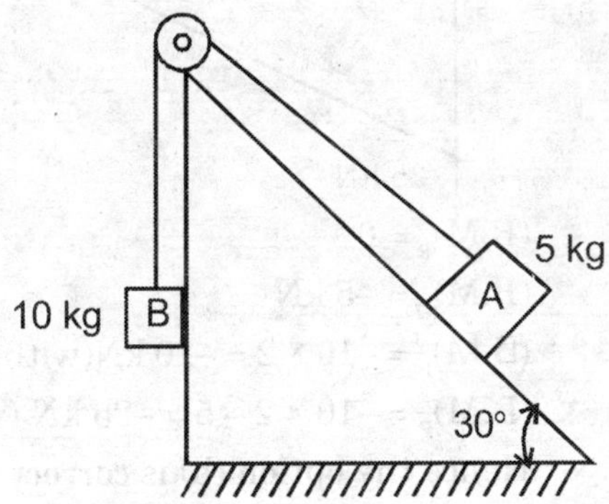

Free Body diagram of Block 'A'

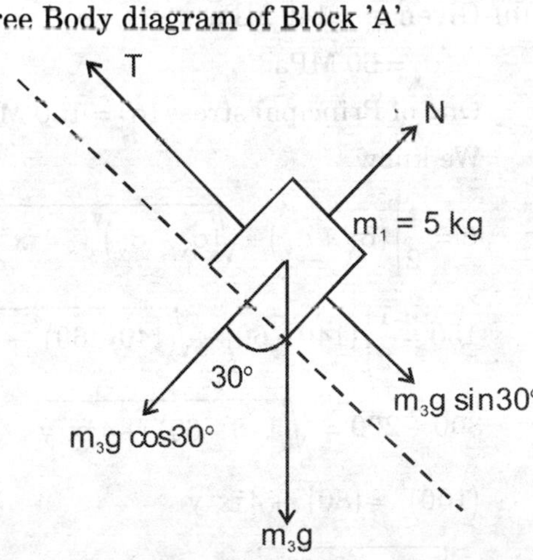

Applying Newton's 2nd low

$$T - m_1 g \sin 30° = 5a$$

$$T = 5a + m_1 g \sin 3a$$

$$= 5a + \left(5 \times 10 \times \frac{1}{2}\right) \quad [\because g = 10 \text{ m/s}^2]$$

$$T = 5a + 25 \quad ...(i)$$

Now FBD of block 'B'

Applying Newton's 2nd law

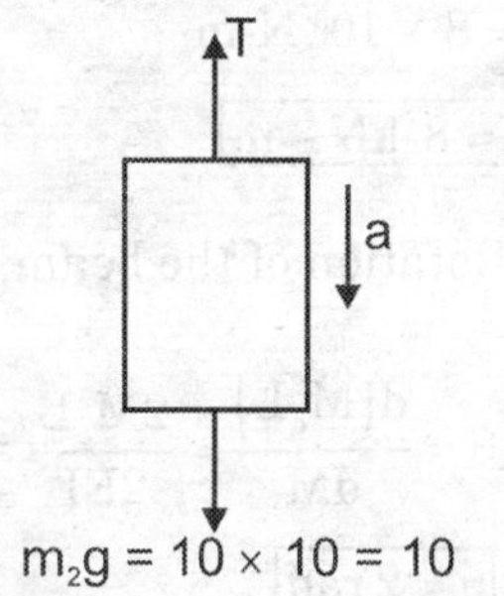

$$m_2 g = 10 \times 10 = 10$$

$$100 - T = 10 \times a \quad ...(ii)$$

Solving equation (i) & (ii) we get

$$a = 5 \text{ m/s}^2$$

We know $v = at = 5 \times 2$

$$\boxed{v = 10 \text{ m/s}}$$

$\therefore$ Velocity of block 'B' = (V_B) = 10 m/s

12. (c)

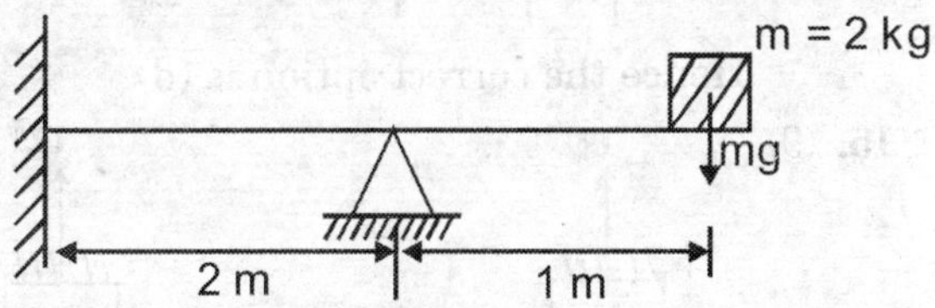

If the system is replaced by an equivalent vibrating spring mass system then

Net force $\quad F_s = mg$

$$kx = mg$$

$$k \times 1 \times 10^{-3} = 2 \times 10 \quad [x = 1 \text{ mm given}]$$

$$\boxed{k = 20 \text{ kN}}$$

Therefore stiffness of the spring k = 20 kN

Now natural frequency $(w_n) = \sqrt{\frac{s}{m}} = \sqrt{\frac{k}{m}}$

$$w_n = \sqrt{\frac{20 \times 10^3}{2}}$$

$$\boxed{w_n = 100 \text{ rad/sec}}$$

Hence the option (c) is correct

13. (b)

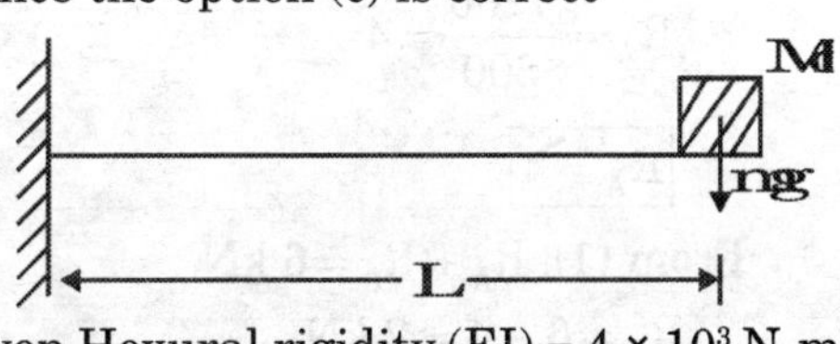

Given Hexural rigidity (EI) = 4×10^3 N-m^2

Length of the beam (L) = 1 m

Applied moment (M_0) = 8 KN –m

$$= 8 \times 10^3 \text{ N-m}$$

We know

$$\text{Strain energy (u)} = \int \frac{M_o^2 dx}{2EI} = \frac{M_o^2 .L}{2EI}$$

$$= \frac{8^2 \times 10^6 \times 1}{2 \times 4 \times 10^3}$$

$U = 8 \times 10^3$ N-m

$\boxed{U = 8\ \text{kN} - \text{m}}$

$\because$ Rotation of the beam $\theta = \dfrac{dU}{dM}$

$$\theta = \frac{d(M_o^2 L)}{dM} = \frac{2M_o L}{2EI} = \frac{28 \times 10^3 \times 1}{4 \times 10^3}$$

$\boxed{\theta = 2\,\text{rad}}$

Hence the option (b) is correct

14. (d) $\varepsilon_x = \dfrac{\Delta l}{l} = \dfrac{-0.002}{1} = -0.002$

$$\varepsilon_y = \frac{\Delta b}{b} = \frac{0.0025}{1} = 0.0025$$

$$\gamma_{xy} = \frac{\delta}{l} = \frac{\Delta l}{l} = \frac{0.002}{1} = -0.002$$

Hence the correct option is (d)

15. (9)

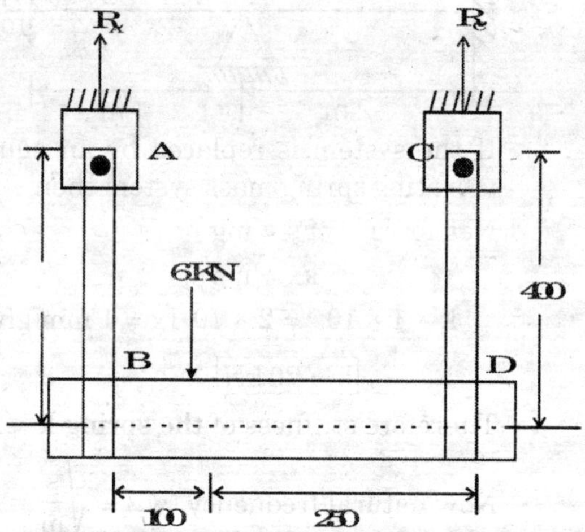

$R_A + R_C = 6$ kN...(i)

Taking moment about 'D'.

$R_A \times 300 - 6 \times 200 = 0$

$RA \times 300 = 1200$

$$R_A = \frac{1200}{300} = 4$$

$\boxed{R_A = 4}$

From (1), $R_A + R_C = 6$ kN

$R_C = 6 - 4 = 2$ kN

$\boxed{R_C = 2\,\text{kN}}$

$\therefore$ elongation of the link AB will be

$$\delta_{AB} = \frac{R_A L}{AE} = \frac{4 \times 10^3 \times 400}{100 \times 100 \times 10^3}$$

$\boxed{\delta_{AB} = 0.16\,\text{mm}}$

18. (b)

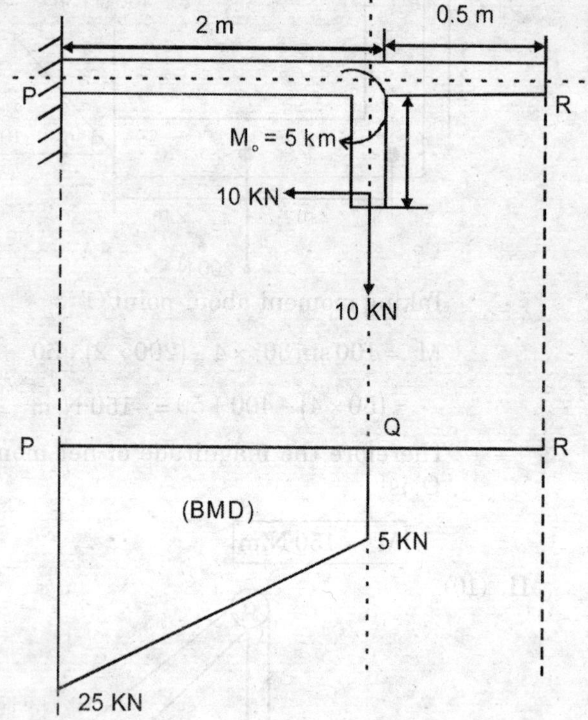

$(B.M)_R = 0$

$(B.M)_Q = -5$ kN

$(B.M)_P = -10 \times 2 = -20$ kN(without M_o)

$(B.M)_P = -10 \times 2 - 5 = -25$ kN (with M_o)

Hence the option (b) is correct

19. (b) Given $\sigma_x = 140$ MPa

$\sigma_y = 60$ MPa

One of Principal stress (σ) = 150 MPa

We know

$$\sigma = \frac{1}{2}\left[(\sigma_x + \sigma_y) + \sqrt{(\sigma_x - \sigma_y)^2 + 4\tau x^2 y}\right]$$

$$150 = \frac{1}{2}\left[(140 + 60) + \sqrt{(140 - 60)^2 + 4\tau x^2 y}\right]$$

$$300 - 200 = \sqrt{(140 - 60)^2 + 4\tau x^2 y}$$

$$(100)^2 = (80)^2 + 4\tau x^2 y$$

$\therefore \boxed{\tau_{xy} = 30\,\text{MPa}}$

Hence the option (b) is correct

20. (80)

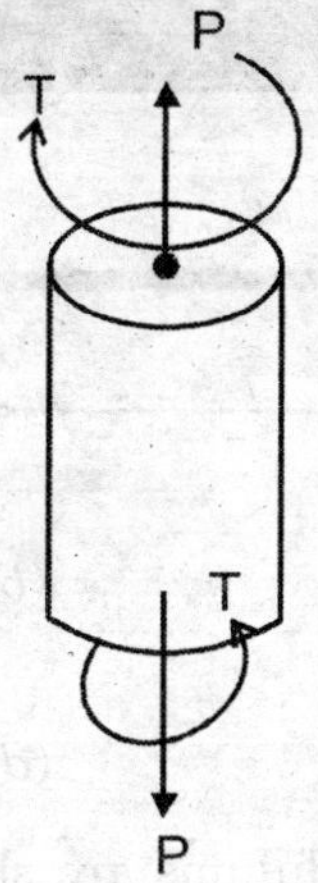

We know

Shear stress developed in the shat due to dorpue 'T'

$$\tau_{max} = \frac{T}{z} = \frac{0.16 \times \pi \times 10^3}{\frac{\pi}{16}(40 \times 10^{-3})^3} \quad \left[z = \frac{\pi}{16}d^3\right]$$

$$\tau_{max} = 40 \text{ MPa}$$

Now axial stress developed in the shaft due to axial load 'P'.

$$\sigma_{max} = \frac{P}{A} = \frac{24 \times \pi \times 10^3}{\frac{\pi}{4}(40)^2}$$

$$= 60 \text{ MPa}$$

Therefore maximum tensile stress induced in the shaft

$$\sigma_1 = \frac{1}{2}\left[\sigma + \sqrt{(\sigma_{axial})^2 + 4\tau^2_{max}}\right]$$

$$= \frac{1}{2}\left[60 + \sqrt{(60)^2 + 4(40)^2}\right]$$

$$\boxed{\sigma_T = 80 \text{ MPa}}$$

21. (3) Given load (P) = 10 kN

Length of block (1) = 400 mm

We know that

Normal stress on an oblique plane

$$\sigma_n = \sigma \cos^2 \theta$$

When θ = 30°, then

$$\sigma_m = \sigma \cos^2(90 - \theta) = \frac{P}{A}\cos^2 60°$$

$$= \frac{10 \times 10^3}{100 \times 50} \times \left(\frac{1}{2}\right)^2 = 0.5$$

When θ = 120°, then

$$\sigma_{x2} = \sigma \cos^2(-30)$$

$$= \frac{10 \times 10^3}{100 \times 50} \times \left(\frac{-\sqrt{3}}{2}\right)^2$$

$$= 2 \times \frac{3}{4} = 1.5$$

$$\boxed{\frac{\sigma_{x2}}{\sigma_{x1}} = \frac{1.5}{0.5} = 3}$$

2017

1. In the truss shown, a mass m = 10 kg is hung from the node J. The magnitude of net force (in Newtons) transferred by the truss EFGHIJ onto the truss JKLMNO at the node J is ______.
Assume acceleration due to gravity, $g = 10\ m/s^2$.

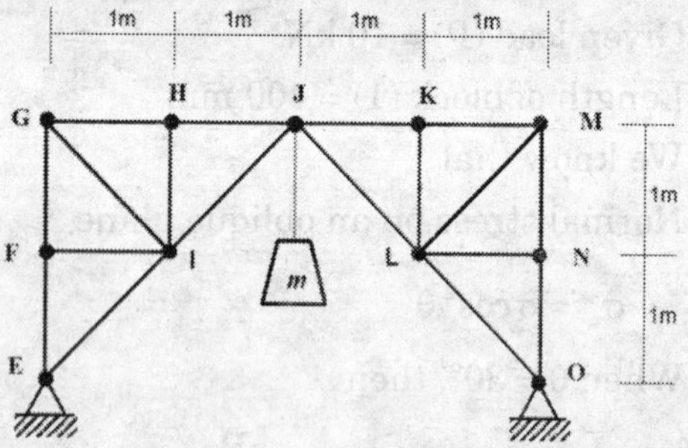

2. A ball moves along a planar frictionless slot as shown. Which one of the paths shown closely matches the path taken by the ball after it exists the slot at E?

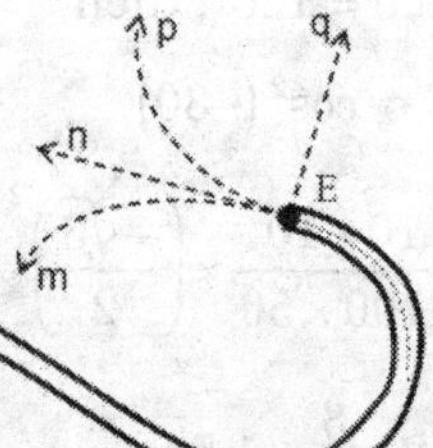

(a) path m (b) path n

(c) path p (d) path q

3. A rod EF moving in a plane has velocity V_E at E and V_F at F that are parallel to each other. Which of the following CANNOT be true?

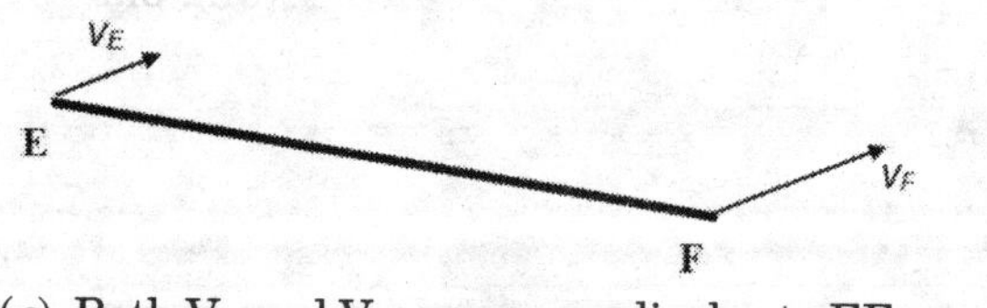

(a) Both V_E and V_F are perpendicular to EF.

(b) Magnitude of V_E is equal to the magnitude V_F and the angular velocity of EF is zero.

(c) The velocity V_E is not perpendicular to EF and the angular velocity of EF is nonzero.

(d) Magnitude of V_E is not equal to magnitude of V_F and the angular velocity of EF is

4. The beam shown below carries two external moments. A counter clockwise moment of magnitude 2M acts at point B and a clockwise moment of magnitude M acts at the free end, C. The beam is fixed at A. The shear force at a section close to the fixed end is equal to

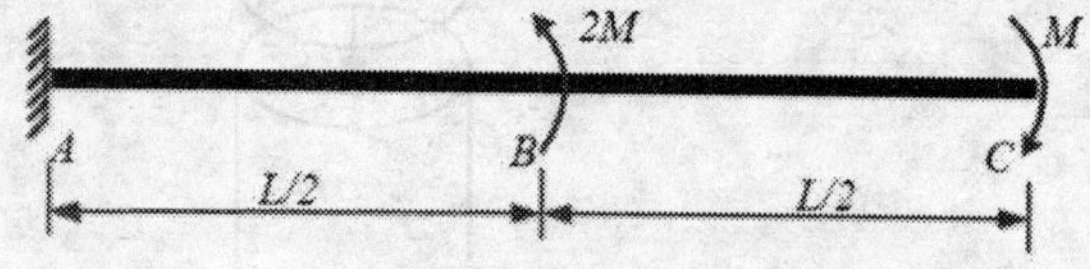

(a) $\frac{2M}{L}$ (b) $\frac{M}{L}$

(c) 0 (d) $-\frac{M}{L}$

5. Two pendulums are shown below. Pendulum-A carries a bob of mass m, hung using a hinged massless rigid rod of length L whereas Pendulum-B carries a bob of mass 4m and length L/4. The ratio of the natural frequencies of Pendulum-A and Pendulum-B is given by

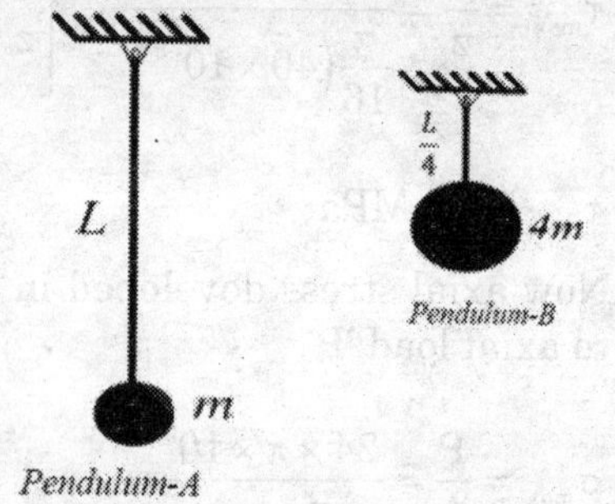

(a) 1 : 2 (b) 1 : 1

(c) $\sqrt{2} : 1$ (d) 2 : 1

6. A closed thin-walled cylindrical steel pressure vessel of wall thickness t = 1 mm is subjected to internal pressure. The maximum value of pressure p (in kPa) that the wall can withstand based on the maximum shear stress failure theory is given by

(Yield strength of steel is 200 MPa and mean radius of the cylinder r = 1 m).

(a) 100 (b) 60200

(c) 300 (d) 400

7. The state of stress at a point in a body is represented using components of stresses along X and Y directions as shown. Which one of the following represents the state of stress along X' and Y' axes? (X' – axis is at 45° clockwise with respect to X - axis)

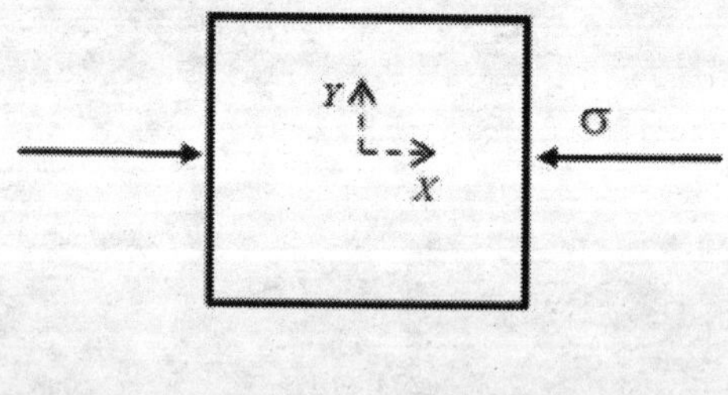

(a) 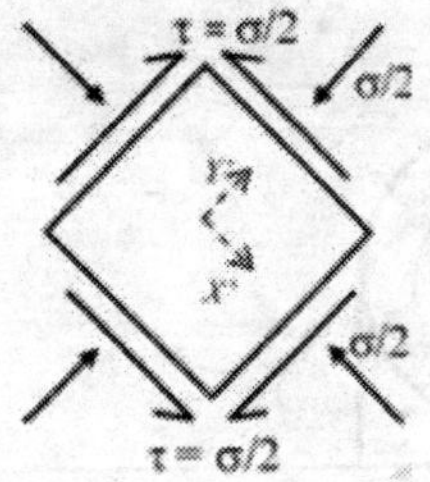

(b) 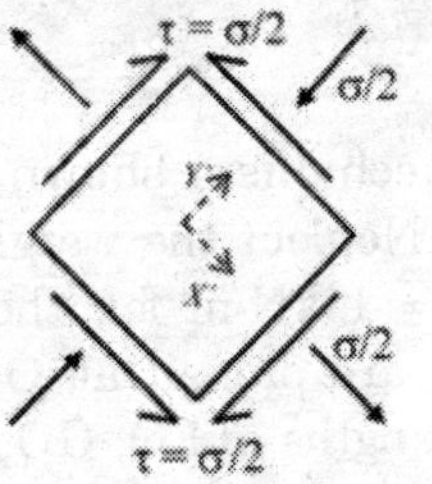

(c) 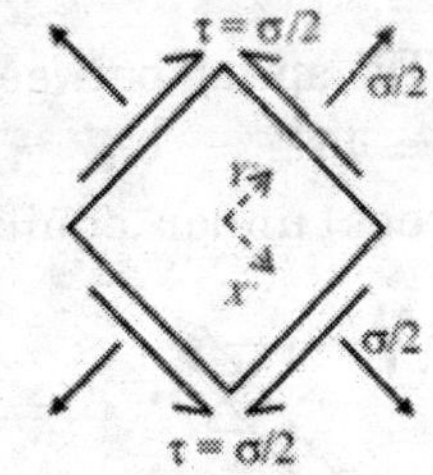

(d) 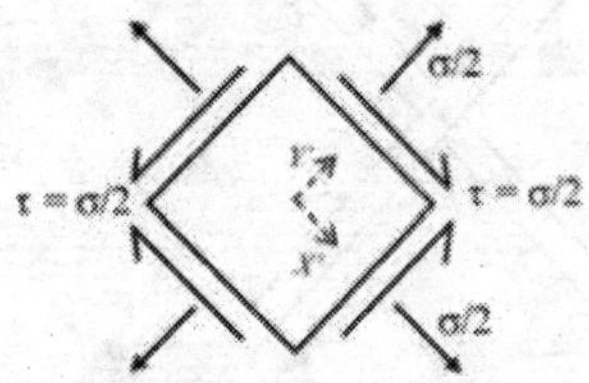

8. An aluminium specimen with a initial gauge diameter d_0 = 10 mm and a gauge length l_0 = 100 mm is subjected to tension test. A tensil force P = 50 kN is applied at the ends of the specimen as shown resulting in an elongation of 1mm in the gauge length. The Poisson's ratio (v) of the specimen is _______.

Shear modulus of the material G = 25GPa. Consider engineering stress-strain conditions.

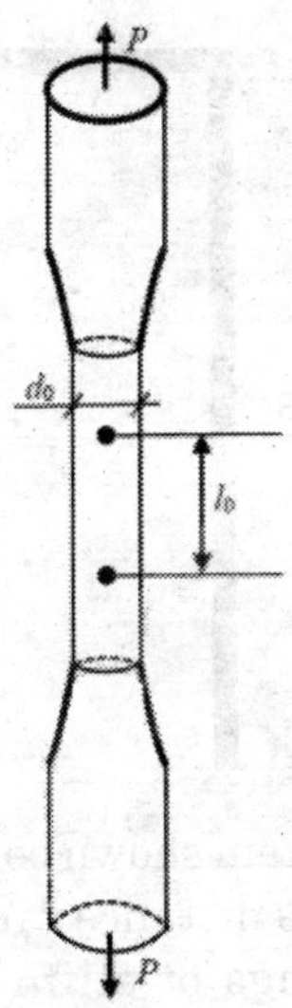

9. A rectangular sheet ABCD of dimensions a and b along X and Y directions, respectively, is stretched to a rectangle AB' C' D', as shown. The maximum principal strain (ε_1) and minimum principal strain (ε_2) due to the stretch are given by

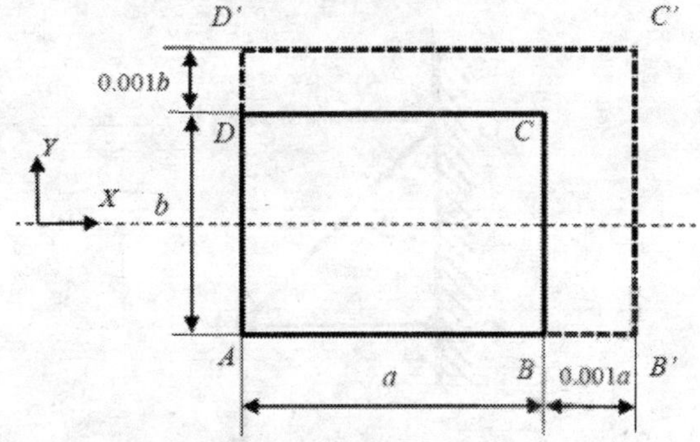

(a) $\varepsilon_1 = 0.001$ and $\varepsilon_2 = 0.001$

(b) $\varepsilon_1 = -0.001$ and $\varepsilon_2 = 0.001$

(c) $\varepsilon_1 = 0.001$ and $\varepsilon_2 = -0.001$

(d) $\varepsilon_1 = -0.001$ and $\varepsilon_2 = -0.001$

10. A solid bar of uniform square cross-section of side b and length L is rigidly fixed to the supports at the two ends. When the temperature in the rod is increased uniformly by Tc, the bar undergoes elastic buckling. Assume Young's modulus E and coefficient of thermal expansion α to be independent of temperature. The coefficient of thermal expansion α is given by

(a) $\dfrac{3\pi^2 b^2}{T_c L^2}$ (b) $\dfrac{\pi^2 b^2}{T_c L^2}$

(c) $\dfrac{\pi^2 b^2}{2T_c L^2}$ (d) $\dfrac{\pi^2 b^2}{3T_c L^2}$

11. Two rigid blocks, of masses 10 kg and 15 kg, are arranged one on top of the other and placed on a horizontal rough surface as shown. The blocks are connected to each other through an inextensible cable passing over a frictionless pulley. The coefficients of static friction between the blocks and also between the bottom block and the surface are all equal to 0.3. The force P (in Newtons) needed to set the blocks in motion towards right is _______.

Assume acceleration due to gravity g = 10 m/s²).

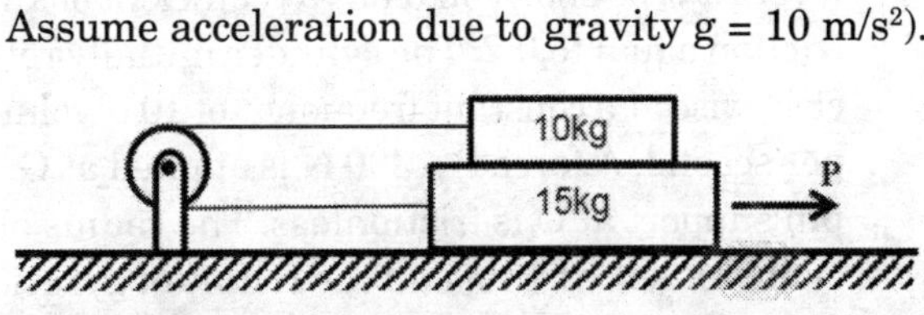

12. A truss system EFGH shown below is built using members EF, GH and FH of the same cross-sectional area 10 mm² and member FG of cross-sectional area 20 mm². The total strain energy stored (in Nm) in the system due to a force P = 1 kN acting at F is ______.

Assume elastic deformations and members are made of steel with elastic modulus of 200 GPa.

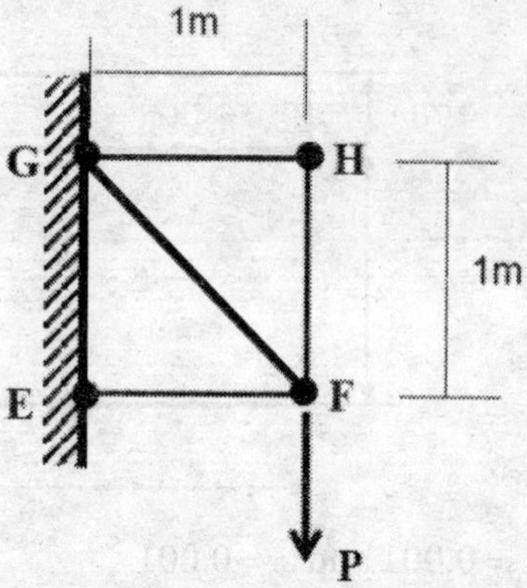

13. A rigid frame grips on to a steel wall as shown using a powerful magnet at the top support G and with a roller support at E. EF is horizontal. A man stands on the platform attached to the frame 1 m away from the wall as shown. Assume the frame and magnet assembly to be a negligible weight and the mass of the man to be 80 kg. The magnitude of the reaction (in Newtons) exerted by the frame onto the steel wall due to the weight of the man is ______.

The magnetic force of attractions of the magnet at no load condition is 1 kN. Magnet can be assumed to be small enough that it offers negligible moment resistance. Assume acceleration due to gravity, $g = 10\ m/s^2$.

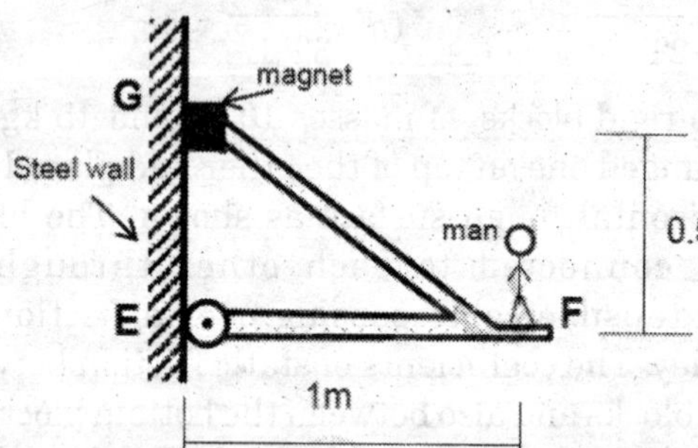

14. A manually operated band brake has a control lever EFG as shown and has a coefficient of kinetic friction equal to 0.2. The cylinder initially rotates clockwise at a constant frequency of 10 revolutions per second. A force P = 300 N is applied at G. The pin support at O is frictionless. The radius of the cylinder is r = 0.15m and the radius of gyration is 0.1m. The mass of the cylinder is 50 kg. Assume acceleration due to gravity $g = 10\ ms/^2$. The time required (in seconds) to reduce the rotational frequency to 5 revolutions per second is ______.

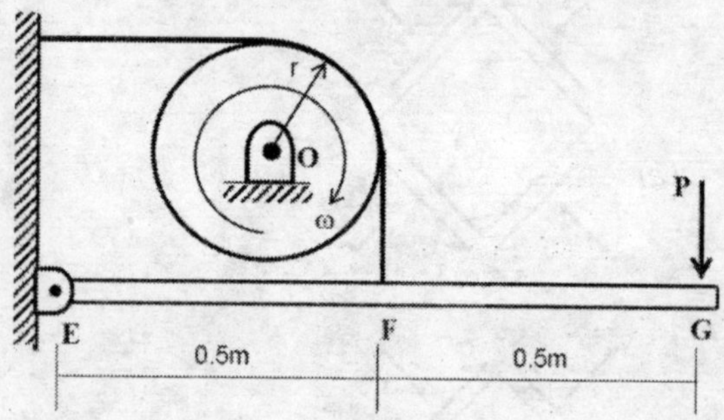

15. In a pin-connected mechanism shown, load P applied at F is 50 N. Neglect the weight of the links and assume k = 1 kN/m for the spring. The bars EH and FG are pinned at O at their centre such that the lengths of EO, GO, HO and FO are

all equal to ℓ= 0.2 m. The spring between G and H is unstretched when θ = 45°.

The angle θ (in degrees) under equilibrium is ______.

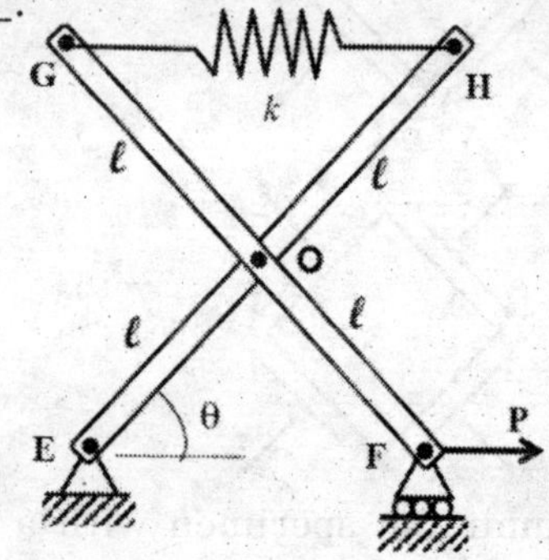

16. The frame shown below carries a vertical load P = 10 kN at its free and D. The frame is fixed at A and has a roller support at B. Magnitude of the reaction force at B (in kN) is______.

Assume tht the effect of the axial force on bending is negligible.

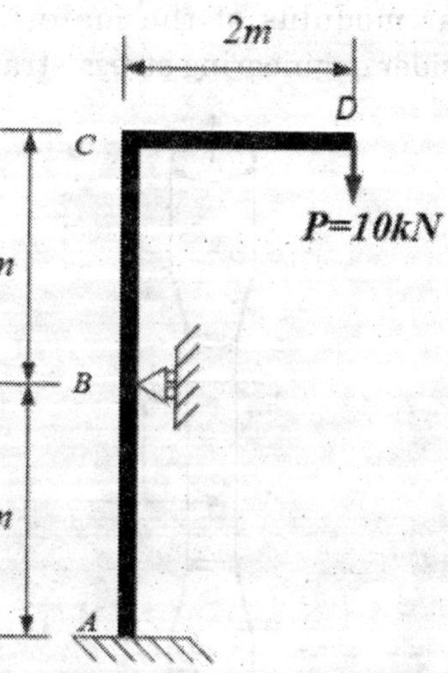

17. Consider the system shown below. Mass M is fixed to the rod AC at a distance x from the hinge point at B. Two springs of stiffness 3K and K are attached to the rod at points A and C, respectively.

The natural frequency of angular oscillation of the system about B is 20 rad/s. Assume the rod to be rigid and massless. Magnitude of x (in metres) is ______.

(M = 30 kg, and K = 1 kN/m).

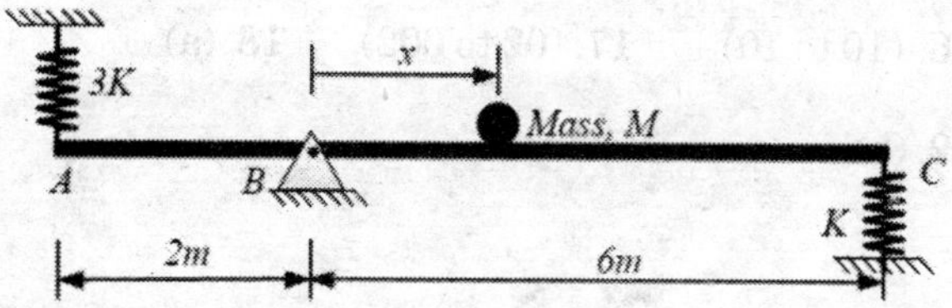

18. The simply supported beam shown below is subjected to a clockwise moment M at point A and two counter clockwise moments 2M and M at points B and C, respectively. Which one of the following is the correct bending moment diagram (tensile at bottom is positive moment) for the beam?

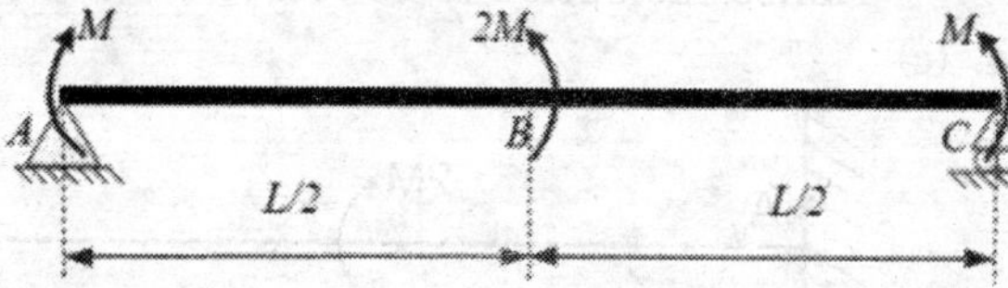

(a) 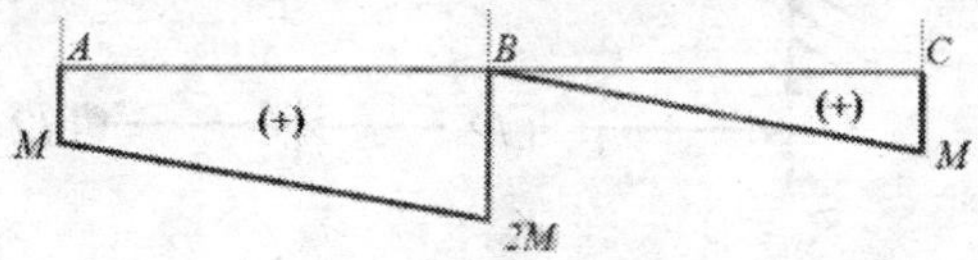

(b) 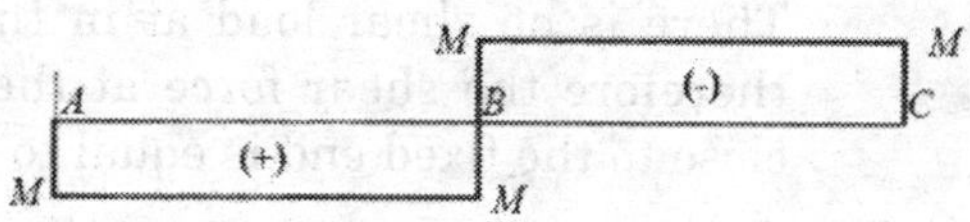

(c) 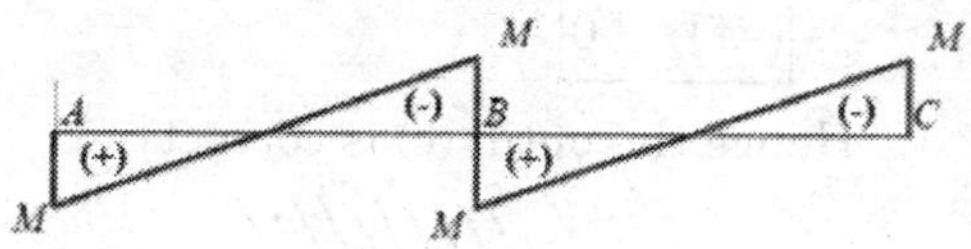

(d) 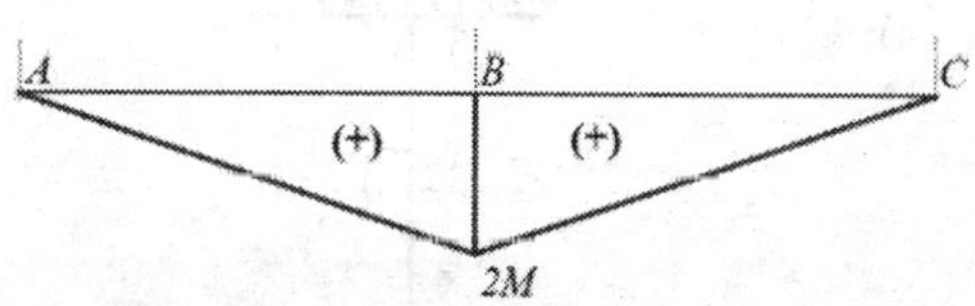

19. The structure shown below is of rectangular cross section and carries a load of 10kN at its free end E. Maximum bending stress (in MPa) developed in the beam due to external load is ______.

The depth of the beam is 300 mm and the width is 150 mm.

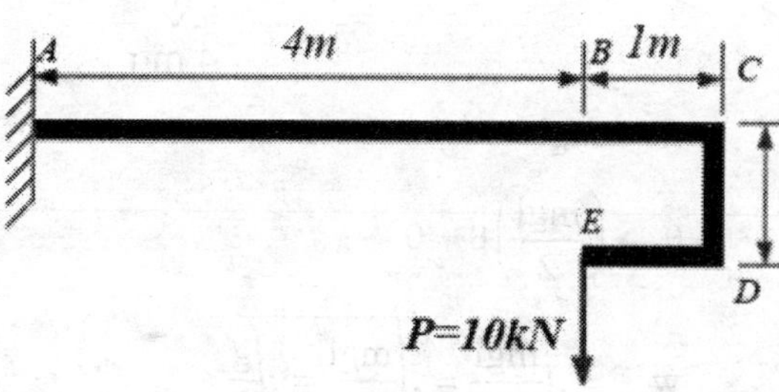

20. Two circular rods shown below carry the same axial load P. The Rod-A has uniform cross-section and the Rod-B has non-uniform cross-section as shown. The ratio of elongation of Rod-A to Rod-B is given by

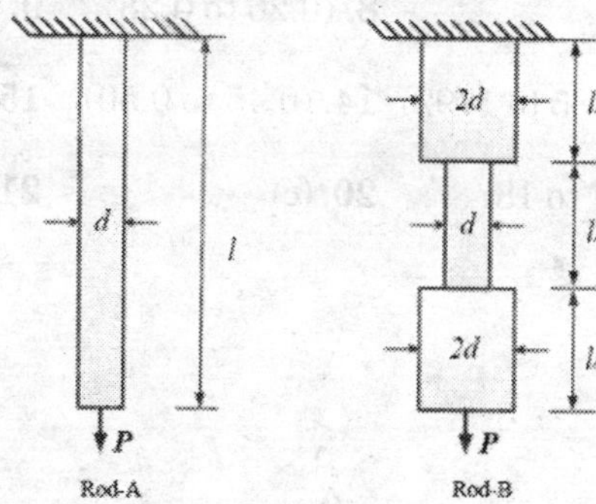

(a) 1 : 1 (b) 1 : 2

(c) 2 : 1 (d) 3 : 1

21. A composite shaft is made of a steel tube with an inner brass core perfectly bonded together as shown. The shaft is fixed at one end and subjected to a torque of 2T at the other end. Shear modulus of steel is G and that of brass is G/2. The outer radius of the steel tube is R = 2r and radius of the inner brass core is r. The magnitude of shear stress at the interface (point X) and in the steel tube is closest to

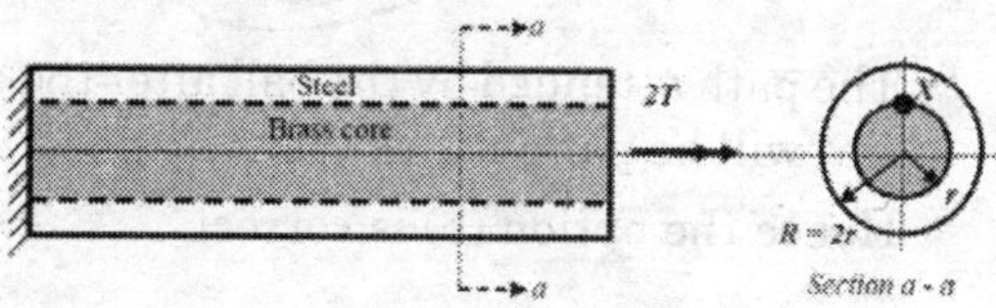

(a) $0.014\frac{T}{r^3}$ (b) $0.082\frac{T}{r^3}$

(c) $0.16\frac{T}{r^3}$ (d) $0.41\frac{T}{r^3}$

22. A massless rod of rectangular cross-section is subjected to a force P at origin O as shown. The expression for the stress σ_{zz} at point Q is given by

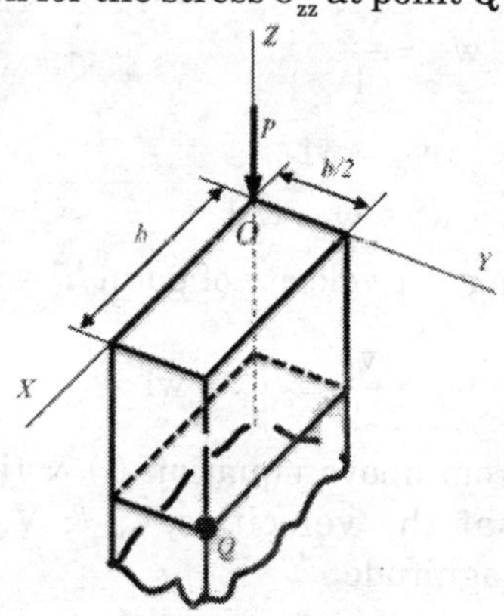

(a) $6\frac{P}{b^2}$ (b) $10\frac{P}{b^2}$

(c) $-14\frac{P}{b^2}$ (d) $-\frac{P}{b^2}$

ANSWERS

1. (69 to 73)	**2.** (b)	**3.** (c)	**4.** (c)	**5.** (a)	**6.** (b)
7. (a)	**8.** (0.26 to 0.28)	**9.** (a)	**10.** (d)	**11.** (134 to 136)	**12.** (0.55 to 0.65)
13. (1785 to 1792)	**14.** (0.45 to 0.50)	**15.** (32 to 35)	**16.** (10 to 10)	**17.** (02 to 002)	**18.** (a)
19. (17 to 18)	**20.** (c)	**21.** (b)	**22.** (b)		

EXPLANATIONS

2. (b)

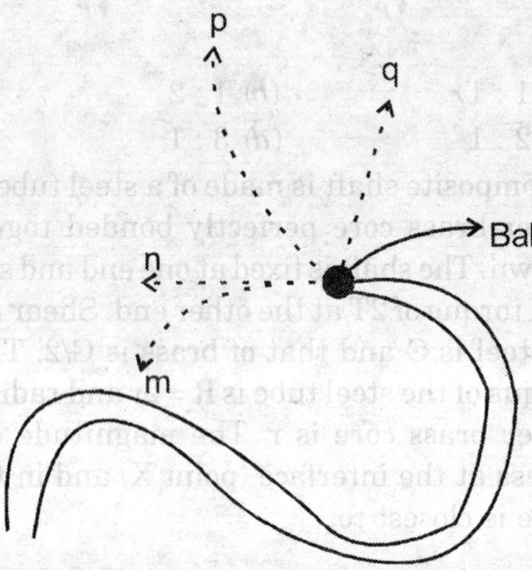

The path attended by the ball after the exit of slot will be path 'n'.

Hence the option (b) is correct.

3. (c)

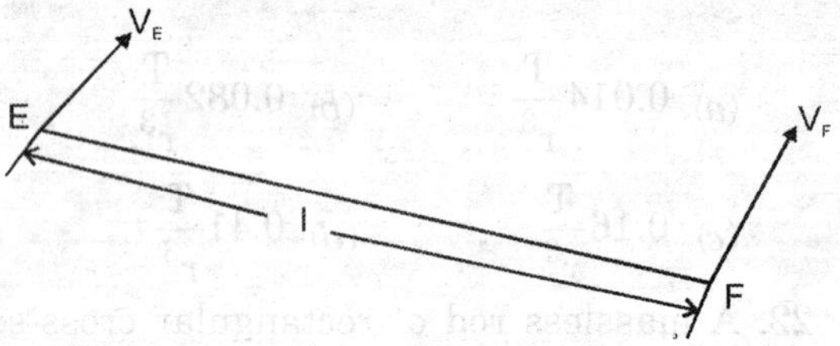

Agular velocity of point 'E' write to f

$$w_E = \frac{V_E}{l}$$

$$\Rightarrow V_E = wl \qquad \text{...(i)}$$

[Let $w_E = w = w_F$]

Angular velocity of point 'f' w.r.t E

$$w_F = \frac{V_F}{l} \Rightarrow V_F = wl \qquad \text{...(ii)}$$

From above equation (i) & (ii), we observe that the velocities V_E & V_F are same in magnitude.

Angular velocity $(W_{EF}) = W_E - E_F$

$$= \frac{V_E}{l} - \frac{V_F}{l}$$

$$= \frac{wl}{l} - \frac{wl}{l} \qquad [\because V_E = wl,\ V_F = wl]$$

$$= 0$$

Therefore Angular velocity of EF is zero.

Hence the option (c) is correct.

4. (c)

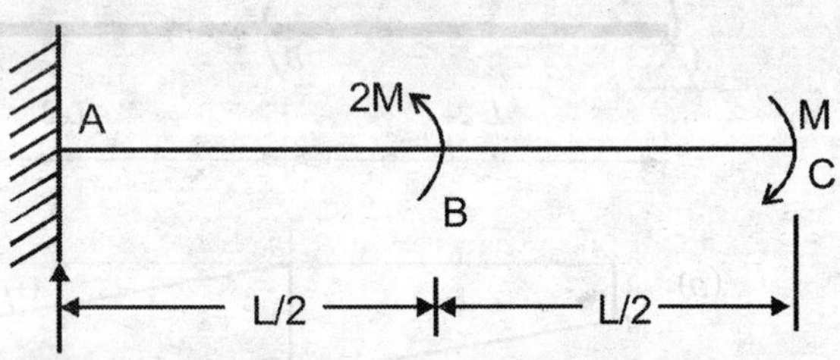

$R_A = 0$

$\because$ There is no shear load at in the beam, therefore the shear force at the section close to the fixed end is equal to zero.

$$\therefore \boxed{(s.f)_A = 0}$$

Hence the optin (c) is correct

5. (a)

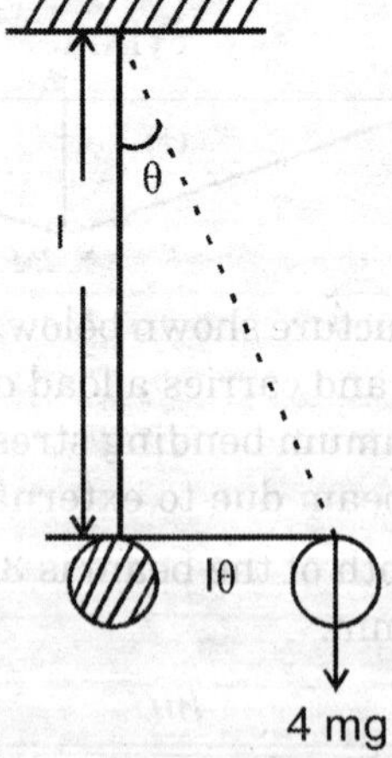

$$2\ddot{\theta} + mglo = \theta$$

$$\ddot{\theta} \to \left(\frac{mgl}{2}\right)\theta = 0$$

$$w_{n_A} = \sqrt{\frac{mgl}{2}} = \sqrt{\frac{mgl}{ml^2}} = \sqrt{\frac{g}{l}} \qquad \text{...(i)}$$

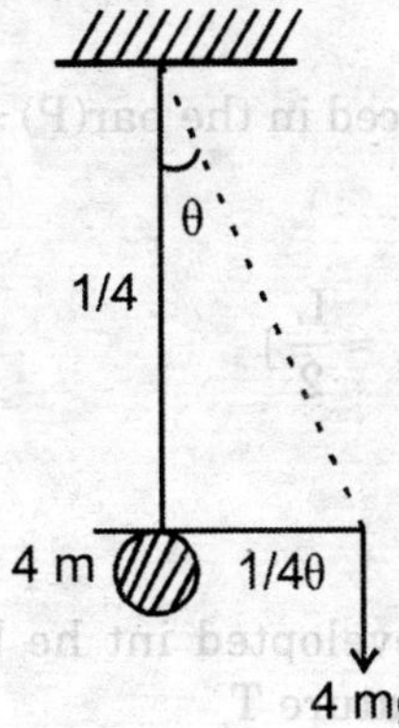

Taking moment about '0', we have

$$2\ddot{\theta}+\left(4mg\frac{1}{4}\theta\right)=0$$

$$\ddot{\theta}+\left(\frac{4mgl}{4I}\right)\theta=0 \qquad \left[\because I=4m\left(\frac{1}{4}\right)^2\right]$$

$$\therefore w_n=\sqrt{\frac{mgl}{I}}=\sqrt{\frac{mgl}{4m\left(\frac{1}{4}\right)^2}}=\sqrt{\frac{gl}{4\times\frac{l^2}{16}}}$$

$$w_{n_B}=\sqrt{\frac{4g}{l}} \qquad \text{...(ii)}$$

$$\frac{w_{n_A}}{w_{n_B}}=\frac{\sqrt{\frac{g}{l}}}{2\sqrt{\frac{g}{l}}}=\frac{1}{2}$$

$$\boxed{\frac{w_{n_A}}{w_{n_B}}=\frac{1}{2}}$$

Hence the option (a) is correct

6. (b) Given that

Thickness of cylinder (t) = 1 mm

Radius of cylinder (r) = 1 m

Yield strength of steel (sys) = 200 MPa

We know, $\sigma_{induced} \le \sigma_{permissible}$

$$\frac{pd}{4t}\le\frac{sys}{N}=\frac{syt}{2N}$$

$$\frac{p\times 2}{4\times 1\times 10^{-3}}\le\frac{200\times 10^6}{2\times 1}$$

$p = 2 \times 10^5$ N/m^2

$$\boxed{p=200\ \text{kN/m}^2}$$

Hence the option (b) is correct

7. (a)

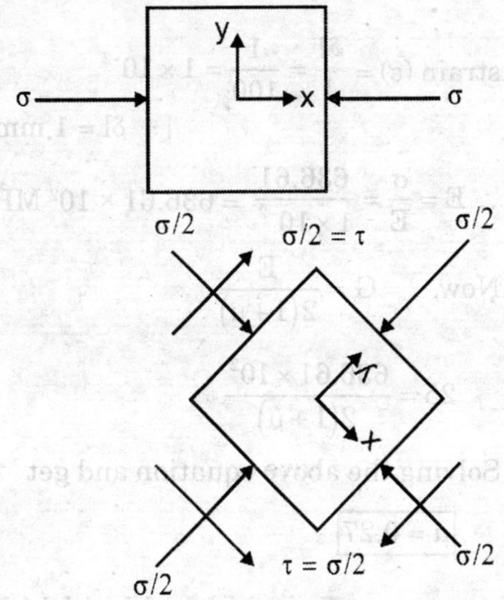

$$\sigma_n=-\sigma\cos^2\theta=-\sigma.\cos^2 45=\frac{-\sigma}{2}=\frac{\sigma}{2}$$

[composive]

$$\tau=\frac{\sigma}{2}\sin 2\theta=\frac{\sigma}{2}\sin(2\times 45)=\frac{\sigma}{2}\sin 90^\circ=\frac{\sigma}{2}$$

Hence the option (a) is correct.

8. (0.27) Given that

gange dialueter (da) = 10 mm

gange length (l) = 100 mm

Tanstle load (P) = 50 kN

Shear modulus (G) = 25 GPa

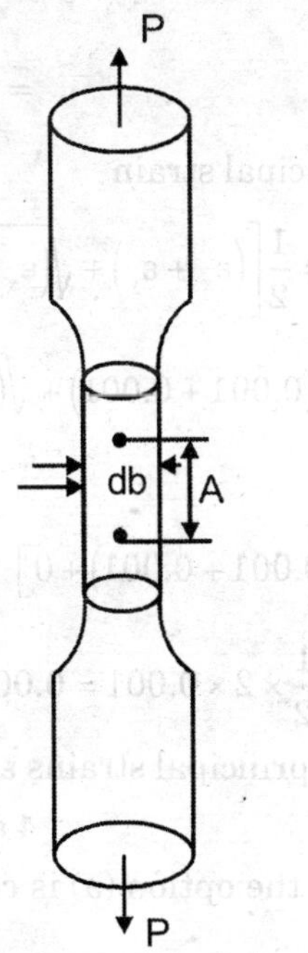

Stress developed in the object

$$\sigma=\frac{P}{A}=\frac{50\times 10^3}{\frac{\pi}{4}(10)^2}$$

σ = 636.61 MPa

$$\text{strain } (\varepsilon) = \frac{\delta l}{l} = \frac{1}{100} = 1 \times 10^{-2}$$

$[\because \delta l = 1 \text{ mm given}]$

$$\therefore E = \frac{\sigma}{E} = \frac{636.61}{1 \times 10^{-2}} = 636.61 \times 10^2 \text{ MPa}$$

$$\text{Now, } = G = \frac{E}{2(1+\mu)}$$

$$25 = \frac{636.61 \times 10^2}{2(1+\mu)}$$

Solving the above equation and get

$$\boxed{\mu = 0.27}$$

9. (a)

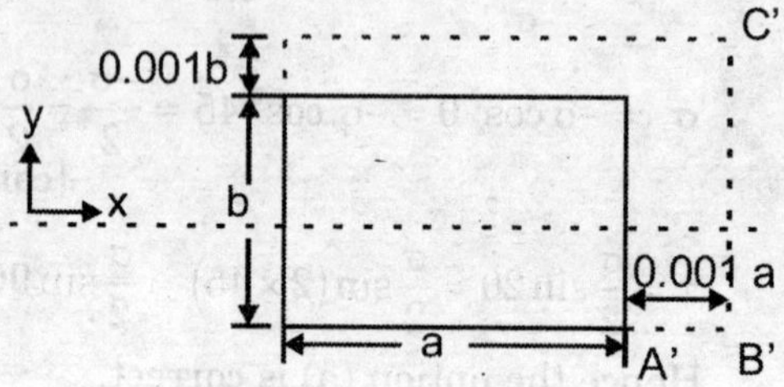

$$\text{Strain along x-direction } (\varepsilon_x) = \frac{\delta l}{l}$$

$$= \frac{0.001a}{a} = 0.001$$

$$\text{Strain along y-direction } (\varepsilon_y) = \frac{\delta b}{b}$$

$$= \frac{0.001b}{b} = 0.001$$

$\therefore$ Principal strain

$$(\varepsilon_{1/2}) = \frac{1}{2}\left[(\varepsilon_x + \varepsilon_y) + \sqrt{(\varepsilon_x - \varepsilon_y)^2 + (r_{xy})^2}\right]$$

$$\varepsilon_{1/2} = \frac{1}{2}\left[(0.001 + 0.001) + \sqrt{(0.001 - 0.001)^2 + 0}\right]$$

$[\because r_{xy} = 0]$

$$= \frac{1}{2}\left[(0.001 + 0.001) + 0\right]$$

$$\varepsilon_{1/2} = \frac{1}{2} \times 2 \times 0.001 = 0.001$$

There principal strains $\varepsilon_1 = 0.001$

$4\ \varepsilon_2 = 0.001$

Hence the option (a) is correct

10. (d)

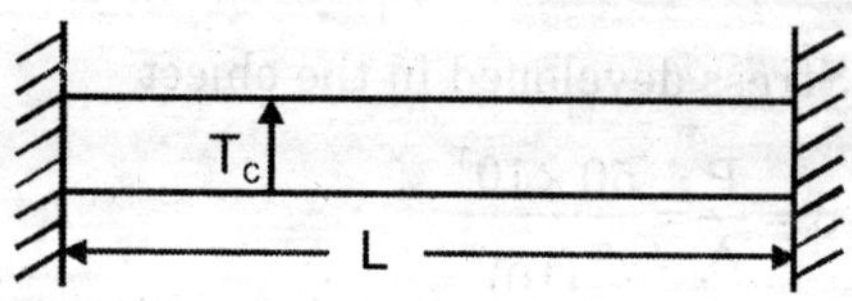

We know

$$\text{Buckling load induced in the bar(P)} = \frac{\pi^2}{L_c^2} EI$$

$$= \frac{\pi^2}{\left(\frac{L}{2}\right)^2} EI \quad [\because L_c = \frac{L}{2}]$$

$$= \frac{4\pi^2 EI}{L^2}$$

Now the stress developted int he bar due increase in temperature T_c

$$\sigma = \frac{P}{A}$$

$$\alpha T_c E = \frac{4\pi^2 EI}{L^2 \times (b^2)}$$

$$\alpha T_c E = \frac{4\pi^2 E b^4}{12 L^2 b^2}$$

$$\alpha\ T_c = \frac{\pi^2 b^2}{3L^2}$$

$$\therefore \boxed{\alpha = \frac{\pi^2 b^2}{3l^2 T_c}}$$

Hence the option (d) is correct

11. (135)

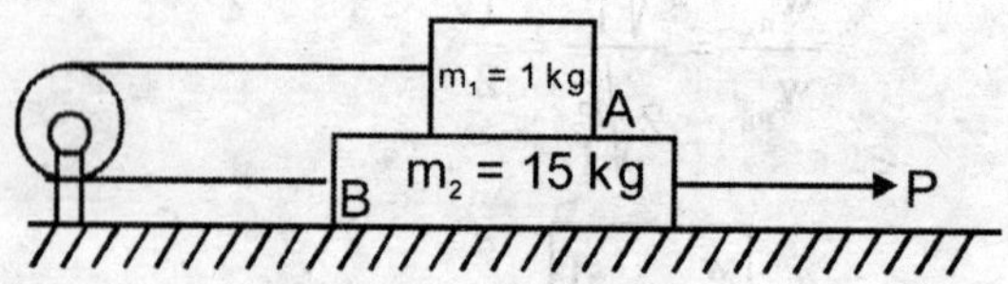

Given Mass of the upper block $(m_1) = 10$ kg

Mass of the lower block $(m_2) = 15$ kg

Coefficient of static friction $(\mu) = 0.3$

Now free body diagram (FBD) for block A

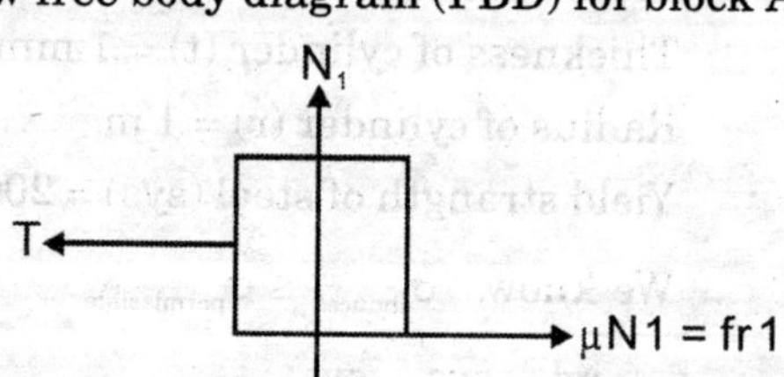

By applying 2nd law of Newton,

$$f_{r1} - T = m_1 a = 0 \, [\because a = 0]$$

$$\mu N_1 - T = 0$$

$$T = 0.3 \times (10 \times 10) = 30 \text{ N} \qquad \text{...(i)}$$

FBD for block B.

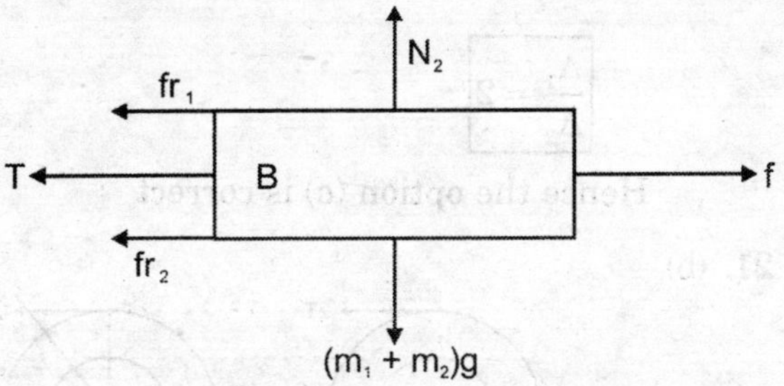

$$f-\left(f_{r1}+f_{r2}+T\right)=\left(m_1+m_2\right)a=0\ [\because a=0]$$

$$f=f_{r1}+f_{r2}+T=\mu N_1+\mu N_2+30\ \text{[from (i)]}$$

$$=\left(\frac{0.3}{10}\times 100\right)+\left(\frac{0.6}{10}\times 25\times 10\right)+30$$

$$=30+75+30$$

$$\boxed{f=135\ N}$$

17. (2)

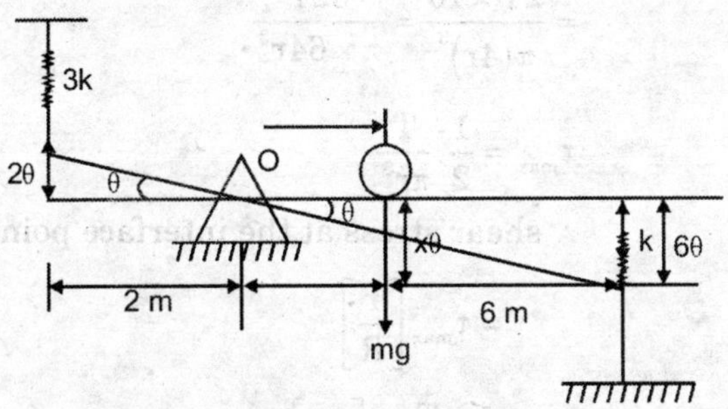

Taking moment about 'O'

$$I\ddot{\theta}+k(6)^2\theta+3k(2)^2\theta=0$$

$$2\ddot{\theta}+48k\theta=0$$

$$\ddot{\theta}+\left(\frac{48k}{2}\right)\theta=0$$

$$\therefore\ w_n^2=\frac{48k}{2}=\frac{48k}{mx^2}$$

$$(20)^2=\frac{48\times 10^3}{30\times x^2}$$

$$\Rightarrow x^2=\frac{48\times 10^3}{30\times 400}=4$$

$$\boxed{x=2m}$$

18. (a)

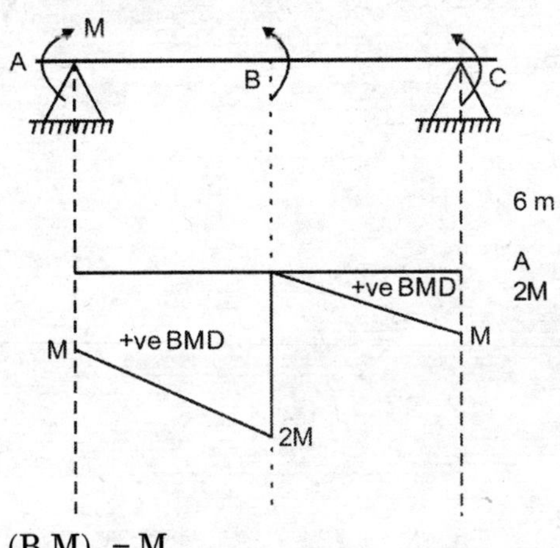

$$(B.M)_C=M$$

$$(B.M)_A=-M$$

$$(B.M)_B=M+2M-M=24$$

$$(BM)_B=M-M=0(\text{without }2M)$$

Hence the option (a) is correct

19. (17.77)

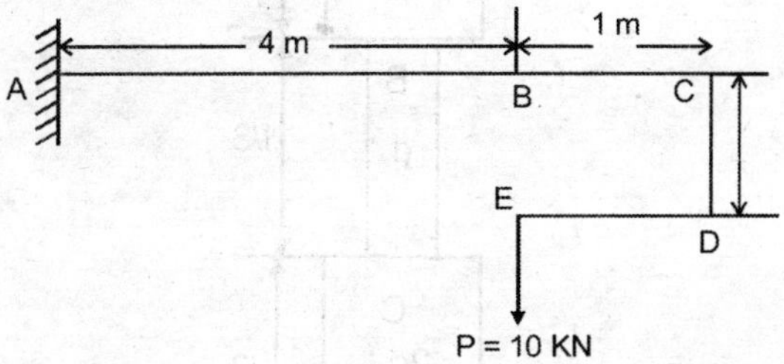

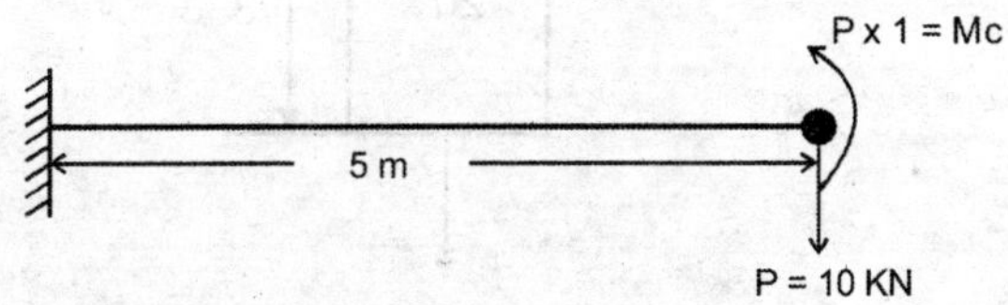

$\therefore$ Total sending stress

$$\sigma_T=\sigma_1+\sigma_2=\frac{M_C}{z_1}+\frac{(-P)\times 5\times 10^3}{z_2}$$

$$=\frac{10\times 10^3\times 1\times 10^3}{\frac{150\times 300^2}{6}}+\frac{\left(-10\times 10^3\times 5\right)\times 10^3}{\frac{150\times 300^2}{6}}$$

$$=-\frac{6\times 4\times 10^3}{150\times 9}\quad \boxed{\sigma_T=17.77\ MPa}$$

20. (c)

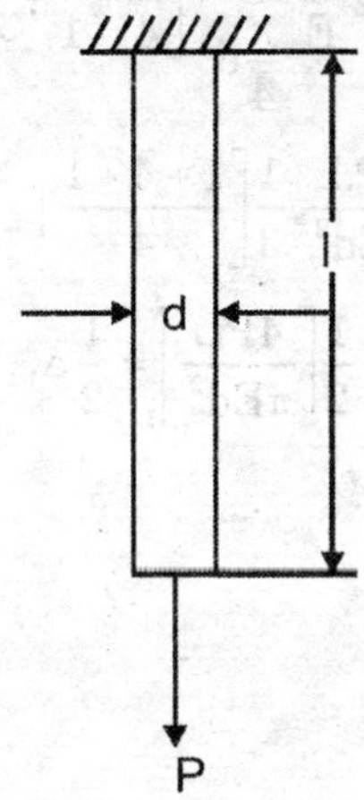

$$\left(\Delta_1\right)=\frac{P_1L_1}{A_1E_1}=\frac{P\times L}{\frac{\pi}{4}d^2E}=\frac{4PL}{\pi d^2E}$$

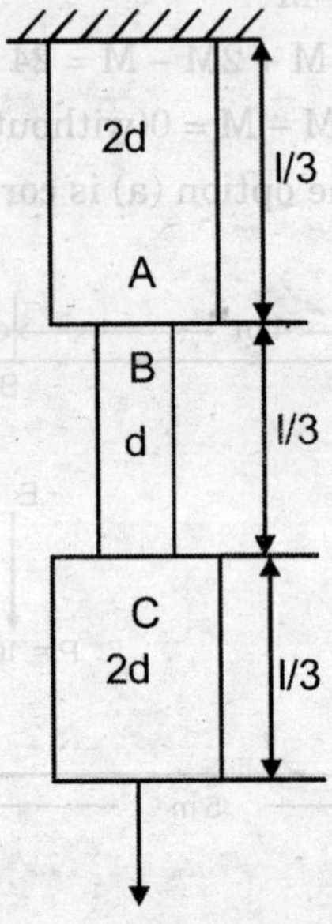

$$\left(\Delta_2\right) = \Delta_A + \Delta_B + \Delta_C = \frac{P_A L_A}{A_A E_A} + \frac{P_B L_B}{A_B E_B} + \frac{P_C L_C}{A_C E_C}$$

$$= \frac{P}{E}\left[\frac{\frac{L}{3}}{\frac{\pi}{4}(2d)^2} + \frac{\frac{L}{3}}{\frac{\pi}{4}d^2} + \frac{\frac{L}{3}}{\frac{\pi}{4}(2d)^2}\right]$$

$$\Rightarrow \frac{P}{E}.\frac{\frac{L}{3}}{\frac{\pi}{4}d^2}\left[\frac{1}{4}+\frac{1}{1}+\frac{1}{4}\right]$$

$$= \frac{4P.L}{\pi E d^2}.\frac{1}{3}\left[\frac{1+4+1}{4}\right]$$

$$\Delta_2 = \frac{1}{2}\left[\frac{4PL}{\pi E d^2}\right] = \frac{1}{2}\Delta_1$$

$$\boxed{\frac{\Delta_1}{\Delta_2} = 2}$$

Hence the option (c) is correct

21. (b)

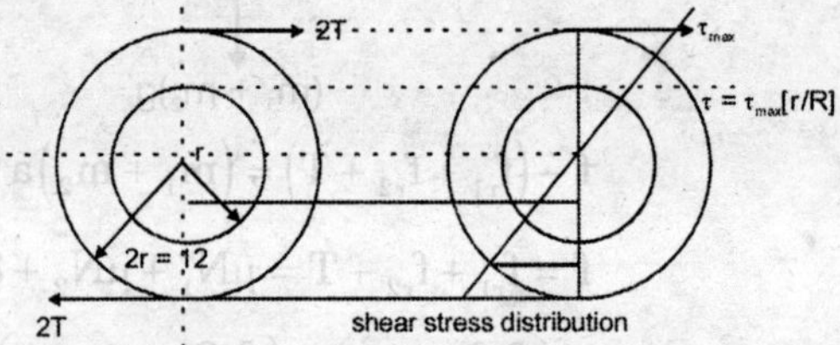

$$\tau_{max} = \frac{T}{z} = \frac{2T}{\frac{\pi}{16}(d)^3}$$

$$= \frac{2T \times 16}{\pi(4r)^3} = \frac{32T}{\pi \times 64 r^3}$$

$$\tau_{max} = \frac{1}{2}.\frac{T}{\pi r^3}$$

$\therefore$ **shear stress at the interface point x is**

$$\tau = \tau_{max}\left[\frac{r}{R}\right]$$

$$= \frac{1}{2}.\frac{T}{\pi r^3} \times \left[\frac{r}{2r}\right]$$

$$= 0.0795\frac{T}{r^3}$$

$$\boxed{\tau \simeq 0.08\frac{T}{r^3}}$$

Hence the option (b) is correct

2018

Q. 1 – Q. 9 carry one mark each.

1. A thin cylinder (thickness, t and diameter, d) with closed ends is subjected to an internal pressure, p The ratio of hoop stress to longitudinal stress developed in the wall of the cylinder is

   (*a*) 1 : 4 (*b*) 1 : 2

   (*c*) 1 : 1 (*d*) 2 : 1

2. A steel rod is fixed at one end and free at the other end. The coefficient of thermal expansion of the steel is α, and modulus of elasticity is E. If the temperature of the rod is increased by ΔT then the stress and strain developed in the rod are respectively

   (*a*) zero, $\alpha\Delta T$ (*b*) $E\alpha\Delta T$, $\alpha\Delta T$

   (*c*) $E\alpha\Delta T$, zero (*d*) zero, zero

3. The effective lengths of the columns with ideal boundary conditions shown in Case-I, Case-II, and Case-III are respectively

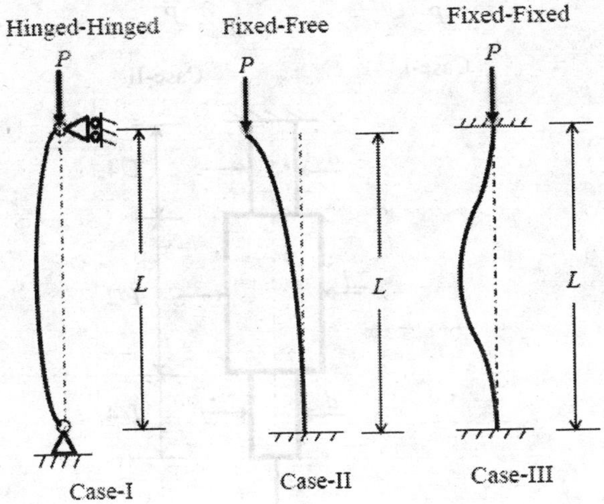

   (*a*) L, 4L, 2L (*b*) L, 2L, L/4

   (*c*) L, L, L (*d*) L, 2L, L/2

4. At a point in a stressed body, sum of the normal stresses acting on perpendicular faces of an arbitrarily oriented plane stress element is always

   (*a*) dependent on the angle of orientation of the element

   (*b*) constant and independent of angle of orientation of the element

   (*c*) one half of the sum of the principal stresses

   (*d*) zero

5. The principal stresses on a plane stress element are shown in the figure. The maximum shear stress (in MPa) is

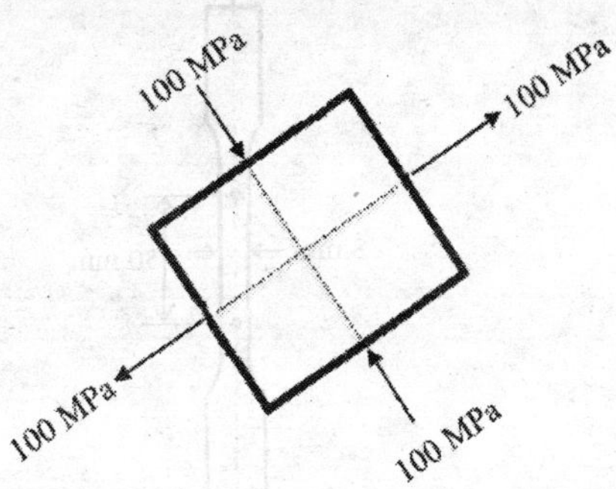

   (*a*) 200 (*b*) 100

   (*c*) 50 (*d*) 0

6. A rigid and thin L-shaped bracket is fixed to the wall at point B, and a force F is applied at point A as shown. For a given force F, the point B experiences the maximum clockwise moment when the inclination θ (in degrees) with the x-axis is .... [up to two decimal places].

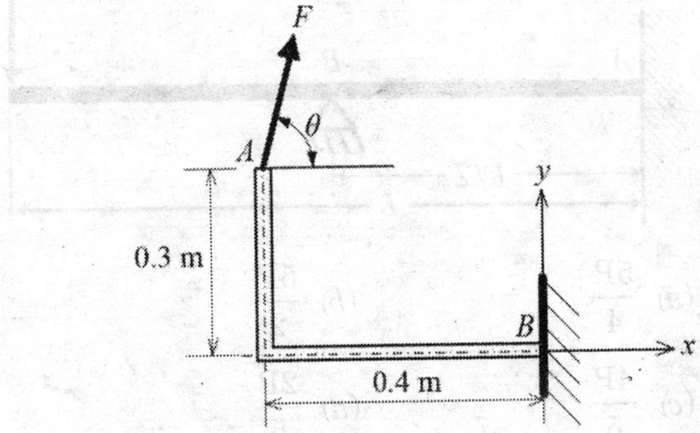

7. A cantilever beam with length, L = 1 m, modulus of elasticity, E = 210 GPa, and area moment of inertia, $I = 1.2 \times 10^{-7} m^4$ carries a concentrated mass m = 100 kg at its freeend. By idealizing it as a single degree-of-freedom system and neglecting the mass of the cantilever beam, the natural frequency (in rad/s) of small transverse oscillations of the mass m is ....[up to two decimal places]

8. For a typical grade of steel, the value of modulus of elasticity (E) and Poisson's ratio (v) are 208 GPa and 0.3 respectively. The value of shear modulus (G) of the steel (in GPa) is ....

9. A tensile test is performed on a metallic specimen of diameter 8 mm and gauge length 50 mm. When the tensile load P reaches a value of 20 kN, the distance between the gauge marks increases by 0.09 mm. If the sample remains within the elastic limit, the modulus of elasticity (in GPa) of the test metal is..........[up to two decimal places]

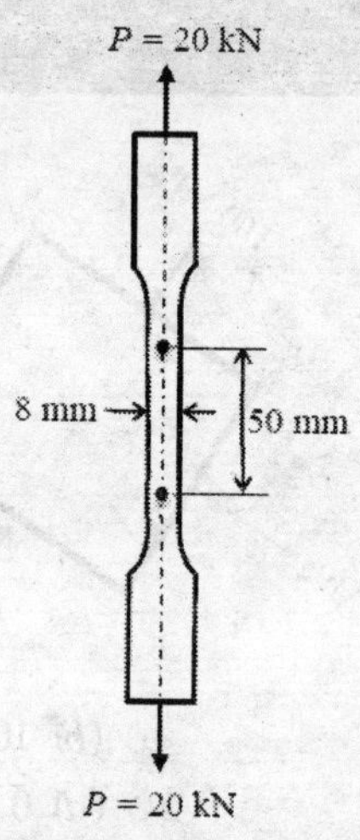

Q. 10 – Q. 22 carry two marks each.

10. A beam ABC is subjected to load P at its free end C as shown in the figure. The flexural rigidity of the beam is ..... The vertical support reaction at point B is

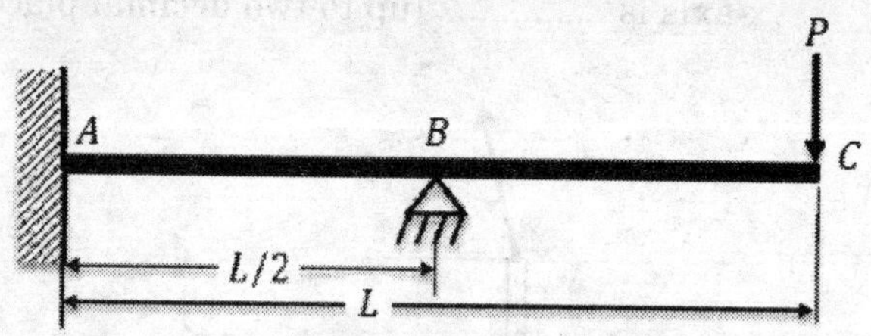

(a) $\frac{5P}{4}$ (b) $\frac{5P}{2}$

(c) $\frac{4P}{5}$ (d) $\frac{2P}{5}$

11. A horizontal effort P is applied to raise a block of weight W on a rough surface inclined at an angle θ with the horizontal. If μ_s is the coefficient of static friction between the block and the surface, the minimum effort P required to impend the upward motion of the block along the surface is

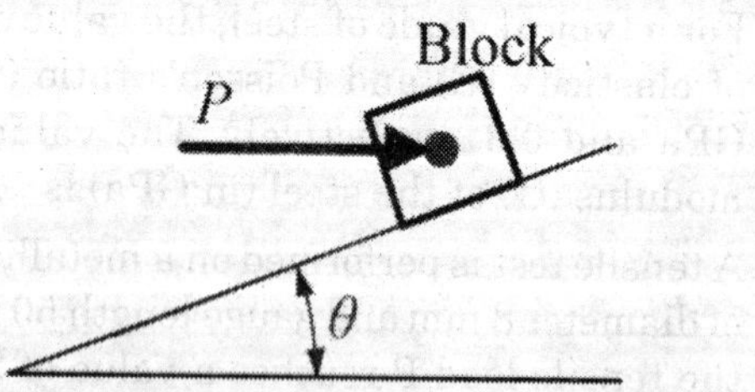

(a) $W\left(\frac{\mu_s - \tan\theta}{1+\mu_s \tan\theta}\right)$ (b) $W\left(\frac{\mu_s + \tan\theta}{1-\mu_s \tan\theta}\right)$

(c) $W\left(\frac{\mu_s - \tan\theta}{1-\mu_s \tan\theta}\right)$ (d) $W\left(\frac{\mu_s + \tan\theta}{1+\mu_s \tan\theta}\right)$

12. Three round bars of same material, equal lengths, and different cross-sectional dimensions are shown in the figures as Case-I, Case-II and Case-III. All the bars are clamped at the upper end, and a concentrated load P is applied at the lower end of each bar. If the elastic strain energy stored in the bar shown in Case-I is U_1 then the elastic strain energy stored in Case-II and Case-III respectively is

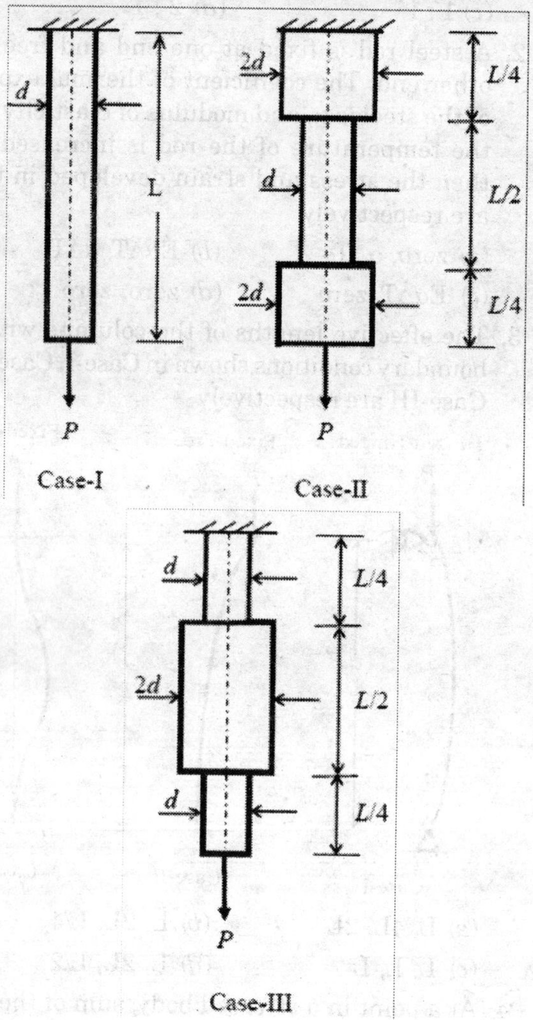

(a) $\left(\frac{8}{5}\right)U_1, \left(\frac{5}{8}\right)U_1$

(b) $\left(\frac{5}{16}\right)U_1, \left(\frac{5}{16}\right)U_1$

(c) $\left(\frac{5}{8}\right)U_1, \left(\frac{8}{5}\right)U_1$

(d) $\left(\frac{5}{8}\right)U_1, \left(\frac{5}{8}\right)U_1$

13. A rigid uniform rod with mass m, length L and center of gravity G is freely suspended from a hinge as shown in the figure. The rod is given a small angular displacement θ in the counter-clockwise direction from the position in which it hangs vertically (θ = 0). If g is the acceleration due to gravity, the natural frequency of oscillations (in rad/s) is

(*a*) $\sqrt{\frac{6g}{L}}$

(*b*) $\sqrt{\frac{2g}{L}}$

(*c*) $\sqrt{\frac{3g}{L}}$

(*d*) $\sqrt{\frac{g}{L}}$

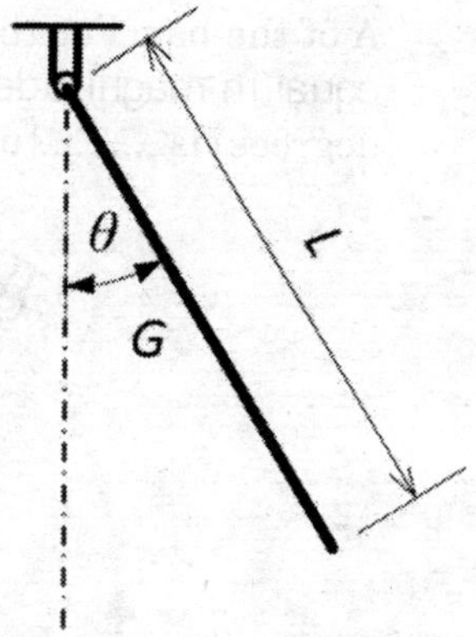

14. A cylindrical member, made up of ductile material, is subjected to pure torsion as shown in the figure. The failure plane (from Option-I to Option-IV) for ductile material as per maximum shear stress theory is represented by

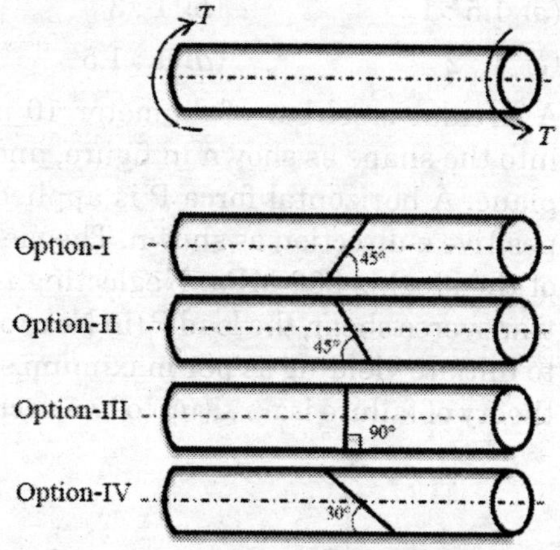

(*a*) Option I (*b*) Option II

(*c*) Option III (*d*) Option IV

15. A simply supported beam of span L is subjected to a couple Mo at a distance a from support A. Among the four options (Option-I to Option-IV) shown, the correct shear force diagram of the beam is

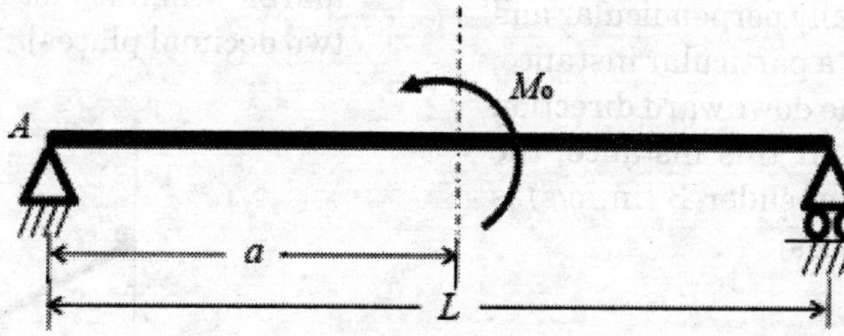

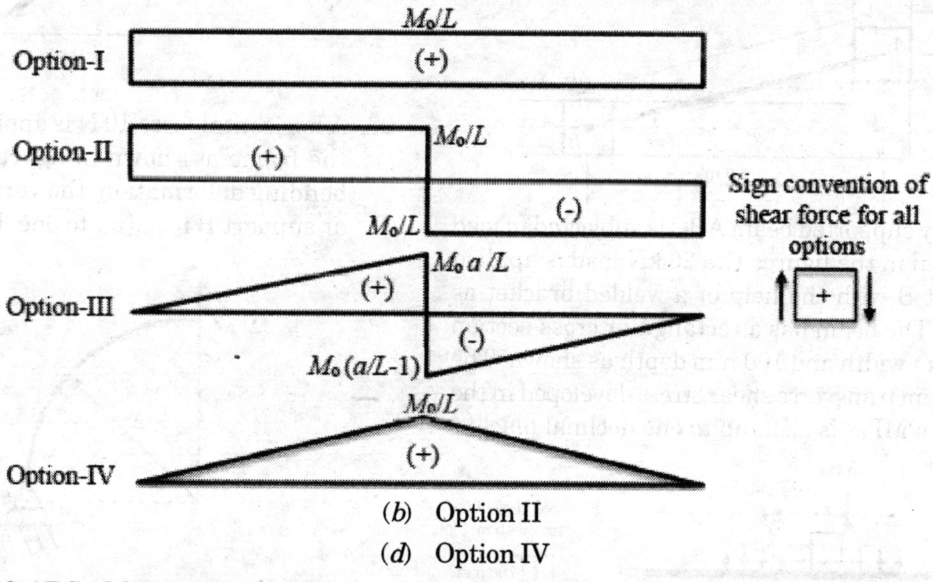

(*a*) Option I (*b*) Option II

(*c*) Option III (*d*) Option IV

16. A circular shaft ABC of diameter, d and length, L is fixed at end A. It is subjected to the torsional moments at point B and point C as shown in the figure. The ratio of angle of twists at point B to point C is

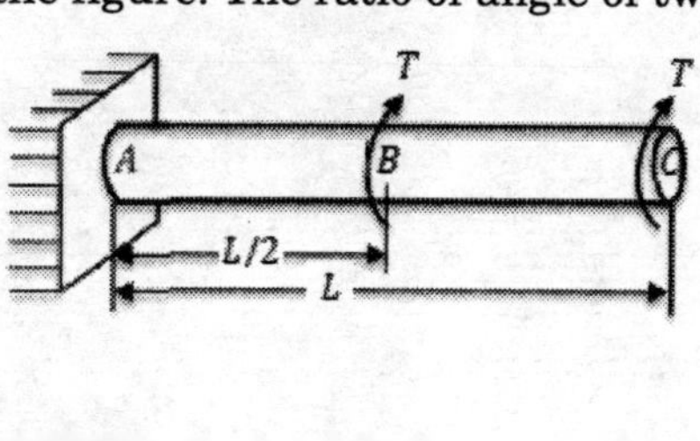

(*a*) 1.5 : 1 (*b*) 1 : 3

(*c*) 1 : 2 (*d*) 1 : 1.5

17. A circular steel bar of diameter 10 mm is bent into the shape as shown in figure, and lies in x-z plane. A horizontal force P is applied along the positive z-direction as shown. The yield strength of the steel is 200 MPa. Neglecting the effect of transverse shear, the load P (in Newton) required to initiate yielding as per maximum shear stress theory of failure is.........[up to two decimal places]

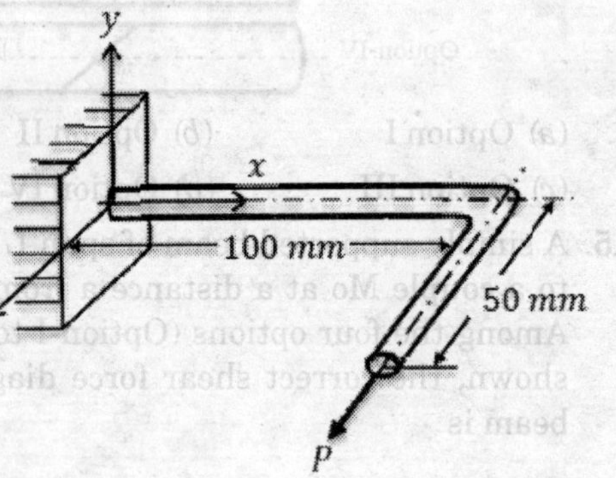

18. Two sliders A and B, connected by a rigid link of length L, slide in two mutually perpendicular and frictionless guide-ways. At a particular instance, the slider A is moving in the downward direction with a speed of 0.05 m/s. At this instance, the magnitude of the velocity of slider B (in m/s) is ....[up to two decimal places]

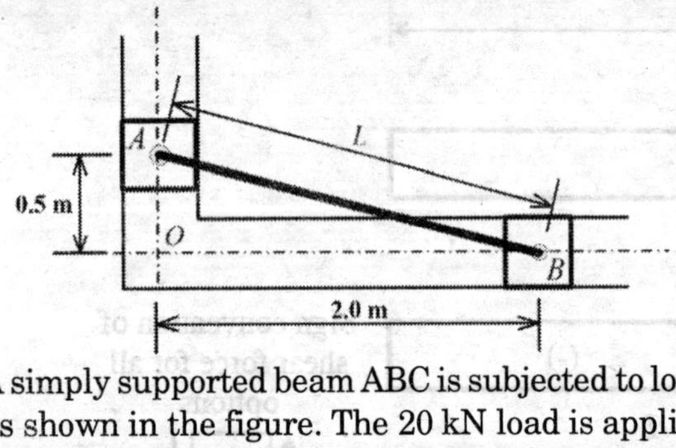

19. A simply supported beam ABC is subjected to load as shown in the figure. The 20 kN load is applied at point B with the help of a welded bracket as shown. The beam has a rectangular cross-section of 15 mm width and 100 mm depth as shown. The maximum transverse shear stress developed in the beam (in MPa) is ....[up to one decimal place]

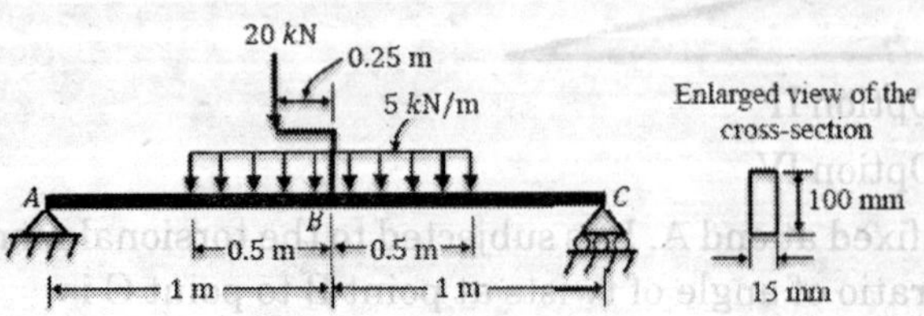

20. A rigid rod ABC, in the form of a quarter-circular arc of radius R, is hinged at C and supported by a roller at B. A vertical force P is applied at the end A of the bar. For the reactions at B and C to be equal in magnitude, the value of the angle θ (in degrees) is ....[up to two decimal places].

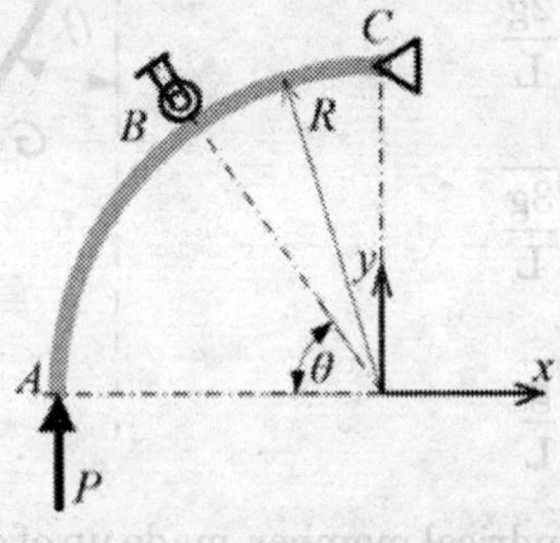

21. A marble of mass m slides along a frictionless linear slide AB kept in a vertical plane. The marble is released from point A with zero initial velocity, and it reaches point B under the action of gravity. Assuming the acceleration due to gravity to be 9.81 m/s^2, the speed (in m/s) of the marble when it just reaches point B is......[up to two decimal places].

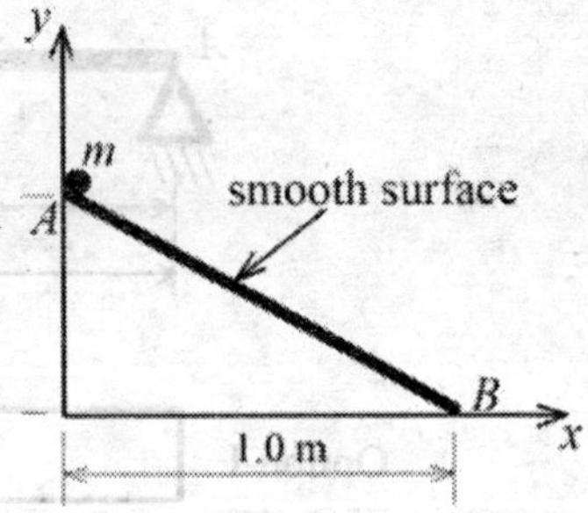

22. A horizontal force 10 N is applied at support B on the frame as shown in figure. Considering only bending deformation, the vertical reaction (in N) at support B is.....[up to one decimal place]

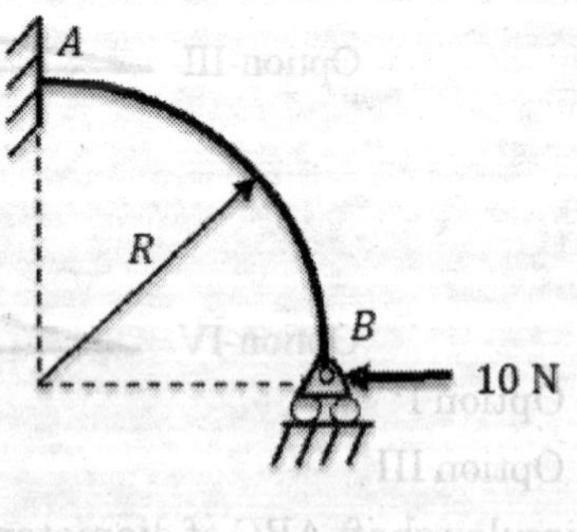

ANSWERS

1. (d)	**2.** (a)	**3.** (d)	**4.** (b)	**5.** (b)	**6.** (52.50 to 53.50)
7. (27.00 to 28.00)	**8.** (80 to 80)	**9.** (220.00 to 222.00)	**10.** (d)	**11.** (c)	**12.** (d)
13. (c)	**14.** (c)	**15.** (a)	**16.** (c)	**17.** (195.00 to 197.00)	
18. (0.01 to 0.02)	**19.** (14.5 to 15.5)	**20.** (26.00 to 27.00)	**21.** (3.10 to 3.20)	**22.** (13.5 to 14.5)	

EXPLANATIONS

1. (d)

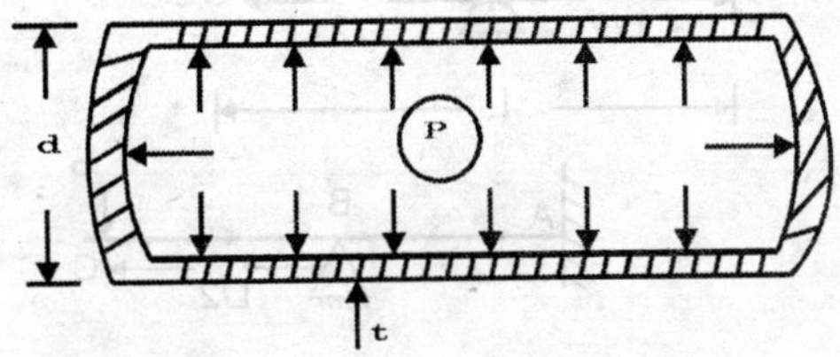

Given that

Thickness of the cylinder = t

Diameter of the cylinder = d

Pressure inside the cylinder = P

We know that

$$\text{Longitudinal stress } (\sigma_{long}) = \frac{Pd}{4t} \quad ...(i)$$

$$\text{Hoop stress } (\sigma_{hoop}) = \frac{Pd}{2t} \quad ...(ii)$$

∴ divide equation (ii) by equation (i)

$$\frac{\sigma_{hoop}}{\sigma_{long}} = \frac{Pd}{2t} \times \frac{4t}{Pd} = 2$$

$$\therefore \boxed{\frac{\sigma_{hoop}}{\sigma_{long}} = \frac{2}{1}}$$

Hence, the option (d) is correct

2. (a)

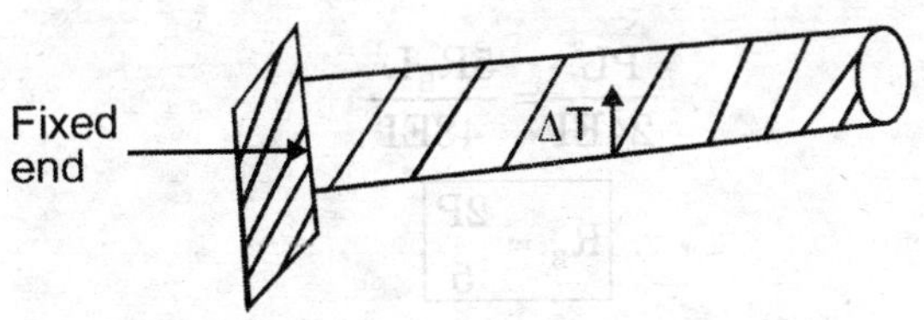

Total stress developed in the rod (σ_T)

$$= \sigma_{Thersvel} + \sigma_{Axial}$$

$$= 0 + 0$$

$$\boxed{\sigma_T = 0}$$

Strain developed in the rod (ε) = $\alpha \Delta T$

Hence the option (a) is correct

3. (d)

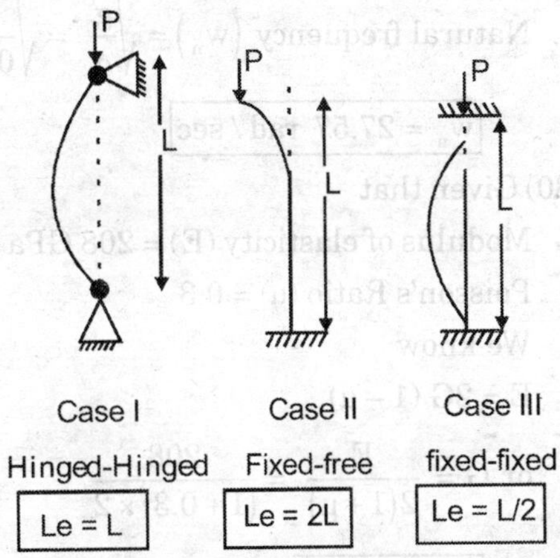

Hence the option (d) is correct

5. (b)

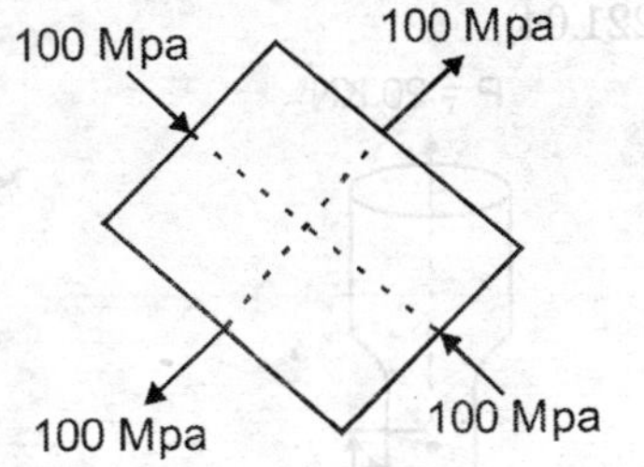

Given that

Principal stresses σ_1 = 100 MPa

σ_2 = −100 MPa

We know

$$\text{Maximum shear stresses } (\tau_{max}) = \frac{\sigma_1 - \sigma_2}{2}$$

$$= \frac{100 - (-100)}{2} = \frac{200}{2}$$

$$\boxed{\tau_{max} = 100 \text{ MPa}}$$

Hence the option (b) is correct

7. (27.57)

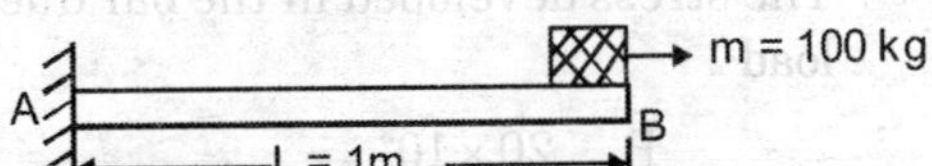

Given that

Length of the beam (L) = 1m

Madulus of elasticity (E) = 210 GPa

$= 210 \times 10^9$ N/s^2

Area Moment of Inertia (I) = 1.2×10^{-7} m^4

We know

Total deflection of the beam due to the given mass at its free end

$$\Delta = \frac{wL^3}{3EI} = \frac{100 \times 9.81 \times 1^3}{3 \times 2w \times 10^9 \times 1.2 \times 10^{-4}}$$

$$\Delta = 0.0129 \text{ m}$$

Now,

$$\text{Natural frequency } (w_n) = \sqrt{\frac{g}{\Delta}} = \sqrt{\frac{9.81}{0.0129}}$$

$$\boxed{w_n = 27.57 \text{ rad / sec}}$$

8. (80) Given that

Modulus of elasticity (E) = 208 GPa

Poisson's Ratio (μ) = 0.3

We know

$E = 2G(1 - \mu)$

$$\text{or } G = \frac{E}{2(1+\mu)} = \frac{208}{(1+0.3) \times 2}$$

$$\boxed{G = 80 \text{ GPa}}$$

9. (221.04)

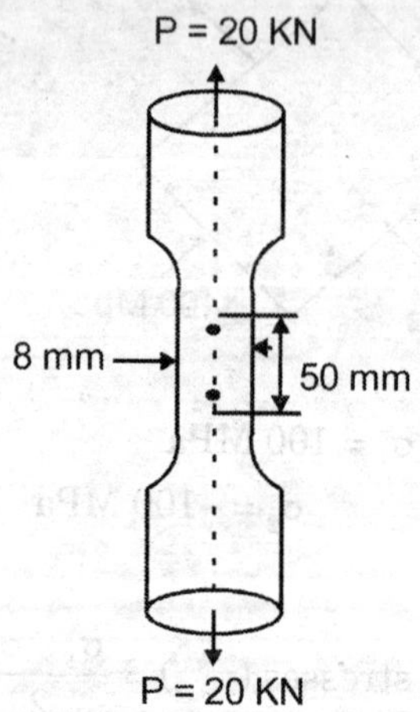

Given that

Specimen diameter (d) = 8 mm

Gange length (L) = 50 mm

Tensile load (P) = 20 kN

Elongation (δl) = 0.09 mm

The stress developed in the bar due to 20 kN load

$$\sigma = \frac{P}{A} = \frac{20 \times 10^3}{\frac{\pi}{4}(8)^2}$$

$$\boxed{\sigma = 397.88 \text{ MPa}}$$

The strain due to load 20 kN

$$E = \frac{\delta L}{L} = \frac{0.09}{50} = 1.8 \times 10^{-3}$$

Therefore, modulus of elasticity as

$$E = \frac{\sigma}{E} = \frac{397.88}{1.8 \times 10^{-3}}$$

$E = 221.04 \times 10^3$ MPa

or $\boxed{E = 221.04 \text{ GPa}}$

10. (d)

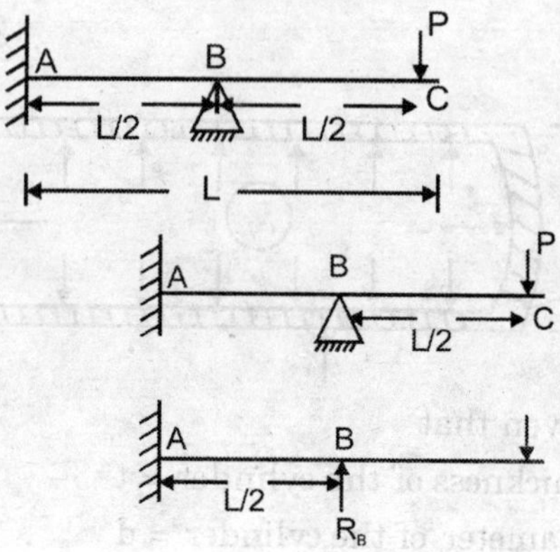

Deflection of the beam due to print load at 'P' at point 'C'

$$y_1 = \frac{-PL^3}{3EI} = \frac{-P\left(\frac{L}{2}\right)^3}{3EI} = \frac{-PL^3}{24EI}$$

Deflection of the beam due Reaction 'R_B' at point 'B'

$$y_2 = \frac{5R_B L^3}{48EI}$$

Net deflection of the beam at $Y_T = y_1 + y_2$

$$y_T = \frac{-PL^3}{24EI} + \frac{5R_B L^3}{48EI}$$

$$0 = \frac{-PL^3}{24EI} + \frac{5R_B L^3}{48EI}$$

$$\frac{PL^3}{24EI} = \frac{5R_B L^3}{48EI}$$

$$\therefore \boxed{R_B = \frac{2P}{5}}$$

Hence the option (d) as correct

11. (c)

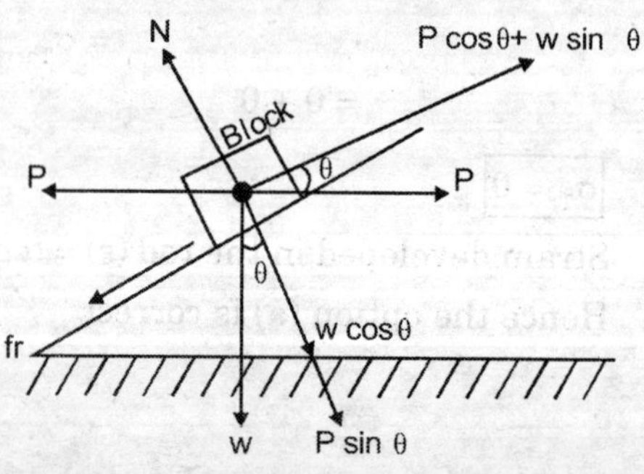

By applying Newton's 2nd Law

$(P\cos\theta + w\sin\theta) - fr = 0 \qquad [\because a = 0]$

$P\cos\theta + w\sin\theta = fr = \mu N \qquad ...(i)$

[fr → friction force, fr = μN, where w = mg]

$P\cos\theta + w\sin\theta = \mu(P\sin\theta + w\cos\theta)$

$[\because N = P\sin\theta + w\cos\theta]$

$P(\cos\theta - \mu\sin\theta) = \mu w\cos\theta - w\sin\theta$

$$P = \frac{\mu w\cos\theta - w\sin\theta}{\cos\theta - \mu\sin\theta}$$

$$\boxed{P = \frac{w(\mu - \tan\theta)}{(1 - \mu\tan\theta)}}$$

Hence the option (c) is correct

12. (d) Case : I

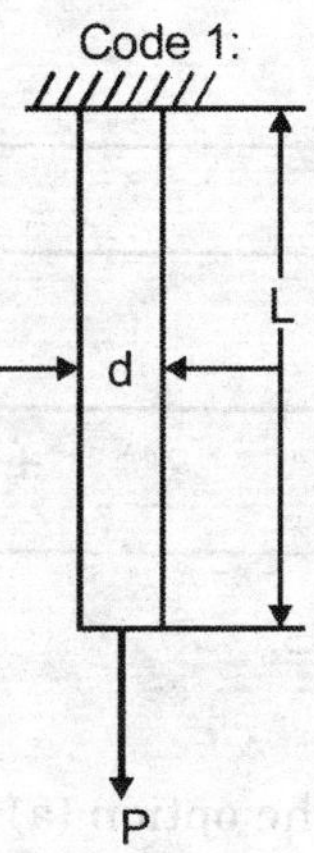

We know

Strain energy $(U)_I = \dfrac{P^2L}{2AE_1} = \dfrac{P_1L}{2.\frac{\pi}{4}d^2E}$

$$T_1 = \frac{2P^2L}{\pi d^2E}$$

Case : II

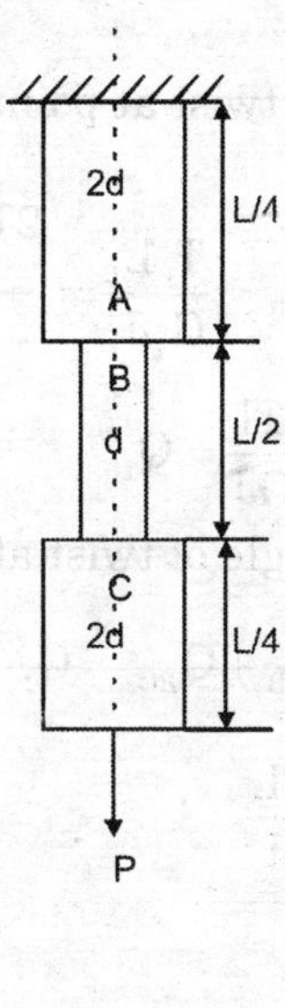

Strain energy $(U)_{II} = U_A + U_B + U_C$

$$U_{II} = \frac{P^2L_A}{2A_AE} + \frac{P^2L_B}{2A_BE} + \frac{P^2L_C}{2A_CE}$$

$$= \frac{P^2}{2E}\left[\frac{\frac{L}{4}}{\frac{\pi}{4}(2d)^2} + \frac{\frac{L}{2}}{\frac{\pi}{4}d^2} + \frac{\frac{L}{4}}{\frac{\pi}{4}(2d)^2}\right]$$

$$U_{II} = \frac{5}{4}\frac{P^2L}{\pi d^2E}$$

$$\boxed{U_{II} = \frac{5}{8}U_I} \qquad \left[\because U_I = \frac{2P^2L}{\pi d^2E}\right]$$

Case–III

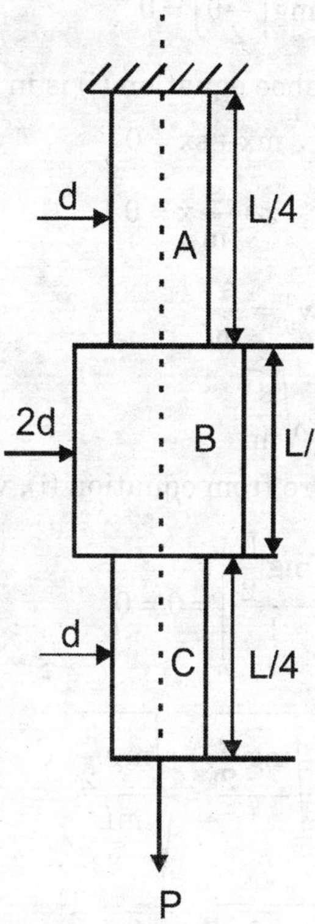

Strain energy $(U)_{III} = U_A + U_B + U_C$

$$(U)_{III} = \frac{P^2L_A}{2A_AE} + \frac{P^2L_B}{2A_BE} + \frac{P^2L_C}{2A_CE}$$

$$= \frac{P^2}{2E}\left[\frac{\frac{L}{4}}{\frac{\pi}{4}(d)^2} + \frac{\frac{L}{2}}{\frac{\pi}{4}(2d)^2} + \frac{\frac{L}{4}}{\frac{\pi}{4}(d)^2}\right]$$

$$U_{III} = \frac{5}{4}\frac{P^2L}{\pi Ed^2}$$

$$\boxed{U_{II} = \frac{5}{8}U_I} \qquad \left[\because U_1 = \frac{2P^2L}{\pi d^2E}\right]$$

Hence the option (d) is correct

13. (c) Taking moment about fixed point 'O'

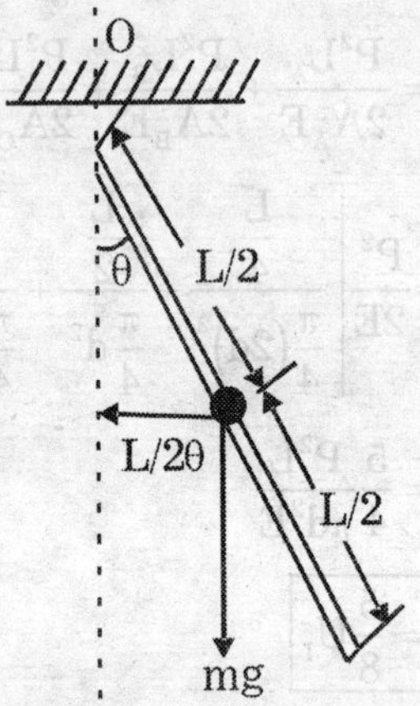

$$2\ddot{\theta} + mg\left(\frac{L}{2}\theta\right) = 0 \qquad \text{...(i)}$$

The aboe equation (i) is in the form of

$$m\ddot{x} + sx = 0$$

$$\ddot{x} + \frac{s}{m}x = 0$$

where $w_n^2 = \frac{s}{m}$

$$\Rightarrow w_n = \sqrt{\frac{s}{m}}$$

Therefore from equation (i), we get

$$\ddot{\theta} + \left(\frac{mg\frac{L}{2}}{I}\right) = \theta = 0$$

$$\therefore w_n = \sqrt{\frac{mg\frac{L}{2}}{I}} = \sqrt{\frac{mg\frac{k}{2}}{\frac{mL^2}{3}}} \qquad [\because I = \frac{mL^2}{3}]$$

$$\boxed{w_n = \sqrt{\frac{3g}{2L}}}$$

Hence the option (c) is correct

14. (c) As a rule of thumb: when brittle matrials are subjected to torsion they fail in the plane. Where tension is at its highest, he at a 45° angle.

Whereas, Ductile Material fail in the plane of Maximum shear stress where shear is maximum. Therefore the plane of failure is perpendicular to the primary axis, i.e. at 90° angle.

Hence the option (c) is correct

15. (a)

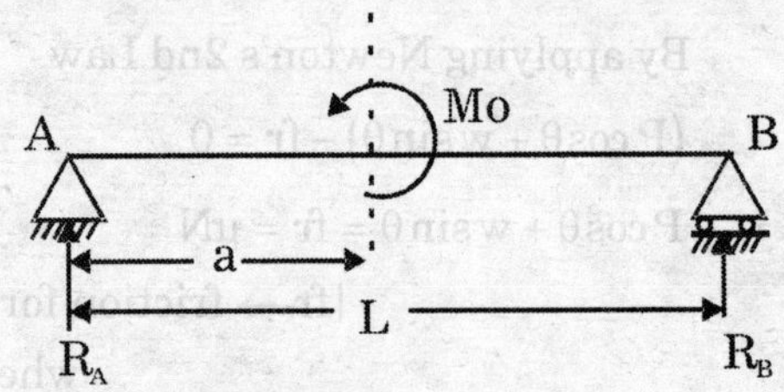

Net vertical force,

$$R_A + R_B = 0 \qquad \text{...(i)}$$

Taking moment about point 'A'

$$R_B \times L + M_o = 0$$

$$\boxed{R_B = \frac{-M_o}{L}}$$

$$\therefore \boxed{R_A = \frac{M_0}{L}}$$

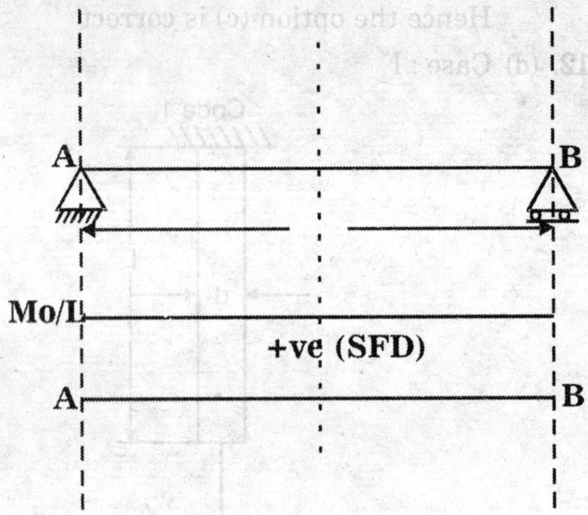

Hence the option (a) is correct

16. (c)

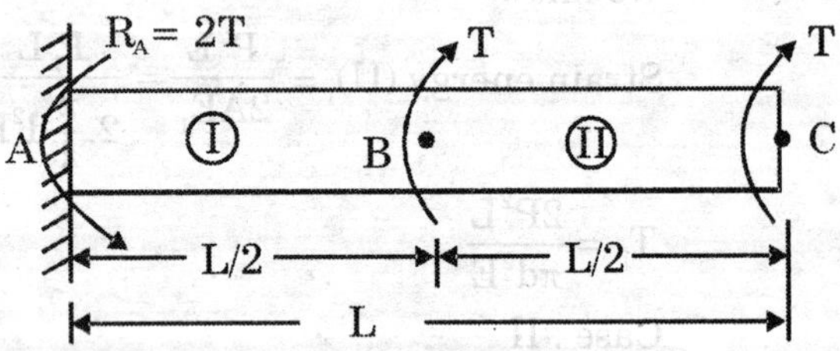

$T_1 = 2T, \qquad T_2 = T$

Angle of twist at point 'B'

$$Q_{AB} = Q_B = \frac{T_1 L_1}{G_1 J_1} = \frac{2T\left(\frac{L}{2}\right)}{GJ} = \frac{2T.\frac{L}{2}}{GJ}$$

$$\therefore Q_B = \frac{Tl}{GJ} = Q_1$$

Now, Angle of twist at point 'C'

$$Q_C = Q_{AB} + Q_{BC} = Q_1 + Q_2 = \frac{TL}{GJ} + \frac{TL}{GJ}$$

$$Q_C = \frac{2TL}{GJ}$$

Therefore $\frac{Q_B}{Q_C} = \frac{TL}{GJ} \times \frac{GJ}{2TL}$

$$\boxed{\frac{Q_B}{Q_C} = \frac{1}{2}}$$

Hence the option (c) is correct

17. (196.34)

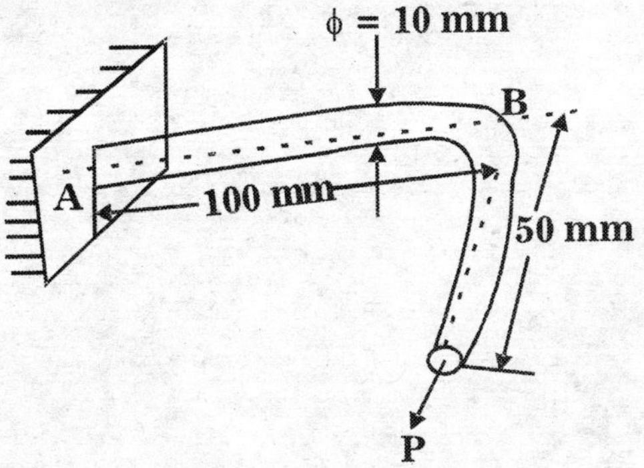

Given that

syt = 200 MPa

diameter (d) = 10 mm

Bending moment at point 'A' due to load 'P' is

$(Bm)_A = P \times 100$ ...(i)

Now, According Maximum shears stress theory

$Abs(\tau_{max}) \le (\tau_s)_{permissible}$

$\frac{T_e}{z_p} \le \frac{sys}{N} = \frac{syt}{2N}$

$\frac{T_e}{z_p} \le \frac{200}{2 \times 1}$ ...(ii)

We have from MSST, $T_e = \sqrt{M_A^2 + T_A^2}$ [from equation (i)]

$T_e = \sqrt{(P \times 100)^2 + 0} = P \times 100$

Now, Put the value in (ii), we get

$\frac{P \times 100}{\frac{\pi}{16}(10)^3} = \frac{200}{2 \times 1}$ $[\because z_p = \frac{\pi}{16} d^3]$

$$\boxed{P = 196.34 \text{ MPa}}$$

18. (0.012)

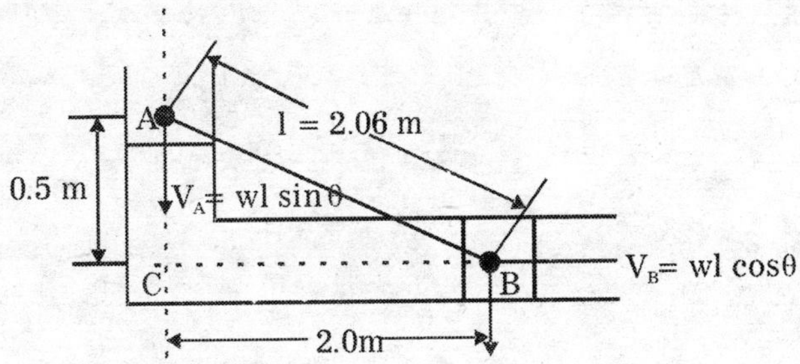

$\because \tan\theta = \frac{BC}{AC} = \frac{20}{0.5} = 4$

$$\boxed{\theta = 75.96}$$

Now $V_A = wl \sin\theta$

$0.05 = w \times 2.06 \times \sin 75.96$

$\therefore \boxed{w = 0.025 \text{ rad/sec}}$

therefore, volocity of slider 'B' is

$V_B = wl \cos\theta$

$= 0.025 \times 2.06 \times \cos(75.96)$

$$\boxed{V_B = 0.012 \text{ m/sec}}$$

21. (3.17)

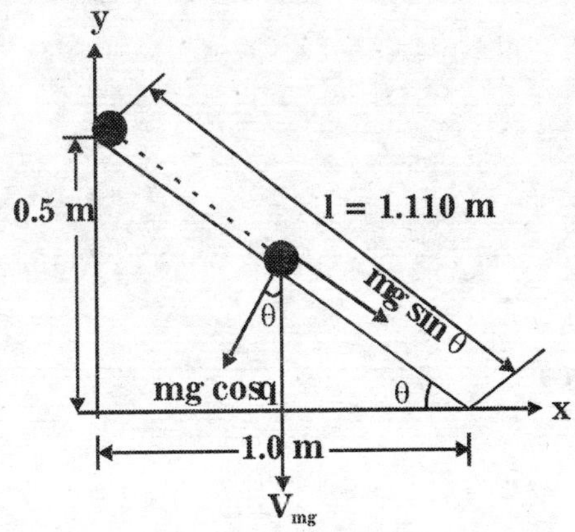

According to Newton's 2nd law

mg sinθ = ma

a = g sin θ ...(i)

$\therefore \tan\theta = \frac{0.5}{1} = 0.5$

$$\boxed{\theta = 25.56}$$

From equation (i).

a = 9.81 × sin 26.56

$$\boxed{a = 4.38 \text{ m/s}^2}$$

Now we know

$V^2 = u^2 + 2as$

$V^2 = 0 + 2 \times 4.38 \times 1.11$

$$\boxed{V = 3.11 \text{ m/s}}$$

XE - E

THERMODYNAMICS

Syllabus

Section 1: Basic Concepts

Continuum and macroscopic approach; thermodynamic systems (closed and open); thermodynamic properties and equilibrium; state of a system, state postulate for simple compressible substances, state diagrams, paths and processes on state diagrams; concepts of heat and work, different modes of work; zeroth law of thermodynamics; concept of temperature.

Section 2: First Law of Thermodynamics

Concept of energy and various forms of energy; internal energy, enthalpy; specific heats; first law applied to elementary processes, closed systems and control volumes, steady and unsteady flow analysis.

Section 3: Second Law of Thermodynamics

Limitations of the first law of thermodynamics, concepts of heat engines and heat pumps/refrigerators, Kelvin-Planck and Clausius statements and their equivalence; reversible and irreversible processes; Carnot cycle and Carnot principles/theorems; thermodynamic temperature scale; Clausius inequality and concept of entropy; microscopic interpretation of entropy, the principle of increase of entropy, T-s diagrams; second law analysis of control volume; availability and irreversibility; third law of thermodynamics.

Section 4: Properties of Pure Substances

Thermodynamic properties of pure substances in solid, liquid and vapor phases; P-vT behaviour of simple compressible substances, phase rule, thermodynamic property tables and charts, ideal and real gases, ideal gas equation of state and van der Waals equation of state; law of corresponding states, compressibility factor and generalized compressibility chart.

Section 5: Thermodynamic Relations

T-ds relations, Helmholtz and Gibbs functions, Gibbs relations, Maxwell relations, Joule-Thomson coefficient, coefficient of volume expansion, adiabatic and isothermal compressibilities, Clapeyron and Clapeyron-Clausius equations.

Section 6: Thermodynamic Cycles

Carnot vapor cycle, ideal Rankine cycle, Rankine reheat cycle, air-standard Otto cycle, air-standard Diesel cycle, air-standard Brayton cycle, vapor-compression refrigeration cycle.

Section 7: Ideal Gas Mixtures

Dalton's and Amagat's laws, properties of ideal gas mixtures, air-water vapor mixtures and simple thermodynamic processes involving them; specific and relative humidities, dew point and wet bulb temperature, adiabatic saturation temperature, psychrometric chart.

Subject-wise Analysis

XE - E : THERMODYNAMICS

Year	1 mark Questions	2 marks Questions	Total Marks
2009	8	16	40
2010	9	13	35
2011	9	13	35
2012	9	13	35
2013	9	13	35
2014	9	13	35
2015	9	13	35
2016	9	13	35
2017	9	13	35
2018	9	13	35

2009

Useful Data : Universal gas constant,

$\bar{R}$ = 8314 J/kmol K; g = 9.8 m/s^2

Q. 1 – 8 carry one mark each

1. The ideal gas law is valid for

(*a*) inert gases

(*b*) gases at high pressure and high temperature

(*c*) gases at low pressure and low temperature

(*d*) gases at low pressure and high temperature

2. During the adiabatic saturation process

(*a*) the relative humidity increases but the specific humidity remains constant

(*b*) both the relative humidity and the specific humidity remain constant

(*c*) both the relative humidity and the specific humidity increase

(*d*) the relative humidity decreases but the specific humidity increases

3. For an ideal gas undergoing at throttling process 1-2, which of the following relationship holds?

(*a*) $T_1 = T_2$ (*b*) $\frac{P_1}{P_2} = \frac{T_1}{T_2}$

(*c*) $\frac{P_1}{P_2} = \left(\frac{T_1}{T_2}\right)^{\gamma/(\gamma-1)}$ (*d*) $\frac{P_1}{P_2} = \frac{T_2}{T_1}$

4. A Carnot refrigerator operating between – 1°C and 33°C has a cooling capacity of 1.6 kW. The power consumed by the refrigerator is

(*a*) 160 W

(*b*) 178 W

(*c*) 200 W

(*d*) 1.8 kW

5. An ideal gas undergoes expansion according to the process $PV^{0.5}$ = constant. The temperature of the gas during the expansion process

(*a*) does not change

(*b*) increases

(*c*) decreases

(*d*) changes depending on the initial condition

6. Air (γ = 1.4) is compressed ideally from an initial state of 1 bar, 300 K to a final temperature of 600 K. The value of the final pressure in bar is

(*a*) 2 (*b*) 3.7

(*c*) 7.2 (*d*) 11.3

7. On a T-s diagram, the slope of the constant volume line for an ideal gas is

(*a*) less than that of constant pressure line

(*b*) more than that of constant pressure line

(*c*) less than that of constant enthalpy line

(*d*) equal to that of constant enthalpy line

8. The thermal efficiency of an ideal Rankine cycle is less than that of Carnot cycle operating between the same maximum and minimum temperature limits, because

(*a*) heat addition does not take place at constant temperature

(*b*) the expansion process is not reversible and adiabatic

(*c*) heat rejection does not take place at constant temperature

(*d*) the compression process is not reversible and adiabatic

Q. 9 – 24 carry two marks each

9. Atmospheric air (R = 287 J/kg; γ = 1.4) at 1 bar and 25 °C is compressed adiabatically to 2 bar and 105°C. Which of the following statements is correct ?

(*a*) The process is possible but irreversible

(*b*) The process is posssible and reversible

(*c*) The process is impossible

(*d*) The process is possible and it is isentropic

10. A pressure cooker contains saturated water-vapour mixture at 100°C with volume of vapour being eight times that of the volume of liquid. The specific volume of saturated liquid and saturated vapour at 100°C are, v_f = 0.001044 m^3/kg and v_g = 1.6729 m^3/kg, respectively. The quality of the mixture is

(*a*) 0.005 (*b*) 0.125

(*c*) 0.889 (*d*) 0.995

11. An ideal gas (γ = 1.39) flows in a pipeline at 450°C and 20 bar. A rigid, insulated and initially evacuated vesssel is connected to the pipeline through a valve. The valve is now opened and the gas is allowed to fill the empty vessel. The final temperature of the gas in the vessel is

(*a*) 247°C (*b*) 450°C

(*c*) 625 °C (*d*) 732 °C

12. An equi-molar mixture of nitrogen ($\gamma = 1.4$) and helium ($\gamma = 1.67$) is initially at 5 bar and 300°C. The mixture is expanded adiabatically to a pressure of 2 bar. The final temperature of the mixture is

(*a*) 149°C (*b*) 200°C
(*c*) 250°C (*d*) 524°C

13. A heat engine E_1 operates between an infinite reservoir at 800°C and a body B. The temperature of the body B remains constant at 550°C. Heat transferred to the engine E_1 is 900 kJ and the work output is 200 kJ. Another engine E_2 operates between the body B and the atmosphere at 27°C. Heat rejected to the atmosphere is 350 kJ. The thermal efficiency of the engine E_2 is

(*a*) 0.39 (*b*) 0.5
(*c*) 0.61 (*d*) 0.635

14. A gas turbine power plant operates with air ($\gamma = 1.4$) between 1 bar and 20 bar. The maximum thermal efficiency (in %) for the corresponding air-standard cycle is

(*a*) 30 (*b*) 36.7
(*c*) 48.2 (*d*) 57.5

15. The saturation pressures of water at 100°C and 105°C are respectively 101.3 kPa and 120.8 kPa. Taking the molecular weight of water as 18, the latent heat of water in kJ/kg at 102.5 °C is approximately equal to

(*a*) 2290 (*b*) 1250
(*c*) 820 (*d*) 330

16. An engine reversibly receives 1200 J of heat at 900 K. After rejecting heat to the ambient at 300 K, it develops 600 J of work. The irreversibility in joules is equal to

(*a*) 600 (*b*) 400
(*c*) 200 (*d*) zero

17. Saturated liquid water at 0.4 MPa and 1000 kg/hr of steam at 0.4 MPa and 300°C enter steadily into an insulated mixing chamber. At 0.4 MPa, the enthalpy of saturated liquid and saturated vapour are 604.73 and 2738.53 kJ/kg respectively; also, the enthalpy of superheated steam at 300°C is 3066.75 kJ/kg. The quality of the water-vapour mixture exiting the chamber is 0.9. The mass flow rate of saturated liquid water in kg/hr is

(*a*) 182 (*b*) 282
(*c*) 382 (*d*) 1000

18. A certain quantity of gas undergoes the polytropic process, $PV^{1.3}$ = constant, from an initial state of 1.5MPa and 0.09 m³ to the final pressure of 7.5 MPa. The work done by the gas in kJ is

(*a*) – 217 (*b*) – 200
(*c*) 200 (*d*) 217

Common Data Questions

Common Data for Questions 19 and 20:

Saturated water vapour enters an adiabatic turbine at 0.8 MPa and leaves at 0.1 MPa. The mass flow rate of water vapour is 25 kg/s. Use the following data table to answer the questions 19 and 20.

Pressure (MPa)	Temperature (°C)	Specific enthalpy		Sepecific entropy	
		h_f(kJ/kg)	h_g(kJ/kg)	s_f(kJ/kg K)	s_g(kJ/kg K)
0.8	170.43	722.11	2769.10	2.0462	6.6628
0.1	99.63	417.46	2675.50	1.3026	7.3594

19. The quality of steam at the exit of the turbine after an isentropic expansion is

(*a*) 0.47 (*b*) 0.72
(*c*) 0.88 (*d*) 0.94

20. If the steam leaves the turbine as saturated vapour, the power produced by the turbine in kW is

(*a*) 1640 (*b*) 2030
(*c*) 2340 (*d*) 8830

Common Data for Question 21 and 22:

The flow rate of refrigerant R-12 in a refrigerator is 0.03 kg/s. The refrigerant enters the compressor as saturated vapour at 150.9 kPa. After adiabatic compression, it enters the condenser as superheated vapour at 500 kPa and 100°C and leaves the condenser as saturated liquid at the same pressure. Use the following data to answer the questions 21 and 22.

Pressure (kPa)	Temperature (°C)	Specific enthalpy	
		h_f(kJ/kg)	h_g(kJ/kg)
150.9	– 20	17.82	178.74
500	15.6	50.64	195.01

For the superheated vapour at 500 kPa and 100°C, h = 252.05 kJ/kg.

21. The refrigeration effect in kW is

(a) 1.71 (b) 3.84

(c) 4.33 (d) 4.83

22. The actual power input to the compressor in kW is

(a) 0.49 (b) 0.99

(c) 1.71 (d) 2.2

Linked Answer Questions

Statement for Linked Answer Questions 23 - 24:

An insulated vertical cylinder encloses 0.1 kg of argon (Ar) with the help of a frictionless non-conducting piston as shown in the figure. The mass of the piston is 5 kg and it initially rests on the bottom of the cylinder. The cylinder is connected to a nitrogen (N_2) tank at 100 bar through a pipeline fitted with a valve. The valve is opened and nitrogen is slowly admitted into the cylinder. During this operation, the piston is lifted through a height of 10 cm by the nitrogen gas. The initial pressure and temperature of argon gas are 100 kPa and 300 K respectively. The final temperature of argon is 320 K. For argon, C_p = 520 J/kgK and C_v = 312 J/kgK.

23. The work done by argon in kJ during the process is

(a) 10 (b) 1.041

(c) –0.624 (d) –1.041

24. The work done by nitrogen in kJ during the process is

(a) 1.046 (b) 0.629

(c) –1.046 (d) –10

ANSWERS

1. (d)	**2.** (b)	**3.** (a)	**4.** (c)	**5.** (b)	**6.** (d)	**7.** (b)
8. (a)	**9.** (c)	**10.** (a)	**11.** (c)	**12.** (b)	**13.** (b)	**14.** (d)
15. (a)	**16.** (c)	**17.** (b)	**18.** (b)	**19.** (c)	**20.** (d)	**21.** (c)
22. (c)	**23.** (c)	**24.** (b)				

EXPLANATIONS

1. (d) Ideal gas law is valid for,

(i) Low pressure

(ii) High Temperature

2. (b) During the adiabatic saturation process, tdb = dwb = tap, so that, both are constant.

$$w = \frac{0.622\,p_v}{p - p_v}$$

$$\phi = \frac{p_v}{p_{vs}}$$

3. (a) An ideal gas undergoing throttling process then,

$h_1 = h_2$

$T_1 = T_2$

4. (c) Given that

$T_1 = 33°C = 306\ K$

$T_2 = -1°C = 272\ K$

$$(COP)_R = \frac{T_2}{T_1 - T_2} = \frac{272}{306 - 272} = 8$$

$$(COP)_R = \frac{Q_2}{W_{in}}$$

$$8 = \frac{1.6 \times 1000}{W_{in}} \qquad [Q_2 = 1.6\ kW = 1600\ W]$$

$$\boxed{W_{in} = 200\ W}$$

5. (b) An ideal gas undergoes expansion process $PV^{0.5} = c$,

the temperature of the gas during the expansion process increases.

$$\frac{T_2}{T_1} = \left(\frac{V_1}{V_2}\right)^{n-1}$$

$$\frac{T_2}{T_1} > 1$$

∴ Temperature increases.

6. (d) Given that

$\gamma = 1.4$

$P_1 = 1\ bar$

$T_1 = 300\ K$

$T_2 = 600\ K$

$$\frac{T_2}{T_1} = \left(\frac{P_2}{P_1}\right)^{\frac{\gamma-1}{\gamma}}$$

$$\left(\frac{T_2}{T_1}\right)^{\frac{\gamma}{\gamma-1}} = \frac{P_2}{P_1}$$

$$\text{Final pressure}(P_2) = \left(\frac{T_2}{T_1}\right)^{\frac{\gamma}{\gamma-1}} \cdot P_1$$

$$P_2 = \left(\frac{600}{300}\right)^{\frac{1.4}{1.4-1}} \times 1$$

$$\boxed{P_2 = 11.31 \text{ bar}}$$

7. (b) On T–S diagram for an ideal gas.

Now

$$T \cdot dS = dU + P \cdot dV$$

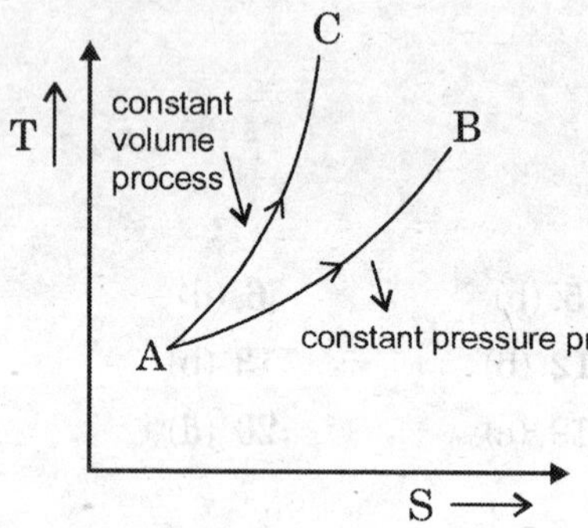

at constant volume

$V = C$

$dv = 0$

$T \cdot ds = du$

$T \cdot ds = Cv \cdot dT$

$\boxed{\frac{dT}{ds} = \frac{T}{C_v}}$ Slope at constant volume

Slope at constant pressure,

$$\left(\frac{dT}{ds}\right)_{p=c} = \frac{T}{C_p}$$

Slope at constant volume is greater that the slope at constant pressure because $C_p > C_v$ for an ideal gas.

8. (a) Thermal efficiency of an ideal Rankine cycle is less than that of carnot cycle operating between the same temperature limits because heat addition and heat rejection don't take place at constant temperature.

9. (c) Given that

$\gamma = 1.4$

$P_1 = 1$ bar, $T_1 = 25°C = 298$ K

$P_2 = 2$ bar, $T_2 = 105°C = 378$ K

$$T_2 = T_1 \left(\frac{P_2}{P_1}\right)^{\frac{\gamma-1}{\gamma}}$$

$$S_2 - S_1 = C_v \ln\left(\frac{T_2}{T_1}\right) - R\ln\left(\frac{P_2}{P_1}\right)$$

$$= .717 \ln\left(\frac{378}{298}\right) - .28 \ln\left(\frac{2}{1}\right)$$

$$S_2 - S_1 = -ve$$

The process is inpossible.

10. (a) Given that

$Vv = 8 \cdot V_L$

$v_f = 0.001044 \text{ m}^3/\text{kg}$

$v_g = 1.6729 \text{ m}^3/\text{kg}$

$$v_f = \frac{V_L}{m_L}$$

$$m_1 = \frac{V_L}{v_f} = \frac{V_L}{0.001044}$$

$$m_g = \frac{V_v}{v_g} = \frac{8 \cdot V_L}{1.6729}$$

$$\text{Dryness fraction}(x) = \frac{m_g}{m_g + m_L}$$

$$x = \frac{\frac{8 \cdot V_L}{1.6729}}{\frac{8 \cdot V_L}{1.6729} + \frac{V_L}{0.001044}}$$

$$\boxed{x = 0.00496}$$

11. (c) Given that

$\gamma = 1.39$

$Ti = 450°C$

$P_i = 20$ bar

From unsteady state condition

Final temperature $(T_f) = \gamma Ti$

$T_f = 1.39 \times 450$

$\boxed{T_f = 625.5 °C}$

12. (b) Given that

$\gamma = 1.67$

$P_1 = 5$ bar

$T_1 = 300°C = 573$ k

$P_2 = 2$ bar

$$\gamma_{mixture} = \frac{\gamma_{N_2} + \gamma_{H_e}}{2} = \frac{1.4 + 1.67}{2} = 1.535$$

For adiabatic process,

$$\frac{T_2}{T_1} = \left(\frac{P_2}{P_1}\right)^{\frac{\gamma-1}{\gamma}}$$

$$T_2 = \left(\frac{2}{5}\right)^{\frac{0.535}{0.5.35}} \times 573$$

$$T_2 = 20°C$$

13. (b) $T_1 = 800°C = 1073\,K$

$T_2 = 550°C = 823\,K$

$Q_1 = 900\,kJ$

$W_1 = 200\,kJ$

$T_2 = 27°C = 300\,K$

$Q_3 = 350\,kJ$

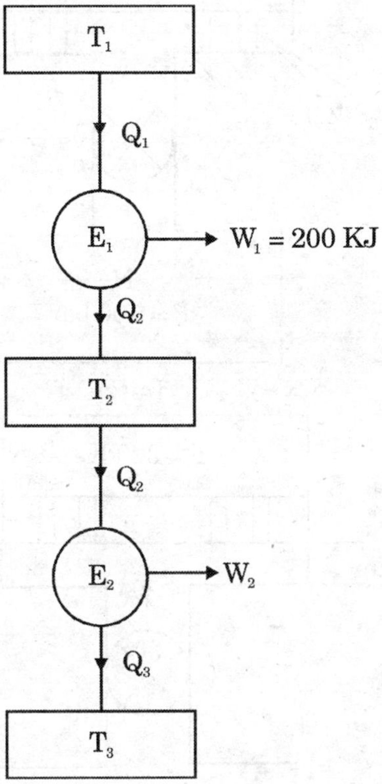

Now

$W_1 = Q_1 - Q_2$

$200 = 900 - Q_2$

$Q_2 = 700\,kJ$

Effeciency of $E_2\left(\eta_{E_2}\right) = \frac{W_2}{Q_2} = \frac{Q_2 - Q_3}{Q_2}$

$$\eta_{E_2} = \frac{700 - 350}{700}$$

$$\boxed{\eta_{E_2} = 0.5}$$

14. (d) Given that

$\gamma = 1.4$

$P_1 = 1\,bar$

$P_2 = 20\,bar$

$$r_P = \frac{P_2}{P_1} = \frac{20}{1} = 20$$

$$\eta = 1 - \frac{1}{(r_p)^{\frac{\gamma-1}{\gamma}}}$$

$$= 1 - \frac{1}{(20)^{\frac{0.4}{1.4}}} = 0.575$$

$$\boxed{\eta = 57.5\%}$$

15. (a) Latent heat of water at 100°C is $2300\frac{kJ}{kg}$

So, as temperature increases, h_{fg} decreases but slowly

$\therefore$ h_{fg} at 102.5°C = 2290 kJ/kg

16. (c) Given that

$Q_1 = 1200\,J$

$T_1 = 900\,K$

$T_0 = 300\,K$

$W = 600\,J$

$Q_2 = Q_1 - W = 1200 - 600 = 600\,J$

$(\Delta s)_{univ} = (\Delta s)_{sys} + (\Delta s)_{surr}$

$$(\Delta s)_{univ} = \frac{-Q_1}{T_1} + \frac{Q_2}{T_0}$$

$$(\Delta s)_{univ} = \frac{-1200}{900} + \frac{600}{300}$$

$$(\Delta s)_{univ} = \frac{2}{3}$$

Irreversibility (I) = $T_0 \cdot (\Delta s)_{univ}$

$$I = 300 \times \frac{2}{3}$$

$$\boxed{I = 200\,J}$$

17. (b) m_v = mass of saturated liquid

m_g = mass of steam

m_m = total mass of mixture

$m_v \times h_v + m_g h_{steam} = m_m (h_m)_{x=0.9}$

$604.73 m_v + 1000 \times 3066.75$

$= [0.9(2738.53 - 604.73) + 604.73]_{(100+x)}$

$m_v = 282$ kg/hr

18. (b) Given that

$PV^{1.3} = C$

$P_1 = 1.5\,MPa$

$P_2 = 7.5\,MPa$

$V_1 = 0.09\,m^3$

$P_1 V_1^n = P_2 V_2^n$

$$V_2 = \left(\frac{P_1}{P_2}\right)^{\frac{1}{n}} \times V_1 = \left(\frac{1.5}{7.5}\right)^{\frac{1}{1.3}} = 0.026\,m^3$$

$$w = \frac{P_1 V_1 - P_2 V_2}{n - 1}$$

$$w = \frac{(1500 \times 0.09) - (7500 \times 0.026)}{1.3 - 1} kJ$$

$$\boxed{w = -200\,kJ}$$

19. (c) From the given table in question, we have

$s_g = s_f + x \cdot s_{fg}$

$6.6628 = 1.3026 + x(sg - sf)$

$6.6628 = 1.3026 + x(7.3594 - 1.3026)$

quality of steam at exit(x) = 0.88

20. (d) Mass flow rate $(\dot{m}) = 25$ kg/s

h_1 at 0.8 MPa = 2769.10 kJ/kg

$h_2 = h_f + x\,h_{fg}$ at 0.1 MPa

$h_2 = 417.46 + 0.88(2675.50 - 417.46)$

$h_2 = 2404.53$ kJ/kg

Power produced by turbine = $\dot{m}(h_1 - h_2)$

$= 25(2769.10 - 2404.53)$

$= 8830$ kW

21. (c) Given that

$\dot{m} = 0.03$ kg/s

At 150.9 kPa, $h_f = 17.82$ kJ/kg, $h_g = 178.74$ kJ/kg

At 500 kPa, $h_f = 50.64$ kJ/kg, $h_g = 195.01$ kJ/kg

For superheated vapour, h = 252.05 kJ/kg

Here,

$h_1 = h_4$

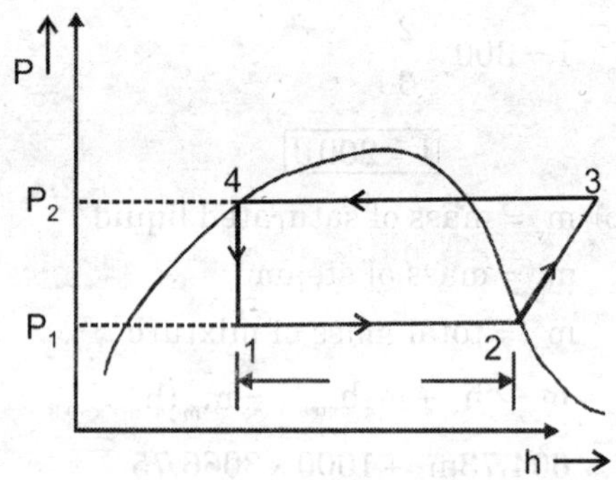

Refrigeration effect (RE) = $\dot{m}(h_2 - h_1)$

$RE = 0.03(195.01 - 50.64)$

$\boxed{RE = 4.33\,kW}$

22. (c) Work input to the compressor (W_c)

$W_c = \dot{m}(h_3 - h_2)$

$W_c = \dot{m}(h - h_2)$ [h – superheated]

$W_c = 0.03(252.05 - 195.01)$

$\boxed{W_c = 1.71\ kW}$

23. (c) Given that

$m_{ar} = 0.1$ kg

$C_p = 520$ kJ/kg k, $C_v = 312$ kJ/kg k

R = 0.208 kJ/ kg k

$$\gamma = \frac{C_p}{C_v} = 1.67$$

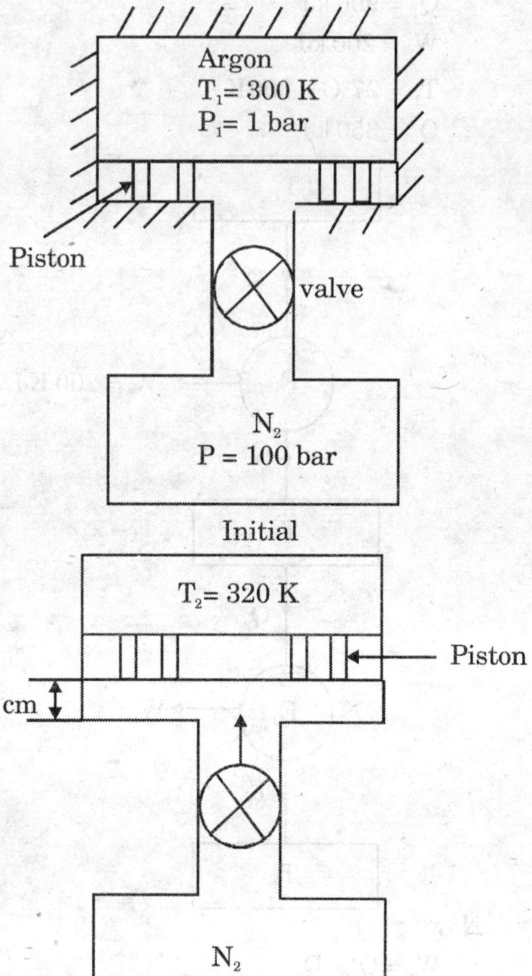

Work done by argon (W_{ar})

$$W_{ar} = \frac{P_1V_1 - P_2V_2}{\gamma - 1} = \frac{mR}{\gamma - 1}(T_1 - T_2)$$

$$W_{ar} = \frac{0.1 \times 0.208}{1.67 - 1}(300 - 320)$$

$$\boxed{W_{ar} = -0.6208\ kJ/kg}$$

24. (b) Work done by Nitrogen$\left(W_{N_2}\right)$

$$\left(W_{N_2}\right) = \frac{mgh}{1000} + 0.6208$$

$$\left(W_{N_2}\right) = \frac{5 \times 9.81 \times 0.1}{1000} + 0.6208$$

$$\boxed{\left(W_{N_2}\right) = 0.626\ kJ}$$

2010

Q.(1 – 9) carry one mark each.

1. Match the items in Group I for their correctness with the corresponding appropriate terms given in Groups II and III.

GROUP-I

P : Pressure

Q : Heat

GROUP-II

1. Path dependent quantity
2. Path independent quantity

GROUP-III

X : Intensive property

Y : Extensive property

(*a*) P, 1, X (*b*) P, 2, X

(*c*) Q, 1, X (*d*) Q, 2, Y

2. An object of mass '*m*' in a wooden box having mass 'M' falls through a height '*h*' under the influence of gravity in vacuum. The work done by the object on the box is

(*a*) 0 (*b*) *mgh*

(*c*) M*gh* (*d*) (*m* + M)*gh*

3. An ideal gas is known to obey following relationships: $u = 200 + 0.718\text{ T}$ and $\text{P}v = 0.287\ (\text{T} + 273)$, where u is specific internal energy (*kJ*/kg), T is temperature (°C), P is pressure (kPa) and v is specific volume (m³/kg). Specific heat (in kJ/kg-K) at constant pressure is

(*a*) 0.287 (*b*) 0.431

(*c*) 0.718 (*d*) 1.005

4. A heat pump, which operates in a cycle, extracts heat energy from the cold reservoir and supplies the same amount of energy to the hot reservoir. Which of the following statements holds for this process?

(*a*) This process violates both the first and the second law

(*b*) This process violates the first law but not the second law

(*c*) This process violates the second law but not the first law

(*d*) This process does not violate both first and second law

5. An insulated rigid container having 1 m³ volume has two compartments having equal volume separated by a thin membrane. Half of the container is filled with helium (R = 2.08 kJ/kg-K, C_p = 5.19 kJ/-Kg-K and C_v = 3.11 kJ/kg-K), while the remaining half is empty. Suddenly the membrane ruptures and helium fills the whole volume of the container. Temperature and pressure of helium before rupture are 500°C and 0.1 MPa respectively. The change in the entropy of helium is

(*a*) 0.019 kJ/K (*b*) 0.045 kJ/K

(*c*) 0.112 kJ/K (*d*) 0.675 kJ/K

6. 1 kg of methane is enclosed in a cylinder having volume 6.4 litres and is maintained at a temperature of 13°C and pressure of 18.56 MPa. If molecular weight of methane is 16 kg/kmol (for methane, critical pressure = 4.64 MPa, critical temperature is 191.1 K; universal gas constant is 8.314 kJ/kmol-K), compressibility factor, Z, is

(*a*) 0.375 (*b*) 0.8

(*c*) 1.25 (*d*) 2.66

7. $\left(\frac{\partial P}{\partial T}\right)_s$ is equal to

(*a*) $\left(\frac{\partial S}{\partial V}\right)_p$ (*b*) $-\left(\frac{\partial S}{\partial V}\right)_p$

(*c*) $\left(\frac{\partial S}{\partial V}\right)_T$ (*d*) $-\left(\frac{\partial S}{\partial V}\right)_T$

8. A rigid spherical vessel contains 1 kg of wet steam of quality x at pressure P_1. This is shown by point A on the T – v diagram. Heat is transferred to the vessel to form superheated steam at pressure P_2 and temperature T_2 as shown by point B.

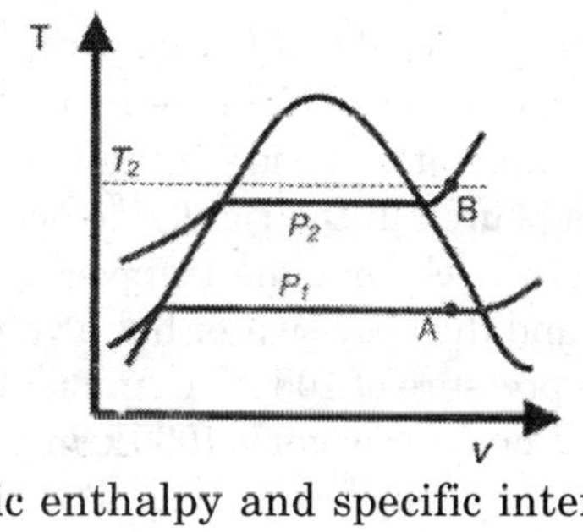

Specific enthalpy and specific internal energy corresponding to the saturated water and saturated vapour at pressures P_1 and P_2 as well as at points A and B are given by

Property	Saturated liquid		Saturated vapour		Point A	Point B
	Pressure P_1	Pressure P_2	Pressure P_1	Pressure P_2		
Specific Enthalpy (kJ/kg)	h_{f1}	h_{f2}	h_{g1}	h_{g2}	h_A	h_B
Specific internal energy (kJ/kg)	u_{f1}	u_{f2}	u_{g1}	u_{g2}	u_A	u_B

Heat transferred to the steam is

(*a*) $h_B - h_A$ (*b*) $h_B - h_{f1}$

(*c*) $u_B - u_A$ (*d*) $u_B - u_{f1}$

9. Determine the correctness or otherwise of the following **Assertion [a]** and the **Reason [r]**

Assertion : Carnot cycle is not used in vapour power cycles.

Reason : Pumping of a two phase mixture is difficult

(*a*) Both **[a]** and **[r]** are true and **[r]** is a reason for **[a]**

(*b*) Both **[a]** and **[r]** are true but **[r]** is NOT a reason for **[a]**

(*c*) **[a]** is true but **[r]** is false

(*d*) Both **[a]** and **[r]** are false

Q.(10 - 22) carry two marks each.

10. A slab of mass 2000 kg and volume 0.2 m^3 is raised slowly in the vertical direction by a massless rope through a height of 2 m from the bottom of a fresh-water lake. Depth of water in the lake is 10 m. If the density of fresh water is 1000 kg/m^3 and acceleration due to gravity is 10 m/s^2, work done by water on the block is

(*a*) – 40 kJ (*b*) 0 kJ

(*c*) 4 kJ (*d*) + 40 kJ

11. A bulb is connected to a U-tube with long vertical column as shown in the figure below. Cross-sectional area of both side of the U-tube is 3 cm^2. Water is poured in the right side open column of the U-tube such that the trapped air in the left column and the above water has a volume of 1000 cc and a pressure of 105 kPa. Ambient pressure is 100 kPa. Density of water is 1000 kg/m^3. Acceleration due to gravity may be taken as 10 m/s^2.

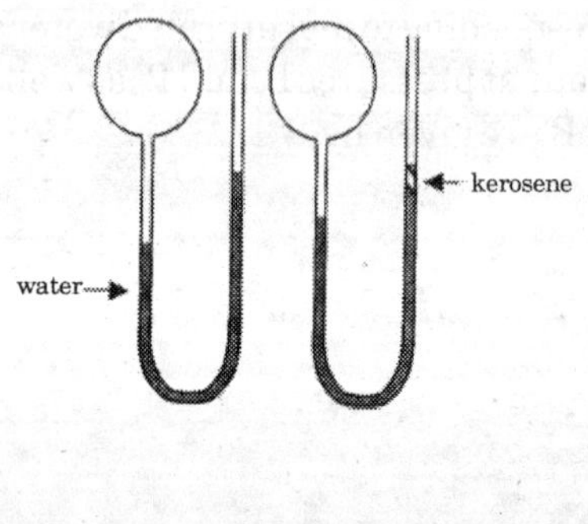

When 90 cc of kerosene oil (specific gravity = 0.7) is poured on the right side of the U-tube, water level in the left column rises by 10 cm. The pressure in the bulb becomes

(*a*) 105.1 kPa (*b*) 108.3 kPa

(*c*) 112.4 kPa (*d*) 119.7 kPa

12.

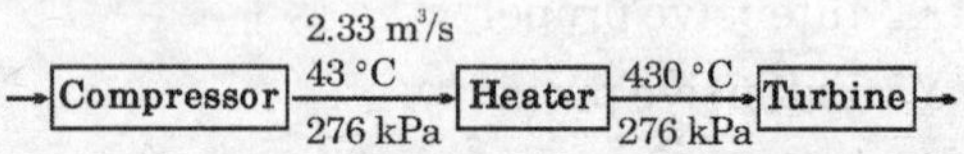

Air (R = 287 J/kg-K C_P = 1005 J/kg-K and γ = 1.4) flows sequentially through a compressor, a heater and a turbine as shown in the figure. Volume flow rate of air coming out from the compressor is 2.33 m^3/s when pressure and temperature are 276 kPa and 43°C respectively. Air is then heated at same pressure to 430°C in a heater. From heater, air flows through a turbine which produces 1860 kW of power. Heat loss from turbine to the surrounding is 90 kW. Air temperature at the turbine exit is

(*a*) 156.4°C (*b*) 181.6°C

(*c*) 223.7°C (*d*) 678.4°C

13. Availability per unit mass associated with air (R = 287 J/kg-K, C_P = 1005 J/kg-K and γ = 1.4) flowing from a reservoir at 10 atm and 25°C when atmosphere is at 1 atm and 25°C is (Neglect changes in the potential and the kinetic energies)

(*a*) 98.4 kJ/kg (*b*) 196.6 kJ/kg

(*c*) 492.3 kJ/kg (*d*) 689.14 kJ/kg

14. In an air standard Diesel cycle, compression ratio is 14. At the beginning of compression of air (R = 287 J/kg-K, C_P = 1005 J/kg-K and γ = 1.4) the temperature is 27°C and the pressure is 1 bar. If the specific volume after heat addition is two times the specific volume after compression, heat added (in kJ per kg of air) at constant pressure is

(*a*) 555.5 (*b*) 622.7

(*c*) 767.8 (*d*) 866.4

15. A mixture of Freon and air is supplied for cleaning a chamber. The mixture contains 70% by volume of air and 30% by volume of Freon. Specific heat ratios for Freon and air are 1.1 and 1.4 respectively.

Molecular mass of Freon is 200 g/mole and that of air is 30 g/mole. Temperature of gas is 300 K. If universal gas constant is 8.314 J/mole-K, specific heat ratio of the mixture is

(*a*) 1.16 (*b*) 1.21

(*c*) 1.25 (*d*) 1.31

16. Air-water vapour mixture having 100% relative humidity at 50°C is heated isobarically to 100°C in a closed system. If saturation pressure at 50°C is 12.352 kPa and at 100°C is 101.42 kPa, final relative humidity is

(*a*) 0% (*b*) 8.2%

(*c*) 12.2% (*d*) 100%

Common Data Questions

Common Data for Q.(17 – 18) :

Saturated vapour enters a turbine at a pressure of 2 bar and leaves the turbine at a pressure of 0.1 bar and a quality of 0.9. After condensation, saturated water at 0.1 bar is pumped into the boiler where it receives heat at a constant pressure of 2 bar. The pumping process can be considered to be isentropic. Use the data given in the following table to answer **Q.17** and **Q.18**.

Pressure (bar)	Saturation temperature (°C)	Specific volume (m^3/kg)		Specific enthalpy (kJ/kg)		Specific entropy (kJ/kg-K)	
		v_f	v_g	h_f	h_g	s_f	s_g
2	120.23	0.001061	0.8857	504.68	2706.6	1.530	7.1271
0.1	45.81	0.001010	14.674	191.81	2584.6	0.6492	8.1501

17. Work done by the turbine is

(*a*) 276.5 kJ/kg (*b*) 303.9 kJ/kg

(*c*) 335.8 kJ/kg (*d*) 361.3 kJ/kg

18. Heat addition in the boiler is

(*a*) 2000.9 kJ/kg (*b*) 2514.6 kJ/kg

(*c*) 3028.2 kJ/kg (*d*) 3554.5 kJ/kg

Common Data for Q.(19 - 20)

An insulated piston-cylinder assembly having a paddle wheel, as shown in the adjacent figure, contains air (R = 287 J/kg-K, and C_v = 718 J/kg-K) of mass 4 kg. Both piston and paddle wheel can be considered as insulated and massless. Temperature and pressure of air inside the cylinder are 300 K and 100 kPa respectively. Ambient pressure is 100 kPa.

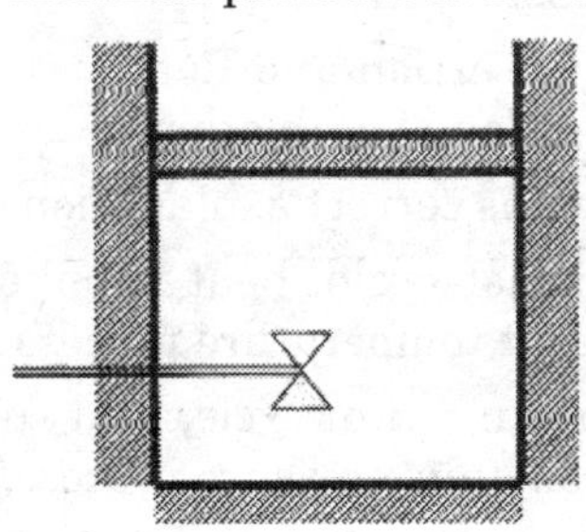

19. If the piston is locked in the fixed position and the paddle wheel delivers 75 kJ of work, final air temperature is

(*a*) 300 K (*b*) 318.7 K

(*c*) 320.6 K (*d*) 326.1 K

20. If the piston is free to slide without any friction when the paddle wheel delivers 75 kJ of work, final temperature of air in the cylinder is

(*a*) 305.2 K (*b*) 309.3 K

(*c*) 312.6 K (*d*) 318.7 K

Linked Answer Questions

Statement for Linked Answer Q.(21 - 22)

In a process industry, two different streams of water (to be considered incompressible) are available at 10°C and 90°C as shown in the figure. Mass flow rates of both the streams are 1 kg/s. Rather than wasting these resources, it is desired to connect a reversible Carnot engine that will continuously extract heat from the hot stream and supply part of it to the cold stream such that the exit temperature of both the streams T_f is identical. Heat capacity of water is 4.18 kJ/kg-K.

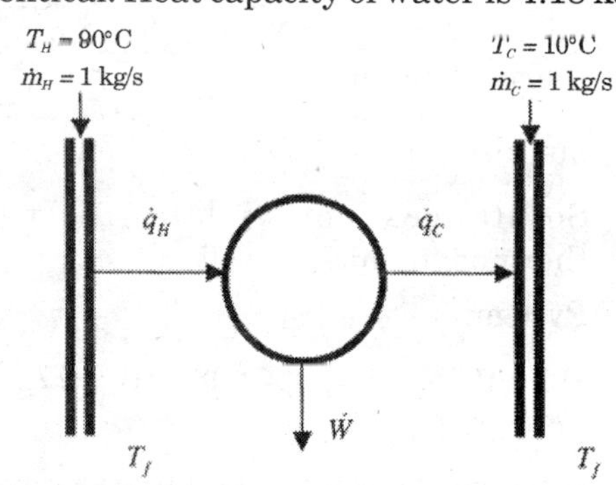

21. Value of T_f is

(*a*) 30°C (*b*) 42.5°C

(*c*) 47.5°C (*d*) 50°C

22. Work output W is

(*a*) 20.8 kW (*b*) 42.5 kW

(*c*) 63 kW (*d*) 167 kW

ANSWERS

1. (b)	**2.** (a)	**3.** (d)	**4.** (c)	**5.** (b)	**6.** (b)	**7.** (a)
8. (a)	**9.** (a)	**10.** (d)	**11.** (b)	**12.** (b)	**13.** (b)	**14.** (d)
15. (a)	**16.** (c)	**17.** (d)	**18.** (b)	**19.** (d)	**20.** (d)	**21.** (d)
22. (d)						

EXPLANATIONS

1. (b) Pressure
   (i) Path independent quantity
   (ii) It is a point function
   (iii) It is an intensive property.
   Heat
   (i) Path function
   P – 2, X

2. (a) During the free expansion process work done is equal to zero.

3. (d) u = 200 + .718 T
   Pv = .287 (T + 273)
   h = u + Pv
   $$C_p = \frac{dh}{dT}$$
   $$= \frac{d}{dT}(200 + .718T + .287T + .287 \times 273)$$
   $$= .718 + .287$$
   $$= 1.005\frac{kJ}{kgK}$$

4. (c) Heat pump is the application of second law of thermodynamics
   $W_{input} = 0$
   So it violates the second law of thermodynamics.
   but also $\oint \delta Q = \oint \delta W$ $\quad \oint \delta Q = 0$
   So it does not violates first law of thermodynamics as well.

5. (b) Pv = mRT
   $$0.1 \times 10^6 \times 0.5 = m \times 2.08 \times 10^3 \times 773$$
   m = .0310 kg
   since container is insulated, temperature remains constant.
   $$ds = mR\ln\left(\frac{V_2}{V_1}\right)$$
   $$= .031 \times 2.08 \times \ln\left(\frac{2}{1}\right)$$
   $$= .045 \text{ kJ/K}$$

6. (b) Compressibility factor $(z) = \frac{P_c}{\delta R\ T_c}$
   $$z = \frac{18.56 \times 10^6}{\frac{8.314 \times 10^3}{16} \times 286 \times \frac{1}{6.4 \times 10^{-3}}}$$
   z = 0.8

7. (a) $T \cdot ds = dH - v \cdot dP$
   $dH = T \cdot ds + v \cdot dP$
   Now,
   $$\left(\frac{\partial T}{\partial P}\right)_s = \left(\frac{\partial v}{\partial s}\right)_p$$
   $$\left(\frac{\partial P}{\partial T}\right)_s = \left(\frac{\partial s}{\partial v}\right)_p$$

8. (a)
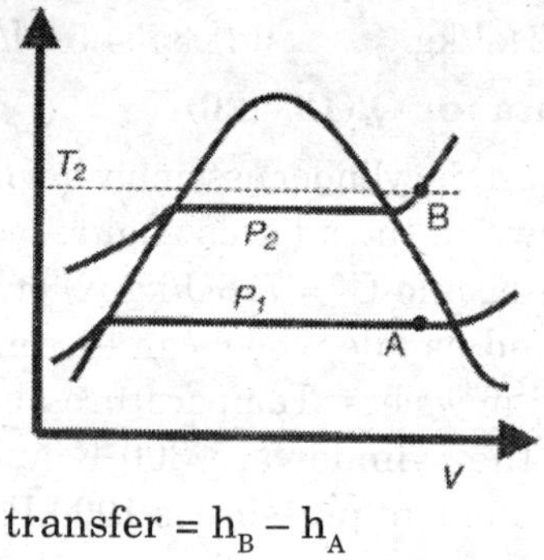

   Heat transfer = $h_B - h_A$

9. (a) Both assertion and Reason are correct, and explantion of eath other.
   Reason is correct explanation of assertion.
   - In Carnot cycle, heat supplied follows the constant temperature process.
   - In vapour power cycle, heat supplied follows the constant pressure process.

10. (d) m = 2000 kg
    h = 2 m
    Work done by water(w) = mgh
    = 2000 × 10 × 2
    w = 40 kJ

11. (b) The rise of the kerosene
    $$A \times h = hK$$
    $$3h = 90$$

$h = 30 \text{ cm} = 0.3 \text{ m}$

$P = \delta gh$

$= 700 \times 10 \times 0.3$

$= 2.1 \text{ kPa}$

Pressure in right limb due to rise of water

$= 1000 \times 0.1 \times 10 = 1 \text{ kPa}$

Total Pressure = 105 + 2.1 + 1 = 108.1 kPa

12. (b) At the exit of compressor,

$Pv = mRT$

$276 \times 10^3 \times 2.33 = m \times 287 \times 316$

$\dot{m} = 7.09 \frac{kg}{sec}$

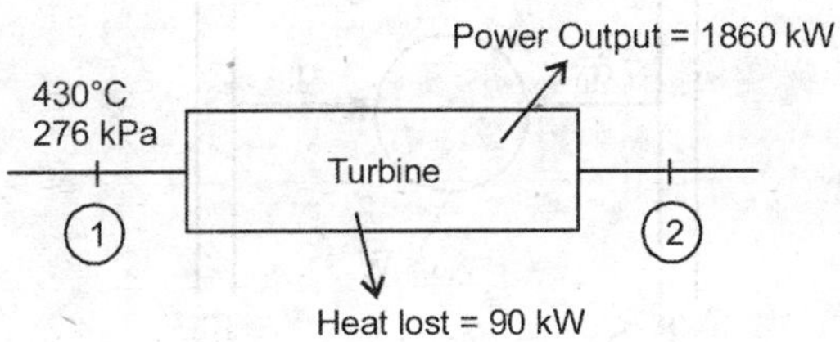

$(\delta\dot{W} - \delta\dot{Q}) = \dot{m}Cp(T_1 - T_2)$

$10^3(1860 - 90) = 7.09 \times 1005 \times (430 - T_2)$

$T_2 = 181.6 \text{ K}$

13. (b) Given that

$R = 0.287$ kJ/kg-k

$T_1 = T_2 = 25°C = 298$ k

$P_1 = 10$ atm

$P_2 = 1$ atm

Change of entropy (ds)

$$= Cp \ln\left(\frac{T_2}{T_1}\right) - R \ln\left(\frac{P_2}{P_1}\right)$$

$$ds = 1.005 \ln\left(\frac{298}{298}\right) - 0.287 \ln\left(\frac{1}{10}\right)$$

$ds = 0.66$ kJ/kg-k

Avaliability $= To \cdot ds$

$= 298 \times 0.66$

$A = 196.93$ kJ/kg

14. (d) For Air standard Diesel cycle.

$\gamma = 1.4$

$r = 14$ (Compression Ratio)

$R = 0.287$ kJ/kg K

$C_p = 1.005$ kJ/kg K

$T_1 = 27°C = 300$ K

$P_1 = 1$ bar

$V_3 = 2V_2$

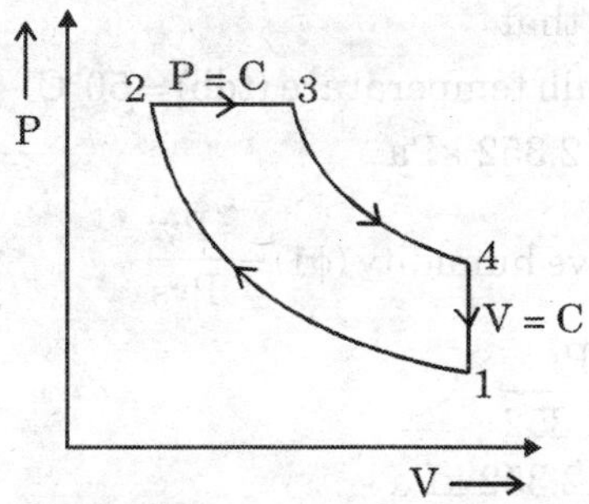

$P_1V_1 = RT_1$

$$V_1 = \frac{0.287 \times 300}{100}$$

$V_1 = 0.861 \text{ m}^3/\text{kg}$

$$V_2 = \frac{0.861}{14} = 0.0615 \text{ m}^3/\text{kg}$$

From 1 – 2 process,

$T_1V_1^{\gamma-1} = T_2V_2^{\gamma-1}$

$T_2 = 300(14)^{1.4-1} = 862.13\text{K}$

$P_1V_1^{\gamma} = P_2V_2^{\gamma}$

$P_2 = 1 \times 14^{1.4} = 40.23$ bar

$V_3 = 2V_2$

$V_3 = 2 \times V_2$ $[V_1 = V_4]$

2 – 3 (constant pressure process)

$P_2V_2 = mRT_2$

$$\frac{V_2}{T_2} = C$$

$$\frac{V_2}{T_2} = \frac{V_3}{T_3}$$

$$T_3 = \left(\frac{V_3}{V_2}\right) \times T_2$$

$T_2 = (2) \times 862.13$

$T_2 = 1724.26$ K

Heat added at constant pressure(Q_p)

$Q_P = mCp(T_3 - T_2)$

$Q_P = 1.005(1724.26 - 862.13)$

$\boxed{Q_P = 866.4 \text{ kJ/kg}}$

15. (a) $$C_{mixture} = \frac{C_{Fr} \times m_{Fr} \times 0.3 + C_{air} \times m_{air} \times 0.7}{m_{Fr} \times 0.3 + m_{air} \times 0.7}$$

$$= \frac{200 \times 1.1 \times 0.3 + 1.4 \times 30 \times 0.7}{200 \times 0.3 + 30 \times 0.7}$$

$= 1.17$

16. (c) Given that

Dry bulb temperature (tdb) = 50°C

$P_{vs} = 12.352$ kPa

$$\text{Relative humidity } (\phi 1) = \frac{Pv}{Pvs}$$

$$1 = \frac{Pv}{12.352}$$

Pv = 12.352 kPa

at $tdb_2 = 100°C$

$P_{vs} = 101.42$ kPa

$$\text{Relative humidity } (\phi 2) = \frac{Pv}{Pvs}$$

$$\phi = \frac{12.352}{101.42}$$

$\phi = 0.1217$

$$\boxed{\phi = 12.17\%}$$

17. (d) From table,

quality (x) = 0.9

$h_2 = h_{f2} + x\, h_{fg2}$

$h_2 = 191.81 + 0.9\,(2584.6 - 191.81)$

$h_2 = 2345.32$ kJ.kg

$h_1 = 2706.60$ kJ/kg

Work done by turbine$(W_T) = h_1 - h_2$

$W_T = 2706.60 - 2345.32$

$$\boxed{W_T = 361.28 \text{ KJ/kg}}$$

18. (b) Heat addition in boiler (Q_b)

$Q_b = h_1 - h_f$ (at 0.1 bar)

$h_1 = 2706.60$ kJ/kg

hf = 191.81 kJ/kg

Qb = 2706.60 – 191.81 = 2514.79 KJ/kg

19. (d) Work supplied to the system will be absorbed as heat in the system, at constant volume

$$\delta Q = dU$$

$$\delta Q = mC_v dT$$

$$75000 = 4 \times 718 \times (T_2 - T_1) \quad (\because T_1 = 300K)$$

$$\therefore \quad T_2 = 326.1 \text{ K}$$

20. (d) When piston is free to move,

$$\delta Q = dH$$

$$\delta Q = mC_v dT$$

$$75000 = 4 \times 1005 \times (T_2 - T_1) \quad (\because T_1 = 300K)$$

$$T_2 = 318.65 \text{ K}$$

21. (d) Given that

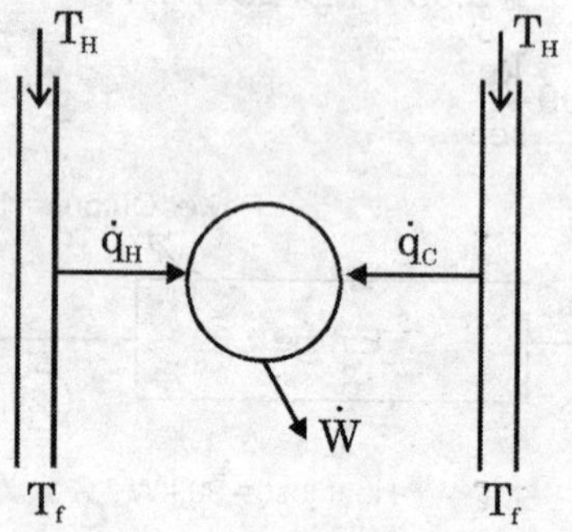

$\dot{m}_h = \dot{m}_c = 1\text{kg/sec}$

$\dot{q}_H = \dot{q}_C, \quad C_{pH} = C_{pc}$

$m_h C_{pH}(T_H - T_f) = m_c C_{pc}(T_f - T_c)$

$(363 - T_f) = T_f - 283$

$2T_f = 646$

$T_f = 323$ k

$$\boxed{T_f = 50\,°C}$$

22. (d) Work output ($\dot{W}$)

$\dot{W} = \dot{m}Cpc\,(T_f - T_c)$

$\dot{W} = 1 \times 4.18(323 - 283)$

$$\boxed{\dot{W} = 167.2 \text{ kW}}$$

2011

Note : Usual notations have been used for thermodynamic variables.

Useful Data

Universal gas constant, $R_0 = 8.314$ kJ/kmol.

K; Acceleration due to gravity, $g = 9.81$ m/s^2.

Molecular mass of air, $M_{air} = 29$ kg/kmol;

Specific heat at constant pressure of air, $c_p = 1.008$ kJ/kgK ;

Ratio of specific heats for air, $\gamma = 1.4$.

Assume air to be a perfect gas unless specified otherwise.

Q.(1-9) carry one mark each.

1. Identify the group containing the appropriate match of items in List-I and List-II.

List-I	List-II
K. A jet engine in flight	P. Closed system
L. Water being heated in a sealed container	Q. Control volume
M. Internal energy	R. Intensive property
N. Specific entropy	S. Extensive property

	K	L	M	N
(a)	P	Q	R	S
(b)	Q	P	R	S
(c)	Q	P	S	R
(d)	P	Q	S	R

2. The mass, internal energy, pressure, and volume of a system are 10 kg, 100 kJ, 1 bar, and 1 m^3, respectively. The value of specific enthalpy, in kJ/kg, of the system is

(a) 10 (b) 20

(c) 30 (d) 40

3. A heat pump extracts heat at the rate of 100 kW from a low temperature reservoir and delivers heat at the rate of 160 kW to a high temperature reservoir. The COP of the heat pump is

(a) 0.60 (b) 1.67

(c) 2.60 (d) 2.67

4. At a certain pressure, the specific volumes (m^3/kg) of saturated liquid, saturated vapor, and wet steam are 1.1565×10^{-3}, 0.1274, and 0.1, respectively. The quality of wet steam is

(a) 0.805 (b) 0.783

(c) 0.776 (d) 0.673

5. Which one of the following expressions represents the Joule-Thomson coefficient?

(a) $\left(\frac{\partial T}{\partial p}\right)_h$ (b) $\left(\frac{\partial T}{\partial v}\right)_h$

(c) $\left(\frac{\partial p}{\partial h}\right)_s$ (d) $\left(\frac{\partial s}{\partial T}\right)_p$

6. Specific heat at constant pressure (C_p) of helium is 5.19 kJ/kgK and its molecular mass is 4 kg/kmol. The specific heat at constant volume of helium, in kJ/kgK, is

(a) 1.11 (b) 2.11

(c) 3.11 (d) 4.11

7. An evacuated, rigid, adiabatic tank is filled slowly with air from a supply line supplying air at a constant pressure, p_L and temperature. T_L. The temperature of air in the tank at the end of the filling process will be

(a) greater than T_L

(b) equal to T_L

(c) less than T_L

(d) the average of ambient temperature and T_L

8. An ideal gas undergoes a cyclic process as shown in p-V diagram below :

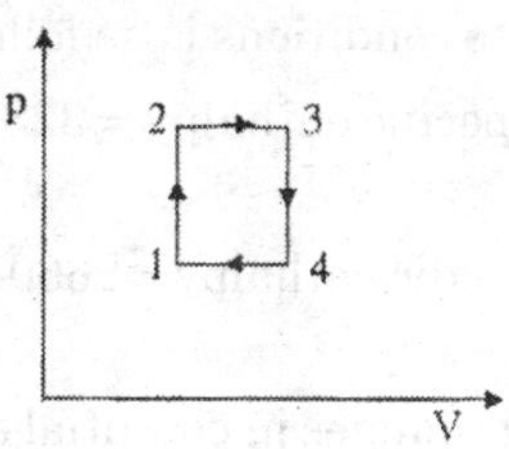

The same cycle, represented in T-S diagram is

(a)

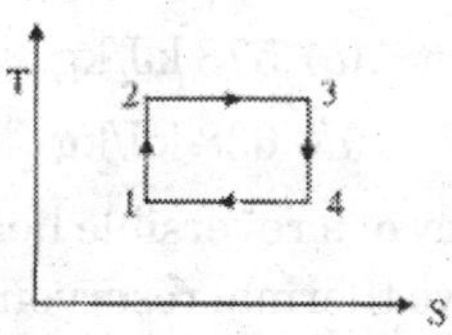

(b)

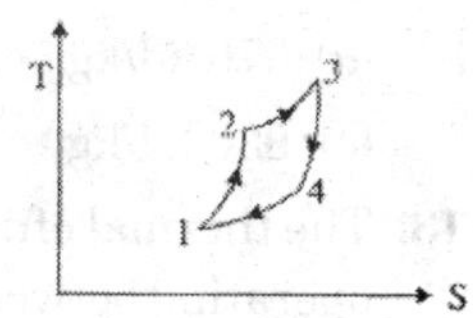

(c)

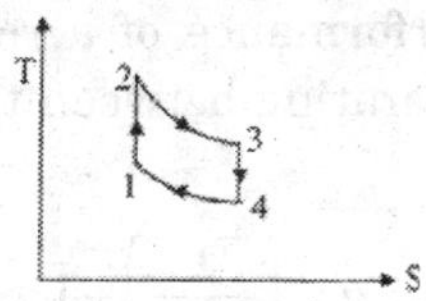

(d)

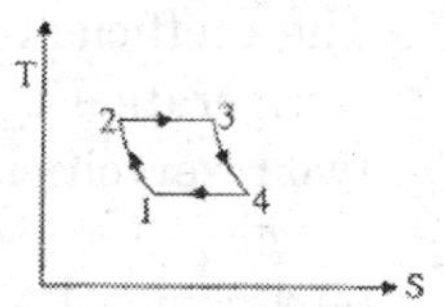

9. Consider three identical tanks A, B and C, shown below. What is the pressure p in tank C?

Tank A	Tank B
1m^3	1m^3
x kg nitrogen	y kg oxygen
1 bar, 300 K	1 bar, 300 K

Tank C
$1m^3$
y kg nitrogen +
y kg oxygen
300 K, p = ?

(a) 1 bar (b) 1.5 bar

(c) 2 bar (d) 2.5 bar

Q.(10-22) carry two marks each.

10. The thermometric property, x, of a thermometer varies with temperature, t, according to the relation $t = ax^2 + b$, where t is in °C, x is in cm, and a & b are constants. At ice point (0°C) and steam point (100°C), the values of x are 5 cm and 20 cm, respectively. When this thermometer is brought in contact with a heated body, the value of x is recorded as 15 cm. The temperature of the heated body in °C is

(a) 83.3 (b) 73.3

(c) 63.3 (d) 53.3

11. Air at 10 bar and 400 K is flowing with a velocity of 20 m/s through a pipe of uniform cross-sectional area of 20 cm^2. The mass flow rate of air in the pipe is approximately

(a) 0.25 kg/s (b) 0.35 kg/s

(c) 2.5 kg/s (d) 3.5 kg/s

12. The condition of steam at inlet and exit of a perfectly insulated steam turbine running under steady-state conditions is as follows :

At inlet : specific enthalpy = 3230 kJ/kg; velocity = 160 m/s

At exit : specific enthalpy = 2660 kJ/kg; velocity = 100 m/s

Neglecting changes in potential energy, the work done by the steam turbine is approximately

(a) 570 kJ/kg (b) 578 kJ/kg

(c) 630 kJ/kg (d) 638 kJ/kg

13. The thermal efficiency of a reversible heat engine operating between two thermal reservoirs is η_{max}. The coefficient of performance of a reversible refrigeration cycle operating between the same two reservoirs is

(a) $\left(\frac{1}{\eta_{max}}\right) - 1$ (b) $\left(\frac{1}{\eta_{max}}\right) + 1$

(c) $\left(\frac{1}{\eta_{max}}\right)$ (d) $1 - \left(\frac{1}{\eta_{max}}\right)$

14. Given for water,

at 190°C, saturation pressure = 12.54 bar

at 210°C, saturation pressure = 19.06 bar

at 200°C, specific volume of saturated liquid = 1.1565×10^{-3} m^3/kg

at 200°C, saturation volume of saturated vapor = 0.1274 m^3/kg

Using Clapeyron equation, the enthalpy of vaporization (kJ/kg) of water at 200°C is approximately

(a) 1857 (b) 1924

(c) 1947 (d) 2025

15. An ideal vapor compression refrigeration cycle uses R-134a as working fluid. The condenser pressure is 5 bar and evaporator pressure is 60 kPa. The refrigerant enters the compressor as saturated vapor and its specific enthalpy at the compressor exit is 267.98 kJ/kg. The other relevant data regarding R-134a are given below :

Pressure, bar	0.6	5
h_f, kJ/kg	3.46	71.33
h_g, kJ/kg	224.72	256.07
s_f, kJ/kg.K	0.0147	0.2723
s_g, kJ/kg.K	0.9520	0.9117

The COP of the cycle is approximately

(a) 2.55 (b) 3.55

(c) 4.11 (d) 5.11

16. One kilomole of hydrogen (M = 2 kg/kmol) is mixed with certain number of kilomoles of argon (M = 40 kg/kmol) such that the mass fraction of argon in the resultant mixture is 0.8. The number of kilomoles of argon in the mixture is

(a) 0.05 (b) 0.10

(c) 0.15 (d) 0.20

COMMON DATA QUESTIONS

Common Data for Questions 17 and 18 :

A steam power plant operates on an ideal Rankine cycle between pressures of 5 MPa (boiler) and 10 kPa (condenser). Water enters the pump at a specific enthalpy of 191.83 kJ/kg, and specific volume 0.00101 m^3/kg, and steam enters the turbine at an enthalpy of 4612.2 kJ/kg, and leaves the turbine as saturated vapor at 10 kPa. The enthalpy of vaporization of water at 10 kPa is 2392.8 kJ/kg.

17. The net work done per cycle (kJ/kg) is approximately

(a) 2023 (b) 2123

(c) 2223 (d) 2323

18. The thermal efficiency of the cycle is

(a) 35.8% (b) 40.8%

(c) 45.8% (d) 50.8%

Common Data for Questions 19 and 20 :

At the beginning of compression process of an air-standard Otto cycle, pressure, temperature, and volume are 1 bar, 290 K, an 400 cm^3, respectively. The compression ratio is 8. The maximum temperature attained in the cycle is 2200 K.

19. The heat transferred (kJ) during the heat addition process is approximately

(*a*) 531 (*b*) 53.1

(*c*) 5.31 (*d*) 0.531

20. The thermal efficiency of the cycle is

(*a*) 41.5% (*b*) 46.5%

(*c*) 51.5% (*d*) 56.5%

LINKED ANSWER QUESTIONS

Statement for Linked Answer Questions 21 & 22 :

An ideal gas undergoes a cyclic process consisting of the following three processes :

Process 1-2 : Compression process with pV = constant

Process 2-3 :Constant pressure

Process 3-1 :Constant volume ; $U_3 - U_1 = 3549$ kJ

Changes in kinetic and potential energies are neglected.

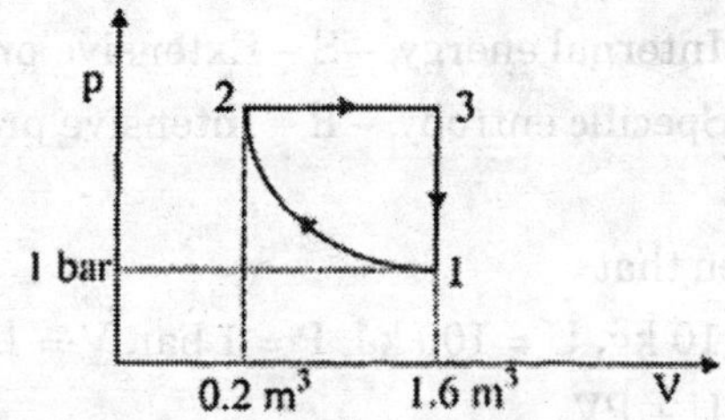

21. The work done (kJ) during the process 2-3 is

(*a*) 1120 (*b*) 1220

(*c*) 1320 (*d*) 1420

22. The heat transferred (kJ) during the process 2-3 is

(*a*) 4269 (*b*) 4469

(*c*) 4569 (*d*) 4669

ANSWERS

1. (c)	**2.** (b)	**3.** (d)	**4.** (b)	**5.** (a)	**6.** (c)	**7.** (a)
8. (b)	**9.** (c)	**10.** (d)	**11.** (b)	**12.** (b)	**13.** (a)	**14.** (c)
15. (b)	**16.** (d)	**17.** (a)	**18.** (c)	**19.** (d)	**20.** (d)	**21.** (a)
22. (d)						

EXPLANATIONS

1. (c)

K – A jet engine in flight, – Q – Control volume

L – Water being heated in a sealed container, – P – Closed system

M – Internal energy, – S – Extensive property

N – Specific entropy, – R – Intensive property

2. (b)

Given that

$m = 10$ kg, $U = 100$ kJ, $P = 1$ bar, $V = 1\ m^3$

$H = U + PV$

$H = 100 + (100 \times 1)$

$H = 200$

$h = \frac{200}{10}$ kJ/kg.

$\boxed{h = 20 \text{ kJ/kg.}}$

3. (d)

$$(COP)_{HP} = \frac{Q_1}{Q - Q_2}$$

$$(COP)_{HP} = \frac{160}{60}$$

$$\boxed{(COP)_{HP} = 2.67}$$

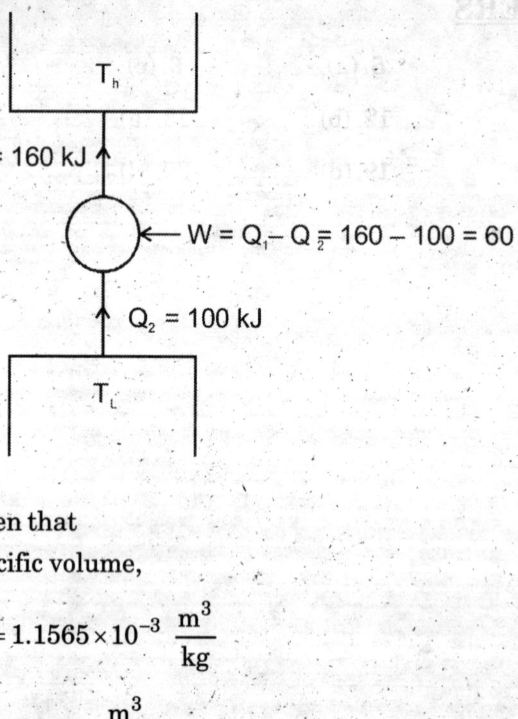

4. (b)

Given that

Specific volume,

$$v_f = 1.1565 \times 10^{-3}\ \frac{m^3}{kg}$$

$$v_g = 0.1274\ \frac{m^3}{kg}$$

$$v_{fg} = 0.1\ m^3/kg$$

Now, $v_g = v_f + x \cdot v_{fg}$

$$0.1274 = (1.1565 \times 10^{-3}) + (x \times 0.1)$$

Dryness fraction (x) = 0.783

5. (a)

The Joule – Thomson coefficient (μ)

$$\mu = \left(\frac{\partial T}{\partial P}\right)_h$$

6. (c)

Given that

Specific heat at constant pressure (C_P) = 5.19 kg/kgK

Molecular mass (M) = 4 kg/k mol.

Characteristics gas constant (R)

$$R = \frac{\bar{R}}{M}$$

$$R = \frac{8.314}{4} = 2.0785 \text{ kJ/kgK}$$

$$C_P - C_V = R$$

$$C_V = C_P - R = 5.19 - 2.0785$$

$$C_V = 3.11 \text{ kJ/kg} \cdot K$$

7. (a)

The temperature of air (T_f)

$$T_f = \gamma \cdot T_i$$

$$\boxed{T_f = \gamma \cdot T_L}$$

8. (b)

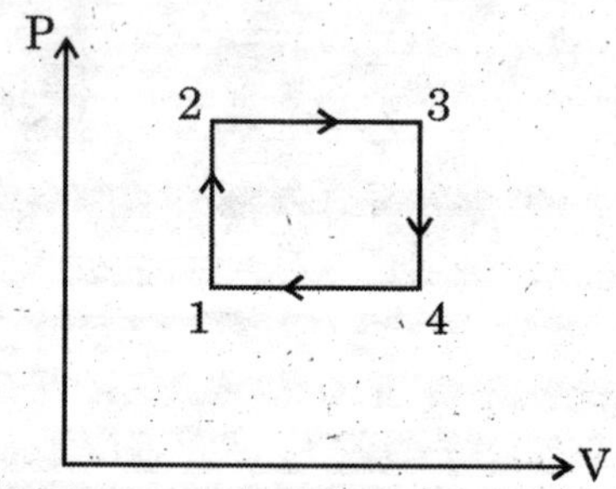

The cycle on T-S diagram.

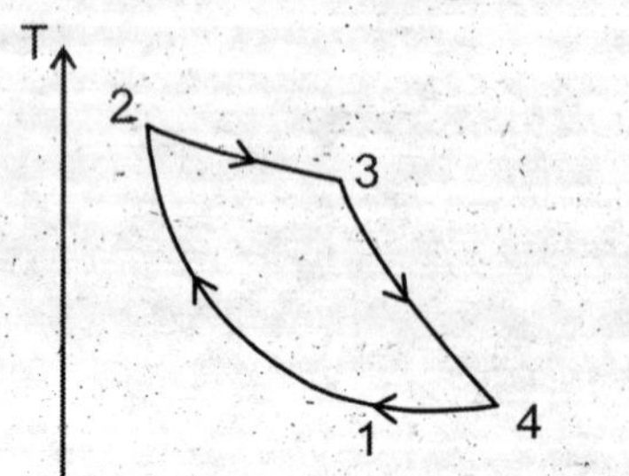

9. (c)

Pressure on the tank C (P_C)

$P_C = P_A + P_B$

$P_C = 1 + 1 = 2$ bar [Given that]

10. (d)

Given that

$t = ax^2 + b$

at $t = 0\,°C$, $x = 5$ cm.

$0 = 25a + b$ ...(1)

at $t = 100\,°C$, $x = 20$ cm.

$100 = 400a + b$ ...(2)

Solving equation (1) & (2)

$$\begin{array}{rcl} 25a + b &=& 0 \\ 400a + b &=& 100 \\ \hline 375a &=& 100 \end{array}$$

$$a = \frac{4}{15}$$

$$b = -25a = -25 \times \frac{4}{15} = -\frac{20}{3}$$

at $x = 15$ cm.

$t = ax^2 + b$

$$t = \frac{4}{15}(15)^2 - \frac{20}{3}$$

$$t = 60 - \frac{20}{3}$$

$$\boxed{t = 63.33°C}$$

11. (b)

P = 10 bar

T = 400 K

V = 20 m/s.

$A = 20\ cm^2 = 20 \times 10^{-4}\ m^2$

PV = mRT

$P = \rho RT$

$$\rho = \frac{10 \times 100}{287 \times 400} = 8.71\ kg/m^3$$

Mass flow rate $(\dot{m}) = 8.71 \times 20 \times 10^{-4} \times 20$

$\dot{m} = 0.3484$ kg/sec.

12. (b)

Given that

$h_1 = 3230$ kJ/kg.

$V_1 = 160$ m/s.

$h_2 = 2660$ kJ/kg.

$V_2 = 100$ m/s.

$$h_1 + \frac{V_1^2}{2} = h_2 + \frac{V_2^2}{2} + W_T$$

$$W_T = (h_1 - h_2) + \frac{V_1^2 - V_2^2}{2000}$$

$$W_T = (3230 - 2660) + \frac{160^2 - 100^2}{2000}$$

$$\boxed{W_T = 577.8\ kJ/kg.}$$

13. (a)

Thermal efficiency of a reversible heat engine (η_{max})

$$\eta_{max} = \frac{T_1 - T_2}{T_1} = 1 - \frac{T_2}{T_1}$$

$$(COP)_R = \frac{T_2}{T_1 - T_2} = \frac{1}{\frac{T_1}{T_2} - 1}$$

$$\boxed{(COP)_R = \left(\frac{1}{\eta_{max}}\right) - 1}$$

14. (c)

Given that

According Clausius Clapeyron equation,

$$\frac{dp}{dT} = \frac{h_{fg}}{T(v_g - v_f)}$$

$$h_{fg} = \frac{dp}{dT} \cdot T(v_g - v_f)$$

$$h_{fg} = \frac{(19.06 - 12.54) \times 10^2}{(210 - 190)} \times 200(0.1274 - 1.1565 \times 10^{-3})$$

Enthalpy of vaporization $h_{fg} = 1947.01$ kJ/kg.

15. (b)

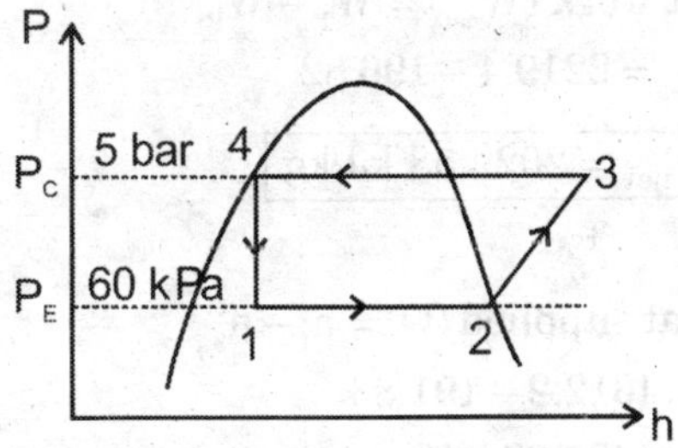

Given that

$h_3 = 267.98$ kJ/kg

$h_4 = 71.33$ kJ/kg.

$h_2 = 224.72$ kJ/kg.

$$COP = \frac{R \cdot E}{W_C}$$

$$COP = \frac{h_2 - h_1}{h_3 - h_2}$$

$$COP = \frac{224.72 - 71.33}{267.98 - 224.72}$$

$$\boxed{COP = 3.55}$$

16. (d)

1 Kilomole for hydrogen, M = 2 kg/K mol.

For Argon, M = 40 kg/K mol.

x kilomole of Argon

$$\text{then mole fraction} = \frac{(0.2 \times 2) + (40 \times 0.8)}{40}$$

(x) mole fraction = 0.81

the kilomole of Argon (x) = 1 – 0.81 = 0.19.

17. (a)

For Rankine cycle

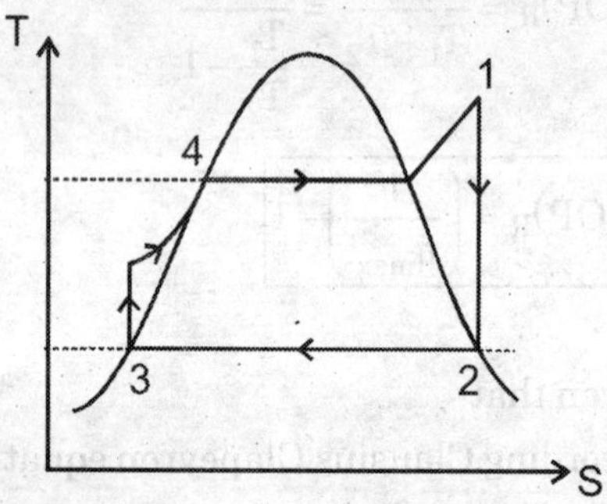

$h_1 = 4612.2$ kJ/kg

$h_2 = 2392.8$ kJ/kg

$h_3 = 191.83$ kJ/kg

$h_4 = (5000 - 10) \times 0.001001$

$h_4 = 4.99$ kJ/kg.

$W_T = h_1 - h_2 = 4612.2 - 2392.8$

Turbine work (W_T) = 2219.4 kJ/kg.

Pump work (W_P) = $h_3 + h_4$ = 196.82 kJ/kg

Net work (W_{net}) = $W_T - W_P$

$W_{net} = 2219.4 - 196.82$

$$\boxed{W_{net} = 2022.58 \text{ kJ/kg.}}$$

18. (c)

Heat supplied (Q) = $h_1 - h_3$

$Q = 4612.2 - 191.83$

$Q = 4420.37$ kJ/kg.

The thermal/efficiency of the cycle

$$\eta_{th} = \frac{W_{net}}{Q}$$

$$\eta_{th} = \frac{2022.58}{4420.37} = 0.4575$$

$$\boxed{\eta_{th} = 45.45\%}$$

19. (d)

For an air standard otto cycle

Given that

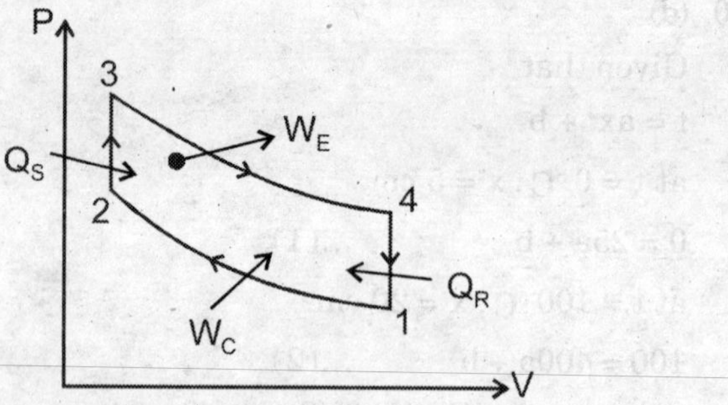

$P_1 = 1$ bar

$T_1 = 290$ K

$V_1 = 400 \text{ cm}^3 = 4 \times 10^{-3} \text{ m}^3$

Compression ratio (r) = 8

$T_{max} = T_3 = 2200$ K

For process 1-2.

$$T_1 V_1^{\gamma-1} = T_2 V_2^{\gamma-1}$$

$$T_2 = T_1 \left(\frac{V_1}{V_2}\right)^{\gamma-1} = 290(8)^{1.4-1}$$

$$T_2 = 666.24 \text{ K}$$

Heat transfer during heat addition process (Q_1)

$$Q_1 = mC_v(T_3 - T_2)$$

$$P_1 V_1 = mRT_1$$

$$m = \frac{P_1 V_1}{RT_1}$$

$$m = \frac{100 \times 4 \times 10^{-4}}{0.287 \times 290}$$

$$m = 4.81 \times 10^{-4} \text{ kg}$$

$$Q_1 = mC_v(T_3 - T_2)$$

$$Q_1 = 4.81 \times 10^{-4} \times 0.718(2200 - 666.24)$$

$$Q_1 = 0.529 \text{ kJ}$$

20. (d)

From P-V diagram

$V_1 = V_4$

$V_2 = V_3$

$$r = \frac{V_1}{V_2} = \frac{V_4}{V_3} = 8$$

From process 3-4

$$T_3 V_3^{\gamma-1} = T_4 V_4^{\gamma-1}$$

$$T_4 = 2200\left(\frac{V_3}{V_4}\right)^{\gamma-1}$$

$$T_4 = 2200\left(\frac{1}{8}\right)^{0.4} = 957.6 \text{ K.}$$

Heat repeated $(Q_2) = mC_v(T_4 - T_1)$

$$Q_2 = 4.81 \times 10^{-4} \times 0.718(957.6 - 290)$$

$$Q_2 = 0.231 \text{ kJ}$$

Thermal efficiency of cycle $(\eta_{th}) = \dfrac{Q_1 - Q_2}{Q_1}$

$$\eta_{th} = \frac{0.529 - 0.231}{0.529}$$

$$\eta_{th} = 0.5633$$

$$\boxed{\eta_{th} = 56.33\%}$$

21. (a)

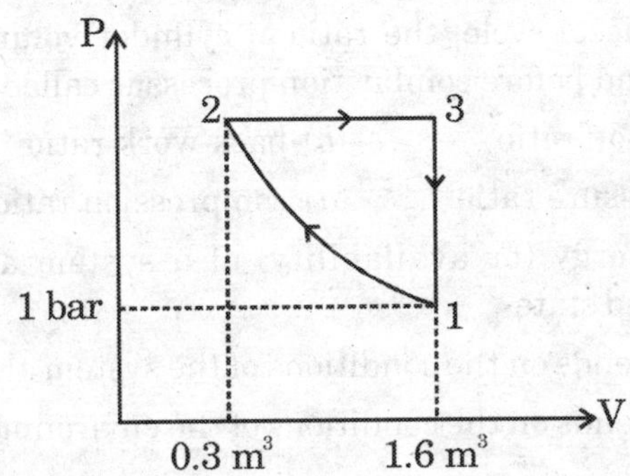

Process1-2, Compression process, PV= C

Process2-3, Constant Pressure

Pressure 3-1, Constant volume, $U_3 - U_1 = 3549$ kJ

$V_2 = 0.2 \text{ m}^3$

$V_1 = V_3 = 1.6 \text{ m}^3$

From process1-2

$$P_1V_1 = P_2V_2$$

$$P_2 = \frac{1 \times 1.6}{0.2} = 8 \text{ bar.}$$

Work done, process (2-3)

$W = P(V_3 - V_2) = 800\,(1.6 - 0.2)$

$W = 1120$ kJ

22. (d)

Heat transfer during process 2-3

$dQ = dU + p\cdot dV$

$dQ = 3549 + 1120$

$$\boxed{dQ = 4669 \text{ kJ}}$$

2012

Note : Usual notations have been used for thermodynamic variables.

Useful Data:

Unless otherwise specified, the following data may be assumed.

Universal gas constant, $\overline{R} = 8.314$ kJ/kmol.K;

Acceleration due to gravity, $g = 9.81$ m/s^2

Molecular mass of air, $M_{air} = 29$ kg/kmol;

Specific heat of air at constant pressure, $C_p = 1.005$ kJ/kg.K

Ratio of specific heats of air, $\gamma = 1.4$. Assume air to be a perfect gas unless specified otherwise.

Q. (1 – 9) carry one mark each.

1. Consider a piston-cylinder arrangement containing a gas. This system is heated by placing it on the top of a burner. The system undergoes

(*a*) a constant volume process

(*b*) a constant pressure process

(*c*) an adiabatic process

(*d*) an isothermal process

2. For a pure substance, at the triple point

(*a*) only solid and liquid phases co-exist in equilibrium

(*b*) only liquid and vapour phases co-exist in equilibrium

(*c*) only solid and vapour phases co-exist in equilibrium

(*d*) solid, liquid and vapor phases co-exist in equilibrium

3. In a saturated liquid-vapour mixture, the property quality, x is defined as

(*a*) $x = \dfrac{m_{vapour}}{m_{liquid} + m_{vapour}}$

(*b*) $x = \dfrac{m_{vapour}}{m_{liquid}}$

(*c*) $x = \dfrac{m_{liquid}}{m_{liquid} + m_{vapour}}$

(*d*) $x = \dfrac{m_{liquid}}{m_{vapour}}$

4. The slope of a Mollier diagram at constant pressure indicates

(*a*) enthalpy (*b*) entropy

(*c*) internal energy (*d*) temperature

5. If Q_L represents the magnitude of heat transfer from a low temperature reservoir to a cyclic device and Q_H represents the magnitude of heat transfer from a cyclic device to a high temperature reservoir, then for the same Q_L and Q_H, the coefficient performance of a refrigerator (COP_R) and the coefficient performance of a heat pump (COP_{HP}) can be related as

(*a*) $COP_R = 1 - COP_{HP}$

(*b*) $COP_{HP} = COP_R + 1$

(*c*) $COP_R . COP_{HP} = 1$

(*d*) $COP_{HP} = COP_R - 1$

6. Clausius inequality is written as

(*a*) $\oint \delta Q > 0$ (*b*) $\oint \delta Q < 0$

(*c*) $\oint \dfrac{\delta Q}{T} \le 0$ (*d*) $\oint \dfrac{\delta Q}{T} > 0$

7. In a Diesel cycle, the ratio of cylinder volumes after and before combustion process is called

(*a*) cut-off ratio (*b*) back work ratio

(*c*) pressure ratio (*d*) compression ratio

8. The exergy (or availability) of a system at a specified state

(*a*) depends on the conditions of the system alone

(*b*) depends on the conditions of the environment alone

(*c*) depends on the conditions of both the system and environment

(*d*) depends neither on the conditions of the system nor the environment

9. In each of the following choices, there are two expressions given. Select the choice that gives, first, the defining expression of volume expansivity and second, the expression of volume expansivity for ideal gases

(*a*) $\dfrac{1}{v}\left(\dfrac{\partial n}{\partial T}\right)_P, \dfrac{1}{T}$ (*b*) $\dfrac{1}{v}\left(\dfrac{\partial n}{\partial P}\right)_T, \dfrac{1}{T}$

(*c*) $-\dfrac{1}{v}\left(\dfrac{\partial v}{\partial P}\right)_T, \dfrac{1}{P}$ (*d*) $\dfrac{1}{v}\left(\dfrac{\partial v}{\partial T}\right)_T, \dfrac{1}{P}$

Q. (10 – 22) carry two marks each.

10. A certain mass of gas at 0°C is expanded to 81 times its original volume under adiabatic conditions. If ratio of specific heats of the gas, $\gamma = 1.25$, the final temperature of the gas is

(*a*) – 235°C (*b*) – 182°C

(*c*) – 91° C (*d*) 0° C

11. A sample of gas with initial average kinetic energy, E is heated from 27°C to 327°C. The average kinetic energy after heating is

(*a*) E (*b*) 2E

(*c*) 27E (*d*) 327E

12. Helium in a piston/cylinder assembly at 20°C and 100 kPa is brought to 400 K in a reversible polytropic process with exponent $n = 1.25$. Assume helium to be an ideal gas. The molecular mass of helium is 4.003 kg/kmol. The specific work in the process is approximately

(*a*) – 800 kJ/kg (*b*) – 788 kJ/kg

(*c*) 788 kJ/kg (*d*) – 888 kJ/kg

13. 32 kg of oxygen is mixed with 28 kg of nitrogen at the same temperature. The gases are at the same pressure of 103 kPa before and after mixing. If $\overline{R}$ is the universal gas constant in kJ/kmol.K, the change in entropy of the mixture is

(*a*) $1.38\ \overline{R}$ (*b*) $0.69\overline{R}$

(*c*) $\overline{R}$ (*d*) $0.34\overline{R}$

14. Consider two Carnot heat engines A and B operating in series. Engine A receives heat from a reservoir at 1750 K and rejects heat to another reservoir at temperature T. Engine B receives an amount of energy same as that rejected by Engine A from the reservoir at temperature T. Engine B then rejects heat to another reservoir at 320 K. In both cases, the engines produce some amount of work. If the thermal efficiencies of both the engines are the same, then the temperature T is approximately

(*a*) 848 K (*b*) 748 K

(*c*) 648 K (*d*) 548 K

15. Joule Thomson coefficient for a gas, μ_j obeying the relation $p(v-b) = RT$ is

(*a*) $\mu_j = \dfrac{c_p}{b}$ (*b*) $\mu_j = \dfrac{b}{c_p}$

(*c*) $\mu_j = -\dfrac{b}{c_p}$ (*d*) $\mu_j = -\dfrac{c_p}{b}$

16. The correct expression representing Z to be a thermodynamic property is

(*a*) $Z = pdv$ (*b*) $Z = vdp$

(*c*) $Z = pdv + vdp$ (*d*) $Z + pdv = vdp$

Common Data Q. (17 – 18) :

The vapour pressure of liquid ammonia (in atmosphere)in the vicinity of the triple point can be expressed as

$$\ln p + \frac{3063}{T} = 15.16$$

where temperature T is expressed in K.

In a similar manner, the vapour pressure of solid ammonia can be expressed as

$$\ln p + \frac{3754}{T} = 18.7$$

Take the molecular mass of ammonia to be 17 kg / kmol.

17. The temperature and pressure at the triple point are

(*a*) 295.2 K, 0.69 atm

(*b*) 295.2 K, 0.59 atm

(*c*) 195.2 K, 0.69 atm

(*d*) 195.2 K, 0.59 atm

18. The latent heat of vaporization is

(*a*) 1298 kJ/kg (*b*) 1398 kJ/kg

(*c*) 1498 kJ/kg (*d*) 1698 kJ/kg

Common Data Q. (19 – 20) :

Consider an ideal reheat cycle utilizing steam. Steam leaves the boiler and enters the turbine at 3 MPa, 400°C (state 3) and then expands to 0.8 MPa (state 4). It is then reheated at constant pressure 0.8 MPa to 400°C (state 5) and expands to 10 kPa in the low pressure turbine (state 6). The entry to the pump corresponds to saturated liquid state (state 1), and state 2 represents inlet to the boiler. The following data are given :

h_1 = 191.8 kJ/kg, h_3 = 3230.82 kJ/kg, h_4 = 2891.6 kJ/kg, h_5 = 3267.97 kJ/kg,

$h_g\big|_{10\,kPa}$ = 2584.63 kJ/kg, x_6 = 0.92285,

$v_f\big|_{10\,kPa}$ = 0.00101m³/kg

19. The heat transfer in the boiler is approximately

(*a*) 4411 kJ/kg (*b*) 3412 kJ/kg

(*c*) 3230 kJ/kg (*d*) 2892 kJ/kg

20. The net workdone in the cycle is approximately

(*a*) 1004 kJ/kg (*b*) 1104 kJ/kg

(*c*) 1204 kJ/kg (*d*) 2004 kJ/kg

Linked Answer Q. (21 – 22) :

A piston-cylinder arrangement as shown in the figure initially contains air at 150 kPa and 400°C. The arrangement is allowed to cool to the ambient temperature of 20°C. The characteristic gas constant for air is 0.287 kJ / kg.K. The cylinder wall has stops of negligible thickness that can prevent the piston from moving down. The stops are 1 m from the inner side of the base surface of the cylinder. At the initial state, the piston is resting 1 m above the stops.

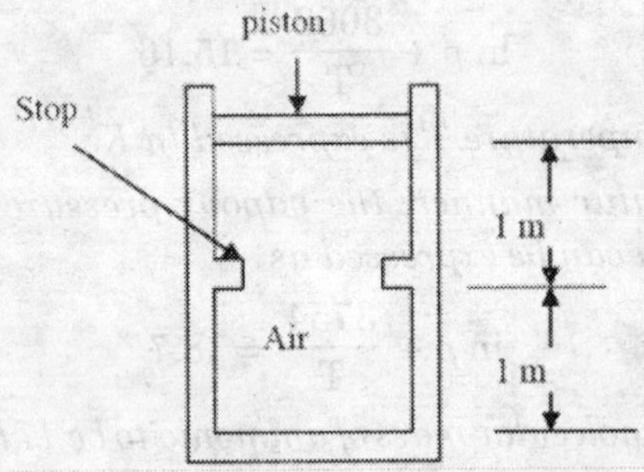

21. The final pressure in the cylinder is

(*a*) 130.7 kPa (*b*) 150 kPa
(*c*) 200.7 kPa (*d*) 230.7 kPa

22. The specific work done by the air during the process is

(*a*) –26.67 kJ/kg (*b*) 26.67 kJ/kg
(*c*) 49.5 kJ/kg (*d*) –96.67 kJ/kg

ANSWERS

1. (b)	**2.** (d)	**3.** (a)	**4.** (d)	**5.** (b)	**6.** (c)	**7.** (a)
8. (c)	**9.** (a)	**10.** (b)	**11.** (b)	**12.** (d)	**13.** (a)	**14.** (b)
15. (c)	**16.** (c)	**17.** (d)	**18.** (c)	**19.** (b)	**20.** (c)	**21.** (a)
22. (d)						

EXPLANATIONS

1. (b)

The piston-cylinder arrangement containing a gas. This system is heated by placing it on the top of a burner. Since the piston in piston cylinder arrangement is free to move therefore the system undergoes constant pressure process.

2. (d)

For a pure substance, at the triple point, three phases solid, liquid and vapour co-exist in equilibrium.

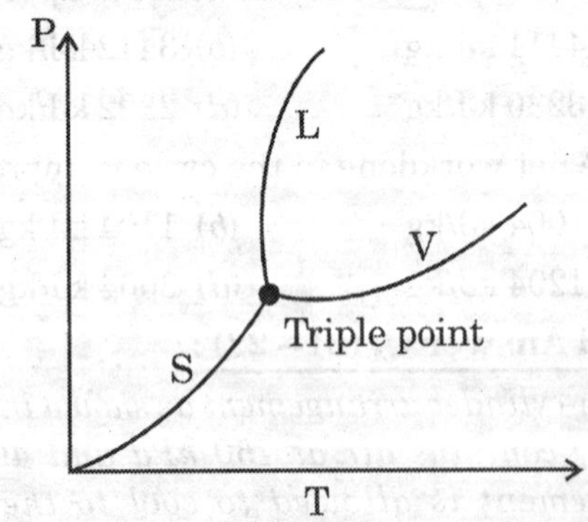

3. (a)

In a saturated liquid vapour-mixture, the quality x.

$$x = \frac{m_{vapour}}{m_{liquid} + m_{vapour}}$$

4. (d)

The slope of a Mollier diagram at a constant pressure indicates the temperature.

The diagram of h-s is called as Mollier diagram.

5. (b)

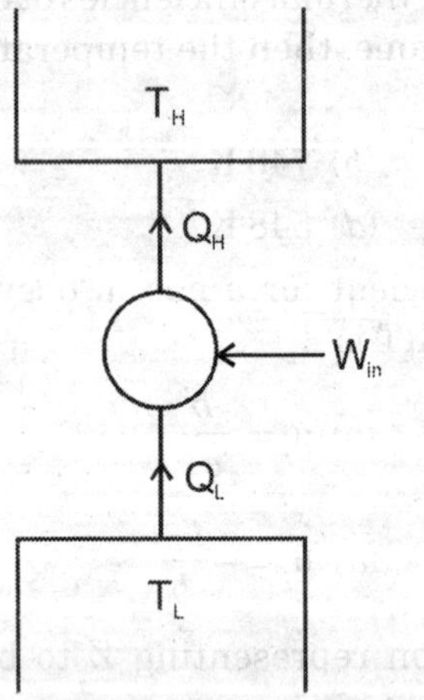

$$(COP)_R = \frac{Q_L}{Q_H - Q_L}$$

$$(COP)_{HP} = \frac{Q_H}{Q_H - Q_L}$$

$$= (COP)_{HP} - (COP)_R$$

$$= \frac{Q_H}{Q_H - Q_L} - \frac{Q_L}{Q_H - Q_L}$$

$(COP)_{HP} - (COP)_R = 1$

$\boxed{(COP)_{HP} = 1 + (COP)_R}$

6. (c)

Clausius Inequality states the cycle integral of $\frac{dQ}{T} \le 0$

$\oint \frac{dQ}{T} = 0$, For Reversible cycle

$\oint \frac{dQ}{T} < 0$, For Irreversible cycle

$\oint \frac{dQ}{T} > 0$, For Impossible cycle

7. (a)

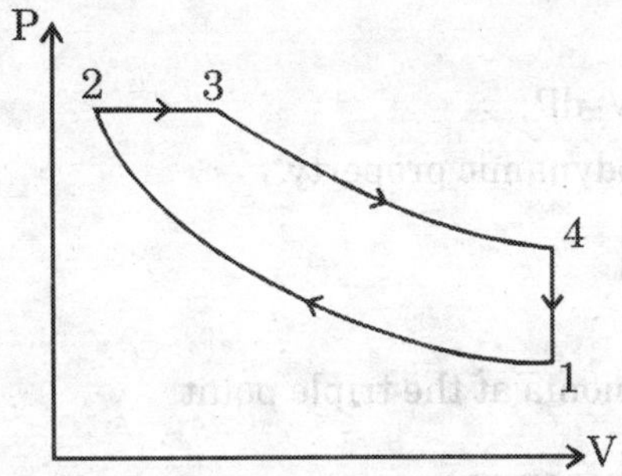

Fig.: P-V diagram of Diesel cycle

1-2, Compression process,

2-3, Heat supply,

3-4, Expansion process.

4-1, Heat rejected

The ratio of cylinder volume after and before combustion is called cut-off ratio.

$\boxed{\text{Cut-off-ratio } (\rho) = \frac{V_3}{V_2}}$

8. (c)

Availability → Availability of a system is defined as the maximum possible work that can be obtained in a cycle when system reaches to dead state.

Availability depends on the condition of both system and environment.

9. (a)

$$\text{Volume expansivity} = \frac{1}{V}\left(\frac{\partial V}{\partial T}\right)_P$$

$$\text{Volume expansivity for ideal gas} = \frac{1}{T}$$

10. (b)

Given that

$T_1 = 0°C = 273 \text{ K}$

$V_2 = 81 \times V_1$

For Adiabatic process,

$$T_1 V_1^{\gamma-1} = T_2 V_2^{\gamma-1}$$

$$T_2 = T_1\left(\frac{V_1}{V_2}\right)^{\gamma-1}$$

$$T_2 = 273\left(\frac{1}{81}\right)^{\gamma-1} \quad \left[V_2 = 81V_1 \Rightarrow \frac{V_1}{V_2} = \frac{1}{81}\right]$$

$$T_2 = 273\left(\frac{1}{81}\right)^{1.25-1}$$

$T_2 = 91 \text{ K}$

$\boxed{T_2 = -182°C}$

11. (b)

$$\text{Kinetic energy (E)} = \frac{3}{2}RT$$

$$E = \frac{3}{2}R \cdot 300 \qquad [T_1 = 27°C = 300 \text{ K}]$$

$$E = \left(\frac{3}{2}R\right) \cdot 300$$

$T_2 = 327°C = 600 \text{ K}$

$$E' = \frac{3}{2}RT_2 = \left(\frac{3}{2}R\right) \cdot 600$$

$$E' = \left(\frac{3}{2}R\right)(300) \times 2$$

$\boxed{E' = 2E}$

12. (d)

Given that

$T_1 = 20°C = 293 \text{ K}$

$P_1 = 100 \text{ kPa}$

$T_2 = 400 \text{ K}$

$n = 1.25$

$M = 4.003 \text{ kg/k mol.}$

$$W = \frac{mR(T_1 - T_2)}{n-1}$$

$$R = \frac{\bar{R}}{M} = \frac{8.314}{4.003} = 2.077 \text{ kJ/kg K}$$

$$W = \frac{2.077(293 - 400)}{1.25 - 1}$$

$\boxed{W = -888.95 \text{ kJ/kg}}$

13. (a)

$$x_{O_2} = \frac{P_{O_2}}{P} = \frac{\frac{32}{32}}{\frac{32}{32}+\frac{28}{28}} = \frac{1}{2} = 0.5$$

$$x_{N_2} = \frac{P_{N_2}}{P} = 0.5$$

Entropy increase due to diffusion,

$$\Delta S = -m_{O_2}R_{O_2}\ln\left(\frac{P_{O_2}}{P}\right) - m_{N_2}R_{N_2}\ln\left(\frac{P_{N_2}}{P}\right)$$

$$= -32\left(\frac{\bar{R}}{32}\right)\ln(0.5) - 28\left(\frac{\bar{R}}{28}\right)\ln(0.5)$$

$$= 1.38\,\bar{R} \quad \left(\because \bar{R} = 8.314\frac{kJ}{kmolK}\right)$$

14. (b)

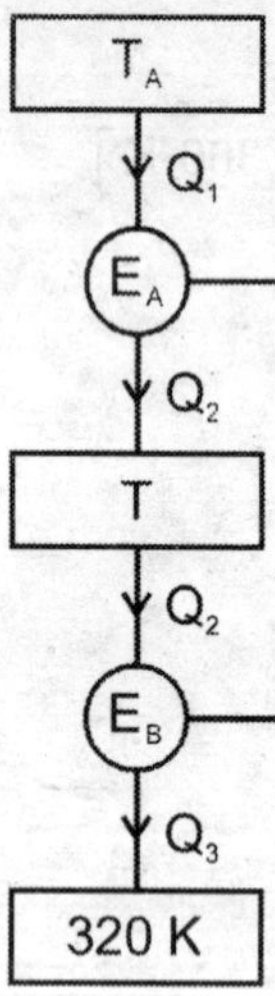

Given that

$T_A = 1750$ K

$W_A = W_B$

$\eta_A = \eta_B$

$$\eta_A = 1 - \frac{T}{T_A}$$

$$\eta_B = 1 - \frac{320}{T}$$

Now, $\eta_A = \eta_B$

$$1 - \frac{T}{T_A} = 1 - \frac{320}{T}$$

$$T^2 = T_A \times 320$$

$$T^2 = 1750 \times 320$$

$$T = \sqrt{1750 \times 320}$$

$$\boxed{T = 748.33\ K}$$

15. (c)

Given relation

$P(V - b) = RT$

Joule-Thomson coefficient for a gas (μ_j)

$$\mu_j = \left(\frac{\partial T}{\partial P}\right)_h$$

$$Pv - Pb = RT$$

$$T = \frac{Pv}{R} - \frac{Pb}{R}$$

$$\left(\frac{\partial T}{\partial P}\right) = \frac{v}{R} - \frac{b}{R}$$

$$\mu = \left(\frac{\partial T}{\partial P}\right)_h = \frac{-b}{C_p}$$

16. (c)

$z = P\cdot dV + V\cdot dP$

is a thermodynamic property.

$$\boxed{z = d(PV)}$$

17. (d)

Liquid ammonia at the triple point

$$\text{In}p + \frac{3063}{T} = 15.16$$

Vapour pressure of solid amonia

$$\text{In}p + \frac{3754}{T} = 18.7$$

Temperature and pressure at triple point

$$\text{In } P + \frac{3754}{T} = 18.7$$

$$\text{In } P + \frac{3063}{T} = 15.16$$

$$\frac{(3754 - 3063)}{T} = 3.54$$

$$\boxed{T = 195.19\ K}$$

$$\text{In } P = 15.16 - \frac{3063}{195.19}$$

$$\boxed{P = 0.587\ atm}$$

18. The correct answer is 1498 kJ/kg

19. (b)

Given that

$h_1 = 191.81$ kJ/kg

$h_3 = 3230.82$ kJ/kg

$h_4 = 2891.6$ kJ/kg

$h_5 = 3267.97$ kJ/kg

$(h_g)_{10kPa} = 2584.63$ kJ/kg

$x = 0.92285$

$(V_f)_{10kPa} = 0.00101$ m^3/kg.

$h_6 = h_1 + xhf_g$

$h_6 = 191.8 + 0.92285(2584.63 - 191.8)$

$h_6 = 2400$ kJ/kg

Heat transfer in the boiler (Q)

$Q = (h_3 - h_1) + (h_3 - h_4)$

$Q = (3230.82 - 191.81) + (3230.82 - 2891.6)$

$Q = 3378.23$ kJ/kg.

20. (c)

Net work done is a cycle $= W_T - W_P$

Work done on pump $= v_{f/10\,kPa} \times (P_2 - P_1)$

$= .00101 \times (3 \times 10^3 - 10)\frac{kJ}{kg}$

$= 3.0199\frac{kJ}{kg}$

Work done by turbine $= W_{T1} + W_{T2}$

$W_{T1} = (h_3 - h_4) = (3230.82 - 2891.6)$

$= 339.22\frac{kJ}{kg}$

$W_{T2} = (h_5 - h_6) = (3267.97 - (191.8 + 0.92285 \times 2392.83)$

$= 867.94$ kJ/kg

$W_T = (W_{T1} + W_{T2} = 1207.16$

$W_{net} = W_T - W_P = 1207.16 - 3.0199$

$= 1204$ kJ/kg

21. (a)

$$\frac{P_1V_1}{T_1} = \frac{P_2V_2}{T_2}$$

$$P_2 = \frac{P_1V_1 \cdot T_2}{T_1 \cdot V_2}$$

$$= \frac{150 \times 2 \times 293}{673 \times 1}$$

$= 130.7$ kPa

22. (d)

$P_1V_1 = mRT_1$

$150 \times V_1 = 1 \times .287 \times 673$

$V_1 = 1.287$ m^3/kg $\quad (\because V_2 = V_1/2)$

$V_2 = .6435$ m^3/kg

Work done by air = PdV

$= 150 \times (.6435 - 1.287)$

$= -96.67\frac{kJ}{kg}$

2013

Notation used :

p-pressure, V- volume, T-temperature, S- entropy, H- enthalpy, U- internal energy G- Gibbs free energy.

Specific properties are designated by lower case symbols.

Useful data :

Universal gas constant (R) = 8.314 J/mol K

Acceleration due to gravity = 9.81 m/s²

Molecular masses in kg/kmol: M_{air} = 29, $M_{nitrogen}$ = 28, M_{water} = 18, M_{helium} = 4

Ratio of ideal gas specific heats : γ_{ai} = 1.4

c_p for water = 4.186 kJ/kg K

Vapour pressure equation for water in the temperature range of 5 to100°C , with p in kPa and T in K

$$\ln(p) = 18.558 - \frac{5190}{T}$$

Q. 1 - 9 carry one mark each.

1. The measured temperature of a system is 30°C. Its exact absolute temperature in K is
   (*a*) 303.00 (*b*) 303.10
   (*c*) 303.15 (*d*) 303.16

2. The fuel air mixture in a petrol engine is ignited with a spark plug at the end of compression stroke. This process
   (*a*) increases the entropy of the fuel air mixture but decreases the entropy of the spark plug
   (*b*) decreases the entropy of the fuel air mixture but increases the entropy of the spark plug
   (*c*) decreases the entropy of the fuel air mixture and of the spark plug
   (*d*) increases the entropy of the fuel air mixture and of the spark plug

3. In the van der Waals equation of state given below :

$$\left(p+\frac{a}{v^2}\right)(v-b) = RT$$

   The constant a represents the effect of
   (*a*) attractive forces between molecules
   (*b*) repulsive forces between molecules
   (*c*) deviation from molecules being spherical
   (*d*) finite size of the molecule

4. For a reversible isothermal expansion of an ideal gas from a state 1 to a state 2,
   (*a*) $s_1 = s_2$ (*b*) $s_1 > s_2$
   (*c*) $s_1 < s_2$ (*d*) $h_1 > h_2$

5. For a pure substance the critical isotherm on the p-v plane exhibits
   (*a*) a maximum (*b*) a minimum
   (*c*) a point of inflection (*d*) a discontinuity

6. For an ideal gas as a working fluid for a given heat input Q, the process that gives the maximum work among the following four processes is
   (*a*) isothermal (*b*) constant volume
   (*c*) constant pressure (*d*) isentropic

7. An air standard Otto cycle has the following shape on a thermodynamic property plane.

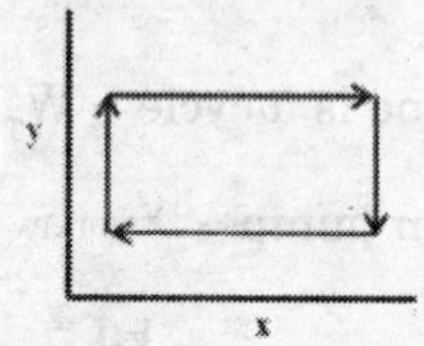

   The x and y coordinates, respectively, are
   (*a*) v and p (*b*) s and v
   (*c*) v and T (*d*) s and p

8. The specific volume of steam after expansion in a turbine is 12 m³/kg. At this pressure the saturated liquid and saturated vapour specific volumes are 0.001 and 15.25 m³/kg respectively. What is the dryness fraction to second decimal place accuracy?

9. Which of the following processes, shown in the figure below, represents the throttling of an ideal gas?

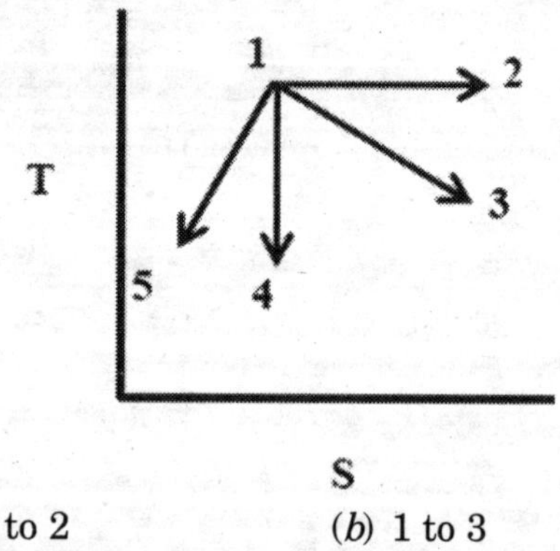

   (*a*) 1 to 2 (*b*) 1 to 3
   (*c*) 1 to 4 (*d*) 1 to 5

Q. 10 - 22 carry two marks each.

10. On a ln *pvsh* coordinate system, where ln *p* is the *y*-coordinate and *h* is the *x* coordinate, the slope of a constant entropy line is
   (*a*) 1/*v* (*b*) *v*
   (*c*) *p*/*v* (*d*) 1/(*pv*)

11. Starting from the definition of Gibbs free energy function $g = h - Ts$, the Maxwell relation that can be derived is

(a) $\left.\frac{\partial p}{\partial T}\right|_s = \frac{v}{s}$ (b) $\left.\frac{\partial p}{\partial T}\right|_s = v$

(c) $\left.\frac{\partial v}{\partial s}\right|_T = \left.\frac{\partial p}{\partial T}\right|_s$ (d) $\left.\frac{\partial v}{\partial T}\right|_p = \left.\frac{\partial s}{\partial p}\right|_T$

12. A thermodynamic cycle operates between one source at a temperature of 600 K, another source at a temperature of 300 K and a sink at a temperature T as shown in the figure below

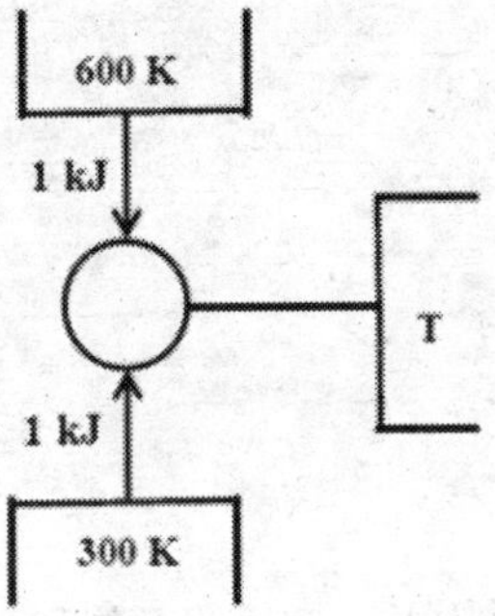

If the First and Second laws of thermodynamics are not violated, what should be the value of T in K? ________

13. A closed system containing an ideal gas undergoes a cycle as shown in the figure shown below. For the process1-2, which one of the following statements is true?

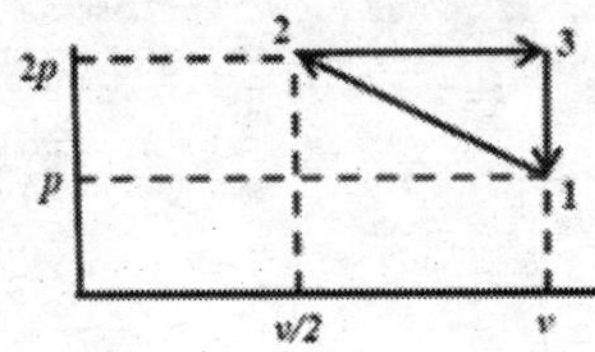

(a) Heat added $= \frac{3}{4} pv$ (b) Heat rejected $= \frac{3}{4} pv$

(c) Heat added $= \frac{1}{2} pv$ (d) Heat rejected $= \frac{1}{4} pv$

14. A well-insulated rigid hot water tank receives steady flow of water from two sources as shown in the figure below

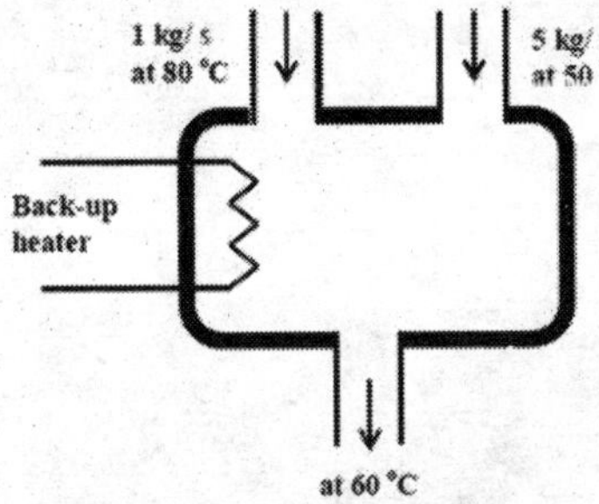

There is no accumulation of water in the tank. A back-up heater is provided to ensure a constant outflow temperature of water at 60°C from the tank under steady state. What is the required capacity of the back-up heater to the nearest kW? ________

15. 1 kg of air in an insulated rigid tank of volume 1 m^3 is churned with a friction-less fan (see figure below) of 600 W capacity for 10 minutes. The fan efficiency is 100%. Treating air as an ideal gas and neglecting kinetic and potential energy changes, what is the increase of pressure, to the nearest kPa? ________

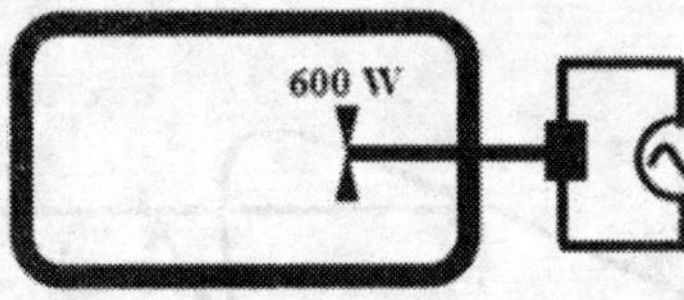

16. The isothermal compressibility of a liquid is 5×10^{-6} /kPa. If it is compressed at constant temperature from 5000 to 10000kPa, what is the ratio of final volume to initial volume, to second decimal place accuracy? ________

COMMON DATA QUESTIONS

Common Data for Questions 17 and 18 :

At a location where the atmospheric pressure is 98 kPa and the ambient temperature is 30°C, the humidity ratio is 0.01 kg/kg of dry air. A high pressure front moves over the location which changes only the atmospheric pressure to 102kPa,while the humidity ratio remains same.

17. What is the partial pressure of water vapour in kPa to the first decimal place accuracy before the high pressure front moves in? ________

18. What is the relative humidity of air under the influence of high pressure front to integer precision in %? ________

Common Data for Questions 19 and 20:

A rigid insulated cylinder is divided into two chambers A and B by a thin rigid insulating barrier as shown in the figure below

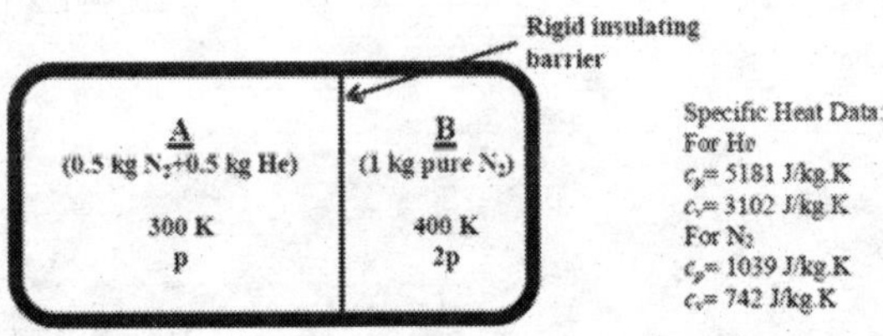

Initially, chamber A contains a mixture of 0.5 kg nitrogen and 0.5 kg helium at 300 K while chamber B contains 1 kg of pure nitrogen at 400 K. The pressure in chamber B is twice that in chamber A. The gases and gas mixtures are assumed to be ideal.

19. What is the ratio of the volumes of chambers A and B, i.e. V_A/V_B, to first decimal place accuracy? ________

20. If the barrier is removed and the gases are allowed to mix and reach thermodynamic equilibrium, what is the final temperature of the mixture, to the nearest K? ________

LINKED ANSWER QUESTIONS

Statement for Linked Answer Questions 21 and 22 :

A combined vapour compression-cum-Brayton cycle is shown below

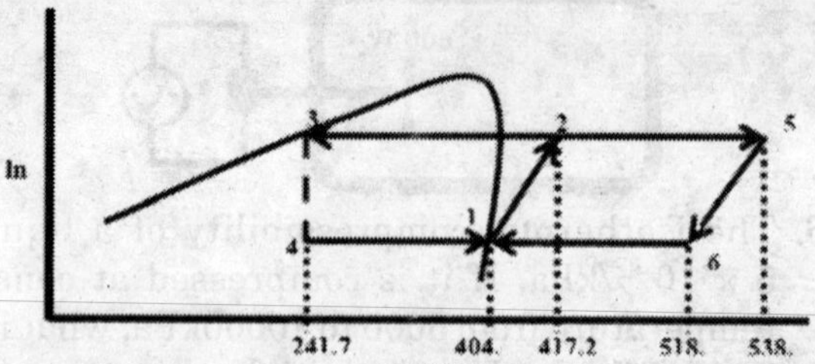

1 – 2 : Isentropic compression, 5 – 6 : isentropic expansion.

The refrigeration system has a cooling capacity of 30 kW and the turbine generates a power of 30 kW.

21. What is the mass flow rate of the working fluid through the turbine, in kg/s, to first decimal place accuracy? ________

22. What is the power required to drive the compressor, to the nearest kW? ________

ANSWERS

1. (c)	**2.** (d)	**3.** (a)	**4.** (c)	**5.** (c)	**6.** (a)	**7.** (b)
8. 0.78 to 0.8	**9.** (a)	**10.** (d)	**11.** (d)	**12.** 399 to 401	**13.** (b)	
14. 120 to 130	**15.** 140 to 150	**16.** 0.97 to 0.98	**17.** 1.5 to 1.7	**18.** 36 to 40	**19.** 5.7 to 6.3	
20. 325 to 330	**21.** 1.4 to 1.6	**22.** 20.5 to 22.5				

EXPLANATIONS

1. (c) $t = 30°C$

Absolute temperature = $t°C + 273.15$
$= 30 + 273.15 = 303.15$ K

2. (d) The fuel air mixture in a petrol engine is ignited with a spark plug at the end of compression stroke. This process increase the entropy of the fuel air mixture and of the spark plug.

3. (a) In Vander Waals equation,

$$\left(P + \frac{a}{V^2}\right)(V - b) = RT$$

The constant 'a' represent the effect of

Attractive force between the molecules

4. (c) For a reversible isothermal expansion of an ideal gas from state 1 to a state 2.

$$S_1 < S_2$$

5. (c) For a pure substance the critical isotherm on the P-v plane having the point of inflection.

6. (a) For an ideal gas as a working fluid for a given heat input Q, the process that gives the maximum work among the following four process is, isothermal.

The maximum work is obtained on isothermal process.

$$\delta Q = dU + \delta W$$

as $\quad dT = 0, dU = 0$

$\therefore \quad \delta Q = \delta W$

7. (b) Otto cycle on P-V diagram,

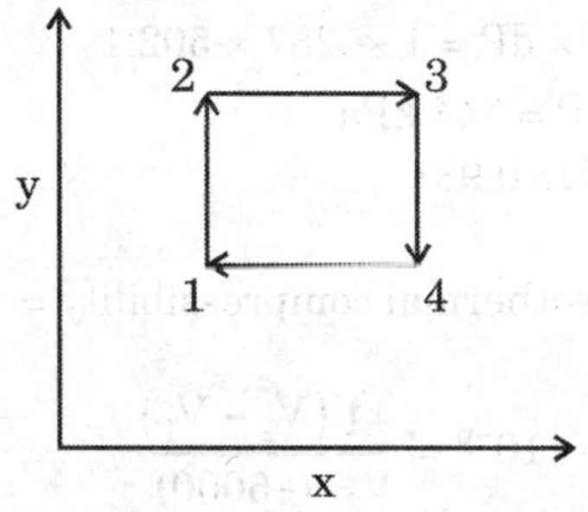

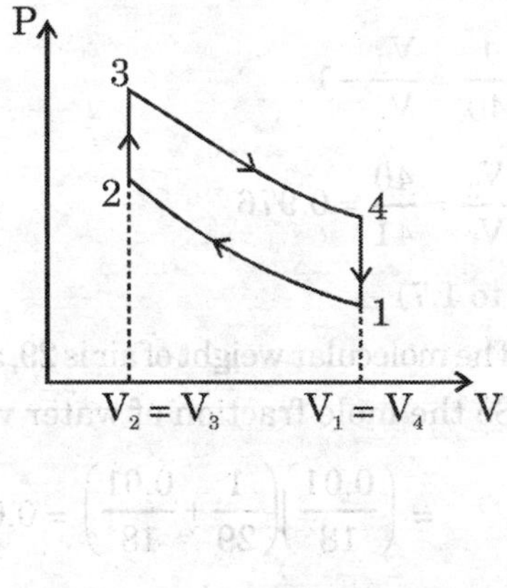

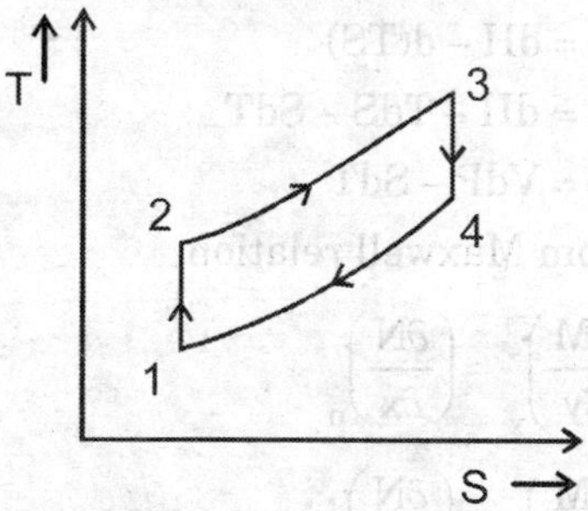

During otto cycle operation, pressure & temperature are continuously changing in nature.

$$Y - S$$
$$X - V$$

8. (0.78 to 0.8)

Specific volume (v) = 12 m^3/kg.

$$v_f = 0.001 \ m^3/kg$$

$$v_g = 15.25 \ m^3/kg$$

$$v_g = v_f + xv_{fg}$$

$$12 = 0.001 + x(15.25 - 0.001)$$

$$x = 0.786$$

Dryness fraction (x) = 0.786

9. (a)

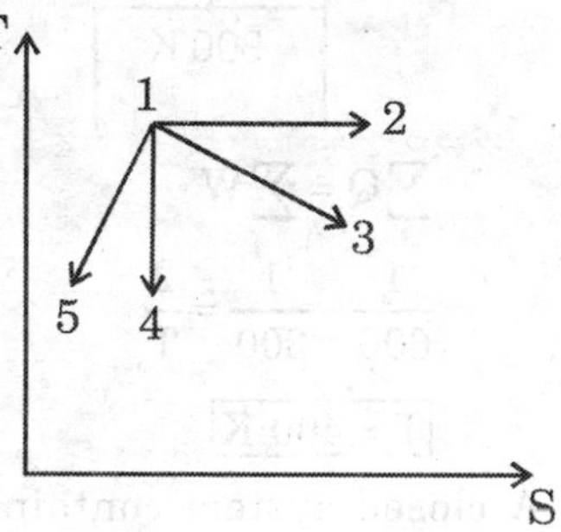

During the throttling of an ideal gas,

$h_1 = h_2$

$C_pT_1 = C_pT_2$

$T_1 = T_2$

$T_1 = T_2$, for process (1-2).

10. (d)

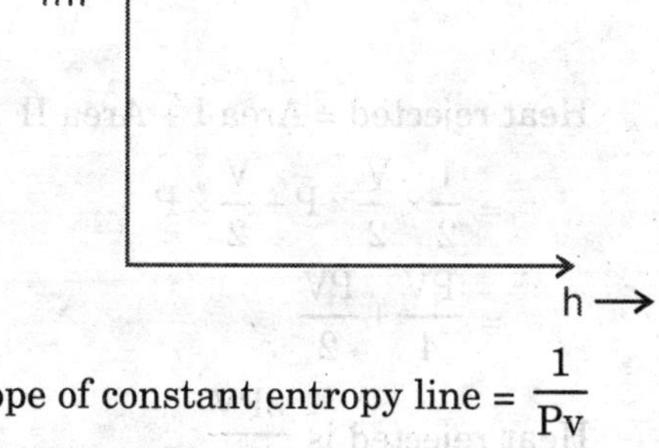

Slope of constant entropy line = $\frac{1}{Pv}$

11. (d) From Gibbs free energy function

$G = H - TS$

$dG = dH - d(TS)$

$dG = dH - TdS - SdT$

$dG = VdP - SdT$

From Maxwell relation

$$\left(\frac{\partial M}{\partial y}\right)_x = \left(\frac{\partial N}{\partial x}\right)_n$$

$$\left(\frac{\partial M}{\partial y}\right)_x = \left(\frac{\partial N}{\partial x}\right)_y$$

$$\left(\frac{\partial v}{\partial T}\right)_P = -\left(\frac{\partial S}{\partial P}\right)_T$$

12. (399 to 401)

First and second laws are not violated then

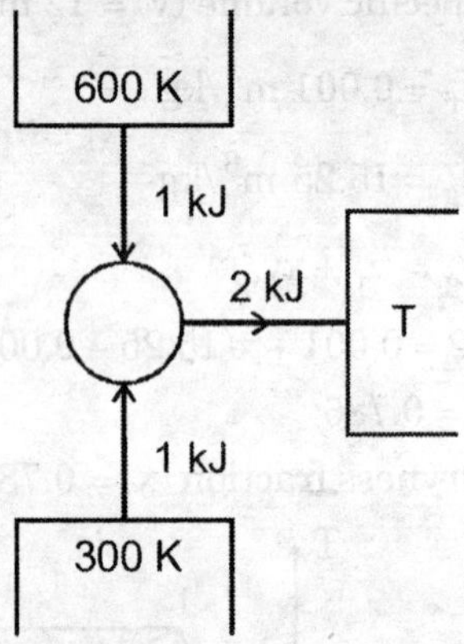

$$\sum Q = \sum W$$

$$\frac{1}{600} + \frac{1}{300} = \frac{2}{T}$$

$$\boxed{T = 400\ K}$$

13. (b) A closed system containing an ideal gas undergoing a cycle.

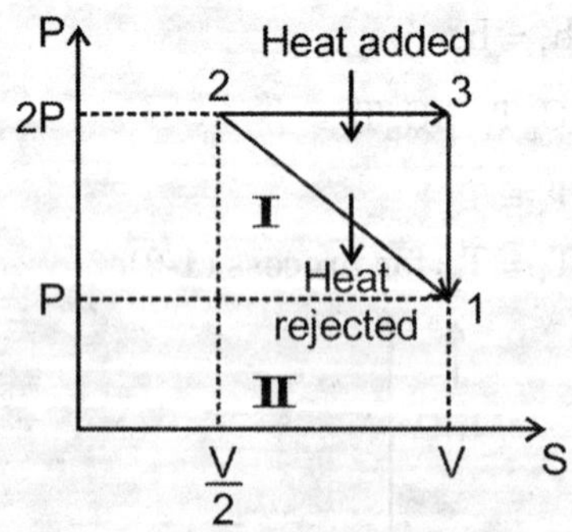

Heat rejected = Area I + Area II

$$= \frac{1}{2} \times \frac{V}{2} \times P + \frac{V}{2} \times P$$

$$= \frac{PV}{4} + \frac{PV}{2}$$

Heat rejected is $\frac{3PV}{4}$.

14. (120 to 130)

Under steady state conditions, $\dot{m}_1 + \dot{m}_2 = \dot{m}_3$

$$\dot{Q} + \dot{m}_1 CpT_1 + \dot{m}_2 CpT_2 = \dot{m}_3 CpT_3$$

$$\dot{Q} + 1 \times 353 \times 4.187 + 5 \times 323 \times 4.187$$

$$= 6 \times 333 \times 4.187$$

$$\dot{Q} = 125.61\ kW$$

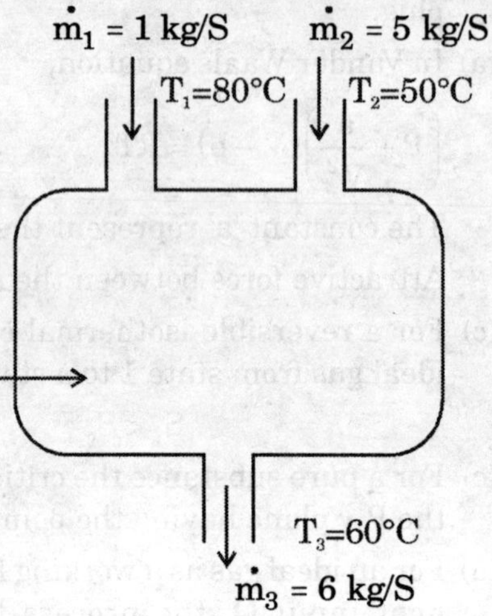

15. (140 to 150)

$\delta Q = dU + \delta W$

$\delta Q = dU$

$600 \times 600 = 1 \times 717 \times dT$

$dT = 502.1\ °C$

$pV = mRT$

For constant volume

$dP.V = mRdT$

$1 \times dP = 1 \times .287 \times 502.1$

$dP = 144\ kPa$

16. (0.97 to 0.98)

$$\text{Isothermal compressibility} = \frac{-1}{V_f}\left(\frac{dV}{dP}\right)_T$$

$$5 \times 10^{-6} = \frac{+1}{V_2}\frac{(V_2 - V_1)}{(+5000)}$$

$$\frac{1}{40} = \frac{V_2}{V_1} - 1$$

$$\frac{V_2}{V_1} = \frac{40}{41} = 0.976$$

17. (1.5 to 1.7)

The molecular weight of air is 29, and of water 18.

So the mole fraction of water vapour

$$= \left(\frac{0.01}{18}\right)\left(\frac{1}{29} + \frac{0.01}{18}\right) = 0.0158$$

Therefore the partial pressure of water vapour $= 0.0158 \times 98 = 1.55$ kPa

18. (36 to 40)

$$\ell n(p) = 18.558 - \left(\frac{5190}{T}\right)$$

Relative humidity (RH) $= \frac{P}{P_S}$

$[\because P = 1.55 \text{ kPa}]$

At T = 303 K,

$$\ell n(P) = 18.558 - \left(\frac{5190}{303}\right)$$

$$RH = \frac{P}{P_S} = \frac{1.55}{4.175}$$

$= 0.3712 = 37.12\%$

19. (a) $R_A = 0.5(5181 - 3102) + 0.5(1039 - 742)$

$$R_A = 1188 \frac{J}{kg\ K}$$

$$P_A V_A = m_A R_A T_A$$

$$V_A = \frac{m_A R_A T_A}{P_A}$$

$$V_A = \frac{1 \times 1188 \times 300}{P}$$

$$V_A = \frac{356400}{P} \quad \ldots(1)$$

(B) $V_B = \frac{m_B R_B T_B}{P_B}$

$$= \frac{1 \times 297 \times 400}{2P} = \frac{59400}{P} \quad \ldots(2)$$

By dividing (1) with (2), we get

$$\frac{V_A}{V_B} = 6$$

20. (325 to 330)

$$1 \times 3110(T - 300) = 1 \times 1039(400 - T)$$

$$\left[\because (C_P)_A = 0.5 \times (C_P)_{He} + 0.5 \times (C_P)_{N_2}\right]$$

$$\boxed{T = 325 \text{ K}}$$

21. (1.4 to 1.6)

$$P_T = \dot{m}_f \cdot dh$$

(P_T = Power developed by the turbine)

$$30 = \dot{m}_f \cdot 20$$

$$\dot{m}_f = 1.5 \text{ kg/sec.}$$

22. (20.5 to 22.5)

$$P_C = \dot{m}_f\, dh$$

$$= 1.5 \times 13.2 = 19.8 \text{ kW} \approx 20 \text{ kW}$$

2014

Notation Used

P-pressure, V-volume, T-temperature,

S-entropy, H-enthalpy, U-internal energy,

c_p-specific heat at constant pressure,

c_v-specific heat at constant volume;

specific properties are designated by lower case symbols.

Subscripts: R-reduced, C-critical, f-saturated liquid, g-saturated vapor.

Properties of air : $c_p = 1.005$ kJ/(kg.K),

specific heat ratio $\gamma = 1.4$,

Gas constant = 0.287 kJ/(kg.K),

Molecular weight = 29 gm/mol.

Universal gas constant = 8.314 kJ/(kmol.K).

Q. (1 – 9) Carry One Mark Each.

1. Entropy is a

(*a*) Path function

(*b*) Point function

(*c*) Property independent function

(*d*) Neither path nor point function

2. A small container has gas at high pressure. It is placed in an evacuated space. If the container is punctured, work done by the gas is

(*a*) Positive (*b*) Negative

(*c*) Zero (*d*) ∞

3. The molecular weight of a mixture is 38.4 gm/mol. The mixture is composed of methane and carbon-dioxide gases. The atomic weights of the elements C, H, and O are 12, 1, and 16 gm/mol, respectively. The mole fraction of methane ($X_{methane}$) is _____ and that of carbon-dioxide ($X_{carbon\text{-}dioxide}$) is _____.

(*a*) $X_{methane} = 0.2$; $X_{carbon\text{-}dioxide} = 0.8$

(*b*) $X_{methane} = 0.8$; $X_{carbon\text{-}dioxide} = 0.2$

(*c*) $X_{methane} = 0.3$; $X_{carbon\text{-}dioxide} = 0.7$

(*d*) $X_{methane} = 0.7$; $X_{carbon\text{-}dioxide} = 0.3$

4. A system undergoes a change from state 1 to state 2. During this process, the change in the internal energy is ΔU. The change in internal energy of the system when executing the cycle 1-2-1 is equal to

(*a*) ΔU (*b*) $2\ \Delta U$

(*c*) Zero (*d*) $-2\ \Delta U$

5. Which among the following plots represents a line joining two states with the same dew point temperature on a standard psychrometric chart, with the dry bulb temperature on the X-axis and the humidity ratio on the Y-axis?

(*a*) 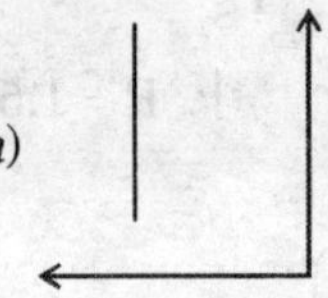 (*b*) 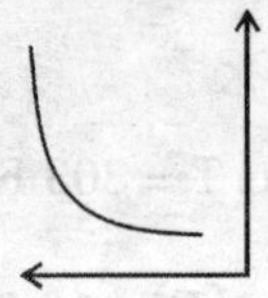

(*c*) 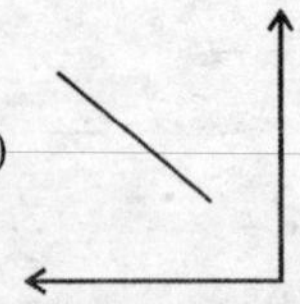 (*d*) 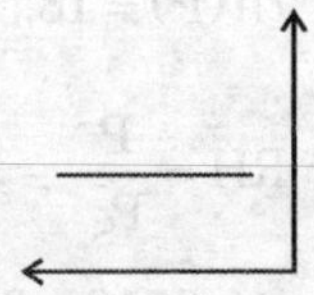

6. The efficiency of a reversible engine operating between two temperatures is 40%. The COP of a reversible refri-gerator operating between the same temperatures is

(*a*) 1.5 (*b*) 2.5

(*c*) 0.4 (*d*) 3.5

7. For a superheated vapor that cannot be approximated as an ideal gas, the expression determining a small change in the specific internal energy is

(*a*) $du = c_p dT + \left.\frac{\partial u}{\partial v}\right|_T dv$

(*b*) $du = c_p dT + \left.\frac{\partial u}{\partial P}\right|_T dP$

(*c*) $du = c_v dT + \left.\frac{\partial u}{\partial v}\right|_T dv$

(*d*) $du = c_v\ dT$

8. The minimum and maximum volumes in an air standard Otto cycle are 100 and 800 cm^3. Its thermal efficiency (%) is

(*a*) 56.47 (*b*) 94.55

(*c*) 54.08 (*d*) 87.50

9. At a saturation temperature T_{sat}, the difference between the entropy of saturated vapor and entropy of saturated liquid can be expressed as

(*a*) $(h_f - h_g)/T_{sat}$ (*b*) $(h_g - h_f)/T_{sat}$

(*c*) $(u_g - u_f)/T_{sat}$ (*d*) $(u_f - u_g)/T_{sat}$

Q. (10 – 22) Carry Two Mark Each.

10. A gas in a closed system is compressed reversibly from an initial volume of 0.2 m^3 to 0.1 m^3 at a constant pressure of 3 bar. During this process, there was a heat transfer of 50 kJ from the gas. The change in internal energy of the gas during this process in kJ is

(a) 20 (b) – 80

(c) 80 (d) – 20

11. In a closed rigid vessel, air is initially at a pressure of 0.3 MPa and volume of 0.1 m^3 at 300 K. A stirrer supplies 100 kJ of work to the air, while 20 kJ of heat is lost to the atmosphere across the container walls. After these processes, the temperature of air changes to _____ K.

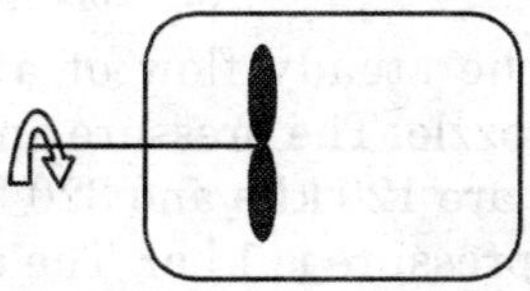

(a) 321.9 (b) 702.4

(c) 782.4 (d) 620.2

12. A reversible heat engine (E) operates using three thermal reservoirs with temperatures as shown in the following figure. If $Q_1 = Q_2$, the efficiency of the engine is __________.

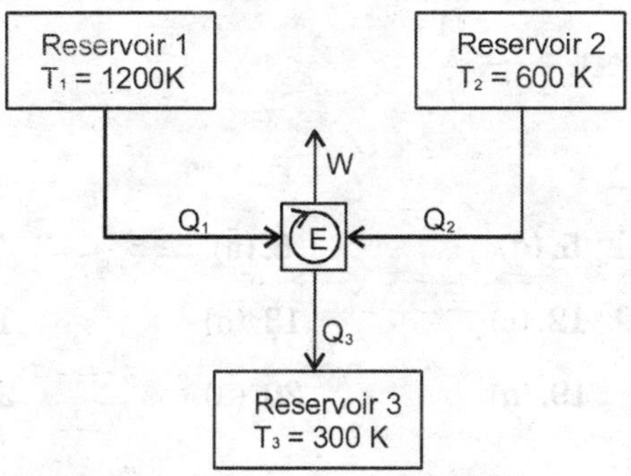

(a) 0.25 (b) 0.125

(c) 0.625 (d) 0.75

13. A metal block of mass 25 kg at 300 K is immersed in an infinitely large liquid nitrogen bath maintained at 77 K. The system comprising of the block and liquid nitrogen attains thermal equilibrium. The average specific heat of the metal is 0.45 kJ/(kg.K). The entropy generated during the process is _____ kJ/K.

(a) 17.28 (b) 32.5

(c) 47.8 (d) – 47.8

14. For a gas obeying the equation of state given by $\left(P+\frac{a}{v^2}\right)v = RT$, the values of the critical volume and the critical temperature are 0.004 m^3/kg and 100°C, respectively. If the value of the gas constant is 250 J/(kg.K), then the value of the constant 'a' is _____ (N.m^4/kg^2). Note that the critical point is the point of inflection on the critical isotherm.

(a) 124.3 (b) 0.75

(c) 186.58 (d) 248.67

15. A rigid closed vessel is initially filled with 2 kg of water which is a mixture of saturated liquid and saturated vapor states at 2 bar. The vessel is placed in an oven which heats the mixture to the critical state. Using the saturated and critical property values from the table given below, the heat transferred from the oven to the vessel is _____ kJ.

Pressure = 2 bar			
v_f(m^3/kg)	v_g(m^3/kg)	u_f(kJ/kg)	u_g(kJ/kg)
0.0010605	0.8857	504.49	2529.5
Critical pressure			
	v_c(m^3/kg)	u_c(kJ/kg)	
	0.003155	2029.6	

(a) 3035.8 (b) 3040.6

(c) 3036.2 (d) 3044.9

16. The equation of state for a certain gas is given by $v = RT/P - C_1/T^2 + C_2$, where C_1 is 50,000 (K^{2}m^3)/kg and C_2 is 0.8 m^3/kg. The relation $\left.\frac{\partial h}{\partial P}\right|_T = v - T\left.\frac{\partial v}{\partial T}\right|_P$ is known for the gas, the inversion temperature, given by the condition, $\left[\left.\frac{\partial h}{\partial P}\right|_T = 0\right]$ is _____ K.

(a) 500.0 (b) 433.0

(c) 353.6 (d) 250.0

17. The maximum pressure and temperature in an air standard diesel cycle are 44 bar and 1600 K, respectively. If the minimum pressure and temperature are 1 bar and 300 K, respectively, then the cut-off ratio (the ratio of the volume at the end of the heat addition process to that at the beginning of the heat addition process) is

(a) 1.000 (b) 14.920

(c) 2.809 (d) 1.809

18. The thermal efficiency of an air standard Brayton cycle 0.35. The pressure ratio across the turbine is

(a) 4.516 (b) 5.232

(c) 7.535 (d) 8.234

19. Steam is isentropically expanded in a turbine from 80 bar to 7 bar. At the inlet of the turbine (state 1) h_1 is 3246 kJ/kg and s_1 is 6.52 kJ/(kg.K).

Pressure = 7 bar			
h_f(kJ/kg)	h_g(kJ/kg)	s_f[kJ/(kg.K)]	s_g[kJ/(kg.K)]
697	2763	2.0	6.7

The enthalpy of the steam exiting the turbine (state 2) in kJ/kg is

(a) 2683.87 (b) 2657.17

(c) 1986.87 (d) 3354.17

20. A thin insulating membrane separates two tanks initially filled with nitrogen [mean c_v = 21.6 J/(mol.K)] and carbon-dioxide [mean c_v = 11.6 J/(mol.K)] as shown below.

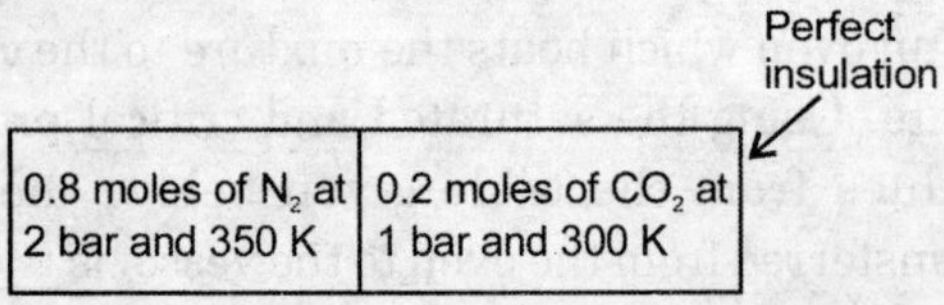

The membrane is ruptured and the gases are allowed to mix to form a homogeneous mixture at equilibrium. During this process there are no heat or work interactions between the tank contents and the surroundings. The final temperature at the equilibrium state in Kelvin is

(*a*) 344.1 (*b*) 306.3

(*c*) 325.0 (*d*) 346.1

21. Two moist air streams MAS1 and MAS2 are mixed adiabatically. The details of MAS1 and MAS2 are given below in the table.

	MAS1	MAS2
h (kJ/kg of dry air)	42	80
v(m³/kg of dry air)	0.85	0.9
Flow rate (m³/min)	85	90

With pressure remaining same and with no work interactions during the mixing process, the enthalpy of the mixed stream is _____ kJ/kg of dry air.

(*a*) 122 (*b*) 61

(*c*) 81 (*d*) 108

22. Consider the steady flow of air through an insulated nozzle. The pressure and temperature at the inlet are 120 kPa and 320 K, respectively. The outlet pressure is 1 bar. The inlet velocity is very small and the air undergoes a reversible adiabatic process. The outlet velocity, in m/s, is

(*a*) 303.7 (*b*) 180.7

(*c*) 5.7 (*d*) 127.3

ANSWERS

1. (b)	**2.** (c)	**3.** (a)	**4.** (c)	**5.** (d)	**6.** (a)	**7.** (c)
8. (a)	**9.** (b)	**10.** (d)	**11.** (d)	**12.** (c)	**13.** (a)	**14.** (c)
15. (b)	**16.** (b)	**17.** (d)	**18.** (a)	**19.** (a)	**20.** (a)	**21.** (b)
22. (b)						

EXPLANATIONS

1. (b) Entropy is a property of system and it is a point function, which is represented on P – V, diagram by a point.

2. (c)

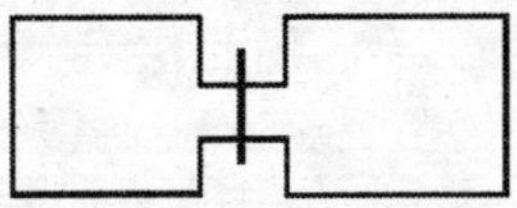

A small container has gas at high pressure, if it is placed in an evacuated space, it represent the free expansion process. In the free expansion process, work done is zero.

3. (a) Molecular weight of CO_2 = 44 gm/mol.

$$\text{Mole fraction of } CO_2\left(X - CO_2\right) = \frac{38.4}{44}$$

$X - CO_2 = 0.8$

X – methane = 1 – 0.8 = 0.2

4. (c) Internal energy is a property of the system and point function. During cyclic process change of property is zero, therefore change of internal energy is zero during the cyclic process.

5.

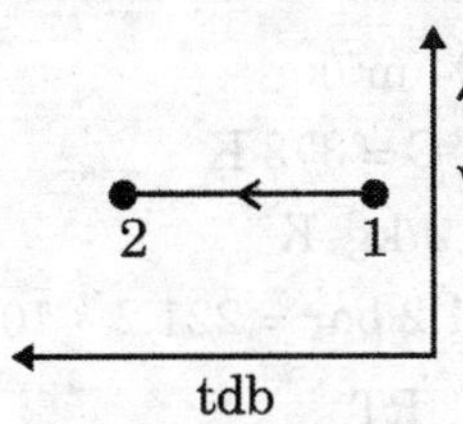

$$\text{Specific humidity (w)} = \frac{Pv}{Pvs}$$

Pv – Vapour pressure at dew point temperature.

If $tdP_1 = tdp_2$

$Pv_1 = Pv_2$

then w = c

6. (a) Given that

$\eta = 40\ \% = 0.4$

$$\eta = 1 - \frac{T_2}{T_1}$$

$$\frac{T_2}{T_1} = 0.6$$

$$COP = \frac{T_2}{T_1 - T_2} = \frac{\frac{T_2}{T_1}}{1 - \frac{T_2}{T_1}} = \frac{0.6}{0.4}$$

COP = 1.5

7. (c) For superheated vapour.

Change in the specific internal energy (du)

$$du = Cv.dT + \left.\frac{\partial u}{\partial v}\right|_T \cdot dv$$

8. (a) Given that

Maximum volume(v_1) = 800 cm^3

Minimum volume(v_2) = 100 cm^3

$$\text{Compression ratio(r)} = \frac{v_1}{v_2}$$

$$r = \frac{800}{100} = 8$$

$\gamma = 1.4$

$$\eta_o = 1 - \frac{1}{(r)^{\gamma-1}}$$

$$\eta_o = 1 - \frac{1}{(8)^{1.4-1}}$$

$$\boxed{\eta_o = 56.47\ \%}$$

9. (b) Change of entropy $= \dfrac{h_{fg}}{T_{sat}}$

$$ds = \frac{h_g - h_f}{T_{sat}}$$

10. (d) Given that

Initial volume(V_1) = 0.2 m^3

Final volume(V_2) = 0.1 m^3

Pressure(P) = 3 bar = 300 kPa

$\delta Q = -50$ KJ

$\delta Q = \delta U + \delta W$

$-50 = dU + P.dV$

$-50 = dU + 300(0.1 - 0.2)$

$-50 = dU - 30$

Change of internal energy (dU) = –20 kJ

11. (d) Given that

$P_1 = 0.3$ MPa = 300 kPa

$v_1 = 0.1$ m^3

$T_1 = 300$ K

$\delta W = -100$ kJ

$\delta Q = -20$ kJ

$$m = \frac{P_1 V_1}{RT_1} = \frac{300 \times 0.1}{0.287 \times 300}$$

m =0.348 kg

$\delta Q = \delta U + \delta W$

$-20 = mCv\,(T_2 - T_1) - 100$

$T_2 - T_1 = \frac{80}{0.348 \times 0.718}$

$T_2 - 300 = 320.17$

$\boxed{T_2 = 620.17\text{ K}}$

12. (c) Given that

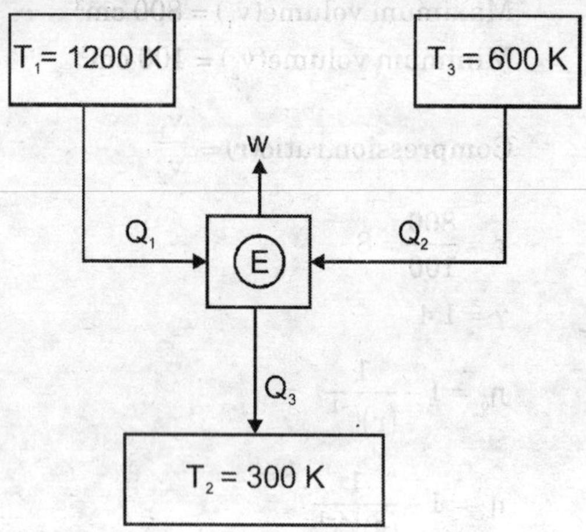

$Q_1 = Q_2$

$\oint \frac{dQ}{T} = 0$ for reversible

$\frac{Q_1}{T_1} + \frac{Q_2}{T_2} - \frac{Q_3}{T_3} = 0$

$\frac{Q_1}{1200} + \frac{Q_2}{600} - \frac{Q_3}{300} = 0$

$\frac{Q_1}{1200} + \frac{Q_1}{600} - \frac{Q_3}{300} = 0$ [$Q_1 = Q_2$ Given]

$\frac{Q_1}{12} + \frac{Q_1}{6} - \frac{Q_3}{3} = 0$

$\frac{Q_1}{4} + \frac{Q_1}{2} = Q_3$

$3Q_1 = 4Q_3$

$Q_1 = \frac{4}{3}Q_3$

$Q_2 = \frac{4}{3}Q_3$

$\eta = \frac{W_{net}}{Q_{in}}$

$Q_{in} = Q_1 + Q_2 = 2Q_1$

$W_{net} = 2Q_1 - Q_3$

$\eta = \frac{2Q_1 - Q_3}{2Q_1}$

$\eta = \frac{\frac{8}{3}Q_3 - Q_3}{\frac{5}{3}Q_3} \quad \left[Q_1 = \frac{4}{3}Q_3\right]$

$\boxed{\eta = 0.625}$

13. (a) Given that

$m = 25$ kg

$T_i = 300$ K

$T_f = 77$ K

$C_f = 0.45$ kJ/kg.k

Entropy change during the process(ds)

$= mC_f \ln\left(\frac{T_f}{T_i}\right)$

$ds = 25 \times 0.45 \ln\left(\frac{77}{300}\right)$

$\boxed{ds = 15.29\text{ kJ}}$

14. (c) Given that

$\left(P + \frac{a}{v^2}\right)v = RT$

$v = 0.004\ m^3/kg$

$T = 100°C = 373$ K

$R = 250$ J/kg·K

$P_C = 221.2$ bar $= 221.2 \times 10^5\ N/m^2$

$P + \frac{a}{v^2} = \frac{RT}{v}$

$\frac{a}{v^2} = \frac{RT}{v} - P$

$a = RTv - Pv^2$

$a = (250 \times 373 \times 0.004) - (221.2 \times 10^5 \times 0.004)^2$

$a = 186.57$

15. (b) Heat transferred from the oven to the vessel is 3040.6 kJ

16. (b) $\left.\frac{\partial h}{\partial P}\right|_T = V - T\left.\frac{\partial V}{\partial T}\right|_P$

For inversion temperature,

$O = V - T\left(\frac{R}{P} + \frac{2C_1}{T^3}\right)$

$V = T\left(\frac{R}{P} + \frac{2C_1}{T^3}\right)$

$\frac{RT}{P} - \frac{C_1}{T^2} + C_2 = \frac{RT}{P} + \frac{2C_1}{T^2}$

$$\frac{3C_1}{T^2} = C_2$$

$$T^2 = \frac{3C_1}{C_2}$$

$$= \frac{3 \times 50000}{0.8}$$

$$= 433 \text{ K}$$

17. (d) Given that

For Diesel cycle.

$P_2 = P_3 = 44$ bar

$T_3 = 1600$ K

$P_1 = 1$ bar

$T_1 = 300$ K

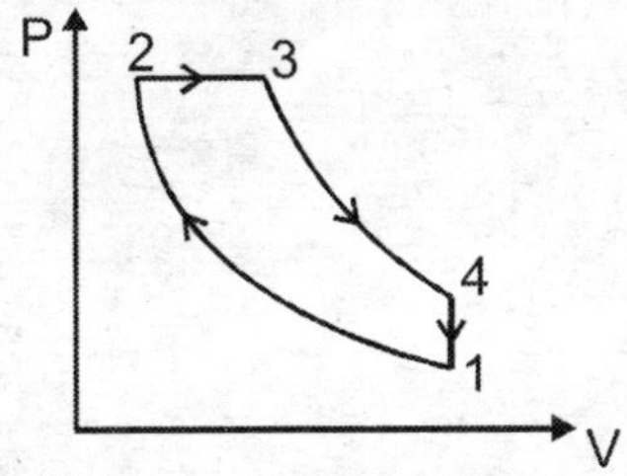

Process 1 – 2 is odiabatic process.

$$\frac{T_2}{T_1} = \left(\frac{P_2}{P_1}\right)^{\frac{\gamma}{\gamma-1}}$$

$$T_2 = 300\left(\frac{44}{12}\right)^{\frac{1.4}{0.4}}$$

$T_2 = 884.46$ K

Process 2 – 3, Heat supplied at Constant pressure.

$$\frac{T_2}{v_2} = \frac{T_3}{v_3}$$

$$\frac{T_3}{T_2} = \frac{V_3}{V_2}$$

$$\text{Cut–off ratio } (\rho) = \frac{V_3}{V_2} = \frac{T_3}{T_2} = \frac{1600}{884.46}$$

$$\boxed{\rho = 1.809}$$

18. (a) Pressure ratio (r_P) = ?

$\eta_B = 0.35$

$\gamma = 1.4$

$$\eta_B = 1 - \frac{1}{(r_P)^{\frac{\gamma-1}{\gamma}}}$$

$$\frac{1}{(r_P)^{\frac{\gamma-1}{\gamma}}} = 1 - 0.35 = 0.65$$

$$(r_P)^{\frac{\gamma-1}{\gamma}} = 1.538$$

$$r_P = (1.538)^{\frac{\gamma}{\gamma-1}}$$

$$r_P = (1.538)^{\frac{1.4}{0.4}}$$

$$\boxed{\text{Pressure ratio}(r_P) = 4.512}$$

19. (a) Given that

At 80 bar

$h_1 = 3246$ kJ/kg

$s_1 = 6.52$ kJ/kg·K

At 4 bar

$s_f = 2$ kJ/kg·k

$s_g = 6.7$ kJ/kg·k

$s_1 = s_f + x\, s_{fg}$

$6.52 = 2 + x(6.7 - 2)$

Dryness fraction(x) = 0.96

Enthalpy at exit (h_2) = $h_f + x\, h_{gf}$

$h_2 = 697 + 0.96(2763 - 697)$

$$\boxed{h_2 = 2680.36 \text{ kJ / kg}}$$

20. (a) dQ = 0

dW = 0

$$C_v(T_1 - T_f) = C_v(T_f - T_2)$$

$$21.6(350 - T_f) = 11.6(T_f - 300)$$

$$11040 = 33.2\, T_f$$

$$\boxed{\text{Final temperature}(T_f) = 332.53 \text{ K}}$$

21. (b) Mixing of two stream.

$h_1 = 42$ kJ/kg

$h_2 = 80$ kJ/kg

$$m_1 = \frac{0.85}{85} \text{ kg/min} = 0.01 \text{ kg/min}$$

$$m_2 = \frac{0.9}{90} \text{ kg/min} = 0.01 \text{ kg/min}$$

$m_1 + m_2 = m_3$

$m_3 = 0.02$ kg/min

$m_1h_1 + m_2h_2 = m_3h_3$

$$h_3 = \frac{m_1h_1 + m_2h_2}{m_3}$$

$$h_3 = \frac{(0.01 \times 42) + (0.01 \times 80)}{0.02}$$

$h_3 = 61$ kJ/kg

22. (b) Given that

$P_1 = 320$ K

$T_1 = 320$ K

$P_2 = 1$ bar $= 100$ kPa

$$T_2 = T_1 \left(\frac{P_2}{P_1}\right)^{\frac{\gamma}{\gamma - 1}}$$

$$T_2 = 320\left(\frac{100}{120}\right)^{\frac{1.4}{0.4}}$$

$T_2 = 303.76$ K

The outlet velocity(v_2) $= 44.72\sqrt{\Delta d}$

$$v_2 = 44.72\sqrt{C_P(T_1 - T_2)}$$

$$v_2 = 44.72\sqrt{1.005(320 - 303.76)}$$

$$\boxed{v_2 = 180.66 \text{ m/s}}$$

2015

Q. (1 – 9) Carry One Mark Each.

1. A gas expands following the relation PV^n = constant, from the initial tate P_1, V_1 to final volume $V_2 = 2V_1$. For the values of '*n*' mentioned below, maximum displacement work is obtained for

(*a*) n = – 1 (*b*) n = 0
(*c*) n = 1 (*d*) n = 1.4

2. A 100 Ω electrical resistor is heated steadily by passing a current of 20 A. If heating is performed in the ambient at 30°C, the rate of increase in entropy of the universe in kW/K is ______.

3. As per Clausius inequality, a system operating on an irreversible cycle transfers

(*a*) more entropy to the sink than what it receives from the source
(*b*) as much entropy to the sink as it receives from the source
(*c*) less entropy to the sink than what it receives from the source
(*d*) less entropy to the sink than that corresponding to a reversible cycle

4. The critical point of a substance corresponds to the state

(*a*) at which the solid, liquid and vapor phases are in equilibrium
(*b*) beyond which liquid will require very large amount of heat to become vapor
(*c*) beyond which solid sublimates directly to the vapor state
(*d*) beyond which the distinction between liquid and vapor phases disappears

5. Consider the process of sensible cooling of air with 60% relative humidity at constant pressure. Which one of the following statements is TRUE for this process?

(*a*) Both humidity ratio and relative humidity increase during process
(*b*) The humidity ratio decreases continuously due to condensation
(*c*) The dry bulb temperature decreases but the wet bulb temperature increases
(*d*) The humidity ratio remains constant

6. The coefficient of performance (COP) of a reversible refrigerator operating between two thermal reservoirs is 4.0. The efficiency (in percentage) of a reversible heat engine operating between the same temperature limits is ______.

7. Differential change in specific enthalpy of a superheated vapor for which ideal gas approximation does not hold, is given by the expression

(*a*) $dh = C_p\, dT$

(*b*) $dh = C_p dT + \left.\frac{\partial h}{\partial v}\right|_T dv$

(*c*) $dh = C_p dT + \left.\frac{\partial h}{\partial p}\right|_T dp$

(*d*) $dh = C_v dT + \left.\frac{\partial h}{\partial p}\right|_T dp$

8. An ideal gas mixture of oxygen (molecular weight = 32 kg/kmol) and carbon dioxide (molecular weight = 44 kg/kmol) has a mass composition of 40% and 60% respectively. If the total pressure is 200 kPa, the partial pressure of oxygen (in kPa) is ______.

9. In an ideal Rankine cycle, increase in superheat of vapor at the exit of boiler leads to

(*a*) decrease in net work output from the cycle
(*b*) increase in cycle efficiency
(*c*) decrease in cycle efficiency
(*d*) decrease in quality of steam at the exit of the turbine

Q. (10 – 22) Carry Two Marks Each.

10. Two moles of air at 1 atm, 21.1°C goes through an adiabatic device and separates into a hot stream of 0.4 moles at 1 atm, 176.3°C and a cold stream of 1.6 moles at 1 atm, – 17.7°C, without any external work. It can be concluded that

(*a*) the total entropy change is zero
(*b*) the total entropy change for the process is positive
(*c*) the device violates Second Law of Thermodynamics
(*d*) the device violates First Law of Thermodynamics

11. For a real gas undergoing volume expansion through a porous plug with $\alpha = \frac{1}{v}\left.\frac{\partial v}{\partial T}\right|_p$, the Joule-Thomson cooling effect is observed if

(*a*) $0 < \alpha T < 1$ (*b*) $\alpha T = 1$
(*c*) $\alpha T > 1$ (*d*) $\alpha T = 0$

12. A lead bullet at 100°C traveling at 500 m/s, strikes a target and adiabatically comes to rest. If the specific heat of lead is 92 J/kg °C, melting temperature is 327.5 °C and heat of fusion is 108 kJ/kg, the percentge mass of lead in the bullet that is melted by collision is ________.

13. An air-water vapor mixture with volume 100 m^3 at 100 kPa, 35°C is at 75% relative humidity. Taking saturation pressure of vapor at 35°C as 5.63 kPa, the mass of vapor in the mixture in kg is ________.

14. An ideal gas of 1 kg mass enclosed inside a rigid vessel at the initial temperature 1200 K is employed as a heat source. The specific heat C_v of the gas is 718 J/kg K. The maximum work in kJ that can be developed by operating a heat engine between the ideal gas and the ambient at 300 K is

(*a*) 646.2 (*b*) 484.7

(*c*) 387.7 (*d*) 347.6

15. The boiling point of water changes from 99.62°C to 105.99°C when the system pressure is changed from 1 bar to 1.25 bar. The boiling point value (in °C) corresponding to a pressure of 1.5 bar is ______.

16. An ideal gas expands in an adiabatic frictionless nozzle from the inlet conditions of 31 bar, 800 K to the ambient pressure of 1 bar at the outlet. The specific heat C_p for the gas is 1 kJ/kg K and the specific heat ratio $\gamma = 1.4$. Neglecting inlet kinetic energy, the velocity of the gas (in m/s) at the nozzle exit is

(*a*) 32 (*b*) 500

(*c*) 707 (*d*) 1000

17. One kmol of hydrogen (molecular weight = 2 kg/kmol, specific heat ratio $\gamma = 1.4$) at 1 bar, 300 K mixes with one kmol of nitrogen (molecular weight = 28 kg/kmol, specific heat ratio $\gamma = 1.4$) at 1 bar, 300 K in an adiabatic vessel. The final mixture is also at 1 bar, 300 K. The entropy change (in kJ/K) for the process is

(*a*) – 5.76 (*b*) zero

(*c*) 5.76 (*d*) 11.53

18. A cycle 1-2-3-1 is proposed with the following processes:

1-2: Constant pressure expansion, 2-3: Reversible adiabatic expansion, 3-1: Irreversible adiabatic compression. Which one of the following statements is TRUE?

(*a*) The net work of the cycle is zero because there is no heat transfer.

(*b*) The cycle is feasible and can deliver net positive work.

(*c*) The cycle is impossible according to Kelvin-Planck statement.

(*d*) The cycle is impossible in accordance with First Law of Thermodynamics.

19. One kg of saturated liquid-vapor mixture of water at 150 kPa (u_f = 467 kJ/kg, v_f = 0.001053 m^3/kg; u_g = 2520 kJ/kg and v_g = 1.159 m^3/kg) with quality of 0.7 is enclosed in a piston cylinder assembly. Heat is added at constant pressure to this system while a paddle wheel transfers a work of 50 kJ. The mixture eventually attains saturated vapor state. The amount of heat added to the mixture (in kJ) is ________.

20. Air enters a pipe at 1 bar and flows isothermally at the rate of 1 kg/s. Due to pipe friction, the pressure drop between two sections of the pipe is 7% of the pressure at inlet section. For ambient temperture, T_0 = 300 K, the rate of irreversibility (in W) between the two sections is ________.

21. A pump raises pressure of saturated liquid water at 100 kPa (density ρ = 959 kg/m^3) to 2 MPa. The isentropic efficiency of the pump is 0.92. The work work done by the pump (in J/kg) is ______.

22. In a Brayton cycle with air ($\gamma = 1.4$) as working fluid, T_{min} = 300 K and T_{max} = 1000 K. The pressure ratio corresponding to maximum net work per cycle is ________.

ANSWERS

1. (a)	**2.** 0.132 – .135	**3.** (a)	**4.** (d)	**5.** (d)	**6.** 19.5 – 20.5	**7.** (c)
8. 92 – 100		**9.** (b)	**10.** (c)	**11.** (c)	**12.** 95 to 97	**13.** 2.9 – 3.0
14. (d)	**15.** 111 – 111.6	**16.** (d)	**17.** (d)	**18.** (c)	**19.** 617 – 619	
20. 6235 – 6245	**21.** 2100 to 2200			**22.** 8.20 to 8.25		

EXPLANATIONS

1. (a) Given that

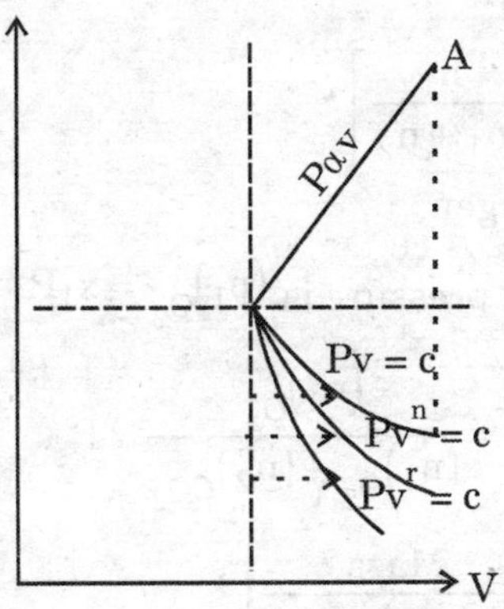

$Pv^n = c$ -----(1)

$V_2 = 2V_1$

For Maximum displacement work, the area under the P-V curve should be maximum.

therefore, if

$n = -1$

$\Rightarrow PV^{-1} = c$ [from eq. (1)

$\Rightarrow \frac{P}{V} = c$

$\Rightarrow P = CV$

$\therefore \boxed{P \alpha V}$ -----(2)

from equation (2), we can observe that the area under the curve 'OA' is maximum and thus the displacement work is maximum.

Hence the option (a) is correct.

2. Given that

Resistance of the resistor (R) = 100 Ω.

Current (I) = 20A

Ambient temperature (T_o) = 30°C = 30 + 273

T_o = 303K

We know

$$(ds)_{universe} = (ds)_{system} + (ds)_{surrounding} = \frac{dQ}{T} + O$$

[∵ No heat interaction b/w system surrounding

$\therefore (ds)_{surrounding} = 0]$

$$= \frac{I^2R}{T} = \frac{(20)^2 \times 100}{303}$$

$(ds)_{universe} = 132.013 W/K$

$\therefore \boxed{(ds)_{universe} = 132.013 W/K}$

therefore,

Rate of increase of entropy of the universe

$$\boxed{\left(\frac{ds}{dt}\right)_{universe} = 0.132 \text{ kW/K}}$$

3. (a) We know that

for irreversible process, clausius inequality will be

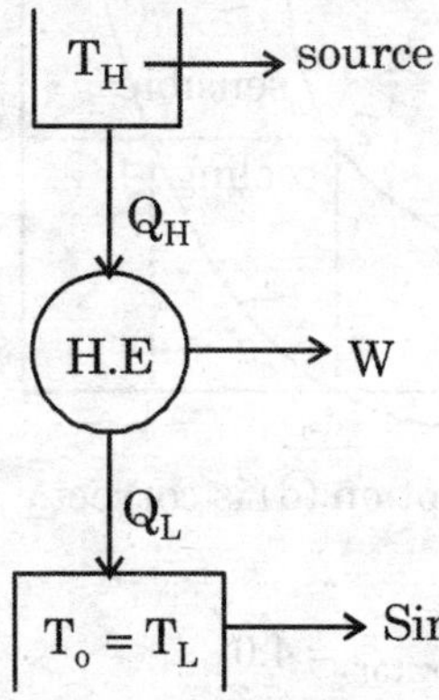

$$\oint \frac{dQ}{T} < 0$$

By applying the above concept to the heat engine

$$\frac{Q_H}{T_H} - \frac{Q_L}{T_L} < 0$$

$$\frac{Q_H}{T_H} < \frac{Q_L}{T_L}$$

$$\boxed{(ds)_{source} < (ds)_{sink}}$$

$$\left[\because ds = \frac{dQ}{T}\right]$$

Hence the option (a) is correct.

4. (d) The critical point of a substance is the state of the substance beyond which the distinction between the liquid and vapour phases disappears.

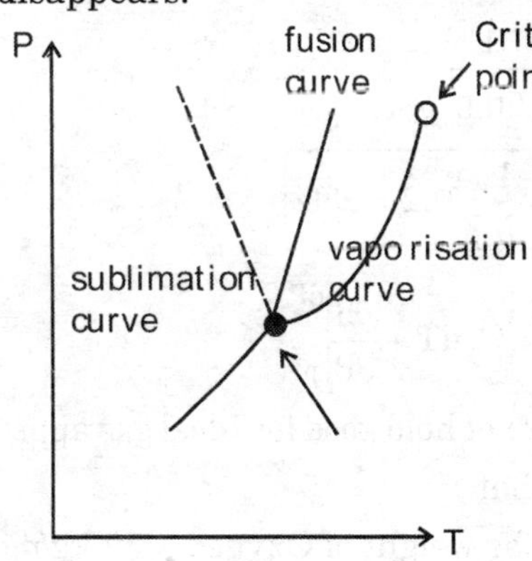

Hence the option (d) is correct.

5. (d) Give that

Relative humidity (ϕ) = 60%

When the process is sensible cooled, then dry bulb temperature (t) decreases.

Humidity ratio (W) remains constant

i.e $\boxed{W_1 = W_2}$

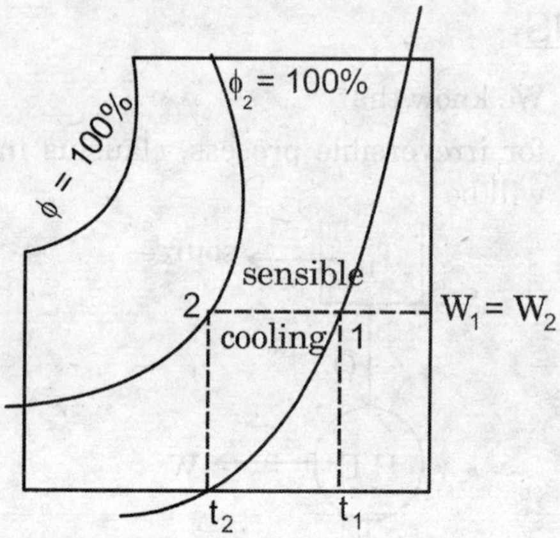

Hence the option (d) is correct.

6. Given that

$(COP)_{refrigerator} = 4.0$

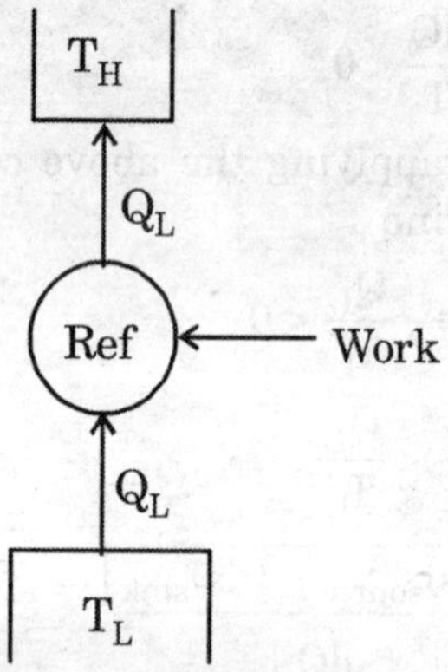

Reversible Refrigerator operating system

We know that

$$1 + (COP)_{refrigerator} = \frac{1}{\eta_{Heat\ engins}}$$

$$1 + 4 = \frac{1}{\eta_{H.E}}$$

$$\boxed{\eta_{H.E} = \frac{1}{5} = 0.2 = 20\%}$$

7. (c) $dh = C_p dT + \left.\frac{\partial h}{\partial P}\right|_T dP$

does not hold good for Ideal gas approximation

8. Given that

Molecular weight of Oxygen = 32 kg/mole

Mol. weight of CO_2 = 44 kg/mole

Mass composition of O_2 = 40%

Mass composition of CO_2 = 60%

Total pressure of the mixture

(P) = 200 kPa

We know that

Partial pressure

$P_1 = x_1 P$

$$\left[x_1 = \frac{n_1}{n_1 + n_2}\right]$$

for oxygen,

partial pressure is $(p_1)_{O_2} = x_1 P$

$$(p_1)_{02} = \frac{(n_1)_{O_2}}{(n_1)_{O_2} + (n_2)_{CO_2}} \times P$$

$$\left[n = \frac{\text{Mass}}{\text{Molecular wt.}}\right]$$

$$= \frac{\left(\frac{0.4}{32}\right)}{\frac{0.4}{32} + \frac{0.6}{44}} \times 200$$

$$\boxed{(p_1)_{O_2} = 95.78 \text{ kPa}}$$

9. (b) Efficiency of the ideal Rankine cycle is

$$\eta_{rankine} = \frac{W_{net}}{Q_s}$$

$$\boxed{\eta_{rankine} = \frac{W_T - W_P}{Q_s}}$$

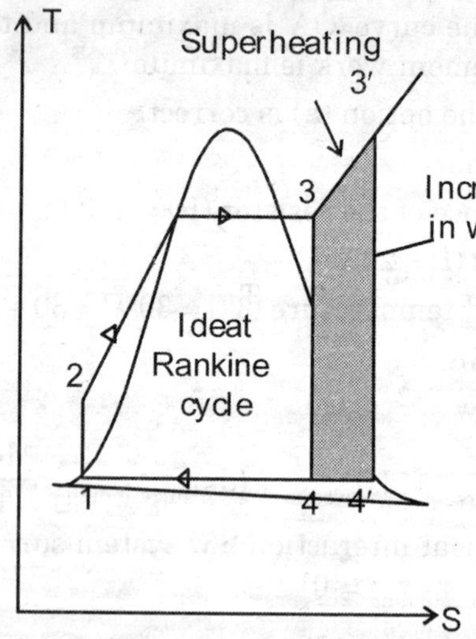

Here, W_{net} increases, the efficiency of the cycle increases.

In another words, superheating the steam to higher temperatures has another very desirable effect, it decreases the moisture content of the steam at the turbine exit.

Hence the option (b) is correct

10. (c) The device violates the second law of thermodynamics which states that a natural process runs only in one sense, and is not reversible for example, heat always flows spontaneously from hotter to colder bodies and never the reverse, unless external work is performed on the system.

11. (c) Given that

$$\alpha = \frac{1}{v}\frac{\partial v}{\partial T}\Big|_P$$

$$\Rightarrow \left(\frac{dv}{dT}\right)_P = \alpha v \quad \text{----(1)}$$

According to Joule thomson coefficient (j)

$$j = \frac{1}{C_P}\left[T.\left(\frac{dv}{dT}\right)_P - v\right] \quad \text{----(2)}$$

Substitute eqn (1) in eqn (2)

$$j = \frac{1}{C_P}[T\,\alpha\,v - v]$$

$$\boxed{j = \frac{v}{C_P}[\alpha\,T - 1]} \quad \text{----(1)}$$

for Joule =thomson cooling effect

$$\boxed{j > 0 \;\&\; \alpha\,T > 1}$$

Hence the option (c) is correct.

12. By applying energy equation, we get

$$h_1 + \frac{1}{2}V_1^2 = h_2 + \frac{1}{2}V_2^2$$

$$CT_1 + \frac{1}{2}V_1^2 = CT_2 + \frac{1}{2}V_2^2$$

But final speed = 0

$\therefore\ V_2 = 0$

$$92 \times 373 + \frac{500^2}{2} = 92 \times T_2$$

$T_2 = 1731.7$ K

T_m, melting temperature = 600 . 5K

Percentage mass of lead melted

$$= \frac{Mc\Delta T}{\text{latent heat of fusion}} = \frac{92 \times 1131.2}{108}$$

= 96.36%

13. Given that, volume of the mixture (V) = 100 m^3.

Total pressure (P) = 100 kPa

Temperature (T) = 35 + 273 = 308 K

Relative humidity (ϕ) = 75%

Saturation pressure of vapoure $\left(P_{v_s}\right)$ = 5.63 KPa

We know that $\phi = \frac{P_v}{P_{v_s}} \Rightarrow 0.75 = \frac{P_v}{5.63}$

$$\therefore\ \boxed{P_v = 4.22 \text{ kPa}}$$

Now, Humidity Ratio $(w) = \frac{0.622\,P_v}{P - P_v}$

$$\frac{m_v}{m_a} = \frac{0.622 \times 4.22}{100 - 4.22}$$

$$\left[\because w = \frac{m_v}{m_a}\right]$$

$$\frac{m_v}{m_a} = 0.0274 \quad \ldots (1)$$

$$\left[\begin{array}{l} M_v \rightarrow \text{mass of vapour} \\ m_a \rightarrow \text{mass of air} \end{array}\right]$$

Total mass (m) = $m_v + m_a$.

$$\frac{PV}{RT} = m_v + m_a \qquad [PV = mRT]$$

$$\frac{100 \times 100}{0.287 \times 303} = m_v + m_a$$

$$115 = m_v + m_a \quad \ldots (2)$$

Solving equation (1) & (2), we get

$$\boxed{m_v = 3.0 \text{ Kg}}$$

14. (d) Given that

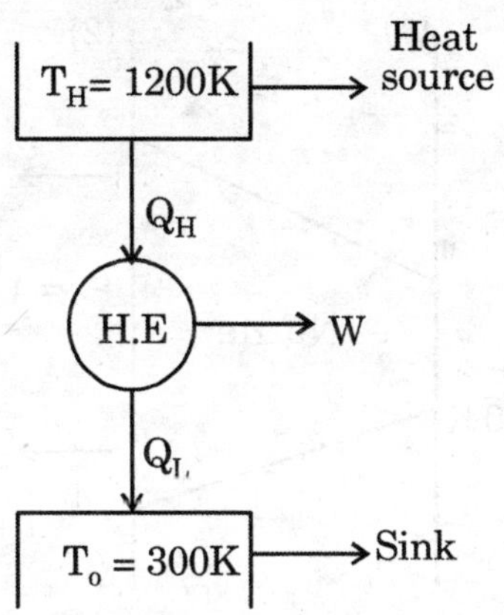

mass of the gas (m) = 1 kg

T_H = 1200 K

T_L = 300 K

Specific heat = 718 J/kg

Heat generated from the heat source

$Q_H = m\,c_v T$

$Q_H = 1 \times 718 \times 1200$

Q_H = 861.6 KJ

By applying clausius inequality

$$\oint \frac{\delta Q}{T} = 0 \Rightarrow \frac{Q_H}{T_H} - \frac{Q_L}{T_L} = O$$

$$\frac{861.6}{1200} = \frac{Q_L}{300}$$

$Q_L = 215.4$ kJ

$\therefore$ Net maximum work (W) = $Q_H - Q_L$.

W = 646.2 kJ

$$\boxed{W = 646.2 \text{ kJ}}$$

Hence the option (a) is correct

15.

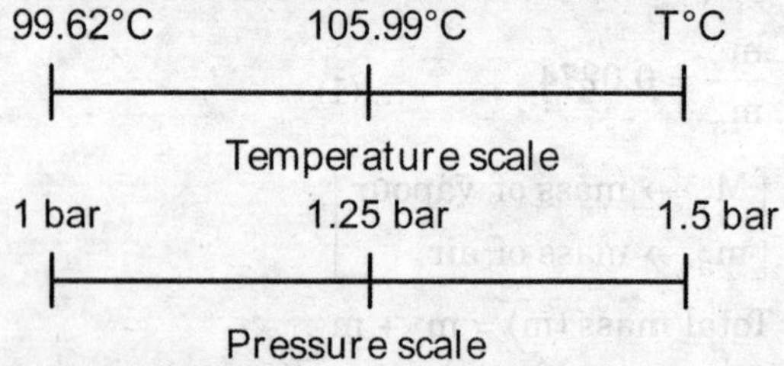

$$\frac{1.25-1}{105.99-99.62} = \frac{1.5-1}{T-99.62}$$

By solving above equation, we get

$$\boxed{T = 112.36°C}$$

16. (d) Given that

inlet pressure (P_1) = 31 bar

inlet temperature (T_1) = 800 K

outlet pressure (P_2) = 1 bar

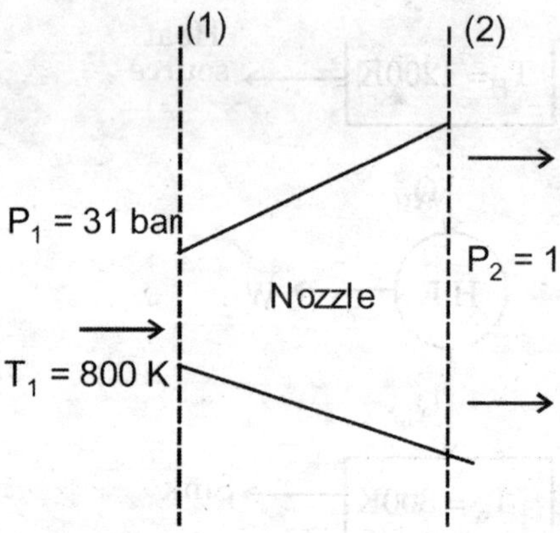

By using steady flow energy equation at (1)- (2)

$$h_1 + \frac{C_1^2}{2000} + \frac{gZ_1}{1000} + Q$$

$$= h_2 + \frac{C_1^2}{2000} + \frac{gZ_2}{1000} + W$$

Here $[Q = 0, C_1 = 0, w = 0, Z_1 = Z_2]$

$$h_2 = h_2 + \frac{C_2^2}{2000}$$

$$\therefore\ C_2 = \sqrt{2000(h_1 - h_2)}$$

$$= \sqrt{2000 \times mc_P(T_1 - T_2)}$$

$$C_2 = \sqrt{2000 \times m \times c_P(800 - T_2)} \quad \text{----(1)}$$

$\because$ the expansion takes place adiabatically

$PV^r = C$

we know

$$\frac{T_2}{T_1} = \left(\frac{P_2}{P_1}\right)^{\frac{r-1}{r}} \left[\frac{T_2}{T_1} = \left(\frac{P_2}{P_1}\right)^{\frac{r-1}{r}} = \left(\frac{V_1}{V_2}\right)^{r-1}\right]$$

$$\frac{T_2}{800} = \left(\frac{1}{31}\right)^{\frac{1.4-1}{1.4}}$$

$$\therefore\ \boxed{T_2 = 300\text{K}}$$

Put T_2 = 300 K in eqn (1), then

$$C_2 = \sqrt{2000 \times 1 \times 1(800-300)}$$

$$\boxed{C_2 = 1000 \text{ m/sec}}$$

Hence the option (d) is correct.

17. (d) $\Delta s = \overline{R}\ell n\left(\frac{V_{f1}}{V_{i1}}\right) + \overline{R}\ell n\left(\frac{V_{f2}}{V_{i2}}\right)$

$$= 8.314 \times 2\ell n(2) = 11.52 \frac{KJ}{K}$$

V_{fi} = Final volume of hydrogen = 2 V_{i1}

V_{i1} = Initial volume of hydrogen

V_{f2} = Final volume of nitrogen = V_{i2}

V_{i2} = Initial volume of nitrogen.

18. (c) The cycle impossible according to the Kelvin-plank statements which states that it impossible to construct a device which operate in a cycle, produces work while interacting with the single thermal reservoir.

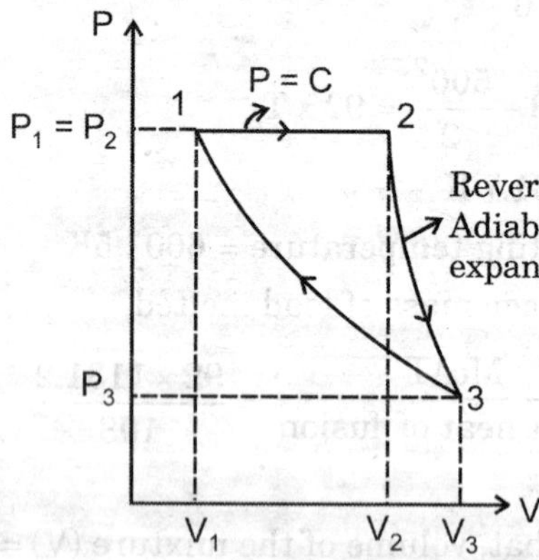

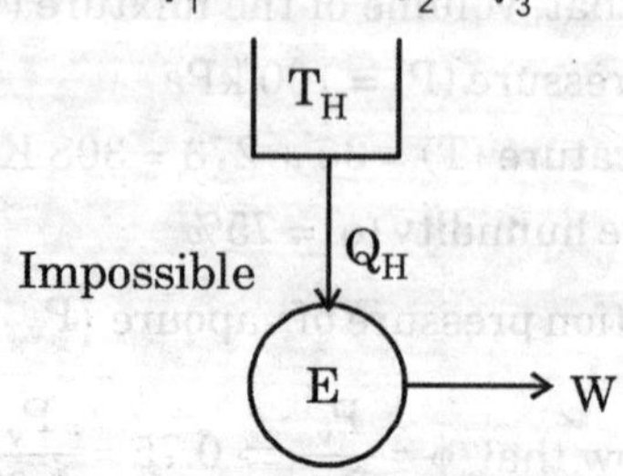

Hence the option (c) is correct.

19. $h_f = u_f + U_f$

$$= 467 + .001053 \times 150 = 467.15 \text{ kJ}$$

$$h_G = u_G + U_G$$

$$= 2520 + 150 \times 1.159 = 02693.85 \text{ kJ}$$

$$h_{fg} = h_g - h_f = 2693.85 - 467.15$$

$$= 2226.59 \text{ kJ}$$

$$\Delta h = h_g - 0.7\, h_{fg} - h_f = 668.007$$

Amount of heat added to the mixture

$$= 668.007\text{–}50$$

$$= 618.007 \text{ kJ}$$

20. Given that

Air flow rate (m) = 1 kg/sec

Inlet pressure (P_1) = 1 bar

Outlet pressure (P_2) = 0.93 bar

Ambient temperature (T) = 300 K

We know

$$ds = C_v \ell n\left(\frac{T_2}{T_1}\right) - R\ell n\left(\frac{P_2}{P_1}\right)$$

$$= 0 - 277\ \ell n\left(\frac{0.93}{1}\right)$$

$[\because\ T_1 = T_2 = T$, Isothermal process$]$

$$ds = 20.0 \text{kJ/kg K}$$

therefore, irrerversibility (I) $= T_0\,(ds)$

$$\therefore\quad I = 300 \times 20.80$$

$$\boxed{I = 6240 \text{ kJ/kgK}}$$

21. Given that

$P_1 = 100$ kPa, $P_2 = 2$ MPa

Density of water (ρ) = 959 kg/m^3

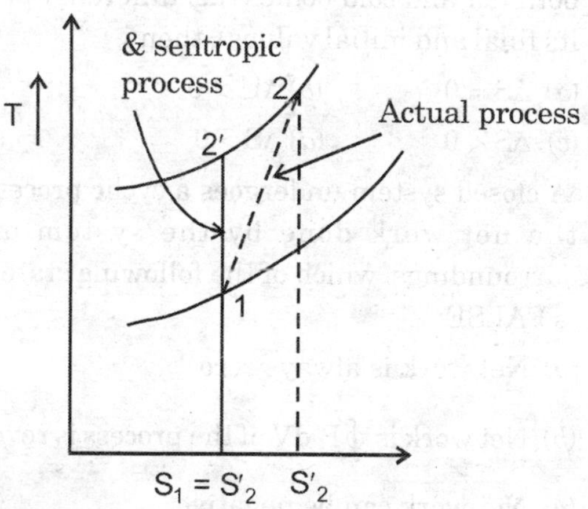

We know

Isentropic efficiency of pump

$$\eta_{Isen} = \frac{h_2^1 - h_2}{(W_{pump})_{Actual}}$$

$$0.93 = \frac{V(P_2 - P_1)}{(W_{pump})_{Actual}}$$

$$\left[\because h_2^2 - h_2 = V(P_2 - P_1)\right]$$

$$\therefore (W_{pump})_{Actual} = \frac{200 - 100}{959 \times 0.93} \text{kJ/kg}$$

$$W_{pump} = 21.30 \text{ kJ/kg}$$

$$\boxed{(W_{pump})_{Actual} = 21.30 \text{ J/kg}}$$

22. Given that

$$T_{min} = 300 = T_1$$

$$T_{max} = 100 \text{ K} = T_3.$$

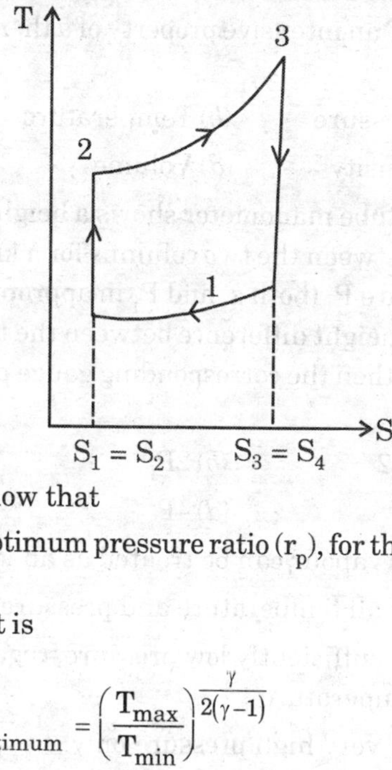

We know that

The optimum pressure ratio (r_p), for the maximum work

output is

$$(r_p)_{optimum} = \left(\frac{T_{max}}{T_{min}}\right)^{\frac{\gamma}{2(\gamma-1)}}$$

$$(r_p)_{optimum} = \left(\frac{1000}{300}\right)^{\frac{1.4}{2(1.4-1)}}$$

$$\therefore \boxed{(r_p)_{optimum} = 8.223}$$

2016

Notation used:

P - pressure, V - volume, T - temperature, S - entropy, H - enthalpy, U - internal energy, A - Helmholtz free energy, C_p - specific heat capacity at constant pressure. Specific properties are designated by lower case symbols.

Useful data:

Universal gas constant R = 8.314 kJ/(kmol.K)

C_p of air = 1.005 kJ/(kg.K)

Ratio of ideal gas specific heats for air: $\gamma = 1.4$

Molecular mass of hydrogen: 2 kg/kmol

(Q. 1 – 9) : Carry One Mark Each.

1. Which of the following thermodynamic properties is NOT an intensive property of a thermodynamic system:

(*a*) Pressure (*b*) Temperature

(*c*) Density (*d*) Volume

2. An U-tube manometer shows a height difference of z_1 between the two columns for a known gauge pressure P_1 (both z_1 and P_1 in appropriate units). If the height difference between the two columns is $2z_1$, then the corresponding gauge pressure will be:

(*a*) $P_1/2$ (*b*) $2P_1$

(*c*) P_1 (*d*) $4P_1$

3. Water vapour can be treated as an ideal gas,

(*a*) for all temperature and pressure

(*b*) for sufficiently low pressure, regardless of its temperature

(*c*) for very high pressure only

(*d*) for sufficiently low temperature, regardless of its pressure

4. The thermal efficiency of a Carnot engine is 0.5. If the temperature of the cold reservoir is 300 K, then the temperature of the hot reservoir is:

(*a*) 600 K (*b*) 1200 K

(*c*) 900 K (*d*) 450 K

5. In a reversible, constant-pressure, non-flow process, heat input is given by:

(*a*) change in internal energy

(*b*) change in enthalpy

(*c*) change in entropy

(*d*) work output

6. Moist air undergoes an adiabatic saturation process such that the relative humidity of air increases. For this process,

(*a*) Dry bulb temperature increases, specific humidity increases

(*b*) Dry bulb temperature increases, specific humidity decreases

(*c*) Dry bulb temperature decreases, specific humidity increases

(*d*) Dry bulb temperature decreases, specific humidity decreases

7. A steadily flowing ideal gas undergoes adiabatic throttling, where

T_1: temperature before throttling

T_2: temperature after throttling

Assuming no change in kinetic and potential energy due to throttling, which of the following is correct:

(*a*) $T_1 = T_2$

(*b*) $T_1 > T_2$

(*c*) $T_1 < T_2$

(*d*) $T_1 = \gamma T_2$, γ: specific heat ratio

8. For irreversible heat transfer from a hot body to a cold body, if Δ denotes the property change of both hot and cold bodies (i.e. difference between its final and initial values), then

(*a*) $\Delta S = 0$ (*b*) $\Delta U > 0$

(*c*) $\Delta S < 0$ (*d*) $\Delta S > 0$

9. A closed system undergoes a cyclic process. For the net work done by the system on the surroundings, which of the following statements is FALSE:

(*a*) Net work is always zero

(*b*) Net work is $\oint P\,dV$ if the process is reversible

(*c*) Net work can be negative

(*d*) Net work can be positive

(Q. 10 – 22) : Carry Two Marks Each.

10. Consider the following statements related to the second law of thermodynamics:

P. A cyclic heat engine cannot produce net work by exchanging heat only with one reservoir.

Q. The efficiency of a reversible heat engine is dependent on the nature and amount of working substance undergoing the cycle.

R. It is impossible to have a cyclic device which will produce no effect other than the transfer of heat from a cold body to a hot body.

S. It is impossible to have heat engines operating between a heat source and sink to have a lower efficiency than that of a reversible heat engine operating between the same source and sink.

For which of the following options, BOTH the statements are inconsistent with the second law of thermodynamics:

(*a*) P and R (*b*) P and Q

(*c*) R and S (*d*) Q and S

11. Consider the following statements related to air-standard Otto, Diesel, and Brayton cycles:

P. Brayton cycle has at least one isentropic and one isobaric process.

Q. Otto cycle has at least one isentropic and one isochoric process.

R. Diesel cycle has at least one isentropic and one isothermal process.

S. At least one of the cycles has an isothermal process.

For which of the following options, BOTH the statements are consistent with the operation of the above cycles:

(*a*) P and R (*b*) P and Q

(*c*) R and S (*d*) P and S

12. Volumetric analysis of a hydrocarbon combustion product shows 8% CO_2, 15% H_2O (vapour), 5.5% O_2 and 71.5% N_2. The combustion product flows steadily through a heat exchanger at 200 kPa pressure. Assume each component in the mixture to be an ideal gas. In order to avoid the condensation of H_2O in the heat exchanger, the minimum allowable temperature (in °C) is _____.

Saturated H_2O Table:

P (kPa)	10	20	30	40	50
T (°C)	45.83	60.09	69.12	75.82	81.35

13. An equimolar mixture of two ideal gases (A, B) expands isentropically in a nozzle. The gas mixture enters the nozzle at 300 kPa, 400 K and exits at 100 kPa. Assuming the mixture to be an ideal gas, the exit temperature of the gas mixture (in K) is _________ .

	Molar mass (kg/kmol)	C_P (kJ/kg-K)
Gas A	28.013	1.04
Gas B	2.016	14.21

14. A rigid vessel of volume 10 m³ is filled with hydrogen at 25 °C and 500 kPa. Due to leakage, some gas has escaped from the vessel until the pressure in the vessel drops down to 200 kPa, and the corresponding temperature of the gas inside the vessel is found to be 15 °C. The amount of gas leaked (in kg) from the vessel is ________.

15. A hot ideal gas (C_p = 1.2 kJ/(kg.K)) steadily flows through a turbine with inlet and exit temperatures of 1500 K and 500 K respectively. The minimum mass flow rate (in kg/s) of the hot gas to achieve a power output of 12 MW is ______.

16. Air pressure inside a spherical balloon is proportional to its diameter. The balloon undergoes a reversible, isothermal, non-flow process. During the process, the balloon maintains its spherical shape, and the air inside the balloon consumes 2 kJ of heat. Initial air pressure inside the balloon was 120 kPa, while the initial balloon diameter was 20 cm. Assuming air to be an ideal gas, the final diameter of the balloon (in cm) is ________.

17. An air-standard diesel engine has a compression ratio of 18 (the ratio of the volume at the beginning of the compression process to that at the end of the compression process), and a cut-off ratio of 2 (the ratio of the volume at the end of the heat addition process to that at the beginning of the heat addition process). The thermal efficiency (in %) of the engine is ________.

18. Compressed air, at 1 MPa pressure, 400 K temperature flows through a large pipe. An evacuated, insulated rigid tank of 0.5 m³ volume is connected to the pipe through a valve. The valve is opened to fill the tank and the valve closes automatically when the tank pressure reaches 1 MPa. Assuming ideal gas behaviour, the final air temperature in the tank (in K) is __________ .

19. A 40 kg metal block (C_p = 0.5 kJ/(kg.K)) at T = 450 °C is quenched in 150 kg oil (C_p = 2.5 kJ/(kg. K)) at T = 25 °C. If the combined (metal block and oil) system is fully isolated from its surroundings, then the net change in the entropy (in kJ/K) of the combined system is ____.

20. For phase change from solid (sol) to liquid (liq) state, if the slope of the solid-liquid coexistence line in the P-T diagram is negative, then:

(*a*) $v_{liq} < v_{sol}$ (*b*) $v_{liq} > v_{sol}$

(*c*) $S_{liq} < S_{sol}$ (*d*) $h_{liq} < h_{sol}$

21. A house-hold refrigerator operates under steady state condition between an evaporator temperature of 263 K and a condenser temperature of 323 K. The heat load to the refrigerator is 3 kW. The actual COP of the refrigerator is half of that of a Carnot refrigerator operating between the same condenser and evaporator temperatures. The power required (in kW) to run the refrigerator is ________.

22. The Maxwell relation that results from the expression for the Helmholtz free energy A = U – TS, is:

(*a*) $\left.\frac{\partial T}{\partial v}\right)_s = -\left.\frac{\partial P}{\partial s}\right)_v$ (*b*) $\left.\frac{\partial T}{\partial P}\right)_s = \left.\frac{\partial v}{\partial s}\right)_P$

(*c*) $\left.\frac{\partial P}{\partial T}\right)_v = \left.\frac{\partial s}{\partial v}\right)_T$ (*d*) $\left.\frac{\partial v}{\partial T}\right)_P = -\left.\frac{\partial s}{\partial P}\right)_T$

ANSWERS

1. (d)	**2.** (b)	**3.** (b)	**4.** (a)	**5.** (b)	**6.** (c)	**7.** (a)
8. (d)	**9.** (a)	**10.** (d)	**11.** (b)	**12.** 67:297	**13.** 286:297	**14.** 2.30:2.40
15. 95:10.5	**16.** 28:36	**17.** 62.0:64.0	**18.** 555:565	**19.** 9.7:9.9	**20.** (a)	**21.** 1.34:1.40
22. (c)						

EXPLANATIONS

1. (d) Pressure, Temperature and density are intensive properties but volume is extensive property. Hence correct option is (D).

2. (b)

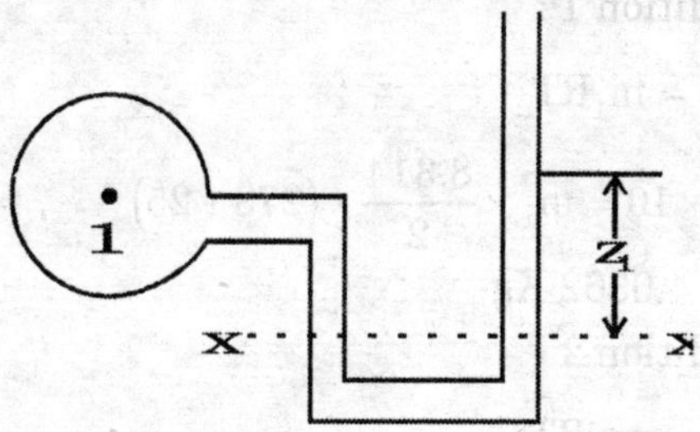

$P_1 = \rho g z_1$

If $z_2 = 2z_1$

$P_2 = \rho g z_2 = \rho g 2z_1 = 2P_1$

3. (b) Any gas behave as an ideal gas under condition:-

(i) Low Pressure

(ii) High Temperature

Hence, correct option is (B)

4. (a) Given that

For carnot engine.

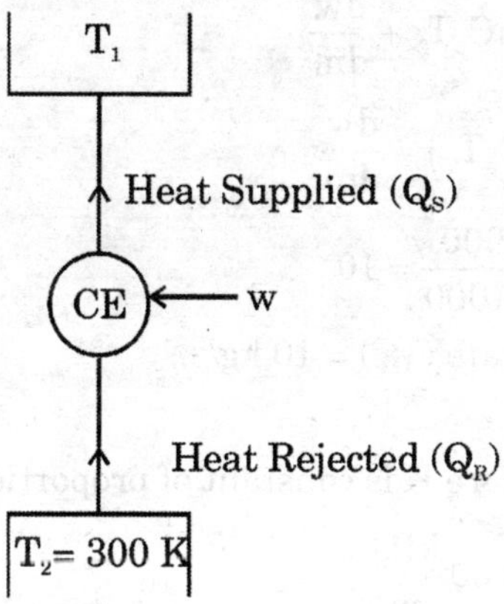

$\eta = 0.5$

$$\eta = 1 - \frac{T_2}{T_1}$$

$$0.5 = 1 - \frac{300}{T_1}$$

$$\boxed{T_1 = 600\,K}$$

5. (b) We know that

For non-flow process

$\delta Q = dU + dW$

$\delta Q = dU + p.dV$

$\delta Q = d(U + PV)$

$\delta Q = dH$

at constant pressure

$$\boxed{\delta Q = dH}$$

Heat transfer at constant pressure is equal to change of enthalpy. Hence correct answer is option (B)

6. (c) Moist air undergoing in an adiabatic saturation process.

$$\text{Relative humidity}(\phi) = \frac{Pv}{Pvs}$$

$Pv \rightarrow$ Dew point temperature.

$Pvs \rightarrow$ Dry Bulb temperature

If $\phi \uparrow \Rightarrow Pvs \downarrow$

If $\phi \uparrow \Rightarrow Pv \uparrow \Rightarrow w \uparrow$

Hence correct answer is option (C).

7. (a) During the adiabatic throttling process.

$h_1 = h_2$

$mCp\ T_1 = mCp\ T_2$

$$\boxed{T_1 = T_2}$$

Hence correct answer is option (A)

8. (d)

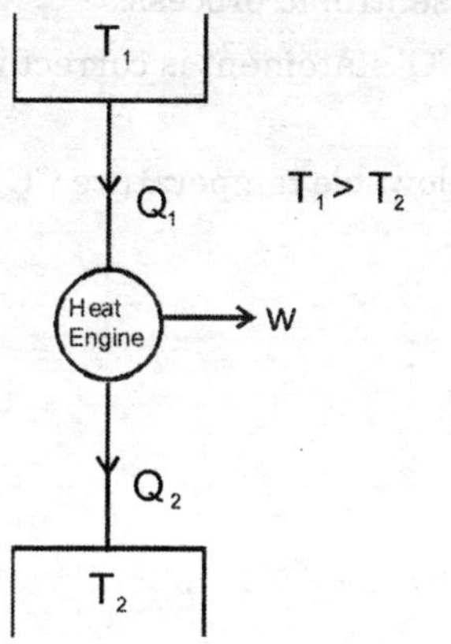

For Irreversible process.

$ds > 0$

$$\boxed{\Delta s > 0}$$

9. (a) A closed system undergoes a cyclic process, then

(i) $W_{net} = \oint P.dv$

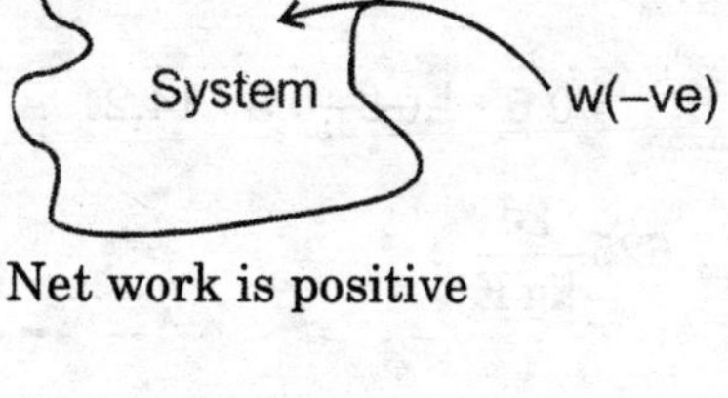

(ii) Net work is positive

(iii) Net work is negative

But Net work is not equal to zero, because work transfer is a path function.

10. (d) According to Second Law of thermodynamics:

(i) The Efficiency of a reversible heat engine is independent of the nature and amount of working substance undergoing the cycle.

(ii) It is impossible to have heat engines operating between a heat source and sink to have a lower efficiency than that of a reversible heat engine operating between the same source and sink.

$$\boxed{\eta_R > \eta_{IR}}$$

So that in the given question 'Q' & 'S' are correct statements according to second law of thermodynamics.

11. (b) (1) Brayton cycle has

Two isobaric and two isentropic process

(2) Otto cycle has two isochoric and two isentropic process.

(3) Diesel cycle has one isobaric, one isochoric and two isentropic process.

So that 'P' & 'Q' statement is correct in given equation.

12. The minimum allowable temperature (°C) is 69°C

13. $T_2 = T_1\left(\frac{P_2}{P_1}\right)^{\frac{\gamma-1}{\gamma}}$

$$C_p = \frac{R\gamma}{\gamma - 1}$$

$$\frac{R}{C_p} = \frac{\gamma - 1}{\gamma}$$

$$R_A = \frac{8.314}{28.013} = .297\frac{kJ}{kg\,K}$$

$$R_B = \frac{8.314}{2.016} = 4.124\frac{kJ}{kg\,K}$$

$$R_{mixture} = 0.5 \times R_A + 0.5\,R_B$$

$$R_{mixture} = .5 \times .297 + .5 \times 4.124 = 2.2105\ \frac{kJ}{kg\,K}$$

$$(C_p)_{mixture} = 0.5 \times C_{PA} + 0.5\,C_{PB}$$

$$(C_p)_{mixture} = 0.5 \times 1.04 + 0.5 \times 14.21$$

$$= 7.625\frac{kJ}{kg\,K}$$

$$T_2 = 400\left(\frac{100}{300}\right)^{\frac{2.2105}{7.625}}$$

$$\boxed{\therefore T_2 = 290.9\,K}$$

14. Condition 1 :

$$P_1 v_1 = m_1 RT_1$$

$$500 \times 10 = m_1 \times \frac{8.314}{2} \times (273 + 25)$$

$$m_1 = 4.0362\ Kg$$

Condition 2

$$P_2 v_2 = m_2 RT_2$$

$$200 \times 10 = m_2 \times 4.157 \times 288$$

$$m_2 = 1.6705\ Kg$$

$$\Delta m = m_1 - m_2 = 2.36\ Kg$$

15. Given that

$C_p = 1.2$ kJ/kg.K

$T_1 = 15000K$

$T_2 = 500K$

Power output = 12 MW.

$$\dot{m}\,C_p T_1 = \dot{m} C_p T_2 + \frac{dw}{dm}$$

$$\dot{m}\left(C_p T_1 - C_p T_2\right) = \frac{dw}{dm}$$

$$\dot{m} = \frac{12 \times 1000}{1.21 \times 1000} = 10$$

Mass flow rate ($\dot{m}$) = 10 kg/s.

16. $P \propto D$

P = K D(where K is constant of proportionality)

$$K = \frac{120}{0.2} = 600\frac{kPa}{m}$$

$$W_{1-2} = \int_{V_1}^{V_2} P\,dV$$

$$= \int_{D_1}^{D_2} KD \times \frac{\pi D^2}{2} \times dD = \frac{K\pi}{2}\left[\frac{D^4}{4}\right]_{D_1}^{D_2}$$

$$2 = \frac{600 \times \pi}{2 \times 4}\left[D_2^4 - .2^4\right]$$

$D_2 = .32$ m or 32 cm

$$V = \frac{\pi D^3}{6}$$

$$dV = \frac{3D^2.\pi dD}{6}$$

$$dV = \frac{\pi D^2}{2}.dD$$

17. Given that

For Diesel cycle.

Compression ratio(r) = 18

Cut – off ratio (ρ) = 2

$\gamma = 1.4$

$$\eta_D = 1 - \frac{1}{\gamma \cdot r^{\gamma-1}}\left[\frac{\rho^\gamma - 1}{\rho - 1}\right]$$

$$\eta_D = 1 - \frac{1}{1.4 \times 18^{0.4}}\left[\frac{2^{1.4} - 1}{2 - 1}\right]$$

$\eta_D = 63.16\ \%$

18. Given that

P = 1 MPa

$T_i = 400$ K

V = 0.5 m^3

From unsteady state condition.

$T_f = \gamma Ti$

$T_f = 1.4 \times 400$

$\boxed{T_f = 560\text{ K}}$

19. By using energy balance equation,

$40 \times 0.5 \times (450 - T_f) = 150 \times 2.5\ (T_f - 25)$

$T_f = 46.51°C$

$\Delta s = 40 \times 0.5$

$$\times \ln\left(\frac{319.51}{723}\right) + 150 \times 2.5 \times \ln\left(\frac{319.51}{298}\right)$$

$$\Delta s = 9.8 \frac{kJ}{kg}$$

20. (a) The melting curve for water is negative on P – T diagram here

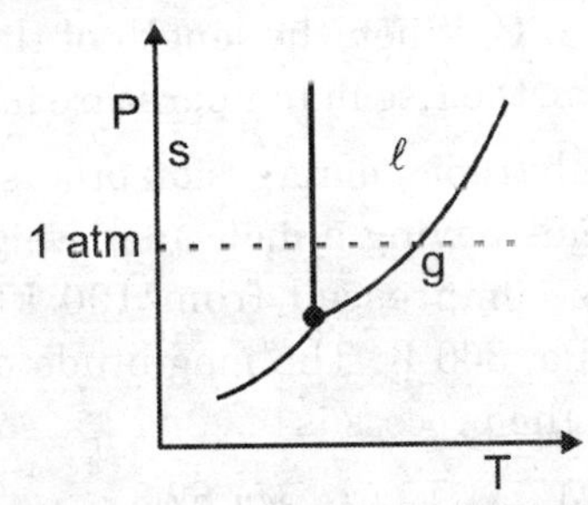

The melting curve for most substance has a positive slope. But water has a negative slope

From clapeyron equation

$$\left(\frac{\partial P}{\partial T}\right)_{\Delta G} = \frac{\Delta S}{\Delta v}$$

For negative slope, $\left(\frac{\partial P}{\partial T}\right)_{\Delta G} = \frac{\Delta S}{\Delta v} < 0$

We know that ΔS for the melting process is positive because the system is going from the highly order crystalline solid state to more disordered liquid state $\left(\Delta S = \frac{\Delta H}{T}\right)$. If ΔV is negative. This mean that

$\boxed{\partial_L < \partial_S}$

21. Given that

$T_E = 263$ K

$T_C = 323$ K

Q = 3 kW

$$(COP)_{CR} = \frac{T_E}{T_C - T_E}$$

$$(COP)_{CR} = \frac{263}{323 - 26.3} = 4.38$$

$$(COP)_{CR} = \frac{1}{2} \times (COP)_{CR} \text{ [From equation]}$$

$$(COP)_{CR} = 2.19$$

$$(COP)_{CR} = \frac{Q}{W_{in}}$$

$$\boxed{W_{in} = \frac{3}{2.19} = 1.369\text{ kW}}$$

22. (c) Helmboltz free energy,

H = U – TS

From Maxwell relation,

H = U – TS

dA = dU – T.ds – s.dT = – Pdv – sdT

dz = m.dx + N.dy

$$\left(\frac{\partial M}{\partial y}\right)_x = \left(\frac{\partial N}{\partial x}\right)_y$$

$$\left(\frac{\partial P}{\partial T}\right)_v = \left(\frac{\partial s}{\partial v}\right)_T$$

2017

1. Given $d\phi = f(T)\,dT + (T/V)$ and $d\Psi = Tdp + (T/p^2)dV$, then
(a) both ϕ and Ψ are properties
(b) neither ϕ nor Ψ is a property
(c) ϕ is a property but Ψ is not a property
(d) Ψ is a property but ϕ is not property

2. A paddle wheel is installed in a rigid insulated tank containing 10 kg air (C_v = 0.718 kJ/kg.K). A torque of 100 N.m is applied on the paddle wheel to rotate it at 60 revolutions per minute for 2 minutes. At the end of the process, the increase in temperature of air in °C is
(a) 0 (b) 5.25
(c) 10.50 (d) 21.50

3. Consider the two systems containing 20 kg of air at the same temperature and pressure. It is desired to increase the temperature of the air in both systems by 10°C. One system undergoes a constant pressure heat addition process and the other undergoes a constant volume heat addition. The difference in the values of heat transferred to the two systems in kJ is
(a) 30.5 (b) 44.2
(c) 57.5 (d) 73.2

4. A refrigerator is used to maintain certain space at 10°C. It pumps 18000 kJ/hour of heat from the space to the atmosphere at 30°C. If the power input to the refrigerator is 2 kW, the ratio of COP of this refrigerator to that of a Carnot refrigerator (up to 2 decimal places) is ________.

5. A thermal cycle receives 2000 kJ of heat from a heat source at 1000 K. It rejects 300 kJ of heat to a heat sink at 300 K and also rejects 250 kJ of heat to another heat sink at 200 K during the cycle. The cycle.
(a) reversible (b) irreversible
(c) impossible (d) work absorbing

6. Saturated liquid water is slowly heated at a constant pressure of 200 kPa to a final state where its quality reaches 0.65. For water at 200 kPa: T_{sat} = 120.23°C, h_f = 504.68 kJ/kg, h_g = 2706.60 kJ/kg. The change in the specific entropy in kJ/kg. K is
(a) 3.04
(b) 3.24
(c) 3.44
(d) 3.64

7. Given the thermodynamic functional relations: p = p(v, T) and T = (T(p, v), the term $\left.\frac{\partial p}{\partial v}\right|_T \times \left.\frac{\partial v}{\partial T}\right|_P$ is equal to
(a) $-\left(\left.\frac{\partial T}{\partial p}\right|_v\right)^{-1}$ (b) $\left(\left.\frac{\partial T}{\partial p}\right|_v\right)^{-1}$
(c) $\left(\left.\frac{\partial T}{\partial p}\right|_v\right)$ (d) 1

8. Two closed cycle gas turbine engines, A and B, operate on air standard Brayton cycle with efficiencies of η_A and η_B, respectively. If they operate between the same maximum and minimum temperatures, but with different pressure ratios of r_{pA} and r_{pB}, ($r_{pA} > r_{pB}$), then,
(a) $\eta_A = \eta_B$
(b) $\eta_A > \eta_B$
(c) $\eta_A < \eta_B$
(d) cannot be determined as the efficiencies are maximum only at the optimal r_p values.

9. The values of density and isentropic compressibility of water at certain pressure and temperature are given as 1000 kg/m³ and 40 × 10^{-10} Pa^{-1}, respectively. The speed at which sound travels in water under these conditions in m/s is equal to ________.

10. Length of a certain metal rod at 0°C is 10 cm. The coefficient of linear expansion of that metal varies with temperature as $10^{-4} + 10^{-5} \times T$ (cm/cm)°C. When the length of the metal rod is 10.2 cm, the rise in temperature in °C is ________.

11. In a polytropic compression process, one kg of an ideal gas having a molecular weight of 40 kg/k mol is compressed from 100 kPa, 300 K to 400 kPa, 360 K. The magnitude of the work in kJ for the process is
(a) 52.3 (b) 62.3
(c) 72.3 (d) 82.3

12. Two streams of air (C_p = 1005 J/kg. K) flow through insulated pipes 1 and 2 with the conditions as shown in figure. They mix in an insulated pipe-3 and the mixture steadily exits with a velocity of 100 m/s at 150 kPa. Neglecting the change in potential energy in all the pipes, the exit area of the pipe-3 in m² (up to 3 decimal places) is ________.

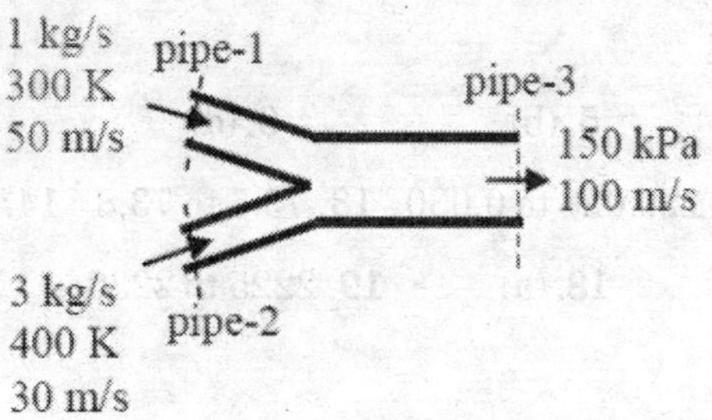

13. A 1 m^3 rigid vessel contains air at 200 kPa. A vacuum pump is connected to the vessel in order to control the pressure inside. The volume flow rate of air through the pump is maintained at a constant value of 0.1 m^3/s. If the pump operates for 10 seconds and the temperature of the air is maintained constant during operation, the pressure in the tank in kPa after 10 seconds (up to 2 decimal places) is ________.

14. A heat engine receives Q_1 kJ of heat from a hot reservoir and rejects Q_2 kJ of heat to a cold reservoir. The work delivered by the heat engine is entirely supplied to a heat pump, which receives Q_3 kJ of heat from another reservoir and rejects Q_4 kJ of heat to the same cold reservoir. If the efficiency of the heat engine is 0.4 and COP of the heat pump is 4.0, the value of $(Q_2 + Q_4)/Q_1$ (up to 1 decimal place) is ______.

15. A block of ice of mass 2 kg at 0°C is dropped into an insulated vessel containing 10 kg of liquid water at 25°C. The latent heat of melting of ice is 330 kJ/kg and specific heat of water is 4.2 kJ/kg.K. The change in the entropy of the universe in kJ/K (up to 3 decimal places) is ______.

16. A pure substance (C_v = 0.733 kJ/kg. K) undergoes a reversible process in which its temperature increases linearly from 40°C to 85°C and its specific entropy increases by 600 J/kg.K. The work done by the system in kJ/kgh is

(*a*) 160.2 (*b*) 164. 3
(*c*) 168.3 (*d*) 172.3

17. An ideal gas having a mass of 0.5 kg is initially at 300 kPa, 80°C and occupies a volume of 0.14 m^3. The gas undergoes an adiabatic process, where 50 kJ of work is transferred to the gas. The pressure and volume at the final state are 300 kPa and 0.20 m^3. The change in the entropy of the gas in J/K is

(*a*) 160.3 (*b*) 175.3
(*c*) 190.3 (*d*) 195.3

18. The van der Waals equation of state is given as, $\left(p+\frac{a}{v^2}\right)(v-b)=R_uT$, where p in bar, v in m^3/kmol and T is in K.

For air, the constants, a and b, are 1.368 bar. (m^3/kmol)^2 and 0.0367 m^3/kmol, respectively, Air is contained in a system at 160 K and 0.08 m^3/kmol. If p_1 is the pressure calculated using ideal gas equation of state and p_2 is pressure calculated using van der Walls equation of state, then p_1/p_2 is equal to

(*a*) 1.78 (*b*) 1.52
(*c*) 1.28 (*d*) 1.0

19. The values of specific volume of H_2O at 100°C for saturated liquid and saturated vapor states are 0.001044 m^3/kg and 1.673 m^3/kg, respectively. The slope of saturation pressure versus temperature curve, i.e., $(dp./dT)_{sat}$ is 3570 Pa/K. The change in enthalpy in kJ/kg between the two saturation states is ______.

20. In a steam power plant, steam is first expanded isentropically in a turbine from an initial condition of 100 bar and 500°C to a pressure of 40 bar. Then the steam is reheated up to 500°C at constant pressure. The steam is then expanded isentropically in another turbine up to a condenser pressure 0.01 bar. For steam, at 100 bar, 500°C: h = 3373.7 kJ/kg, s = 6.5966 kJ/kgK; at 40 bar, 500°C: h = 3445.3 kJ/kg, s = 7.0901 kJ/kg. K and at 0.01 bar: h_f = 29.3 kJ/kg, h_g = 2514.2 kJ/kg, s_f = 0.1059 kJ/kg. K, s_g = 8.9756 kJ/kg.K. The dryness fraction at the condenser inlet (up to 2 decimal places) is ______.

21. Air contains by volume 79% N_2 (molecular weight = 28 kg/kmol) and 21% O_2 (molecular weight = 32 kg/kmol). A stream of air flows at 32°C, 1 bar, at a rate of 2 m^3/s and is mixed with another stream of O_2 flowing at 0.4 kg/s. The molecular weight of the mixture (up to 2 decimal places) is ________.

22. Moist air enters a duct at a rate of 3 kg/s at 10°C, 80% relative humidity. The air heated as it flows through the duct and exits at 30°C. No moisture is added or removed and the pressure of air in the duct is constant at 1 bar. The saturation vapor pressure (p_g) of H_2O at 10°C 0.01228 bar. Specific enthalpy values of dry air at inlet and outlet of the duct are respectively 283.1 kJ/kg and 303.2 kJ/kg. The corresponding specific enthalpy values for water vapor are 2519.8kJ/kg and 2556.3 kJ/kg. For steady state operation the amount of heat added to the moist air in kW (up to 2 decimal places) is ______.

ANSWERS

1. (b) **2.** (c) **3.** (c) **4.** 0.17 to 0.18 **5.** (b) **6.** (d) **7.** (a)

8. (b) **9.** 499 to 501 **10.** 53 to 55 **11.** (d) **12.** .026 to 0.030 **13.** 73.5 to 73.8 **14.** 2.2 to 2.2

15. .140 to .150 **16.** (c) **17.** (a) **18.** (a) **19.** 2225 to 2230

20. .76 to 0.82 **21.** 29.0 to 30.0 **22.** 60.5 to 61.5

EXPLANATIONS

1. (b) Both ϕ and Ψ are not a property. It is a scalar function of T, P and v.

2. (c) m = 10 kg,

   Cv = 0.718 kJ/kg k

   T = 100 N.m, N = 60 × 2 = 120 rpm.

   $P = 2\pi NT$ J

   P = 75360 J

   $P = mCv\,\Delta T$

   $75360 = 10 \times 718.\ \Delta t$

   Increase in temperature = 10.495°C

3. (c) m = 20 kg.

   Increase of temperature = 10°C

   Difference of heat transferred

   $= m\,\Delta T\,(C_p - C_v)$

   $= 20 \times 10(R)$

   $= 20 \times 10 \times 0.287$

   = 57.4 kJ

4. T_1 = 10°C = 283 K

   T_2 = 30°C = 303 K

   $$(COP)_{Carnot} = \frac{T_2}{T_1 - T_2} = \frac{283}{303 - 283} = 14.15$$

   $$(COP)_R = \frac{\text{Desired effect}}{W_{input}}$$

   $$= \frac{5}{2} = 2.5 \quad \frac{(COP)_R}{(COP)_{carnot}} = \frac{2.5}{14.15} = 0.176$$

5. (b)

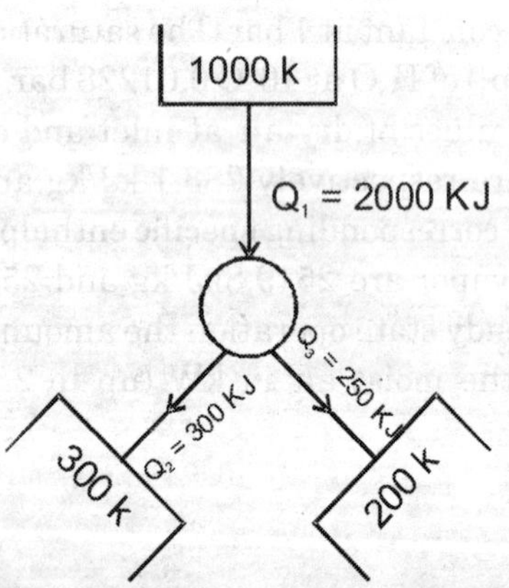

   $$\text{Now, } \oint \frac{dQ}{T} = \frac{Q_1}{T_1} - \frac{Q_2}{T_2} - \frac{Q_3}{T_3}$$

   $$\oint \frac{dQ}{T} = \frac{2000}{1000} - \frac{300}{300} - \frac{250}{200}$$

   $$\oint \frac{dQ}{T} = -0.25$$

   If $\oint \frac{dQ}{T} < 0$, it means that given cycle is irreversible.

6. (d) x = 0.65

   T_{sat} = 120.23°C = 393.23 K

   h_f = 504.68 kJ/kg

   h_g = 2706.60 kJ/kg

   $h_{fg} = h_g - h_f$

   h_{fg} = 2201.92

   $$= \frac{xh_{gf}}{T_{sat}} = \frac{0.65 \times 2201.92}{393.23}$$

   $$\boxed{\text{Specific entropy (s)} = 3.64 \text{ KJ / kg} \cdot \text{k}}$$

7. (a) P = P(v, T)

   T = T(P, v)

   According to thermodynamic relation,

   $$\left.\frac{\partial P}{\partial v}\right|_T \times \left.\frac{\partial v}{\partial T}\right|_P = -\left(\frac{\partial T}{\partial P}\right)_v^{-1}$$

8. (b) $$\eta = 1 - \frac{1}{(r_P)^{\frac{\gamma-1}{\gamma}}}$$

   r_P – pressure ratio,

   if $(r_{PA} > r_{PB})$, then

   $r_{PA} \uparrow \rightarrow \eta \uparrow$

   So that $\eta_A > \eta_B$

9. Density (ρ) = 1000 Kg/m³

   $\beta = 40 \times 10^{-10}$ Pa^{-1}

   $$\text{Bulk modulus(K)} = \frac{1}{\beta} = \frac{1}{40 \times 10^{-10}}$$

   $= 2.5 \times 10^8$ Pa.

Velocity of sound(C) = $\sqrt{\frac{k}{\rho}} = \sqrt{\frac{2.5\times10^{8}}{1000}}$

$\boxed{C = 500\ m/sec}$

10. $L_0 = 10$ cm
$L = 10.2$ cm
$\alpha = 10^{-4} + 10^{-5}T$ (cm/cm) /°C
$L = L_0 + \alpha\,\Delta T$

$$10.2 = 10 + 10^{-4}(1 + \frac{T}{10})\cdot(T-0)$$

$$0.2 = 10^{-4}\left(T + \frac{T^2}{10}\right)$$

$\boxed{T = 54.99\,°C}$

11. (d) M = 40 kg/k mol
$P_1 = 100$ KPa, $T_1 = 300$ k
$P_2 = 400$ KPa, $T_2 = 360$ k
n = 1.2

$$W_C = \frac{mR(T_1 - T_2)}{n-1} = \frac{1\times\frac{8.34}{40}(300-360)}{1.2-1}$$

$W_C = -82.37$ KJ

Work done in compression process.

$$\frac{T_2}{T_1} = \left(\frac{P_2}{P_1}\right)^{\frac{\gamma-1}{\gamma}}$$

$$\frac{300}{360} = \left(\frac{400}{100}\right)^{\frac{\gamma-1}{\gamma}}$$

By solving above equation, we get $\gamma = 1.1514$

12.

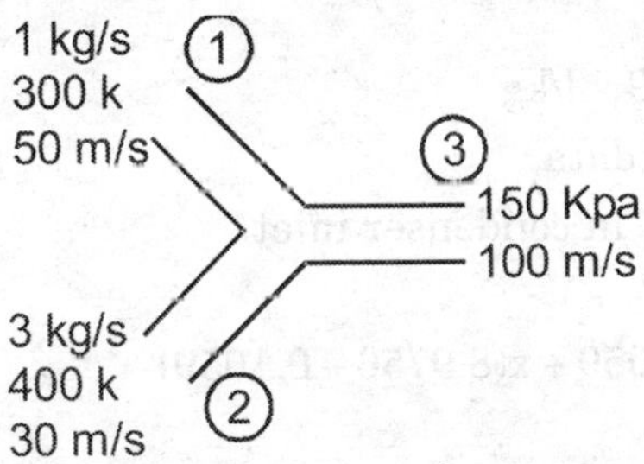

$\dot{m}_1 + \dot{m}_2 = \dot{m}_3$

$\dot{m}_3 = 1 + 3 = 4$ Kg / sec

We know that

$$\dot{m}_3 = \rho_3 \times A_3 \times V_3$$

$$A_3 = \frac{\dot{m}_3}{\rho_3 \times V_3}$$

$$A_3 = \frac{4}{1.21\times100}$$

$\boxed{A_3 = 0.03\ m^2}$

13. The correct answer is 73.5 kPa

14. Given that
$Q_1 - Q_2 = Q_3 - Q_4$

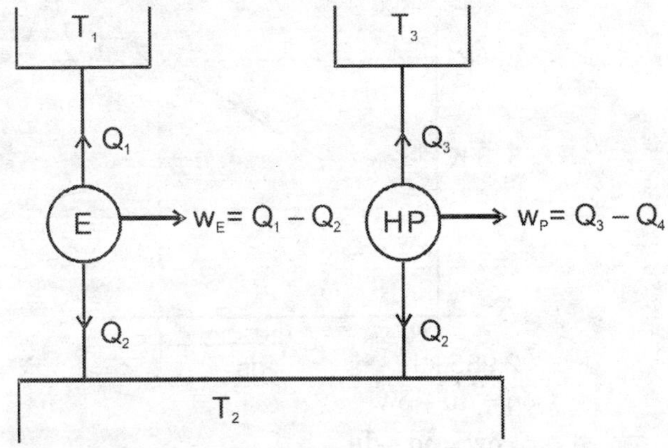

$$\eta_E = \frac{Q_1 - Q_2}{Q_1}$$

$$0.4 = \frac{Q_1 - Q_2}{Q_1}$$

$0.4Q_1 = Q_1 - Q_2$, $\boxed{Q_2 = 0.6Q_1}$

$$(COP)_P = \frac{Q_4}{Q_3 - Q_4}$$

$$4 = \frac{Q_4}{Q_3 - Q_4}$$

$$4Q_3 - 4Q_4 = Q_4$$

$$4(Q_1 - Q_2) = Q_4 \quad \frac{Q_2 + Q_4}{Q_1} = \frac{Q_2 + 4Q_1 - 4Q_2}{Q_1}$$

$$= \frac{0.6Q_1 + 4(Q_1 - 0.6Q_1)}{Q_1}$$

$$\frac{Q_2 + Q_4}{Q_1} = 0.6 + (4\times0.4)$$

$$\boxed{\frac{Q_2 + Q_4}{Q_1} = 2.2}$$

15. $S_1 = \frac{h_{fg}}{T} = \frac{330}{373}$
$S_1 = 0.88$ kJ/K

$$S_2 = mC\ell n\left(\frac{T_f}{T_1}\right) = 2\times4.2\ell n\left(\frac{298}{273}\right)$$

= 0.73 kJ/k

Change of entropy = 0.15 kJ/K

16. (c) $\delta q = Tds = 313\times600 + \frac{1}{2}\times45\times600$

$$= 201300\frac{J}{kg} = 201.3\frac{kJ}{kg}$$

$du = mCv\ dT$

$du = C_v\ dT = .733 \times 45 = 32.985\ \frac{kJ}{kg}$

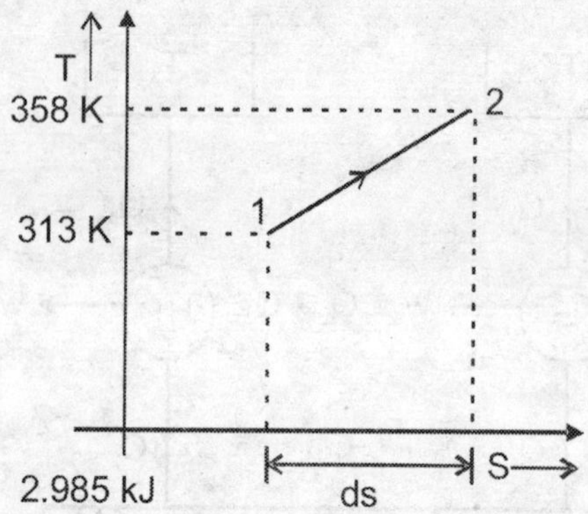

$\delta q = du + \delta w$

$\Rightarrow \delta w = \delta q - du$

$= 201.3 - 32.985$

$= 168.3\ \frac{kJ}{kg}$

17. (a) $Pv = mRT$

$.14 \times 300 \times 10^3 = 0.5 \times R \times 353$

$R = 237.97 \frac{J}{kg\ K}$

By $\frac{v_1}{T_1} = \frac{v_2}{T_2}$

$\frac{.14}{353} = \frac{.2}{T_2}$

$T_2 = 504.286$

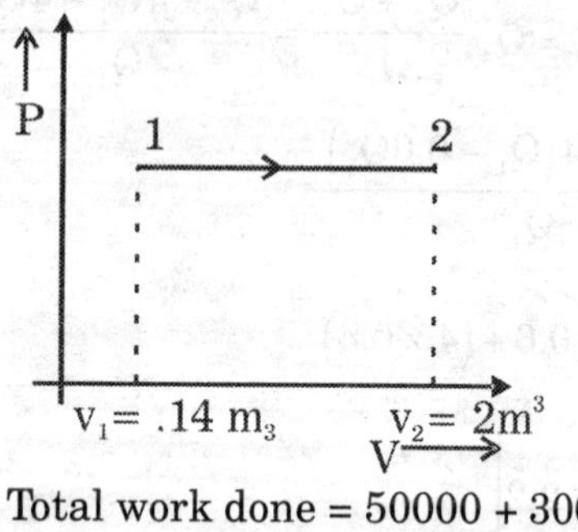

Total work done = 50000 + 300 × 103(.06)

= 68000 J

$dH = mC_p\ dT$

$68000 = 0.5 \times Cp \times 151.286$

But $Cp = \frac{R\gamma}{\gamma - 1}$

$68000 = 0.5 \times \frac{237.97 \times \gamma}{\gamma - 1} \times 151.286$

$\gamma = 1.3538$

$ds = mR\ In\left(\frac{T_2}{T_1}\right) + mC_v \ln\left(\frac{v_2}{v_1}\right)$

$= 0.5 \times 237.94 \times In\left(\frac{504.286}{353}\right) + 0.5 \times$

$672.611 \times In\left(\frac{.2}{.14}\right)$

$d\rho = 160.3\ \frac{J}{K}$

18. (a) $\left(P + \frac{a}{v^2}\right)(v - b) = R_u T$

$T = 160k$

$V = 0.08\ m^3/K\ mol$

$P_2V = nR_uT$

$P_2 = \frac{8.314 \times 160}{0.08} = 166.28\ bar$

$P_1 = \frac{R_uT}{v - b} - \frac{a}{v^2}$

$P_1 = \frac{8.314 \times 160}{0.08 - 0.0367} - \frac{1.368}{0.082}$

$P_1 = 305.077\ bar$

$\boxed{\frac{P_1}{P_2} = 1.81}$

19. Given that

$T_{sat} = 100°C = 373\ K$

$v_f = 0.001044\ m^3/kg$

$\left(\frac{dP}{dT}\right)_{sat} = 3570\ Pa/K$

$\left(\frac{dP}{dT}\right)_{sat} = \frac{h_{fg}}{T_{sat}\left(v_g - v_f\right)}$

$h_{fg} = \left(\frac{dP}{dT}\right)_{sat} .T_{sat}\left(v_g - v_f\right)$

$= 2226.39\ kJ/kg$

20. From Given data

$s_g = s_f + x\ s_{fg}$ in condenser inlet

$s_g = s_f + x(s_g - s_f)$

$7.0901 = 0.1059 + x(8.9756 - 0.1059)$

$\boxed{x = 0.787}$

21. Molecular weight of Air = 28 × .79 + 32 × .21

$= 28.84\ \frac{kg}{k\ mol}$

For air : $P\dot{V} = \dot{m}RT$

$\dot{m} = \frac{P\dot{V}}{RT} = \frac{1 \times 10^5 \times 2}{287 \times 305} = 2.284\ \frac{kg}{sec}$

Molecular weight of the mixture

$= \frac{(2.284 \times 28.84) + (0.4 \times 32)}{2.684} = 29.3195\ \frac{kg}{k\ mol}$

22. Correct answer is 61 KW

2018

Q. 1 – Q. 9 carry one mark each.

Notation used : *P : pressure, V : volume, T : temperature, s : specific entropy, h : specific enthalpy, C_p : molar heat capacity at constant pressure, C_v : molar heat capacity at constant volume, R : universal gas constant*

1. When a fixed mass of air-water vapour mixture is heated at constant pressure,

(*a*) both relative and specific humidity decrease.

(*b*) relative humidity decreases, but specific humidity remains unchanged.

(*c*) specific humidity decreases, but relative humidity remains unchanged.

(*d*) both relative and specific humidity increase.

2. For a reversible isothermal expansion of one mole of an ideal gas from state 1 to state 2, the magnitude of work done is

(*a*) $RT\ln\left(\frac{P_1}{P_2}\right)$ (*b*) $P_2V_2 - P_1V_1$

(*c*) $R\ln\left(\frac{V_1}{V_2}\right)$ (*d*) 0

3. The statement which is NOT a consequence of the first law of thermodynamics is

(*a*) Heat is a path function

(*b*) Energy is a property of a system

(*c*) Energy of an isolated system is not conserved

(*d*) A perpetual motion machine of the first kind is not possible

4. For a refrigerator absorbing heat Q_L from a cold region and rejecting heat Q_H to a hot region, the coefficient of performance is written as

(*a*) $\frac{Q_L}{Q_H - Q_L}$ (*b*) $\frac{Q_H}{Q_H - Q_L}$

(*c*) $\frac{Q_H - Q_L}{Q_L}$ (*d*) $\frac{Q_L}{Q_H}$

5. The value of the compressibility factor at the critical point evaluated using the van der Waals equation of state is

(*a*) $\frac{2}{7}$ (*b*) $\frac{5}{8}$

(*c*) $\frac{3}{8}$ (*d*) $\frac{1}{7}$

6. The vapour pressure of a liquid at 8°C is 2.7 kPa. Its enthalpy of vaporization is constant and equal to 42700 kJ/kmol. Take R = 8.314 kJ/kmol.K. The temperature (in °C) at a vapour pressure of 13.5 kPa is

(*a*) 58.7 (*b*) 51.4

(*c*) 44.3 (*d*) 35.2

7. One kmol of an ideal gas (C_p = 21 kJ/kmol.K) undergoes a constant pressure process from 300 K to 500 K. The molar entropy of the gas at 300 K is 150 kJ/kmol.K. The molar entropy (in kJ/kmol.K) at 500 K (up to 1 decimal place) is _____.

8. A spring, having a spring constant of 350 kN/m, is initially compressed by 0.4 cm. The work required (in J) to compress it by another 0.6 cm (up to 1 decimal place) is ________.

9. An ideal gas has a molar mass of 40 kg/kmol. Take R = 8.314 kJ/kmol.K. At a pressure of 2 bar and a temperature of 300 K, the volume (in m³) of 1 kg of this gas (up to 2 decimal places) is _____.

Q. 10 – Q. 22 carry two marks each.

10. Consider the following statements for an ideal gas undergoing a reversible non-flow process:

P. If the process is adiabatic, the change in enthalpy of the gas is necessarily zero.

Q. If the process is adiabatic, the change in entropy of the gas is necessarily zero.

R. If the process is isothermal, the change in enthalpy of the gas is necessarily zero.

S. If the process is isothermal, the change in entropy of the gas is necessarily zero.

Which one of the following options is valid?

(*a*) Only P is correct

(*b*) Only S is correct

(*c*) Only Q and R are correct

(*d*) Only P and S are correct

11. An ideal Otto cycle (O) and an ideal diesel cycle (D) have the same maximum temperature and reject equal amount of heat. Also, the working fluid enters at the same state before compression. One of the following statements always true about their efficiencies (η) is

(*a*) $\eta_O > \eta_D$ (*b*) $\eta_O = \eta_D$

(*c*) $\eta_O < \eta_D$ (*d*) $\eta_O = 1 - \eta_D$

12. A reversible engine receives 75 kJ/s of energy from a reservoir at 750 K and does 12 kJ/s of work. The heat is rejected to two reservoirs at 650 K and 550 K. The rate of heat rejection (in kJ/s) to the reservoir at 650 K is

(*a*) 11 (*b*) 31

(*c*) 41 (*d*) 52

13. A gas obeys the following equation of state:

$$P(\bar{v}-b) = RT + \frac{aP^2}{T},$$

where $\bar{v}$ is molar volume, and a, b are constants with values $a = 10^{-5}$ J.K/Pa².kmol and $b = 8 \times 10^{-2}$ m³/ kmol. Take Cp = 30 kJ/kmol.K. At 10 bar and 500 K, the value of the Joule-Thomson coefficient (in K/Pa) is

(*a*) -2×10^{-6} (*b*) -4×10^{-6}

(*c*) 2×10^{-6} (*d*) 4×10^{-6}

14. In an ideal Rankine cycle, steam enters the turbine at 10 MPa and 500°C (h = 3375.1 kJ/kg, s = 6.5995 kJ/kg.K). It is cooled in the condenser at a pressure of 10 kPa. At 10 kPa, h_f = 191.81 kJ/kg, s_f = 0.6492 kJ/kg.K, h_g = 2583.9 kJ/kg and s_g = 8.1488 kJ/kg.K. The heat rejected in the condenser (in kJ/kg) is

(*a*) 1898 (*b*) 3796

(*c*) 949 (*d*) 2847

15. Methane has compressibility factor value of 0.9 at reduced pressure of 1.0 and reduced temperature of 1.5. For propane, the critical temperature and pressure are 369.8 K and 42.48 bar, respectively. Take R = 8.314 kJ/kmol.K. Applying the principle of corresponding states, the molar volume of propane (in m³/kmol) at the same reduced pressure and temperature is

(*a*) 0.355 (*b*) 0.526

(*c*) 0.791 (*d*) 0.977

16. A rigid insulated vessel is divided into two compartments by a partition. One compartment contains 12 kg of oxygen at 200 kPa and 280 K. The other compartment contains 26 kg of carbon dioxide at 400 kPa and 360 K. The specific heats at constant volume in kJ/kg.K for oxygen and carbon dioxide are 0.662 and 0.653, respectively. The partition is removed and the gases are allowed to mix. Considering both gases are ideal, the final temperature (in K) of the mixture (up to 1 decimal place) is _______.

17. Moist air having 60% relative humidity, enters a steady-flow air-conditioning unit at 102 kPa and 30°C. The volume flow rate of the moist air entering the unit is 0.1 m³/s. The moist air leaves the unit at 95 kPa and 15°C with a relative humidity of 100%. Liquid condensate leaves the unit at 15°C.

For water: at 15°C, h_f = 62.982 kJ/kg, h_g = 2528.3 kJ/kg, P_{sat} = 1.7057 kPa.

at 30°C, h_f = 125.74 kJ/kg, h_g = 2555.6 kJ/kg, P_{sat} = 4.2469 kPa.

For air, specific heat at constant pressure is 1.004 kJ/kg.K and the specific gas constant is 0.287 kJ/kg.K.

Neglecting heat leakage to the surrounding, the magnitude of heat extracted (in kW) from the air stream (up to 2 decimal places) is _________.

18. Air at 150 kPa and 323 K is filled in a rigid vessel of 0.05 m³ capacity. For air, assumed as an ideal gas, specific heat at constant volume is 0.7163 kJ/kg.K and the specific gas constant is 0.287 kJ/kg.K. Neglect kinetic and potential energy changes. If 30 kJ of heat is added, the final temperature (in K) of air (up to 1 decimal place) is _________.

19. Superheated steam at 2 bar and 300°C, with an enthalpy of 3072.1 kJ/kg, enters a horizontal adiabatic nozzle with negligible velocity and leaves at 0.2 bar as saturated vapour with an enthalpy of 2609.9 kJ/kg. Assuming steady flow and neglecting the potential energy changes, the exit velocity (in m/s) of the steam (up to 1 decimal place) is _________.

20. A given mass of a simple compressible substance undergoes a reversible cycle, as shown in the P-V diagram. The magnitude of the net work done during the cycle is 3 kJ. The pressure (in bar) at point C (up to 1 decimal place) is _________.

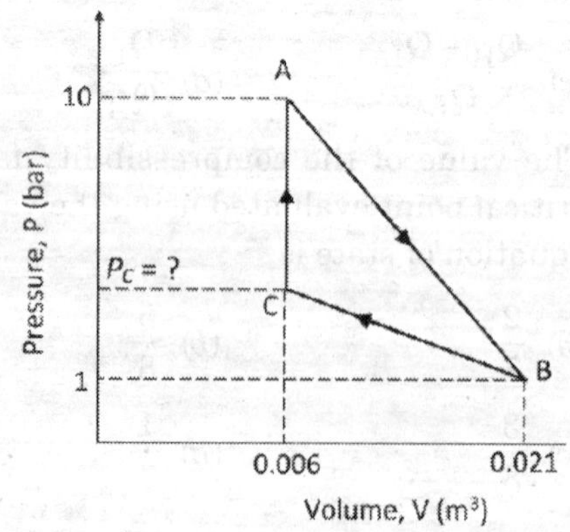

21. One kmol of an ideal gas at 300 K and 10 bar is reversibly heated in a constant volume process to 500 K. It is then reversibly and isothermally expanded to 2 bar. Take C_v = 20.8 kJ/kmol.K and R = 8.314 kJ/kmol.K. The total heat supplied (in kJ) to the gas (up to 1 decimal place) is ________.

22. A rigid container is completely filled with a liquid having a constant isothermal compressibility of 1.09×10^{-4} bar^{-1} and a constant coefficient of volume expansion of 1.12×10^{-3} K^{-1}. The liquid is initially at 300 K and 1 bar. Heat is supplied to the liquid to raise its temperature to 350 K. Assuming that no phase change occurs, the final pressure (in bar) of the liquid (up to 1 decimal place) is ________.

ANSWERS

1. (b)	**2.** (a)	**3.** (c)	**4.** (a)	**5.** (c)	**6.** (d)
7. 160.0 to 161.0	**8.** 14.5 to 15.0	**9.** 0.30 to 0.32	**10.** (c)	**11.** (c)	**12.** (d)
13. (b)	**14.** (a)	**15.** (d)	**16.** 332.5 to 336.5	**17.** 3.01 to 3.07	**18.** 838.0 to 844.0
19. 961.0 to 962.0	**20.** 5.7 to 6.1	**21.** 12950.0 to 13000.0		**22.** 513.0 to 517.0	

EXPLANATIONS

1. (b) Specific humidity (w) $= 0.622\dfrac{Pv}{P - Pv}$

P–Total pressure, Pv – Partial Vapour Pressure

w = f(Pv), so that specific humidity is Constant, and Relative humidity decreases, because it depends on the saturation pressure.

2. (a) We know that,

$$W = \int_1^2 P.dv$$

Pv = mRT for ideal gas

$$P = \frac{mRT}{V}$$

$$W = \int_{V_1}^{V_2} \frac{mRT}{V}.dV = mRT = \int_{V_1}^{V_2} \frac{1}{V}.dV$$

$$W = mRT \ \ln\left[\frac{V_2}{V_1}\right], \quad \boxed{W = RT\ln\left(\frac{V_2}{V_1}\right)}$$

3. (c) Following are the consequences of first law of thermodynamics

(i) Heat is a path function.

(ii) Work is a path function.

(iii) PMM–1 is not possible.

(iv) Energy of an isolated system is constant.

(v) Energy is a property of the system.

4. (a) $Q_H = W + Q_L$

$W = Q_H - Q_L$

$$(COP)_R = \frac{\text{Cooling effect}}{\text{Net work input}}$$

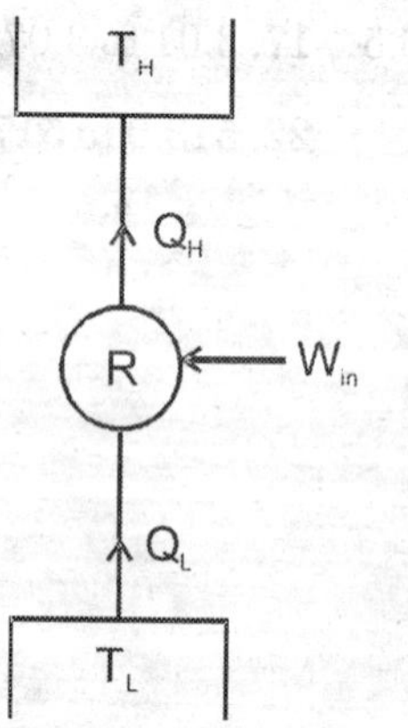

$$\boxed{(COP)_R = \frac{Q_L}{Q_H - Q_L}}$$

5. (c) Compressibility factor (z) $= \dfrac{PV}{RT}$

For critical point

$$Z_c = \frac{P_c V_c}{R \cdot T_c}$$

We know that, from Vandar Waal's equation.

$$T_c = \frac{8a}{27bR},\ P_c = \frac{a}{27b^2} \text{ and } V_c = 3b.$$

$$Z_c = \frac{\dfrac{a}{27b^2}}{R\cdot\dfrac{8a}{27bR}}$$

$$\boxed{Z_c = \frac{3}{8}}$$

6. (d) $T_1 = 8°C = 281$ K, $P_1 = 2.7$ kPa.

$\Delta H = 42700$ kJ/K mol, R = 8.314 kJ/K mol.K

$P_2 = 13.5$ kPa, T_2 = ?

We know that

$$\text{In}\left(\frac{P_1}{P_2}\right) = -\frac{\Delta H}{R}\left[\frac{1}{T_2} - \frac{1}{T_1}\right]$$

$$\text{In}\left(\frac{2.7}{13.5}\right) = -\frac{42700}{8.314}\left[\frac{1}{T_1} - \frac{1}{281}\right]$$

$T_2 = 308.2$K, $\boxed{T_2 = 35.2°C}$

7. (*) $C_p = 21$ kJ/k mol.K

$T_1 = 300$K, $\quad T_2 = 500$ K

$S_1 = 150$ kJ/k mol.K, $\quad S_2$ = ?

At constant pressure

$$ds = C_p \ln\left[\frac{T_2}{T_1}\right]$$

$$s_2 - s_1 = C_p \ln\left[\frac{T_2}{T_1}\right]$$

$$s_2 - 150 = 21 \ln\left[\frac{500}{300}\right]$$

$$\boxed{S_2 = 160.73 \text{ kJ/k mol.K}}$$

8. Spring constant (K) = 350 KN/m

$x_1 = 0.4$ cm $= 0.4 \times 10^{-2}$ m

$$W_1 = \frac{1}{2}Kx_1^2 = \frac{1}{2} \times 350 \times 1000 + \left(0.4 \times 10^{-2}\right)^2$$

$W_1 = 2.8$ J

$x_2 = 0.4 + 0.6 = 1$ cm $= 10^{-2}$ m

$$W_2 = \frac{1}{2}Kx_2^2 = \frac{1}{2}\times 350\times 1000\times 10^{-4}$$

$W_2 = 14.5$ J

The work required to compress it by another 0.6 cm is

$W = W_2 - W_1$

$W = 17.5 - 2.8$

$\boxed{W = 14.75}$ J

9. Molar mass (M) = 40 kg/k mol.

R = 8.314 kJ/k mol.K

P = 2 bar = 2×10^5 N/m²

$T_1 = 300$ K

PV = mRT

$$V = \frac{mRT}{P} = \frac{1\times\left(\frac{8.314}{40}\right)\times 300}{2\times 10^5}$$

$\boxed{V = 0.31\,m^3}$

10. (c) According to question,

For an ideal gas undergoing a reversible non-flow process, then,

$Q > dS = \frac{dQ}{T}$, for adiabatic process,

$\boxed{ds = 0}$, change of entropy is equal to zero.

R > We know that for ideal gas.

$dh = C_p dT$

For isothermal process.

T = C

dT = 0

$\boxed{dh = 0}$

Change of enthalpy is equal to zero.

Hence, 'Q' and 'R' is correct, according to given condition.

11. (c)

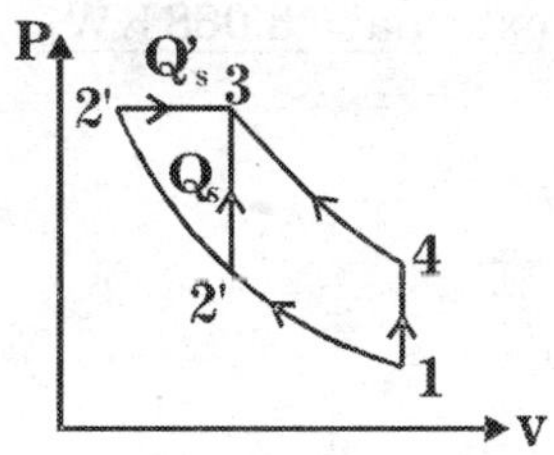

2'–3 – Heat supplied at Constant pressure (Q's)

2 – 3 – Heat supplied at Constant volume (Qs)

$$\eta_O = 1 - \frac{Q_R}{Q_S}, \quad \eta_D = 1 - \frac{Q_R}{Q_{S'}}$$

Here, Q's > Q_s, therefore the $\boxed{\eta_D > \eta_O}$

12. (d) $w = Q_1 - (Q_2 + Q_3)$

$12 = 75 - (Q_2 + Q_3)$

$Q_2 + Q_3 = 63$

$Q_3 = 63 - Q_2$

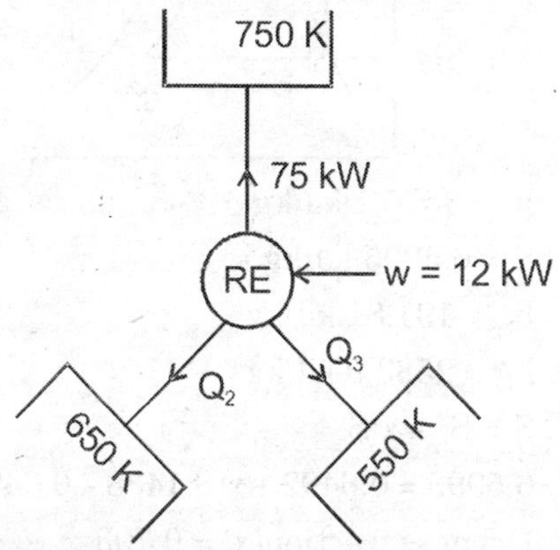

From Clausius inequality

$$\oint \frac{\delta Q}{T} = 0$$

$$\frac{Q_1}{T_1} - \frac{Q_1}{T_2} - \frac{Q_3}{T_3} = 0$$

$$\frac{75}{750} - \frac{Q_2}{650} - \frac{(63 - Q_2)}{550} = 0$$

$\boxed{Q_2 = 52\,kJ/s}$

13. (b) Given that

$$P(\overline{V} - b) = RT + \frac{aP^2}{T}$$

$$\overline{V} - b = \frac{RT}{P} + \frac{aP}{T}$$

$$\overline{V} = \frac{RT}{P} + \frac{aP}{T} + b$$

$$\overline{V} = \frac{8.314\times 500}{10\times 100} + \frac{10^{-5}\times 10\times 105}{500} + 8\times 10^{-2}$$

$\overline{V} = 4.257\,m^3/mol$

$$\left(\frac{\partial \overline{V}}{\partial T}\right)_P = \frac{R}{P} + \frac{aP}{T^2} = \frac{8.314}{1000} + \frac{10^{-5}\times 10\times 10^5}{500^2}$$

$$\frac{\partial \overline{V}}{\partial T} = 8.354\times 10^{-3}$$

$$\mu_j = \left(\frac{\partial T}{\partial P}\right)_h = \frac{-1}{C_p}\left[\overline{V} - \left(\frac{\partial \overline{V}}{\partial T}\right)_P \cdot T\right]$$

$$\mu_j = -\frac{1}{30\times 1000}\left[4.257 - (8.354\times 10^{-3}\times 500)\right]$$

$\boxed{\mu_j = -2.86\times 10^{-6}}$

14. (a)

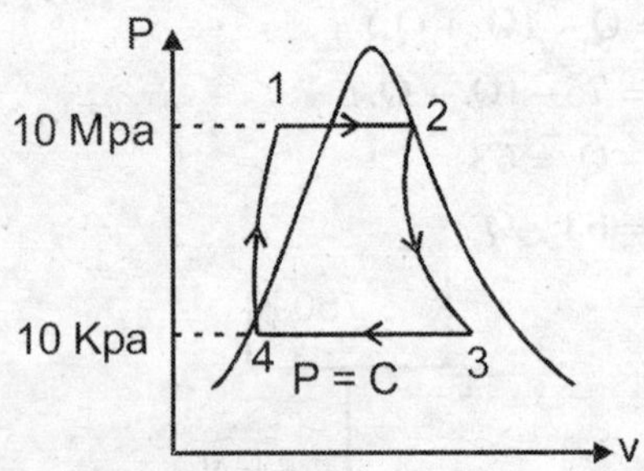

$h_2 = 3375.1$ kJ/kg

$S = 6.5995$ kJ/kg k

$h_{f4} = 191.81$ kJ/kg

hg = 2583.9 kJ/kg

$S = S_f + x\, S_{fg}$

$6.5995 = 0.6492 + x(8.1488 - 0.6492)$

Dryness fraction(x) = 0.793

$h_3 = h_{f4} + xh_{fg}$

$= 191.8 + 0.793(2583.9 - 191.81)$

$h_3 = 2088.727$

Heat rejected in condenser(Q_r) = $h_3 - h_{f4}$

$Q_r = 2088.727 - 191.8$

$\boxed{Q_r = 1897 \text{ kJ / kg}}$

15. (d) $Z = 0.9,\ P_r = 1,\ T_r = 1.5$

$$Pr = \frac{P}{Pc}$$

$P = Pc \cdot Pr = 42 \cdot 48 \times 1 = 42.48$ bar

$T = Tc \cdot Tr = 369.8 \times 1.5 = 554.7$ K

$$Z = \frac{vP}{RT}$$

$$0.9 = \frac{v \times 42.48 \times 1000}{8.314 \times 554.7}$$

$\boxed{v = 0.977}$ m^3

16.

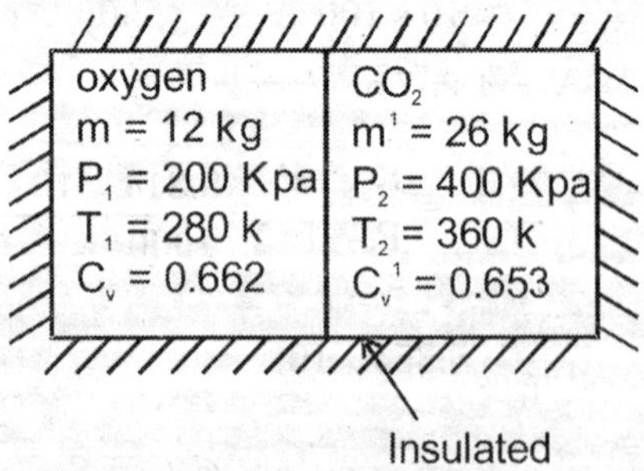

Now $mc_v\,(T_1 - T_f) = m'c_v\left(T_f - T_2\right)$

Now,

$12 \times 0.662\,(280 - T_f) = 0.653 \times 26(T_f - 360)$

Final temperature (T_f) = 334.497 K

17. $\phi_1 = 60\% = 0.6$

P = 102 kPa, at 30°C

$P_{sat} = 4.2469$ kPa

$$\phi_1 = \frac{Pv_1}{Pv_{s2}}$$

$Pv_1 = 0.6 \times 4.2469 = 2.548$ KPa

$$w_1 = \frac{0.622\,Pv_1}{P - Pv_1} = \frac{0.622 \times 2.548}{102 - 2.548}$$

$w_1 = 0.0159$ kJ/kg of dry air.

$h_1 = 1.005\, td_1 + w_1(2500 + 1.9\, td_1)$

$h_1 = (1.005 \times 30) + 0.0159(2500 + 1.9 \times 30)$

$h_1 = 40.806$ kJ/kg

$$\phi_1 = \frac{Pv_2}{Pv_{s2}}$$

$$1 = \frac{Pv_2}{1.7057}$$

$Pv_2 = 1.7057$ kPa

$$w_2 = \frac{0.622 \times 1.7057}{95 - 1.7057}$$

$= 0.0114$ kJ/kg of dry air

$h_2 = (1.005 \times 15) + 0.0114(2500 + 1.9 \times 15)$

$h_2 = 43.89$ kJ/kg

$$\dot{m}_a = \frac{v_1}{v_{s1}}$$

$$v_{s1} = \frac{R_a td_1}{Pa} = \frac{287 \times 303}{99.452 \times 1000}$$

$$= \frac{874.40}{1000} = 0.874 \text{ m}^3 / \text{kg}$$

$$\dot{m}_a = \frac{v_1}{v_{s1}} = \frac{0.1}{0.874} = 0.114 \text{ kg/see}$$

Rate of heat = $\dot{m}_a\left(h_1 - h_2\right)$

$= 0.114(70.806 - 43.89)$

$\boxed{\text{Rate of heat external} = 3.068 \text{ KW}}$

18. $P_1 = 150$ KPa

$T_1 = 323$ K

$v_1 = 0.05\ m^3$

$P_1V_1 = mRT_1$

$$m = \frac{P_1 V_1}{RT_1} = \frac{150 \times 0.05}{0.287 \times 323} = 0.081 \text{ kg}$$

$dQ = mCv\ dT$

$30 = 0.081 \times 0.7163 \times (T_2 - 323)$

$\boxed{T_2 = 840.06 \text{ K}}$

19. Given that

$h_1 = 3072.1$ KJ/kg

$h_2 = 2609.9$ KJ/kg

$v_2 = 44.72\sqrt{\Delta h}$

$v_2 = 44.72\sqrt{(3072.1 - 2609.9)}$

$v_2 = 961.43 \text{ m/s}$

20. $\frac{1}{2} \times .015 \times 90 - \frac{1}{2} \times .015 \times (P_c - 100)$

= 3[Area (ΔAOB – Area(ΔCOB)

= Network done during cycle]

$P_c = 60.0$ kPa

or

$P_c = 6$ bar 10

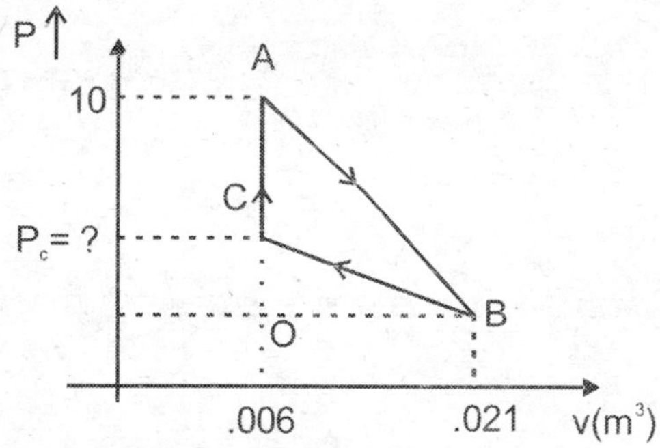

21. $T_1 = 300$ K, $P_1 = 10$ bar

$T_2 = 500$K, $P_2 = 2$ bar

$C_v = 20.8$ kJ/k mol.K

R = 8.314 kJ/k mol.K

At constant volume

$Q_1 = mc_v \Delta T = 4160$ kJ

At constant temperature

$$Q_2 = mRT \ln\left(\frac{V_2}{V_1}\right) = mRT \ln\left(\frac{P_1}{P_2}\right)$$

$$= 8.314 \times 500 \ln\left(\frac{16.6}{2}\right)$$

$Q_2 = 8814.76$ kJ

$\boxed{Q = Q_1 + Q_2 = 12975}$

22. Isothermal compressibility :

$$K_T = \frac{-1}{V}\left(\frac{\partial V}{\partial P}\right)_T \quad ...(i)$$

Volume expansivity :

$$\beta = \frac{1}{V}\left(\frac{\partial V}{\partial T}\right)_P \quad ...(ii)$$

By dividing (i) & (ii), we get

$$\frac{K_T}{\beta} = -\left(\frac{\partial T}{\partial P}\right)$$

$$\frac{1.09 \times 10^{-4}}{1.12 \times 10^{-3}} = \frac{T_2 - T_1}{P_1 - P_2}$$

$$\frac{109}{1120} = \frac{50}{1 - P_2}$$

$\boxed{P_2 = 514.76 \text{ bar}}$

XE - F

POLYMER SCIENCE AND ENGINEERING

Syllabus

Section 1: Chemistry of high polymers

Monomers, functionality, degree of polymerizations, classification of polymers, glass transition, melting transition, criteria for rubberiness, polymerization methods: addition and condensation; their kinetics, metallocene polymers and other newer techniques of polymerization, copolymerization, monomer reactivity ratios and its significance, kinetics, different copolymers, random, alternating, azeotropic copolymerization, block and graft copolymers, techniques for copolymerizationbulk, solution, suspension, emulsion.

Section 2: Polymer Characterization

Solubility and swelling, concept of average molecular weight, determination of number average, weight average, viscosity average and Z-average molecular weights, polymer crystallinity, analysis of polymers using IR, XRD, thermal (DSC, DMTA, TGA), microscopic (optical and electronic) techniques.

Section 3: Synthesis and properties

Commodity and general purpose thermoplastics: PE, PP, PS, PVC, Polyesters, Acrylic, PU polymers. Engineering Plastics: Nylon, PC, PBT, PSU, PPO, ABS, Fluoropolymers Thermosetting polymers: PF, MF, UF, Epoxy, Unsaturated polyester, Alkyds. Natural and synthetic rubbers: Recovery of NR hydrocarbon from latex, SBR, Nitrile, CR, CSM, EPDM, IIR, BR, Silicone, TPE.

Section 4: Polymer blends and composites

Difference between blends and composites, their significance, choice of polymers for blending, blend miscibility-miscible and immiscible blends, thermodynamics, phase morphology, polymer alloys, polymer eutectics, plastic-plastic, rubber-plastic and rubber-rubber blends, FRP, particulate, long and short fibre reinforced composites.

Section 5: Polymer Technology

Polymer compounding-need and significance, different compounding ingredients for rubber and plastics, cross-linking and vulcanization, vulcanization kinetics.

Section 6: Polymer rheology

Flow of Newtonian and non-Newtonian fluids, different flow equations, dependence of shear modulus on temperature, molecular/segmental deformations at different zones and transitions. Measurements of rheological parameters by capillary rotating, parallel plate, cone-plate rheometer. Viscoelasticity-creep and stress relaxations, mechanical models, control of rheological characteristics through compounding, rubber curing in parallel plate viscometer, ODR and MDR.

Section 7: Polymer processing

Compression molding, transfer molding, injection molding, blow molding, reaction injection molding, extrusion, pultrusion, calendaring, rotational molding, thermoforming, rubber processing in two-roll mill, internal mixer.

Section 8: Polymer testing

Mechanical-static and dynamic tensile, flexural, compressive, abrasion, endurance, fatigue, hardness, tear, resilience, impact, toughness. Conductivity-thermal and electrical, dielectric constant, dissipation factor, power factor, electric resistance, surface resistivity, volume resistivity, swelling, ageing resistance, environmental stress cracking resistance.

Subject-wise Analysis

XE - F : Polymer Science and Engineering

Year	1 mark Questions	2 marks Questions	Total Marks
2009	8	16	40
2010	9	13	35
2011	9	13	35
2012	9	13	35
2013	9	13	35
2014	9	13	35
2015	9	13	35
2016	9	13	35
2017	9	13	35
2018	9	13	35

2009

1. Which of the following trends is the most appropriate for a thixotropic fluid?
   (*a*) Viscosity increases with increase in the rate of shear.
   (*b*) Viscosity increases with increase in the time of application of shear.
   (*c*) Viscosity decreases with increase in the time of application of shear.
   (*d*) Viscosity increases with decrease in the rate of shear.

2. The temperature at which thermoforming is best carried out is
   (*a*) Softening temperature
   (*b*) melting temperature
   (*c*) Glass transition temperature
   (*d*) 10 % above melting temperature

3. Which of the following blends is immiscible?
   (*a*) SAN/ PMMA
   (*b*) PE/ PP
   (*c*) PC / PS
   (*d*) PET /PBT

4. A flexible garden hose pipe made of PVC was observed to get hardened after a length of time. The observation is most likely due to
   (*a*) Chain scission
   (*b*) loss of plasticizer
   (*c*) Loss of UV stabilizer
   (*d*) loss of thermal stabilizer

5. A doped polymer that conducts electricity is
   (*a*) poly(vinyl chloride) (*b*) polyethylene
   (*c*) polypropylene (*d*) polypyrrole

6. During chain growth polymerization, the molecular weight of the polymer
   (*a*) increases with conversion
   (*b*) decreases with conversion
   (*c*) does not change with conversion
   (*d*) first increases and then decreases with conversion

7. Based on the solubility parameter (δ) the best solvent for polyethylene ($\delta = 16.2\ MPa^{1/2}$) is
   (*a*) tetrahydrofuran ($\delta = 20.3\ MPa^{1/2}$)
   (*b*) toluene ($\delta = 18.3\ MPa^{1/2}$)
   (*c*) acetone ($\delta = 19.9\ MPa^{1/2}$)
   (*d*) methanol ($\delta = 29.7\ MPa^{1/2}$)

8. For any polymer, the number average molecular weight (Mn). weight average molecular weight (Mw) and viscosity average molecular weight (Mv). in general, obey the following relationship :
   (*a*) Mn > Mw > Mv
   (*b*) Mv > Mn > Mw
   (*c*) Mw > Mv > Mn
   (*d*) Mv > Mw > Mn

9. Pair the items in the Column I with those in the Column II.

Processing step	***Item***
P. rotational molding	1. polyurethane
Q. extrusion	2. use of a gas
R. reaction injection molding	3. centrifugal force
S. blow molding	4. twin screw

   (*a*) P-3, Q-1, R-2, S-4
   (*b*) P-2, Q-4, R-3, S-1
   (*c*) P-4, Q-2, R-1, S-3
   (*d*) P-3, Q-4, R-1, S-2

10. Strain γ, in a polymer melt varies with time on application of stress *s* by the following relation :

$$\eta\frac{d\gamma}{dt} + G\gamma = s$$

If a steady shear stress, s_o, is applied, the strain at the steady state, γ_o is given by
   (*a*) s_o/G (*b*) s_o/η
   (*c*) s_oG (*d*) $s_o\eta$

11. Match the polymerization initiator with the respective process.

Initiator	***Process***
P. benzyl lithium	1. coordination polymerization
Q. tropylium chloride	2. anionic polymerization
R. AIBN	3. cationic polymerization
S. $TiCl_4/\ Al(Et)_3$	4. radical polymerization

   (*a*) P-2, Q-3, R-4, S-1
   (*b*) P-2, Q-3, R-1, S-4
   (*c*) P-3, Q-1, R-2, S-4
   (*d*) P-4, Q-2, R-1, S-3

12. Arrange the following polyamides (PA) in decreasing order of their melting points:

I. PA66

II. PA6

III. PA10

IV. PA12

(*a*) IV > I > II > III

(*b*) I > II > III > IV

(*c*) III > II > IV > I

(*d*) II > IV > III > I

13. Match the characterization technique with the most appropriate property

Characterization technique	***Property***
P. infrared spectroscopy	1. melting point
Q. thermo-gravimetric analysis	2. functional group
R. transmission electron microscopy	3. degradation temperature
S. differential scanning calorimetry	4. morphology

(*a*) P-3, Q-2, R-4, S-1

(*b*) P-3, Q-4, R-2, S-1

(*c*) P-2, Q-1, R-4, S-3

(*d*) P-2, Q-3, R-4, S-1

14. Match the rubber ingredients with their appropriate function.

Rubber ingredient	***Function***
P. ZnO	1. tackifier
Q. salicylic acid	2. extender
R. ester gum	3. accelerator
S. paraffin oil	4. retarder

(*a*) P-3. Q-4, R-1, S-2

(*b*) P-3, Q-4, R-2, S-1

(*c*) P-4. Q-3, R-2, S-1

(*d*) P-4, Q-3, R-1, S-2

15. At the start of a step growth polymerization there are N_0 moles of monomer A (molecular weight M_A) and N_0 moles of monomer B (molecular weight M_B). At the end of the polymerization there are N moles of polymer chains. Assuming no condensation product, the number of average molecular weight is

(*a*) $2 N_0 (M_A + M_B)/ N$

(*b*) $N_0 (M_A + M_B)/ N$

(*c*) $N_0 (M_A + M_B)/2 N$

(*d*) $N_0^2 (M_A + M_B)/ N^2$

16. The ratio of the complex dynamic modulus to the storage modulus of a polymer system with a phase angle of 45° is

(*a*) 0 (*b*) 1 – i

(*c*) 1 + *i* (*d*) 1 ± *i*

17. Match the additive to its most common function.

Additive	***Function***
P. talc	1. plasticizer
Q. carbon fibre	2. flame retardant
R. dioctyl phthalate	3. filler
S. antimony trioxide	4. reinforcement

(*a*) P-3, Q-4, R-2, S-1

(*b*) P-4, Q-3, R-1, S-2

(*c*) P-4, Q-3, R-2, S-1

(*d*) P-3, Q-4, R-1, S-2

18. Match the polymer mechanical property with the appropriate testing method.

Mechanical property	***Testing method***
P. flexural strength	1. notched Izod
Q. impact strength	2. Shore-D
R. hardness	3. ASTM D 638
S. tensile strength	4. three-point bending

(*a*) P-4, Q-1, R-2, S-3

(*b*) P-3, Q-2, R-1, S-4

(*c*) P-3, Q-1, R-2, S-4

(*d*) P-4, Q-1, R-3, S-2

Common Data for Questions 19 and 20:

An aligned short carbon fibre reinforced polyester composite has a fibre content of 40% by volume. The elastic moduli of carbon fibre and polyester resin are 250 GPa and 35 GPa respectively. The fibre diameter is 5 m and the ultimate tensile strength of the fibre is 1240 MPa.

19. The modulus of the composite is

(*a*) 121 GPa (*b*) 215 GPa

(*c*) 285 GPa (*d*) 142.5 GPa

20. The fibre-matrix bond strength, assuming a critical fibre length of 12 mm, is

(*a*) 258 MPa (*b*) 2.58 MPa

(*c*) 25.8 MPa (*d*) 0.258 MPa

Common Data for Questions 21 and 22:

A plasticating screw of an injection molding unit injects 0.1 L/s of polymer through a mold, which is a cylindrical tube having a diameter of 20 mm and a length of 100 mm. The pressure drop across the mold is 100 MPa.

21. The shear stress exerted by the polymer on the wall of the mold is

(*a*) 2.5 MPa (*b*) 10 MPa

(*c*) 5 MPa (*d*) 1 MPa

22. The power consumed by the plasticizing screw is

(*a*) 5 kW (*b*) 1 kW

(*c*) 2.5 kW (*d*) 10 kW

Statement for Linked Answer Questions 23 and 24:

The density of a poly (ethylene terephthalate) (PET) sample is 1.407 g/cm^3, and the heat of fusion of the sample obtained from differential scanning calorimetry (DSC) is 54.6 J/g. The density of the PET crystalline phase is 1.515 g/cm^3 and of the PET amorphous phase is 1.335 g/cm^3

23. The fractional crystallinity of the sample is

(*a*) 0.23 (*b*) 0.36

(*c*) 0.40 (*d*) 0.43

24. The heat of fusion of the PET crystalline phase is

(*a*) 21.8 J/g (*b*) 136.5 J/g

(*c*) 68.2 J/g (*d*) 158.3 J/g

ANSWERS

1. (*c*)	**2.** (*b*)	**3.** (*b*)	**4.** (*c*)	**5.** (*d*)	**6.** (*a*)	**7.** (*b*)	**8.** (*c*)	**9.** (*d*)	**10.** (*a*)
11. (*a*)	**12.** (*b*)	**13.** (*d*)	**14.** (*a*)	**15.** (*c*)	**16.** (*c*)	**17.** (*d*)	**18.** (*a*)	**19.** (*a*)	**20.** (*d*)
21. (*c*)	**22.** (*d*)	**23.** (*d*)							

24. Marks to all (*heat of fusion of the PET amorphous phase is not given)

EXPLANATIONS

1.

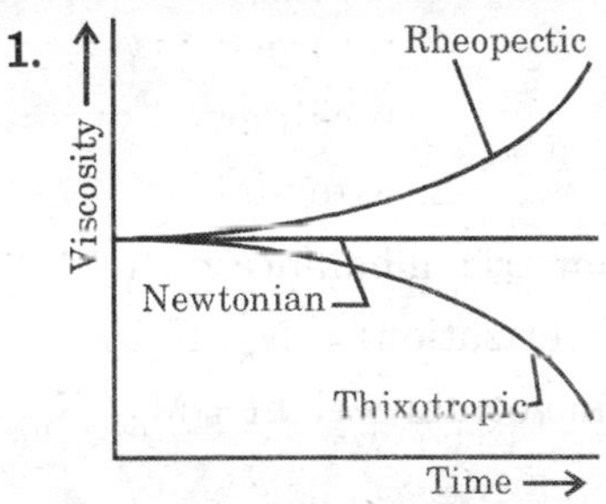

Thixotropic Fluids are fluids that show a shear thinning property. .. In more technical language: some non-Newtonian pseudoplastic fluids show time dependent change in viscosity; the longer the fluid undergoes shear stress, the lower its **viscosity**.

3. Immiscible polymer blends (heterogeneous polymer blends): This is by far the most populous group. If the blend is made of two polymers, two glass transition temperatures will be observed.

4. Exposure to ultraviolet (UV) radiation may cause the significant degradation of many materials. UV radiation causes photooxidative degradation which results in breaking of the polymer chains, produces free radical and reduces the molecular weight, causing deterioration of mechanical properties and leading to useless materials, after an unpredictable time.

5.

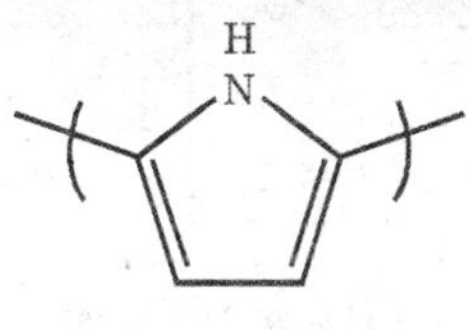

polypyrrole (PPy)

Conducting Polymers are organic polymers that conduct electricity. Conductive polymers are generally not thermoplastics, i.e., they are not thermoformable. But, like insulating polymers, they are organic materials. They can offer high electrical conductivity but do not show similar mechanical properties to other commercially available polymers. The electrical properties can be fine-tuned using the methods of organic synthesis and by advanced dispersion techniques. All conducting polymer system is alternating single bond and double bond along the backbone.

6.

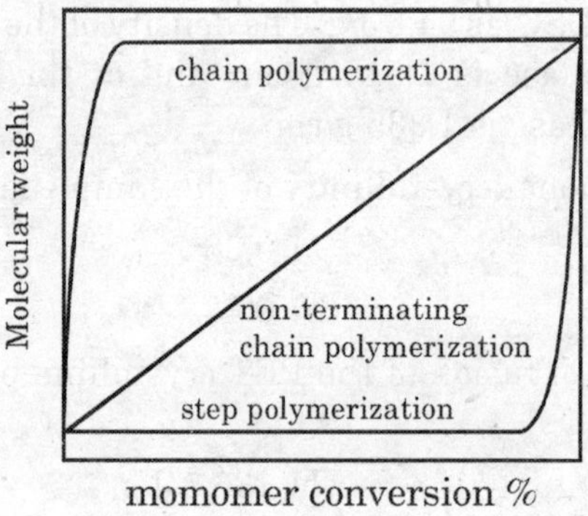

7. The above equation may be rewritten;

$$\frac{\Delta H_m}{V} = \phi, \phi_2[\delta_1 - \delta_2]^2$$

This equation is valid only when ΔHm is zero or greater. It is invalid for exothermic mixing, that is, when ΔHm is negative.

The square of the difference between the solubility parameters of two liquids decides the magnitude of the heat of mixing. Hence, for the positive heat of mixing to be at a minimum value so as to let mixing occur, the difference between the solubility parameters ($\delta_1 - \delta_{2)}$ of the two liquids must be as low as possible.

Difference of the solubility parameters is low between toluene and polyethylene. So best solvent is toluene for polyethylene.

8.

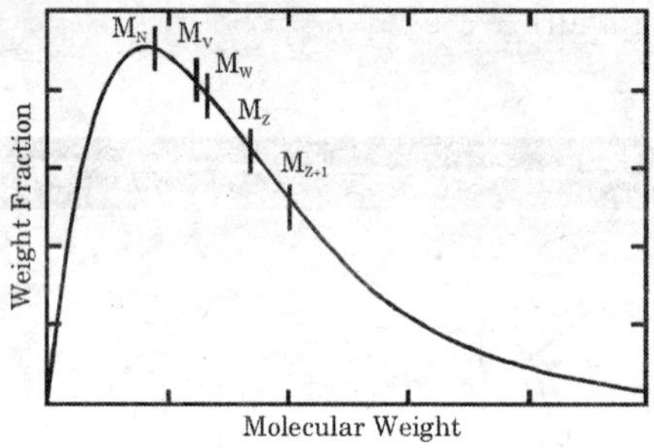

9.

	Processing step	Item
P.	rotational molding	3. centrifugal force
Q.	extrusion	4. twin screw
R.	reaction injection molding	1. polyurethane
S.	blow molding	2. use of a gas

10. At steady state condition means process unchanged with the time.

$$\eta \frac{d\gamma_o}{dt} + G\gamma_0 = s_o$$

at steady state $\frac{d\gamma_o}{dt} = 0$ hence

$$\gamma_0 = s_o/G$$

11. 1. Ziegler-Natta catalyst: These are complex catalyst systems derived from a transition metal compound from groups IVB to VIIIB of the periodic table and an organometallic compound usually from a group IA or IIIA metal. Aluminium alkyls (AlEt or $AlEt_2Cl$) acts as electron acceptor in combination with the electron donor titanium chloride ($TiCl_3$& $TiCl_4$) are used most commonly.

2. Cationic catalyst: Strong Lewis acid like boron trifluoride (BF_3) or Friedel-Crafts catalysts such as $AlCl_3$, H_2SO_4, $SnCl_4$ and $TiCl_4$

3. Anionic catalyst: Strong base Butyl lithium, triphenylmethyl potassium, ethyl sodium, alkali metal amides, alcoxides and hyderoxides.

4. Radical catalyst: AIBN, Benzoyl Peroxides,

12.

PA66	PA6	PA10	PA12
268.6°C	220°C	185°C	180°C

13.

	Characterization technique	***Property***
P.	infrared spectroscopy	2. functional group
Q.	thermo-gravimetric analysis	3. degradation temperature
R.	transmission electron microscopy	4. morphology
S.	differential scanning calorimetry	1. melting point

14.

	Rubber ingredient	***Function***
P.	ZnO	3. accelerator
Q.	salicylic acid	4. retarder
R.	ester gum	1. tackifier
S.	paraffin oil	2. extender

15. Where no stoichiometric imbalance $r = 1$

X_n (degree of polymerization) = N_o / N

Number average molecular weight = $M_o * X_n$

$M_o = M_A + M_B / 2$

So Number average molecular weight

$$= N_0 (M_A + M_B)/2N$$

16.

$$\delta = 45°$$

$$E^* = E' + iE''$$

$$= E'(1 + i \tan \delta)$$

$$E^* / E' = (1 + i \tan 45°)$$

$$= 1 + i$$

17. Talc-filler, Carbon black- reinforcing filler, Phthalate-plasticizer and antimony oxide-flame retardant.

19. Modulus of the composite

$$E_{cl} = E_m V_m + E_f V_f$$

$$E_{cl} = 0.6 * 35 + 0.4 * 100 = 121 \text{ GPa}$$

20. Critical fibre length

$$L_c = \frac{\sigma d}{2\tau_c} \qquad \tau_c = \frac{\sigma d}{2L_c},$$

$$\tau_c = \frac{1240 \times 5 \times 10^{-6}}{2 \times 12 \times 10^{-3}}$$

$$= 0.258 \text{ MPa}$$

21. Shear stress distribution in the Hagen-Poiseuille equation;

$$\tau = \frac{R}{2}\frac{dp}{dx}$$

$$\tau = \frac{10}{2} \times \frac{100}{100} = 5\text{MPa}$$

22. Power = volumetric rate * pressure drop

$$\dot{W} = \dot{V}\Delta P$$

Power = 0.1 * 10^{-3} * 100 * 10^{6} = 10 kW

23. Mass fraction of the crystallinity

$$X_c = \frac{1/d_a - 1/d}{1/d_a - 1/d_c}$$

$$= \frac{1/1.335 - 1/1.407}{1/1.335 - 1/1.515}$$

$$= 0.43$$

(d_a = density of amorphous part, d_c = density of crystallinity part, d = density of the polymer)

24. Degree of crystallinity = $X_c = \dfrac{H_a - H}{H_a - H_c}$

(H_a = enthalpy of amorphous part, H_c = enthalpy of crystallinity part, H = enthalpy of the polymer)

2010

1. Which one of the following molecules undergoes ring opening polymerization?

(a)  (b) 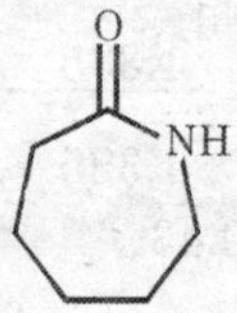

(c) 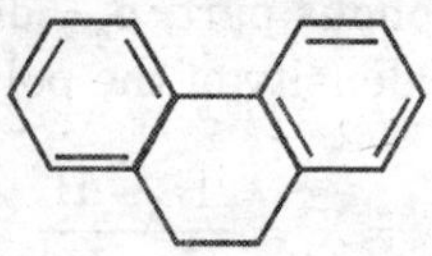 (d) 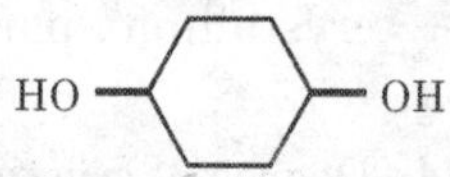

2. Out of the following polymers, which one shows the highest melting temperature?
   (a) Poly (ethylene terephthalate)
   (b) Poly(propylene terephthalate)
   (c) Poly (butylene terephthalate)
   (d) Poly (hexylene terephthalate)

3. Which one of the following reagents is used to prevent coagulation of natural rubber latex?
   (a) Ammonia
   (b) Acetic acid
   (c) Tolyl mercaptan
   (d) Sodium chloride

4. From the following four groups of polymers, identify the group in which all four polymer are semi crystalline
   (a) HDPE, PP, PS, UF
   (b) PET, PVC, PF, ABS
   (c) Nylon 6. EPDM, PMMA. SBR
   (d) Nylon 66, PP, HDPE. PET

5. If η represents viscosity of polymer solution and η_o represents viscosity of pure solvent, then the specific viscosity (η_{sp})of the polymer solution is expressed as
   (a) η/η_o
   (b) $\eta/\eta_o - 1$
   (c) $\eta_o/\eta - 1$
   (d) η_o/η

6. A polymer blend is developed by mixing
   (a) A polymer and a monomer
   (b) A polymer and a stabilizer
   (c) A polymer with another polymer
   (d) A polymer and a filler

7. In shear deformation of a polymer melt, the unit of shear rate is
   (a) $m^3\ sec^{-1}$ (b) $m^2\ sec^{-1}$
   (c) $m\ sec^{-1}$ (d) sec^{-1}

8. A rubber compound is made by mixing functional additives with the rubber using
   (a) Two-roll mill
   (b) Compression molding machine
   (c) Three-roll calender
   (d) Thermoforming machine

9. The group of polymer consisting of LDPE. PP, PS and PVC is best categorized as
   (a) Engineering polymers
   (b) Biodegradable polymers
   (c) Commodity polymers
   (d) Natural polymers

10. A small molecule is eliminated as a byproduct during the synthesis of
    (a) Polycaprolactone
    (b) Poly (ethylene terephthalate)
    (c) Styrene butadiene copolymer
    (d) Polytetrafluoroethylene

11. For free radical copolymerization of monomers M_1 and M_2, if the reactivity ratios r_1 and r_2 are both found to be zero, then the resulting copolymer is
    (a) Random
    (b) Branched
    (c) Block
    (d) Alternating

12. Pair each item in Column I with the appropriate one in Column II

Column I	Column II
P. Shampoo bottles	1. Injection molding
Q. Overhead tank	2. Extrusion
R. Helmet	3. Blow molding
S. Insulation for cables	4. Rotomolding

    (a) P-1; Q-2; R-4; S-3
    (b) P-2; Q-1: R-3; S-4
    (c) P-3: Q-4; R-1; S-2
    (d) P-4; Q-3; R-2; S-1

13. Toughness of a plastic material can be judged from the area under the stress-strain curve obtained from tensile test. The plastic having the highest toughness exhibits

(*a*) High tensile strength and low elongation

(*b*) Low tensile strength and high elongation

(*c*) High tensile strength and high elongation

(*d*) Low tensile strength and low elongation

14. Match the following additives for plastics with their respective functions

Additives	***Functions***
P. Iron oxide	1. Blowing agent
Q. Azodicarbonamide	2. Coloring ageut
R. Phenyl salicylate	3. Filler
S. Calcium carbonate	4. UV stabilizer

(*a*) P-1; Q· 2; R-3; S-4

(*b*) P-2; Q-1; R-4; S-3

(*c*) P-3; Q-4; R-1; S-2

(*d*) P-4; Q-3; R-2; S-1

15. The volume fraction of epoxy resin in a glass fibre/ epoxy composite is 0.48. The densities of glass fibre and composite are 2540 kg/m^3 and 1950 kg/m^3, respectively. The weight fraction of the fibre in the composite is:

(*a*) 0.68 (*b*) 0.52

(*c*) 0.48 (*d*) 0.32

16. The change of shear stress with shear rate of a polymer melt as shown in the figure below indicates

(*a*) Viscosity increase with increase in shear rate

(*b*) Viscosity decrease with increase in shear rate

(*c*) Viscosity remaining independent of shear rate

(*d*) Viscosity oscillation with increase in shear rare

Common Data for Questions 17 and 18 :

For polyesterification of HO-$(CH_2)_{14}$COOH, the number average degree of polymerization, X_n, is related to the stoichiometric imbalance *r*between the functional group and the extent of polymerization *p* by the equation

$$X_n = \frac{1+r}{1+r-2Pr}$$

17. For 100% polyesterification, the X_n will be

(*a*) 100 (*b*) 1000

(*c*) 10000 (*d*) infinity

18. The percentage Conversion of functional groups required to obtain the polyester with a molecular weight of 24000 g/mol will

(*a*) 99 (*b*) 96

(*c*) 93 (*d*) 90

Common Data for Questions 19 and 20 :

The glass transition temperatures of PVC, plasticized PVC and nitrile rubber are 81°C, –3°C and –50°C respectively. For making a plastic raincoat, 40 mass percent of polymeric plasticizer is added to PVC

19. The glass transition temperature of the polymeric plasticizer will be

(*a*) –84°C (*b*) –74°C

(*c*) 74 °C (*d*) 84°C

20. If the polymeric plasticizer is replaced by nitrile rubber to make the same raincoat, the mass percent of rubber to be blended with PVC is

(*a*) 40 (*b*) 47

(*c*) 53 (*d*) 60

Question no.19 and 20 marks to all due to technical error in the questions.

Linked Answer Questions

Statement for Linked Answer Questions 21 and 22:

A polydisperse polymer consists of the following three different fractions

	Fraction 1	Fraction 2	Fraction 3
Mass of Polymer (%)	30	20	50
Molecular weight (g/mol)	30,000	80,000	150,000

21. The number average molecular weight, M_n(g/mol) of the polymer is

(*a*) 1.02×10^5

(*b*) 0.99×10^5

(*c*) 0.63×10^5

(*d*) 0.55×10^5

22. The polydispersity index of the polymer is

(*a*) 2.59 (*b*) 1.87

(*c*) 1.72 (*d*) 1.59

ANSWERS

1. (*b*)	2. (*a*)	3. (*a*)	4. (*d*)	5. (*b*)	6. (*c*)	7. (*d*)	8. (*a*)	9. (*c*)	10. (*)
11. (*d*)	12. (*c*)	13. (*c*)	14. (*b*)	15. (*a*)	16. (*c*)	17. (*d*)	18. (*a*)	19. (*)	20. (*)
21. (*c*)	22. (*d*)								

EXPLANATIONS

1. Structure in Option B is caprolactum. Nylon 6 is polymerization by caprolactum by ring opening polymerization.

4. Materials such as polystyrene (PS), polycarbonate (PC), acrylics (PMMA), acrylonitrile-butadiene styrene (ABS) and polyvinyl chloride (PVC) are said to be amorphous thermoplastics. This signifies that in the solid state their molecular structure is random and disordered the long chain molecules being all entangled rather like solidified spaghetti.
Materials such as most of the nylons (PA), polyacetal (POM), polypropylene (PP), polyethylene (PE) and the thermoplastic polyesters (PET) have a much more ordered structure in the solid state, a considerable proportion of the long chain molecules being closely packed in regular alignment. These materials are known as semi crystalline thermoplastics.

5. $\dfrac{t-t_0}{t_0} = \dfrac{\eta-\eta_o}{\eta_o} = \eta_r - 1 = \eta_{sp}$

6. Polymer blends are physical mixtures of two or more polymers with/without any chemical bonding between them.

7. $\sigma = \eta\dfrac{d\gamma}{dt}$

σ stress (Pa) and η viscosity (Pa-s)

$$\frac{d\gamma}{dt} = \frac{pa}{(pa\text{-}s)} = 1/\text{sec}$$

8. Two roll is used for the mixing purpose. Option B, C & D is used to shape the product.

9.

Engineering polymers	ABS, Nylon, PA, PBT, PEEK , PET etc.
Biodegradable polymers	Poly(α-esters), Polylactide
Commodity polymers	PE, PP, PS, PVC
Natural polymers	Silk, wool , DNA, cellulose and NR

10. Only step growth polymerization small molecule is eliminated as a byproduct.
Option B is polymerized by the step growth polymerization. Option A is synthesis by the ring opening polymerization. Option C & D is synthesis by freeradical polymerization.

11. $r_1 = r_2 = 0$
   a. Monomer 1 reacts exclusively with monomer 2 and viceversa
   b. Alternating copolymer
   c. Example: styrene: maleic anhydride $r_1 = 0.041$: $r_2 = 0.01$

13.

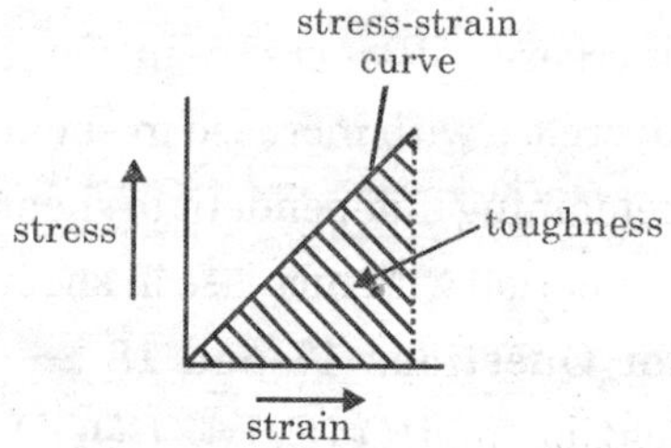

Toughness = /12*stress * strain

High tensile and high elongation both will give high value of the ear under the curve by stress and strain.

14.

carbonamide	Blowing agent	Play a role in the formation of air bubbles by breaking down into gas at high temperature.
Calcium carbonate	Filler	Used as fillers in rubber industry
Iron oxide	Coloring agent	Using as pigment
Phenyl salicylate	UV stabilizer	The primary function is to protect the substance from the long term UV degradation effects from ultraviolet radition.

15. Mass fraction of the fiber in the composite

= mass of fiber / total composite mass

Assume total composite volume is 1

$$V_c = V_m + V_f$$

$$V_f = 1 - 0.48 = 0.52$$

Mass of fiber = density * volume

weight fraction of the fiber

$$= \frac{(2540*0.52)}{(1950*1)} = 0.6777$$

16. Stress is proportional to the strain rate. According to Newton's law, the response of a fluid to a shearing stress σ is viscous flow, given by

$$\sigma \propto \frac{d\gamma}{dt}$$

$$\sigma = \eta \frac{d\gamma}{dt}$$

Where η is viscosity and $d\gamma/dt$ is strain rate.

Viscosity constant with the shear rate and stress.

17. $P = 100\%$

$$X_n = \frac{1+r}{1+r-2Pr}$$

At 100% of the extent reaction number average degree of polymerization equal to infinity

18. Molecular weight of the

$$HO–(CH_2)_{14}COOH = 258 = M_o$$

$$M_n = \frac{M_o}{1-p}$$

$$M_n = 24000 \text{ g/mol}$$

$$24000 = \frac{258}{(1-p)}$$

$$P = 99\%$$

19. $$\frac{1}{T_g} = \frac{W_1}{T_{g1}} + \frac{W_2}{T_{g2}}$$

$$\frac{1}{T_g} = \frac{0.4}{270} + \frac{0.6}{354}$$

$$T_g = 41.8°C$$

20. $$\frac{1}{T_g} = \frac{W_1}{T_{g1}} + \frac{W_2}{T_{g2}}$$

21. We know that weight of the sample is

$$W = \sum w_i = \sum N_i M_i$$

Mass in terms of percentage given, we use mass as w_i, and divide w_i by the molecular weight and get the number of chains. Number average molecular weight:

$$M_n = \frac{\sum N_i M_i}{\sum N_i}$$

$$M_n = \frac{30+20+50}{\frac{30}{3000}+\frac{20}{80000}+\frac{50}{150000}}$$

$$M_n = 0.633 \times 10^5$$

22. Weight average molecualr weight:

$$M_w = \frac{\sum N_i M_i^2}{\sum N_i M_i}$$

$$M_w = \frac{(30 \times 30000 + 20 \times 80000 + 50 \times 150000)}{30+20+50}$$

$$M_w = 10^5$$

$$\text{Poly Dispersity} = \frac{M_w}{M_n}$$

$$\text{Poly Dispersity} = \frac{10^5}{0.633 \times 10^5}$$

$$PD = 1.59$$

2011

1. Monomers that undergo free radical polymerization have
   (*a*) one hydroxyl and one carboxylic group
   (*b*) double bonds
   (*c*) one amine group and one carboxylic group
   (*d*) two amine groups
2. In the polymerization of phthalic anhydride (3 moles) and glycerols (3 moles), the average functionality of the reacting system is
   (*a*) 1.2 (*b*) 4.2
   (*c*) 3.8 (*d*) 2.5
3. A mixture consists of 10 moles of dimer, 15 moles of trimer, 20 moles of tetramer and 5 moles of pentamer. The average chain length of the mixture is
   (*a*) 3.4
   (*b*) 8.5
   (*c*) 6.8
   (*d*) 1.7
4. Polyurethane is formed by
   (*a*) self condensation of polyols
   (*b*) self condensation of diisocyane
   (*c*) reaction of polyol and diisocyanate
   (*d*) reaction of polyol with adipic acid
5. The glass transition temperature (T g) is governed by
   (*a*) translational motion of entire molecule
   (*b*) long cooperative wriggling motion of 40 to 50 C-C bonds
   (*c*) short cooperative motion of 5 to 6 bonds of the molecules
   (*d*) vibration of carbon atoms of the polymer molecules
6. The miscibility in binary polymer blends of poly vinyl fluoride and poly acrylate is governed by which one of the following interactions?
   (*a*) dipole-dipole (*b*) acid-base type
   (*c*) hydrogen bonding (*d*) ion-dipole
7. The melting temperature of poly tetrafluoroethylene is higher than its degradation temperature due to
   (*a*) hydrogen bonding
   (*b*) n-hydrogen bonding
   (*c*) van der Waal's interaction
   (*d*) dipole moment interaction
8. As the molecular weight (M) increases beyond the critical molecular weight (Mc), the zero shear viscosity of a polymer melt becomes proportional to
   (*a*) $M^{0.5}$
   (*b*) $M^{3.4}$
   (*c*) $M^{1.0}$
   (*d*) M^{0}
9. In the step-growth polymerization of phenol with formaldehyde the functionality of phenol is
   (*a*) 2 (*b*) 5
   (*c*) 4 (*d*) 3
10. During the processing of polymers, degradation occurs. Pair each item in Column-I with the appropriate one in Column-II.

Column-I	Column-II
P. Poly vinyl chloride	1. Elimination of low molecular weight compound
Q. Poly acrylonitrile	2. Unzipping
R. Poly methyl methacrylate	3. Cyclization
S. Polyethylene	4. Random scission

   (*a*) P-1, Q-3, R-2, S-4 (*b*) P-1, Q-2, R-3, S-4
   (*c*) P-2, Q-1, R-3, S-4 (*d*) P-4, Q-2, R-3, S-1
11. The solubility parameter of polystyrene is 9.1 $(cal/cm^3)^{1/2}$, the solubility parameter of n-hexane is 7.3 $(cal/cm^3)^{1/2}$ and that of benzene is 9.2 $(cal/cm^3)^{1/2}$. Then polystyrene will
   (*a*) dissolve inn-hexane
   (*b*) not dissolve in 80:20 mixture ofn-hexane and benzene
   (*c*) not dissolve in benzene
   (*d*) dissolve in benzene
12. Fiber glass composites are prepared by coating unidirectional fiber glass with epoxy prepolymer. If slippage **DOES NOT** occur at the interface in the loading in the direction of fiber, then
   (*a*) force sheared by fiber glass (Pr) and matrix (P m) should be equal
   (*b*) strain as well force in fiber (Er and Pr) are both equal to strain and the force in matrix (Er and Pm)
   (*c*) strain at the interface between fiber glass (Er) and matrix (Em) should be equal
   (*d*) None of the above is true

13. The power factor (PF) and dissipative factor (DF) of polymeric materials are related by

(a) $PF = DF/(1-DF^2)^{1/2}$ (b) $DF = PF/(1-PF^2)^{1/2}$

(c) $DF = 1/(1-PF^2)^{1/2}$ (d) $PF = 1/(1-DF^2)^{1/2}$

14. Poly methyl methacrylate (PMMA) and silicone can be used respectively in

(a) medical syringe and hip joint

(b) soft tissue replacement and contact lenses

(c) hip joint and medical syringe

(d) contact lenses and soft tissue replacement

15. In the log (viscosity) versus log (shear rate) behaviour of polymer melts, the upper (μ_{upper}) and lower (μ_{lower}) Newtonian viscosities are related as

(a) μ_{upper} and μ_{lower} versus shear rate are connected by discontinuous curve

(b) $\mu_{upper} > \mu_{lower}$

(c) $\mu_{upper} < \mu_{lower}$

(d) $\mu_{upper} = \mu_{lower}$

16. Match the following **additives** for plastic with their respective **functions.**

Additives	**Functions**
P. Cetyl Palmitate	1. Fire retardant
Q. Asbestos	2. Heat stabilizer
R. $BaSO_4$	3. Lubricant
S. SbO_2	4. Filler

(a) P-1, Q-2, R-3, S-4 (b) P-3, Q-4, R-2, S-1

(c) P-3, Q-2, R-1, S-4 (d) P-4, Q-3, R-2, S-1

Common Data for Questions 17 and 18:

In a single-screw extruder the dimensions of the channel are Diameter (D)= 5.03 cm, Width (W) = 1.1 cm, Height (H)= 0.36 cm, Screw Angle (8) = 6.3° and Speed of Rotation= 10 revolution per second (rps)

17. Assume a flat plate model in the melt zone of a single-screw extruder $\partial P/\partial z = \eta \partial^2 v_z/\partial y^2$; $v_z = 0$ at $y = 0$ and $v_z = V_{bz}$ at $y = H$

For fully developed flow in the z direction v_z is found to be

(a) $v_z = V_{bz}(y/H) - [y(H-y)\,\partial P/\partial z]/(2\eta)$

(b) $v_z = (y/H) + [y(H-y)\,\partial P/\partial z]/(2\eta)$

(c) $v_z = (y^2/H) + [y(H-y)\,\partial P/\partial z]/(2\eta)$

(d) $v_z = (y^2/H) - [y(H-y)\,\partial P/\partial z]/(2\eta)$

18. In the absence of a die in a single screw extruder $\Delta P = 0$ and Q is equal to

(a) 0.31 cm^3/sec

(b) 3.11 cm^3/sec

(c) 311 cm^3/sec

(d) 31.1 cm^3/sec

Common Data for Questions 19 and 20:

In the polymerization of styrene at 60□C, it is assumed that all steps are irreversible and

propagation rate constant (kp) = 145 lit/mol-sec

mutual termination rate constant (k1) = 0.13×10^7 lit/mol-sec

initiator rate constant (ki) = 4.38×10^{-6} sec–1

initiator concentration ([I2])= 1.65×10^{-2} mol/lit

monomer concentration([M])= 7.69 mol/lit

19. Assuming the rate constant for combination of initiator radical with monomer, $k_1 = 10k_p$ in the polymerization of styrene, the primary radical concentration ([I·]) can be calculated as

(a) [I·] = 0.364 X 10^{-8} mol/lit

(b) [I·] = 1.65 X 10^{-10} mol/lit

(c) [I·] = 6.48 X 10^{-12} mol/lit

(d) [I·] = 2.74 X 10^{-4} mol/lit

20. The total molar concentration of the growing polymer radicals (λp) in the polymerization of styrene is given by

(a) $\lambda_p = 4.72 \times 10^{-8}$ mol/lit

(b) $\lambda_p = 2.36 \times 10^{-7}$ mol/lit

(c) $\lambda_p = 7.08 \times 10^{-6}$ mol/lit

(d) $\lambda_p = 9.44 \times 10^{-5}$ mol/lit

Statement for Linked Answer Questions 21 and 22:

For semi-crystalline polymeric materials, free of voids and impurities, we define the following:

d = average density of a semi-crystalline polymer

dc = density of the crystalline phase

da = density of the amorphous phase

21. The relation between the average density and the densities of crystalline and amorphous phases can be related to mass fraction of the crystalline phase (X) as

The degree of crystallinity of the sample can be calculated by given equation:

$$\text{Degree of crystallinity by mass} = X_c = \frac{1/d_a - 1/d}{1/d_a - 1/d_c}$$

$1/d = (X_C/d_c) + \{(1 - X_C)/d_a$

22. When a fiber is made of PET, it has crystalline as well as amorphous phases and experiments have revealed that the average density of the overall material is 1.375 g/cm^3 .It is known that the density of the crystalline phase is 1.455 g/cm^3 and that of amorphous phase is 1.335 g/cm^3. The mass fraction crystallinity (X) in % is given by

(a) 35.3% (b) 50.5%

(c) 80.1% (d) 5.1%

ANSWERS

1. (*b*)	2. (*d*)	3. (*a*)	4. (*c*)	5. (*b*)	6. (*a*)	7. (*c*)	8. (*b*)	9. (*d*)	10. (*a*)
11. (*d*)	12. (*c*)	13. (*b*)	14. (*d*)	15. (*c*)	16. (*b*)	17. (*a*)	18. (*d*)	19. (*c*)	20. (*b*)
21. (*d*)	22. (*a*)								

EXPLANATIONS

1. Monomer has double bond will undergo free radical polymerization. Monomer has functional group will undergo step growth polymerization.

2. The functionality of a molecule is simply its interlinking capacity, or the number of sites available for bonding(presence of reactive functional group like %OH, %COOH,% NH_2, % SH, % NCO, etc.) with other molecules under the specific polymerization conditions

Average functionality

$$= \frac{\text{Functionality of compound 1*no.of mole} + \text{Functionality of compound 2*no.of mole}}{\text{Total mole}}$$

Figure. Structure of Glycerol

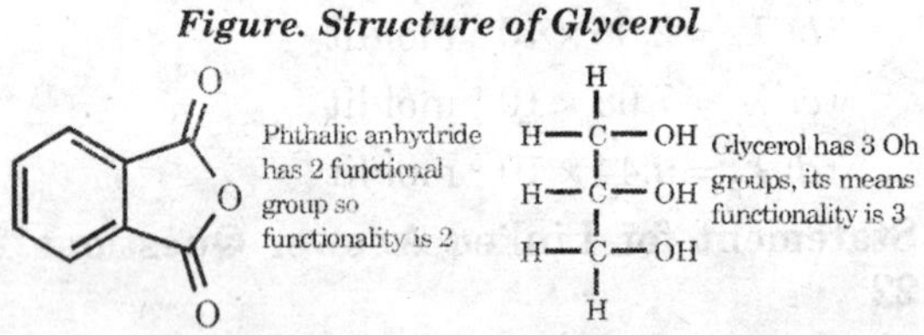

Average functionality = 2*3+ 3*3 / (3+3) = 2.5

3. Average chain length

$$= \frac{10\times2+15\times3+20\times4+5\times5}{(10+15+20+5)} = 3.4$$

4. 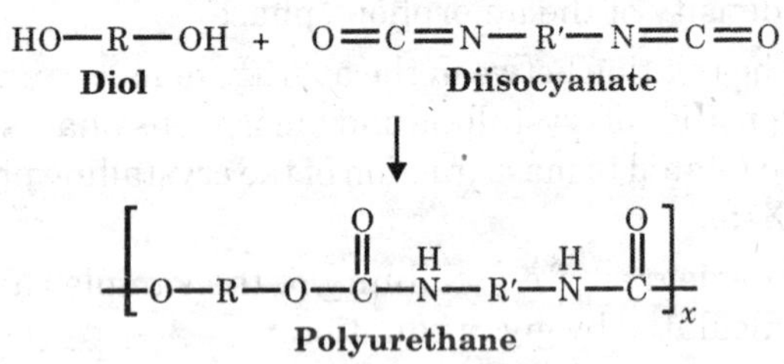

7. PTFE is very crystalline in the sense that there are large areas where the molecules are lying in a very regular arrangement. Remember that PTFE molecules can be thought of as long thin rods. These rods will pack very closely together. Due to weak van der waal's interaction its degrade before to melt.

8. At low molar mass values

$$\eta_0 = K_L M^{1.0}$$

melt viscosity is directly proportional to molar mass. For M values larger than a critical molar mass, M_c, there is a sharp increase in melt viscosity since

$$\eta_0 = K_H M^{3.4}$$

K_L and K_H in the equations reported above represent constants that are temperature dependent.

9. 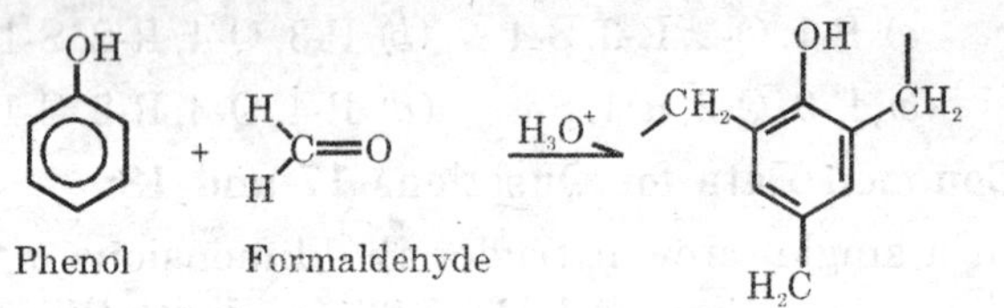

Formaldehyde attach on the 3 position on the phenol. At every place accommodate the formaldehyde. So functionality o phenol is 3.

10. 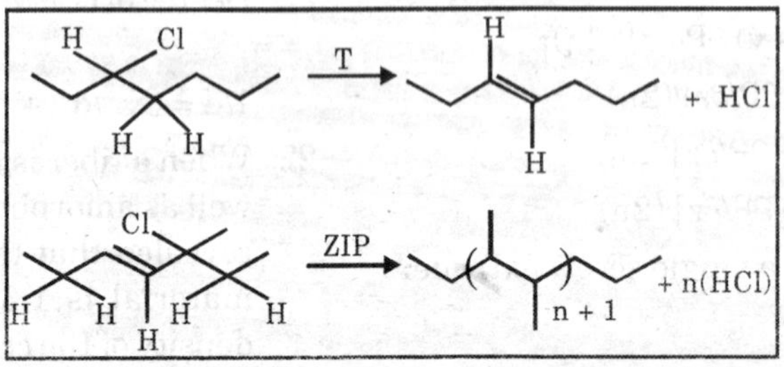

Scheme-1 - dehydrochlorination reactions in PVC macromolecules (Veronelli et al., 19999)

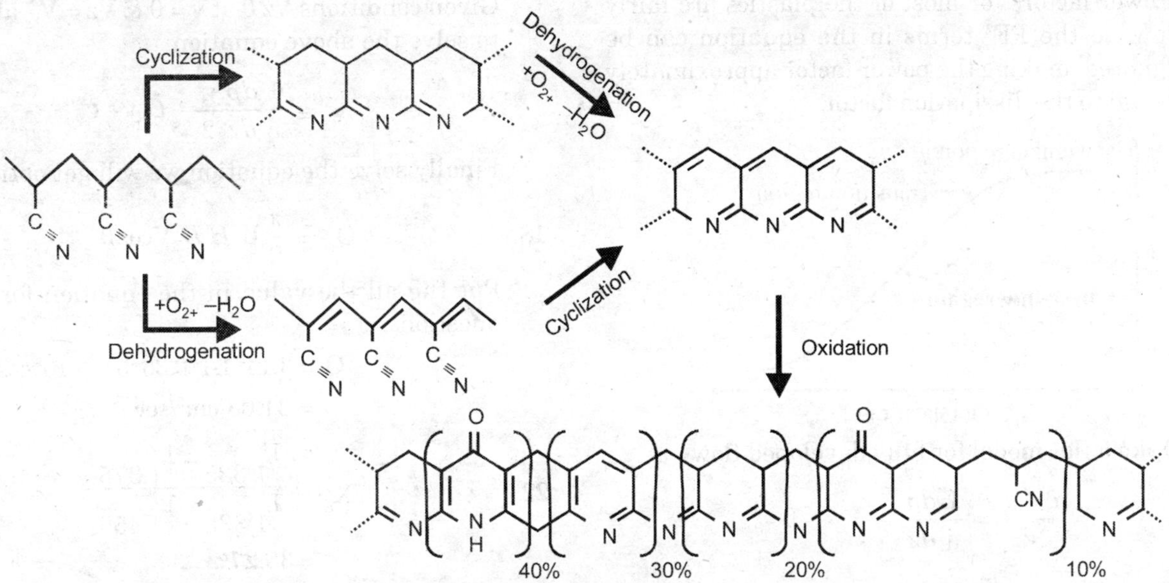

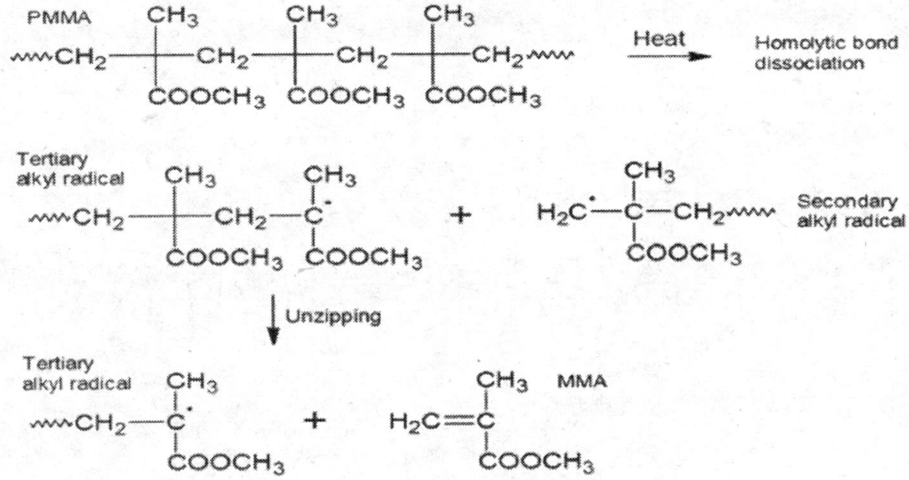

11. The above equation may be rewritten;

$$\frac{\Delta H_m}{V} = \phi_1\phi_2\left[\delta_1 - \delta_2\right]^2$$

This equation is validi only when ΔH_m is zero or greater. It is invalid for exothermic mixing, that is, when ΔHm is negative.

The square of the difference between the solubility parameters of two liquids decides the magnitude of the heat of mixing. Hence, for the positive heat of mixing to be at a minimum value so as to let mixing occur, the difference between the solubility parameters (δ_1-δ_2) of the two liquids must be as low as possible.

Difference of the solubility in the polystyrene and benzene are low compare to the hexane, so polystyrene will dissolve in benzenes.

13. The power factor is a measure of the energy absorbed by the material as the alternating current constantly changes direction and the dipoles try to align themselves with the field.

The dissipation factor (D) is the ratio of the current of the current of the resistive component (I_r) to the current of the capacitive component (I_c) and is also equal to the tangent to the dielectric loss angle δ, i.e.

$$D = \frac{I_r}{I_c} = \tan\delta$$

Typical values of the dissipation factor vary from 01 to less than 0.0005.

The power factor (PF) is the ratio of power dissipated to the product of the effective power output (volts × amps) and is a measure of the dielectric loss in the insulating plastic (which is effectively acting as capacitor). The power factor and dissipation factor are related by the equation:

$$DF = \frac{PF}{\sqrt{\left[1-(PF)^2\right]}}$$

Power factors for most of the plastics are fairly low, so the PF^2 terms in the equation can be ignored, making the power factor approximately equal to the dissipation factor.

15.

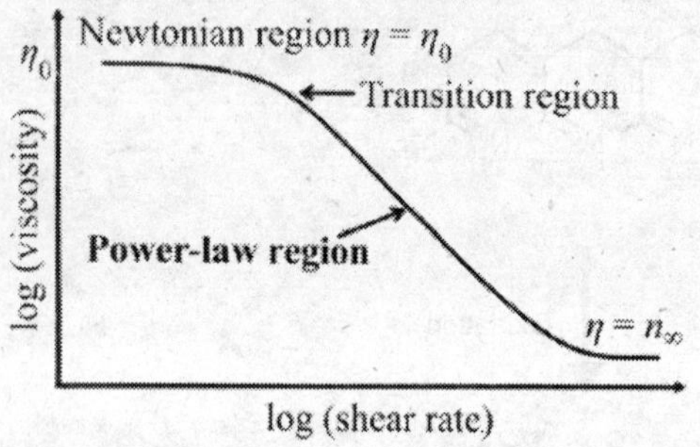

17. Take a flat model for fully developed flow

$$\frac{\partial^2 V_z}{\partial y^2} = \frac{1}{\eta}\frac{dp}{dz}$$

Given conditions Vz 0 at y = 0 & Vz = V_{bz} at y = H to solve the above equation

$$V_z = \frac{1}{\eta}\frac{dp}{dz}\frac{y^2}{2} + C_1 y + C_2$$

Finally solve the equation we will get option A.

18. $$Q_D = \frac{\pi}{2} WHDN\cos\theta$$

Put the all the value in the equation form the question

$$Q = 3.14*1.1*0.36*5.03*10*\cos(6.3)/2$$
$$= 31.08 \text{ cm}^3/\text{sec}$$

22. $$X_c = \frac{\frac{1}{1.335} - \frac{1}{1.375}}{\frac{1}{1.335} - \frac{1}{1.455}}$$
$$= 35.27\%$$

2012

1. The diamine used for the synthesis of nylon 6,4 is
   (*a*) tetramethylene diamine
   (*b*) hexamethylene diamine
   (*c*) phenylene diamine
   (*d*) cyclohexane-1,4-diamine
2. A small amount of hydroquinone is added to methyl methacrylate monomer in order to
   (*a*) improve its thermal stability
   (*b*) improve its hydrolytic stability
   (*c*) inhibit polymerization
   (*d*) regulate molecular
3. Which of the following compounds is used in Zeigler-Natta catalyst?
   (*a*) $TiCl_3$
   (*b*) $CaCl_2$
   (*c*) $ZnCl_2$
   (*d*) NaCl
4. The z-average molecular weight, Mz can be estimated using
   (*a*) End group analysis
   (*b*) Osmometry
   (*c*) Viscometry
   (*d*) Ultracentrifugation
5. The unit of intrinsic viscosity is
   (*a*) dl/g
   (*b*) g/dl
   (*c*) poise
   (*d*) g/mol
6. Heat dissipation is most inefficient in
   (*a*) Emulsion polymerization
   (*b*) Solution polymerization
   (*c*) Bulk polymerization
   (*d*) Suspension polymerization
7. The crystalline melting temperature of the polymers high density polyethylene (HDPE), isotactic polypropylene (iPP), nylon 6 (PA6) and poly(ethylene terephthalate) (PET) can be arranged as
   (*a*) THDPE > TPET > TiPP > TPA6
   (*b*) TPET > TPA6 > TiPP > THDPE
   (*c*) TPA6 > THDPE > TPET > TiPP
   (*d*) TiPP > TPA6 > TPET > THDPE
8. Which of the following polymers is sensitive to moisture absorption?
   (*a*) Polyethylene
   (*b*) Polystyrene
   (*c*) Polyamide 6
   (*d*) Polybutadiene
9. Most engineering polymers have heat distortion temperature (HDT) in excess of
   (*a*) –80°C
   (*b*) 0°C
   (*c*) 20°C
   (*d*) 80°C
10. Two miscible polymers A and B are blended in weight ratio of 30:70. If the glass transition temperature, T_g of polymer A is –50°C and that of polymer B is 100°C, then the T_g of the blend is
    (*a*) –10.5°C
    (*b*) 25.0°C
    (*c*) 37.5°C
    (*d*) 74.8°C
11. If the flow curves of two viscous fluids are represented by

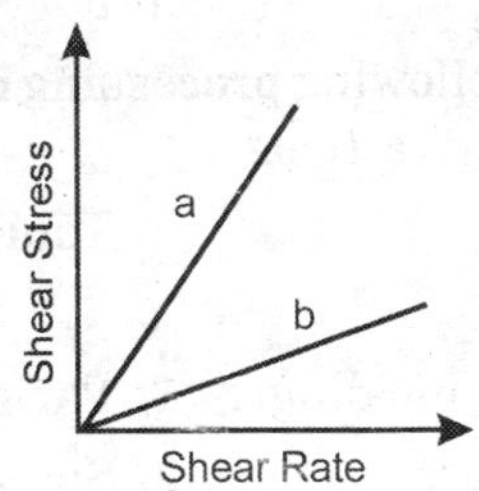

then, the viscosity plots of the fluids will be

(*a*)
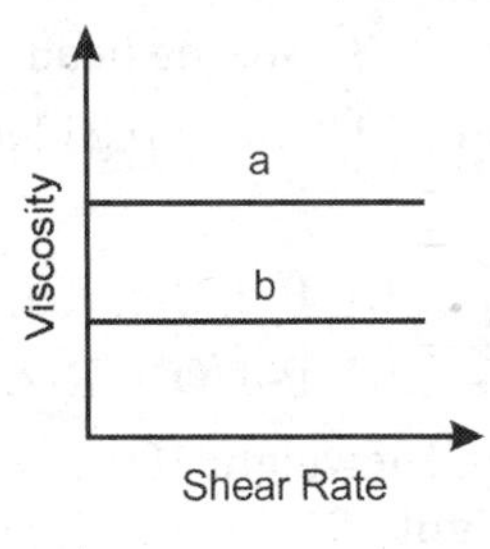

(*b*)
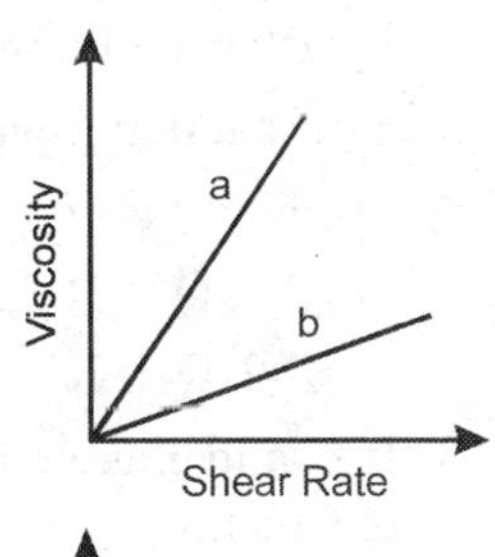

(*c*)
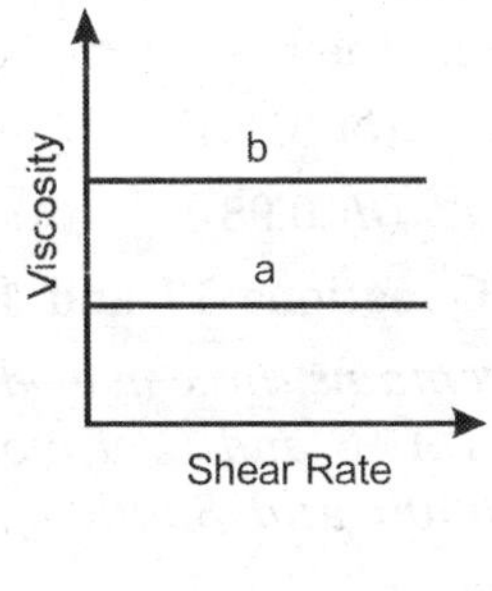

(*d*)
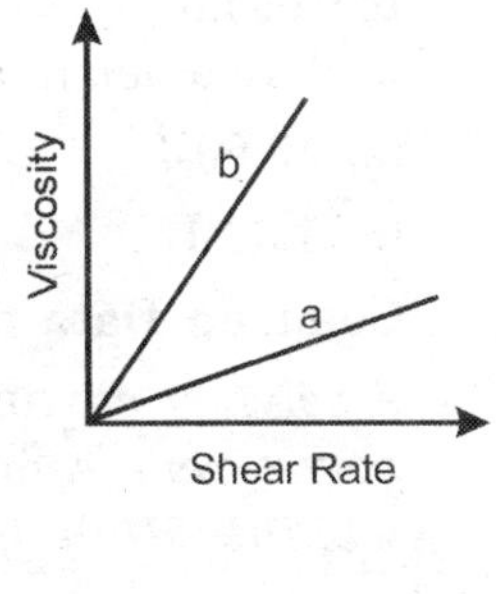

12. Match the following **compounding ingredients** in plastic technology with their respective **functions**:

Compounding ingredients	**Functions**
P. Tricresyl phosphate	1. Filler
Q. Calcium carbonate	2. UV stabilizer
R. Azodicarbonamide	3. Plasticizer
S. *o*-Hydroxybenzophenone	4. Blowing agent

(*a*) P-3, Q-1, R-4, S-2 (*b*) P-1, Q-4, R-3, S-2
(*c*) P-4, Q-1, R-2, S-3 (*d*) P-3, Q-4, R-1, S-2

13. In a three-point bending mode for flexural test of a polymer sample with the following data:

load applied in the mid-span = 80 kg
width of the specimen = 3 cm
depth of the specimen = 2 cm
length of the specimen = 30 cm

the flexural strength will be

(*a*) 17.5 MPa (*b*) 29.4 MPa
(*c*) 35.7 MPa (*d*) 41.3 MPa

14. How many stereo isomers are possible in total on polymerization of butadiene?

(*a*) 2 (*b*) 3
(*c*) 4 (*d*) 5

15. Match the following **processing operations** with their respective **tools:**

Processing operations	**Tools**
P. Injection molding	1. Parison mold
Q. Twin screw extrusion	2. Sprue-runner system
R. Blow molding	3. Mixing head
S. Reaction injection molding	4. Kneading blocks

(*a*) P-1, Q-2, R-3, S-4 (*b*) P-2, Q-4, R-1, S-3
(*c*) P-2, Q-1, R-4, S-3 (*d*) P-4, Q-1, R-2, S-3

16. If a solid elastomer ball of weight 100 g is allowed to free fall from a height of 10 m and it rebounds back to a height of 8 m, the hysteresis loss is

(*a*) 19.60 J (*b*) 9.80 J
(*c*) 1.96 J (*d*) 0.98 J

Common Data for Questions 17 and 18:

A polymer mixture contains three monodisperse polystyrene samples A, B and C of molecular weights 10000, 20000 and 30000 g mol^{-1}*, respectively.*

17. If the mixture contains equal number of molecules of A, B and C, the weight average molecular weight, *Mw* will be

(*a*) 1.93×10^4 g mol^{-1}
(*b*) 2.13×10^4 g mol^{-1}
(*c*) 2.33×10^4 g mol^{-1}
(*d*) 2.53×10^4 g mol^{-1}

18. If the mixture contains equal masses of A, B and C, then the number average molecular weight, *Mn* will be

(*a*) 1.64 * 10^4 g mol^{-1}
(*b*) 1.74 * 10^4 g mol^{-1}
(*c*) 1.84 *10^4 g mol^{-1}
(*d*) 1.94 * 10^4 g mol^{-1}

Common Data for Questions 19 and 20:

The stress relaxation equation for polymers is given by, $\sigma = \sigma_0 e^{-t/\tau}$, where σ is the stress at any time instant *t*, σ_0 is the initial stress and τ is the relaxation time. The relaxation process is temperature dependent and follows the Arrhenius law. For a rubber sample, the relaxation time is 60 days at 25°C.

19. If the above sample is stressed to 2 MPa initially, then the time required to relax the stress to 1 MPa will be

(*a*) 31.6 days (*b*) 41.6 days
(*c*) 51.6 days (*d*) 61.6 days

20. If the activation energy for relaxation is 30 KJ/mol, the relaxation time at 35°C will be

(*a*) 30.5 days (*b*) 35.5 days
(*c*) 40.5 days (*d*) 45.5 days

Statement for Linked Answer Questions 21 and 22:

In tensile testing of carbon fiber reinforced epoxy composite samples, the following data were recorded:

gauge length = 4 cm
cross-section = 0.8 cm × 0.3 cm
increase in gauge length at the break point = 0.03 cm
breaking load = 50 kg

21. Considering the stress-strain curve to be linear up to the break point, the tensile strength is

(*a*) 191.2 kg/cm^2 (*b*) 208.3 kg/cm^2
(*c*) 312.1 kg/cm^2 (*d*) 535.5 kg/cm^2

22. The Young's modulus of the composite is

(*a*) 1.77×10^4 kg/cm^2 (*b*) 1.99×10^4 kg/cm^2
(*c*) 2.34×10^4 kg/cm^2 (*d*) 2.77×10^4 kg/cm^2

ANSWERS

1. (*b*)	**2.** (*c*)	**3.** (*a*)	**4.** (*d*)	**5.** (*a*)	**6.** (*c*)	**7.** (*b*)	**8.** (*c*)	**9.** (*d*)	**10.** (*c*)
11. (*a*)	**12.** (*a*)	**13.** (*b*)	**14.** (*d*)	**15.** (*b*)	**16.** (*c*)	**17.** (*c*)	**18.** (*a*)	**19.** (*b*)	**20.** (*c*)
21. (*b*)	**22.** (*d*)								

EXPLANATIONS

1.

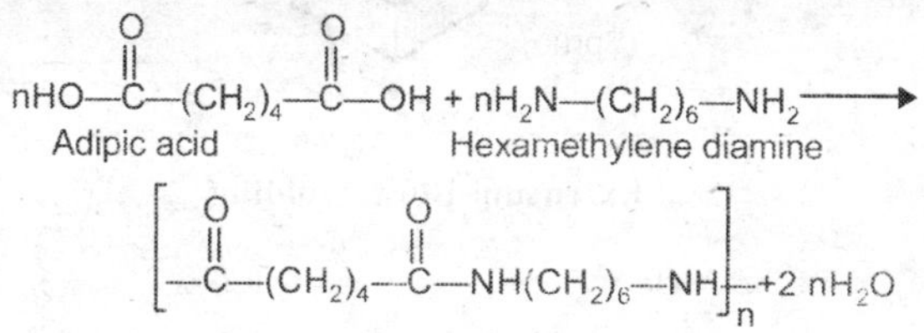

2. Monomers are often stabilized by the addition of inhibitors to prevent polymerization initiation by light, heat and air. For example, stable radical compounds which can trap free radicals are used to inhibit radical polymerization.

3. Zeigler-Natta complex catalyst systems derived from a transition metal compound from groups IVB to VIIIB of the periodic table and an organometallic compound usually from a group IA or IIIA metal. Aluminium alkyls(AlEt or $AlEt_2Cl$) acts as electron acceptor in combination with the electron donor titanium chloride ($TiCl_3$& $TiCl_4$) are used most commonly. The complex formed is insoluble in the solvent and is, hence, heterogeneous in nature.

4.

Technique	Measures	Range, g/mol
End Group	M_n	up to 2500
Osmometry	M_n	15000 – 750000
Ebulliometry	M_n	up to 100000
Light scattering	M_w	20000 to 10^7
Ultra centrifuge	M_w, M_z, MWD	2000 to 10^7
Solution viscosity	M_v, M_w	15000 – 10^6
Vapour-phase osmometry	M_n	up to 25000
Gel-permiation chromatography	M_n, M_w, M_v, M_z, MWD	up to 10^6

6. Bulk polymerization of pure liquid monomer is the simplest process and carried out by using initiator in the absence of diluent or solvent. For this process reaction rate is high due to high monomer concentration and result in high yield per volume of reactor. Another advantage is that the relatively pure product is produced. However, control of the bulk polymerization, exothermic in nature, is difficult. The viscosity of the reaction system increases rapidly even at relatively low conversion. The heat removal is difficult due to high viscosity and low thermal conductivity of the polymer melt. Consequently local hot spots may occur resulting in degradation and discoloration of the polymer product. Bulk polymerization requires careful temperature control and strong elaborate stirring equipment. Though, bulk polymerization is commercially less used, polymerization of ethylene, styrene and methyl methcrylate are carried out by this method. The heat dissipation and viscosity problem are reduced by carrying out polymerization at low conversion. Bulk polymerization can be carried out in conventional stirred tank reactor, long tubular reactor with high surface to volume ratio and screw extruder reactors.

8. Polyamide is more sensitive to moisture absorption due to presence of the functional group $-NH_2$.

10. Convert all the temperature in the Kelvin from centigrade. Total weight fraction is equal to 1.

$$\frac{1}{T_g} = \frac{W_1}{T_{g_1}} + \frac{W_2}{T_{g_2}}$$

$$\frac{1}{T_g} = \frac{0.3}{223} + \frac{0.7}{373}$$

$$T_g = 37.36°C$$

11. Shear stress α strain rate

$$\sigma = \eta \, d\varepsilon/dt$$

Viscosity is constant

From the question if strain rate is same for the both the situation and stress is different than viscosity is higher in the case where stress is higher.

12.

carbonamide	Blowing agent	Play a role in the formation of air bubbles by breaking down into gas at high temperature.
Calcium carbonate	Filler	Used as fillers in rubber industry
Tricresyl phosphate	plasticizer	Additives increase plasticity
o-Hydroxyben-zophenone	UV stabilizer	The primary function is to protect the substance from the long term UV degradation effects from ultraviolet radition.

13. Flexural strength $\sigma_f = \frac{3FL}{2bd^2}$

where, **F** is the force at midspan, **L** is the support span, **h** the specimen thickness, b the width and **s** is the selection of the tensile surface of the specimen at midpoint

Force = mg = 80*9.8

Flexural strength $\sigma_f = \frac{3 \times 80 \times 9.8 \times 0.3}{2 \times 0.03 \times 0.02^2}$

$= 29.4$ MPa

14. In stereochemistry, stereoisomers are isomeric molecules that have the same molecular formula and sequence of bonded atoms (constitution), but differ in the three-dimensional orientations of their atoms in space.

15.

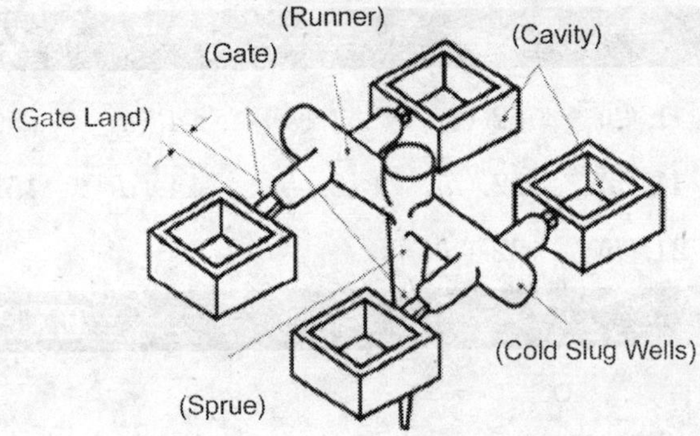

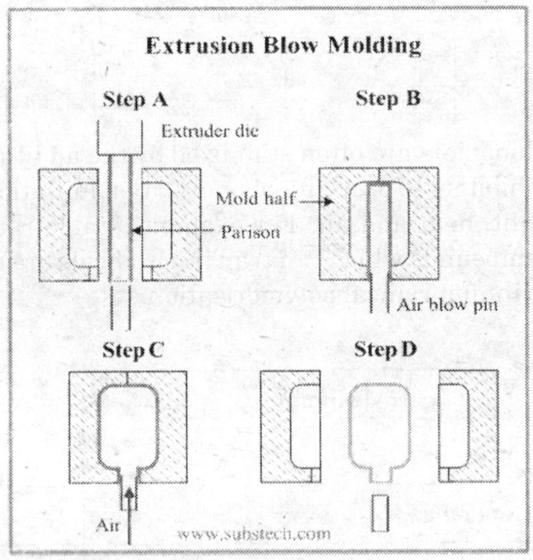

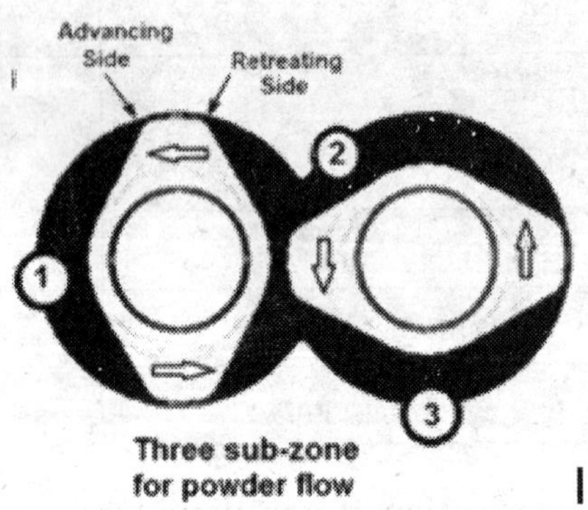

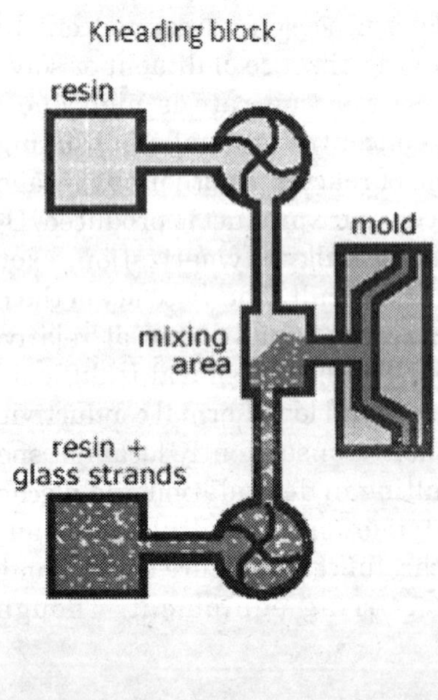

16. Hysteresis loss = $Mg(H_2-H_1)$, H_1= initial height, H_2 = height after rebound,

loss = 0.1* 9.8*(10-8) = 1.96 J

17. We know that weight of the sample is

$$W = \sum w_i = \sum N_i M_i$$

Mass in terms of percentage given, we use mass as w_i and divide w_i by the molecular weight and get the number of chains. Weight average molecular weight:

$$M_w = \frac{\sum N_i M_i^2}{\sum N_i M_i}$$

N_i = x for all samples given condition

$$M_w = \frac{X\times 10000^2 + X\times 20000^2 + X\times 30000^2}{X+X+X}$$

weight average molecular weight, M_w will be $= 2.33 \times 10^4$ g/mol

18. $$M_n = \frac{\Sigma N_i M_i}{\Sigma N_i M_i}$$

$$W = \sum w_i = \sum N_i M_i$$

w_i is equal for all sample given condition = x

$$M_n = \frac{X+X+X}{\dfrac{X}{10000}+\dfrac{X}{20000}+\dfrac{X}{30000}}$$

The number average molecular weight, *Mn* will be $= 1.64 \times 10^4$ g/mol

19. $$\sigma = \sigma_o e^{(-t/60)}$$

Where τ **(relaxation time)**=η/E, σ stress at time t and σ_o initial stress

τ = 60 days σ = 1 MPa and σ_o = 2MPa

$$1 = 2e^{(-t/60)}$$

$$t = 41.8 \text{ days.}$$

20. $\ln \frac{\tau_2}{\tau_1} = \frac{E_a}{R}\left(\frac{1}{T_1}-\frac{1}{T_2}\right)$, T_1 = 273 + 25 = 298, T_2 = 273 + 35 = 308, τ_1 = 60 days

$\ln \frac{\tau_2}{90} = \frac{30000}{8.317}\left(\frac{1}{298}-\frac{1}{318}\right)$, τ_2 = 40.5 days

21. Tensile strength = force/ Area = 50/0.8*0.3 = 208.3 kg/cm²

22. Young modulus = tensile stress/ strain = 208.3/ (0.03/4) = 2.77×10^4 kg/cm²

Strain = increase in length / original length

2013

1. In free radical polymerization, one of the following techniques permits simultaneous increase in rate of polymerization and polymer molecular weight.
   (a) Solution polymerization.
   (b) Suspension polymerization.
   (c) Bulk polymerization.
   (d) Emulsion polymerization.
2. The shear modulus, G, of plastic is related to the elastic modulus, E, and the Poisson ratio, as
   (a) E = 2(1 –) G (b) G = 2(1 +) E
   (c) E = 2(1 +) G (d) E = (1 +) G
3. LLDPE is obtained by
   (a) Ziegler-Natta polymerization of ethylene.
   (b) free-Radical polymerization of ethylene.
   (c) free-Radical polymerization of ethylene and alpha-olefins.
   (d) Ziegler-Natta copolymerization of ethylene and alpha-olefins.
4. A hindered phenol is added to a polyolefin
   (a) to increase ozone resistance
   (b) to increase foamability
   (c) to increase oxidation resistance
   (d) to increase crosslinkability
5. Stretching of rubber leads to
   (a) decrease in alignment of polymer chains
   (b) increase in alignment of polymer chains
   (c) no change in alignment of polymer chains
   (d) decrease in strength of rubber
6. In a cone and plate viscometer, the rate of strain is related to the speed of rotation of the cone, ? (radian/second), and the angle between the cone and the plate, ? (radian), by the following relation
   (a) $\omega\alpha$ (b) $\omega \cos\omega\alpha$
   (c) α/ω (d) ω/α
7. The tensile breaking strength of polycarbonate (I), low density polyethylene (II), polystyrene (III) and polypropylene (IV) can be arranged as
   (a) IV > II > I > III (b) I > II > IV > III
   (c) I > III > IV > II (d) III > I > II > IV
8. High molecular weight polymers could be obtained even at low monomer conversion in case of
   (a) Step growth polymerization
   (b) Living polymerization
   (c) Chain growth polymerization
   (d) Solid state polymerization
9. A reinforced polymer composite is made by the incorporation of
   (a) elastomers into the polymer.
   (b) fibers into the polymer.
   (c) plasticizers into the polymer.
   (d) gaseous additives into the polymer.
10. Match the following for free-radical copolymerization of two monomers with reactivity ratios, r1 and r2.

Reactivity Ratios	Copolymer Nature
P. r1 = r2 = 0 Q. r1 = r2 = 1 R. r1 >1, r2 > 1 S. 0 < r1r2 < 1	1. Random copolymer 2. Alternate copolymer 3. Block copolymer 4. Random-Block copolymer

   (a) P-2; Q-1; R-3; S-4 (b) P-3; Q-1; R-2; S-4
   (c) P-2; Q-4; R-3; S-1 (d) P-2; Q-3; R-1; S-4
11. The relative viscosity of a 1% solution (weight/volume) of a given polymer was found to be 1.1. The inherent viscosity of this polymer will be
   (a) 0.065 dl/g (b) 0.075 dl/g
   (c) 0.085 dl/g (d) 0.095 dl/g
12. Match the following in case of step-growth polymerization, where A reacts only with B, and reacts only with A (Note: A – A is expressed as A_2, and A

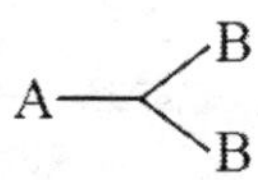
is expressed as AB_2).

Monomers	Polymer
P. $A_2 + AB_3$	1. Hyperbranched Polymer
Q. AB_2	2. Crosslinked Polymer
R. $AB + B_3$	3. Star Polymer
S. $A_2 + B_2$	4. Linear Polymer

   (a) P–2: Q–3; R–1; S–4
   (b) P–2: Q–1; R–3; S–4
   (c) P–1: Q–2; R–3; S–4
   (d) P–2: Q–4; R–1; S–3

13. Match each of the following additives for plastics with its function

Additive	Function
P. α-Cellulose	1. Flame retarder
Q. Zinc chromate	2. Plasticizer extender
R. Alumina trihydrate	3. Organic fibrous filler
S. Chlorinated paraffin wax	4. Colorant

(a) P-1; Q-2; R-3; S-4 (b) P-2; Q-3; R-4; S-1
(c) P-3; Q-4; R-1; S-2 (d) P-4; Q-1; R-2; S-3

14. The length of a glass fiber reinforced polymer increased by 0.03mm, from its initial length of 100mm, when the temperature was changed from -30°C to +30°C. The coefficient of linear thermal expansion is

(a) 1.03×10^{-5} °C^{-1} (b) 9.82×10^{-6} °C^{-1}
(c) 5.00×10^{-6} °C^{-1} (d) 14.4×10^{-5} °C^{-1}

15. A 40mm × 40mm square polymer composite sample with 5mm thickness (heat transfer distance) exhibited a heat flow rate of 60W, when the temperatures of the warm and cold surfaces were 90 °C and 25 °C respectively. The thermal conductivity of the sample in W.m^{-1}.K^{-1}

(a) 5.67 (b) 15.3
(c) 2.88 (d) 0.667

16. An extruder is supplied with 40 kW of power. The mass flow rate of a polymer through the extruder is 240 kg h^{-1} and the specific heat capacity of the polymer is 4 kJ kg^{-1} K^{-1}. The maximum possible temperature rise in the polymer is

(a) 150 K (b) 100 K
(c) 600 K (d) Zero

Common Data Questions

Common Data for Questions 17 and 18:

For a given free-radical polymerization, the only mode of termination is the bimolecular termination and there is no chain transfer. The final polymer produced was analyzed to contain an average of 1.60 initiator fragments per polymer chain.

17. Percentage of final polymer chains containing one initiator fragment per chain is

(a) 40 % (b) 50 %
(c) 60 % (d) 70 %

18. Percentage of polymer radicals terminated by coupling is

(a) 65 % (b) 75 %
(c) 85 % (d) 95 %

Common Data for Questions 19 and 20:

For the synthesis of polyester, 1.5 mole of pentaerythritol (tetra-ol) was reacted with 1.0 mole of a tricarboxylic acid.

19. The extent of reaction when the number average degree of polymerization of the reaction mixture approaches infinity is

(a) 80.33% (b) 83.33%
(c) 84.33% (d) 86.33%

20. The number average degree of polymerization of the reaction mixture when the polymerization was stopped at 80% conversion, is

(a) 1000 (b) 100
(c) 50 (d) 25

Linked Answer Questions

Statement for Linked Answer Questions 21 and 22:

A viscoelastic fluid is modeled as a spring and two dashpots, all connected in series. The spring has elastic modulus G and the fluids in two dashpots have viscosities η_1 and η_2.

21. The constitutive equation (relation between stress α and γ in which overdot represents the time derivative) for the fluid is:

(a) $\sigma = G\gamma + (\eta_1 + \eta_2)\gamma$ (b) $\sigma = G\gamma + (\eta_1 - \eta_2)\gamma$

(c) $\gamma = \frac{\sigma}{G} + (\frac{1}{\eta_1} + \frac{1}{\eta_2})\sigma$ (d) $\gamma = \frac{\sigma}{G} + \frac{\sigma}{\eta_1 + \eta_2}$

22. For a periodic stress $_0e^{i\ t}$ the strain is given by

(a) $\gamma = \sigma_0[\frac{1}{G} + \frac{\iota}{\omega}(\frac{1}{\eta_1} + \frac{1}{\eta_2})]e^{i\omega t}$

(b) $\gamma = \sigma_0[\frac{1}{G} - \frac{\iota}{\omega}(\frac{1}{\eta_1} + \frac{1}{\eta_2})]e^{i\omega t}$

(c) $\gamma = [\sigma_0 + (\eta_1 + \eta_2)\omega i]\frac{e^{i\omega t}}{G}$

(d) $\gamma = [\sigma_0 - (\eta_1 + \eta_2)\omega i]\frac{e^{i\omega t}}{G}$

ANSWERS

1. (*d*)	2. (*c*)	3. (*d*)	4. (*c*)	5. (*b*)	6. (*d*)	7. (*c*)	8. (*c*)	9. (*b*)	10. (*a*)
11. (*d*)	12. ()	13. (*c*)	14. (*c*)	15. (*c*)	16. (*a*)	17. (*a*)	18. (b)	19. (*b*)	20. (*d*)
21. (*c*)	22. (*b*)								

EXPLANATIONS

1. Emulsion polymerization

2. E = 2(1 +) G

3.

(A) Ziegler-Natta polymerization of ethylene.	HDPE
(B) free-Radical polymerization of ethylene.	LDPE
(C) free-Radical polymerization of ethylene and alpha-olefins.	------
(D) Ziegler-Natta copolymerization of ethylene and alpha-olefins.	LLDPE

4. Phenols (ArOH) are known to reduce the rates of oxidation of organic matter by transferring a H atom (from their OH groups) to the chain-carrying ROO* radicals, a mechanism that most likely involves a concerted transfer of the hydrogen as a proton and of one electron between the two oxygen atoms, O-H---O* (proton-coupled electron transfer mechanism). The antioxidant capabilities of phenols are strongly reduced by hydrogen-bond accepting solvents since the hydrogen-bonded molecules ArOH---S are virtually unreactive toward ROO* radicals.

5. Rubber is composed of long hydrocarbon chains that are interlinked to one another at very distant intervals along the polymer backbone. These chains tend to form random coils -- you might think of them as very long chains of spaghetti. When you pull on the rubber band, the chains tend to uncoil and partially align with one another. However, because the chains are linked together by these bridging bonds, there is a limit to how much the coils can be stretched out. When that point is reached it gets harder to stretch the rubber band any further.

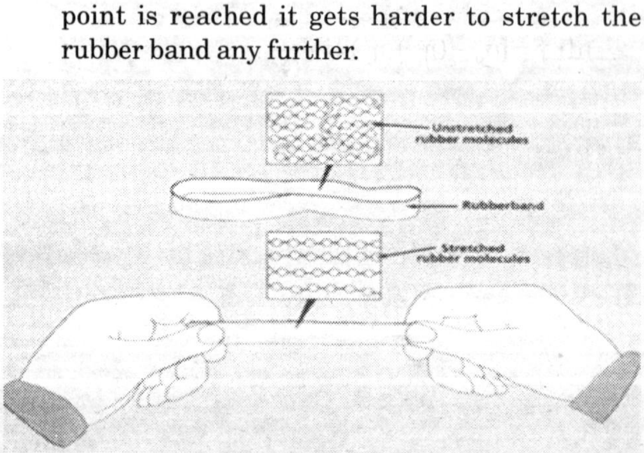

6. Rate of strian G(1/sec) =

Speed of rotation (radian/sec)/the angle between the cone and the plate (radian)

G = ω/α

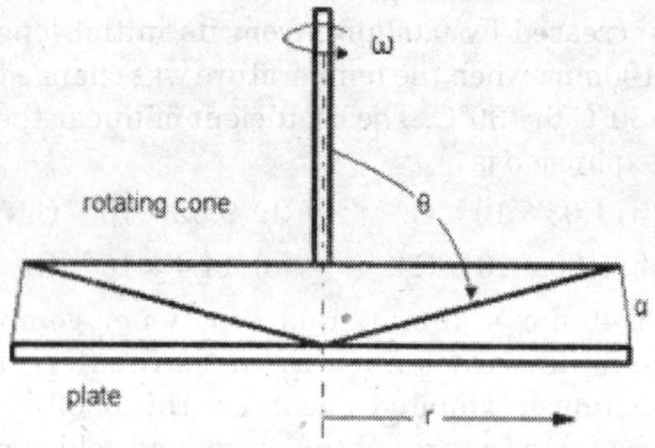

7. Crystallinity makes a material strong, but it also makes it brittle. A completely crystalline polymer would be too brittle to be used as plastic. The amorphous regions give polymer toughness, that is, the ability to bend without breaking. Higher the crystallinity in the polymer region required more force to break it. In other words, Lower the glass transition temp lower the tensile strength.

Polymer	Glass Transition temp
PC	147?C
PS	100?C
PE	-125?C
PP	5-15?C

8. Chain-growth polymerization or chain polymerization is a polymerization technique where unsaturated monomer molecules add onto the active site of a growing polymer chain one at a time. Growth of the polymer occurs only at one (or possibly more) ends. Addition of each monomer unit regenerates the active site. This type of polymerization results in high molecular weight polymer being formed at low conversion. This final weight is determined by the rate of propagation compared to the rate of individual chain termination, which includes both chain transfer and chain termination steps. Above a certain ceiling temperature, no polymerization occurs.

9. Polymer Matrix Composites combine a resin system and reinforcing fibres, the properties of the resulting composite material will combine something of the properties of the resin on its own with that of the fibres on their own.

10. Effect of reactivity ratios on copolymer composition

Case 1: $r_1 = r_2 = 1$ (Random Copolymer)

a. No preference for monomer 1 or 2

b. Truly random copolymer

c. Example: ethylene: vinyl acetate $r_1 = 0.97$: $r_2 = 1.02$

Case II: $r_1 = r_2 = 0$

a. Monomer 1 reacts exclusively with monomer 2 and vice versa

b. Alternating copolymer

c. Example: styrene: maleic anhydride $r_1 = 0.041$: $r_2 = 0.01$

Case III: $r_1, r_2 < 1$

a. Copolymer has a tendency toward a pattern (like alternation)

$r_1 = 0.5 \Rightarrow M_1$ reacts twice as fast with M_2 as with M_1.

b. Composition of polymer is close to "azeotropic" concentration

i. Azeotrope - composition of copolymer is equal to concentration of feed

$$\frac{d[M_1]}{d[M_2]} \quad \frac{[M_1]}{[M_2]}$$

ii. Compare to azeotrope in liquid-vapour phase diagram where composition of vapour is equal to composition of liquid.

c. Example: styrene: methyl methacrylate $r_1 = 0.52$: $r_2 = 0.46$

Case IV: $r_1 > 1, r_2 < 1$

a. Both M_1 and M_2 prefer to react with M_1

b. Composition of copolymer changes significantly during copolymerization (heavy in M_1 initially, heavy in M_2 at the end).

c. Example: styrene: vinyl chloride $r_1 = 17$: $r_2 = 0.02$.

Case V: $r_1.r_2 = 1$

a. An "ideal polymerization"

b. Relative rates of monomer consumption independent of identity of radical

c. Result is a random copolymer

d. Example: methyl methacrylate: vinyl chloride $r_1 = 10$: $r_2 = 0.1$

Case VI: $r_1 > 1, r_2 > 1$

a. It forms either a mixture of the two homopolymers or, under certain favourable condition, block copolymers having long sequences of M_1 followed by a long sequence of M_2.

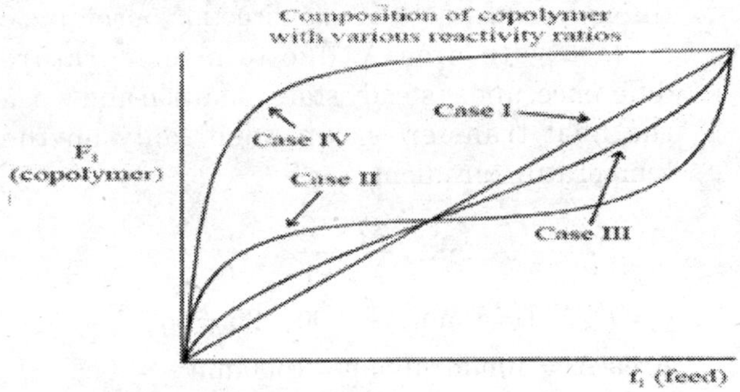

11. Concentration: 0.01 g/ml

Inherent viscosity: Ln(r / C)

where, r is relative viscosity & C is the concentration

Inherent Viscosity = ln(1.1/(0.01*100) = 0.095dl/g

12. [Question has some technical error.][Marks to all]

13.

Additive	Function
P. α-Cellulose	3. Organic fibrous filler
Q. Zinc chromate	4. Colorant
R. Alumina trihydrate	1. Flame retarder
S. Chlorinated paraffin wax	2. Plasticizer extender

Wax is used as plasticizer in the rubber or plastic industries to reduce the viscosity and friction.

14. Linear expansion

$$\frac{L}{L_0} \quad T$$

$\Delta L = 0.03$mm, L =100mm $\Delta T = [30 - (-30)]$

= 60 °C

$$\frac{L}{L_0 \quad T} \quad \frac{0.03}{100 \quad 60}$$

$= 5.0 \times 10^{-6}$/°C

15.

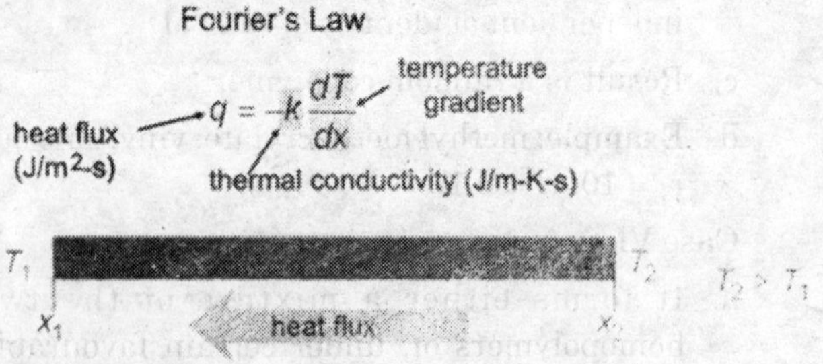

Thermal conductivity can be calculated from a number of measured quantities, but normally is defined as: the quantity of Heat Q, transmitted through a thickness L, in a direction normal to a surface with area A, due to a temperature difference, under steady state condition and when the heat transfer is dependent only on the temperature gradient.

$$k \quad \frac{Q \times L}{A \times \Delta T}$$

$Q = 60W$, $L = 5mm$, $\Delta T = 90 - 25 = 65\ °C$,

$Area(A) = 40mm*40mm = 1600mm^2$

$k = 60 \times 5*10^{-3} / (1600*10^{-6} \times 65)$

$k = 2.88$

16. Power = energy/time Energy = $mc_p\ \Delta T$

$P = mc_p\ \Delta T/T$

$\Delta T = P \times T/(m\ C_)$, $m/T = 240\ kg/h$

$\Delta T = 40*10^3 \times 3600/(240*4*103)$

$\Delta T = 150K$

19. $\overline{X_n} \quad \frac{1+r}{1+r-2rp_A}$

$r = N_A/N_B = 1.5/1 = 1.5$

When the number average degree of polymerization of the reaction mixture approaches infinity it means [1 + r – 2rp] goes to zero.

$1 + 1.2 - 2*1.5*P = 0$

$P = 0.8333$

$P = 83.33\%$

20. $\overline{X_n} \quad \frac{1+r}{1+r-2rp_A}$

r =1.5 & p=0.8

$X_n = [1+1.5]/[1+1.5-2*1.5*0.8]$

$X_n = 25$

21. In the series stress is constant.

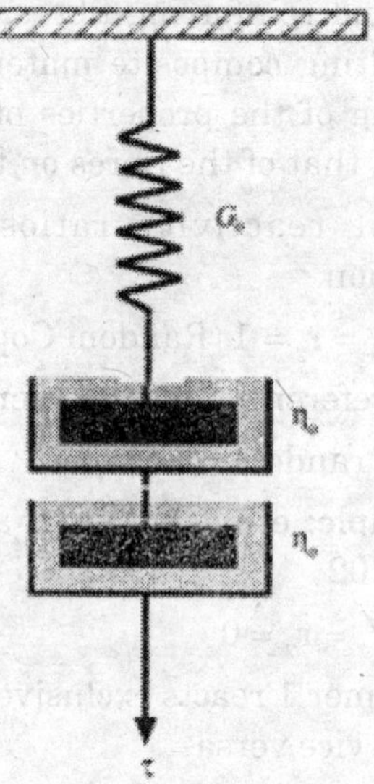

For spring strain = $\varepsilon_1 = \frac{\sigma}{G}$

Total strain = $\varepsilon_1 + \varepsilon_2 + \varepsilon_3$

For dashpot 1 & 2 strain rate $\frac{d\varepsilon}{dt} = \frac{\sigma}{\eta}$

Take differentiation both sides:

$$\frac{d\varepsilon_t}{dt} = \frac{d\varepsilon_1}{dt} + \frac{d\varepsilon_2}{dt} + \frac{d\varepsilon_3}{dt}$$

$$= \frac{d\sigma}{dt} + \frac{\sigma_1}{\eta_1} + \frac{\sigma_2}{}$$

$$= \frac{d\sigma}{dt} + \sigma\left(\frac{1}{\eta_1} + \frac{1}{\eta_2}\right)$$

22. Total strain = $\varepsilon_1 + \varepsilon_2 + \varepsilon_3$

$\varepsilon_1 = \frac{\sigma_0 e^{i\omega t}}{G}$ (for spring),

For Dashpots

$\sigma = \frac{d\varepsilon}{dt}$, $\sigma_0 e^{i\omega t} = \frac{d\varepsilon}{dt}$

take both sides integrations

$\varepsilon = \frac{\sigma_0 \varepsilon^{i\omega t}}{\eta . i\omega}$ put all the values in the total strain

$$\varepsilon = \sigma_0 [\frac{1}{G} - \frac{\iota}{\omega} \frac{1}{\eta_1} + \frac{1}{\eta_2}]$$

2014

1. The estimation of the molecular weight of a polymer by gel permeation chromatography (GPC) is based on its
   (*a*) polarity
   (*b*) size
   (*c*) adsorption to stationary phase
   (*d*) crystallinity
2. Elastomers are characterized by
   (*a*) high modulus and high elongation at break
   (*b*) high modulus and low elongation at break
   (*c*) low modulus and high elongation at break
   (*d*) low modulus and low elongation at break
3. Thermodynamically, two polymers with enthalpy of mixing (ΔH) and entropy of mixing (ΔS) form a miscible blend at temperature T when
   (*a*) $\Delta H/\Delta S = 0.5T$ (*b*) $\Delta H/\Delta S = T$
   (*c*) $\Delta H/\Delta S = 1.5T$ (*d*) $\Delta H/\Delta S = 2T$
4. The tensile strain of a uniformly extending plastic specimen of initial length L_0 and extended length L is
   (*a*) L_0/L (*b*) L / L_0
   (*c*) $L_0/ (L - L_0)$ (*d*) $(L - L_0) / L_0$
5. In natural rubber compounding, a peptizer is added
   (*a*) at the beginning of the compounding cycle
   (*b*) after the addition of filler
   (*c*) at the end of the compounding cycle
   (*d*) after the addition of antioxidant
6. A continuous annular product is produced by
   (*a*) compression molding (*b*) extrusion
   (*c*) blow molding (*d*) injection molding
7. Relate the three varieties of polyethylene in the left column with their chain structures given in the right column.

   P. HDPE 1. long as well as short branches
   Q. LDPE 2. only short branches
   R. LLDPE 3. no branches

   (*a*) P-1, Q-3, R-2 (*b*) P-3, Q-2, R-1
   (*c*) P-2, Q-3, R-1 (*d*) P-3, Q-1, R-2
8. Match the following changes observed in the calorimetric analysis of a polymer sample when heat flow (y-axis) is plotted against temperature (x-axis):

   P. step increase in heat flow 1. crystallization
   Q. exothermic peak 2. melting
   R. endothermic peak 3. glass transition

   (*a*) P-1, Q-2, R-3
   (*b*) P-2, Q-1, R-3
   (*c*) P-3, Q-1, R-2
   (*d*) P-3, Q-2, R-1
9. A Bingham plastic fluid is flowing under gravity, down a vertical plate, as a film. Find the appropriate match for the fully developed velocity profile of the fluid in the film, from among those shown below.

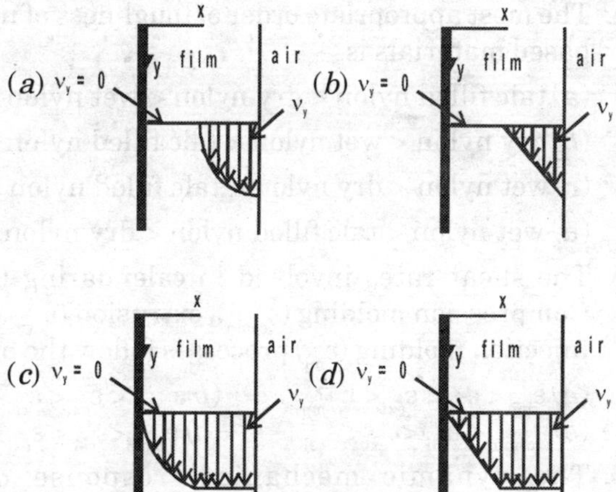

10. Calculate the mass percent of the crystalline phase in a polymer sample of density 975 kg/m^3. The density of amorphous phase is 866 kg/m^3 and that of the crystalline phase is 996 kg/m^3.
11. Find the rate of initiation (molL^{-1}s^{-1}) of a polymerization reaction using a peroxide initiator with a half life of 0.1 s and efficiency of 70%, if the concentration of the initiator is 0.05 molL^{-1}.
12. The constitutive equation of a shear thinning polymeric fluid is given by

$$\sigma = \frac{\mu_0 \gamma}{1 + (\gamma / \gamma_0)}$$

   Where σ represents shear stress (Pa), and γ, the corresponding shear rate (s^{-1}). The quantities μ_0 = 20 Pas and γ_0 = 10 s^{-1} are constants. Find the apparent viscosity of the sample (Pas) when the applied shear rate is 40 s^{-1}.
13. Identify the monomer for the polymer shown below prepared by ring opening metathesis polymerization.

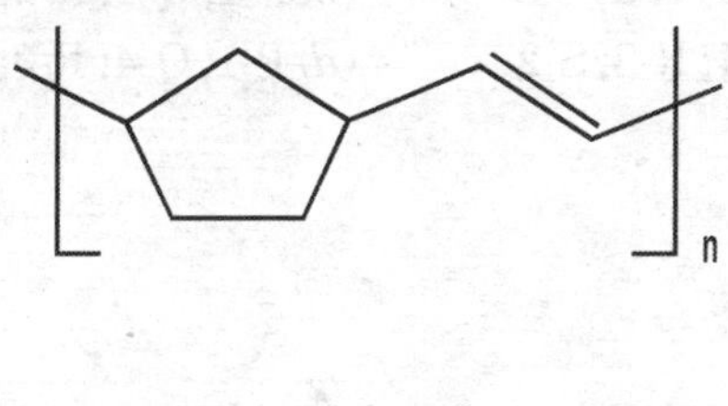

(a) 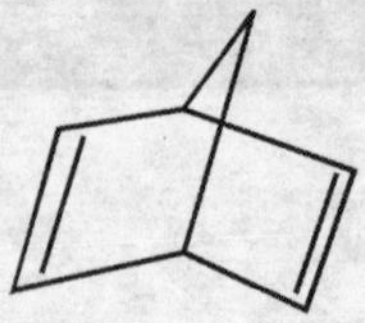 (b) 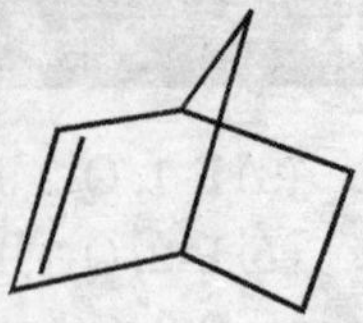

(c) 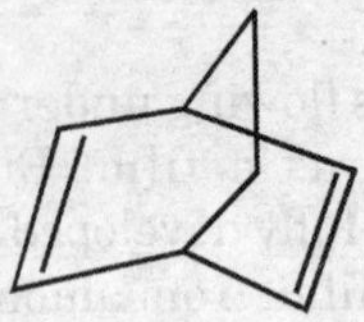 (d) 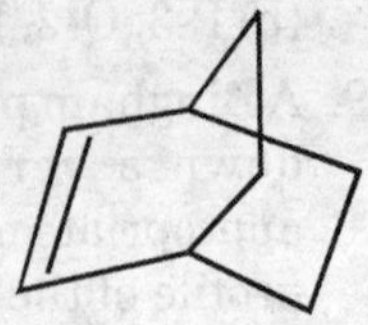

14. The most appropriate order of toughness of nylon based materials is

(a) talc filled nylon < dry nylon < wet nylon

(b) dry nylon < wet nylon < talc filled nylon

(c) wet nylon < dry nylon < talc filled nylon

(d) wet nylon < talc filled nylon < dry nylon

15. The shear rates involved in calendaring (ε_{cal}), Compression molding (ε_{comp}), extrusion (ε_{ext}), and injection molding (ε_{inj}) processes follow the order

(a) $\varepsilon_{inj} < \varepsilon_{cal} < \varepsilon_{ext} < \varepsilon_{comp}$ (b) $\varepsilon_{comp} < \varepsilon_{ext} < \varepsilon_{cal} < \varepsilon_{inj}$

(c) $\varepsilon_{comp} < \varepsilon_{inj} < \varepsilon_{ext} < \varepsilon_{cal}$ (d) $\varepsilon_{comp} < \varepsilon_{cal} < \varepsilon_{ext} < \varepsilon_{inj}$

16. The dynamic mechanical response of a thermoplastic automotive component has shown a loss angle of 45° and storage modulus of 3500 MPa. Calculate the loss modulus (MPa) of the component.

17. Match the terms in Column A with the appropriate terms in Column B:

Column A	Column B
P. processability	1. Rockwell scale
Q. moisture permeation	2.rubber modification
R. hardness	3. melt flow index
S. fracture toughening	4. Fick's law

(a) P-3; Q-4; R-2; S-1 (b) P-3; Q-4; R-1; S-2

(c) P-4; Q-3; R-1; S-2 (d) P-4; Q-3; R-2; S-1

18. Match the following additives for plastics with their respective functions:

Column A	Column B
P. dilaurylthiodipropionate	1. solid layer lubricant
Q. graphite	2. flame retardant
R. antimony trioxide	3. reinforcement
S. carbon fibre	4. antioxidant

(a) P-4; Q-1; R-2; S-3 (b) P-1; Q-4; R-2; S-3

(c) P-4; Q-1; R-3; S-2 (d) P-1; Q-4; R-3; S-2

19. Match the following catalyst/initiator with the type of polymerization reaction:

Catalyst/initiator	Polymerization reaction
P. butyl lithium	1. Ziegler-Natta
Q. $TiCl_4 + Et_3Al$	2. cationic
R. CuBr	3. anionic
S. H_2SO_4	4. atom transfer radical polymerization

(a) P-2; Q-1; R-4; S-3 (b) P-2; Q-4; R-1; S-3

(c) P-3; Q-1; R-4; S-2 (d) P-3; Q-4; R-1; S-2;

20. For AIBN (mol. wt. = 164 gmol-1) initiated free radical polymerization of methyl methacrylate (mol. wt. = 100 gmol-1), where the termination is only by radical coupling, the M_n of PMMA is found to be 4636 gmol-1. Calculate the degree of polymerization.

21. Polymer solution is made by dissolving 5 g of polymer in 1000 ml of solvent. The flow time of the solvent and that of the polymer solution between two appropriate marks in a viscometer are 40 s and 60 s, respectively. The reduced viscosity (in dLg-1) of the polymer solution is:

(a) 0.25 (b) 0.50

(c) 1.0 (d) 1.5

22. The volume resistivity of a polymeric material is 107 Ωm. Find the resistance (in MΩ) of a cube of the material of side 1 cm. The direction of current flow is as shown in the figure below.

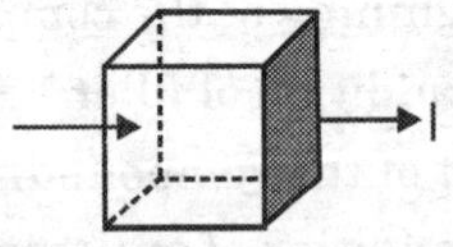

ANSWERS

1. (*b*) 2. (*c*) 3. (*a*) 4. (*d*) 5. (*a*) 6. (*b*) 7. (*d*) 8. (*c*) 9. (*c*) 10. (85.15 to 86.15)
11. (0.48 to 0.50) 12. (4) 13. (*b*) 14. (*a*) 15. (*d*) 16. (3500 Mpa) 17. (*b*) 18. (*a*) 19. (*c*) 20. (45)
21. (*c*) 22. (1000)

EXPLANATIONS

1. Gel permeation chromatography (GPC) is a type of size exclusion chromatography (SEC) that separates analytes on the basis of size. The technique is often used for the analysis of polymers.

2. **Elastomer**, any rubbery material composed of long chainlike molecules, or polymers, that are capable of recovering their original shape after being stretched to great extents—hence the name *elastomer*, from "elastic polymer." Under normal conditions the long molecules making up an elastomeric material are irregularly coiled. With the application of force, however, the molecules straighten out in the direction in which they are being pulled. Upon release, the molecules spontaneously return to their normal compact, random arrangement. A small force on the elastomer gives the 400-500% strain to it.

3. Generally, polymer blends can be completely miscible, partially miscible or immiscible, depending on the value of ΔGm. The free energy of mixing is given by

$\Delta G_m = \Delta H_m - T.\ \Delta S_m$

For miscibility (binary blend), the following two conditions must be satisfied: the first condition ΔGm < 0; and the second condition

$$\left(\frac{\partial(\Delta Gm)}{\partial\phi_i^2}\right)_{T.p} > 0$$

$\Delta Gm < 0$

$\Delta H_m - T.\ \Delta S_m < 0$

$\Delta H/\Delta S < T$

Now we have to wisely choose the condition amongst the options. Option a is only satisfied the condition of miscibility.

4. Elongation: final length –initial length

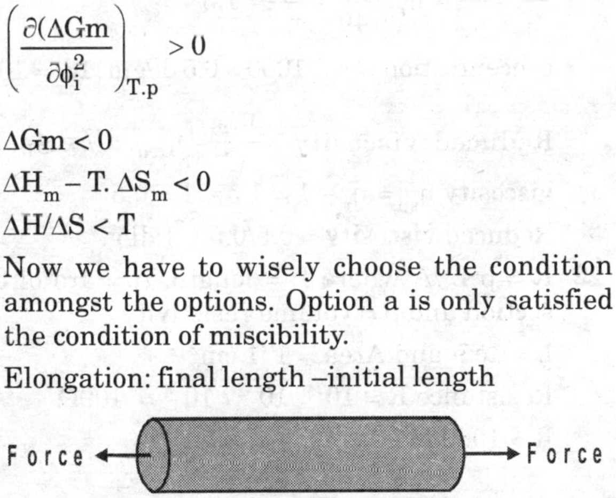

$$\text{Strain} = \frac{\text{Elongation}}{\text{Orininal Length}} = \frac{\Delta L}{L_0}$$

5. Peptizer: Mastication and peptization are processing stages, on the open mill or in the internal mixer, in which the viscosity of the rubber is reduced to a level facilitating further processing or even making processing feasible. Natural molecular weight is very high; Processing of the NR rubber is very difficult due to high molecular weight. Peptizer add at the beginning of the compounding cycle to lower the molecular weight and viscosity and rubber become easy processable.

6. Compression molding, blow molding and injection molding are the batch process. Extrusion is the continuous process. Continuous annular product(Pipes) is produced by only extrusion process.

7. Density α1/branches

Density order: HDPE > LDPE

LDPE (Low Density Polyethylene) has the most excessive branching. This causes the low density to have a less compact molecular structure which is what makes it less dense. It has a density of 0.910 – 0.925 g/cm^3.

HDPE (High Density Polyethylene) has minimal branching of its' polymer chains. Because it is denser it is more rigid and less permeable then the LDPE. It has a density of 0.941 – 0.965 g/cm^3.

LLDPE (Linear Low Density Polyethylene) has a significant number of short branches. Because it has shorter and more branches its chains are able slide against each other upon elongation without becoming entangled like LPDE which has long branching chains that would get caught on each other. This gives LLDPE higher tensile strength and higher impact and puncture resistance than the LDPE. It has a density of 0.91 – 0.94 g/cm^3.

8. 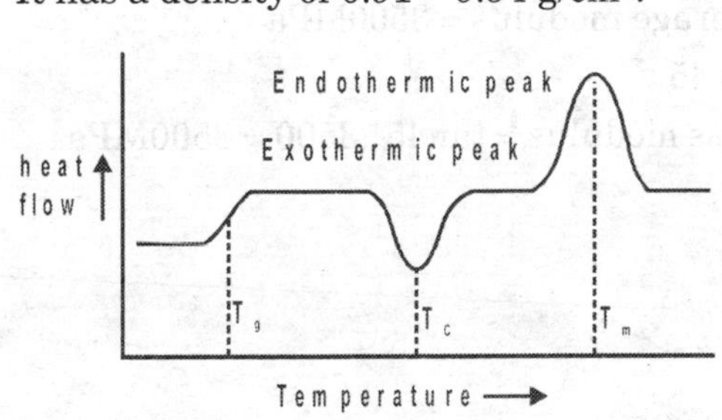

The entire DSC plot, right before your very eyes!

10. Mass fraction of the crystallinity

$$= X_c = \frac{\frac{1}{d_a} - \frac{1}{d}}{\frac{1}{d_a} - \frac{1}{d_c}} = \frac{\frac{1}{866} - \frac{1}{975}}{\frac{1}{866} - \frac{1}{996}} = 0.8565$$

X_c = 85.65% (da= density of amorphous part, dc = density of crystallinity part, d = density of the polymer)

11. The rate expression for initiation may be written as:

$$Ri = \frac{d[M]}{dt} = 2\,fkd[1]$$

Where [I] represents the initiator concentration and f is the efficiency of the initiator.

Free radical polymerization is the first order reaction;

$$K_d = \frac{0.693}{half\ life\frac{t_1}{2}} = \frac{0.963}{0.1} = 6.93$$

$$Ri = 2 * 0.7 * 0.05 * 6.93 = 0.4851\ \frac{mol}{L.s}$$

12. μ_0 = 20 Pas, γ_0 = 10s $^{-1}$ & γ = 40 s^{-1} Put all the values in the above equation

$$\sigma = \frac{40 \times 20}{1 + \frac{40}{10}} = 160$$

$\sigma = \eta\gamma$ where η = viscosity

η = 160/40 = 4Pas

14. Toughness is really a measure of the energy a sample can absorb before it breaks.

15.

Processing	**Shear rate (1/s)**
Compression molding	1 to 10
Calendering	10 to 100
Extrusion	100 to 1000
Injection molding	10^3 to 10^4
Fiber spining	> 10^4

16. $$\tan\delta = \frac{\text{loss modulus}}{\text{storage modulus}}$$

Loss modulus = ?

Storage modulus = 3500MPa

$\delta = 45$

Loss modulus = tan45* 3500 = 3500MPa

17.

Column A	Column B
P. processability	3. melt flow index
Q. moisture permeation	4. Fick's law
R. hardness	1. Rockwell scale
S. fracture toughening	2.rubber modification

18.

Additive	Function
P.dilaurylthiodipropionate	4. antioxidant
Q. graphite	1. solid layer lubricant
R. antimony trioxide	2. flame retardant
S. carbon fibre	3. reinforcement

19. 1. Ziegler-Natta catalyst: These are complex catalyst systems derived from a transition metal compound from groups IVB to VIIIB of the periodic table and an organometallic compound usually from a group IA or IIIA metal. Aluminium alkyls (AlEt or $AlEt_2Cl$) acts as electron acceptor in combination with the electron donor titanium chloride ($TiCl_3$ & $TiCl_4$) are used most commonly.

2. Cationic catalyst: Strong Lewis acid like boron trifluoride (BF_3) or Friedel–Crafts catalysts such as $AlCl_3$, H_2SO_4, $SnCl_4$ and $TiCl_4$

3. Anionic catalyst: Strong base Butyl lithium, triphenylmethyl potassium, ethyl sodium, alkali metal amides, alcoxides and hyderoxides

21. The time of flow of liquid depends on the viscosity and composition. If η and η_o are the absolute viscosities of a solution and the pure solvent, respectively, and t and to are their corresponding ratio of flow times.

$$\frac{t}{t_0} = \frac{\eta}{\eta_o} = \eta_r \quad \frac{60}{40} = \frac{\eta}{\eta_o} = 1.5$$

concentration C = 5/1000 = 0.5 dl/gm [1dl = 100ml]

Reduced viscosity $\frac{\eta_{sp}}{C} = \eta_{red}, dl/g$ Specific viscosity $\eta_{sp} = \eta_r - 1$ = 1.5 – 1 = 0.5

Reduced viscosity = 0.5/0.5 = 1 dl/g.

22. R = ρ L /A where L = length, A = are of cross section and ρ = volume resistivity

L = 1cm and Area = 1*1 cm^2

Resistance R = 10^{7*} 10^{-2} / 10^{-4} = 109Ω

R = 1000 MΩ

2015

Q. 1 – 9 carry one mark each.

1. The biodegradable polymer among the following polymers is
   (*a*) poly(lactic acid)
   (*b*) poly(butyleneterephthalate)
   (*c*) polystyrene
   (*d*) polypropylene

2. Notched impact strength of a plastic decreases with
   (*a*) increase in notch tip radius
   (*b*) increase in notch depth
   (*c*) increase in temperature
   (*d*) decrease in notch depth

3. The compound used as a reactive diluent in unsaturated polyester resins is
   (*a*) benzene (*b*) cresol
   (*c*) styrene (*d*) adipic acid

4. The diameter of a die of an extruder producing extrudate of diameter 2.4 mm with a die-swell ratio of 2 is ________ mm.

5. The degree of polymerization of Nylon 6 (ignore end-groups) with molar mass of 1,00,000 g mol^{-1} is ________ .

6. The Polymer synthesized by ring opening polymerization is
   (*a*) Poly(acrylic acid)
   (*b*) Nylon 6
   (*c*) Nylon 66
   (*d*) Poly (ethylene terephthalate)

7. Which among the following are used as initiators for free radical polymerization?
   P. K_2SO_4
   Q. $K_2S_2O_8$
   R. AIBN
   S. t-Butyl hydroperoxide + Fe^{2+}
   (*a*) P, Q & R only (*b*) Q, R & S only
   (*c*) P, R & S only (*d*) P, Q, R & S

8. Weight average molecular weight can be determined by
   (*a*) Osmometry (*b*) Ebulliometry
   (*c*) End group analysis (*d*) Light scattering

9. Butyl rubber is a copolymer of
   (*a*) Isobutylene and butadiene
   (*c*) Isobutylene and isoprene
   (*b*) Butadiene and 1-butene
   (*d*) Isoprene and 1-butene

Q. 10 – 22 carry two marks each.

10. Match the characterization technique with the polymer property it is used to determine

Technique	Property
P. X-ray diffraction	1. Melting temperature
Q. Differential scanning calorimetry	2. Crystallinity & crystal size
R. Thermogravimetric analysis	3. Glass transition temperature
S. Dynamic mechanical analysis	4. Ash content

   (*a*) P-3; Q-1; R-4; S-2 (*b*) P-3; Q-4; R-2; S-1
   (*c*) P-2; Q-4; R-1; S-3 (*d*) P-2; Q-1; R-4; S-3

11. Match the following plastic additives with their function

Additive	Function
P. Di-isooctyl phthalate	1. Antioxidant
Q. 4-Methyl-2,6-t-butylphenol	2. Plasticizer
R. Dicumyl peroxide	3. Antistatic agent
S. Quaternary ammonium compound	4. Cross-linking agent

   (*a*) P-2; Q-4; R-1; S-3 (*b*) P-4; Q-1; R-3; S-2
   (*c*) P-2; Q-1; R-4; S-3 (*d*) P-3; Q-1; R-4; S-2

12. The correct statement with respect to electrical property of polymeric materials is
   (*a*) For non-polar materials, dielectric constant is independent of frequency & temperature
   (*b*) For polar materials, dielectric constant depends on frequency but not on temperature
   (*c*) For non polar materials, power losses are high and depend on temperature & frequency
   (*d*) For polar materials, power losses are low and independent of frequency

13. The order of melting point for the given polymers is
   (*a*) Nylon 66 > PTFE > Nylon 6 > PP
   (*b*) Nylon 66 > Nylon 6 > PTFE > PP
   (*c*) PTFE > Nylon 66 > Nylon 6 > PP
   (*d*) PTFE > Nylon 6 > Nylon 66 > PP

14. Match the processing technique used to manufacture the appropriate product

Processing Technique	Product
P. Calendering	1. Pipes
Q. Extrusion	2. Disposable cups
R. Injection Molding	3. Sheets
S. Thermoforming	4. Nylon gears

(*a*) P-3; Q-2; R-1; S-4 (*b*) P-3; Q-1; R-2; S-4
(*c*) P-3; Q-1; R-4; S-2 (*d*) P-3; Q-2; R-4; S-1

15. Match the thermosetting resins to the raw materials they are synthesized from

Resin	Raw material
P. Epoxy	1. Cresol + furfural
Q. Phenolic	2. Diethylene glycol + diallyl phthalate
R. Unsaturated polyester	3. Bisphenol A + epichlorohydrin
S. Allyl	4. Maleic acid + 1,2-propylene glycol

(*a*) P-4; Q-2; R-3; S-1 (*b*) P-3; Q-1; R-2; S-4
(*c*) P-3; Q-2; R-1; S-4 (*d*) P-3; Q-1; R-4; S-2

16. The damping factor of a solid polymer under sinusoidal loading in single cantilever mode showing 80 percent recovery in modulus is ______.

17. A styrene-butadiene random copolymer with equal weight fraction of polystyrene (T_g = 100 °C) and polybutadiene (T_g = – 100 °C) shows a single glass transition peak. The T_g of the copolymer is ______ °C.

18. In a unidirectional carbon fibre reinforced epoxy composite, the ratio of fibre-to-matrix moduli is 30 and the fibres take up 50% of the cross-section. The percentage of applied force taken up by the fibres is ______.

19. The viscoelastic behavior of a plastic is represented by spring and dashpot elements having constants of 2 GN m^{-2} and 90 GN s m^{-2}, respectively. If a constant stress of 12 MN m^{-2} is applied, the strain predicted by Maxwell model after 50 s is ______ %.

20. Match the elastomers given below to their suitable application

Elastomer	Application
P. EPDM	1. Golf balls
Q. Polyurethane	2. Fuel transfer hose
R. Nitrile rubber	3. Cable insulation
S. Polybutadiene	4. Footwear

(*a*) P-3; Q-4; R-2; S-1 (*b*) P-4; Q-3; R-2; S-1
(*c*) P-3; Q-2; R-4; S-1 (*d*) P-1; Q-4; R-2; S-3

21. Match the following reagents to their function in natural rubber latex technology

Reagent	Function
P. Ammonia	1. Prevent storage hardening
Q. Hydroxylamine	2. Delay plugging mechanism
R. Formic acid	3. Stabilizer
S. Ethephone	4. Coagulating agent

(*a*) P-3; Q-1; R-2; S-4 (*b*) P-3; Q-2; R-4; S-1
(*c*) P-3; Q-1; R-4; S-2 (*d*) P-3; Q-4; R-1; S-2

22. 1.0 g of a polybutadiene sample with carboxylic acid groups at both the ends requires 2.5 mL of 0.1 M KOH for complete neutralization. The molecular weight of the polymer in g mol^{-1} is______.

ANSWERS

1. (*) **2.** (*) **3.** (*) **4.** (*) **5.** (883.7) **6.** (*) **7.** (*) **8.** (*d*) **9.** (*c*) **10.** (*d*)
11. (*c*) **12.** (*a*) **13.** (*c*) **14.** (*c*) **15.** (*d*) **16.** (0.25) **17.** (–37 to –36) **18.** (96.77%)
19. (1.267%) **20.** (*a*) **21.** (*c*) **22.** (8000)

EXPLANATIONS

1. Biodegradable polymers are a specific type of polymer that breaks down after its intended purpose to result in natural byproducts such as gases (CO_2, N_2), water, biomass, and inorganic salts. These polymers are found both naturally and synthetically made, and largely consist of ester, amide, and ether functional groups. PP and PS and PBS are not breaks into natural byproduct such as gas, water, biomass and inorganic salts. Poly(lactic acid) is a biodegradable and bioactive thermoplastic aliphatic polyester derived from renewable resources, such as corn starch, cassava roots, chips or starch, or sugarcane.
2. Notch depth α 1/ impact strength
3. Reactive diluents are substances which reduce the viscosity of a lacquer for processing and become part of the lacquer during its subsequent curing via copolymerization. Reactive diluents used are low-viscosity, mono-, bi- or polyfunctional monomers or oligomers. Styrene used as reactive diluents in unsaturated polyester resin.
4. Die swell ratio = D/D_o,

   $D_o = 2.4/2 = 1.2$ mm

   D = extrudate diameter,

   D_o = Die diameter
5. Nylon 6 molecular weight : 113.16

   Molar mass : 10000g/mol

   Degree of polymerization :

   $$DP = \frac{\overline{M}_n}{m}$$

   $$\frac{100000}{113.16} = 883.7$$
6. Monomer having a ring structure can be opened and polymerised through ring opening polymerization method. Nylon 6 monomer has ring structure only.

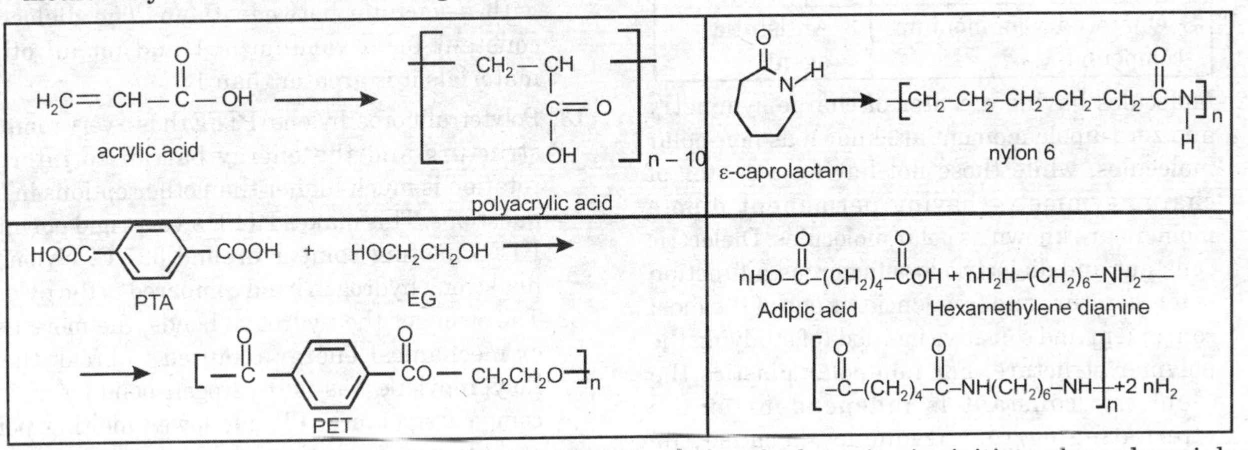

7. Solution: In a free radical addition polymerization, the choice of polymerization initiator depends mainly on two factors: a) its solubility and b) its decomposition temperature. If the polymerization is performed in an organic solvent, then the initiator should be soluble in that solvent and the decomposition temperature of the initiator must be at or below the boiling point of the solvent. Commonly, 2,2'-azo-bis-isobutyrylnitrile (AIBN) and Benzoyl peroxide suit these requirements. Initiator can also induce to decompose into radicals by using suitable catalyst. An electron transfer mechanism resulting in a reduction-oxidation reaction and peroxides and hydroperoxides are more susceptible to this kind of decomposition. Potassium persulfate gives free radical instead of potassium sulfate. Answer is B
8.

Technique	Measures	Range, g/mol
End Group	M_n	Up to 2500
Osmometry	M_n	15000 – 750000
Ebulliometry	M_n	Up to 100000
Light scattering	M_w	20000 to 10^7
Ultra centrifuge	M_w, M_z, MWD	2000 to 10^7
Solution viscosity	M_v, M_w	15000 – 10^6
Vapour-phase osmometry	M_n	Up to 25000
Gel-permiation chromatography	M_n, M_w, M_v, M_z, MWD	Up to 10^6

9.

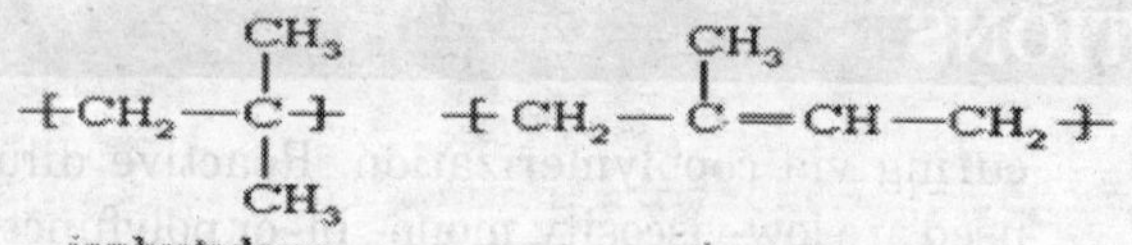

isobutylene isoprene

10.

Technique	Property
P. X-ray diffraction	2. Crystallinity & crystal size
Q. Differential scanning calorimetry	1. Melting temperature
R. Thermogravimetric analysis	4. Ash content
S. Dynamic mechanical analysis	3. Glass transition temperature

11.

Additive	Function
P. Di-isooctyl phthalate	2. Plasticizer
Q. 4-Methyl-2,6-t-butylphenol	1. Antioxidant
R. Dicumyl peroxide	4. Cross-linking agent
S. Quaternary ammonium compound	3. Antistatic agent

12. Molecules having a center of charge symmetry and zero dipole moment are known as non-polar molecules, while those not having a center of charge symmetry having permanent dipole moment are known as polar molecules. Dielectric constant and dielectric loss factor, as a function of temperature and frequencies is one of the most convenient and sensitive methods of studying the polymer structure. For non-polar plastics the dielectric constant is independent of the alternating current frequency because the electron polarization is effectively instantaneous. Non-polar plastics always have dielectric constants of less than 3. Raising the temperature of a polar plastic allows faster movement of the polymer chains and faster alignment of the dipoles. This is particularly true if the temperature is raised above Tg (see Newsletter on Low Temperature Plastics) because above Tg much more molecular movement is possible. Raising the temperature inevitably raises the dielectric constant of polar plastics. Non-polar plastics, such as the fluoropolymers, are not as affected by the water because they tend not to absorb water and temperature effects are not generally as severe because increased temperature does not affect the electronic polarization.

The dielectric constant is a measure of the influence of a particular dielectric on the capacitance of a condenser. It measures how well a material separates the plates in a capacitor and is defined as the ratio of the capacitance of a set of electrodes with the dielectric material between them to the capacitance of the same electrodes with a vacuum between them. The dielectric constant for a vacuum is 1 and for all other materials it is greater than 1.

13. Polytetrafluoroethylene (PTFE) has a very compact structure and the energy barrier to internal rotation is much higher than other options in the questions. This makes PTFE a very rigid polymer. PTFE melting point is around 327°C. Nylon 66 has strong hydrogen bond compared to the nylon6. The stronger the hydrogen bonds, the more heat or mechanical energy required to break them. Polypropylene has no hydrogen bonding and no compact structure. PP has lower melting point other than available options.

14.

Calendering	Calendering is a finishing process used on cloth, paper, or plastic film. A calender is employed, usually to smooth, coat, or thin a material. With textiles, fabric is passed under rollers at high temperatures and pressures
Extrusion	Extrusion is a process used to create objects of a fixed cross-sectional profile. A material is pushed through a die of the desired cross-section.
Injection molding	The plastic is melted in the injection molding machine and then injected into the mold, where it cools and solidifies into the final part
Thermoforming	Thermoforming describes the process of heating a thermoplastic sheet to its softening point, stretching it over or into a single-sided mold, and holding it in place while it cools and solidifies into the desired shape.

15.

Resin	Raw material
P. Epoxy	3. Bisphenol A + epichlorohydrin
Q. Phenolic	1. Cresol + furfural
R. Unsaturated polyester	4. Maleic acid + 1,2-propylene glycol
S. Allyl	Diethylene glycol + diallyl phthalate

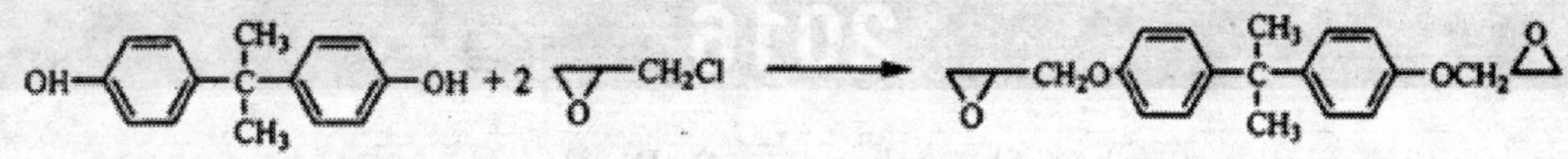

Bisphenol A *Epicholorohydrin* *"Uncured Epoxy Resin"*

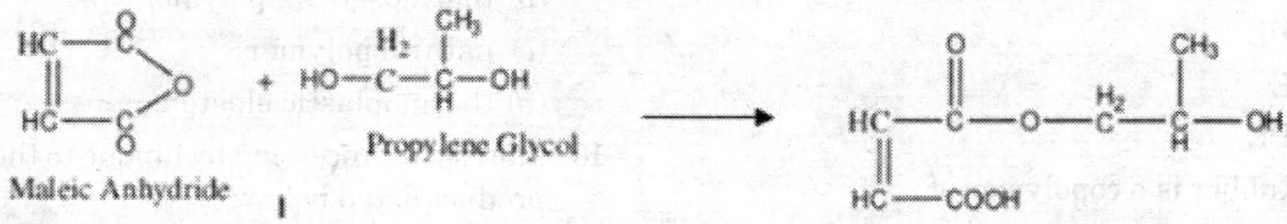

Unsaturated Polyester

16. $\tan\delta = \dfrac{\text{loss modulus}}{\text{storage modulus}}$

Loss modulus = 20% (loss in the recovery)
Storage modulus = 80 (stored in the recovery)
Damping Factor = tanδ
Damping factor = 20/80 = 0.25

17. Convert all the temperature in the Kelvin from centigrade. Total weight fraction is equal to 1.

$$\frac{1}{T_g} = \frac{W_1}{T_{g_1}} + \frac{W_1}{T_{g_2}}$$

$$\frac{1}{T_g} = \frac{0.5}{173} + \frac{0.5}{373}$$

$$T_g = -36.59°C$$

18. $$\frac{P_f}{P_c} = \frac{\sigma_f A_f}{\sigma_f A_f + \sigma_m A_m}$$

P_f = force by fiber & P_m = force by matrix
$P_C = P_f + P_m$
Unidirectional direction applied strain on the composite is equal to fiber and equal to the matrix. So, Stress is equal to the modulus.
Ratio of the fiber and matrix modulus = 30
Ratio of the fiber and matrix area= 0.5/0.5=1

$$\frac{P_f}{P_c} = \frac{1}{1 + \dfrac{\sigma_m A_m}{\sigma_f A_f}}$$

$$\frac{P_f}{P_c} = \frac{1}{1 + \dfrac{1}{30}} = \frac{30}{31}$$

The percentage of applied force taken up by the fibres is 0.9677*100 = 96.77%

19. Maxwell model equation at constant stress is :

$$\gamma_{\text{maxwell}} = \frac{\sigma_0}{E} + \frac{\sigma_0}{\eta}t$$

$E = 2\ GN/m^2$, $\eta = 90\ GNs/m^2$, $\sigma_o = 12\ MN/m^2$, t = 50s

$$\gamma = \frac{12\times10^3}{2\times10^6} + \frac{12\times10^3}{90\times10^6}\times50$$

$$\gamma = 1.267\%$$

20.

Elastomer	**Application**
P. EPDM	3. Cable insulation
Q. Polyurethane	4. Footwear
R. Nitrile rubber	2. Fuel transfer hose
S. Polybutadiene	1. Golf balls

21.

Reagent	**Function**
P. Ammonia	3. Stabilizer
Q. Hydroxylamine	1. Prevent storage hardening
R. Formic acid	2. Delay plugging mechanism
S. Ethephone	4. Coagulating agent

22. Let the Volume of KOH consumed = 2.5ml
Normality of KOH solution = 0.1
Weight of the sample taken = 1.0g

Carboxyl Vlaue = $\dfrac{VN(56.1)}{W}$ mg/KOH

Carboxy equivalent/100g

$$= \frac{VN(56.1)}{W\times1000}\times\frac{100}{56.1} = \frac{VN}{10W}$$

Carboxyl equvalent/100g

$$= \frac{2.5\times0.1}{10\times1} = 0.025$$

$$M_n = \frac{\text{functionality}}{\text{functional group equivalent}}$$

$$M_n = \frac{2\times100}{0.025} = 8000$$

2016

1. The polymer with minimum number of branches is
   (a) HDPE
   (b) VLDPE
   (c) LDPE
   (d) LLDPE

2. Nitrile Rubber is a copolymer of
   (a) isoprene and acrylonitrile
   (b) butadiene and acrylonitrile
   (c) cyclopentadiene and acrylonitrile
   (d) isobutylene and acrylonitrile

3. The functionality of 1,4-divinylbenzene in reactions involving addition across carbon-carbon double bond is
   (a) 1 (b) 2
   (c) 3 (d) 4

4. The comonomer common to Nylon 66 and Nylon 46 is
   (a) hexamethylene diamine
   (b) butylene diamine
   (c) adipic acid
   (d) octane dicarboxylic acid

5. Polyethylene and polypropylene form an immiscible blend mainly due to
   (a) entropy factor
   (b) enthalpy factor
   (c) crystallinity
   (d) solubility

6. Rubber modulus is
   (a) ratio of stress to strain
   (b) same as Young's modulus
   (c) stress at specified strain
   (d) stress at break

7. The solubility parameter is determined by using
   (a) Bragg's equation
   (b) Fox equation
   (c) Hildebrand equation
   (d) Carother's equation

8. 'Roller die' consists of a combination of
   (a) a two-roll calender with internal mixer feeding
   (b) a two-roll calender with open mill feeding
   (c) a three-roll vertical calender with two-roll mixer feeding
   (d) a two-roll calender with extruder feeding

9. Resole is an example of
   (a) thermoplastic polymer
   (b) thermosetting polymer
   (c) natural polymer
   (d) thermoplastic elastomer

10. Match the processing technique to the appropriate product listed below:

Processing Technique	Product
P. Blow molding	1. Bucket
Q. Co-extrusion	2. Blister packaging
R. Injection molding	3. Bottles
S. Thermoforming	4. Multilayered sheets

   (a) P-3; Q-4; R-2; S-1
   (b) P-3; Q-1; R-4; S-2
   (c) P-3; Q-4; R-1; S-2
   (d) P-3; Q-2; R-1; S-4

11. For a high molecular weight polymer sample with a viscosity of 6×10^{11} Poise and a stress relaxation modulus of 3×10^6 dyne cm^{-2} at a given temperature, the relaxation time will be _____ hours.

12. Match the following polymer additives to their function :

Additive	Function
P. Azocarbonamide	1. Chemical plasticizer
Q. Antimony trioxide	2. Accelerator
R. Pentachlorothiophenol	3. Flame retardant
S. Mercaptobenzothiazole	4. Blowing agent

   (a) P-4; Q-1; R-3; S-2
   (b) P-4; Q-2; R-1; S-3
   (c) P-4; Q-3; R-2; S-1
   (d) P-4; Q-3; R-1; S-2

13. Tensile force of 165 N is applied to a piece of vulcanized rubber of dimension 4 mm × 4 mm × 30 mm. If the sample is elongated by 50% of its original length under the same applied force, the true stress will be _______ MPa.

14. The order of glass transition temperature for the given polymers is [NR = natural rubber; PP = polypropylene; PE = polyethylene; PMMA = poly (methyl methacrylate)]
   (a) NR < PE < PP < PMMA
   (b) PE < NR < PP < PMMA
   (c) PE < PP < NR < PMMA
   (d) NR < PP < PE < PMMA

15. Dynamic mechanical analysis of polystyrene (T_g = 100 °C) measured at a frequency of 1 Hz shows the damping peak at 110°C. If the measurement is made at 10^4 Hz, then the peak temperature (°C) will be

(*a*) 123.2 (*b*) 133.2
(*c*) 143.2 (*d*) 153.2

16. Match the product to the most suitable plastic listed below :

Product	Plastic
P. Baby feeding bottle	1. Polypropylene
Q. Tiffin box	2. Poly(ethylene terephthalate)
R. Water bottle	3. Poly(vinyl chloride)
S. Blood bag	4. Polycarbonate

(*a*) P-1; Q-4; R-2; S-3 (*b*) P-4; Q-1; R-2; S-3
(*c*) P-1; Q-3; R-2; S-4 (*d*) P-4; Q-3; R-2; S-1

17. The number average molecular weight for the polymerization of adipic acid and ethylene glycol (feed ratio 1 : 1) at 99 percent conversion is _______ g mol^{-1}.

18. A composite material contains 30 % by volume of uniaxially aligned glass fibres in a matrix of alkyl resin. The tensile modulus of the glass fibre and alkyl resin are 76 GPa and 3 GPa, respectively. If a tensile stress of 100 MPa is applied parallel to the fibres, the percentage longitudinal strain is _____ .

19. Match the Elastomers listed below to the appropriate curing agent :

Elastomer	Curing Agent
P. Silicone rubber	1. Zinc oxide + ethylene thiourea
Q. Natural rubber	2. Diamine
R. Chloroprene rubber	3. Sulfur
S. Acrylate elastomer	4. Dicumyl peroxide

(*a*) P-4; Q-3; R-1; S-2
(*b*) P-3; Q-4; R-1; S-2
(*c*) P-4; Q-1; R-3; S-2
(*d*) P-2; Q-3; R-4; S-1

20. The weight of graphite fiber (density = 1800 kg m^{-3}) that should be added to 1.00 kg of vinyl ester resin (density = 1250 kg m^{-3}) to produce a composite with a density of 1600 kg m^{-3} is _____ kg.

21. If the values of K and *a* in the Mark-Houwink equation are 1.5×10^{-4} dL g^{-1} and 0.60 respectively, the viscosity average molecular weight of a polymer having an intrinsic viscosity of 0.05 dL g^{-1} is _______ kg mol^{-1}

22. A rectangular polymer bar of length 40 mm fits exactly into a steel mold cavity and the entire assembly was heated from 20° to 100 °C. The linear thermal expansion coefficients of the polymer and steel are 80×10^{-6} °C^{-1} and 11×10^{-6} °C^{-1}, respectively. The strain encountered by the polymer sample along the length will be _____ %.

ANSWERS

1. (*a*) **2.** (*b*) **3.** (*d*) **4.** (*c*) **5.** (*a*) **6.** (*c*) **7.** (*c*) **8.** (*d*) **9.** (*b*) **10.** (*c*)
11. (55–56 hr) **12.** (*d*) **13.** (15.42:15.52) **14.** (*b*) **15.** (*) **16.** (*b*) **17.** (8590–8610)
18. (0.35–0.45) **19.** (*a*) **20.** (2.3–2.7) **21.** (14.5–16.5) **22.** (5.4–5.6)

EXPLANATIONS

1. Density α 1/branches

   HDPE : High density polyethylene

   VLDPE : Very Low density polyethylene

   LDPE : Low density polyethylene

   LLDPE : Liner Low density polyethylene

   HDPE has highest density among the 4 polymer. Density is inversely proportionate to the branches in the main Chain. So the right answer is HDPE.

2. Nitrile rubber, also known as Buna-N, Perbunan, acrylonitrile butadiene rubber, and NBR, is a synthetic rubber copolymer of acrylonitrile (ACN) and butadiene.

   BUTADIENE $CH_2 = CH - CH - CH$ ACRYLONITRILE $CH_2 = CH - CN$

   NITRILE RUBBER (NBR)

   $$(\text{---}(CH_2 - CH = CH - CH_2) \text{---} (CH_2\text{--}\underset{\underset{CN}{1}}{CH}) \text{---})_n$$

3. The functionality of a molecule is simply its interlinking capacity, or the number of sites available for bonding (presence of reactive functional group like —OH, —COOH, —NH_2, — SH, —NCO, etc.) with other molecules under the specific polymerization conditions

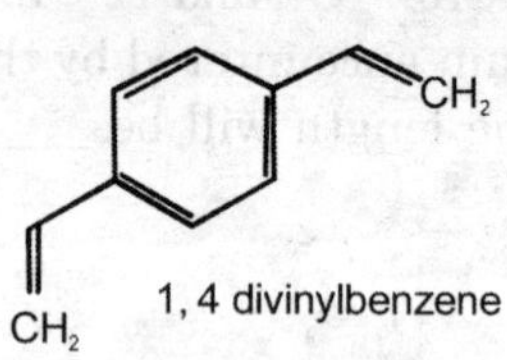

1, 4 divinylbenzene

   1, 4 divinylbenzene has 2 nos. of double bond on the chain ends. Each bond gives bifunctionality to the system. The functionality to the system. The functionality of 1, 4-divinylbenzene in reactions involving addition across carbon-carbon double bond is 4 due to presence of the two double bond.

   Some compounds, however, do not contain any reactive functional group but the presence of double bond or triple bonds in the molecule bestows polyfunctionality on them.

   Example :

   (*i*) Ethylene (bifunctional because of the presence of the double bond).

   (*ii*) Acetylene (tetrafunctional because of the presence of the triple bond).

4. Nylon 66

   $$n\text{HO}-\overset{\overset{O}{\|}}{C}-(CH_2)_4-\overset{\overset{O}{\|}}{C}-\text{OH} + nH_2N-(CH_2)_6-NH_2 \longrightarrow$$

   Adipic acid Hexamethylene diamine

   $$\left[-\overset{\overset{O}{\|}}{C}-(CH_2)_4-\overset{\overset{O}{\|}}{C}-NH(CH_2)_6-NH-\right]_n + 2\,nH_2O$$

   Nylon 46 : Nylon 46 was synthesized from the salt of 1, 4-diaminobutane and adipic acid.

   Common monomer in both nylon is adipic acid.

5. Entropy Factor. ***Entropy* is the name we scientists call disorder. Entropy is proportional to the amorphous or disorder in the chains. PE has lower disorder than PP.** Polyethylene and polypropylene form an immiscible blend mainly due to the entropy factor.

6. Young's modulus represents the factor of proportionality in Hooke's law, which relates the stress and the strain. However, Hooke's law is only valid under the assumption of an elastic and linear response. Rubber behavior is not like a linear material when applied the stress on it. Rubber modulus or elastic modulus calculate at different specific strain such as 100%, 200% or 300% due to non-linear behavior of the rubber material when applied stress.

7. (*a*) Bragg's equation — The Bragg law is useful for measuring wavelengths and for determining the lattice spacing of crystals

   (*b*) Fox equation — The Flory-Fox equation is a simple empirical formula that relates molecular weight to the glass transition temperature of a polymer system.

   (*c*) Hildebrand equation — The Hildebrand solubility parameter (δ) provides a numerical estimate of the degree of interaction between materials, and can be a good indication of solubility, particularly for nonpolar materials such as many polymers. Materials with similar values of δ are likely to be miscible.

   (*d*) Carother's equation — Carothers equation measures the degree of the polymerization

8. 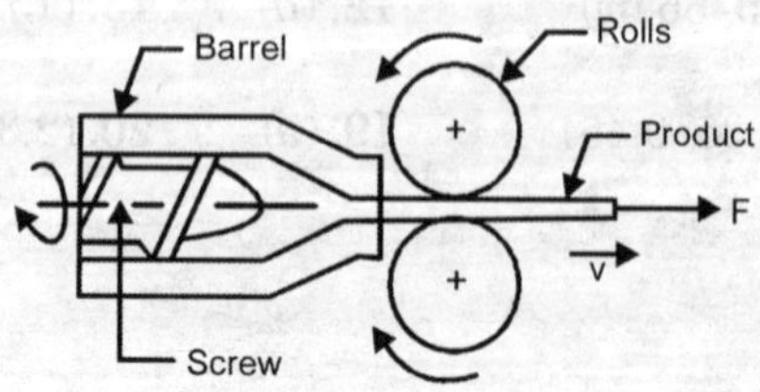

   Combination of extrusion and calendaring that result in better quality product than either extrusion or calendering alone.

9.

Thermoplastic polymer	Become soft when heat is applied and have a smooth, hard finish when cooled.
Thermosetting polymer	The plastic which when molded once, cannot be softed by heating
Natural polymer	**Natural polymers** occur in nature and can be extracted
Thermoplastic elastomer	Copolymers or a physical mix of polymers (usually a plastic and a rubber) which consist of materials with both **thermoplastic** and **elastomeric** properties.

Resol : Base-catalysed phenol-formaldehyde resins are made with formaldehyde to phenol ratio of greater than one (usually around 1.5). These resins are called resoles. hydroxymethyl phenols will crosslink on heating to around 120 °C to form methylene and methyl ether bridges through the elimination of water molecules. At this point the resin is a 3-dimensional network, which is typical of polymerised phenolic resins. The high crosslinking gives this type of phenolic resin its hardness, good thermal stability, and chemical imperviousness. Presence of the 3-dimesnional network resol comes under the thermosetting polymers.

10.

Blow molding	**Blow molding** is a manufacturing process by which hollow plastic parts are formed
Co-extrusion	**Coextrusion** is the process of pressing two or more materials through the same die to produce a single piece
Injection molding	The plastic is melted in the injection molding machine and then injected into the mold, where it cools and solidifies into the final part
Thermoforming	Thermoforming describes the process of heating a thermoplastic sheet to its softening point, stretching it over or into a single-sided mold, and holding it in place while it cools and solidifies into the desired shape.

11. Use Maxwell model

$$\text{Relaxation time} = \frac{\eta}{E}$$

$$\eta\ (\text{Viscosity}) = 6 \times 10^{11}\ \text{Poise}$$

$$E\ (\text{Modulus}) = 3 \times 10^{6}\ \text{dyne cm}^{-2}$$

$$1\ \text{Poise} = 1\ \text{dyne/cm}^2.\text{S}$$

$$\text{Relaxation time} = \frac{6\times10^{11}}{3\times10^{6} * 3600} = 55.5\ \text{hours}$$

12.

carbonamide	Blowing agent	Play a role in the formation of air bubbles by breaking down into gas at high temperature.
Antimony trioxide	Flame retardant	reducing the devastating impact of fires
Pentachlorothiophenol	plasticizer	Additives increase plasticity
Thiazole	Accelerator	Accelerators the curing system in rubber curing

13. true stress, S, will actually occer, γ = strain

$$S = \frac{S_o L}{L_o} = S_o(1+\gamma)$$

$$\gamma = \frac{L - L_o}{L_o}$$

$$S_o = \frac{F}{A} = \frac{165}{4*10^{-3}*4*10^{-3}} = 10.31 * 10^6\ \text{Pa}$$

$$S = 10.31 * 10^6 * (1 + 0.5) = 15.47\ \text{MPa}$$

14. The glass transition occurs in amorphous (or glassy) and semi-crystalline polymers and is due to a reduction in motion of large segments of molecular chains with decreasing temperature.

PE	-125°C
NR	-70 °C
PP	0-5 °C
PMMA	105°C

16.

Baby Feeding Bottle	Polycarbonate is transparent and better Tensile strength and no harm to body
Tiffin box	Generally made from PE & PP
Water bottle	Bottles are made from PET because Pet is recyclable easily
Blood Bag	PVC is the best material that satisfies the strict criteria in terms of performance, safety and cost for medical applications,

17. Product of the adipic acid and ethylene glycol

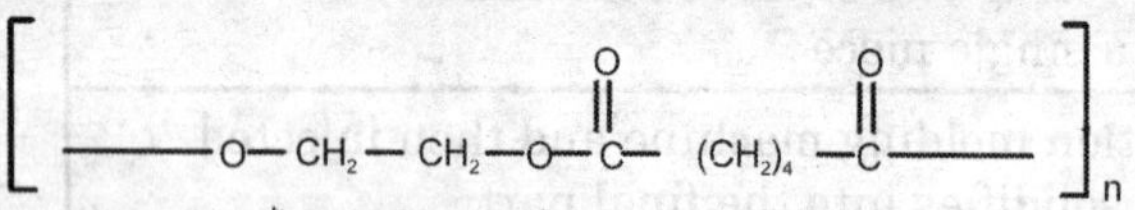

Number average molecular weight :

$$M_n = \frac{M_o}{(1-p)}$$

$p = 0.99$

& M_o = molecular weight of the repeated unit/2

$$= \frac{172}{2} = 86$$

$$M_n = \frac{86}{(1-0.99)} = 8600$$

18. Modulus of the composite

$$E_{cl} = E_m V_m + E_f V_f$$

E_m & E_f are the elastic modulus for matrix and fiber.

$$V_m + V_f = 1$$

V_m and V_f are volume fraction for matrix and fiber.

$V_m = 0.7$

& $V_f = 0.3$

$E_m = 3\text{GPa}$

$E_r = 76\text{ GPa}$

$E_{cl} = 0.7 * 3 + 0.3 * 76$

$= 24.9\text{ GPa}$

$$\text{Elastic modulus} = \frac{\text{tensile stress}}{\text{strain}}$$

$$\text{Strain }(\%) = \frac{\text{Tensile stress}}{\text{Elastic modulus}}$$

$$= \frac{100\text{MPa}}{24.9*1000\text{MPa}}$$

$$= 0.40\%$$

20. $$\text{Density} = \frac{\text{mass}}{\text{volume}}$$

	Mass (Kg)	Density (kg m^{-3})	Volume m^3
Fiber	x	1800	x/1800
Resin	1	1250	1/1250
Total	x + 1		x/1800 + 1/1250

$$\text{Composite density} = \frac{\text{total mass}}{\text{total volume}}$$

$1600 = x + 1/x/1800 + 1/1250$

$X = 2.54$ kg;

21. Mark-Houwink equation the relationship among the molecular weight and viscosity are given below

$$[\eta] = K M^a$$

Where $[\eta]$ is the intrinsic viscosity, 'M' is Molecular weight, 'K' and 'a' are constants for a particular polymer solvent system.

$K = 1.5 \times 10^{-4}\text{ dL g}^{-1}$

$a = 0.60$

and $\eta = 0.05\text{ dL g}^{-1}$

$0.05 = 1.5 \times 10^{-4} * [M]^{0.6}$

$M = 16024.995$ g/mol

$= 16.02$ Kg/mol

2017

1. Poly(ethylene terephthalate) is synthesized from
   (*a*) Ethylene + dimethyl terephthalate
   (*b*) Ethylene + terephthalate
   (*c*) Glycerol + terephthalate
   (*d*) Ethylene Glycol + terephthalate

2. Poly(vinyl chloride) has a higher T_g than polypropylen due to presence of
   (*a*) Bulky side groups
   (*b*) Polar interactions
   (*c*) Restriction of bond rotation
   (*d*) Non-polar interaction

3. The filler which would impart electrical conductivity to a polymer is
   (*a*) Carbon Black
   (*b*) Talc
   (*c*) Glass beads
   (*d*) Calcium carbonate

4. Which one of the following catalysts is used to prepare 'isotactic' polypropylene?
   (*a*) Alkyl lithium (*b*) BF_3
   (*c*) Ziegler-Natta (*d*) AIBN

5. Novolac and Resole are A–stage low molecualr weight phenolic resin products that are
   (*a*) soluble and fusible
   (*b*) Insoluble but fusible
   (*c*) Insoluble and infusible
   (*d*) soluble and infusible

6. Which of the following reagents can act as an initiator at room temperature?
   (*a*) AIBN (*b*) dicumyl peroxide
   (*c*) Dibenzoyl peroxide (*d*) $Fe^{2+} + H_2O_2$

7. The impact strength of polystyrene can be enhanced by blending/mixing with
   (*a*) Carbon black
   (*b*) PMMA
   (*c*) Polybutadiene
   (*d*) Glass fibre

8. The melt processing temperature of a semicrystalline thermoplastic polymer is
   (*a*) Between T_g and T_m
   (*b*) Equal to T_m
   (*c*) Lower then T_m
   (*d*) Higher than T_m

9. The unit of viscosity of a polymer is expressed as
   (*a*) Pa.s
   (*b*) Pa/s
   (*c*) Pa/s^2
   (*d*) Pa/s^3

10. Based on the graph 1-5, which option best described the stress-strain behaviour of materials listed as P, Q, R, S and T?

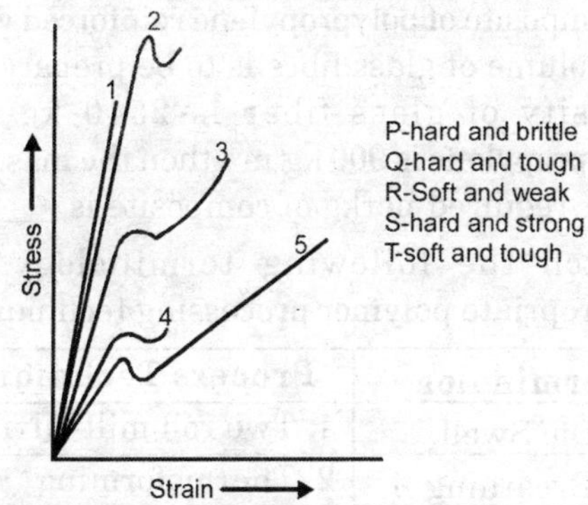

   (*a*) P-2; Q-1; R-5; S-4; T-3
   (*b*) P-1; Q-3: R-4; S-2; T-5
   (*c*) P-1: Q-2; R-5; S-3; T-4
   (*d*) P-2; Q-3; R-1; S-4; T-5

11. The two characterization techniques which can be used to determine degree of crystallinity of a polymer are
   P. Scanning Electron and Microscopy
   Q. Thermogravimetric Analysis
   R. Wide Angle X-Ray Diffreaction
   S. Differential Scanning Calorimetry.
   (*a*) P&R
   (*b*) Q&R
   (*c*) R&S
   (*d*) Q&S

12. The density of polyethylene crystals is 998 kg/m^3 and that of totally amorphous polyethylene is 886 kg/m^3. If the density of a polyethylene sample is 949 kg/m^3, the crystallinity in volume fraction is ___%.

13. The polydispersity index of a polymer sample containing 200 molecules each of molecular weight 10,000g/mol, 300 molecules each of molecular weight 30,000g/mol and 500 molecules each of molecualr weight 50,000 g/mol is_____.

14. Match the following rubber additives to their function:

Additives	Function
P. Dicumyl Peroxide	1.Ultra fast accelerator
Q. Pentachlorothiophenol	2. Activator
R. Zno with stearic acid	3. Curing Agent
S. Zinc diethyldithiocarbamate	4. Peptizer

(*a*) P-3; Q-1; R-2; S-4
(*b*) P-3; Q-1: R-4; S-2
(*c*) P-3: Q-4; R-2; S-1
(*d*) P-3; Q-4; R-1; S-2

15. A composite of polypropylene reinforced with 20% by volume of glass fibre is to be prepared. If the density of glass fiber is 2540 kg/m^3 and polypropylene is 900 kg/m^3, then the mass of glass fiber required perkg of composite is ______g.

16. Match the following terminology to the appropriate polymer processing technique :

Terminology	Process Technique
P. Die Swell	1. Two roll mill mixing
Q. Breathing	2. Thermoforming
R. Plug-assisted	3. Extrusion
S. Mastication	4. compressor moulding

(*a*) P-1; Q-2; R-3; S-4 (*b*) P-3; Q-4: R-2; S-1
(*c*) P-2: Q-3; R-4; S-1 (*d*) P-2; Q-1; R-4; S-3

17. Match the polymer in Column A to its application in Column B:

Column A	Column B
P. Nylon	1.Television cabinet
Q. Polyethylene	2. Tyre
R. Cis-1,4 polyisoprene	3. Mechanical gear
S. Acrylonitrile-butadiene -styrene	4. Packaging

(*a*) P-3; Q-4; R-2; S-1 (*b*) P-4; Q-3: R-2; S-1
(*c*) P-4: Q-2; R-3; S-1 (*d*) P-3; Q-4; R-1; S-2

18. For the polycondensation of equimolar amounts of adipic acid with hexamethylene diamine, if the number average degree of polymerization is 100, then the extent of reaction is___%.

19. The relaxation time for a rubber band at 23°C is 60 days. If it is stressed to 2MPa initially, then the time required before the dtress relaxation to 1 MPa is_____days.

20. Match the processing technique in Column A to the corresponding shear rate(s^{-1}) in Column B

Column A	Column B
P. Injection Moulding	1.1-10
Q. Extrusion	2. 10-100
R. Calendering	3. 100-1000
S. Compression molding	4. 1000-10000

(*a*) P-1; Q-3; R-2; S-4 (*b*) P-4; Q-2: R-3; S-1
(*c*) P-4: Q-3; R-1; S-2 (*d*) P-4; Q-3; R-2; S-1

21. Given T_g of Polymer A is 100°C and that of polymer B is -100°C, then the T_g of a miscible blend of A and B contaning 30wt% of A is______°C.

22. Match plots 1-4 given in the figure below with the the correct flow behavior of polymeric fluid listed as P, Q, R & S

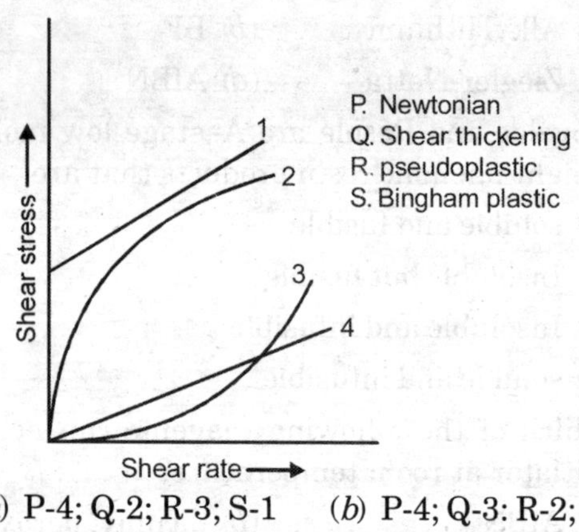

(*a*) P-4; Q-2; R-3; S-1 (*b*) P-4; Q-3: R-2; S-1
(*c*) P-4: Q-3; R-1; S-2 (*d*) P-1; Q-3; R-2; S-4

ANSWERS

1. (*d*) **2.** (*b*) **3.** (*a*) **4.** (*c*) **5.** (*a*) **6.** (*d*) **7.** (*c*) **8.** (*d*) **9.** (*a*) **10.** (*b*)
11. (*c*) **12.** (55.95 to 56.55) **13.** (1.14 to 1.24) **14.** (*c*) **15.** (410 to 418) **16.** (*b*) **17.** (*a*)
18. (98.5 to 99.5) **19.** (41.00 to 42.20) **20.** (*d*) **21.** (– 67.3 to – 66.3) **22.** (*b*)

EXPLANATIONS

1.

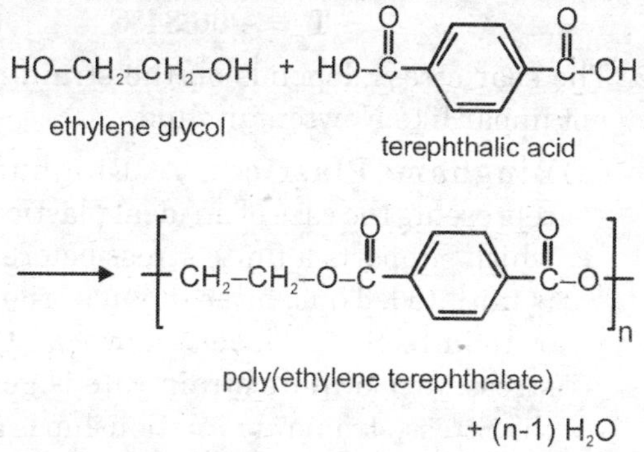

2. Stronger intermolecular forces lead to a higher T_g. PVC has stronger intermolecular forces than polypropylene because of the dipole-dipole forces from the C-Cl bond.

   Polypropylene $T_g = -20°C$

   Poly(vinyl chloride) = 81°C

3. Carbon black when used to impart electrical conductivity to plastics exhibits a phenomenon known as percolation - where the level of carbon black is sufficient to cause a significant and abrupt increase in electrical conductivity. As the loading of the carbon black in the compound increases, the plastic compound remains initially insulating, as the loading increases the conductivity passes through a sharp and abrupt rise over a very narrow black concentration (loading) range. Further increases in loading part this threshold cause little increase in the conductivity. This narrow range is known as the percolation threshold.

6. Redox reaction has occurred at room temperature.

7. Polystyrene is toobrittle for many applications. To solve this problem we incorporate some rubber in the matrix of the polystyrene. Butadiene increase the toughness of the polymer and reduce the sofetening point and imparts the impact strength.

9. $$\sigma = \eta \frac{dy}{dt}$$

   σ stress (Pa) and dy/dt (shear rate) = sec^{-1}

   $$\text{viscosity} = \text{Pa/sec}^{-1} = \text{Pa.s}$$

10. Hard means required force is highet to break it. Soft material requires less force to break the material and give higher elongation. Tough means higher the stress strain curve. Brittle materials is similar to glass which requires the higher force to elongate a small percentage.

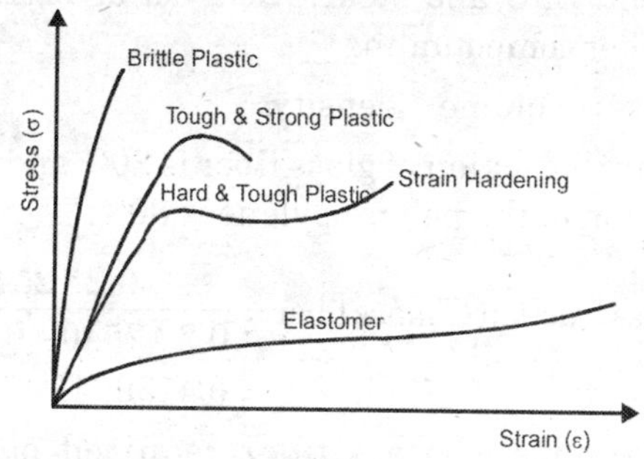

12. Degree of crystallinity = X_v

$$= \frac{d_a - d}{d_a - d_c} = \frac{886 - 949}{886 - 998}$$

$$= 0.5625$$

$$X_v = 56.25\%$$

(d_a = density of amorphous part, d_c = density of crystallinity part, d = density of the polymer)

13.

	1	2	3
Molecules	200	300	500
Molecular weight (g/mol)	10,000	30,000	50,000

We know that weight of the sample is

$$W = \sum w_i = \sum N_i M_i$$

Mass in terms of percentage given, we use mass as w_i, and divide w_i by the molecular weight and get the number of chains. Number average molecular weight :

$$M_n = \frac{\Sigma N_i M_i}{\Sigma N_i}$$

$$M_n = \frac{200 * 10^4 + 300 * 3 * 10^4 + 500 * 5 * 10^4}{200 + 300 + 500}$$

$$M_n = 36000$$

Weight average molecular weight :

$$M_w = \frac{\Sigma N_i M_i^2}{\Sigma N_i M_i}$$

$$M_w = \frac{200 * (10^4)^2 + 300 * (3 * 10^4)^2 + 500 * \left(5 * 10^4\right)^2}{200 * 10^4 + 300 * 3 * 10^4 + 500 * 5 * 10^4}$$

$$M_w = 4277$$

$$\text{Poly Dispersity} = \frac{M_w}{M_n}$$

$$\text{Poly Dispersity} = \frac{42777}{36000}$$

$$PD = 1.18$$

14. Phenol generally used as peptizer for Natural rubber in rubber industry. Peroxide used as curing agent. ZnO and stearic acid act as activator in rubber compounding.

15. Mass = volume * density,

Volume fraction of glass fiber is 20% and Volume faction of the polypropylene is 80%.

$$\text{Mass fraction of glass fiber} = \frac{0.2 * 2540}{0.2 * 2540 + 0.8 * 900}$$

$$= 0.4136;\ 41.4\%$$

The mass of glass fiber required per kg of composite is 0.414 * 1000 = 414 g.

17. Polyethylene world wide used in the packaging. C is 1, 4 polyisoprene is the structure of the natural rubber and natural rubber is used in the tire. ABS is used where the hard and durable plastic compound required.

18.

$$X_n = \frac{1+r}{1+r-2Pr}$$

For euimolar $r = 1$

$$X_n = \frac{1}{(1-p)} = 100$$

$$P = 0.99 = 99\%$$

19.

$$\sigma = \sigma_o e^{(-t/\tau)}$$

Where τ (**relaxation time**) = η/E,

σ stress at time t and σ_o initial stress

$$\tau = 60 \text{ days}$$

$$\sigma = 1 \text{ MPa}$$

and $\sigma_o = 2$ MPa

$$1 = 2e^{(-t/60)}$$

$$t = 41.8 \text{ days}$$

20.

Processing	Shear rate (1/s)
Compression molding	1 to 10
Calendering	10 to 100
Extrusion	100 to 1000
Injection molding	10^3 to 10^4
Fiber spinning	> 10^4

21. Convert all the temperature in the Kelvin from centrigrade. Total weight fraction is equal to 1.

$$\frac{1}{T_g} = \frac{W_1}{T_{g_1}} + \frac{W_2}{T_{g_2}}$$

$$\frac{1}{T_g} = \frac{0.3}{373} + \frac{0.7}{173}$$

$$T_g = -66.84°C$$

22. The sear stress depends on the strain rate, but not linear like Newtonian fluid.

(*a*) **Bingham Plastic** : A Bingham body represents the case of an ideal plastic material which supports a finite stress before a strain is initiated. Thus, after the material reaches a definite Stress level known as the yield stress, (σ_o), a finite strain rate is generated; the Stress-strain rate relationship is assumed to be linear, as in Newtonian fluids, but only after attainment of the yield stress (σ_o). Shear stress equation for bingham plastic is given by :

$$\sigma = \sigma_o + \eta \frac{dy}{dt}$$

Examples : clay suspensions, drilling mud, toothpaste, mayonnaise, chocolate and mustard.

(*b*) **Pseudoplastic** : The behavior of most real plastics is, however, characterized by non-linear relationships which are concave downward as shown by the pseudoplastic behavior. Most plastic melts, solutions or dispersions exhibit pseudoplastic flow patterns. At rest, long chain molecules of a plastic melt, solution or dispersion are believed to engage each other into stable associations as a consequence of chain entanglements and due to molar cohesion. At low shear rates, the stress required to overcome or undo the effect of chain entanglements is relatively high, but proportionately lower stress is required to allow the melt, solution or dispersion flow at a higher strain rate once the molecules are set in motion with respect to each other. The entanglement effect becomes virtually non-existent and almost Newtonian behavior assumes prominence at high shear rates depending on the exact nature of the fluid materials. If allowed to stand at rest thereatter, the fluid materials set again. Such isothermal reversible sol-gel transformations are known as thixotropy. The equation for Pseudoplastic is given by :

$$\sigma = \eta \frac{dy^n}{dt}$$

$n < 1$ pseudoplastics

Examples : Nail polish, whipped cream, ketchup, molasses, syrups, paper pulp in water, latex paint, ice, blood, some silicone oils, some silicone coatings, sand in water.

(*c*) **Dilatancy** : Concave upward curve in the shear stress vs. shear rate curve represents the behavior of dilatants substance. Such behaviour is exemplified by some highly concentrated suspensions of certain fine powders in appropriate liquids and by some melts in rare cases where crystallization or structure formation occurs at high rate of shearing. Such systems usually resemble a Newtonian fluid at low shear rates, but beyond a specific shear rate, they sharply gain in viscosity which may finally reach a very high value. When set at rest again, they return to the flow behavior of a near-normal viscous liquid. This reversible flow pattern is called *dilatancy.* The equation for Pseudoplastic is given by :

$$\sigma = \eta \frac{dy^n}{dt}$$

$n > 1$ dilatant

Examples : Suspensions of corn starch in water

Pseudoplastic are shear thinning while dilatants are shear thickening.

2018

Q. 1 – Q. 9 carry one mark each.

1. Which one of the following polymers occurs naturally?
   (*a*) Bakelite (*b*) Teflon
   (*c*) Cellulose (*d*) Perspex
2. The order of average molecular weights of a polymer is
   (*a*) $M_z > M_w > M_v > M_n$
   (*b*) $M_w > M_z > M_n > M_v$
   (*c*) $M_n > M_w > M_v > M_z$
   (*d*) $M_z > M_v > M_n > M_w$
3. Rubbers are a class of polymer known for
   (*a*) High intermolecular forces
   (*b*) High T_g polymers
   (*c*) Crystalline polymers
   (*d*) Low intermolecular forces
4. Nylon 6 is manufactured from
   (*a*) Sebacic acid and hexamethylene diamine
   (*b*) Caprolactam
   (*c*) Adipic acid and hexamethylene diamine
   (*d*) Caprolactone
5. Storage modulus and tan δ of a polymer are experimentally measured by
   (*a*) Differential scanning calorimetry
   (*b*) Thermogravimetric analysis
   (*c*) Thermomechanical analysis
   (*d*) Dynamic mechanical thermal analysis
6. A plastic bucket is manufactured by
   (*a*) Compression moulding
   (*b*) Injection moulding
   (*c*) Extrusion
   (*d*) Blow moulding
7. The monomers, A and B with reactivity ratios r_A and r_B, form alternate copolymers when
   (*a*) $r_A = r_B = 0$ (*b*) $r_A = r_B = 1$
   (*c*) $r_A > 1, r_B > 1$ (*d*) $r_A < 1, r_B < 1$
8. The degree of polymerization of a poly(methyl methacrylate) sample having number average molecular weight of 1,50,000 g/mol is__________. (C = 12, H = 1, O = 16 g/mol).
9. If the heat of fusion of 100 % crystalline polyethylene is 290 mJ/mg, a sample of polyethylene with heat of fusion of 141 mJ/mg will have _______ % crystallinity.

Q. 10 – Q. 22 carry two marks each.

10. Match the following:

P. Butyl rubber	1. Metallocene polymerization
Q. Cold SBR	2. Cationic polymerization
R. Poly(ethylene terephthalate)	3. Redox polymerization
S. Polypropylene	4. Condensation polymerization

   (*a*) P-3; Q-1; R-2; S-1 (*b*) P-2; Q-3; R-1; S-4
   (*c*) P-4; Q-3; R-1; S-2 (*d*) P-2; Q-3; R-4; S-1
11. Match the following:

P. Polyaramid	1. Baby-feeding nipple
Q. Polytetrafluoroethylene	2. Optical glasses
R. Polycarbonate	3. Non-stick cookware
S. Poly(dimethyl siloxane)	4. Bullet-proof jacket

   (*a*) P-4; Q-3; R-2; S-1 (*b*) P-2; Q-3; R-4; S-1
   (*c*) P-4; Q-1; R-2; S-3 (*d*) P-3; Q-4; R-2; S-1
12. Flexible PVC tubes are used for watering. If some organic solvents are passed through this tube, it becomes stiff. This is due to the fact that the organic solvents
   (*a*) plasticize PVC and raise T_g.
   (*b*) remove plasticizer and raise T_g.
   (*c*) remove plasticizer and lower T_g.
   (*d*) react with PVC and increase T_g
13. Match the following:

P. Plastic egg container	1. Injection moulding
Q. Water tank	2. Extrusion
R. Chair	3. Rotational moulding
S. Cable	4. Thermoforming

   (*a*) P-3; Q-1; R-4; S-2 (*b*) P-4; Q-3; R-2; S-1
   (*c*) P-2; Q-3; R-4; S-1 (*d*) P-4; Q-3; R-1; S-2
14. Match the following:

P. Flame retardant	1. 4-Methyl-2,6-di-t-butyl phenol
Q. UV absorber	2. Azocarbonamide
R. Blowing agent	3. Phenyl salisylate
S. Antioxidant	4. Aluminium trihydrate

   (*a*) P-4; Q-1; R-2; S-3 (*b*) P-4; Q-3; R-2; S-1

(*c*) P-3; Q-4; R-2; S-1 (*d*) P-2; Q-4; R-1; S-3

15. A plot of strain (%) versus time of a polymer is given below. Based on this plot and the properties as mentioned below, find out the correct combination.

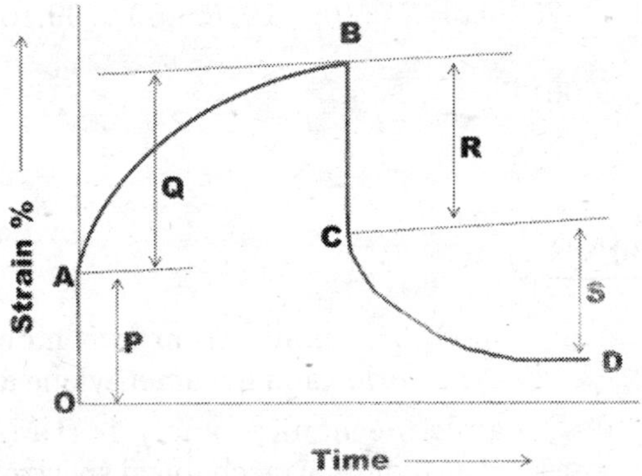

1 = Viscoelastic deformation;

2 = Elastic deformation

3 = Viscoelastic recovery;

4 = Elastic recovery

(*a*) P-1; Q-4; R-2; S-3

(*b*) P-2; Q-3; R-4; S-1

(*c*) P-3; Q-1; R-2; S-4

(*d*) P-2; Q-1; R-4; S-3

16. The plot shows apparent viscosity versus shear rate of Newtonian, Dilatent and Pseudoplastic fluids. Based on this plot and the fluid behaviour as mentioned below, find out the correct combination.

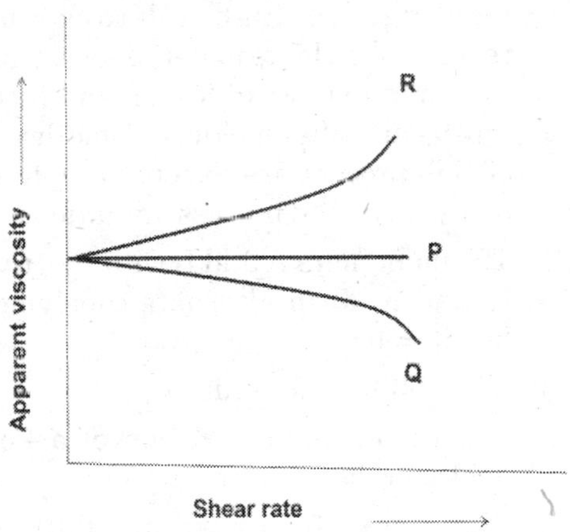

1 = Dilatent fluid;

2 = Newtonian fluid,

3 = Pseudoplastic fluid

(*a*) P-1; Q-2; R-3

(*b*) P-3; Q-1; R-2

(*c*) P-2; Q-3; R-1

(*d*) P-2; Q-3; Q-1

17. Plot of the modulus versus temperature of different types of polymers is given below.

Based on this plot and the nature of the polymers as mentioned below, find out the correct combination.

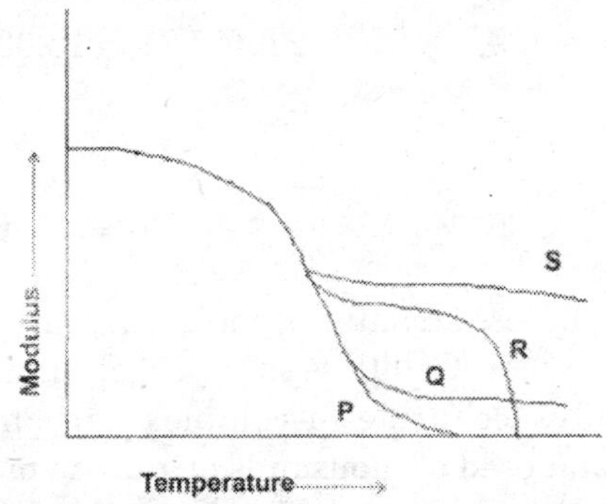

1 = An amorphous polymer of high molecular weight having entanglements

2 = An amorphous polymer of moderate molecular weight

3 = Highly crosslinked polymer

4 = Semi-crystalline polymer

(*a*) P-2; Q-1; R-3; S-4

(*b*) P-1; Q-2; R-3; S-4

(*c*) P-2; Q-1; R-4; S-3

(*d*) P-1; Q-3; R-4; S-2

18. The T_g of homopolymers of A and B are +100°C and –70°C respectively. The T_g of a random copolymer of A and B having 40 wt% A and 60 wt% B is ________°C.

19. The number average molecular weight of a polymer prepared from $HO(CH_2)_{14}COOH$ is 24,000 g/mol. The conversion of the monomer required to reach the above molecular weight is ________%. (C = 12, H = 1, O = 16 g/mol).

20. Glass fibers in nylon provide reinforcement. The modulus of elasticity for each component of the composite is; $E_{glass} = 10.5 \times 10^6$ psi; $E_{nylon} = 0.4 \times 10^6$ psi. If the nylon contains 30 vol % E-glass, the fraction of the applied force is carried by the glass fiber is________. (Assume that both glass fiber and nylon have equal strain).

21. The solubility parameter of a polymer having cohesive energy density (E_{coh}) 43870 J/mol and molar volume (V) 136 cm³/mol is ________ $(J/cm^3)^{1/2}$.

22. The heat of polymerization of styrene is 20 Kcal/mol. Heat of 5×10^5 Kcal will be released on polymerization of ________ Kg of styrene

ANSWERS

1. (c) **2.** (a) **3.** (d) **4.** (b) **5.** (d) **6.** (b) **7.** (a)

8. (1500 to 1500) **9.** (48.52 to 48.72) **10.** (d) **11.** (a) **12.** (b) **13.** (d)

14. (b) **15.** (d) **16.** (c) **17.** (c) **18.** (–25.00 to –24.50) **19.** (98.80 to 99.10)

20. (0.90 to 0.93) **21.** (17.91 to 18.00) **22.** (2600 to 2600)

EXPLANATIONS

1. (c) Cellulose : A naturally occurring carbohydrate polymer. Cellulose is a largely indigestible polysaccharide (a complex carbohydrate), composed of thousands of glucose molecules. Cellulose is naturally present in plants. It is an important structural component of the primary cell wall of green plants, many forms of algae and the oomycetes.

2. (a) Most polymers are poly disperse, they contain more than one chain length.

The average distribution of chain masses can be described in more than one way:

- M_n, the number-average molecular weight
- M_w, the weight-average molecular weight
- M_z, the z-average molecular weight
- M_v, the viscosity-average molecular weight

$M_z > M_w > M_v > M_n$

Each value is determined by an aspect of polymer structure.

3. (d) Rubber is a natural polymer of Isoprene (2-Methyl -1, 3 – Butadiene). It is a linear, 1, 4 – addition polymer of Isoprene.

Natural rubber has elastic properties and it undergoes long range of reversible extension even if relatively small forces are applied on it. Therefore, it is also known as Elastomer. Natural rubber is prepared from latex which is a Colloidal Solution of Rubber in Water.

4. (b) Nylon 6 is only made from one kind of monomer, a monomer called caprolactam.

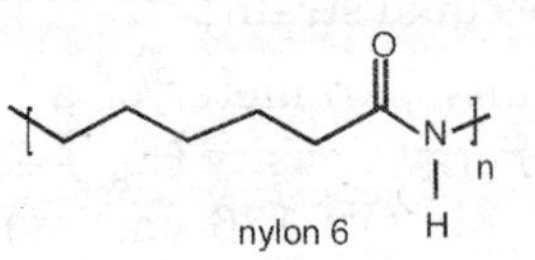

nylon 6

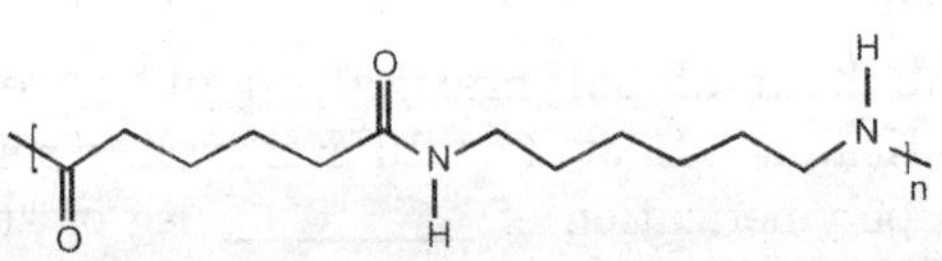

nylon 6,6

Nylon 6, 6 is made from two monomers, adipoyl chloride and hexamethylene diamine.

5. (d) Dynamic mechanical analysis (DMA), also called as dynamic mechanical spectroscopy is a technique used to study and characterize materials. It is most useful for studying the viscoelastic behavior of polymers. A sinusoidal stress is applied and the strain in the material is measured, allowing one to determine the complex modulus.

6. (b) Plastic injection molding is the process used for manufacturing plastic parts. Injection molding is used to develop many things such as electronic housings, containers, bottle caps, combs, automotive interiors, and most other plastic products available today. It is ideal for producing high volumes of plastic parts due to the reason that several parts can be produced in each cycle by using multi-cavity injection molds. Some of the advantages of injection molding are high tolerance precision, repeatability, large material selection, low labor cost, minimal scrap losses, and little need to finish parts after molding. Some disadvantages of this process are expensive upfront tooling investment and process limitations.

7. (a) The monomers A and B with reactivity ratios R_A and R_B form alternate copolymers when, $R_A = R_B = 0$.

8. (1500 to 1500) To find:

Degree of Polymerization of a Polymethyl (methacrylate)

Chemical formula of Poly (methyl methacrylate) is $(C_5O_2H_8)_n$

Degree of Polymerization

$$= \frac{\text{Average molecular weight}}{(\text{Mol.wt})_{C_5} + (\text{Mol.wt})_{H_8} + (\text{Mol.wt})_{O_2}}$$

$$= \frac{15000}{12 \times 5 + 1 \times 8 + 16 \times 2} = 1500$$

9. (48.52 to 48.72)

% crystallinity

$$= \frac{\text{(Heat of fusion) polyethylene}}{\text{(Heat of fusion) crystallite polyethylene}}$$

$$= \frac{141}{290} \times 1 = 48.62\%$$

Correct matches are as follows:

10. (d)

P. Butyl rubber	2. Cationic polymerization
Q. Cold SBR	3. Redox polymerization
R. Poly(ethylene terephthalate)	4. Condensation polymerization
S. Polypropylene	1. Metallocene polymerization

Condensation Polymerization: In This type of polymerization, polymers are generated with the help of condensation of two monomer units with the loss of small molecules such as H_2O, HCl, and NH_3 etc. Here the monomer units must have two functional groups in order to condensation reaction took place. Dacron®, nylon 6 and nylon 66 are the few examples for condensation polymers.

Cationic polymerization is a type of chain growth polymerization in which a cationic initiator transfers charge to a monomer which then becomes reactive. This reactive monomer goes on to react similarly with other monomers to form a polymer.

Redox Polymerization: Reduction of hydrogen peroxide or an alkyl hydrogen peroxide by iron. Other reductants such as Cr^{2+}, V^{2+}, Ti^{3+}, Co^{2+}, and Cu^{+} can be employed in place of ferrous ion in many instances.

Metallocene polymerization: The substitution pattern of the "-Cp ligands of metallocene polymerization catalysts is the major factor influencing both propagation rate and type and extent of chain transfer reactions, that in turn determine the molecular weight of the poly(a-olefin) produced.

12. (b) Flexible PVC tubes are used for watering. If some organic solvents are passed through this tube, it becomes stiff. This is due to the fact that the organic solvents remove plasticizer and raise Tg.

13. (d) P-4; Q-3; R-1; S-2

Applications of :

1. **Injection molding** : wire spools, packaging, bottle caps, automotive dashboards, pocket combs
2. **Extrusion :** Trim parts as used in automobile and construction equipment, railings, window frame members, structural parts etc
3. **Rotational Molding** : storage tanks, furniture, road signs and bollards, planters, pet houses, toys, bins and refuse containers, doll parts, road cones, footballs, helmets, canoes, rowing boats, tornado shelters, kayak hulls and playground slides
4. **Thermoforming** : disposable cups, containers, lids, trays, blisters, clamshells, and other products for the food, medical, and general retail industries

14. (b) P-4; Q-3; R-2; S-1

15. (d) P-2; Q-1; R-4; S-3

1 = Viscoelastic deformation;

2 = Elastic deformation

3 = Viscoelastic recovery;

4 = Elastic recovery

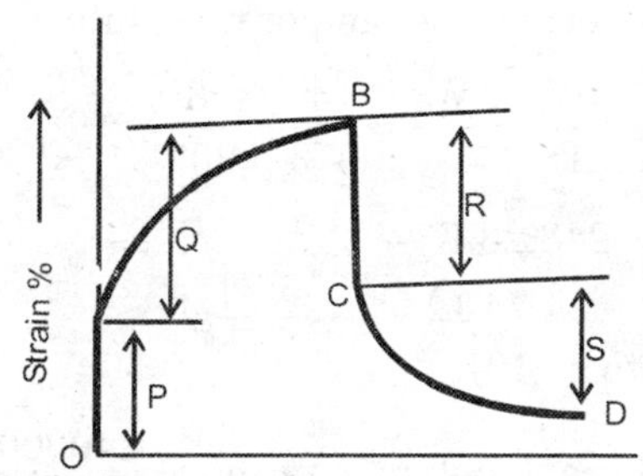

16. (c) P-2; Q-3; R-1

1 = Dilatent fluid;

2 = Newtonian fluid,

3 = Pseudoplastic fluid

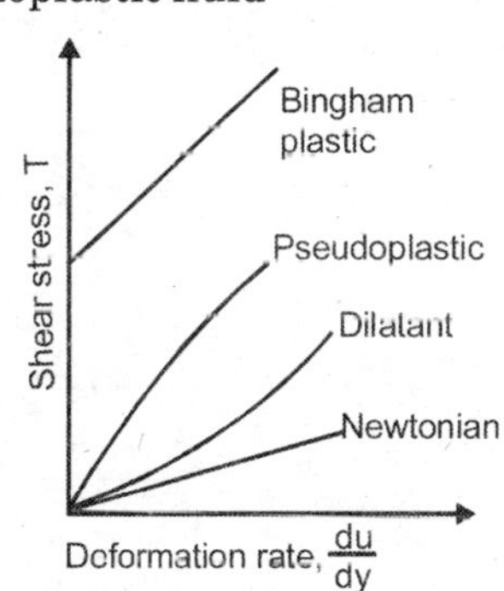

(a)

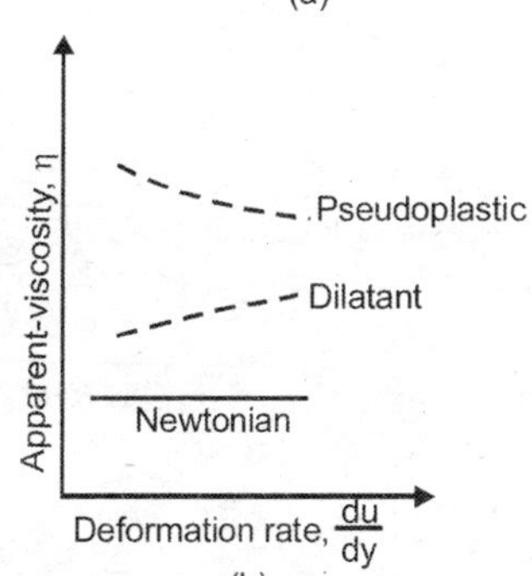

(b)

17. (c) P-2; Q-1; R-4; S-3

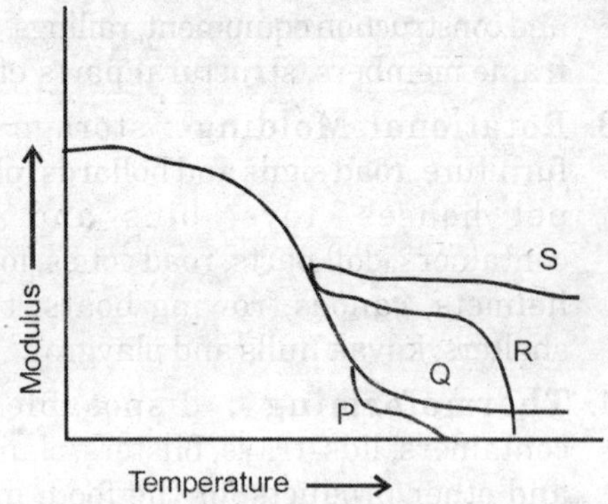

1 = An amorphous polymer of high molecular weight having entanglements

2 = An amorphous polymer of moderate molecular weight

3 = Highly crosslinked polymer

4 = Semi-crystalline polymer

18. (–25.00 to –24.50)

Convert all the temperature into Kelvin scale

$$\frac{1}{T_g} = \frac{W_1}{T_{g1}} = \frac{W_2}{T_{g2}} = \frac{.4}{373} + \frac{.6}{203}$$

$T_g = 248.26$ K

$T_g = -24.74°C$

19. (98.80 to 99.10)

$$\text{Degree of polymerization} = \frac{240000}{258} \times 100$$

$$= 93.023\%$$

20. (0.90 to 0.93)

$$E_{Total} = 10.5 \times 10^6 \times 0.3 + 0.4 \times 10^6 \times 0.7$$

$$= 3.43 \times 10^6$$

$$\delta_{glass} = \delta_{nylon}$$

$$\frac{P_{glass} l_{glass}}{E_{glass} A_{glass}} = \frac{P_{nylon} \times l_{nylon}}{E_{nylone} \times A_{nylone}}$$

$$\frac{P_{glass}}{P_{nylon}} = \frac{10.5 \times 10^6 \times 0.3}{0.4 \times 10^6 \times 0.7}$$

$$\% \text{ fraction} = \frac{P_{glass}}{P_{total}} = \frac{11.25}{11.25 + 1} = \frac{11.25}{12.25} = 91.8\%$$

21. (17.91 to 18.00)

$$\{\text{Solubility Parameter of a polymer}\} = \sqrt{\frac{E_{coh}}{V}}$$

$$= \sqrt{\frac{43870}{136}} = 17.96(\text{J/cm}^3)^{1/3}$$

22. (2600 to 2600)

Styrene chemical formulas is $C_6H_5CH = CH_2$

Mass of styrene

$$= \frac{\text{Total heat released} \times \text{Total mol.wt of styrene}}{\text{Heat of Polymerization of styrene}}$$

$$\text{Mass of styrene required} = \frac{5 \times 10^5 \times 104}{20}$$

$= 2600000$ gms

$= 2600$ kgs.

XE-G

FOOD TECHNOLOGY

Syllabus

Section 1: Food Chemistry and Nutrition

Carbohydrates: structure and functional properties of mono-, oligo-, & polysaccharides including starch, cellulose, pectic substances and dietary fibre, gelatinization and retrogradation of starch. Proteins: classification and structure of proteins in food, biochemical changes in post mortem and tenderization of muscles. Lipids: classification and structure of lipids, rancidity, polymerization and polymorphism. Pigments: carotenoids, chlorophylls, anthocyanins, tannins and myoglobin. Food flavours: terpenes, esters, aldehydes, ketones and quinines.

Enzymes: specificity, simple and inhibition kinetics, coenzymes, enzymatic and nonenzymatic browning. Nutrition: balanced diet, essential amino acids and essential fatty acids, protein efficiency ratio, water soluble and fat soluble vitamins, role of minerals in nutrition, co-factors, anti-nutrients, nutraceuticals, nutrient deficiency diseases. Chemical and biochemical changes: changes occur in foods during different processing.

Section 2: Food Microbiology

Characteristics of microorganisms: morphology of bacteria, yeast, mold and actinomycetes, spores and vegetative cells, gram-staining. Microbial growt h: growth and death kinetics, serial dilution technique. Food spoilage: spoilage microorganisms in different food products including milk, fish, meat, egg, cereals and their products. Toxins from microbes: pathogens and non-pathogens including Staphylococcus, Salmonella, Shigella, Escherichia, Bacillus, Clostridium, and Aspergillus genera. Fermented foods and beverages: curd, yoghurt, cheese, pickles, soya-sauce, sauerkraut, idli, dosa, vinegar, alcoholic beverages and sausage.

Section 3: Food Products Technology

Processing principles: thermal processing, chilling, freezing, dehydration, addition of preservatives and food additives, irradiation, fermentation, hurdle technology, intermediate moisture foods. Food pack aging and storage: packaging materials, aseptic packaging, controlled and modified atmosphere storage. Cereal processing and products: milling of rice, wheat, and maize, parboiling of paddy, bread, biscuits, extruded products and ready to eat breakfast cereals. Oil processing: expelling, solvent extraction, refining and hydrogenation. Fruits a nd vegetables p rocessing: extraction, clarification, concentration and packaging of fruit juice, jam, jelly, marmalade, squash, candies, tomato sauce, ketchup, and puree, potato chips, pickles. Plantation crops processing and products: tea, coffee, cocoa, spice, extraction of essential oils and oleoresins from spices. Milk and milk products processing: pasteurization and sterilization, cream, butter, ghee, icecream, cheese and milk powder. Processing of animal products: drying, canning, and freezing of fish and meat; production of egg powder. Waste utilization: pectin from fruit wastes, uses of by-products from rice milling. Food standards and qu ality maintenance: FPO, PFA, Agmark, ISI, HACCP, food plant sanitation and cleaning in place (CIP).

Section 4: Food Engineering

Mass and energy balance; Momentum transfer: Flow rate and pressure drop relationships for Newtonian fluids flowing through pipe, Reynolds number. Heat transfer: heat transfer by conduction, convection, radiation, heat exchangers. Mass transfer: molecular diffusion and Fick's law, conduction and convective mass transfer, permeability through single and multilayer films. Mechanical operations: size reduction of solids, high pressure homogenization, filtration, centrifugation, settling, sieving, mixing & agitation of liquid. Thermal operations: thermal sterilization, evaporation of liquid foods, hot air drying of solids, spray and freeze-drying, freezing and crystallization. Mass transfer oper ations: psychrometry, humidification and dehumidification operations.

Subject-wise Analysis

XE -G : FOOD TECHNOLOGY

Year	1 mark Questions	2 marks Questions	Total Marks
2009	8	16	40
2010	9	13	35
2011	9	13	35
2012	9	13	35
2013	9	13	35
2014	9	13	35
2015	9	13	35
2016	9	13	35
2017	9	13	35
2018	9	13	35

2009

Q. 1 – 8 carry one mark each

1. Among the following amion acids, the one that has a disulfide linkage is

(*a*) (–)–proline (*b*) (–)–cystine

(*c*) (–)–cysteine (*d*) (–)–histidine

2. The method of packaging of food under sterile enviroment, after independently sterilizing the food and packing material, is termed as

(*a*) active packaging (*b*) vacuum packaging

(*c*) flexible packaging (*d*) aseptic packaging

3. Mild heat treatment of food to inactivate enzymes that would otherwise cause its deterioration during frozen storage is termed as

(*a*) stewing (*b*) blanching

(*c*) boiling (*d*) pasteurization

4. The most suitable evaporator for concentration of fruit juices is

(*a*) agitated film evaporator

(*b*) falling film evaporator

(*c*) long tube evaporator

(*d*) short tube evaporator

5. Souring of milk is promarily due to the conversion of lactose to

(*a*) lactobionic acid (*b*) lactic acid

(*c*) lactol (*d*) lactonic acid

6. The selective media used for isolating *Escherichia coli* is

(*a*) blood agar

(*b*) mannitol salt agar

(*c*) eosin methylene blue agar

(*d*) rose bengal malt extract agar

7. A method in which continuous electric current is passed through food to heat it rapidly while maintaining quality is called

(*a*) microwave cooking (*b*) irradiation

(*c*) ohmic heating (*d*) sonication

8. A cyclone separator is used for the separation of

(*a*) particles from liquid

(*b*) liquid droplets from gas

(*c*) fine particles from gas

(*d*) fine particles from solids

Q. 9 – 24 carry two mark each

9. Match the items in Group I with the most appropriate items in Group II.

GROUP I	GROUP II
P. Tocopherol	1. Oxygen binding
Q. Myoglobin	2. Yellow pigment
R. Crocetin	3. Antioxidant
S. Catechin	4. Green pigment
	5. Tanning agent

(*a*) P-3, Q-1, R-2, S-5 (*b*) P-1, Q-3, R-4, S-5

(*c*) P-3, Q-1, R-5, S-2 (*d*) P-1, Q-3, R-5, S-4

10. Two key reactions involved in enzymatic browning of food are

(*a*) hydroxylation of phenol to *p*-dihydroxybenzene followed by its oxidation to *p*-quinone

(*b*) oxidation of phenol to *p*-quinone followed by its reduction to *p*-dihydroxybenzene

(*c*) oxidation of phenol to *o*-quinone followed by its reduction to *o*-dihydroxybenzene

(*d*) hydroxylation of phenol to *o*-dihydroxybenzene followed by its oxidation to *o*-quinone

11. The correct stucture of synthetic antioxidant BHT (butylated hydroxy toluene) is

(*a*) 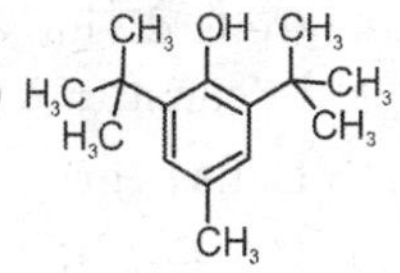

(*b*) 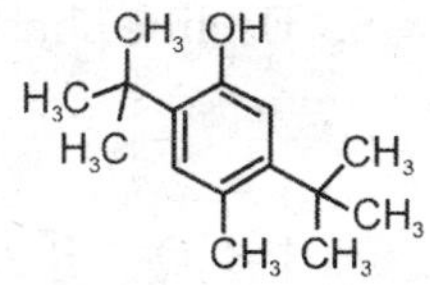

(*c*) 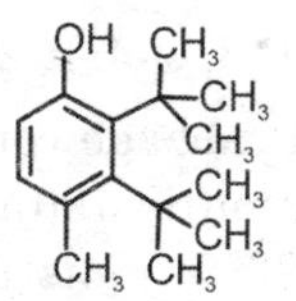

(*d*) 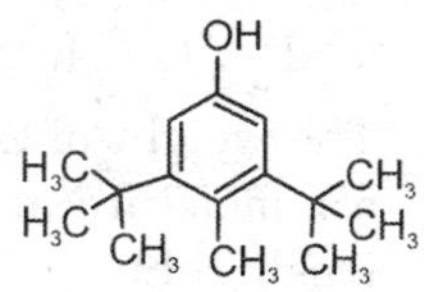

12. Wet grain was dried from an initial moisture content of 50% to a final moisture content of 20% (on wet basis). The amount.of moisture removed to get 1000 kg of the final product is

(*a*) 800 kg (*b*) 200 kg

(*c*) 300 kg (*d*) 600 kg

13. The correct pair of food borne disease and its causative microorganism is

(*a*) Hemorrhagic inflammation of intestinal wall – *Camplylobacter jejuni*

(*b*) Paratyphoid fever – *Staphylococcus aureus*

(*c*) Typhoid fever – *Solmonella typhimurium*

(*d*) Listerellosis – *Leptospira biflexa*

14. Fermentation process of vinegar production involves

(*a*) ethanolic fermentation followed by reduction of ethanol (*b*) direct acetic production without ethanolic fermentation

(*c*) anaerobic fermentation of acetone

(*d*) ethanolic fermentation followed by oxidation of ethanol

15. In a double pipe heat exchanger the outer diameter of the inner pipe is d_1 and the inner diameter of the outer pipe is d_2. The equivalent diameter of the annulus for heat transfer is

(*a*) $(d_1+d_2)/2$ (*b*) $(d_2^2-d_1^2)/d_1$

(*c*) (d_2-d_1) (*d*) $(d_2^2-d_1^2)/d_1d_2$

16. Match various phases of a typical bacterial growth cycle in Group I with most appropriate bacterial activity in Group II.

GROUP I	GROUP II
P. Lag phase	1. Number of viable cells decreases
Q. Exponential phase	2. Growth ceases and population remains constant
R. Stationary phase	3. Preparatory phase for cell division
S. Decline phase	4. Cells divide steadily at constant rate
	5. Cells aggregate

(*a*) P-4, Q-3, R-2, S-1 (*b*) P-5, Q-4, R-1, S-2

(*c*) P-2, Q-1, R-3, S-4 (*d*) P-3, Q-4, R-2, S-1

17. The weight of 20 g of dried cabbage containing 5% moisture after rehydration is 190 g. If the fresh cabbage contained 93 % moisture, the coefficient of rehydration is

(*a*) 0.70 (*b*) 0.75

(*c*) 0.07 (*d*) 0.57

18. At atmospheric pressure, the solubilities of CO_2, in a beverage at 15.5°C and 0°C are1.0 volume and 1.7 volume respectively. The pressure (in atm.) required to carbonate the beverage at 4.5 °C so as to maintain a gas volume of 4.0 is

(*a*) 1.04 (*b*) 1.47

(*c*) 1.6 7 (*d*) 1.76

Common Data Questions

Common Data for Questions 19 and 20:

The partial pressure and vapour pressure of water vapour in air at 27°C and 1 atm. are 0.028 and 0.035 atm. respectively. (Molecular weight of air is 29.)

19. The humidity of air (kg water/kg air) is

(*a*) 0.496 (*b*) 0.082

(*c*) 0.018 (*d*) 0.046

20. The percentage relative humidity of air is

(*a*) 46 (*b*) 80

(*c*) 20 (*d*) 35

Common Data for Questions 21 and 22 :

In an ice-cream manufacturing plant, 1450 litres of ice-cream was obtained from 1000 litres of icecream mix. The composition of ice-cream mix was as follows:

Fat : 12.0%, Sugar : 15.0%, Milk solids not fat : 11.0%, Stabilizer & emulsifier : 0.3%

21. Specific gravity of ice-cream mix at 16°C is

(*a*) 1.096 (*b*) 0.196

(*c*) 1.906 (*d*) 0.916

22. Percent over run in the ice-cream was

(*a*) 35 (*b*) 50

(*c*) 40 (*d*) 45

Linked Answer Questions

Statement for Linked Answer Questions 23 - 24:

In an experiment, the thermal death time (TDT) values for a microorganism were obtained as 2.78 minutes and 9.98 minutes at 121.1 °C and 115.5 °C, respectively.

23. The z-value (°C) of the microorganism is

(*a*) 9.91 (*b*) 9.19

(*c*) 1.99 (*d*) 0.19

24. The TDT value (minutes) at 110°C is

(*a*) 35.1 (*b*) 25.8

(*c*) 12.9 (*d*) 21.9

ANSWERS

1. (c)	**2.** (d)	**3.** (b)	**4.** (b)	**5.** (b)	**6.** (c)	**7.** (c)
8. (a)	**9.** (b)	**10.** (a)	**11.** (a)	**12.** (b)	**13.** (b)	**14.** (d)
15. (a)	**16.** (d)	**17.** (b)	**18.** (a)	**19.** (c)	**20.** (d)	**21.** (a)
22. (d)	**23.** (a)	**24.** (a)				

EXPLANATIONS

1. (c) Cysteine is a proteinogenic amino acid with the formula $HO_2CCH(NH_2)CH_2SH$. The thiol side chain in cysteine participates in enzymatic reactions, as a nucleophile. The thiol is susceptible to oxidation to give the disulfide derivative cystine, which serves vital role in many of the proteins.

2. (d) It is defined as the filling of a commercially sterile product into a sterile container under aseptic conditions and hermetically sealing the containers so that reinfection can be temperature & pressure conditions prevented.This results in a product which is shelf stable at ambient conditions.The term aseptic is derived from the Greek word "septicos" which means the absence of putrefactive micro organisms.

3. (b) Blanching operations are designed to expose the entire product to high temperatures for a given short period of time, by using steam heating indirectly for fruit juices and directly for vegetables. The primary function of this operation is to inactivate or retard bacterial and enzyme action, which could otherwise cause rapid degeneration of quality. Two other desirable effects include the expelling of air and gases in the product, and a reduction in the product volume.

4. (b) Falling film evaporator : The liquid to be concentrated is supplied to the top of the heating tubes and distributed in such a way as to flow down the inside of the tube walls as a thin film. The liquid film starts to boil due to the external heating of the heating tubes and is partially evaporated as a result. Residual film liquid and vapor is separated in the lower part of the calandria and in the downstream centrifugal droplet separator. It is essential that the entire film heating surface, especially in the lower regions, be evenly and sufficiently wetted with liquid.

5. (b) *Lactic acid* fermentation is performed by *lactic acid* bacteria, which convert simple carbohydrates such as glucose, sucrose, or galactose to *lactic acid*. These bacteria can also grow in the mouth. So, Lactose gets converted into lactic acid.

6. (c) EMB is the selective and differential medium for ccliforms. It is a blend of two stains, eosin and methylene blue in the ratio of 6:1.

7. (c) *Ohmic heating* is a food processing process where an electric current will be passed through the food with the main purpose of heating the food. In OH, there is no need to transfer *heat* through solid-liquid interfaces or inside solid particles once the energy is dissipated directly into the foods.

8. (a) Cyclone separator is a device which can separate particulates from gas or liquid. But, in food industry ,we mainly deal with liquid. So,it can separate particles from liquid.

9. (b) 1. Tocopherols (TCP) are a class of organic chemical compounds (more precisely, various methylated phenols), many of which have vitamin E activity.

   2. *Myoglobin* (symbol Mb or MB) is an iron- and oxygen-binding protein found in the muscle tissue of vertebrates and in all mammals. It is related to hemoglobin, which is the iron- and oxygen-binding protein in blood, to be specific in the red blood cells.

   3. *Crocetin* is a natural apocarotenoid dicarboxylic acid that is present in the crocus flower and Gardenia jasminoides (fruits). It forms brick red crystals with a melting.

   Melting point : 285 °C (545 °F; 558 K)

   Chemical formula : $C_{20}H_{24}O_4$

   EC Number : 248-708-0

   Acidity (pK_a) : 4.39

   4. Catechin : *Catechin* is a flavan-3-ol, a type of natural phenol and antioxidant. It is a plant secondary metabolite. It belongs to the group of flavan-3-ols (or simply flavanols), part of the chemical family of flavonoids.

10. (a) Enzymatic browning takes place when the enzyme polyphenol oxidase or other enzymes catalyze the oxidation of phenols in the fruit to form compounds called quinones. The quinones can then polymerize to form melanins, which cause the brown pigments. Enzymatic browning of fruits and vegetables is usually not desirable and creates heavy economic losses for growers. Enzymatic browning is beneficial for: Developing flavor in tea (here the reaction is incorrectly called fermentation) Developing color and flavor in dried fruit such as figs and raisins.Enzymatic browning is detrimental to:Fresh fruit and vegetables, in particular apples and potatoes Seafood such as shrimp.

11. (a) Butylated hydroxytoluene (BHT), also known as dibutylhydroxytoluene, is a lipophilic organic compound, chemically a derivative of phenol, that is useful for its antioxidant properties.U.S.

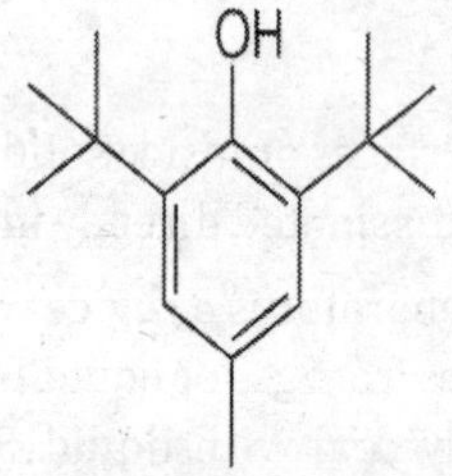

12. (b) 1000 kg of the final product consists of 800kg of product without moisture and 200 kg of moisture. Initial Product consists of 800 kg of product without moisture and 400 kg of moisture. Difference of Initial moisture and final moisture will be 200 kg.

13. (b) Only option (C) is correct. *Typhoid fever* is an acute illness associated with *fever* caused by the Salmonella enterica serotype Typhimurium bacteria. It can also be caused by Salmonella paratyphi, a related bacterium that usually causes a less severe illness.

14. (d) During vinegar production, ethanolic fermentation followed by oxidation of ethanol takes place in fermentation process.

15. (a) For heat transfer the wetted perimeter is the outer circumference of the inner pipe with diameter d_1, and for heat transfer in

$$D_e = 4r_h = \frac{4\,X\,\text{flow area}}{\text{wetted perimeter}}$$

$$= \frac{4\pi(d_2^2 - d_1^2)}{4\pi d_1} = \frac{(d_2^2 - d_1^2)}{d_1}$$

In pressure-drop calculations the friction not only results from the resist-ance of the outer pipe but is also affected by the outer surface of the inner ripe. The total wetted perimeter is $\Pi(d_2 + d_1)$, and for the pressure drop in annuli

$$D_e' = \frac{4\,X\,\text{flow area}}{\text{frictional wetted perimeter}}$$

$$= \frac{4\pi(d_2^2 - d_1^2)}{4\pi(d_2 + d_1)} = (d_2 - d_1)$$

16. (d) The growth of bacteria (or other microorganisms, as protozoa, microalgae or yeasts) in batch culture can be modeled with four different phases: lag phase (A), log phase or exponential phase (B), stationary phase (C), and death phase(D).

1. During lag phase, bacteria adapt themselves to growth conditions. It is the period where the individual bacteria are maturing and not yet able to divide.
2. The log phase (sometimes called the logarithmic phase or the exponential phase) is a period characterized by cell doubling.

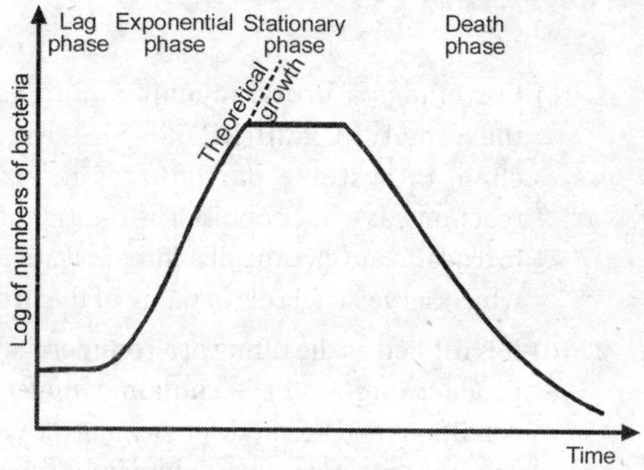

3. The stationary phase is often due to a growth-limiting factor such as the depletion of an essential nutrient, and/or the formation of an inhibitory product such as an organic acid. Stationary phase results from a situation in which growth rate and death rate are equal.
4. At death phase (decline phase), bacteria die. This could be caused by lack of nutrients, environmental temperature above or below the tolerance band for the species, or other injurious conditions.

17. (b) Rehydration coefficient. The weight of rehydrated sample is 60 g (W_r); the weight of dried sample is 10 g (W_d) and its moisture is 5% (W_u); raw material before drying had 87% water (A); rehydration coefficient is :

$$\frac{W_r}{\frac{W_d - W_u x100}{100 - A}} = \frac{190}{\frac{20 - .05x100}{100 - 93}} = .75$$

19. (c) Humidity of air,

$$w = \frac{.622P_v}{P = P_v} = \frac{.622\,X.028}{(1 - .028)} = .018$$

Where, Partial pressure of water vapour,

$P_v = .028$ atm, $P = .035$ atm

20. (d) $\varphi = \frac{wP}{(.622 + w)P_g}$

$$= \frac{.018X100}{(.622 + .018).035}$$

$= 80.35\%$

21. (a)
1. Fat : 12.0%
2. Sugar : 15.0%
3. MSNF : 11%
4. Stabilizer and emulsifier = 0.3%
5. Water : (100-38.3)% = 61.7%
6. Sum of 1,2,3 and 4 is called as TS(Total Solids) = 38.3%

$$\rho = \frac{1}{(\frac{\%fat}{100}X1.075) + \left[(\frac{\%TS}{100} - \frac{\%fat}{100})X.63\right] + \frac{\%Water}{100}}$$

$= 1.096\,kg/L$ of mix

22. (d) % over – run =

$$= \frac{(\text{Volume of ice cream} - \text{volume of mix used})}{\text{Volume of mix used}}$$

$$= \frac{(1450-1000)}{1000} X100 = 45\%$$

23. (a) $z - value(^{\circ}C) = \frac{(T_2 - T_1)}{\log D_1 - \log D_2}$

$$= \frac{(115.5-121.1)}{\log 2.78 - \log 9.98} = 10.0$$

24. (a) $9.91 = \frac{(T_2 - T_1)}{\log D_1 - \log D_2}$

$$9.91 = \frac{(110-115.5)}{\log 9.98 - \log D_2}$$

By Solving above equation, we get D_2=35.1 minutes

Q.(1 – 9) carry one mark each.

1. A food material contains 70% moisture (wet basis). The food is dried for 3 hours at 80°C air temperature in a tray dryer such that 80% of its initial moisture is removed. Final moisture content (wet basis) of the dried food is

(*a*) 31.82% (*b*) 46.67%

(*c*) 56.00% (*d*) 20.01%

2. Liquid A obeys power law equation $\sigma = k\,\dot{\gamma}^n$ (as shown in the attached figure) where σ is shear stress, $\dot{\gamma}$ is shear rate, k is consistency index and n is flow behavior index.

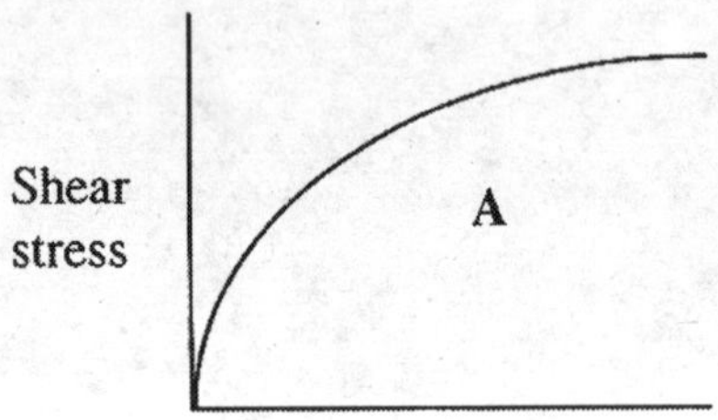

The correct unit of consistency index and nature of liquid are

(*a*) Pa s and shear thinning

(*b*) Pa s^n and shear thickening

(*c*) Pa s and shear thickening

(*d*) Pa s^n and shear thinning

3. Identify the amino acid tyrosine from the following structures

(*a*) CH_3-CH-CH-COOH (with OH and NH_2 on the two CH groups)

(*b*) 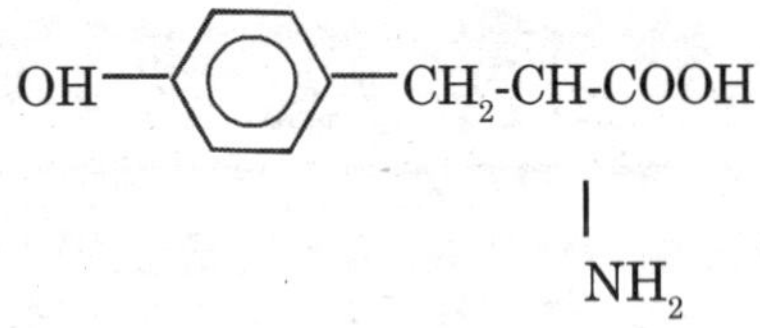

(*c*) $(CH_3)_2$-CH-CH_2-CH-COOH (with NH_2 on the CH)

(*d*) 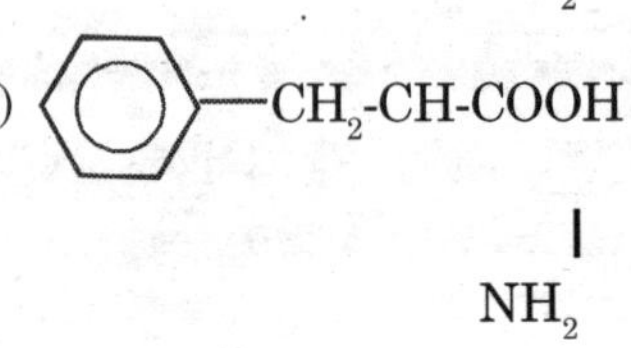

4. Saponification number of a fat is the milligrams of KOH required to saponify 1 g of fat. The correct statement on saponification is

(*a*) fat with high amount of low molecular weight fatty acids will have high saponification number

(*b*) butter has low saponification number

(*c*) fatty acids with long carbon chains have high saponification number

(*d*) fat with low *Reichert-Meissel* number has very high saponification number

5. The expansion of the terms HACCP and GRAS are

(*a*) Hygienic Associated Critical Control Point; Grossly Recommended As Safe

(*b*) Hazard Analysis and Critical Control Point; Generally Recognized As Safe

(*c*) Hygienic and Aesthetic Concept of Critical Products; Generally Recognized As Safe

(*d*) Hazard Analysis and Critical Control Point; Grossly Recommended As Safe

6. Which two of the following statements are NOT the objectives of homogenization of milk?

(*i*) Counteracting segregation for the most part of creaming thus avoiding sedimentation or phase separation

(*ii*) Arresting rancidity of fat globules in milk

(*iii*) Increasing fluidity of milk by lowering viscosity

(*iv*) Improving the colour of the milk (more whitish)

(*v*) Improving milk stability by preventing partial coalescence of fat globules

(*a*) (*i*) and (*ii*) (*b*) (*ii*) and (*iii*)

(*c*) (*iii*) and (*iv*) (*d*) (*iv*) and (*v*)

7. Shelf-life of fish can be extended by chilling as it

(*a*) reduces chemical activity of food constituents and increases biochemical activity

(*b*) reduces water activity and increases biochemical reaction rate

(*c*) reduces chemical and biochemical reactions in fish cells

(*d*) destroys pathogenic microbes

8. Major spoilage organisms of poultry meat at low temperatures are

(*a*) *Candida* and *Staphylococcus*

(*b*) *Torula* and *Clostridium*

(*c*) *Pseudomonas* and *Acinetobacter*

(*d*) *Flavobacteria* and *Lactobacillus*

9. The appropriate explanation for spoilage of egg, stored at low temperature, might be due to

(*a*) shell of egg is porous and only fungal hyphae can enter and contaminate the egg liquid

(*b*) shells are non-porous and the spoilage is mainly attributed to chemical decomposition (*c*) shell of egg is porous and microor-ganisms contaminating the shell penetrate it and cause the spoilage

(*d*) eggs are contaminated before they are laid by hen

Q.(10–22) carry two marks each.

10. Two faces of a metal plate having thermal conductivity 17 W m^{-1} K^{-1} and thickness 10 mm are maintained at 80°C and 100°C. If the thickness of the plate is increased by 20% and the temperature of the hotter face is increased to 120°C, then the percent increase in heat flux under steady state heat transfer is

(*a*) 20.67 (*b*) 40.00

(*c*) 59.99 (*d*) 66.67

11. Match the items in Group I with the most appropriate items in Group II

Group I	Group II
P. Freeze concentration	1. Triple point of water
Q. Reverse osmosis	2. Heat transfer by conduction
R. Drum drying	3. Eutectic point
S. Freeze drying	4. Radiation heat transfer
	5. Concentration polarization

(*a*) P-4, Q-5, R-2, S-1 (*b*) P-3, Q-2, R-5, S-1

(*c*) P-3, Q-5, R-2, S-1 (*d*) P-1, Q-2, R-4, S-3

12. Match the following items in Group I and Group II in relation to nutritional requirement of human body

Group I	Group II
P. Calcium and Phosphorus	1. Elements not needed in diet
Q. Vitamin D	2. Promotes absorption of iron
R. Manganese and Chromium	3. Elements that are required required in small quantities
S. Vitamin K	4. Promotes the absorption of Calcium
	5. Essential for normal clotting of blood
	6. Elements that are required in large quantities

(*a*) P-6, Q-2, R-1, S-5 (*b*) P-5, Q-2, R-6, S-4

(*c*) P-6, Q-4, R-3, S-5 (*d*) P-2, Q-5, R-1, S-4

13. 9.5 g of corn flakes containing 5% moisture (wet basis) is oxidized completely to CO_2 and H_2O by ignition in a Bomb Calorimeter (as shown in the figure). The combustion increases the temperature of 2500 g of water from 15°C to 27°C. Assume that the heat capacity and latent heat of vaporization of water are 4.187 kJ $kg^{-1}K^{-1}$ and 2257 kJ kg^{-1}, respectively. Neglect any sensible heat gain by water vapour. The calorific value of the flake is

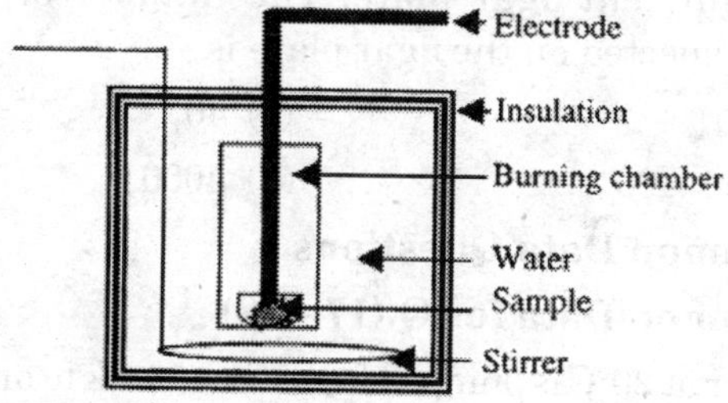

(*a*) 18.28 kJ g^{-1} (4.37 kcal g^{-1})

(*b*) 9.79 kJ g^{-1} (2.34 kcal g^{-1})

(*c*) 14.04 kJ g^{-1} (3.36 kcal g^{-1})

(*d*) 22.43 kJ g^{-1} (5.36 kcal g^{-1})

14. Match the following items in Group I and Group II in relation to permitted food additives/preservatives in India

Group I	Group II
P. Jelly	1. Calcium propionate
Q. Edible oil	2. Monosodium glutamate
R. Meat flavour enhancer	3. Sodium benzoate
S. Bread	4. Butylated hydroxylated anisole
	5. Tricalcium silicate

(*a*) P-3, Q-4, R-2, S-1 (*b*) P-5, Q-3, R-2, S-4

(*c*) P-1, Q-3, R-4, S-5 (*d*) P-2, Q-3, R-1, S-5

15. Preparation of sweet coated breakfast cereals like corn flakes includes several major processing steps, like

P : Soaking in water followed by steaming of corn grits

Q : Coating of sugar followed by drying of flakes

R : Breaking the whole corn into large grits

S : Flaking of cooked grits

T : Packaging of finished product

U : Toasting of flakes

V : Cleaning of whole corn

The correct sequence for the preparation of sugar coated corn flake is

(*a*) V → U → Q → P → S → R → T

(*b*) V → R → S → P → U → Q → T

(*c*) V → U → P → Q → S → R → T

(*d*) V → R → P → S → U → Q → T

16. A bacterial strain isolated from meat is inoculated in a growth medium at a cell density of 2×10^5 cell/ml. Then, 0.2 ml of the culture broth is withdrawn immediately and mixed with 0.8 ml of sterile saline. This sample is diluted by mixing 0.1 ml of it with 99.9 ml sterile water. Then 0.1

ml of this diluted solution is spread on appropriate nutrient agar plate. The number of colonies expected on the agar plate is

(*a*) 4 (*b*) 40

(*c*) 400 (*d*) 4000

Common Data Questions

Common Data for Q.(17 – 18)

Water at 20°C is pumped from a base tank to an elevated tank, 15 m above the base tank. Water flows at a rate of 5.0×10^{-3} m^3 s^{-1} through a pipe having internal diameter of 0.1023 m. Frictional energy loss in the pipe is 6.837 J kg^{-1}. The pump has an efficiency of 65%. Density and viscosity of water are 998.2 kg m^{-3} and 1.005×10^{-3} Pa s, respectively.

17. Reynolds number for water flowing through the pipe is

(*a*) 5.286×10^4 (*b*) 6.180×10^4

(*c*) 2.285×10^4 (*d*) 1.252×10^5

18. Power needed for pumping water in kW is

(*a*) 1.182 (*b*) 3.334

(*c*) 0.985 (*d*) 2.226

Common Data for Q.(19–20)

True density and bulk density of rice grain are 1230 kgm^{-3} and 740 kg m^{-3}, respectively, and that of wheat grain are 1360 kg m^{-3} and 650 kg m^{-3}, respectively.

19. The void fractions of a bed of rice and that of wheat are respectively,

(*a*) 0.331 and 0.546 (*b*) 0.662 and 0.261

(*c*) 0.398 and 0.480 (*d*) 0.398 and 0.522

20. Assume the bulk volume for the mixture of two grains follows additive rule. If the bulk density of a *mixture* of rice and wheat is 700 kg m^{-3}, weight percentage of wheat in the mixture is nearly

(*a*) 41 (*b*) 50

(*c*) 24 (*d*) 36

Linked Answer Questions

Statement for Linked Answer Q.(21–22)

D_0 value of bacterium is determined by using two thin walled glass capillary tubes filled with same bacterial suspension in distilled water. The sealed capillaries are dipped in an oil bath maintained at 121°C and kept for 60 s and 135 s, respectively. These capillaries are cooled immediately in ice water. Number of survivals remained in the respective tubes are 2000 and 300.

21. D_0 value (in minutes) of the bacterium is

(*a*) 1.32 (*b*) 0.52

(*c*) 1.52 (*d*) 2.52

22. The processing time (in minutes) to kill 99.999% of the bacteria in any food at 121°C will be

(*a*) 7.60 (*b*) 6.60

(*c*) 12.62 (*d*) 2.60

ANSWERS

1. (a)	**2.** (c)	**3.** (b)	**4.** (b)	**5.** (b)	**6.** (c)	**7.** (c)
8. (c)	**9.** (d)	**10.** (d)	**11.** (c)	**12.** (c)	**13.** (c)	**14.** (a)
15. (d)	**16.** (a)	**17.** (b)	**18.** (a)	**19.** (d)	**20.** (a)	**21.** (c)
22. (a)						

EXPLANATIONS

1. (a) Let 1 kg be the initial mass which has 0.7 kg of moisture in it.80% of the initial moisture 0.7 kg has been removed which is 0.56 kg. So , total mass left in the product is 0.44 kg. So moisture left in the product will be 0.7 kg – 0.56 kg=.14 kg. So, % moisture will be .14*100/.44=31.82%

2. (c) It's a dilatant fluid which has consistency index Pa-s because curve is originating from origin. And nature of liquid will be shear thickening since slope of the curve is decreasing in nature.

3. (b) Tyrosine (symbol Tyr or Y) or 4-hydroxyphenylalanine is one of the 20 standard amino acids that are used by cells to synthesize proteins. It is a non-essential amino acid with a polar side group. The word "tyrosine" is from the Greek tyros, meaning cheese, as it was first discovered in 1846 by German chemist Justus von Liebig in the protein casein from cheese. It is called tyrosyl when referred to as a functional group or side chain. While tyrosine is generally classified as a hydrophobicamino acid, it is more hydrophilic than phenylalanine.

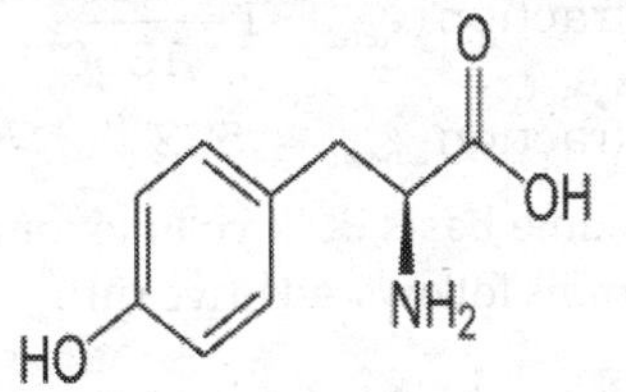

4. (b) All statements are incorrect except (b).Butter has more fat content and therefore has low saponification number.

5. (b) Hazard Analysis Critical Control Point (HACCP): HACCP is a management system in which food safety is addressed through the analysis and control of biological, chemical, and physical hazards from raw material production, procurement and handling, to manufacturing, distribution and consumption of the finished product.

Generally recognized as safe (GRAS) is an American Food and Drug Administration (FDA) designation that a chemical or substance added to food is considered safe by experts.

6. (c) Homogenization does not change color of the milk. And it does not lower the viscosity.

7. (c) Shelf-life of fish can be extended by deep freezing as it minimizes chemical and biochemical reaction in fish cells. Biochemical reactions are almost reduced.

8. (c) The organisms most commonly found growing on poultry carcases stored at 1° are pigmented and non pigmented strains of Pseudomonas, Pseudomonas putrefaciens and strains of Acinetobacter.The microflora of fresh meat is composed primarily of:

1. Gram negative aerobic rods such as Pseudomonas, Acinetobacter and Moraxella.
2. Bacillus and clostridia (e.g. C. perfringens) are also common on all types of meat.

9. (d) Eggs may get spoiled at even low temperatures if eggs are contaminated before they are laid by hen.

10. (d) $Q = \dfrac{(T_1 - T_2)}{\dfrac{L}{kA}}$,

where L is thickness of the plate , k is thermal conductivity and A is area of the plate.

$$Q_1 = \frac{20}{\dfrac{10}{kAX1000}} = 2000kA$$

$$Q_2 = \frac{40}{\dfrac{12}{kAX1000}} = 3333.33kA$$

$$\%\,\text{increase in heat flux} = \frac{Q_2 - Q_1}{Q_1}$$

$$\%\,\text{increase in heat flux} = \frac{3333.33 - 2000}{2000} X100$$

$$= 66.67\%$$

11. (c) Reverse osmosis is a process by which a solvent passes through a porous membrane in the direction opposite to that for natural osmosis when subjected to a hydrostatic pressure greater than the osmotic pressure.

Drum drying is a method used for drying out liquids from raw materials with drying drum. In the drum-drying process, pureed raw ingredients are dried at relatively low temperatures over rotating, high-capacity drums that produce sheets of drum-dried product. It uses conduction phenomenon.

Freeze drying, also known as lyophilisation or cryodessication, is a low temperature dehydration process which involves freezing the product, lowering pressure, then removing the ice by sublimation. This is in contrast to dehydration by most conventional methods that evaporate water using heat.

Freeze concentration processes consist of a crystallization section where part of the water is converted into solid ice crystals using a refrigeration system.

12. (c) Vitamin K helps in blood clotting system.Vitamin D helps in absorption of Calcium.Calcium and Phosphorous are always required in large quantities.These are macro nutrients.Manganese and Chromium are micro nutrients.

13. (c) m_{corn}XCalorific Value = m_{water}Xc_pX(27 – 15)

9.5XC.V. = 2.5X4.187X12

By solving above equation, we get calorific value = 14.04 kJ/g

14. (a)

Product	Preservatives
Jelly	Sodium benzoate
Edible Oil	Butylated hydroxylated anisole
Meat flavor enhancer	Monosodium glutamate
Bread	Calcium propionate

15. (d) V→R→P→S→U→Q→T

16. (a) 1 ml of sample contains 2 X 10^5 cells0.2 ml of the culture withdrawn will have 40000 cells, this is added to 0.8 ml of sterile saline.0.1 ml withdrawn of the immediate culture will have 4000 cells and added to 99.9 ml sterile water.0.1 ml of above culture withdrawn from the sample having 100 ml will have 4 cells.

17. (b) $$V = \frac{Q}{A} = \frac{5X10^{-3}}{\frac{\pi}{4}X.1023^2} = .6083\,m/s$$

Where, V = Velocity of fluid in pipe,m/s

A = Area of the pipe, m^2

$$Re = \frac{\rho VD}{\mu}$$

$$= \frac{998.2X.6083X.1023}{.001005} = 61808.04$$

18. (a) Water Power = $\dot{m}gh = \rho AVgh$

$$\dot{m}gh = \rho AVgh$$

$$= \frac{998.2X22X.1023^2X.6083X9.81X15}{7X4}$$

= 734.4 W

Mass flow rate = .005 X 998.2

Frictional Power = $\dot{m}$XFrictional loss

= 4.991X6.837

= 34.12 W

Total Power required to,

Pump water = Water Power + Frictional Power

= 734.4 + 34.12

= 768.52 W

= .76852 kW

Power required by pump for pumping water

= .76852/.65

= 1.182 kW

19. (d) $$\text{void fraction} = 1 - \frac{\rho_{bulk}}{\rho_{true}}$$

$$\text{void fraction}_{rice} = 1 - \frac{740}{1230}$$

$$\text{void fraction}_{rice} = .398$$

$$\text{void fraction}_{wheat} = 1 - \frac{650}{1360}$$

$$\text{void fraction}_{wheat} = .522$$

20. (a) On Volume basis,Bulk volume for the mixture of two grains follows additive rule.

$$\frac{1}{700} = \frac{1-x}{740} + \frac{x}{650}$$

By solving above equation , we get x = 41%

21. (c) $$D_0 \text{ value} = \frac{135-60}{\log 2000 - \log 300}$$

$$D_0 \text{ value} = 1.52$$

22. (a) $$\frac{135-60}{\log 2000 - \log 300} = \frac{60-x}{\log .2 - \log 2000}$$

.001% of 2000 is 0.2

By solving above equation, we get Processing time = 7.60 minutes

2011

Q.(1-9) carry one mark each.

1. The protein responsible for spongy structure in bread is

(*a*) Albumin (*b*) Zein

(*c*) Gluten (*d*) Gliadin

2. The factor most responsible for making a good ice cream is

(*a*) Water content (*b*) Homogenization

(*c*) Emulsifying agent (*d*) Mixing index

3. Listed below are some of the functions of fats in the human nutrition. Identify the INCORRECT function.

(*a*) Concentrated source of energy

(*b*) Transport of oxygen to various organs

(*c*) Absorption of fat soluble vitamins

(*d*) Synthesis of cell membranes and hormones

4. During ripening of cheese by *Penicillium roqueforti* the characteristic aroma is because of

(*a*) Methyl ketones (*b*) Aceto acetic acid

(*c*) Diacetyl (*d*) Acetoin

5. Which of the following statements is NOT TRUE in case of oxidative rancidity of fatty foods?

(*a*) Peroxides and hydroperoxides are formed during auto-oxidation

(*b*) Auto-oxidation is a complex chain reaction

(*c*) The final breakdown products of auto-oxidation are aldehydes, ketones and alcohols

(*d*) The reaction is brought about by an enzyme, called lipase

6. Which of the following group of characteristics is CORRECT in respect of *Shigella* species found as food pathogen?

(*a*) Gram positive, motile by gliding, spore forming cocci and transmitted by contaminated food

(*b*) Gram negative, motile by flagella, spore forming bacilli and transmitted by contaminated water

(*c*) Gram positive, non-motile, non-spore forming cocci and transmitted by contaminated air and water both

(*d*) Gram negative, non-motile, non-spore forming and transmitted by fecal-oral route

7. Relate the vitamins listed below (left hand side) with the associated diseases (right hand side)

P. Thiamin	1. Pellagra
Q. Nocotinic acid	2. Beriberi
R. Folic acid	3. Scurvy
S. Ascorbic acid	4. Anemia

	P	Q	R	S
(*a*)	1	2	3	4
(*b*)	4	3	2	1
(*c*)	2	1	4	3
(*d*)	3	4	1	2

8. Which of the following conditions for the heat resistance of microorganisms is CORRECT?

(*a*) Psychrophiles < Mesophiles < Thermophiles

(*b*) Psychrophiles > Mesophiles > Thermophiles

(*c*) Thermophiles > Psychrophiles > Mesophiles

(*d*) Mesophiles < Thermophiles < Psychrophiles

9. The solubility of sodium bicarbonate in water is 9.6 g/100 g at 20°C and 16.4 g/100 g at 60°C. If a saturated solution of sodium bicarbonate at 60°C is cooled to 20°C, the percentage of the dissolved salt crystallized out will be

(*a*) 20.5 (*b*) 25.4

(*c*) 41.5 (*d*) 45.2

Q.(10-22) carry two marks each.

10. A sugar syrup (density = 1040 kg/m^3 and viscosity = 1600×10^{-6} Pa.s) is required to be pumped into a tank (1.5 m diameter and 3 m height) by a 3 cm inside diameter pipe. If the liquid is required to flow under laminar flow conditions, the minimum time to fill the tank with the syrup will be

(*a*) 192.9 h (*b*) 19.3 h

(*c*) 38.6 h (*d*) 57.9 h

11. Match the following sauerkraut defects for their causative agents

P. Soft kraut	1. Due to growth of bacteria, mold and/or yeast
Q. Slimy kraut	2. Due to surface growth of *Torula* yeast
R. Rotted kraut	3. Bacterial growth does not initiate till last stage
S. Pink kraut	4. Rapid growth of Lactobacillus cucumens and L plantarum specially at elevated temperature

	P	Q	R	S
(*a*)	4	2	3	1
(*b*)	3	4	1	2
(*c*)	3	1	2	4
(*d*)	4	3	1	2

12. Match the following carbohydrates with their use in the food processing

P. High amylose starch	1. White sauces in cook freeze operations

Q. Pectin	2. Edible film for wrapping candies
R. Starch phosphates	3. As humectant in confectionary
S. Glucose	4. Setting agent in jams and jellies

	P	Q	R	S
(*a*)	1	2	4	3
(*b*)	2	4	1	3
(*c*)	3	1	2	4
(*d*)	4	3	1	2

13. Match the food items and their principal flavouring agents given in the two columns below

P. Butter	1. Menthol
Q. Orange	2. Limonene
R. Cloves	3. Eugenol
S. Mint	4. Diacetal

	P	Q	R	S
(*a*)	3	2	4	1
(*b*)	2	3	1	4
(*c*)	4	1	3	2
(*d*)	4	2	3	1

14. Match the food items on left hand side with their colloidal nature on right hand side

P. Curd	1. Foam
Q. Butter	2. Emulsion
R. Vegetable soup	3. Sol
S. Whipped egg white	4. Gel

	P	Q	R	S
(*a*)	2	1	3	4
(*b*)	4	3	2	1
(*c*)	4	2	3	1
(*d*)	3	4	1	2

15. In an actively growing (exponential phase) yeast culture, the cell concentration increased from 10^3 cells per ml to 10^7 cells per ml in 4 h. The doubling time of the yeast is

(*a*) 120 minutes (*b*) 30 minutes

(*c*) 18 minutes (*d*) 60 minutes

16. The steps followed in Gram's staining of microorganisms are

P. Washing with neutral organic solvent

Q. Counter staining with a contrast dye

R. Staining with basic dye

S. Fixing the colour with a suitable mordant

Identify the CORRECT sequence.

(*a*) $Q \rightarrow S \rightarrow R \rightarrow P$ (*b*) $P \rightarrow Q \rightarrow R \rightarrow S$

(*c*) $Q \rightarrow P \rightarrow S \rightarrow R$ (*d*) $R \rightarrow S \rightarrow P \rightarrow Q$

COMMON DATA QUESTIONS

Common Data for Questions 17 and 18 :

A continuous dryer was used to dry 12 kg/min of a blanched vegetable containing 50% moisture (wet weight basis) to give a product containing 10% moisture. As the dryer could handle feed material with moisture content not more than 25%, a part of dried material was recycled and mixed with the fresh feed.

17. The evaporation rate in the dryer would be

(*a*) 2.08 kg/min (*b*) 5.33 kg/min

(*c*) 3.33 kg/min (*d*) 2.93 kg/min

18. The recycle ratio to achieve the drying requirement would be

(*a*) 2.00 (*b*) 1.25

(*c*) 1.67 (*d*) 4.16

Common Data for Questions 19 and 20 :

An enzyme has a K_m of 4.7×10^{-5} M and V_m is 22 micro moles per litre per min. The enzyme reaction is carried out at a substrate concentration of 2×10^{-4} M.

19. The initial reaction velocity for this enzyme catalyzed reaction will be

(*a*) 6.5 micro moles per litre per min

(*b*) 17.8 micro moles per litre per min

(*c*) 13.0 micro moles per litre per min

(*d*) 8.9 micro moles per litre per min

20. Adition of a competitive inhibitor ($K_i = 3 \times 10^{-4}$ M) at concentration of 5×10^{-4} M to the above reaction system will result in the inhibition of enzymatic reaction by

(*a*) 24.0% (*b*) 62.5%

(*c*) 76.0% (*d*) 57.5%

LINKED ANSWER QUESTIONS

Statement for Linked Answer Questions 21 & 22 :

The F – Value at 121.1 °C, equivalent to 99.9999 per cent destruction of a strain of *Clostridium botulinum*. is 1.8 min.

(D_0 and F_o represent the decimal reduction time and lethality of the destruction process at reference temperature, respectively)

21. The D_0 value of the organism will be

(*a*) 10.8 min (*b*) 0.3 min

(*c*) 6.0 min (*d*) 0.2 min

22. The F_0 value, based on 12D concept using the D_0 value of the above organism and a most likely spore load in the product of 100, will be

(*a*) 3.0 min (*b*) 1.2 min

(*c*) 1.5 min (*d*) 4.2 min

ANSWERS

1. (c)	**2.** (b)	**3.** (b)	**4.** (a)	**5.** (d)	**6.** (d)	**7.** (c)
8. (a)	**9.** (c)	**10.** (b)	**11.** (b)	**12.** (b)	**13.** (d)	**14.** (c)
15. (c)	**16.** (d)	**17.** (b)	**18.** (c)	**19.** (b)	**20.** (a)	**21.** (b)
22. (d)						

EXPLANATIONS

1. (c) Gliadin and Glutenin then combine to form the protein called GLUTEN. Gluten is a tough, rubbery and elastic substance, which has the capacity to stretch and rise due to the action of baking powder or yeast. When flour is mixed with water, the gluten swells to form a continuous network of fine strands.

2. (b) After pasteurization, the mix is put through a **homogenizer**. During this process the mix is pushed at high pressure through a nozzle to break up any clumps in the **ice cream**, **creating** a creamy, smooth texture.

3. (b) Fat does not transport oxygen to various organs of human body.

4. (a) A **methyl ketone** is a **ketone** in which a **methyl** group is attached to the carbonyl group.The mold Penicillium roqueforti which is involved in the ripening of blue cheese can oxidize fatty acids to the corresponding methyl ketones with one less carbon atom. Therefore aroma is produced due to Methyl Ketones

5. (d) Lipase is an enzyme that breaks down dietary fats into smaller molecules called fatty acids and glycerol. A small amount of lipase, called gastric lipase, is made by cells in your stomach. This enzyme specifically digests butter **fat** in your food.

6. (d) **Shigella spp**. are **bacteria** that cause **shigellosis**, also known as bacillary dysentery. They are a highly infectious organism, with foodborne outbreaks often involving infected **food** handlers. Unlike other common foodborne **pathogens**, humans are the only natural hosts of **Shigella spp**.

7. (c) P-2,Q-1,R-4,S-3

Vitamin A deficiency can cause keratomalacia.

Thiamine (vitamin B1) deficiency causes **beriberi** and Wernicke–Korsakoff syndrome.

Riboflavin (vitamin B2) deficiency causes ariboflavinosis.

Niacin (vitamin B3) deficiency causes **pellagra**. Pantothenic acid (vitamin B5) deficiency causes chronic paresthesia.

Vitamin C (ascorbic acid) short-term deficiency can lead to weakness, weight loss and general aches and pains. Longer-term depletion may affect the connective tissue. Persistent vitamin C deficiency leads to scurvy.

8. (a) Psychrophiles < Mesophiles < Thermophiles

The cold-loving organisms are **psychrophiles** defined by their ability to grow at 0 degrees. A variant of a **psychrophile** (which usually has an optimum T of 10-15 degrees) is a psychrotroph, which grows at 0 degrees but displays an optimum T in the **mesophile** range, nearer room temperature.

Organisms with an optimum T between about 45 degrees and 70 degrees are **thermophiles**. Some Archaea with an optimum T of 80 degrees or higher and a maximum T as high as 115 degrees, are now referred to as **extreme thermophiles** or **hyperthermophiles**.

9. (c) % of dissolved salt precipitated $= \frac{16.4 - 9.6}{16.4} \text{x} 100$

$= 41.5\ \%$

10. (b) For laminar flow, Re = 2000

$$Re = \frac{\rho VD}{\mu}$$

$$2000 = \frac{1040 x V x .03}{1600 x 10^{-6}}$$

$$V = .10256\, m/s$$

$$Q = AxV$$

$$Q = \frac{22 x .03^2 x .10256}{7 x 4}$$

$$Q = 7.2495x10^{-5}\ m^3/s$$

$$t = \frac{\text{Capacity of Tank}}{Q}$$

$$t = \frac{3.14x1.5^2x3}{4x7.2495x10^{-5}}$$

T = 20 hours or 73000 seconds.

11. (b)

Product Defect	Causative organisms
Sauerkraut	
Pinking	Rbodolorula
Ropy/slimy	Leuconostoc, Lactobacillus
Softening	Enterobacter, Flavobacterium. Pseudomonas

P-3,Q-4,R-1,S-2

12. (b) P-2,Q-4,R-1,S-3

Highly amylose starch acts as an edible film for wrapping candies.Pectin is used as an agent in jams and jeliies. Starch phosphates is used for white sauces Glucose is used as humectant in confectionary items

13. (d) P-4,Q-2,R-3,S-1

Chemical	Odour
Diacetyl, acetylpropionyl, acetoin	Butter/
Isoamyl acetate	Banana
Benzaldehyde	Bitter almond, cherry
Cinnamaldehyde	Cinnamon
Ethyl propionate	Fruity
Methyl anthranilate	Grape
Limonene	Orange
Ethyl decadienoate	Pear
Allyl hexanoate	Pineapple

14. (c) P-4,Q-2,R-3,S-1

Curd is Gel type.

Butter is Emulsion type.

Vegetable soup is Sol type

Whipped egg white is Gel type

15. (c) $$\frac{4}{\ln 10^7 - \ln 10^3} = \frac{T}{\ln(2x10^3) - \ln 10^3}$$

T = .3 hours

T = 18 minutes

16. (d) Correct sequence is as follows :

R. Staining with basic dye

S. Fixing the color with a suitable mordant

P. Washing with neutral organic solvent

Q. Counter staining with a contrast dye

19. (b) $$V = \frac{V_{max}.[S]}{Km + [S]}$$

$$V = \frac{22x2x10^{-4}}{.47x10^{-4} + 2x10^{-4}}$$

$$= 17.8\ \mu mol/l\ min$$

20. (a) $$V = \frac{V_{max}.[S]}{Km + [S]}$$

$$V = \frac{22x5x10^{-4}}{3.47x10^{-4} + 5x10^{-4}}$$

$$= 13\ \mu mol/l\ min$$

$$\%reduction = \frac{17.8 - 13}{17.8} x100$$

$$= 27\%$$

2012

Q. (1 – 9) carry one mark each.

1. Among the following fatty acids, which group is known as essential fatty acids?
   (*a*) 9,11-Octadecadienoic and 9,11,13-Octadecatrienoic
   (*b*) 9,12-Octadecadienoic and 9,12,15-Octadecatrienoic
   (*c*) 9-Octadecenoic and 9,11-Octade-cadienoic
   (*d*) 9,11-Octadecadienoic and 9-Eicosenoic

2. Cellulose, the structural polysaccharide of plant, is a polymer of
   (*a*) β–D-Glucose (*b*) β–D-Glucose
   (*c*) β–D-Galactose (*d*) β–D-Galcturonic acid

3. The important role of carotenoids in the human diet is their ability to serve as precursors of
   (*a*) Vitamin C (*b*) Vitamin D
   (*c*) Vitamin A (*d*) Vitamin K

4. Which one of the following microorganisms is used in the preparation of bread?
   (*a*) *Candida utilis*
   (*b*) *Saccharomyces cerevisiae*
   (*c*) *Saccharomyces cevarum*
   (*d*) *Aspergilus niger*

5. Which one of the microorganisms given below is NOT RESPONSIBLE for ropy or stringy fermentation of milk?
   (*a*) *Alcaligenes viscolactis*
   (*b*) *Enterobacter aerogenes*
   (*c*) *Streptococcus cremoris*
   (*d*) *Streptococcus lactis*

6. A mild heat treatment of foods that destroys pathogens and extends its shelf life is called
   (*a*) Baking (*b*) Blanching
   (*c*) Sterilization (*d*) Pasteurization

7. The most common and least expensive plastic film used for packaging of solid food materials is
   (*a*) Polyethylene (*b*) Polystyrene
   (*c*) Polypropylene (*d*) Polyvinylchloride

8. Reassociation of amylose and formation of crystalline structure upon cooling of cooked starch solution is termed as
   (*a*) Synersis
   (*b*) Gelatinization
   (*c*) Retrogradation
   (*d*) Denaturation

9. Thermal destruction of microorganisms follows a kinetics of
   (*a*) Zero order (*b*) First order
   (*c*) Second order (*d*) Fractional order

Q. (10 – 22) carry two marks each.

10. Which one of the following is NOT A CORRECT statement?
    (*a*) Meatiness is the taste produced by compounds such as glutamate in products like cheese, soy sauce.
    (*b*) Astringency is a dry mouth feel in the oral cavity that is most associated with phenolic compounds.
    (*c*) Saltiness is a taste that is mainly produced by chloride ions.
    (*d*) Sourness is related to acidity and is sensed by hydrogen ion channels in the human tongue.

11. The following plot represents the *Lineweaver-Burk* equation of an enzymatic reaction both in the presence and the absence of inhibitor. Here, V is the velocity of reaction and S is the substrate concentration.

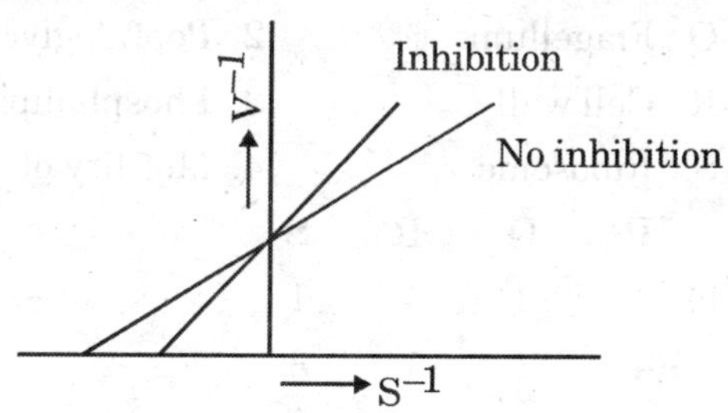

The nature of inhibition shown in the plot is
    (*a*) Non-competitive
    (*b*) Anti-competitive
    (*c*) Competitive
    (*d*) Mixed type

12. Make the correct match of the food constituents in **Group I** with their nature given in **Group II**.

Group I	Group II
P. Ascorbic Acid	1. Sugar
Q. Phenyl alanine	2. Chelate
R. Dextrose	3. Amino Acid
S. Haemoglobin	4. Antioxidant

	P	Q	R	S
(*a*)	4	3	1	2
(*b*)	4	1	3	2
(*c*)	3	4	2	1
(*d*)	4	2	1	3

13. Make the correct match of the fermented food products in **Group I** with the microorganisms in **Group II.**

Group I	**Group II**
P. Yoghurt	1. *Lactobacillus acido-philus and Lactoba-cillus delbrueckii*
Q. Cheese	2. *Leuconostoc mesent-eroides and Lactoba-cillus plantarum*
R. Sauerkraut	3. *Lactobacillus delbru-eckii and Streptococcus thermophillus*
S. Kefir	4. *Lactobacillus casei and Streptococcus thermo-phillus*

	P	Q	R	S
(a)	1	4	2	3
(b)	4	3	1	2
(c)	3	4	2	1
(d)	3	2	4	1

14. Match the following between organelle or cellular components of a bacterium cell in **Group I** with the constituents and functionalities in **Group II**.

Group I	**Group II**
P. Cytoplasmic membrane	1. Protein synthesis
Q. Flagellum	2. Peptidoglycan
R. Cell wall	3. Phospholipid bilayer
S. Ribosome	4. Motility of cell

	P	Q	R	S
(a)	4	2	4	1
(b)	4	2	1	3
(c)	3	4	2	1
(d)	2	3	4	1

15. Thermal death time (TDT) of *Clostridium botulinum* at 121°C is 2.78 min with a z-value of 10°C. The TDT of the microorganism at 116°C (in min) is

(a) 5.270 (b) 8.791

(c) 1.390 (d) 0.712

16. Make the correct match between specific food processing operations in **Group I** with their mechanism of action in **Group II.**

Group I	**Group II**
P. Ball Mill	1. Compression and shear
Q. Roller Mill	2. Pressure bursting
R. Flash Peeling	3. Friction and shear
S. Abrasive Peeling	4. Impact and shear

	P	Q	R	S
(a)	4	2	1	3
(b)	4	1	2	3
(c)	4	3	2	1
(d)	3	1	4	2

Common Data Q. (17 – 18) :

650 g of a wet food containing 405 g water is dried in a tray dryer to a final moisture content of 6.8 % (dry basis). It is observed that the drying process occurs under constant rate period and it takes 8 h.

17. Initial moisture content (in percentage) of the food on wet basis is

(a) 62.31 (b) 70.45

(c) 162.31 (d) 165.31

18. The rate of drying (in kg/h) is

(a) 128.79 (b) 126.35

(c) 77.81 (d) 0.0485

Common Data Q. (19 – 20) :

Air at 1 atmospheric pressure (101.325 kPa) and 30°C with absolute humidity of 0.0218 kg/kg of dry air is flowing in a drying chamber. The saturated vapor pressure of water (p_w^0, in kPa) is related to temperature (T, in °C) as given below

$$\text{in } p_w^0 = 18.6556 - \frac{5217.635}{T + 273}$$

Heat capacities of dry air (average molecular weight 29) and that of water vapor (molecular weight 18) are 1.005 and 1.884 kJ/kg.K, respectively. Latent heat of vaporization of water at reference temperature

19. The relative humidity of air (in percentage) is

(a) 62.82 (b) 68.22

(c) 86.62 (d) 81.80

20. The enthalpy (in kJ/kg) of moist air is

(a) 85.93 (b) 54.55

(c) 31.38 (d) 99.38

Linked Answer Q. (21 – 22) :

The total solids content in a milk sample is 18 %. It is desired to produce 1000 kg of sweetened condensed milk (SCM) having 40 % sugar, 25 % moisture and rest milk solids.

21. What is the 'Sugar Ratio' (in percentage) in the SCM in terms of sugar and water content in the final product?

(a) 48.19 (b) 61.54

(c) 54.16 (d) 56.14

22. If the 'Concentration Degree' is 2.5, the amount of sugar added in kg in the milk sample is

(a) 246.16 (b) 216.64

(c) 192.76 (d) 224.56

ANSWERS

1. (b)	**2.** (a)	**3.** (c)	**4.** (b)	**5.** (d)	**6.** (d)	**7.** (a)
8. (c)	**9.** (b)	**10.** (c)	**11.** (c)	**12.** (a)	**13.** (c)	**14.** (c)
15. (b)	**16.** (b)	**17.** (a)	**18.** (d)	**19.** (d)	**20.** (a)	**21.** (b)
22. (a)						

EXPLANATIONS

1. (b) Among the following fatty acids, 9,12-Octadecadienoic and 9,12,15-Octadecatrienoic group is known as essential fatty acids .

2. (a) Cellulose, the structural polysaccharide of plant, is a polymer of β-D-Glucose.

3. (c) Carotenoids (Crts) are structurally and functionally a very diversing group of natural pigments of the polyene type . They occur ubiquitously in all organisms capable of conducting photosynthesis, a process in which sun light is effectively converted into chemical energy. Carotenoids are important parts of photosynthetic organelles of all higher plants, mosses, ferns and algae. They are also found in photosynthetic membranes of phototropic bacteria and cyanobacteria. These not synthesized by humans and animals, but they are also present in their blood and tissues. They are important precursors of retinol (vitamin A);

4. (b) Saccharomyces cerevisiae, the yeast commonly used as baker's yeast. Gradation marks are 1 m apart.

5. (d) Streptococcus lactis.

6. (d) Pasteurization, heat-treatment process that destroys pathogenic microorganisms in certain foods and beverages. It is named for the French scientist Louis Pasteur, who in the 1860s demonstrated that abnormal fermentation of wine and beer could be prevented by heating the beverages to about 57° C (135° F) for a few minutes.

7. (a) The most common and least expensive plastic film used for packaging of solid food materials is Polyethylene.

8. (c) Retrogradation is a reaction that takes place when the amylose and amylopectin chains in cooked, gelatinized starch realign themselves as the cooked starch cools.

9. (b) The decrease of a microbial population, under thermal processing, also follows a first order kinetics and all the relationships deduced for chemical kinetics can be applied.

10. (c) Saltiness is a taste that is mainly produced by sodium ions.

11. (c)

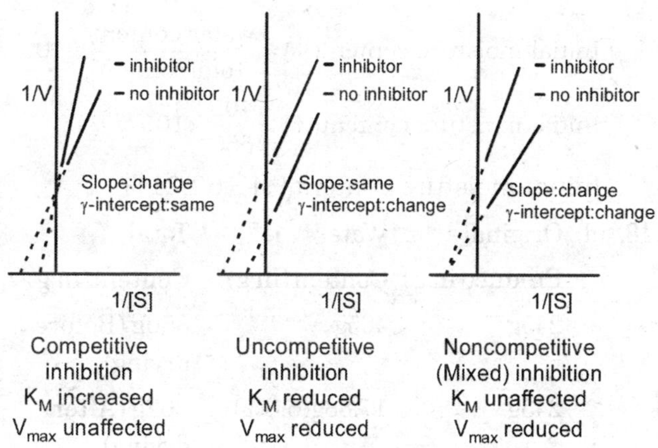

Lineweaver-Burk equation of an enzymatic reaction

12. (a) P-4, Q-3, R-1, S-2

P) Ascorbic Acid	4) Antioxidant
Q) Phenyl alanine	3) Amino Acid
R) Dextrose	1) Sugar
S) Haemoglobin	2) Chelate

13. (c) P-3, Q-4, R-2, S-1

P) Yoghurt	3) Lactobacillus delbrueckii and Streptococcus thermophillus
Q) Cheese	4) Lactobacillus casei and Streptococcus thermophillus
R) Sauerkraut	2) Leuconostoc mesenteroides and Lactobacillus plantarum
S) Kefir	1) Lactobacillus acidophilus and Lactobacillus delbrueckii

14. (c) P-3, Q-4, R-2, S-1

P) Cytoplasmic membrane	3) Phospholipid bilayer
Q) Flagellum	4) Motility of cell
R) Cell wall	2) Peptidoglycan
S) Ribosome	1) Protein synthesis

15. (b) $z = \dfrac{T_2 - T_1}{\log D_1 - \log D_2}$

$$10 = \frac{121-116}{\log D_1 - \log(2.78 \times 60)}$$

$$D_1 = 527.5 \text{ seconds}$$

$$D_1 = 8.791 \text{ minutes}$$

16. (b) P-4, Q-1, R-2, S-3

P) Ball Mill	4) Impact and shear
Q) Roller Mill	1) Compression and shear
R) Flash Peeling	2) Pressure bursting
S) Abrasive Peeling	3) Friction and shear

17. (a)

$$\text{Initial moisture content(\%)} = \frac{\text{water content}}{\text{Total content}} \times 100$$

$$\text{Initial moisture content(\%)} = \frac{405}{650} \times 100$$

$$\text{Initial moisture content(\%)} = 62.31\%$$

18. (d)

Original Product(in g)	Water Content(in g)	Total Content(in g)
245g	405g	650g (Before drying)
245g	17.88g(6.8%)	262g (After drying)

$$\text{Rate of drying(in kg/h)} = \frac{405 - 17.88}{8 \times 1000}$$

$$\text{Rate of drying(in kg/h)} = .0485$$

19. (d)

$$w = .0218$$

$$w = \frac{.622P_v}{P - P_v}$$

$$.0218 = \frac{.622P_v}{101.325 - P_v}$$

$$2.2088 - 0218P_v = .622P_v$$

$$Pv = 3.309 \text{ kPa}$$

$$\ln p^\circ_w = 18.6556 - \frac{5217.635}{T+273}$$

$$\ln p^\circ_w = 18.6556 - \frac{5217.635}{30+273}$$

$$\ln p^\circ_w = 18.6556 - \frac{5217.635}{303}$$

$$p^\circ_w = e^{1.4356} = 4.2022 \text{ kPa}$$

$$\text{(Relative Humidity)}\varphi = \frac{3.4309}{4.2022} \times 100 = 81.64\%$$

20. (a) **Enthalpy of moist air,**

$$h = 1.005t_d + w(2502.3 + 1.9t_d)$$

$$h = 1.005 \times 30 + (.0218)(2502.3 + 1.9 \times 30)$$

$$h = 30.15 + 55.79$$

$$h = 85.94 \text{ kJ/kg}$$

21. (b) **1000 kg of Sweetened Condensed milk(SCM) has 180 kg of total solid content which means SCM will contain 72 kg of sugar(40%),45 kg of moisture(25%) and 63 kg of rest milk solids.**

$$\text{Sugar ratio} = \frac{\text{Sugar content(in kg)}}{\text{Sugar content(in kg)} + \text{moisture content(in kg)}}$$

$$\text{Sugar ratio} = \frac{72}{72+45} \times 100$$

$$\text{Sugar ratio} = 61.54\%$$

22. (a)

$$\text{Amount of sugar added} = \frac{\text{Sugar ratio} \times 1000}{\text{Concentration degree} \times 100}$$

$$\text{Amount of sugar added} = \frac{61.54 \times 1000}{2.5 \times 100}$$

2013

Q. 1 - 9 carry one mark each.

1. Kawashiorkor disease is caused due to the deficiency of

   (*a*) lysine (*b*) unsaturated fatty acids

   (*c*) vitamin K (*d*) protein

2. Which of the following statements is TRUE in case of oxidative rancidity of vegetable oils and fats?

   (*a*) It is caused by the reaction of saturated fatty acids and oxygen

   (*b*) It involves polymerization of fatty acids

   (*c*) It is caused by the reaction of unsaturated fatty acids with oxygen

   (*d*) It is caused by oxidative enzymes

3. The food borne disease, Q fever is caused by the organism,

   (*a*) Clostridium perfringens

   (*b*) Coxiella burnetti

   (*c*) Bacillus cereus

   (*d*) Staphylococcus aureus

4. The primary bacterial spoilage of poultry meat at low temperature, with characteristic sliminess at outer surface, is caused by

   (*a*) Pseudomonas spp. (*b*) Aspergillus spp.

   (*c*) Bacillus spp. (*d*) Candida spp.

5. The weight gain (in gram) per gram protein consumed is called

   (*a*) Net Protein Ratio (NPR)

   (*b*) Biological Value (BV)

   (*c*) Protein Efficiency Ratio (PER)

   (*d*) Chemical Score (CS)

6. Which of the following carbohydrates is NOT classified as dietary fibre?

   (*a*) Agar (*b*) Pectin

   (*c*) Sodium alginate (*d*) Tapioca starch

7. In the extruder barrel, the compression is achieved by back pressure created by the die and by

   (*a*) increasing pitch and decreasing diameter of the screw

   (*b*) using the tapered barrel with constant pitch

   (*c*) increase in the clearance between barrel surface and screw

   (*d*) opening of the die

8. The brown colour of bread crust during baking is due to Maillard reaction between

   (*a*) aldehyde groups of sugars and amino groups of proteins

   (*b*) aldehyde groups of sugars and vitamins

   (*c*) aldehyde groups of sugars and salt

   (*d*) starch and yeast

9. Blanching influences vegetable tissues in terms of

   (*a*) enzymes production

   (*b*) alteration of cytoplasmic membrane

   (*c*) stabilization of cytoplasmic proteins

   (*d*) stabilization of nuclear proteins

Q. 10 - 22 carry two marks each.

10. Match the toxicants of plant foods in Group I with their main plant source given in Group II.

Group I	*Group II*
P. Gossypol	1. Khesari Dahl (*Lathyrus sativus*)
Q. Vicine	2. Cotton seeds
R. Glucosinolates	3. Fava beans
S. BOAA (beta-N-Oxalyl Amino L-Alanine)	4. Rapeseeds

   (*a*) P-2, Q-3, R-4, S-1

   (*b*) P-2, Q-4, R-3, S-1

   (*c*) P-3, Q-1, R-2, S-4

   (*d*) P-4, Q-3, R-1, S-2

11. Match the products in Group I with the enzymes used for their preparation given in Group II.

Group I	*Group II*
P. Aspartame	1. Lipase
Q. Cocoa butter substitute	2. Glucose isomerase
R. High fructose corn syrup	3. Thermolysin
S. Lactose free milk	4. Invertase
	5. Beta galactosidase

   (*a*) P-2, Q-1, R-4, S-3 (*b*) P-3, Q-1, R-2, S-5

   (*c*) P-1, Q-3, R-2, S-4 (*d*) P-1, Q-2, R-4, S-5

12. Match the food items in Group I with the type of colloidal dispersion given in Group II.

Group I	*Group II*
P. Mayonnaise	1. Sol
Q. Tomato ketchup	2. Emulsion
R. Cake	3. Gel
S. Curd	4. Solid foam

   (*a*) P-4, Q-1, R-2, S-3 (*b*) P-3, Q-1, R-2, S-4

   (*c*) P-2, Q-3, R-4, S-1 (*d*) P-2, Q-1, R-4, S-3

13. [a] **Assertion:** In the presence of sucrose, the temperature and time for gelatinization of starch increases .

[r] **Reason:** Sucrose, due to its hygroscopic nature, competes with starch for water needed for gelatinization.

(*a*) Both [a] and [r] are true and [r] is the correct reason for [a]

(*b*) Both [a] and [r] are true but [r] is not the correct reason for [a]

(*c*) Both [a] and [r] are false

(*d*) [a] is true but [r] is false

14. Thermal death of viable spores of *Bacillus subtilis* in a food sample follows a first order kinetics with a specific death rate constant of 0.23 min^{-1} at 100°C. The time (in minutes) required to kill 99% of spores in the food sample at 100°C will be

(*a*) 10 (*b*) 20
(*c*) 23 (*d*) 60

15. How much skim milk (in kg) containing 0.1% fat should be added to 500 kg of cream containing 50% fat to produce standardized cream containing 36% fat?.

(*a*) 140 (*b*) 165
(*c*) 195 (*d*) 210

16. Which of the following statements is NOT CORRECT in relation to muscle proteins?

(*a*) Actin and myosin interact to form actomyosin which is responsible for muscle contraction

(*b*) Collagen contributes to the toughness of muscles due to its abundant presence

(*c*) Elastin, a constituent of ligaments, is tougher than collagen

(*d*) Actomyosin is not the main state of actin and myosin in post-mortem muscles

COMMON DATA QUESTIONS

Common Data for Questions 17 and 18 : A cold storage plant is used for storing 50 tonnes of apples in perforated plastic crates. During the storage, apples are cooled down from 28°C to storage temperature of 2°C. (Specific heat of the apple = 0.874 kCal kg^{-1} $°C^{-1}$)

17. If the required cooling is attained in 16 hours, the refrigeration plant capacity (in Tons) is

(*a*) 19 (*b*) 24
(*c*) 29 (*d*) 32

18. If the cooling is to be achieved in 8 hours, the power required (in Horse Power) to operate the plant having a Coefficient of Performance (COP) of 2.5 will be

(*a*) 47 (*b*) 65
(*c*) 89 (*d*) 96

Common Data for Questions 19 and 20 : An actively growing culture of Acetobacter aceti is added to the vigorously aerated fermented fruit juice medium containing 10 g l^{-1} ethanol to produce vinegar. After some time, the ethanol concentration in the medium is 0.8 g l^{-1} and acetic acid produced is 8.4 g l^{-1}.

19. What is the conversion efficiency of the process with respect to theoretical yield?

(*a*) 30 (*b*) 50
(*c*) 70 (*d*) 90

20. The concentration of fermentable sugars (g l^{-1}) required in the fruit juice to produce 10 gl^{-1} ethanol, based on 90% fermentation efficiency is

(*a*) 20.0 (*b*) 21.7
(*c*) 22.8 (*d*) 25.1

LINKED ANSWER QUESTIONS

Statement for Linked Answer Questions 21 and 22 : An enzyme catalyzed reaction (following Michaelis-Menten kinetics) exhibits maximum reaction velocity (V_m) of 75 nmol l^{-1} min^{-1}. The enzyme at a substrate concentration of 1.0×10^{-4} M shows the initial reaction velocity of 60 nmol l^{-1} min^{-1}.

21. The K_m value of the enzyme in molar concentration (M) is

(*a*) 2.5×10^{-5} (*b*) 5.0×10^{-5}
(*c*) 2.5×10^{-4} (*d*) 5.0×10^{-4}

22. If the enzyme concentration for the reaction is doubled at a substrate concentration of 5.0×10^{-5} M, the initial reaction velocity in nmol l^{-1} min^{-1} will be

(*a*) 37.5 (*b*) 50
(*c*) 60 (*d*) 100

ANSWERS

1. (d)	**2.** (a,b,c,d)	**3.** (b)	**4.** (a)	**5.** (c)	**6.** (d)	**7.** (b)
8. (a)	**9.** (b)	**10.** (a)	**11.** (b)	**12.** (d)	**13.** (a)	**14.** (b)
15. (c)	**16.** (d)	**17.** (b)	**18.** (c)	**19.** (c)	**20.** (c)	**21.** (a)
22. (d)						

EXPLANATIONS

1. (d) Kwashiorkor, also known as "edematous malnutrition" because of its association with edema (fluid retention), is a nutritional disorder most often seen in regions experiencing famine. It is a form of malnutrition caused by a lack of protein in the diet.

2. (a,b,c,d) All options are correct.

3. (b) Q fever is a disease caused by infection with Coxiella burnetii, a bacterium that affects humans and other animals. This organism is uncommon, but may be found in cattle, sheep, goats, and other domestic mammals, including cats and dogs.

4. (a) The primary bacterial spoilage of poultry meat at low temperature, with characteristic sliminess at outer surface, is caused by Pseudomonas spp.

5. (c) Protein efficiency ratio (PER) is based on the weight gain of a test subject divided by its intake of a particular food protein during the test period.

6. (d) Tapioca starch is made from the root of the cassava plant. Although it can be used as a flour for baking, its mainly used as a thickener.

7. (b) In the extruder barrel, the compression is achieved by back pressure created by the die and by using the tapered barrel with constant pitch.

8. (a) The brown colour of bread crust during baking is due to Maillard reaction between aldehyde groups of sugars and amino groups of proteins.

9. (b) Blanching is a cooking process wherein a food, usually a vegetable or fruit, is scalded in boiling water, removed after a small, timed interval, and finally plunged into iced water or placed under cold running water to halt the cooking process.

10. (a) P-2, Q-3, R-4, S-1

Gossypol is a natural phenol derived from the cotton plant (genus Gossypium). Gossypol is a phenolic aldehyde that permeates cells and acts as an inhibitor for several dehydrogenase enzymes. It is a yellow pigment.

Vicine is an alkaloid glycoside found in fava beans. Vicine is toxic, causing the disease favism, in individuals who have a hereditary loss of the enzyme glucose-6-phosphate dehydrogenase. Divicine is the name of the aglycone of vicine.

The glucosinolates are natural components of many pungent plants such as mustard, cabbage, and horseradish. The pungency of those plants is due to the release of oils containing glucosinolates when the plant material is chewed, cut, or otherwise damaged.

Lathyrism or neurolathyrism is a neurological disease of humans and domestic animals. This problem is mainly associated with Lathyrus sativus (also known as Grass pea, Kesari Dal, Khesari Dal or Almorta). BOAA causes paralysis, characterized by lack of strength in or inability to move

11. (b) (B) P-3, Q-1, R-2, S-5

12. (d) P-2, Q-1, R-4, S-3

P) Mayonnaise-Emulsion

Q) Tomato ketchup-Sol

R) Cake-Solid foam

S) Curd- Gel

13. (a) Both [a] and [r] are true and [r] is the correct reason for [a]

14. (b) 20

15. (c) $500 \times 0.5 = .001x + (500-x) \times .36$

By solving above equation,

we get x = 195 kg

16. (d) Actomyosin is the main state of actin and myosin in post-mortem muscles

17. (b) m = 50000

kgc = 3.657 kJ/°kg C

dT = 26 °C

$$\text{Refrigerating effect(R.E.)} = \frac{\text{mc.dT}}{16 \times 3600 \times 3.5}$$

R.E. = 23.58 Ton

18. (c) $\text{COP} = \frac{\text{R.E.}}{\text{P}}$

By putting respective values,

we get P = 89 hp

19. (c) 70

20. (c) 21.7

21. (a) $V = Vm\frac{[S]}{K_m + [S]}$ $60 = 75\frac{10^{-4}}{K_m + 10^{-4}}$

By solving above equation,

we get Km = 2.5×10^{-5}

22. (d) V_m is doubled to 150

[S] = 5×10^{-5}

Km = 2.5×10^{-5}

$$V = Vm\frac{[S]}{K_m + [S]}$$

By putting above values,

we get V = 100 nmol l^{-1} min^{-1}

2014

Q. (1 – 9) Carry One Mark Each.

1. The systematic name of sucrose is

(a) α-D-Fructofuranosyl (1 → 2) β-D-Glucopyranoside

(b) α-D-Glucopyranosyl (1 → 2) β-D-Fructofuranoside

(c) α-D-Glucopyranosyl (2 → 1) β-D-Fructofuranoside

(d) α-D-Fructofuranosyl ((2 → 1) β-D-Glucopyranoside

2. A non-hydrolyzable lipid is

(a) Lecithin (b) Arachidic acid

(c) Tocopherol (d) Tristearin

3. The respiratory quotient (RQ) for the reaction

$2\,C_{57}H_{110}O_6 + 163\,O_2 \rightarrow 114\,CO_2 + 110\,H_2O$ is

(a) 0.70 (b) 1.14

(c) 1.43 (d) 0.14

4. Liver necrosis may be caused by the deficiency of

(a) Vitamin A (b) Vitamin D

(c) Vitamin K (d) Vitamin E

5. Which of the following non-nutritive sweeteners contains similar calories per gram as that of sucrose?

(a) Saccharin (b) Aspartame

(c) Sucralose (d) Cyclamate

6. The objective of heating milk to about 65°C before homogenization is to inactivate

(a) Glucose oxidase (b) Lipases

(c) Lactases (d) Invertases

7. Make the correct match of the processes in Column I with the suitable materials/products in Column II

Column I	Column II
1. Rendering	P. Lecithin
2. Hydrogenation	Q. Fullers' earth
3. Degumming	R. Lard
4. Bleaching	S. Margarine

(a) 1-R, 2-P, 3-Q, 4-S (b) 1-P, 2-Q, 3-S, 4-R

(c) 1-R, 2-P, 3-S, 4-Q (d) 1-R, 2-S, 3-P, 4-Q

8. A fruit juice of viscosity and density ρ is agitated using an impeller of diameter D at a speed of N revolutions per minute. The terms $X = \frac{P}{\rho N^3 D^5}$, $Y = \frac{D^2 N \rho}{\mu}$, $Z = \frac{N^2 D}{g}$ represent three process related numbers, where P is power imparted by impeller and g is acceleration due to gravity. Which of the following is correct representation of these numbers?

(a) X = Power, Y= Froude, Z = Reynolds

(b) X = Power, Y = Reynolds, Z = Froude

(c) X = Froude, Y= Reynolds, Z = Power

(d) X = Reynolds, Y = Power, Z = Froude

9. The energy required to reduce the size of a food material from a mean diameter of 12 mm to 4 mm is 10 kJ kg^{-1}. From Rittingers' law, the energy needed to reduce the same material from a diameter of 1.2 mm to 0.4 mm in kJ kg^{-1} is ______

Q. (10 – 22) Carry Two Mark Each.

10. *Saccharomyces cerevisiae* (mean doubling time 3.2 *h*) is grown in a batch fermenter with an operating volume of $12 m^3$. A 2% (v/v) inoculum, which contains 5 kg cells per $100 m^3$ is mixed with the substrate. The residence time in the fermenter is 24h and the density of broth is 1010 kg m^{-3}. The mass of *S. cerevisiae* obtained from the fermenter, in kg, is ______

11. Make the correct combination of operations in Column I with the machines in Column II

Column I	Column II
1. Rice milling	P. Pin mill
2. Wheat milling	Q. Rubber rolls
3. Mustard oil expelling	R. Break rolls
4. Pepper grinding	S. Screw press

(a) 1-Q, 2-R, 3-S, 4-P

(b) 1-R, 2- Q, 3-S, 4-P

(c) 1-Q, 2-P, 3-S, 4-R

(d) 1-Q, 2-R, 3- P, 4- S

12. The correct order for D_{121} values of the spores of food spoilage bacteria in aqueous medium is

(a) *B. stearothermophilus* > *C. sporogenes* > *C. botulinum type A* > *B. coagulans*

(b) *C. sporogenes* > *B. stearothermophilus* > *C. botulinum type A* > *B .coagulans*

(c) *C. botulinum type A* > *B. stearothermophilus* > *C. sporogenes* > *B. coagulans*

(d) *B. stearothermophilus* > *C. botulinum type A* > *C. sporogenes* > *B. coagulans*

13. Make the correct combination of pigments/ microorganisms in Column I with the process/ products in Column II

Column I	Column II
1. Anthocyanin	P. Ropiness
2. Chlorophyll	Q. Koji
3. *Bacillus subtilis*	R. Glycosides
4. *Aspergillus oryzae*	S. Porphyrins

(*a*) 1-S, 2- R, 3-P, 4-Q (*b*) 1-R, 2-S, 3-Q, 4- P
(*c*) 1-Q, 2-S, 3-P, 4-R (*d*) 1-R, 2-S, 3-P, 4-Q

14. Make the correct combination of underlying principles in Column I with the processes in Column II

Column I

1. Carbonyl derivatives react with free amino acids to yield aldehydes
2. Starch aggregates and forms micro-crystals
3. Starch granules swell and leach amylose
4. Pyranose or furanose rings open up by pyrolytic reactions to form furfural derivatives

Column II

P. Gelatinization
Q. Strecker degradation
R. Caramelization
S. Retrogradation

(*a*) 1-Q, 2-R, 3-P, 4- S (*b*) 1-Q, 2-S, 3-P, 4-R
(*c*) 1-R, 2-S, 3-P, 4- Q (*d*) 1-Q, 2-P, 3- S, 4-R

15. Which one of the following statements is FASLE?

(*a*) The peptide bond is planar offering restricted rotation around its axis
(*b*) Full range of water activity is $0 \le a_w \le 1$ and it has well defined unit.
(*c*) The autooxidation of lipids proceeds via free radical mechanism
(*d*) The carbonyl group of sugar reacts with nucleophilic amino group of amino acids in Amadori rearrange-ment

16. Which one of the following statements is TRUE?

(*a*) Pectate lyase hydrolyzes methyl ester bond of pectin
(*b*) α-Solanine is a non-toxic compound found in solanaceae plants
(*c*) Egg proteins have lower digestibility than pea proteins
(*d*) Lipoxygenase catalyses the conversion of cis, cis-1,4-pentadiene to hydroperoxides

17. Fish fillet having 84% moisture (wet basis) is frozen from top using an air blast freezer maintained at – 32°C. The initial temperature of the fillet (density 1050 kg m^{-3}) is – 2°C (freezing point). Convective heat transfer coefficient of air is 25 Wm^{-2} K^{-1}, thermal conductivity of frozen fish is 1.0 Wm^{-1} K^{-1} and latent heat of crystallization is 340 kJ kg^{-1}. The freezing time, in min, for a 20 mm thick block of fish fillet weighing 1 kg is _____.

18. Make the correct combination of properties in Column I with their dimensions in Column II

Column I	Column II
1. Dynamic viscosity	P. m^2 s^{-2} K^{-1}
2. Thermal conductivity	Q. kg m s^{-2}
3. Specific heat	R. kg m^{-1} s^{-1}
4. Force	S. kg m s^{-3} K^{-1}

(*a*) 1-R, 2-S, 3-Q, 4- P (*b*) 1- Q, 2-S, 3-P, 4- R
(*c*) 1-R, 2-S, 3-P, 4-Q (*d*) 1- S, 2-R, 3-P, 4-Q

19. The viscosity and density of a fruit juice at 21°C are 6.3 × 10^{-3} Pa s and 1029 kg m^{-3}, respectively. The juice flows at the rate of 0.12 m^3 min^{-1} in a 2.54 cm inner diameter steel pipe. Correct combination of the Reynolds number (N_{Re}) and the nature of flow of juice is

(*a*) N_{Re} = 1048, Laminar
(*b*) N_{Re} = 2056, Laminar
(*c*) N_{Re} = 16375, Turbulent
(*d*) N_{Re} = 28656, Turbulent

20. For a typical food sorption isotherm curve (Figure 1), which one of the following statements is CORRECT?

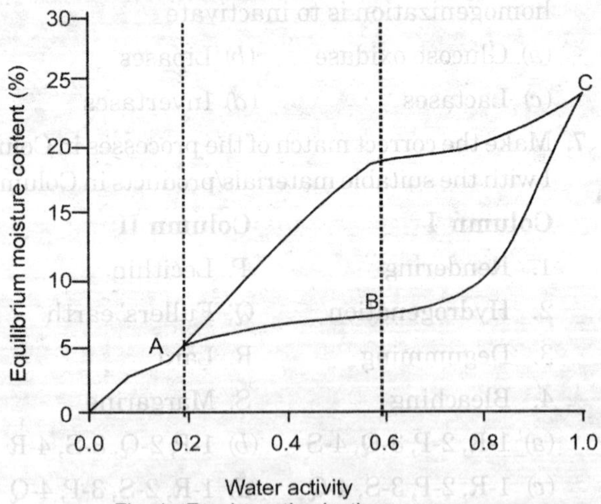

Fig. 1 : Food sorption isotherm curve

(*a*) Y-coordinate of A represents monolayer water content of food, A-B represents water absorbed in the multilayer within the food and B-C represents free water within the capillary network of the food.

(*b*) Y-coordinate of B represents monolayer water content of food, A-B represents water absorbed in the multilayer within the food and B-C represents free water within the capillary network of the food.

(*c*) Y-coordinate of A represents monolayer water content of food, Y-coordinate of B represents water absorbed in the multilayer within the food and B-C represents free water within the capillary network of the food.

(*d*) Y-coordinate of A represents monolayer water content of food, A-C represents water absorbed in the multilayer within the food and Y-coordinate of C represents free water within the capillary network of the food.

21. 10,000 kg milk (7% fat) is passed through a cream separator to obtain cream (40% fat) and skim milk (0.1 % fat). The cream, thus obtained, is churned to make butter of 80.5% fat. If a loss of 0.5% of initial milk fat occurs during the manufacturing process, the % overrun is ________.

22. A 50 mm thick pack of farm fresh berries is cooled at one side from 24°C to 7°C. The relevant properties of berries are: density 1025 kg m^{-3}, specific heat 3.78 kJ kg^{-1} K^{-1}, convective heat transfer coefficient 30 Wm^{-2} K^{-1}, and thermal conductivity 0.3 Wm^{-1} K^{-1}. The Fourier number for a cooling span of 30 min is ________

ANSWERS

1. (b)	**2.** (c)	**3.** (a)	**4.** (d)	**5.** (b)	**6.** (b)	**7.** (d)
8. (b)	**9.** (100 to 100)	**10.** (2.1 to 2.2)	**11.** (a)	**12.** (a)	**13.** (d)	**14.** (b)
15. (b)	**16.** (d)	**17.** (74 to 76)	**18.** (c)	**19.** (c)	**20.** (a)	**21.** (22.0 to 22.2)
22. (.055 to .057)						

EXPLANATIONS

1. (b) Sucrose (common name : table sugar, also called saccharose) is a disaccharide (glucose + fructose) with the molecular formula $C_{12}H_{22}O_{11}$. Its systematic name is α-D-glucopyranosyl-(1→2)-β-D-fructofuranose. It is best known for its role in human nutrition and is formed by plants but not by higher organisms.

2. (c) Tocopherols (TCP) are a class of organic chemical compounds (more precisely, various methylated phenols), many of which have vitamin E activity.

3. (a) $$\text{Respiratory Quotient(RQ)} = \frac{CO_2 \text{ produced}}{O_2 \text{ consumed}}$$

RQ = 114/163 = 0.7

4. (d) The clinical course of acute hepatic necrosis resembles an acute, toxic injury to the liver with sudden and precipitous onset, marked elevations in serum aminotransferase levels, and early signs of hepatic (or other organ) dysfunction or failure despite minimal or no jaundice. It takes place due to deficiency of Vitamin E.

5. (b) Aspartame has same calories per gram as that of sugar but Aspartame is almost 200 times sweeter than table sugar.

6. (b) To inactivate lipases.

7. (d) 1-R,2-S,3-P,4-Q

Lard is pig fat in both its rendered and unrendered forms. It is obtained from any part of the pig where there is a high proportion of adipose tissue. It can be renderedby steaming it or boiling it in water and then separating the insoluble fat from the water, or by the use of dry heat.

Fuller's Earth is also known by the following other names : "Bleaching clay", probably because fulling whitenes the cloth.

Water degumming is the oldest degumming treatment and also forms the basis of the production of commercial lecithin.Vegetable oils are commonly referred to as "polyunsaturated". This simply means that there are several double bonds present. Vegetable oils may be converted from liquids to solids by the hydrogenation reaction. Margarines and shortenings are "hardened" in this way to make them solid or semi-solids.

8. (b) Y and Z are dimensionless numbers. Whereas, X has unit of Power.

9. (100 to 100)

By Rittinger's equation,

$$E = K\left(\frac{1}{x_2} - \frac{1}{x_1}\right)$$

$$10 = K\left(\frac{1}{4} - \frac{1}{12}\right) \text{—(1)}$$

$$P = K\left(\frac{1}{.4} - \frac{1}{1.2}\right) \text{—(2)}$$

By solving above equations, we get

P = 100 kJ/kg

11. (a) 1-Q, 2-R, 3-S, 4-P

1. Rubber rolls are used in Rice milling.
2. Pin mill is used in pepper grinding
3. Break rolls are used in Wheat milling.
4. Screw press is used in mustard oil expelling

12. (a) B. siearoihermophilus > C. sporogenes > C. botulinum type A > B. coagulans

13. (d) 1-R, 2-S, 3-P, 4-Q

Koji, a masculine Japanese given name; Koji, the common name of the fungus Aspergillus oryzae.

As chlorophyll they are at the heart of photosynthesis. The role of porphyrins in electron transport appears to be as handmaidens to the redox properties of iron ions, but their role in photosynthesis is far more unique.

They are involved in the critical energy conversion step of photosynthesis. A glucoside is a glycoside that is derived from glucose. Glucosides are common in plants, but rare in animals. Glucose is produced when a glucoside is hydrolysed by purely chemical means, or decomposed by fermentation or enzymes.

Ropiness in bread is usually due to bacterial growth and is considered more prevalent in home made breads. The chief causative organism is Bacillus subtilis or B. licheniformis.

14. (b) 1-Q, 2-R, 3-P, 4-S

1) Strecker degradation : Carbonyl derivatives react with free amino acids to yield aldehydes

2) Retrogradation : Starch aggregates and forms micro-crystals

3) Gelatinization : Starch granules swell and leach amylose

4) Carainelization :Pyranose or furanose rings open up by pyrolytic reactions to form furfural derivatives

15. (b) It means the relative product equilibrium humidity, the freely available water for micro organisms to grow. Its measuring value is called "water activity" (a_w-value) in a range of 0.03 - 1.000 a_w.

16. (d) Lipoxygenase catalyses the conversion of cis, cis-1,4-pcntadienc to hydroperoxides.

17. (74 to 76)

$$t_f = \frac{dH.\rho}{dT}\left(\frac{P.a}{h_s} + \frac{R.a^2}{k}\right)$$

$$t_f = \frac{340x1000x1050}{30}\left(\frac{20}{50000} + \frac{20^2}{8x10^6}\right)$$

$$t_f = 5357.7 \text{ seconds}$$

But for 84% moisture content,

$$t_f' = (0.84x5357.7) \text{ seconds}$$

$$t_f' = 75 \text{ minutes}$$

18. (c) 1-R, 2-S, 3-P, 4-Q

Column I	Column II
1) Dynamic viscosity	R) kg m^{-1} s^{-1}
2) Thermal conductivity	S) kg m s^{-3} K^{-1}
3) Specific heat	P) $m^2s^{-2}K^{-1}$
4) Force	Q) kg m s^{-2}

19. (c) $N_{Re} = \frac{\rho VD}{\mu}$

$$= \frac{1029x2.54x10^{-2}x3.95}{6.3x10^{-3}}$$

$$= 16375$$

Since Reynolds number is greater than 4000, therefore flow will be turbulent in nature.

20. (a)

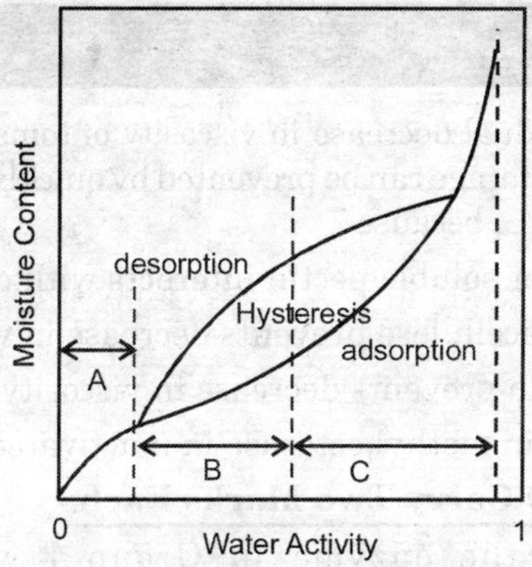

Figure 1. Sorption isotherm for a typical food product, showing the hysteresis.

Y-coordinate of A represents monolayer water content of food, A-B represents water absorbed in the multilayer within the food and B-C represents free water within the capillary network of the food.

22. (055 to .057)

$$Fo = \frac{\alpha t}{L^2}$$

$$\alpha = \frac{k}{\rho c_p}$$

$$Fo = \frac{kt}{\rho c_p L^2}$$

$$Fo = \frac{0.3x60x60}{1025x3.78x10^3x2500x10^{-6}} = .0558$$

Where, K is thermal conductivity T is time in secondsr is the density in kg/m^{3}c$_p$ is the specific heat in kJ kg^{-1} K^{-1}L is the thickness of the berries

2015

Q. (1 – 9) Carry One Mark Each.

1. Standard pasteurization protocol for milk is adequate for destroying

(*a*) *Clostridium sporogenes*

(*b*) *Bacillus cereus*

(*c*) *Clostridium botulinum*

(*d*) *Listeria monocytogenes*

2. Which one of the following is NOT a component of an evaporator?

(*a*) Heat exchanger (*b*) Vacuum separator

(*c*) Condenser (*d*) Cyclone separator

3. Among the following animal foods, the fat content is least in

(*a*) Beef (*b*) Chicken meat

(*c*) Pork (*d*) Lamb flesh

4. The enzyme that hydrolyzes starch to maltose is

(*a*) α-amylase

(*b*) β-amylase

(*c*) glucoamylase

(*d*) cyclodextrin glucanotransferase

5. Which one of the following is NOT enriched in endosperm during parboiling of paddy?

(*a*) Thiamine (*b*) Niacin

(*c*) Iron (*d*) Fat

6. Heat-treated legume seed proteins are more digestible than those of untreated legume seed proteins due to

(*a*) reaction of reducing sugars with ε-amino group of lysine

(*b*) increased binding of lectins to intestinal mucosal cells

(*c*) thermolabile nature of lectins and Kunitz-type protease inhibitors

(*d*) thermolabile nature of Bowman-Birk type of inhibitor

7. What is the percent relative humidity at which both the dry bulb and wet bulb thermometers would record equal temperatures?

(*a*) 0 (*b*) 10

(*c*) 50 (*d*) 100

8. How many fold would the *g*-number of a centrifuge increase by doubling both the spinning speed and bowl diameter?

(*a*) 2 (*b*) 4

(*c*) 8 (*d*) 16

9. The gradual decrease in viscosity of tomato paste during storage can be prevented by quickly heating it to 82 °C, because

(*a*) water soluble pectin interacts with calcium

(*b*) hemicellulose prevents decrease in viscosity

(*c*) lignin prevents decrease in viscosity

(*d*) pectin methyl esterase in inactivated

Q. (10 – 22) Carry Two Marks Each.

10. Match the enzyme in Group I with its corresponding application in Group II

Group I

P. Chymosin

Q. Sulfhydryl oxidase

R. β-Galactosidase

S. Microbial proteases

Group II

1. Removal of cooked flavor from milk
2. Soybean milk coagulation
3. For rennet puddings
4. Lactose removal

(*a*) P-3; Q-2; R-1; S-4

(*b*) P-3; Q-1; R-4; S-2

(*c*) P-1; Q-3; R-4; S-2

(*d*) P-4; Q-3; R-2; S-1

11. Milk is flowing at 0.12 m^3/min in a 2.5 cm diameter pipe. The temperature of the milk is 21 °C and the corresponding viscosity and density are 2.1×10^{-3} Pas and 1029 kg/m^3, respectively. If the flow is found to be turbulent under the given conditions, the Reynolds number is __________.

12. Whole milk (34, 950 kg) containing 4% fat is to be separated in 6 h period into skim milk with 0.45% fat and cream with 45% fat. The flow rate of cream stream (kg/h) from the separator is __________.

13. Match the edible plant tissue in Group I with the type of carotenoid given in Group II

Group I	Group II
P. Corn	1. Lycopene
Q. Red pepper	2. β-Carotene
R. Pumpkin	3. Capsanthin
S. Tomato	4. Lutein

(*a*) P-3; Q-4; R-2; S-1

(*b*) P-2; Q-1; R-3; S-4

(*c*) P-4; Q-3; R-2; S-1

(*d*) P-1; Q-2; R-4; S-3

14. Undesirable bitterness frequently encountered in cured cheese is due to the

(*a*) presence of naringen

(*b*) formation of limonin

(*c*) overall hydrophobicity of amino acid side-chains in peptide

(*d*) conversion of humulone to isohumulone

15. Green tea is considered to be a more healthy option than black tea because it

(*a*) has high content of polyphenols

(*b*) is richer in thearubigin

(*c*) does not require any sweetener during tea preparation

(*d*) has no microbial load

16. Multiple effect evaporation leads to

(*a*) reduction in operating cost and reduction in capital cost

(*b*) increase in operating cost and increase in capital cost

(*c*) increase in operating cost and reduction in capital cost

(*d*) reduction in operating cost and increase in capital cost

17. A dilute pineapple juice is heated in a double pipe heat exchanger from 28 °C to 75 °C heat exchanging with hot water flowing in shell in counter current direction. Hot water is entering the shell at 95 °C and leaving at 85 °C. The log mean temperature difference (°C) is ____________.

18. Heat is transferred by radiation to a loaf of bread in an oven at a uniform temperature of 177 °C. The total surface area and temperature of the loaf are 0.0645 m^2 and 100 °C, respectively. The surface emissivity of the loaf is 0.85 and the value of Stefan-Boltzmann constant is 5.67×10^{-8} $W.m^{-2}.K^{-4}$. The net heat transfer (W) is ________.

19. Granulated sugar, having an average particle size of 500 μm, is milled to produce icing sugar having an average particle size of 25 μm. The power requirement was 10 kW as obtained by Rittinger's law. If the same mill were to be used to produce fondant sugar having an average particle size of 20 μm at the same capacity, the power requirement (kW) would be ________.

20. One ton of soybean containing 18% oil, 35% protein, 27.1% carbohydrates, 9.4% of fibre and ash, and 10.5% moisture is crushed and pressed. The residual oil content in the pressed cake is 6%. Assuming that there is no loss of protein and water with oil, the amount of oil (kg) obtained from the crusher is ________.

21. Match the processing method in Group I with the operation carried out in Group II

Group I

P. Degumming

Q. Deacidifying

R. Bleaching

S. Winterizing

Group II

1. Crystallization of triacylglycerol by cooling to remove fat crystals
2. Passing heated oil over charcoal
3. Using alkaline solution to remove fatty acids
4. Wetting with water to remove lecithin

(*a*) P-3; Q-1; R-4; S-2

(*b*) P-4; Q-3; R-1; S-2

(*c*) P-4; Q-3; R-2; S-1

(*d*) P-3; Q-1; R-2; S-4

22. The order of succession of microbes in the spoilage of milk, involving (P) *Lactobacillus*, (Q) protein digesting bacteria, (R) *Lactococcus lactis*, (S) yeasts and molds, is

(*a*) S > R > Q > P (*b*) S > Q > R > P

(*c*) R > P > S > Q (*d*) Q > S > P > R

ANSWERS

1. (d)	**2.** (d)	**3.** (b)	**4.** (b)	**5.** (d)	**6.** (c)
7. (d)	**8.** (c)	**9.** (d)	**10.** (b)	**11.** (49000 to 50225)	
12. (455 to 475)	**13.** (c)	**14.** (c)	**15.** (a)	**16.** (d)	**17.** (35 to 36)
18. (67.8 to 68.2)	**19.** (12.4 to 12.8)	**20.** (127 to 128)	**21.** (c)	**22.** (c)	

EXPLANATIONS

2. (d) Cuelonic separator is a method of removing particulates from an air, gas or liquid steam, without the use of filters, through vortex separation

Hence the option (d) is correct

3. (b)

Sr. No.	Name of Product	% fat
1	Beet	1.80%
2	Chicken Meat	0.90%
3	Pork	0.90%
4	Lamb flesh	8-10%

Hence the option (b) is correct

4. (b) β-Amylase is the enzyme that hydrolyzes starch to Maltose.

Hence the option (b) is correct

5. (d) Paraboiling rice is also an economical way to improve the storage stability & to increase nutritional value of the rice. During this, vitamins, minerals such as Iron seep into the endosperm from their natural location in the bran and aleurone layers.

6. (c) Thermolabile nature of luteins & Kunitz-type protease inhibitors make heat treated legume seed proteins more digestible & reduce risks of contracting diseases such as goitre.

7. (d) When the Relative Humidity is 100%

i.e $\phi = 100\%$

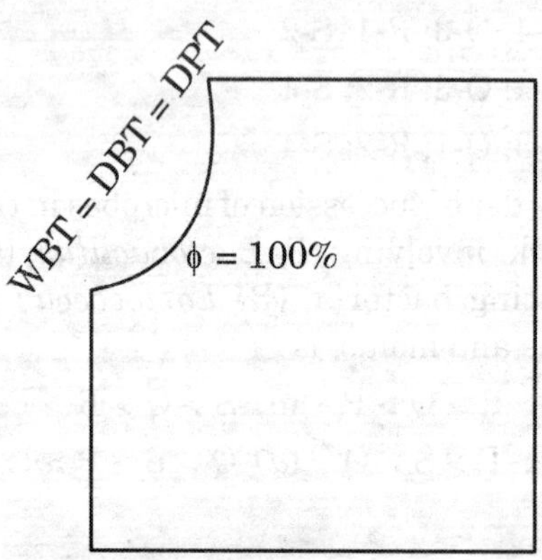

then $\boxed{DBT = WBT}$

Hence the option (d) is correct

8. (c) $F = \dfrac{MV^2}{R} = 4\pi^2 Mn^2 R$

F is the centrifugal force

m is the mass

R is the radius

V is the speed

n is the number of revolutions/sec

$$F_1 = 4\pi^2 mn^2 R \quad ....(1)$$

$$F_2 = 4\pi^2 m(2n)^2 2R$$

$$= 4\pi^2 \times m \times 8n^2 R \quad ....(2)$$

$$\boxed{F_2 = 4F_1}$$

9. (d) By heating tomato paste to 82°C, pectin methyl esterase is inactivated due to which gradual decrease in viscosity can be prevented.

10. (b) Chymosin → Rennet is used during the cheese making process to coagulate or thicken milk into curd. Rennet traditionally comes from animals specially it is an enzyme called chymosin or remain that is taken from the stomach living of calf, ewe, or kid (baby goat)

Sulfhydryl oxidase → Removal of cooked flavour from milk.

β-Galactosidase → is important for organisms as it is a beg provider in the product of energy and source of carbon through the break down of lactose to galactose and glucose.

Microbeal proteases → soybean milk coagulation

Hence the option (b) is correct

11. (49000 to 50225)

Cross sectional area of pipe

$$= \left(\frac{\pi}{4}\right) D^2 = \frac{\pi}{4} \times .025^2$$

$$= 4.9 \times 10^{-4}\, m^2$$

Rate of flow $= .12 m^3/min$

$$= 2 \times 10^{-3} m^3/sec$$

$$\text{Velocity of flow} = \frac{2510^{-3}}{4.9 \times 10^{-4}} = 4.1 m/s$$

$$V = \frac{Q(\text{Discharge rate})}{\text{Area}}$$

$$\text{Reynold's number, } Re = \frac{SVD}{}$$

$$= \frac{.025 \times 4.1 \times 1029}{2.1 \times 10^{-3}} = 50230$$

& this is greater than 4000 so that flow is turbulent.

12. (455 to 475)

Basis 1 hour's flow of whole milk

Mass in

Total mass $= \frac{24950}{6} = 5825$ kg

Fat $= 5825 \times 0.4 = 233$ kg

And so water plus solids-not-fat = 5592 kg

Mass out

Let the mass of cream by x kg then its total fat content will be 0.45x. The mass of skim milk (5825–x) & its total fat content is (.30045(5825–x))

Material balance on fat:

Fat in = Fat out

$5825 \times 0.4 = .0045(5825 - x) + 0.45x$

& x = 464.17 kg

Flow of cream = 464.17 kg/hour

Flow of skim milk = 5360.83 kg/hour

13. (c) Corn → Lutein

Red pepper → capsanthin

pumpkin → b-carotene

Tomato → Lycopene

Hence the option (c) is correct

14. (c) Undesirable batteries frequently enloubtered in cured cheese is due to the overall hydrophobicity of amino acid side-chains in peptide.

Hence the option (c) is correct

15. (a) Because green tea leaves are young and have not been oxidized.

Green tea has upto 40% polyphenots, while black tea contains only about 10% polyphenols

Hence the option (a) is correct.

16. (d) Multiple effect evaporation leads to reduction in operating cost and increases in capital cost

Hence the option (d) is correct

17. (35 to 36)

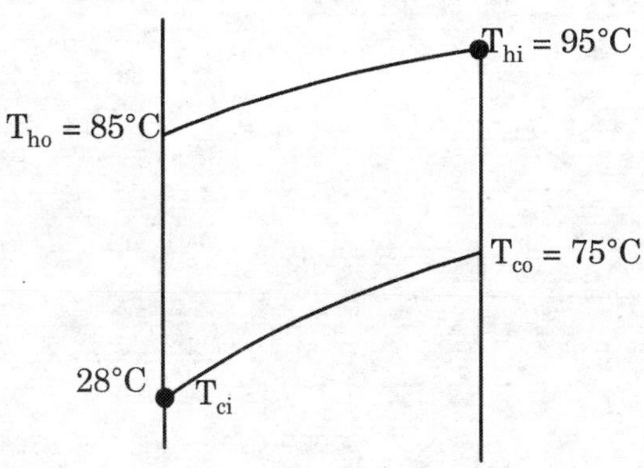

x = 0

x = l

$\theta_1 = T_{ho} - T_{ci} = 57°C$

$\theta_2 = T_{ho} - T_{co} = 20°C$

Log mean temperature difference:

$$\theta_m = \frac{\theta_1 - \theta_2}{l_n\left(\frac{\theta_1}{\theta_2}\right)} = \frac{57 - 20}{l_n\left(\frac{57}{20}\right)}$$

$\theta_m = 35.3283°C$

18. (67.8 to 68.2)

$\sigma A\left(T_1^4 - T_2^4\right). \in$

$= 5.67 \times 10^{-18} \times .0645\left(450^4 - 373^4\right) \times .85$

= 67.29 watts

19. (12.4 to 12.8)

$P \alpha \Delta$ (original size + final size)

$P \alpha$ K.Δ S (where K is constant)

10 kW = K. 25

P = K. 20

$10kW = \frac{K}{25}$

$P = \frac{K}{20}$

P = 12.5 KW

$$\frac{dE}{dL} = KL^n$$

dE is the differential energy

dL is the change in typical dimension

L = Magnitude of typical length dimension

K, n are constants.

20. (127 to 128)

Basis : 1000 kg

Mass in oil $= 1000 \times \frac{18}{100} - 180$ kg

Protein $= 1000 \times \frac{35}{100} - 350$ kg

Total non oil constituents = 820 kg

Carbohydrate, ash, fibre & water are calculated in a similar manner to fat as well as protein.

Mass out : In original beans, 820 kg of protein, water etc, are associated with 180 kg of oil. In pressed material, 94 parts of protein, water, etc., are associated with 6 parts oil.

Total oil in expressed material

$$820 \times \frac{6}{94} = 52.3 \text{ kg oil exacted impress}$$

$$= 180 - 52.3 = 127.7 \text{ kg}$$

21. (c) Degumming → water refining usually called degumming to remove lecithin.

Deacidifying → Process of removing fatty acid by using alkaline solutions.

Bleaching → Passing heated oil over charcoal.

Winterizing → Crystallization of triacylglycerol by cooling to remove fat crystals.

Hence the option (c) is correct

22. (c) Raw milk undergoes four steps succession of microorganism during spoilage acid product by lactococcus lactis is followed by additional acid product associated with growth of more acid tolerant Lactobacillus.

At this point yeasts and molds becomes dominant and degrade the lactic acid gradually decreasing the acidity

Eventually protein digesting bacteria becomes active resulting in putrid odor and bitter flavour making the milk clear.

Hence option (c) is correct.

2016

(Q. 1 – 9) : Carry One Mark Each.

1. Bread staling is caused by ________

(*a*) Caramelisation (*b*) Gelatinisation

(*c*) Retrogradation (*d*) Aggregation

2. Arrange the grades of tea in the increasing order of their leaf size __ , __ and ______ .

(*a*) Souchang, pekoe and orange pekoe

(*b*) Pekoe, souchang and orange pekoe

(*c*) Orange pekoe, souchang, and pekoe

(*d*) Orange pekoe, pekoe, and souchang

3. Fruit juice is being pasteurized in a tubular heat exchanger. The retention time in holding tube of 0.2 m^2 cross sectional area is 3 seconds. If the flow rate of juice is 0.4 $m^3\ s^{-1}$, the length of the holding tube in m, is _____ .

4. The oil, which experiences flavor reversion even at the lower peroxide value is _____.

(*a*) Mustard (*b*) Soybean

(*c*) Palm (*d*) Sesame

5. 80 kg of wheat containing 10 kg of moisture has been dried to a moisture content of 8% wet basis in 3 hours under constant rate period of drying. The drying rate in kg h^{-1} is _____.

6. The rate of cream separation in a disc bowl centrifuge can be increased by ______.

(*a*) Increasing the size of the bowl

(*b*) Lower viscosity of fluid

(*c*) Increasing RPM of the bowl

(*d*) All of these

7. Which one of the following is not used in mass transfer analysis?

(*a*) Biot number (*b*) Peclet number

(*c*) Schmidt number (*d*) Sherwood number

8. Oxygen is permeating through an EVOH film of thickness 't' and solubility coefficient 'S'. If diffusivity of oxygen through the film is 'D', then permeability of oxygen through the film will be __________.

(*a*) D/t (*b*) D/s

(*c*) D × S (*d*) S/D

9. Condensing steam is used to heat vegetable oil in a double pipe co-current heat exchanger. If the inlet and outlet temperature of steam are T_{hi} and T_{ho}, and for vegetable oil T_{ci} and T_{co}, respectively, the log mean temperature difference (ΔT_{LM}) will be_________.

(*a*) $\dfrac{T_{hi} - T_{co}}{\ln \dfrac{T_{hi} - T_{ci}}{T_{hi} - T_{co}}}$

(*b*) $\dfrac{(T_{ho} - T_{co}) - (T_{hi} - T_{co})}{\ln \dfrac{T_{ho} - T_{ci}}{T_{ho} - T_{co}}}$

(*c*) $\dfrac{(T_{hi} - T_{co}) - (T_{ho} - T_{ci})}{\ln \dfrac{T_{hi} - T_{ci}}{T_{co} - T_{ci}}}$

(*d*) $\dfrac{T_{co} - T_{ci}}{\ln \dfrac{T_{hi} - T_{ci}}{T_{hi} - T_{co}}}$

(Q. 10 – 22) : Carry Two Marks Each.

10. Match the food spoilage organisms given in Column I with the associated foods given in Column II

Column I

P. Clostridium botulinum

Q. Salmonella spp.

R. Vibrio parahaemolyticus

S. Bacillus cereus

Column II

1. Fish
2. Cooked starch foods
3. Meat, egg and poultry
4. Canned foods

(*a*) P-4, Q-3, R-1, S-2

(*b*) P-3, Q-4, R 2, S-1

(*c*) P-2, Q-1, R-3, S-4

(*d*) P-4, Q-3, R-2, S-1

11. Fluid is flowing inside a pipe of radius 'R' in fully developed laminar flow. If the velocity of the fluid at the centre at a distance 'L' is 'v_{max}', velocity at radial distance of ¾ (R) will be _____ times v_{max}

(*a*) 9/16 (*b*) 7/16

(*c*) 16/9 (*d*) 16/7

12. A meat ball with a radius of 25.4 mm at a temperature of 700 K, is suddenly plunged into a medium whose temperature is held at 395 K. Assume a convective heat transfer coefficient of 11.5 W $m^{-2}\ K^{-1}$ and take the average physical properties as: K = 44 W $m^{-1}K^{-1}$, ρ = 7850 kg m^{-3} and c_p = 0.4606 kJ $kg^{-1}K^{-1}$. The temperature (K) of the meat ball after one hour is _______.

13. a) **Assertion:** Acidulates are added in soft drinks to provide a buffering action.

r) **Reason:** Buffers tend to prevent changes in pH and prevent excessive tartness.

Choose the correct answer from the following

(*a*) Both a) and r) are true but r) is not the correct reason

(*b*) Both a) and r) are true and r) is the correct reason for a)

(*c*) a) is true but r) is false

(*d*) Both a) and r) are false

14. The D_{121} and Z values for C. **botulinum** spores in canned food are 0.2 min and 10 °C respectively. Total time required in min, to reduce the spores from 10^2 to 10^{-6} at 111 °C is ________.

15. a) **Assertion:** Olestra is used as a zero calorie fat replacer

r) **Reason:** It is a sucrose polyester with 6-8 acyl group and is not absorbed in the human digestive system.

Choose the correct answer from the following

(*a*) Both a) and r) are false

(*b*) Both a) and r) are true, but r) is not the correct reason for a)

(*c*) a) is true but r) is false

(*d*) Both a) and r) are true and r) is the correct reason for a)

16. Match the enzymes in Column I with their functions in Column II

Column I

P. Amylase

Q. Invertase

R. Phosphatase

S. Protease

Column II

1. Conversion of sucrose to glucose and fructose
2. Softening of dough
3. Effectiveness of pasteurization
4. Conversion of starch to maltose

(*a*) P-1, Q-2, R-3, S-4

(*b*) P-4, Q-1, R-3, S-2

(*c*) P-1, Q-4, R-2, S-3

(*d*) P-2, Q-4, R-3, S-1

17. Match the terms in Column I with their most appropriate description in Column II

Column I

P. Enrichment

Q. Fortification

R. Supplementation

S. Complementation

Column II

1. Overcome the deficiency of nutrient by mixing of two plant sources
2. Overcome the deficiency of nutrient from a synthetic source
3. Restoration of nutrient which is lost during processing
4. Addition of nutrient which may or may not originally present

(*a*) P-3, Q-4, R-2, S-1

(*b*) P-4, Q-3, R-1, S-2

(*c*) P-1, Q-2, R-3, S-4

(*d*) P-2, Q-3, R-1, S-4

18. Match the products in Column I with their Original Phase in Column II

Column I	**Column II**
P. Milk	1. Colloidal
Q. Butter	2. Solution
R. Lactose	3. Water in oil emulsion
S. Casein	4. Oil in water emulsion

(*a*) P-3, Q-4, R-1, S-2

(*b*) P-3, Q-4, R-2, S-1

(*c*) P-4, Q-3, R-2, S-1

(*d*) P-4, Q-3, R-1, S-2

19. a) **Assertion:** Presence of low sulphur containing amino acids make casein in milk to boil, sterilize and concentrate without coagulation even at higher temperature.

r) **Reason:** This is due to the restricted formation of di-sulphide bonds resulting in increased stability.

Choose the correct answer from the following

(*a*) Both a) and r) are true and r) is the correct reason for a)

(*b*) Both a) and r) are true but r) is not the correct reason for a)

(*c*) Both a) and r) are false

(*d*) a) is true but r) is false

20. In a typical Psychrometric Chart shown below, the processes OP, OQ and OR related to air water vapor mixture are _____ , _____ and _____ .

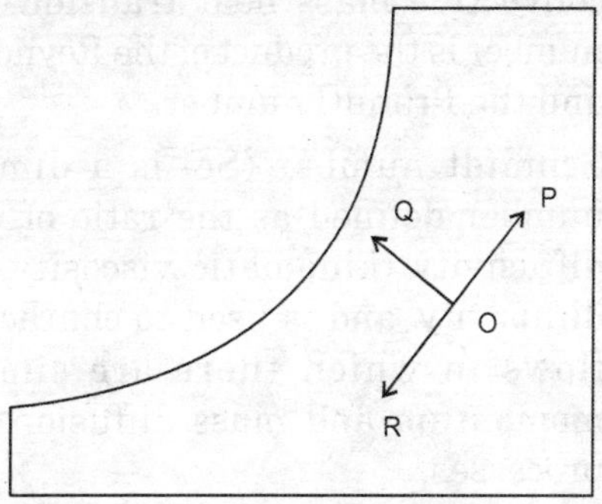

(*a*) Cooling & dehumidification, cooling & humidification, heating & humidification

(*b*) Cooling & dehumidification, heating & humidification, drying

(*c*) Heating & humidification, cooling & humidification, cooling & dehumidification

(*d*) Heating & humidification, cooling & dehumidification, drying

21. A fruit juice with a negligible boiling point rise is being evaporated using saturated steam at 121.1 °C in a triple effect evaporator having equal area in each effect. The pressure of the vapor in the last effect is 25.6 kPa absolute and the corresponding saturation temperature is 65.7 °C. The heat transfer coefficients are U_1 = 2760, U_2 = 1875 and U_3 = 1350 W m^{-2} K^{-1}. The boiling point (°C) in the first effect is _________.

22. In an aeration system, 520 kg of wheat grains having average size of 0.15 mm, shape factor of 0.88 and density of 1040 kg m^{-3} are fluidized using air at 2 atm absolute and 25 °C. If the cross section of empty bed is 0.4 m^2, the minimum height (m) of the fluidized bed, with voidage of 0.45 will be _____.

ANSWERS

1. (c)	**2.** (a)	**3.** (5.9 to 6.1)	**4.** (b)	**5.** (1.3 to 1.4)	**6.** (d)	**7.** (a)
8. (c)	**9.** (d)	**10.** (a)	**11.** (b)	**12.** (472 to 475)	**13.** (b)	**14.** (15 to 17)
15. (d)	**16.** (b)	**17.** (a)	**18.** (c)	**19.** (a)	**20.** (c)	**21.** (108 to 109)
22. (2.0 to 2.3)						

EXPLANATIONS

1. (c) Retrogradation : Retrogradation is a reaction that takes place when the amylose and amylopectin chains in cooked, gelatinized starch realign themselves as the cooked starch cools.

2. (a) Grades of tea in the increasing order of their life Souchang,pekoe and orange pekoe.

3. (5.9 to 6.1)

$$L_{\text{holding tube}} = \frac{Qxt}{A} = \frac{.4x3}{2} = 6\,m$$

Q is discharge or volume flow rate.
A is the cross sectional area of tube.

4. (b) Detection of peroxide gives the initial evidence of rancidity in unsaturated fats and oils. Peroxide value is the most widely used. It gives a measure of the extent to which an oil sample has undergone primary oxidation, extent of secondary oxidation may be determined from p-anisidine test.

The double bonds found in fats and oils play a role in autoxidation. Oils with a high degree of unsaturationare most susceptible in autoxidation. The best test for autoxidation (oxidative rancidity) is determination of the peroxide value. Peroxides are intermediates in the autoxidation reaction.

5. (1.3 to 1.4)

80 kg of wheat contains 10 kg moisture.

It means wheat has 12.5% moisture.,70 kg wheat solid and 10 kg moisture.

When drying takes place, moisture remains at 8%.

Which leaves 70 kg wheat solid and 6.087 kg moisture.

Drying rate = (10-6.087)/3 = 1.3 kh/hour

6. (d) Rate of cream separation can be increased by

1. Increasing size of bowl.
2. By lowering viscosity of fluid.
3. By increasing RPM

7. (a) Biot Number : The Biot number (Bi) is a dimensionless quantity used in heat transfer calculations. It is named after the eighteenth century French physicist Jean-Baptiste Biot (1774–1862), and gives a simple index of the ratio of the heat transfer resistances inside of and at the surface of a body.

The Peclet number is a dimensionless number used in calculations involving convective mass heat transfer. The Peclet number is the product of the Reynolds number and the Prandtl number.

Schmidt number (Sc) is a dimensionless number defined as the ratio of momentum diffusivity (kinematic viscosity) and mass diffusivity, and is used to characterize fluid flows in which there are simultaneous momentum and mass diffusion convection processes.

The Sherwood number (Sh) (also called the mass transfer Nusselt number) is a dimensionless number used in mass-transfer operation. It represents the ratio of the convective mass transfer to the rate of diffusive mass transport.

8. (c) Permeability of oxygen = DxS

9. (d)
$$LMTD = \frac{T_{co} - T_{ci}}{\ln\left(\frac{T_{hi} - T_{ci}}{T_{hi} - T_{co}}\right)}$$

10. (a) C. botulinum is commonly associated with bulging canned food; bulging, misshapen cans are due to an internal increase in pressure caused by gas produced by the bacteria.

Salmonella growth kinetics have been developed for chicken, pork, tomatoes, and melons. Salmonella reproduce asexually with a cell division interval of 40 minutes.

Vibrio parahaemolyticus : Outbreaks tend to be concentrated along coastal regions during the summer and early fall when higher water temperatures favor higher levels of bacteria.

Bacillus cereus is found in Cooked starch foods.

11. (b) For Laminar flow,

$$V_{\max at 0.75R} = [1 - (\frac{3^2}{4^2})]V_{\max at R}$$

$$V_{\max at 0.75R} = (\frac{7}{16})V_{\max at R}$$

13. (b) Buffers tend to prevent changes in pH and prevent excessive tartness. And in order to have this, acidulates are added in soft drinks.

15. (d) Olestra is used as a zero calorie fat replacer because it is a sucrose polyester with 6-8 acyl group and is not absorbed in the human digestive system.

16. (b) P-4,Q-1,R-3,S-2

An amylase is an enzyme that catalyzes the hydrolysis of starch into sugars. Amylase is present in the saliva of humans and some other mammals, where it starts the chemical process of digestion.

Invertase is an enzyme that catalyzes the hydrolysis (breakdown) of sucrose (table sugar) into fructose and glucose.

A phosphatase is an enzyme that uses water to cleave a phosphoric acid monoester into a phosphate ion and an alcohol. Because a phosphatase enzyme catalyzes the hydrolysis of its substrate, it is a subcategory of hydrolases.

Protease is an enzyme which breaks down proteins and peptides.

17. (a) P. Enrichment-3. Restoration of nutrient which is lost during processing

Q. Fortification-4. Addition of nutrient which may or may not originally present

R. Supplementation-1. Overcome the deficiency of nutrient by mixing of two plant sources

S. Complementation-2. Overcome the deficiency of nutrient from a synthetic source

18. (c) P-4,Q-3,R-2,S-1

Case in is a family of related phosphoproteins. These proteins are commonly found in mammalian milk, comprising c. 80% of the proteins in cow's milk and between 20% and 45% of the proteins in human milk.

19. (a) Both a) and r) are true and r) is the correct reason for a)

20. (c) OP represents heating and humidification in which DBT increases and humidity also increases.

OQ represents cooling and humidification in which DBT decreases and Humidity increases.

OR represents cooling and dehumidification in which DBT decreases and Humidity also decreases.

2017

1. Indicate the correct group that contains a monosaccharide, a disaccharide and a trisaccharide.
   (a) Glucose, sucrose, mannose
   (b) Ribose, lactose, raffinose
   (c) Mannose, jmaltose, lactose
   (d) Raffinose, stachyose, glucose

2. In which of the following products, 'must is used as the substrate for fermentation?
   (*a*) Beer (*b*) Wine
   (*c*) Idli (*d*) Tempeh

3. Identify the food borne illness which is not caused by bacteria.
   (*a*) Botulism (*b*) Listeriosis
   (*c*) Vibriosis (*d*) Cysticerocosis

4. Nutrient composition of wheat flour changes with extent of extraction from whole wheat grain. Which of the following statements is true if the extraction rate increased from 50% to 90%?
   (*a*) Starch increases, protein increase, fat increases, mineral increases
   (*b*) Starch decreases, protein increases, fat increases, mineral increases
   (*c*) Starch decreases, protein deceases, fat increases, mineral decreases
   (*d*) Starch decreases, protein increases, fat decreases, mineral decreases

5. You have two samples of milk one (X) with 3.8% fat and another (Y) with 0.5% fat. In order to produce a milk with 3.5% fat, 100 ml of Y should be mixed with ____ ml of X.

6. Match the items in column I with the items in column II in relation to food safety and standards.

Column I	Column II
P. HACCP	1. International food standards
Q. FSSAI	2. Quality control protocol
R. CIP	3. Food plant sanitation and hygiene protocol
S. CODEX	4. Indian food standards

   (*a*) P-2; Q-4; R-3; S-1
   (*b*) P-2; Q-3; R-1; S-1
   (*c*) P-1; Q-4; R-2; S-3
   (*d*) P-4; Q-2; R-3; S-1

7. A 50% sucrose solution at 20°C is flowing at a rate of $3.5 m^3/h$ through a pipe with an inside diameter of 0.0475 m and length of 12 m. The viscosity and the density of the solution are 15.43 cp and $1232 kg/m^3$, respectively. The Reynolds number of the flows is ______.

8. In a pineapple juice, fibre particles having mean diameter of 160 m and density of $1075 kg/m^3$ are settling by gravity. If the density and viscosity of the juice are $1015 kg/m^3$ and 0.98 cp, respectively, terminal velocity of the fibre particles is ___ mm/s.

9. Power consumption in liquid mixing is proportional to ______.
   (*a*) Power number × liquid density × $(\text{rotational speed})^3$ × $(\text{impeller diameter})^5$.
   (*b*) Power number × liquid density × $(\text{rotational speed})^2$ × $(\text{impeller diameter})^3$.
   (*c*) Liquid density × viscosity of the liquid × $(\text{rotational speed})^2$ × $(\text{impeller diameter})^3$.
   (*d*) Acceleration due to gravity × liquid density × $(\text{rotational speed})^3$ × $(\text{impeller diameter})^5$.

10. Match the following items of group I with the items of group II in relation to the quality of fat.

Group I	Group II
P. Saponification number	1. Unsaturation of fatty acid
Q. Iodine number	2. Volatile water soluble fatty acid
R. Reichert Meissl number	3. Hydroxy fatty acid
S. Acetyl value	4. Molecular weight of fatty acid

   (*a*) P-1; Q-2; R-3; S-4 (*b*) P-1; Q-3; R-4; S-2
   (*c*) P-4; Q-1; R-2; S-3 (*d*) P-2; Q-1; R-3; S-4

11. Match the following metabolic product (Column I) that indicates the quality of food (Column II)

Column I	Column II
P. Ethanol	1. Canned vegetable
Q. Lactic acid	2. Fish
R. Trimethylamine	3. Butter
S. Volatile fatty acid	4. Apple juice

   (*a*) P-3; Q-2; R-4; S-1
   (*b*) P-4; Q-1; R-2; S-3
   (*c*) P-4; Q-3; R-2; S-1
   (*d*) P-3; Q-4; R-2; S-1

12. Correlate the vitamins in column I with their role in promoting reaction/process in column II.

Column I	Column II
P. Riboflavin	1. Visual cycle
Q. Vitamin D	2. Acyl group transfer
R. Pantothenic Acid	3. Regulation of Ca^{2+} metabolism
S. Vitamin A	4. Oxidation-reduction reaction

(*a*) P-1; Q-2; R-4; S-3
(*b*) P-2; Q-1; R-3; S-4
(*c*) P-3; Q-4; R-1; S-2
(*d*) P-4; Q-3; R-2; S-1

13. A pure strain with generation time of 60 min is used in a fermentation process. Following inoculation (0 h), the strain takes 2 h for adaptation. 10 h to achieve maximum growth and 12 h to arrive at the point where the death rate is higher than the growth rate. If the inoculation load is 100 cells, the total population at the end of 10 h will be ______.

14. Refer the shear stress - shear rate plot shown in the figure below. Match the lines (Column I) with appropriate rheological behaviour (Column II).

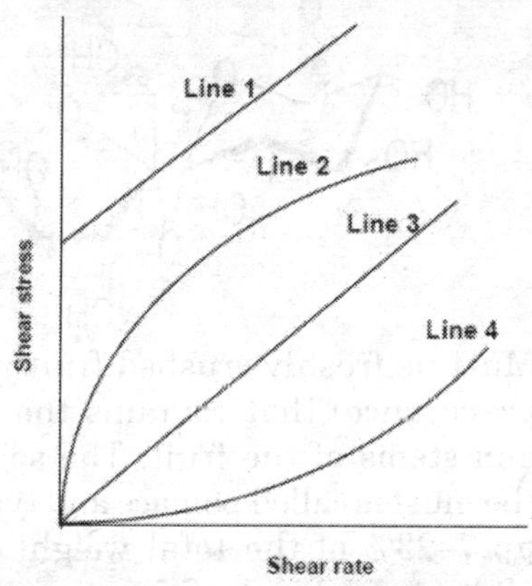

Column I	Column II
P. Line 1	1. Dilatant
Q. Line 2	2. Newtonian
R. Line 3	3. Pseudoplastic
S. Line 4	4. Bingham plastic

(*a*) P-2; Q-3; R-4; S-1 (*b*) P-1; Q-3; R-4; S-2
(*c*) P-2; Q-4; R-3; S-1 (*d*) P-4; Q-3; R-2; S-1

15. Water flowing at a rate of 1 kg/min is heated from 12 to 80°C with flue gas supplied at a rate of the flue gas are 180°C and 1.05 kJ/kg. K, respectively. If specific heat of water is 4.2 kJ/kg/ K and the flow is parallel, then the logarithmic mean temperature difference will be ______°C.

16. The Lineweaver-Burk plot of an enzymatic reaction shows V_{max} of 160 mol/l.min and k_m of 60 mol/l. For a substrate concentration of 40 mol/l, the velocity of the reaction is estimated to be ____ mol/l. min.

17. Bread is wrapped in 0.1 mm thick cellophane film having water vapour permeability of 1.82×10^{-10} m^3 water (STP)/s.m^2. atm/m at 38°C. If the surface area of pack, vapour pressure of water inside and outside of the pack is 0.20 m^2, 10 mm Hg and 5 mm Hg, respectively, the loss of water vapour at 38 °C in g/day is ____.

18. Match the following methods/system (column I) with appropriate operations (column II).

Column I	Column II
P. Parboiling	1. Sugarcane juice extraction
Q. Pearling	2. Hydrothermal treatment
R. Wet milling	3. Corn milling
S. Degerming	4. Wheat milling
T. Break rolls	5. Barley processing
U. Crushing rolls	6. Pulse milling

(*a*) P-4 Q-1; R-3; S-6, T-2, U-5
(*b*) P-4 Q-5; R-2; S-6, T-1, U-3
(*c*) P-3 Q-5; R-2; S-1, T-3, U-4
(*d*) P-2 Q-5; R-6; S-3, T-4, U-1

19. A 12 mm thick fish fillet having 80% moisture content (wet basis) is to be frozen using a plate freezer. The plates are maintained at –35°C. Assume the heat transfer coefficient; initial freezing temperature and latent heat of fusion are 2.0 W/m^2K,–2°C and 330 kJ/kg, respectively. If the density and thermal conductivity of frozen fish fillet are 1050 kg/m^3 and 1.48 W/m-K, respectively, the time required to freeze the fillet from the initial freezing temperature is ______ h.

20. A suspension containing 2×10^4 spores of organism A having a $D_{121.1°C}$ value of 1.5 min and 8×10^5 spores of organism B having a $D_{121.1°C}$ value of 0.8 min is heated at a constant temperature of 121.1°C. The heating time needed to obtain a probability of spoilage '1 in 1000' is ____ min.

21. In a evaporation process, a compressor picks up 0.05 m^3 air in each revolution and compresses 500 kg of air per minute. If the specific volume of air is 0.9 m^3/kg, then the compressor speed is ______ rpm.

22. For a soybean oil extraction system solvent: soy ratio is maintained at 0.5 : 1 (w/w). Original seed contains 18% oil (w/w). If the meal (soy solid) after final desolventization has 0.01 kg oil per kg oil free meal then, the effectiveness of the solvent (kg oil/kg solvent) in the extraction process is ____.

ANSWERS

1. (b)	**2.** (b)	**3.** (d)	**4.** (b)	**5.** (1000 to 1000)	**6.** (a)
7. (2078 to 2086)	**8.** (0.8 to 0.9)	**9.** (a)	**10.** (c)	**11.** (b)	**12.** (d)
13. (2550 to 25650)	**14.** (d)	**15.** (53.3 to 55.25)	**16.** (64.0 to 64.0)	**17.** (.14 to .18)	**18.** (d)
19. (6.8 to 7.2)	**20.** (10.8 to 11.2)	**21.** (9000 to 9000)	**22.** (.03 to 0.4)		

EXPLANATIONS

1. (b) Lactose is a disaccharide. It is a sugar composed of galactose and glucose. Lactose makes up around 2–8% of milk (by weight).The compound is a white, water-soluble, non-hygroscopic solid with a mildly sweet taste. It is mainly used in the food industry.

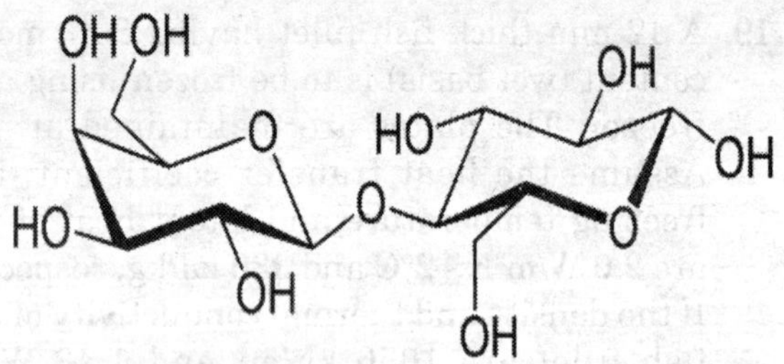

Ribose is a carbohydrate with the formula $C_5H_{10}O_5$; specifically, it is a pentose monosaccharide (simple sugar) with linear form H–(C = O)–$(CHOH)_4$–H, which has all the hydroxyl groups on the same side in the Fischer projection.

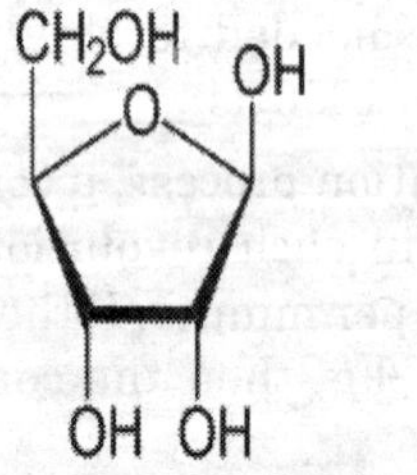

Raffinose is a trisaccharide composed of galactose, glucose, and fructose. It can be found in beans, cabbage, brussels sprouts, broccoli, and whole grains.

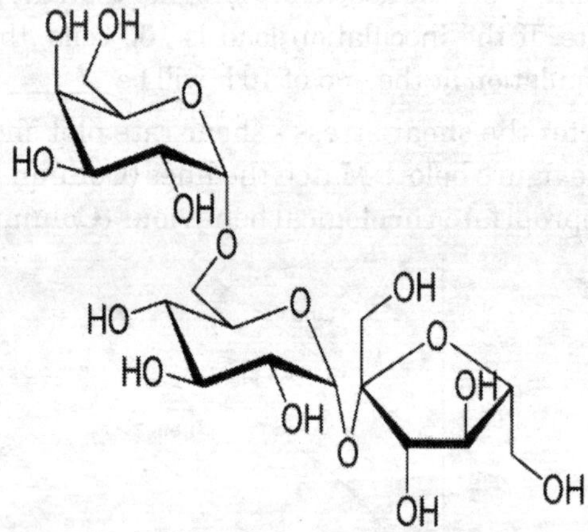

2. (b) Must is freshly crushed fruit juice (usually grape juice) that contains the skins, seeds, and stems of the fruit. The solid portion of the must is called pomace and typically makes up 7–23% of the total weight of the must. Making must is the first step in winemaking.

3. (d) Cysticercosis is a tissue infection caused by the young form of the pork tapeworm.

Vibrio is a genus of Gram-negative bacteria, possessing a curved-rod shape (comma shape), several species of which can cause food borne infection, usually associated with eating undercooked seafood. Typically found in salt water,

Listeriosis is a bacterial infectionmost commonly caused by Listeria monocytogenes,

Botulism is a rare and potentially fatal illness caused by a toxin produced by the bacterium Clostridium botulinum.

4. (b) Starch content decreases. But, as extraction rate increases, protein, fat and mineral increases.

5. (1000 to 1000)

$$100X0.5 + 3.8x = 3.5(100 + x)$$

By solving above equation, we get

x = 1000 ml

6. (a) Hazard Analysis Critical Control Point (HACCP): HACCP is a management system in which food safety is addressed through the analysis and control of biological, chemical, and physical hazards from raw material production, procurement and handling, to manufacturing, distribution and consumption of the finished product.

Generally recognized as safe (GRAS) is an American Food and Drug Administration (FDA) designation that a chemical or substance added to food is considered safe by experts.

Clean-in-place (CIP) is a technique of cleaning the interior surfaces of pipes, vessels, process equipment, filters and associated fittings, without disassembly them

The Codex Alimentarius is a collection of internationally recognized standards, codes of practice, guidelines, and other recommendations relating to foods, food production, and foodsafety. Its name is derived from the Codex Alimentarius Austriacus..

7. (2078 to 2086)

$$V = \frac{Q}{A} = \frac{\frac{3.5}{3600}}{\frac{\pi}{4}X.0475^2} = .5486\,m/s$$

Where, V = Velocity of fluid in pipe,m/s

A = Area of the pipe, m2

$$Re = \frac{\rho VD}{\mu} = \frac{1323X.5486X.0475}{15.43X10^{-3}} = 2080.7$$

8. (0.8 to 0.9)

$$\frac{\pi D^3 g(\rho_p - \rho_f)}{6} = \rho_f V^2 \frac{\pi D^2}{8}$$

By solving above equation, we get terminal velocity = 0.85 m/s

9. (a) Power Consumption is proportional to

Power number X liquid density X (rotational speed)3 X (impeller diameter)5

10. (c) The saponification number is the number of milligrams of potassium hydroxide required to neutralize the fatty acids resulting from the complete hydrolysis of 1g of fat.

The iodine value (or iodine adsorption value or iodine number or iodine index) in chemistry is the mass of iodine in grams that is consumed by 100 grams of a chemical substance.

The Reichert value (or more fully, the Reichert-Meissl-Wollny value or Reichert-Meissl-Wollny number) is a value determined when examining fat. The Reichert value is an indicator of how much volatile fatty acid can be extracted from fat through saponification. This number is a useful indicator of non-fat compounds in edible fats, and is especially high in butter.

The acetyl value is determined by the milligrams of potassium hydroxide require to neutralize the acetic acid produced when 1 gram of fat or oil is acetylated with acetic anhydride.

11. (b) P-4,Q-1,R-2,S-3

Trimethylamine (TMA) is an organic compound with the formula $N(CH_3)_3$. This colorless, hygroscopic, and flammable tertiary amine has a strong "fishy"odor in low concentrations and an ammonia-like odor at higher concentrations.

12. (d) P-4,Q-3,R-2,S-1

Riboflavin, also known as vitamin B2, is a vitamin found in food and used as a dietary supplement. Food sources include eggs, green vegetables,.

Riboflavin is important biochemically because it is vital for proper utilization of carbohydrates, fats, and proteins as energy sources. It is a component of two coenzymes, flavin adenine dinucleotide (FAD) and flavin mononucleotide (FMN). (Coenzymes are molecules that must be added to certain polypeptides to make them functional enzymes.) In general, FAD and FMN, when tightly bound to specific enzymes, easily lose or gain one or two electrons, or hydrogen atoms, and so drive oxidation/ reduction reactions. In the 1930s.

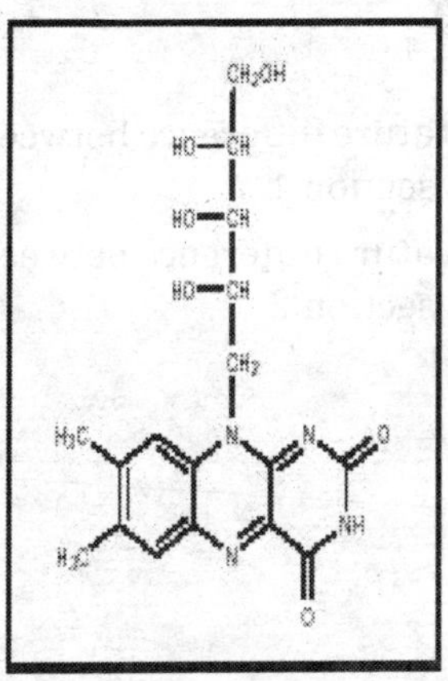

Figure 1.

The effects of active vitamin D are primarily to increase intestinal Ca^{2+} absorption, facilitate PTH-mediated calcium reabsorption in the distal renal tubules, and to suppress the synthesis and release of PTH from the parathyroid gland. Vitamin D also plays a role in regulation of bone resorption and formation.

Pantothenic acid, also called "pantothenate" or "vitamin B5" (a B vitamin), is a water-soluble vitamin. For many animals, pantothenic acid is an essential nutrient. Animals require pantothenic acid to synthesize coenzyme-A (CoA), as well as to synthesize and metabolize proteins, carbohydrates, and fats

Vitamin A helps in Visual Cycle. The conversion of dietary vitamin A (all- trans - retinol) into the light-sensitive visual chromophore (11- cis -retinal), and its regeneration following photobleaching, is accomplished through a series of enzymatic reactions known as the visual cycle.

14. (d)

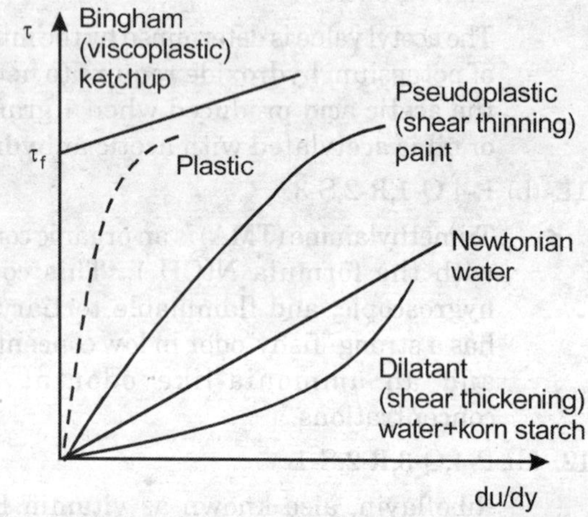

Type of behaviour	Description	Example
Thixotropic	Viscosity decreases with stress over time	Honey–keep stirring, and solid honey becomes liquid
Rheopetic	Viscosity increases with stress over time	Cream–the longer you whip it the thicker it gets
Shear thinning	Viscosity decreases with increased stress	Tomato sauce
Dilatent or shear thickening	Viscosity increases with increased stress	Oobleck

15. (53.3 to 55.25)

$$\theta_m = \frac{\theta_1 - \theta_2}{\ln\left(\frac{\theta_1}{\theta_2}\right)}$$

θ_1 = Temperature difference between hot and cold fluid at section 1

θ_2 = Temperature difference between hot and cold fluid at section 2

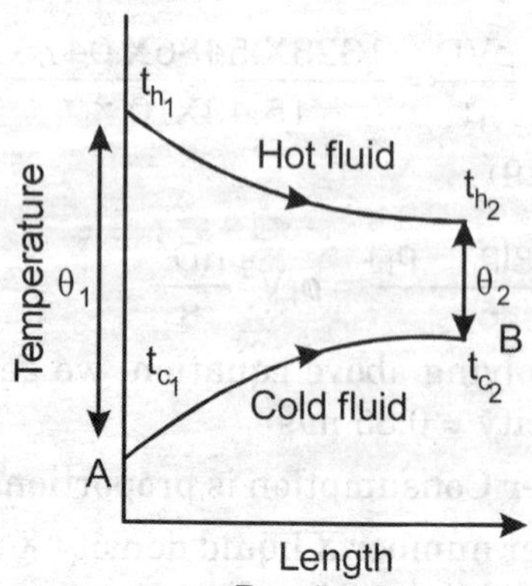

Parallel flow

$\theta_1 = 180 - 12 = 168$ °C

By using energy balance equation,we get

Energy lost by hot fluid = Energy gained by cold fluid

$3X1.05X(180-x) = 4.2X1X68$

By solving temperature at the exit of hot fluid, $x = 89.33°C$

$\theta_2 = 89.33 - 80 = 9.33\ °C$

By substituting θ_1 and θ_2 in θ_m, we het

$\theta_m = 54.9\ °C$

16. (64.0 to 64.0)

$$V = \frac{V_{max} \cdot [S]}{Km + S}$$

$= 64\ \mu mol/l\ min$

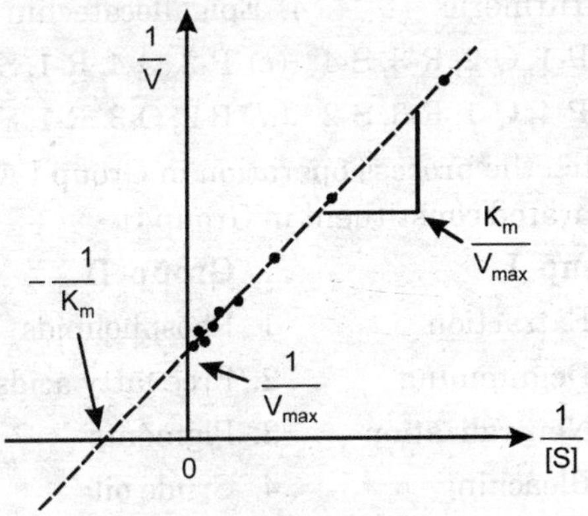

18. (d) P-2,Q-5,5-6,S-3,T-4,U-1

Parboiling: The food items are added to boiling water and cooked until they start to soften, then removed before they are fully cooked. Parboiling is usually used to partially cook an item which will then be cooked another way such as braising, grilling, or stir-frying.

Pearling is the namc of barley processing method.

Degerming is a process in which the germ or embryo from (a kernel of grain) is removed, usually through milling.

19. (6.8 to 7.2)

$$t_f = \frac{dH.\rho}{dT}\left(\frac{P.a}{h_s} + \frac{R.a^2}{k}\right)$$

$$t_f = \frac{330X1050}{33}\left(\frac{12}{4} + \frac{12^2}{8X1.48}\right)$$

$$t_f = 31627.7\ \text{seconds}$$

But for 80% moisture content,

$$t_f' = (0.8x31627.7)\ \text{seconds}$$

$$t_f' = 25302.16\ \text{seconds}$$

$$t_f' = 7.02\ \text{hours}$$

20. (10.8 to 11.2)

$$t = -D\log\left(\frac{N}{N_0}\right)$$

$$t_A = 1.5\log\left(\frac{20000}{.001}\right) = 10.95\ \text{minutes}$$

$$t_B = 0.8\log\left(\frac{800000}{.001}\right) = 7.12\ \text{minutes}$$

We select maximum out of the above values, therefore heating time = 10.95 min

21. (9000 to 9000)

$$N = \frac{v.\dot{m}}{V_{rev}}$$

$$= \frac{0.9X500}{.05}$$

$= 9000\ rpm$

2018

Q. 1 – Q. 9 carry one mark each.

1. Which of the following is oil soluble pigment present in fruits and vegetables?

   (*a*) Flavonoids (*b*) Carotenoids

   (*c*) Anthocyanins (*d*) Tannins

2. Which of the following represent the group of saturated fatty acids?

   (*a*) Lauric, Myristic, Arachidic

   (*b*) Palmitic, Linoleic, Linolenic

   (*c*) Capric, Stearic & Oleic

   (*d*) Behenic, Caprylic, Arachidonic

3. The anti-nutritional factor present in fava bean is

   (*a*) Gossypol (*b*) Curcine

   (*c*) Vicine (*d*) Cyanogen

4. Irradiation carried out to reduce viable non-spore forming pathogenic bacteria using a dose between 3 to 10 kGy is called

   (*a*) Radurization (*b*) Thermoradiation

   (*c*) Radappertization (*d*) Radicidation

5. Identify the correct statement related to the viscosity of Newtonian fluids from the following

   (*a*) It is not influenced by temperature

   (*b*) It increases with shearing rate

   (*c*) It decreases with shearing rate

   (*d*) It is not influenced by shearing rate

6. Adult male Wistar rats were fed with a protein based diet. Total 150 g of protein was ingested per animal. If the average weight increased from 110 g to 350 g after the end of the experiment, the Protein efficiency ratio of the given protein would be _______. (up to two decimal points)

7. The initial moisture content of a food on wet basis is 50.76%. Its moisture content (%) on dry basis is _______.(up to two decimal points)

8. The oxygen transmission rate through a 2.54×10^{-3} cm thick low density polyethylene film with air on one side and inert gas on the other side is 3.5×10^{-6} mL cm^{-2} s^{-1}. Oxygen partial pressure difference across the film is 0.21 atm. The permeability coefficient of the film to oxygen is _______ x 10^{-11} mL (STP) cm cm^{-2} s^{-1} (cm Hg)$^{-1}$.

9. Ambient air at 30°C dry bulb temperature and 80% relative humidity was heated to a dry bulb temperature of 80°C in a heat exchanger by indirect heating. The amount of moisture gain (g kg^{-1} dry air) during the process would be _____.

Q. 10 – Q. 22 carry two marks each.

10. Match the commodity in Group I with the bioactive constituent in Group II

Group I	Group II
P. Ginger	1. Lutein
Q. Green tea	2. Gingerol
R. Spinach	3. Curcumin
S. Turmeric	4. Epigallocatechin gallate

   (*a*) P-1, Q-2, R-3, S-4 (*b*) P-2, Q-4, R-1, S-3

   (*c*) P-4, Q-1, R-3, S-2 (*d*) P-2, Q-3, R-1, S-4

11. Match the process operation in Group I with the separated constituent in Group II

Group I	Group II
P. Extraction	1. Phospholipids
Q. Degumming	2. Free fatty acids
R. Neutralization	3. Pigments
S. Bleaching	4. Crude oil

   (*a*) P-3, Q-2, R-4, S-1 (*b*) P-4, Q-3, R-1, S-2

   (*c*) P-4, Q-1, R-2, S-3 (*d*) P-4, Q-1, R-3, S-2

12. Match the spoilage symptom in Group I with the causative microorganism in Group II

Group I	Group II
P. Green rot of eggs	1. *Micrococcus spp.*
Q. Putrid swell in canned fish	2. *Serratia marcescens*
R. Red bread	3. *Pseudomonas fluorescens*
S. Yellow discoloration of meat	4. *Clostridium sporogenes*

   (*a*) P-4, Q-3, R-2, S-1 (*b*) P-2, Q-1, R-4, S-3

   (*c*) P-3, Q-4, R-2, S-1 (*d*) P-1, Q-4, R-3, S-2

13. Match the fermented product in Group I with the base material in Group II

Group I	Group II
P. Sake	1. Milk
Q. Chhurpi	2. Cabbage
R. Natto	3. Rice
S. Sauerkraut	4. Soybean

   (*a*) P-3, Q-1, R-4, S-2 (*b*) P-1, Q-3, R-4, S-2

   (*c*) P-4, Q-1, R-3, S-2 (*d*) P-3, Q-2, R-1, S-4

14. Match the operation in Group I with the process in Group II

Group I	**Group II**
P. Cleaning	1. Quality separation
Q. Grading	2. Clarification
R. Size reduction	3. Screening
S. Filtration	4. Comminution

(*a*) P-1, Q-3, R-4, S-2 (*b*) P-4, Q-1, R-3, S-2
(*c*) P-2, Q-4, R-1, S-3 (*d*) P-3, Q-1, R-4, S-2

15. Out of 7 principles of HACCP system, 4 are listed below. Arrange these principles in the order in which they are applied.

(P) Conduct a hazard analysis
(Q) Establish monitoring process
(R) Establish critical limit
(S) Establish record keeping and documentation process

(*a*) P, R, Q, S (*b*) Q, R, P, S
(*c*) P, Q, R, S (*d*) R, S, P, Q

16. Apple juice of 10% total solids (TS) is being concentrated in a single effect evaporator working with a surface condenser to 40% TS under a vacuum of 20 kPa. After some time the vacuum pump stops but the evaporation process continued. Choose the combination of possible implications from the following.

(P) Product quality is affected
(Q) Substantial increase in thermal energy requirement
(R) Decrease in the rate of evaporation

(*a*) P & Q
(*b*) Q & R
(*c*) R & P
(*d*) P,Q & R

17. Identify an example of a classical diffusional mass transfer process without involving heat, among the following.

(*a*) Drying of food grains
(*b*) Carbonation of beverages
(*c*) Distillation of alcohol
(*d*) Concentration of fruit juice

18. For an enzyme catalyzed reaction $S \rightarrow P$, the kinetic parameters are:
$[S] = 40$ M, $V_0 = 9.6$ M s^{-1} and $V_{max} = 12.0$ M s^{-1}.
The K_m of the enzyme in M will be _______.
(up to one decimal points)

19. A microbial sample taken at 10 AM contained 1×10^5 CFU/mL. The count reached to 1×10^{10} CFU/mL at 8 PM of the same day. The growth rate (h^{-1}) of the microorganism would be _______.
(up to two decimal points)

20. Black pepper is ground from an equivalent particle size of 6 mm to 0.12 mm using a 10 hp motor. Assuming Rittinger's equation and that 1 hp = 745.7 W, the power (hp) of motor required to fine grind black pepper to 0.08 mm would be _______.(up to two decimal points)

21. Green pea (average diameter 0.8 cm) is frozen in a blast freezer operating at –40°C and with a surface heat transfer coefficient of 30 W m^{-2} K^{-1}. The thermal conductivity of pea is 2.5 W $m^{-1}K^{-1}$, and latent heat of crystallization is 2.74×10^2 kJ kg^{-1}. If the freezing point of pea is –1°C and the density is 1160 kg m^{-3}, the freezing time in minutes will be _______. (up to two decimal points)

22. The rate of heat transfer from a metal plate is 1000 W m^{-2}. The surface temperature of the plate is 120°C and ambient temperature is 20°C. The convective heat transfer coefficient (W m^{-2} °C^{-1}) using the Newton's law of cooling will be _______.

ANSWERS

1. (b)	**2.** (a)	**3.** (c)	**4.** (d)	**5.** (d)
6. (1.55 to 1.65)	**7.** (103.0 to 103.2)	**8.** (54 to 56)	**9.** (0 to 0)	**10.** (b)
11. (c)	**12.** (c)	**13.** (a)	**14.** (d)	**15.** (a)
16. (c)	**17.** (b)	**18.** (9.8 to 10.2)	**19.** (1.1 to 1.8)	**20.** (14.5 to 15.5)
21. (6 to 6.4)	**22.** (10 to 10)			

EXPLANATIONS

1. (b) Carotenoids also called tetraterpenoids, are organic pigments that are produced by plants and algae, as well as several bacteria and fungi.

2. (a) List of saturated fatty acids :

Common Name	Systematic Name	Structural Formula	Lipid Numbers
Caprylic acid	Octanoic acid	$CH_3(CH_2)_6COOH$	C8:0
Pelargonic acid	Nonanoic acid	$CH_3(CH_2)_7COOH$	C9:0
Capric acid	Decanoic acid	$CH_3(CH_2)_8COOH$	C10:0
Undecylic acid	Undecanoic acid	$CH_3(CH_2)_9COOH$	C11:0
Lauric acid	Dodecanoic acid	$CH_3(CH_2)_{10}COOH$	C12:0
Tridecylic acid	Tridecanoic acid	$CH_3(CH_2)_{11}COOH$	C13:0
Myristic acid	Tetradecanoic acid	$CH_3(CH_2)_{12}COOH$	C14:0
Pentadecylic acid	Pentadecanoic acid	$CH_3(CH_2)_{13}COOH$	C15:0
Palmitic acid	Hexadecanoic acid	$CH_3(CH_2)_{14}COOH$	C16:0
Margaric acid	Heptadecanoic acid	$CH_3(CH_2)_{15}COOH$	C17:0
Stearic acid	Octadecanoic acid	$CH_3(CH_2)_{16}COOH$	C18:0
Nonadecylic acid	Nonadecanoic acid	$CH_3(CH_2)_{17}COOH$	C19:0
Arachidic acid	Eicosanoic acid	$CH_3(CH_2)_{18}COOH$	C20:0
Cerotic acid	Hexacosanoic acid	$CH_3(CH_2)_{24}COOH$	C26:0

3. (c) Vicine is an alkaloid glycoside found in fava beans. Vicine is toxic, causing the disease favism, in individuals who have a hereditary loss of the enzyme glucose-6-phosphate dehydrogenase. Divicine is the name of the aglycone of vicine.

4. (d) Radicidation is a specific case of food irradiation where the dose of ionizing radiation applied to the food is sufficient to reduce the number of viable specific non-spore-forming pathogenic bacteria to such a level that none are detectable when the treated food is examined by any recognized method. The required dose is in the range of 3 - 10 kGy.

5. (d) Newtonian fluids are those fluid which follows newton's law of viscosity. It is not influenced by shearing rate.

6. (1.55 to 1.65)

$$\text{Protein efficiency ratio} = \frac{\text{Weight increased in grams}}{\text{Protein ingested in grams}}$$

$$\text{Protein efficiency ratio} = \frac{240}{150} = 1.6$$

7. (103.0 to 103.2)

$$MCdb = \frac{100 \times MC_{wb}}{100 - MC_{wb}}$$

$$= \frac{100 \times 50.76}{100 - 50.76}$$

$$= 103.09$$

8. (54 to 56)

Permeability coeficient

$$= \frac{2.54 \times 10^{-3} X 3.5 \times 10^{-6}}{.21 \times 760}$$

$$= 5.7 \times 10^{-11} \text{ mL(STP) cm cm}^{-2}\text{ s}^{-1}(\text{cmHg})^{-1}$$

Formula used :

Permeability coeficient

$$= \frac{\text{Thickness.OTR}}{P_{atm} \times 760}$$

9. (0 to 0)

In indirect heating process, amount of moisture added will be zero.

10. (b) P-2,Q-4,R-1,S-3

The term "bioactive food component" refers to nonessential biomolecules that are present in foods and exhibit the capacity to modulate one or more metabolic processes, which results in the promotion of better health. Bioactive food components are usually found in multiple forms such as glycosylated, esterified, thiolyated, or hydroxylated.

Summary listing of various bioactive food components, common food sources, and biological functions

Bioactive component	Food source	Function
Glucosinolates, diallyl sulfides, isothiocyanates	Broccoli, cauliflower, brussels sprouts, garlic, onions	Induction of detoxifying enzyme systems, antimicrobial, immunomodulator, anticancer
Tocopherols and tocotrienols	Vegetable oil, nuts, seeds	Antioxidant, immunomodulator
Isoflavonoids and polyphenols	Grapes, red wine, tea, fresh fruit, and vegetables	Antioxidant, lipid- lowering, immunomodulator, antiosteoporotic, anticancer
Phytoestrogens (genistein, daidzein)	Soybean and other soy-based products, flaxseed, cabbage, legumes, tea	Antiestrogen, anti- osteoporotic, antiproliferative
Phytosterols	Vegetable oils, nuts	Lipid-lowering
Dietary fiber	Whole grains, oats, fresh fruit with skin	Lipid-lowering
γ-linolenic acid, α-linolenic acid, and omega-3 fatty acids	Evening primrose or borage oil, walnuts, rapeseed, flaxseed, fish, microalgae	Anti-inflammatory, lipid- lowering
Lutein	Green leafy vegetables	Reduction in age- related macular degeneration
Cartenoids	Carrots, corn, squash, green leafy vegetables, oranges, papaya, red palm oil	Antioxidant immunomodulators
Lycopene	Tomatoes	Antiproliferative, anticancer
Bioactive peptides: lactoferrin, glycomacropeptide	Milk and fermented milk products	Immuno system enhancing, antiproliferative, antimicrobial
Probiotics	Fermented milk products	Immunomodulators, anticancer , gastrointestinal health modulators

11. (c) P-4,Q-1,R-2,S-3

Degumming is the first stage in the refining process. It is sued to separate the gums, phospholipids, proteins etc

Bleaching is applied in the refining of edible oils and fats and also in fruit and vegetable canning.

12. (c) Micrococcus spp. are gram-positive, oxidase-positive, and strictly aerobic cocci belonging to the family Micrococcaceae(2,3). They usually occur in irregular clusters, tetrads, and pairs(2,3), where individual cells are about 1 to 1.8 mm in diameter(2) and are usually non-motile and non-spore-forming(3).

Serratia marcescens is a species of rod-shaped gram-negative bacteria in the family Enterobacteriaceae.

Pseudomonas fluorescens is a common Gram-negative, rod-shaped bacterium. It belongs to the Pseudomonas genus.

Clostridium sporogenes is a species of Gram-positive bacteria that belongs to the genus Clostridium.

13. (a) P-3,Q-1,R-4,S-2

Sake also referred to as a Japanese rice wine, is an alcoholic beverage made by fermenting rice that has been polished to remove the bran. Unlike wine, in which alcohol is produced by fermenting sugar that is naturally present in fruit, typically grapes, sake is produced by a brewing process more akin to that of beer, where starch is converted into sugars which ferment into alcohol.

Chhurpi or durkha is a traditional cheese, similar to ricotta, consumed in the Himalayan regions of Nepal, Sikkim, Darjeelinghills, Bhutan, and Tibet. The two varieties of chhurpi are a soft variety (consumed as a side dish with rice) and a hard variety (chewed like a betel nut).

Natto is a traditional Japanese foodmade from soybeans fermented with Bacillus subtilis var. natto. Some eat it as a breakfast food.It is served with soy sauce.

Sauerkraut is finely cut cabbage that has been fermented by various lactic acid bacteria.

14. (d) P-3,Q-1,R-4,S-2

Comminution is the reduction of solid materials from one average particle size to a smaller average particle size, by crushing, grinding, cutting, or vibrating.

15. (a) **Seven Principles of HACCP in order are as follows :**

Principle 1 - Conduct a Hazard Analysis : The application of this principle involves listing the steps in the process and identifying where significant hazards are likely to Occur. The HACCP team will focus on hazards that can be prevented, eliminated or controlled by the HACCP plan. A justification for including or excluding the hazard is reported and possible control measures are identified.

Principle 2 - Identify the Critical Control Points : A critical control point (CCP) is a point, step or procedure at which control can be applied and a food safety hazard can be prevented, eliminated or reduced to acceptable levels. The HACCP team will use a CCP decision tree to help identify the critical control points in the process

Principle 3 - Establish Critical Limits : A critical limit (CL) is the maximum and/or minimum value to which a biological, chemical, or physical parameter must be controlled at a CCP to prevent, eliminate, or reduce to an acceptable level the occurrence of a food safety hazard. The critical limit is usually a measure such as time, temperature, water activity (Aw), pH, weight, or some other measure that is based on scientific literature and/or regulatory standards.

Principle 4- Monitor CCP : The HACCP team will describe monitoring procedures for the measurement of the critical limit at each critical control point. Monitoring procedures should describe how the measurement will be taken, when the measurement is taken, who is responsible for the measurement and how frequently the measurement is taken during production.

Principle 5 - Establish Corrective Action : Corrective actions are the procedures that are followed when a deviation in a critical limit occurs. The HACCP team will identify the steps that will be taken to prevent potentially hazardous food from entering the food chain and the steps that are needed to correct the process.

Principle 6 - Verification : Those activities, other than monitoring, that determine the validity of the HACCP plan and that the system is operating according to the plan. The HACCP team may identify activities such as auditing of CCP's, record review, prior shipment review, instrument calibration and product testing as part of the verification activities.

Principle 7 - Recordkeeping : A key component of the HACCP plan is recording information that can be used to prove that the a food was produced safely. The records also need to include information about the HACCP plan. Record should include information on the HACCP Team, product description, flow diagrams, the hazard analysis, the CCP's identified, Critical Limits, Monitoring System, Corrective Actions, Recordkeeping Procedures, and Verification Procedures.

16. (c) Decrease in the rate of evaporation

Product quality is affected.

17. (b) Carbonated Beverages are drinks that have had carbon dioxide added to them. Carbon dioxide is colourless and flavourless.

Carbonated Beverages have as their base either carbonated water or soda water - which are considered Carbonated Beverages in their own right. Usually flavour and sweetener are added. Such beverages are mostly water, 90% and up (diet ones can be 99% water.)

18. (9.8 to 10.2)

$$V = \frac{V_{max} \cdot [S]}{K_m + [S]}$$

$$9.6 = 12 \frac{40}{K_m + 40}$$

By solving above equation, we get

Km = 10 μM

19. (1.1 to 1.8)

$$\frac{\ln(1 \times 10^{10}) - \ln(1 \times 10^{5})}{10} = 1.15/h$$

20. (14.5 to 15.5)

By Rittinger's equation,

$$E = K(\frac{1}{x_2} - \frac{1}{x_1})$$

$$10 = K(\frac{1}{.12} - \frac{1}{6}) \text{ ---(1)}$$

$$P = K(\frac{1}{.08} - \frac{1}{6}) \text{ ---(2)}$$

By solving above equations, we get

P = 15 hp

21. (6 to 6.4)

$$t_f = \frac{dH.\rho}{T - T_a}(\frac{P.a}{h_s} + \frac{R.a^2}{k})$$

P = 1/6 for sphere, R = 1/24 for sphere

P = 1/2 for slab, R = 1/8 for slab

$$t_f = \frac{274 \times 1000}{39}(\frac{.004}{6 \times 30} + \frac{.004^2}{24 \times 2.5})$$

$$t_f = 6.24 \text{ minutes}$$

22. (10 to 10)

By Newton's law of cooling, we get

$$Q = hA.dT$$

$$1000 = h \times 1 \times 100$$

By solving above equation, we get

$$h = 10 \frac{W}{m^2 C}$$

XE-H

ATMOSPHERIC AND OCEANIC SCIENCES

Syllabus

Section A: Atmospheric Science

Fundamental of Meteorology, Thermal structure of the atmosphere and its composition, Radiation Balance and Laws, Wind Belts, Monsoon, Climate. Atmospheric Thermodynamics. Hydrostatic equilibrium and: Hydrostatic equation, variation of pressure with height, geopotential, Tropical convection. Atmospheric Electricity. Cloud Physics. Observation Techniques of the Atmosepheric Properties.

Fundamental equations. Pressure, gravity, centripetal and Corolis forces, continuity equation in Cartesian and isobaric coordinates, Scale analysis, inertial flow, geostrophic and gradient winds, thermal wind, vorticity. Atmospheric turbulence, baroclinic instabiltiy. Atmosphreric Waves. Tropical meteorology: Trade wind inversion, ITCZ; monsoon trough tropical cyclones, their structure and development theory; monsoon depressions; Climate variability and forcings; Madden-Julian oscillation(MJO), ENSO, QBO (quasi-biennial oscillation) and sunspot cycles. Primitive equations of Numerical Weather Prediction. General Circulation and Climate Modelling. Synoptic weather forecasting, prediction of weather elements such as rain, maximum and minimum temperature and fog. Data Assimilation.

Section B: Ocean Sciences

Seawater Properties, T-S diagrams, Ocean Observations, Ocean Tide and Waves and their properties. Coastal processes and Estuary Dynamics. coastal zone management. Wind Driven Circulation: Ekman, Sverdrup, Stommel and Munk theories, Inertial currents; geostrophic motion; barotropic and baroclinic conditions; Oceanic eddies. Global conveyor belt circulation. Subtropical gyres; Western boundary currents; equatorial current systems; Current System in the Indian Ocean.

Momentum equation, mass conservation, vorticity. Ocean and Wave Modeling, Ocean State Forecasting. Data Assimilation. Ocean Turbulence. Chemical Property of seawater, major and minor elements, their behavior and chemical exchanges across interfaces and residence times in seawater, Element chemistry in atypical conditions-estuaries, Biochemical cycling of nutrients, trace metals and organic matter. Air-sea exchange of important biogenic dissolved gases; carbon dioxide-carbonate system; alkalinity and control of pH; biological pump. Marine Pollution. Primary and secondary production; factors controlling phytoplankton and zooplankton abundance and diversity; nekton and fisheries oceanography.

Subject-wise Analysis

XE - H : Atmospheric and Oceanic Sciences

Year	1 mark Questions	2 marks Questions	Total Marks
2009	8	16	40
2010	9	13	35
2011	9	13	35
2012	9	13	35
2013	9	13	35
2014	9	13	35
2015	9	13	35
2016	9	13	35
2017	9	13	35
2018	9	13	35

2017

1. Rossby Number is the ratio of
   (a) Coriolis Force to Inertial Force
   (b) Inertial Force to Coriolis Force
   (c) Gravitational Force to Coriolis Force
   (d) Viscous Force to Inertial Force

2. Kuroshio Current and Gulf Stream are
   (a) EBC, WBC (b) EBC, EBC
   (c) WBC, WBC (d) WBC, EBC
   [WBC : Western Boundary Current, EBC: Eastern Boundary Current]

3. The velocity of a tsunami wave is in ocean basin of depth 1 km is ______ ms^{-1}.
   [Density of seawater : 1025 kg m^{-3}, g : 10 ms^{-2}]

4. A thin iceberg is observed to move southeastward in the Arctic Ocean. If the surface current is wind driven, the prevailing wind is
   (a) Easterly (b) Northerly
   (c) Southerly (d) Westerly

5. Equatorial Kelvin and Rossby waves respectively propagate
   (a) Westward and Eastward
   (b) Eastward and Westward
   (c) Westward and Westward
   (d) Eastward and Eastward

6. The largest contributor to the atmospheric greenhouse effect is
   (a) CO_2 (b) N_2
   (c) CH_4 (d) H_2O

7. If T_v, T, T_w and T_d denote virtual, dry bulb, wet bulb and dew point temperatures of a moist air parcel, then the correct order of their values is
   (a) $T_v > T > T_w > T_d$ (b) $T_v \geq T \geq T_w \geq T_d$
   (c) $T_v > T \geq T_w \geq T_d$ (a) $T > T_v > T_w > T_d$

8. Burning of fossil fuel is increasing the concentration of CO_2 in the atmosphere. A consequence of this is
   (a) Ocean water which is presently basic will drift towards pH neutral
   (b) Ocean water which is presently acidic will become more acidic
   (c) No effect on ocean pH
   (d) Ocean water which is presently slightly basic will become more basic

9. Mixed layer depths measured in the Pacific Ocean in two different years are schematically shown in the figure below.

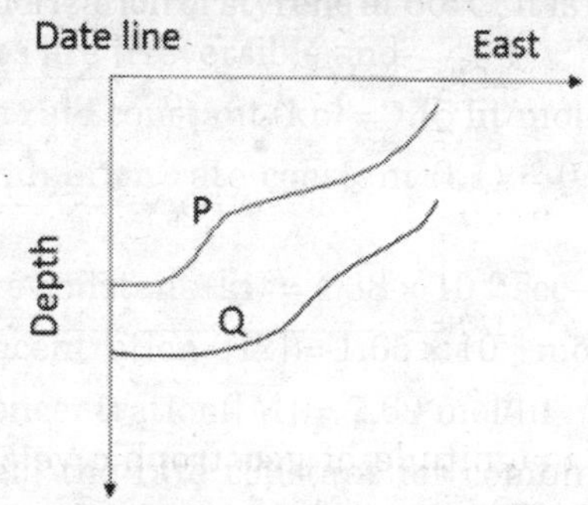

   Years P and Q belongs to
   (a) P: El-Nino, Q: La-Nina
   (b) P: La-Nino, Q: El-Nino
   (c) P: El-Nino, Q: QBO
   (d) P: QBO, Q: La-Nina

10. Average surface temperatures of the Sun and the Earth are 6300 K and 285 K, respectively. The ratio of the wavelength of peak radiation of the Earth to that of the Sun is ______.

11. In the month of April, the mixed layer in the Arabian Sea received a net heat flux of 50 W m^{-2}. If the mixed layer depth is 50 m, the increase in temperature at the end of April is ______ °C.
   [Density of seawater: 1025 kg m^{-3}, Density of freshwater: 1000 kg m^{-3}, Specific heat of seawater: 4200 J kg^{-1} K^{-1}, Latent heat of evaporation: 2.45×10^6 J kg^{-1}]

12. The thickness of an atmospheric layer between 600 hPa and 500 hPa is 1.5 km. If the layer is isothermal, then its temperature is ______ K.
   [Gas constant of air: 287 J kg^{-1} K^{-1}, g : 10 ms^{-2}]

13. At 17°N, a mass of fluid is moving under geostrophic balance at 0.3 m s^{-1} towards east. Suddenly the pressure gradient force becomes zero. Then the fluid will
   (a) continue to move towards the east at 0.3 ms^{-1}.
   (b) undergo circular motion with radius of about 17 km
   (c) undergo circular motion with radius of about 7 km
   (d) move southward
   [Angular velocity of the Earth: 7.27×10^{-5} rad s^{-1}]

14. At 45°N wind is blowing northward and its magnitude decreases eastward from 10 m s^{-1} to 1 m s^{-1} over a distance of 18 km. The absolute velocity of the flow is ______ $\times 10^{-4}$ s^{-1}.
   [Angular velocity of the Earth: 7.27×10^{-5} rad s^{-1}]

15. Sea surface height anomalies at the locations A, B, C and D are –10, –15, 5 and 0 cm respectively.

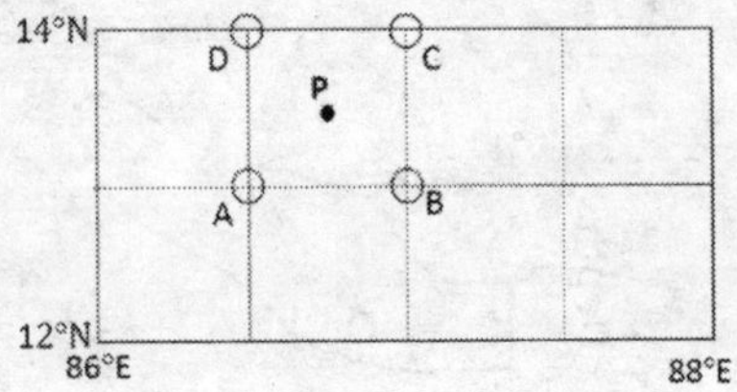

The magnitude of geostrophic velocity at P is _____ ms^{-1}.

[Take 1° = 100 km, g = 10 m s^{-2}, Angular velocity of the Earth: 7.27×10^{-5} rad s^{-1}]

16. In a severe tropical cyclone, 250 mm of rainfall occurs in an area having a radius of 200 km. If the energy supplied to the system from this rainfall is N times the energy of one atomic bomb ($=1.5 \times 10^{15}$ kJ), then the value of N is _______.

[Density of freshwater: 1000 kg m^{-3}, specific heat of seawater: 4200 J kg^{-1} K^{-1}, Latent heat of evaporation: 2.45×10^6 J kg^{-1}]

17. A student wants to numerically solve the linear 1-D advection equation $\frac{\partial\varphi}{\partial t}+C\frac{\partial\varphi}{\partial x}=0$, where c = 300 m s^{-1}. The value of the maximum time-step the student can consider according to CFL criterion for a spatial resolution of 3 km is

(*a*) 15 s (*b*) 10 s

(*c*) 25 s (*d*) 20 s

18. Planets in the solar system are in radiative equilibrium. Let S_0, α, T_0 and R denote solar constant, albedo, average temperature and radius of a planet, respectively, and σ is Stefan's constant. Then the energy balance of this planet is given by the expression.

(*a*) $(1-\alpha)S_0 = 4\,\sigma T_0^4$ (*b*) $(1-\alpha)S_0 = 2\,\sigma T_0^4$

(*c*) $\alpha S_0 = \sigma T_0^4$ (*d*) $\pi R^2(1-\alpha)S_0 = 4\sigma T_0^4$

19. A cumulonimbus cloud forms by an air parcel rising from the sea level with an initial temperature and specific humidity of 27°C and 20 gm kg^{-1}, respectively. Assume that moist static energy is conserved in this cloud. Then the cloud temperature at an altitude of 15 km is _____ K.

[Specific heat of dry air at constant pressure: 1005 J kg^{-1} K^{-1}, Specific heat of water vapour at constant pressure: 1850 J kg^{-1} K^{-1}, g = 10 m s^{-2}, Latent heat of evaporation: 2.45×10^6 J kg^{-1}]

20. If u_g and v_g are respectively zonal and merideonal components of a flow field in geostrophic balance, then the divergence of this flow is

(*a*) 0 (*b*) $\frac{u_g}{f}\frac{\partial f}{\partial x}$

(*c*) $-\frac{1}{\rho f}\frac{\partial^2 p}{\partial y^2}$ (*d*) $-\frac{v_g}{f}\frac{\partial f}{\partial y}$

[x, y, f, p, ρ are zonal distance, meridional distance, Coriolis parameter, pressure and density, respectively]

21. During the Indian summer monsoon season, depressions do not intensify to tropical cyclones because

P: Indian sub-continent is very hot and large land-sea temperature difference pulls depressions quickly to land before they can intensify into cyclones.

Q: southwesterly winds at low level are not conductive for the formation of tropical cyclones.

R: SST cooling due to strong monsoonal winds prevents cyclone formation.

S: strong zonal wind shear during the monsoon season does not allow warm core formation.

Which of the above statement(s) is (are) correct.

(*a*) P & Q (*b*) Only R

(*c*) Only S (*d*) R & S

22. Which among the following statement(s) (are) correct,

P: ENSO and El-Nino are the same and refer to the warming of Equatorial Eastern Pacific SST.

Q: ENSO is an atmosphere-ocean coupled phenomenon and El-Nino is its oceanic part.

R: ENSO is an atmospheric phenomenon and El-Nino is an oceanic phenomenon

S: ENSO is the oscillatory component of El-Nino having a period of 4.7 years.

(*a*) P & R (*b*) Only Q

(*c*) P, Q and S (*d*) R & S

ANSWERS

1. (b)	**2.** (c)	**3.** (98.9 to 100.5)	**4.** (d)	**5.** (b)
6. (d)	**7.** (c)	**8.** (a)	**9.** (b)	**10.** (2.05 to 22.15)
11. (.59 to .61)	**12.** (286.55 to 287.55)	**13.** (c)	**14.** (–4.10 to –3.90)	**15.** (.42 to .46)
16. (51 to 52)	**17.** (b)	**18.** (a)	**19.** (199 to 205)	**20.** (d)
21. (c)	**22.** (b)			

EXPLANATIONS

1. (b) Rossby number – The ratio between Nonlinear or Inertial terms of equation of motion and Coriolis forces is termed as Rossby number. It is

$$R_0 = \frac{\text{Inertial terms}}{\text{coriolis}}$$

$$R_0 = \frac{U^2}{L} \times \frac{1}{fU}$$

$$R_0 = \frac{U}{2\Omega L \sin\theta}$$

2. (c) Kuroshio Current and Gulf Stream are Western Boundary Current, which are warm, faster, and narrower.

Gulf Stream current is located in east coast of North America in Northern western Atlantic Ocean and Kuroshio Current is located in east coast of Asia in Northern western Pacific Ocean.

3. (98.9 to 100.5)

The velocity of tsunami wave in ocean base of depth 1 km is

Given, $\rho_\omega = 1025$ kg/m^3

$g = 10$ m/s^2

Depth = 1km = 1000m

Velocity = $\sqrt{gd}$

$= \sqrt{10 \times 1000}$

$= 100$ m/s.

4. (d) As the iceberg is observed to move southeastward. Therefore, as per open ocean surface current the prevailing winds will be westerlies.

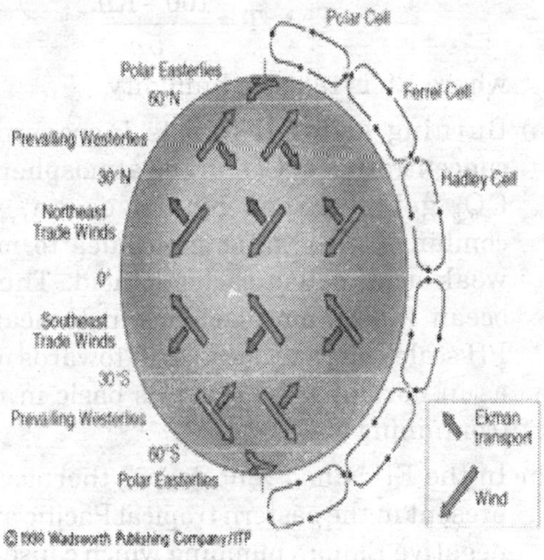

5. (b) Equatorial kelvin waves are a special type of kelvin wave that balances the Coriolis force in northern hemisphere against its southern hemisphere counterpart. This wave always propagates eastward and only exists on the equator.

Equatorial Rossby waves often called planetary waves are very long low frequency waves found near the equation and propagates westward.

6. (d) Water vapour (H_2O) is the largest contribution of greenhouse effect on average, it probably accounts for about 60% of the warming effect.

7. (c) The virtual temperature (T_v) of a moist air parcel is the temperature at which a theoretical dry air parcel would have a total pressure and density equal to the moist parcel of air.

$$T_v = \frac{T}{1-\frac{e}{p}(1-\epsilon)}$$

$$e = \rho_v R_v T$$

$$P = \rho R_d T_v$$

where T = dry bulb temperature

ρ_d and ρ_v densities the dry air and water vapour respectively.

$$\epsilon = \frac{R_d}{R_v}$$

Thus, T_v is greater than T

T_w is the wet bulb temperature which smaller than equal to T for moist air

and T_d is the dew point temperature which is smaller than all the temperature

$$T_d = T - \frac{100 - Rh}{5}$$

where Rh is relative humidity.

8. (a) Burning of fossil fuel is increasing the concentration of CO_2 in the atmosphere. The CO_2 get absorbed by the ocean, which combines with water molecules to make a weak acid, called carbonic acid. Thus, the ocean will be more acidic which means the PH scale will reduce and drift towards neutral as in normally the ocean is basic in nature (containing NaCl).

9. (b) In the El- Nino event deeper thermocline is present in the eastern tropical Pacific and has negative Ekman pumping which cause mixed layer depth (MLD) to increase hence positive MLD anomalies occurs and in La-Nina event the thermocline is shallow and hence the MLD decrease.

10. (2.05 to 22.15) $T_{sun} = 6300$ K

$T_{earth} = 285$ K

All the radiation travel with speed of 3×10^8 m/s in vacuum.

According the Wien's law

$$\lambda_{max} = \frac{\text{Constant}}{T} = \frac{3000\mu mK}{T}$$

$$\frac{\lambda_{max,earth}}{\lambda_{max,sun}} = \frac{3000\mu m}{285} \times \frac{6300}{3000\mu m}$$

$$= 22.10$$

11. (.59 to .61)

Net heat flux, $Q_{net} = 50$ W/m²

Depth, $d = 50$ m

$\rho_{seawater} = 1025$ kg/m³

Specific heat of seawater = 4200 J/kg/K

Latent heat of evaporation = 2.45×10^6 J/kg

Here April contains 30 days

Also 1 W = 1 J/s

We know, $Q_{net} = \frac{Q_{net}}{\rho c_p h}$

$$= \frac{50 \times 30 \times 86400}{1025 \times 4200 \times 50}$$

$$= 0.60$$

12. (286.55 to 287.55)

If the layer is isothermal

Therefore, $\frac{dT}{dz} = 0$

$$P = P_0 e^{\frac{z}{H}}$$

where, $H = \frac{RT}{g}$

$$\ln\left(\frac{P}{P_0}\right) = \frac{zg}{RT}$$

$$T = \frac{zg}{R\ln\left(\frac{P}{P_0}\right)}$$

$$T = \frac{1.5 \times 1000 \times 10}{287 \times 0.182}$$

$$T = 287.16 \text{ K}$$

13. (c) In this case, as per geostrophic flow

$$\frac{V^2}{R} - fV = \nabla P$$

Here pressure gradient is equal to zero

Therefore, $\frac{V^2}{R} = fV$

$$R = \frac{V}{f}$$

$$R = \frac{V}{2\Omega \sin\varphi}$$

$$R = \frac{0.3}{2 \times 7.27 \times 10^{-5} \sin(17)}$$

$$R = 7057 \text{ m}$$

$$R = 7 \text{ km}$$

and the fluid particle will undego circular motion.

14. (–4.10 to 3.90)

Absolute vorticity is the sum of planetary and relative vorticity

$$\zeta_a = \zeta + f$$

$$\zeta = \frac{\partial v}{\partial x} - \frac{\partial u}{\partial y}$$

$$f = 2\Omega \sin\phi$$

$$f = 2 \times 7.27 \times 10^{-5} \times \sin(45°)$$

$$f = 1.028 \times 10^{-4}$$

$$\because \quad v = 0$$

$$\zeta = \frac{-(10-1)}{18 \times 1000} = -5 \times 10^{-4}$$

$$\zeta_a = (-5 + 1.028) \times 10^{-4}$$

$$\zeta_a = -3.97 \times 10^{-4}$$

15. (.42 to .46)

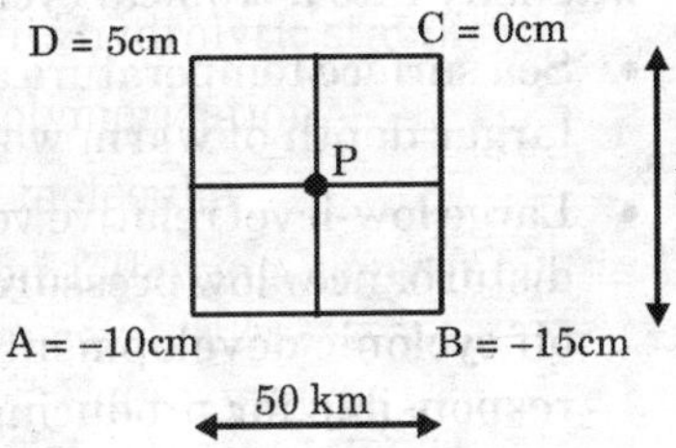

For x direction,

Pressure at mid-point of A and B are given as

$$\left(\frac{\partial P}{\partial x}\right)_{A,B} = \frac{0 - 5 \times 10^{-2}}{50 \times 10^3}$$

$$= -1 \times 10^{-6}$$

$$\left(\frac{\partial P}{\partial x}\right)_{D,C} = \frac{(-10 - \{-15\}) \times 10^{-2}}{50 \times 10^3}$$

$$= 1 \times 10^{-6}$$

Therefore, at the center point P of the area, taking the mean of the values of midpoint of A, B and D, C we get,

$$\left(\frac{\partial P}{\partial x}\right)_P = 0$$

For y direction,

$$\left(\frac{\partial P}{\partial y}\right)_{A,D} = \frac{(0 - \{-10\}) \times 10^{-2}}{50 \times 10^3}$$

$$= 1 \times 10^{-6}$$

$$\left(\frac{\partial P}{\partial y}\right)_{B,C} = \frac{(5 - \{-15\}) \times 10^{-2}}{50 \times 10^3}$$

$$= 2 \times 10^{-6}$$

$$\left(\frac{\partial P}{\partial y}\right)_P = \frac{1 \times 10^{-6} + 2 \times 10^{-6}}{2}$$

$$= 1.5 \times 10^{-6}$$

$$f = 2\Omega \sin\phi$$

$$f = 2 \times 7.27 \times 10^{-5} \times \sin(13.5°)$$

$$f = 3.39 \times 10^{-5}$$

$$\phi = 13.5° \text{ (from figure)}$$

$$\Delta P = \left(\frac{\partial P}{\partial x}\right)_P + \left(\frac{\partial P}{\partial y}\right)_P$$

$$\Delta P = 1.5 \times 10^{-6}$$

Now, $V_g = \frac{g}{f} \times \Delta P$

$$\Rightarrow \quad V_g = \frac{10}{3.39 \times 10^{-5}} \times 1.5 \times 10^{-6}$$

$$\Rightarrow \quad V_g = 0.442$$

16. (51 to 52)

As per question

density of freshwater = ρ = 1000 kg/m^3

Energy of atomic bomb = E_{bomb} = 1.5×10^{15} kJ

Energy of atomic bomb = E_{bomb} = 1.5×10^{15} kJ

$= 1.5 \times 10^{18}$ J

Radius = r = 200km = 200×10^3 m

Height of rainfall column = h = 250mm

$= 250 \times 10^{-3}$ m

Latent heat of evaporation = L = 2.45×10^6 J/kg

Energy of Tropical cyclone = N × Energy of atomic bomb

$$m \times L = N \times E_{bomb}$$

$$\text{mass} = m = V \times \rho$$

Volume of area, $V = \pi \times r^2 h$

$$N = \frac{\pi \times r^2 h \times \rho \times L}{E_{bomb}}$$

$$N = \frac{\pi \times (200 \times 10^3)^2 \times 250 \times 10^{-3} \times 1000 \times 2.45 \times 10^6}{1.5 \times 10^{18}}$$

$$N = 51$$

17. (b) As per CFL criteria

$$C\frac{\partial t}{\partial x} < 1$$

$$C\partial t < \partial x$$

$$\partial t < \frac{\partial x}{C}$$

$$\partial t < \frac{3 \times 1000}{100}$$

$$\partial t < 10\text{s}$$

18. (a) Let S_o = Solar Constant

α = albedo

T_o = Avg Temperature

R = Radius of planet

σ = Stefan's Constant

$$\text{Energy}_{\text{intercepted}} = S_o \times \pi R^2$$
$$\text{Energy}_{\text{absorbed}} = S_o(1-\alpha) \times \pi R^2$$

As per Stefan-Boltzmann law,

$$\text{Energy}_{\text{flux}} = 4\sigma T^4$$
$$\text{Energy}_{\text{emitted}} = 4\sigma T^4 \times \pi R^2$$

As per conservation of energy

$$\text{Energy}_{\text{absorbed}} = \text{Energy}_{\text{emitted}}$$
$$S_o(1-\alpha) \times \pi R^2 = 4\sigma T^4 \times \pi R^2$$
$$S_o(1-\alpha) = 4\sigma T^4$$

19. (199 to 205)

As per conservation of equation of moist static energy (S)

Given, $T_0 = 273 + 27 = 300$ K

$$S_0 = S_{15}$$
$$c_p T_o + gz_o + Lq = c_p T_{15} + gz_{15} + L_q$$
$$T_{15} = \frac{(c_p T_o + gz_o + Lq) - (gz_{15})}{c_p}$$
$$\Rightarrow T_{15} = \frac{(1005 \times 300 + 2.45 \times 10^6 \times 2 \times 10^{-3}) - (10 \times 15 \times 10^3)}{1005}$$
$$\Rightarrow T_{15} = \frac{(1005 \times 300 + 2.45 \times 10^6 \times 20 \times 10^{-3}) - (10 \times 15 \times 10^3)}{1005}$$
$$\Rightarrow T_{15} = 199.5 \text{ K}$$

20. (d) As per geostrophic flow

$$fV_g = \alpha \nabla P$$
$$\Rightarrow f(u_g + v_g) = \frac{1}{\rho}\frac{\partial P}{\partial x} + \frac{1}{\rho}\frac{\partial P}{\partial y}$$

Assuming ρ is constant

Divergence of V_g

$$\Rightarrow \nabla((f(u_g + v_g)) = -\frac{1}{\rho}\frac{\partial^2}{\partial x \partial y} + \frac{1}{\rho}\frac{\partial^2 P}{\partial x \partial y}$$

$$f\frac{\partial u_g}{\partial x} + u_g\frac{\partial f}{\partial x} + f\frac{\partial v_g}{\partial y} + v_g\frac{\partial f}{\partial y} = 0$$

$u_g \frac{\partial f}{\partial x}$ is neglected Coriolis parameter does not change in x direction.

$$f\left(\frac{\partial u_g}{\partial x} + \frac{\partial v_g}{\partial y}\right) = v_g \frac{\partial f}{\partial y}$$
$$\nabla V_g = \frac{v_g}{f}\frac{\partial f}{\partial y}$$

21. (*c*) Factors responsible for a depression to intensify into a tropical cyclone

- Sea surface temperature above 26.5°C with larger depth of warm water
- Large low-level relative vorticity : Incipient disturbance / low pressure region to trigger off cyclonic development [Convergence is responsible for producing rotation / spin-up in developing cyclone]
- Large middle tropospheric relative humidity
- Strong convective instability over a larger vertical column of lower troposphere
- Weak vertical shear of horizontal wind
- Formation of warm core, combined with an area of low level convergence an area of upper level divergence should be present (between 9-15 km), so that outflow of air in upper atmosphere is maintained.

22. (b) El-Nino and Southern Oscillation (ENSO) is a coupled phenomenon of atmosphere and ocean where periodic fluctuation in sea surface temperature (El Niño) and the air pressure of the overlying atmosphere (Southern Oscillation) across the equatorial Pacific Ocean.

2018

Q. 1 – Q. 9 carry one mark each.

1. The most abundant gas in the atmosphere among inert gases is

   (*a*) Helium (*b*) Argon

   (*c*) Neon (*d*) Krypton

2. The pair of variables that always exhibit monotonic decrease with height in the atmosphere is

   (*a*) Pressure, Temperature

   (*b*) Pressure, Ozone concentration

   (*c*) Air Density, Pressure

   (*d*) Temperature, Water Vapour

3. Correct order of the maximum vertical extent of atmospheric circulation cells is

   (*a*) Hadley > Ferrel > Polar

   (*b*) Polar > Hadley > Ferrel

   (*c*) Hadley > Polar > Ferrel

   (*d*) Ferrel > Hadley > Polar

4. Analysis of an atmospheric variable shows prominent modes at 5,40 and 1460 days. These modes correspond respectively to

   (*a*) Tidal, MJO and ENSO

   (*b*) Synoptic, MJO and ENSO

   (*c*) Synoptic, MJO and Decadal

   (*d*) Tidal, Milankhovich and ENSO

5. Atmospheric vertical profile of temperature is measured by radiosonde. Equivalent instruments for measuring ocean temperature profile among the following are:

   (P) drifting buoy (Q) ARGO float

   (R) current mooring (S) XBT

   (*a*) Q, R, S (*b*) Q, S

   (*c*) R, S (*d*) P, R, S

6. When deep water sinks in the North Atlantic and moves away from where it formed, it gets

   (*a*) richer in oxygen and nutrients

   (*b*) less acidic and richer in metals

   (*c*) richer in CO_2 and poorer in O_2

   (*d*) richer in CO_2 and O_2

7. The speed of sound in the ocean depends on

   (*a*) temperature alone

   (*b*) temperature and pressure

   (*c*) temperature and salinity

   (*d*) temperature, salinity and pressure

8. In a numerical weather prediction model with a horizontal grid resolution of 50 km, convective cloud processes are parameterized, because

   (*a*) Cloud physics is not known for modelling

   (*b*) Models cannot handle phase change

   (*c*) Cloud size is larger than the grid size

   (*d*) Cloud size is much smaller than the grid size

9. Figure below shows SST anomaly (in°C). It is associated with the phenomenon known as

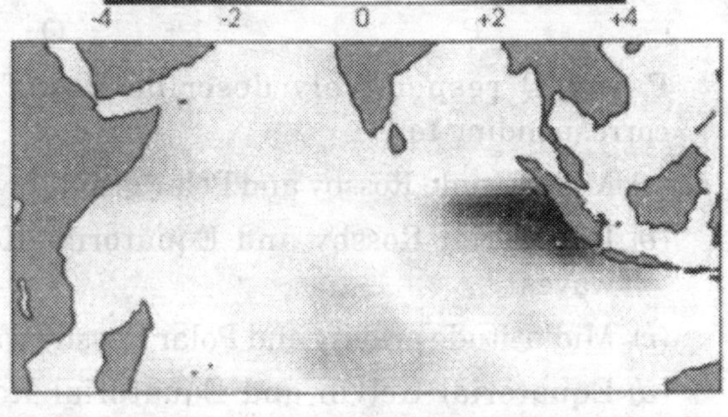

   (*a*) El Nino (*b*) Indian Ocean dipole

   (*c*) La Nina (*d*) MJO

Q. 10 – Q. 22 carry two marks each.

10. For an inviscid and barotropic ocean of constant depth (D), a water parcel with initial vorticity 2Ω is displaced from the equator to the north pole. Latitudinal variation of the parcel vorticity (ζ) is well represented by the curve

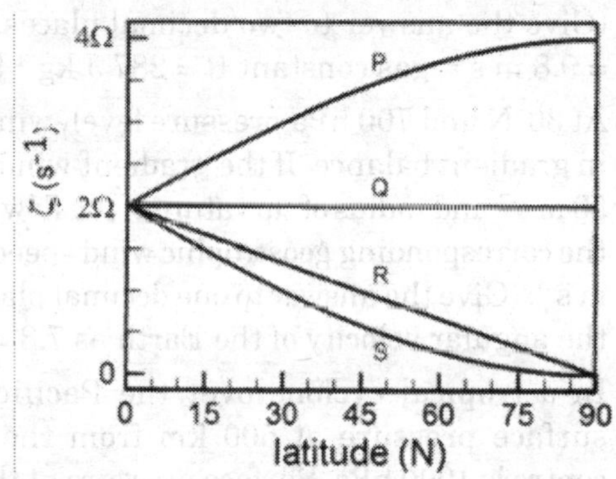

   (*a*) S (*b*) Q

   (*c*) P (*d*) R

11. A wave progresses up an estuary of decreasing water depth. If friction is neglected, then

   (*a*) wave amplitude decreases and wave length increases

   (*b*) wave amplitude increases and wave length decreases

   (*c*) wave amplitude decreases and wave length decreases

   (*d*) wave amplitude increases and wave length increases

12. In the Ekman flow limit, directions of ocean surface current and the geostrophic wind are

(*a*) the same

(*b*) surface current is 45° to the left of the geostrophic wind

(*c*) surface current is 45° to the right of the geostrophic wind

(*d*) exactly opposite to each other

13.

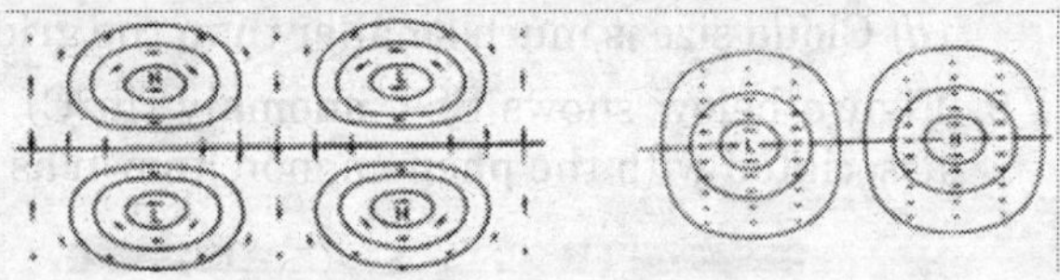

P Q

P and Q respectively describe flow fields corresponding to

(*a*) Mid latitude Rossby and Polar gravity waves

(*b*) Equatorial Rossby and Equatorial Kelvin waves

(*c*) Midlatitude gravity and Polar Rossby waves

(*d*) Equatorial Kelvin and Equatorial Rossby waves

14. On the summer solstice day, the maximum incident shortwave radiation at the top of the atmosphere over the equator (up to one decimal place) is _____ $W\ m^{-2}$. (Take solar constant as 1368 $W\ m^{-2}$).

15. In an isothermal atmosphere having a temperature of 15°C, the height at which pressure decreases to 1/10 of its value at the surface is _______ km. (Give the answer to two decimal places.) Take g = 9.8 $m\ s^{-2}$, gas constant R = 287 $J\ kg^{-1}\ K^{-1}$

16. At 30°N and 700 hPa pressure level, wind field is in gradient balance. If the gradient wind speed is 50 $m\ s^{-1}$ and radius of curvature of the flow is 50 km, the corresponding geostrophic wind speed is _____ $m\ s^{-1}$. (Give the answer to one decimal place.) Take the angular velocity of the Earth as $7.3 \times 10^{-5}\ s^{-1}$

17. In a tropical cyclone over the Pacific Ocean, surface pressure at 500 km from the cyclone centre is 1000 hPa. Surface pressure at the centre is 900 hPa. Sea surface temperature and surface air temperature remain constant at 28°C and 27°C, respectively. Difference in potential temperature between 500 km and cyclone centre is __________ K. (Give the answer to two decimal places.) Take g = 9.8 $m\ s^{-2}$, Cp = 1005 $J\ kg^{-1}\ K^{-1}$, gas constant R = 287 $J\ kg^{-1}\ K^{-1}$.

18. A cloud forms by the lifting of moist air from the surface with the initial conditions T_o = 30°C, RH = 80% and P_o = 1005 hPa. If the vapour pressure of this parcel at 500 hPa is 6.5 hPa, the liquid water content of the parcel if no precipitation takes place is ________ $gm\ kg^{-1}$. (Give the answer to one decimal place.) Saturation vapour pressure of water at 30°C is 42.43 hPa.

19. A numerical model of the atmosphere uses sigma (σ) coordinate system in vertical. At locations P and Q, surface pressures are 1005 hPa and 500 hPa, respectively. Absolute difference in the heights of σ = 0.9 level between these locations is ____________ meters.

(Give the answer to one decimal place.) Layer mean temperatures at P and Q are 300 K and 270 K, respectively. (Take g = 9.8 $m\ s^{-2}$ gas constant R = 287 $J\ kg^{-1}\ K^{-1}$)

20. If difference in sea surface elevation is 1 m in 100 km at 30°N latitude, the corresponding geostrophic current is ______ $m\ s^{-1}$. (Give the answer to one decimal place.) Take g = 9.8 $m\ s^{-2}$ and angular velocity of the Earth = $7.3 \times 10^{-5}\ s^{-1}$.

21. If wind speed over ocean surface is 10 $m\ s^{-1}$, air-sea interface momentum flux is _____ $N\ m^{-2}$. (Give the answer to two decimal places.) Surface air temperature and pressure are 27°C and 1000 hPa, respectively. Take drag coefficient as 0.001 and gas constant R = 287 $J\ kg^{-1}K^{-1}$.

22. Let L_x, L_y be length scales in x- and y-directions and corresponding mass transports are M_x and M_y. The ratio of M_x and M_y (to nearest integer) is ___________ , if the ratio of L_x and L_y is 10 and vertical velocity is zero.

ANSWERS

1. (b)	**2.** (c)	**3.** (a)	**4.** (b)	**5.** (b)
6. (c)	**7.** (d)	**8.** (d)	**9.** (b)	**10.** (a)
11. (b)	**12.** (a)	**13.** (b)	**14.** (1254 to 1256)	**15.** (19.42 to 19.44)
16. (734.8 to 735)	**17.** (9.16 to 9.18)	**18.** (12.8 to 13.2)	**19.** (92.4 to 92.8)	**20.** (1.3 to 1.4)
21. (0.11 to 0.12)	**22.** (9.99 to 10.01)			

EXPLANATIONS

1. (b) Argon, an inert gas, is the third most abundant gas in the atmosphere which is around 0.934% (9340 ppmv). It is more than twice as abundant as water vapor, 23 times as abundant as carbon dioxide (400 ppmv), and more than 500 times as abundant as neon (18 ppmv). Argon is the most abundant noble gas in Earth's crust, comprising 0.00015% of the crust.

2. (c) Temperature at altitude h m above sea level is given by

$$T = T_o - Lh$$

The pressure at altitude h is given by

$$P = P_o\left(\frac{1 - Lh}{T_o}\right)^{\frac{gM}{RL}}$$

Density ρ can be calculated as

$$\rho = \frac{PM}{RT}$$

where P = absolute pressure

T = absolute temperature

R = ideal gas constant

M = molar mass

T_o = mean sea level temperature

P_o = mean sea atmospheric pressure

g = acceleration due to gravity

L = Lapse rate

As per equation

$$P \propto \rho$$

Also the pressure equation states that pressure decreases with height so does air density.

As per International standard atmosphere (ISA), it first decreases then increases whereas temperature which depends on lapse rate.

3. (a)
- Hadley cell is the largest atmospheric cell extend from the equator to between 30° and 40° north and south. The trade winds blow towards the equator, then ascend near the equator as a broken line of thunderstorms, which forms the Inter-Tropical-Convergence Zone (ITCZ).
- Ferrel cell is the middle cell in which air converges at low altitudes to ascend along the boundaries between cool polar air and the warm subtropical air that generally occurs between 60 and 70 degrees north and south. The Ferrel cell moves in the opposite direction to the two other cells (Hadley cell and Polar cell).
- The smallest and weakest cells are the Polar cells, which extend from between 60 and 70 degrees north and south, to the poles. Air in these cells sinks over the highest latitudes and flows out towards the lower latitudes at the surface.

4. (b)
- Synoptic scale like cyclones occurs for 5-10 days
- In the tropical atmosphere the Madden-Julian Oscillation (MJO) is the largest element of the intra seasonal (30 to 90 days) variability.
- El-Nino Southern Oscillation (ENSO) is an irregularly periodic variation in winds and sea temperature over the equatorial Pacific Ocean. The ENSO occurs in a duration of 4-8 days.
- Whereas Tides are the rise and fall of sea levels caused by the combined effects of the gravitational forces exerted by the Moon and the Sun and the rotation of Earth. The duration of the tidal mode is few hours.
- Decadal mode is stated for any phenomenon to occur for 10 years or 3652 days.
- Milankovitch Cycles occurs about 100,000 years. This cycles are constantly fluctuating cycles whose orbital shape ranges between more and less elliptical.

5. (b) Atmospheric vertical profile of temperature is measured by radiosonde.

Similarly, for measuring ocean temperature profile (vertical)

The expendable bathythermograph (XBT)- It is a device that holds a temperature sensor and a transducer to detect changes in water temperature versus depth down to a depth of approximately 285 meters (935 feet).

Argo floats- These floats drift at a depth of 1000 m and, every 10 days, by changing their buoyancy, dive to a depth of 2000 m and then move to the sea-surface, measuring conductivity and temperature profiles as well as pressure. From these, salinity and density can be calculated.

whereas

Drifting buoys and current buoys are floating devices that measure the sea surface temperature, significant wave height, surface pressure, wind stress and direction etc. They are anchored (stationary) or allowed to drift with ocean currents. These buoys are used only for surface measurements and are not suitable of vertical profile of ocean measurements.

6. (c) In the oceanic conveyor belt, the best way to map out the oxygen content of deep water is where the water is old i.e. where it has been deep for a long time much of its oxygen has been spent in bacterial decay of organic matter and its respiration. Also it is rich in CO_2 content whereas, the surface water sinks in North Atlantic is young and rich in oxygen content. So, when deep water sinks in North Atlantic and moves away from where it formed it (like to Pacific Ocean) losses its oxygen content and its CO_2 content gets richer.

7. (d) The speed of sound through water has been found to be function of three factors .They are temperature (T), pressure or depth (z) and salinity (S).

Speed of sound, *c* is given as

$$c(T,z,S) = 1449.2 + 4.6T - 5.5 \times 10^{-2}T^2 + 2.9 \times 10^{-4} T^3 + (1.34 \times 10^{-2}T)(S - 35) + 1.6 \times 10^{-2}z$$

with limits

$0 \leq T \leq 35°C$

$0 \leq S \leq 45$ p.s.u

$0 \leq z \times 1000$ m

8. (d) The global model grid size is typically about 100 kilometers which is too coarse to capture clouds. For Numerical weather prediction CFL criteria is followed which states that the time step must be less than a certain time in many explicit time-marching computer simulations, otherwise the simulation produces incorrect results. In case of convective clouds

$$C\frac{\partial t}{\partial x} \ll 1$$

Thus, cloud size is smaller so it is parameterized in numerical weather model wherein a convective cloud is replaced as the processes are too small-scale or complex to be physically represented in the model by a simplified process.

9. (b) The figure is indicated in Equatorial Indian Ocean where Indian Ocean Dipole (IOD) phenomenon is prominent whereas El-Nino and La-Nina occurs in tropical Pacific Ocean and MJO is a more influenced atmospheric phenomenon rather than oceanic.

10. (a) As per conservation of potential vorticity

$$\Pi = \frac{f+\zeta}{h} = \text{constant}$$

where ζ is absolute vorticity, f is coriolis parameter or planetary vorticity and h is height

here h is negligible which reduces the equation to

$$\pi = f + \tau = \text{constant}$$

As per given question, at equator

$$\zeta = 2\Omega$$

and $f = 2\Omega \sin f$

$$f = 2 \times \Omega \times \sin(0°) = 0$$

$\therefore \quad \pi_{equator} = \zeta = 2\Omega$

At poles, $f = 2 \times \Omega \times \sin(90°) = 2\Omega$

$\therefore \quad \pi_{poles} = f = 2\Omega$

and at poles τ is zero.

So, the graph shows clearly the curve S satisfy as per above equation. The curvature of S and not R is due to the presence of $\zeta \to 0$ at poles.

11. (b) In case of wave progress up to estuary of decreasing depth as per question which is a case of Shallow Water Wave.

For Shallow Water Wave propagation the phase speed = $\sqrt{gD}$,

where D = depth of water

As the water depth decreases the phase speed/wave speed also decreases.

Also, equation of wave motion is given as,

$$c = \lambda \times f$$

where, c is wave speed, λ is wavelength of wave and f is frequency

Thus, from above equation the wavelength decreases

As per conservation of Energy in a wave,

Total Energy = K.E + P.E = constant

$$\text{Total Energy} = \frac{1}{2}\rho c^2 + \rho g a$$

So, as c is decreasing then to conserve the total energy amplitude a increases.

12. (a) The governing equation of Ekman flow is given as

$$fv + K_z \frac{\partial^2 u}{\partial z^2} = 0$$

$$-fu + K_z \frac{\partial^2 u}{\partial z^2} = 0$$

where K_z is constant.

$$(u, v) = (u_g + v_g, u_{ag} + v_{ag})$$

where u and v are velocity components of ocean current,

u_g and v_g are velocity components of geostrophic wind

u_{ag} and v_{ag} are velocity components of ageostrophic wind respectively

At Ekman limit i.e $z = -\delta$ (Ekman layer limit/depth)

$$(u_{ag}, v_{ag}) \to 0$$

$$\therefore \quad (u, v) = (u_g + v_g)$$

Thus, the ocean surface current and geostrophic wind are in same direction.

13. (b) A second order differential equation in the single unknown υ is given as

$$\frac{\partial^2 v}{\partial z^2} + \left[\left(\frac{v^2}{gh_e} - k^2 - \frac{k}{v}\beta\right) - \frac{\beta^2 y^2}{gh_e}\right]$$

where $\beta \equiv \frac{2\Omega}{a}$, a is radius of Earth, Ω is angular velocity of earth and h_e is the mean depth.

It turns out that solutions to the above equation, satisfies the condition of decay far from the equator, and exists only when the constant part of coefficient in square bracket satisfies the following relationship

$$\frac{\sqrt{gh_e}}{\beta}\left(\frac{v^2}{gh_e} - k^2 - \frac{k}{v}\beta\right) = 2n+1; n = 0, 1, 2, ...$$

which is a cubic dispersion equation determining the frequencies of permitted equatorially trapped free oscillations for zonal wave number k and meridional mode number n. In general the three solutions of the above equation can be interpreted as eastward and westward moving equatorially trapped gravity waves, and westward moving equatorial Rossby waves. The case $n = 0$ (for which the meridional velocity perturbation has a Gaussian distribution centered at the equator) must be treated separately. In this case the dispersion relationship factors as

$$\left(\frac{v}{\sqrt{gh_e}} - \frac{\beta}{v} - k\right)\left(\frac{v}{\sqrt{gh_e}} + k\right) = 0$$

The roots given by first term of parentheses of this equation is

$$v = k\sqrt{gh_e}\left[\frac{1}{2} \pm \frac{1}{2}\left(1 + \frac{4\beta}{k^2\sqrt{gh_e}}\right)^{1/2}\right]$$

The positive root corresponds to an eastward propagating equatorial inertia-gravity wave, whereas the negative root corresponds to a westward propagating wave, which resembles an inertia-gravity wave for long zonal scale ($k \to 0$) and resembles a Rossby wave for zonal scales characteristic of synoptic-scale disturbances. This mode is generally referred to as a Rossby-gravity wave. The horizontal structure of the westward propagating $n = 0$ solution is shown in Figure (P).

The perturbation wind and geopotential structure for the Kelvin wave are shown in plan view in Figure (Q). The meridional force balance for the Kelvin mode is an exact geostrophic balance between the zonal velocity and the meridional pressure gradient. It is the change in sign of the Coriolis parameter at the equator that permits this special type of equatorial mode to exist.

14. (1254 to 1256)

Given, $S_o = 1368 \text{ W/m}^2$

As per question, summer solstice day occurs in 21 June and the Sun's declination from the celestial equator is 23.44°.

Therefore, θ = 23.44°

The maximum incident shortwave radiation is given by,

$$S = S_o \cos\theta$$

$$\Rightarrow \quad S = 1368 \times \cos(23.44°)$$

$$S = 1254 \text{ W/m}^2$$

15. (19.42 to 19.44)

If the layer is isothermal

Therefore, $\frac{dT}{dz} = 0$

$$P = P_o e^{\frac{-z}{H}}$$

where, $H = \frac{RT}{g}$

$$T = 273 + 15 = 288 \text{ K}$$

$$\ln\left(\frac{P}{P_o}\right) = \frac{-zg}{RT}$$

$$z = \frac{-RT\ln\left(\frac{P}{P_o}\right)}{g}$$

$$\Rightarrow \quad z = \frac{-287 \times 288 \times -2.302}{9.8}$$

$$\Rightarrow \quad z = 194206 \text{ m} = 19.4 \text{ km}$$

16. (734.8 to 735)

As per given question,

Gradient wind speed, V = 50 m/s

Radius of curvature, R = 50 km = 50×10^3 m

Latitude, $\phi = 30°N$

Coriolis parameter $f = 2\Omega \sin\phi$

$$= 2 \times 7.3 \times 10^{-5} \times \sin(30°)$$

$$f = 7.3 \times 10^{-5}$$

Now, as per geostrophic flow

$$fV_g = \nabla P$$

and equation of gradient wind states that,

$$\frac{V^2}{R} + fV = \nabla P$$

Also, from geostrophic equation,

$$\frac{V^2}{R} + fV = fV_g$$

Dividing by fV we get,

$$\frac{V_g}{V} = 1 + \frac{V}{fR}$$

$$\Rightarrow \quad V_g = \left(1 + \frac{V}{fR}\right) \times V$$

$$\Rightarrow \quad V_g = \left(1 + \frac{50}{7.3 \times 10^{-5} \times 50 \times 10^3}\right) \times 50$$

$$V_g = 734.9 \text{ m/s}$$

17. (9.16 to 9.18) $P_{500} = 1000$ hPa

$P_{center} = 900$ hPa

$R = 287$ J/kg/K

$c_p = 1005$ J/kg/K

$T = 273 + 27 = 300$ K

Potential temperature at 500 km is given by,

$$\theta = T\left(\frac{P_{500}}{P}\right)^{\frac{R}{c_p}}$$

$$\Rightarrow \quad \theta = 300 \times \left(\frac{1000}{900}\right)^{\frac{287}{1005}}$$

$$\theta = 309.16 \text{ K}$$

Difference in Potential temperature

$$\doteq 309.16 - 300$$

$$= 9.16 \text{ K}$$

18. (12.8 to 13.2)

Given,

$T_o = 30°C = 273 + 30 = 303$ K

RH = 80% at $P_o = 1005$ hPa

$e_{s0} = 42.43$ hPa at $T_o = 30°C$

$P_v = 6.5$ hPa at P = 500 hPa

$$RH = \frac{\text{Actual pressure}}{\text{Saturation vapour pressure}} \times 100$$

$$\Rightarrow \quad RH = \frac{e_0}{e_{s0}} \times 100$$

$$\Rightarrow \quad e_0 = \frac{RH}{100} \times e_{s0}$$

$$= \frac{80}{100} \times 42.43$$

$$= 33.94 \text{ hPa}$$

At $P_o = 1005$ hPa

$$\in = 0.61$$

Mixing ratio is given as,

$$\omega_1 = \in \times \frac{e_0}{P - e_0}$$

$$\Rightarrow \quad \omega_1 = 0.61 \times \frac{33.94}{1005 - 33.94}$$

$$= 0.0213 \frac{\text{kg}}{\text{kg}} = 21.3 \text{ g/kg}$$

At Pv = 6.5 hPa at P = 500 hPa

$$\Rightarrow \quad \omega_2 = 0.61 \times \frac{6.5}{500 - 6.5}$$

$$= 0.00819 \frac{\text{kg}}{\text{kg}} = 8.02 \text{ g/kg}$$

Liquid water content

$$LWC = \omega_1 - \omega_2$$

$$\Rightarrow \quad LWC = 21.3 - 8.02$$

$$\Rightarrow \quad LWC = 13.26 \text{ g/kg}$$

19. (92.4 to 92.8)

Given,

Surface pressure at P = P_{PS} = 1005 hpa

Surface pressure at Q = P_{QS} = 450 hpa

Sigma coordinate = σ = 0.9

Mean Temperature at P = T_P = 300 K

Mean Temperature at Q = T_Q = 270 K

$$g = 9.8 \text{ m/s}^2$$

Gas constant = R = 287 J/kg/ K

As per formula, $\sigma = \frac{P}{P_S}$

$$P = \sigma \times P_S$$

where P is vertical coordinate pressure and P_S is the surface pressure.

Therefore,

$$P_P = 0.9 \times 1005$$
$$= 904.5 \text{ hpa}$$
$$P_Q = 0.9 \times 500$$
$$= 450 \text{ hpa}$$

Now, applying the following formula we find the height of the pressure levels

$$Z = \frac{-RT \ln\left(\frac{P}{P_S}\right)}{g}$$

$$Z_P = \frac{-RT_p \ln\left(\frac{P_P}{P_{PS}}\right)}{g}$$

$$= \frac{-287 \times 300 \times \ln\left(\frac{904.5}{1005}\right)}{9.8}$$

$$= 925.66 \text{ m}$$

$$Z_Q = \frac{-RT_Q \ln\left(\frac{P_Q}{P_{QS}}\right)}{g}$$

$$= \frac{-287 \times 270 \times \ln\left(\frac{450}{500}\right)}{9.8}$$

$$= 833.10 \text{ m}$$

Absolute difference in heights is given as

$$= Z_P - Z_Q$$
$$= 925.66 - 833.10$$
$$= 92.56 \text{ m}$$

20. (1.3 to 1.4)

Given,

Latitude, $\phi = 30°N$

$g = 9.8 \text{ m/s}^2$

Coriolis parameter $f = 2\Omega \sin \phi$

$$= 2 \times 7.3 \times 10^{-5} \times \sin(30°)$$
$$= 7.3 \times 10^{-5}$$

Sea surface elevation,

$$x = 100 \text{ km}$$
$$= 100 \times 10^3 \text{ m}$$

As per equation of geostrophic flow,

$$V_g = \frac{g}{f}\frac{\partial \Phi}{\partial x}$$

$$\Rightarrow \quad V_g = \frac{9.8}{7.3 \times 10^{-5}} \frac{1}{100 \times 10^3}$$

$$\Rightarrow \quad V_g = 1.34 \text{ m/s}$$

21. (0.11 to 0.12)

Given,

Gas constant = R = 287 J/kg/ K

Drag coefficient $C_d = 0.001$

Pressure, P = 1000 hpa

wind speed U = 10 m/s

Temperature, T = 273 + 27 = 300 K

As per Bulk aerodynamic formula,

$$\tau = \rho C_d U^2$$

Now, $P = \rho RT$

$$\rho = \frac{P}{RT}$$

$$\rho = \frac{1000 \times 10^3}{287 \times 300}$$

$$= 11.61 \text{ kg/m}^3$$

Therefore, $\tau = 11.61 \times 0.001 \times 10^2$

$$= 1.16 \text{ N/m}^2$$

22. (9.99 to 10.01)

The mass transport M for a fluid flow is given by

$$M = \int_{L_2}^{L_1}\int_{-h}^{0} \rho v dz\, dx$$

$$\Rightarrow \quad M = \rho \frac{g}{f} \Delta h L$$

where h = steric height (is the surface relative to some depth)

Let, L_x and L_y be length scale in x and y direction

Given, $\frac{L_x}{L_y} = 10$ and $w = 0$

Now, as per above formula

Also, $u = v$

$$\frac{M_x}{M_y} = \frac{L_x \times \rho \times u \times \Delta h}{L_y \times \rho \times v \times \Delta h}$$

$$\frac{M_x}{M_y} = 10$$

SOLVED PAPER 2019
XE : ENGINEERING SCIENCES

GENERAL APTITUDE (GA)

Q. 1 – 5 carry one mark each

1. A final examination is the _____ of a series of evaluations that a student has to go through.

(a) culmination

(b) consultation

(c) desperation

(d) insinuation

2. Once the team of analysts identify the problem, we ______ in a better position to comment on the issue.

Which one of the following choices CANNOT fill the given blank?

(a) are going to be (b) might be

(c) were to be (d) will be

3. Are there enough seats here? There are _____ people here than I expected.

(a) most (b) least

(c) many (d) more

4. If IMHO = JNIP; IDK = JEL; and SO = TP, then IDC =

(a) JDC (b) JED

(c) JCD (d) JDE

5. The product of three integers X, Y and Z is 192. Z is equal to 4 and P is equal to the average of X and Y. What is the minimum possible value of P?

(a) 6 (b) 9.5

(c) 7 (d) 8

Q. 6 – 10 carry two marks each

6. While teaching a creative writing class in India, I was surprised at receiving stories from the students that were all set in distant places: in the American West with cowboys and in Manhattan penthouses with clinking ice cubes. This was, till an eminent Caribbean writer gave the writers in the once-colonised countries the confidence to see the shabby lives around them as worthy of being "told".

The writer of this passage is surpised by the creative writing assignments of his students because________.

(a) None of the students had written about ice cubes and cowboys

(b) Some of the students had written about ice cubes and cowboys

(c) None of the students had written stories set in India

(d) Some of the students had written stories set in foreign places

7. X is an online media provider. By offering unlimited and exclusive online content at attractive prices for a loyalty membership, X is almost forcing its customers towards its loyalty membership. If its loyalty membership continues to grow at its current rate, within the next eight years more housholds will be watching X than cable television.

Which one of the following statements can be inferred from the above paragraph?

(a) Most households that subscribe to X's loyalty membership discontinue watching cable television

(b) Non-members prefer to watch cable television

(c) The X is cancelling accounts of non-members

(d) Cable television operators don't subscribe to X's loyalty membership

8. Fiscal deficit was 4% of the GDP in 2015 and that increased to 5% in 2016. If the GDP increased by 10% from 2015 to 2016, the percentage increase in the actual fiscal deficit is

(a) 25.00 (b) 37.50

(c) 35.70 (d) 10.00

9. Two pipes P and Q can fill a tank in 6 hours and 9 hours respectively, while a third pipe R can empty the tank in 12 hours. Initially, P and R are open for 4 hours. Then P is closed. The total time taken to fill the tank (in hours) is _____.

(a) 13.50 (b) 16.50

(c) 14.50 (d) 15.50

10. Mola is a digital platform for taxis in a city. It offers three types of rides – Pool, Mini and Prime. The table below presents the number of rides for the past four months. The platform earns one US dollar per ride. What is the percentage share of revenue contributed by Prime to the total revenues of Mola, for the entire duration?

	Month			
Type	January	February	March	April
Pool	170	320	215	190
Mini	110	220	180	70
Prime	75	180	120	90

(a) 23.97 (b) 16.24

(c) 38.74 (d) 25.86

ENGINEERING MATHEMATICS

Q. 1 – 7 carry one mark each & Q. 8 – 11 carry two marks each.

1. Let X be the Poisson random variable with parameter A = 1. Then, the probability $P(2 \le X \le 4)$ equals

(a) $\frac{19}{24e}$ (b) $\frac{17}{24e}$

(c) $\frac{13}{24e}$ (d) $\frac{11}{24e}$

2. For the series $\sum_{n=1}^{\infty} \frac{(x+1)^n}{n2^n}, -\infty < x < \infty$, which of the following statements is NOT correct?

(a) The series converges at x = –3

(b) The series converges at x = –1

(c) 'llie series converges at x = 0

(d) The series converges at x = 1

3. Let $f(z) = \bar{z}e^{-|z|^2}$, where $\bar{z}$ is tlie complex conjugate of z. Then, it is differentiate on

(a) $|z| > 1$ (b) $|z| < 1$

(c) $|z| = 1$

(d) the entire complex plane $\mathbb{C}$

4. If the transformation $u(x, t) = e^x v(x, t)$ reduces the partial differential equation $\frac{\partial^2 u}{\partial x^2} - 2\frac{\partial u}{\partial x} - \frac{\partial u}{\partial t} + u = 9$ u = 9 to the equation $\frac{\partial v}{\partial t} - \frac{\partial^2 v}{\partial x^2} = 9f(x)$, then f(x) equals

(a) $-e^{-x}$ (b) e^{-x}

(c) $-2e^{-x}$ (d) $2e^{-x}$

5. The value of α for which the system of equations

$$x - y - 3z = 3$$
$$2x + z = 0$$
$$-2y - 7z = a$$

has a solution is ________.

6. The value of the line integral $\frac{2}{\pi}\oint_\gamma (-y^3 dx + x^3 dy)$, where γ is the circle $x^2 + y^2 = 1$ oriented counter clockwise, is ________.

7. Let $y_1(x)$ and $y_2(x)$ be two linearly independent solutions of the differential equation $x^2 \frac{d^2y}{dx^2} + x\frac{dy}{dx} - 4y = 0$, $x > 0$. If $y_1(x) = x^2$, then $\lim_{x \to \infty} y_2(x)$ is ________.

8. If $Q = \begin{pmatrix} 3 & 2 & 4 \\ 2 & 0 & 2 \\ 4 & 2 & 3 \end{pmatrix}$ and $P = (v_1\ v_2\ v_3)$ is the matrix where v_1, v_2 and v_3 are linearly independent eigenvectors of the matrix Q, then the sum of the **absolute** values of all the elements of the matrix $P^{-1}QP$ is

(a) 6 (b) 10

(c) 14 (d) 22

9. If $P(x) = ax^3 + bx^2 + cx + d$ is the polynomial obtained by Lagrange interpolation satisfying P(0) = –8, P(1) = –7, P(2) = –6 and P(4) = 20, then the value of a – b + c + c is

(a) 1 (b) 3

(c) 5 (d) 7

10. The number of critical points of the function $f(x, y) = x^3 + 3xy^2 - 15x - 12y$ at which there is neither maximum nor minimum is ________.

11. Let $I = \frac{10^5 i}{2\pi}\oint_\gamma \frac{dz}{(z-4)(z^7-1)}$, where $i = \sqrt{-1}$ and g is the circle |z| = 2 oriented counter clockwise. Then, the value of I rounded off to one decimal place is ________.

FLUID MECHANICS

Q. 1 – 9 carry one mark each

1. For stable equilibrium of a "floating body, which one of the following statements is correct?

(a) Centre of gravity must be located below the centre of buoyancy.

(b) Centre of buoyancy must, be located below the centre of gravity.

(c) Metacentre must be located below the centre of gravity.

(d) Centre of gravity must be located below the metacentre.

2. If u and v are the velocity components in the x- and y-directions respectively, the z-component of vorticity ω_z at a point in a flow field is

(*a*) $\frac{\partial v}{\partial x}+\frac{\partial u}{\partial y}$ (*b*) $\frac{\partial v}{\partial x}-\frac{\partial u}{\partial y}$

(*c*) $\frac{\partial v}{\partial y}+\frac{\partial u}{\partial x}$ (*d*) $\frac{\partial v}{\partial y}-\frac{\partial u}{\partial x}$

3. In which one of the following devices the difference between static and total pressure is used to determine the flow velocity?

(*a*) Piezometer (*b*) Pitot static tube

(*c*) Orificemeter (*d*) Venturimeter

4. A golf ball is dimpled to make the flow turbulent and consequently to reduce the drag. Turbulent flow reduces the drag on the golf ball because

(*a*) skin friction coefficient is lower in a turbulent flow.

(*b*) skin friction coefficient is higher in a turbulent flow.

(*c*) turbulent flow has a lower tendency to separate.

(*d*) turbulent flow has a higher tendency to separate.

5. For a steady laminar incompressible boundary layer flow over a sharp-edged flat plate at zero incidence,

(*a*) the edge of the boundary layer is a streamline.

(*b*) the edge of the boundary layer is a pathline.

(*c*) the skin friction coefficient decreases as the distance from the leading edge increases.

(*d*) the skin friction coefficient remains constant all along the plate.

6. Hie power input P to a centrifugal pump is a function of the volume flow rate Q, impeller diameter D, rotational speed Ω, fluid density ρ, dynamic viscosity μ, and surface roughness ∈. To carry out a dimensional analysis using Buckingham's π theorem, which one of the following sets can be taken as the set of repeating variables?

(*a*) Q, Ω, D (*b*) Q, ∈, D

(*c*) ∈, D, ρ (*d*) D, ρ, Ω

7. Consider the two-dimensional laminar flow of water ($\mu = 0.001$ N.s/m^2) between two infinitely long parallel plates 0.1 m apart as shown in the figure below. The velocity profile at any location is given by $u(y) = 100(0.1y - y^2)$ m/s where y is in m. The magnitude of shear stress (in N/m^2, rounded off to 2 decimal places) acting on the bottom plate is

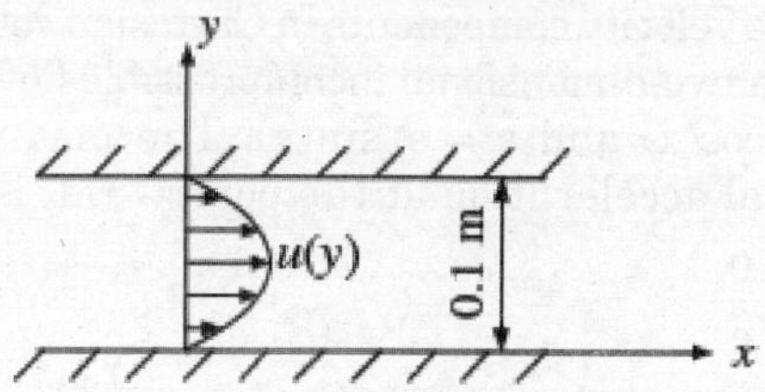

8. The maximum velocity in a fully developed laminar incompressible flow through a circular pipe of constant cross-sectional area is 6 m/s. Hie average velocity (in m/s) of the flow is _____

9. The theoretical discharge for the flow through an orificemeter is 40 m^3/s. If the measured discharge in an experiment is 32 m^3/s, then the discharge coefficient (rounded off to one decimal place) is _____

Q. 10 – 22 carry two marks each.

10. Consider the flow between two infinitely long parallel plates of large width separated by a distance 2H. The upper plate is moving with a constant velocity U while the lower plate is stationary. The volumetric flow rate per unit width of the plate is

(*a*) 0.25 UH (*b*) 0.5 UH

(*c*) UH (*d*) 2 UH

11. The velocity field in Cartesian coordinates in a two-dimensional steady incompressible flow of a fluid with density ρ is V = xi — yj. Assuming no body and line forces, the magnitude of pressure gradient ∇p at point (1,1) is

(*a*) $\sqrt{2}\rho$ (*b*) ρ

(*c*) $\rho/\sqrt{2}$ (*d*) ρ/2

12. A two-dimensional velocity field in Cartesian coordinates is defined by V = yi – xj. This flow is

(*a*) compressible and rotational

(*b*) compressible and irrotational

(*c*) incompressible and rotational

(*d*) incompressible and irrotational

13. Assertion [a] : The streamlines in a free vortex flow are concentric circles.

Reasoning [r] : 'There exists only radial component for the velocity field in a free vortex flow.

(*a*) Both [a] and [r] are true and [r] is the correct reason for [a]

(*b*) Both [a] and [r] are true but [r] is not the correct reason for [a]

(*c*) [a] is true but [r] is false

(*d*) [a] is false but [rj is true

14. The velocity components in Cartesian coordinates in a two-dimensional incompressible flow are $u = e^y \cos(x)$ and $v = e^y \sin(x)$. The magnitude of total acceleration at the point (–1, 1) is

(*a*) 0 (*b*) 1

(*c*) e (*d*) e^2

15. For steady laminar flow at zero incidence over a flat plate, the component of velocity parallel to the plate in the boundary layer is given by $u(y) = a + by + cy^2$, where y is the distance measured normal to the flat plate. If μ is the coefficient of dynamic viscosity, U is the velocity parallel to the wall at the edge of the boundary layer and S is the boundary layer thickness, the wall shear stress is given by

(*a*) $\mu U / \delta$ (*b*) $2\mu U / \delta$

(*c*) $2\mu(U / \delta)^2$ (*d*) $3\mu U / \delta$

16. A fluid with constant density of 1 kg/m³ flows past a semi-cylindrical structure with a freestream velocity of 2 m/s as shown in the figure below. The difference in static pressure between points P and Q is 10 N/m². If the gravitational acceleration g is 10 m/s² and the flow is assumed to be potential, what is the radius r (in m) of the semi-cylindrical structure?

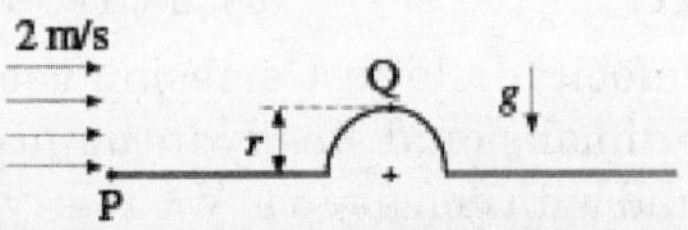

(*a*) 1 (*b*) 0.8

(*c*) 0.6 (*d*) 0.4

17. The mercury manometer shown in the figure below is connected to a water pipe at one end while the other end is open to the atmosphere. The density of water is 1000 kg/m³, the specific gravity of mercury is 13.6 and the gravitational acceleration g is 10 m/s². The gauge pressure p_w (in kN/m², rounded off to 2 decimal places) in the water pipe is ________

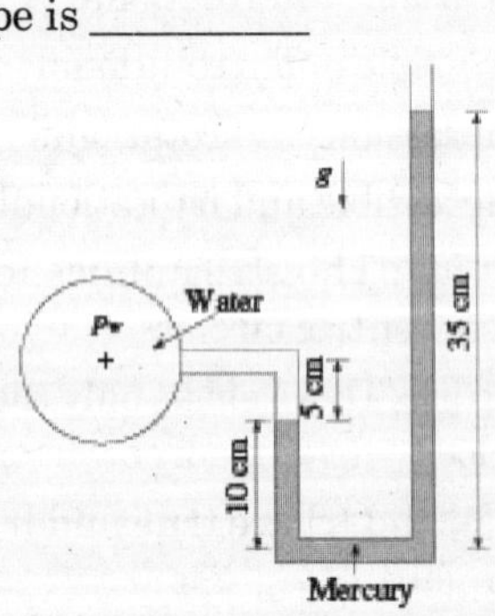

18. Water (ρ = 1000 kg/m³, μ = 0.001 N.s/m²) flows through a smooth circular pipe of radius 0.05 m. If the flow Reynolds number is 1000, then the pressure drop (in N/m², rounded off to 2 decimal places) over a length of 5 m will be ________

19. A uniform flow with a velocity of 2 m/s in the x-direction approaches a line source placed on the x-axis at a distance of 0.1 m from the origin. If the origin is the stagnation point in the resulting flow, the strength of the source (in m²/s, rounded off to 2 decimal places) is ________

20. In a steady incompressible flow of a fluid past a smooth stationary sphere, the drag force F depends on the flow velocity U, diameter D, and the dynamic viscosity μ and density ρ of the fluid. Experiments are conducted on the same sphere at the same flow velocity using two different fluids. The density of the second fluid is two times that of the first fluid. The dynamic viscosity of the second fluid is n times that of the first fluid. If the non-dimensional force $F/(\rho U^2 D^2)$ remains the same in both the experiments, the value of n is ________

21. An incompressible fluid flows past a flat plate as shown in the figure below with a uniform inlet velocity profile u = U and a parabolic exit velocity profile $u = U(2\eta - \eta^2)$, where u is the component of velocity parallel to the wall, y is the normal distance from the plate and $\eta = y/\delta$. If the volume flow rate across the top surface of the control volume (CV) is $Q = pU\delta$ per unit width (perpendicular to the x-y plane) of the plate, the value of p (rounded off to 2 decimal places) is

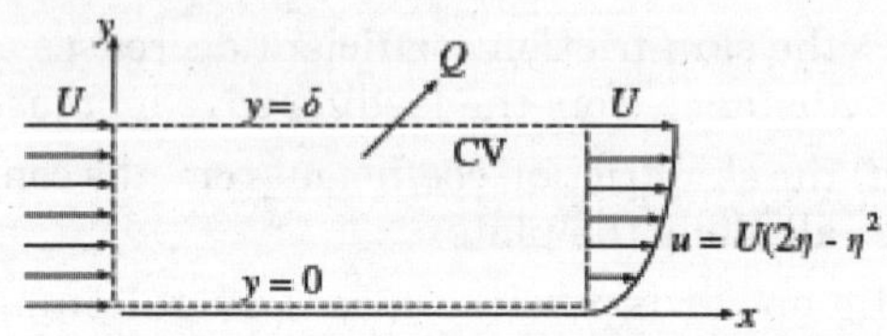

22. A jet engine is io be tested on a thrust stand as shown in the figure below. The conditions prevailing in a typical test are as follows: Axial intake air velocity = 100 m/s; axial exhaust gas velocity = 250 m/s; intake cross-sectional area = 1 m²; intake static pressure = –22 kPa (gauge); exhaust static pressure = 0 kPa (gauge); mass flow rate through the engine = 100 kg/s. The anchoring force (in kN) in axial direction on the thrust stand is ________

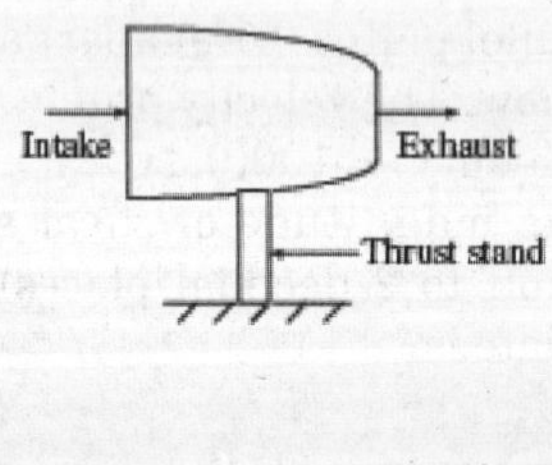

MATERIALS SCIENCE

Q. 1 – 9 carry one mark each & Q. 10 – 22 carry two marks each.

1. On decreasing the objective aperture size in an optical microscope

(*a*) the spherical aberration increases

(*b*) the depth of field increases

(*c*) the diffraction-limited resolution increases

(*d*) the astigmatism increases

2. Piliing-Bedworth ratios for oxides of some metals are given in the table.

Metal	Pilling-Bedworth Ratio
Li	0.57
Ce	1.16
Ta	2.33
W	3.40

Based on the criterion of Piliing-Bedworth ratio alone, which one of the following metals will be most protected from high temperature oxidation?

(*a*) Li (*b*) Ce

(*c*) Ta (*d*) W

3. In NaCl the substitution of a Na^+ ion by a Ca^{2+} ion would most probably lead to

(*a*) the formation of aNa^+ vacancy

(*b*) the creation of a Cl^- interstitial

(*c*) the formation of a Cl^- vacancy

(*d*) the formation of a Na^+ and Cl^- vacancy pair

4. Which one of the following is time-independent?

(*a*) Elastic deformation

(*b*) Aneiaslie deJbrmation

(*c*) Viscoelastic deformation

(*d*) Creep deformation

5. Copper is diffused into aluminium at 400 °C for 100 hours to obtain a certain concentration at a given depth. In another experiment conducted at 500 °C, to achieve the same concentration of copper at the same depth, the time required in hours is

(*Given:* Diffusion coefficients of copper in aluminium at 400 °C and 500 °C are 5×10^{-14} m^2 s^{-1} and 6×10^{-13} m^2 s^{-1} respectively)

(*a*) 7.33 (*b*) 8.33

(*c*) 9.33 (*d*) 10.33

6. If carbon (*c*) in iron (Fe) is 6 percent by weight, then its atomic percent is approximately

{Given: atomic weight C = 12.. Fe = 56)

(*a*) 13 (*b*) 23

(*c*) 30 (*d*) 50

7. GaAs has advantage over silicon when used in intergrated circuits at low power because it has

(*a*) larger band gap

(*b*) more than one element

(*c*) higher electron mobility

(*d*) higher hole mobility

8. Glass transition temperature of a polymer can be determined by

(*a*) Thermo-gravimetric analysis

(*b*) Raman spectroscopy

(*c*) NMR spectroscopy

(*d*) Differential scanning caiorimetry

9. The maximum wavelength of radiation to which Germanium (Ge) is opaque will be

{Given: energy gap of Ge = 0.67 eV, Planck's constant h = 6.63^{-34} J s, velocity of light c = 3×10^8 m s^{-1} and 1 eV = 1.6×10^{-19}J)

(*a*) 0.8 μm (*b*) 1.8 μm

(*c*) 2.8 μm (*d*) 4.8 μm

10. An alternating copolymer has number-averaged molecular weight of 10^5 gmol^{-1} and degree of polymerization of 2210. If one of the repeat units is ethylene, the other one is

(Given: atomic weight of H = 1, C = 12, F = 19 and Cl = 35.5)

(*a*) $–CH_2–CH(CH_3)–$ (*b*) $–CH_2–CHCl–$

(*c*) $–CF_2–CF_2–$ (*d*) $–CH_2–CH(C_6H_5)–$

11. Match the sintering processes in **column I** with the most suitable products in **column II.**

Column I

(P) Solid state sintering

(Q) Liquid phase sintering

(R) Spark plasma sintering

(S) Laser sintering

Column II

(1) Carbon nanotube products

(2) Mixture of Cu and Zn powder products

(3) Iron powder products

(4) 3D printed products

(*a*) P-4: Q-1; R-2; S-3 (*b*) P-3; Q-2; R-1; S-4

(*c*) P-3; Q-2; R-4; S-1 (*d*) P-2; Q-3; R-1; S-4

12. Which one of the following conditions will NOT favour the separation of impurities in zone refining process?

(*a*) Increase in the gap between solidus and liquidus lines

(*b*) Increase in the solubility of impurities in solid as compared to that in liquid phase

(*c*) Agitation of melt

(*d*) Low cooling rate of melt

13. A monochromatic X-ray beam of wavelength 0.154 nm is incident on a cubic crystal having lattice parameter a = 0.245 nm. The diffraction angle (26) for the first order reflection from a set of planes represented by the schematic plane below is _______ degrees.

(round off to 1 decimal place)

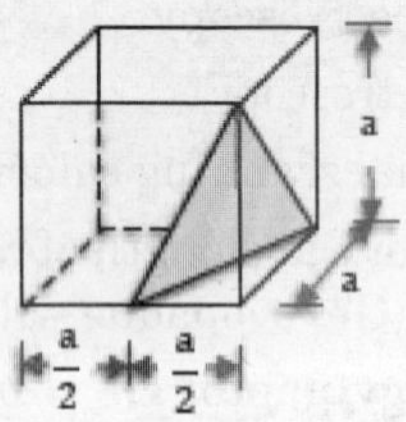

14. Nickel corrodes at 298 K in a solution of 0.06 M nickel chloride having pH 4. Assuming complete dissociation of nickel chloride, the partial pressure of hydrogen required to stop the corrosion of nickel is _______ atm. (round off to the nearest integer)

{Given: Standard reduction potential of nickel = – 0.25 V,
Faraday's constant = 96500 C mol^{-1}, Universal gas constant = 8.314 $JK^{-1} mol^{-1}$)

15. The potential energy, U(r), of a pair of atoms spaced at a distance r in a solid is given by

$$U(r) = -\frac{A}{r^3} + \frac{B}{r^7}.$$

The equilibrium distance between the atom pair is _______ nm.

(round off to 2 decimal places)

{Given: Constants A = 6 × 10^{-20} Jnm^3, B = 2.1 × 10^{-22} Jnm^7)

16. Tensile true stress - true strain curve for plastic region of an alloy is given by $\sigma(MPa) = 600\ \varepsilon^n$.

When true strain is 0.05, the true stress is 350 MPa. For the same alloy, when engineering strain is 0.12 then the engineering stress is _____ MPa. (round off to the nearest integer)

17. Box-S has 2 white and 4 black balls and box-T has 5 white and 3 black balls. A ball is drawn at random, from the hox-S and put in box-T. Subsequently, the probability of drawing a white ball from box-T is _______ (round off to 2 decimal places)

18. The zero point energy of an electron in a box of 0.2 mn width is _______ eV.

(round off to 1 decimal place)

(Given: Planck's constant = 6.63 × 10^{-34} J s, electron mass = 9.11 × 10^{-31} kg and 1 eV = 1.6 × 10^{-19} J)

19. The de Broglie wavelength of an electron accelerated across a 300 kV potential in an electron microscope is _____ × 10^{-12} m. Ignore relativistic effects.

(round off to 2 decimal places)

(Given: Planck's constant = 6.63 × 10^{-34} J s, electron rest mass = 9.11 × 10^{-31} kg, electron charge = 1.6 × 10^{-19} C)

20. A stress of 17 MPa is applied to a polymer serving as a fastener in a complex assembly. At constant strain the stress drops to 16.6 MPa after 100 hours. The stress on the polymer must remain above 14.5 MPa in order for the assembly to function properly. The expected life of the assembly is _____ hours, (round off to the nearest integer)

21. A piezoelectric material has a Young's modulus of 72 GPa. The stress required to change the polarization from 640 to 645 C m m^{-3} is _____ MPa. (round off to the nearest integer)

22. An iron bar magnet having coercivity of 7000 A m^{-1} is to be demagnetized. The bar is introduced fully inside a 0.25 ra long solenoid having 150 turns of wire. The electric current required to generate the necessary magnetic field is _____ A.

(round off to 1 decimal place)

SOLID MECHANICS

Q. 1 – 9 carry one mark each

1. For a pinned-pinned slender column of length L, with uniform circular cross section, and with second moment of cross sectional area /, the critical buckling load is Per. For a similar column, made of same material, having length '1L and second moment of cross sectional area 4/, the critical buckling load is

(*a*) 4 P_{cr} (*b*) 2 P_{cr}

(*c*) P_{cr} (*d*) 0.5P_{cr}

2. Two rods P and Q of uniform circular cross section are made of same material and are subjected to identical uniaxial tensile load. The length of rod P is twice the length of rod Q and the diameter of rod P is also twice the diameter of rod Q. The ratio of elastic strain energy stored in rod P to that stored in rod Q is

(*a*) 4 : 1 (*b*) 2 : 1

(*c*) 1 : 1 (*d*) 1 : 2

3. The frame comprises members ABC, CD, and BEDF. This frame is fixed at A, and the connections between the three members are pins. Load P is applied at F. Which of the following statements is TRUE?

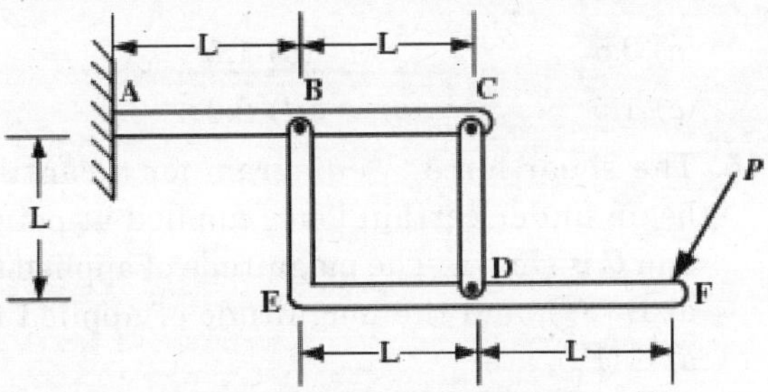

(*a*) ABC and CD are two-force members

(*b*) CD is the only two-force member

(*c*) BEDF and CD are two-force members

(*d*) ABC is the only two-force member

4. When two spheres moving in a horizontal plane collide obliquely, which one of the following statements is ALWAYS TRUE?

(*a*) Energy of the two-sphere system is conserved

(*b*) Linear momentum of the two-sphere system is conserved along the normal to the plane of impact

(*c*) Linear momentum of each individual sphere is conserved along the normal to the plane of impact

(*d*) Lineal' momentum of each individual sphere is NOT conserved along the horizontal tangent to the plane of impact

5. In a bending moment diagram of a simply supported beam, discontinuity in the bending moment occurs

(*a*) at the point of application of a couple

(*b*) at the point of application of a concentrated force

(*c*) at a point where the cross section of the beam changes abruptly

(*d*) at a point where shear force is zero

6. For a solid circular shaft fixed at one end and a torque applied at the other end, which one of the following is TRUE?

(*a*) Every cross section of the shaft is subjected to the same internal torque

(*b*) The internal torque gradually increases from one end of the shaft to the other end

(*c*) Every cross section of the shaft is subjected to the same bending moment

(*d*) No cross section of the shaft rotates relative to another cross section

7. Flexural stress occurs in a straight beam when

(*a*) it is subjected to pure torsion

(*b*) it is subjected to axial loads

(*c*) it is subjected to transverse loads

(*d*) it is subjected to combined axial and torsional loads

8. The bending moment at a point in a beam changes its sign. Which one of the following is ALWAYS TRUE?

(*a*) The slope of the beam changes its sign at that point

(*b*) The deflection of the beam changes its sign at that point

(*c*) The shear force of the beam changes its sign at that point

(*d*) The curvature of the beam changes its sign at that point

9. The state of stress at a point in a body under plane stress condition is given as $\sigma_x = 5$ MPa, $\sigma_y = -1$ MPa and $\tau_{xy} = 4$ MPa. If one of the principal stresses at that point is 7 MPa, then the other non-zero principal stress fin MPa) is ________

Q. 10 – 22 carry two marks each,

10. A massless rod of length L having a concentrated mass m attached at its mid-point is held at rest between the smooth ground and the smooth wall when it makes an acute angle θ_0 with the ground as shown. The rod is then released. The acceleration due to gravity is g. At an instant when the angle between the rod and ground is θ, the velocity of the mass m is

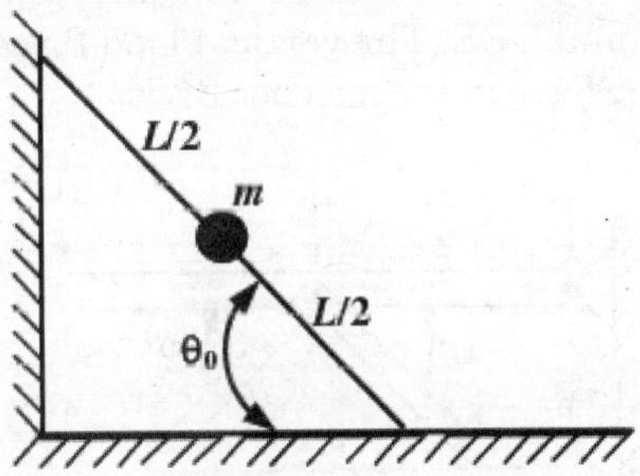

(*a*) $\sqrt{gL(\cos\theta - \cos\theta_0)}$ (*b*) $\sqrt{gL(\sin\theta - \sin\theta)}$

(*c*) $\sqrt{gL(\cos\theta + \cos\theta)}$ (*d*) $\sqrt{gL(\sin\theta_0 + \sin\theta)}$

11. A cantilever beam of constant cross section is subjected to a positive moment M and a positive axial force F as shown. The normal stress due to the axial force F is 200 MPa and the maximum compressive stress due to the moment M is 100 MPa. If the point A is located on the top surface and the point B is located on the centroidal axis of the beam, the stress elements at these points are (with stresses shown in MPa)

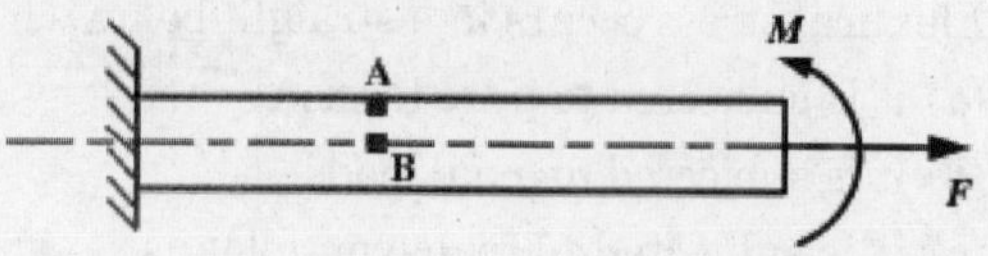

(a) 100 ← A → 100 (b) 100 ← A → 100

200 ← B → 200 300 ← B → 300

(c) 100 → A ← 100 (d) 300 → A ← 300

0 ← B → 0 200 ← B → 200

12. The block of mass m = 10 kg is held in place along the incline by the force F applied via a pulley arrangement as shown. The coefficient of static friction between the block and incline is 0.65. The range of F (in N) for which the block will remain at rest is (use acceleration due to gravity g = 10 ni/s^2)

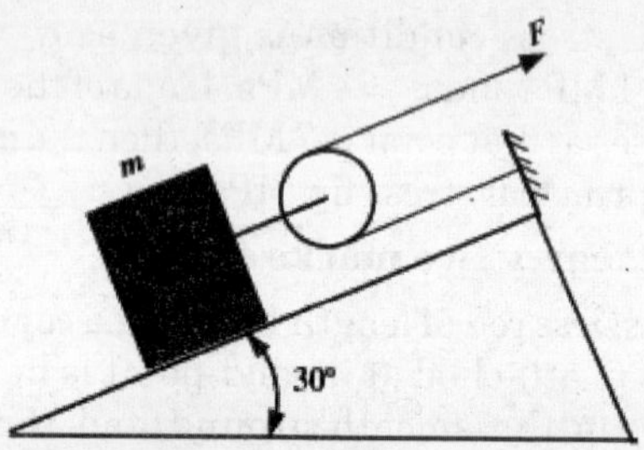

(a) 63 to 106.30 (b) 0 to 10630
(c) 0 to 53.15 (d) 53.15 to 106.30

13. The frame shown comprises members ABC, BE, BF, and DBF. It is pin supported at A and D. Members BE and BF are pin connected to ABC and DEE. The vertical load P is applied at C. The force in the member BF is

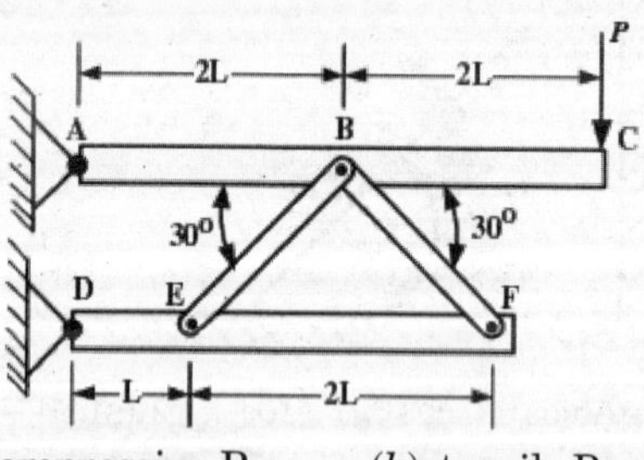

(a) compressive P (b) tensile P
(c) compressive 2P (d) tensile 2P

14. A simply supported beam is subjected to a concentrated force P = 48000 N at the mid-span as shown. The flexural rigidity of the beam is El = 11 × 10^6 N-m^2. The curvature of the beam between points Q and R is given as κ = (Px/ 2EI). The magnitude of deflection (in mm) of the beam at x = 0,5 m is

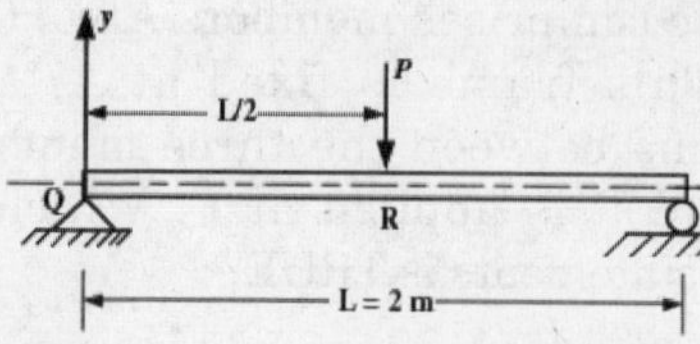

(a) 0.5 (b) 1.0
(c) 1.5 (d) 2.0

15. The shear force (V) diagram for a cantilever beam under certain loads applied at points B and C is shown. The magnitude of applied force at B (F_B), and the magnitude of applied force at C (F_C), are

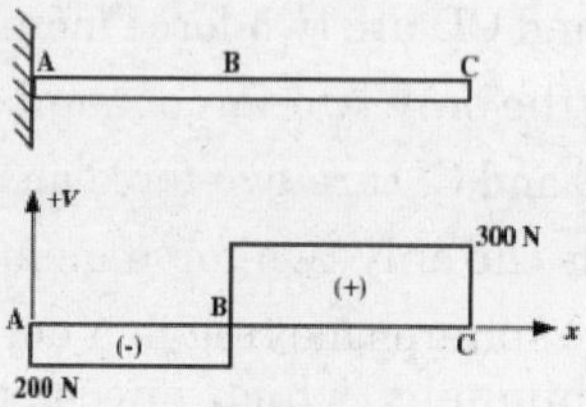

(a) F_B = 500 N F_C = 300 N
(b) F_B = 200 N F_C = 300 N
(c) F_B = 300 N F_C = 200 N
(d) F_B = 300 N F_C = 500 N

16. A bar of length 1 m, having a uniform area of cross section 1000 mm^2, is fixed at one end and subjected to a uniaxial tensile load of 200 kN at the other end. Along with the tensile load, the rod also experiences a uniform temperature rise ΔT = 100 °C. The material of the rod has a Young's modulus E = 200 GPa and coefficient of thermal expansion a = 1 × 10^{-5} / °C. The elongation of the rod (in mm) is ______

17. A solid rod, with uniform circular cross section, of length 1000 mm and diameter 100 mm is subjected to a uniaxial tensile load. As a result, the increase in length of the rod is observed to be 1 mm while the reduction in diameter of the rod is observed to be 0.03 mm. If the Young's modulus of the material of the rod is 260 GPa, the shear modulus (in GPa) of the material of the rod is ______

18. A spherical pressure vessel has inner diameter of 4 m and wall thickness of 10 mm. It is made of a steel having tensile yield strength of 200 MPa. Neglecting radial stress and using the maximum shear stress theory, the maximum value of gauge pressure (in MPa) that the wall of the pressure vessel can withstand is ______ (rounded off to two decimal places)

19. Two solid shafts P and Q made of same material transmit equal torque. Shaft P has a uniform circular cross section of area A and shaft Q has a uniform circular cross section of area 4A. If the maximum torsional shear stress developed in shaft P is 160 MPa, the maximum torsional shear stress developed (in MPa) in shaft Q is ________

20. The slotted arm AB rotates in the horizontal plane about point A at a constant angular speed ω = 10 rad/sec. The particle P is released from rest, in the frictioniess slot, at a distance of 2 m from end A. The magnitude of the velocity of the particle (in m/s) just before it hits end B is ________ (rounded off to two decimal places)

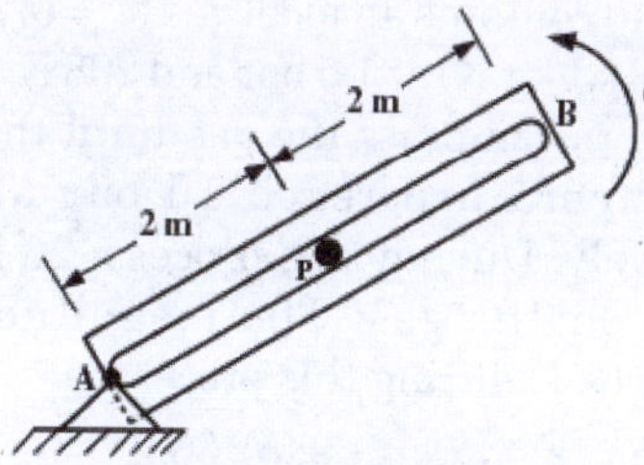

21. The uniform pendulum rod, having mass 10 kg and length L = 5 m, is attached to a viscous damper having damping coefficient c. Use acceleration due to gravity g = 10 m/s². The least value of c (in N-s/m) such that small motions of the pendulum rod decay without oscillations is ______

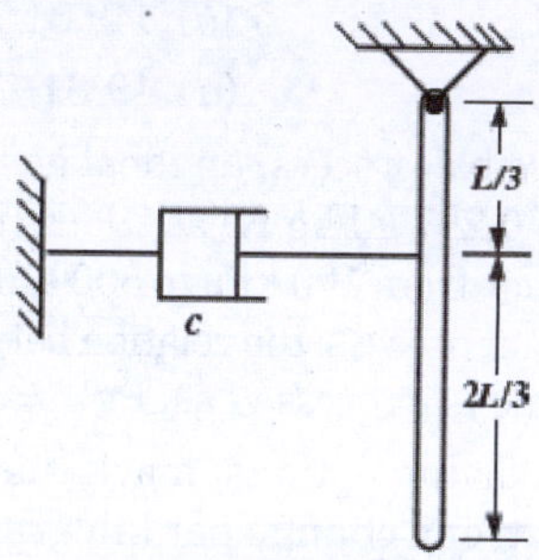

22. A beam having a constant solid rectangular cross section is subjected to pure bending. The longitudinal strain at a point A on the cross section is 3.0×10^{-5}. If the Young's modulus of the material of the beam is 200 GPa, the maximum tensile stress (in MPa) is ______

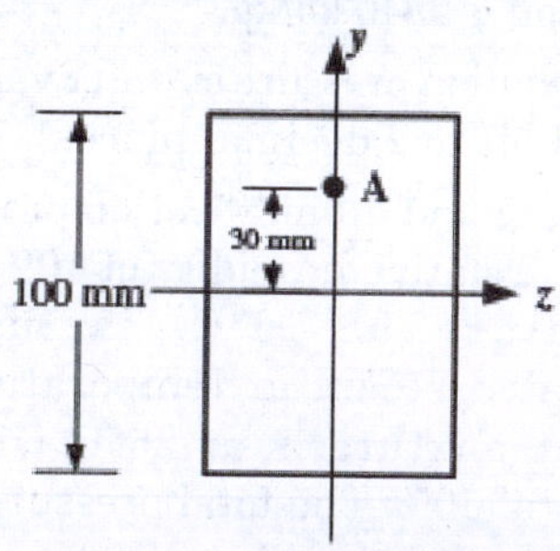

THERMODYNAMICS

Q. 1 – 9 carry one mark each & Q. 10 – 22 carry two marks each.

Notations used :

c_p : Specific heat at constant pressure;

c_v : Specific heat at constant volume;

h : Specific enthalpy;

s : Specific entropy;

v : Specific volume;

P : Pressure;

V : Volume;

T : Temperature;

R : Gas constant;

R_u : Universal Gas constant

1. One kg of an ideal gas (molecular weight = X) occupies a certain volume V at temperature Jand pressure P_1. Four kgs of another ideal gas (molecular weight = 2X) is added to the first gas keeping the volume V and temperature I'same as before. The final pressure is

(*a*) 2 P_1 (*b*) 3 P_1

(*c*) 4 P_1 (*d*) 5 P_1

2. Which of the following is an intensive thermodynamic property?

(*a*) Enthalpy (*b*) Internal energy

(*c*) Entropy (*d*) Pressure

3. A thermodynamic process for a substance is represented as a constant pressure process on a P-V diagram, and a constant temperature process on a J-V diagram. Which of the following statements is TRUE?

(*a*) The substance is an ideal gas.

(*b*) Thc substance is a van dcr Waals gas.

(*c*) The substance undergoes a phase change.

(*d*) Such a process is not possible.

4. Addition of a reheat process to a simple Rankine cycle will always

(*a*) increase both efficiency and work output of the cycle.

(*b*) increase efficiency but not necessarily work output of the cycle.

(*c*) increase work output but not necessarily efficiency of the cycle.

(*d*) decrease both efficiency and work output of the cycle.

5. A Carnot cycle implemented using an ideal gas in a piston-cylinder system will have

(*a*) work transfer in two parts of the cycle and heat transfer in the other two parts of the cycle.

(*b*) work transfer in all four parts of the cycle and heat transfer in two parts of the cycle.

(*c*) work transfer in two parts of the cycle and heat transfer in all four parts of the cycle.

(*d*) work transfer in all four parts of the cycle and heat transfer in all four parts of the cycle.

6. The "degrees of freedom" for a pure substance at its triple point is

(*a*) 3 (*b*) 2

(*c*) 1 (*d*) 0

7. Hie differential of the Gibbs function (g) for a simple compressible system can be represented as dg = v dP — s dT. Using the appropriate Maxwell relation derived from the above equation, the quantity, $\left[-\left(\frac{\partial s}{\partial P}\right)_T\right]$ for an ideal eas is

(*a*) $\frac{R}{T}$ (*b*) $\frac{R}{P}$

(*c*) $\frac{P}{R}$ (*d*) $\frac{T}{R}$

8. A Carnot engine receives 2000 kJ heat from a source at 1200 K and rejects 500 kJ heat to sink. The sink temperature (K) is ________.

9. A closed vessel contains a pure substance with 0.1 m^3 of saturated liquid and 0.9 m^3 of saturated vapor. The specific volume of the liquid and vapor phases are 0.000843 m^3/kg and 0.02671 m^3/kg respectively. The percentage of vapor in the mixture on mass basis (rounded off to two decimal places) is ________.

10. A gas obeys the van der Waais equation of state $\left(P+\frac{a}{v^2}\right)(v-b)=RT$, where a and b are van der Waals constants. The conpressibility factor $\left[Z=\frac{Pv}{RT}\right]$ in the limit of high specific volume (b << v) is

(*a*) $1+\frac{1}{v}\left(b-\frac{a}{RT}\right)$ (*b*) $1+\frac{1}{v}\left(b+\frac{a}{RT}\right)$

(*c*) $1-\frac{1}{v}\left(b+\frac{a}{RT}\right)$ (*d*) $1-\frac{1}{v}\left(b-\frac{a}{RT}\right)$

11. Consider the thermodynamic relation, $\left(\frac{\partial c_p}{\partial P}\right)_T=-T\left(\frac{\partial^2 v}{\partial T^2}\right)_P$. For a certain gas, if the isobaric thermal expansion coefficient, $\alpha=\frac{1}{T}$ (where T is the temperature measured in K), then the value of $\left(\frac{\partial c_p}{\partial P}\right)_T$ is,

(*a*) 0 (*b*) 1

(*c*) 2 (*d*) $\frac{v}{T}$

12. A piston-cylinder system has an initial volume of 0.1 m^3 and contains an ideal gas (c_v = 0.74 kJ/kg.K, R = 0.288 kJ/kg.K) at 1.5 bar and 298 K. The piston is moved to compress the gas until the pressure and temperature reach 10 bar and 423 K respectively. During this process, 20 kJ of work is done on the gas. The magnitude of heat transfer (**kJ**) during this process is

(*a*) 134.8 (*b*) 36.2

(*c*) 8.2 (*d*) 3.8

13. A reversible cyclic device absorbs 400 kJ/s heat from a reservoir maintained at 265 K and rejects heat to another reservoir at 298 K. The magnitude of the rate of work done (kW) by the device is

(*a*) 849.81 (*b*) 449.81

(*c*) 355.70 (*d*) 49.81

14. In a reversible process, an ideal gas (c_p = 1.04 kJ/kg.K, R = 0,297 kJ/kg.K) at 293 K, is compressed in a cylinder from 100 kPa to 500 kPa. During this compression process, the relation between pressure and volume is expressed as, $Pv^{1.3}$ = constant.

In the following options, mark the closest value for the entropy change per kilogram (J/kg.K).

(*a*) –91 (*b*) 91

(*c*) –864 (*d*) 864

15. The sublimation pressure of water vapor at 233 K is 13 Pa. Assume the water vapor to behave as an ideal gas (R = 0.46 kJ/kg.K), and the latent heal of sublimation is 2840 kJ/kg.

The sublimation pressure of water vapor at 213 K, (rounded off to 2 decimal places), is ______ **Pa.**

16. In a heating and humidification process, dry air (d.a) with relative humidity of 10% at ambient temperature of 10°C, is mixed with superheated steam at temperature 'T'. The resultant mixture is at 25°C with relative humidity of 50%. The total pressure before and after the process remains the same at 102 kPa.

The saturated pressure of water vapor (v) at 10°C and 25°C are "1.2281 kPa, and 3/1698 kPa, respectively. The specific humidity in grams of water vapour per kg of dry air, can be calculated using $\omega = 622 \frac{p_v}{p_{d.a}}$, where p_v and $p_{d.a}$ are the partial pressures of water vapour and dry air respectively. The quantity of steam in grams (rounded off to 2 decimal places) added per kg of dry air is ______

17. In a vapor compression system, a refrigerant leaves the evaporator and enters the compressor in saturated vapor condition at 0°C. The specific enthalpies of the saturated liquid and the saturated vapor at 0°C are 50 kJ/kg and 250 kJ/kg, respectively. The refrigerant leaves the compressor with a specific enthalpy of 300 kJ/kg. From the exit of the condenser, the refrigerant is throttled to the evaporator pressure. If the coefficient of performance (COP) of the refrigerator is 2.8 then the **dryness fraction** of the refrigerant entering the evaporator is_______.

18. Hot air, assumed as an ideal gas (c_p = 1000 J/kg.K, γ = 1.4) enters a gas turbine at 10 bar, 1000 K and leaves at a pressure of 5 bar. Subsequently it expands in a nozzle to a pressure of 1 bar. Assume both these processes to be reversible and adiabatic. If the inlet velocity of the air to the nozzle is negligible, the final velocity (m/s, rounded off to 1 decimal place) of air at the exit of the nozzle is ______.

19.

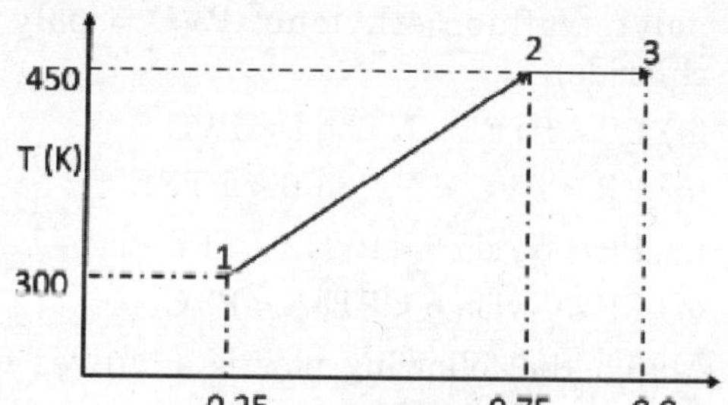

The "T-5 diagram" for a thermodynamic process is shown in the figure. The heat transferred **(kJ/kg)** during the reversible process 1 -3 is ______.

20. In a simple Rankine cycle, superheated steam enters the turbine at 100 bar, 500°C (h = 3375.1 kJ/kg, s = 6.5995 kJ/kg.K) at a rate of 1000 kg/s. It expands isentropically in the turbine to the condenser pressure. The temperature in the condenser is 30°C (h_f = 125.74 kJ/kg, s_f = 0.4368 kJ/kg.K, h_g = 2555.6 kJ/kg, s_g = 8.4520 kJ/kg.K). Saturated water from the condenser is pumped back to the boiler. Neglecting the pump work, the efficiency (in percentage, rounded off to 1 decimal) of the cycle is _______.

21. A rigid, adiabatic container has two parts (*a*) and (*b*) separated by a piston. Initially, part (*a*) is filled with 1 kg of an ideal gas (c_v = 750 J/kg.K, R = 500 J/kg.K) at 1 bar, 600 K. The part (*b*) is filled with 1 kg of the same ideal gas at 2 bar, 400 K. The piston is now removed. In the final state, the pressure (bar, rounded off to two decimal places) will be ________

22. In a steady flow process, superheated steam at 70 bar and 450°C, is throttled to 30 bar in an insulated valve, at a rate of 2 kg/s. The changes in the kinetic energy and potential energy during the process are assumed to be zero. The rate of entropy generation **(kJ/s.K** rounded off to three decimal places) during the throttling process is ________.

The thermodynamic property data for superheated steam at 70 bar and 450°C:

h = 3288.3 kJ/kg, s - 6.6353 kJ/kg.K

The thermodynamic property data for superheated steam at 30 bar are given in the following table.

T (°C)	h (kJ/kg)	s (kJ/kg K)
400	3231.7	6.9235
450	3344.9	7.0856

POLYMER SCIENCE AND ENGINEERING

XE-A : Q. 1 – 7 carry one mark each & Q. 8 – 11 carry two marks each.

XE (B to II) : Q. 1 – 9 carry one mark each & Q. 10 – 22 carry two marks each.

1. The functionality of allyl alcohol ($CH_2 = CH - CH_2OH$) for condensation reaction with terephlhalic acid is

(*a*) 0 (*b*) 1
(*c*) 2 (*d*) 3

2. Which of the following polymers can be synthesized by ring opening polymerization?

(*a*) Poly(vinyl alcohol)
(*b*) Nylon 66
(*c*) Poly(ethylene terephthalate)
(*d*) Poly(ε-caprolactone)

3. The weather resistant polymer among the following is

(*a*) natural rubber
(*b*) styrene butadiene rubber
(*c*) nitrile rubber
(*d*) silicone rubber

4. If a combination of sodium and naphthalene is used to initiate polymerization of styrene, then the most likely mechanism of polymerization is

(*a*) free radical (*b*) cationic

(*c*) anionic (*d*) olefin metathesis

5. Hypalon is the trade name for

(*a*) chlorosulfonatcd polyethylene

(*b*) chlorinated polyethylene

(*c*) ultra high molecular weight polyethylene

(*d*) cross-linked polyethylene

6. The copolymer(s) among high density polyethylene (HDPE), low density polyethylene (LDPE) and linear low density polyethylene (LLDPE) is/are

(*a*) LDPE only (*b*) LLDPE only

(*c*) LDPE and LLDPE (*d*) HDPE and LDPE

7. The polymer which lacks the ability to exhibit, tacticity among the following is

(*a*) polypropylene

(*b*) polystyrene

(*c*) polyisobutylene

(*d*) poly(methyl methacrylate)

8. The correct order of glass transition temperature (T_g) for the polymers listed below is (PC = polycarbonate; PET = poly(ethylene terephthalate); PE = polyethylene; PP = polypropylene)

(*a*) PC > PET > PP > PE (*b*) PC > PET > PE > PP

(*c*) PET > PC > PE > PP (*d*) PET > PC > PP > PE

9. Polystyrene coffee cup can be most economically manufactured by

(*a*) thermoforming

(*b*) injection molding

(*c*) compression molding

(*d*) blow molding

10. Match the following rubber additives to their function:

Additive

P. Tetramethyl thiuram disulfide

Q. Xylyl mercaptan

R. Sulfur

S. Microcrystalline wax

Function

1. Peptizer
2. Antiozonant
3. Accelerator
4. Cross-linking agent

(*a*) P-4,Q-1,R-3,S-2 (*b*) P-3, Q-1, R-4, S-2

(*c*) P-3, Q-2, R-4, S-1 (*d*) P-4, Q-2, R-3, S-1

11. Match the following polymers with their characteristic infrared (IK) stretching frequency:

Polymer	IR stretching frequency (cm^{-1})
P. Polystyrene	1. ~2242
Q. Polyacrylonitrile	2. ~1733
R. Poly(elhylene terephthalate)	3. ~3392
S. Poly(vinyl alcohol)	4. ~3100

(*a*) P-2, Q-3, R-1, S-4 (*b*) P-2, Q-1, R-4, S-3

(*c*) P-4, Q-2, R-3, S-1 (*d*) P-4, Q-1, R-2, S-3

12. Match the following products to Hie most suitable polymer for their manufacture:

Product

P. Electrical cables

Q. Electrical switches

R. Optical lenses

S. Shoe soles

Polymer

1. Polyurethane
2. Poly(methyl methacrylate)
3. Cross-linked polyethylene
4. Phenol formaldehyde resin

(*a*) P-3, Q-2, R-1, S-4 (*b*) P-3, Q-4, R-2, S-1

(*c*) P-4, Q-3, R-1, S-2 (*d*) P-2, Q-3, R-1, S-4

13. The order of limiting oxygen index for the following polymers is (polypropylene = PP; PTFE = polytetrafluoroethylene; PVC = poly (vinyl chloride))

(*a*) PP < PTFE < Nylon 6 < PVC

(*b*) PP < PVC < Nylon 6 < PTFE

(*c*) PP < Nylon 6 < PVC < PTFE

(*d*) PP < Nylon 6 < PTFE < PVC

14. Match the following plastic additives to their function:

Additive

P. Molybdenum disulfide

Q. Glycerol monostearate

R. Tribasic lead sulphate

S. 2-Hydroxybenzophenone

Function

1. Heat stabilizer
2. UV-absorber
3. Antistatic agent
4. Solid layer lubricant

(*a*) P-3, Q-1, R-2, S-4 (*b*) P-3, Q-4, R-1, S-2

(*c*) P-4, Q-3, R-2, S-1 (*d*) P-4, Q-3, R-1, S-2

15. Match the material classification in Column A with the appropriate one in Column B: (PS = polystyrene, PPO = polyphenylene oxide, PDMS = poly(dimethyl siloxane), PP = polypropylcnc, PE = polyethylene, PP-g-MA- malcic anhydride grafted PP)

Column A

P. Miscible blend

Q. Immiscible blend

R. Compatibilizcd blend

S. Polymer composite

Column B

1. PDMS + Fumed silica
2. Nylon 6 + PP-g-MA + PP
3. PS + PPO
4. PE + PP

(*a*) P-3, Q-4, R-2, S-1

(*b*) P-4, Q-3, R-2, S-1

(*c*) P-3, Q-4, R-1, S-2

(*d*) P-4, Q-3, R-1, S-2

16. A lineal' amorphous polymer has a T_g of + 10 °C. At 28 °C, it has a melt viscosity of 4×10^8 poise. The viscosity of the polymer at its T_g is ______ $\times 10^{13}$ poise (round off to one decimal place).

17. A polypropylene (PP) bar with a 10 mm × 10 mm square section is 225 mm long. The modulus of PP bar is 861 MN m^{-2}. It is pinned at both ends and an axial compressive load of 140 N is applied. The strain due to the applied load experienced by the PP bar is ______ % (round off to two decimal places).

18. In a unidirectional carbon fibre/vinyl ester composite, the ratio of the moduli of the carbon fibrc to that of vinyl ester is 35 and the fibres take up 50% of the cross-section. The percentage of applied force taken up by thc fibros is _____ % (round off to one decimal placc).

19. A 3 mm thick layer of softened poly(methyl methacrylate) at 190 °C is sandwiched between two flat parallel plates. A shear stress of 100 kPa is applied to the softened polymer. Assuming the softened polymer as a Newtonian fluid with an apparent viscosity of 3.9×10^4 Pa s, the relative sliding velocity between the two plates is ______ mm s^{-1} (round off to one decimal place).

20. For AIBN initialed polymerization of styrene, if both the monomer and initiator concentration are doubled, then the rate of polymerization increases by a factor of _____ (round off to two decimal places).

21. If 49 moles of hcxamcthylene diamine is reacted with 50 moles of adipic acid to prepare Nylon 66, then the number average molecular weight, $\overline{M}_n$ of the resulting polymer at 99.5% conversion (ignoring contribution from end groups) is g mol^{-1} (round off to 2 decimal places).

22. A single screw extruder is to be used to manufacture a nylon rod of 0.5 cm diameter at a production rate of 2.5 cm s^{-1}. The density of solid nylon and nylon melt are 1.140 g cm^{-3} and 0.790 g cm^3, respectively. The melt flow rate through the die is ______ cm^3 s^{-1} (round off to two decimal places).

FOOD TECHNOLOGY

Q. 1 – 9 carry one mark each & Q. 10 – 22 carry two marks each.

1. Colloidal stability of milk casein is because of the highly hydrated carbohydrate residues in ______.

(*a*) α_{s1} casein (*b*) α_{S2} casein

(*c*) β casein (*d*) κ casein

2. Rice bran is stabilized prior to oil extraction to protect it from the activity of ________.

(*a*) Poiyphenoi oxidase (*b*) Peroxida.se

(*c*) lipase (*d*) Lipoxygenase

3. Sticking of powder to wall of the chamber during spray drvins of fruit juice is due to ______.

(*a*) Low glass transition temperature of the compounds in juice

(*b*) High glass transition temperature of the compounds in juice

(*c*) Improper processing parameters of spray dryer

(*d*) Presence of gums in feed material

4. Thearubigins and thea.flav.ins in black tea are formed by the oxidation and dimerization of_____.

(*a*) Quercetin (*b*) Catechins

(*c*) Gallic acid (*d*) Kaempferol

5. Ratio of Schmidt number to Lewis number is____.

(*a*) Prandtl number (*b*) Reynolds number

(*c*) Nusselt number (*d*) Sherwood number

6. 'Red dog' is one of the byproducts during milling of

(*a*) Corn (*b*) Rice

(*c*) Ragi (*d*) Wheat

7. An ice cream mix of 870 g L^{-1} has been used to prepare ice cream which yielded a finished product of 490 g L^{-1}. The per cent over run is _________ (round off to 1 decimal place).

8. Impeller in a fruit juice mixing tank is rotating at 200 rpm with a Reynolds number $>10^4$. Density of juice is 1045 kg m^3. If diameter of the impeller is doubled and other conditions remained constant, the power requirement of mixing will increase by a factor of ______.

9. Paddy consisting of 20% husk has been milled to remove 6% bran during polishing. Assuming no other losses, yield (per cent) of polished rice from the paddy is ______ (round off to 1 decimal place).

10. Match the following laws in Column I with corresponding phenomenon in Column II.

Column I

P. Newton's law

Q. Hertz constant stress theory

R. Fick's law

S. Bond's law

Column II

1. Size reduction
2. Substance cooling
3. Damage of fruits and vegetables
4. Molecular diffusion

(*a*) P-2, Q-3, R-4, S-1 (*b*) P-3, Q-2, R-4, S-1

(*c*) P-3, Q-1, R-4, S-2 (*d*) P-4, Q-3, R-2, S-1

11. Match the mold in Column I with its asexual/ sexual spore shown in Column II

Column I	Column II
P. Aspergillus	1. Arthrospore
Q. Geotrichum	2. Oospores
R. Rhizopus	3. Conidia
S. Oomycetes	4. Sporangiospores

(*a*) P-3, Q-1, R-4, S-2

(*b*) P-1, Q-4, R-3, S-2

(*c*) P-4, Q-3, R-1, S-2

(*d*) P-4, Q-1, R-2, S-3

12. Match the foods given in Column I with their specific usage given in Column II.

Column I	Column II
P. Egg yolk	1. Ice cream
Q. Pregelatinised starch	2. Mayonnaise
R. Gum	3. Baking powder
S. Starch	4. Baby food

(*a*) P-2, Q-4, R-1, S-3

(*b*) P-4, Q-1, R-2, S-3

(*c*) P-2, Q-3, R-1, S-4

(*d*) P-1, Q-4, R-1, S-3

13. Match the bioactive compounds in Column I with their botanical source given in Column II.

Column I	Column II
P. Isoflavones	1. Corn
Q. Resistant starch	2. Grapes
R. Xanthophyll	3. Soybean
S. Resveratrol	4. Plantain (culinary banana)

(*a*) P-2, Q-4, R-1, S-3 (*b*) P-3, Q-4, R-1, S-2

(*c*) P-4, Q-1, R-2, S-3 (*d*) P-4, Q-3, R-2, S-1

14. Match the following raicrobial species in Column I with related disease caused by them as listed in Column II.

Column I	Column II
P Vibrio sp.	1 Gastroenteritis
Q Shigella sp.	2 Typhoid
R E. coli	3 Cholera
S Salmonella typki	4 Bacillary dysentery

(*a*) P-1, Q-3, R-4, S-2 (*b*) P-2, Q-3, R-4, S-1

(*c*) P-3, Q-1, R-4, S-2 (*d*) P-3, Q-4, R-1, S-2

15. Buffalo milk having density of 1030 kg m^{-3} is homogenized with a pressure of 30 MPa. Given, acceleration due to gravity as 9.81 m s^{-2} and assuming no pressure loss, the velocity (m s^{-1}) of the milk flowing through the homogenizer valve will be ______ (round off to 2 decimal places).

16. Potato slices have been dehydrated from an initial solid content of 12% to a final solid content of 94%. If the peeling and other losses are to the tune of 10%, final yield (per cent) of the dried chips per ton of fresh potato taken is ______ (round off to 2 decimal places).

17. A mixed fruit beverage with 12 °Brix having specifc heat of 4298 J kg^{-1} K^{-1} is being heated from 30 °C to 95 °C for pasteurization at a flow rate of 1000 L h^{-1} in a tubular heal exchanger. Steam at 100 °C is used as heating medium which is converted into condensate at 100 °C. If the density of beverage is 1075 kg m^{-3} and the latent heat of steam at the given temeparture is 2257 kJ kg^{-1}, the mass flow rate of steam (kg min^{-1}) is ______ (round off to 2 decimal places).

18. Fruit juice was cooled in a tubular heal exchanger from 50 °C to 7 °C using water at 2 °C, which gets heated to 5 °C. Assume, Pr = 0.72, Re = 20000 and thermal conductivity = 0.6 W m^{-1} $°C^{-1}$ and no viscous effect. If pipe diameter was 10 cm, the convective heat transfer coefficient (W m^2 $°C^{-1}$) is ______ (round, off to 1 decimal place).

19. Room air is at 40 °C with 60%, relative humidity. Saturated vapour pressure of water at 40 °C is 7.375 kPa. Volume of humid air (m^3 per kg of dry air) is(round off to 3 decimal places).

20. A shallow-bed horizontal belt type solvent extractor is operating on 0.3 mm thick soybean flakes with 0.5 m bed depth and a forward speed of 0.8 m min^{-1} with miscelia flux rate of 0.25 m min^{-1}. If porosity of the flakes is 60%, the distance between washing nozzle and miscclla collecting receptacle (cm) is ________ (round, off to 1 decimal place).

21. An extruded snack food is packed in a barrier film having water vapour transmission rate of 0.02 mL m^{-2} day^{-1}. Pack surface area is 0.0012 m^2 per gram of dry food solids, EMC of the food is 6% (d.b.), initial moisture content is 2% (d.b.), critical moisture content is 5% (d.b.) and slope of moisture sorption isotherm is 3.4% (d.b.) per unit water activity (a_w). Sealed pack is stored at 30 °C. Assume that the vapor pressure of pure water at 30 °C is 31.7 torr. Time required for the food to reach the critical moisture content (days) is ______ (round off to 1 decimal place).

22. Freezing of 100 mm spherical meat ball with 60% moisture at 35 °C is being done in an air blast freezer maintained at –45 °C. Given, latent heat of fusion for water is 333.2 kJ kg^{-1}, thermal conductivity of meat is 1.5 W m^{-1} °C^{-1}, convective heat transfer coefficient is 40 W m^2 °C^{-1} density of frozen meat is 980 kg m^{-3} and initial freezing temperature of meat ball is –10°C. Using Plank's equation, freezing time (h) is ______ (round off to 2 decimal places).

ATMOSPHERIC AND OCEANIC SCIENCES

XE-A : Q. 1 – 7 carry one mark each & Q. 8 – 11 carry two marks each.

XE (B to II) : Q. 1 – 9 carry one mark each & Q. 10 – 22 carry two marks each.

XL-P : Q. 1 – 5 carry one mark each & Q. 6 – 15 carry two marks each

XL (Q to U) : Q. 1 – 10 carry one mark each & Q. 11 – 20 carry two marks each.

1. In the equatorial belt, the Madden-Julian Oscillation (MJO) propagates

(*a*) northward

(*b*) southward

(*c*) eastward

(*d*) westward

2. Potential temperature increases with height in an atmosphere. Upon adiabatic vertical displacement, a parcel of dry air is

(*a*) stable

(*b*) absolutely unstable

(*c*) neutral

(*d*) conditionally unstable

3. Oxygen minimum zones in the ocean arc formed in the layers where

(*a*) Photosyntliesis is dominant

(*b*) Sunlight is maximum

(*c*) Decomposition of organic matter is dominant

(*d*) iron availability is maximum

4. Water level in an estuary rises and falls in response to tidal forces. Assume that the only tidal period in an estuary is 12 hours. The speed of the ebb current in that estuary would be maximum

(*a*) At high tide

(*b*) At low tide

(*c*) About 3 hours after high tide

(*d*) About 6 hours after low tide

5. The most abundant ions in seawater are

(*a*) Sodium and chloride ions

(*b*) Magnesium and chloride ions

(*c*) Calcium and chloride ions

(*d*) Magnesium and sulphide ions

6. Which among the following statements is true with respect to the axis of the Indian monsoon trough

(*a*) Relative vorticity is cyclonic at 850 hPa

(*b*) Relative vorticity is anticyclonic at 850 hPa

(*c*) It shifts to the south during break monsoon

(*d*) It moves to the head Bay of Bengal when a monsoon depression is present over the Indian subcontinent

7. A student considered the following possibilities regarding trade wind inversion:

P. Zonal component of wind is westerly

Q. Zonal component of wind is casterly

R. Meridional component of wind is poleward

S. Meridional component of wind is equatorward

The only combination of correct statements is

(*a*) P, S

(*b*) Q, S

(*c*) Q, R

(*d*) P, R

8. Which of the following statements about thermal wind is NOT correct:

(*a*) Flow is in geostrophic balance

(*b*) Horizontal temperature gradient is non-zero

(*c*) Wind speed varies with height

(*d*) Flow is qua si-geostrophic

9. A numerical weather forecasting model follows sigma (σ) coordinate system in the vertical. At two locations A and B, the surface pressure is 1000 hPa and 500 hPa respectively. The values of pressure at A and B (in that order) for $\sigma = 0.5$ are:

(*a*) 2000 hPa, 1000 hPa

(*b*) 500 hPa, 250 hPa

(*c*) 999.5 hPa, 499.5 hPa

(*d*) 500 hPa, 0 hPa

10. At a given temperature and pressure, equilibrium vapor pressure over the surface of a rain droplet compared to that over a flat surface of pure water is

(*a*) Always less

(*b*) Always more

(*c*) Always equal

(*d*) Can be less, equal to or more depending on the size and chemical composition of the cloud condensation nuclei, and droplet radius

11. Velocity field in a geophysical flow is given by u = –5x, v = 4y, w = 0 where u, v, w are zonal, meridional and vertical components of velocity, and x and y refer to eastward and northward distance. Which among the following statements is **NOT** true for the flow at 45°N:

(*a*) Row is incompressible

(*b*) Relative vorticity is zero

(*c*) Potential vorticity is positive

(*d*) Row is compressible

12. The distance between Mars and Earth is 0.52 times the distance between Sun and Earth. If the solar constant on Earth is 1370 W m^{-2}, then the solar constant on Mars is W m^2. (Round off to one decimal place.)

13. Ideal gas equation for dry air is expressed as $p = \rho_d R_d T$, where p is pressure, ρ_d is density, T is temperature and R_d is gas constant. If CO_2 concentration in Earth's atmosphere quadruples due to fossil fuel burning, then the new gas constant is ______ J kg^{-1} K^{-1}. (Round off to two decimal places.)

Take the universal gas constant (R^*) = 8.314 J K^{-1} $mole^{-1}$ molecular weight of CO_2 = 44, and present value of Rd = 287 J kg^{-1} K^{-1}, present concentration of CO_2 = 500 mg kg^{-1}.

14. On a calm evening under a clear sky, the lowest 20 m of the atmosphere is continuously losing energy at a rate of 10 W m^{-2} due to radiative cooling. Sensible and latent heat flux exchange with the ground is zero. If air temperature and dew point depression at 6 p.m. are 25°C and 8°C, respectively, the probability of fog formation at midnight is ______ %. (Round off to the nearest integer.)

Take C_p = 1005 J kg^{-1} K^{-1}, density of air = 1.15 kg m^{-3}.

15. The dynamical core of a general circulation model uses 256 spherical harmonic waves to cover the globe in east-west direction. If two grids of the model exactly resolve the shortest of those waves, then the grid spacing at the equator is ______ km. (Round off to one decimal place.)

Take radius of the earth (R_E) = 6400 km.

16. It rains 10 cm in a day over an ocean with a mixed layer depth of 5 m and initial salinity 30 g kg^{-1}. The rain water uniformly mixes with the existing mixed layer water. The final salinity after the rain is ______ g kg^{-1}. (Round off to one decimal place.)

17. A moist air parcel is lifted adiabatically to 800 hPa level. Initial temperature and pressure of the parcel are 300 K and 1010 hPa, and the partial pressure of water vapor is 10 hPa. The partial pressure of the water vapor at the new location is ______ hPa. (Round off to one decimal place.)

Take saturation vapor pressure of water at temperature T as

$$e_s(T) = Ae^{-\frac{B}{T}},$$

where A = 2.53 × 10^9 hPa, B = 5420 K, gas constant (R) = 287 J kg^{-1} K^1, and C_p = 1005 Jkg^{-1} K^{-1}.

18. In steady state, global mean precipitation P is equal to global mean evaporation E from the surface of the earth. If P is 2.8 mm day^{-1}, then global mean surface latent heat flux is ______ W m^2. (Round off to the nearest integer).

Take latent heat of evaporation of water as 2.5 × 10^6 J kg^{-1}.

19. A column of seawater from surface to 1000 m depth is initially at rest at the equator. The column is moved to 30°N with no change in column depth. In the absence of friction, the vertical component of relative vorticity of the water in the column at 30°N is ______ × 10^{-5} s^{-1}. (Round off to one decimal place.)

20. Consider two north-south (i.e. meridional) sections A and B in the ocean, separated by 100 km; section B is to the east of section A (as shown in figure below). The steady northward wind stress at A and B are 0.1 N/m^2 and 0.2 N/m^2, respectively; zonal wind stress is zero everywhere. Density of seawater is 1000 kg m^{-3}. At 30°N, the vertical velocity at the base of the Ekman layer is______ meters per day. (Round off to one decimal place.)

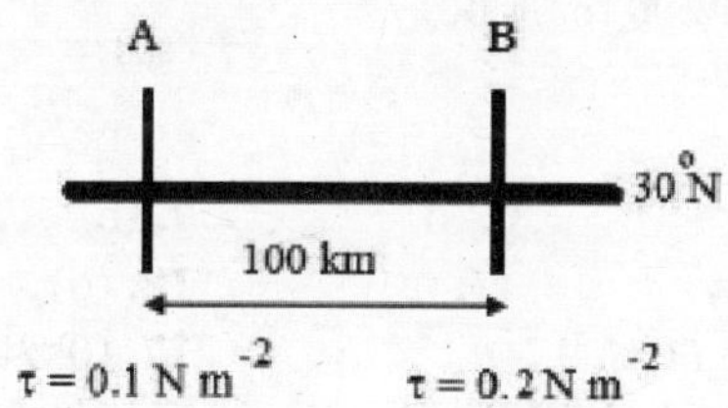

21. A tsunami is a long gravity wave, i.e. its wavelength is much larger than the depth of the ocean. A tsunami moves from point A to point B (see figure below); water depth at A is 2000 in, and water depth at B is 500 m. Assume no frictional losses. By energy conservation, C_g <E> is constant, where C_g is group speed and <E> is the total (potential plus kinetic) energy density of the wave field. If the amplitude of the tsunami at point A is 1 m, its amplitude at B is ______ m. (Round off to two decimal places.)

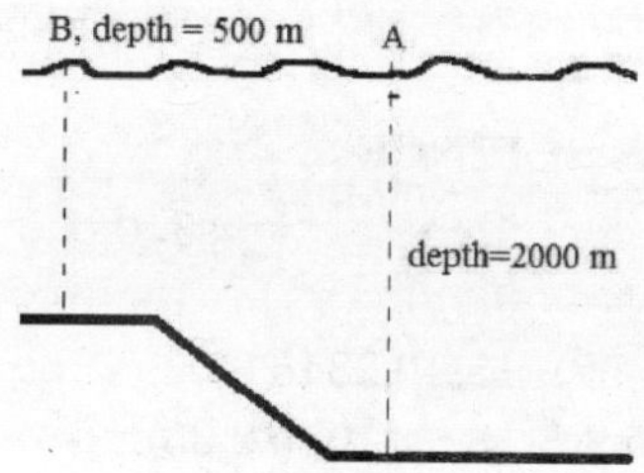

22. At mid-latitudes, long Rossby waves obey the dispersion relation $\sigma = -\beta k / L^{-2}$, where σ is frequency, k is zonal wavenumber, β is rate of change of Coriolis parameter with northward distance (units m^{-1} s^{-1}), and L is Rossby radius. If L = 100 km, the speed of the long Rossby wave at45°N is ________ m s^{-1}. (Round off to two decimal places.) Take radius of the earth as 6400 km.

ANSWERS

GENERAL APTITUDE (GA)

1. (*a*) **2.** (*c*) **3.** (*d*) **4.** (*b*) **5.** (*c*) **6.** (*d*) **7.** (*a*)
8. (*b*) **9.** (*c*) **10.** (*a*)

ENGINEERING MATHEMATICS

1. (d) **2.** (d) **3.** (c) **4.** (a) **5.** (6 to 6) **6.** (3 to 3) **7.** (0 to 0)
8. (b) **9.** (d) **10.** (2 to 2) **11.** (6.0 to 6.2)

FLUID MECHANICS

1. (d) **2.** (b) **3.** (b) **4.** (c) **5.** (c) **6.** (d)
7. (0.01 to 0.01) **8.** (3 to 3) **9.** (0.8 to 0.8) **10.** (c) **11.** (a) **12.** (c)
13. (c) **14.** (d) **15.** (b) **16.** (d) **17.** (33.25 to 33.75)
18. (0.15 to 0.17) **19.** (1.25 to 1.27) **20.** (2 to 2) **21.** (0.32 to 0.34)
22. (37 to 37)

MATERIALS SCIENCE

1. (b) **2.** (b) **3.** (a) **4.** (a) **5.** (b) **6.** (b) **7.** (c)
8. (d) **9.** (b) **10.** (b) **11.** (b) **12.** (b) **13.** (100.0 to 102.0)
14. (45 to 52) **15.** (0.28 to 0.32) **16.** (360 to 364) **17.** (0.58 to 0.61)
18. (9.2 to 9.6) **19.** (2.10 to 2.35) **20.** (666 to 670) **21.** (560 to 565)
22. (11.0 to 12.5)

SOLID MECHANICS

1. (c) **2.** (d) **3.** (b) **4.** (b) **5.** (a) **6.** (a) **7.** (c)
8. (d) **9.** (–3.0 to –3.0) **10.** (b) **11.** (a) **12.** (c) **13.** (d)
14. (a) **15.** (a) **16.** (2.0 to 2.0) **17.** (100 to 100)
18. (1.90 to 2.10) **19.** (19.0 to 21.0) **20.** (52.00 to 54.00)
21. (102.0 to 106.0) **22.** (10.0 to 10.0)

THERMODYNAMICS

1. (b) **2.** (d) **3.** (c) **4.** (c) **5.** (b) **6.** (d) **7.** (b)
8. (300 to 300) **9.** (22.00 to 22.30) **10.** (a) **11.** (a) **12.** (d)
13. (d) **14.** (a) **15.** (1.05 to 1.15) **16.** (8.50 to 9.50) **17.** (0.3 to 0.3)
18. (770.0 to 785.0) **19.** (255 to 255) **20.** (41.0 to 43.5)
21. (1.25 to 1.25) **22.** (0.700 to 0.750)

POLYMER SCIENCE AND ENGINEERING

1. (b) **2.** (d) **3.** (d) **4.** (c) **5.** (a) **6.** (b) **7.** (c)
8. (a) **9.** (a) **10.** (b) **11.** (d) **12.** (b) **13.** (c) **14.** (d)
15. (a) **16.** (1.2 to 1.4) **17.** (0.15 to 0.17) **18.** (97.0 to 97.5)
19. (7.6 to 7.8) **20.** (2.80 to 2.84) **21.** (7497.72 to 7517.72)
22. (0.69 to 0.73)

FOOD TECHNOLOGY

1. (d) **2.** (c) **3.** (a) **4.** (b) **5.** (d) **6.** (d)
7. (77.4 to 77.6) **8.** (32 to 32) **9.** (74.8 to 75.6) **10.** (a) **11.** (a)
12. (a) **13.** (b) **14.** (d) **15.** (240.00 to 242.00) **16.** (11.45 to 11.55)
17. (2.16 to 2.25) **18.** (339.1 to 342.9) **19.** (0.924 to 0.930)
20. (95.0 to 100.0) **21.** (58.1 to 62.9) **22.** (1.06 to 1.09)

ATMOSPHERIC AND OCEANIC SCIENCES

1. (c) **2.** (a) **3.** (c) **4.** (c) **5.** (a) **6.** (a) **7.** (b)
8. (d) **9.** (b) **10.** (d) **11.** (a) **12.** (592.8 to 593.2)
13. (286.40 to 287.30) **14.** (99 to 100) **15.** (78.4 to 78.6)
16. (29.3 to 29.5) **17.** (7.8 to 8.1) **18.** (75 to 85) **19.** (–7.6 to –7.0)
20. (1.1 to 1.3) **21.** (1.30 to 1.50) **22.** (0.13 to 0.19)

EXPLANATIONS

General Aptitude

4. (b) For IMHO = JNIP

$I + 1 = J$

$M + 1 = N$

$H + 1 = I$

$O + 1 = P$

For

SO = TP

$S + 1 = T$

$O + 1 = P$

Similarly, for

IDC = JED

$I + 1 = J$

$D + 1 = E$

$C + 1 = D$

5. Given product of three integers,

$X.Y.Z = 192$

Given : Z = 4

$\therefore X.Y = \frac{192}{4} = 48$

$\because P = \frac{X+Y}{2}$

Now,

Possible value of (X, Y) are

(48, 1), (24, 2), (16, 3), (12, 4), (8, 6)

So, $(P)_{min} = \frac{8+6}{2} = 7$

So, min value of P is 7.

8. Let GDP in 2015 be x

Now, in 2016, GDP will be 1.1x

Fiscal deficit in 2015 = $0.04 \times x$

Fiscal deficit in 2016 = $0.05 \times 1.1\,x$

Here percentage increase in the actual fiscal deficit

$$= \left(\frac{0.05 \times 1.1x - 0.04 \times x}{0.04x}\right) \times 100 = 37.5\%$$

9. Here,

In 1 hr, part of tank filled by P alone is = $\frac{1}{6}$

In 1 hr, part of tank filled by Q alone is = $\frac{1}{9}$

In 1 hr, part of tank emptied by R is = $\frac{1}{12}$

Initially, P and R opened for 4 hr,

So, part of tank filled = $4\left(\frac{1}{6} - \frac{1}{12}\right) = \frac{4}{12} = \frac{1}{3}$

After 4 hr, P is closed and for 6 more hr Q and R are opened.

So, part of tank filled = $6\left(\frac{1}{9} - \frac{1}{12}\right) = \frac{1}{6}$

So, total tank filled in 10 hr = $\frac{1}{6} + \frac{1}{3} = \frac{2+4}{12} = \frac{1}{2}$

So, remaining 1/2 part of the tank is filled by Q alone

Now, Q will take 9/2 = 4.5 hr to fill remaining half of the tank.

Hence, total time taken = 4 + 6 + 4.5 = 14.5 hr

10. Here,

Revenue shared by pool = 170 + 320 + 215 + 190 = 895$

Similarly,

Revenue shared by mini = 110 + 220 + 180 + 70 = 580$

and,

Revenue shared by prime = 75 + 180 + 120 + 90 = 465$

Now, percentage share of revenue contributed by prime

$$= \frac{465}{895 + 580 + 465} = 23.97\%$$

Engineering Mathematics

1. (d) $P(2 \le x \le 4)$

$= p(x = 2) - p(x = 4)$

$$= \frac{e^{-1}(1^2)}{2!} - \frac{e^{-1}(1^4)}{4!}$$

$$= e^{-1}\left(\frac{1}{2} - \frac{1}{24}\right)$$

$$= \frac{1}{e}\left(\frac{11}{24}\right) = \frac{11}{24e}$$

2. We have

$$a_n = \frac{1}{n2^n}$$

$$a_{nH} = \frac{1}{(nH)2^{(nH)}}$$

$$\therefore \lim_{n\to\infty}\left|\frac{a_{nH}}{a_n}\right|$$

$$= \lim_{n\to\infty}\left|\frac{\dfrac{(n+1)^{n+1}}{(n+1)2^{(nH)}}}{\left(\dfrac{(n+1)^n}{n2^n}\right)}\right|$$

$$= \lim_{n\to\infty}\left|\frac{(n+1)}{2}\cdot\frac{n}{n+1}\right|$$

$$= \left|\frac{n+1}{2}\right| \le 1$$

$\Rightarrow |nH| \le 2$

$\Rightarrow -2 \le nH \le 2$

$\Rightarrow -3 \le n \le 1$

5. The given system is

$x - y - 32 = 3$

$2x + z = 0$

$-2y - 72 = \rightarrow \alpha$

The angumented matrix is

$$A = \left[\begin{array}{ccc|c} 1 & -1 & -3 & 3 \\ 2 & 0 & 1 & 0 \\ 0 & -2 & -7 & \alpha \end{array}\right]$$

$$\sim \left[\begin{array}{ccc|c} 1 & -1 & -3 & 3 \\ 0 & 2 & 7 & -6 \\ 0 & -2 & -7 & \alpha \end{array}\right]$$

$$\sim \left[\begin{array}{ccc|c} 1 & -1 & -3 & 3 \\ 0 & 2 & 7 & -6 \\ 0 & 0 & 0 & -6+\alpha \end{array}\right]$$

$\therefore$ The given system has solution if

$-6 + \alpha = 0 \Rightarrow \alpha = 6.$

7. $\dfrac{x^2d^2y}{dx^2} + x\dfrac{dy}{dx} - 4y = 0$

By couchy euler eqn,

Let $z = \log x$ $[x = e^z]$

$\theta(\theta - 1)y + \theta y - 4y = 0$

$(\theta^2 - \theta)y + \theta y - 4y = 0$

$(\theta^2 - \theta + \theta - 4)y = 0$

$(\theta^2 - 4)y = 0$

Auxiliary equation,

$m^2 - 4 = 0 \Rightarrow m = \pm 2,$

$\therefore \theta = \pm 2$

General solution,

$y = CF$

$y = C_1e^{2z} + C_2e^{-2z}$

$= C_1e^{2\log x} + C_2e^{-2}\log x$

$= C_1e\log x^2 + C_2e\log x^{-2}$

$= C_1x^2 + C_2(x^{-2})$

$$\lim_{x\to\infty}(x-2) = \frac{1}{x^2} = \frac{1}{(\infty)^2} = 0$$

9. (d) We have

$P(x) = ax^3 + bx^2 + cx + d$

$\therefore$ $P(0) = -8$

$\Rightarrow d = -8$ ...(1)

$P(1) = -7$

$a + b + c + d = -7$

$\Rightarrow a + b + c = 1$

$P(2) = -6$

$\Rightarrow 8a + 4b + 2c + d = -6$

$\Rightarrow 8a + 4b + 2c = 2$ $[\because d = -8]$

$\Rightarrow 4a + 2b + c = 1$ ..(3)

$P(4) = 20$

$\Rightarrow 64a + 16b + 4c + d = 20$

$\Rightarrow 64a + 16b + 4c = 28$

$\Rightarrow 16a + 4b + c = 7$ ...(4)

On solving (2), (3) and (4), we get

$a = 1, b = -3, c = 3$

$\therefore$ $a - b + c = 1 + 3 + 3 = 7$

$\therefore$ option (d) is correct.

10. $f(x, y) = x^3 + 3xy^2 - 15x - 12y$

Now, $\frac{\partial f}{\partial x} = 3x^2 + 3y^2 - 15 = 0 \Rightarrow x^2 + y^2 = 5$

$\frac{\partial f}{\partial y} = 6xy - 12 = 0 \Rightarrow xy = 2$

Now,

$(x + y)^2 = x^2 + y^2 + 2xy = 5 + 2 \times 2 = 9$

and

$(x - y)^2 = (5 - 4)^2 = 1$

Here, $(x + y)^2 = 9$ Þ $x + y = \pm 3$ ....eqn (i)

and

$(x - y)^2 = 1 \Rightarrow x - y = \pm 1$ ....eqn (ii)

Solving eqn (i) and (ii), we get,

$x = 2, -2$

$y = 1, -1$

Now, critical point (2, 1) and (–2, –1)

Now, number of critical points = 2

Fluid Mechanics

1.

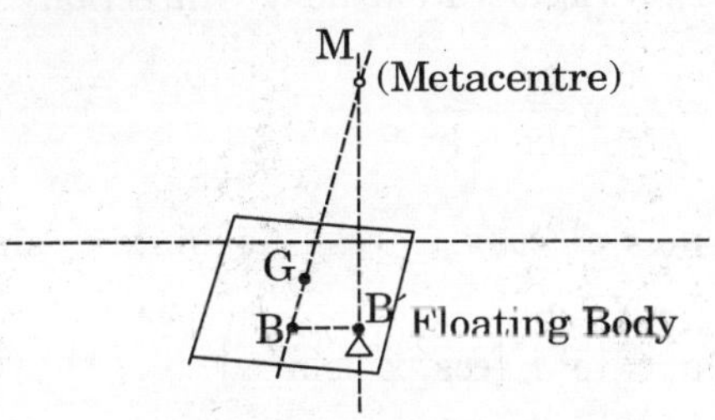

For stable equilibrium condition of a floating body is that, the metacentre must be located above centre of gravity.

2. We know vorticity (w_z) at a point in a flow field along x and y-direction

$$\boxed{w_z = \frac{dv}{dx} - \frac{du}{dy}}$$

u $\rightarrow$ velocity component

v $\rightarrow$ velocity component in y-direction

Hence the option (b) is correct

3. We know by Bernoulli's equation.

Static pressure + dynamic pressure = Total pressure

$$\boxed{P_s + \frac{\rho v^2}{2} = P_t}$$

The above equation used to derive the flow velocity (V) and this equation is derived from pitot-static tube.

4. Dimples on a golf ball create a thin turbulent boundary layer of air that clings to ball's surface This allows the smoothly flowing air to follow the ball's surface a little farther around the back side of the ball, thereby decreasing the size of the wake.

5. We know for laminar flow, skin friction coefficient

$$C_f = \frac{0.664}{\sqrt{R_{ex}}} = \sqrt{\frac{0.664}{\frac{\rho vx}{u}}}$$

$$\boxed{C_f \propto \frac{1}{\sqrt{x}}}$$

As the distance of the leading edge increases the skin-friction coefficient decreases.

7. (0.01 to 0.01)

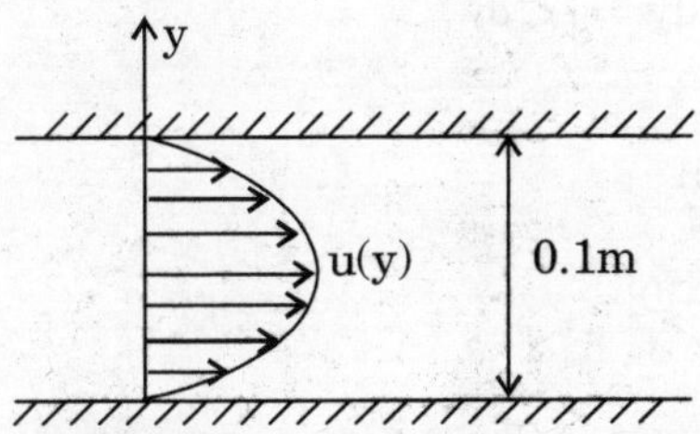

Given that

Velocity equation $u(y) = 100\,(0.1y - y^2)$

$\therefore \frac{du}{dy} = 100(0.1 \times 1 - 2y)$

at $y = 0$

$\frac{du}{dy} = 100(0.1 - 0) = 10$ ...(1)

Now shear stress acting on the bottom plate

$\tau = \mu \frac{du}{dy}$ $[\because \mu = 0.001\ \text{N.s/m}^2]$

$= 0.001 \times 10$ [from equ. 1]

$\boxed{\tau = 0.01\ \text{N/m}}$

8. We know that the ratio of average velocity to the maximum velocity in a fully developed laminar incompressible flow through a circular pipe is

$$\frac{\overline{U}}{U_{max}} = \frac{1}{2}$$

$$\frac{\overline{U}}{6} = \frac{1}{2}$$ [U_{max} = 6 m/s given]

$\therefore$ $\boxed{\overline{U} = 3\ \text{m/s}}$

9. We know

$$Q_{actual} = C_d \times Q_{theoretical}.$$

$$\therefore C_d = \frac{Q_{Act}}{Q_{theo}} = \frac{32}{40}$$

$$\boxed{C_d = 0.8}$$

therefore coefficient of discharge through an orifice is $C_d = 0.8$

10.

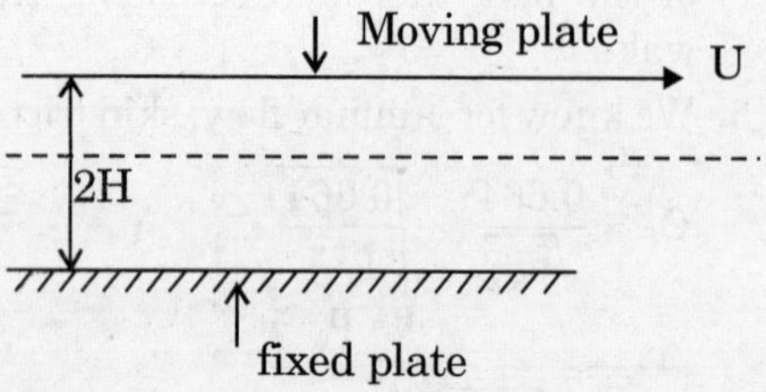

We know

Volumetric flow rate per unit width of the plate

$$Q = \int_0^H V_x \, dy = \int_0^H U \, dy$$

$$Q = u[y]_0^H$$

$$Q = U[H - 0]$$

$$\boxed{Q = UH}$$

12. Given velocity field

$V = yi - xj$

Here $u = y$

$v = -x$

For Irrotational flow

$$w_z = \frac{1}{2}\left[\frac{dv}{dx} - \frac{du}{dy}\right] = 0 = \frac{1}{2}[-1-1]$$

$$w_z = -1 \neq 0$$

therefore the flow is rotational. Now, for incompressible flow

$$\frac{du}{dx} + \frac{dv}{dy} = 0$$

$$0 + 0 = 0$$

$$0 = 0$$

therefore the flow is incompressible.

Hence the option (c) is correct.

13.

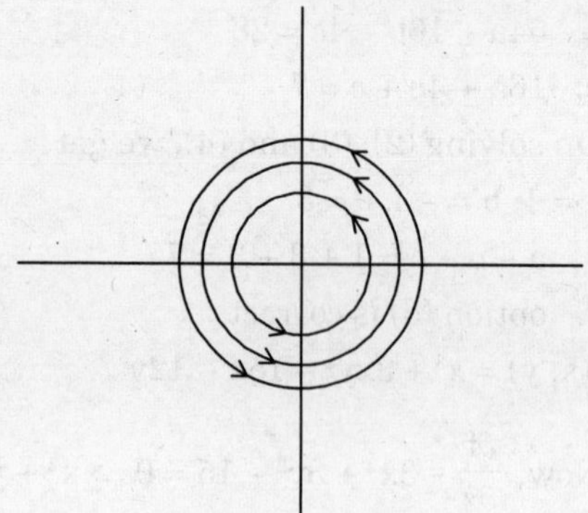

free vortex flow

For a free vortex flow:

(a) fluid particles move in a circle about a point

(b) There only non trivial velocity component is tangential

14. Given velocity components

$u = e^y \cos x$

$v = e^y \sin x$

$\because$ Acceleration along x-direction

$$a_x = u\frac{du}{dx} + v\frac{du}{dy}$$

$$= \left[e^{\psi} \cos x.\, e^y (-\sin x)\right] + \left[e^y \sin.\, e^y \cos x\right]$$

$$\boxed{a_x = 0}$$

Now Acceleration along y-direction

$$a_y = u\frac{dv}{dx} + \frac{dv}{dy}$$

$$a_y = e^y \cos x.\, e^y \cos x + e^y \sin x.\, e^y \sin x$$

$$a_y = (e^y)^2 [\cos^2 x + \sin^2 x]$$

$$a_y = (e^y)^2$$

$$\therefore (a_y)_{(-1,1)} = e^2$$

$\therefore$ Magnitude of total acceleration at point (–1, 1) is

$$a_T = \sqrt{(a_x)^2 + (a_y)^2}$$

$$a_T = \sqrt{0 + (e^2)^2}$$

$$\boxed{a_T = e^2}$$

17.

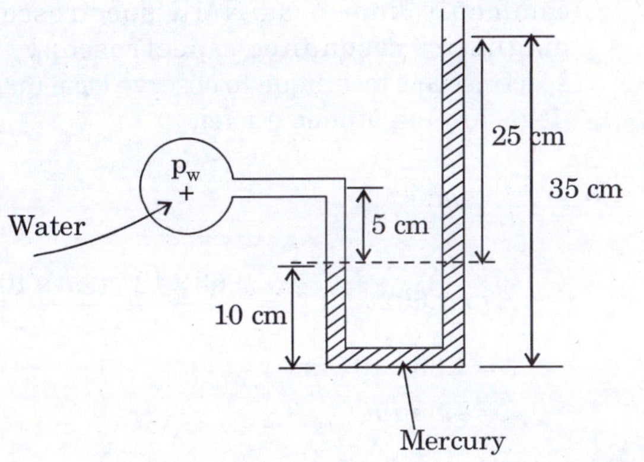

$$P_w + (\rho_w g.h_w) - (\rho_m .g.h_m) = 0$$

$$P_w + \rho_m .g.h_w - \rho_m .g.h_m$$

= (13.6 × 1000 × 10 × 0.25) – (1000 × 10 × 0.05)

= 34000 – 500

P_w = 33500 N/m².

$$\boxed{P_w = 33.50\ KN/m^2}$$

therefore gauge pressure, P_w = 33.50 KN/m².

18. Given that

density of water (ρ) = 1000 Kg/m³.

dynamic viscosity (μ) = 0.001 N.S/m².

Reynolds no. (Re) = 1000

We know

$$R_e = \frac{\rho v d}{\mu}$$

$$R_e = \frac{1000 \times V \times (0.05 \times 2)}{0.001} \quad \text{[Given r = 0.05 m]}$$

$$1000 = \frac{1000 \times V \times 0.1}{0.001}$$

V = 10^{-2} m/s

Now pressure drop in pipe flow

$$\Delta P = \frac{32 \mu V L}{D^2}$$

$$\Delta P = \frac{32 \times 0.001 \times 10^{-2} \times 5}{(0.1)^2} \quad [\because L = 5m]$$

$$\boxed{\Delta P = 0.160\ N/m^2}$$

Materials Science

1 On decreasing the objective aperture size in an optical microscope, the depth of field increases.

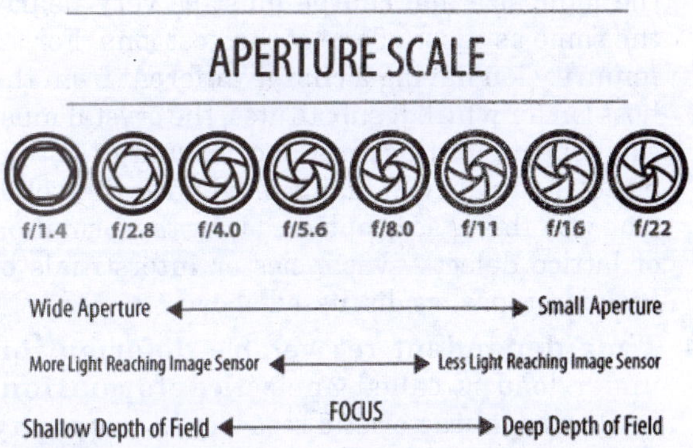

2. The Pilling–Bedworth ratio, in corrosion of metals, is the ratio of the volume of the elementary cell of a metal oxide to the volume of the elementary cell of the corresponding metal. On the basis of the P–B ratio, it can be judged if the metal is likely to passivate in dry air by creation of a protective oxide layer.

3. IMPURITIES IN CERAMICS

Impurity atoms can form solid solutions in ceramic materials much as they do in metals. Solid solutions of both substitutional and interstitial types are possible. For an interstitial, the ionic radius of the impurity must be relatively small in comparison to the anion. Since there are both anions and cations, a substitutional impurity will substitute for the host ion to which it is most similar in an electrical sense: if the impurity atom normally forms a cation in a ceramic material, it most probably will substitute for a host cation.

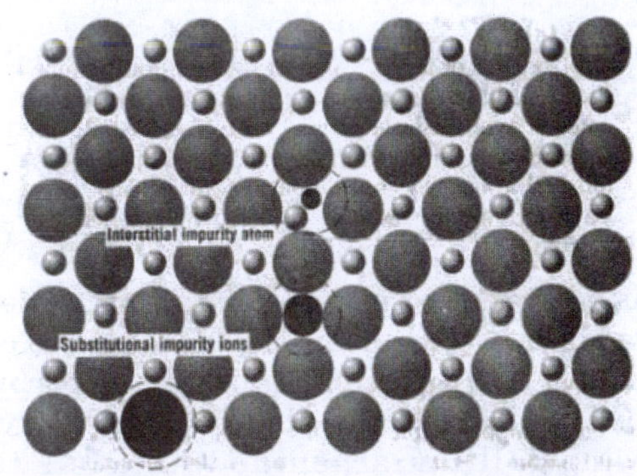

For example, in sodium chloride, impurity Ca^{2+} and O^{2-} ions would most likely substitute for Na^+

and Cl^- ions, respectively. Schematic representations for cation and anion substitutional as well as interstitial impurities are shown in Figure. To achieve any appreciable solid solubility of substitut-ing impurity atoms, the ionic size and charge must be very nearly the same as those of one of the host ions. For an impurity ion having a charge different from the host ion for which it substitutes, the crystal must compensate for this difference in charge so that electroneutrality is maintained with the solid. One way this is accomplished is by the formation of lattice defects—vacancies or interstitials of both ion types, as discussed above.

4. **Time dependent** recoverable **deformation** under load is called anelastic**deformation**, while the characteristic recovery of temporary **deformation** after removal of load as a function of **time** is called **elastic** aftereffect. **Time dependent**i.e. progressive permanent **deformation** under constant load/stress is called creep.

6. **From wt% to at%**

First we must consider weight in terms of grams. To do this, consider a 100 grams sample of the alloy. For example, 100g of Fe-6wt%C contains 6g of Carbon and 94g of Iron.

Atomic weight of Fe is 56

Atomic weight of C is 12

Avogadro's Number is 6.022×10^{23}

Number of atoms in 6 grams of Carbon

= 6 × Avogadro's Constant / atomic weight of C

= $6 \times 6.022 \times 10^{23} / 12$

= 3.011×10^{23}

Similarly, number of atoms in 94 grams of Iron

= 94 × Avogadro's Constant / atomic weight of Fe

= $94 \times 6.022 \times 10^{23} / 56$

= 10.1084×10^{23}

Therefore, at% of Carbon

= C atoms × 100% / (total of C + Fe atoms)

= $3.011 \times 10^{23} \times 100 / (3.011 + 10.1084) \times 10^{23}$

= 23%

So, Fe-6wt%C is equivalent to Fe-23at%C

7. **GaAs advantages**: Some **electronic** properties **of gallium arsenide** are superior **to** those **of silicon**. It **has** a **higher** saturated **electron** velocity **and higher electron mobility**, allowing **gallium arsenide** transistors **to** function at frequencies in excess **of** 250 GHz.**Because of** its wide band gap, pure **GaAs is** highly resistive.

8. Nuclear magnetic resonance spectroscopy, most commonly known as NMR spectroscopy or magnetic resonance spectroscopy, is a spectroscopic technique to observe local magnetic fields around atomic nuclei.

9.
$$E = \frac{hc}{\lambda}$$
$$.67 \times 1.602 \times 10^{-19} = \frac{6.63 \times 10^{-34} \times 3 \times 10^{8}}{\lambda}$$
$$\lambda = 1.85 \times 10^{-6}\, m$$
$$\lambda = 1.85 \mu m$$

10.
$$2210 = \frac{10^5}{x}$$
$$x = 46(approx)$$

11. (P) Solid state sintering- (3) Iron powder products

(Q) Liquid phase sintering - (2) Mixture of Cu and Zn powder products

(R) Spark plasma sintering- (1) Carbon nanotube products

(S) Laser sintering- (4)3D printed products

Solid-state sintering is the bonding and, usually, densification of particles by the application of heat below the melting point of a material.

Liquid phase sintering (LPS) is a process for forming high performance, multiple-*phase* components from powders. It involves *sintering* under conditions where solid grains coexist with a wetting *liquid*.

Spark plasma sintering (SPS), also known as field assisted sintering technique (FAST) or pulsed electric current sintering (PECS), is a sintering technique.

Selective laser sintering is an additive manufacturing technique that uses a laser as the power source to sinter powdered material, aiming the laser automatically at points in space defined by a 3D model, binding the material together to create a solid structure.

12. Increase in the solubility of impurities in solid as compared to that in liquid phase will not favour the separation of impurities in zone refining process

13 100.0 to 102.0

Given data are:

Wavelength (ë) of X-rays = 0.154 nm

Lattice constant (a) = 0.245 nm

Plane (hkl) = (121)

Order of diffraction = 1

Glancing angle $\theta = ?$

Bragg's law is $2d \sin \theta = n\lambda$

$$d = \frac{a}{\sqrt{b^2 + k^2 + l^2}}, \text{ because rock salt is FCC}$$

$$= \frac{0.245 \times 10^{-9}}{\sqrt{1^2 + 2^2 + 1^2}} m = \frac{0.245 \times 10^{-9}}{\sqrt{6}} m = .1 \times 10^{-9} \text{ m}$$

Substitute in Bragg's equation

$$2 \times .1 \times 10^{-9} \sin\theta = 1 \times .154 \times 10^{-9}$$

$$\sin\theta = \frac{77}{100} = .77$$

$$\theta = \sin^{-1}(.77) = 50.35$$

$$2\theta = 100.7°$$

15. $\frac{\partial U(r)}{\partial r} = 0$

$$\frac{3A}{r^4} - \frac{7B}{r^8} = 0$$

$$r^4 = \frac{7B}{3A}$$

$$r^4 = \frac{7 \times 2.1 \times 10^{-22}}{3 \times 6 \times 10^{-20}}$$

$$r = .3nm$$

18. $Zero\ point\ energy = \frac{h^2}{2ma^2}$

$$Zero\ point\ energy = \frac{(6.63 \times 10^{-34})^2}{2 \times 9.11 \times 10^{-31} (.2 \times 10^{-9})^2} J$$

$$Zero\ point\ energy = \frac{(6.63 \times 10^{-34})^2 \times 1.6 \times 10^{-19}}{2 \times 9.11 \times 10^{-31} (.2 \times 10^{-9})^2} eV$$

$$Zero\ point\ energy = 9.6eV$$

19. Well, since an electron is a particle *with mass*, it can be described by the **de Broglie relation**:

$\lambda = h/mv$

where:

- $h = 6.626 \times 10^{-34}$ J.s is Planck's constant.
- Remember that 1 J = 1 kg.m²s.
- me = 9.109×10^{-31}kg is the rest mass of an electron.
- v is its speed.

And since it is also moving with a certain speed, it will have a kinetic energy of:

$K = mv^2/2 = p^2/2m$

where:

- p = mv is the linear momentum of the particle.
- m is the mass of the particle.
- v is the speed of the particle.

And so, we can get the forward momentum in terms of the kinetic energy:

$$p = \sqrt{2mK} = mv$$

Lastly, note that in 1 eV, there is 1.602×10^{-19}J, i.e. one needs 1 eV to push one electrons' worth of charge through a potential difference of 1V in an electric field.

Therefore, the de Broglie wavelength of the electron is:

$$\lambda = \frac{h}{\sqrt{2m_e K_e}}$$

by Substituting all the values, we get

$\lambda = 2.225 \times 10^{-12}$ m

21. $Strain = \frac{5}{640}$

$$Strain = \frac{1}{128}$$

$$\sigma = E\varepsilon$$

$$\sigma = 72 \times 10^9 \times \frac{1}{128}$$

$$\sigma = 562.5 \times 10^6\ Pa$$

$$\sigma = 562.5\ MPa$$

22. Intensity of magnetizing field -

$$H = \frac{B}{\mu}$$

- where in

H is coercivity

B is magnetic field

Solid Mechanics

1. For first pinned-pinned slender column,

Length = L

Moment of Inertia = I

Critical buckling load = P_{Cr}

For second column,

Length = 2 L

Moment of Inertia = 4 I

Critical buckling load = ?

Now,

$$P_{Cr_1} = P_{Cr} = \frac{\pi^2 EI_{min}}{(L)^2} \quad [L_e = \text{Effective length} = L]$$

$$P_{Cr_2} = \frac{\pi^2 E(4I_{min})}{(2L)^2}$$

Now,

$$\frac{N_{Cr_2}}{P_{Cr_1}} = \frac{\pi^2 E(4I_{min})}{4L^2} \times \frac{L^2}{\pi^2 EI_{min}} = 1$$

$$\Rightarrow \frac{P_{Cr_2}}{P_{Cr}} = 1 \Rightarrow P_{Cr_2} = P_{Cr}$$

Hence, option (c) is correct.

2. For rod P,

Length = 2 L

Diametre = 2 D

For rod Q,

Length = L

Diametre = D

Now,

Strain Energy (SE) for Q = SE_Q

$$= \frac{1}{2} \times P \times \delta = \frac{1}{2} \times P \times \frac{PL}{AE}$$

$$= \frac{P^2 L}{2AE}$$

$$= \frac{2P^2 L}{\pi D^2 E} \quad \left[\because \text{Area} = \frac{\pi}{4} D^2\right]$$

Similarly,

Strain Energy (SE) for P = SE_P

$$= \frac{2P^2 (2L)}{\pi (2D)^2 E} = \frac{P^2 L}{\pi D^2 E}$$

Now, Ratio of elastic strain energy stored in rod P to that stored in rod Q is:

$$\frac{SE_P}{SE_Q} = \frac{P^2 L}{\pi D^2 E} \times \frac{\pi D^2 E}{2P^2 L} = \frac{1}{2} = 1:2$$

Hence, option 'D' is correct.

4. Linear momentum of the two sphere system is conserved along the normal to the plane of impact, if two spheres moving in horizontal plane collids obliquely.

5. Discontinuity in the bending moment occurs at the point of application of a couple.

6. If the shaft is fixed at one end and torque applied at the other end, then every cross-section of the shaft is subjected to the same bending moment.

7. When straight beam is subjected to transverse loads, then fluxural stress occurs.

8. If bending moment at a point in a beam changes its sign, then curvature of the beam changes its sign at the point.

9. Given that,

$$\sigma_x = 5\text{mPa}, \ \sigma_y = -1\text{mPa}$$

$$z_{xy} = 4\text{mPa}$$

One of the principal stresses $(\sigma_1) = 7\text{mPa}$

Let, other principal stress = σ_2

We know that, Principal stresses is

$$\sigma_{1,2} = \left(\frac{\sigma_x + \sigma_y}{2}\right) \pm \sqrt{\left(\frac{\sigma_x - \sigma_y}{2}\right)^2 + z_{xy}^2}$$

Also, we know

$$\sigma_x + \sigma_y = \sigma_1 + \sigma_2$$

$$\Rightarrow 5 + (-1) = 7 + \sigma_2$$

$$\Rightarrow \sigma_2 = 4 - 7 = -3$$

$$\therefore \sigma_2 = 3\text{mPa}$$

11. 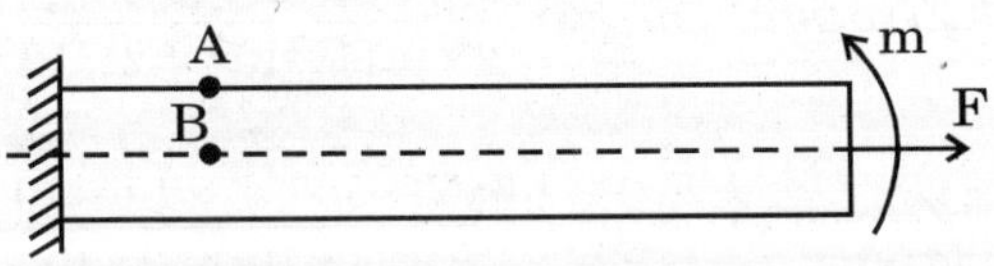

Stress at point B = $\sigma_B = \frac{\text{Force}}{\text{Area}} = 200$ mPa

[Since, point B is located at the neutral axis of the beam, so at point B only axial stress is applicable.]

Now,

Stress at point A = $\sigma_A = \left(\frac{-m}{2} + \frac{F}{A}\right)$

= (–100 + 200) mPa = +100 mPa

So, we can conclude,

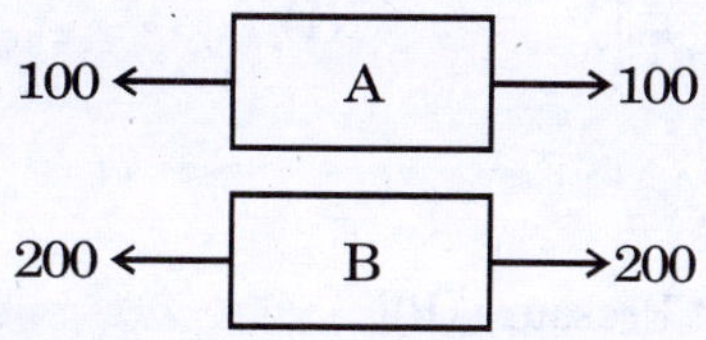

12.

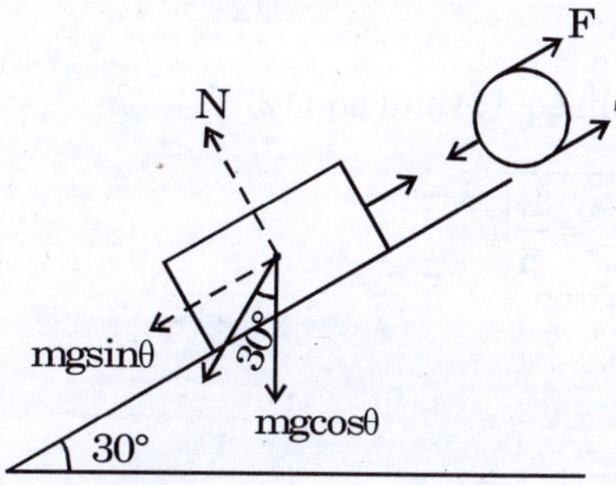

From FBD,

$= (mg\sin\theta + F) = F + T$

$mg\sin\theta + \mu N = F + T$

$(mg\sin\theta + \mu mg\cos\theta) = F + T = 2F$

[Let us assume F = T, for maximum value of F]

$\Rightarrow mg(\sin\theta + \mu\cos\theta) = 2F$

$\Rightarrow 10 \times 10(\sin 30° + 0.65 \times \cos 30°) = 2F$

$\Rightarrow F = 53.14$ N

15.

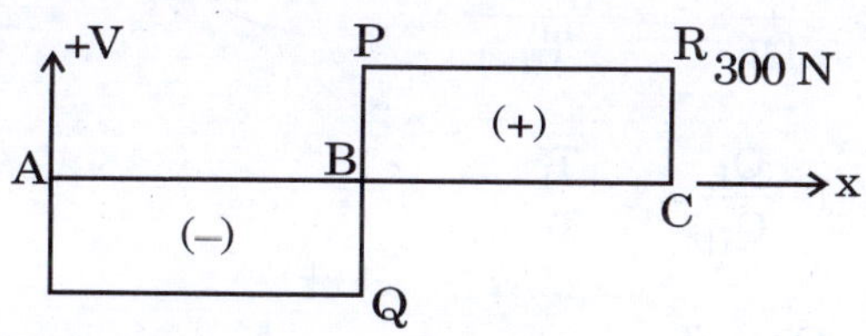

We know that,

$$\text{Concentrate load } (-W) = \frac{d(SF)}{d(x)}$$

By diagram,

Applied force at point B = Shear force at point P – Shear force at point Q

= [300 – (–200)] N

= 500 N

Applied force at point C = Shear force at point R – Shear force at point C

= [300 – 0] N = 300 N

Hence, $F_B = 500$ N and $F_C = 300$ N

16. Elongation of the rod is given by,

(Thermal Elongation + Mechanical Elongation)

$$= \alpha \Delta TL + \frac{PL}{AE}$$

$$= 10^{-5} \times 100 \times 1 + \frac{200 \times 10^3 \times 1}{1000 \times \frac{1}{(1000)^2} \times 200 \times 10^9}$$

$$= 10^{-3} + \frac{1000 \times 1000}{10^9}$$

$$= 10^{-3} + 10^{-3} = (2 \times 10^{-3}) \text{ m}$$

$$= (2 \times 10^{-3} \times 10^{-3}) \text{ mm}$$

$$= 2 \text{ mm}$$

Thermodynamics

1. We know

$$P_{Total} = P_1 + P_2$$

$$= \frac{n_1\bar{R}\uparrow}{V} + \frac{n_2\bar{R}\uparrow}{V}$$

$$[\therefore Pv = n\bar{R}T]$$

$$= \left(\frac{m_1}{\text{Mol.wt.}} \times \frac{\bar{R}T}{V}\right) + \frac{m_2}{\text{Mol.wt.}} \times \frac{\bar{R}T}{V}$$

$$= \left(\frac{1}{1} \times \frac{\bar{R}T}{V}\right) + \left(\frac{4}{2} \times \frac{\bar{R}T}{V}\right)$$

$$P_{Total} = \frac{3\bar{R}T}{V}$$

$$\left[\therefore P_1 = \frac{\bar{R}T}{V}\right]$$

$$\boxed{P_{Total} = 3P_1}$$

2. The property which is independent of mass is called intensive properties.

For example :- Temperature

Pressure

Specific heat

Specific Volume etc.

3.

Case I

Case II

In both cases, we observe that both the temperature and pressure remains unchanged but the volume is changing which shows that the substance undergoes a phase change.

Hence the option (c) is correct.

4. Cycle efficiency improves with reheat, however the cycle efficiency in a single reheat plant is influened by pressure at which steam is reheated. The efficiency increases as the reheat pressure is lowered and reaches peak at a pressure ratio between 0.20 and 0.25.

Reheating steam also increases the net work output of turbine.

6. We know

At triple point all the three phases liquid, gas and solid exist of single component.

Therefore 'degree of freedom (F)

$F = C - P + 2$

$= 1 - 3 + 2$

$\boxed{F = 0}$

where C → No. of component

P → No. of phases

Hence the option (d) is correct.

7. Given that

Gibb's function $dg = vdp - sd\tau$

By using Maxwell relation we get

$$-\left(\frac{\partial s}{\partial p}\right)_T = \left(\frac{\partial v}{\partial T}\right)_P \quad ...(1)$$

$\therefore \; PV = RT$

$P\,dv = R\,dT$

(At constant Pressure (P)]

$$\frac{dv}{dT} = \frac{R}{P} \quad ...(2)$$

Now from eqⁿ (1) and eqn (2)

$$\boxed{-\left(\frac{\partial s}{\partial p}\right)_T = \frac{R}{p}}$$

8. We have

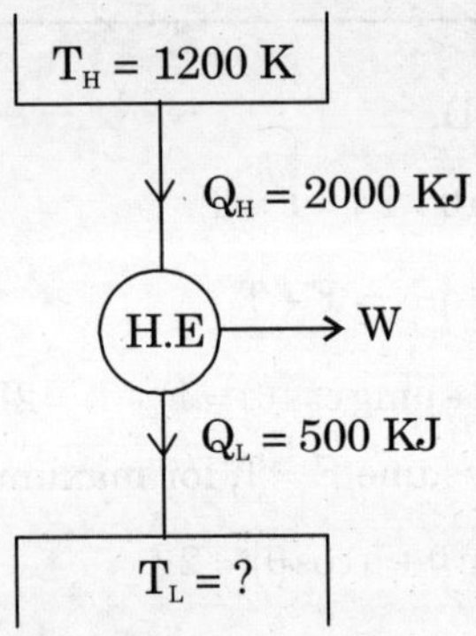

$$\text{FIG}\;\frac{Q_H - Q_L}{Q_H} = \frac{T_H - T_L}{T_H}$$

$$1 - \frac{Q_L}{Q_H} = 1 - \frac{T_L}{T_H}$$

$$\frac{Q_L}{Q_H} = \frac{T_L}{T_H}$$

$$\frac{500}{2000} = \frac{T_L}{1200}$$

$$\therefore \; \boxed{T_L = 300\text{ K}}$$

Hence the lower temperature (or)

Sink temperature is $T_L = 300$ K

9. Given that

Volume of liquid $(V_l) = 0.1\text{ m}^3$

Volume of Vapour. $(V_g) = 0.9\text{ m}^3$

Specific Volume of liquid $(v_l) = 0.000843\text{ m}^3/\text{kg}$

Specific Volume of gas $(v_g) = 0.02671\text{ m}^3/\text{kg}$

We know

$$v_g = \frac{V_g}{m_g}$$

$$\Rightarrow mg = \frac{V_g}{v_g} = \frac{0.9}{0.02871} = 33.69 \text{ kg}$$

Similarly

$$m_l = \frac{V_l}{v_i} = \frac{0.1}{0.000843} = 118.620 \text{ kg}$$

$\therefore$ Total mass of substance $(m_T) = m_g + m_l$

$m_T = 152.31$

$\therefore$ % of Vapour in the Mixture on

mass basis $'x' = \frac{m_g}{m}$

$$x = \frac{33.69}{152.31}$$

$\boxed{x = 22.11\%}$

12. (d) Given that

Initial Volume $(V_1) = 0.1 \text{ m}^3$

Initial Pressure $(P_1) = 1.5$ bar

Initial Temperature $(T_1) = 29$ K

Final Pressure $(P_2) = 10$ bar

Final Temperature (T2) = 423 K

Work given to the system (w) = – 20 KJ

We know

$P_1V_1 = mRT_1$

$1.5 \times 10^5 \times 0.1 = m \times 0.288 \times 10^3 \times 298$

$\therefore \boxed{m = 0.174 \text{ kg}}$

$\therefore$ dQ = du + dw

$= mC_v (T_2 - T_1) + dw$

$dQ = 0.174 \times 0.74 \times 10^3 (423 - 298) - 20$

dQ = – 3.83 KJ

therefore magnitude of heat transfer

$\boxed{Q = 3.83 \text{ KJ}}$

13. (d) We know

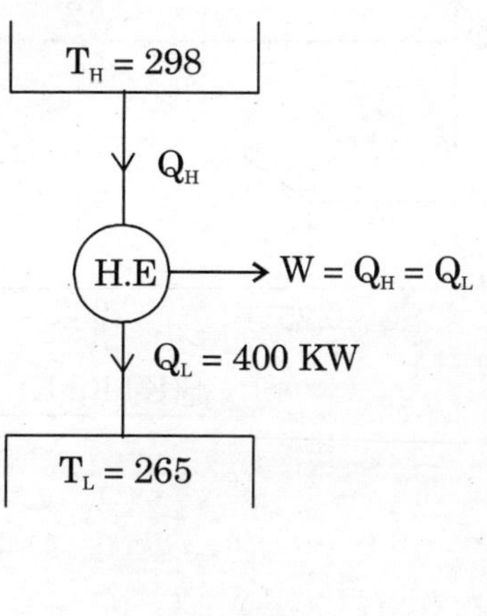

$$\frac{Q_L}{Q_H - Q_L} = \frac{T_L}{T_H - T_L}$$

$$\frac{400}{W} = \frac{265}{298 - 265}$$

$\boxed{W = 49.81 \text{ Kw}}$

Therefore magnitude of the rate of work done

$\boxed{W = 49.81 \text{ Kw}}$

14. (a) Initial pressure $(P_1) = 100$ KPa

Final pressure $(P_2) = 500$ KPa

Initial temperature $(T_1) = 293$ K

Given process is $PV^{1.3}$ = constant

We know

$$\frac{T_2}{T_1} = \left(\frac{P_2}{P_1}\right)^{\frac{\gamma-1}{\gamma}}$$

$$\Rightarrow \frac{T_2}{293} = \left(\frac{500}{100}\right)^{\frac{1.3-1}{1.3}}$$

$\Rightarrow \boxed{T_2 = 424.70 \text{ K}}$

$\therefore$ Entropy change $(ds) = \left(\frac{T_2}{T_1}\right) - R\ell_n\left(\frac{P_2}{P_1}\right)$

$$ds = 1.04 \times 10^3 \ell n\left(\frac{424.70}{293}\right) - 0.297 \times 10^3 \ell n\left(\frac{500}{100}\right)$$

ds = 386.05 – 478.0

$\boxed{\cdot\ ds = -91 \text{ J/Kg. K}}$

17. (0.3 to 0.3)

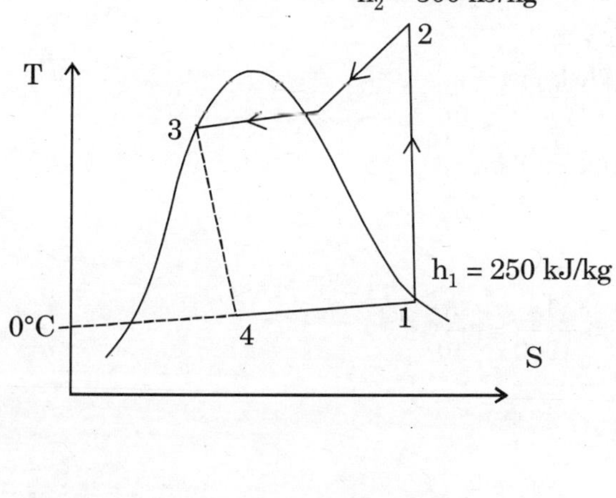

Given that specific enthalpy of saturated liquid at 0°C

$h_f = 50$ kJ/kg

specific enthalpy of saturated vapour at 0°C

$h_g = 250$ kJ/k $= h_1$.

$h_2 = 300$ kJ/kg

COP = 2.8

We know

$$COP = \frac{h_1 - h_4}{h_2 - h_1}$$

$$2.8 = \frac{250 - 4_y}{300 - 250}$$

$$\boxed{h_4 = 110 \text{ kJ}}$$

Now

$$h_4 = h_{f_4} + x_4\left(h_{g_4} - h_{f_4}\right)$$

$$110 = 50 + x_4 (250 - 50)$$

Therefore dryness fraction of the refrigerant entering the evaporator is $\boxed{x_4 = 0.3}$

18. (770 to 785)

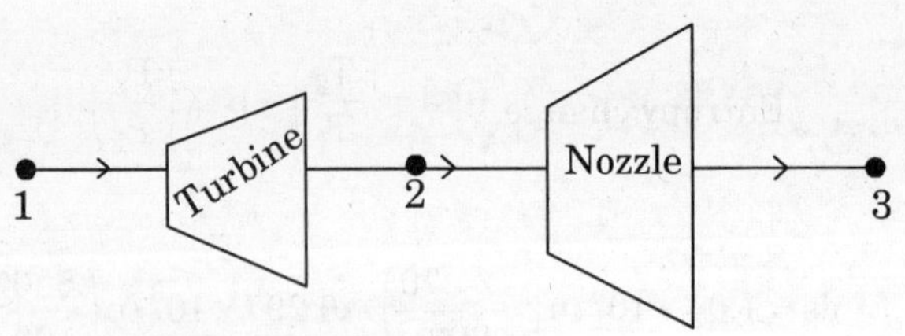

Given that

$P_1 = 10$ bar

$T_1 = 1000$ k

$P_2 = 5$ bar

$P_3 = 1$ bar

$C_p = 1000$ J/kg

r = 1.4

We know

$$\frac{T_2}{T_1} = \left(\frac{P_2}{P_1}\right)^{\frac{\gamma-1}{\gamma}}$$

$$\Rightarrow \frac{T_2}{1000} = \left(\frac{5}{10}\right)^{\frac{1.4-1}{1.4}}$$

$$\boxed{T_2 = 820.33 \text{ k}}$$

Similarly for process 2-3

$$\frac{T_3}{T_2} = \left(\frac{P_3}{P_1}\right)^{\frac{\gamma-1}{\gamma}}$$

$$T_3 = 820.33\left(\frac{1}{5}\right)^{\frac{1.4-1}{1.4}}$$

$$\boxed{T_3 = 517.94 \text{ k}}$$

By Applying energy conservation in nozzle

$$h_2 + \frac{(C_2)^2}{2} + Q = H_3 + \frac{(C_3)^2}{2} + W$$

for adiabatic process Q = 0

No work in the nozzle W = 0

$$h_2 - h_3 = \frac{(C_3)^2 - (C_2)^2}{2}$$

$$C_p(T_2 - T_3) = \frac{(C_3)^2 - 0}{2}$$

Given inlet velocity at nozzle is negligible

$$1000(820.33 - 517.94) = \frac{C_3^2}{2}$$

$$(C_3)^2 = 604765.06$$

$$\therefore \boxed{C_3 = 777.66 \text{ m/s}}$$

therefore exit velocity of nozzle is 777.66m/s.

19. (255 to 255)

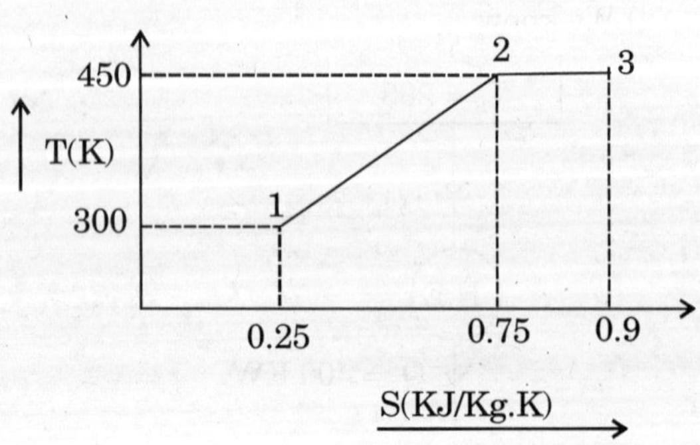

We know that

Area under curve on T-S diagram will provide heat.

Therefore heat transferred during the reversible process 1-3 is

Q_{1-3} = Area under (1 – 2) + Area under (2 – 3)

$$= \frac{1}{2}(0.50 \times 150) + (0.50 \times 300) + 450\,(0.90 - 0.75)$$

$$= 37.5 + 150 + 67.5$$

$$\boxed{Q_{1-3} = 255 \text{ kJ/kg}}$$

21. (1.25 to 1.25)

(A)	(B)
$P_1 = 1$ bar	$P_2 = 2$ bar
$T_1 = 600$k	$T_2 = 400$ k
$m_1 = 1$ kg	$m_2 = 1$ kg

When the piston is removed the final pressure of the mixture will be

$$P_3 = \frac{M_1T_1 + M_2T_2}{\frac{M_1T_1}{P_1} + \frac{M_2T_2}{P_2}}$$

$$= \frac{(1\times 600) + (1 \times 400)}{\left(\frac{600}{1}\right) + \left(\frac{400}{2}\right)}$$

$$P_3 = \frac{1000}{800}$$

$$\boxed{P_3 = 1.25 \text{ bar}}$$

therefore final pressure of the mixture will be $P_3 = 1.25$ bar

Polymer Science and Engineering

1. Ally 1 compounds $CH_2 = CH\text{-}CH_2Y$ with Y = OH, $OCHOCH_3$. can be polymerized free radically only to low degrees of polymerization. Examples arc di- and triallyl ester monomers produced by the reaction of allyl alcohol with acids, acid anhydrides, or acid chlorides . Examples of this are the reaction of phthalic anhydride with allyl alcohol to diallyl phthalate, and the conversion of trichloro-s-triazine (by trimerization of C1CN) to triallyl cyanuratc.

   The monomers are polymerized free radically up to yields of about 25% via the vinyl group give products of 10,000-25,000 g/m. Then, the prepolymers are cross-linked or cured. Polydiethylene glycol bisallyl carbonate is used for sunglass lenses, and as molding resin for related optical articles (the transparency is similar to that of polymethyl methacrylate, but the abrasion resistance is 30-40 times greater). The cured resins have an electrical resistivity between polytetra- fluoroethylene and porcelain, which makes them useful for electrical insulation.

2. A polymerization in which a cyclic monomer yields a monomeric unit which is acyclic or contains fewer cycles than the monomer.

   Note : If the monomer is polycyclic, the opening of a single ring is sufficient to classify the reaction as ring-opening polymerization.

   Modified from the earlier definition.

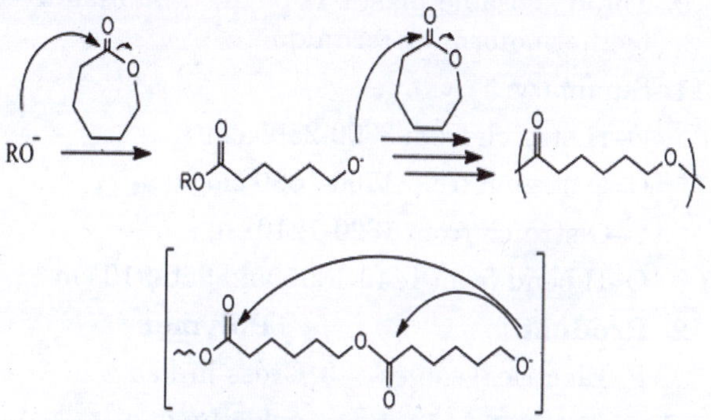

3. Silicone elastomers are ideal for high and low temperature applications. They have high heat and oxidative stability, outstanding low temperature flexibility, and are resistant to many chemicals, weathering, ozone and sunlight (UV). VMQ rubbers are extremely heat resistant.

4. If a combination of sodium and naphthalene is used to initiate polymerization of styrene, then the most likely mechanism of polymerization is anionic.

5. In polymer science, Hypalon is a trademark for chlorosulfonated polyethylene (CSPE) synthetic rubber (CSM) noted for its resistance to chemicals, temperature extremes, and ultraviolet light. It was a product of DuPont Performance Elastomers, a subsidiary of DuPont.

6. Linear low-density polyethylene (**LLDPE**) is a substantially linear polymer (polyethylene), with significant numbers of short branches, commonly made by copolymerization of ethylene with longer-chain olefins.

7. Polyisobutylene lacks the ability to exhibit tacticity.

8.

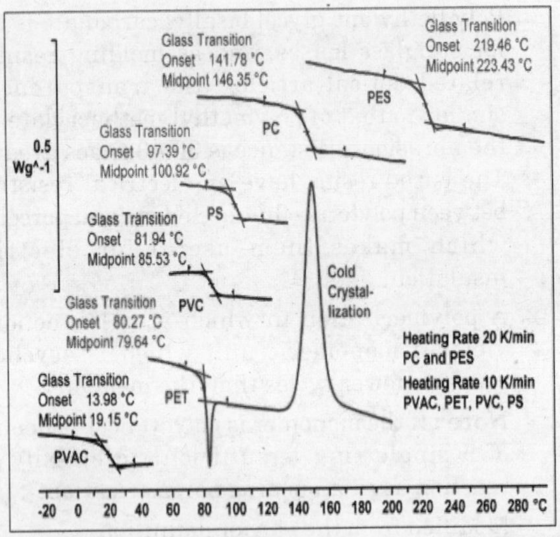

9. The disposable plastic cups are manufactured by thermoforming technique.

11. Summary:

O–H stretch from 3300-2500 cm^{-1}

C=O stretch from 1760-1690 cm^{-1}

C–O stretch from 1320-1210 cm^{-1}

O–H bend from 1440-1395 and 950-910 cm^{-1}

12.

Product	**Polymer**
P. Electrical cables	3. Cross-linked polyethylene
Q. Electrical switches	4. Phenol formaldehyde resin
R. Optical lenses	2. Poly(methyl methacrylate)
S. Shoe soles	1. Polyurethane

Cross-linked polyethylene, commonly abbreviated PEX, XPE or XLPE, is a form of polyethylene with cross-links. It is used predominantly in building services pipework systems, hydronic radiant heating and cooling systems, domestic water piping, and insulation for high tension (high voltage) electrical cables.

Then later, Bakelite, which was an early form brittle plastic, made from formaldehyde and phenol. Baekeland developed the plastic by combining phenolic resin, formaldehyde, and heat. It is a polymer which becomes irreversibly hardened upon being cured. A cured thermosetting polymer is called a thermoset.

Poly(methyl methacrylate) (PMMA), also known as acrylic, acrylic glass, or plexiglass as well as by the trade names Crylux, Plexiglas, Acrylite, Lucite, and Perspex among several others (see below), is a transparent thermoplastic often used in sheet form as a lightweight or shatter-resistant alternative to glass.

Polyurethane (PU) is one of the most common polymers used in shoes, and for good reason - it's lightweight, durable, and waterproof. PU is also remarkably oil-resistant, which makes it a great ingredient in boots designed for construction or industrial use. Thermoplastic polyurethane (TPU) is a subset of PU.

13.

Polymer	LOI
Polypropylene	18
Poly (butylene terephthalate)	20
Poly (ethylene terephthalate)	21
Nylon-6,6	24
Nylon-6	21
Cotton	16
Polyester fabric	21
Wool	24
Polyacrylonitrile	18
Polyaramid	38

14.

Additive	**Function**
P. Molybdenum disulfide	4. Solid layer lubricant
Q. Glycerol monostearate	3. Antistatic agent
R. Tribasic lead sulphate	1. Heat stabilizer
S. 2-Hydroxybenzophenone	2. UV-absorber

15. Polyphenylene oxide is completely miscible with polystyrene in all proportions. The glass transition temperature of these blends change smoothly with composition. These blends with 25-60% (weight) of PPO, designed for injection or blow moulding, extrusion calendering, thermoforming and reinforced with 0-30 weight percentage glass fiber. They have good processability, toughness, flame retardance, dimensional stability, resistance to hot water, economy, HD T (95-150°C), low density and moisture absorption.

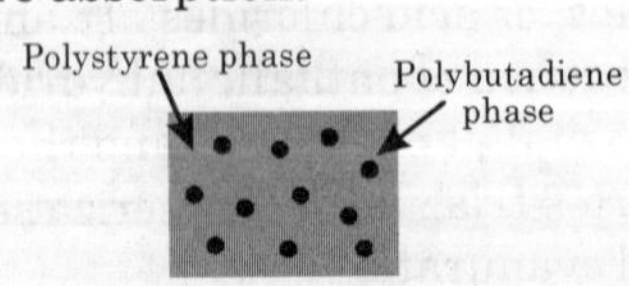

Compatibilization in polymer chemistry is the addition of a substance to an immiscible blend of polymers that will increase their stability.

16. From WLF equation, we have

$$\log\left(\frac{\eta}{\eta_g}\right) = -\left(\frac{17.44(T - T_g)}{51.6 + T - T_g}\right)$$

$$\log\left(\frac{\eta}{\eta_g}\right) = -\left(\frac{17.44(18)}{51.6 + 18}\right)$$

$$\log\left(\frac{4\times10^8}{\eta_g}\right) = -\left(\frac{17.44(18)}{51.6 + 18}\right)$$

$$\eta_g = 1.2954\times10^{13}$$

17. $\varepsilon = \frac{\delta l}{l} = \frac{P}{AE}$

$$\varepsilon = \frac{140}{861\times10^6\times100\times10^{-6}}$$

$$\varepsilon = .001626$$

$$\varepsilon = .1626\,(\text{in}\,\%)$$

18. $\frac{E_{CF}}{E_{CF}} = 35$

$$A_{CF} = 50\%$$

$$F_{CF} = \frac{E_{CF}\times A_{CF}}{\Sigma EA}\times F$$

$$F_{CF} = \frac{E_{CF}\times A_{CF}}{(E_{CF}\times A_{CF} + E_{VE}\times A_{VE})}\times F$$

$$\frac{F_{CF}}{F} = \frac{35E_{VE}}{36E_{VE}}\times100$$

$$\frac{F_{CF}}{F} = 97.22\%$$

19. $\tau = \mu\frac{du}{dy}$

$$100\times10^3 = 3.9\times10^4\times\frac{du}{3\times10^{-3}}$$

$$du = .007692\,\text{m/s}$$

$$du = 7.692\,\text{mm/s}$$

20. The kinetic schemes for free-radical polymerizations in bulk monomer, solution, or suspension (but not in emulsion) are the same, the rate of polymerization (R_p) being given by

$$-d[\text{M}]/dt = R_p = \left(k_p/k_t^{1/2}\right)\left(fk_d[\text{I}]\right)^{1/2}[\text{M}]$$

The above equation above shows that the rate of polymerization depends directly on the monomer concentration and on the square root of the rate of initiation. Thus, doubling the monomer concentration doubles the polymerization rate, while doubling the rate of initiation or initiator concentration increases the polymerization rate only by the factor $\sqrt{2}$.

So, doubling both, we get $2\sqrt{2} = 2.82$

21. $M_{HMD} = 116.21\,\text{g/mol}$

$$M_{AA} = 146.14\,\text{g/mol}$$

$$M_{N66} = \frac{116.21\times49 + 146.14\times50}{99}$$

$$M_{N66} = 131.32\,\text{g/mol}$$

At 99.5% conversion, $M_{N66} = 130.57\,\text{g/mol}$

22. The melt flow rate through the die is 0.71 $cm^3\,s^{-1}$.

Food Technology

1. **κ-casein :** Here the molecules are normally located on the external region of casein micelles. Steric interactions between the highly hydrated carbohydrate residues of K-casein protect milk casein micelles from coagulation and ensure their colloidal stability. After treating dr-casein with the enzyme chymosin, the "carbohydrate" covering is removed allowing milk casein micelles to coagulate. Chymosin induced coagulation of κ-casein is crucial for cheese manufacture.

2. In the production process of rice bran oil (RBO), the bran itself requires stabilization immediately after milling to protect against oil degradation from lipase.

3. In recent years, the stickiness problem of sugar-rich products has been related to their low glass-transition temperatures.

4. The green lea leaves are healed to inaetivale the enzymes and thus preserve the constituents in the dried tea leaves. A typical brewed green tea beverage (e.g., 2.5 g tea leaves in 250 ml of hot water) contains 240 to 320 mg of catechins and 20 to 50 mg of caffeine. The catechins include (-)-cpigallocatcchin 3-gallatc (HGCG), (-)-cpigallocalechin (EGC), (-)-epicatechin-3-gallate (ECG), and (-)-epicatechin (EC), with HGCG as the major catcchin in tea (Figure 11.1). Tea leaves also contain flavonols such as quercetin and myricclin as well as the alkaloids caffeine and theobromine. During the production of black tea, the tea leaves are withered, crushed, and allowed to undergo enzyme-mediated oxidation in a process commonly referred to as "fermentation." During this process, much of the catcchin content of the tea leaves is converted to theaflavins and thearubigins. Theaflavins are produced from the oxidation and dimerization of two catcchin molecules (Figure 11.1). The structures of the thearubigins, polymers formed by catechins, arc poorly characterized. In brewed black tea, catechins, theaflavins, and thearubigins account for 3 to 10%, 2 to 6%, and greater than 20% of the dry weight, respectively. Theaflavins are responsible for the red-brown color and contribute to the taste of black tea.

5. The Lewis number is the ratio of mass diffusion to thermal effects. It can be calculated as the ratio of Prandtl to Schmidt number. It varies slightly with temperature and is greater than one.

6. 'Red dog' is one of the byproducts during milling of Wheat.

Wheat milling by-products according to AAFCO (1996) involve wheat bran, wheat middlings, wheat shorts, wheat red dog, and wheat feed flour, representing almost 25% – 30% of the total. The wheat milling by-products are used as animal feed because they provide a source of energy, amino acids, and phosphorus. The chemical composition of wheat by-products varies due to differences in the variety of the wheat being processed, environmental factors, and differences in the processing.

7. $\text{Percent over run} = \frac{(870-490)}{490} \times 100$

Percent over run = 77.55%

8. $P \propto D^5$

If diameter is doubled, then Power will increase by a factor of 32.

9. Yield (per cent) of polished rice from the paddy

$$= \frac{(100-20)\times(100-6)}{100}\%$$

Yield = 75.2%

10.

P. Newton's law	2. Substance cooling
Q. Hertz constant stress theory	3. Damage of fruits and vegetables
R. Fick's law	4. Molecular diffusion
S. Bond's law	1. Size reduction

Newton's law of cooling states that the rate of heat loss of a body is directly proportional to the difference in the temperatures between the body and its surroundings provided the temperature difference is small and the nature of radiating surface remains same.

Hertz constant stress theory -The classical theory of contact focused primarily on non-adhesive contact where no tension force is allowed to occur within the contact area, i.e., contacting bodies can be separated without adhesion forces.

Fick's laws of diffusion describe diffusion and were derived by Adolf Fick in 1855. They can be used to solve for the diffusion coefficient, D. Fick's first law can be used to derive his second law which in turn is identical to the diffusion equation.

Bond's law : The total work input represented by a given weight of crushed product is inversely proportional to the square root of the diameter of the product particles.

11.

P. Aspergillus	3. Conidia
Q. Geotriehum	1. Arthrospore
R. Rhizopus	4. Sporangiospores
S. Oomycetes	2. Oospores

Aspergillus fumigatus is a species of fungus in the genus Aspergillus. A large portion of inhaled conidia are cleared by the mucociliary action of the respiratory epithelium.

Geotrichum has septate mycelia which readily fragment into arthrospores (Figure 5), which are the organism's primary means of reproduction.

Rhizopus species grow as filamentous, branching hyphae that generally lack cross-walls (i.e., they are coenocytic). They reproduce by forming asexual and sexual spores. In asexual reproduction, sporangiospores are produced inside a spherical structure, the sporangium.

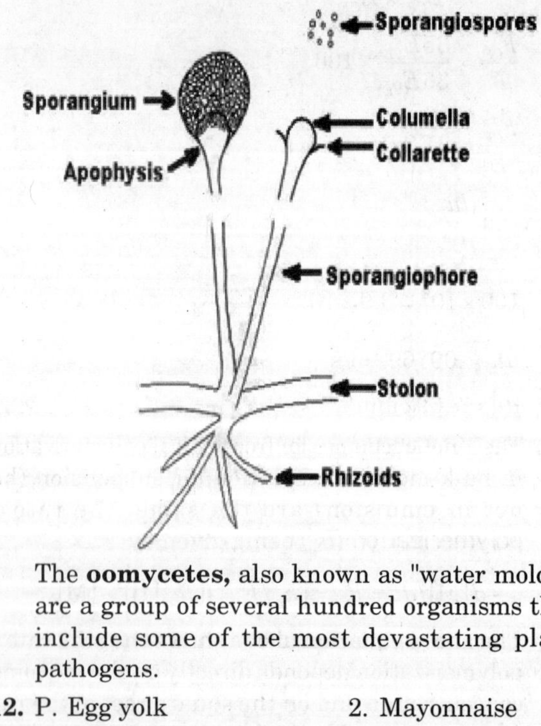

The **oomycetes,** also known as "water molds", are a group of several hundred organisms that include some of the most devastating plant pathogens.

12.

P. Egg yolk	2. Mayonnaise
Q. Pregelatinised starch	4. Baby food
R. Gum	1. Ice cream
S. Starch	3. Baking powder

13. **Isoflavones** are a class of phytoestrogens - plant-derived compounds with estrogenic activity. Soybeans and soy products are the richest sources of **isoflavones** in the human diet.

Resistant starch is starch, including its degradation products, that escapes from digestion in the small intestine of healthy individuals. Resistant starch occurs naturally in foods but is also added to foods by the addition of isolated or manufactured types of resistant starch.

Xanthophylls are yellow pigments that occur widely in nature and form one of two major divisions of the carotenoid group; the other division is formed by the carotenes. The name is from Greek xanthos and phyllon, due to their formation of the yellow band seen in early chromatography of leaf pigments.

Resveratrol is part of a group of compounds called polyphenols. They're thought to act like antioxidants, protecting the body against damage that can put you at higher risk for things like cancer and heart disease.It's in the skin of red grapes, but you can also find it in peanuts and berries.

14. Vibrio cholerae is a Gram-negative, comma-shaped bacterium. The bacterium's natural habitat is brackish or saltwater. Some strains of V. cholerae cause the disease cholera. V. cholerae is a facultative anaerobe and has a flagellum at one cell pole as well as pili.

Shigella is a genus of Gram-negative, facultative aerobic, non-spore-forming, nonmotile, ... Photomicrograph of Shigella sp. in a stool specimen.

Escherichia coli (E. coli) bacteria normally live in the intestines of healthy people and animals. Most varieties of E. coli are harmless or cause relatively brief diarrhea. But a few particularly nasty strains, such as E. coli O157:H7, can cause severe abdominal cramps, bloody diarrhea and vomiting.

Salmonella Typhi (S. Typhi) are bacteria that infect the intestinal tract and the blood. The disease is referred to as **typhoid** fever. **S.** Paratyphi bacteria cause a similar, but milder illness, which comes under the same title.

15. $$\frac{P}{\rho g} = \frac{V^2}{2g}$$

$$V = \sqrt{\frac{2P}{\rho}}$$

$$V = \sqrt{\frac{2 \times 30 \times 10^6}{1030}}$$

$$V = 241.36 \text{ m/s}$$

16. Let the total mass be 1000 kg.

Due to 10% loss, mass will be 900 kg

Out of 900 kg, 108 kg will be solid content (12%) and 792 kg will be water content (88%).

After dehydration 108 kg becomes 94% solid and rest 6 % will be water content.

Rest 6% means 6.89 kg.

Total mass will be 108 + 6.89 = 114.89 kg

Total mass after dehydration will be 11.489 % of initial mass 1000 kg.

17. By applying Energy conservation principle, we get

$$\left(\rho \dot{\upsilon} c_p \Delta T\right)_{\text{Beverage}} = \left(\dot{m} h_{fg}\right)_{\text{steam}}$$

$$1015 \times \frac{1000 \times 10^{-3}}{60} \times 4298 \times 65 = 2257 \times 1000 \times \dot{m}_{\text{steam}}$$

$$\dot{m} = 2.218 \text{ kg/min}$$

18. Relation between Nusselt number, Reynolds number and Prandtl number is given by,

$$Nu = 0.023 Re^{0.8} Pr^{0.4}$$

$$\frac{hD}{k} = 0.023(20000)^{0.8}(0.72)^{0.4}$$

$$\frac{h \times 0.1}{0.6} = 0.023(20000)^{0.8}(0.72)^{0.4}$$

$$h = 340 W/m^2 {}^{\circ}C$$

22. • Plank's equation was derived based on energy balance principle. The general form of plank equation can be written as :

$$t_f - \frac{\rho L_v}{T_F - T_\infty}\left[\frac{P * a}{h_c} + \frac{R * a^2}{k_1}\right]$$

• where TF is the initial freezing point of the product, kI the thermal conductivity of frozen food, hc the convective heat transfer coefficient, T¥ is the ambient temperature, Lv the latent heat of freezing, and r the food density. The P, R values can be given as :

	Infinite slab	Infinite cylinder	Sphere
P	1/2	1/4	1/6
R	1/8	1/16	1/24

$$1 t_f = \frac{\rho L_v}{T_F - T_\infty}\left[\frac{P \times a}{h_c} + \frac{R \times a^2}{k_1}\right]$$

Bu substituting all the values, we get

$$t_f = 1.075 \text{ hours}$$

Atmospheric and oceanic sciences

1. The Madden–Julian Oscillation (MJO) is the greatest source of variability on intra-seasonal time scales throughout the tropics. It consists of alternate large-scale envelopes of active and suppressed convection propagating slowly **eastwards** from the Indian Ocean to the Pacific Ocean along the equator.

2. Potential temperature is very useful. It is a method of comparing the temperature of air at two different heights. And if potential temperature is constant or increasing with height then the atmosphere is stable. The atmosphere is mostly stable for dry air.

3. The Oxygen minimum zone (OMZ), sometimes referred to as the shadow zone, is the zone in which oxygen saturation in seawater in the ocean is at its lowest. This zone occurs at depths of about 200 to 1,500 m, depending on local circumstances. These layers are formed where decomposition of organic matter is dominant.

4. Water level in an estuary rises and falls in response to tidal forces. Assume that the only tidal period in an estuary is 12 hours. The speed of the ebb current in that estuary would be maximum about 3 hours after high tide.

5. Seawater in the world's oceans has a salinity of about 3.5%. This means that every kilogram of seawater has approximately 35 grams of dissolved salts predominantly sodium (Na+) and chloride (Cl") ions.

6. With respect to Indian monsoon trough, relative vorticity is cyclonic at 850 hPa as in northern hemisphere is cyclonic.

7. In general, Zonal component of trade winds are westerly i.e., west to east directional wind flow and meridional component of trade winds flow equatorward. Thus, in case of trade wind inversion, Zonal component of wind is easterly and Meridional component of wind is equatorward.

8. For thermal wind,

   Flow is in geostrophic balance

   Horizontal temperature gradient is non-zero

   Wind speed varies with height

   Thus, for thermal wind, flow is not quasi-geostrophic balance.

9. $\frac{\partial P}{\partial \sigma} = \rho g H$

   $\therefore \partial P = \rho g H \partial \sigma$

   Integrating both sides

   $P = \sigma \rho g H$

   For, $P_A = 1000\ hPa$ *and* $\sigma = 1$

   $\rho g H = \frac{P_A}{\sigma} = \frac{1000}{1} = 1000$

   Now, for $\sigma = 0.5$

   $P_A = \sigma \rho g H = 1000 \times 0.5 = 500\ hPa$

   For, $P_B = 500\ hPa$ *and* $\sigma = 1$

   $\rho g H = \frac{P_B}{\sigma} = \frac{500}{1} = 500$

   for $\sigma = 0.5$

   $P_B = \sigma \rho g H = 500 \times 0.5 = 250\ hPa$

10. At a given temperature and pressure, equilibrium vapor pressure over the surface of a rain droplet compared to that over a flat surface of pure water can be less, equal to or more depending on the size and chemical composition of the cloud condensation nuclei, and droplet radius.

11. *Given,*

    $u = -5x$

    $v = 4y$

    Test for vorticity

    If $\nabla \times V = 0$

    The flow has zero relative vorticity

    $\therefore \frac{\partial v}{\partial x} - \frac{\partial u}{\partial y} = 0$

    Here,

    $\Rightarrow 0 - 0 = 0$

    Thus flow has zero relative vorticity

    Test for incompressibility

    $\nabla . V = 0$

    $\therefore \frac{\partial u}{\partial x} + \frac{\partial v}{\partial y} = 0$

    Here,

    $-5 + 4 = -1 \neq 0$

    Thus, flow is compressible

    Now, According to the question flow is incompressible is not true.

12. Given,

Solar constant, $S_E = 1370\ W/m^2$

Let, the distance between Sun and Earth, R_{ES}

Distance between Mars and Earth, $R_{ME} = 0.52 \times R_{ES}$

Therefore, Distance between Mars and Sun is

$R_{MS} = R_{ES} + 0.52R_{ES} = 1.52R_{ES}$

Now,

As per formula,

$$S = \frac{P_{sun}}{4\pi R^2}$$

where, P_{sun} is the power output of sun

S is the solar constant of the planet

R is the distance *between sun and planet*

Here,

$$\frac{S_M}{S_E} = \frac{P_{sun}}{4\pi(1.52R^2)} \times \frac{4\pi R^2}{P_{sun}}$$

$$\Rightarrow S_M = \frac{S_E}{1.52^2} = \frac{1370}{2.3104} = 592.97\ W/m^2$$

13. For the change in the gas concentration, the gas constant does not change.

Thus the new gas constant will be $287\ J/kg/K$

14. *Given,*

Rate of loosing energy = 10 W/m^2

$\Delta T = 25 - 8 = 17$

$\therefore$ *Total Heat energy* $= \rho C_P \Delta T$

$= 1.15 \times 1005 \times 17 = 19647.75\ W/m^2$

As per question

Probability of fog formation at midnight

$$= \frac{19647.75}{10 \times 20} = 98.2 \approx 99$$

15. *Given,*

Radius, $R = 6400\ km = 6400 \times 10^3\ m$

Total Distance $= 2\pi R = 2\pi \times 6400 \times 10^3$

$= 40.21 \times 10^6\ m$

The model exactly resolve the shortest of those waves = 2 grid

$\therefore$ The grid spacing at the equator

$$= \frac{\textit{Total Distance}}{256 \times 2}$$

$$= \frac{40.21 \times 10^6}{256 \times 2} = 78535.15 = 78.5\ km$$

16. *Given,*

Initial mixed layer depth, $h_1 = 5\ m$

Initial Salinity, $\rho_1 = 30\ g/kg$

The new mixed layer depth, $h_2 = 5 + 0.1 = 5.1\ m$

As per question

$\rho_1 V_1 = \rho_2 V_2$

$$\rho_2 = \frac{\rho_1 V_1}{V_2} = \frac{30 \times 5}{5.1} = 29.4\ g/kg$$

17. *Given,*

Initial pressure, $P_1 = 1010\ hPa$

Final pressure, $P_2 = 800\ hPa$

Initial Temperature, $T_1 = 300\ K$

$A = 2.53 \times 10^9\ hPa$

$B = 5420\ K$

$C_P = 1005\ J/kg/K$

$R = 287\ J/kg/K$

Initial Partial pressure, $P_{a1} = 10\ hPa$

$$\gamma = \frac{C_P}{C_P - R} = \frac{1005}{1005 - 287} = 1.4$$

As the process is adiabatic in nature

$P^{1-\gamma}T^{\gamma}$ = Constant

Therefore,

$$P_1^{1-\gamma}T_1^{\gamma} = P_2^{1-\gamma}T_2^{\gamma}$$

$$T_2 = \sqrt[\gamma]{\frac{P_1^{1-\gamma}T_1^{\gamma}}{P_2^{1-\gamma}}} = \sqrt[1.4]{\frac{1010^{0.4}300^{1.4}}{800^{0.4}}} = 320.65\ K$$

Now as per given formula

$$e_s(T) = Ae^{-\frac{B}{T}}$$

$$e_s(T_1) = 2.53 \times 10^9 \times e^{-\frac{5420}{300}} = 36.04\ hPa$$

$$e_s(T_2) = 2.53 \times 10^9 \times e^{-\frac{5420}{320.65}} = 111.49\ hPa$$

$\therefore$ The partial pressure of the water vapor at the new location *is*

$$P_{a2} = \frac{e_s(T_2)}{e_s(T_1)} - P_{a1} = \frac{111.49}{36.04} - 10 = 3 - 10 = -7$$

18. Given,

Global mean Precipitation = Global mean evaporation

P = E

And, $P = E = 2.8\ mm/day$

Latent heat of evaporation of water, $l_v = 2.5 \times 10^6\ J/kg$

Now, according to formula,

Global mean surface latent heat flux, $Q_l = \rho \times l_v \times E$

$$Q_l = \frac{1000 \times 2.8 \times 1 \times 2.5 \times 10^6}{1000 \times 86400} = 81.01\ W/m^2$$

19. *Given,*

Column depth = 0 *to* 1000 *m*

$\phi = 30°N$

We know,

$\Omega = 7.29 \times 10^{-5}\ rad/s$

In the absence of friction, the vertical component of relative vorticity of the water in the column at $30°N$.

Absolute vorticity = Relative vorticity + Planetary vorticity

$\varsigma_a = \varsigma + f$

here $\varsigma_a = 0$

$$\varsigma = -f = -2\Omega \sin\phi = -2 \times 7.29 \times 10^{-5} \times \sin 30 = -7.29 \times 10^{-5}$$

20. *Given,*

Wind stress at A, $\tau_A = 0.1\ N/m^2$

Wind stress at B, $\tau_B = 0.2\ N/m^2$

Density of seawater, $\rho_{sea} = 1000\ kg/m^3$

$\phi = 30°\ N$

$f = 2\Omega \sin\phi = 2 \times 7.29 \times 10^{-5} \times \sin 30 = 7.29 \times 10^{-5}$

As per question,

$d\tau = \rho f v_E dz$

Here,

$dz = 1$

$$\therefore v_E = \frac{d\tau}{\rho f dz} = \frac{0.2 - 0.1}{1000 \times 7.29 \times 10^{-5} \times 1} = 1.37$$

21. *Given,*

Water depth at A, $h_A = 2000\ m$

Water depth at B, $h_B = 500\ m$

Amplitude of tsunami at point A = 1 *m*

As per question

$$\frac{a_B}{a_A} = \left(\frac{h_B}{h_A}\right)^{\frac{1}{4}}$$

$$\Rightarrow a_B = 1 \times \left(\frac{2000}{500}\right)^{\frac{1}{4}} = (4)^{\frac{1}{4}} = 1.41\ m$$

22. *Given,*

$\sigma = -\beta k / L^{-2}$

k is zonal wavenumber

σ is frequency

L is Rossby radius

$L = 100\ km = 100 \times 10^3\ m$

$a = 6400\ km = 6400 \times 10^3\ m$

$\phi = 45°\ N$

For, at mid-latitudes, long Rossby

$\sigma = -ck$

where, c is the phase speed

As per question,

$-ck = -\beta k / L^{-2}$

$c = \beta / L^{-2}$

$$c = \frac{2\Omega \cos\phi \times L^2}{a}$$

$$c = \frac{2 \times 7.29 \times 10^{-5} \times \cos(45) \times (100 \times 10^3)^2}{6400 \times 10^3} = 0.16$$

SOLVED PAPER 2020

XE : ENGINEERING SCIENCES

GENERAL APTITUDE

Q 1 – 5 carry one mark each.

1. Rajiv Gandhi Khel Ratna Award was conferred ______ Mary Kom, a six-time world champion in boxing, recently in a ceremony ______ the Rashtrapati Bhawan (the President's official residence) in New Delhi.

(*a*) with, at (*b*) on, in

(*c*) on, at (*d*) to, at

2. Despite a string of poor performances, the chances of K.L. Rahul's selection in the team are______

(*a*) slim (*b*) bright

(*c*) obvious (*d*) uncertain

3. Select the word that fits the analogy:

Cover : Uncover : : Associate : ______

(*a*) Unassociate (*b*) Inassociate

(*c*) Misassociate (*d*) Dissociate

4. Hit by floods, the kharif (summer sown) crops in various parts of the country have been affected. Officials believe that the loss in production of the kharif crops can be recovered in the output of the rabi (winter sown) crops so that the country can achieve its food-grain production target of 291 million tones in the crop year 2019-20 (July-June). They are hopeful that good rains in July-August will help the soil retain moisture for a longer period, helping winter sown crops such as wheat and pulses during the November-February period.

Which of the following statements can be inferred from the given passage?

(*a*) Officials declared that the food-grain production target will be met due to good rains.

(*b*) Officials want the food-grain production target to be met by the November-February period.

(*c*) Officials feel that the food-grain production target cannot be met due to floods.

(*d*) Officials hope that the food-grain production target will be met due to a good rabi produce.

5. The difference between the sum of the first 2n natural numbers and the sum of the first n natural odd numbers is ______

(*a*) $n^2 - n$ (*b*) $n^2 + n$

(*c*) $2n^2 - n$ (*d*) $2n^2 + n$

Q 6 - Q 10 carry two marks each.

6. Repo rate is the rate at which Reserve Bank of India (RBI) lends commercial bank, and reverse repo rate is the rate at which RBI borrows money from commercial banks.

Which of the following statements can be inferred from the above passage?

(*a*) Decrease in repo rate will increase cost of borrowing and decrease lending by commercial banks

(*b*) Increase in repo rate will decrease cost of borrowing and increase lending by commercial banks

(*c*) Increase in repo rate will decrease cost of borrowing and decrease lending by commercial banks

(*d*) Decrease in repo rate will decrease cost of borrowing and increase lending by commercial banks

7. P, Q, R, S, T, U, V and W are seated around a circular table.

(*i*) S is seated opposite to W.

(*ii*) U is seated at the second place to the right of R.

(*iii*) T is seated at the third place of the left of R.

(*iv*) V is a neighbour of S.

Which of the following must be true ?

(*a*) P is neighbour of R.

(*b*) Q is neighbour of R.

(*c*) P is not seated opposite to Q.

(*d*) R is the left neighbour of S.

8. The distance between Delhi and Agra is 233km. A car p started travelling from Delhi to Agra and another car Q started from Agra to Delhi along the same road 1 hour after the car P started. The two cars crossed each other 75 minutes after the car Q started. Both cars were travelling at constant speed. The speed of car P was 10 km/hr more than the speed of car Q. How many kilometers the car Q had travelled when the cars crossed each other?

(*a*) 66.6 (*b*) 75.2

(*c*) 88.2 (*d*) 116.5

9. For a matrix $M = [m_{ij}]$; i, j = 1,2,3,4 the diagonal elements required to fully specify the matrix is

(*a*) 0 (*b*) 6

(*c*) 12 (*d*) 16

10. The profit shares of two companies P and Q are shown in the figure. If the two companies have invested a fixed and equal amount every year, then the ratio of the total revenue of company P to the total revenue of company Q, during 2013 - 2018 is ______.

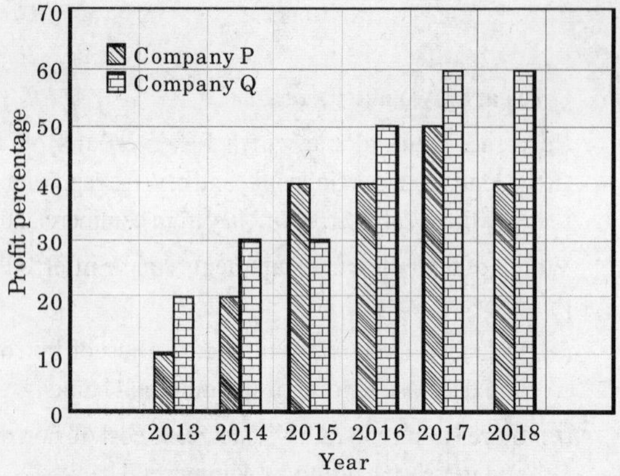

(*a*) 15 : 17 (*b*) 16 : 17

(*c*) 17 : 15 (*d*) 17 : 16

ENGINEERING MATHEMATICS (COMPULSORY)

Q 1 – 7 carry one mark each.

1. Let *A* be a 4 x 3 non - zero matrix and let be a 4 x 1 column vector. Then $Ax = b$

(*a*) a solution for every b.

(*b*) no solution for some *b*.

(*c*) a solution only when b = 0

(*d*) a solution if *b* and the coloumns of *A* from a linearly independent set.

2. Let $x_0, x_1, x_2......$ be the sequence generated by the Newton-Raphson applied to the function $f(x) = x^3 - 2x + 2$ with $x_0 = 1$. Then the sequence

(*a*) converges to 0.

(*b*) becomes unbounded

(*c*) converges to a root of $f(x)$.

(*d*) does not converge.

3. Let $z(t)$ be the solution of the initial value problem

$$\frac{d^2z}{dt^2} = bz\,,\ z(0) = 0,\ \frac{dz}{dt}(0) = 1,\ \text{for } t \geq 0.$$

If the planar curve paramaterized by t having x–coordinate $z(t)$ and y–coordinate $\frac{dz}{dt}$ is closed, necessarily

(*a*) b > 0

(*b*) b < 0

(*c*) b = 0

(*d*) b is non zero rational number.

4. Let z be a complex number. Then the series $\sum_{n=0}^{\infty} \frac{z^{2n}}{(2n)!}$

(*a*) converges for all z.

(*b*) converges for $|z| \leq 1$ and diverges for $|z| > 1$.

(*c*) converges for z = 0 and diverges for any $z \neq 0$.

(*d*) converges for $|z| < 1$ and diverges for $|z| \geq 1$.

5. Let $\vec{V}(x,y,z) = ax\hat{i} - bz\hat{j} + cy\hat{k}$ be a vector field whose curl is zero. Then necessarily

(*a*) a = b = c (*b*) a = – b = c

(*c*) b = c (*d*) b = – c

6. Let f(x) be a continuous function on the real line such that for any x,

$$\int_0^{x^2} f(t)dt = x^2(1 + x^2).$$ Then f(2) is ______.

7. The number of points at which the function f(x, y) = $\frac{x^2}{2} + \frac{y^4}{4} - \frac{y^2}{2}$ has local minima is ______.

Q 8 – 11 carry two marks each.

8. Let $f(t)$ be a real -valued differentiable function on (–1,1) such that $f(0) = 0$ and

$$\left|\frac{df}{dt}\right| < 1 \text{ for } 0 < t < 1.$$

Then the series $\sum_{n=0}^{\infty} f(0.5)^n$

(*a*) converges but not absolutely.

(*b*) is unbounded.

(*c*) converges absolutely.

(*d*) is bounded but does not converge.

9. Let *X* be a random variable with probability density function

$$f(t) = \begin{cases} \exp(-t) & \text{for } t < 0, \\ 0 & \text{for } t \geq 0. \end{cases}$$

Let $0 < a < b$. Then the probability $P(X \leq b \mid X \geq a)$ depends only on

(*a*) b – a. (*b*) b.

(*c*) a. (*d*) a + b.

10. Let A be a 3×3 matrix such that $A^4 = A$. Then it is necessary that

(*a*) A is identity matrix or the zero matrix.

(*b*) the determinant of A^2 is either 0 or 1.

(*c*) the rank of A is 3.

(*d*) A has one imeginary eigenvalue.

11. Players A and B take turns to throw a fair dice with six faces. If A is the first player to throw, then the probability of B being the first one to get a six is ___(round off to two decimal places).

FLUID MECHANICS

Q 1 – 9 carry one mark each.

1. Figure given below show the velocity and shear stress profiles for the flow in a duct. In each option, '1' represents shear stress profile.

Choose the correct option that closely represents the turbulent flow condition.

(*a*)
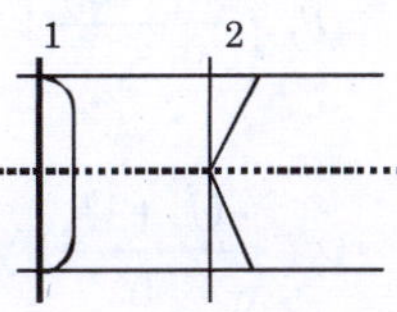

(*b*)
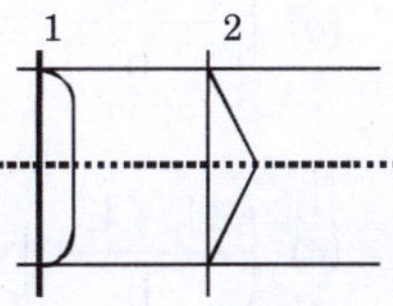

(*c*)
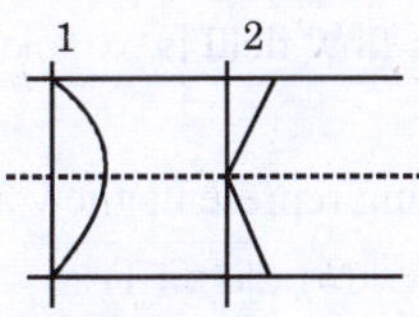

(*d*)
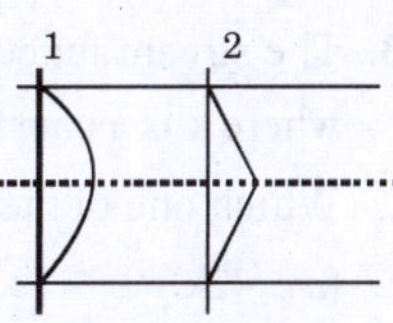

2. The variation of the shear stress (τ) against strain rate (du/dy) is given in the Figure. Identify the line / curve among P, Q, R and S, that represents an ideal fluid.

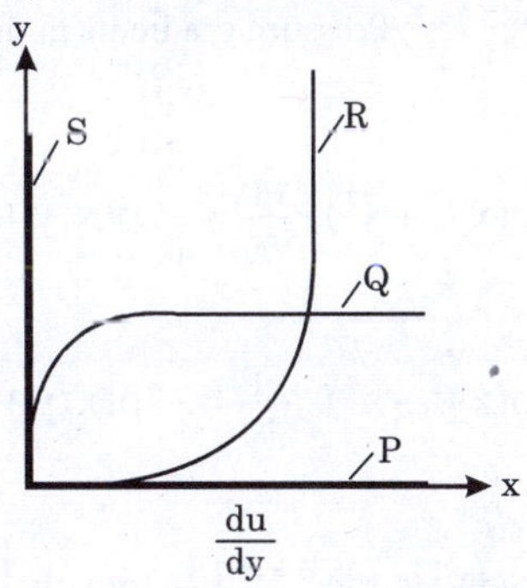

(*a*) S (*b*) P

(*c*) Q (*d*) R

3. A body is under stable equilibrium in a homogeneous fluid, where CG and CB are centre of gravity and centre of buoyancy, respectively.

Two statements 'P' and 'Q' are given below:

P : For a fully submerged condition, CG should always be below CB

Q : For a floating body, CG need not be below CB

Choose the option that is valid for the present situation.

(*a*) **P** is False; **Q** is True when metacentre is below CG

(*b*) **P** is False; **Q** is True when metacentre is above CG

(*c*) **P** is True; **Q** is True when metacentre is below CG

(*d*) **P** is True; **Q** is True when metacentre is above CG

4. A laminar hydrodynamic boundary layer over a smooth flat plate is shown in the Figure. The shear stress at the wall is denoted by τ_w. Which one of the following conditions is correct.

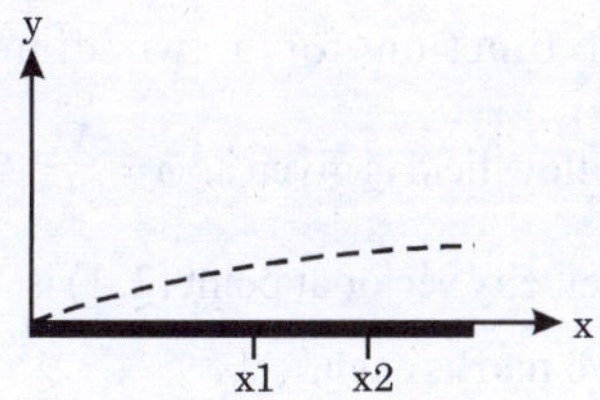

(*a*) Pressure is varying along 'x' and $(\tau_w)_{x1} > (\tau_w)_{x2}$

(*b*) Pressure is constant along 'x' and $(\tau_w)_{x2} > (\tau_w)_{x1}$

(*c*) Pressure is constant along 'x' and $(\tau_w)_{x1} > (\tau_w)_{x2}$

(*d*) Pressure is varying along 'x' and $(\tau_w)_{x2} > (\tau_w)_{x1}$

5. A non-dimensional number known as **Weber** number is used to characterize which one of following flows,

(*a*) Motion of fluid in open channel

(*b*) Motion of fluid droplets

(*c*) Motion of fluid at high velocity

(*d*) Motion of fluid through a pipe

6. A uniform approach flow is subjected to an unsteady and periodic flapping plate as shown in the Figure. Tracer is released to obtain flow visualization lines, which are marked as 'P', 'Q' and 'R'.

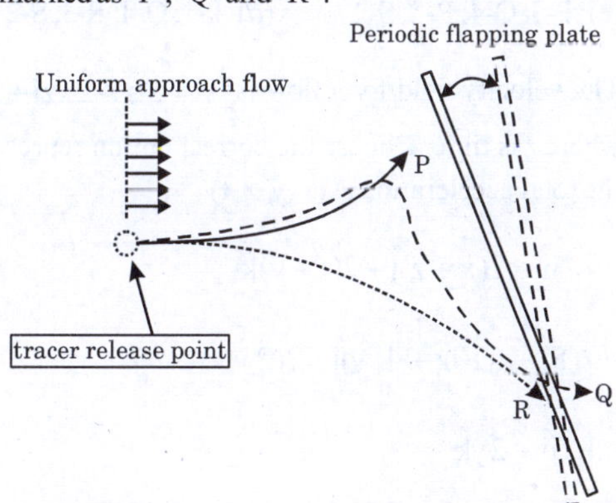

Choose the correct option that line 'R' represents

(*a*) Streakline (*b*) Streamline

(*c*) Pathline (*d*) Timeline

7. The volume flow between any two points not laying on the same steamline in a flow field is equal to

(*a*) Change in strain rate between the points

(*b*) Change in vorticity between the points

(*c*) Change in potential function between the points

(*d*) Change in stream function between the points

8. A liquid flow through a horizontal smooth pipe of diameter 5 cm and discharges into a collection tank of dimension 50 cm *50 cm* 50 cm. Time taken for a 10 cm rise of liquid level in the collection tank is 40s.

The flow velocity in the pipe is ____ m/s (round off to two decimal places).

9. The potential function for a two dimensional incompressible flow field is given as $\varphi = \frac{x^3}{3} - xy^2$.

Magnitude of velocity vector at point (2, 1) is ____ m/s.

Q 10 – Q 22 carry two marks each.

10. Column I represents a list of elementary plane flows and Column II represents flow past geometry obtained by superposition of these elementary plane flows.

Column I	Column II
P. Source, sink, uniform flow	1. Rankine half body
Q. Doublet, Uniform flow	2. Rotating Cylinder
R. Source, Uniform flow	3. Rankine oval
S. Doublet, Free vortex, Uniform flow	4. Cylinder

The correct match betweem columns I and II is,

(*a*) P-3; Q-2; R-1; S-4 (*b*) P-1; Q-2; R-3; S-4

(*c*) P-3; Q-4; R-1; S-2 (*d*) P-1; Q-4; R-3; S-2

11. The velocity field for a flow is $\vec{V} = 5t\hat{i} + 2xz\hat{j} + 2ty\hat{k}$, where *t* is time. Choose the correct option representing the total acceleration at (x, y, z, t)

(*a*) $5\hat{i} + 2(x+z)\hat{j} + 2(t+y)\hat{k}$

(*b*) $5\hat{i} + t(10z + 4xy)\hat{j} + 2(2y + 4xzt)\hat{k}$

(*c*) $5\hat{i} + 2y\hat{k}$

(*d*) $2t(2xy + 5z)\hat{j} + 4xzt\hat{k}$

12. An incompressible viscous fluid is placed between two infinite horizontal parallel plates as shown in figure. The plates move in opposite direction with constant velocities U_1 and U_2. The pressure gradient in the x-direction is zero and the only body force is due to the fluid weight. The flow is steady, laminar and two-dimensional. Assume velocity component in 'y' direction to be zero.

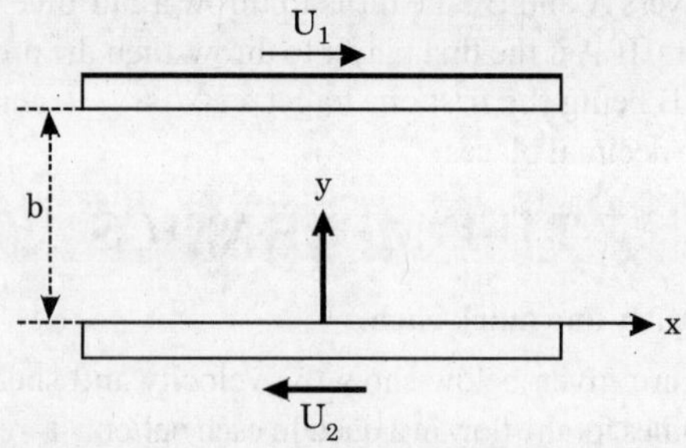

The correct expression for the velocity distribution between the plates is :

(*a*) $\left(\frac{U_1 + U_2}{b}\right)y - U_2$ (*b*) $\left(\frac{U_1 - U_2}{b}\right)y - U_1$

(*c*) $\left(\frac{U_1 + U_2}{b}\right)y + U_2$ (*d*) $\left(\frac{U_1 + U_2}{b}\right)y + U_1$

13. The stream function of a flow field is $\psi = k(x^2 - y^2x)$ where k is a constant.

Which one of the following represents the vorticity?

(*a*) $-2k$ (*b*) $2k(x+1)$

(*c*) $2k(x-1)$ (*d*) $-2k(x+1)$

14. Consider a two dimensional, incompressible steady flow of a Newtonian fluid in which the velocity field is $u = -2xy$, $v = y^2 - x^2$. Pressure gradients in the x-and y-directions are

(*a*) $\frac{\partial p}{\partial x} = -2\rho(xy^2 + x^3)$, $\frac{\partial p}{\partial y} = -2\rho(x^2y + y^3)$

(*b*) $\frac{\partial p}{\partial x} = -2\rho(xy^2 - x^3)$, $\frac{\partial p}{\partial y} = -2\rho(x^2y + y^3)$

(*c*) $\frac{\partial p}{\partial x} = -2\rho(xy^2 + x^3)$, $\frac{\partial p}{\partial y} = -2\rho(x^2y - y^3)$

(*d*) $\frac{\partial p}{\partial x} = -2\rho(xy^2 - x^3)$, $\frac{\partial p}{\partial y} = -2\rho(x^2y - y^3)$

15. A hydroelectric power plant takes in 30 m^3/s of water through its turbine and discharges it to the atmosphere with $V_2 = 2$ m/s. The total head loss in the turbine and penstock system is 20 m. (Assume turbulent flow with kinetic energy correction factor as 1.1. Density of water is 1000 kg/m^3 and acceleration due to gravity, g is 10 m/s^2)

The net head available to the turbine for power generation is_________ m. (rounded off to one decimal place).

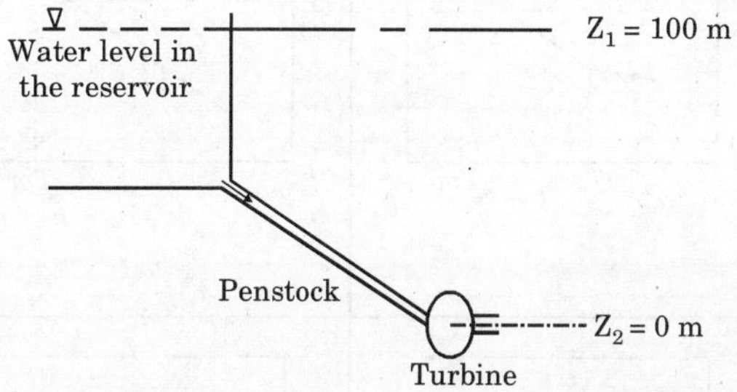

16. Water flows at an average velocity, v of 10 m/s through a horizontal smooth tube of diameter, d 5 cm. The friction factor, f is 0.02. Head loss is obtained using Darcy-Weisbach relation $\frac{flv^2}{2gd}$. The fluid pressure, p measured at various stations are reported in the table below. The length of the pipe, l between station 0 and station 6 is 6m.

Station	0	1	2	3	4	5	6
p, kpa	304	273	255	240	226	213	–

If acceleration due to gravity, g = 10 m/s^2 and density of water = 1000 kg/m^3, then the fluid pressure at station 6 is________ kPa (rounded off to one decimal place).

17. A sphere model of 10 cm diameter is tested in water flowing at 2 m/s. The drag force is measured as 5 N. Prototype of 1.5 m in diameter is tested in air with dynamic similarity conditions. (Density of water is 1000 kg/m^3, density of air is 1.2 kg/m^3, viscosity of water is 0.001 Ns/m^2 and viscosity of air is 1.78×10^{-5} Ns/m^2).

Drag force experienced by the prototype is ______ N (rounded off to two decimal places).

18. A liquid of viscosity 1.74×10^{-3} Ns/m^2 is flowing through a horizontal capillary tube of diameter 0.5 mm. The flow in the tube is steady, incompressible, and fully developed laminar flow. The pressure drop across two locations spaced 1 m apart in the tube is 1.0 MPa.

The flow rate in the tube ________ mm^3/s.

19. A venturimeter with 75 mm diameter throat is placed in a 150 mm diameter pipeline carrying water at 25°C. The pressure drop between the upstream tap and the venturi throat is 40 kPa. (Density of water = 1000 kg/m^3).

The flow rate is ________ m^3/s (rounded off to three decimal places).

20. A water jet with velocity $\vec{V}_{jet}$ impinges normal to a moving flat plate with velocity $\vec{V}_{plate}$ such that the jet splits equally into two halves as shown in the figure. The jet cross-sectional area is 2 cm^2, $\vec{V}_{jet}$ is 20 m/s and $\vec{V}_{plate}$ is 10 m/s and density of water is 1000 kg/m^3. Consider steady flow and neglect weight of the jet, weight of the plate and frictional losses.

The absolute value of the force required to keep the plate moving at constant velocity $\vec{V}_{plate}$ is ________ N.

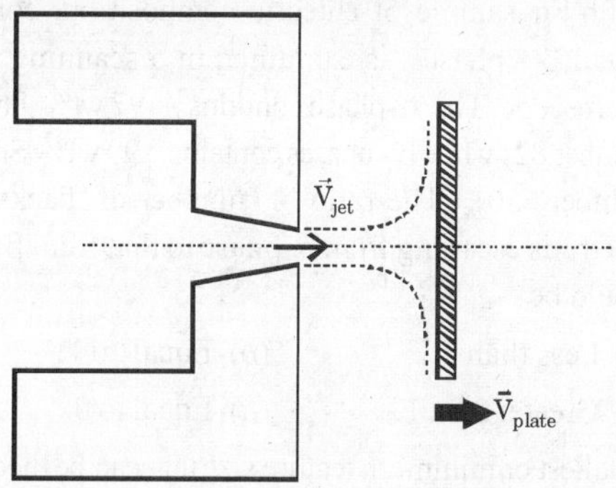

21. In an inverted manometer (as shown in Figure), the pressure difference,

$p_a - p_b$ is 100 kPa.

Use specific gravity of oil as 0.8, density of water as 1000 kg/m^3, density of mercury as 13600 kg/m^3 and acceleration due to gravity as 10m/s^2.

The height of the water column, *H* is_____cm. (rounded off to one decimal place).

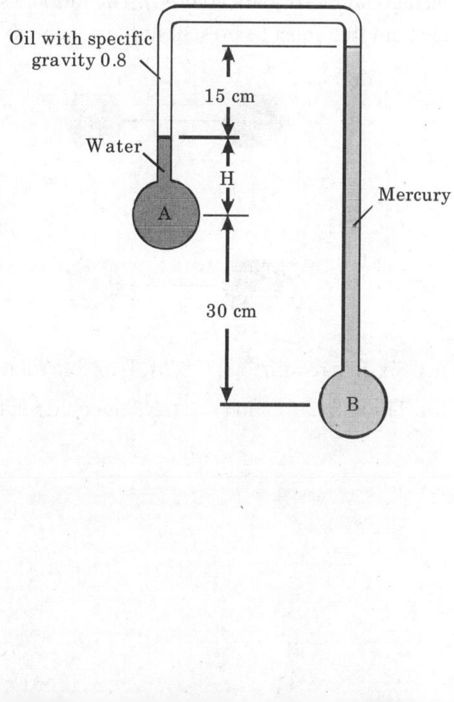

22. An incompressible, steady flow with uniform velocity condition at the inlet between parallel plates is shown in Figure. The flow develops into a parabolic laminar profile with $u = ay\,(y_0 - y)$ at the downstream end, where 'a' is a constant. Assume unit depth of the plate. For $U_0 = 7.5$ cm/s, $y_0 = 3cm$ and the fluid with density, $\rho = 800 kg/m^3$ The value of 'a' is ____.

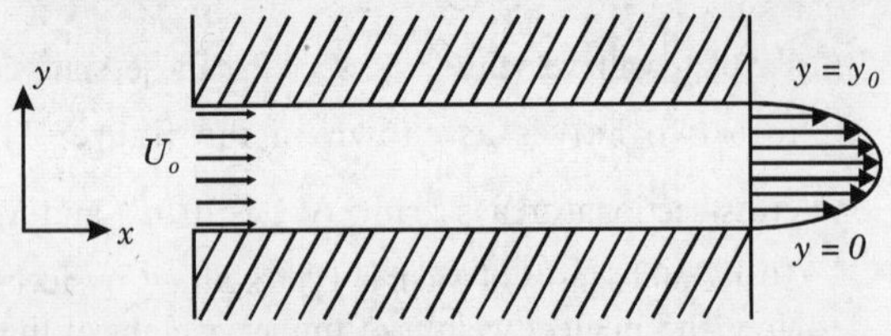

MATERIALS SCIENCE

Q 1 – Q 9 carry one mark each.

1. A Pb-Sn sample of eutectic composition, containing α and β - phases, is examined in a scanning electron microscope. The α-phase contains ~ 97 wt% Pb (atomic number 82) while β - phases contains ~ 99 wt% Sn (atomic number 50). The ratio of number of backscattered electrons escaping from α-phase to that from β - phases would be :

(*a*) Less than 1 (*b*) Equal to 1
(*c*) Greater than 1 (*d*) Equal to 0

2. Smallest or minimum feature size that can be theoretically resolved in an optical microscope does NOT depend on:

(*a*) Refractive index of the medium between the lens and the focal point
(*b*) Intensity of radiation
(*c*) Wavelength of radiation
(*d*) Numerical aperture of the objective lens

3. Following diagram shows a square 2-D lattice with a hexagonal motif (dark colored). The rational symmetry element that must be present in the system is:

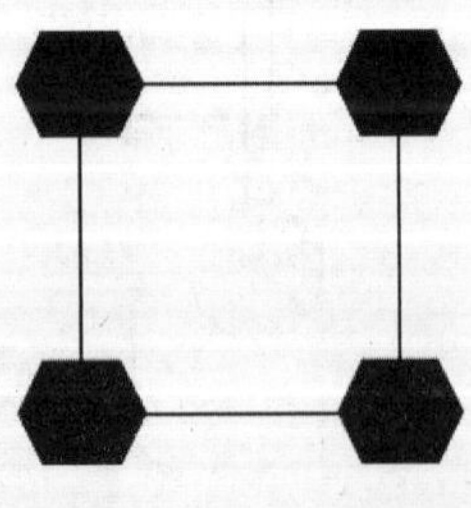

(*a*) Six-fold rotation (*b*) Two-fold rotation
(*c*) Three-fold rotation (*d*) Four-fold rotation

4. Density of states, D(E), in a three dimensional solid varies with energy (E) as

(*a*) $E^{1/2}$ (*b*) E^0
(*c*) $E^{-1/2}$ (*d*) $E^{3/2}$

5. The variation of molar volume (Vm) of a liquid showing glass transition temperature (Tg) while cooling from its melting temprature (Tm) is depicted by:

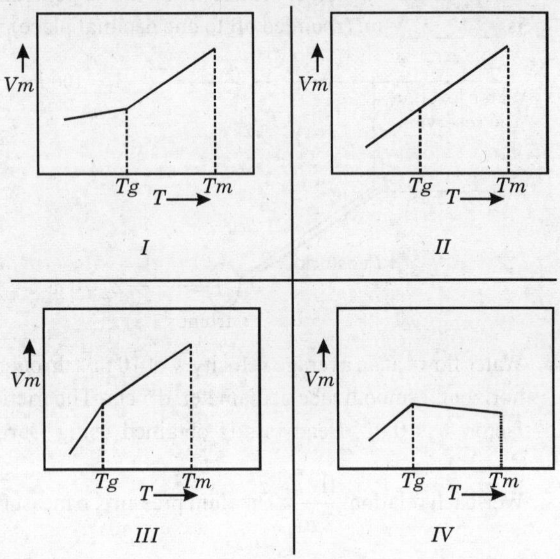

(*a*) I (*b*) II
(*c*) III (*d*) IV

6. Find the correct match between polymer name in column I and the monomer type in column II.

Column I	Column II
I. Polyethylene	(P) 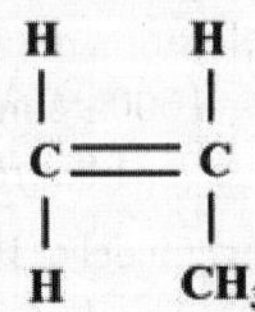
II. Polypropylene	(Q) 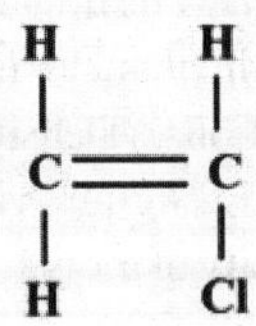
III. Polyvinyl chloride	(R) 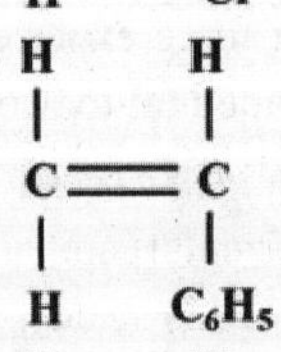
IV. Polystrene	(S) 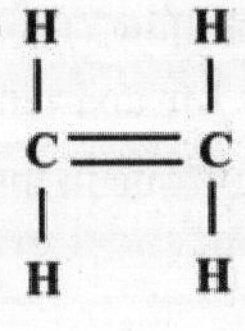

(*a*) I-P, II-S, III-R, IV-Q

(*b*) I-R, II-Q, III-S, IV-P

(*c*) I-S, II-P, III-Q, IV-R

(*d*) I-S, II-R, III-Q, IV-P

7. A ceramic has a fracture toughness (K_{Ic}) of 1 MPa.m$^{1/2}$. If this ceramic is to be exposed to a maximum stress (σ) of 200 MPa, the maximum value of half crack length 'a' (in micrometer, μm), below which the material does not fail, is ________ μm, (round off to one decimal place). Loading condition for the sample is shown in the schematic. Assume geometrical factor f = 1.2.

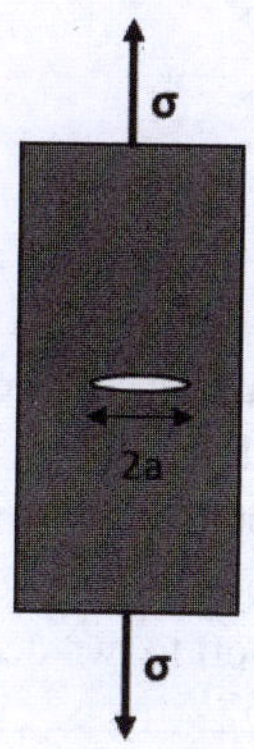

8. A ceramic material is periodically heated and cooled between 25°C and a higher temperature, T_f. During thermal cycling, the material remains dimensionally constrained. The material can withstand a maximum compressive stress of 200 MPa without failure. Material's coefficient of thermal expansion is 7.5×10^{-6} °C^{-1} and modulus of elasticity (E) is 200 GPa. The lowest value of T_f(in °C) at which material will fail is _______ °C (round off to the nearest integer). Assume that there is no plastic deformation during thermal cycling.

9. During homogeneous solidification of a liquid metal, the radius of critical nucleus (in nanometer, nm) at a temperature T_S which is below the melting point (T_m), is _______ nm (round off to one decimal place). Given that γ_{SI} (solid liquid interfacial energy) is 0.18 J.m^{-2} and Δ_{Gv} (change in volume free energy upon transformation from liquid to solid) at T_S is 0.18×10^9 J.m^{-3}.

Q 10 - Q 22 carry two marks each.

10. Read the two statements related to sintering and select the correct option.

Statement-I : Sintering in vacuum leads to improved densification as compared to sintering under ambient (at atmospheric pressure) condition.

Statement-II : Closed pores formed during sintering inhibit full densification.

(*a*) Both Statement-I and Statement-II are FALSE

(*b*) Both Statement-I and Statement-II are TRUE

(*c*) Statement-I is TRUE but Statement-II is FALSE

(*d*) Statement-I is FALSE but Statement-II is TRUE

11. Select the correct option that appropriately matches the process to the material/product that can be fabricated using them.

Process	Material/Product
(I) Powder processing	(P) Organic semiconductor thin films
(II) Spin coating	(Q) Single crystal silicon
(III) Czocharalski process	(R) Poly-silicon
(IV) Chemical vapour deposition	(S) Porous bronze bearings

(*a*) I-S, II-P, III-R, IV-Q (*b*) I-S, II-R, III-Q, IV-P

(*c*) I-S, II-P, III-Q, IV-R (*d*) I-P, II-R, III-Q, IV-S

12. Consider a FCC structured metal with lattice parameter a = 3.5 Å. If the material is irradiated using X-rays of wavelength λ = 1.54056 Å, the Bragg angle (2θ) corresponding to the fourth reflection will be :

(*a*) 88.21° (*b*) 76.99°

(*c*) 99.35° (*d*) 93.80°

13. The number of Schottky defects per mole of KCl at 300°C under equilibrium condition will be :

Given :

Activation energy for the formation of Schottky defect = 250 k.J. mol^{-1}

Avogadro number = 6.023×10^{23} mol^{-1}

Universal Gas Constant = 8.314 J.K^{-1}.mol^{-1}

(*a*) 1.21×10^{18}

(*b*) 1.52×10^{16}

(*c*) 9.75

(*d*) 2.42×10^{12}

14. In an industry, the probability of an accident occuring in a given month is $\frac{1}{100}$. Let P(n) denote the probability that there will be no accident over a period of 'n' months. Assume that the events of individual months are independent of each other. The smallest integer value of 'n' such that P(n) $\leq \frac{1}{2}$ is _______ (round off to the nearest integer).

15. For a FCC metal, the ratio of surface energy of {1 1 1} surface to {100} surface is _______ (round-off to two decimal places). Assume that only the nearest neighbor broken bonds contribute to the surface energy.

16. Pure silicon (Si) has a band gap (E_g) of 1.1 eV. This Si is doped with 1 ppm (parts per million) of phosphorous atoms. Si contains 5×10^{28} atoms per m^3 in pure form. At temperature T = 300 K, the shift in Fermi energy upon doping with respect to intrinsic Fermi level of pure Si will be ______ eV (with appropriate sign and round-off to two decimal places).

Intrinsic carrier concentration of Si, n_i, is given as :

$$n_i = 2\left(\frac{2\pi m k_B T}{h^2}\right)^{3/2} \exp\left(-\frac{E_g}{2k_B T}\right)$$

Given :

Mass of an electron, $m = 9.1 \times 10^{-31}$ kg

Charge of an electron, $e = 1.6 \times 10^{-19}$ C

Boltzmann constant, $k_B = 1.38 \times 10^{-23}$ J.K^{-1}

Planck's constant, $h = 6.6 \times 10^{-34}$ J.s^{-1}

17. The schematic diagram shows the light of intensity I_0 incident on a material (shaded grey) of thickness, x, which has an absorption coefficient, α and reflectance, R. The intensity of transmitted light is I. The reflection of light (of a particular wavelength) occurs at both the surfaces (surfaces indicated in the diagram). The transmittance is estimated to be ______ (round-off to three decimal places).

Given that for the wavelength used, $\alpha = 10^3$ m^{-1} and R = 0.05.

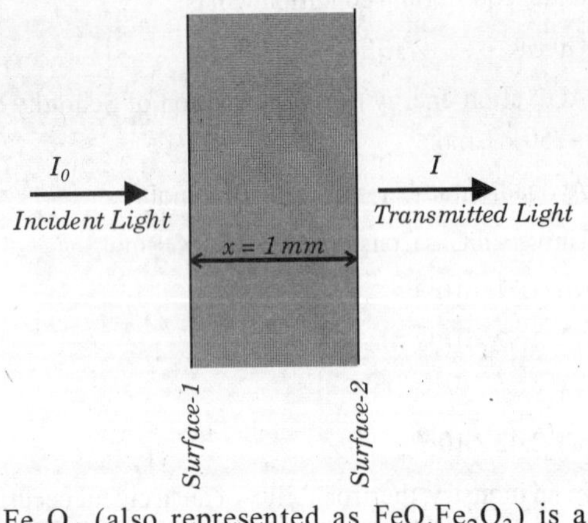

18. Fe_3O_4 (also represented as $FeO.Fe_2O_3$) is a FCC structured inverse spinel (AB_2O_4) material where 1/8 of tetrahedral sites are occupied by half of B cations and 1/2 of the octahedral sites are occupied by remaining B and A cations. The magnetic moments of cations on octahedral sites are antiparallel with respect to those on tetrahedral sites. Atomic number of Fe is 26 and that of O is 8. The saturation magnetic moment of Fe_3O_4 per formula unit in terms of Bohr magnetons (μ_B) will be ______ μ_B. Ignore contribution from orbital magnetic moments.

19. A piezoelectric ceramic with piezoelectric coefficient (d_{zz}) value of 100×10^{-12} C.N^{-1} is subjected to a force, Fz, of 10 N, applied normal to its x-y face, as shown in the figure. If relative dielectric constant (ε_r) of the material is 1100, the voltage developed along the z-direction of the sample will be ______ Volts (round-off to two decimal places). Ignore any nonlinear effects.

Given : Permittivity of free space (ε_o) is 8.85×10^{-12} F.m^{-1}.

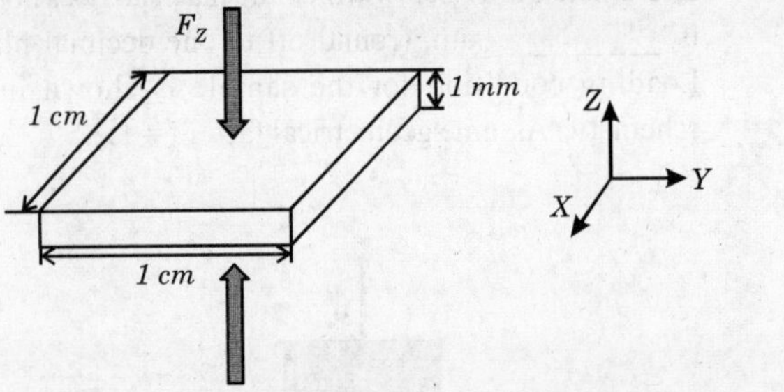

20. Silicon carbide (SiC) particles are added to Aluminum (Al) matrix to fabricate particle reinforced Al-SiC composite. The resulting composite is required to possess specific modulus (E/ρ; E : elastic modulus, ρ : density) three times that of pure Al. Assuming iso-strain condition, the volume fraction of SiC particles in the composite will be ______ (round-off to two decimal places).

Material	E (GPa)	ρ(g.cm^{-3})
Al	69	2.70
SiC	379	2.36

21. Isothermal weight gain per unit area (ΔW/A, where ΔW is the weight gain (in mg) and A is the area (in cm^2) during oxidation of a metal at 600°C follows parabolic rate law, where, ΔW/A = 1.0 mg.cm^{-2} after 100 min of oxidation. The ΔW/A after 500 min at 600°C will be ______ mg.cm^{-2} (round-off to two decimal places).

22. A plain carbon steel sample containing 0.1 wt% carbon is undergoing carburization at 1100°C in a carbon rich surroundings with fixed carbon content of 1.0 wt% all the time. The carburization time necessary to achieve a carbon concentration of 0.46 wt% at a depth of 5 mm at 1100 °C is ______ hour (round off to the nearest integer).

Given : Diffusivity of carbon in iron at 1100 °C is 6.0×10^{-11} m^2.s^{-1} and

erf(z)	z
0.56	0.55
0.60	0.60
0.64	0.65
0.68	0.70

SOLID MECHANICS

Q 1 - Q 9 carry one mark each.

1. Which among the following statements is true for a body moving on a dry surface under the action of applied forces?

 (*a*) Kinetic-friction force is zero.

 (*b*) Kinetic-friction force is equal to the static-friction force.

 (*c*) Kinetic-friction force is greater than the static-friction force.

 (*d*) Kinetic-friction force is lower than the static-friction force.

2. Consider an isotropic material with Young's modulus E and Poisson's ratio *v*. The bulk modulus of this material is given by ________.

 (*a*) $\dfrac{E}{(1-v^2)}$ (*b*) $\dfrac{E}{2(1+v)}$

 (*c*) $\dfrac{E}{3(1-2v)}$ (*d*) $\dfrac{Ev}{(1+v)(1-2v)}$

3. A body subjected to ________ does not undergo change in volume.

 (*a*) uniform tension

 (*b*) pure shear

 (*c*) pure bending

 (*d*) hydrostatic pressure

4. The angular momentum of a particle moving under a central force is

 (*a*) zero

 (*b*) constant in both magnitude and direction.

 (*c*) constant in magnitude but not direction.

 (*d*) constant in direction but not magnitude.

5. According to Euler-Bernoulli beam theory, which one of the following statements best describes the state of a beam subjected to pure bending ?

 (*a*) Transverse shear stress and transverse shear strain are zero.

 (*b*) Transverse shear stress is not zero but transverse shear strain is zero.

 (*c*) Transverse shear stress is zero but transverse shear strain is not zero.

 (*d*) Transverse shear stress and transverse shear strain are not zero.

6. A rigid square ABCD is subjected to planar forces at the corners as shown.

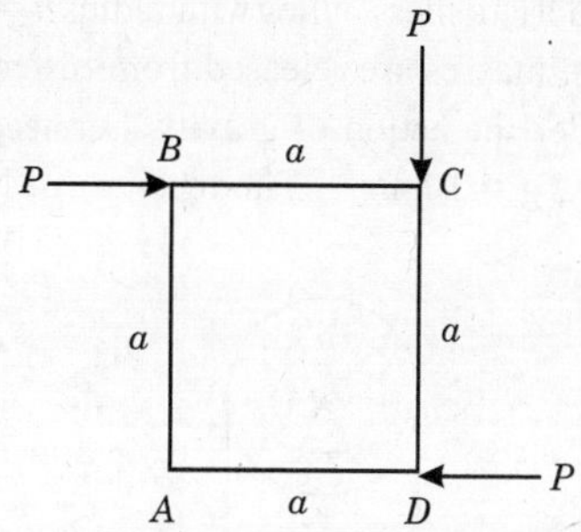

For this planar force system, the equivalent force couple system at the corner A can be represented as

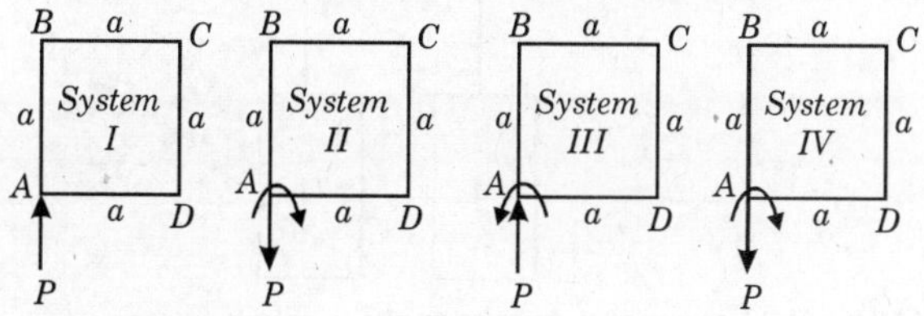

 (*a*) System I

 (*b*) System II

 (*c*) System III

 (*d*) System IV

7. A particle of mass 0.1kg, which is released from rest, falls vertically downward under gravity in a fluid. The fluid offers a resistive force, which is linearly proportional to the particle velocity with 0.1Ns/m as the constant of proportionality. The uniform gravitational acceleration is 10 m/s^2 throughout the trajectory of particle. The magnitude of the particle velocity (in m/s) at time 1s after release (rounded off to two decimal places) is________

8. The state of two-dimensional plane stress at a point in a body is shown on the triangular element ABC, where $\cos\theta = \dfrac{3}{5}$ and $\sin\theta = \dfrac{4}{5}$. The normal stress (in MPa) on the plane AC is ______.

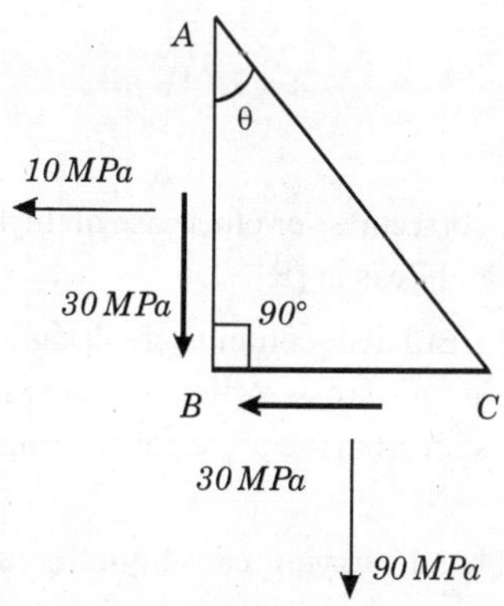

9. Consider two point masses $m = 10$ kg and $M = 30$ kg connected by a massless inextensible string passing over a massless and frictionless pulley with radius $a = 100$ mm as shown. The masses are released from rest and move vertically under the action of gravity. Let acceleration due to gravity, $g = 10 m/s^2$. The tension (in N) in the string is______.

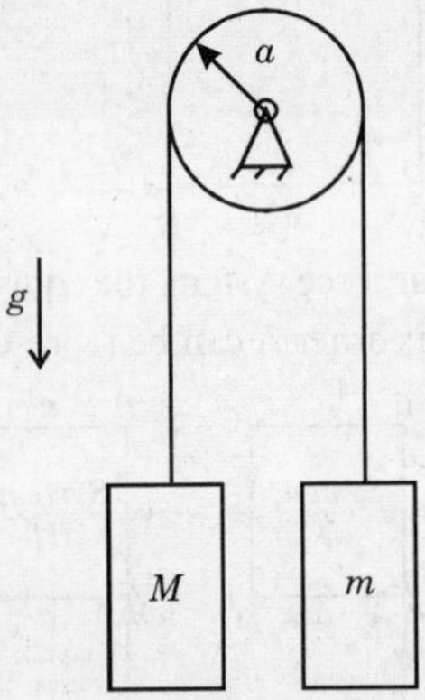

Q 10 - Q 22 carry two marks each.

10. The cantilever beam AC is composed of two segments AB and BC that are rigidly connected at B. The flexural rigidity of the segment AB is EI, whereas, the flexural rigidity of the segment BC is assumed to be infinite. Determine the magnitude of slope at B due to a force P applied at C.

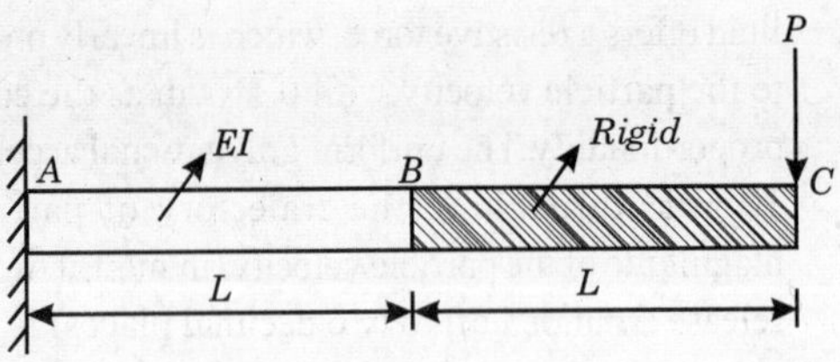

(*a*) $\frac{5PL^2}{6EI}$

(*b*) $\frac{PL^2}{2EI}$

(*c*) $\frac{PL^2}{EI}$

(*d*) $\frac{3PL^2}{2EI}$

11. Determine the correctness or otherwise of the following Assertion [A] and Reason [R].

Assertion [A] : Efficient columns are designed so that most of the column's cross-sectional area is located as far away as possible from the principal centroidal axes of the section.

Reason [R] : Load carrying capacity of columns will increase as the moment of inertia of the cross-section increases.

(*a*) Both (A) and (R) are true and (R) is the correct reason for (A).

(*b*) Both (A) and (R) are true but (R) is not correct reason for (A).

(*c*) Both (A) and (R) are false.

(*d*) (A) is true but (R) is false.

12. Consider the structure consisting of two massless elastic bars AB and BC, each of length L, cross-sectional area A, and Young's modulus E. Connections at A, B, C are all pinned. A horizontal force P acts on the joint B as shown. Calculate the horizontal deflection of the joint B.

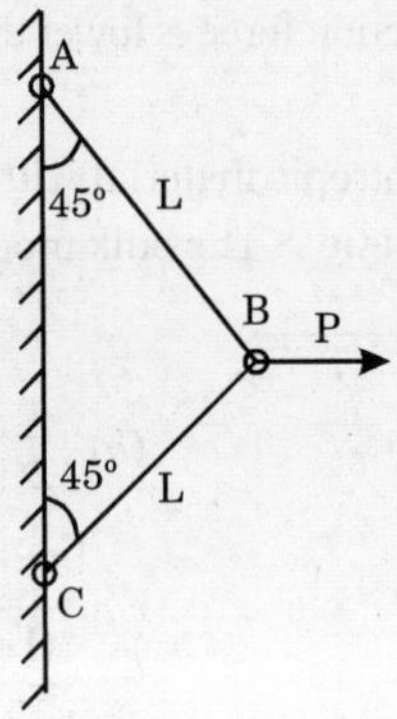

(*a*) $\frac{PL}{2AE}$ (*b*) $\frac{PL}{AE}$

(*c*) $\frac{\sqrt{2}PL}{AE}$ (*d*) $\frac{PL}{\sqrt{2}AE}$

13. A rigid bar ABC of mass m and length L is hinged at A and has a point mass M attached at C. An elastic spring with linear stiffness k is attached at B as shown. Ignore the effect of gravity and damping. The natural frequency of small oscillations of this system is_____.

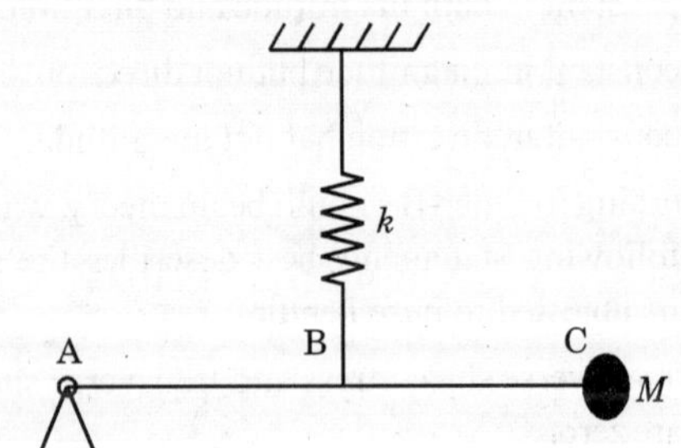

(*a*) $\sqrt{\frac{k}{4M+m}}$ (*b*) $\sqrt{\frac{k}{2(M+2m)}}$

(*c*) $\sqrt{\frac{3k}{3M+m}}$ (*d*) $\sqrt{\frac{3k}{4(3M+m)}}$

14. A beam of flexural rigidity EI is fixed at A and supported by a linear spring of stiffness $k = EI/L^3$ at B. Determine the compressive force developed in the spring, when the beam is subjected to a uniformly distributed load of w per unit length.

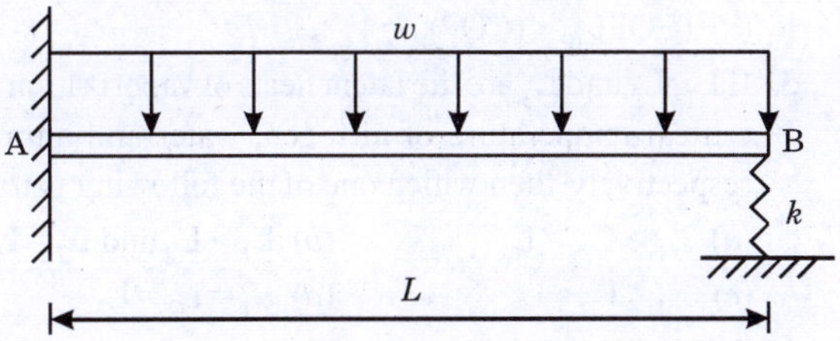

(*a*) $\dfrac{3wL}{32}$ (*b*) $\dfrac{3wL}{16}$

(*c*) $\dfrac{wL}{2}$ (*d*) $\dfrac{wL}{8}$

15. The bar AB is fixed at A and is separated by a gap of 0.005 mm from wall at C as shown. The temperature of the bar is E = 200 GPa and coefficient of thermal expansion is $\alpha = 10 \times 10^{-6}$/°C, then the magnitude of the compressive stress (in MPa) developed in the bar is _____.

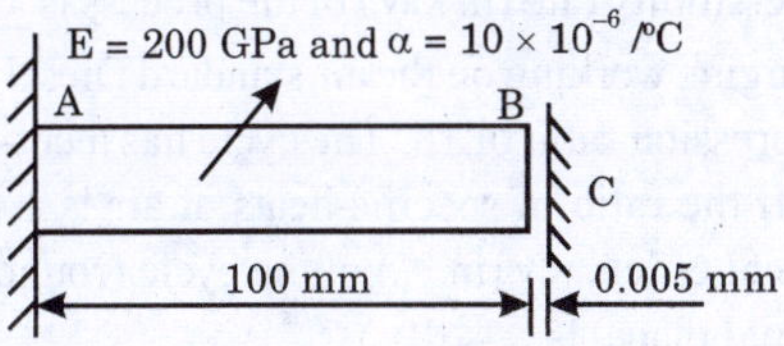

16. A thin walled spherical pressure vessel has mean radius 1000 mm and wall thickness 10 mm. The material has Young's modulus 200 GPa and Poisson's ratio 0.25. If the internal pressure is 100 MPa, the radial displacement (in mm) of the spherical pressure vessels (rounded off to two decimal places) is

17. A particle of mass m = 100 kg is released from rest and falls under gravity through a height of H = 1 m directly onto an upright massless elastic bar of length L = 200 mm, Young's modulus 200 GPa, and cross- sectional area 100 m m^2. Assume the following during impact :

(a) particle mass sticks to the bar,

(b) the bar does not buckle, and

(c) no energy is lost. Use gravitational acceleration $g = 10\ m/s^2$. The maximum axial compression (in mm) of the bar due to the impact (rounded off to three decimal places) is _______.

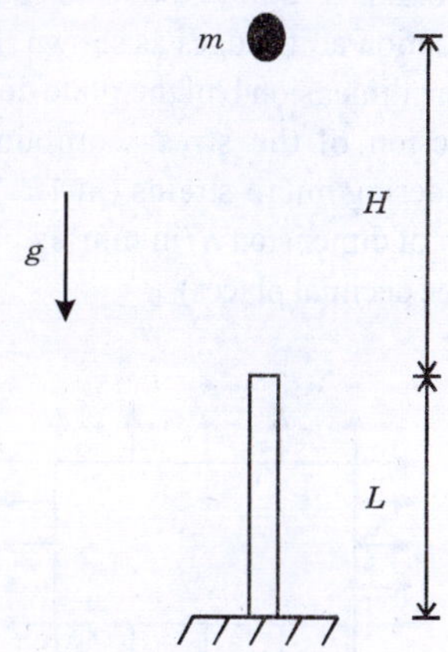

18. A pin-jointed truss has a pin support at A and a roller support at C. All the members are made of same material and have the same cross-section. Neglect the self-weight of the members. Due to the applied loading shown, the total number of zero force members is____.

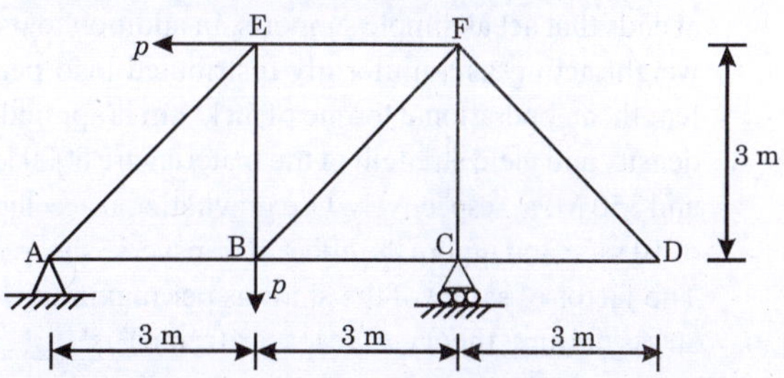

19. Two beams AB and BC having diameter of 100 mm are connected by an internal hinge at B. The structure is fixed at A and roller supported at C. Load of P = 1 kN is applied at B. Ignoring the effect of any transverse shear stress, the tensile stress (in MPa) developed at A due to bending (round off to three decimal places) is _______.

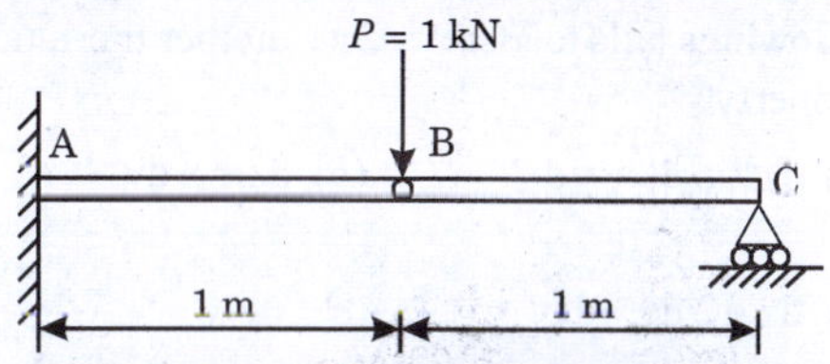

20. The shear force diagram for a beam AD, which is simply supported at A and D, is shown. The magnitude of the maximum bending moment (in kN.m) is ________ (rounded off to three decimal places).

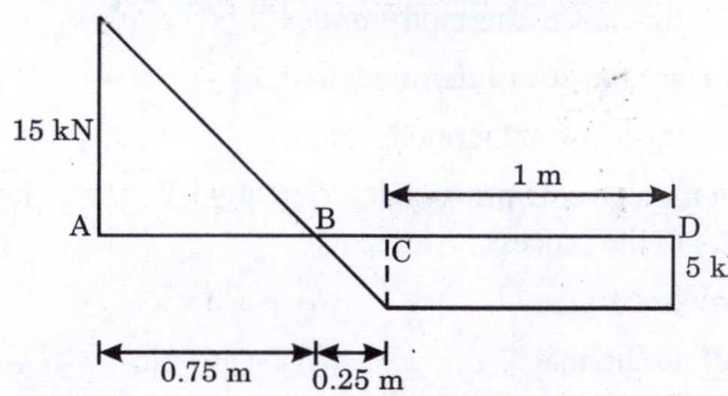

21. A rectangular thin plate with Young's modulus 200 GPa and Poisson's ratio 0.30 is subjected to uniform stress distribution at its edges as shown. However, it is stated that the dimension b of the plate does not change under the action of the stress components σ_{xx} and σ_{yy}. Considering micro-strains (in 10^{-6}), the change in the length of dimension *a* (in mm) is ______ (rounded off to three decimal places).

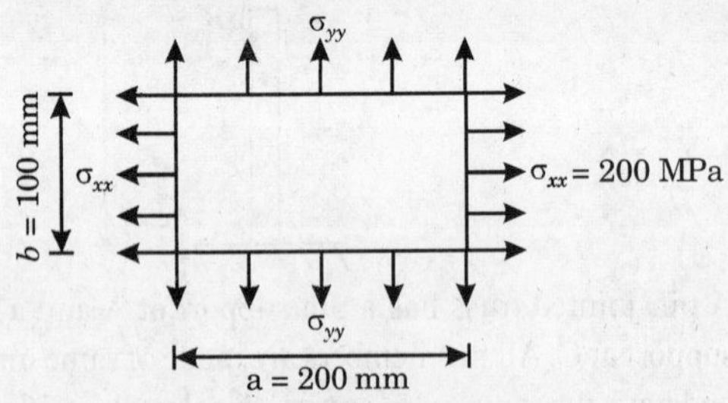

22. A solid transmission shaft has length 10 m and diameter 100 mm. The shaft is supported by frictionless bearings at ends that act as simple supports. In addition to its self-weight acting as a uniformly distributed load per unit length, an operational torque of 5π kN.m is applied. The density and yield strength of the material are 8000 kg/m^3 and 350 MPa, respectively. Use gravitational acceleration of 10 m/s^2 and ignore the effect of transverse shear stress. The factor of safety of the shaft as per maximum shear stress failure theory (Tresca criterion) is ______ (rounded off to two decimal places).

THERMODYNAMICS

Q 1 - Q 9 carry one mark each.

1. If x and y are two independent intensive properties of a thermodynamic system, then which relation among the followings fails to identify z as another thermodynamic property?

(*a*) $dz = x\,dy + y\,dx$ (*b*) $dz = x\,dy - y\,dx$

(*c*) $dz = 2\,dy + dx$ (*d*) $dz = \frac{dy}{x} - \frac{y\,dx}{x^2}$

2. Internal energy of a thermodynamic system is defined by the

(*a*) zeroth law of thermodynamics

(*b*) first law of thermodynamics

(*c*) second law of thermodynamics

(*d*) third law of thermodynamics

3. In a polytropic process described by PV^n = constant, if n = 0, the process is called as

(*a*) isobaric (*b*) isochoric

(*c*) isothermal (*d*) isentropic

4. The relation between the coefficient of performance of a refrigerator $(COP)_R$ and the coefficient of performance of a heat pump $(COP)_{HP}$ is

(*a*) $(COP)_{HP} = (COP)_R + 1$

(*b*) $(COP)_{HP} = (COP)_R - 1$

(*c*) $(COP)_{HP} = 1 - (COP)_R$

(*d*) $(COP)_{HP} \times (COP)_R = 1$

5. If L_1, L_2 and L_3 are the latent heats of vaporization at the critical temperature of nitrogen, water and ammonia, respectively, then which one of the following is true?

(*a*) $L_1 > L_2 > L_3$ (*b*) $L_1 > L_2$ and $L_2 = L_3$

(*c*) $L_1 < L_2 < L_3$ (*d*) $L_1 = L_2 = L_3$

6. A new temperature scale (°N) has been proposed where the normal freezing and normal boiling points of water are marked as 500 °N and 100 °N, respectively. If the temperature of a system is measured to be 0 °N, its temperature according to the Celsius scale (in °C) is ______.

7. Let Z_1 represents the compressibility factor of air at 2 bar and 600 K, and Z_2 represents the compressibility factor of air at 1 bar and 300 K. If air is assumed to be an ideal gas having gas constant of 0.287 kJ/kg.K, then Z_1/Z_2 is ______.

8. The rate of heat received by a heat engine from a source at 900 K is 600 kJ/s. The engine rejects heat to the sink of 300 K. The heat engine produces a power of 200 kW. The irreversibility rate (in kW) of the process is ______.

9. An engine working on the air standard Diesel cycle has a compression ratio of 18. The cycle has a cut-off ratio of 1.7. If the ratio of specific heats of air is 1.4, then the thermal efficiency (in %) of the cycle (rounded off to 1 decimal place) is ______.

Q 10 - Q 22 carry two marks each.

10. A system with rigid walls is initially at a temperature of T_1. It is used as the heat source for a heat engine, which rejects heat to a reservoir maintained at T_0 ($T_0 < T_1$). The specific heats of the system are constant. If the temperature of the system finally reduces to T_0, then the maximum work recoverable from the heat engine per unit mass of the system is

(*a*) $c_v\left[(T_1 - T_0) - T_0 \ln\left(\frac{T_1}{T_0}\right)\right]$

(*b*) $c_v(T_1 - T_0)$

(*c*) $c_v T_0 \ln\left(\frac{T_1}{T_0}\right)$

(*d*) $c_v \frac{T_1^2}{T_0}$

11. A reversible heat engine is operating between two reservoirs maintained at T_1 and T_2, where $T_1 > T_2$. Which one of the following is the most effective option for increasing its thermal efficiency?
(*a*) increasing, T_1, while keeping T_2 constant
(*b*) decreasing T_1, while keeping T_2 constant
(*c*) increasing T_2, while keeping T_1 constant
(*d*) decreasing T_2, while keeping T_1 constant

12. A 4-m^3 reservoir contains 10 kg of a real gas at 200 K. If this gas follows the vander Waal's equation of state with a = 0.0687 m^6.kPa/kg^2, b = 0.00657 m^3/kg and R = 0.187 kJ/kg.K, then the reservoir pressure (in kPa) is
(*a*) 93.5 (*b*) 94.6
(*c*) 95.7 (*d*) 101.3

13. Air at a pressure of 86 kPa and specific volume of 1 m^3/kg is heated at constant pressure till it reaches 627 °C. Air is assumed to be an ideal gas with constant specific heats. It has the gas constant of 0.287 kJ/kg.K and ratio of specific heats of 1.4. The change in specific entropy of air (in kJ/kg.K) during this process will be
(*a*) 1.104 (*b*) 0.740
(*c*) 0.788 (*d*) 0.529

14. An air standard Otto cycle has compression ratio of 4. The compression ratio of this cycle is changed to 6. If the ratio of specific heats is 1.4, the percentage increase in its thermal efficiency will be
(*a*) 20.2 (*b*) 27.2
(*c*) 42.6 (*d*) 51.2

15. In a mixture of gas there are 0.1 kmol of oxygen (O_2), 0.1 kmol of nitrogen (N_2) and 0.8 kmol of methane (CH_4). If the molar mass of O_2, N_2 and and CH_4 are 32 kg/kmol, 28 kg/kmol and 16 kg/kmol, respectively, then the mass fraction of N_2 in the gas mixture is
(*a*) 0.100 (*b*) 0.170
(*c*) 0.148 (*d*) 0.680

16. A particular gas sample is initially maintained at 6000 cm^3 and 100 kPa. It is compressed during a quasistatic process following the relation PV^2 = constant. The compression continues till the volume becomes 2000 cm^3. The magnitude of the corresponding work transfer (in kJ) (rounded off to 2 decimal places) is ________

17. Carbon dioxide (CO_2) enters an adiabatic rigid nozzle steadily at 1 MPa and 500 °C with a mass flow rate of 1.5 kg/s. The inlet area of the nozzle is 40 cm^2 and the exit velocity is 10 times of that at the inlet. If CO_2 can be considered as an ideal gas with gas constant of 0.19 kJ/kg.K and the ratio of specific heats of 1.29, the exit temperature (in K) (rounded off to 1 decimal place) is ________

18. A closed system containing 8 kg of gas undergoes an expansion process following the relation $PV^{1.2}$ = constant. The initial and final pressures are 1 MPa and 5 kPa, respectively, while the initial volume is 1 m^3. If the specific internal energy of the gas decreases by 40 kJ/kg during the process, the heat transfer (in kJ) associated with the process (rounded off to 1 decimal place) is________

19. Saturation pressure of water at 5 °C is 0.8725 kPa. If the latent heat of vaporization is 2489.1 kJ/kg and gas constant is 0.4615 kJ/kg.K, then the saturation pressure at 10°C (in kPa) (rounded off to 2 decimal places) is________

20. The turbine inlet conditions of a Rankine cycle are 10 MPa and 500°C, while the condenser pressure is 10 kPa. The enthalpy and entropy of saturated liquid at 10 kPa are 191.8 kJ/kg and 0.6492 kJ/kg.K, respectively, while the enthalpy and entropy of vapourization at 10 kPa are 2392.1 kJ/kg and 7.4996 kJ/kg.K, respectively. The enthalpy and entropy at the inlet to the turbine are 3375.1 kJ/kg and 6.5995 kJ/kg.K, respectively. The condenser outlet has saturated liquid. Neglecting the pump work, the thermal efficiency (in %) of the cycle (rounded off to 1 decimal place) is________

21. The minimum and maximum temperatures of an air standard Brayton cycle are 300 K and 1100 K, respectively. The pressure ratio of this cycle is 6. The ratio of specific heats is 1.4 and the specific heats are constant. For this cycle, the ratio of network output to the turbine work (rounded off to 2 decimal places) is________

22. The specific humidity of air at 100 kPa is 0.015 kg of vapour per kg of dry air. The partial pressure of vapour (in kPa) in the existing state (rounded off to 2 decimal places) is ________

POLYMER SCIENCE AND ENGINEERING

Q 1 - Q 9 carry one mark each.

1. The solvent in which chain transfer is maximum in a radical polymerization is
(*a*) Benzene (*b*) Chloroform
(*c*) Carbon tetrachloride (*d*) Toluene

2. The monomer that can NOT be polymerized by anionic polymerization is
(*a*) Styrene (*b*) Ethyl vinyl ether
(*c*) Butadiene (*d*) Methyl methacry late

3. The elastomer retaining flexibility at the lowest temperature is
(*a*) Styrene butadine rubber
(*b*) Nitrile rubber
(*c*) Silicone rubber
(*d*) Butyl rubber

4. The polymer with minimum number of branches is
(*a*) LDPE (*b*) LLDPE
(*c*) HDPE (*d*) VLDPE

5. The nearest value of conductivity of Nylon 6 is
(*a*) 10^6 S/m (*b*) 10^0 S/m
(*c*) 10^{-13} S/m (*d*) 10^{-21} S/m

6. Aramid is
(*a*) Polyamide (*b*) Polyether
(*c*) Polyester (*d*) Polyimide

7. A miscible blend in 1 : 1 (by weight) composition is formed with
(*a*) Polystyrene and polybutadiene
(*b*) Polystyrene and poly (phenylene oxide)
(*c*) Polystyrene and poly (methyl methacrylate)
(*d*) Polystyrene and poly (dimethyl siloxane)

8. Dicumyl peroxide is
(*a*) Plasticizer
(*b*) Cross-linking agent
(*c*) Mold release agent
(*d*) Peptizer

9. The change in stress of a polymer as a function of time at a fixed strain is known as
(*a*) Fatigue
(*b*) Creep
(*c*) Stress relaxation
(*d*) Fracture toughness

Q 10 - Q 22 carry two marks each.

10. Match the polymers in Column A with their corresponding polymerization methods in Column B

Column A	Column B
P. Bisphenol A polycarbonate	1. Cationic
Q. Polyethylene	2. Step-growth
R. Poly (Styrene-b-butadiene)	3. Coordination
S. Butyl rubber	4. Anionic

(*a*) P-3, Q-2, R-4, S-1
(*b*) P-3, Q-3, R-2, S-1
(*c*) P-2, Q-3, R-4, S-1
(*d*) P-1, Q-3, R-4, S-2

11. Match the appropriate processing technique in Column A to fabricate the product in Column B

Column A	Column B
P. Blow molding	1. Plastic buckets
Q. Rotational molding	2. PVC sheet
R. Injection molding	3. Basket ball
S. Calendering	4. Plastic bottles

(*a*) P-2, Q-3, R-4, S-1 (*b*) P-3, Q-4, R-2, S-1
(*c*) P-4, Q-2, R-1, S-3 (*d*) P-4, Q-3, R-1, S-2

12. Match the appropriate characterization technique in Column A used to determine the polymer attributes in Column B

Column A	Column B
P. Gel permeation chromatography	1. Functional group
Q. FT-IR spectroscopy	2. Crystal structure
R. Differential scanning calorimetry	3. Glass transition temperature
S. X-ray diffraction	4. Molecular weight

(*a*) P-2, Q-3, R-4, S-1 (*b*) P-3, Q-4, R-1, S-2
(*c*) P-2, Q-4, R-1, S-3 (*d*) P-4, Q-1, R-3, S-2

13. Match each additive in Column A with its function given in Column B

Column A	Column B
P. Azodiformamide	1. Curing agent
Q. Di-octyl phthalate	2. UV stabilizer
R. Benzoyl peroxide	3. Blowing agent
S. 2-(2′-hydroxy phenyl) benzotriazole	4. Plasticizer

(*a*) P-2, Q-3, R-1, S-2 (*b*) P-3, Q-4, R-1, S-2
(*c*) P-4, Q-3, R-2, S-1 (*d*) P-1, Q-3, R-4, S-2

14. Plot of shear stress against shear rate for various types of fluids is given below. The appropriate assignment for P, Q, R and S is

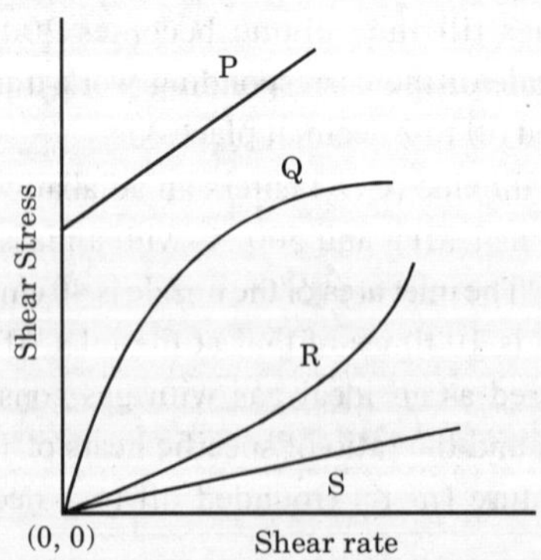

(*a*) P-Dilatant, Q-Bingham plastic, R-Pseudoplastic, S-Newtonian

(*b*) P-Bingham plastic, Q-Pseudoplastic, R-Dilatant, S-Newtonian

(*c*) P-Pseudoplastic, Q-Bingham plastic, R-Newtonian, S-Dilatant

(*d*) P-Newtonian, Q-Dilatant, R-Pseudoplastic, S-Bingham plastic

15. The number average molecular weight of a polyster formed from equimolar mixture of adipic acid and ethylene glycol at a conversion of 99.5% will be (*round off to nearest integer*).

16. For a freely jointed linear polyethylene chain with molar mass of 1.4×10^5 $gmol^{-1}$, the value of root mean square end-to-end distance in nanometer is ________ (*round off to 1 decimal place*). [Given : C-C bond length = 0.154 nanometer]

17. Viscosity measurements were performed for a set of PMMA solutions of different concentrations in toluene at 25°C. The plot of reduced viscosity against concentration (c) of the PMMA solutions produced an intercept of 21.0 cm^3 g^{-1} on the ordinate at $c = 0$. The value of viscosity average molecular weight of PMMA in toluene at 25°C is ________ (*round off to nearest integer*). [Given : Mark-Houwink constants K = 7.5×10^{-3} cm^3 g^{-1} and $a = 0.72$ for PMMA in toluene at 25°C]

18. Glass fibre reinforced PP composite is to be prepared with 20 volume % of glass fibre. The densities of glass fibre and PP are 2540 kg m^{-3} and 900 kg m^{-3}, respectively. The mass of glass fibre required to produce 1 kg of the composite in kg is _____ (*round off to 2 decimal places*).

19. A polymer solution flows through a cylindrical tube with a diameter of 4 mm at a volumetric flow rate of 10^{-9} m^3 s^{-1}. Under laminar flow condition and assuming the polymer solution to be a Newtonian fluid with viscosity 10^2 N s m^{-2}, the value of pressure drop per unit length of the tube in N m^{-3} is ________ (*round off to nearest integer*). [Consider the value of π as 3.14]

20. A molten polymer with a bulk modulus of 1 GPa is pressurized to 200 MPa during injection molding. The fractional decrease in volume of the molten polymer at this pressurized condition is ________ (*round off to 1 decimal place*).

21. Assume that each cross-link produced by vulcanization of polyisoprene contains an average of two sulphur atoms and that the sulphur is present only in the cross-links. If 40% of the isoprene units are cross-linked, the sulphur content in weight percentage is ________ (*round off to 2 decimal places*).

22. A tensile force of 160 N is applied to a piece of vulcanized rubber of dimension 30 mm × 4 mm × 4 mm. Assuming the vulcanized rubber to be incompressible, if the sample is elongated to 150% of its original length under the same applied force, the true stress in N mm^{-2} will be ________ (*round off to 1 decimal place*).

FOOD TECHNOLOGY

Q 1 - Q 9 carry one mark each.

1. The enzyme majorly involved in postmortem degradation of muscle proteins is

(*a*) Trypsin

(*b*) Calpin

(*c*) Transglutaminase

(*d*) Pepsin

2. Which of the following is the corret pair of essential fatty acids ?

(*a*) Oleic acid and Lenoleic acid

(*b*) Lenoleic acid and Linolenic acid

(*c*) Linolenic acid and Lauric acid

(*d*) Linolenic acid and Oleic acid

3. Nisin A is produced by

(*a*) *Aspergillus niger*

(*b*) *Acetobacter aceti*

(*c*) *Lactobacillus lactis*

(*d*) *Clostridium perfringens*

4. Which of the following bacteria will stain purple color after Gram staining?

(*a*) *Bacillus subtilis*

(*b*) *Escherichia coli*

(*c*) *Pseudomonas aeruginaosa*

(*d*) *Yersinia pestis*

5. The enzyme system used for removal of glucose from egg white prior to its drying consists of

(*a*) Glucose oxidase and Catalase

(*b*) Glucosidase and Glucoisomerase

(*c*) Glucoisomerase and Catalase

(*d*) Glucoamy lase and Glucose oxidase

6. The **INCORRECT** pair of food borne illness and its causative microorganism is

(*a*) Brucellosis – Brucella Sp.

(*b*) Peptic ulcers – Bacillus subtilis

(*c*) Bubonic plague – Yersinia pestis

(*d*) Q fever – Coxiella burnatii

7. Which of the following is commonly used as a preservative in the tomato sauce ?

(*a*) Sodium sulphite
(*b*) Potassium sorbate
(*c*) Potassium sulphite
(*d*) Sodium benzoate

8. The velocity of 2.2 μm diameter fat particles inside a centrifuge, running at 6000 rpm and 20°C, is 0.25 mm s^{-1}. The velocity of 1.5 μm diameter fat particles inside the same centrifuge running at 7500 rpm and same temperature (round off to 2 decimal places) will be ______ mm s^{-1}.

9. The initial population of a bacterial strain increases from 1×10^4 cells per mL to 1×10^6 cells per mL in 120 minutes. The generation time for this strain (round off to 2 decimal places) is ______ minutes.

Q 10 - Q 22 carry two marks each.

10. Match the protein in **Column I** with its food source in **Column II**.

Column I	Column II
P. Zein	1. Soybean
Q. Gluten	2. Maize
R. Glycinin	3. Egg
S. Ovalbumin	4. Wheat

(*a*) P-4, Q-1, R-2, S-3 (*b*) P-4, Q-3, R-1, S-2
(*c*) P-2, Q-3, R-1, S-4 (*d*) P-2, Q-4, R-1, S-3

11. Match the carbohydrate in **Column I** with corresponding enzyme used for its hydrolysis in **Column II**.

Column I	Column II
P. Pectin	1. Xylanase
Q. Lactose	2. β-galactosidase
R. Hemicellulose	3. Poly galacturonase
S. Inulin	4. β-fructofuranosidase

(*a*) P-3, Q-2, R-1, S-4 (*b*) P-2, Q-4, R-1, S-3
(*c*) P-1, Q-2, R-3, S-4 (*d*) P-4, Q-3, R-1, S-2

12. Match the edible oil refining stage in **Column I** with its purpose in **Column II**.

Column I	Column II
P. Degumming	1. Separation of triglycerides
Q. Neutralization	2. Removal of pigments
R. Bleaching	3. Removal of phosphatides
S. Winterization	4. Removal of free fatty acids

(*a*) P-3, Q-1, R-2, S-4
(*b*) P-1, Q-4, R-2, S-3
(*c*) P-4, Q-3, R-1, S-2
(*d*) P-3, Q-4, R-2, S-1

13. Match the food material in **Column I** with its related term in **Column II**.

Column I	Column II
P. Coffee	1. Wort
Q. Cocoa	2. Must
R. Beer	3. Arabica
S. Wine	4. Theobroma

(*a*) P-4, Q-2, R-1, S-3 (*b*) P-3, Q-4, R-1, S-2
(*c*) P-3, Q-4, R-2, S-1 (*d*) P-1, Q-3, R-4, S-2

14. Match the component/system in **Column I** with the peeling method for fruits and vegetables in **Column II**.

Column I	Column II
P. Lye solution	1. Flash peeling
Q. Carborundum rollers	2. Flame peeling
R. Pressure vessel	3. Abrasion peeling
S. Conveyor belt	4. Caustic peeling

(*a*) P-4, Q-3, R-2, S-1 (*b*) P-3, Q-4, R-1, S-2
(*c*) P-4, Q-3, R-1, S-2 (*d*) P-3, Q-4, R-2, S-1

15. Which among the given options correctly explains the nature of the microbial culture represented by curves 1, 2 and 3 in the following figure ?

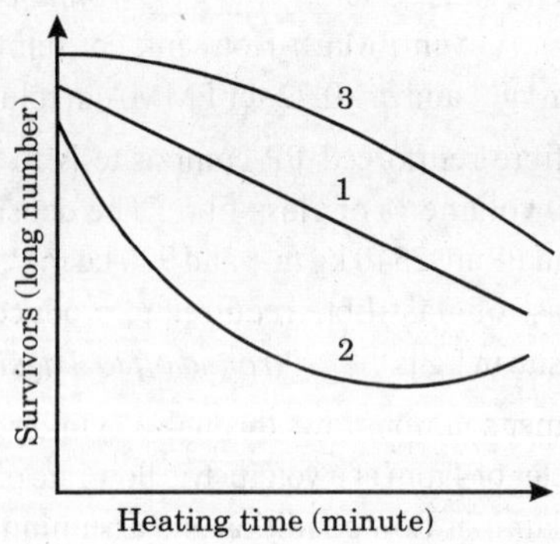

(*a*)
1. Germination of spores
2. Homogeneous population
3. Mixed population of spores and vegetative cells

(*b*)
1. Homogeneous population
2. Mixed population of heat sensitive and heat resistant microbes
3. Germination of spores

(*c*)
1. Composite population
2. Spores activated by short exposure to heat
3. Thermo sensitive and thermo resistant microbes

(*d*)
1. Mixed population
2. Microorganisms activated by short exposure to heat
3. Germination of spores

16. Match the equation/law in **Column I** with its application in **Column II**.

Column I	Column II
P. Plank's equation	1. Terminal velocity
Q. Arrhenius equation	2. Freezing time
R. Guggenheim-Anderson -de Boer equation	3. Activation energy
S. Stoke's law	4. Monolayer moisture content

(*a*) P-1, Q-3, R-4, S-2
(*b*) P-2, Q-3, R-1, S-4
(*c*) P-2, Q-3, R-4, S-1
(*d*) P-4, Q-3, R-1, S-2

17. Match the absorber used in modified atmosphere packaging and storage in **Column I** with the scavenger in **Column II**.

Column I	Column II
P. Oxygen absorber	1. Calcium chloride
Q. Carbon dioxide absorber	2. Magnesium oxide
R. Ethylene absorber	3. Ferric oxide
S. Moisture absorber	4. Potassium permanganate

(*a*) P-3, Q-2, R-4, S-1
(*b*) P-1, Q-2, R-4, S-3
(*c*) P-2, Q-3, R-4, S-1
(*d*) P-3, Q-2, R-1, S-4

18. During extrusion cooking, food materials are generally subjected to a combination of
(*a*) high shear and low pressure
(*b*) high temperature and high shear
(*c*) low shear and high temperature
(*d*) low shear and low pressure

19. The whole milk at 22 °C is pumped through a stainless steel pipe at a flow rate of 3 L s^{-1}. The length and inner diameter of the pipe are 40 m and 4 cm, respectively. If viscosity and density of the milk at the pumping temperature of 0.2 Pa s and 1032 kg m^{-3}, respectively, the Reynolds number (rounded off to nearest integer) will be ________.

20. A hammer mill, operating at a feed rate of 108 ton h^{-1}, consumes 10 kW power for reducing size of wheat grain from 3.92 mm to 1.25 mm. If Bond's law holds good, the feed rate (round off to 2 decimal places) for reducing the size of the wheat grain to 0.75 mm at the same power consumption level is ______ ton h^{-1}.

21. During spray drying of a milk sample, inlet and outlet temperatures are maintained at 132 °C and 80 °C, respectively. If the ambient temperature is 29 °C, the thermal efficiency (round off to 2 decimal places) of the dryer will be ________%.

22. An orange juice flowing at 0.80 kg s^{-1} enters a counter current double pipe heat exchanger at 20°C and leaves at 72 °C. Inlet and outlet temperatures of the hot water used as heating medium in the exchanger are 81 °C and 74 °C, respectively. The specific heat of the orange juice is 3.74 kJ kg^{-1} K^{-1} and overall heat transfer coefficient is 492 W m^{-2} K^{-1}. The heat transfer surface area (round off to 2 decimal places) will be ________ m^2.

ATMOSPHERIC AND OCEANIC SCIENCES

Q 1 - Q 9 carry one mark each.

1. In the northern hemisphere, the flow in the middle depths of the ocean is geostrophic. As we go down from that level and start approaching the bottom of the ocean, the flow deflects to the left of the geostrophic current because
(*a*) friction decreases and Coriolis force increases
(*b*) friction decreases and Coriolis force decreases
(*c*) friction increases and Coriolis force increases
(*d*) friction increases and Coriolis force decreases

2. Which one of the following is the definition of a monsoon?
(*a*) Seasonal reversal of wind direction
(*b*) High rainfall
(*c*) Occurs in the summer
(*d*) Occurs in the tropics

3. Anthropogenic emission of _______ is the main contributor to the ongoing ocean acidification
(*a*) Methane
(*b*) Carbon dioxide
(*c*) Nitrous oxide
(*d*) Sulphuric acid

4. What are phytoplankton?
(*a*) Microscopic animal life floating on surface of water bodies
(*b*) Pollen floating freely on surface of water bodies
(*c*) Microscopic plant life floating on surface of water bodies
(*d*) Microscopic plant life living on the floor of water bodies

5. Consider the two atmospheric virtual temperature profiles observed in Delhi given in Figures (i) and (ii) below.

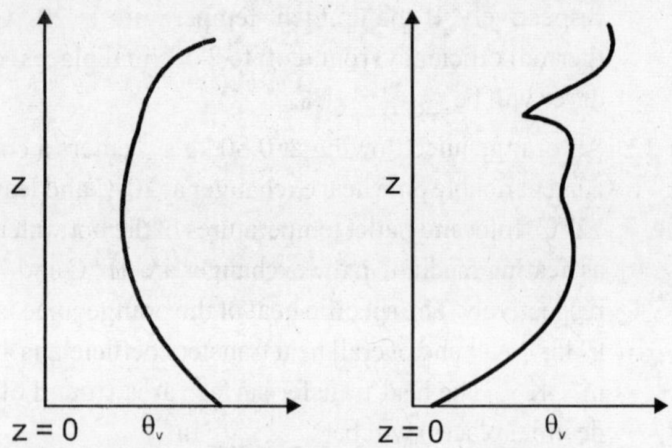

(i) (ii)

At what times of the day are you most likely to see such profiles?

(a) (i) Midnight and (ii) noon

(b) (i) 3 pm and (ii) 3 am

(c) (i) Sunrise and (ii) sunset

(d) (i) 3 am and (ii) 3 pm

6. A south-easterly wind is blowing towards which direction?

(a) 135°

(b) 157.5°

(c) 315°

(d) 225°

7. Consider a dry parcel at 30°C in an isothermal environment at 25°C. The parcel rises adiabatically by 1 km. Assuming $g = 10\ ms^{-2}$ and air density = $1\ kgm^{-3}$, the buoyancy force at the new location (rounded off to 2 decimal places) is ________ ms^{-2}.

8. The emissivity of polluted air that reflects and transmits 20% and 60% of the incoming solar radiation, respectively, at a given wavelength (correct up to 1 decimal place) is ________.

9. Given that the angular velocity of rotation of the Earth = $7.3 \times 10^{-5}\ s^{-1}$, the period of inertial oscillations generated in the oceans by surface winds at 30°N latitude (rounded off to the nearest integer) is ________ hours.

Q 10 - Q 22 carry two marks each.

10. Consider a high pressure centre in the northern hemisphere with tangential winds of $10\ ms^{-1}$ at a distance of 500 km from the centre. Assuming solid body rotation principles, what is the relative vorticity of the flow?

(a) $2 \times 10^{-5}\ s^{-1}$

(b) $-2 \times 10^{-5}\ s^{-1}$

(c) $4 \times 10^{-5}\ s^{-1}$

(d) $-4 \times 10^{-5}\ s^{-1}$

11. Which one of the following statements is true for atmospheric and oceanic general circulation models?

(a) Vertical velocity is ignored in oceanic models but not in atmospheric models.

(b) Boussinesq approximation is adequate in oceanic models but not in atmospheric models.

(c) Atmospheric models need a longer spin-up and integration time than oceanic models.

(d) Atmospheric models need parameterizations for subgrid scale processes but oceanic models do not.

12. Consider a scenario where air temperature increases by 2°C. We know that saturation vapour pressure for water increases with temperatures. As a result of this effect, the water vapour content of the atmosphere will ________ and the net warming will be ________ than 2°C. The correct pair of words to fill in the blanks (in the right order) is

(a) increase, more (b) increase, less

(c) decrease, more (d) decrease, less

13. The prevailing Trade winds over the Equator in the Pacific Ocean result in piling up of waters in the ________ part of the ocean. As a result, the gradients of the thermocline and the ocean surface have ________ signs. The correct pair of words to fill in the blanks (in the right order) is

(a) western, opposite

(b) western, same

(c) eastern, opposite

(d) eastern, same

14. Consider two different cases, shown in Figures (i) and (ii) below, with two layers of water of same density on top of each other.

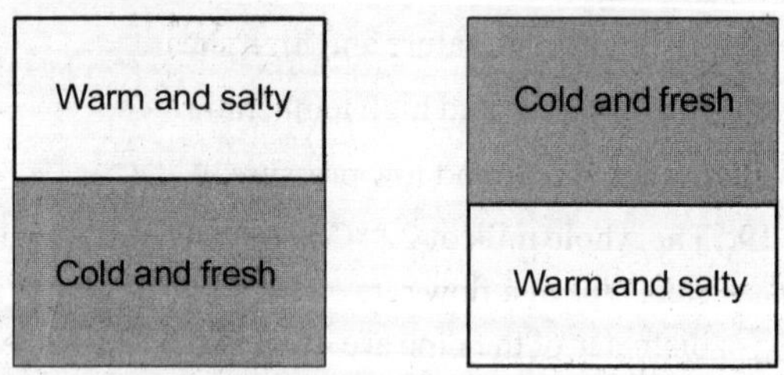

(i) (ii)

Which one of the following statements is true about convective plumes across the interface of the two layers?

(a) Upward convective plumes in (i) and downward convective plumes in (ii)

(b) Downward convective plumes in (i) and upward convective plumes in (ii)

(c) No convective plume in (i) and (ii)

(d) No convective plume in (i) but upward convective plumes in (ii)

15. During the Indian summer monsoon, surface outgoing longwave radiation (OLR) over the Arabian Sea is often observed to be low because

(*a*) of enhanced convection

(*b*) monsoon winds advect the OLR away into the Indian subcontinent

(*c*) monsoon clouds limit incoming solar radiation

(*d*) surface Bowen ratio is low

16. A tsunami wave in the ocean is approaching the coast. Assuming $g = 10$ ms^{-2}, the correct group speed of the wave at a depth of 1 km is

(*a*) 1 ms^{-1} (*b*) 10 ms^{-1}

(*c*) 100 ms^{-1} (*d*) 1000 ms^{-1}

17. A tornado is in cyclostrophic balance where the horizontal pressure gradient and centrifugal forces balance each other. Consider a tornado with 100 m radius and a tangential velocity of 100 ms^{-1} at the edge. Assuming air density = 1 kgm^{-3}, the magnitude of the pressure-drop between the centre and the edge of the tornado is ______ kgm^{-1}s^{-2}.

18. While driving south a distance of 1000 km, the temperature outside your car increases from 10 °C to 20°C. Assuming the air is completely dry, g = 10 ms^{-2} and Coriolis parameter = 10^{-4} s^{-1}, the vertical gradient of the geostrophic wind (rounded off to 2 decimal place) is _________ms^{-1}km^{-1}

19. A westerly wind of 10 ms^{-1} is blowing at a location in Pacific Ocean in the northern hemisphere. Assuming density of sea water = 1000 kgm^{-3}, Coriolis parameter = 10^{-4} s^{-1} and drag coefficient for sea water = 10^{-6}, the Ekman transport due to the wind at that location is_________ kgm^{-1}s^{-1}.

20. M_x and M_y represent the ocean mass transport in the x and y directions, respectively. L_x and L_y are the corresponding east-west and north-south length scales. For a typical equatorial ocean gyre, if the ratio of zonal to meridional mass transport ≈ 10, then $L_x \approx$ _______ L_y.

21. Assume that pressure varies exponentially with height : $p(z) = p_0 e^{-z/H}$, where $p(z)$ is the pressure at a height z above the surface, p_0 is the surface pressure, and scale height H = 7.5 km. Under these conditions, one-fourth of the total mass of the atmosphere lies above a height (round off to 1 decimal place) of _______ km above the surface.

22. Consider an atmospheric column of depth 300 m at the Earth's surface with an average temperature of 300 K. If the temperature of the layer rises by $\Delta T = 10$°C, the layer depth h will increase by Δh. Assuming $\Delta T/T \approx \Delta h/h$, air density remains unchanged at 1 kgm^{-3} and g = 10 ms^{-2}, the change in surface pressure is _______ kgm^{-1}s^{-2}.

ANSWERS

GENERAL APTITUDE

1. (*c*)	**2.** (*b*)	**3.** (*d*)	**4.** (*d*)	**5.** (*b*)	**6.** (*d*)	**7.** (*c*)
8. (*b*)	**9.** (*b*)	**10.** (*b*)				

ENGINEERING MATHEMATICS (COMPULSORY)

1. (*b*)	**2.** (*d*)	**3.** (*b*)	**4.** (*a*)	**5.** (*d*)	**6.** (5 to 5)	**7.** (2 to 2)
8. (*c*)	**9.** (*a*)	**10.** (*b*)	**11.** (0.43 to 0.47)			

FLUID MECHANICS

1. (*a*)	**2.** (*b*)	**3.** (*d*)	**4.** (*c*)	**5.** (*b*)	**6.** (*a*)	**7.** (*d*)
8. (0.30 to 0.33)	**9.** (5 to 5)	**10.** (*c*)	**11.** (*b*)	**12.** (*a*)	**13.** (*c*)	**14.** (*a*)
15. (79.60 to 79.90)		**16.** (183.5 to 184.5)		**17.** (1.21 to 1.35)	**18.** (857 to 924)	**19.** (0.035 to 0.045)
20. (20 to 20)	**21.** (30.5 to 34.5)	**22.** (5 to 5 or 500 to 500)				

MATERIALS SCIENCE

1. (*c*)	**2.** (*b*)	**3.** (*b*)	**4.** (*a*)	**5.** (*a*)	**6.** (*c*)	**7.** (5.3 to 5.7)
8. (157 to 159)	**9.** (1.9 to 2.1)	**10.** (*b*)	**11.** (*c*)	**12.** (*d*)	**13.** (*d*)	**14.** (69 to 69)
15. (0.84 to 0.90)	**16.** (0.33 to 0.45)	**17.** (0.330 to 0.334)		**18.** (4 to 4)	**19.** (0.97 to 1.09)	**20.** (0.36 to 0.45)
21. (2.20 to 2.28)	**22.** (77 to 83)					

SOLID MECHANICS

1. (*d*)	**2.** (*c*)	**3.** (*b*)	**4.** (*b*)	**5.** (*a*)	**6.** (*b*)	**7.** (6.25 to 6.40)
8. (90 to 90)	**9.** (150 to 150)	**10.** (*d*)	**11.** (*a*)	**12.** (*b*)	**13.** (*d*)	**14.** (*a*)
15. (9.99 to 10.00)		**16.** (18.60 to 18.90)		**17.** (4.477 to 4.487)		**18.** (7 to 7)
19. (10.160 to 10.340)		**20.** (5.600 to 5.650)		**21.** (0.175 to 0.190)		**22.** (1.90 to 2.85)

THERMODYNAMICS

1. (*b*) **2.** (*b*) **3.** (*a*) **4.** (*a*) **5.** (*d*) **6.** (125 to 125) **7.** (1 to 1)

8. (198 to 202) **9.** (63.5 to 65.5) **10.** (*a*) **11.** (*d*) **12.** (*b*) **13.** (*a*) **14.** (*a*)

15. (*c*) **16.** (1.15 to 1.25) **17.** (590 to 600) **18.** (2610 to 2614) **19.** (1.20 to 1.26) **20.** (39 to 42)

21. (0.52 to 0.57) **22.** (2.30 to 2.40)

POLYMER SCIENCE AND ENGINEERING

1. (*c*) **2.** (*b*) **3.** (*c*) **4.** (*c*) **5.** (*c*) **6.** (*a*) **7.** (*b*)

8. (*b*) **9.** (*c*) **10.** (*c*) **11.** (*d*) **12.** (*d*) **13.** (*b*) **14.** (*b*)

15. (17150 to 17250) **16.** (15.3 to 15.5) **17.** (60000 to 62000) **18.** (0.40 to 0.42)

19. (15800 to 16000) **20.** (0.2 to 0.2) **21.** (15.75 to 15.95) **22.** (14.9 to 15.1)

FOOD TECHNOLOGY

1. (*b*) **2.** (*b*) **3.** (*c*) **4.** (*a*) **5.** (*a*) **6.** (*b*) **7.** (*d*)

8. (0.15 to 0.21) **9.** (17.00 to 19.00) **10.** (*d*) **11.** (*a*) **12.** (*d*) **13.** (*b*)

14. (*c*) **15.** (*b*) **16.** (*c*) **17.** (*a*) **18.** (*b*) **19.** (491.00 to 495.00)

20. (63.00 to 67.00) **21.** (49.00 to 52.00) **22.** (11.00 to 14.00)

ATMOSPHERIC AND OCEANIC SCIENCES

1. (*d*) **2.** (*a*) **3.** (*b*) **4.** (*c*) **5.** (*b*) **6.** (*c*) **7.** (–0.17 to –0.16)

8. (0.2 to 0.2) **9.** (24 to 24) **10.** (*d*) **11.** (*b*) **12.** (*a*) **13.** (*a*) **14.** (*b*)

15. (*c*) **16.** (*c*) **17.** (10000 to 10000) **18.** (3.40 to 3.50) **19.** (–1000 to –1000)

20. (10 to 10) **21.** (10.3 to 10.5) **22.** (–100 to –100 or 100 to 100)

EXPLANATIONS

GENERAL APTITUDE

1. Rajiv Gandhi Khel Ratana Award was conferred on Mary Kom, a six time world champion in boxing, recently in ceremony at the Rashtrapati Bhawan (the President's official residence) in New Delhi.

2. Despite means without being affected by something.

3. Dissociate is the opposite of Associate.

4. Officials hope that the food grain production target will be met due to a good rabi produce.

5. Sum of first n natural numbers

$$= \frac{[n(n+1)]}{2}$$

Sum of first 2n natural numbers

$$= \frac{[2n(2n+1)]}{2} \quad ....(1)$$

Sum of first n odd numbers = n^2 ...(2)

Difference of (1) and (2)

$D = 2n^2 + n - n^2 = n^2 + n$

6. Decrease in repo-rate will decrease cost of borrowing and increase lending by commercial banks is the correct inference.

7.

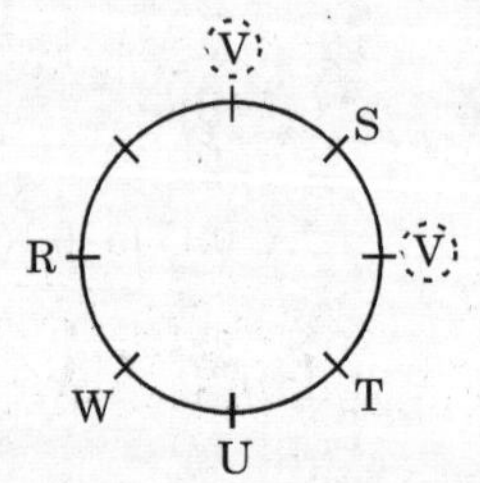

V can take any of the two positions.

In above diagram, P can not be seated opposite to Q.

8.

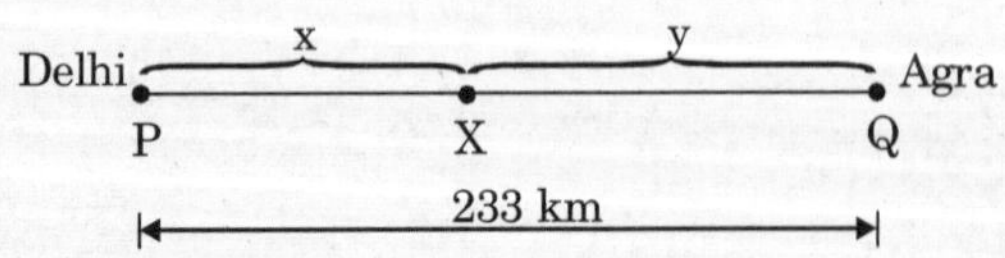

$V_P = V_Q + 10$ ...(i)

$x = V_P \times 2.25$

$y = V_Q \times 1.25$

$x + y = 233$

$V_P \times 2.25 + V_Q \times 1.25 = 233$ ...(ii)

$$V_P = \frac{491}{7} \text{ km/hr}$$

$$V_Q = \frac{421}{7} \text{ km/hr}$$

Distance travelled by Q

$\Rightarrow y = V_Q \times 1.25$

$$= \frac{421}{7} \times 1.25$$

$y = 75.17$ km

9. Total no. of elements in in 4×4 matrix is 16; of which four diagonal elements are zero. Remaining 12 elements can be divided in to two equal groups of 6 elements each. Both groups are opposite signed elements to meet the given condition.

10. Let the fixed investment annually be 'A'.

Total revenue of company P (2013 to 2018) = 1.1A + 1.2A + 1.4A + 1.4A + 1.5A + 1.4A = 8A

Total revenue of Company Q (2013 to 2018) = 1.2A + 1.3A + 1.3A + 1.5A + 1.6A + 1.6A = 8.5 A

FLUID MECHANICS

1. $\tau \propto \frac{du}{dy}$

2. An ideal fluid is icompressible and have no viscocity.

3. P is true, since a submerged body will be stable if its centre of gravity is below centre of buoyancy. For stable equilibrium of floating bodies met a centre has to be above centre of gravity.

4. Pressure is constant in laminar boundary layer and shear stress = $\mu \frac{u}{y}$ hence $(\tau_\omega)_{x_1} > (\tau_\omega)_{x_2}$

5. Weber number is the ratio of inertial and surface tension forces, hence its used to characteristic motion of fluid drops.

6. Streak lines the loci of points of all fluid particles that have passed continuously through a particular spatial point in the past.

7. The difference between stream function values at any two points gives volumetric flux.

8. By continuity equation

$A_1V_1 = A_2V_2$

$A_1 = \frac{\pi}{4} \times 0.05^2 = 0.00196m^2$

$A_2 = 0.5 \times 0.5 = 0.25\ m^2;\ V_2 = \frac{0.1}{40}$

$A_2V_2 = \frac{0.5 \times 0.5 \times 0.1}{40} = 0.000625\ m^3/s$

$V_1 = \frac{A_2V_2}{A_1} = 0.318\ m/s$

9. $\Phi = \frac{x^3}{3} - xy^2$

$\frac{d\Phi}{dx} = x^2 - y^2; \frac{\Phi\varphi}{\varphi y} = -2xy$

$u = \frac{-\Phi\varphi}{\varphi x}$ and $v = \frac{-\Phi\varphi}{\varphi y}$

at point (2, 1)

$u = -[4 - 1]$; $v = -[-2 \times 2 \times 1]$

$u = -3; v = 4$

Resultant velocity magnitude $= \sqrt{(-3)^2 + 4^2}$

$= \sqrt{25} = 5m/s$

10. Source, sink, uniform flow – Rankine oval

Doublet, uniform flow – Cylinder

Source, uniform flow – Rankine half body

Doublet, free vorflex – Rotating cylinder uniform flow

11. $\vec{V} = 5t\hat{i} + 2x2\hat{j} + 2ty\hat{k}$

$a_x = u \times \frac{\varphi u}{dx} + v \times \frac{\varphi u}{dy} + w \times \frac{\varphi u}{dz} + \frac{\varphi u}{dt}$

$= 5t \times 0 + 2xz \times 0 + 0 + 5$

$a_y = u \times \frac{\varphi v}{dx} + v \times \frac{\varphi v}{dy} + w \times \frac{\varphi v}{dz} + \frac{\varphi v}{dt}$

$= 5t \times 2z + 2xz \times 0 + 2ty \times 0 + 0$

$a_z = u \times \frac{\varphi w}{dx} + v \times \frac{\varphi w}{dy} + w \times \frac{\varphi w}{dz} + \frac{\varphi w}{dt}$

$0 + 2xz \times 2t + 0 + 2y$

Total acceleration $= 5\hat{i} + t(10z + 4xy)\hat{j} + (2y + 4xzt)\hat{k}$

13. $\psi = k(x^2 - y^2x)$

vorticity $= 2\omega$

$\frac{\Phi\varphi}{dx} = v = k \times 2x - ky^2 = 2xk - ky^2$

$\frac{-\Phi\varphi}{dy} = u = -[0 - 2yk] = 2yk$

$\omega z = \frac{1}{2}\left[\frac{dv}{dx} - \frac{du}{dy}\right] = \frac{-1}{2}[2k - 2xk]$

$\omega x = 0, \omega y = 0$

Hence vorticity $\varepsilon_p = 2 \times \frac{-1}{2}[2k - 2xk] = 2k(x - 1)$

14. $u = -2xy,\ v = y^2 - x^2$

1. $\frac{-1}{\rho}\frac{\Phi\rho}{\Phi x} = u \times \frac{\Phi u}{dx} + v \times \frac{\varphi u}{dy}$

$= -2xy \times -2y + (y^2 - x^2)(-2x)$

$\frac{\varphi p}{\varphi x} = -2\rho(xy^2 + x^3)$

2. $\frac{-1}{\rho}\frac{\varphi\rho}{\Phi y} = u \times \frac{\varphi v}{\varphi x} + v \times \frac{\varphi v}{\varphi y}$

$= -2\rho(x^2y + y^3)$

15. $v_1 = 0,\ p_1 = patm,\ z_1 = 100m$

$\frac{P_1}{\nu} + \frac{\propto_1 v_1^2}{2g} + z_1 = \frac{P_2}{\nu} + \frac{\propto_2 v_2^2}{2g} + z_2 + h_t + h_f$

$\frac{Pa}{\nu} + \frac{1.06 \times 0}{2g} + 100 = \frac{Pa}{\nu} + \frac{1.06 \times 2^2}{2g} + h_t + 20m$

head available $h_t = 100 - 20 - 0.2 = 79.8$ m

16. $h_f = \frac{flv^2}{2gd} = \frac{0.02 \times 6 \times 10^2 \times 10}{2 \times 9.81 \times 5}$

$= 12.33$ m

by darcy-weisbach equation

$h_f = \frac{P_1 - P_2}{\rho g}$

$$12.33 = \frac{P_1 - 304\times10^3}{1000\times9.81}$$

$P_1 = 183{,}042$ Pa

= 183.042KPa

17. Model | Proto type

Model	Proto type
Fm = 5M	Fp = ?
ρm = 1000	ρp = 1.5
Lm = 0.1	Lp = 1.5
Vm = 2	Vp = ?
$\upsilon_m = 0.001$	$\upsilon_m = 1.78\times10^{-5}$

$$\frac{VpLp}{\upsilon p} = \frac{VmLm}{\upsilon m}$$

$$Vp = \frac{2\times0.1\times1.78\times10^{-5}}{0.001\times1.5}$$

= 0.0024 m/s

$$\frac{Fp}{Fm} = \frac{\rho_p L_p^2 V_p^2}{\rho m L_m^2 V_m^2}$$

$$Fp = 5\times\frac{1.2}{1000}\times\left(\frac{1.5}{0.1}\right)^2\left(\frac{0.024}{2}\right)^2$$

= 1.25 N

18. By Hagen-Poiseuille formula

$$P_1 - P_2 = \frac{32\mu VL}{D^2}$$

$$1\times10^6 = \frac{32\times1.74\times10^{-3}\times V\times1}{(0.5\times10^{-3})^2}$$

$$V = \frac{1\times10^6\times(0.5\times10^{-3})^2}{32\times1.74\times10^{-3}\times1}$$

$$\frac{Q}{\frac{\pi}{4}d^2} = \frac{(1\times10^6\times0.5\times10^{-3})^2}{32\times1.74\times10^{-3}}$$

Q = 862

19. $$Q = cd\times\frac{a_1a_2}{\sqrt{a_1^2 - a_2^2}}\times\sqrt{2gh}$$

$$h = \frac{P_1 - P_2}{\rho g}\ \frac{40\times10^3}{1000\times9.81} = 4.077\text{m}$$

ad = 0.98

$$a_2 = \frac{\pi}{4}\times\left(\frac{7.5}{100}\right)^2,\ a_1 = \frac{\pi}{4}\times\left(\frac{15}{100}\right)^2$$

= 0.0044, 0.0177

$$Q = 0.98\times\frac{0.0177\times0.0044}{\sqrt{(0.0177)^2 - (0.0044)^2}}\times\sqrt{2\times9.81\times4.077}$$

= 0.0385 m^3/s

20. $Fx = \rho a[Vplete - Vjet]^2$

$\rho = 1000$, $a = 2\times10^{-4}\ m^2$

Vplete = 20 m/s Vjet = 10 m/s

Hence $F = 1000\times2\times10^{-4}[20-10]^2 = 20N$

21. Taking Datum line at the point where mercury and oil coincide at naming the section X-X.

Pressure below X-X in left limb

$= P_A - 1000\times10\times H - 800\times10\times0.15$

Pressure below X-X in right limb

$= P_B - 13600\times10\times(0.45 + H)$

Equating both pressures:

$P_A - 10000H - 1200 = P_B - 61200 - 136000H$

$P_B - P_A = 612\,00 - 1200 + 136000\,H - 10000H$

$100\times1000 = 60000 + 135000H$

H = 33.75 cm ≈ 33.8 cm

22. $u = ay\,(y_0 - y)$

Qout – Qin = 0

$$\int udA - u_0\int dA = 0$$

$$\int_0^{y_0} ay(u_0 - y)bdy - u_0\int_0^{y_0} bdy = 0$$

$$aby_0\left[\frac{y^2}{2}\right]_0^{y_0} - ab\left[\frac{y^3}{3}\right]_0^{y_0} - u_0b[y]_0^{y_0}$$

$$a = \frac{6u_0}{y_0^2} = \frac{6\times7.5}{9} = 5$$

MATERIALS SCIENCE

2. The **optical microscope**, also referred to as a **light microscope**, is a type of microscope that commonly uses visible light and a system of lenses to generate magnified images of small objects. so for the smallest or minimum feature size that can be theoretically resolved in an optical microscope and it does not depend on intensity of radiation.

3. As we know that

Normal crystals contain only five kinds of improper rotational symmetry:

1. One fold, $\phi = 360°$
2. Two fold, $\phi = 180°$
3. Three fold, $\phi = 120°$
4. Four fold, $\phi = 90°$
5. Six fold, $\phi = 60°$

The given square 2-D lattice with a hexagonal motif (dark colors):

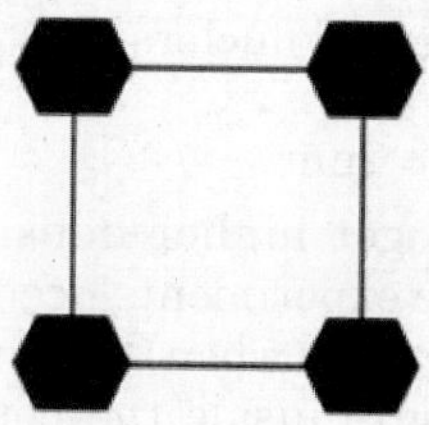

the rotational symmetry element that must be present with 180° in the system is two- fold rotation.

4. The density of states (DOS) is essentially the number of different states at a particular energy level that electrons are allowed to occupy, i.e. the number of electron states per unit volume per unit energy. Bulk properties such as specific heat, paramagnetic susceptibility, and other transport phenomena of conductive solids depend on this function. DOS calculations allow one to determine the general distribution of states as a function of energy and can also determine the spacing between energy bands in semi-conductors.

In three dimensional solid, the DOS can be calculated for most of the energy band (E), using:

$$g(E) = \frac{1}{2\pi^2}\left(\frac{2m^*}{\hbar^2}\right)^{3/2} E^{1/2}$$

5. The **glass-transition temperature** T_g of a material characterizes the range of temperatures over which this glass transition occurs. It is always lower than the melting temperature T_m, of the crystalline state of the material, if one exists.

So, the variation of molar volume (V_m) of a liquid showing glass transition temperature (Tg) while cooling from its melting temperature (T_m) is depicted by:

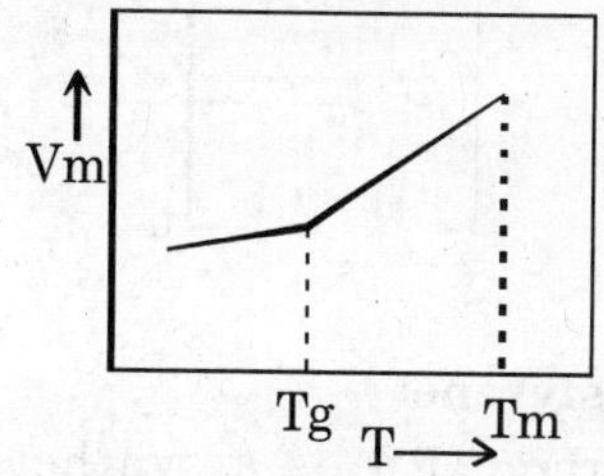

6. (i) Polyethylene:

Polyethylene is made from **the** polymerization of **ethylene** (or ethene) **monomer. Polyethylene** chemical formula is (C2H4)n. Content.

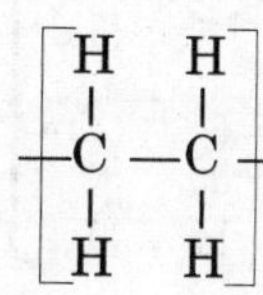

(ii) Polypropylene:

Polypropylene (PP), also known as polypropene, is a thermoplastic polymer used in a wide variety of applications. It is produced via chain-growth polymerization from the monomer propylene. **Polypropylene** belongs to the group of polyolefins and is partially crystalline and non-polar.

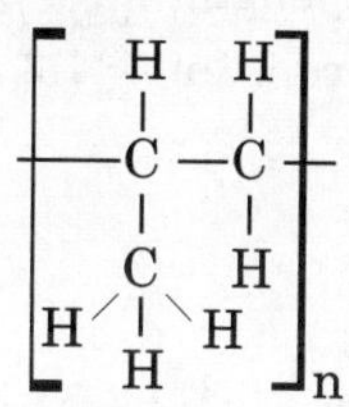

(iii) Polyvinyl chloride:

Poly(vinyl chloride) is the plastic known at the hardware store as PVC. This is the PVC from which pipes are made, and PVC pipe is everywhere. The plumbing in your house is probably PVC pipe, unless it's an older house. Polyvinyl Chloride (PVC or Vinyl) is a high strength thermoplastic material widely used in applications, such as pipes, medical devices, wires and cable insulation, etc.

Vinyl chloride is an organo chloride with the formula H_2C = CHCl that is also called **vinyl chloride monomer** (VCM) or chloro-ethene. This colorless compound is an important industrial chemical chiefly used to produce the polymer **polyvinyl chloride (PVC).**

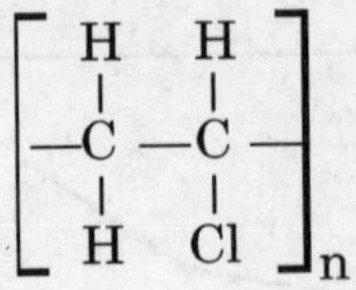

(iv) **Polystyrene:**

Polystyrene (PS) is a synthetic aromatic hydrocarbon polymer made from the monomer known as styrene.

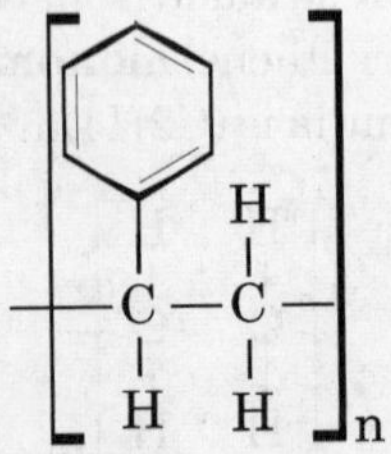

7. Fractive tonghness

$K_1 = y\sigma\sqrt{\pi a}$

K_1 = stress intensity factor = $1 MPa.m^{\frac{1}{2}}$

σ = Applied stress = 200 MPa

a = edge crack length

y = geometric constant = 1.2

$$a = \frac{1}{\pi}\left(\frac{K_1}{y\sigma}\right)^2$$

$$a = \frac{1}{3.14}\left(\frac{1}{1.2 \times 200}\right)^2$$

$a = 5.52 \times 10^{-6}$m

$a = 5.52\mu$m

8. $\frac{\Delta L}{L} = \propto (T_f - 25)$

$$\frac{\Delta L}{L} = \frac{\text{Stress}}{\text{Modulus}} = \frac{200}{200 \times 10^3} = 10^{-3}$$

$10^{-3} = \propto (T_f - 25)$

$10^{-3} = 7.5 \times 10^{-6} (T_f - 25)$

$T_f = 158.33°C$

9. For cretical size of nucleus

$$\frac{d(\Delta GT)}{dr} = 0 \text{ when } \gamma = \gamma^*$$

$$\gamma^* = \frac{2Y_s}{\Delta G_v}$$

γ_s = Solid lequid interfacial energy = 0.18 J/m^2

ΔG_v = Change in volume free energy upon transformation from lequid to solid

= 0.18×10^9 J/m^3

$$\text{Radius of cretical nuclers} = \frac{2 \times 0.18}{0.18 \times 10^9}$$

= 2×10^{-9}m = 2nm

10. In order to get higher densification, higher temperature experiments need to be run. Since the sintering stops when the driving force is equal to the pressure inside the pores, real vacuum condition can be applied to extract the air out of the furnace and closed pores formed during sintering inhibit until it gets full densification. Sintering in vacuum leads to improved densification as compared to sintering under ambient (at atmospheric pressure) condition. Hence both statements are true.

11. 1. **Powder processing – porous bronze bearing:**

Powder processing normally involves cold compaction followed by a high temperature sintering stage, in which heat is applied with or without pressure.

Starting materials may be mixed elemental powders of copper and tin, fully pre-alloyed bronze powder, or mixtures of the three.

If pre-alloyed bronze is used shrinkage on sintering normally results, while with elemental powder a significant growth can be obtained.

In addition to the normal lubricant graphite is often added to the mixes used for **porous bronze bearings.**

2. **Spin coating – organic semiconductor thin films:**

Spin coating is a technique that allows a highly controlled and reproducible deposition of organic semiconductor thin films on the desired substrates over large areas with high structural homogeneity.

3. **Czochralski process – single crystal silicon:**

The **Czochralski method** or **Czochralski process**, is a method of crystal growth used to obtain **single crystals** of semiconductors (e.g. silicon, germanium and gallium arsenide), metals (e.g. palladium, platinum, silver, gold), salts and synthetic gemstones.

4. **Chemical vapor deposition – poly- silicon:**

Low pressure **chemical vapour deposition** of polysilicon from silane is an industry-standard technique. In this technique silane gas is passed through a hot walled deposition vessel containing the substrates on which it is required to deposit the polysilicon, such that the reaction of the silane with the heated substrates causes decomposition of the silane such that respective polysilicon films are deposited on the substrates.

12. For fourth reflection and FCC structured

$h^2 + k^2 + l^2 = 11$

λ = wavelength

d = Distance between atoms in a crystal

$$\lambda = 2d\sin\theta = \frac{2\times a}{\sqrt{h^2+k^2+l^2}}.\sin\theta$$

θ = angle of scattering

$$d = \frac{a}{\sqrt{h^2+k^2+l^2}}$$

$$1.54056 = 2\times\frac{3.5}{\sqrt{11}}\times\sin\theta$$

$\theta = 46.755$

Brag angle $2\theta = 93.51$

13. Number of schottky defects

$$N_s = N\exp\left(-\frac{Q}{2KT}\right)$$

$N \approx Na$ = Number of lattice sites per cubic meters

$= 6.023 \times 10^{23}$

$$N_s = 6.023\times10^{23}\exp\left(-\frac{250000}{2\times8.314\times573}\right)$$

$N_s = 2.42 \times 10^{12}$

14. Given, probability of accident happens in a particular month $= \frac{1}{100}$.

It is given that each events are independent from each other = binomial distribution

$\Rightarrow$ Number of trials = Number of months = n

failure probability $q = 1 - p = 1 - \frac{1}{100} = 0.99$.

$\Rightarrow$ The probability that no accident happens over a period of n months

$= P(x = 0)$

$= {}^nC_0 \times p^0 \times q^n$

$= 1 \times 1 \times (0.99)^n$

$= (0.99)^n$

This probability has to less than or equal to $\frac{1}{2}$

$$(0.99)^n \geq \frac{1}{2}$$

$n \geq 68.9675$.

Smallest integer value of n = 69.

15. Surface energy of

$$\gamma_{111} = \frac{2\Delta H_s}{\sqrt{3}N_A{a_0}^2}.$$

ΔH_s = molar enthalpy of sublimation

$$\gamma_{100} = \frac{4\Delta H_s}{3N_A a_0 2}$$

The ratio of the two surface energy

$$\frac{\gamma_{111}}{\gamma_{100}} = \frac{\frac{2\Delta H_s}{\sqrt{3}N_A{a_0}^2}}{\frac{4\Delta H_s}{3N_A{a_0}^2}} = 0.87$$

16. $n_i = 2\left(\frac{2\pi m K_b T}{h^2}\right)^{\frac{3}{2}}\exp\left(-\frac{E_g}{2KT}\right)$

$KT = 1.38 \times 10^{-23} \times 300K = 0.026ev$

$$n_i = 2\left(\frac{2\times3.14\times9.1\times10^{-31}\times1.38\times10^{-23}\times300}{\left(6.64\times10^{-34}\right)^2}\right)^{\frac{3}{2}}$$

$$\exp\left(\frac{-1.1}{2\times0.026}\right)$$

$= 2 \times 1.242 \times 10^{25} \times \exp(-21.153)$

$= 2 \times 1.242 \times 10^{25} \times 6.50 \times 10^{-10}$

$n_i = 1.615 \times 10^{16}$

$$N = n_i \exp\left(-\frac{\Delta E}{KT}\right)$$

$$\Delta E_f = KT \ln\left(\frac{N}{Ni}\right)$$

$$= 0.026 \ln\left(\frac{5\times10^{22}}{1.615\times10^{16}}\right)$$

$= 0.026 \times 14.944$

$= 0.3885$

$\Delta E_f = 0.39 ev$

17. $$\frac{I_T}{I_O} = (1-R)^2 e^{-\propto x}$$

$$= (1-0.05)^2 \times e^{-10^3 \times 10^{-3}}$$

$= (0.95)^2 \times e^{-1}$

$= 0.33$

18. In compounds with inverse spinel structure e.g. Fe_3O_4, $N.fe_2O_4$, half of B cations occupy tetrahedral sites and all A and the remaining 50% B cations occupy octahedral sites.

So, if we take Fe_3O_4 which is actually $fe^{2+}fe_2^{3+}O_4$, then the arrangement of spin is like what is shown below:

½ of fe^{3+} cations (Tetrahedral)	½ of fe^{3+} All fe^{2+} Cations (Octahedral)	
↓ ↓ ↓ ↓ ↓	↑↑↑↑↑	↑↑↑↑↑
fe^{3+} : 5 μB	fe^{3+} : 5 μB	fe^{2+} : 4 μB

μ_{net} (Per fe_3O_4 formula unit) $= \mu_{oct} - \mu_{tet}$

$= 9\mu_B - 5\mu_B = 4\mu_B$

Hence the net magnetization of fe_3O_4 is $4\mu_B$ per formula unit.

19. Voltage developed in z-direction

$$= \frac{d_{zz} f_z . z}{E_0 E_r . A_z}$$

$$= \frac{100\times10^{-12}\times10\times1\times10^{-3}}{8.85\times10^{-12}\times1100\times10^{-2}\times10^{-2}}$$

$= 1.027$ volt

20. Rule of mixute

$E_C = E_A V_A + E_S V_S$

$\rho_C = \rho_A V_A + \rho_B V_B$

C – Compsete

A – Al

S – Sic

V – Vomule factor

$$\frac{E_C}{\rho_C} = \frac{E_A V_A + C_S V_S}{\rho_A V_A + \rho_S V_S}$$

$V_A + V_S = 1$

$$3\left(\frac{E_A}{\rho_A}\right) = \frac{E_A(1-V_S) + E_S V_S}{\rho_A(1-V_S) + \rho_S V_S}$$

$$3\left(\frac{69}{2.7}\right) = \frac{69(1-V_S) + 379.V_S}{2.7(1-V_S) + 2.36.V_S}$$

$V_S = 0.4106$

21. The oxidation of metal is believed to be diffusion controlled and can be described by the parabole rate law

$$\left(\frac{\Delta W}{A}\right)^2 = K{\cdot}t$$

Case 1: $(1)^2 = K{\cdot}100$ min

Case 2: $\left(\frac{\Delta W}{A}\right)_{500}^2 = K{\cdot}500$ min

Case 1 ÷ Case 2

$$\frac{1}{\left(\frac{\Delta W}{A}\right)_{500}^2} = \frac{K{\cdot}100}{K{\cdot}500}$$

$$\left(\frac{\Delta W}{A}\right)_{500} = 2.2361 \text{ mg/cm}^2$$

22. $$\frac{C_x - C_o}{C_s - C_o} = 1 - \text{erf}\left(\frac{x}{2\sqrt{Dt}}\right)$$

$C_x = 0.46$, $C_o = 0.1$, $C_s = 1$

$$\frac{0.46-0.1}{1-0.1} = 1 - \text{erf}\left(\frac{5\times10^{-3}}{2\times\sqrt{6\times10^{-11}}\times\sqrt{t}}\right)$$

$$0.6 = \text{erf}\left(\frac{322.99}{\sqrt{t}}\right)$$

From the table given in question

$0.6 = \text{erf}(0.6)$

$$0.6 = \frac{322.99}{\sqrt{t}}$$

$t = 289798$

$t = 80.49$ hrs.

SOLID MECHANICS

1. Kinetic friction is less than static friction because in kinetic friction one object is in moving conditon and less inter molecular attraction between the objects where as in static friction there is muck more intermolecular attraction.

2. Bulk modulers $K = \frac{P}{e_v}$

$$e_v = e_x + e_y + e_z$$

$$= \frac{3p}{E}(1 - 2\mu)$$

$$K = \frac{p}{\frac{3p(1-2\mu)}{E}}, \text{ i.e., } K = \frac{E}{3(1-2\mu)}$$

3. Shear strain length of the element's sides as no dimension is affected, volume donot change.
4. Under a central force torque about C.M is zero, therefore angular momentum is conserved.
5. Pure bending is a condition of stress where a bending moment is applied to a beam without simultaneous presence of axial, shear, or torsional forces. Under constant bending moment S.F is zero.
6.

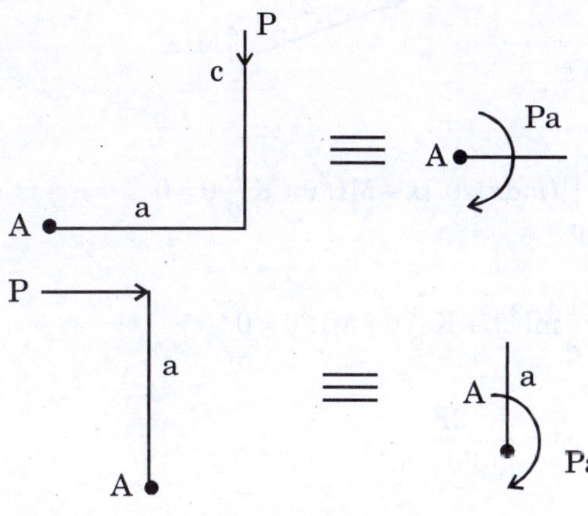

Thus, option B is the answer.

7.

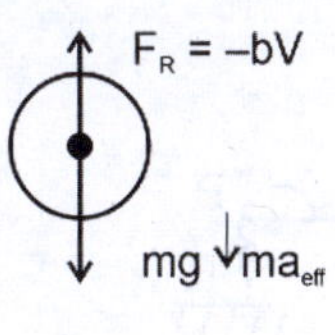

$$a_{eff} = \frac{F_{net}}{m}$$

$$a = \frac{mg - bv}{m}$$

$$\frac{dv}{dt} = g - \frac{b}{m}V$$

$$\frac{dv}{g - \frac{b}{m}V} = dt$$

Integrating

$$\int_a^V \frac{dv}{g - \frac{b}{m}V} = \int_a^t dt$$

$$\ln\left|\frac{g - \frac{b}{m}V}{g}\right| = ta - \frac{b}{m}$$

$$e^{\frac{-b}{m}t} = \frac{g - \frac{b}{m}V}{g}$$

$$ge^{\frac{-b}{mt}} = g^{\frac{-b}{m}}V$$

$$V = \frac{mg}{b} - \frac{mg}{b}e^{\frac{-b}{m}t}$$

Now we get a relation between velocity and time.

Velocity at t = 1 sec. when m = 0.1 kg

$g = 10\ m/s^2$

$b = 0.1$ N s/m

$$V = \frac{mg}{b}\left[1 - e^{\frac{-b}{mt}}\right]$$

$$= \frac{0.1 \times 10}{0.1}\left[1 - e^{-1}\right]$$

$$= 6.32$$

8.

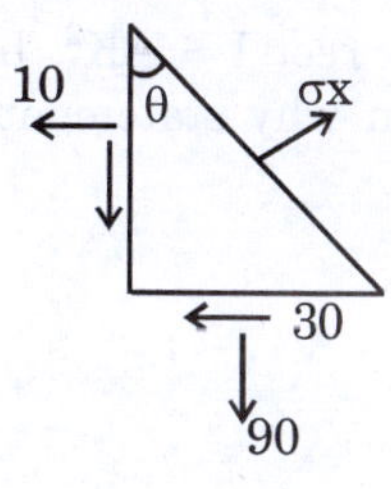

$$\sigma_{x_1} = \sigma_x \cos^2\theta + \sigma_y \sin^2\theta + 2\tau xy \sin\theta\cos\theta$$

$$= 30 \times \frac{9}{25} + 90 \times \frac{16}{25} + 2 \times 30 \times \frac{12}{25}$$

$$= 90 \text{ MPa}$$

9.

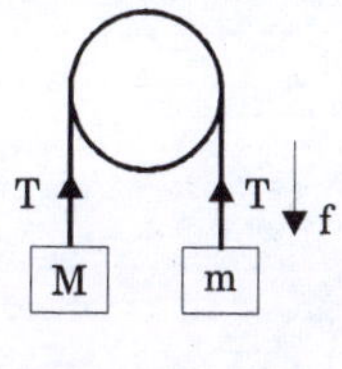

Let 'f' be the accelration of the system.

Downward force on M = Mg – T = M × f

Upward force on m = T – mg = m × f

adding both $f = \frac{M-m}{M+m}$

but T – mg = mf

$$T = \frac{2Mm}{M+m} \times g = \frac{2\times 30\times 10\times 10}{40} = 150N.$$

10.

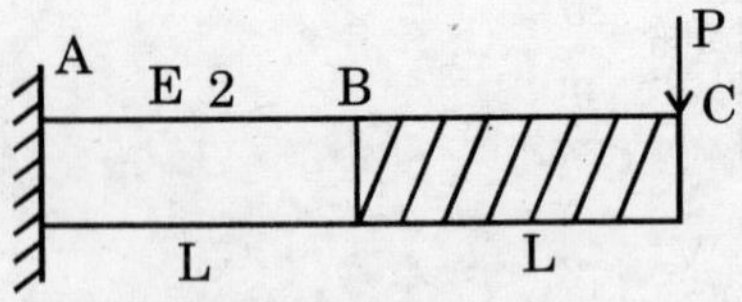

$$\theta_b = \frac{P\times L^2}{2EI} + \frac{PL^2}{EI}$$

$$\frac{3}{2}\frac{PL^2}{EI}$$

11. According to Euler's theory

$$Pcr = \frac{\pi^2 E1}{L^2} \text{ hence } Pcr \propto 1,$$

thus statement 2 is correct

$$Pcr = \frac{\pi^2 EAK^2}{L^2} \; Pcr \propto K^2$$

Hence statement 1 is correct I = AK^2, hence statement 2 is the reason why statement 1 is correct.

12.

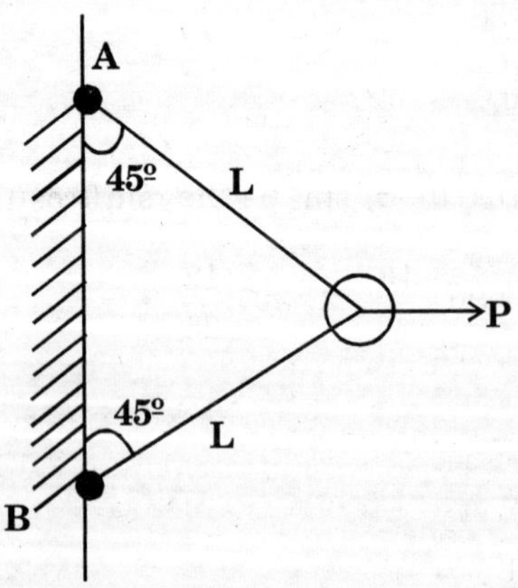

Let the four in each will be S

Resolving the fores at 'O' vertically

25 sin 45 = P.

$$S = \frac{P}{\sqrt{2}}.$$

$$\text{Total S.E. stored} = 2\times\left(\frac{P}{\sqrt{2}}\right)^2 \times \frac{1}{2AE}$$

$$Wi = \frac{P^2 1}{2AE}.$$

Horizontal deffertion at O

$$= \frac{dWi}{dP} = \frac{Pl}{AE}.$$

13.

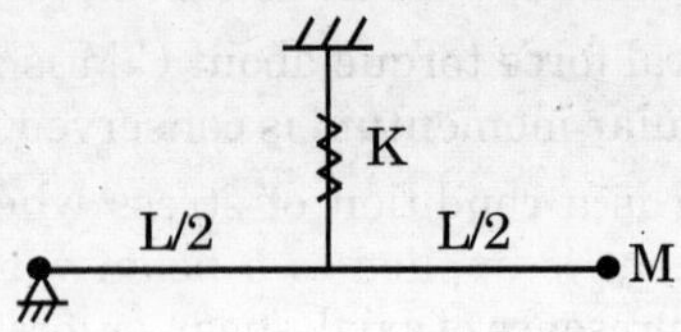

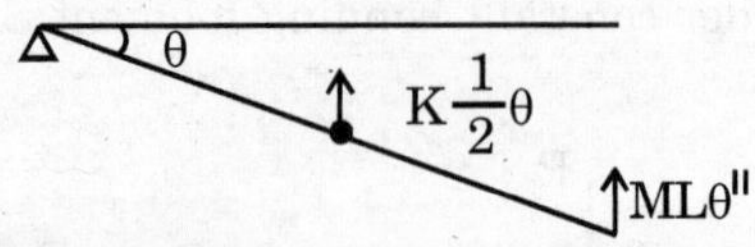

$$\int_0^1 [(mdx)x\theta'']x + ML^2\theta + K\frac{L}{2}\theta = 0$$

$$\frac{1}{3}ml^3\theta'' + K\frac{L}{2}\theta + ML^2\theta = 0$$

$$p^2 = \frac{3K}{4[3M+m]}$$

$$p = \sqrt{\frac{3K}{4[3M+m]}}$$

14.

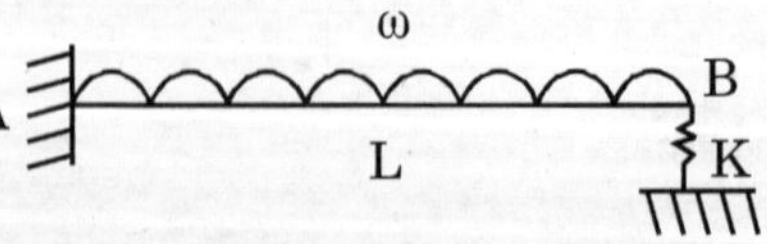

Here, both spring and beam are in series.

$$\text{without spring } \delta = \frac{\omega l^4}{8E1}$$

with spring reaction $\delta - \delta_{spring} = \delta_R$.

$$\frac{\omega l^4}{8E1} - \frac{R}{K} = \frac{RL^3}{3E1}$$

$$\frac{\omega l^4}{8E1} = \frac{R}{K} + \frac{RL^3}{3E1};$$

but $K = \frac{EI}{L^3}, \frac{Wl^4}{8E1} = \frac{RL^3}{E1} + \frac{RL^3}{3E1}$

R=3WL/32

15. Free extensions under temperature increase

$= l \propto \Delta \in$

$= 100 \times 10x \times 10^{-6} \times 10$

$= 10^{-2}$ mm

But there is an extension of 0.005 mm hence compression is caused by

$10^{-2} - 0.005\text{mm} = 5 \times 10^{-3}$ mm

Stress produced by strain of 5×10^{-3} mm

$$= \frac{\text{stress}}{\frac{5 \times 10^{-3}}{100}} = E$$

stress $2 \times 10^5 \times \frac{5 \times 10^{-3}}{100} = 10$ MPa

16. Radial displacement of spherical thin vessel

$\delta r = r\varepsilon_c$

$$\frac{1-\upsilon}{E} \times \frac{Pr^2}{2t}$$

$$= \frac{1-0.25}{2 \times 10^5} \times \frac{100 \times (1000)^2}{2 \times 10}$$

= 18.75 mm

17. σ due to impact $= \frac{P}{A} + \sqrt{\left(\frac{P}{A}\right)^2 + \frac{2Eph}{AL}}$

$$= \frac{100 \times 10}{100} + \sqrt{\frac{(10)^2 + 2 \times 2 \times 10^5 \times 1000 \times 100}{1000 \times 200}}$$

$= 447.72$ N/mm^2

$\frac{\sigma}{\frac{\Delta l}{l}} = E$ hence $\Delta l = \frac{\sigma \times l}{E}$

$$\Delta l = \frac{447.72 \times 200 \times 10}{2 \times 10^5}$$

= 4.477 mm

18.

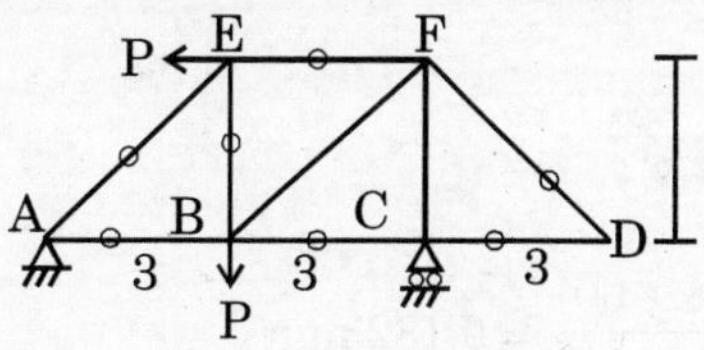

7 members are zero four members.

19.

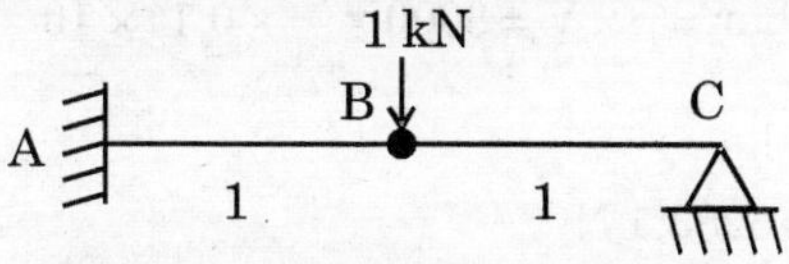

The internal hinge makes AB a contileres. These

MA = 1 kN × 1m = 1000 Nm

Now by simple bending

Stress $f = \frac{M}{1}$ y max;

here $\frac{1}{y\,max} = \frac{\pi d^3}{32}$

$$f = \frac{1000 \times 32}{\pi \times (0.1)^3} = 10.191 \text{ MPa}$$

20.

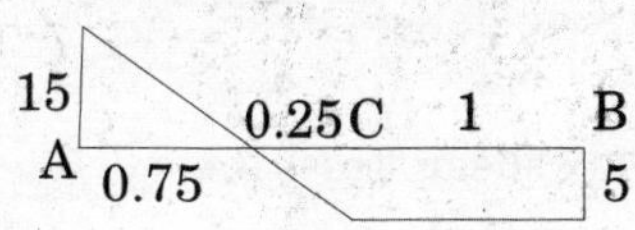

S.F.D.

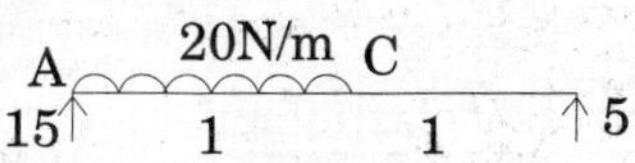

Load dig

BM is man where SF = 0 hence, at 0.75m from A.

$$BM = 15 \times 0.75 - \frac{20 \times 0.75^2}{2}$$

= 5.625 N/m.

21. $ey = \frac{-\mu\sigma_{xx}}{E} - \frac{\sigma_{yy}}{E} = 0$

$$\frac{-0.3 \times 200}{E} = \frac{\sigma_{yy}}{E}$$

$\sigma_{yy} = -60MPa$

$$e_x = \frac{\sigma_{xx}}{E} + \frac{M\sigma_{yy}}{E}$$

$$= \frac{200}{E} + \frac{(0.3x - 60)}{E} = 0.182 \text{ mm}$$

22. Mass of bar $= \rho \times V = 8000 \times \frac{\pi}{4} \times 0.1^2 \times 10$

$= 628.31$ kg

height = 6283.1 N

$$\text{Max bending moment } M = \frac{wl^2}{8} = \frac{6283.1 \times 10^2}{8}$$

$= 78537.5$ N/m

$$\text{Now q max} = \frac{P_1 - P_2}{2}$$

$$P_1 = \frac{16}{\pi d^3}\left[M + \sqrt{M^2 + T^2}\right] \qquad T = 5\pi \text{ KNm}$$

$$P_2 = \frac{16}{\pi d^3}\left[M - \sqrt{M^2 + T^2}\right]$$

$$q_{max} = \frac{16}{2.\pi d^3} \times \left[\sqrt{78537.5^2 + (5\pi \times 1000)^2}\right]$$

$= 253 \times 10^9$

According to Max shear stress theory.

$$q_{max} = \frac{\text{Shear stress at elastic limit}}{\text{FOS}}$$

$$\text{FOS} = \frac{\text{Shear stress at elastic limit}}{q_{man}} = 2.01$$

THERMODYNAMICS

1. dz = xdy – ydx, fails to identify 'z' as another thermodynamics property
2. Internal energy of a thermodynamic system is defined by the first law of thermodynamics

$\boxed{\Delta U = Q - W}$

ΔU = Change in internal energy

Q = Heat added to the system

W = Work done by the system

3. If PV^n = Constant and n = 0, then the process is called Isobaric process (constant pressure process).
4. The relation between $(COP)_{HP}$ and $(COP)_R$ is given as:

$(COP)_{HP} = 1 + (COP)_R$

5. $L_1 = L_2 = L_3$
6.

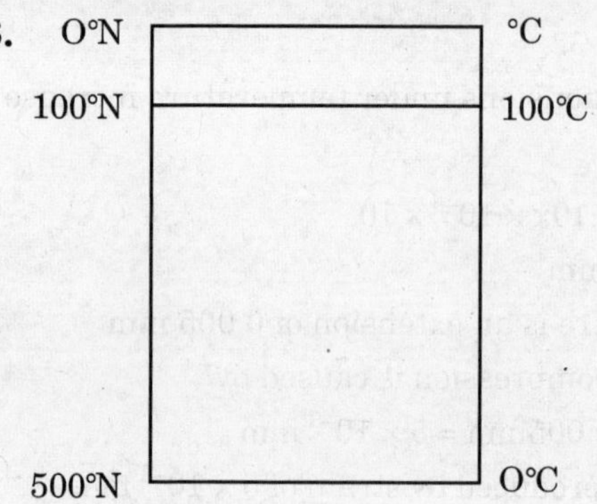

By temperature scales:

$$\frac{100 - 500}{0 - 500} = \frac{100 - 0}{C - 0}$$

$$\text{or } °C = \frac{100 \times 500}{400} = 125\,°C$$

$\therefore$ Required answer is 125 °C

7. Compressibility factor $(Z_1) = \frac{P_1 V}{RT_1}$

Compressibility factor $(Z_2) = \frac{P_2 V}{RT_2}$

Now,

$$\frac{Z_1}{Z_2} = \frac{P_1 V}{RT_1} \times \frac{RT_2}{P_2 V}$$

$$= \frac{2 \times 300}{1 \times 600} = 1$$

Hence, $\frac{Z_1}{Z_2} = 1$

8.

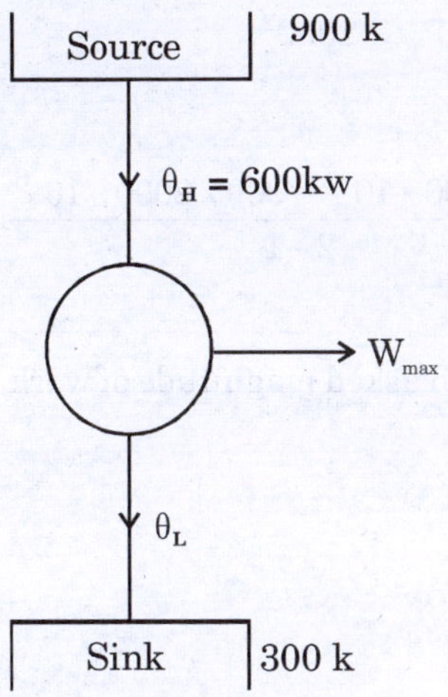

Here, $\eta = \frac{1 - T_L}{T_H} = \frac{W_{max}}{\theta_H}$

or $\frac{1-300}{900} = \frac{W_{max}}{600}$

or $\frac{2}{3} = \frac{W_{max}}{600}$

$\Rightarrow W_{max} = \frac{600 \times 3}{2} = 400$ kw

Now,

we know that,

Irreversibility rate = $W_{max} - W_{actual}$

= (400 – 200)kw

= 200 kw

9. Given that:

Compression ratio (r) = 18

$r_c = 1.7$

$\upsilon = 1.4$

Now,

η (Thermal efficiency)

$$1 - \frac{1}{\upsilon(r)^{\upsilon-1}} \times \frac{r_c^{\upsilon} - 1}{r_c - 1}$$

$$= 1 - \frac{1}{1.4(18)^{0.4}} \times \frac{(1.7)^{1.4} - 1}{1.7 - 1}$$

= 0.6462

= 64.62%

Hence, η = 64.62%

10. $C_V\left[(T_1 - T_0) - T_0 \ln\left(\frac{T_1}{T_0}\right)\right]$

11. By decreasing T_2 and keeping T_1 constant, the thermal efficiency of given reversible heat engine can be increased.

12. Vander Waals Equation:

$$\left(P + \frac{an^2}{v^2}\right)(V - nb) = nRT$$

P, V, T, n are the pressure, volume, temperature and mole of the gas. 'a' and 'b' constant specific to each gas.

Sometimes use n as weight of gas by seeing the unit of a and b.

$$P = \frac{nRT}{V - nb} - \frac{an^2}{V^2}$$

$$= \frac{10 \times 0.187 \times 200}{4 - 10 \times 0.00657} - \frac{0.0687 \times 100}{16}$$

$\boxed{P = 94.6 \text{ kPa}}$

Answers [B]

13. At constant pressure specific entropy of air is given by :

$$\Delta Sm = mCp \ln\left(\frac{T_2}{T_1}\right)$$

ideal gas Pv = mRT

$$T_1 = \frac{86 \times 10^3 \times 1}{0.287} = 299.652$$

$T_2 = 627 + 273 = 900$ k is given

Specific Heat ratio $r = \frac{C_P}{C_V}$ ... (1)

$R = C_P - C_V$

$0.287 = C_P - C_V$

$1.4 = \frac{C_P}{C_V} : C_P = 1.4 \times C_V$

$0.287 = 0.4 \times C_V$

$$C_V = \frac{0.287}{0.4}$$

$$C_P = 1.4 \times \frac{0.287}{0.4} = 1.0045$$

$$m = 1kg$$

$$\Delta Sm = 1 \times 1.0045 \ln\left(\frac{900}{299.652}\right)$$

$$\Delta Sm = 1.10472$$

Option [A]

14. Otto Cycle – Thermal efficiency is given by

$$\eta = 1 - \frac{1}{r^{k-1}}$$

r = compression ratio

k = ratio of specific heat

$$\text{Case I: } \eta_1 = 1 - \frac{1}{4^{1.4-1}} = 0.42565$$

$$\text{Case II: } \eta_2 = 1 - \frac{1}{6^{1.4-1}} = 0.51164$$

Increase in Thermal efficiency

$$\frac{\eta_2 - \eta_1}{\eta_1} \times 100$$

$$= \frac{0.51164 - 0.42565}{0.42565} \times 100 = 20.20\%$$

15. Molecular weight $= \frac{\text{Mass}}{\text{no. of moles}}$

	O_2	N_2	CH_4
Mass	32×0.1	28×0.1	0.8×16
	3.2 kg	2.8 kg	12.8kg

Total weight of the mixture 3.2 + 2.8 + 12.8

= 18.8 kg

Mass fraction of $N_2 = \frac{2.8}{18.2} = 0.148$

Option [c]

16. An adiabatic process pressure and volume of a sample of gas change from (P_1, V_1) to (P_2, V_2)

Work done is given by

$$W = \frac{P_1V_1 - P_2V_2}{\gamma - 1}$$

r = specific heat ratio

= 2 $\quad P_1V_1^2 = P_2V_2^2$

$100 \times 600^2 = P_1 \times 2000^2$

$P_2 = 900$ kPa.

$$W = \frac{100 \times 6000 \times 10^{-2} - 900 \times 2000 \times 16^{-6}}{2-1}$$

W = – 1.2 kJ

In the question asked magnitude of work transfer so

$W = |-1.2|$

= 1.2 kJ

17. P_1= 1MPa

T_1= 773 k

A_1= 40 cm ^{2} $\longrightarrow$ T_2= ?

V_1= ? $\quad$ V_2= 10V $_1$

Since the flow is steady the mass flow rate through the entire nozzle well remain constant

specific volume at inlet $v_1 = \frac{RT}{P_1} = \frac{0.19 \times 773}{1000}$

= 0.14887 m³/kg

Mass flow rate $\dot{m} = \frac{A_1V_1}{v_1}$

$$15 = \frac{40 \times 10^{-4} \times V_1}{0.14687}$$

$V_1 = 55.08$ m/s

$$\frac{C_P}{C_V} = \gamma = 1.29$$

$C_r - C_V = 0.19$ k J/kg K

$1.29\ C_p - C_V = 0.19$

$$C_V = \frac{0.19}{0.29} \text{ kJ/kg k}$$

$$C_P = 1.29 \times \frac{0.19}{0.29} \text{ kJ/kg.k}$$

= 0.8452 kJ/kg k

= 8452 J/kg. k

Exit temp T_2 at end of the nozzle given by

$$T_2 = T_1 - \frac{V_2^2 - V_1^2}{2C_P}$$

$$= 773 - \frac{100V_1^2 - V_1^2}{2C_P}$$

$$= 773 - \frac{99 \times (5508)^2}{2 \times 8452}$$

$$= 773 - 177.68$$

$$\boxed{T_2 = 595.32k}$$

18. Heat transfer = Work done – internal energy

Internal energy = 40 kJ/kg × 8kg – 320 kJ

Work done by the system

$$W = \frac{P_1V_1}{\gamma - 1}\left[1 - \left(\frac{P_2}{P_1}\right)^{\frac{\gamma-1}{\gamma}}\right]$$

$P_1 = 1$ MPa = 1000 k pa

$V_1 = 1\ m^3$

$\gamma = 1.2$

$$W = \frac{1000 \times 1}{1.2 - 1}\left[1 - \left(\frac{5}{1000}\right)^{\frac{1.2-1}{1.2}}\right]$$

W = 2932.407 kJ

U = 320 kJ

Heat transfer = 2932.41 – 320

= 2612.41 kJ

19. $\ln\left[\frac{P_2}{P_1}\right] = \frac{\Delta H_y}{R}\left[\frac{1}{T_1} - \frac{1}{T_2}\right]$

$P_1 = 0.8725$

$T_1 = 278k$

$T_2 = 283k$

$\Delta H_v = 24891$

$R = 0.4615$

$$\ln\left[\frac{P_2}{0.8725}\right] = \frac{24891}{0.4615} \times \left[\frac{1}{278} - \frac{1}{283}\right]$$

$P_2 = 1.229$ kPa

20. Rankine Cycle

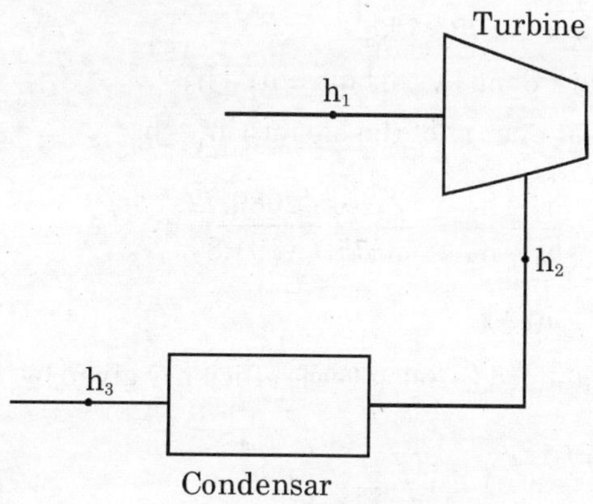

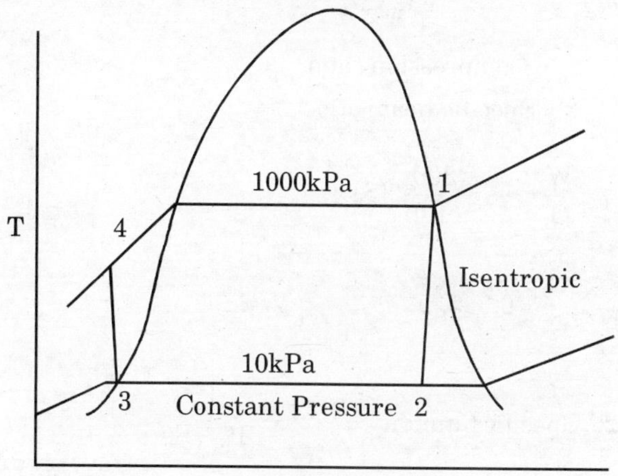

η = Work done by the steam turbine heat energy of steam.

$$\eta = \frac{\text{Work done by the steam turbine}}{\text{heat energy of steam}}$$

$P_1 = 10000$ kPa

$T_1 = 500\,°C$

$h_1 = 3375.1$ kJ/kg

$S_1 = 6.5995$ kJ/kg.k

$P_2 = 10$ kPa   hf = 191.8 kJ/kg   Sf = 0.6492

Sat. hq,   hfg = 2392.1 kJ/kg,   stg = 7.4996

f = for saturated liquid

fg = fg saturated vapor

Point 1 – 2 process in isentropic

$S_1 = S_2 = S_f + x\,S_{fg}$

$6.5995 = 0.6492 + x \times 7.4996$

$x = 0.79341$

Similar value of enthalpy at h_2

$h_2 = h_f + xh_{fg}$

$= 191.8 + 0.79341 \times 2392.1$

$h_2 = 2089.72$ kJ/kg

Work done by turbine $= h_1 - h_2$

Heat energy of the steam $= h_1 - h_3$

$$\eta = \frac{h_1 - h_2}{h_1 - h_3} = \frac{3375.1 - 2089.72}{3375.1 - 191.8}$$

$\eta = 40.4\%$

21. Standard Braylon Cycle efficiency given by

$$\frac{W}{Q} = \eta = 1 - \left(\frac{T_{min}}{T_{max}}\right)(r_p)^{\frac{K-1}{K}}.$$

r_p = Compression ratio

k = Specific heat ratio

$$\frac{W}{Q} = 1 - \left(\frac{300}{1100}\right)(6)^{\frac{1.4-1}{1.4}}$$

$$\frac{W}{Q} = 0.545.$$

22. Specific humidity $= 0.622 \times \frac{P_V}{P_b - P_V}$

P_V = Partial pressure of Vapour

P_b = Barometric pressure = 100 kPa

Specific humidity = 0.015 kg of vapour per kg of dry an.

$$0.015 = 0.622 \times \frac{P_V}{100 - P_V}$$

$P_V = 2.354$ kPa

POLYMER SCIENCE AND ENGINEERING

1. **Chain transfer** is a polymerization reaction by which the activity of a growing polymer chain is transferred to another molecule. Chain transfer reactions occur in most forms of addition polymerization including radical polymerization, ring-opening polymerization, coordination polymerization, and cationic polymerization, as well as anionic polymerization. Common chain transfer agents include thiols, especially dodecyl mercaptan (DDM), and halocarbons such as carbon tetrachloride.
2. Anionic polymerization is a form of chain-growth polymerization that encompasses the polymerization of vinyl monomers with strong electronegative groups. This type of polymerization is often employed to produce synthetic polydiene rubbers, solution styrene-butadiene rubbers (SBR), and thermoplastic styrenic elastomers.

   All monomers with (strong) electronegative substituents polymerize readily in the presence of carbanions. Some electron-withdrawing substituents that stabilize the negative charge through charge delocalization, and hence permit anionic polymerization include – CN, – COOR, $-C_6H_5$, and $-CH=CH_2$, to name only a few. Therefore, monomers such as styrenes, dienes, acrylates and methacrylates, aldehydes, epoxides, acrylonitriles and cyanoacrylates readily undergo anionic polymerization reactions.
3. When it comes to selecting the most suitable base elastomer when low temperature flexibility must be a key characteristic of a product or component, we can clearly see that, silicone is the most attractive option, offering an excellent level of flexibility when operating at low temperatures.

   Silicone stays flexible at the lowest temperatures of all elastomers, without noticeable change in its properties. When silicone is compared to other elastomers, it is clearly seen that it stays elastic at lower operating temperatures whilst other such materials become brittle.
4. HDPE (High Density Polyethylene) has minimal branching of its' **polymer** chains. Because it is denser it is more rigid and less permeable than the LDPE. It has a density of 0.941-0.965g/cm^3.
5. The nearest value of conductivity of nylon 6 is $10^{(-13)}$ S/m.
6. **Aramid**, also known as **aromatic polyamide**, any of a series of synthetic polymers (substances made of long chainlike multiple-unit molecules) in which repeating units containing large phenyl rings are linked together by amide groups. Amide groups (CO-NH) form strong bonds that are resistant to solvents and heat.
7. A miscible blend in 1:1 (by weight) composition is formed with polystyrene and poly(phenylene oxide) .
8. Dicumyl peroxide is mainly used as cross-linking agent for polymers and elastomers. Polymers which can be cross-linked with organic peroxides are used to produce hose, wires, tires, rubber seals, etc. Dicumyl peroxide can also be used as flame-retardant synergist in expanded polystyrene(EPS). In that case, the peroxide is incorporated in small quantities in EPS, and will catalyse the flame retardant action in case of heating.

9. Stress relaxation is a time-dependent decrease in stress under a constant strain. This characteristic behaviour of the polymer is studied by applying a fixed amount of deformation to a specimen and measuring the load required to maintain it as a function of time.

10.

Polymers	Polymerization Method
Bisphenol A Polycarbonate	Step- growth
Polyethylene	Coordination
Poly(styrene-b-butadiene)	Anionic
Butyl rubber	Cationic

Step-growth polymerization refers to a type of **polymerization mechanism** in which bi-functional or multifunctional monomers react to form first dimers, then trimers, longer oligomers and eventually long **chain polymers**.

Coordination polymerisation is a form of **polymerization** that is catalyzed by transition metal salts and complexes.

Anionic polymerization is a form of chain-growth **polymerization** that encompasses the **polymerization** of vinyl monomers with strong electronegative groups.

Cationic polymerization is a type of chain growth **polymerization** in which a **cationic** initiator transfers charge to a monomer which then becomes reactive.

11.

Processing Technique	**Products**
Blow molding	Plastic bottles
Rotational Molding	Basket ball
Injection molding	Plastic buckets
Calendering	PVC sheet

Blow molding is a molding process used in the manufacturing industry to create hollow objects made of plastic.

Rotational Molding is used mainly for producing hollow parts by adding plastic powder to a shell-like mold, and then rotating and heating it at the same time.

Injection molding is a manufacturing process commonly used for fabricating items from plastic trinkets and toys to automotive body parts, cell phone cases, water bottles, and containers.

Calendering is a speciality process for high-volume, high quality plastic film and sheet, mainly used for PVC as well as for certain other modified thermoplastics.

12.

Characterization Technique	**Polymer Attributes**
Gel permeation chromatography	Molecular weight
FT- IR spectroscopy	Functional group
Differential Scanning calorimetry	Glass transition temperature
X- ray diffraction	Crystal structure

13.

Additive	**Functions**
Azodiformamide	Blowing agent
Di- octyl phthalate	Plasticizer
Benzoyl peroxide	Curing agent
2-(2'-hydroxy phenyl) benzotriazole	UV Stabilizer

14. P-Bingham Plastic, Q- Pseudoplastic, R- Dilatant, S- Newtonian

The four basic types of time-independent fluids are shown in the figures below:

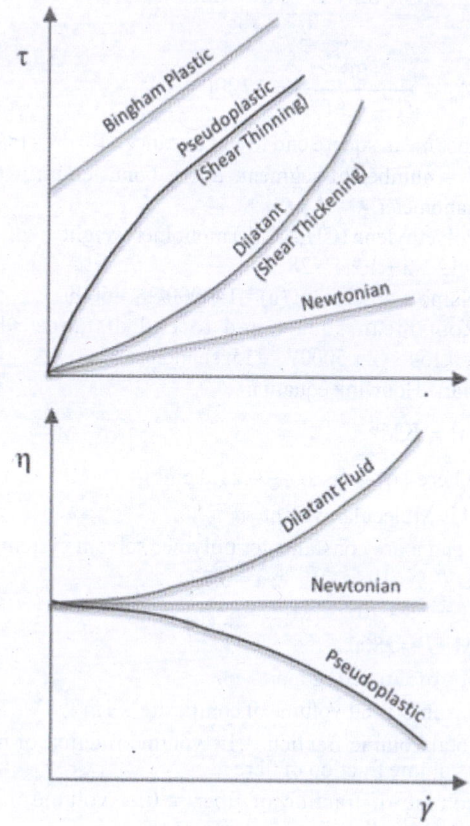

15. The reaction 46of adipic acid and ethylene glycol

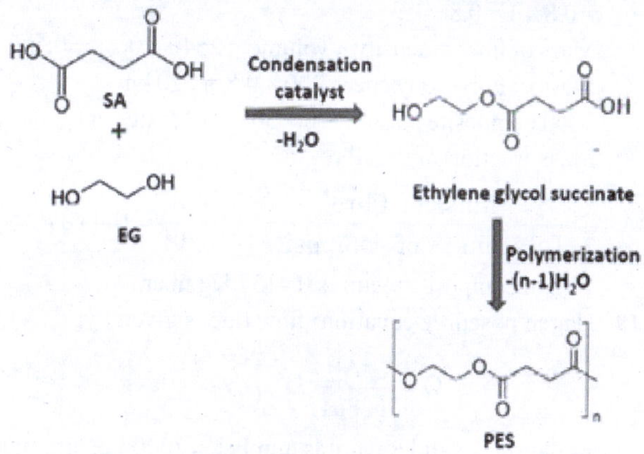

Number average molecular weight:

$$M_n = \frac{M_o}{(1-p)}$$

$$P = 0.995$$

M_0 = molecuar weight of the monomer

$$= \frac{\text{Mole of monomer 1} \times \text{Mol. wt. monomer 1} + \text{mole of monomer 2} \times \text{Mol. wt. of monoer 2} - \text{water molecules weight}}{\text{Total mole}}$$

$$M_o = \frac{1 \times 62 + 1 \times 146 - 36}{2} = 86$$

$$M_n = \frac{86}{(1 - 0.995)} = 17200$$

16. Root mean square end to end distance $\langle R^2 \rangle^{1/2} = l \langle 2n \rangle^{1/2}$

n = number of segment, & l = bond length = 0.154 nanometer.

Polyethylene ($CH_2 - CH_2$) monomer weight $= 12 * 2 + 1 * 4 = 28$

Number of segment (n) = 140000/28 = 5000

Root mean square end to end distance $\langle R^2 \rangle^{1/2} = 0.154 \times (2 \times 5000)^{1/2} = 15.4$ nanometer

17. Mark-Houwink equation

$[\eta] = KM^a$

Where $[\eta]$ = viscosity = 21.0 cm³/g

M is Molecular weight

k and a are constants for polymer solvent system.

$K = 7.5 \times 10^{-3} cm^3/g$ & $a = 0.72$

$21 = 7.5 \times 10^{-3} \times [M]^{0.72}$

$[M]^{0.72} = 2800$

M = 61336.5703 g/mol

18. Assume Total volume of composite = 1m³

Total volume fraction = 1 = volume fraction of matrix + volume fraction of fibre

Volume of fraction of fiber = 0.2; volume of fiber $= 0.2 \times 1 = 0.2$

Volume fraction of matrix = 0.8; volume of matrix $= 0.8 \times 1 = 0.8$

Mass of fiber = density × volume = 2540 × 0.2 = 508 kg

Mass of Polypropylene = 900 × 0.8 = 720 kg

Total composite mass = 720 + 508 = 1228 kg

Mass fraction of the fibre

$$= \frac{\text{Mass of fibre}}{\text{Total mass of composite}} = \frac{508}{1228} = 0.4137$$

1 kg of composite requires 0.4137 Kg fiber.

19. Hagen poseuille equation: flow rate is given by:

$$Q = \frac{\pi \Delta P}{128 \mu L} D^4$$

Equation is valid Reynolds number < 2000 i.e. laminar flow.

Q = flow rate;

ΔP = pressure drop;

μ = viscosity of the fluid;

L = length of the path

D = diameter of pipe = 2 × radius

$Q = 10^{-9} m^3/s$;

$D = 2R = 2 \times 2 \times 10^{-3} m$;

$\mu = 10^2 Nsm^{-2}$

$$\frac{\Delta P}{L} = \frac{128 Q_{\mu}}{\pi D^4} = \frac{128 \times 10^{-9} \times 10^2}{3.14 \times (2 \times 2 \times 10^{-3})^4} = 15923.57$$

20. Bulk modulus (β)

$$\beta = \frac{\Delta P}{\frac{\Delta V}{V}}$$

Bulk modulus (β) = 1 GPa; applied pressure = 200 MPa

$$\frac{\Delta V}{V} = \frac{\text{pressure applied}}{\beta} = \frac{200}{1 \times 10^3} = 0.2$$

21. Isoprene monomer weight = 68 g/mol & Sulphur molecular weight = 32 g/mol

Assume weight of isoprene = 100gm & Number of moles of isoprene = 100/68 = 1.47 mol

The mass of sulphur required for 40 percent cross-linking is therefore

Mass of sulphur = 0.40 × 1.47 × 32 = 18.816 gm

Total weight of isoprene & sulphur = 100+ 18.816 = 118.816 gm

Weight fraction of the sulphur = (18.816/118.816) × 100 = 15.84%

22. True stress, S, will actually occur at y = strain

$$S = \frac{S_o L}{L_o} = S_o(1 + \gamma);$$

$$\gamma = \frac{L - L_o}{L_o} = \frac{150 - 100}{100} = 0.5$$

$F = 160N; A = 4 * 4 = 16 mm^2$

$S_o = F/A = 160/(16) = 10 N/mm^2$

$S = 10 \times (1 + 0.5) = 15 N/mm^2$

FOOD TECHNOLOGY

1. The enzyme majorly involved in postmortem degradation of muscle proteins is calpain.

Enzymes of the calpain family are known to play a major role in muscle protein degradation.

They are responsible for proteolytic degradation of a variety of proteins, including desmin and tropomyosin.

2. Essential fatty acids should not be confused with essential oils, which are "essential" in the sense of being a concentrated essence.

Only two fatty acids are known to be essential for humans: alpha-linolenic acid (an omega-3 fatty acid) and linoleic acid (an omega-6 fatty acid).

Hence, lenoleic acid and linioletic acid are the correct pair of essential fatty acids.

3. Production of Nisin by Lactococcus lactis in Media with Skimmed Milk. Nisin is a bacteriocin that inhibits the germination and growth of Gram-positive bacteria.

Nisin production was improved under nutritional limitation in fed-batch culture of Lactococcus lactis.

4. Bacillus subtilis is a bacteria will stain purple color after gram staining. Bacillus subtilis is a rod-shaped, Gram-positive bacterium that forms spores.

Subtilis to become a paradigm for the study of spore formation and of low GC Gram-positive bacteria. Gram-positive bacteria have a thick mesh-like cell wall made of peptidoglycan (50-90% of cell envelope), and as a result are stained purple by crystal violet, whereas gram-negative bacteria have a thinner layer (10% of cell envelope), so do not retain the purple stain and are counter-stained pink by safranin.

5. Glucose (0.3-0.5%) is removed from the liquid egg white prior to drying by fermentation using a yeast or bacterial culture or by oxidation to gluconic acid using a glucose oxidase-catalase enzyme system.

With glucose removed, dried egg whites are completely stable. So the enzyme system used for removal of glucose from egg white prior to its drying consists of glucose oxidase and catalase.

6. The most common causes of peptic ulcers are infection with the bacterium Helicobacter pylori (H. pylori) and long-term use of aspirin and nonsteroidal anti-inflammatory drugs (NSAIDs) (Advil, Aleve, others).

Stress and spicy foods do not cause peptic ulcers. However, they can make your symptoms worse. Where as bacillus subtilis is a bacteria will stain purple color after gram staining.

Hence, Peptic ulcers- Bacillus subtilis is incorrect pair of food illness and its causative microorganism.

7. Sodium benzoate is a substance which has the chemical formula C_6H_5COONa. It is a widely used food preservative, with an E number of E211.

Sodium benzoate inhibits the growth of potentially harmful bacteria, mold, and other microbes in food, thus deterring spoilage. It's particularly effective in acidic foods.

Therefore, it's commonly used in foods, such as soda, bottled lemon juice, pickles, jelly, salad dressing, soy sauce, tomato sauce, and other condiments.

Tomato juice also prepared with sodium benzoate was more stable than others and spoiled at the end of the 60 days storage period at room temperature.

8. Settling velocity in centrifuge is given by:

$$V_1 = \frac{\omega_1^2 r d_1^2}{18\mu}(\rho_p - \rho_s) \quad \text{... (1)}$$

ω = angular velocity

r = roter radius

d_1 = particle diameter

ρ_p = density of particle

ρ_s = density of solution

$$V_2 = \frac{\omega_2^2 r d_2^2}{18\mu}(\rho_p - \rho_s) \quad \text{... (2)}$$

(1) ÷ (2)

$$\frac{V_1}{V_2} = \frac{\omega_1^2 d_1^2}{\omega_2^2 d_2^2}$$

$$\frac{0.25}{V_2} = \frac{6000^2 \times 2.2^2}{7500^2 \times 1.5^2}$$

$$V_2 = 0.181 \text{ mms}^{-1}$$

9. (18 min)

The initial population of 10^4 cells/mL of a bacterial species increases to 10^6 cells/mL in 120 minutes, its generation time will be

$$G = \frac{0.3 \times 120}{6-4} = 18 \text{ min}$$

10. 1. **Zein - Maize:** Zein is a prolamine and an alcohol soluble protein present in maize endosperm cells.

2. **Gluten - Wheat:** Gluten is a family of proteins found in grains, including wheat, rye, spelt, and barley. Of the gluten-containing grains, wheat is by far the most common.

3. **Glycinin - Soybean:** Glycinin (11S globulin) and β-conglycinin (7S globulin) are the most important soybean proteins.
4. **Ovalbumin - Egg:** Ovalbumin (abbreviated OVA) is the main protein found in egg white, making up approximately 55% of the total protein.

11. 1. **Pectin - polygalacturonase:** Polygalacturonase is an important pectin degrading enzyme. The western honey bee (Apis mellifera L.) collects pollens from different flowers which are rich source of pectin. The microbiota in the gut of honey bee, release polygalacturonase enzyme and help in pectin digestion.
2. **Lactose - β galactosidase**: β-galactosidase, also called lactose, beta-gal or β-gal, is a glycoside hydrolase enzyme that catalyzes the hydrolysis of β-galactosides into monosaccharides through the breaking of a glycosidic bond.
3. **Hemicellulose - xylanase:** Xylanase (CAS number: 9014-63-5) is a group of hemicelluloses that represents the third most abundant biopolymer on Earth.
4. **Inulin - β fructofuranosidase:** Growth on inulin included one encoding a β-fructo-furanosidase protein that was prominent in the proteome of inulin-grown cells.

12. 1. **Degumming- Removal of phosphatides:** Degumming is a process for removal of phosphatides from crude soybean and other vegetable oils to improve physical stability and facilitate further processing. The phosphatides are also called gums and lecithin.
2. **Neutralization- Removal of free fatty acids:** Neutralization is the process of removing free fatty acids from crude oils by using caustic soda or lime. After addition of alkaline solutions, soap is formed in the oil and removed by washing with hot water. Then, the oil is dried in order to remove the remaining water in the neutralized oil.
3. **Bleaching - Removal of phosphatides:** For palm oil, often dry degumming is applied, in which concentrated acid is added to the oil, removing the phosphatides in the subsequent bleaching process. It means that the added acid is not removed by centrifuging, but in the bleaching step.
4. **Winterization - Removal of free fatty acids:** Winterization is an oil refinement technique commonly used in biotechnology for removal of free fatty acids. The process involves dissolving the extract in ethanol, then placing the mixture in a freezer to chill. The cold allows for the separation of compounds by differences in their melting and precipitation points.

13. 1. **Coffee - arabica:** Arabica-Coffea, also known as the Arabian coffee, "coffee shrub of Arabia", "mountain coffee" or "arabica coffee", is a species of Coffea.
2. **Cocoa - theobroma:** Theobroma-cacao, also called the cacao tree and the cocoa tree, is a small (4-8 m (13-26 ft) tall) evergreen tree in the family Malvaceae, native to the deep tropical regions of Mesoamerica. Its seeds, cocoa beans, are used to make chocolate liquor, cocoa solids, cocoa butter and chocolate.
3. **Beer - wort:** Wort is "beer starter" – comprised of malt extract (from grain mash) and water.
4. **Wine - must:** Must (from the Latin vinum mustum, "young wine") is freshly crushed fruit juice (usually grape juice) that contains the skins, seeds, and stems of the fruit.

14. 1. **Lye solution - caustic peeling:** A method of removing skins from vegetables by immersion in hot caustic soda solution (lye) followed by 'tumbling' in a wash to remove the skin and chemicals is called lye solution peeling.
2. **Carborundum rollers - abrasion peeling:** In abrasion peeling, the material to be peeled is fed onto carborundum rollers or fed into a rotating bowl, which is then lined with carborundum. The abrasive carborundum surface removes the skin, which is then washed away with a copious supply of water. The process is normally carried out at ambient temperature.
3. **Pressure vessel - flash peeling:** Flash steam peeling or flash peeling fruits/vegetables are fed in batches into a pressure vessel, which is rotated at 4-6-rev min".
4. **Conveyor belt - flame peeling:** Flame peeling is used in onions, garlic and brinjal. The peeler consists of a conveyor belt which carries and rotates food through a furnace heated to more than 1000°C. The outer layer and root hairs of onion are burnt off and charred skin is

removed manually.

15. 1. **Homogeneous population:** Using a continuous microbial culture system which provides homogeneous population culture environment. The homogeneous population model assumes equal expected value and variance for the variable of interest for all population units.

2. **Mixed population of heat sensitive and heat resistant microbes:** The nature of a microbial population culture is influenced by biological, chemical, and physical factors. In mixed populations, the end products of metabolism of one organism may be unfavorable to the extreme for certain other forms. Perfringens can be isolated from mixed population cultures. Perfringens differ widely in their resistance to heat, and heat-sensitive and heat-resistant strains have been observed.

3. **Germination of spores:** Germination of spore and outgrowth progression are often very heterogeneous and therefore, predictions of nature of microbial culture stability of food products are exceedingly difficult. Vegetative cells are formed by germination of spores followed by a more extended outgrowth phase.

16. 1. **Plank's equation - Freezing time:** Planck's equation for predicting freezing times ignores sensible heat and the gradual character of the phase-change process in foodstuffs.

2. **Arrhenius equation - Activation energy:** The Arrhenius equation allows us to calculate activation energies if the rate constant is known, or vice versa.

As well, it mathematically expresses the relationships we established earlier: as activation energy term Ea increases, the rate constant k decreases and therefore the rate of reaction decreases.

3. **Guggenheim - Anderson de Boer equation - Monolayer moisture content:**

Moisture isotherms model

The Guggenheim- Anderson de Boer equation (GAB) equation was used to describe the dry basis moisture content (X) as a function of water activity (a_w):

$$X = \frac{CkX_0a_w}{\left[(1-ka_w)(1-ka_w+Cka_w)\right]}$$

Where, C and k are constants and X_0 is described in the Literature as the monolayer moisture content on dry basis (kg/kg).

4. **Stoke's law- Terminal velocity:** Stokes law is the basis of the falling-sphere viscometer, in which the fluid is stationary in a vertical glass tube. A sphere of known size and density is allowed to descend through the liquid. If correctly selected, it reaches terminal velocity, which can be measured by the time it takes to pass two marks on the tube. Electronic sensing can be used for opaque fluids.

Knowing the terminal velocity, the size and density of the sphere, and the density of the liquid, Stokes' law can be used to calculate the viscosity of the fluid.

17. 1. **Oxygen absorber - ferric oxide:** Oxygen absorbers are made of a chemical compound, the active ingredient of which is a powdered iron oxide or ferric oxide. Oxygen absorbers are applied to a wide variety of food types for preservation including bread and biscuits, fruit and vegetables, nut products, fish and seafood products, and meat products.

2. **Carbon dioxide - magnesium oxide:** The carbonate layer formed on the surface during the initial stage of adsorption hinders further adsorption of carbon dioxide molecules. This results in low performance of pure magnesium oxide.

3. **Ethylene absorber - Potassium Permanganate:** Ethylene scavenger is important to control the presence of ethylene in food packages to improve the shelf life of food product. Silica embedded potassium permanganate is the most common ethylene absorbing agent in use currently. now silica absorbs ethylene which in turn is converted to ethylene glycol by potassium permanganate.

4. **Moisture absorber - Calcium Chloride:** Calcium chloride ($CaCl_2$) effectively absorbs moisture from the air. It can attract several times its own weight in water, dissolving into a liquid brine if the air is humid enough and the temperature is high enough.

18. During extrusion cooking, food materials are generally subjected to high temperature and pressure in combination with high shearing stress.

Physical and chemical changes include starch gelatinization, protein denaturization and many other macro- and micronutrient interactions.

19. Reynolds Number (Re) = $\frac{\rho VD}{\mu}$

ρ = density of milk

V= velocity of milk

D = diameter of pipe

μ = Visocisty

Volume= A_1V_1

$$\text{Velocity } V = \frac{\text{Volume}}{A} = \frac{3\times10^{-3}}{\frac{\pi}{4}\times0.04\times0.04}$$

V = 2.39 m/s

$$Re = \frac{1032\times2.39\times4\times10^{-2}}{0.2}$$

$$\boxed{Re = 493.296}$$

20. Bond's Law:

$$\frac{P}{m} = Kb\left[\frac{1}{\sqrt{x_2}} - \frac{1}{\sqrt{x_1}}\right]$$

P = Power required

m = Feed Rate

x_1 = larger particle size – feed

x_2 = smaller particle size – product

Case I:

$$\frac{10}{108} = Kb\left[\frac{1}{\sqrt{1.25}} - \frac{1}{\sqrt{3.92}}\right] \quad \ldots (1)$$

Case II:

$$\frac{10}{f} = Kb\left[\frac{1}{\sqrt{0.75}} - \frac{1}{\sqrt{3.92}}\right] \quad \ldots (2)$$

(1) ÷ (2)

$$\frac{f}{108} = \frac{\frac{1}{\sqrt{1.25}} - \frac{1}{\sqrt{3.92}}}{\frac{1}{\sqrt{0.75}} - \frac{1}{\sqrt{3.92}}}$$

$$\frac{1}{108} = 0.599$$

Feed rate = 0.599 × 108 = 64.73 ton/h

21. Thermal efficiency of spray drying η can be approximately expressed as

$$\eta = \frac{T_1 - T_2}{T_1 - T_0}\times100\%$$

T_1 = inlet temperature (°C) = 132°C

T_2 = outlet temperature (°C) = 80°C

T_0 = ambient temperature = 29°C

$$\eta = \frac{132-80}{132-29}\times100 = 50.48\%$$

22.

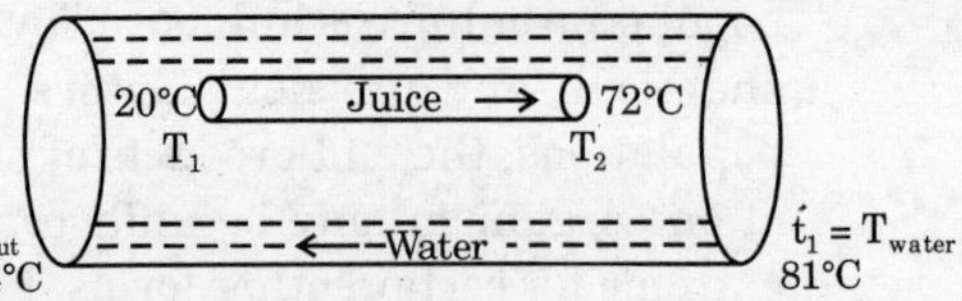

$$A = \frac{Q}{U\times\Delta T_m}$$

A = Heat Transfer Surface Area = ?

Q = Heat Transfer Rate = ?

U = Overall Heat transfer difference

ΔT_m = log mean temp. difference = ?

(1) Calculate Q. heat transfer rate

$Q = mC_p(T_2 - T_1)$

m = 0.80 kg/s

C_p = 3.74 kJ/kg.K

= 0.80 × 3.74 × (72 – 20)

= 155.584 kJ/sec.

$$A = \frac{155.584\times1000}{492\times25.14} = 12.57 \text{ m}^2$$

(2) ΔT_m

$$\Delta T_m = \frac{(T_1 - t_2) - (T_2 - t_1)}{\ell n\left[\frac{T_1 - t_2}{T_2 - t_1}\right]}$$

$T_1 = 20$ $\quad t_1 = 72$

$t_1 = 81$ $\quad t_2 = 74$

$$\Delta T_m = \frac{(20-74)-(72-81)}{\ell n\frac{20-74}{72-81}}$$

$\Delta T_m = -25.14$

$\Delta T_m = |-25.14| = 25.14$

ATMOSPHERIC AND OCEANIC SCIENCES

1. In the northern hemisphere, before friction begins to act we have a geostrophic current with Coriolis force acting to the right and pressure gradient force to the left. With a barotropic case i.e., near the bottom of the ocean the pressure gradient is independent of depth. As bottom is approached, friction slows the flow, the Coriolis force decreases and the pressure gradient to the left is not completely balanced. The flow swings to the left until the vector sum of the Coriolis and friction forces can balance the pressure gradient force.

2. Monsoon is traditionally defined as a seasonal reversing wind accompanied by corresponding changes in precipitation, but is now used to describe seasonal changes in atmospheric circulation and precipitation associated with the asymmetric heating of land and sea.

3. Anthropogenic emission of **Carbon Dioxide** is the main contributor to the ongoing ocean acidification.

   Ocean acidification is the ongoing decrease in the pH of the Earth's oceans, caused by the uptake of carbon dioxide (CO_2) from the atmosphere. Seawater is slightly basic (meaning pH > 7), and ocean acidification involves a shift towards pH-neutral conditions rather than a transition to acidic conditions (pH < 7).

4. **Phytoplankton,** also known as microalgae, are similar to terrestrial plants in that they contain chlorophyll and require sunlight in order to live and grow. Most phytoplankton are buoyant and float in the upper part of the ocean, where sunlight penetrates the water. Phytoplankton also require inorganic nutrients such as nitrates, phosphates, and sulfur which they convert into proteins, fats, and carbohydrates.

5. As we know Potential virtual temperature is given as

$$\theta_v = T\left(\frac{P_s}{P}\right)^{\frac{R}{C_p}}$$

$$\therefore \quad \theta_v \propto T$$

So, at $z = 0$

Temperature is less before sunrise which is almost at 3 am thus, θ_v is also near to zero at z = 0 depicted in graph (*ii*)

Similarly,

In graph (*i*) temperature is highest at ground around 3 pm and also θ_v.

6. Angle of south-easterly wind blowing towards 315°

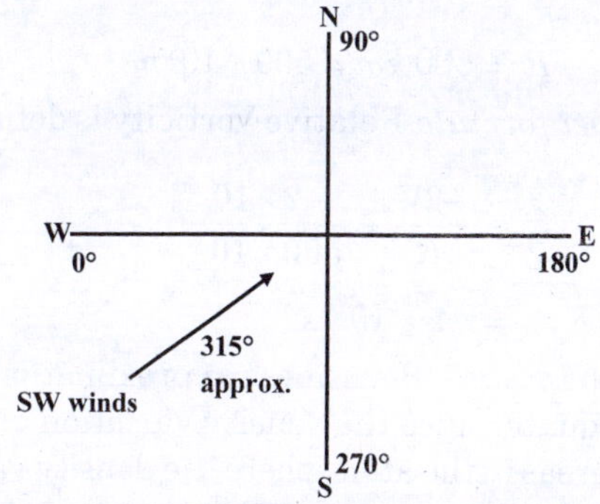

7. *Given,*

Temperature of parcel,

$$T_p = 30°C$$
$$= 273 + 30 = 303 \text{ K}$$

Temperature of environment,

$$T_e = 25°C$$
$$= 273 + 25 = 298 \text{ K}$$

Buoyancy force per mass is given as

$$\frac{F}{m} = \left[\frac{T_p - T_e}{T_e}\right] \times -g$$

$$= \left[\frac{303 - 298}{298}\right] \times -10 = -0.16\ m/s^2$$

8. Given,

Transmiting Energy = 20%

Reflective Enery = 60%

∴ As per formula

Emissivity + Transmiting Energy + Reflective Energy = 100%

⇒ Emissivity = 100 – (60+20)

= 20% = 0.2

9. Given,

$$\Omega = 7.3 \times 10^{-5}\ S^{-1}$$
$$\theta = 30°\ N$$

Now,

$$\text{Time} = \frac{\text{Distance}}{\text{Velocity}} = \frac{2\pi}{\Omega}$$

$$= \frac{2\pi}{7.3 \times 10^{-5}} = 86071.03\ s$$

$$= 23.9 \approx 24 \text{ hrs.}$$

10. *Given,*

$$V = 10\ m/s$$

$$R = 500\ km = 500 \times 10^3\ m$$

As per formula Relative Vorticity is defined

$$\varsigma = \frac{-2V}{R} = \frac{-2 \times 10}{500 \times 10^3}$$

$$= -4 \times 10^{-5}\ s^{-1}$$

11. In the oceans, Boussinesq approximation is quite adequate, since the density variation are small, whereas in the atmosphere the density variations are large, thus methods like anelastic approximation, which filters out acoustic waves but retains density changes are often used in atmospheric models.

12. The saturation vapor pressure sets a limit on the maximum possible water vapor that can be in the air at a given air temperature. The actual amount of water vapor in the air (measured by the vapor pressure) is often less than the saturation amount.

Therefore, the water vapour content of the atmosphere will **increase** and the net warming will be **more** than 2°C.

13. The trade winds blowing from east to west along the equator allowing the upwelling of cold, nutrient rich water from deeper levels off the northwest coast of South America. In addition, these same trade winds push water west which piles higher in the Western Pacific. The trade winds, in piling up water in the Western Pacific, make a deep 150 meters warm layer in the west that pushes the thermocline down, while it rises in the east.

14. Since, the two layers of the water of same density on top of each other. Thus, the convective plumes across the interface of the two layers flows are based on the temperature difference only. Therefore, Downward convective plumes in figure (i) and upward convective plumes in figure (ii).

15. During the Indian summer monsoon surface Outgoing longwave radiation over the Arabian Sea is often observed to be low because monsoon clouds limit the incoming solar radiation.

16. *Given,*

$$g = 100\ m/s$$

$$h = 1\ km = 1000\ m$$

$$c = \sqrt{gh} = \sqrt{10 \times 1000}$$

$$= 100\ m/s$$

17. *Given,*

$$R = 100\ m$$

$$V = 100\ m/s$$

$$\rho = 1\ kg/m^3$$

As per formula

$$-\frac{\partial P}{\partial x} = \frac{V^2}{R}$$

$$\Rightarrow -\left(\frac{P_2 - P_1}{x_2 - x_1}\right) = \frac{V^2}{R}$$

$$\Rightarrow P_1 - P_2 = \frac{V^2}{R} \times (x_2 - x_1)$$

$$= \frac{100^2}{100} \times 100 = 10000$$

18. *Given,*

$$T_1 = 10°C$$

$$= 273 + 10 = 283\ K$$

$$T_2 = 20°C$$

$$= 273 + 20 = 293\ K$$

$$g = 10\ m/s$$

$$f = 10\ m/s$$

As per formula

$$\Rightarrow V_g = \frac{g}{f}\left(\frac{T_2 - T_1}{T_1}\right)$$

$$\Rightarrow V_g = \frac{10}{10^{-4}}\left(\frac{293 - 283}{283}\right) = 353.56\ m/s$$

The vertical gradient of geostropic wind is

$$V_T = \frac{V_g}{R} = \frac{353.56}{1000}$$

$$= 3.53\ m/s/km$$

19. *Given,*

$$v = 10\ m/s$$

$$\rho_w = 1000\ kg/m^3$$

$$f = 10^{-4} s^{-1}$$

$$C_d = 10^{-6}$$

As per formula

$$fM_{yE} = -T_x$$

$$\therefore \quad T_x = \rho C_d V^2$$

$$M_{yE} = \frac{-T_x}{f} = \frac{\rho C_d V^2}{f}$$

$$= \frac{1000 \times 10^{-6} \times 10^2}{10^{-4}}$$

$$= \frac{0.1}{10^{-4}} = -1000$$

20. The mass transport M for a fluid flow is given by

$$M = \int_{L_2}^{L_1}\int_{-h}^{0} \rho v dz dx$$

$$\Rightarrow \quad M = \rho \frac{g}{f} \Delta h l$$

where h = steric height(is the surface relative to some depth)

Let, L_x and L_y be length scale in x and y direction

Given

$$\frac{M_x}{M_y} = 10 \text{ and } w = 0$$

Now, as per above formula

Also, $u = v$

$$\frac{M_x}{M_y} = \frac{L_x \times \rho \times u \times \Delta h}{L_y \times \rho \times u \times \Delta h}$$

$$\therefore \quad \frac{L_x}{L_y} = 10$$

21. *Given,*

$$P(z) = P_o e^{\frac{-z}{H}}$$

$$H = 7.5\ km$$

Mass of earth, $m = 4.808 \times 10^{-26}\ kg$

$$\rho = 1.22\ kg/m^3$$

$$T = 288\ K$$

$$P_o = 1.01 \times 10^5\ Pa$$

Botlzmann's Constant $= 1.38 \times 10^{-23}\ J/K$

As per formula

$$P = \frac{\rho k T}{m}$$

$$\text{here } m_{\frac{1}{4}} = \frac{lm}{4}$$

$$= \frac{4.808 \times 10^{-26}}{4}$$

$$= 1.202 \times 10^{-26}\ kg$$

$$\Rightarrow \quad P = \frac{1.22 \times 1.38 \times 10^{-23} \times 288}{1.202 \times 10^{-26}}$$

$$= 4.033 \times 10^5\ Pa$$

Now,

$$P(z) = P_o e^{\frac{-z}{H}}$$

$$\Rightarrow \text{in}\left(\frac{P}{P_o}\right) = \frac{|-z|}{H}$$

$$\Rightarrow \quad z = H \times \ln\left(\frac{P}{P_o}\right)$$

$$\Rightarrow \quad z = 7.5 \times \ln\left(\frac{4.033 \times 10^5}{1.01 \times 10^5}\right) = 10.3 km$$

22. *Given,*

$$h = 300\ m$$

$$T = 300\ K$$

$$\rho = 1\ kg/m^3$$

$$g = 10\ m/s^2$$

Also,

$$\frac{\Delta T}{T} \approx \frac{\Delta h}{h}$$

$$\Delta h = \frac{\Delta T}{T} \times h$$

$$= \frac{10}{300} \times 300 = 10$$

Now, as per formula

$$(P_2 - P_1) = \rho g \Delta h$$

$$= 1 \times 10 \times 10 = 100\ kg/m/s^2$$

SOLVED PAPER 2021

XE : ENGINEERING SCIENCES

GENERAL APTITUDE

Q. 1 – Q. 5 carry one mark each

1. Gauri said that she can play the keyboard ____ her sister.

(*a*) as well as (*b*) as better as

(*c*) as nicest as (*d*) as worse as

2. As transparent square sheet shown above is folded along the dotted line. The folded sheet will look like ____.

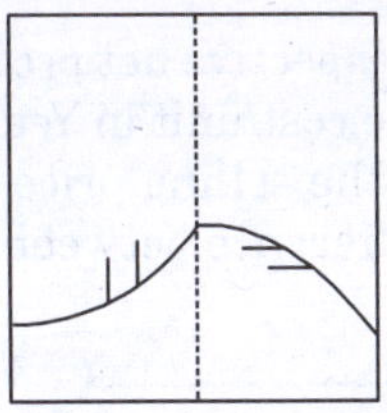

(*a*) 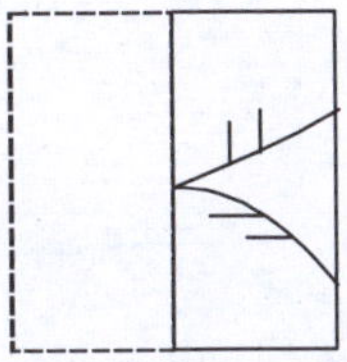 (*b*) 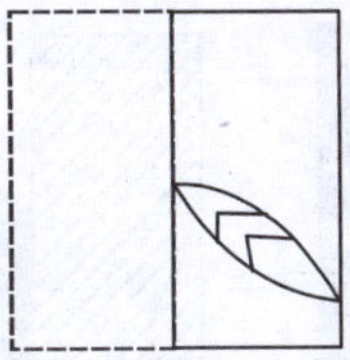

(*c*) 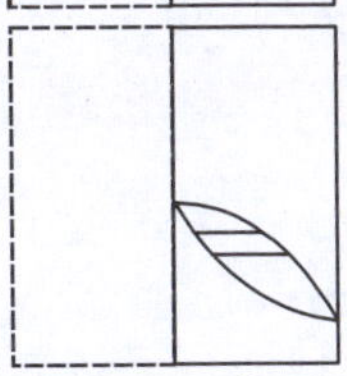 (*d*) 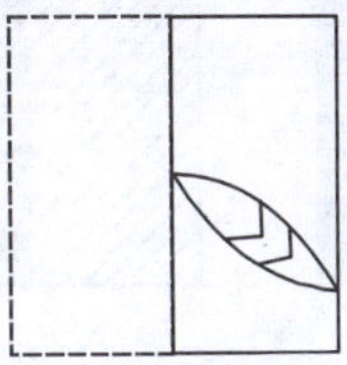

3. If θ is the angle, in degrees, between the longest diagonal of the cube and any one of the edges of the cube, then, cos θ =

(*a*) $\frac{1}{2}$ (*b*) $\frac{1}{\sqrt{3}}$

(*c*) $\frac{1}{\sqrt{2}}$ (*d*) $\frac{\sqrt{3}}{2}$

4. If $\left(x-\frac{1}{2}\right)^2-\left(x-\frac{3}{2}\right)^2=x+2$, then the value of x is :

(*a*) 2 (*b*) 4

(c) 6 (*d*) 8

5. Pen : Write :: Knife : ____.

Which one of the following options maintains a similar logical relation in the above ?

(*a*) Vegetables

(*b*) Sharp

(*c*) Cut

(*d*) Blunt

Q. 6 to Q. 10 : carry two marks each.

6. Listening to music during exercise improve exercise performance and reduces discomfort. Scientists researched whether listening to music while studying can help students learn better and the results were inconclusive. Students who needed external stimulation for studying fared worse while students who did not need any external stimulation benefited from music.

Which one of the following statements is the CORRECT inference of the above passage ?

(*a*) Listening to music has no effect on learning and a positive effect on physical exercise.

(*b*) Listening to music has a clear positive effect both on physical exercise and on learning.

(*c*) Listening to music has a clear positive effect on physical exercise. Music has a positive effect on learning only in some students.

(*d*) Listening to music has a clear positive effect on learning in all students. Music has a positive effect only in some students who exercise.

7. A jigsaw puzzle has 2 pieces. One of the pieces is shown below. Which one of the given options for the missing piece when assembled will form a rectangle ? The piece can be moved, rotated or flipped to assemble will the below piece.

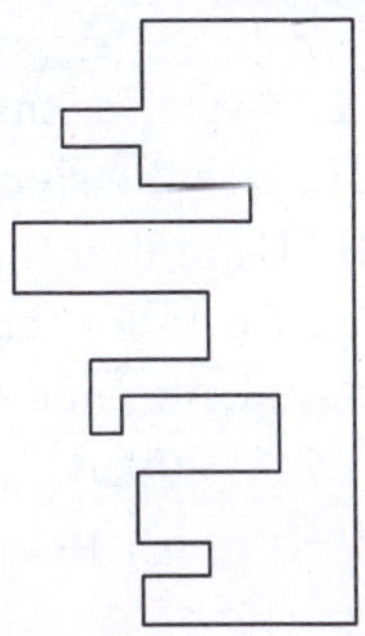

(a) 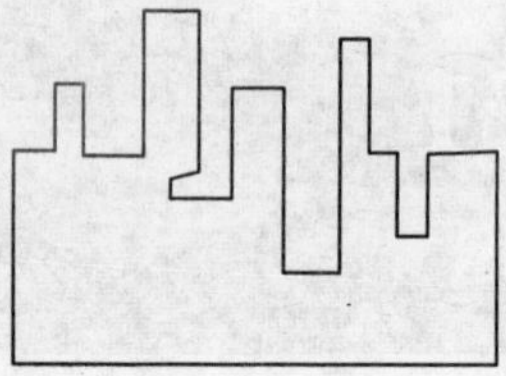

(b) 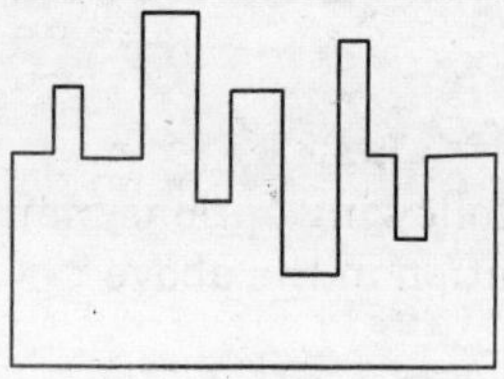

(c) 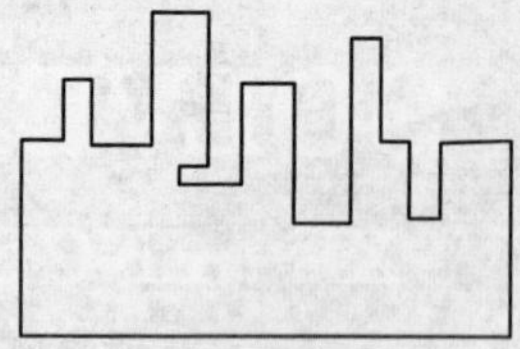

(d) 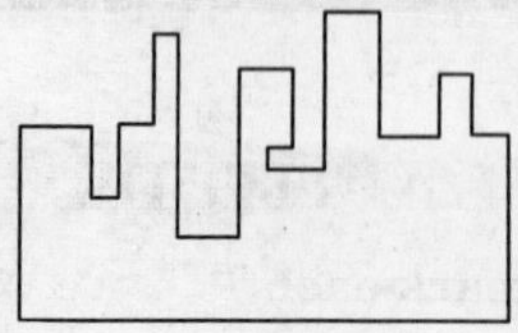

8. The number of students in three classes is in the ratio 3 : 13 : 6. If 18 students are added to each class, the ratio changes to 15 : 35 : 21.

The total number of students in all the three classes in the beginning was :

(a) 22 (b) 66

(c) 88 (d) 110

9. The number of units of a product sold in three different years and the respective net profits are presented in the figure below. The cost/unit in Year 3 was Rs. 1, which was half the cost/unit in Year 2. The cost/unit in Year 3 was one-third of the cost/unit in Year 1. Taxes were paid on the selling price at 10%, 13% and 15% respectively for the three years. Net profit is calculated as the difference between the selling price and the sum of cost and taxes paid in that year.

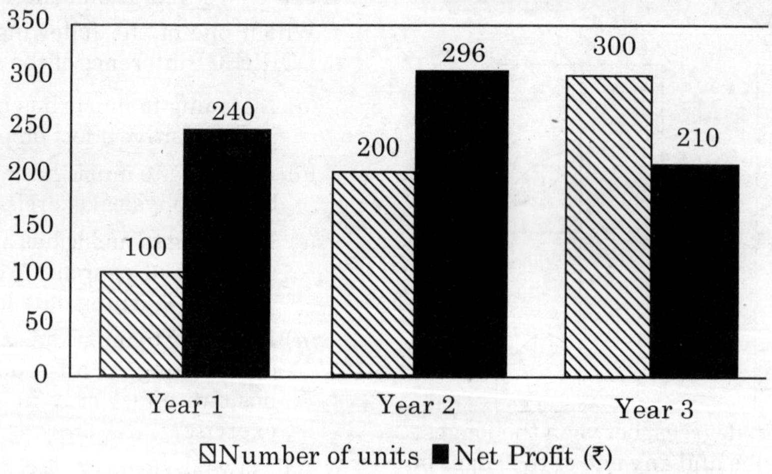

The ratio of the selling price in Year 2 to the selling price in Year 3 is ______.

(a) 4 : 3 (b) 1 : 1

(c) 3 : 4 (d) 1 : 2

10. Six students P, Q, R, S, T and U, with distinct heights, compare their heights and make the following observations.

Observation I : S is taller than R.

Observation II : Q is the shortest of all.

Observation III : U is taller than only one student.

Observation IV : T is taller than S but is not the tallest.

The number of students that are taller than R is the same as the number of students shorter than ____.

(a) T (b) R

(c) S (d) P

ENGINEERING MATHEMATICS

Q. 1 – Q. 3 carry one mark each

1. Let

$$S = \left\{ AX : A = \begin{bmatrix} 2 & -4 \\ 1 & 1 \\ 1 & -1 \end{bmatrix} \text{ and } X = \begin{bmatrix} x_1 \\ x_2 \end{bmatrix} \right\}.$$

If $\begin{bmatrix} -1 \\ \alpha \\ 1 \end{bmatrix} \in S$, then the value of α is

(a) –4 (b) –2

(c) 2 (d) 4

2. Let C be the boundary of the region $R: 0 \le x \le \pi$, $0 \le y \le \sin x$ in the xy-plane and α be the area of the region R. If C traverses once in the counter clockwise direction, then the value of the line integral $\oint_c (2y\,dx + 5x\,dy)$ is equal to

(a) α (b) 2α

(c) 3α (d) 4α

3. Given that $i = \sqrt{-1}$. The value of

$$\lim_{z \to e^{\frac{\pi i}{3}}} \frac{z^3 + 1}{z^4 + z^2 + 1} \text{ is}$$

(a) $\frac{3}{4} + i\frac{\sqrt{3}}{4}$ (b) $\frac{3}{4} - i\frac{\sqrt{3}}{4}$

(c) $\frac{-3}{4} + i\frac{\sqrt{3}}{4}$ (d) $\frac{-3}{4} - i\frac{\sqrt{3}}{4}$

Q. 4 – Q. 7 carry one mark each

4. Let f(x) be a non-negative continuous function of real variable x. If the area under the curve y = f(x) from x = 0 to x = a is

$\frac{a^2}{2} + \frac{a}{2}\sin a + \frac{\pi}{2}\cos a - \frac{\pi}{2}$, then the value of

$f\left(\frac{\pi}{2}\right)$ is ________ (round off to one decimal place).

5. If the numerical approximation of the value of the integral $\int_0^4 2^{\alpha x}\,dx$ using the Trapezoidal rule with two subintervals is 9, then the value of the real constant α is ________ (round off to one decimal place).

6. Let the transformation $y(x) = e^x v(x)$ reduce the ordinray differential equation

$$x\frac{d^2y}{dx^2} + 2(1-x)\frac{dy}{dx} + (x-2)y = 0;\ x > 0$$

to

$$\alpha x\frac{d^2v}{dx^2} + 2\beta\frac{dv}{dx} + 3\gamma v = 0,$$

where α, β, γ are real constants. Then the arithmetic mean of α, β, γ is ________ (round off to three decimal places).

7. A person, who speaks the truth 3 out of 4 times, throws a fair dice with six faces and informs that the outcome is 5. The probability that the outcome is really 5 is ________ (round off to three decimal places).

Q. 8 – Q. 9 carry two marks each

8. Let $f(x, y) = x^4 + y^4 - 2x^2 + 4xy - 2y^2 + \alpha$ be a real valued function. Then, which one of the following statements is TRUE for all α ?

(a) (0, 0) is not a stationary point of f

(b) f has a local maxima at (0, 0)

(c) f has a local minima at (0, 0)

(d) f has a saddle point at (0, 0)

9. Let $u(x, y) = (x^2 - y^2)v(x, y)$ be such that both u(x, y) and v(x, y) satisfy the Laplace equation in a domain Ω of the xy-plane. Then, which one of the following is TRUE in Ω ?

(a) $x\frac{\partial v}{\partial x} - y\frac{\partial v}{\partial y} = 0$ (b) $x\frac{\partial v}{\partial x} + y\frac{\partial v}{\partial y} = 0$

(c) $x\frac{\partial v}{\partial y} - y\frac{\partial v}{\partial x} = 0$ (d) $x\frac{\partial v}{\partial y} + y\frac{\partial v}{\partial x} = 0$

Q. 10 – Q. 11 carry two marks each

10. Let I denote the identity matrix of order 7, and A be a 7 × 7 real matrix having characteristic polynomial $C_A(\lambda) = \lambda^2 (\lambda - 1)^\alpha (\lambda + 2)^\beta$, where α and β are positive integers. If A is diagonalizable and rank (A) = rank (A + 2I), then rank (A – I) is ________ (in integer)

11. Let C_1 be the line segment from (0, 1) to $\left(\frac{4}{5}, \frac{3}{5}\right)$, and let C_2 be the arc of the circle $x^2 + y^2 = 1$ from (0, 1) to $\left(\frac{4}{5}, \frac{3}{5}\right)$. If

$$\alpha = \int_{C_1} \left(\frac{2x}{y}\hat{i} + \frac{1-x^2}{y^2}\hat{j}\right) \cdot d\vec{r}$$

$$\text{and } \beta = \int_{C_2} \left(\frac{2x}{y}\hat{i} + \frac{1-x^2}{y^2}\hat{j}\right) \cdot d\vec{r},$$

where $\vec{r} = x\hat{i} + y\hat{j}$, then the value of $\alpha^2 + \beta^2$ is ________ (round off to two decimal places).

FLUID MECHANICS

Q. 1 – Q. 8 carry one mark each

1. The general relationship between shear stress, τ, and the velocity gradient $\left(\frac{du}{dy}\right)$ for a fluid is given by $\tau = k\left(\frac{du}{dy}\right)^n$, where k is a constant with appropriate units. The fluid is Newtonian if

(a) $n > 1$ *(b)* $n < 1$

(c) $n = 1$ *(d)* $n = 0$

2. Which one of the following options is TRUE?

(a) Pathlines and streaklines are the same in an unsteady flow, and streamlines are tangential to the local fluid velocity at a point.

(b) Streamlines are perpendicular to the local fluid velocity at a point, and streamlines and streaklines are the same in a steady flow.

(c) Pathlines and streaklines are the same in an unsteady flow, and streamlines and streaklines are the same in a steady flow.

(d) Streamlines are tangential to the local fluid velocity at a point, and streamlines and streaklines are the same in a steady flow.

3. If P_{in} = 1.2 Pa and P_{out} = 1.0 Pa are the average pressures at inlet and outlet respectively for a fully-developed flow inside a channel having a height of 50 cm, then the absolute value of average shear stress (in Pa) acting on the walls of the channel of length 5 m is

(a) 0.005 *(b)* 0.02

(c) 0.01 *(d)* 0.05

4. Consider the fully-developed flow of a Newtonian fluid (density ρ; viscosity μ) through a smooth pipe of diameter D and length L. The average velocity of the flow is V. If the length of the pipe is doubled, keeping V, D, ρ, μ constant, the friction factor

(a) increases by two times

(b) remains the same

(c) decreases by two times

(d) increases by four times

5. The absolute value of pressure difference between the inside and outside of a spherical soap bubble of radius, R, and surface tension, γ, is :

(a) $\frac{2\gamma}{R}$ *(b)* $\frac{\gamma}{R}$

(c) $\frac{\gamma}{2R}$ *(d)* $\frac{4\gamma}{R}$

6. Which one of the following statements is TRUE about the continuity equation $\frac{\partial u}{\partial x} + \frac{\partial v}{\partial y} + \frac{\partial w}{\partial z} = 0$ (where u, v, w are the velocity components along the x, y and z coordinates respectively) :

(a) The equation is valid only for steady incompressible flows.

(b) The equation is valid for both steady and unsteady incompressible flows.

(c) The equation is valid only for steady compressible flows.

(d) The equation is valid only for unsteady compressible flows.

7. The head loss (K_L) associated with the flow entry of water to an internal passage depends on the shape of the entry. The following figure shows three different types of flow entry into a pipe. Which one of the following relationships correctly represents the head loss associated with the three different flow entries ?

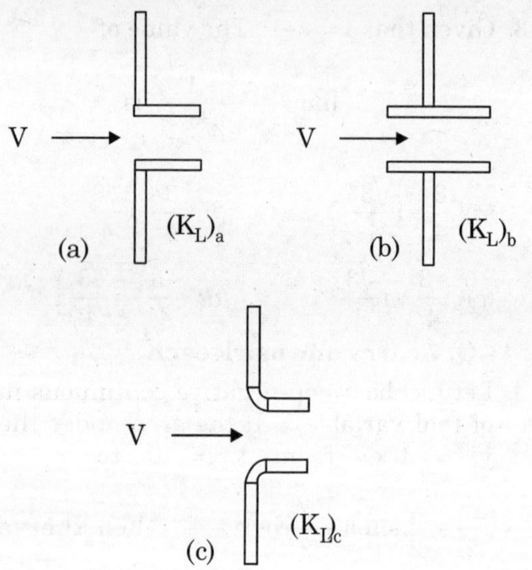

(a) $(K_L)_a > (K_L)_b > (K_L)_c$

(b) $(K_L)_b > (K_L)_a > (K_L)_c$

(c) $(K_L)_b \le (K_L)_a = (K_L)_c$

(d) $(K_L)_b < (K_L)_a < (K_L)_c$

8. The form and friction drags together contribute to the total drag when flow of air occurs past any object. Two orientations of a finite flat plate are shown in the figure. In Orientation-1, the plate is placed perpendicular to the flow while in Orientation-2, the plate is placed parallel to the flow. If the velocity (V) of air in both orientations is the same, which one of the following options is TRUE ?

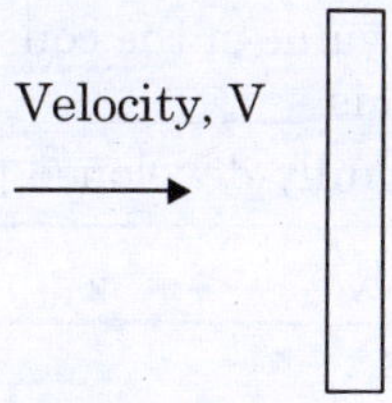

(Orientation-1)

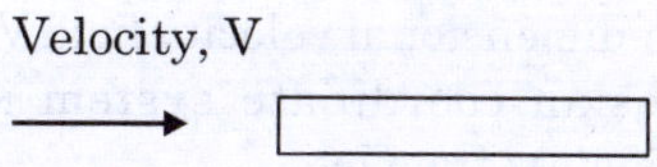

(Orientation-2)

(*a*) Orientation-1 has higher form drag and lower friction drag and Orientation-2 has lower form drag and higher friction drag

(*b*) Orientation-1 has lower form drag and lower friction drag and Orientation-2 has higher form drag and higher friction drag

(*c*) Orientation-1 has lower form drag and higher friction drag and Orientation-2 has higher form drag and lower friction drag

(*d*) Orientation-1 has higher form drag and higher friction drag and Orientation-2 has lower form drag and lower friction drag

Q. 9 carry one marks

9. A spherical ball is steadily supported against gravity by an upward air jet as shown in the figure. Take acceleration due to gravity to be $g = 10\ m/s^2$. The mass flow rate of air, reaching the ball, is 0.01 kg/s and the air reaches the ball at an upward velocity of 3 m/s. Neglecting the buoyancy force, and using the principle of integral momentum balance, the mass (in grams, up to one decimal place) of the ball is ________ .

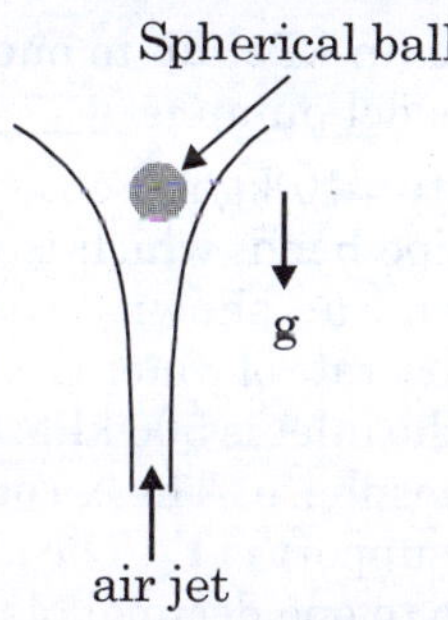

Q. 10 – Q. 12 carry two marks each

10. The incompressible flow of air over a curved surface having possible flow separation is schematically shown in the figure. Two zones P and Q are indicated in the figure. Which one of the following combinations is TRUE for zones P and Q ?

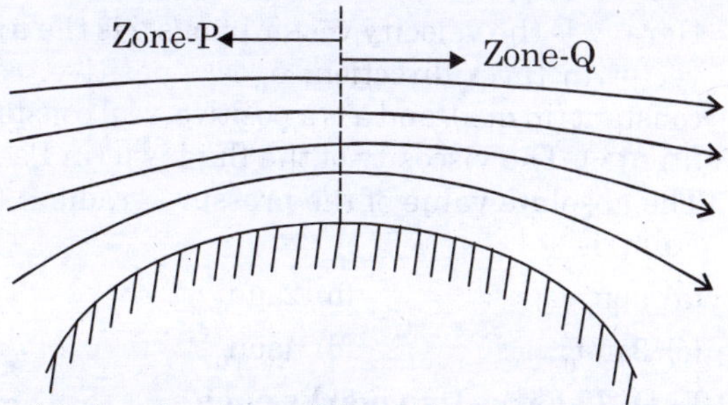

(a) Acceleration of flow, (b) Deceleration of flow, (c) Adverse pressure gradient, (d) Favorable pressure gradient, (e) No flow separation, (f) Possible flow separation.

(*a*) P : (a), (d), (e) and Q : (b), (c), (f)

(*b*) P : (a), (d), (f) and Q : (b), (d), (f)

(*c*) P : (a), (c), (f) and Q : (a), (d), (e

(*d*) P : (a), (c), (e) and Q : (a), (d), (f)

11. A spherical metal ball (of density ρ_s and diameter D), attached to a string, is exposed to a crossflow (of velocity U_∞) of a viscous fluid (of viscosity μ and density ρ_f). Due to the crossflow, the string makes an angle of inclination, θ, with the top surface as shown in the figure. The acceleration due to gravity is denoted by g. For this flow, Reynolds number, $Re = \frac{\rho_f U_\infty D}{\mu} << 1$ and buoyancy force in the fluid is negligible compared to viscous force. Assuming the string to be weightless and offering negligible drag, the expression for θ is

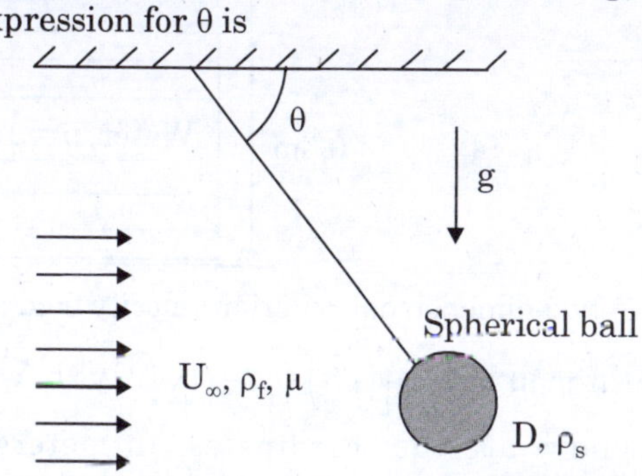

(*a*) $\tan^{-1}\left[\frac{1}{18}\frac{D^2\rho_s^2 g}{\mu U_\infty \rho_f}\right]$ (*b*) $\tan^{-1}\left[\frac{1}{18}\frac{D^2\rho_f g}{\mu U_\infty}\right]$

(*c*) $\sin^{-1}\left[\frac{2}{9}\frac{D^2\rho_s g}{\mu U_\infty}\right]$ (*d*) $\tan^{-1}\left[\frac{1}{18}\frac{D^2\rho_s g}{\mu U_\infty}\right]$

12. In a Cartesian coordinate system, a steady, incompressible velocity field of a Newtonian fluid is given by

$$V = u_0\left(1 - ay^2\right)i$$

Here, V is the velocity vector in m/s, i is the unit vector in the x-direction, u_0 is a positive, real constant (in m/s), and a is a positive, real constant (in m^{-2}). The viscosity of the fluid is μ (in Pa-s). The absolute value of the pressure gradient (in Pa/m) is

(a) $a\mu u_0$ (b) $2a\mu u_0$

(c) $3a\mu u_0$ (d) $4a\mu u_0$

Q. 13 – Q. 22 carry two marks each

13. In a laminar, incompressible, fully-developed pipe flow of a Newtonian fluid, as shown in the figure, the velocity profile over a cross-section is given by $u = U\left(1 - \frac{r^2}{R^2}\right)$, where U is a constant. The pipe length is L and the fluid viscosity is μ. The power P required to sustain the flow is expressed as $P = c\mu LU^2$, where c is a dimensionless constant. The value of the constant c (up to one decimal place) is ______.

Laminar, fully-developed pipe flow

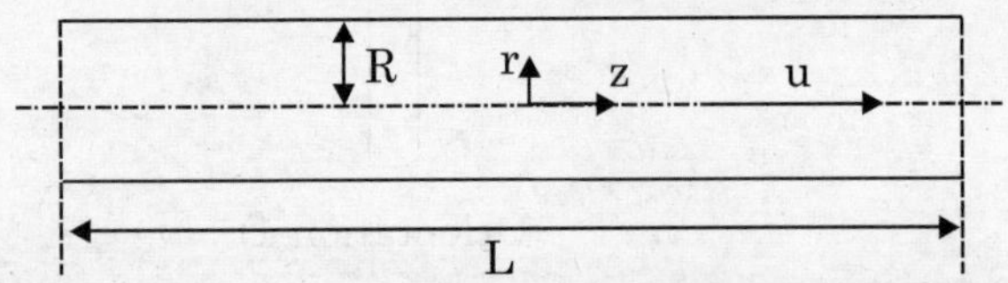

14. The two-dimensional velocity field V of a flow in a Cartesian coordinate system is given in dimensionless from by

$$V = \left(x^2 - axy\right)i + \left(bxy - \frac{y^2}{2}\right)j.$$

Here, i and j are the unit vectors along the x and y directions respectively, a and b are independent of x, y and time. If the flow is incompressible, then the value of (a – b), up to one decimal place, is ______.

15. For the configuration shown in the figure, oil of density 800 kg/m³ lies above water of density 1000 kg/m³. Assuming hydrostatic conditions and acceleration due to gravity g = 10 m/s², the length L (in meters, up to one decimal place) of water in the inclined tube is ______.

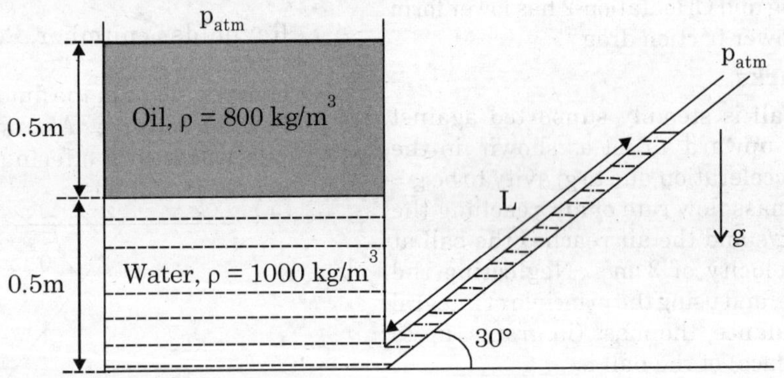

16. A two-dimensional Eulerian velocity field is given (in m/s) by $V = \left[\left(\sqrt{5}\right)x\right]i - \left[\left(\sqrt{12}\right)y\right]j$, where x and y are the coordinates (in meters) in a Cartesian coordinate system. The magnitude of the acceleration (in m/s², up to one decimal place) of a fluid particle at x = 1 m and y = –1 m is ______.

17. A large pump is to deliver oil at an average velocity (V) of 1.5 m/s. The pump has an impeller diameter (D) of 40 cm and the pressure rise across the pump is 400 kPa. To design this pump, a lab-scale model pump with an impeller diameter of 4 cm is to be used with water as the fluid. The viscosity (μ) of the oil is 100 times that of water, an the densities (ρ) of oil and water are identical. A complete geometric similarity is maintained between the model and prototype. If the pressure rise is a function only of V, D, ρ and μ, the pressure rise (in kPa, up to one decimal place) across the model pump is ______.

18. Water (density = 10³kg/m³) enters steadily into a horizontal pipe bend, which is part of a larger piping system, as shown in the figure. The volumetric flow rate of water is 0.1 m³/s. The gage pressure at the inlet is 500 kPa, while the exit is open to atmosphere. The x-component of the force on the support is F_x. The absolute value of F_x (in kN, up to one decimal place) is ______.

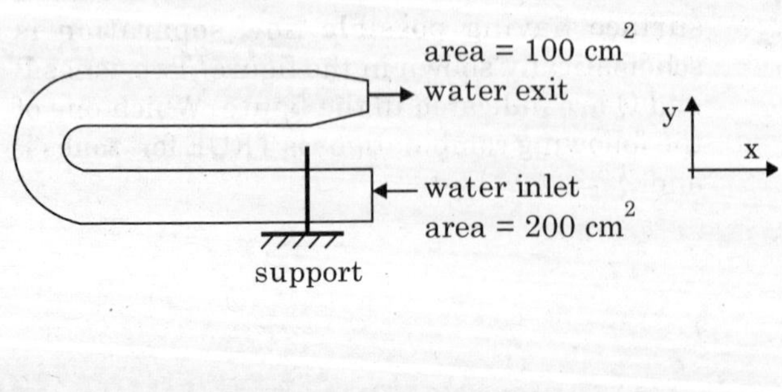

19. Air (of density 0.5 kg/m^3) enters horizontally into a jet engine at a steady speed of 200 m/s through an inlet area of 1.0 m^2. Upon entering the engine, the air passes through the combustion chamber and the exhaust gas exits the jet engine horizontally at a constant speed of 700 m/s. The fuel mass flow rate added in the combustion chamber is negligible compared to the air mass flow rate. Also neglect the pressure difference between the inlet air and the exhaust gas. The absolute value of the horizontal force (in kN, up to one decimal place) on the jet engine is _______.

20. Water discharges from a cylindrical tank through an orifice, as shown in the figure. The flow is considered frictionless. Initially, the water level in the tank was h_1 = 2 m. The diameter of the tank is D = 1 m, while the diameter of the jet is d = 10 cm, and the acceleration due to gravity is g = 10 m/s^2. The time taken (in seconds, up to one decimal place) for the water level in the tank to come down to h_1 = 1 m is _______.

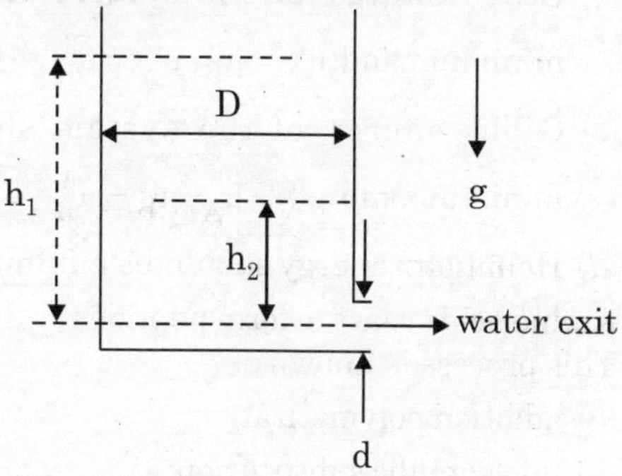

21. Water discharges steadily from a large reservoir through a long pipeline, as shown in the figure. The Darcy friction factor in the pipe is 0.02. The pipe diameter is 20 cm and the discharge of water is 360 m^3/h. Water level in the reservoir is 10 m and acceleration due to gravity g = 10 m/s^2. If minor losses are negligible, the length L (in meters, up to one decimal place) of the pipeline is _______.

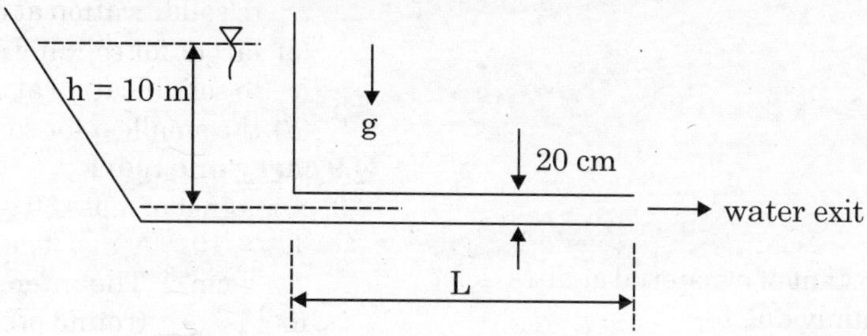

22. Water is flowing with a flow rate Q in a horizontal circular pipe. Due to the low pressure created at the venturi section (Section –1 in the figure), water from a reservoir is drawn upward using a connecting pipe as shown in the figure. Take acceleration due to gravity g = 10m/s^2. The flow rate Q = 0.1 m^3/s, D_1 = 8 cm, and D_2 = 20 cm. The maximum height (h, in meters, up to one decimal place) of the venturi from the reservoir just sufficient to raise the liquid upto Section–1 is _______.

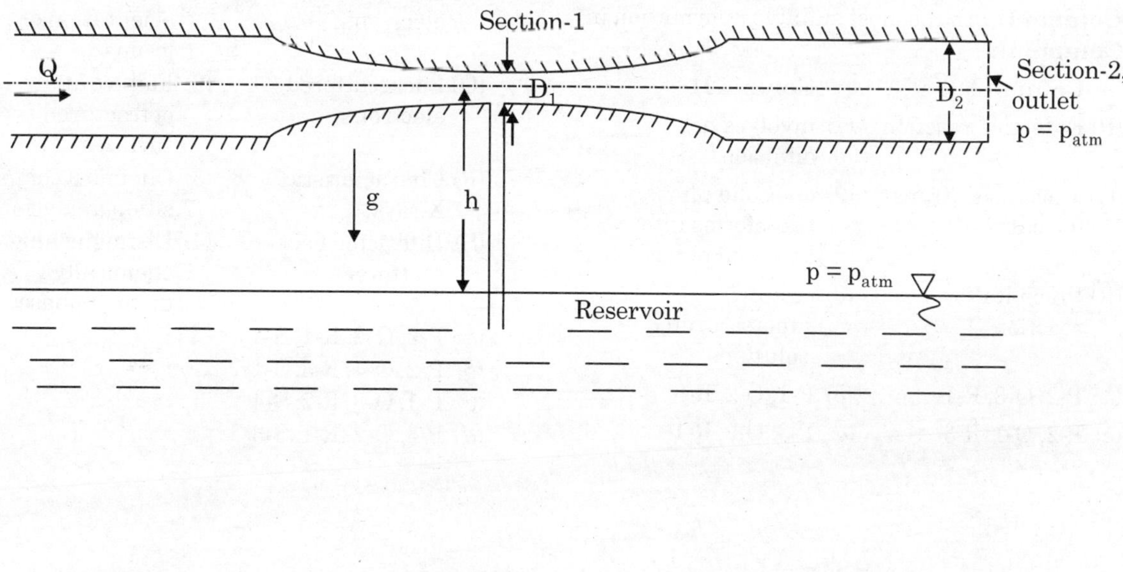

MATERIALS SCIENCE

Q. 1 – Q. 7 carry one mark each

1. Condition to be satisfied for α and β phases to be in equilibrium in a two-component (A and B) system at constant temperature and pressure is

(Given: μ is the chemical potential)

(*a*) entropy of the system should be maximum

(*b*) Gibbs energy of the system should be minimum and $\mu_A^\alpha = \mu_B^\alpha, \mu_A^\beta = \mu_B^\beta$

(*c*) Gibbs energy of the system should be minimum and $\mu_A^\alpha = \mu_A^\beta, \mu_B^\alpha = \mu_B^\beta$

(*d*) Helmholtz energy should be minimum

2. Amino acids react to form peptides and proteins. This process is known as

(*a*) addition polymerization

(*b*) nucleophilic substitution

(*c*) condensation polymerization

(*d*) hydration

3. The most favoured slip system in face centered cubic metal is

(*a*) (111) [110]

(*b*) (110) $[1\bar{1}1]$

(*c*) $(11\bar{1})$ [112]

(*d*) (111) $[1\bar{1}0]$

4. The dielectric constant of a material at ultraviolet frequencies is mainly due to

(*a*) dipolar polarizability

(*b*) ionic polarizability

(*c*) electronic polarizability

(*d*) interfacial polarizability

5. Match the different transformations/reactions in **Column I** with the most suitable information in **Column II**.

Column I	Column II
(P) Eutectoid reaction	(1) involves no diffusion
(Q) Martensitic transformation	(2) one solid phase transforms into two solid phases
(R) Precipitation reaction	(3) occurs in supersaturated solutions

(*a*) P-2, Q-3, R-1 (*b*) P-1, Q-2, R-3

(*c*) P-2, Q-1, R-3 (*d*) P-3, Q-2, R-1

6. In scanning electron microscopy, the resolution of backscattered electron (BSE) image is poorer compared to that of secondary electron (SE) image, because

(*a*) energy of BSE is lower

(*b*) sampling volume of BSE is larger

(*c*) yield of BSE is lower

(*d*) sampling volume of SE is larger

7. Which of the following deposition conditions favour the formation of larger grains in thin film?

(*a*) Low deposition rate and low substrate temperature

(*b*) Low deposition rate and high substrate temperature

(*c*) High deposition rate and low substrate temperature

(*d*) High deposition rate and high substrate temperature

Q.8 carry one mark

8. A metal has a melting point of 600°C. By rapid cooling, liquid metal can be made to solidify either at 500°C or 400°C or 300°C. Critical size of the solid nuclei is

(*a*) same for solidification at 400°C and 500°C

(*b*) smaller for solidification at 400°C as compared to solidification at 500°C

(*c*) larger for solidification at 400°C as compared to solidification at 500°C

(*d*) the smallest for solidification at 300°C

Q.9 carry one mark

9. A magnet of mass 50 g has a magnetic moment of 4.2×10^{-7} A m^2. The density of the magnet is 7.2 g cm^{-3}. The intensity of magnetization in A m^{-1} is ____ (round off to 3 decimal places)

Q. 10 – Q. 12 carry two marks each

10. In the context of scanning electron microscopy, match the information in **Column I** with most appropriate information in **Column II**.

Column I	Column II
(P) Secondary electrons	(1) Crystallographic orientation of grains
(Q) Backscattered electrons	(2) Failure analysis of fractured surfaces
(R) Characteristic X-rays	(3) Chemical composition analysis
(S) Diffracted back-scattered	(4) Distinguishing chemically distinct phases

(*a*) P-3, Q-2, R-1, S-4

(*b*) P-2, Q-4, R-3, S-1

(*c*) P-1, Q-3, R-2, S-4

(*d*) P-4, Q-2, R-1, S-3

11. Match the heat treatment processes given in **Column I** with the most suitable outcomes in **Column II**.

Column I	Column II
(P) Quenching	(1) hardens the steel
(Q) Annealing	(2) softens the cold worked steel
(R) Tempering	(3) toughens the steel
(S) Carburizing	(4) hardens the surface of steel

(*a*) P-3, Q-2, R-1, S-4
(*b*) P-2, Q-4, R-3, S-1
(*c*) P-1, Q-2, R-3, S-4
(*d*) P-1, Q-3, R-4, S-2

12. A co-joined cross-ply laminate composite, as shown in figure, is distorted upon heating. What are the resultant shapes of edges XY and YZ ?

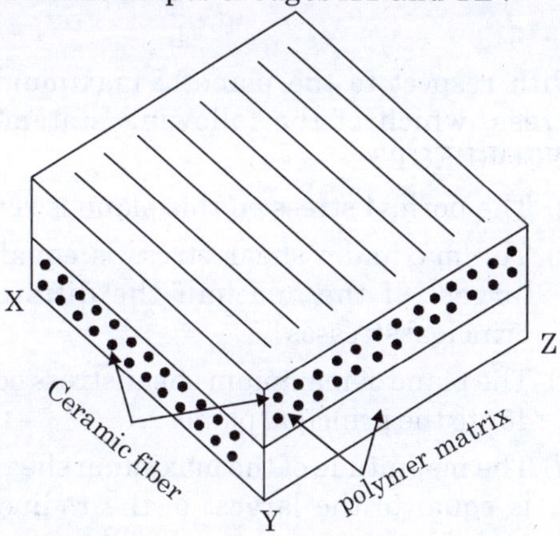

(*a*) X —— Y, Y ⌣ Z
(*b*) X ⌣ Y, Y —— Z
(*c*) X ⌣ Y, Y ⌢ Z
(*d*) X ⌢ Y, Y ⌣ Z

Q 13. carry two marks

13. X-ray diffraction peak broadening enables the estimation of
(*a*) crystallite size of the material
(*b*) microstrain in the material
(*c*) precise lattice parameter
(*d*) residual macrostress acting on the material

Q. 14 – Q. 22 carry two marks each

14. Fe - 10 atom % C austenite (fcc), having no Fe vacancies, has a lattice parameter of 4 Å. The density of austenite in g cm^{-3} is ____ (round off to 2 decimal places)
(Given: atomic weight of Fe = 55.8; atomic weight of C = 12.0; Avogadro's number = 6.023×10^{23})

15. An element transforms from α to β at 773 K and 1 atm pressure with 912 J mol^{-1} as enthalpy of transformation. The molar volumes of α and β phases are 7.377 cm^3 and 7.317 cm^3, respectively. Assume that the difference in molar volumes of α and β is independent of pressure. The pressure (in atm) required for α to β transformation to occur at 723 K is ____ (round off to nearest integer)
(Given: 1 atm = 1.01325×10^5 Pa)

16. A binary A-B alloy has α and β phases at equilibrium. The ratio of weight percentages (wt.%) of α to β is 4. The wt.% of A in α and β phases is 70 and 20, respectively. The wt.% of B in the alloy is ____ (round off to nearest integer)

17. During heating, Ti undergoes allotropic transformation from hcp to bcc at 882°C. The percent volume change accompanying this transformation is ____ (round off to 1 decimal place)
(Given: atomic weight of Ti = 47.9; lattice parameter of bcc Ti = 0.332 nm; density of hcp Ti = 4.51 g cm^{-3}; Avogadro's number = 6.023×10^{23})

18. Vickers hardness test is performed with an indenter of square-base diamond pyramid having an included angle of 136° between the opposite faces of the pyramid. If the applied load is 10 kg and the average length of diagonals of square indentation is 0.5 mm, the Vickers hardness in kg mm^{-2} is ____ (round off to nearest integer)

19. The drift mobility of electron in an n-type Si crystal doped with 10^{16} cm^{-3} phosphorous atoms is 1350 cm^2 V^{-1} s^{-1}. The electrical conductivity in Ω^{-1} m^{-1} is ____ (round off to nearest integer)
(Given: intrinsic charge concentration of Si = 1.45×10^{10} cm^{-3}; Charge of an electron, e = 1.6×10^{-19} C)

20. At 1000 K, the linear thermal expansion coefficients of graphite, parallel and perpendicular to the graphite layers, are 0.8×10^{-6} K^{-1} and 29×10^{-6} K^{-1}, respectively. The percentage increase in the volume of graphite when heated from 900 K to 1100 K is ____ (round off to 2 decimal places)

21. A certain ceramic has a theoretical density and sintered density of 6.76 g cm^{-3} and 6.60 g cm^{-3}, respectively. The green compact has 18 volume percent porosity. For a sintered cube of side 2 cm, the required side of the cubic green compact in cm is ____ (round off to 2 decimal places)

22. When a metal (M) is immersed in de-aerated acid electrolyte, it polarizes anodically by 0.4V. The $\frac{M}{M^{n+}}$ exchange current density is 10^{-5} A m^{-2} and Tafel slope is 0.1 V/decade for the anodic reaction. Assume that corrosion is uniform and, anodic and cathodic reactions are under activation control. The rate of metal dissolution in A m^{-2} is ____ (round off to 1 decimal place)

SOLID MECHANICS

Q. 1 – Q. 8 carry one mark each

1. A force F = 40 kN is applied on the hook as shown. The equivalent force-couple system at B is

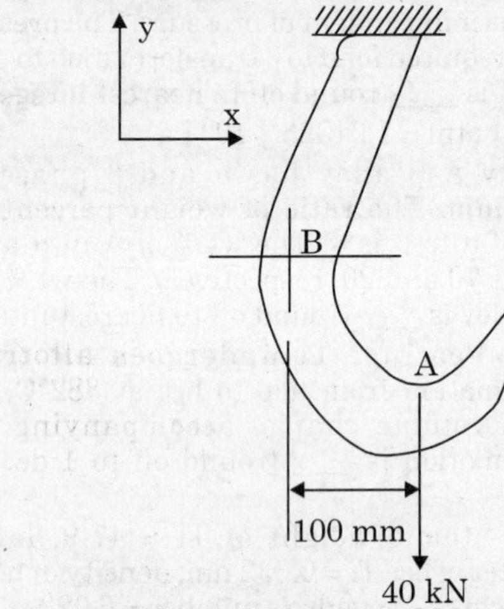

(*a*) 40 kN in +y direction and M = 0
(*b*) 40 kN in –y direction and M = 0
(*c*) 40 kN in +y direction and M = 4000 Nm counter clockwise
(*d*) 40 kN in –y direction and M = 4000 Nm clockwise

2. A rigid rod OA rotates clockwise at an angular velocity of 10 rad/s. A bead B (OB = 1 m) translates outward on the rod at a speed of 5 m/s and acceleration 2.5 m/s^2 (both quantities with respect to the rod). The Coriolis component of acceleration is

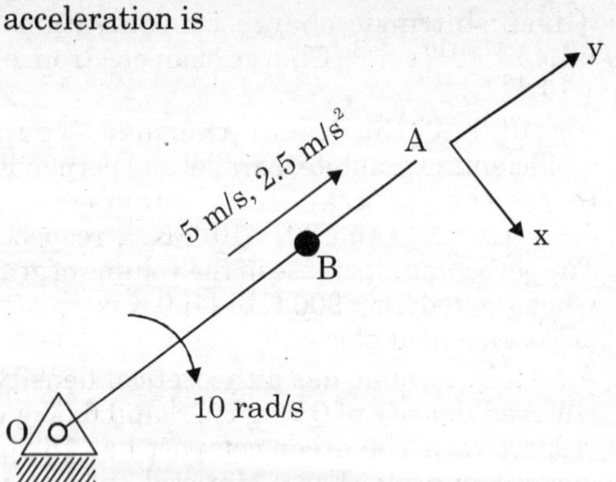

(*a*) 2.5 m/s^2 in +x direction
(*b*) 100 m/s^2 in + x direction
(*c*) 100 m/s^2 in –y direction
(*d*) 25 m/s^2 in +y direction

3. A two force member in equilibrium is one in which
(*a*) Forces act at two points and forces are collinear
(*b*) Forces act at two points and member is always straight
(*c*) Forces act at two points but the member is free to carry moment at any point
(*d*) Force acts at one point and moment acts at second point

4. If the yield point shear stress obtained from the torsion test of a cylindrical specimen is τ_y, then what is the maximum value of principal strain at yielding? (μ is Poisson's ratio and E is Young's modulus)

(*a*) $\frac{\tau_y}{E}$ (*b*) $\frac{(1+\mu)\tau_y}{E}$

(*c*) $\frac{\tau_y}{2E}$ (*d*) $\frac{(1-\mu)\tau_y}{E}$

5. If the ratio of Young's modulus to bulk modulus of a material is $\frac{3}{2}$, then the ratio of shear modulus to the Young's modulus of the material is

(*a*) 1 (*b*) $\frac{2}{5}$

(*c*) $\frac{1}{3}$ (*d*) $\frac{3}{5}$

6. With respect to the plane of maximum shear stress, which of the following statements is INCORRECT?
(*a*) The normal stress on this plane is zero.
(*b*) The maximum shear stress is equal to the largest of the one half the difference of principal stresses
(*c*) The plane of maximum shear stress occurs at 45° to the principal planes.
(*d*) The magnitude of the maximum shear stress is equal to the largest of the radius of the Mohr's circles.

7. A simply supported beam of length L is loaded by two symmetrically applied point loads P at $\frac{L}{3}$ from each support. Both the loads are then shifted to new points which are at a distance $\frac{L}{4}$ from each support. The bending moments at the mid-section of the beam in both the cases are same. The magnitude of P1 in terms of P is

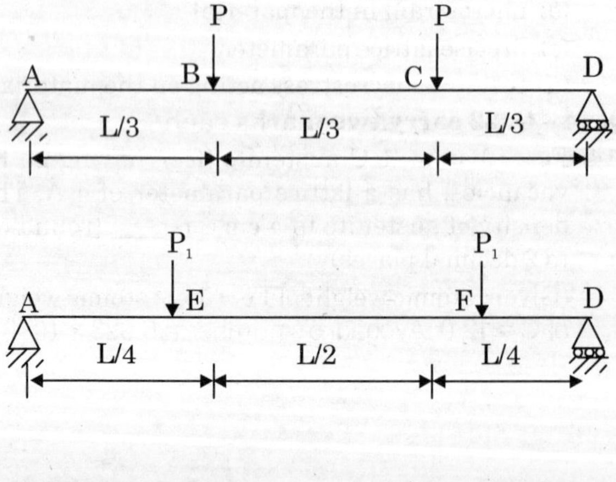

(a) $\frac{P}{4}$ (b) $\frac{8P}{3}$

(c) $\frac{4P}{3}$ (d) $\frac{P}{3}$

8. A beam having rectangular cross section is subjected to transverse loading. The ratio of maximum shear stress developed in the beam to the average shear stress is

(a) 1.50 (b) 1.25

(c) 1.33 (d) 1.66

Q. 9 carry one mark

9. During an earthquake, a structure vibrates and the vibration can be assumed to be in simple harmonic motion at 5 Hz. At a measurement point, the RMS value of acceleration is 10 m/s^2. The approximate amplitude of motion (in mm) at this point (rounded off to two decimal places) is

____

Q. 10 – Q. 15 carry Two marks each

10. For the state of plane stress shown, the components of normal and shear stresses are given in terms of stress σ and unknown constants m and n. If the normal and shear components of stress on a 45° plane are 2σ and zero, the values of m and n would be :

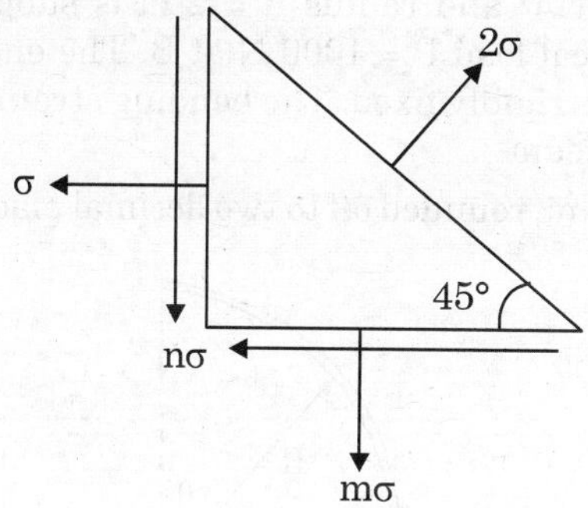

(a) m = 1, n = 2 (b) m = 2, n = 1

(c) m = 1, n = 1 (d) m = 2, n = 2

11. For a state of plane strain, the normal strains are given by $\varepsilon xx = 100 \times 10^{-6}$, $\varepsilon yy = 200 \times 10^{-6}$ and the maximum shear strain is $\gamma max = 1000 \times 10^{-6}$. The value of shear strain γxy for this strain state is

(a) 600×10^{-6} (b) 183×10^{-6}

(c) 1000×10^{-6} (d) 800×10^{-6}

12. A thin cylinder (closed at its ends) of radius r and thickness t ($r \gg t$) is subjected to internal pressure p. The maximum shear stress in the wall of the cylinder is

(a) $\frac{pr}{t}$ (b) $\frac{pr}{2t}$

(c) $\frac{pr}{4t}$ (d) $\frac{3pr}{2t}$

13. The truss shown is subjected to a force P. All members of the truss have the same length L. The reaction at A and force in member AB are

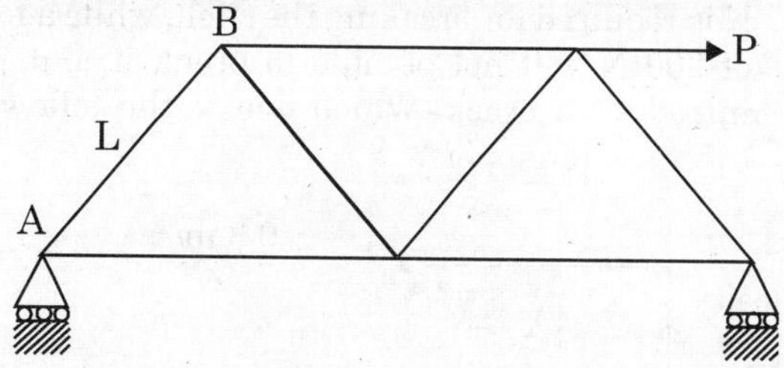

(a) $\frac{P\sqrt{3}}{4}$ and $\frac{P}{2}$

(b) $\frac{P\sqrt{3}}{8}$ and $\frac{P\sqrt{3}}{4}$

(c) $\frac{P\sqrt{3}}{4}$ and $\frac{P}{4}$

(d) P and $\frac{P}{4}$

14. The figure shows a structure with supports. The correct free body diagram when the supports are removed is

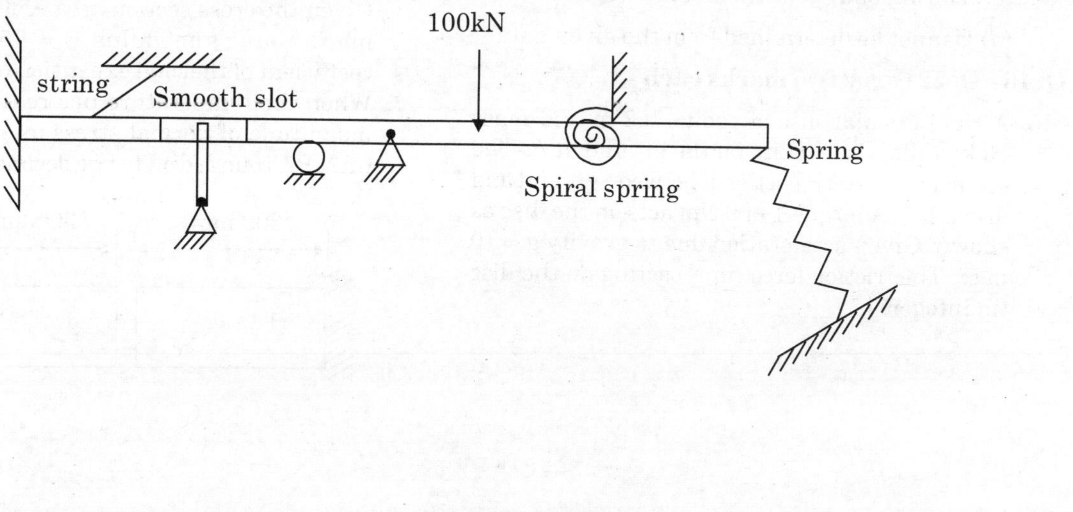

(a) 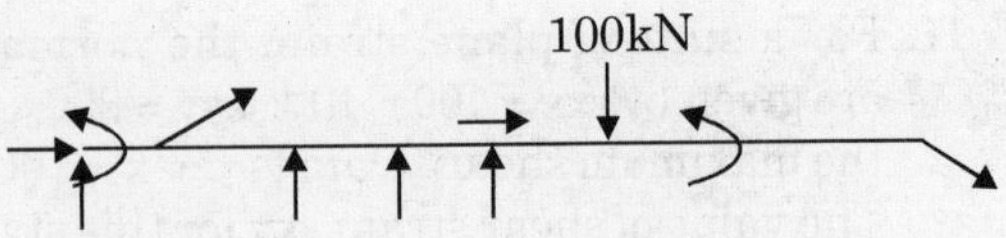

(b) 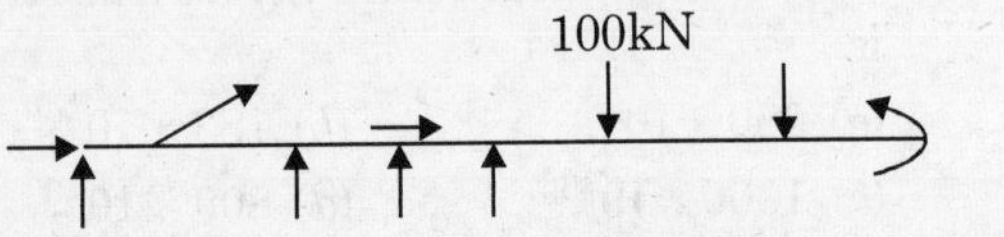

(c) 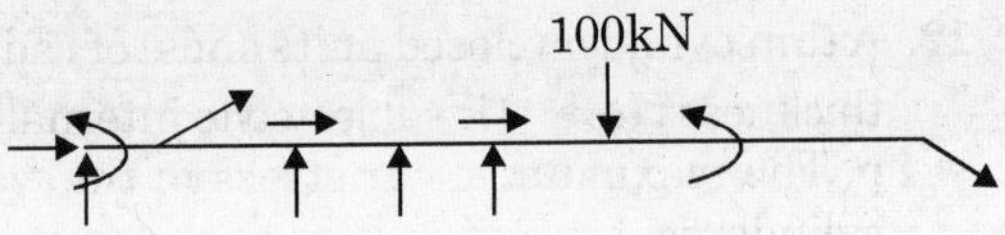

(d) 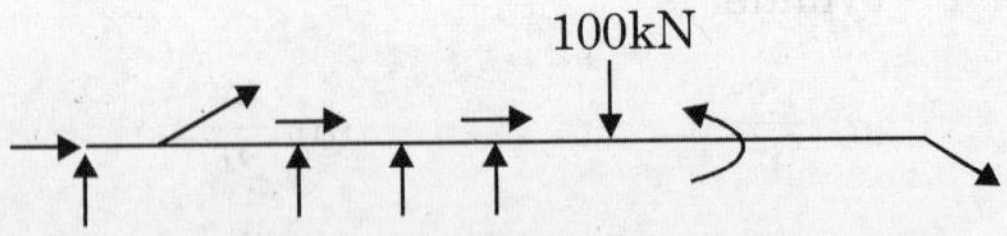

15. A hammer of mass 1 kg is used to break an almond shell. The velocity time graph of the hammer during the impact duration is shown in the figure. The shape of force time graph is also given, which can be approximated as a triangle. A force of 300 N is required for breaking the shell, while a force of 200 N will not be able to break it, but just introduce a crack. Which one of the following events will happen ?

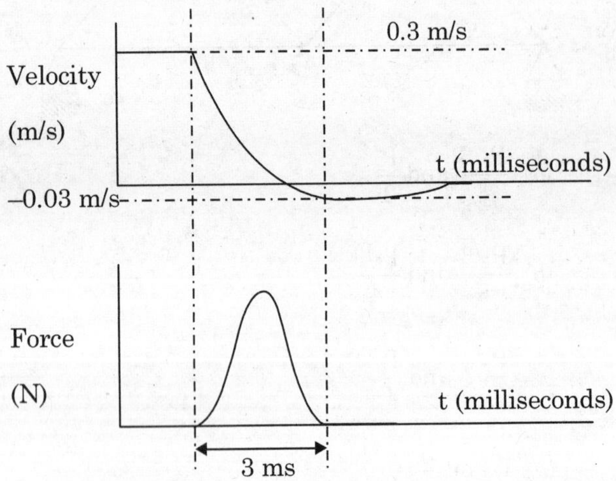

(a) The almond shell will crack but not break

(b) The almond shell will not crack.

(c) The almond shell will break

(d) Cannot be determined from the given data

Q. 16 – Q. 22 carry two marks each

16. A rigid circular disc of radius 0.2 m and mass 10 kg rolls without slip on the ground at A. The coefficient of static friction μ between ground and disc is 0.7. A torque T of 9 Nm acts on the disc as shown. Given acceleration due to gravity g = 10 m/s^2. The friction force (in N) acting on the disc (in integer) is ____

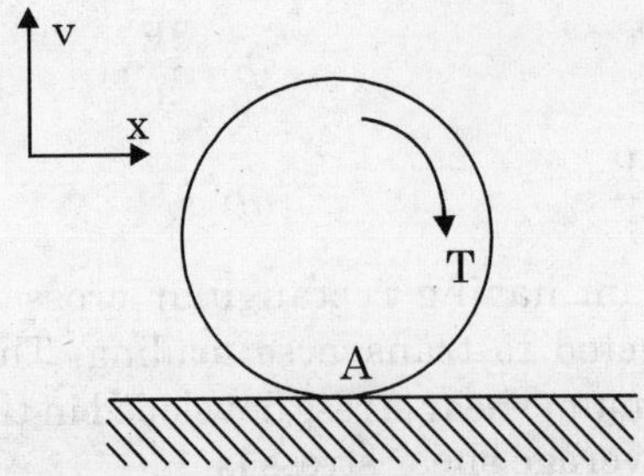

17. A prismatic solid circular rod of diameter d is bent to introduce an offset s = d as shown. The rod is further subjected to an axial load P. If the maximum longitudinal stress at a section A-B in the rod (with offset) is n times the longitudinal stress in the straight rod, the value of n (in integer) would be ____

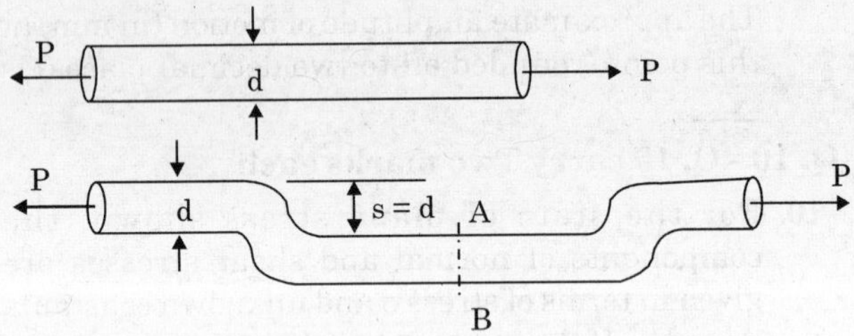

18. A naturally curved steel beam AB having Young's modulus 208 GPa, area moment of inertia I = 26.7 cm^4 and radius R = 2 m is subjected to a vertical load P = 1000 N at B. The end A at θ = 900 is rigidly fixed. The bending strain energy of the beam

(in Nm, rounded off to two decimal places) is ____

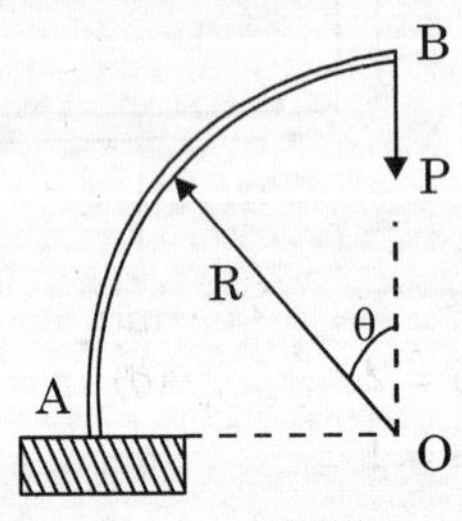

19. At room temperature of 25°C, a gap of 1 mm exists between the ends of the rods 1 and 2 as shown. Given the cross section area A of the rod is 1500 mm^2, Young's modulus E = 75 GPa and the coefficient of thermal expansion $\alpha = 23 \times 10^{-6}$/°C. When the temperature has reached 150°C, the magnitude of normal stress in each of the rods (in MPa, rounded off to two decimal places) is ____

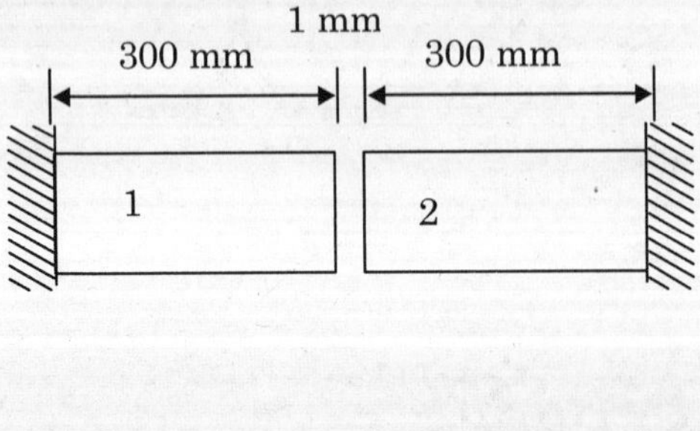

20. A tube of inner radius 4 cm and outer radius 5 cm can carry a maximum torque of T. This tube is now replaced by a solid circular shaft of the same material. The minimum radius of the solid circular shaft (in cm, rounded off to two decimal places) to carry the same amount of torque T is ____

21. In System A, a rectangular block of mass M is centrally supported on a spring of stiffness K as shown. In the System B, the mass is hinged at one of its ends and is supported centrally by the spring. The ratio of natural frequency of System B to that of System A (rounded off to two decimal places) is ____

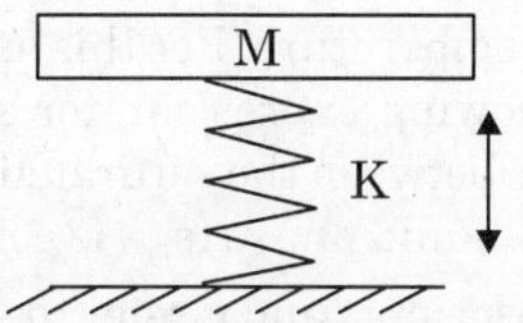

System A

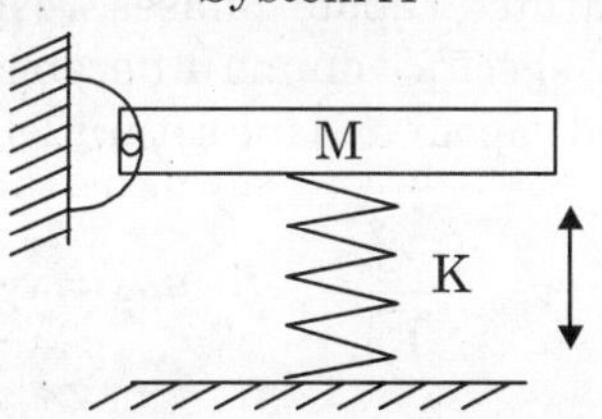

System B

22. A coronavirus droplet of mass 1 microgram ejects from the mouth of a patient with a velocity of 0.7 m/s and travels through air. The gravitational force experienced by it can be neglected due to the buoyancy effect. However, the droplet experiences air drag force proportional to its velocity and the drag coefficient is given as 1.0 μN-s/m. The distance travelled by the droplet before its velocity drops to 10% of its initial velocity (in m, rounded off to two decimal places) is ____.

THERMODYNAMICS

Q. 1 – Q. 7 carry one mark each

1. A refrigerator working on a reversed Carnot cycle has a Coefficient of Performance (COP) of 4. If it works as a heat pump and consumes work input of 1 kW, the heating effect will be :

(*a*) 1 kW (*b*) 4 kW

(*c*) 5 kW (*d*) 6 kW

2. The liquid phase of a pure substance is termed as ________, if its temperature is lower than the saturation temperature corresponding to its pressure P.

(*a*) super-heated liquid

(*b*) sub-cooled liquid

(*c*) metastable liquid

(*d*) flashing liquid

3. Two air streams of mass flow rates $\dot{m}_1$ and $\dot{m}_2$ enter a mixing chamber and exit after perfect mixing. The corresponding temperatures of the inlet streams are T_1 and T_2, respectively. Heat loss rate from the mixing chamber to the surrounding is Q. Assume that the process is steady, specific heat capacity is constant, and air behaves as an ideal gas. Identify the correct expression for the final exit temperature T_3 after mixing. The mass specific heat capacities of the gas at constant volume and constant pressure are c_v and c_p, respectively. Neglect the bulk kinetic and potential energies of the streams.

(*a*) $T_3 = \dfrac{\dot{m}_1 T_1 + \dot{m}_2 T_2}{\dot{m}_1 + \dot{m}_2} - \dfrac{\dot{Q}}{c_v(\dot{m}_1 + \dot{m}_2)}$

(*b*) $T_3 = \dfrac{\dot{m}_1 T_1 + \dot{m}_2 T_2}{\dot{m}_1 + \dot{m}_2} + \dfrac{\dot{Q}}{c_p(\dot{m}_1 + \dot{m}_2)}$

(*c*) $T_3 = \dfrac{\dot{m}_1 T_1 + \dot{m}_2 T_2}{\dot{m}_1 + \dot{m}_2} - \dfrac{\dot{Q}}{c_p(\dot{m}_1 + \dot{m}_2)}$

(*d*) $T_3 = \dfrac{\dot{m}_1 T_1 + \dot{m}_2 T_2}{\dot{m}_1 + \dot{m}_2} + \dfrac{\dot{Q}}{c_v(\dot{m}_1 + \dot{m}_2)}$

4. If h is the mass specific enthalpy,

s is the mass specific entropy,

P is the pressure,

T is the temperature,

C_V is the mass specific heat at constant volume,

C_P is the mass specific heat at constant pressure,

β is the coefficient of thermal expansion,

ν is the mass specific volume,

κ is the isothermal compressibility,

then the partial derivative $\left(\dfrac{\partial h}{\partial s}\right)_P =$

(*a*) $\left(T - \dfrac{1}{\beta}\right)\left(\dfrac{C_P}{C_V}\right)$ (*b*) $\left(T - \dfrac{1}{\beta}\right)$

(*c*) $T\left(1 - \dfrac{\nu\beta}{\kappa C_V}\right)$ (*d*) T

5. If v is the mass specific volume,

s is the mass specific entropy,

P is the pressure,

T is the temperature,

then using Maxwell relations, $\left(\frac{\partial s}{\partial P}\right)_T =$

(a) $\left(\frac{\partial v}{\partial T}\right)_P$ (b) $-\left(\frac{\partial v}{\partial T}\right)_P$

(c) $\left(\frac{\partial v}{\partial T}\right)_s$ (d) $-\left(\frac{\partial v}{\partial T}\right)_s$

6. A closed system consists of a solution of liquid water and ethanol in equilibrium with its vapours. Using the Gibbs phase rule, the degree of freedom of the system is :

(a) 0 (b) 1

(c) 2 (d) 3

7. For a real gas passing through an insulated throttling valve, the outlet temperature of the gas _________ with respect to the inlet temperature.

(a) is always higher

(b) is always lower

(c) may be higher, lower or same

(d) is always same

Q. 8 carry one mark

8. Atmospheric air with Dry Bulb Temperature (DBT) of 24°C and Relative Humidity of 35%, entering in a circular duct (assume no pressure drop in the duct) is heated by an electrical resistance arrangement inside the duct. The DBT of air measured at the outlet of the duct is equal to 30°C. Considering the flow to be steady, which of the following statement(s) is (are) correct as regards to the outlet air, with respect to the inlet air ?

(a) There is no change in the Relative Humidity

(b) There is no change in the Dew Point Temperature

(c) There is no change in the Specific Humidity

(d) There is no change in the Specific Enthalpy

Q. 9 carry one mark

9. A cylinder of volume 1 m^3 contains a mixture of CO_2 (20% by mol) and O_2 (80% by mol) at 100 kPa and 300 K. This cylinder is connected to a 1 MPa pressure line carrying N_2 at 300 K. The cylinder is filled isothermally till the pressure of gas mixture inside it becomes 500 kPa, and then the filling is stopped. The amount of N_2 gas that has entered the cylinder is _________ (in mole, 2 decimal places).

The universal gas constant is 8.3145 J/(mol K).

Q. 10 to Q. 13 carry two marks

10. The saturation pressure P_{sat} of a pure liquid is represented by an equation of the form :

$$\ln P_{sat} = A - \left(\frac{B}{T}\right),$$

where, A and B are constants, and T is the absolute temperature. For this substance, which of the following expression for specific entropy difference between the saturated vapour and the saturated liquid phase (s_{fg}) is correct ?

Note: Subscripts f and g refer to saturated liquid and saturated vapour phases, respectively, and v_{fg} is the specific volume difference between the saturated vapour and the saturated liquid phases.

(a) $s_{fg} = v_{fg}\frac{B P_{sat}^2}{T^2}$ (b) $s_{fg} = v_{fg}\frac{B P_{sat}}{T^2}$

(c) $s_{fg} = v_{fg}\frac{B P_{sat}}{T^3}$ (d) $s_{fg} = v_{fg}\frac{B P_{sat}^3}{T^3}$

11. For a refrigeration cycle, the ratio of actual COP to the COP of a reversible refrigerator operating between the same temperature limits is 0.8. The condenser and evaporator temperatures are 51°C and –30 °C, respectively. If the cooling capacity of the plant is 2.4 kW, then the power input to the refrigerator is :

(COP: Coefficient of Performance)

(a) 1.00 kW

(b) 1.33 kW

(c) 1.25 kW

(d) 2.08 kW

12. Two identical pressure cookers, Cooker A and Cooker B, each having a total internal capacity of 6 litres are available. Cooker A is filled with 2 litres of liquid water at 110°C and Cooker B is filled with 4 litres of liquid water at 110°C. The remaining space in both the cookers is filled with saturated water vapour in equilibrium with the liquid water. If g represents the specific Gibbs free energy, and subscripts v and l represent the saturated vapour and the saturated liquid phases, respectively, which of the following expressions is correct ?

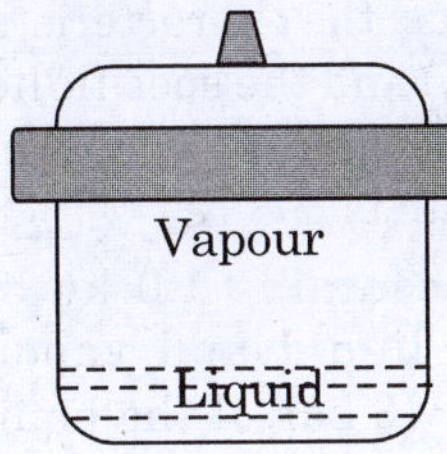

Cooker A

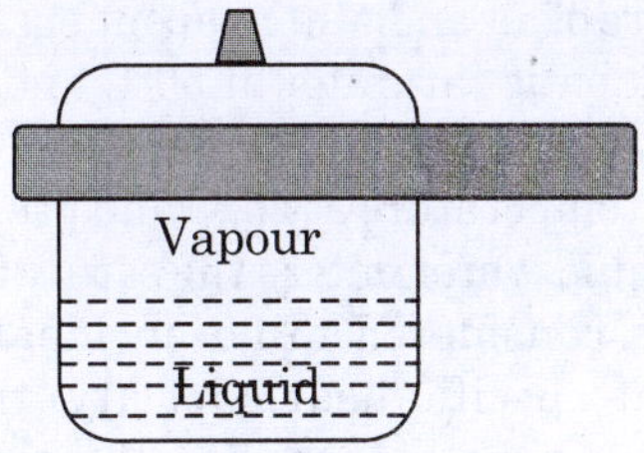

Cooker B

(*a*) $g_{v,A} > g_{l,B}$ (*b*) $g_{v,A} < g_{l,B}$

(*c*) $g_{v,A} = g_{l,B}$ (*d*) $g_{l,B} = 2\,g_{l,A}$

13. Four different Entropy (S) – Temperature (T) diagrams, representing liquid to vapour phase transition process of a pure substance in a closed system under constant pressure are shown. The diagram, which correctly represents the process, is :

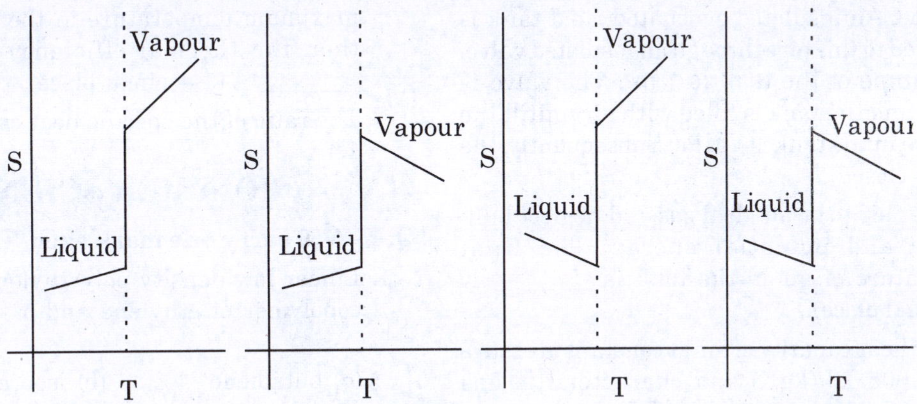

(1) (2) (3) (4)

(*a*) 1 (*b*) 2 (*c*) 3 (*d*) 4

Q. 14 to Q. 22 carry two marks each

14. Air having a mass flow rate of 2 kg/s enters a diffuser at 100 kPa and 30°C, with a velocity of 200 m/s. Exit area of the diffuser is 400 cm^2 while the exit temperature of the air is 45°C. The rate of heat loss from the diffuser to the surrounding is 8 kJ/s. The pressure at the diffuser exit is ________ kPa (2 decimal places).

For air, the characteristic gas constant is 287 J/(kgK) and specific heat capacity at constant pressure is 1005 J/(kgK). Assume air to be an ideal gas and the flow in the diffuser is steady.

15. For the Refrigerant R–134 (at 1 MPa and 50°C), the difference between the specific volume computed by assuming it to be an ideal gas and its actual specific volume is : $v_{ideal} - v_{actual} = 4.529 \times 10^{-3}$ m^3/Kg. If the compressibility factor associated with this state is Z = 0.84, then $v_{com} - v_{actual}$ = ______ $\times 10^{-3}$ m^3/kg (3 decimal places).

Here v_{com} is the specific volume calculated using the compressibility factor.

For Refrigerant R – 134 (at 1 MPa and 50°C):

The characteristic gas constant : 0.0815 J/(kgK), The critical pressure and temperature are, respectively, P_{cr} = 4.059 MPa and T_{cr} = 374.2 K.

16. A mixture of air and water vapour enters a steady-flow adiabatic saturator at 50°C and 100 kPa. It leaves the saturator in a completely saturated state at temperature at 25°C and pressure of 100 kPa. Liquid water enters the saturat or at 25°C. If air is considered to be an ideal gas, with constant specific heat capacity, the relative humidity of the air entering the saturator is% (1 decimal place).

Use the following data :

at 25°C h_f = 104.87 kJ/kg, h_g = 2547.17 kJ/kg, P_{sat} = 3.161 kPa

at 50°C h_f = 209.31 kJ/kg, h_g = 2592.06 kJ/kg, P_{sat} = 12.335 kPa

Specific heat capacity of air at constant pressure C_P = 1.005 kJ/(kgK)

17. Air at a pressure of 1 MPa and 300 K is flowing in a pipe. An insulated evacuated rigid tank is connected to this pipe through an insulated valve. The volume of the tank is 1 m^3. The valve is opened and the tank is filled with air unitill the pressure in the tank is 1 MPa. Subsequently, the valve is closed.

Consider air to be an ideal gas and neglect bulk kinetic and potential energy. The final temperature of air in the tank is _______ K (1 decimal place).

Specific heat capacity of air at constant pressure C_P = 1.005 kJ/(kgK) and characteristic gas constant for air = 0.287 kJ/(kgK)

18. A cylinder of volume 0.1 m^3 is filled with 100 mol of propane (C_3H_8) at 2 MPa. If propane is assumed to obey the van der Waals equation of state, then its temperature is ________ K (1 decimal place).

The van der Waals constants for propane are: a = 939.2 kPa $(m^3/kmol)^2$ and b = 0.0905 m^3/kmol. The universal gas constant is 8.3145 J/(mol K).

19. A frictionless piston cylinder device contains 1 kg of an ideal gas. The gas is compressed according to $Pv^{1.3}$ = constant (P is pressure and v is mass specific volume), from 100 kPa, 250 K, till it reaches a temperature of 500 K. The heat transfer from the piston cylinder device to its surroundings is ________ kJ (2 decimal places).

The characteristic gas constant is 287 J/(kgK) and the ratio of specific heat capacities is 1.4.

20. A 0.8 m^3 insulated rigid tank contains 1.5 kg of an ideal gas at 100 kPa. Electric work is done on the system until the pressure in the tank rises to 135 kPa. The loss in availability (exergy) associated with the process is ________ kJ (2 decimal places)

For the ideal gas, the characteristic gas constant is 188.9 J/(kgK) and the specific heat capacity at constant volume is 680 J/(kgK). The temperature of the dead state is 298 K.

21. A rigid tank contains 1.0 kg of pure water consisting of liquid and vapour phases in equilibrium at 10 bar. If the liquid and vapour phase each occupies one half of the volume of the tank, then the net enthalpy of the contents of the tank is ________ kJ (1 decimal place).

For saturated liquid and vapour at 10 bar, the thermodynamic data table provides the following values:

$v_f = 1.127 \times 10^{-3}$ m^3/kg, $v_g = 194.3 \times 10^{-3}$ m^3/kg, h_f = 762.6 kJ/kg, h_g = 2776.2 kJ/kg

22. An air-standard Diesel cycle with a compression ratio of 16 takes air at 1 bar and 300 K. If the maximum temperature in the cycle is 2100 K, then the thermal efficiency of the cycle is ________ % (1 decimal place).

The ratio of the specific heat capacities of air is 1.4.

POLYMER SCIENCE

Q. 1 – Q. 9 carry one mark each

1. Linear low density polyethylene (LLDPE) is a copolymer of ethylene and a small fraction of ________ .

(*a*) butadiene (*b*) isoprene
(*c*) butene (*d*) hexadiene

2. Binary polymer blends of polypropylene and polyamide 6 are immiscible. From a thermodynamic viewpoint this is due to ________.

(*a*) low enthalpy of mixing
(*b*) high entropy of mixing
(*c*) high enthalpy of mixing
(*d*) low entropy of mixing

3. Which one of the following is an elastomer ?

(*a*) Polyamide 6, 6
(*b*) Poly(ethylene terepthalate)
(*c*) Vulcanized polybutadiene
(*d*) High density polyethylene

4. Compression moulded isotropic polypropylene film exhibits ________ in X-ray diffraction analysis.

(*a*) spot pattern
(*b*) circular ring pattern
(*c*) circular ring and spot pattern
(*d*) arc pattern

5. Which one of the following is an example of a biodegradable polymer ?

(*a*) Polyethylene (*b*) Polyamide 6, 6

(c) Polypropylene (*d*) Polylactic acid

6. Polymer crystals show a range of melting points in contrast to single melting point of crystals of small molecules, because ______.

(*a*) there is an absence of intermolecular interactions

(*b*) there is an absence of long range ordering

(*c*) the polymer chains are not in thermodynamic equilibrium in a metastable state

(*d*) the melting behavior of polymer crystal is independent of sample thermal history

7. When the rate of cooling is increased during the solidification process, the glass transition temperature of a polymer ______.

(*a*) decreases

(*b*) increases

(*c*) stays unaltered

(*d*) shows a non-monotonic dependence

8. Equal and opposite forces of a constant magnitude F are applied at the two ends of a thin elastomeric rod, which is held at a temperature $T_1 (T_g < T_1 < T_m)$, where T_g and T_m are the glass transition temperature and melting temperature respectively. If the temperature is increased to T_2 $(T_g < T_2 < T_m$ and $T_2 > T_1)$, the rod will

(*a*) expand along the loading direction and the transverse direction

(*b*) shrink along the loading direction

(*c*) remain dimensionally unaltered

(*d*) expand only along the loading direction

9. The size of coiled polymer chain in a dilute solution is R_G in a good solvent, R_I in an ideal solvent and R_P in a poor solvent. Select the correct ordering of sizes.

(*a*) $R_G > R_I > R_P$ (*b*) $R_G < R_I < R_P$

(*c*) $R_P > R_G > R_I$ (*d*) $R_P < R_G < R_I$

Q. 10 – Q. 12 carry two marks

10. Match the Additive of its Function.

Additive	Function
P. Tritolyl phosphate	1. Coupling Agent
Q. Triethoxy vinyl silane	2. Antioxidant
R. Azoisobutyronitrile	3. Plasticizer
S. 4-Methyl-2, 6-di-t-butyl phenol	4. Blowing Agent

(*a*) P-3, Q-2, R-1, S-4

(*b*) P-3, Q-1, R-4, S-2

(*c*) P-4, Q-1, R-3, S-2

(*d*) P-1, Q-2, R-4, S-3

11. Match the polymer processing operation with respect to its typical range of shear rate.

Processing Operation	Shear rate (s^{-1})
P. Compression Moulding	1. 1000 - 10000
Q. Extrusion	2. 100 - 1000
R. Calendering	3. 1 - 10
S. Injection Moulding	4. 10 - 100

(*a*) P-3, Q-4, R-2, S-1 (*b*) P-1, Q-3, R-2, S-4

(*c*) P-2, Q-4, R-3, S-1 (*d*) P-3, Q-2, R-1, S-4

12. Shear stress (σ) and shear viscosity (η) are plotted as functions of the shear rate, $\dot{\gamma}$, for idealized "solid-like with yielding (1)" and "liquid-like (2)" materials.

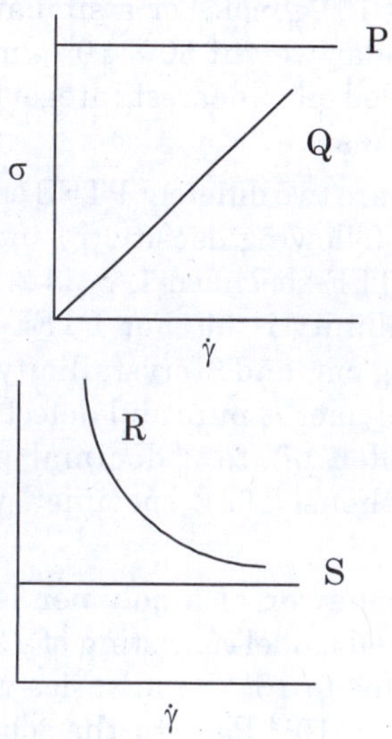

Associate the shear stress and viscosity plots with the appropriate material responses.

(*a*) P-2, Q-1, R-2, S-1

(*b*) P-1, Q-2, R-1, S-2

(*c*) P-1, Q-2, R-2, S-1

(*d*) P-2, Q-1, R-1, S-2

Q. 13 – Q. 22 carry two marks each

13. The plateau modulus of polystyrene has a value of 0.2×10^6 Pa at 150°C. Given, the density of polystyrene is 1.05 g/cm³, the universal gas constant, R = 8.3 J K^{-1} mol^{-1}, and the monomer molecular weight is 104 g/mol. The molecular weight between entanglements (rounded off to the nearest integer) of polystyrene chains is ______ g/mol.

14. A unidirectional composite of epoxy and carbon fiber of 50% by volume is made. The elastic modulus of epoxy and carbon fiber are 3.5 GPa and 350 GPa, respectively. The ratio (rounded off to one decimal place) of the modulus of the composite to the matrix modulus is ________.

15. A single screw extruder is operating at a rotational speed of 2 revolutions per second for the extrusion of a Newtonian polymer under open-discharge conditions (in absence of a die, the pressure drop, $\Delta p = 0$). The extruder has a screw diameter, D = 5 cm, a channel depth, H = 0.4 cm, distance between flights, W = 1 cm, and a helix angle, $\theta = 20°$. Assume the value of $\pi = 3.14$. The volumetric flow rate (rounded off to 2 decimal places) is ________ cm^3/s.

16. At 215°C, the viscosity of a polystyrene of molecular weight 250×10^3 g/mol is 8.0×10^3 Pa.s. The critical molecular weight of polystyrene, $M_c = 35 \times 10^3$ g/mol. For a similar polystyrene of molecular weight 50×10^3 g/mol, the viscosity (rounded off to nearest integer) will be ________ $\times 10^3$ Pa.s.

17. There are two different PTFE polymer specimens of the following density (ρ) and % crystallinity. For PTFE-specimen-1, ρ is 2.144 g/cm^3 and % crystallinity is 50. For PTFE-specimen-2, ρ is 2.215 g/cm^3 and % crystallinity is 75. Assuming the polymer is pure and defect free, the density (rounded off to 3 decimal places) of 100% amorphous PTFE specimen will be ________ g/cm^3.

18. The behavior of a polymer is described by a Maxwell model consisting of a spring element of modulus to 10^{10} Pa in series with a dashpot of viscosity 10^{12} Pa.s. In the solid, 50 s after the sudden application of a fixed strain of 1%, the strees (rounded off to 2 decimal places) will be ________ $\times 10^7$ Pa.

19. A particular free radical polymerization process yields a polymer with a number averaged degree of polymerization, $\bar{x}_n = 100$. The monomer concentration is doubled and the initiator concentration is increased by four times. Assuming that all rate coefficients and other parameters remain unchanged, the value of $\bar{x}_n$ (rounded off to the nearest integer) is ________.

20. A polymer is synthesized from 2 moles of terephthalic acid (molecular weight of the repeat unit, ($-OCC_6H_4CO-$), is 132 g/mol), 1 mol of ethylene glycol (molecular weight of the repeat unit, ($-OCH_2CH_2O-$), is 60 g/mol), and 1 mol of butylene glycol (molecular weight of the repeat unit, ($-O(CH_2)_4O-$), is 88 g/mol). The reaction is terminated at 99% conversion of the acid. The reaction is terminated at 99% conversion of the acid. The number averaged molecular weight, $\overline{M}_n$ (rounded off the nearest integer) is ________ g/mol.

21. A sample of natural rubber (cis-1,4-polyisoprene) is vulcanized such that one of every 240 chain carbon atoms is cross-linked. The formula unit of the isoprene monomer is C_5H_8 (molecular weight = 68 g/mol). The average molecular weight (rounded off to the nearest) between cross-links is ________ g/mol.

22. A sample of an oriented semi-crystalline polymer is subjected to uniaxial tensile stress, σ, in an X-ray diffractometer. The wavelength of X-ray radiation (Cu K_α) is $\lambda = 1.542$ Å. The position of the (002) peak, which was found initially at a Bragg angle of 37.50° at σ = 0Mpa stifted to 37.45° at σ = 160 MPa. Assuming elastic deformation, the strain (rounded off to three decimal places) in the sample along the direction of applied stress is ________ $\times 10^{-3}$.

FOOD TECHNOLOGY

Q. 1 – Q. 9 carry one mark each

1. In a typical bacterial growth curve, the first order kinetics for growth rate is observed in
   (*a*) Lag phase (*b*) Log phase
   (*c*) Stationary phase (*d*) Death phase

2. Which one of the following microorganism is NOT a causative agent for food borne diseases?
   (*a*) *Campylobacter jejuni*
   (*b*) *Clostridium perfingens*
   (*c*) Norovirus
   (*d*) *Borrelia burgdorferi*

3. Which of the following is NOT a fermented food product ?
   (*a*) Tofu (*b*) Vinegar
   (*c*) Sauerkraut (*d*) Tempeh

4. The Protein Efficiency Ratio (PER) is defined as
   (*a*) Percentage of absorbed nitrogen retained in the body
   (*b*) Weight gain in body mass (in gram) per gram protein intake
   (*c*) Ratio of essential and non-essential amino acids in a protein
   (*d*) Percent *in vitro* digestibility of a protein

5. Which one of the following enzymes sequentially releases maltose from starch ?
   (*a*) α–Amylase (*b*) β–Amylase
   (*c*) Glucoamylase (*d*) Pullulanase

6. Which one of the following enzymes is involved in proteolysis of casein in cheese during aging?
(*a*) Myrosinase (*b*) Alliinases
(*c*) Cathepsin (*d*) Plasmin

7. Which one of the following compounds is present in soybean and acts as phytoesterogen ?
(*a*) Tangeretin (*b*) Lutin
(*c*) Quercetin (*d*) Genistein

8. Ultra high temperature (UHT) process of pasteurization of milk is achieved by Heating at
(*a*) 145°F for 30 minutes
(*b*) 161°F for 15 seconds
(*c*) 280°F for 2 seconds
(*d*) 400°F for 15 seconds

9. Bittering agent in grape fruit formed after juice extraction under acidic conditions is
(*a*) Quinine (*b*) Theobromine
(*c*) Isohumulone (*d*) Limonin

Q. 10 – Q. 12 carry two marks each

10. The conversion of pyruvate to lactic acid in homolactic fermentation is catalyzed by
(*a*) Lactate dehydrogenase
(*b*) Pyruvate dehydrogenase
(*c*) Lactase
(*d*) Pyruvate decarboxylase

11. Which one of the following statements is INCORRECT with respect to Controlled Atmosphere Package (CAP) and Modified Atmosphere Package (MAP) of agro-produce?
(*a*) CAP and MAP limit microbial as well as biochemical activities.
(*b*) Gas composition inside a MAP during the storage is continuously monitored and regulated.
(*c*) CAP implies a greater degree of precision than MAP in maintaining specific levels of the gas composition.
(*d*) Modification of the atmosphere inside a MAP is achieved by natural interplay between respiration of products and permeation of gases through the packaging film.

12. Match unit operation in **Column I** with its application in food processing in **Column II**.

Column I	**Column II**
P. Hydrogenation	1. Removal of soft wax
Q. Blanching	2. Shortening of fat
R. Leaching	3. Inactivation of enzyme
S. Winterization	4. Separation of dye

(*a*) P-2, Q-4, R-2, S-1 (*b*) P-2, Q-4, R-3, S-1
(*c*) P-4, Q-1, R-2, S-3 (*d*) P-4, Q-2, R-1, S-3

Q. 13 – Q. 19 carry two marks each

13. Which of the following are correct pair of GRAS chemical food preservative, affected organism and given food matrix ?
(*a*) Sodium lactate-Bacteria-Pre-cooked meats
(*b*) Caprylic acid-Insects-Cheese wraps
(*c*) Dehydroacetic acid-Molds-Squash
(*d*) Sodium nitrite-Clostridia-Meat curing preparations

14. Choose the correct pair of pigment and their corresponding color in given plant product.
(*a*) Carotene-Yellow-orange-Peppers
(*b*) Betanin-Purple/red-Cactus pear
(*c*) Lycopene-Red-Red beets
(*d*) Flavanols-Orange-Cauliflower

15. Which of the following compounds act as anti-nutritional factors ?
(*a*) Phytate (*b*) Isoflavones
(*c*) Trypsin Inhibitor (*d*) Resveratrol

16. Which of the followings is/are commonly used medium/media in the supercritical fluid extraction of spices and tea ?
(*a*) Water
(*b*) Carbon dioxide
(*c*) Dichloromethane
(*d*) Carbon dioxide with Ethanol

17. Which of the following expressions represent the Reynolds number of a fluid flowing through a uniform circular cross section pipe?

(*a*) $$\frac{(\text{density of the fluid})\times(\text{average velocity of the fluid})\times(\text{internal diameter of the pipe})}{(\text{dynamic viscosity of the fluid})}$$

(*b*) $$\frac{(\text{average velocity of the fluid})\times(\text{internal diameter of the pipe})}{(\text{kinematic viscosity of the fluid})}$$

(*c*) $$\frac{(\text{dynamic viscosity of the fluid})}{(\text{average velocity of the fluid})\times(\text{density of the fluid})(\text{internal diameter of the pipe})}$$

(*d*) $$\frac{(\text{kinematic viscosity of the fluid})}{(\text{average velocity of the fluid})\times(\text{internal diameter of the pipe})}$$

18. Which of the following combinations of analytical equipment, property measured and food property are correct ?

(*a*) Particle size analyzer - particle size distribution - span value

(*b*) Texture profile analyzer - morphology - chewiness

(*c*) Differential scanning calorimeter - glass transition temperature - degree of caking

(*d*) Capillary viscometer - viscosity - sensory

19. Choose the correct pair(s) of Govering Law and corresponding application(s)

(*a*) Hagen Poiseuille law - Pressure drop

(*b*) Rittinger's law - Vapour pressure

(*c*) Stefan Boltzmann law - Radiation heat transfer

(*d*) Raoult's law - Size reductin

Q. 20 – Q. 22 carry two marks each

20. An orange juice sample is concentrated from 10% to 40% (by weight) total soluble solids in a single effect evaporator with a feed rate of 3600 kg hr^{-1} at 25°C. The evaporator operates at sufficient vacuum to allow the product moisture to evaporate at 55°C. The specific heat of both feed and concentrated juice is 4.0 kJ kg^{-1} °C^{-1}. If enthalpy of water vapour at 55°C is 2600 kJ kg^{-1}, heat transfer rate through the heating surface area of the evaporator in kilowatt (in integer) will be ____.

21. Dry air is fed into a tray dryer. The percentage relative humidity of the air leaving the dryer is 60% at 70°C and 101.35 KPa. If saturated vapour pressure of water at 70°C is 31.2 kPa, the humidity of the air leaving the dryer in kg water per kg dry air (round off to 3 decimal places) will be ____.
(Given : Molecular weight of water and air are 18.02 g mol^{-1} and 28.97 g mol^{-1} respectively)

22. In a cold storage plant, 5000 kg potato having a constant specific heat capacity of 3.65 kJ^{-1}°C^{-1} are cooled from 28°C to 2°C in 24 hours. The heat of respiration of potato per 24 hour is 3.12 kJ kg^{-1} during the storage. Assuming the efficiency of the storage plant to be 70%, the capacity of the plant in ton of refrigeration (round off to 2 decimal places) is ____.
(Given : 1 ton of refrigeration = 3.517 kilowatt)

ATMOSPHERIC AND OCEANIC SCIENCES

Q. 1 – Q. 5 carry one mark each

1. Western Boundary Current in the ocean is primarily due to

(*a*) Ekman pumping.

(*b*) rotation of the earth.

(*c*) river water forcing

(*d*) ocean floor topography

2. The relevant nondimensional number in deciding deepeing of the thermocline driven by instability of ocean currents is

(*a*) Rossby number. (*b*) Reynolds number.

(*c*) Richardson number. (*d*) Ekman number.

3. During July-August, the highest number of monsoon low pressure systems form over.

(*a*) Arabian Sea. (*b*) Bay of Bengal.

(*c*) South India. (*d*) Himalayan foothills.

4. CO_2 concentration in the Earth's atmosphere is increasing because 50% of the annual anthropogenic emissions are retained in the atmosphere. If nations agree to reduce annual CO_2 emissions by one Giga ton every year starting from 2021, then in which year will the CO_2 concentration in the atmosphere stop rising due to anthropogenic emissions ?
Take the anthropogenic CO_2 emissions in 2020 as 40 Giga tons.

(*a*) 2020 (*b*) 2050

(*c*) 2060 (*d*) 2100

5. The figure shows a schematic of Indian Ocean surface circulation. This pattern is representative of the circulation in which month of the year ?

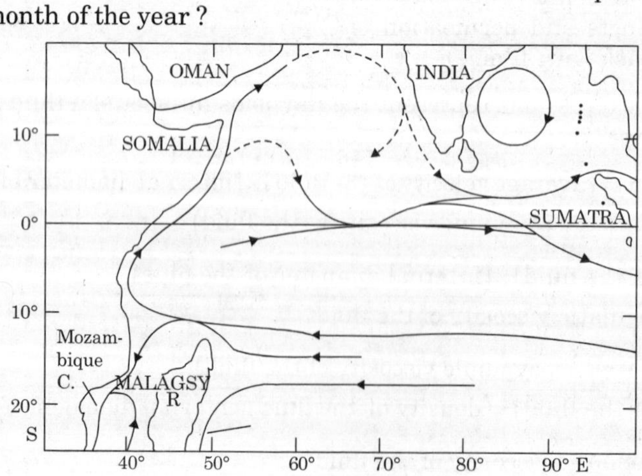

(*a*) January (*b*) July (*c*) May (*d*) November

Q. 6 – Q. 7 carry one mark each

6. Over the open ocean, if the air sea temperature difference is zero, then which of the following statements is/are always true ?
   (*a*) Sensible heat flux is zero.
   (*b*) Latent heat flux is zero.
   (*c*) Momentum flux is zero.
   (*d*) Net energy flux is zero.

7. The psychrometric equation, which is useful in measuring humidity, is derived assuming the following process(es).
   (*a*) Isobaric process
   (*b*) Isothermal process
   (*c*) Adiabatic process
   (*d*) Isentropic process

Q. 8 – Q. 9 carry one mark each

8. The water vapour mixing ratio of an air parcel increases from 10 g kg^{-1} to 20 g kg^{-1} at a constant pressure of 1010 hPa and temperature of 300 K. The change in virtual temperature is ____ K (to one decimal place).

9. The Ekman layer thickness, if turbulent diffusivity is 0.01 $m^2\ s^{-1}$, is ____ m. Take Coriolis parameter to be $10^{-4}\ s^{-1}$. Calculate to the nearest integer.

Q. 10 carry Two mark each

10. The figure shows vertical variation of two chemicals P and Q measured in the Pacific Ocean. Identify the correct combination showing (P, Q) pair from the list below

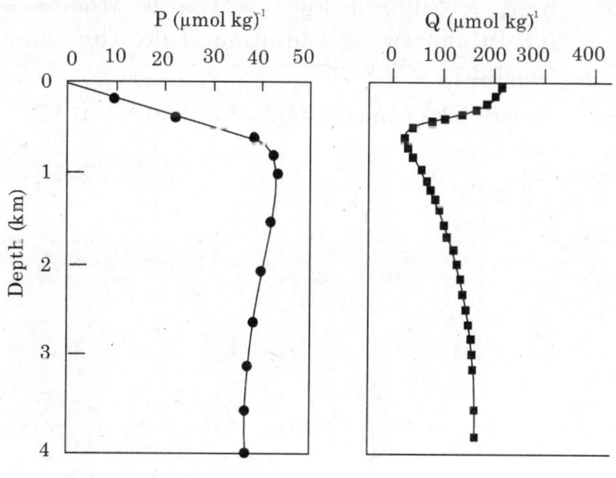

   (*a*) Oxygen, Nitrate
   (*b*) Oxygen, Neon
   (*c*) Nitrate, Oxygen
   (*d*) Neon, Nitrate

Q. 11 – Q. 15 carry Two marks each

11. Consider tropical high-level clouds and low-level stratus clouds with bases at 12 km and 1 km above the surface of the Earth, respectively. Which of the following statement(s) is/are correct?
   (*a*) High clouds are composed of ice crystals.
   (*b*) High clouds have a larger albedo than low clouds.
   (*c*) High clouds have a net warming effect on climate.
   (*d*) Low clouds have a net warming effect on climate.

12. Which of the following statement(s) is/are correct in the context of Sverdrup transport?
   (*a*) Sverdrup transport is always in the meridional direction.
   (*b*) Sverdrup transport is always orthogonal to the wind direction.
   (*c*) Sverdrup transport depends on the variation of the Coriolis parameter.
   (*d*) Sverdrup transport is only due to ageostrophic currents.

13. Which of the following statement(s) is/are true with regard to the Hadley circulation ?
   (*a*) The ascending branch is narrower than its descending branch.
   (*b*) Thunderstorms are more frequent in the subsiding region of the Hadley cell than in its ascending region.
   (*c*) The lower level winds between the ascending and descending branches of the Hadley cell are north-westerly.
   (*d*) Latent heat is transported from the subsiding to the ascending region of the Hadley cell.

14. Which of the following statements(s) is/are true about the ocean circulation ?
   (*a*) Large-scale ocean surface currents are driven by winds
   (*b*) Cold, dense and salty water forms in the North Atlantic Ocean.
   (*c*) Upwelling currents bring warm nutrient deficient water to the surface of the ocean.
   (*d*) Thermohaline circulation does not transport energy in the meridional direction.

15. Coral reefs are found primarily is tropical and subtropical shallow seawaters. Which of the following statements.
   (*a*) Corals require plenty of sunlight for photosynthesis and sunlight is abundant in the tropical and subtropical latitudes.

(*b*) Corals grow optimally in seawater unsaturated in carbonate, which is found only in the tropical and subtropical oceans.

(*c*) Corals grow optimally in fresh low-salinity water.

(*d*) Corals grow optimally in water temperatures between 23°C and 29°C.

Q. 16 – Q. 22 carry Two marks each

16. In an incompressible fluid, the horizontal divergence is $-0.01\ s^{-1}$. Then, the vertical velocity at 50 m above a flat surface is ____ ms^{-1}. Round off to one decimal place.

17. In an atmosphere, temperature (T) decreases linearly with height above the ground (z), i.e., $T(z) = T_o - \gamma z$, where γ is a constant. Surface pressure is 900 hPa. If the atmosphere is at rest, then the value of z at which the pressure decreases to half of that at the surface is ____ m (round off to the nearest integer).

Take acceleration due to gravity $g = 10\ ms^{-2}$, gas constant $R = 300\ J\ kg^{-1}\ K^{-1}$, $T_o = 300$ K and $\gamma = \frac{1}{30} K\ m^{-1}$, and the atmosphere behaves as an ideal gas.

18. In a local Cartesian system, a zonal jet has a form $u(y) = u_o \left(\frac{1 - y^2}{L^2}\right)$, for $-L \le y \le L$. Here, y is the meridional distance measured from the axis of the jet and is positive northward. The vertical component of vorticity of this flow at $y = \frac{L}{2}$ is ____ s^{-1}. Round off to 3 decimal places.

Take $u_o = 50\ ms^{-1}$ and L = 5 km.

19. An eastward flow with a speed to $10\ ms^{-1}$ goes from station M to station N, which are separated by a distance of 1 km. The temperature at station N is always higher than at station M by 10 K. The absolute change in temperature due to advection at the mid-point between the stations is 50 s is ____ K (round off to nearest integer).

20. Suppose, because of the doubling of atmospheric CO_2 concentration, an ocean water column receives an additional net energy input of $4\ Wm^{-2}$. If the entire water column of depth 1 km heats up uniformly, the water temperature will increase by 1 K in ____ years (round off to the nearest integer).

Assume all the additional heat added is retained and not lost. Take density of seawater = 1000 kg m^{-3}, specific heat capacity of seawater = $4200\ J\ kg^{-1}\ K^{-1}$.

21. Consider a layer of atmosphere between 5 and 6 km height. The downwelling longwave radiation at 5 and 6 km is 240 and 230 Wm^{-2}, respectively. The upwelling longwave radiation at these heights is 260 and 240 Wm^{-2}, respectively. The longwave heating rate in this layer is ____ K per day. (Round off to one decimal place.)

Take the average density of air in this layer to be 0.5 kg m^{-3}, Specific heat capacity of air at constant pressure = $1000\ J\ kg^{-1}\ K^{-1}$

22. A spherical asteroid, revolving around the sun in a circular orbit, is in radiative balance. Suddenly, the asteroid enters the shadow of a planet and solar radiation is cut off. Assuming that the asteroid emits as a blackbody in the longwave regime, the time taken to reduce the average temperature of the asteroid by 0.5 K is ____ seconds (round off to the nearest integer). Ignore the temporal change in radiation emitted by the asteroid during this cooling period.

The physical properties of the asteroid are: diameter = 2m, density = 300 kg m^{-3}, specific heat = $2000\ J\ kg^{-1}\ K^{-1}$ and albedo = 0.8 in shortwave radiation. Take the solar constant = $500\ W\ m^{-2}$,

Stefan-Boltzmann constant = $5.67 \times 10^{-8}\ W\ m^{-2}\ K^{-4}$.

ANSWERS

General Aptitude

1. (*a*) **2.** (*b*) **3.** (*b*) **4.** (*b*) **5.** (*c*) **6.** (*c*) **7.** (*a*)
8. (*c*) **9.** (*a*) **10.** (*c*)

Engineering Mathematics

1. (*d*) **2.** (*c*) **3.** (*b*) **4.** (0.5 to 0.5) **5.** (0.5 to 0.5) **6.** (0.665 to 0.668)
7. (0.375 to 0.375) **8.** (*d*) **9.** (*a*) **10.** (4 to 4) **11.** (0.32 to 0.32)

Fluid Mechanics

1. (*c*) **2.** (*d*) **3.** (*c*) **4.** (*b*) **5.** (*d*) **6.** (*b*) **7.** (*b*)
8. (*a*) **9.** (2.9 to 3.1) **10.** (*a*) **11.** (*d*) **12.** (*b*) **13.** (6.0 to 6.5)
14. (0.9 to 1.1) **15.** (1.7 to 1.9) **16.** (12.9 to 13.1) **17.** (3.9 to 4.1)
18. (10.5 to 12.0) **19.** (48.0 to 52.0) **20.** (17.5 to 19.5)
21. (182.0 to 192.0) **22.** (18.5 to 20.0)

Materials Science

1. (*c*) **2.** (*c*) **3.** (*d*) **4.** (*c*) **5.** (*c*) **6.** (*b*) **7.** (*b*)
8. (*b, d*) **9.** (0.058 to 0.063) **10.** (*b*) **11.** (*c*) **12.** (*c*) **13.** (*a, b*)
14. (5.85 to 6.00) **15.** (9600 to 9800) **16.** (38 to 42) **17.** (3.4 to 4.8)
18. (71 to 77) **19.** (210 to 220) **20.** (0.60 to 0.62) **21.** (2.10 to 2.14)
22. (0.1 to 0.1)

Solid Mechanics

1. (*d*) **2.** (*b*) **3.** (*a*) **4.** (*b*) **5.** (*b*) **6.** (*a*) **7.** (*c*)
8. (*a*) **9.** (14.30 to 14.35) **10.** (*c*) **11.** (*a*) **12.** (*b*) **13.** (*a*)
14. (*a*) **15.** (*a*) **16.** (30 to 30) **17.** (9 to 9) **18.** (56.50 to 56.60)
19. (90.60 to 90.65) **20.** (4.17 to 4.21) **21.** (0.80 to 0.90) **22.** (0.58 to 0.68)

Thermodynamics

1. (*c*) **2.** (*b*) **3.** (*c*) **4.** (*d*) **5.** (*b*) **6.** (*c*) **7.** (*c*)
8. (*b, c*) **9.** (159.00 to 162.00) **10.** (*b*) **11.** (*a*) **12.** (*c*) **13.** (*a*)
14. (100.79 to 111.39) **15.** (0.300 to 0.330) **16.** (12.1 to 13.0) **17.** (418.0 to 422.0)
18. (320.0 to 323.0) **19.** (56.80 to 62.78) **20.** (86.66 to 95.78) **21.** (773.0 to 775.0)
22. (59.0 to 61.0)

Polymer Science

1. (*c*) **2.** (*d*) **3.** (*c*) **4.** (*b*) **5.** (*d*) **6.** (*c*) **7.** (*b*)
8. (*b*) **9.** (*a*) **10.** (*b*) **11.** (*a*) **12.** (*b*) **13.** (18000 to 19000)
14. (50.4 to 50.6) **15.** (5.00 to 7.00) **16.** (62 to 99)
17. (1.980 to 2.040) **18.** (6.00 to 7.00) **19.** (100 to 100)
20. (10000 to 10500) **21.** (4000 to 4200) **22.** (1.000 to 1.500)

Food Technology

1. (*b*) **2.** (*d*) **3.** (*a*) **4.** (*b*) **5.** (*b*) **6.** (*d*) **7.** (*d*)
8. (*c*) **9.** (*d*) **10.** (*a*) **11.** (*b*) **12.** (*b*) **13.** (*a*, *d*) **14.** (*a*, *b*)
15. (*a*, *c*) **16.** (*a*, *b*, *d*) **17.** (*a*, *b*) **18.** (*a*, *c*) **19.** (*a*, *c*) **20.** (1900 to 1910)
21. (0.135 to 0.150) **22.** (2.20 to 2.40)

Atmospheric and Oceanic Sciences

1. (*b*) **2.** (*c*) **3.** (*b*) **4.** (*c*) **5.** (*c*) **6.** (*a*) **7.** (*a*, *c*)
8. (1.7 to 1.9) **9.** (9 to 11) **10.** (*c*) **11.** (*a*, *c*) **12.** (*a*, *c*) **13.** (*a*, *d*) **14.** (*a*, *b*)
15. (*a*, *d*) **16.** (0.4 to 0.6) **17.** (4450 to 4550) **18.** (0.009 to 0.011)
19. (4 to 6) **20.** (32 to 34) **21.** (1.6 to 1.8) **22.** (39990 to 40010)

EXPLANATIONS

General Aptitude

1. 'As well as' is used in positive degree.
2. The folded sheet is look like option (b).

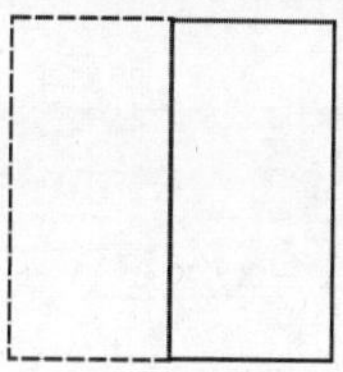

3.

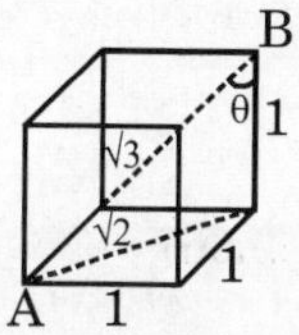

$$\cos\theta = \frac{BX}{AB} = \frac{1}{\sqrt{3}}$$

4. As we know that

$(a-b)^2 = a^2 + b^2 - 2ab$

As given expression

$$\left(x-\frac{1}{2}\right)^2 - \left(x-\frac{3}{2}\right)^2 = (x+2)$$

So,

$$\Rightarrow \left[x^2 + \left(\frac{1}{2}\right)^2 - 2(x)\left(\frac{1}{2}\right)\right] - \left[x^2 + \left(\frac{9}{4}\right) - 2(x).\left(\frac{3}{2}\right)\right]$$

$= [x + 2]$

$$\Rightarrow \left[x^2 + \frac{1}{4} - x - x^2 - \frac{9}{4} + 3x\right] = [x+2]$$

$\Rightarrow [-2 + 2x] = [x + 2]$

$x = 4$

5. Pen is to write just as knife is to cut.
6. Listening to music improves exercise performance and reduces discomfort. Music as positive effort only on few students during learning.
7. When the piece can be moved, rotated or flipped to assemble with the given piece is option (a)
8. The total number of students in three classes in the beginning.

Let us consider:

$3a + 13a + 6a = 22a = n$

after addition of 18 student total number of students.

$15b + 35b + 21b = 71b$

$3a + 18 + 13a + 18 + 6a + 18 = 22a + 54$

$22a + 54 = 71b$

$n + 54 = 71b$

$n = 71b - 54$

Let $(b = 2)$

So, $n = 71 \times 2 - 54$

$= 88$

9. Let us consider, the selling price in year '2' be 'x'.

$\therefore$ Net profit = Selling price – (Cost + Taxes)

$296 = x - (400 + 0.13x)$

$296 = 0.87x - 400$

$$x = \frac{696}{0.87} = 800$$

So, selling price in year '2' = Rs.800

In year 3, total cost = 300 × 1 = Rs.300

Taxes paid = 0.15% of selling price.

Now, let us consider, the selling price in year 3 be 'y'

$\therefore 210 = y - (300 + 0.15y)$

$\Rightarrow 210 = (0.85y - 300)$

$$y = \frac{510}{0.85} = 600$$

$\therefore$ Selling price in year '3' = 600

$$\text{So, } \frac{x}{y} = \frac{800}{600} = \frac{4}{3}.$$

10. I: S > R

II: Q < P, R, S, T, U

From III & IV observation

P > T > S > R > U > Q

The number of students that are taller than 'R' is same as the number of students shorter than S.

Engineering Mathematics

1. $$AX = \begin{bmatrix} 2 & -4 \\ 1 & 1 \\ 1 & -1 \end{bmatrix}\begin{bmatrix} x_1 \\ x_2 \end{bmatrix} = \begin{bmatrix} 2x_1 - 4x_2 \\ x_1 + x_2 \\ x_1 - x_2 \end{bmatrix}$$

$\therefore 2x_1 - 4x_2 = -1$

$x_1 + x_2 = \alpha$ &

$x_1 - x_2 = 1$

$$\therefore x_1 = \frac{5}{2} \text{ \& } x_2 = \frac{3}{2}$$

$\therefore \alpha = x_1 + x_2 = 4$

2.

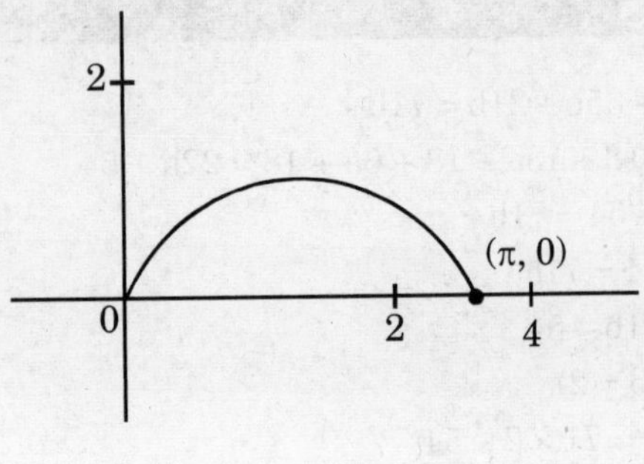

u = 2y & v = 5x. $\frac{\partial v}{\partial x} - \frac{\partial v}{\partial y} = 5 - 2 = 3$

∴ By Green's theorem, the line integral =

$\iint 3d_x d_y = 3\iint 1d_x d_y = 3\alpha$

3. Let us consider,

$(2^3 + 1)_e \frac{i\pi}{3} = 0$

$z^4 + z^2 + 1 = -z + z^2 + 1$, as $z^3 = -3$

$= -\frac{1}{2} - i\frac{\sqrt{3}}{2} + \frac{-1}{2} + i\frac{\sqrt{3}}{2} + 1$

$= 0$

Now

By L'Hôpital's rule, we have

$$L = \lim_{z \to e^{i\pi/3}} \frac{3z^2}{4z^3 + 2z} = \lim_{z - e^{i\pi/3}} \frac{3z}{4z^2 + 2} = \frac{3}{4} - i\frac{\sqrt{3}}{4}$$

4. $\int_0^a f(x)dx = \frac{a^2}{2} + \frac{a}{2}\sin(a) + \frac{\pi}{2}\cos(a) - \frac{\pi}{2}$

By the fundamental theorem of integral caluclus,

$\frac{d}{da}\int_0^a f(x)dx = f(a)$

$= a + \frac{1}{2}[\sin(a) + a\cos(a)] + \frac{\pi}{2} \times -\sin(a) - 0$

$\therefore f\left(\frac{\pi}{2}\right) = \frac{\pi}{2} + \frac{1}{2}[1 + 0] + \frac{\pi}{2} \times -1 = \frac{1}{2} = 0.5$

5. As given ghat

$f_{(x)} = 2^{\alpha x}$

a = 0, b = 4, n = 2 ⇒ h =

$\therefore\ 9 = \frac{2}{2} \times [1 + 2 \times 2^{2\alpha} + 2^{4\alpha}]$

$\therefore\ 8 = 2^{2\alpha+1} + 2^{4\alpha}$

If $\alpha = \frac{1}{2}, 4\alpha = 2$ & $2\alpha + 1 = 2$

∴ RHS = 4 + 4 = 8 = LHS!

6. As given that,

$y = e^x v$

$\therefore y' = e^x(v + v')$

& $y'' = e^x(v + 2v' + v'')$

Given ODE transforms to

$x \times e^x \times [v'' + 2v' + v] + 2 \times (1 - x) \times e^x (v + v') + (x - 2) \times e^x \times v = 0$

$\therefore v'' \times x \times 1 + v' \times [2x + 2 - 2x] + v \times [x + 2 - 2x + x - 2] = 0$

$xv'' + 2v' + ov = 0$

$\Rightarrow \alpha = 1,$ $\frac{\alpha + \beta + \gamma}{3} = \frac{2}{3} = 0.667$

$\beta = 1$

$\gamma = 0$

7. As given that

$P(\text{speaks truth}) = \frac{3}{4}$

$P(\text{lies}) = \frac{1}{4}$

Required probability $= \frac{1}{6} \times \frac{3}{4} = \frac{3}{24} = \frac{1}{8} = 0.375$

8. As given that,

$\frac{\partial f}{\partial x} = 4x^3 - 4x + 4y = 0$ &

$\frac{\partial f}{\partial y} = 4y^3 + 4x - 4y = 0$

Clearly (0, 0) satisfies both equations

$r = \frac{\partial^2 f}{\partial x^2} = 12x^2 - 4 = -4$ $s = \frac{\partial^2 f}{\partial y \partial x} = 4$

$t = \frac{\partial^2 f}{\partial y^2} = 12y^2 - 4 = -4$ $rt - s^2 = 16 - 16 = 0$

∴ no conclusion

We let y = 0} $f_1(x) = x^4 - 2x^2 + \alpha$

$\frac{df_1}{dx} = 4x^3 - 4x$

$\frac{d^2 f_1}{dx^2} = 12x^2 - 4$ & at x = 0, this is –ve

∴ local max

we let y = x} $f_2(x) = 2x^4$ clearly local min

function has a saddle point!

9. As given that,

u & v are harmonic

$\therefore$ f(2) = u + iv is analytic

$u = (r^2 - y^2) \times v$

Due to analyticity, $u_x = v_y$ & $u_y = -v_x$

$u_x = 2_{xv} + (x^2 - y^2)v_x = v_y$ & $4_y = -2_{yv} + (x^2 - y^2)v_y = -v_x$

$u = (x^2 - y^2) \times v$

$u_x = 2_x.v + (x^2 - y^2).v_x$

$u_{xx} = 2.v + 2x.v_x + 2x.v_x + (x^2 - y^2).v_{xx}$

$4_y = -2y.v + (x^2 - y^2).v_y$

$4_{yy} = -2.v + -2_y.v_y + (-2y).v_y + (x^2 - y^2).v_{yy}$

$4_{xx} + 4_{yy} = 2v - 2v + 4xv_x - 4yvy + (x^2 - y^2)(v_{xx} + v_{yy})$

$\therefore 0 = O + 4[xv_x - yv_y] + (x^2 - y^2)(0)$

$\therefore xv_x - yv_y = 0$

10. As given that,

$\therefore \rho(A) = \rho(A + 2I) = 5$

Geometric multiplicity of an eigenvalue

$\lambda = n - \rho(A - \lambda I_n)$

$\therefore$ for $\lambda = -2, (6 - M)_{-2} = 7 - 3(A + 2I_7)$

$\therefore (6 - M)_{-2} = 7 - 5 = 2$

Now, $(6 - M)_{\lambda=1} + (6 - M)_{\lambda=-2} + (6 - M)_{\lambda=0} = 7$

$(6 - M)_{\lambda=1} + 2 + 2 = 7$

$(6\text{–}M)_{\lambda=1}$ is 3.

11. As given that,

$$\alpha = \int_{c_1} \frac{2x}{y}dx + \frac{1-x^2}{y^2}dy$$

$$\beta = \int_{c_2} \frac{2x}{y}dx + \frac{1-x^2}{y^2}dy$$

x = cos(θ), dx = – sin(0)dθ

y = sin(θ), dy = cos(θ)dθ

& for c_2, θ starts at $\frac{\pi}{2}$ & ends at $\tan^{-1}\left(\frac{3}{5}\right)$

$$\therefore \beta = \int_{\pi/2}^{\tan^{-1}\left(\frac{3}{5}\right)} \left[\frac{2\cos(0)}{\sin(\theta)}[-\sin(\theta)] + \frac{1-\cos^2(\theta)}{\sin^2(\theta)}[\cos(\theta)]\right] d\theta$$

$$\therefore \beta = \int_{\pi/2}^{\tan^{-1}\left(\frac{3}{5}\right)} \left[\frac{2\cos(\theta)}{\sin(\theta)}[-\sin(\theta)] + \frac{1-\cos^2(\theta)}{\sin^2(\theta)}[\cos(\theta)]\right] d\theta$$

$$= [(-2)\sin(\theta) + \sin(\theta) = (-1)\sin(\theta)]\tan^{-1}\left(\frac{3}{5}\right)$$

$$= (-1)\left[\frac{3}{5} - 1\right] = \frac{2}{5}$$

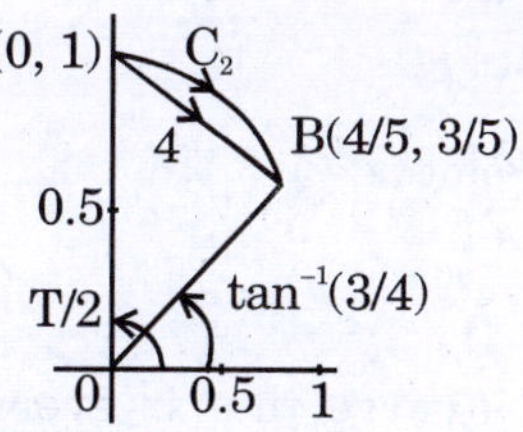

$$\nabla \times \vec{F} = \begin{vmatrix} \hat{i} & \hat{j} & \hat{k} \\ \frac{\partial}{\partial x} & \frac{\partial}{\partial y} & \frac{\partial}{\partial z} \\ \frac{2x}{y} & \frac{1-x^2}{y^2} & 0 \end{vmatrix}$$

$$= 0\hat{i} - 0\hat{j} + \left[\frac{-2x}{y^2} - \frac{-2x}{y^2} = 0\right]\hat{k} = \vec{0}$$

$\therefore$ our field is conservative

$\therefore$ line integral is path independent

$$\therefore \beta = \alpha = \frac{2}{5}$$

$$\therefore \alpha^2 + \beta^2 = 2\beta^2 = 2\left(\frac{4}{25}\right) = \frac{8}{25}$$

Fluid Mechanics

1.

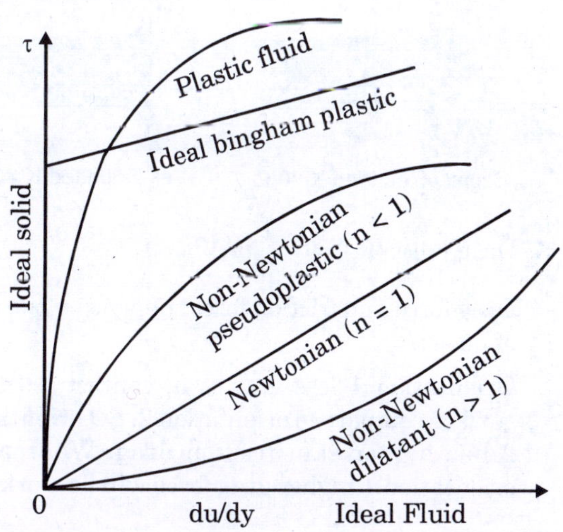

2. 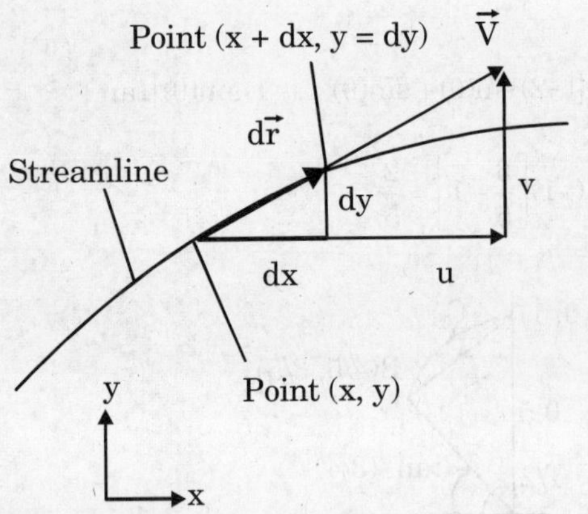

A streamline is a curve that is everywhere tangent to the *instantaneous* local velocity vector.

For steady flow, streamlines, pathlines, and streaklines are identical. For unsteady flow, they can be very different.

In steady flow (when the velocity vector-field does not change with time), the streamlines, pathlines, and streaklines coincide.

6. Continuity equation is valid for steady or unsteady incompressible flow. $\frac{\partial p}{\partial t} = 0$ regardless of whether the flow is steady or unsteady. Continuity equation is valid for both steady and unsteady flow.

7. 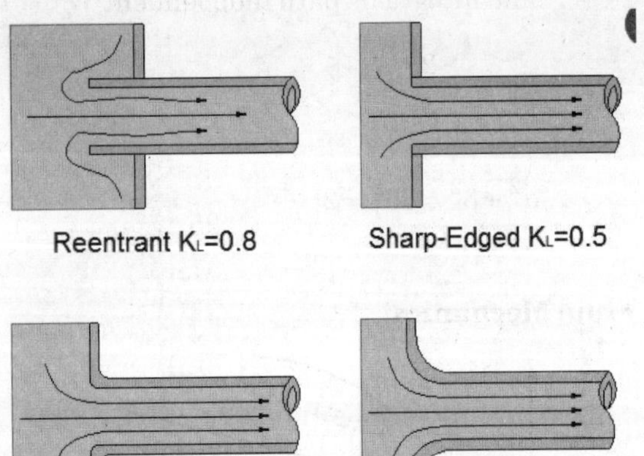

8. Drag Force$(F_D) = C_D \frac{1}{2} \rho A V^2$

Drag force and friction forces depend on the area of obejct.

Orientation 1 less area is in contact with the surface compare to orientation 2. SO orientation 2 has higher skin friction drag. Whereas in orientation 1 higher drag force will be work.

10.

Zone P	Zone Q
Area decreasing due to that velocity increasing, Pressure increasing So Acceleration of flow, favorable pressure gradient, No flow separation	Area increasing due to that velocity decreasing, Pressure decreasing So deceleration of flow, Adverse pressure gradient, Possible flow separation

Materials Science

1. (c)

Chemical potential $\mu = \frac{\partial G}{\partial n} = \frac{\text{Gibbs energy}}{\text{No. of particles}}$

A system is in equilibrium if its free energy is at a minimum.

Substance is in equilibrium between two phase then its chemical potential must have the same value in both the phase.

$\mu_A^{\alpha} = \mu_A^{B}, \mu_B^{\alpha} = \mu_B^{B}$

2. (c)

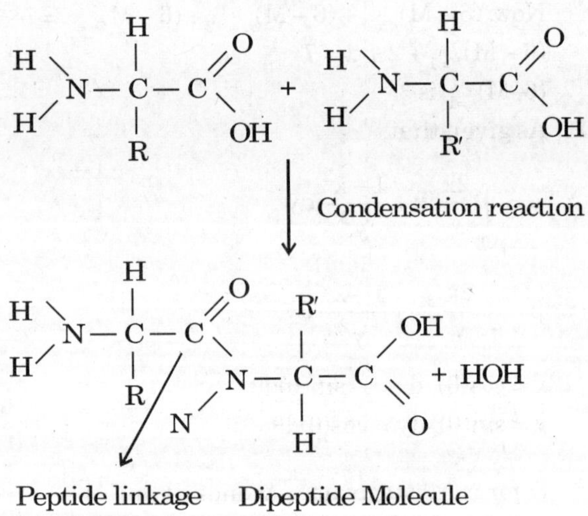

3. (D) For FCC   Sup direction

{111}   <110>

4. (c)

Electronic polarization loses its response in the ultraviolet region around 10^{15}Hz

Total polarization $P = P_0 + P_i + P_e$

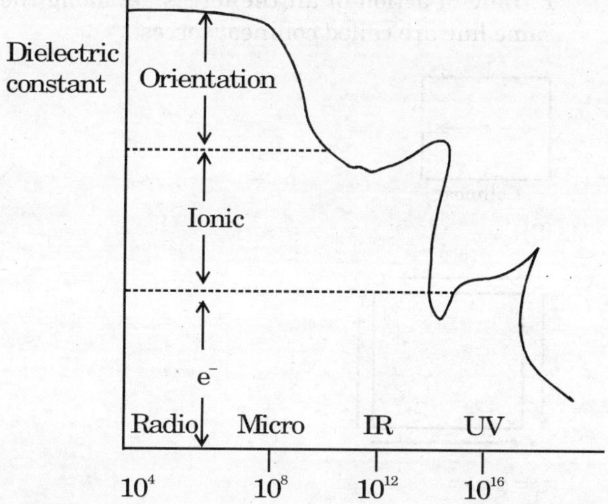

5. (c)

Martensite → Forms when austentized iron - carbon alloy is rapidly cooled (quenching)

→ Martensite is a non-equilibrium single phase structure

→ Formed by diffusion - less transformation

→ FCC austenite changes to BCT Maten side

Eutectoid → "eutectic- like" chemical reaction in which solid transforms into two solid phase at the same time upon cooling

→ Solid, $\underset{\text{Heating}}{\overset{\text{Cooling}}{\rightleftharpoons}}$ $Solid_2 + Solid_3$

→ Solid convert into two solid

Precipitation → It is a heat treatment technique used to increase the yield strength of malleable materials, including most structural alloy of aluminum, Magnesium and stainless steels.

reaction occurs in supersaturated solutions

6. (b)

Sampling volume of BSE is larger than secondary electrons.

BSE originated from much deeper within the sample (a few μm below the surface) and interact much more strongly with the sample. They therefore provide compositional information but give lower resolution image.

7. (B)

→ High deposition rate means more number of nuclei → No space for grow to nuclei

If low deposition rate means low number of nucler → space available for grow to nuclei

→ grain size will increase.

→ Transformation taking place close to 1m has low nucleation rate & high diffusion rate. Hence, few nuclei growing over large size leading to coarse grain microstructure.

→ Transformation at low temp corresponds to high nucleation rate means fine particles.

10. (B)

→ Secondary electron → Topography, fractography & texture

→ Backsscattered electron → Compositional information, distinguishing chemically distinct phase

→ Characteristics x-Ray → Compositional analysis

→ Diffracted backscaltered → Crystallographic information electron

P - 2 , Q – 4, R – 3, S – 1

11. (1) Annealing → Relieve stress & increase softness, ductility Austenite $\xrightarrow[\text{Cooling}]{\text{Slow}}$ pearlite

(2) Quenching → To impart high hardness to steel

Austenite $\xrightarrow[\text{Cooling}]{\text{Rapid}}$ Martensite

(3) Tempering → To provide toughness by again reheating the quenched steel (Martensite)

(4) Carburing → Carbon diffuses on to the steel surface which increase the surface hardness

P – 1, Q – 2, R – 3, S – 4

12. Polymer

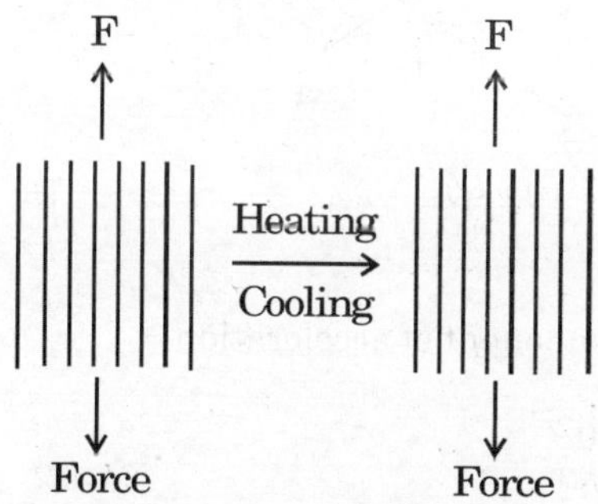

Polymer shrink on heating entropy will increase

Ceramic : Low coefficient of thermal expansion.

Mo change on heating. It is a thermal insulator.

Polymer get heat only three direction.

So polymer show inward bending

X ◡ Y, Y ◠ Z

13. X-ray diffraction peak broadening enables the estimation of crystallite size & micro strain.

Solid Mechanics

1. [D]

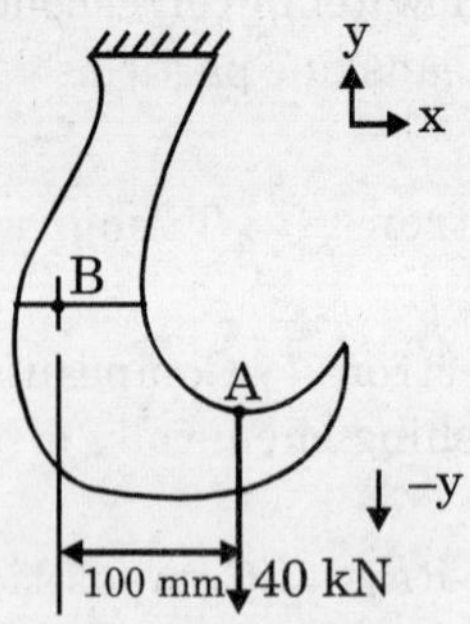

The equivalent force-couple system at B is
Any set of forces on a body can be replaced by a single force and a single couple acting that is statically equivalent to the original set of forces and moments.
This set of an equivalent force and a couple is known as the equivalent force couple system.
The force applied on B point replaced by A point & the force on A is 40 kN in negative y direction.
Moment of force (M) = F.d
$= 40000 \times 100 \times 10^{-3}$
M = 4000 Nm clockwise
Force on a well shift towards B.

2. [B]

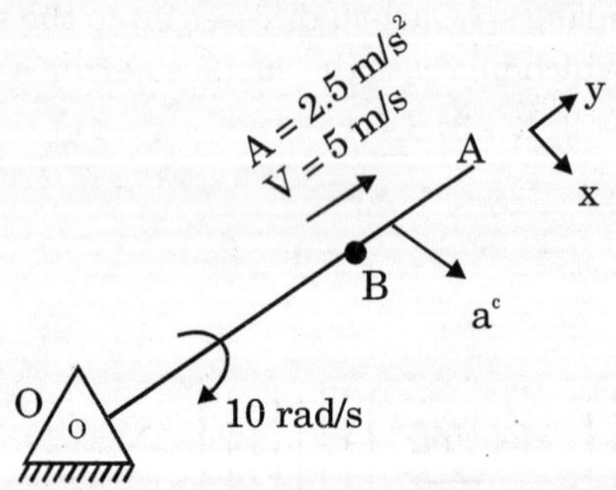

Coriolis component of acceleration
$a^c = 2V\omega$
$= 2 \times 5 \times 10$
$a^c = 100 \text{ m/s}^2$
Coriolis component of acceleration direction always lead by 90° in the direction of rotation (w).
So Coriolis in the direction of x.
$a^c = 1 \text{ w m/s}^2$ in + x direction.

3. [A]

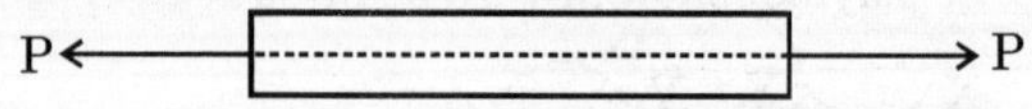

Forces are in equilibrium. Both the side force is P. Line of action of all the forces act along the same line are called collinear forces.

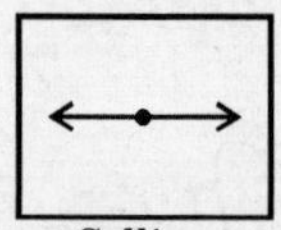

Collinear

4. [B]

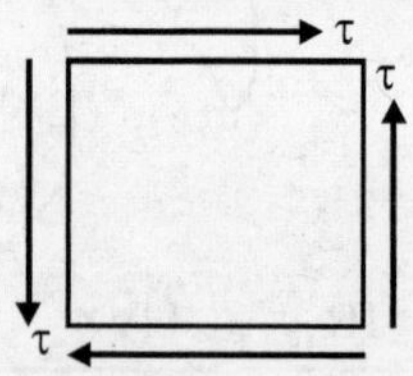

$\tau_1 = \tau = -\tau_2$

Maximum strain $\varepsilon = \frac{\tau_1}{E} - \mu\frac{\tau_2}{E}$

$\varepsilon = \frac{\tau_1}{E} + \mu\frac{\tau}{E}$

$\varepsilon = \frac{\tau}{E} + (1 + \mu)$

5. [B]

$E = 3K(1 - 2\mu)$ ...(1)

E = young modulus $\frac{E}{K} = \frac{3}{2}$ given

K = bulk modulus $E = \frac{3}{2}K$

μ = Poisson ratio
Put in equation (1)

$\frac{3}{2}K = 3K(1 - 24)$

μ = 0.25
G = Shear Modulus
$E = 2G(1 + \mu)$

$\frac{G}{E} = \frac{1}{2(1+\mu)}$

Put m = 0.25 in equation (2)

$\frac{G}{E} = \frac{1}{2(1+0.5)} = \frac{1}{2.5} = \frac{2}{5}$

6. [A]

The value of normal stress in the plane of maximum shear stress may or **may not be zero**. When the shaft is subjected to pure torsion the normal stress in the plane of maximum shear will be zero. While for other cases the normal stress will exist in the plane of maximum shear stress.

1.Principal stresses occur on mutually perpendicular planes. 2. Shear stresses are zero on principal planes. 3. Planes of maximum shear stress occur at 45° to the principal planes. 4. The maximum shear stress is equal to one half the differences of the principal stresses. Magnitude of maximum shear stress is equal to the largest radius of the Mohr's Circle.

7. [C]

Momentum = Force × distance

In case I: Moment of force at Point A

$$= P \times \frac{L}{3}$$

In case II: Moment of force at Point A

$$= P_1 \times \frac{L}{4}$$

Moment of force at Point A

$$= P \times \frac{L}{3} = P_1 \times \frac{L}{4}$$

$$\boxed{P_1 = \frac{4P}{3}}$$

8. [A]

$$\frac{\tau_{max}}{\tau_{avg}} = \frac{3}{2} \text{ for rectangular}$$

Rectangular Circle Cylinder

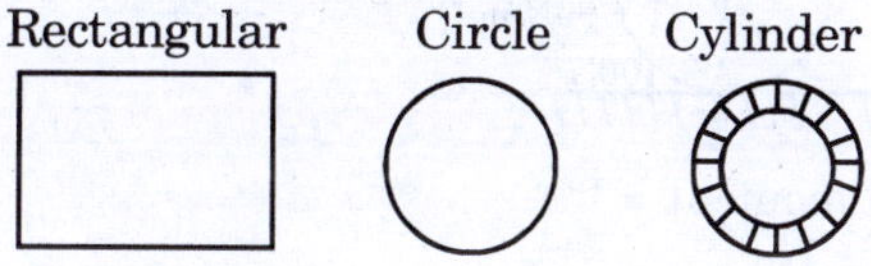

$$\frac{\tau_{max}}{\tau_{avg}} = \frac{3}{2} \quad \frac{\tau_{max}}{\tau_{avg}} = \frac{4}{3} \quad \frac{\tau_{max}}{\tau_{avg}} = 2$$

9. For a simple Harmonic motion

$a_{max} = Aw^2$

a_{max} = Maximum acceleration

A = Amplitude of motion

$a_{max} = Aw^2$

w = Vibration/frequency = $2\pi f$

$a_{max} = \sqrt{2} \times a_{RMS}$

a_{RMS} = R wt mean square of accelration

$Aw^2 = \sqrt{2} \times a_{RMS}$

$$A = \sqrt{2} \times \frac{10}{(2\pi f)^2}$$

$$= \sqrt{2} \times \frac{10}{(2 \times \pi \times 5)^2} = 0.0143 \text{ m}$$

A = 14.34 mm

10. [C]

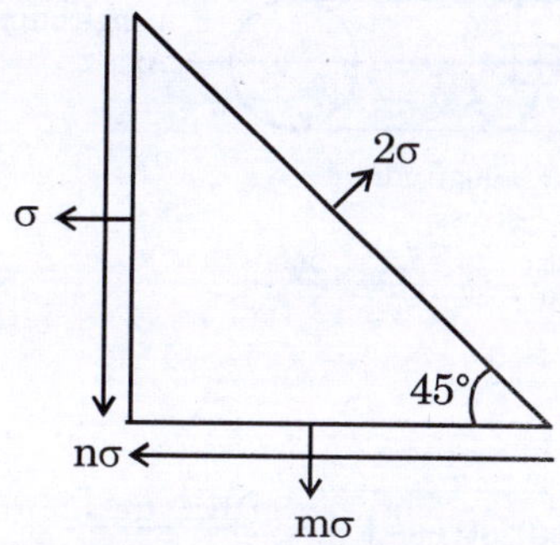

Normal stress at angle θ

$$(\sigma_n)_\theta = \left(\frac{\sigma_x + \sigma_y}{2}\right) + \left(\frac{\sigma_x - \sigma_y}{2}\right)\cos 2\theta + t_{xy} \sin 2\theta$$

Shear stress at angle θ

$$\tau_\theta = \left(\frac{\sigma_x - \sigma_y}{2}\right)\sin 2\theta - \tau_{xy} \cos 2\theta$$

Given: $\sigma_x = m\sigma$ $\sigma_n = 2\sigma$

$\sigma_y = \sigma$

$\theta = 45$

$\tau = 0$

$\tau_{xy} = n\sigma$

Put the value in the equation (1) & (2)

$$2\sigma = \left(\frac{m\sigma + \sigma}{2}\right) + n\sigma \quad ...(3)$$

$$0 = \left(\frac{m\sigma - \sigma}{2}\right) \quad ...(4)$$

m = 1 put the value is 3

$$2\sigma = \frac{\sigma + \sigma}{2} + n\sigma$$

n = 1 m = 1 & n = 1

11. [A]

$$\gamma_{max} = 1000 \times 10^{-6}$$

$$\frac{\gamma_{max}}{2} = \sqrt{\left(\frac{\varepsilon_{xx} - \varepsilon_{yy}}{2}\right)^2 + \left(\frac{\gamma_{xy}}{2}\right)^{-2}}$$

$\varepsilon_{xx} = 1000 \times 10^{-6}$

$\varepsilon_{yy} = 2\omega \times 10^{-}$

$$\frac{10^3 \times 10^{-6}}{2} = \sqrt{\left(\frac{1000 \times 10^{-6}}{2}\right)^2 + \left(\frac{\gamma \times y}{2}\right)^2}$$

$\gamma_{xy} = 600 \times 10^{-6}$

12. [B]

Their cylinder mean ratio of $\frac{Di}{b} > 20$

[Di = internal diameter & t = thickness]

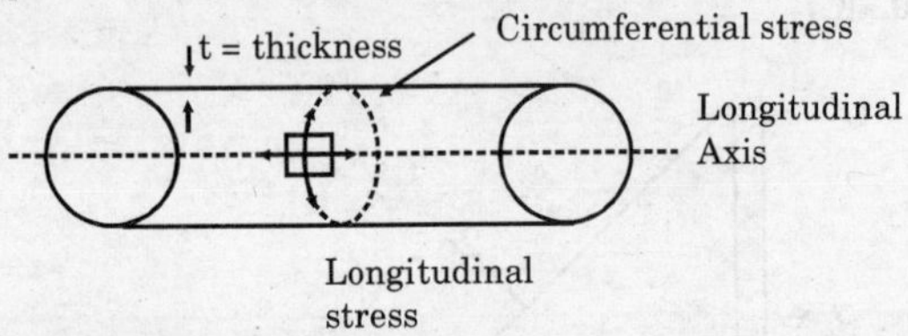

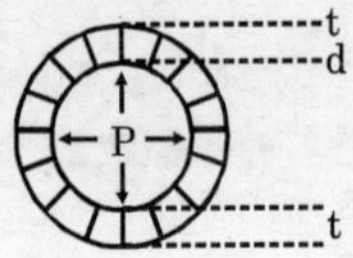

Maximum shear stress $t_{max} = \frac{\tau_{HwPstress}}{2}$

[On the wall of cylinder]

For thin wll cylinder $\tau_{Hwpstress} = \frac{pd}{2t} = \frac{p \times \alpha \times r}{2t}$

d = 2r $\qquad = \frac{Pr}{t}$

$$T_{max} = \frac{Pr}{2 \times t} = \frac{Pr}{2t}$$

13. [A]

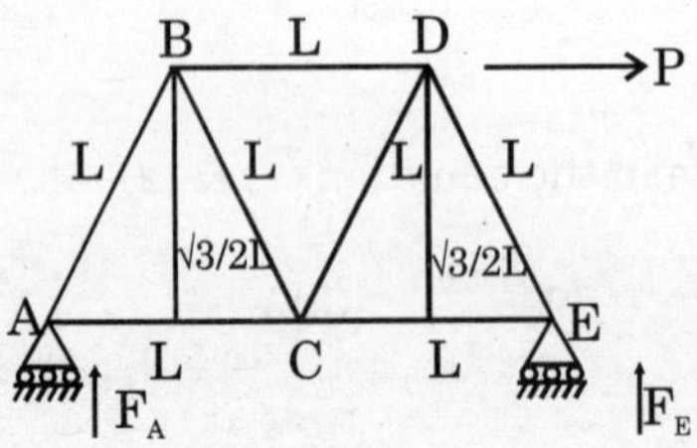

Total force $F_A + F_E = 0$ $\quad \boxed{F_A = -F_E}$

Moment at P and E

$$F_E \times 2L = P \times L \times \frac{\sqrt{3}}{2}$$

$$FE = \frac{\sqrt{3}}{4}P$$

$$F_A = -\frac{\sqrt{3}}{4}P$$

Direction of A is in opposite side

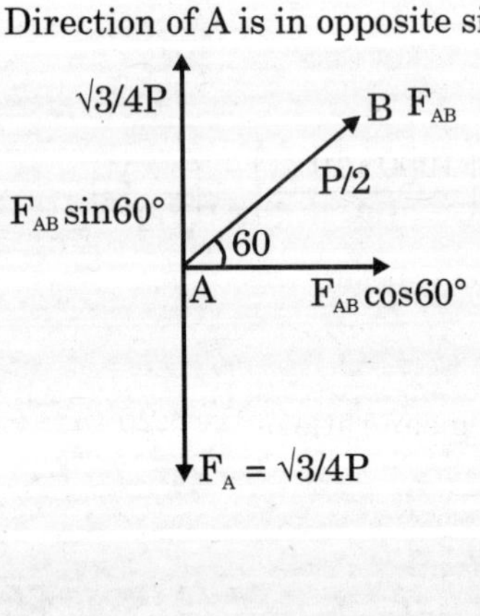

Force $F_{AB} = F_{AB} \times \text{Sin}60° = \frac{\sqrt{3}}{4}P$

$$F_{AB} \times \frac{\sqrt{3}}{4} = \frac{\sqrt{3}}{4}P$$

$$F_{AB} = \frac{P}{2}$$

14. [A]

We start from left to right then

Spring gives –3 moments

String – 1 moment

So option A is most suitable answer

15. [A] Read both curve given in question

$$\text{Acceleration (a)} = \frac{dv}{dt} = \frac{\text{change in velocity}}{\text{change in time}}$$

$$= \frac{0.3-(0.03)}{3 \times 10^{-3}} = 110 m/s^2$$

Force = m × a

= 1 × 110

Force = 110N < 200N

So, the almmd shell will crack but not break

16. [30]

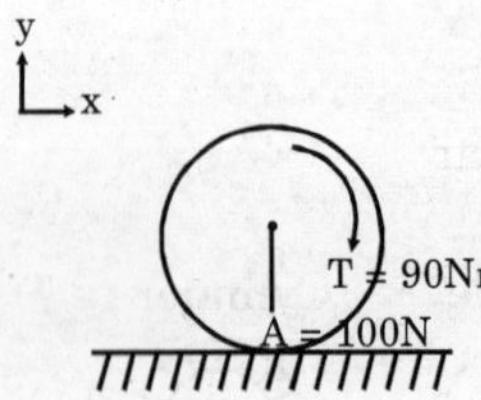

Total moment = Iα

ΣMo = Tα

$$c_j = \left[\frac{10 \times (0.2)^2}{2} + 10 \times (0.2)^2\right]\alpha$$

$$I = \frac{MR^2}{z}$$

Friction force F = m γ α

= 10 × 0.2 × 15 = 30 N

17. [9]

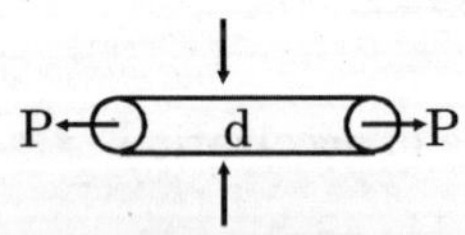

$$\sigma = \frac{P}{\frac{\pi}{4}d^2} = \frac{4P}{\pi d^2}$$

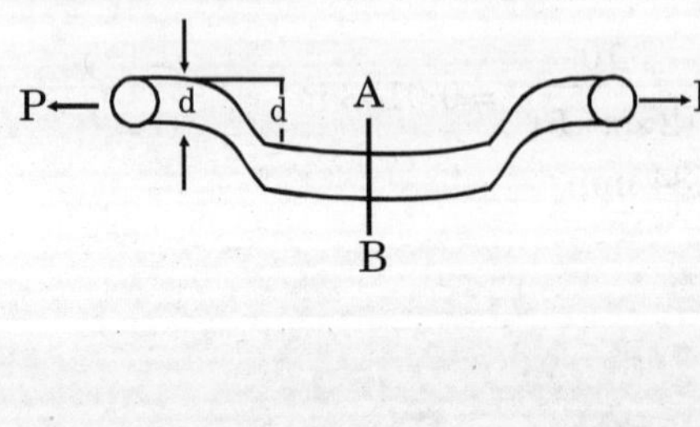

$M = P \times d$ $\quad \sigma$ = Direct stress + Bending stress

$Z = \frac{I}{y}$ $\quad = \frac{4P}{\pi d^2} + \frac{My}{I}$

$$= \frac{4P}{\pi d^2} + \frac{P \times d}{\frac{\pi \times d^3}{32}}$$

$$= \frac{36P}{\pi d^2}$$

I = Moment of inerta = $\frac{\pi r^4}{4} = \frac{\pi d^4}{64}$

y = distance of the outer most length from the neutralance = $\frac{d}{2}$

The maximum longitudinal stress at a section

A – B in the rod = $\frac{\frac{36P}{\pi d^2}}{\frac{4P}{\pi d^2}}$ = 9 times the longitudinal stress the staight rod.

Value of n = 9

18. [56.54]

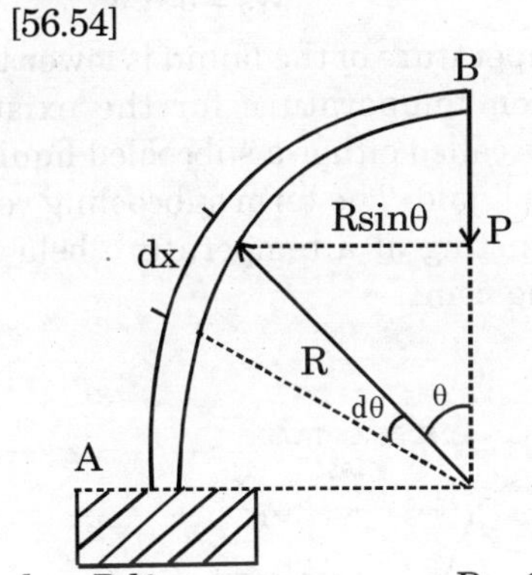

$dx = Rd\theta$ $\quad R = 2m$

$dx = 2d\theta$ $\quad P = 1000$ N

Normal = PR sinθ

= 1000 × 2 × sinθ

M = 2000 sinθ

Bending strain energy $\quad U = \int_0^{\pi/2} \frac{M^2 dx}{2 \in I}$

E = 208 GPa = 208 × 10^9 Pa

I = 26.7 cm^4 = 26.7 × 10^{-8} m^4

$$U = \int_0^{\pi/2} \frac{2000^2 \sin^2\theta \times 2d\theta}{2 \times 208 \times 10^9 \times 26.7 \times 10^{-6}}$$

$$U = 72.03 \underbrace{\int_0^{\pi/2} \sin^2\theta d\theta}_{\frac{\pi}{4}}$$

$$U = 72.03 \times \frac{3.14}{4}$$

U = 56.54 Nm

19. [90.63]

In the given diagram two stress are exits

Between the road: Mechanical stress + Thermal stress

Maximum strain by stress = 1 mm(gap between the road)

Strain by thermal stress $(E)_T = (L\alpha\Delta T)_1 + (L\alpha\Delta T)_2$

Strain by compressive (Mechanical) stress = $\left(\frac{PL}{AE}\right)_1 + \left(\frac{PL}{AE}\right)_2$

1 = rod 1

2 = rod2

$$(L\alpha\Delta T)_1 + (L\alpha\Delta T)_2 + \left(\frac{PL}{AE}\right)_1 + \left(\frac{PL}{AE}\right)_2 = 1\text{mm}$$

$$= 2\left[(300 \times 23 \times 10^{-6} \times 125) + \frac{\sigma \times 300}{75 \times 10^3}\right] = 1\text{mm}$$

σ = 90.625 MPa

20. [4.19]

$T_{holow\ tube} = T_{solid}$

$\tau \times Z_H = \tau \times Z_s$

$Z_H = Z_s$

τ = Maxikum stress at surface

Z = Polar modulus

$r_i = 4$

$r_0 = 5$

$$\frac{\pi}{16}\left(\frac{d_0^4 - d_i^4}{d_0}\right) = \frac{\pi d^3}{16} \qquad d = \frac{r}{2} \text{ Put in both side}$$

$$\frac{5^4 - 4^4}{5} = r^3$$

r = 4.19 cm

21. [0.87]

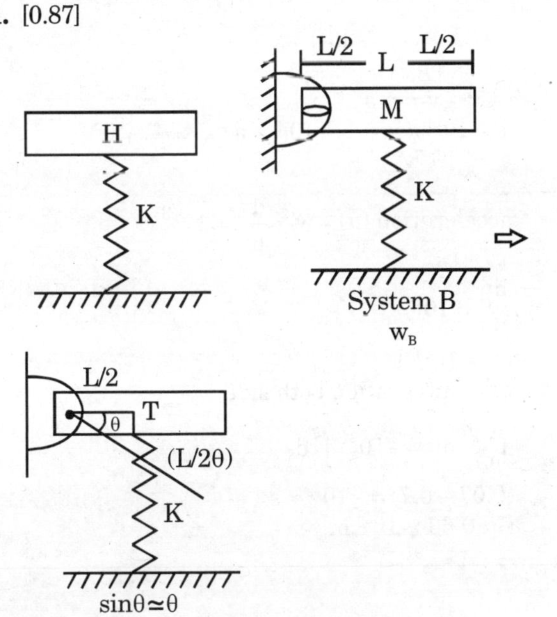

$$w_A = \sqrt{\frac{k}{m}}$$

k = spring constant
m = mass
w_A = natural frequency

x = deflection system B = $\frac{L}{2}$

I = for rod = $\frac{ML^2}{3}$

In the system B mass is hinged at one of its end due to this it will bend at angle of θ
Using of d'alembert's principle

$$I\ddot{\theta} + kx\left(\frac{L}{2}\theta\right) = 0$$

$$\frac{ML^2}{3}\ddot{\theta} + k.\frac{L}{2}.\frac{L}{2} - \theta = 0$$

$$\frac{ML^2}{3}\ddot{\theta} + k\frac{L^2}{4}.\theta = 0$$

$$\ddot{\theta} + \frac{KL^2}{4}.\frac{3w^2}{HL^2}\theta = 0$$

$$w_B = \sqrt{\frac{3}{4}\frac{k}{14}}$$

$$\frac{w_B}{w_A} = \frac{\sqrt{\frac{3}{4}\frac{k}{M}}}{\frac{k}{M}} = \sqrt{\frac{3}{4}} = 0.87$$

22. [0.63]
drag coefficient Cd = 1 μN⁻ˢ/m = 1 × 10^{-6} NS/m
initial velocity v_i = 0.7 m/s

final velocity $v_f = 0.7 \times \frac{10}{100} = 0.07$ m/s

Use Newton's law
ΣF = ma
–Cd × v = ma
$1 \times 10^{-6} \times v = 1 \times 10^{-9} \times a$
$a = -10^{+3}v$

acceleration (a) = $v = \frac{dv}{ds} = 10^{+3}v$

$$\frac{dv}{ds} = 10^{+3}$$

$dv = 10^{+3}\, ds$
take integration both side

$$\int_{0.7}^{0.07} dv = -10^{+3}\int_0^s ds$$

$(0.07 - 0.7) = -10^{+3} \times s$
$s = 0.63 \times 10^{-3}$ m

Thermodynamics

1. (c)

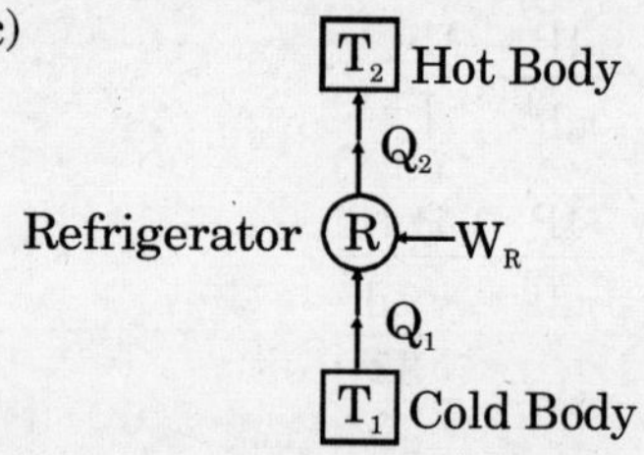

$$(COP)\text{Refrigerator} = 4 = \frac{Q_1}{W_R}$$

$(COP)_{RHP} = (COP)_{R\,Ref.} + 1$
RHP = Reversed heat pump
R Ref = Reversed Refrigerates
Q_1 = desired output at time of Refrigerator

$(COP)_{RHP} = 4 + 1 = 5$

Q_2 = desired output at time of Heat pump

$$\frac{Q_2}{W_R} = 5$$

$$\frac{Q_2}{1\,KW} = 5$$

$Q_2 = 5$ KW

2. (b) If the temperature of the liquid is lower than the saturation temperature for the existing pressure, it is called either a subcooled liquid or a compressed liquid. The term subcooling refers to a liquid existing at a temperature below its normal boiling point.

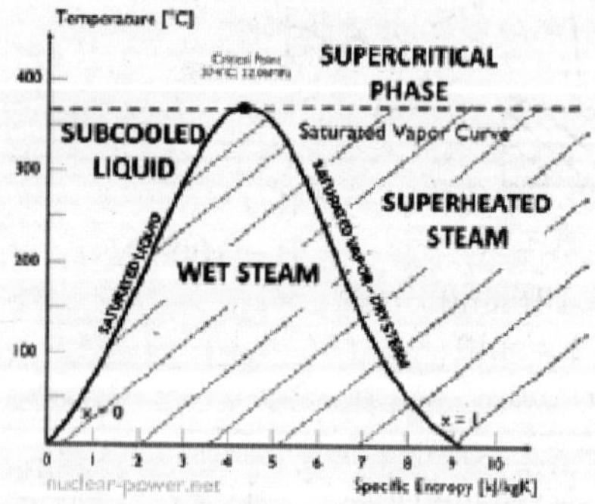

3. (b)

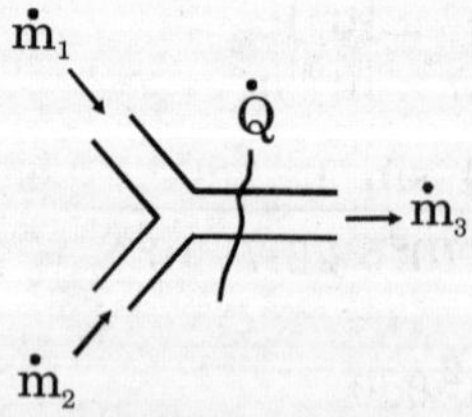

$\dot{m}_3 = \dot{m}_1 + \dot{m}_2$

To express the mixture temperature as a function of the mass flow rates of the initial components and the heat transfer we start from the nergy relation.

$\dot{E}_{in} = \dot{E}_{out}$

$\dot{m}_1 h_1 + \dot{m}_2 h_2 + \dot{Q} = \dot{m}_3 h_3 \qquad \dot{m}_3 = \dot{m}_1 + \dot{m}_2$

$-\dot{Q} = \dot{m}_1 (h_3 - h_1) + \dot{m}_2 (h_3 - h_2)$

$-\dot{Q} = \dot{m}_1 C_P (T_3 - T_1) + \dot{m}_2 C_P (T_3 - T_2)$

$h = CpT$

"–" Negative sign used because of heat loss in chamber

$$T_3 C_P (\dot{m}_1 + \dot{m}_2) = C_P (\dot{m}_1 T_1 + \dot{m}_2 T_2) - \dot{Q}$$

$$T_3 = \frac{\dot{m}_1 T_1 + \dot{m}_2 T_2}{\dot{m}_1 + \dot{m}_2} - \frac{\dot{Q}}{Cp(\dot{m}_1 + \dot{m}_2)}$$

4. (d) We apply Tds equation

$T.ds = dh - v.dp$

$$T = \frac{dh}{ds} - v\frac{dp}{ds}$$

at constant pressure $\frac{dp}{ds} = 0$

$$\left(\frac{dh}{ds}\right)_p = T$$

5. (b) Maxwell equation

$dG = -SdT + Vdp$

In mathematics

$dz = Mdx + Ndy$

$$M = \left(\frac{\partial z}{\partial x}\right)_y ; \; N = \left(\frac{\partial z}{\partial y}\right)_x$$

The differential equation has an exact solution in only situation

$$\left(\frac{\partial M}{\partial y}\right)_x = \left(\frac{\partial N}{\partial x}\right)_y$$

Applied this rule on Maxwell equation

$$-\left(\frac{\partial S}{\partial P}\right)_T = \left(\frac{\partial V}{\partial T}\right)_P$$

$$\left(\frac{\partial S}{\partial P}\right)_T = -\left(\frac{\partial V}{\partial T}\right)_P$$

6. (c) Gibbs phase rule

$F = C - P + 2$

C = No. of component = liquid water & ethanol = 2

P = No. of phase = liquid & vapor = 2

F = degree of freedom $\qquad F = 2 - 2 + 2 = 2$

7. (c)

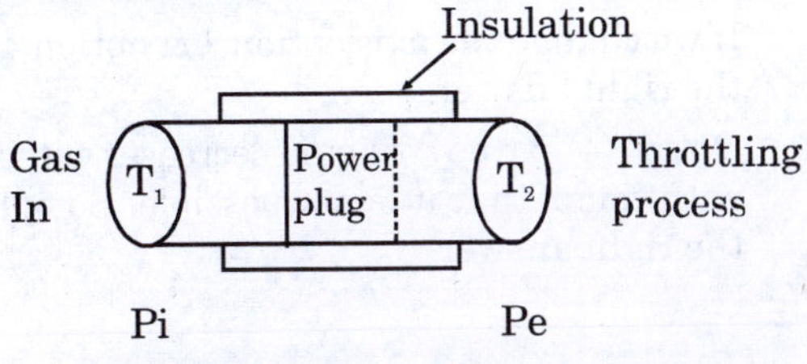

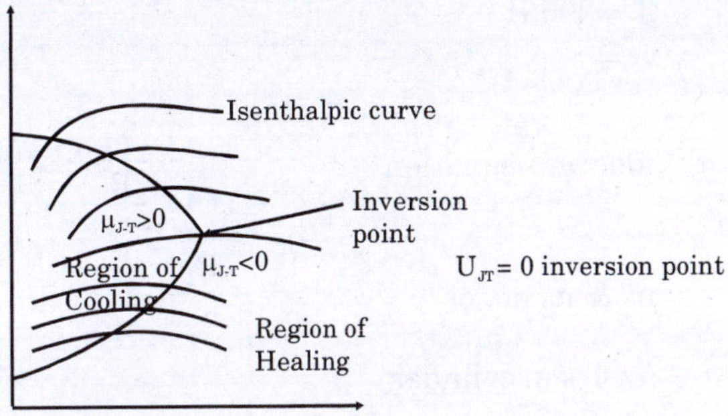

1. $\mu_{JT} > 0$

   So Pi > Pe

   In case whenever a real gas is subjected to throttling, the temperature of the gas decreased if initial state lies in the region to the left of the isenthalpic curve

2. $\mu_{JT} > 0$

   Pi > Pe

   gas lies in the Right region of isenthalpic curve temperature of the gas increases upon throttling.

8. (b,c)

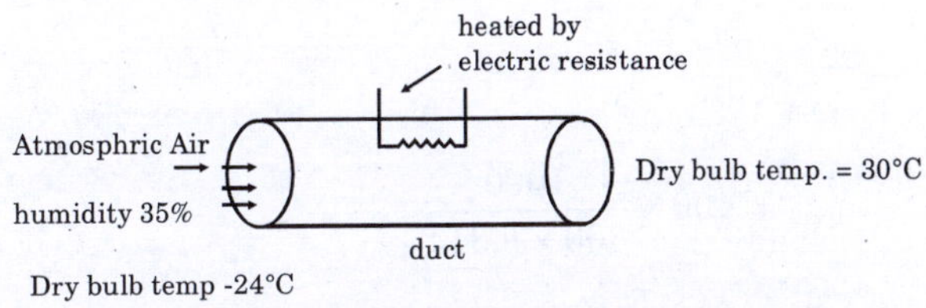

In the circular duct, there is no water vapour passes through the duct. It is a case of sensible heating. Due to electric heating temp increase at exit point. In the sensible heating humidity writ be changed and there is no change in the Dew point temperature.

9. (160.37)

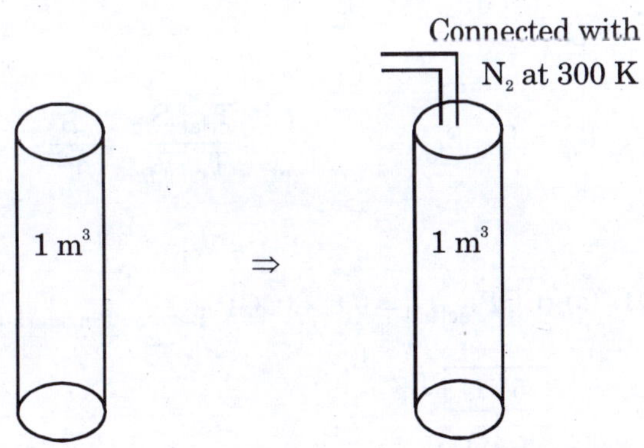

20% CO_2	$P_2 = 500$ kPa
80% O_2	$V_2 = 1\ m^3$
$P_1 = 100$k Pa	T = 300 K
$V = 1\ m^3$	$n_2 = 500$ x

T = 300 K

$n_1 = 100\,x$

ideal gas equation $\frac{P_1 V_1}{n_1 R} = \frac{P_2 V_2}{n_2 R}$ $V_1 = V_2$

n_1 & n_2 no. of $\frac{P_1}{n_1} = \frac{P_2}{n_2}$

misles in cylinder

$$\frac{n_1}{n_2} = \frac{100}{500} = \frac{1}{5}$$

$n_2 = 5 \times n_1 = 5 \times 100x = 500x$

Mole of $N_2 = n_2 - n_1 = 500x - 100x = 400x$

Dalton's law on CO_2

$P_{CO_2} \times V = n_{CO_2} RT$

Partial pressure of P_{CO_2} = Mole fraction × P

$0.2 \times 100 \times 10^3 \times 1 = 20 \times x \times R \times 300$

$$x = \frac{1000}{300R}$$

$$\eta_{N_2} = 400 \times \frac{1000}{300 \times 8.314}$$

$\eta_{N_2} = 160.37$ moles

10. (b) Using the clapeyron Equation

$$\left.\frac{dP}{dT}\right|_{sat} = \frac{h_{fg}}{T_{sat} V_{fg}} \quad ...(1)$$

equation given $\ln P_{sat} = A - \left(\frac{B}{T}\right)$ Put in eq. (1)

$$\frac{1}{P_{sat}} \times \left(\frac{dP}{dT}\right)_{sat} = \frac{B}{T^2}$$

$h_{fg} = T_{sat}.S_{fg}$ $\frac{1}{P_{sat}} \times \frac{T_{Sat}.S_{fg}}{T_{sat} V_{fg}} = \frac{B}{T^2}$

$$S_{fg} = V_{fg} \times \frac{B_{sat}}{T^2}$$

11. (a) $(COP)_{actual} = 0.8 \times (COP)_{Reverse\ Refrigerator}$

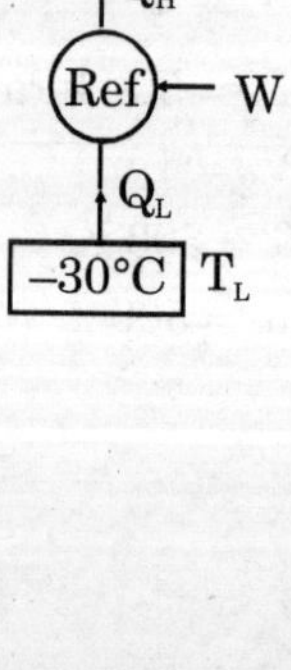

$$(COP)_{Reverse\ Refrigerator} = \frac{T_L}{T_H - T_L}$$

$$= \frac{273 - 30}{(273 + 50) - (273 - 30)} = \frac{243}{81} = 3$$

$(COP)_{actual} = 0.8 \times 3$

$(COP)_{atual} = 2.4$

$$\frac{\text{Output Capacity}}{\text{Power input(W)}} = 2.4$$

$$\frac{2.4}{W} = 2.4$$

$W_{input} = 1KW$

12. (c)

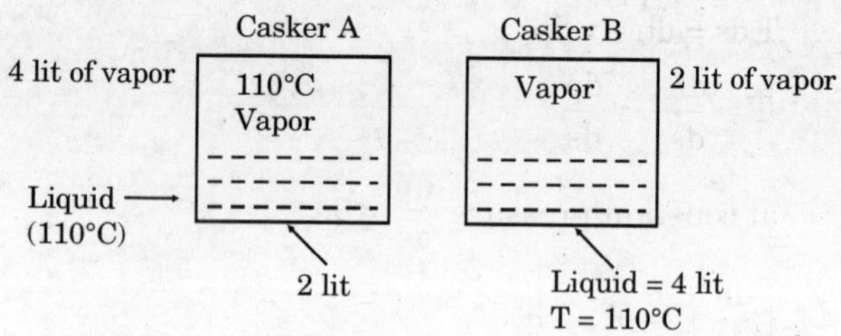

Free energy G = h – Ts

dg = dh – Tds – sdT

TdS = dh – vdP put in equation

dg = vdP – sdT

for a pure substance change in pressure remain constant dP = 0

There is no change in both looker temp

dT = 110 – 110 = 0

dg = 0

It means there is no change ingibbs energy

13. (a)

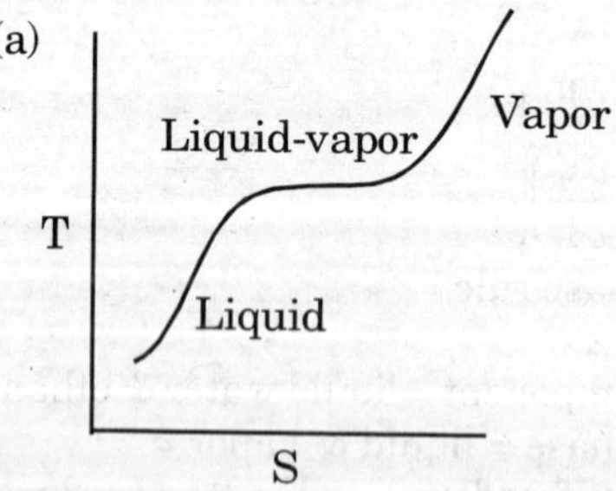

If we change the axis option 1 & option 4 may be the right answer.

Option (4) extropy of liquid decreases with increase with temp. that it is not possible. So option 2 is the right answer.

14. (106.09)

$\dot{m}$ = 2kg/s
P_1 = 1wkPa
T_1 = 30°C
V_1 = 2wm/s

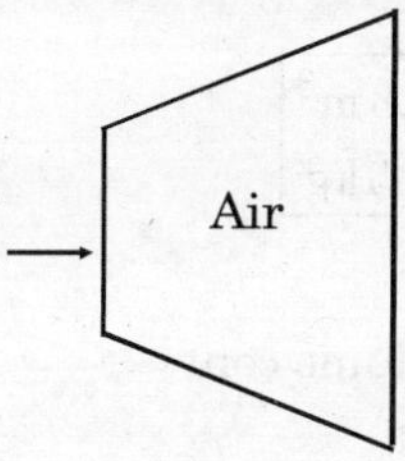

A_2 = 4wcm²=0.04 m²
T_2 = 45°C
P_2 = ?

Heat loss = 8 kJ/s = 8000 J/s

⇒ Flow in the diffuses is steady:

Stedy flow Energy equation

$\dot{Q} - \dot{W} = \Delta KE + \Delta PE + \Delta \dot{H}$

In the process there is no external work $\dot{W} = 0$

There is no change in height (datun) = $\Delta PE = 0$

$\dot{Q} = \Delta KE + \Delta \dot{H}$

$\dot{Q} = -8kJ/s$

"–" because of heat loss $-8000 = \Delta KE + \Delta \dot{H}$

$\Delta KE = \frac{1}{2}mv^2$ $-8000 = \frac{1}{2} \times 2(V_{exit}{}^2 - 2w^2)$
$+ 2 \times 1005 \times 15$

$\Delta H = \dot{m}\, C \Delta DT$ $V_{exit} = 43.011$ m/s

Mass flow rate is constant

$\dot{m}_1 = m_{exit}$

$2 \times 1000 = \rho_{exit} \times A_2 \times V_{exit}$

$\rho_{exit} = \frac{2 \times 1000}{0.04 \times 43.011} = 1162.4 \text{ g/m}^4$

$\rho_{exit} = 1.1624 \text{ kg/m}^3$

Pressure at the diffuses exit

$P_{exit}\ \rho_{exit}\ R\ T_{exit}$

$= 11624 \times 0.087 \times (273 + 45)$

$\rho_{out} = 106.096$ kPa

15. (0.3171)

Compressibility factor $Z = 0.84 = \frac{Vcom}{Videal}$

Vcom = 0.84 Videal

$Videal = \frac{RT}{P} = \frac{0.0815 \times 323}{10^3}$

$= 26.3245 \times 10^{-3}$

$Videal - Vactual = 4.529 \times 10^{-3}$

$Vcom - Vactual = Vcom - Videal + Videal - Vactual$

$= 0.84\ Videal - Videal + 4.529 \times 10^{-3}$

$= 0.16 \times 26.3245 \times 10^{-3} + 4.529 \times 10^{-3}$

$= [4.529 - 4.21192] \times 10^{-3}$

$Vcom - Vactual = 0.3171 \times 10^{-3} \text{ m}^3/\text{kg}$

16. (12.61)

Sec 1
T = 50°C
P_1 = 100 kPa
d_1 = ?

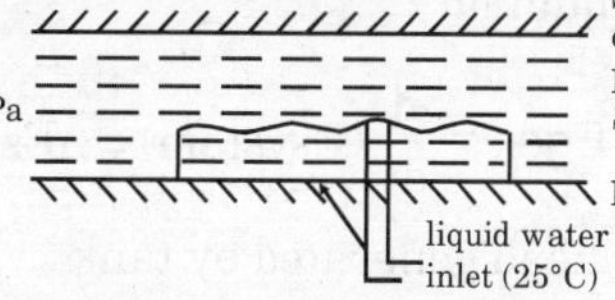

Sec 2
d_2 = 100%
Relative humidity = 100%
T = 25°C
P_1 = 1w kPa

→ Relative humidity 1 w% at the exit due to saturation (complete)

→ for complete saturation. Some liquid water situated in the saturation.

Specific humidity (w) = $\frac{0.622Pv}{P - Pv}$

Relative humidity (ϕ) = $\frac{pv}{pv, sat}$

→ At the exit $\phi = 100 = \frac{Pv}{Pv, sat} \times 100$

Pv,2 = Pv, sat,2 = 3.161 kPa(given)

speedfit humdity $(w_2) = \frac{0.622 \times 3.161}{1w - 3.161}$

Total pressure P = 1wkPa $= 0.0203 \frac{\text{kg of vapour}}{\text{kg of dry air}}$

→ Energy Balance at sec 1 & sec 2

$h_1 + (w_2 - w_1)\, h_{f@25} = h_2$

$1.005(50°C) + w_1(2592.06) + (w_2 - w_1)\, 104.87$
$= 1.005(25) + w_2\,(2547.17)$

$w_2 = 0.0203$ Put in the equation & get w,

⇒ $w_1 = 0.0098$ kg of vapor/kg of dry air

$w_1 = \frac{0.622 \times Pv,1}{P - Pv,1}$

$0.0098 = \frac{0.622 \times Pv,1}{1w - Pv_1}$

$Pv_1 = 1.556$ kPa

Relative humidity $(\phi_1) = \frac{Pv_1}{Pv_1, sat} = \frac{1.556}{12.335} \times 100$

$= 12.61\%$

17. (419.91k)

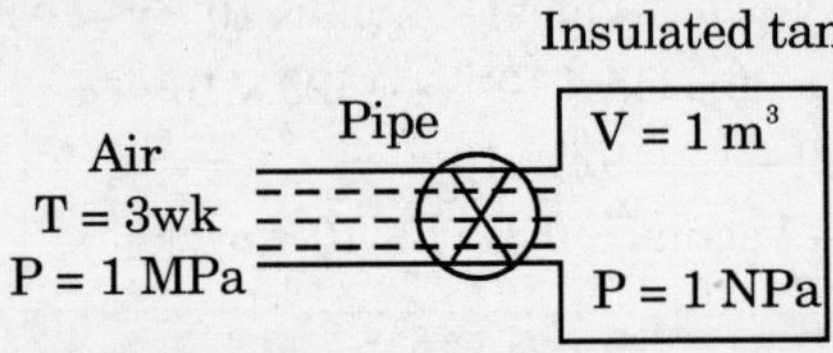

Steady flow equation

$$Ein - Eout + Egen = \frac{d}{dt}(Esystem) = \Delta Esystem$$

Egen = 0; No head generated by tank

Eout = 0; Energy out going to tank zero

$Cp = Cv + R$

$Cv = Cp - R$

$= 1.005 - 0.287$

$= 0.718$

$Ein = \Delta Esystem$

$m\ Cp\ Tin = m\ Cv\ T_{final}$

$$T_{final} = \frac{Cp}{Cv}.Tin = \gamma Tin$$

$$T_{final} = \frac{1.005}{0.718} \times 300$$

$= 419.91$ K

18. (321.51K) Vander Waals equation

$$\left(P + \frac{n^2a}{V^2}\right)(V - nb) = nRT$$

$V = 0.1\ m^3$

n = 1 w mol = 0.1 k mol

P = 2 MPa = 2000 k Pa

Put in the equation

$$\left(2000 + \frac{0.1^2 \times 939.2}{0.1^2}\right) \times (0.1 - 0.1 \times 0.0905)$$

$= 0.1 \times 8.3145 \times T$

T = 321.51 K

19. (59.79) Ideal gas

PV^n = const.

then

$$\text{heat transfer},\ Q = \left(\frac{\gamma - n}{\gamma - 1}\right) \times \frac{M \times R \times (T_2 - T_1)}{1 - n}$$

n = 1.3, $\gamma = 1.4$; R = 287 J/kg.k = 0.287 kJ/kg k

M = 1 kg $T_1 = 250$ k $T_2 = 500$ k

$$Q = \left(\frac{1.4 - 1.3}{1.4 - 1}\right) \times \frac{1 \times 0.287 \times (500 - 250)}{(1.0 - 1.3)}$$

Q = – 589.79 kJ

heat out from the system = 59.79 kJ

20. (91.219 kJ) $V = 0.8\,m^3$, $m = 1.5\,kg$

Insulated tank

Rigid tank i.e. volume cont

$P_1 = 100$ k Pa

After electric work $P_2 = 135$ k Pa

loss in availability i.e. Irrepressibility (IR)

$IR = To\ (\Delta S_{univ})$

To = 298 k

$\Delta S_{univ} = \Delta S_{system} + \Delta S_{surrounding}$

$\Delta S_{univ} = \Delta S_{system}$

$IR = 298 \times \Delta S_{system}$

$$\Delta S_{univ} = \Delta S_{system} = mCv\ln\left(\frac{T_2}{T_1}\right)$$

$$= 1.5 \times 0.680 \times \ln\left(\frac{135}{100}\right)$$

$$IR = 298 \times 1.5 \times 0.680 \times \ln\left(\frac{135}{100}\right)$$

= 91.219 kJ

21. (774.28)

liquid	Vapour
$\frac{V}{2}$	$\frac{V}{2}$

M = 1 kg

P = 10 ba

Specific enthalphy; $h_f + x(h_g - h_f)$

Mass of liquid × specific volume of liquid = $m_e.v_e$

Mass of vapour × specific volume of vapour = $m_v.\ v_v$

$$m_e.v_e = m_v.v_v = \frac{v}{2}$$

$$1 + \frac{m_e}{m_\gamma} = \frac{v_v}{v_e} + 1$$

$$\frac{m_\gamma + m_e}{m_v} = \frac{v_v + v_e}{v_e}$$

$$\frac{1}{x} = \frac{v_v + v_e}{v_e}$$

$$\frac{1}{x} = \frac{194.3 \times 10^{-3} + 1.127 \times 10^{-3}}{1.127 \times 10^{-3}}$$

$$\frac{1}{x} = 173.41$$

$x = 0.00576$

Now

$h = h_f + x(hg - h_f)$

$= 762.6 + 0.0058 \times (2776.2 - 762.6)$

$h = 774.28$ kJ/kg

Enthalpy $= m \times h$

$= 1 \times 774.28$

$= 774.28$ kJ

22. (59.91)

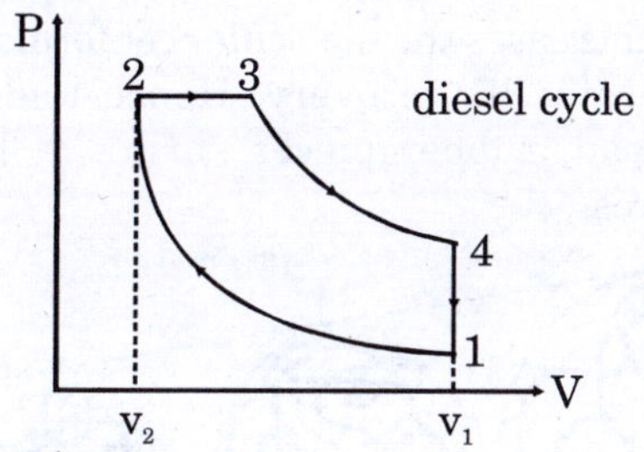

$$\frac{V_1}{V_2} = 16 = \text{Compression ration}$$

$P_1 = 1$ bar $= 100$ k Pa

$T_1 = 300$ k

$T_3 = 2100$ k

efficiency of diesel cycle

$$\eta = 1 - \frac{1}{r^{\gamma-1}}\left[\frac{\rho^{\gamma} - 1}{\gamma(\rho - 1)}\right] \quad ...(1)$$

$$\rho = \frac{V}{V_2} = \frac{T_3}{T_2} \quad \& \; T_1 V_1^{\gamma-1} = T_2 V_2^{\gamma-1}$$

Cut off ratio

$r = 16$ $$\frac{T_2}{T_1} = \left(\frac{V_1}{V_2}\right)^{\gamma-1}$$

Compression ratio $$\frac{T_2}{T_1} = (16)^{\gamma-1}$$

$$T_2 = 300 \times (16)^{1.4-1}$$

$$T_2 = 909.429$$

$$\Rightarrow \rho = \frac{2100}{909.429} = 2.309$$

Put all the value in eq. 1

$$\eta = 1 - \frac{1}{16^{1.4-1}} \times \left[\frac{2.309^{1.4} - 1}{1.4(2.309 - 1)}\right]$$

$$\eta = 59.91\%$$

Polymer Science

1. **Linear low-density polyethylene** (LLDPE) is a substantially linear polymer (polyethylene), with significant numbers of short branches, commonly made by copolymerization of ethylene with longer-chain olefins.

2. $$\Delta G = \Delta H - T\Delta S$$

   ΔG = change in Gibbs Free Energy

   ΔH = change in enthalpy

   T = temperature in Kelvin

   ΔS = change in entropy

   $\Delta G_m > 0$ miscible

   $\Delta G_m > 0$ immiscible

   $$\Delta H - T\Delta S > 0$$

   $$\frac{\Delta H}{T.\Delta S} > 0$$

   Low entropy and low temperture is the favourable condition for immiscibility of blends.

3. Elastomer – Polybutadiene

   Plastics – PA66, PET and HDPE

4. 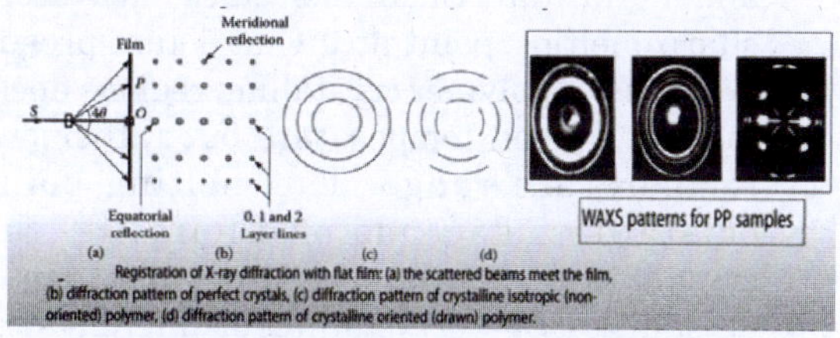

   X-ray diffraction patterns of un-oriented semi-crystallinc polymers are characterized by a series of concentric rings. As the specimen is oriented, these ring break up into arcs and spots. From the intensity and the size of these arcs, the degree of orientation of the crystalline can be determined.

   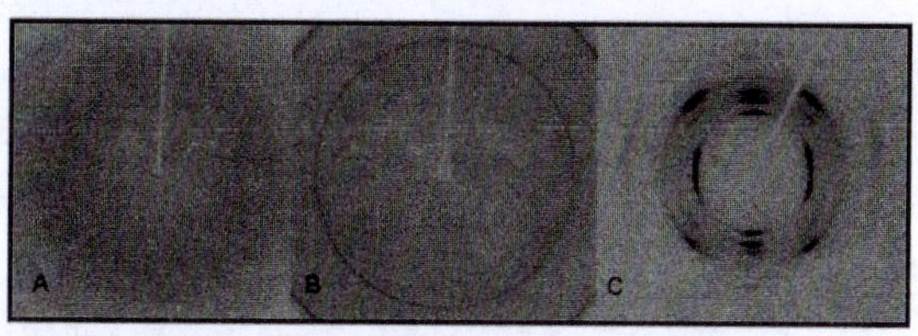

   A. amorphous scattering,

   B. unoriented polycrystalline scattering,

   C. oriented polycrystalline and amorphous scattering

5. Esdegradable polymers (BPs) disposed in bioactive environments degrade by the enzymatic action of microorganisms such as bacteria, fungi, and algae. Their polymer chains may also be broken down by nonenzymatic processes such as chemical hydrolysis. BPs are often derived from plant processing of atmospheric CO_2. Biodegradation converts them to CO_2, CH_4, water, biomass, humic matter, and other natural substances. These polymers are found both naturally and synthetically made, and largely consist of ester, amide, and ether functional groups. An example of biodegradable polymers are: Poly β-hydroxybutyrate–co-β-hydroxy valerate (PHBV). It is formed by combining 3-hydroxy butanoic acid and 3-hydroxy pentanoic acid. Other examples include polyglycolic acid (PGA), Polyhydroxy butyrate (PHB), Polycaprolactone (pcl), Nylon-2-nylon-6.

List of Biodegradable Polymers

- Polylactic acid (PLA)
- Polycarprolactones (PCL)
- Polyglycolic acid (PGA)
- Polyhydroxybutarate (PHB)
- Polyethylene oxide (PEO)
- Polyanhydride (PA)
- Polyvinyl alcohol (PVA)
- Polyalkyls

6. Melting is a transition between crystalline solid and liquid state. Small molecules like water has sharp melting point at 0°C at 1 tam presuure. Whereas in polymer crystalline regions does not melt at a fixed temperature but rather over a temperature range. The melting point of crystallites depends on their size. Small crystallites melt at lower temperatures than large crystallites. Since crystallites of different size are present, polymers always melt over a temperature range.

7. During very fast cooling the chain does not get enough time to rearrange themselves in an efficient manner and tends to appear hard/glassy at a earlier stage and showing Higher Tg.

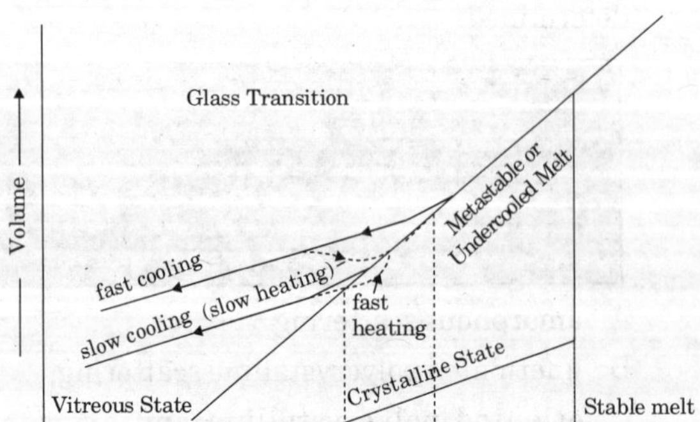

8.

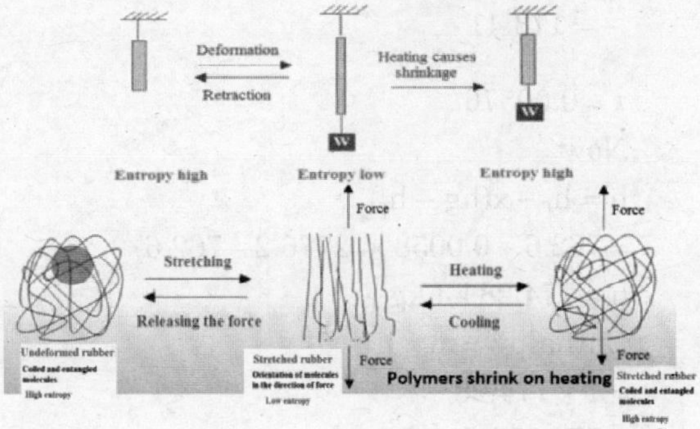

When stretched rubber is heated, molecules undergo chaotic movement, and thus a random state is formed. The enhanced thermal energy of the chains increases the molecular conformation and accelerates its recovery. Hence under a constant load, rubber pieces shrink at high temperatures.

9.

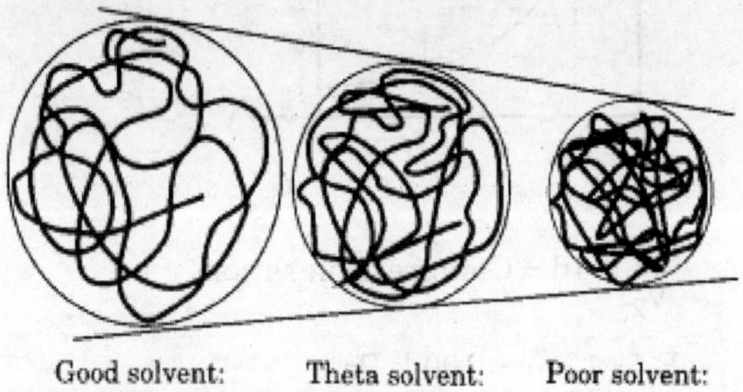

Good, Bad and θ solvents;

Solvent quaity : Interaction monomer – solvent

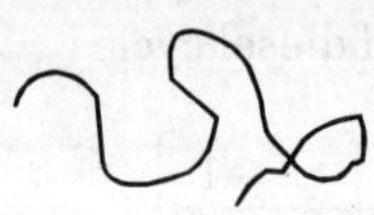

Good solvent : maximization of monomer-solvent contacts

→ chain expansion, "swollen chain"

Theta solvent : balanced interactions

→ unperturbed (ideal chain

Poor solvent : minimization of monomer-solvent contact

→ chain collapse, clustering, precipitation

10. 1. Antioxidants are used to prevent thermal degradation and inhibit oxidation (i.e. degradation). Hindered Amine, Hindered Phenol, Thioesters, Phosphites, Hydroxylamines, dithiopropionate, phenyl salisylate etc.

2. Chemical blowing agents are added to the polymer either prior to or during plasticization. They consist of organic or inorganic compounds that dissociate gas due to a reaction during processing. Chemical blowing agents include isocyanate and water for polyurethane, azodicarbonamide for vinyl, hydrazine and other nitrogen-based materials for thermoplastic and elastomeric foams, and sodium bicarbonate for thermoplastic foams.
3. Plasticizer is necessary in rubber formulations for improving the processing of the rubber by lowering the compound's viscosity, and also for aiding in low temperature flexibility of the formulated rubber. The most common plasticisers include esters such as adipates, azelates, citrates, benzoates, ortho-phthalates, terephthalates, sebacates, and trimellitates.
4. Coupling agents are chemicals which improve the interfacial properties of mineral fillers and polymers (they reduce the interfacial tension which is diadvnatageous rather than advantageous, but simultaneously they reduce the agglomeration tendency of filler particles, thus improving their accessibility to polymer molecules). Example -Silane

11.

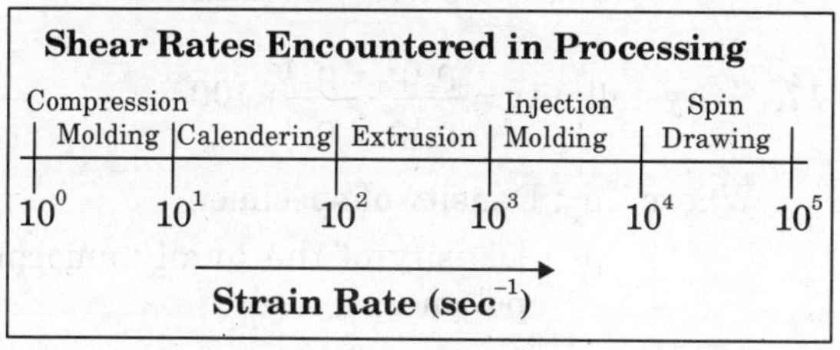

12. Viscous fluids follow the Newton's law; Stress is proportional to the strain rate.

$$\sigma = \alpha\frac{dy}{dt}, \qquad \sigma = \eta\frac{dy}{dt}$$

Where η is viscosity and $\frac{dy}{dt}$ is strain rate.

Viscosity is independent on the shear rate in case of Newtonian fluid. Liquid generally comes in the category of Newtonian fluid. Curve Q & S represents the liquid like fluid.

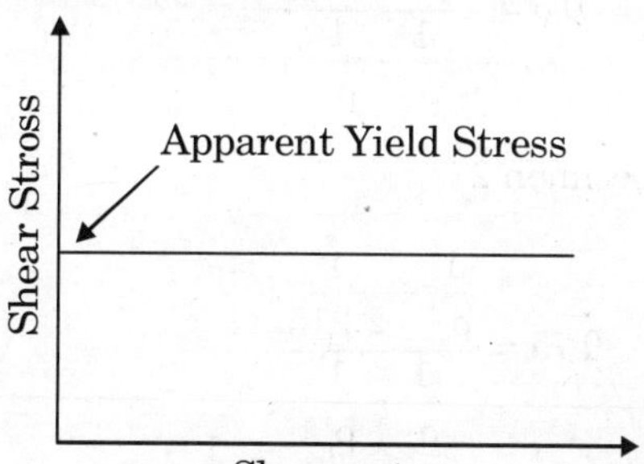

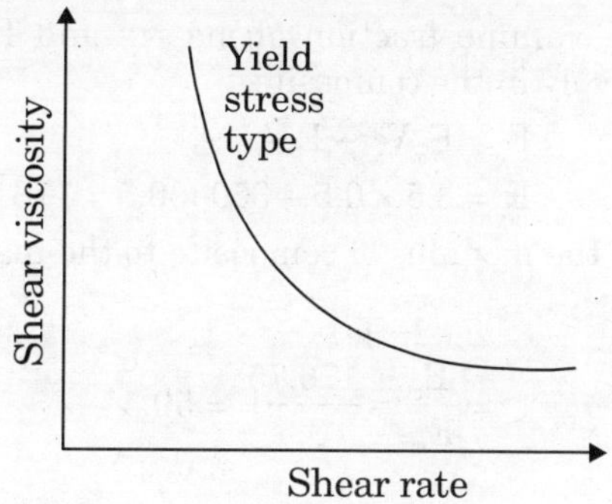

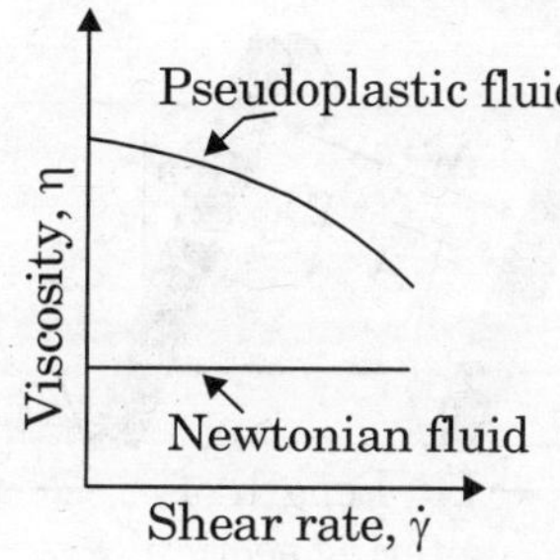

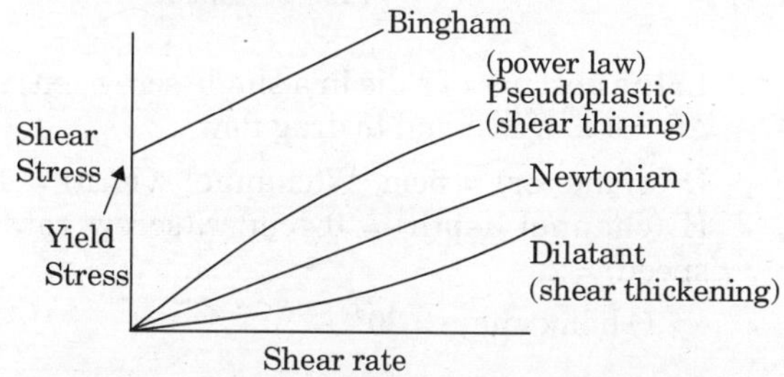

Viscosity is independent on the shear rate in case of Newtonian fluid. Liquid generally comes in the category of Newtonian fluid. Curve Q & S represents the liquid like fluid.

13.

$$G_P = \frac{\rho RT}{M_e}$$

$$M_e = \frac{\rho RT}{G_p}$$

G_p = Plateau modulus

M_c = molecular weight between entanglements of polystyrene chains

$G_P = 0.2 \times 10^6$, $p = 1.05\frac{g}{cm^3}$, $R = 8.314\,jK^{-1}mol^{-1}$,

T = 423 K

$$= \frac{1.05 \times 10^6 \times 8.314 \times 423}{0.2 \times 10^6} = 18360\ \frac{g}{mol}$$

14. Ratio of the modulus of composite to the matrix modulus = $\frac{E_c}{E_m}$

E_c – modulus of composite; E_f – modulus of filler & E_m – modulus of matrix

V_m & V_f volume fraction of matrix and filler respectively in the composite.

$$E_c = E_m V_m + E_f V_f$$
$$E_c = 3.5 \times 0.5 + 350 \times 0.5 = 176.75$$

Ratio of the modulus of composite to the matrix modulus

$$= \frac{E_c}{E_m} = \frac{176.75}{3.5} = 50.5$$

15.

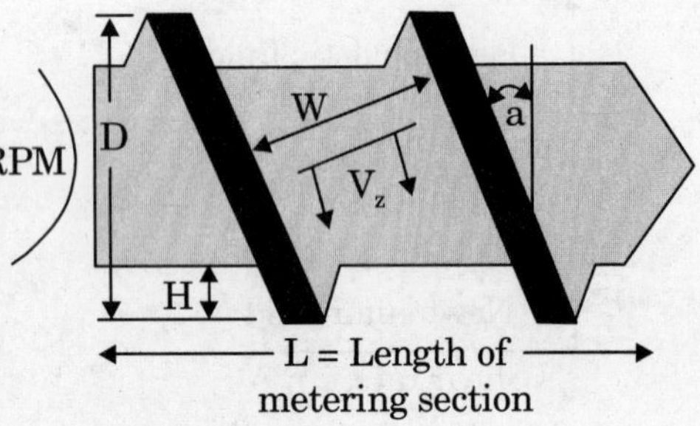

In the absence of a die in a single screw extruder $\Delta P = 0$ & Q is equal to drag flow

D (diameter) = 5cm; W(channel width) = 1cm; H (channel depth) = 0.4cm; N(screw rotation speed) = 2;

Θ helix angle = 20°

$$Q = \frac{1}{2}\pi \text{ WHND} \cos\theta$$

$$Q_D = \frac{1}{2} \times 3.14 \times 1 \times 0.4 \times 2 \times 5 \times \cos 20$$

$$Q_D = 5.90 \text{ cm}^3\text{/s}$$

16.

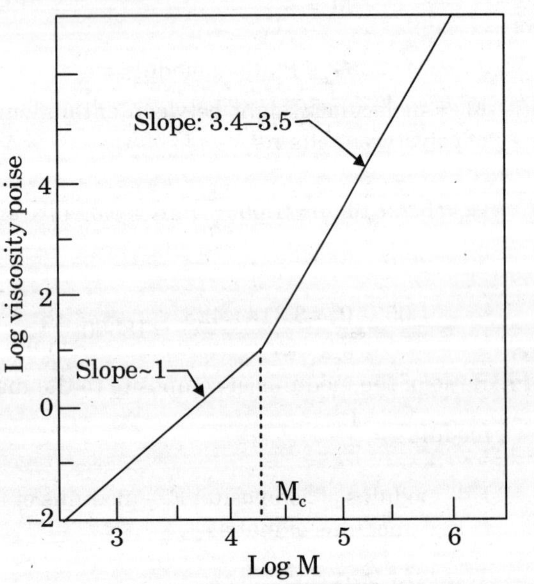

Dependence of polymer (melt) viscosity on molecular weight (M): a typical plot of log viscosity against log M.

$[\eta] = K[M]^1$ when $M < M_c$

$[\eta] = K[M]^{3.4}$ when $M > M_c$

M_c = critical molecular weight = 35×10^3 g/mol

$M_1 = 250 \times 10^3$ g/mol

$M_2 = 500 \times 10^3$ g/mol

Both the molecular weight is higher than the critical molecular weight. Hence, we have to use:

$$[\eta] = K[M]^{3.4} \text{ when } M > M_c$$

Condition 1 :

$$8 \times 10^3 = K[250 \times 10^3]^{3.4}$$

$$K = \frac{8 \times 10^3}{[250 \times 10^3]^{3.4}}$$

Condition 2 :

$$[\eta] = K[500 \times 10^3]^{3.4}$$

K value put in the equation from condition 1

$$[\eta] = \frac{8 \times 10^3}{[250 \times 10^3]^{3.4}} \times [500 \times 10^3]^{3.4}$$

$$= 84.45 \times 10^3 \text{ Pa.s}$$

17. $$\%\text{Crystallinity} = \frac{\rho_c(\rho_s - \rho_a)}{\rho_s(\rho_c - \rho_a)} \times 100$$

Where, ρ_s : Density of specimen

ρ_a : Density of the totally amorphous polymer.

ρ_c : Density of the perfectly crystalline polymer

$$\% \text{ crysatllinity} = \frac{\frac{1}{\rho_a} - \frac{1}{\rho_s}}{\frac{1}{\rho_a} - \frac{1}{\rho_c}}$$

PTFE - Specimen 1

$$0.5 = \frac{\frac{1}{\rho_a} - \frac{1}{2.215}}{\frac{1}{\rho_a} - \frac{1}{\rho_c}} \quad ...(1)$$

PTFE Specimen 2

$$0.75 = \frac{\frac{1}{\rho_a} - \frac{1}{2.215}}{\frac{1}{\rho_a} - \frac{1}{\rho_c}} \quad ...(2)$$

Put the value of $\frac{1}{\rho_a} - \frac{1}{\rho_c}$ from equation 1 to 2

$$0.75 = \frac{1/\rho_a - 1/2.215}{\dfrac{1/\rho_a - 1/2.144}{0.5}}$$

Density of 10% amorphous = 2.013 g/cm³.

18. Use Maxwell model equation

$\therefore$ Given, $E = 10^{10}$, $\eta = 10^{12}$, $\varepsilon = 1\%$ $t = 50$ s

$$\boxed{\varepsilon(t) = \sigma_0\left[\frac{1}{E} + \frac{t}{\eta}\right]}$$

$$0.01 = \sigma\left[\frac{1}{10^{10}} + \frac{50}{10^{12}}\right]$$

$$\Rightarrow \quad \sigma = 6.66 \times 10^7 \text{ Pa}$$

19. $$v = \frac{K_p[M]}{2(K_d K_t f[I])^{1/2}}$$

Degree of polymerization is equal to kinetic chain length, Hence

$$x_n \alpha \frac{[M]}{[I]^{1/2}}$$

Condition 1:

$$100 = \frac{[M]}{[I]^{1/2}}$$

Condition 2:

$$x_n = \frac{[2M]}{[4I]^{1/2}} = \frac{[M]}{[I]^{1/2}} = 100$$

$$x_n = 100$$

20. p = extent of reaction = 0.99; then degree of polymerization is given by

$$X_n = \frac{1}{1-p} = \frac{1}{1-0.99} = 100$$

$$M_n = M_o X_n$$

M_n = number average molecular weight of Polymer & M_0 = monomer weight

Average monomer weight

$$M_0 = \frac{\sum M_i N_i}{\sum N_i}$$

$$= \frac{132 \times 2 + 60 \times 1 + 88 \times 1}{2+1+1} = 103$$

$$M_n = 103 \times 100 = 10300$$

21. 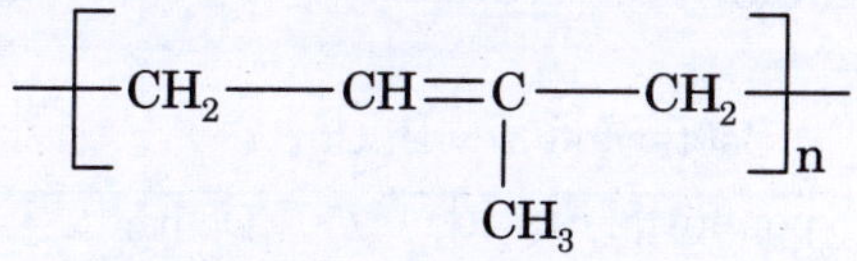

As there are 240 chain c atoms and in one isoprene unit there are 5 c atoms but out of these 4 C atoms are contributing to chain only and remaining 1 C atoms remains in side.

240 number of carbon atom requires how much

monomer $= \frac{240}{4} = 60.$

Molecular weight between crosslinks

$$= 60 \text{ units} \times 68\frac{g}{mol} = 4080\frac{g}{mol}$$

22. strain $= \frac{\text{change in lattice parameter}}{\text{original lattice parameter}} = \frac{a_2 - a_1}{a_1}$

According to Bragg s law.

$$\Rightarrow \quad n\lambda = 2d\sin\theta$$

$$\Rightarrow \quad \lambda = \frac{2a}{\sqrt{h^2+k^2+l^2}}\sin\theta$$

$$\Rightarrow \quad a = \frac{\left(\lambda\sqrt{h^2+k^2+l^2}\right)}{2\sin\theta}$$

where θ is the usual Bragg angle, λ is the radiation wave length = 1.542, d = Inter planner spacing & n = an integer, order of reflection

Lattice planes is determined by three integers h, k, and the Miller indices. Peak = (002) mean; h = 0, k = 0 and l = 2

$$\text{strain} = \frac{\text{change in lattice parameter}}{\text{original lattice parameter}}$$

$$= \frac{a_2 - a_1}{a_1}$$

$$= \frac{\dfrac{\left(\lambda\sqrt{h^2+k^2+l^2}\right)}{2\sin\theta_2} - \dfrac{\left(\lambda\sqrt{h^2+k^2+l^2}\right)}{2\sin\theta_1}}{\dfrac{\left(\lambda\sqrt{h^2+k^2+l^2}\right)}{2\sin\theta_1}}$$

$$\Rightarrow \text{Strain} = \frac{\dfrac{1}{\sin 37.45°} - \dfrac{1}{\sin 37.50°}}{\dfrac{1}{\sin 37.50°}}$$

$$= 1.138 \times 10^{-3}$$

Food Technology

1.

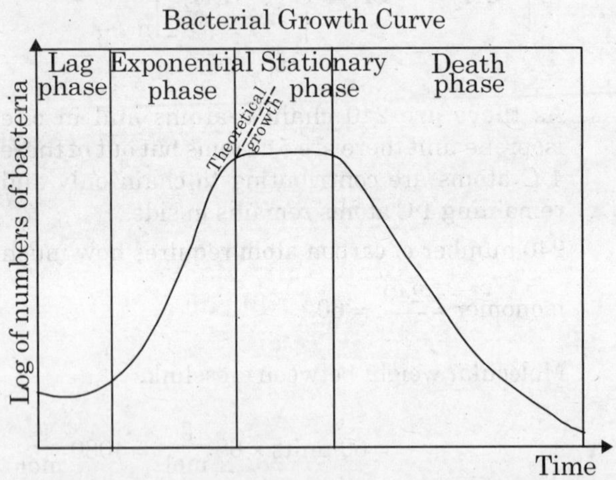

The bacterial growth curve represents the number of live cells in a bacterial population over a period of time.

- There are four distinct phases of the growth curve: lag, exponential (log), stationary, and death.
- The initial phase is the lag phase where bacteria are metabolically active but not dividing.
- The exponential or log phase is a time of exponential growth.
- In the stationary phase, growth reaches a plateau as the number of dying cells equals the number of dividing cells.
- The death phase is characterized by an exponential decrease in the number of living cells.

In the log phase growth is exponential follows the first order kinetics.

2. (d) The top five germs that cause illnesses from food eaten in the United States are:
   1. Norovirus
   2. Salmonella
   3. Clostridium perfringens
   4. Campylobacter
   5. Staphylococcus aureus (Staph)
3. (a) Fermented foods are foods and beverages that have undergone controlled microbial growth and fermentation 1. Fermentation is an anaerobic process in which microorganisms like yeast and bacteria break down food components (e.g. sugars such as glucose) into other products (e.g. organic acids, gases or alcohol). This gives fermented foods their unique and desirable taste, aroma, texture and appearance.

   There are thousands of different types of fermented foods, including:
   - cultured milk and yoghurt
   - wine
   - beer
   - cider
   - tempeh
   - miso
   - kimchi
   - sauerkraut
   - fermented sausage
   - vinegar
4. (B) Protein Efficiency Ratio is the ratio of grams of body weight gain (in specified time) to the grams of protein consumed.
5. (b) Maltose is produced by hydrolysis of starch using the enzyme β-amylase, which specifically releases disaccharide (maltose) units sequentially from the non-reducing ends of the starch polysaccharides. Maltose occurs only rarely in nature and only in plants, mostly in germinating seeds as a result of partial hydrolysis of starch by β-amylase.
6. (d) Proteolysis is the breakdown of protein. Since casein protein is the main structure of cheese, proteolysis can cause cheese to soften/breakdown over time. Proteases are enzymes that cause proteolysis to occur. These can come about through various ways. A few ways include:
   - Coagulation enzymes (e.g. rennet) can break down protein. Rennet, after all, is a protease. Rennet activity can also cause bitterness; as protein breaks down, bitter peptides can form.
   - Proteases from the starter bacteria can break down proteins with time.
   - Proteases from surface mold/bacteria can break down proteins near the surface of certain cheeses. This is partly responsible for Brie's texture as we already discussed.
   - Proteases from non-starter lactic acid bacteria (NSLAB) can break down proteins.
   - Endogenous proteases found in the milk from the get-go, like Plasmin. However most of these don't survive pasteurization, so this may be seen more in raw milk cheeses.

7. (d) Genistein, 4′,5,7-trihydroxyisoflavone, is a major isoflavone in soybean, which is known as phytestrogen having known benefit to brain functions. Categories of phytoestrogens:

Category	Phytoestrogen Examples	Dietary Sources	Basic Chemical Structure
Isoflavones	Daidzein Genistein Fomononetin Biochanin A Glycitein	Clovers (red and white) Soybeans Beans Split peas	
Flavanones	Naringenin	Hops (8-prenylnaringenin) Apples, red onions	
Flavonoids	Apigenin Luteolin	Parsley Capsicum pepper Alfalfa	
	Quercetin Kaempferol	Tomatoes Broccoli Apples Onions	
Plant sterols	B-sitosterol	Corn Soybeans Sugar beet forage Saw palmetto (*Serenoa repens*) Avocados Pistachios and almonds Wood	
Coumestans	Coumestrol	Legumes (alfalfa, clover) Spinach Split peas, lima beans Soy bean sprouts	
Stilbenes	Trans-resveratrol (trans-3,5,4′-trihydroxystilbene)	Grape skin (red wine) peanuts	
Lignans	Secoisolariciresinol, matairesinol	Flaxseed (linseed) Squash, pumpkin seeds Tea (black and green) Sunflower seeds Strawberries Cranberries Brans	

8. (c) Ultra-high-temperature (UHT) pasteurization involves heating milk or cream to 138-150 °C (280-302 °F) for one or two seconds. Packaged in sterile, hermetically sealed containers, UHT milk may be stored without refrigeration for months. Ultrapasteurized milk and cream are heated to at least 138°C for at least two seconds, but, because of less stringent packaging, they must be refrigerated. Shelf life is extended to 60–90 days. After opening, spoilage times for both UHT and ultrapasteurized products are similar to those of conventionally pasteurized products.

9. (d) Limonin is formed as a result of heat treatment of the juice during processing and a chemical reaction in the acidic juice medium.

10. (a) Lactate dehydrogenase is an enzyme found in nearly all living cells. LDH catalyzes the conversion of lactate to pyruvate and back, as it converts NAD^+ to NADH and back.

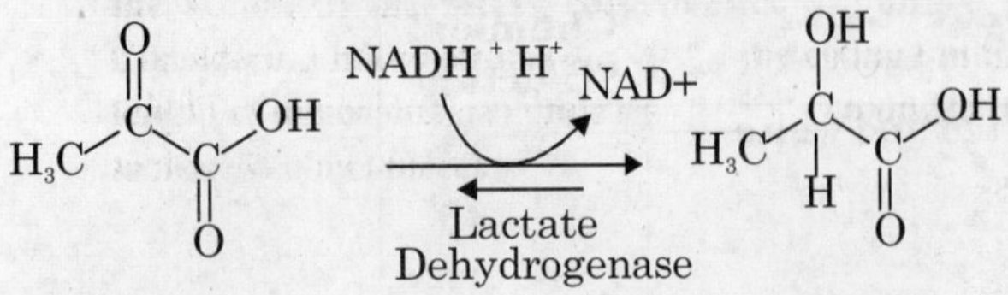

Pyruvate Lactic Acid

11. (b) In CA storage, inside a food storage room is the gas composition that is continually monitored and adapted to maintain the optimum concentration within completely close tolerances.

12. (b)

Column I	Column II
P. Hydrogenation	1. Shortening of fat
Q. Blanching	2. Inactivation of enzyme
R. Leaching	3. Separation of dye
S. Winterization	4. Removal of soft wax

Hydrogenation converts liquid vegetable oils into solid or semi-solid **fats**, such as those present in margarine. **Changing** the degree of saturation of the **fat changes** some important physical properties, such as the melting range, which is why liquid oils become semi-solid.

Blanching is a process that deactivatess the enzyme to slow down the browning process so the fruit or vegetable will remain fresher for longer. Basically is it **heating the food to 75-100 degrees Celsius for a short period of time** to deactivate the polyphenol oxidase.

Wax is also found in edible oils. Before the oil is consumed, wax is usually separated from the oil by placing the oil at a cold place where the temperature is low enough to make wax crystallize and precipitate at the bottom of the container. This process is called winterization, after which wax is easily removed by filtration or centrifugation.

13. (a, d) Sodium lactate-Bacteria-Pre-cooked meats

Sodium nitrite-Clostridia-Meat curing preparations

14. (a, b) Carotene - Yellow-orange-Peppers

Betanin - Purple/red-Cactus pear

Lycopene - Red- red colored fruits and vegetables such as tomato, papaya, pink grapefruit, pink guava and watermelon

Flavanols - yellow- lemons, oranges, and grapefruit

15. (a, c) Major anti-nutritional factors, which are found in edible crops include **saponins, tannins, phytic acid, gossypol, lectins, protease inhibitors**, amylase inhibitor, trypsin inhibitors and goitrogens. Anti-nutritional factors combine with nutrients and act as the major concern because of reduced nutrient bioavailability.06

16. (a, b, d) Supercritical fluid extraction (SFE) is the process of separating one component (the extractant) from another (the matrix) using supercritical fluids as the extracting solvent. Water and Carbon dioxide (CO_2) is the most used supercritical fluid, sometimes modified by co-solvents such as ethanol or methanol.

17. Reynold's number from kinematic viscosity:

$$Re = \frac{V.d}{v}$$

where:

Re is the Reynolds number

V is the pipe velocity (m/s)

d is the pipe diameter (m)

v is the kinematic viscosity (m^2/s)

Reynold's number from absolute (dynamic) viscosity:

$$Re = \frac{V.d.\rho}{\mu}$$

where:

Re is the Reynolds number

V is the pipe velocity (m/s)

d is the pipe diameter (m)

ρ is the fluid density (kg/m^3)

μ is the dynamic viscosity (Ns/m^2)

18. (a, c) Particle size analyzer - particle size distribution - span value

Texture profile analyzer - hardness or cohesiveness - chewiness

Differential scanning calorimeter - glass transition temperature - degree of caking

Capillary viscometer - viscosity - thickness of liquid

19. (a, c) Hagen Poiseuille law - Pressure drop

Rittinger's law - Size reduction

Stefan Boltzmann law - Radiation heat transfer

Raoult's law - Vapour pressure

20. (2070)

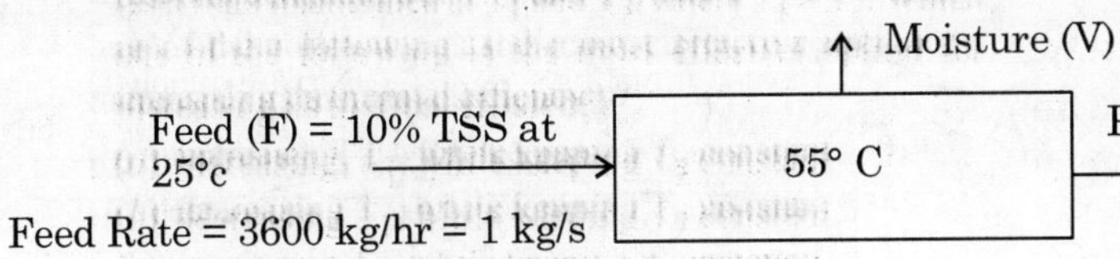

Total soluble solid content is equal in mass.

F = V + P

0.1F = 0.4P

P = 0.25F

F = V + 0.25F

V = 0.75F = 0.75 X 1 = 0.75 kg/s

Total heat Transfer = MCpΔT + mL

= 1 x 4 × 30 + 0.75 × 2600 = 2070 KJ/s

21. (0.14) Relative humidity

$$= \frac{\text{Partial pressure of water vapor}}{\text{Saturation vapor pressure}} \times 100$$

RH = 60% & Saturated vapor pressure = 31.2 kPa

Partial Pressure of water vapor = 0.6 *7.375

= 18.72 kPa

$$\boxed{\omega = 0.622\left(\frac{P_v}{P_a}\right) = 0.622\left(\frac{P_v}{P - P_v}\right)\left[\frac{\text{kg vapour}}{\text{kg dry air}}\right]}$$

P total Pressure = 101.325 KPa

Partial Pressure of water vapor (Pv) = 18.72 KpA

$$X = 0.622\left[\frac{18.72}{(101.325 - 18.72)}\right] = 0.140 \text{ kg of water/}$$

kg of air

22. (2.3) Total energy in the storage tank = $Q_1 + Q_R$

Q_1 is the energy in the cold storage = mCpΔT = 5000 × 3.65 × (28 – 2) = 474500 KJ

Q_R is the energy in respiration of potato = mass × heat = 5000 × 3.12 = 15600 KJ

Total energy in the storage tank = 474500 + 15600 = 490100 KJ

$$\text{Capacity (Power) of plant} = \frac{\text{Energy}}{\text{time}}$$

$$= \frac{490100}{(24 \times 3600)} = 5.67 \text{ KW}$$

$$\text{Efficiency of plant} = 0.7\ ;\ Q_{refrigeration} = \frac{5.67}{0.7}$$

= 8.103 KW

Capacity of plant in ton = 8.103/3.517 = 2.30 tonn

Atmospheric and Oceanic Sciences

1. (B) Surface currents located on the western side of the subtropical gyres, so-called western boundary currents, are faster than their eastern counterparts. In fact, they are among the fastest surface currents in the ocean. One reason for the westward intensification of boundary currents has to do with the strengthening of the Coriolis effect with latitude. The Coriolis effect is stronger in the latitudes of the westerlies than in the latitudes of the trade winds.

Coriolis effect- Our planet's rotation produces a **force on all bodies moving relative to theEarth.** Due to Earth's approximately spherical shape, this force is greatest at the poles and least at the Equator. The force, called the "Coriolis effect," causes the direction of winds and ocean currents to be deflected

2. (C) Dynamic instability in which a stable fluid is made unstable by velocity shear. Another example of dynamic instability, the KelvinHelmholtz instability, occurs when the density contrast in a sheared flow is much less than at the sea surface, such as in the thermocline or at the top of a stable, atmospheric boundary layer. The relative importance of static stability and dynamic instability is expressed by the Richardson Number :

$$\boxed{R_i = \frac{g\,E}{(\partial U / \partial z)^2}}$$

3. (D) During July-August, the highest number of monsoon low pressure systems form over-bay of Bengal.

There were five low pressure systems (LPS) in the north Bay of Bengal during August 2020, that accounted for 27 low pressure days against a normal of 15 days, according to the India Meteorological Department (IMD).

4. (C) 2020 - 40 tonn

Government reduces every year 1 giga ton

So it will take 40 years.

5. (C) Diagram shows circulation pattern in may month.

6. (A) In meteorology, the term 'sensible heat flux' means **the conductive heat flux from the Earth's surface to the atmosphere.** It is an important component of Earth's surface energy budget. Sensible heat flux is commonly measured with the eddy covariance method.

It represents the loss of energy by the surface by heat transfer to the atmosphere. It is positive when directed away from the surface into the atmosphere. Sensible heat flux is zero if the temperature gradient or wind speed is zero.

7. (A,C) Humidity ratio, $\omega = 0.622\dfrac{p_v}{p_b - p_v}$

Partial pressure vapour,

$$p_v = p_{sw} - \frac{(p_b - p_{sw})(t_d - t_w)}{1527.4 - 1.3t_w}$$

Where,

p_{sw} - saturation pressure corresponding to WBT

p_b - barometric pressure = 1bar

t_d - Dry bulb temperature

t_w - Wet bulb temperature

Ways of reaching saturation

- Isobaric cooling (dew point temperature)
- Isobaric and adiabatic cooling by evaporation of water (wet bulb temperature) Temperatura wilgotnego termometru)
- Adiabatic and isobaric mixing

8. [1.86] Virtual temperature $(T_v) = (1+.61r)\,T$

Case 1: r = water vapor mixture = 10 kg/kg

$T_{V1} = (1+.61 \times 10) \times 300$

Case 2: r = water vapor mixture = 20 kg/kg

$T_{V2} = (1+.61 \times 20) \times 300$

Divide both equations

$T_{V2} = 1.86\ T_{v1}$

9. [14.14] $d = \sqrt{2K_m / |f|}$

$$d = \sqrt{\frac{2 \times 0.01}{10^{-2}}} = 14.14$$

10. (C)

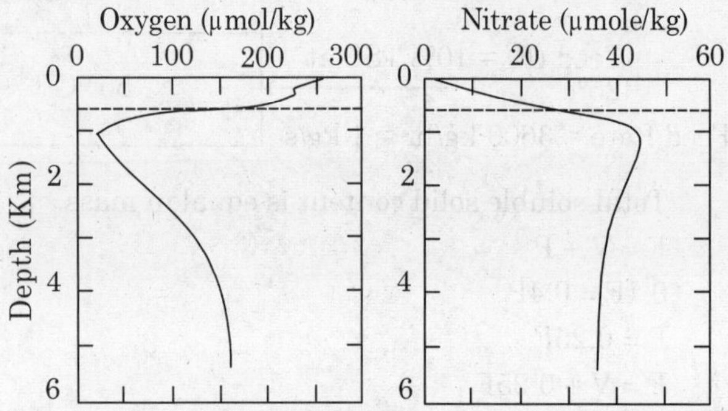

11. [A,C] High clouds are **cirrus**, cirrostratus, and cirrocumulus. Cirrus clouds are wispy, feathery, and composed entirely of ice crystals.

High Clouds warm the Earth by absorbing infrared radiation emitted from the surface and reradiating it back down. The process traps heat like a blanket and slows the rate at which the surface can cool. The blanketing effect warms Earth's surface by some 7°C.

12. [A,C] The Sverdrup transport is **the net meridional transport diagnosed in both the subtropical and subpolar gyres,** resulting from planetary vorticity changes that balance Ekman pumping or Ekman suction.

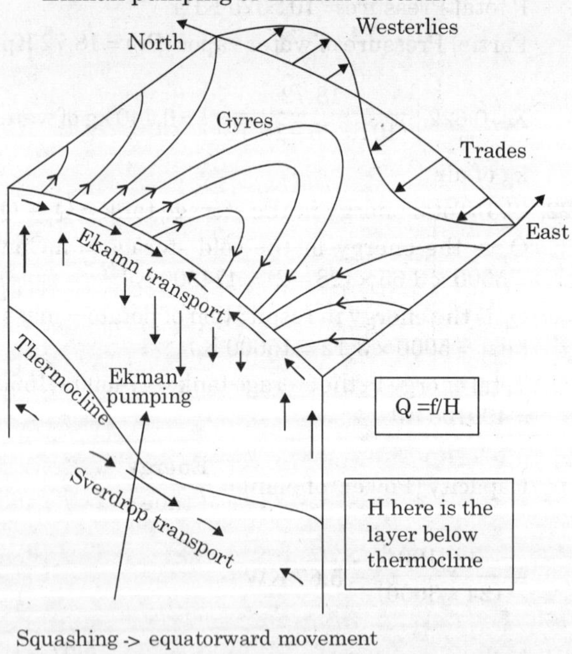

Squashing -> equatorward movement

transport

- Ekman pumping provides the squashing or stretching.

- The water columns musst respond. They do this by changing latitude.
- (They do not spin up in place.)

  Stretching -> poleward

  The Coriolis effect strongly affects the large-scale oceanic and atmospheric circulation, leading to the formation of robust features like jet streams and western boundary currents. Such features are in geostrophic balance, meaning that the Coriolis and pressure gradient forces balance each other. Coriolis acceleration is also responsible for the propagation of many types of waves in the ocean and atmosphere, including Rossby waves and Kelvin waves. It is also instrumental in the so-called Ekman dynamics in the ocean, and in the establishment of the large-scale ocean flow pattern called the Sverdrup balance.

13. [A,D] The Hadley circulation, or Hadley cell-a worldwide tropical atmospheric circulation pattern that occurs due to uneven solar heating at different latitudes surrounding the equator-causes air around the equator to rise to about 10-15 kilometers, flow poleward (toward the North Pole above the equator, the South Pole below the equator), descend in the subtropics, and then flow back to the equator along the Earth's surface. The location of the ascending branch has moved south of the equator into the summer hemisphere. In the southern hemisphere winter the southern Hadley cell dominates the tropical circulation, the northern Hadley cell is weak and very narrow.

In the Hadley cell, both **sensible and latent heat are transported equatorward near the surface.**

14. [A,B] Ocean circulation is the large scale movement of waters in the ocean basins. Winds drive surface circulation, and the cooling and sinking of waters in the polar regions drive deep circulation. So option A is right.

In cold regions, such as the North Atlantic Ocean, ocean **water loses heat to the atmosphere** and becomes cold and dense. When ocean water freezes, forming sea ice, salt is left behind causing surrounding seawater to become saltier and denser. Dense-cold-salty water sinks to the ocean bottom.

15. [A,D] Corals need to grow in shallow water where sunlight can reach them. Corals depend on the zooxanthellae (algae) that grow inside of them for oxygen and other things, and since these algae needs sunlight to survive, corals also need sunlight to survive. Corals rarely develop in water deeper than 165 feet (50 meters).

Reef-building corals require warm water conditions to survive. Different corals living in different regions can withstand various temperature fluctuations. However, corals generally live in water temperatures of 68-90° F or 20-32° C.

Option A & D are correct.

16. [0.5] $\Delta . V = \frac{\partial v}{\partial y}$

$0.01 = v/50$

$V = 0.01 * 50 = 0.5$ m/s

17. [4500] If temperature profile follow the equation $T = T_o - Lh$

$$P = P_0\left(1 - \frac{Lh}{T_0}\right)^{\frac{g}{RL}}$$

$$P = \frac{1}{2}P_o\ ;\ L = \frac{1}{30};\ T_o = 300K$$

$$\frac{1}{2}P_o = P_o\left(1 - \frac{\frac{1}{30}Z}{300}\right)^{\frac{10}{300\times\frac{1}{30}}}$$

Z = 4500 m

18. [0.01] Vertical component of vorticity vector (rotation in a horizontal plane)

$$\zeta = \hat{k}\bullet\bar{v}\bullet = \hat{k}\bullet\Delta\times\bar{V} = \frac{\partial v}{\partial x} - \frac{\partial u}{\partial y}$$

$$u(y) = u_o\left(1 - \frac{y^2}{L^2}\right),$$

$$\zeta\quad \frac{\partial v}{\partial x} - \frac{\partial u}{\partial y} = 0 - \frac{\partial\left(u_0\left(1 - \frac{y^2}{L^2}\right)\right)}{\partial y} = \frac{2u_oy}{L^2}$$

$$\zeta = \frac{2u_oy}{L^2} = \frac{2\times\frac{50}{1000}\times\frac{5}{2}}{5^2} = 0.011/s$$

19. [5] The magnitude of the temperature gradient between the stations of M & N

$$\frac{\Delta T}{\Delta d} = \frac{10}{1000} = \frac{1}{100}k/m$$

Magnitude of temperature is constant form M to N point; Temperature at mid point

$$\frac{1}{100}\frac{k}{m} = \frac{\Delta T}{500}$$

$\Delta T = 5K$

20. [33.29] $\dot{Q} = \frac{m\, c_P\, \Delta T}{r}$

$\dot{Q}$ = Mean heat transfer rate (kW (kJ/s)

m = Mass of the fluid (kg)

c_p = Specific heat capacity of the fluid (kJ/kg °C)

ΔT = Increase in fluid temperature (°C)

t = Time for the heating process (seconds)

Energy in put = 4 w/m² = 4 J/m²s.

$$\frac{\text{Power}}{\text{Area}} = mC_P \frac{\Delta T}{A.T} = 4J/m^2s$$

Mass = volume × Density = A × Depth × Density

$$\frac{\text{Power}}{\text{Area}} = A \times \text{Depth} \times \text{Density}\; C_P \frac{\Delta T}{A.T} = 4J/m^2s$$

$$\text{Depth} \times \text{Density}\; C_P \frac{\Delta T}{t} = 4$$

$$t = 1000 \times 1000 \times 4200 \times \frac{1}{4} = 1.05 \times 10^9 \text{ sec}$$

t = 33.29 years

21. [1.73]

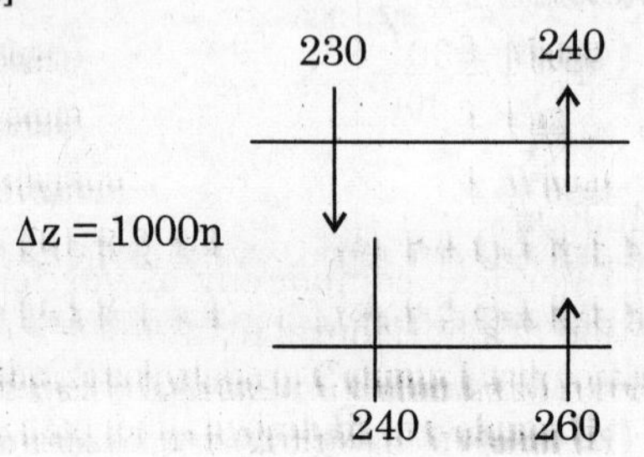

Total net energy radiation = (260 – 240) – (240 – 230) = 10 W/m²

$$\frac{\text{Power}}{\text{Area}} = mC_P \frac{\Delta T}{A.t} = 10J/m^2s$$

Mass = volume × Density = A × Depth × Density

$$\frac{\text{Power}}{\text{Area}} = A \times \text{Depth} \times \text{Density}\; C_P \frac{\Delta T}{A.\Delta t} = 10J/m^2s$$

$$\text{Depth} \times \text{Density}\; C_P \frac{\Delta T}{\Delta t} = 10$$

$$\frac{\Delta T}{\Delta t} = \frac{10}{\text{Depth} \times \text{Density}} C_P = \frac{10}{(1000 \times 0.5 \times 1000)}$$

= 0.00002 K/s = 1.728 K/Day

22. [40000] $E_{absorbed} = E_{emitted}$

$$K_S \times (1 - \text{albedo}) \times \pi R_E^2 = \sigma T^4 \times 4\pi R_E^2$$

$K_s \times (1 - \text{albedo}) = 4\sigma T^4$

T equivalent = 204.93 K

SOLVED PAPER 2022

XE : ENGINEERING SCIENCES

GENERAL APTITUDE

Q.1 to Q. 5 : Carry One Mark Each.

1. The movie was funny and I _______.

(a) could help laughing

(b) couldn't help laughed

(c) couldn't help laughing

(d) could helped laughed

2. $x : y : z = \frac{1}{2} : \frac{1}{3} : \frac{1}{4}$.

What is the value of $\frac{x + z - y}{y}$?

(a) 0.75 (b) 1.25

(c) 2.25 (d) 3.25

3. Both the numerator and the denominator of $\frac{3}{4}$ are increased by a positive integer, x, and those of $\left(\frac{15}{17}\right)$ are decreased by the same integer. This operation results in the same value for both the fractions.

What is the value of x?

(a) 1 (b) 2

(c) 3 (d) 4

4. A survey of 450 students about their subjects of interest resulted in the following outcome.

- 150 students are interested in Mathematics.
- 200 students are interested in Physics.
- 175 students are interested in Chemistry.
- 50 students are interested in Mathematics and Physics.
- 60 students are interested in Physics and Chemistry.
- 40 students are interested in Mathematics and Chemistry.
- 30 students are interested in Mathematics, Physics and Chemistry.
- Remaining students are interested in Humanities.

Based on the above information, the number of students interested in Humanities is

(a) 10 (b) 30

(c) 40 (d) 45

5.

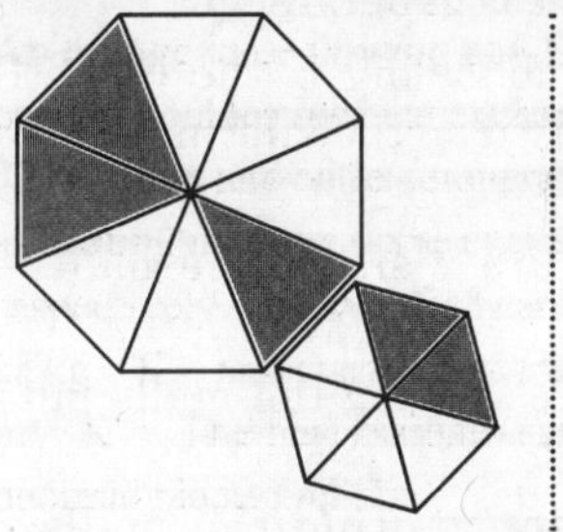

For the picture shown above, which one of the following is the correct picture representing reflection with respect to the mirror shown as the dotted line?

(a)

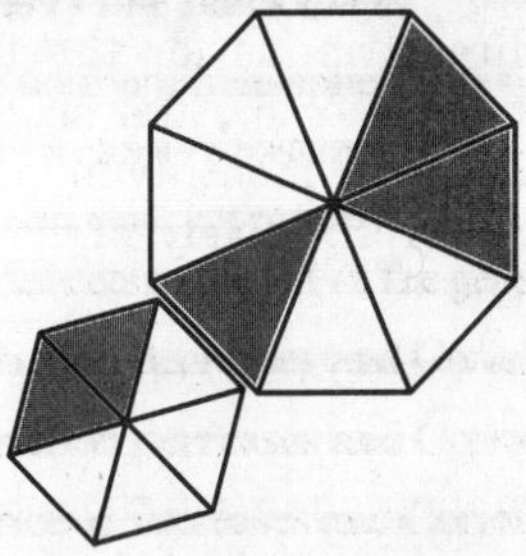

(b)

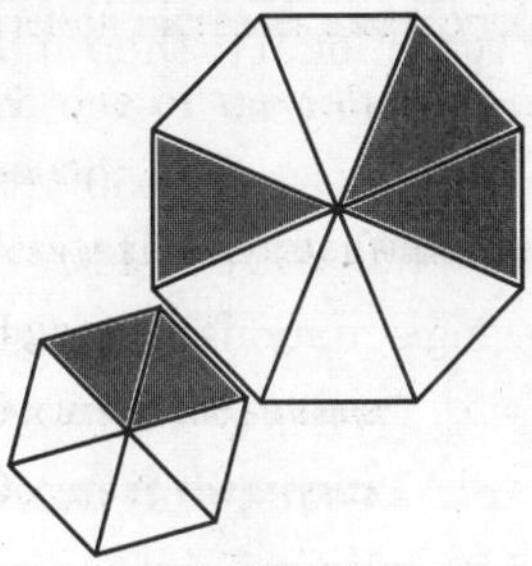

(c)

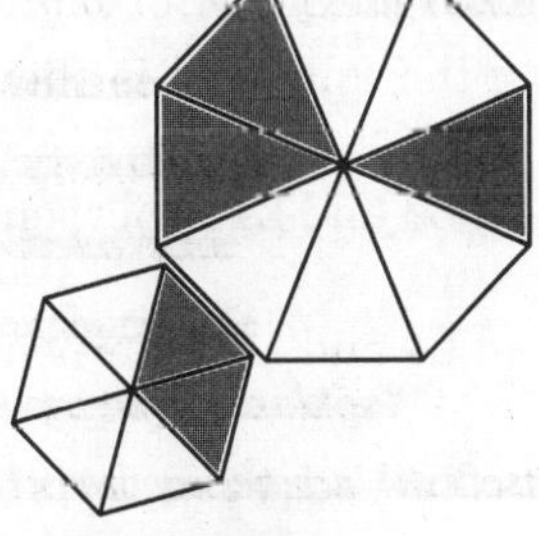

(d)

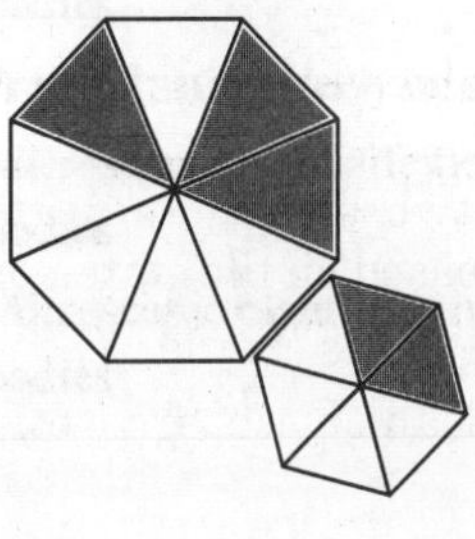

Q. 6 to Q. 10 : Carry Two Marks Each.

6. In the last few years, several new shopping malls were opened in the city. The total number of visitors in the malls is impressive. However, the total revenue generated through sales in the shops in these malls is generally low.

Which one of the following is the CORRECT logical inference based on the information in the above passage?

(a) Fewer people are visiting the malls but spending more

(b) More people are visiting the malls but not spending enough

(c) More people are visiting the malls and spending more

(d) Fewer people are visiting the malls and not spending enough

7. In a partnership business the monthly investment by three friends for the first six months is in the ratio 3 : 4 : 5. After six months, they had to increase their monthly investments by 10%, 15% and 20%, respectively, of their initial monthly investment.

The new investment ratio was kept constant for the next six months.

What is the ratio of their shares in the total profit (in the same order) at the end of the year such that the share is proportional to their individual total investment over the year?

(a) 22 : 23 : 24 (b) 22 : 33 : 50

(c) 33 : 46 : 60 (d) 63 : 86 : 110

8. Consider the following equations of straight lines:

Line L1: $2x - 3y = 5$

Line L2: $3x + 2y = 8$

Line L3: $4x - 6y = 5$

Line L4: $6x - 9y = 6$

Which one among the following is the correct statement?

(a) L1 is parallel to L2 and L1 is perpendicular to L3

(b) L2 is parallel to L4 and L2 is perpendicular to L1

(c) L3 is perpendicular to L4 and L3 is parallel to L2

(d) L4 is perpendicular to L2 and L4 is parallel to L3

9. Given below are two statements and four conclusions drawn based on the statements.

Statement 1: Some soaps are clean.

Statement 2: All clean objects are wet.

Conclusion I: Some clean objects are soaps.

Conclusion II: No clean object is a soap.

Conclusion III: Some wet objects are soaps.

Conclusion IV: All wet objects are soaps.

Which one of the following options can be logically inferred?

(a) Only conclusion I is correct

(b) Either conclusion I or conclusion II is correct

(c) Either conclusion III or conclusion IV is correct

(d) Only conclusion I and conclusion III are correct

10. An ant walks in a straight line on a plane leaving behind a trace of its movement. The initial position of the ant is at point P facing east.

The ant first turns 72° anticlockwise at P, and then does the following two steps in sequence exactly FIVE times before halting.

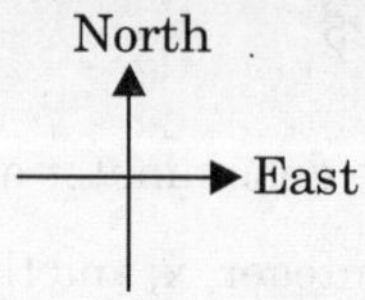

1. Moves forward for 10 cm
2. Turns 144° clockwise.

The pattern made by the trace left behind by the ant is

(a)

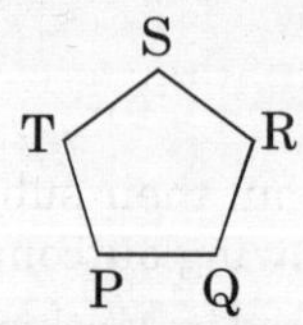

PQ = QR = RS = ST = TP = 10cm

(b)

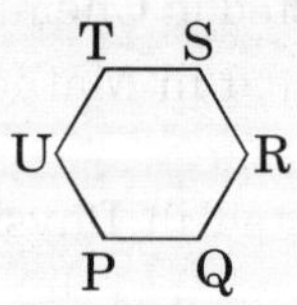

PQ = QR = RS = ST = TU = UP = 10cm

(c)

SQ = QT = TR = RP = PS = 10 cm

(d)

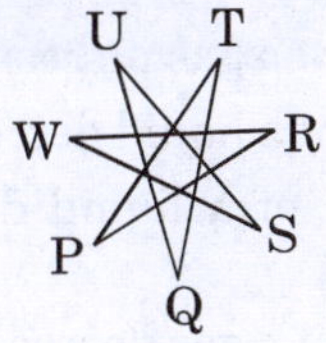

SW = WR = RP = PT = TQ = QU = US = 10cm

A: ENGINEERING MATHEMATICS

Q.11 – Q.17 Carry ONE mark Each

11. The value of

$$\lim_{x\to 0}\frac{1}{x}\int_2^{2+x}(t+\sqrt{t^2+5})dt$$

is

(a) 0 (b) 4

(c) 5 (d) 6

12. Let $\mathbb{C} = \{z = x + iy : x$ and y are real numbers, $i = \sqrt{-1}\}$ be the set of complex numbers. Let the function $f(z) = u(x, y) + i\, v(x, y)$ for $z = x + iy \in \mathbb{C}$ be analytic in $\mathbb{C}$, where

$$u(x, y) = x y^3 - y x^3 \text{ and } v(x, y) = \frac{x^4}{4}+\frac{y^4}{4}-\frac{3}{2}x^2y^2.$$

If $f'(z)$ denotes the derivative of $f(z)$, then

(a) $|f'(-1+i)|^2 = 1$

(b) $|f'(-1+i)|^2 = 7$

(c) $|f'(-1+i)|^2 = 8$

(d) $|f'(-1+i)|^2 = 10$

13. If the partial differential equation

$$(x+2)\frac{\partial^2 u}{\partial x^2}+2(x+y)\frac{\partial^2 u}{\partial x\partial y}+2(y-1)\frac{\partial^2 u}{\partial y^2}-3y^2\frac{\partial u}{\partial y}=0$$

is parabolic on the circle $(x - a)^2 + (y - b)^2 = r^2$, then the values of a, b and r are given by

(a) $a = 1, b = 2, r = 1$

(b) $a = -1, b = 2, r = 1$

(c) $a = 1, b = -2, r = 1$

(d) $a = -1, b = -2, r = 1$

14. Let Γ be the positively oriented circle $x^2 + y^2 = 9$ in the xy-plane. If

$$\oint_\Gamma (3y+e^{x\sin x})dx+(7x+\sqrt{e^y+2})dy = \alpha\pi,$$

where α is real constant, then α is equal to _____.

15. Let $y_1(x)$ and $y_2(x)$ be two linearly independent solutions of

$$x^2\frac{d^2y}{dx^2}-2x\frac{dy}{dx}+2y=0,\ x>0$$

Let $W(y_1, y_2)(x)$ denote the Wronskian of $y_1(x)$ and $y_2(x)$ at x.

If $W(y_1, y_2)(1) = 1$ then $W(y_1, y_2)(2)$ is equal to _____.

16. Let $A = \begin{bmatrix} 2 & 0 & 1 & 1 \\ 1 & 2 & 5 & -5 \\ 0 & 0 & 3 & 0 \\ 0 & 0 & 1 & 3 \end{bmatrix}$. Then the sum of the geometric multiplicities of the distinct eigenvalues of A is equal to ________.

17. In a cosmopolitan city, the population comprises of 30% female and 70% male. Suppose that 5% of female and 30% of male in the population are foreigners. A person is selected at random from this population. Given that the selected person is a foreigner, the probability that the person is a female is ________*(round off to three decimal places).*

Q.18 - Q.21 Carry TWO marks Each

18. Let $f: (0, \infty) \to \mathbb{R}$ be the continuous function such that $f(x) = 2 + \frac{g(x)}{x}$ for all $x > 0$, where $g(x) = \int_1^x f(t)\,dt$ for all $x > 0$. Then $f(2)$ is equal to

(a) $2 + \log_e 2$ (b) $2 - \log_e 2$

(c) $2 + \log_e 4$ (d) $2 - \log_e 4$

19. Let A and B be $n \times n$ matrices with real entries. Consider the following statements:

P: If A is symmetric then rank(A) = Number of nonzero eigenvalues (counting multiplicity) of A.

Q: If AB = 0 then rank(A) + rank(B) $\le$ n. Then

(a) both P and Q are TRUE

(b) P is TRUE and Q is FALSE

(c) P is FALSE and Q is TRUE

(d) both P and Q are FALSE

20. Let $f : \mathbb{R}^2 \to \mathbb{R}$ be given by $f(x, y) = 4xy - 2x^2 - y^4 + 1$. The number of critical points where f has local maximum is equal to ______.

21. If the quadrature rule

$$\int_{-1}^{1} f(x)dx \approx f(\alpha)+\gamma f(\beta),$$

where α, β and γ are real constants, is exact for all polynomials of degree ≤ 3, then $\gamma + 3(\alpha^2 + \beta^2) + (\alpha^3 + \beta^3)$ is equal to ______.

B : FLUID MECHANICS

Q.22 – Q.30 Carry ONE mark Each

22. A heavy horizontal cylinder of diameter D supports a mass of liquid having density ρ as shown in the figure. Find out the vertical component of force exerted by the liquid per unit length of the cylinder if g is the acceleration due to gravity.

(a) $\frac{\pi D^2}{4}\rho g$ (b) $\frac{\pi D^2}{8}\rho g$

(c) $\frac{\pi D^2}{2}\rho g$ (d) $\frac{\pi D^2}{3}\rho g$

23. The figure shows the developing zone and the fully developed region in a pipe flow where the steady flow takes place from left to right. The wall shear stress in the sections A, B, C, and D are given by τ_A, τ_B, τ_C, and τ_D, respectively. Select the correct statement.

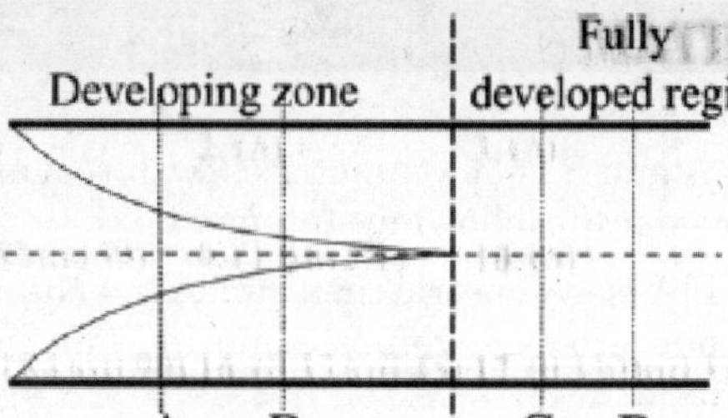

(a) $\tau_A > \tau_B$ (b) $\tau_B > \tau_A$

(c) $\tau_C > \tau_B$ (d) $\tau_C > \tau_D$

24. The left hand column lists some non-dimensional numbers and the right hand column lists some physical phenomena. Indicate the correct combination

1. Reynolds number	i. Wave drag
2. Froude number	ii. Compressible flow
3. Mach number	iii. Viscous drag
4. Weber number	iv. Spray fromation

(a) 1-iii, 2-i, 3-ii, 4-iv

(b) 1-i, 2-ii, 3-iv, 4-iii

(c) 1-iv, 2-iii, 3-iv, 4-iii

(d) 2-iv, 1-iii, 3-ii, 4-i

25. As temperature increases

(a) the dynamic viscosity of a gas increases.

(b) the dynamic viscosity of a liquid decreases.

(c) the dynamic viscosity of a liquid does not change.

(d) the dynamic viscosity of a gas decreases.

26. Which of the following statement(s) regarding a venturimeter is/are correct?

(a) In the direction of flow, it consists of a converging section, a throat, and a diverging section.

(b) In the direction of flow, it consists of a diverging section, a throat, and a converging section.

(c) It is used for flow measurement at a very low Reynolds number.

(d) Pressure tappings are provided just upstream of the venturimeter and at the throat.

27. Which of the following statement(s) is/are true for streamlines in a steady incompressible flow?

(a) Two streamlines cannot intersect each other.

(b) Flow rate increases between two diverging streamlines.

(c) Flow rate decreases between two diverging streamlines.

(d) Stream function has a constant value along a streamline.

28. A flow has a velocity potential given by $\phi = Ax^3$ where 'A' is a non-zero constant. Which of the following statement(s) is/are true about the flow?

(a) The flow is incompressible.

(b) The flow is irrotational.

(c) The flow has local acceleration.

(d) The flow has convective acceleration.

29. A boundary layer develops due to a two-dimensional steady flow over a horizontal flat plate. Consider a vertical line away from the leading edge which extends from the wall to the edge of the boundary layer. Which of the following quantity/quantities is/are not constant along the vertical line? u and v represent the components of velocity in the direction along the plate and normal to it, respectively and x is taken along the length of the plate while p is the pressure.

Neglect body forces.

(a) u (b) $\frac{\partial u}{\partial x}$

(c) v (d) p

30. A 10 kg mass placed on an infinitely long horizontal massless flat platform is to be supported by a steady vertical water jet as shown in the figure. The diameter of the jet is 5 cm. What minimum average velocity is required to hold the mass in place?

Assume $\rho_{water} = 1000\ kg/m^3$, $g = 10\dfrac{m}{s^2}$ and $\pi = 3.14$. Neglect friction.

(Round off to two decimal places)

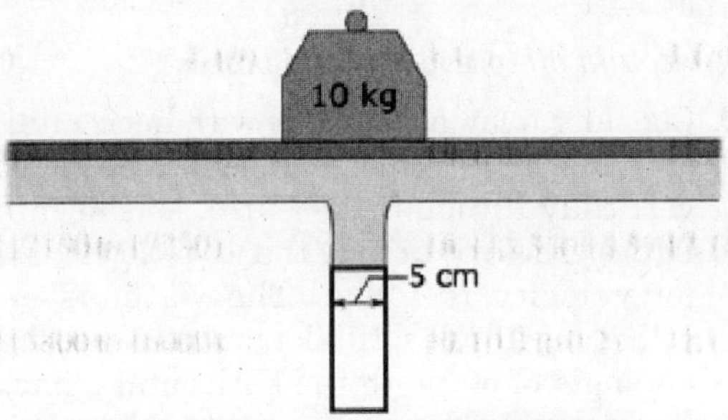

Q.31 – Q.43 Carry TWO marks Each

31. Consider an inviscid flow through a smooth pipe which has a pitot-static tube arrangement as shown. Find the centre-line velocity in the pipe.

Consider that the density of the fluid is 1000 kg/m^3, acceleration due to gravity is 10 m/s^2, and the specific gravity of the manometric fluid is 11.

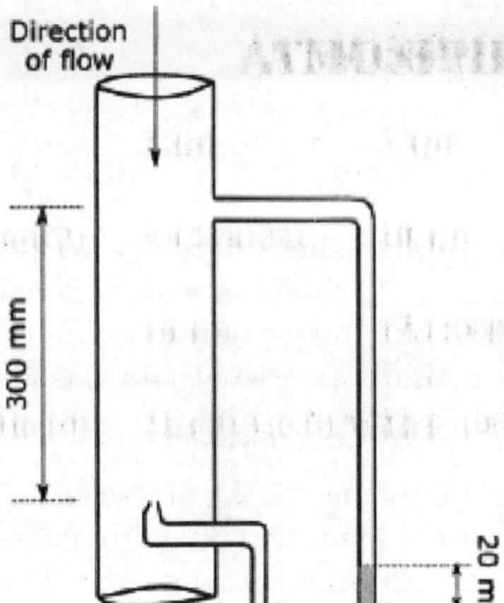

(a) 2 m/s
(b) 3 m/s
(c) 5 m/s
(d) 7 m/s

32. The speed of propagation, c, of a capillary wave depends on the density of the fluid, ρ the wavelength of the wave, λ, and the surface tension, σ. If the density and wavelength remain constant, halving the surface tension would lead to a new velocity, c', given by

(a) $c' = 2c$

(b) $c' = \sqrt{2}c$

(c) $c' = \dfrac{c}{\sqrt{2}}$

(d) $c' = c$

33. A two-dimensional flow field is described by a combination of a source of strength m at the origin and a uniform flow, U, in the positive x-direction such that the velocity potential is given by

$$\phi = Ux + \frac{m}{2\pi}\ln\sqrt{x^2 + y^2}$$

The stagnation streamline is shown in the figure. Find the distance aa′.

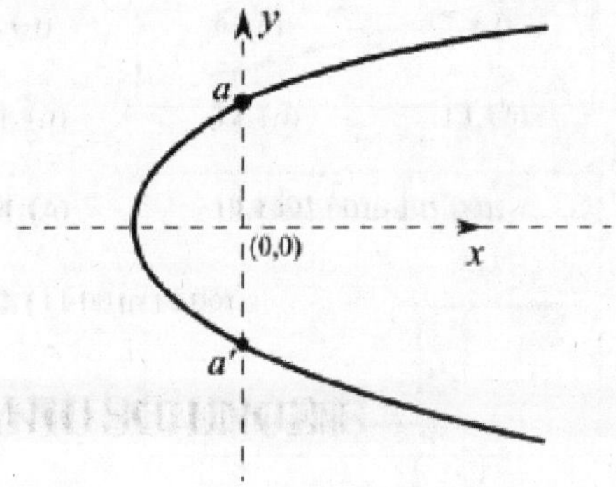

(a) $\dfrac{m}{U}$ (b) $\dfrac{2m}{U}$

(c) $\dfrac{8m}{U}$ (d) $\dfrac{m}{2U}$

34. A typical boundary layer over a flat plate has a linear velocity profile with zero velocity at the wall and freestream velocity, U_∞, at the outer edge of the boundary layer. What is the ratio of the momentum thickness to the thickness of the boundary layer?

(a) $1/2$ (b) $1/4$

(c) $1/6$ (d) $1/3$

35. Identify the configuration(s) in which steady two-dimensional internal flow may show boundary layer separation if the flow direction is left to right.

(a)

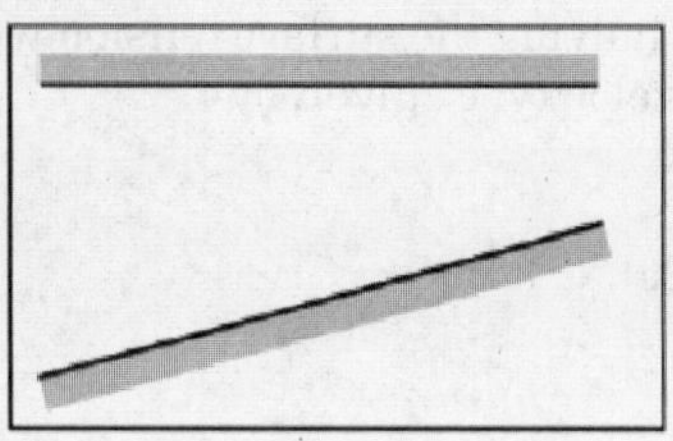

(b)

(c)

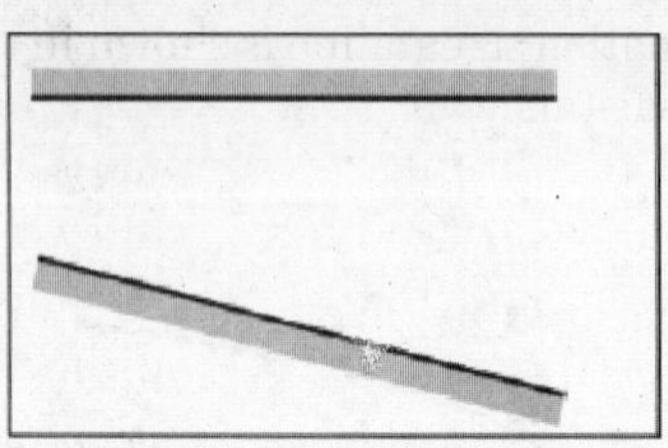

(d)

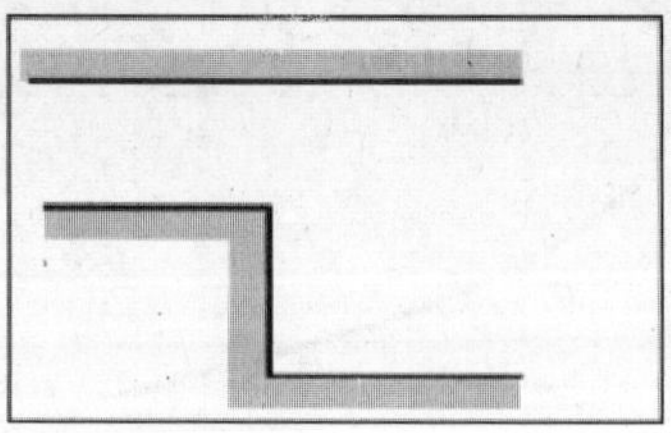

36. Consider steady fully developed flow of a liquid through two large horizontal flat parallel plates separated by a distance of 2 mm. One of the plate is fixed and the other plate moves at a speed of 0.5 m/s. What is the magnitude of the pressure gradient (in Pa/m) in the direction of the flow required to ensure that the net flow through the plates is zero?

Dynamic viscosity of the liquid is 5×10^{-4} Ns/m^2

(Round off to the nearest integer)

37. Consider two-dimensional turbulent flow of air over a horizontal flat plate of length 1 m. Skin friction coefficient at a length x from the leading edge of the plate is obtained as:

$$Z_f = \frac{0.06}{(Re_x)^{0.2}}$$

where, Re_x is the local Reynolds number.

Find out the drag force per unit width (in N/m^2) on the plate if the free stream air velocity is 10 m/s.

Density and dynamic viscosity of air are given as 1.2 kg/m^3 and 1.83×10^{-5} N–s/m^2, respectively.

(Round off to three decimal places)

38. For an inviscid fluid with density 1 kg/m^3, the Cartesian velocity field is given as:

$u = (-2x + y)ti + (2x + y)tj$ m/s

Neglecting the body forces, find the magnitude of pressure gradient in (Pa/m) at (x, y) = (1 m, 1 m) at t = 1 s.

(Round off to two decimal places)

39. Consider a lawn sprinkler with horizontal arms of radius, a = 10 cm which has water introduced vertically through the centre, as shown in the figure. The exit area of the jet is 25 cm^2 and the jet velocity is 1 m/s. The water is ejected orthogonal to the sprinkler arm and the jet makes an angle of 60° with the horizontal plane. Find the torque (in N-m) required to hold the sprinkler stationary.

Consider water density 1000 kg/m^3. Neglect the effects of friction and gravity.

(Round off to two decimal places)

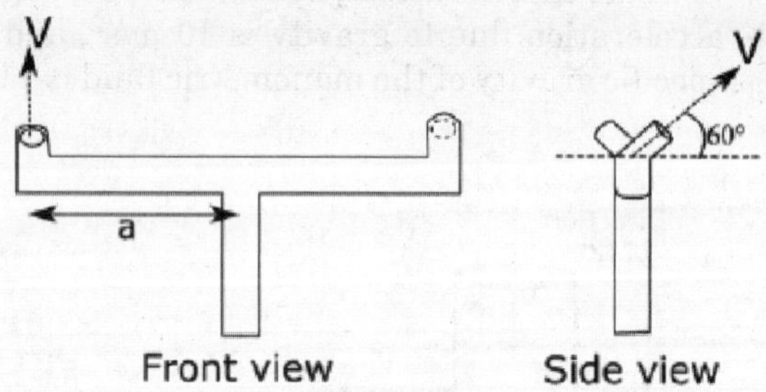

40. A wooden cylinder (specific gravity = 0.6) of length L and diameter D floats in water (density 1000 kg/m^3). Find out the minimum value of D/L for which the cylinder floats with its axis vertical.

(Round off to three decimal places)

41. Consider a cart of mass 10 kg placed on an inclined plane (angle of inclination 60° with horizontal) as shown in the figure. A turning vane of negligible weight is mounted on the cart. A horizontal steady water jet is issued from a stationary nozzle of area 0.1 m^2 and strikes the turning vane as shown in the figure. The vane turns the jet downward parallel to the inclined plane. Find out the minimum jet velocity (in m/s) which will not allow the cart to come down. Neglect friction, consider density of water = 1000 kg/m^3 and acceleration due to gravity = 10 m/s^2. *(Round off to two decimal places)*

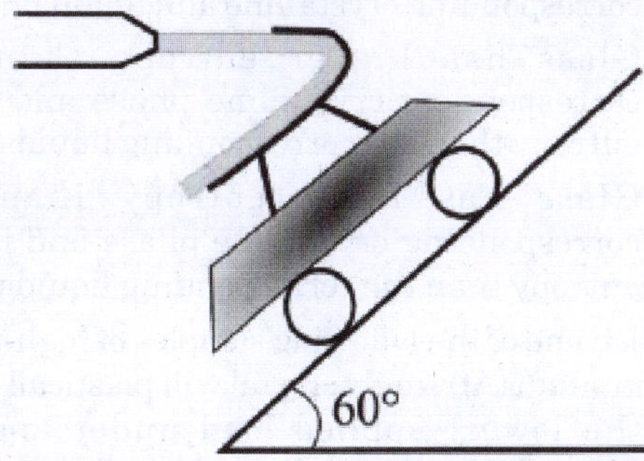

42. A siphon is used to drain out water (density 1000 kg/m^3) from a tank as shown in the figure. What can be the maximum height z (in meter) of the point C?

Consider acceleration due to gravity = 10 m/s^2, pressure at point A = 101 kPa, vapour pressure of water = 29.5 kPa and neglect friction. *(Round off to two decimal places)*

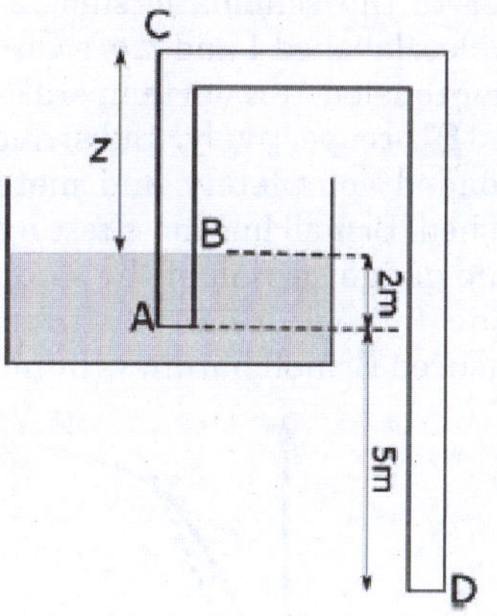

43. The horizontal belt of negligible weight shown in the figure moves with a steady velocity (V) of 2.5 m/s and skims over the top surface of an oil-film of depth h = 3 cm. The length (L) and width (b) of the belt are, respectively, 2 m and 60 cm. Find the viscosity of the oil (in Pa-s), given that the minimum power required to move the belt is 100 W. Neglect the end effects. *(Round off to two decimal places)*

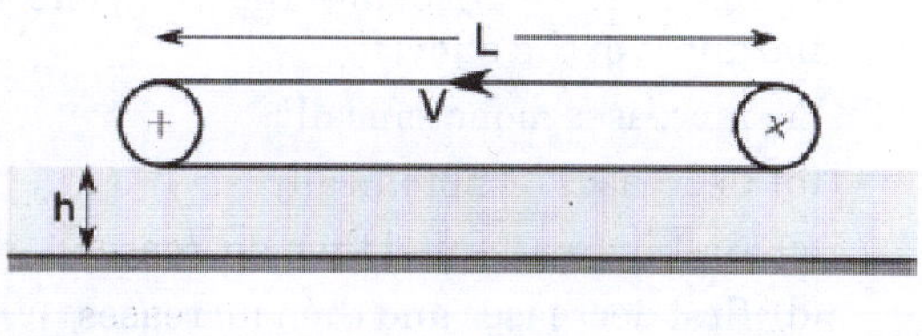

C : MATERIALS SCIENCE

Q.44 – Q.52 Carry ONE mark Each

44. Number of atoms per unit area of the (110) plane of a body centered cubic crystal, with lattice parameter 'a', is

(a) $\frac{1}{a^2}$ (b) $\frac{\sqrt{2}}{a^2}$

(c) $\frac{1}{\sqrt{3}a^2}$ (d) $\frac{1}{\sqrt{2}a^2}$

45. Match the following materials with their corresponding bonding types.

Material	Bonding
P: $Cu_{0.5}Al_{0.5}$	1: Ionic
Q: ZnS	2: Covalent
R: Na_2O	3: Metallic
S: Li_4SiO_4	4: Mixed

(a) P - 4; Q - 2; R - 3; S - 1
(b) P - 3; Q - 4; R - 2; S - 1
(c) P - 3; Q - 2; R - 1; S - 4
(d) P - 3; Q - 1; R - 4; S - 2

46. In an ideal rubber, the primary factor responsible for elasticity up to small strains is

(a) Change in both enthalpy and entropy
(b) Change in enthalpy, but no change in the entropy
(c) No change in enthalpy, but change in the entropy
(d) Neither a change in enthalpy, nor a change in the entropy

47. Which one of the following statements is true for an intrinsic semiconductor?

(a) Electrical conductivity increases with increasing temperature and pressure
(b) Electrical conductivity increases with increasing temperature and decreasing pressure
(c) Electrical conductivity increases with decreasing temperature and increasing pressure
(d) Electrical conductivity increases with decreasing temperature and pressure

48. A differential scanning calorimetry (DSC) experiment tracks the heat flow into or out of a system as a function of temperature. If the experiments given in the options below are performed at 1 atmospheric pressure, then in which case will the DSC thermogram exhibit a spike, either upward or downward?

(a) Heating 10 mg of pure Cu from 323 K to 673 K
(b) Cooling pure water from 323 K to 278 K
(c) Heating pure ice from 263 K to 284 K
(d) Cooling a Pb-Sn alloy at the eutectic composition from 323 K to 273 K

49. Which one of the following solvent environments will likely result in swelling of solid polystyrene?
(a) 0.1 M NaOH in H_2O
(b) HCl (aq.) of pH = 6
(c) Distilled water
(d) Benzene

50. Vickers microhardness (HV) of a ductile material A is higher than another ductile material B. Which of the following is/are true?
(a) Young's modulus of A is greater than B
(b) Yield strength of A is greater than B
(c) Scratch resistance of A is greater than B
(d) Ductility of A is greater than B

51. The enthalpy required to create an oxygen vacancy in CeO_2 is 4 eV. The number of oxygen vacancies present per mole of CeO_2 at 1000 K is ______.
(Round off to the nearest integer)
Given:
N_A: Avogadro's number = 6.02×10^{23} mole^{-1}
k_B: Boltzmann's constant = 8.62×10^{-5} e V/K

52. An electrochemical reaction is known to occur at +4.50 V against a Li^+/Li reference electrode. The potential of the same reaction against a Zn^{2+}/Zn reference electrode is ______V. (Round off to two decimal places).
Given:
E^0 (Li^+/Li) = –3.04 V versus Standard Hydrogen Electrode
E^0 (Zn^{2+}/Zn) = –0.77 V versus Standard Hydrogen Electrode

Q.53 - Q.65 Carry TWO marks Each

53. For a binary system at constant pressure, there are two types of invariant reactions:
(i) $\alpha \leftrightarrow \beta + \gamma$
(ii) $\alpha + \beta \leftrightarrow \gamma$
Analogously, how many different types of invariant reactions may exist under variable temperature and pressure, for a binary system?
(a) 1 (b) 2
(c) 3 (d) 4

54. For a glass marginally below its glass transition temperature, which one of the following statements is true?
(a) Glass has higher enthalpy than both the corresponding crystalline and liquid phases
(b) Glass has lower enthalpy than both the corresponding crystalline and liquid phases
(c) Glass has higher entropy than the corresponding crystalline phase and lower entropy than the corresponding liquid phase
(d) Glass has lower entropy than the corresponding crystalline phase and higher entropy than the corresponding liquid phase

55. Which one of the following samples of high-purity aluminium (Al) single crystal will plastically yield at the lowest applied load under ambient conditions? Loading axis is along the direction shown in the schematic.

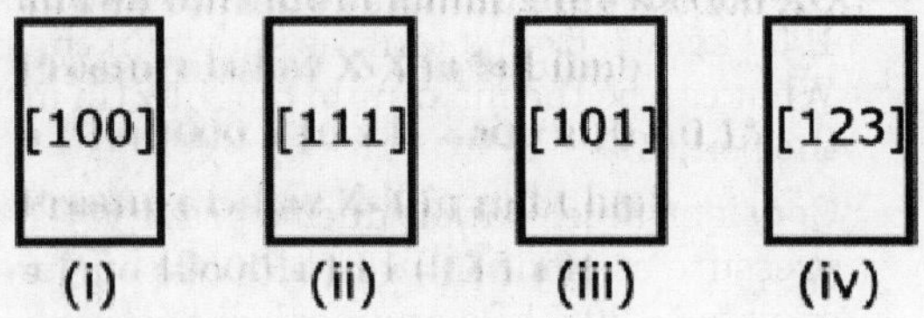

(a) (i) (b) (ii)
(c) (iii) (d) (iv)

56. Refer to the schematic shown. Two dog-bone samples, labelled 1 and 2, of a Cu- alloy are tested under tension at room temperature to points "E" and "P", respectively. Subsequently, they are unloaded completely and metallographically polished. Brinell hardness testing was performed in the gauge section of the samples. Which one of the following can be inferred about the measured Brinell hardness number (BHN)?

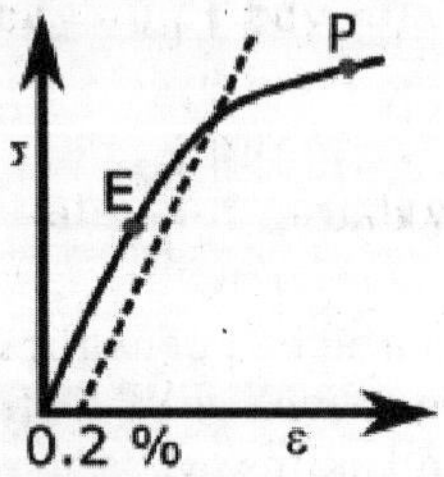

(a) BHN of 1 > BHN of 2
(b) BHN of 1 = BHN of 2
(c) BHN of 1 < BHN of 2
(d) A conclusion about BHN of samples 1 and 2 cannot be made, with the provided information

57. During the ageing of a homogenized Al-Cu alloy (1 to 4 wt.% Cu) below the GP zone solvus, hardness of the alloy:
(a) increases monotonically
(b) decreases monotonically
(c) first increases and then decreases
(d) first decreases and then increases

58. A student aims to deposit a thin metallic film on SiO_2 substrate, with an adhesion layer between the metal film and substrate, in a contiguous planar fashion. Island type of growth must be avoided. The student performs an extensive optimization exercise. Which one of the following steps is in the right direction?

(a) Choose a metallic adhesion layer with very low interfacial energy with the deposited thin film

(b) Choose a metallic adhesion layer with very low interfacial energy with SiO_2, irrespective of its interaction with metal film to be deposited

(c) Increase the substrate temperature and decrease the deposition rate

(d) Use intermittent stages of deposition followed by annealing

59. For a diffusional transformation (i.e., growth of β precipitates in an α matrix), which of the following is/are true with increasing degree of undercooling?

(a) Rate of transformation first increases and then decreases

(b) Rate of transformation first decreases and then increases

(c) Thermodynamic driving force increases monotonically

(d) Mobility of atoms in α matrix remains unchanged

60. A two-phase (α + β) mixture of an A-B binary system has the following properties:

(i) Phase α has equal weight percentages of A and B.

(ii) Phase β has twice the mole fraction of A compared to B.

(iii) The two-phase mixture has equal amounts of α and β.

(iv) Atomic mass of A is twice that of B.

The mole fraction of A in the resultant two-phase mixture is ______. (Round off to one decimal)

61. It is known that component A diffuses into a solid to a depth of 10 μm in 1 hour at 300 K. Treat diffusion in one dimension. The time taken for A to diffuse to the same depth at 600 K is ______ seconds. (Round off to 1 decimal).

Diffusivity of A in the solid is given by

$$D_A = D_A^0 \exp\left(-\frac{E_a}{k_B T}\right)$$

D_A^0: Diffusivity coefficient

E_a: Activation energy = 0.3 eV

k_B: Boltzmann's constant = 8.62×10^{-5} eV/K

T: Absolute temperature

62. A spherical β particle nucleates from the α matrix on a non-deformable substrate δ, forming a contact angle of θ as shown in the schematic.

The value of $\frac{\Delta G^*_{het}}{\Delta G^*_{hom}}$ is ______. (Round off to three decimal places)

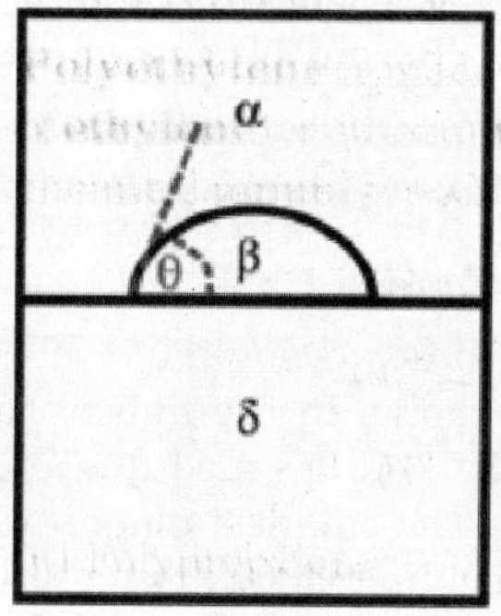

ΔG^*_{hom} = Gibbs free energy change at the critical radius for homogeneous nucleation

ΔG^*_{het} = Gibbs free energy change at the critical radius for heterogeneous nucleation

α–β interfacial energy = 0.4 J/m²

α–δ interfacial energy = 0.3 J/m²

β–δ interfacial energy = 0.02 J/m²

63. The resistivity of a pure semiconductor at 298 K is 3000 Ωm. Assume that the number of electrons excited (n_e) across the band gap is given by the relation

$$n_e = N_A \exp\left(-\frac{E_g}{k_B T}\right)$$

N_A: Avogadro's number = 6.02×10^{23} mole^{-1}

k_B: Boltzmann's constant = 8.62×10^{-5} eV/K

T: Absolute temperature

Mobility of electrons in the semiconductor = 0.14 m²/(V s)

Mobility of holes in the semiconductor = 0.06 m²/(V s)

Absolute charge of an electron = 1.60×10^{-19} C

The band gap (E_g) of the semiconductor is ____eV. (Round off to two decimals)

64. A new glass material is developed to minimize the transmission of the light through the window with glass panel of thickness 5 mm. The refractive index of the glass material is 1.5 and the absorption coefficient can be changed from

0.3 cm^{-1} to 1 cm^{-1}. In the given range of absorption coefficients, the ratio of the maximum to the minimum fraction of the light coming out of the other side of the glass panel is _______. (Round off to two decimal places)

65. The third peak in the X-ray diffraction pattern of a face-centered cubic crystal is at 2θ value of 45°, where 2θ is the angle between the incident and reflected rays. The wavelength of the monochromatic X-ray beam is 1.54 Å. Considering first-order reflection, the lattice parameter of the crystal is _______Å. (Round off to two decimal places)

D : SOLID MECHANICS

Q.66 - Q.74 Carry ONE mark Each

66. A force F is applied at an angle θ = 30° on an elastic column as shown in the figure. E and I are respectively the Young's modulus and area moment of inertia. The smallest magnitude of F needed to cause buckling is

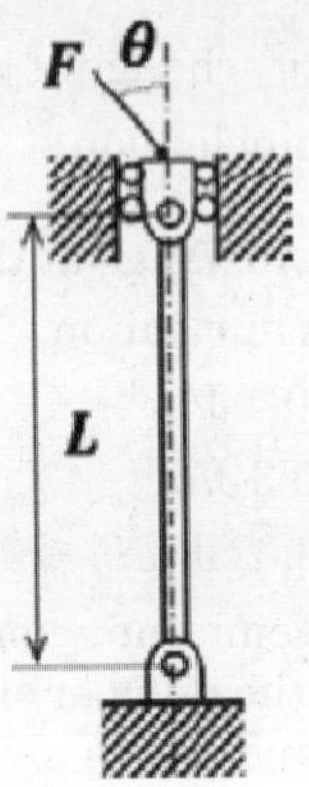

(a) $\frac{2}{\sqrt{3}}\frac{\pi^2 EI}{L^2}$ (b) $\frac{\sqrt{3}}{2}\frac{\pi^2 EI}{L^2}$

(c) $\frac{\pi^2 EI}{2L^2}$ (d) $\frac{2\pi^2 EI}{L^2}$

67. The shear stress due to a transverse shear force in a linear elastic isotropic beam of rectangular cross-section

(a) varies linearly along the depth in the transverse direction of the beam

(b) is zero at the neutral axis

(c) is maximum at the neutral axis

(d) remains constant along the depth in the transverse direction of the beam

68. A massless semicircular rod held fixed at end A is in the xy-plane, as shown in the figure. A force P along the negative z direction is acting at point B on the rod. The unit vectors along x, y and z directions are denoted respectively as **i**, **j** and **k**. Due to the applied force P, the cross-section of the rod at point D will be subjected to

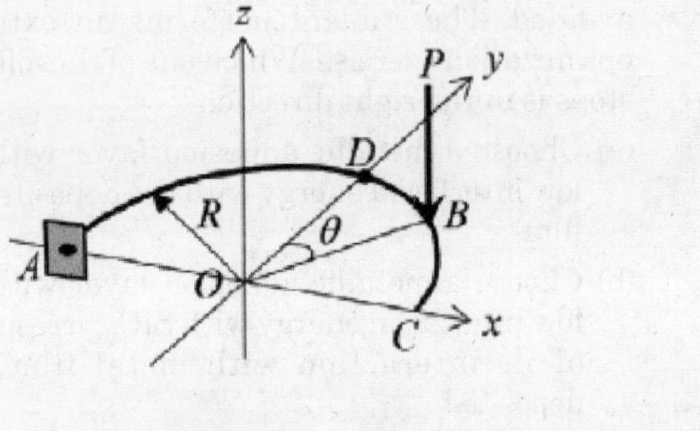

(a) a twisting moment PR(1 – cos θ) **i**, a bending moment PR sin θ **j**, and a shear force –P **k**

(b) a twisting moment PR(1 – sin θ) **i**, a bending moment PR cos θ **j**, and a shear force P **k**

(c) a twisting moment PR(cos θ – 1) **i**, a bending moment –PR sin θ **j**, and a shear force –P **k**

(d) a twisting moment PR sinθ **i**, a bending moment PR (1 – cos) θ **j**, and a shear force P **k**

69. In the truss shown in the figure, all the members are pin jointed to each other. The members AB, BD, DE and DC have the same length. For the given loading, which of the following is the correct statement?

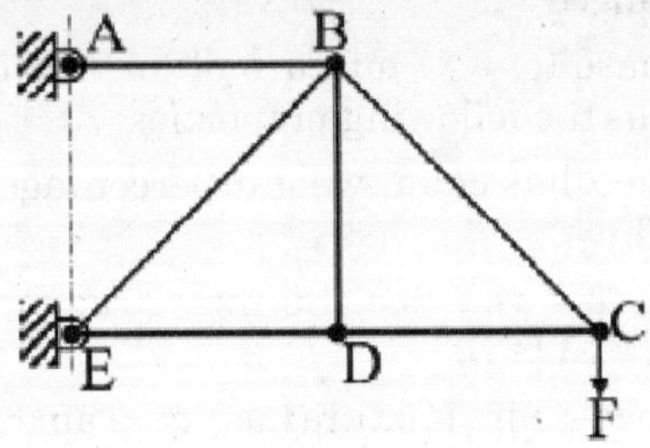

(a) BD is a zero-force member, and AB and ED are in compression

(b) AB is in tension, ED is in compression, and BD is a zero-force member

(c) AB and DC are in tension, and BC is in compression

(d) ED is in tension, and DC and BC are in compression

70. End B of the 2 m long rigid rod AB is constrained to move horizontally in the slot as shown in the figure and has a velocity of 1.0 **i** m/s. The angular velocity of the rod at the instant shown is 2 rad/s. The unit vectors along x and y directions are denoted respectively as **i** and **j**. The velocity of point A in m/s is then given by

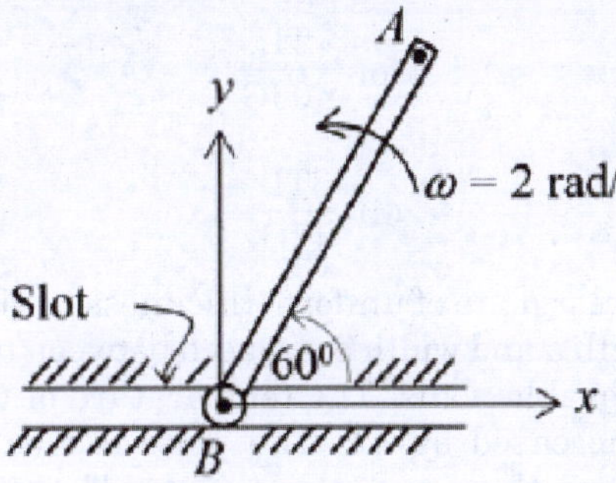

(a) $(1-2\sqrt{3})\mathbf{i}+2\mathbf{j}$ (b) $(1+2\sqrt{3})\mathbf{i}-2\mathbf{j}$

(c) $-2\sqrt{3}\mathbf{i}+2\mathbf{j}$ (d) $2\sqrt{3}\mathbf{i}-2\mathbf{j}$

71. The assembly of four masses connected by rigid mass-less rods is kept on a smooth horizontal floor as shown in the figure. Under the applied force 2F, the magnitude of angular acceleration of the assembly at the instant shown is

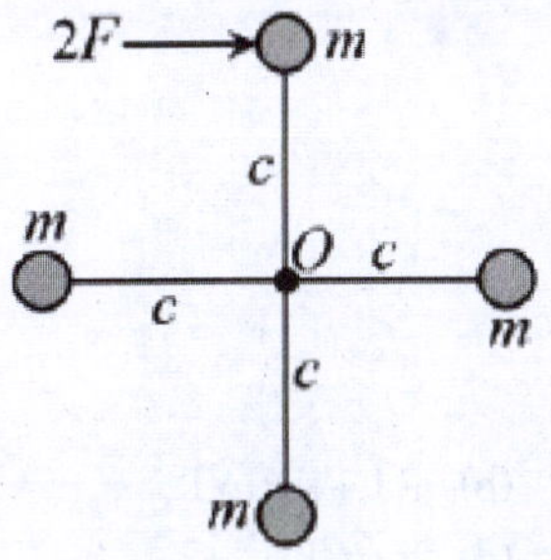

(a) $\frac{F}{mC}$ (b) $\frac{F}{2mC}$

(c) $\frac{2F}{3mC}$ (d) $\frac{F}{3mC}$

72. A particle is constrained to move at a constant speed on an inclined plane (ABCD) along the curved path shown in the figure. Edges AD and BC are parallel to the y axis. The inclined plane makes an angle θ with the xy-plane. The velocity vector of the particle makes an angle ϕ with the dotted line which is parallel to edge AB. If the speed of the particle is 2 m/s, ϕ = 30°, and θ = 40°, then the z-component of the velocity of the particle in m/s is ______

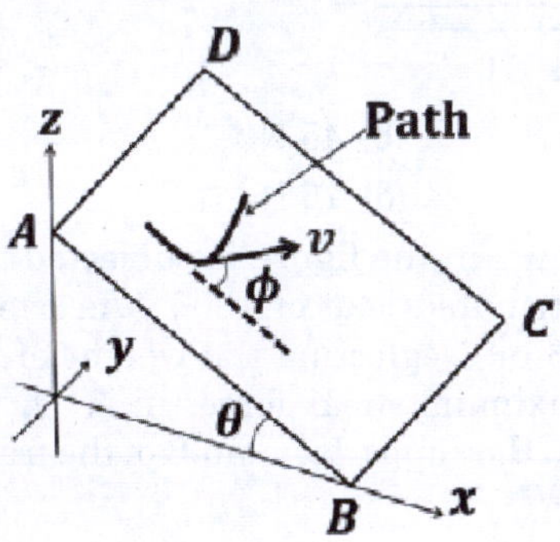

(a) –1.32 (b) –1.00

(c) –1.11 (d) –1.50

73. A uniform elastic rod of constant cross-section is fixed at its left end as shown in the figure. An axial force P is acting as shown. Assume that plane sections remain plane during deformation. The ratio of axial displacements at point A (x = 4L) to that at point B (x = L) is ______(rounded off to one decimal place)

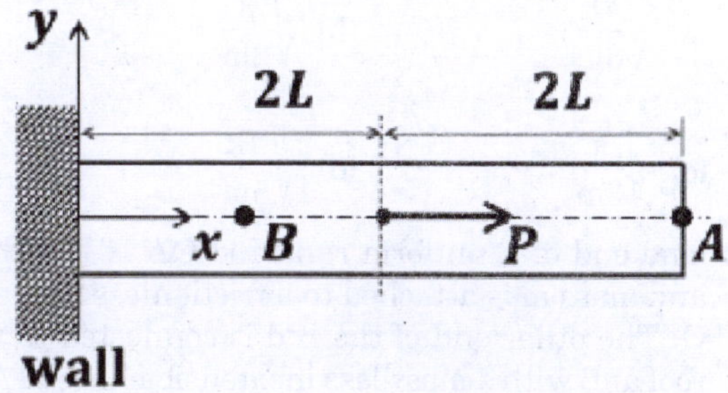

74. A thin-walled spherical pressure vessel has a mean radius of 500 mm and wall thickness of 10 mm. The yield strength of the material is 500 MPa. The internal pressure in MPa at which the spherical pressure vessel will yield according to the Tresca criterion is ______(rounded off to one decimal place)

Q.75 – Q.87 Carry TWO marks Each

75. The beam in the figure is subjected to a moment M_0 at mid span as shown. Which of the following is the vertical reaction at B?

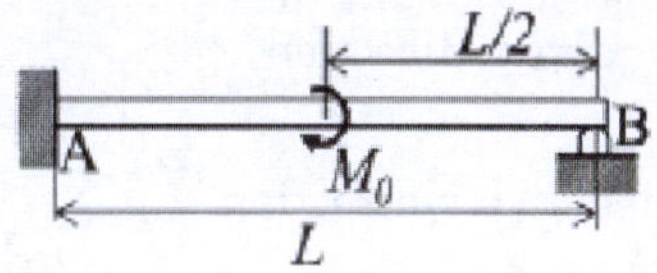

(a) $\frac{9M_0}{8L}$ (b) $\frac{15M_0}{8L}$

C) $\frac{3M_0}{4L}$ (d) $\frac{9M_0}{4L}$

76. A spring-mass system having a mass m and spring constant k, placed horizontally on a foundation, is connected to a vertically hanging mass m with the help of an inextensible string. Ignore the friction in the pulleys and also the inertia of pulleys, string and spring. Gravity is acting vertically downward as shown. The natural frequency of the system in rad/s is

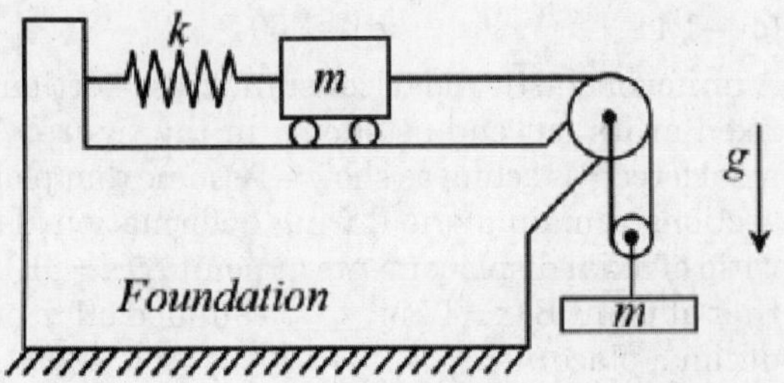

(a) $\sqrt{\frac{4k}{3m}}$ (b) $\sqrt{\frac{k}{2m}}$

(c) $\sqrt{\frac{k}{3m}}$ (d) $\sqrt{\frac{4k}{5m}}$

77. One end of a uniform rigid rod OA of length L and mass m is attached to a frictionless hinge at O. The other end of the rod is connected to the roof at B with a mass-less inextensible thread AB. Initially the rod is horizontal and at rest. The gravity is acting vertically downward as shown. Immediately after the thread AB is cut, the reaction on the rod at O is

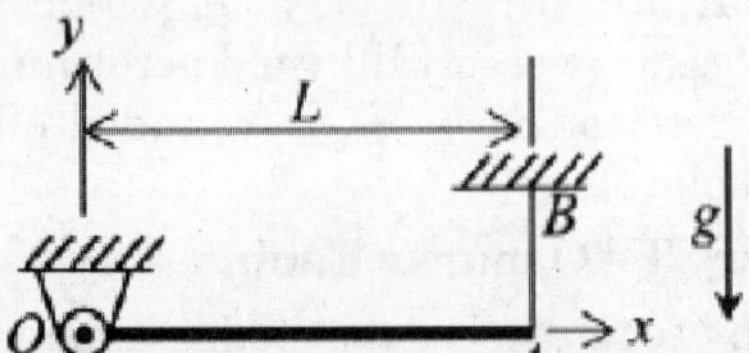

(a) $\frac{mg}{4}$ in the positive y-direction

(b) $\frac{mg}{2}$ in the negative y-direction

(c) $\frac{3mg}{4}$ in the negative y-direction

(d) mg in the positive y-direction

78. A circular shaft is rigidly connected to a wall at one end. The shaft has a solid portion and a hollow portion as shown in the figure. The length of each portion is L and the shear modulus of the material is G. The polar moment of inertia of the hollow portion is J and that of the solid portion is 50J. A torque T is applied at the right most end as shown. The rotation of the section PQ is

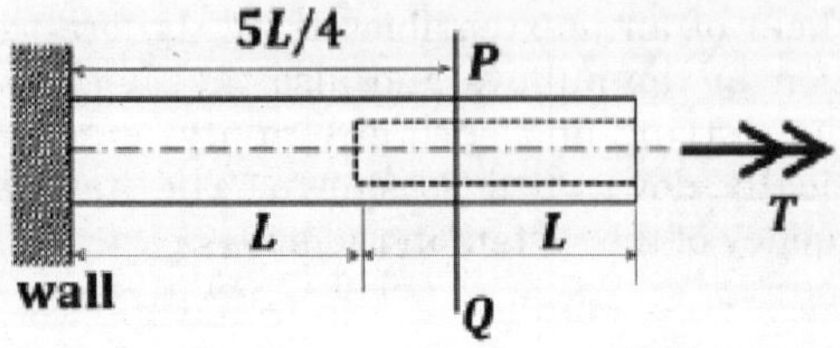

(a) $\frac{27TL}{100JG}$ (b) $\frac{TL}{40JG}$

(c) $\frac{5TL}{4JG}$ (d) $\frac{3TL}{4JG}$

79. A rectangular plate of uniform thickness having initial length a and width b is placed between two rigid immovable walls. The temperature of the plate is increased by ΔT. The plate is free to expand along the y and z directions. The mid-surface of the plate remains in the xy-plane. The Poisson's ratio is v and the coefficient of thermal expansion is α. Assuming that the plate is initially free of stresses, the change in length of the plate after the increase in temperature is given by

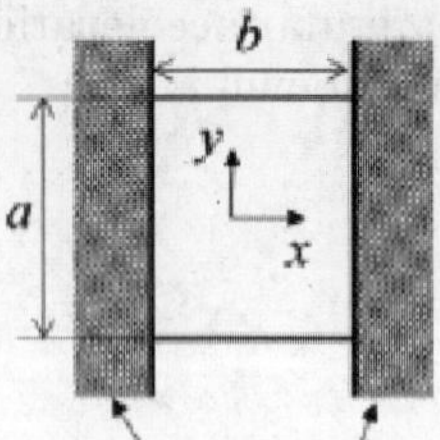

Rigid walls

(a) $a(1-v)\alpha\Delta T$ (b) $a(1+v)\alpha\Delta T$

(c) $a\alpha\Delta T$ (d) $2a\alpha\Delta T$

80. A mass m = 10 kg is attached to a spring as shown in the figure. The coefficient of friction between the mass and the inclined plane is 0.25. Assume that the acceleration due to gravity is 10 m/s^2 and that static and kinematic friction coefficients are the same. Equilibrium of the mass is impossible if the spring force is

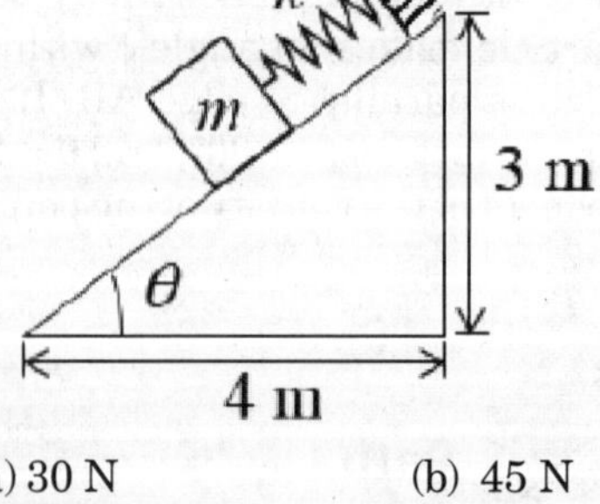

(a) 30 N (b) 45 N

(c) 60 N (d) 75 N

81. The frame shown in the figure is subjected to a uniformly distributed load of 1000 N/m over a distance of 0.6 m. Neglecting the weight of the frame, the maximum shear force (in N) in the region between the supports A and B of the frame is ______.

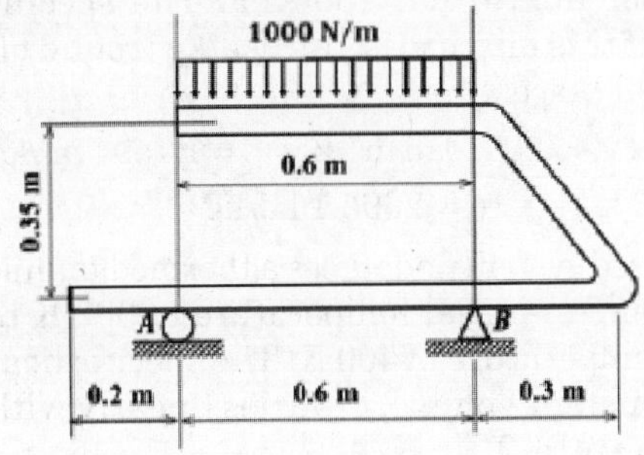

82. A slider moving in a frictionless slot is connected with a linear spring OA as shown in the figure. The following is known: stiffness of the spring = 2 kN/m, mass of the slider = 10 kg, and the unstretched length of the spring = 1 m. If the slider is released from rest at A the magnitude of its velocity (in m/s) when it passes through point B is ________(rounded off to the nearest integer)

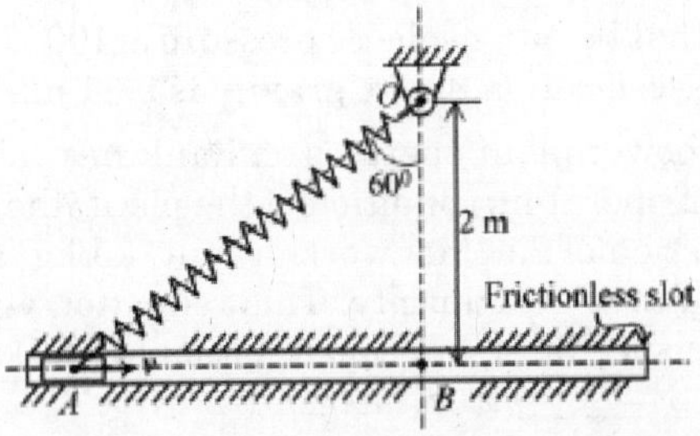

83. A sphere A of mass m is thrown into the air at 50 m/s along a direction $\tan^{-1}(3/4)$ up from the horizontal. At the topmost point of its trajectory, it has a central (non-oblique) collision with another sphere B which is at rest on top of a vertical pole. Sphere B has a mass of 3m. Acceleration due to gravity is 10 m/s^2. Neglect contact friction and air-resistance. If the coefficient of restitution is 0.3, then the speed (in m/s) of sphere A immediately after the collision is ________(rounded off to one decimal place)

84. The truck shown in the figure is moving on a horizontal road at a speed of 20 m/s. It is carrying a box of mass 1000 kg, which is simply placed on the truck platform. The coefficient of friction between the truck platform and the box is 0.25. Take acceleration due to gravity as 10 m/s^2. Assume uniform deceleration during braking, and the coefficients of static and kinetic friction to be the same. The shortest distance in meters in which the truck can be brought to rest without the box slipping is ______(round off to the nearest integer)

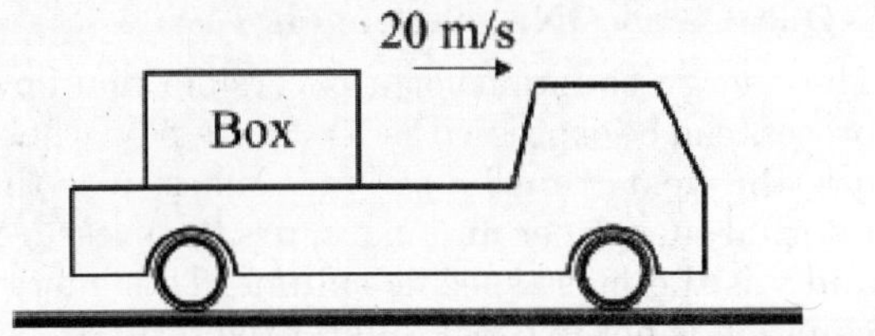

85. The stepped rod of length 2 m, shown in the figure, is fixed at both ends (A and C). The area of cross-section of portion AB is 200 mm^2 and that of portion BC is 100 mm^2. Force F is applied at section B such that the section is displaced by 0.1 mm in the direction of the force. Young's modulus of the rod is 200 GPa. The applied force F in N is______(round off to the nearest integer)

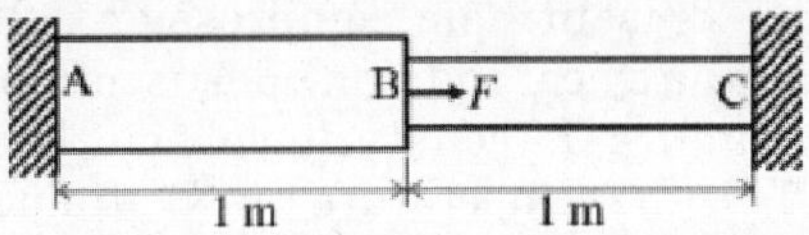

86. A cantilever beam has a span of 1 m and carries a uniformly distributed load of q = 1250 N/m over a portion as shown. A force F = 1000 N acts at a distance L from the fixed end. The distance L is such that the bending moment at the fixed end is zero. The beam has a rectangular cross-section of depth 20 mm and width 24 mm. For this loading, the magnitude of the maximum bending stress in the beam in MPa is ______(round off to the nearest integer)

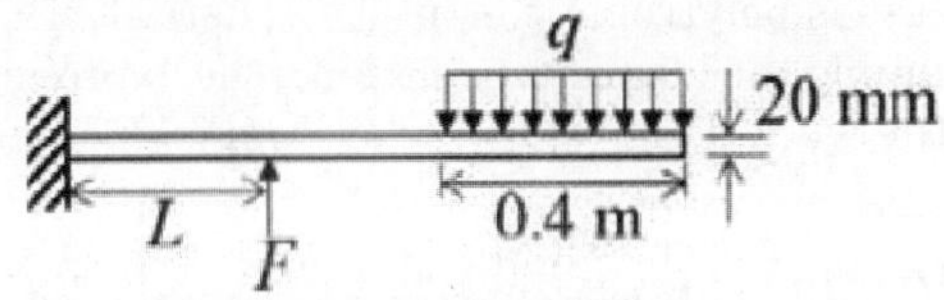

87. The figure shows two identical mass-less beams AB and CD, each clamped at one of their ends. The left end of beam CD rests on the right end of beam AB such that the ends of the beams are just in contact. The beams are unstressed before the application of load. Assume no friction at the contact. Now, if a uniformly distributed load of 800 N/m is applied on beam CD, the bending moment at the end B of beam AB in N.m is ________(rounded off to the nearest integer)

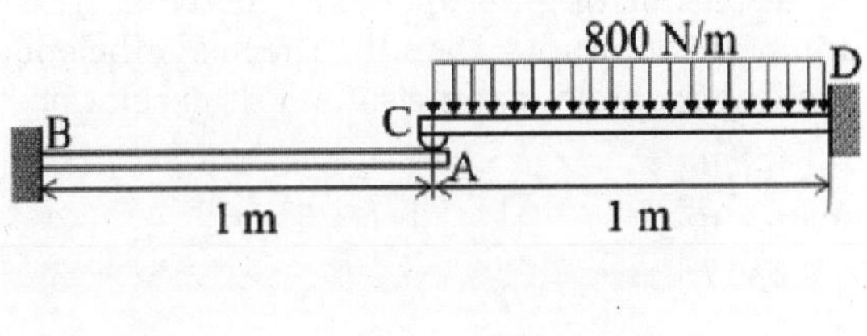

E : THERMODYNAMICS

Q.88 – Q.96 Carry ONE mark each

88. The energy equation for a reversible non-flow process can be expressed as $\delta q = du + pdv$, where q is the heat transfer per unit mass, u is the internal energy per unit mass, p is the pressure, and v is the mass specific volume. This energy equation is not in exact differential form. It can be made exact differential by multiplying with the following integrating factor:

(T is the absolute temperature)

(a) $\frac{1}{p}$ (b) $\frac{1}{v}$

(c) $\frac{1}{T}$ (d) $\frac{1}{uT}$

89. An air standard Diesel cycle consists of four processes: 1-2 (isentropic compression), 2-3 (constant pressure heat addition), 3-4 (isentropic expansion) and 4-1 (constant volume heat rejection). T_4 is the temperature (in K) attained at the end of isentropic expansion (3-4) before constant volume heat rejection. The constant volume heat rejection process (4-1) is replaced by a constant pressure heat rejection process (4a-1) such that T_{4a} is the temperature (in K) reached at the end of isentropic expansion (3-4a), and the state point 1 remains the same. Then

(a) $T_{4a} < T_4$ (b) $T_{4a} > T_4$

(c) $T_{4a} = T_4$ (d) $T_{4a} = 2T_4$

90. Gas in a cylinder-piston device expands from state $1(p_1, V_1, T_1)$ to state $2(p_2, V_2, T_2)$. The expansion process is polytropic, i.e., pV^n = constant, $n \neq 1$. Assuming the ideal gas behaviour, the expression for the work done, W by the system is given by

(a) $W = p_1V_1 \ln\left(\frac{T_2}{T_1}\right)$ (b) $W = \frac{p_2V_2 - p_1V_1}{1-n}$

(c) $W = p_1V_1 \ln\left(\frac{V_1}{V_2}\right)$ (d) $W = p_2V_2 \ln\left(\frac{p_2}{p_1}\right)$

91. The temperature of the working fluid in a real heat engine cycle changes during heat addition and heat rejection processes. The maximum and minimum temperatures of the cycle are T_{max} and T_{min}, respectively. If η_C is the thermal efficiency of a Carnot engine operating between these temperature limits, then the thermal efficiency, η of the real heat engine satisfies the relation

(a) $\eta > \eta_C$ (b) $\eta < \eta_C$

(c) $\eta = \eta_C$ (d) $\eta = 1 + \eta_C$

92. A 1.2 m^3 rigid vessel contains 8 kg of saturated liquid-vapor mixture at 150 kPa. The specific enthalpy of this mixture is _____kJ/kg (round off to 2 decimal places).

At 150 kPa: $v_f = 0.001053$ m^3/kg, $v_g = 1.1594$ m^3/kg
$h_f = 467.13$ kJ/kg, $h_g = 2693.1$ kJ/kg

93. Air in a closed system undergoes a thermodynamic process from an initial temperature of 300 K to the final temperature of 400 K. The specific heat of air at constant volume, c_v varies linearly with the temperature, T (in K) as

$c_v = (0.7 + 0.27 \times 10^{-3}T)$ kJ/(kg K)

Change in the specific internal energy of the air in the system is _______kJ/kg (*round off to 2 decimal places*).

94. A vertical cylinder-piston device contains a fixed mass of gas in equilibrium. The cross-sectional area of the piston is 0.05 m^2. For 150 kPa pressure of the gas in the cylinder, the mass of the piston is ______kg (*round off to 2 decimal places*).

Assume that the atmospheric pressure is 100 kPa and the acceleration due to gravity is 9.81 m/s^2.

95. A steam power plant operates on Rankine cycle. At certain operating condition of the plant, there is a reduction of 20% net work output (kJ/kg) as compared to 100% capacity. This reduction will cause the specific steam consumption (kg/kJ) to increase by ______% (*in integer*).

96. A Carnot heat pump extracts heat from the environment at 250 K, and supplies 6 kW of heat to a room which is maintained at a constant temperature T_H. The heat pump requires a power input of 1 kW for its operation. Then, the temperature of the room T_H is _______K (*round off to nearest integer*).

Q.97 – Q.109 Carry TWO marks each

97. One of the Maxwell equations is expressed as $\left(\frac{\partial s}{\partial v}\right)_T = \left(\frac{\partial p}{\partial T}\right)_v$, where s is the entropy per unit mass, v is the mass specific volume, p is the pressure, and T is the temperature. In this expression, s is a continuous function of T and v. The derivatives of s are also continuous. Let c_v be specific heat capacity at constant volume for a gas. Then, $\left(\frac{\partial c_v}{\partial v}\right)_T$ can be written as

(a) $\frac{p}{T}\left(\frac{\partial^2 p}{\partial T^2}\right)_v$ (b) $\frac{v}{T}\left(\frac{\partial^2 p}{\partial v^2}\right)_T$

(c) $T\left(\frac{\partial^2 p}{\partial T^2}\right)_v$ (d) $\frac{1}{T}\left(\frac{\partial^2 p}{\partial v^2}\right)_T$

98. The general relation among the properties x , y and z at any state point can be expressed as $\left(\frac{\partial x}{\partial y}\right)_z\left(\frac{\partial y}{\partial z}\right)_x\left(\frac{\partial z}{\partial x}\right)_y = -1$. If p, T and h are continuous functions and $c_p = \left(\frac{\partial h}{\partial T}\right)_p$, μ is the Joule-Thomson coefficient, then $\left(\frac{\partial h}{\partial p}\right)_T$ is

(a) $-\mu c_p$ (b) c_pT

(c) $-\frac{c_p}{T}$ (d) μc_p

99. An air-conditioning system consists of an insulated rigid mixing chamber designed to supply air at 24°C to a building. The mixing chamber mixes two air streams: (i) a cold air stream at 10 °C and mass flow rate $\dot{m}_c$ (kg/s), and (ii) a stream of fresh ambient air at 30 °C and mass flow rate $\dot{m}_a$ (kg/s). Assume air to be an ideal gas with constant specific heat (c_p) 1.005 kJ/(kg K), $\gamma = c_p / c_v = 1.4$). Neglect change in kinetic and potential energies as compared to change in enthalpy. Under the steady state condition, the ratio of the mass flow rates of the two streams $(\dot{m}_c / \dot{m}_a)$ is

(a) $\frac{7}{3}$ (b) $\frac{3}{7}$

(c) $\frac{2}{7}$ (d) $\frac{4}{7}$

100. An ideal gas mixture consists of 80% N_2 and 20% O_2 on mass basis. If the total pressure is 300 kPa, then the partial pressure of N_2 (in kPa) is (Molecular weights of N_2 = 28 kg/kmol and O_2 = 32 kg/kmol)

(a) 246.15 (b) 230.34

(c) 254.78 (d) 213.54

101. On the basis of the ideal gas equation and van der Waals equation, the temperatures of a gas at pressure 10 MPa and specific volume 0.005 m³/kg would be, respectively

(Assume gas constant R = 0.3 kJ/(kg K), a = 0.18 m⁶ kPa/kg² and b = 0.0014 m³/kg)

(a) 166.67 K and 235.89 K

(b) 166.67 K and 206.40 K

(c) 166.67 K and 267.21 K

(d) 166.67 K and 240.90 K

102. An ideal Brayton cycle operates between maximum and minimum temperatures of T_3 and T_1 , respectively. For constant values of T_3 and T_1 , the pressure ratio (r_p) for maximum work output is

(γ is the specific heat ratio of air)

(a) $\left(\frac{T_3}{T_1}\right)^{\frac{\gamma}{(\gamma-1)}}$ (b) $\left(\frac{T_3}{T_1}\right)^{\frac{2\gamma}{(\gamma-1)}}$

(c) $\left(\frac{T_3}{T_1}\right)^{\frac{\gamma}{2(\gamma-1)}}$ (d) $\left(\frac{T_3}{T_1}\right)^{\frac{2}{(\gamma-1)}}$

103. An insulated rigid tank of volume 10 m³ contains air initially at 1 MPa and 600 K. A valve connected to the tank is opened, and air is allowed to escape until the temperature inside the tank drops to 400 K. The temperature of the discharged air can be approximated as the average of the initial and final temperatures of the air in the tank. Neglect kinetic and potential energies of the discharged air. Assume that air behaves as an ideal gas with constant specific heat so that internal energy $u = c_vT$ and enthalpy $h = c_pT$. Then, the final pressure of the air in the tank is ______ MPa (*round off to 2 decimal places*).

Assume c_p = 1.005 kJ/(kg K), $\gamma = c_p / c_v = 1.4$

104. Steam enters a steam turbine at 5 MPa and 600 °C, and exits as saturated vapor at 50 kPa. Under steady state condition, the turbine loses heat to the surroundings at the rate of 50 kJ per kilogram of steam flowing through the turbine. The ambient temperature is 300 K, and the heat transfer to the surroundings takes place at the outer surface of the turbine at a temperature of 450 K. The irreversibility per unit mass of steam flowing through the turbine is______kJ/kg (*round off to 2 decimal places*).

Neglect the change in kinetic and potential energies of the steam, and use the following property values:

Superheated steam at 5 MPa, 600 °C

v = 0.07870 m³/kg, u = 3273.3 kJ/kg, h = 3666.9 kJ/kg, s = 7.2605 kJ/(kg K)

Saturated vapour at 50 kPa

v_g = 3.2403 m³/kg, u_g = 2483.2 kJ/kg, h_g = 2645.2 kJ/kg, s_g = 7.5931 kJ/(kg K)

105. A heat engine receives heat at 1000 K and rejects heat to the environment at 300 K. The efficiency of the heat engine is half of the efficiency of a Carnot engine operating between the above mentioned temperature limits. The work output from the heat engine is completely used to drive a refrigerator that steadily removes heat from a cold space at 260 K at a rate of 5.2 kW, and rejects the heat to the same environment at 300 K. The COP (coefficient of performance) of the refrigerator is half of the COP of the Carnot refrigerator operating between the same temperature limits as that of the refrigerator. Then, rate of heat supplied to the heat engine is ________kW (*round off to 2 decimal places*).

106. A room contains air at 25 °C, 100 kPa and 80% relative humidity. If the saturation pressure of water vapor at 25 °C is 3.1698 kPa, then the specific humidity of air is ________1 kg of water vapor / kg of dry air (round off to 4 decimal places).

107. An insulated rigid container is divided into two parts by a thin partition. One part of the container contains 6 kg of saturated liquid-vapor mixture with a dryness fraction of 0.7 at 0.3 MPa. The other part contains 12 kg of saturated liquid at 0.6 MPa of the same substance. When the partition is removed and the system attains equilibrium, the final specific volume of the mixture is________m^3/kg (*round off to 2 decimal places*).

Use the following property values:

At 0.3 MPa : $v_f = 0.001073\ m^3/kg$, $v_g = 0.60582\ m^3/kg$

At 0.6 MPa : $v_f = 0.001101\ m^3/kg$, $v_g = 0.31560\ m^3/kg$

108. During a steady state air-conditioning process, air enters a heating section at 15 °C with 40% relative humidity and leaves at 30 °C. Assuming the heating process takes place at 100 kPa, the relative humidity of the air at exit is ____% (*round off to nearest integer*).

Saturation pressures of water vapor at 15 °C and 30 °C are 1.7057 kPa and 4.2469 kPa respectively.

109. Steam enters a steam turbine at 10 MPa and 600°C with a mass flow rate of 16 kg/s. The steam exits the turbine as saturated vapor at 10 kPa. Under steady state condition, the turbine generates 16.2 MW power. If the ambient temperature is 25 °C, the rate of entropy generation in the turbine is ______kW/K (*round off to 2 decimal places*).

Neglect the change in kinetic and potential energies of the steam, and use the following property values:

Superheated steam at 10 MPa, 600 °C

v = 0.03837 m^3/kg, u = 3241.68 kJ/kg,

h = 3625.34 kJ/kg, s = 6.9028 kJ/(kg K)

Saturated vapour at 10 kPa

v_g = 14.67355 m^3/kg, u_g = 2437.89 kJ/kg,

h_g = 2584.63 kJ/kg, s_g = 8.1501 kJ/(kg K)

F : POLYMER SCIENCE AND ENGINEERING

Q.110 – Q.118 Carry ONE mark Each

110. Interfacial polymerization can be used to prepare

(a) Nylon 6 (b) Nylon 66

(c) Polyacrylonitrile (d) Poly(butyl acrylate)

111. In a rubber sample with a Mooney viscosity of 60 ML(1 + 4) 100 °C, the number 4 signifies

(a) Applied shear rate in s^{-1}

(b) Number of samples tested

(c) Time in minutes after starting the motor when the measurement is taken

(d) Preheating time in minutes

112. The initiator system which can be used for free radical polymerization at 5 °C is

(a) $FeSO_4$ + t-butyl hydroperoxide

(b) Azobisisobutyronitrile

(c) Potassium persulfate

(d) Benzoyl peroxide

113. Weather resistance of high impact polystyrene can be improved by blending polystyrene with

(a) Styrene butadiene rubber

(b) Natural rubber

(c) Ethylene propylene rubber

(d) Nitrile rubber

114. Which of the following is a discontinuous polymer processing operation?

(a) Calendering

(b) Extrusion

(c) Film blowing

(d) Thermoforming

115. The blend of polyethylene and polypropylene is

(a) Immiscible due to enthalpic constraints

(b) Immiscible due to entropic constraints

(c) Miscible as they are polyolefins

(d) Miscible due to comparable solubility parameters

116. Toughness in a polymer can be inferred from

(a) Izod impact strength

(b) Depth of indentation

(c) Area under the stress-strain curve

(d) Charpy impact strength

117. Which of the following polymers are polyesters?

(a) Poly(acrylic acid)

(b) Poly(lactic acid)

(c) Polyhydroxybutyrate

(d) Poly(ε-caprolactone)

118. The functionality of adipic acid for condensation reaction with glycerol is ______ *(in integer)*.

Q.119 - Q.131 Carry TWO marks Each

119. From the dynamic mechanical analysis of a polymer sample with a phase angle of 30°, the relationship between storage modulus (E') and loss modulus (E'') can be expressed as

(a) $E' = \sqrt{3}\,E''$ (b) $2E' = \sqrt{3}\,E''$

(c) $E'' = \sqrt{3}\,E'$ (d) $2E'' = \sqrt{3}\,E'$

120. Match the properties in **Column A** with their respective unit in **Column B**

Column A	**Column B**
P. Surface resistivity	1. Ohm-cm
Q. Volume resistivity	2. $S\ cm^{-1}$
R. Coefficient of thermal expansion	3. Ohm
S. Electrical conductivity	4. K^{-1}

(a) P-1; Q-3; R-4; S-2

(b) P-2; Q-3; R-1; S-4

(c) P-3; Q-1; R-4; S-2

(d) P-3; Q-1; R-2; S-4

121. Match the following polymer product with its most appropriate processing technique

Polymer product	**Processing technique**
P. Bottle	1. Extrusion
Q. Blister packaging	2. Pultrusion
R. Reinforced electric poles	3. Injection blow molding
S. Sewage pipes	4. Thermoforming

(a) P-3; Q-4; R-1; S-2

(b) P-3; Q-4; R-2; S-1

(c) P-4; Q-3; R-2; S-1

(d) P-3; Q-2; R-4; S-1

122. Match the following additives to their respective functions

Additive	**Function**
P. p-phenylenediamine	1. Flame retardant
Q. Trixylyl phosphate	2. Impact modifier
R. Polybutadiene rubber	3. Nucleating agent
S. Talc	4. Antiozonant

(a) P-4; Q-2; R-1; S-3

(b) P-3; Q-1; R-2; S-4

(c) P-4; Q-1; R-3; S-2

(d) P-4; Q-1; R-2; S-3

123. Match the polymers with their characteristic infrared (IR) stretching frequency

Polymer	**IR stretch (cm^{-1})**
P. Polyurethane	1. ~2234
Q. Polyethylene	2. ~1151
R. Polysulfone	3. ~1720
S. Acrylonitrile-butadiene-styrene copolymer	4. ~2914

(a) P-4; Q-3; R-2; S-1

(b) P-3; Q-4; R-2; S-1

(c) P-3; Q-4; R-1; S-2

(d) P-3; Q-2; R-4; S-1

124. Match the following polymers to the most appropriate product

Polymer	**Product**
P. Expanded polystyrene	1. Motor bearings
Q. Polyether ether ketone	2. TV cabinet
R. Polycarbonate	3. Sound proof walls
S. Poly(butylene terephthalate)	4. Safety glass

(a) P-2; Q-1; R-4; S-3

(b) P-2; Q-4; R-1; S-3

(c) P-3; Q-1; R-2; S-4

(d) P-3; Q-1; R-4; S-2

125. Match the polymers to the polymerization method used for their synthesis

Polymer	**Polymerization method**
P. Linear low density polyethylene	1. Ring opening
Q. Nylon 6	2. Ziegler-Natta
R. Styrene-butadiene rubber	3. Condensation
S. Aromatic polyamide	4. Emulsion

(a) P-2; Q-1; R-4; S-3 (b) P-2; Q-1; R-3; S 4

(c) P-2; Q-3; R-4; S-1 (d) P-2; Q-4; R-1; S-3

126. If 5 g of a monodisperse polystyrene sample of molecular weight 10,000 g mol^{-1} is mixed with 15 g of another monodisperse polystyrene sample of molecular weight 20,000 g mol^{-1}, then the polydispersity of the resulting mixture is ______ *(rounded off to two decimal places)*.

127. For a polymer sample with a viscosity of 6×10^{11} poise, if the apparent plateau modulus of 3×10^{6} dyne cm^{-2} drops to zero above a certain temperature, the relaxation time of the polymer is ______ days (rounded off to one decimal place).

128. The thermal conductivity values of glass fiber and epoxy resin are 1.05 W m^{-1} K^{-1} and 0.25 W m^{-1} K^{-1}, respectively. The thermal conductivity of a glass fiber reinforced epoxy composite with a fiber content of 60% by volume along the fiber direction is ______W m^{-1} K^{-1} *(rounded off to two decimal places)*.

129. The tensile modulus of a thermosetting polyester resin and glass fiber are 3 GPa and 80 GPa, respectively. If a tensile stress of 110 MPa is applied along the fiber direction on a continuous uniaxially aligned glass fiber reinforced thermosetting polyester composite with a fiber content of 60% by volume, the resulting strain will be ______ $\times 10^{-3}$ *(rounded off to one decimal place)*.

130. The amount of low molecular weight plasticizer with a T_g of –60 °C that must be added to nylon 6 to reduce its T_g from 50 °C to 30 °C is ______% *(rounded off to nearest integer)*.

131. The enthalpy of fusion for a polymer is found to decrease from 135.6 J g^{-1} to 120 J g^{-1} after five years of use. If the enthalpy of fusion of the same polymer with 100% crystallinity is 290 J g^{-1}, then the loss in crystallinity after five years is ______ % *(rounded off to one decimal place)*.

G : FOOD TECHNOLOGY

Q.132 – Q.140 Carry ONE mark Each

132. Which among the given options truly depict the lines 1 and 2 in the figure below with respect to the effect of heat processing on food?

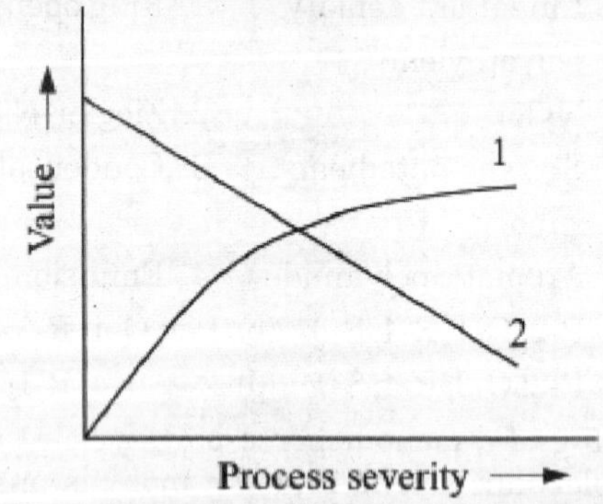

(a) 1-Safety, 2-Quality (b) 1-Yield, 2-Safety
(c) 1-Yield, 2-Quality (d) 1-Quality, 2-Safety

133. Homogenization of milk leads to disintegration of fat globules by
(a) Turbulence and pasteurization
(b) Pasteurization and cavitation
(c) Pasteurization and pressurization
(d) Turbulence and cavitation

134. The lowest water activity (a_w) supporting the growth of Staphylococcus aureus in food under aerobic condition is
(a) 0.98 (b) 0.91
(c) 0.89 (d) 0.86

135. Cultures used in industrial production of yogurt are
(a) *Lactococcus lactis subsp. lactis*
(b) *Streptococcus thermophilus*
(c) *Leuconostoc mesenteroides subsp. cremoris*
(d) *Lactobacillus delbrueckii subsp. bulgaricus*

136. In a dairy plant, spray drying technology is used to produce whey powder. The rate of spray drying depends on
(a) Temperature of the incoming air
(b) Shape of the cyclone separator
(c) Diameter of the whey droplet
(d) Heat transfer coefficient of hot air

137. The parboiling of paddy results into
(a) Increase in the milling losses
(b) Increase in the nutritional value of rice
(c) Increase in the head rice recovery
(d) Increase in the broken rice percentage

138. One hundred kg paddy is dried from 18% wet basis to 13% wet basis moisture content. The amount of water removed (in kg) from the paddy is ______ *(round off to one decimal place)*.

139. The radius of a centrifuge bowl is 0.1 m and is rotating at 850 revolutions per minute. The centrifugal force developed in terms of gravity force (g-force) is ______ *(round off to two decimal places)*.

Given: Acceleration of gravity (g) = 9.81 m s^{-2} and $\pi = 3.14$

140. In a canning industry, the total process time (F_0) was calculated as 3 min. If each can contains 20 spores having decimal reduction time of 1.6 min, the probability of spoilage would be ______ in 100 cans *(round off to the nearest integer)*.

Q.141 - Q.153 Carry TWO marks Each

141. Match the edible oil refining stages given in Column I with their respective functions in Column II

Column I	Column II
P. Degumming	1. Separation of waxes
Q. Neutralization	2. Removal of pigments
R. Bleaching	3. Removal of phosphatides
S. Winterization	4. Removal of free fatty acids

(a) P-3, Q-2, R-1, S-4
(b) P-2, Q-1, R-3, S-4
(c) P-3, Q-4, R-2, S-1
(d) P-3, Q-1, R-2, S-4

142. Make the correct pair of food packaging technology given in Column I with operating principle or description in Column II.

Column I	Column II
P. Aseptic packaging	1. Control of the concentration of O_2 and CO_2 inside the package
Q. Active packaging	2. Create a skin tight package wall
R. Modified atmosphere packaging	3. Independent sterilization of food and packaging material and packaging under sterile environment
S. Vacuum packaging	4. Makes non-passive contribution to product development

(a) P-3, Q-4, R-1, S-2
(b) P-3, Q-2, R-1, S-4
(c) P-1, Q-4, R-3, S-2
(d) P-3, Q-1, R-4, S-2

143. Which of the following is not a caramel flavour producing compound?
(a) 3-Hydroxy-2-methylpyran-4-one
(b) 2H-4-Hydroxy-5-methylfuran-3-one
(c) 3-Hydroxy-2-acetylfuran
(d) p-Amino benzoicacid

144. Match the size reduction equipment in Column I with the method of operation in Column II.

Column I	Column II
P. Hammer mill	1. Compression
Q. Burr mill	2. Impact
R. Crushing rolls	3. Cutting
S. Rotary knife	4. Attrition

(a) P-2, Q-4, R-1, S-3
(b) P-3, Q-1, R-2, S-4
(c) P-4, Q-1, R-2, S-3
(d) P-3, Q-4, R-2, S-1

145. Most commonly used refrigerant in direct immersion freezing of food is
(a) Monochlorodifluoromethane
(b) Dichlorodifluoromethane
(c) Liquid nitrogen
(d) Freon

146. Which among the following are ω-6 poly unsaturated essential fatty acids?
(a) 18:2 Linoleic acid
(b) 18:3 α -Linolenic acid
(c) 18:3 α -Linolenic acid
(d) 20:4 Arachidonic acid

147. Which among the following statements are true with respect to protein denaturation?
(a) There may be an increase in α-helix and β-sheet structure
(b) It is an irreversible process
(c) When fully denatured, globular proteins resemble a random coil
(d) The peptide bonds are broken

148. Identify the correct pair(s) of milling equipment and the grain for which it is used.
(a) Mist polisher-Rice
(b) Break roll-Wheat
(c) Rubber roll-Pigeon pea
(d) Beall degermer-Maize

149. Which among the following expression(s) is/are correct?

(a) Reynolds number = $\dfrac{\text{Density} \times \text{Velocity} \times \text{Characteristic dimension}}{\text{Viscosity}}$

(b) Nusselt number = $\dfrac{\text{Convective heat transfer coefficient} \times \text{dimension}}{\text{Thermal conductivity of solid}}$

(c) Schmidt number = $\dfrac{\text{Kinematic viscosity of fluid}}{\text{Diffusivity}}$

(d) Biot number = $\dfrac{\text{Convective heat transfer coefficient} \times \text{Characteristic dimension}}{\text{Thermal conductivity of fluid}}$

150. In sieve analysis of coffee powder, the particle size distribution is given below

Number of particles	Mean particle size (μm)
5	40
8	30
50	20
90	17.5
148	12.5
10	10

The Sauter mean diameter (in μm) of the coffee powder is ______
(*round off to one decimal place*).

151. In a dairy processing plant, milk enters a 30 m long and 2 cm diameter tube at 60 °C and leaves at 57 °C. The total heat loss over the tube length is 381.15 W. The specific heat capacity, density, and viscosity of milk are 3.85 kJ kg^{-1} K^{-1}, 1020 kg m^{-3}, and 1.20 cP, respectively. The Reynolds number for the flow is ______ *(round off to the nearest integer)*.

Given: $\pi = 3.14$

152. Apple juice flows through a steel pipe having thermal conductivity of 50 W m^{-1} K^{-1}. The outer surface of pipe is exposed to ambient environment. The inside diameter and thickness of the pipe are 3 cm and 1.5 cm, respectively. The overall heat transfer coefficient based on inside area is 25 W m^{-2} K^{-1}. If the internal convective heat transfer coefficient is 30 W m^{-2} K^{-1}, the external convective heat transfer coefficient (in W $m^{-2}K^{-1}$) will be ______ *(round off to two decimal places)*.

153. The dry bulb temperature and relative humidity of air inside a storage chamber are 37 °C and 50%, respectively. The saturation pressure of water vapour at 37 °C and barometric pressure are 6.28 kPa and 101.32 kPa, respectively. The humidity ratio of air inside the chamber is ______ kg water (kg dry air)$^{-1}$ *(round off to three decimal places)*.

Given: Molecular weight of water vapour and dry air are 18.02 g mol^{-1} and 28.97 g mol^{-1}, respectively.

H : ATMOSPHERIC AND OCEANIC SCIENCES

Q.154 - Q.162 Carry ONE mark Each

154. The figure shows a schematic of vertical profiles of concentrations of two gases P and Q in the atmosphere near a coastal station. The correct pair representing P and Q, respectively, is

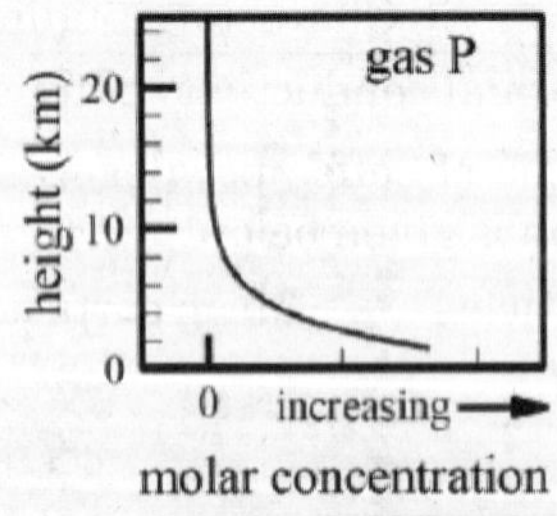

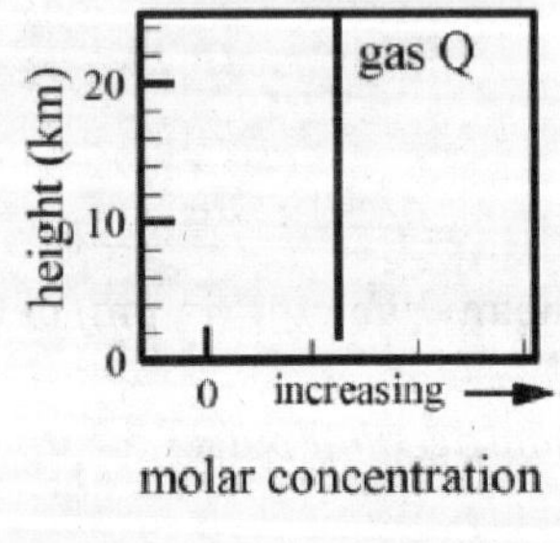

(a) water vapor and CO_2

(b) O_3 and water vapor

(c) CO_2 and O_3

(d) N_2 and O_2

155. A form of momentum equation for an incompressible fluid is

$$\underset{(i)}{\rho\frac{DV}{Dt}} = \underset{(ii)}{-\nabla p} + \underset{(iii)}{\mu\nabla^2 V} + \underset{(iv)}{B}$$

where ρ is density, V is velocity, t is time, p is pressure, μ is viscosity and B represents body force per unit volume. The dimension of term (iii) is (M, L and T stand for mass, length and time, respectively).

(a) $[L]^1 [T]^{-2}$

(b) $[M]^1 [L]^{-2} [T]^{-2}$

(c) $[M]^1 [L]^1 [T]^{-2}$

(d) $[M]^1 [L]^1 [T]^{-1}$

156. Tropical cyclones usually do not form close to the Equator primarily because

(a) sea surface temperature at the Equator is too cold.

(b) beta effect dissipates clouds.

(c) Coriolis force is too weak.

(d) vertical shear of the zonal wind is weak.

157. Which one of the following statements regarding equatorial under current (EUC) in the Pacific Ocean is correct?

(a) EUC flows from west to east.

(b) EUC flows from east to west.

(c) EUC flows from north to south.

(d) EUC flows from south to north.

158. Which one of the following statements is correct regarding the dominant energy balance in the troposphere in a tropical convergence zone?

(a) Shortwave heating balances longwave radiative cooling.

(b) Compressional heating balances radiative cooling.

(c) Radiative cooling balances heating due to viscous dissipation of kinetic energy.

(d) Condensational heating balances adiabatic cooling.

159. Which one of the following processes is primarily responsible for the poleward transport of energy in the midlatitude troposphere?

(a) atmospheric tides

(b) baroclinic waves

(c) gravity waves

(d) turbulence in the boundary layer

160. Which of the following feature(s) characterize the seasonal mean flow in the upper troposphere near 200 hPa level over the Tibetan Plateau during the boreal summer?

(a) cyclonic (b) anticyclonic

(c) irrotational (d) divergent

161. The Rossby number of a synoptic system with a length scale of 1000 km, characteristic velocity scale of 10 m s^{-1} at a latitude where the Coriolis parameter equals 10^{-4} s^{-1}, is _____. (Round off to two decimal places)

162. The ratio of scattering efficiency of red light of wavelength 0.65 μm to blue light of wavelength 0.45 μm by air molecules in the atmosphere is_____. (Round off to two decimal places)

Q.163 – Q.175 Carry TWO marks Each

163. An unsaturated moist air parcel undergoes adiabatic ascent in atmosphere without mixing with surrounding air. Air is so clean that there is no possibility for heterogeneous nucleation. Which one of the following plots depicts the vertical variation of water vapor pressure (shown as continuous line) and saturation water vapor pressure (shown as dotted/dashed line) of the parcel?

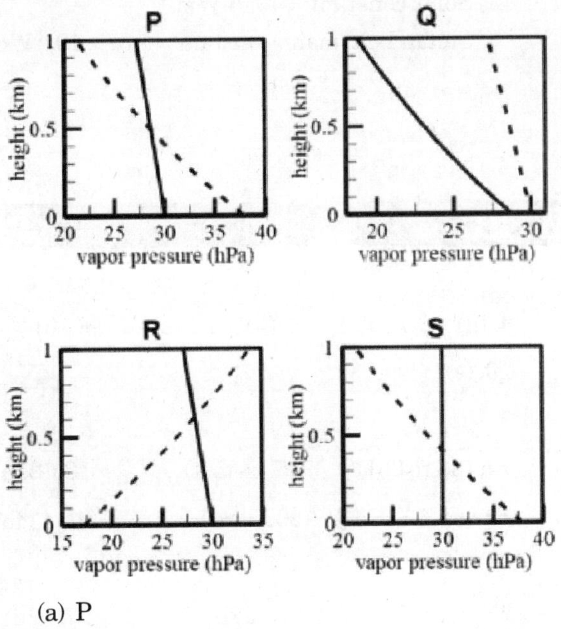

(a) P

(b) Q

(c) R

(d) S

164. A fluid is in solid body rotation in a cylindrical container of radius R rotating with an angular velocity $\Omega = (0, 0, \Omega)$. The circulation per unit area around a circular loop in the horizontal plane of radius r (r < R), whose center coincides with the axis of rotation is

(a) 2Ω (b) Ω^2

(c) $\Omega/2$ (d) $\Omega/4$

165. Consider a layer of atmosphere where temperature increases with height. If the concentration of a vertically well-mixed greenhouse gas suddenly increases in this layer, then an immediate consequence is that

(a) infrared radiation leaving the top of the layer decreases.

(b) infrared radiation leaving the top of the layer increases.

(c) infrared radiation leaving the top of the layer remains unchanged.

(d) the layer becomes optically thinner to infrared radiation.

166. Consider an atmosphere where the mole fractions of N_2, Ar and CO_2 are 7.81×10^{-1}, 9.34×10^{-3} and 4.05×10^{-4}, respectively. This atmosphere exchanges gases with sea water below having temperature and salinity of 20°C and 35 psu, respectively. In the absence of biological and chemical activity, relative concentrations of dissolved gases in the surface sea water at equilibrium are ordered as

(a) $[N_2] > [Ar] > [CO_2]$ (b) $[CO_2] > [N_2] > [Ar]$

(c) $[N_2] > [CO_2] > [Ar]$ (d) $[Ar] > [CO_2] > [N_2]$

167. Gravitational forces exerted by the Sun and the Moon are mainly responsible for ocean tides. Which of the following statement(s) regarding ocean tides is/are correct?

(a) Tidal amplitude corresponding to diurnal period is larger than that of the semi-diurnal period.

(b) Diurnal time period of lunar forced tides is longer than that of the solar forced tides.

(c) Tidal amplitudes are larger during a solar eclipse compared to that during a lunar eclipse.

(d) Tides are absent during equinoxes.

168. Which of the following statement(s) is/are true about northern hemisphere tropical cyclones?

(a) They have a warm core.

(b) Their low-level flow is cyclonic.

(c) Strong wind shear in the vertical is required for their intensification.

(d) They are characterized by upper-level divergence.

169. In gradient wind balance, which of the following statement(s) is/are true for flow around a region of low pressure in the northern hemisphere?

(a) The flow is clockwise.

(b) The flow is anti-clockwise.

(c) The wind speed is faster than the geostrophic wind.

(d) The wind speed is slower than the geostrophic wind.

170. Which of the following statement(s) is/are true regarding biogeochemical cycle in the ocean?

(a) Shutdown of the biological pump in the ocean would have resulted in higher CO_2 concentration in the atmosphere compared to present-day.

(b) If atmospheric CO_2 concentration increases, solubility pump would lead to a decrease in dissolved inorganic carbon in the ocean.

(c) All carbon sequestered by marine photosynthesis settles down on the ocean floor as organic matter.

(d) Calcification (the process of making shells and skeletons) by marine organisms in the surface ocean layer would lead to an increase in the surface ocean CO_2.

171. Consider the atmosphere to be a heat engine, which converts absorbed radiation to kinetic energy of winds. Let the global mean radiation absorbed be 200 Wm^{-2}.

In steady-state, if the global mean kinetic energy dissipation is 10 Wm^{-2}, then the efficiency of the atmospheric heat engine is ______%.

(Round off to one decimal place)

172. A drifter on the surface of the ocean performs inertial oscillation. The speed of the drifter is 2 m s^{-1} and the Coriolis parameter at the latitude is 2×10^{-4} s^{-1}. The radius of the inertial oscillation is ______ km. (Round off to the nearest integer)

173. Consider a tornado in cyclostrophic balance. The tangential wind speed at a radial distance of 500 m from the center of the tornado is ______ m s^{-1}, if the pressure gradient at that location in the radial direction is 5 N m^{-3}. Assume the density of air to be 1 kg m^{-3}. (Round off to the nearest integer)

174. Consider two weather stations A and B having the same altitude. Station B is 5 km north of Station A and is always 2 K warmer than Station A. A steady northerly wind blows at 1 m s^{-1}. The change in temperature at Station A in 2 hours is ______ K. (Round off to one decimal place)

175. Assume the Earth is in radiative equilibrium with effective radiative temperature of 255 K. If the planetary albedo increases by 0.05, then the effective radiative temperature of the planet will be ______ K. (Round off to the nearest integer)

Given:

Solar constant = 1370 Wm^{-2}

Stefan Boltzmann constant = 5.67×10^{-8} $Wm^{-2} K^{-4}$

ANSWERS

General Aptitude (GA)

1. (c)	**2.** (b)	**3.** (c)	**4.** (d)	**5.** (a)	**6.** (b)
7. (d)	**8.** (d)	**9.** (d)	**10.** (c)		

A : Engineering Mathematics

11. (c)	**12.** (c)	**13.** (b)	**14.** (36 to 36)	**15.** (4 to 4)	**16.** (2 to 2)
17. (0.066 to 0.068)		**18.** (c)	**19.** (a)	**20.** (2 to 2)	**21.** (3 to 3)

B : Fluid Mechanics

22. (b)	**23.** (a)	**24** (a)	**25.** (a,b)	**26.** (a,d)	**27.** (a,d)
28. (b,d)	**29.** (a,b,c)	**30.** (6.95 to 7.20)	**31.** (a)	**32.** (c)	**33.** (d)
34. (c)	**35.** (c,d)	**36.** (374 to 376)	**37.** (MTA)	**38.** (5.60 to 5.70)	
39. (0.24 to 0.26)	**40.** (1.380 to 1.390)		**41.** (0.74 to 0.78)	**42.** (5.1 to 5.2)	
43. (0.39 to 0.41)					

C : Materials Science

44. (b) **45.** (c) **46.** (c) **47.** (a) **48.** (c) **49.** (d)
50. (b,c) **51.** (4232 to 4249) **52.** (2.20 to 2.30) **53.** (c) **54.** (c) **55.** (d)
56. (c) **57.** (c) **58.** (a) **59.** (a,c) **60.** (0.5 to 0.5) **61.** (9.7 to 11.2)
62. (0.057 to 0.062) **63.** (0.45 to 0.47) **64.** (1.40 to 1.43) **65.** (5.64 to 5.73)

D : Solid Mechanics

66. (a) **67.** (c) **68.** (a) **69.** (b) **70.** (a) **71.** (b)
72. (c) **73.** (1.9 to 2.1) **74.** (19 to 21) **75.** (a) **76.** (d) **77.** (a)
78. (a) **79.** (b) **80.** (a) **81.** (295 to 305) **82.** (39 to 41) **83.** (0.9 to 1.1)
84. (79 to 81) **85.** (5990 to 6010) **86.** (123 to 127) **87.** (148 to 152)

E : Thermodynamics

88. (c) **89.** (a) **90.** (b) **91.** (b) **92.** (750.00 to 756.00)
93. (79.00 to 80.00) **94.** (254.00 to 255.00) **95.** (25 to 25) **96.** (300 to 300)
97. (c) **98.** (a) **99.** (b) **100.** (a) **101.** (b) **102.** (c)
103. (0.20 to 0.24) **104.** (132.00 to 134.00) **105.** (4.50 to 4.65) **106.** (0.0160 to 0.0165)
107. (0.13 to 0.15) **108.** (15 to 17) **109.** (21.00 to 22.00)

F : Polymer Science and Engineering

110. (b) **111.** (c) **112.** (a) **113.** (c) **114.** (d) **115.** (b)
116. (a,c,d) **117.** (b,c,d) **118.** (2 to 2) **119.** (a) **120.** (c) **121.** (b)
122. (d) **123.** (b) **124.** (d) **125.** (a) **126.** (1.04 to 1.14) **127.** (2.1 to 2.5)
128. (0.71 to 0.75) **129.** (2.1 to 2.4) **130.** (12 to 14) **131.** (5.2 to 5.6)

G : Food Technology

132. (a) **133.** (d) **134.** (d) **135.** (b,d) **136.** (a,c,d)
137. (b,c) **138.** (5.5 to 6.0) **139.** (79.50 to 81.50) **140.** (25 to 28) **141.** (c)
142. (a) **143.** (d) **144.** (a) **145.** (c) **146.** (a,c,d) **147.** (a,c)
148. (a,b,d) **149.** (a,c) **150.** (19.0 to 20.5) **151.** (1700 to 1800) **152.** (73.00 to 84.00)
153. (0.017 to 0.021)

H : Atmospheric and Oceanic Sciences

154. (a) **155.** (b) **156.** (c) **157.** (a) **158.** (d) **159.** (b)
160. (b,d) **161.** (0.08 to 0.12) **162.** (0.22 to 0.24) **163.** (a) **164.** (a) **165.** (b)
166. (c) **167.** (b,c) **168.** (a,b,d) **169.** (b,d) **170.** (a,d)
171. (4.9 to 5.1) **172.** (9 to 11) **173.** (49 to 51) **174.** (2.8 to 3.0) **175.** (249 to 252)

EXPLANATIONS

GENERAL APTITUDE

1. Used to show a reason or cause:
The movies was funny and I could not help laughing.
The movie was so funny that he couldn't control his/her laughing , means he laughed.

2. As given that,

$$x : y : z = \frac{1}{2} : \frac{1}{3} : \frac{1}{4} = 6 : 4 : 3$$

Let us consider,

$$\left.\begin{aligned} x &= 6K \\ \text{i.e } y &= 4K \\ z &= 3K \end{aligned}\right\}$$

Then $\frac{x+z-y}{y} = \frac{9K-4K}{4K} = \frac{5}{4} = 1.25$

3. According to the question :

$$\left(\frac{3+x}{4+x}\right) = \left(\frac{15-x}{17-x}\right)$$

$\Rightarrow 51 - x^2 + 14x = 60 - x^2 + 11x$

By putting,
$x = 1, 2, 3, 4$
So, only $x = 3$ is satisfying.

4. According to the question,

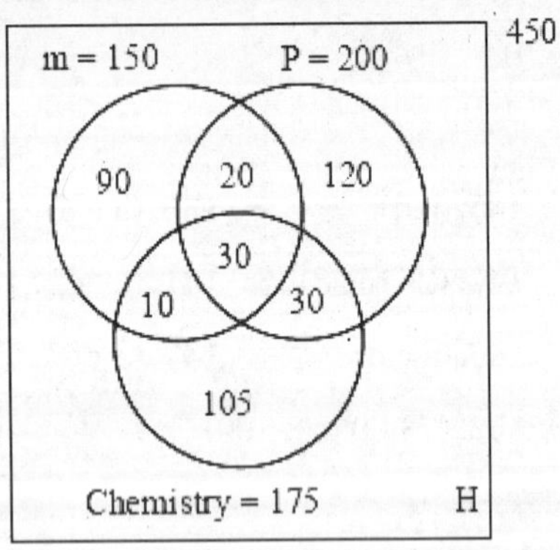

As we know that,

$$n(M \cup P \cup C) = n(M) + n(P) + n(C) - n(M \cap P) - n(P \cap C) - n(C \cap M) + n(M \cap P \cap C)$$

$= 150 + 200 + 175 - 50 - 60 - 40 + 30$

$= 405$

n(Humanities) $= 450 - 405 = 45$

5. Mirror image of the figure is in options (a).

6. More people visit but don't spend that is why the low revenue generation through sales in shops in these malls.

7. Let capitals invested by three friends for 6 months are;

$c_1 = 3x$
$c_2 = 4x$
$c_3 = 5x$

For next 6 minutes

300,	400,	500(say)
↓ 10%	↓ 15%	↓ 20%
330,	460,	600

Total investment;

$\Rightarrow [3 + (3 \times 1.1)] : [4 + (4 \times 1.15)] : [5 + (5 \times 1.2)]$

$= 6.3 : 8.6 : 11$

So, ratio is 63 : 86 : 110.

8. As given that,

Line $L_1 : 2x - 3y = 5$

Slope $= m_1 = -\frac{b}{a} = \frac{3}{2}$

Line $L_2 : 3x + 2y = 8$

Slope $m_2 = -\frac{2}{3}$

Line $L_3 : 4x - 6y = 5$

Slope $m_3 = \frac{6}{4} = \frac{3}{2}$

Line $L_4 : 6x - 9y = 6$

Slope $= m_4 = \frac{9}{6} = \frac{3}{2}$

$\Rightarrow \frac{-3}{2} \times \frac{2}{3} = -1$

L_4 is parallel to L_3 and L_4 is perpendicular to L_2

$$= \frac{3}{2} \times \frac{2}{3} = -1$$

Hence, slopes are equal.

9. Statement 1; probability (1) :

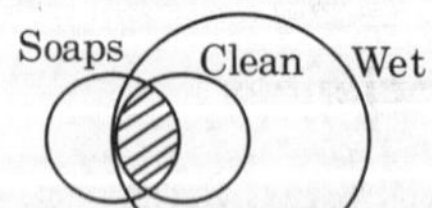

Statement 2; probability (2) :

According to conclusion (I) & (III) followed in both possibilities.

10. According to the question :

P faces east and turned 72°C Angle.

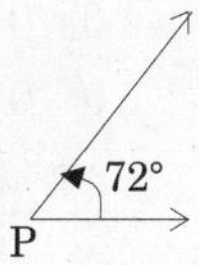

Travelled 10 cm then 144° CW

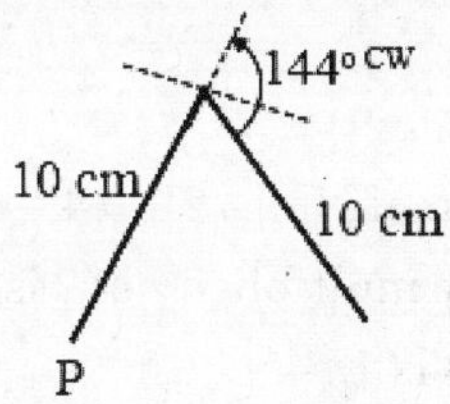

From above two steps, we can conclude

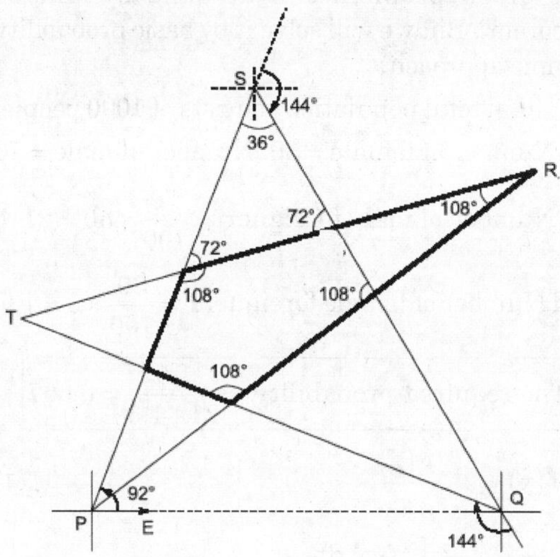

$\Rightarrow$ PS = SQ = QT = TR = RP = 10 cm.

Hence, option (c) is correct

A: ENGINEERING MATHEMATICS

11. The given limit is of the indeterminate form $\left(\frac{0}{0}\right)$

$\therefore$ By using L' hospital rule.

$$= \lim_{x\to 0} \frac{\frac{d}{dx}\int_2^{2+x}\left(t+\sqrt{t^2+5}\right)dt}{\frac{d}{dx}(x)}$$

By applying Leibtiz rule :

$$= \lim_{x\to 0} \frac{\left[(2+x)+\sqrt{(2+x)^2+5}\right]\times\frac{d}{dx}(2+x) + \left[2+\sqrt{4+5}\right]\times\frac{d}{dx}(2) - \int_2^{2+2x} 0.dt}{1}$$

$$= \lim_{x\to 0}\left[2+x+\sqrt{(2+x)^2+5}\right]\times 1 = 2+3 = 5$$

12. We need to find $|f'(-1+i)|^2$.

Given that f(x) is analytic. So it is differentiable (infinitely) and 'along any path'

Let us take y = k (constant)

$$\therefore\ f'(z) = f'(u+iv) = \left(\frac{\partial u}{\partial x} + i\frac{\partial v}{\partial x}\right)$$

$$= (y^3 - 3x^2y) + i(x^3 + 0 - 3xy^2)$$

Now $f'(-1+i) = (1-3) + i(-1+3) = -2 + 2i$

$\therefore\ u(x, y) = xy^3 - yx^3$

$$\therefore\ v(x,y) = \frac{x^4}{4} + \frac{y^4}{4} - \frac{3}{2}x^2y^2$$

$$\therefore\ |f'(-1+i)|^2 = \left(\sqrt{(-2)^2+2^2}\right) = 8$$

13. The given PDE is a homogenous liner differential equation of 2nd order.

The standard form is

$$A\frac{d^2u}{dx} + B\frac{\partial^2 u}{\partial x\,\partial y} + C\frac{\partial^2 u}{\partial y^2} + D\frac{\partial u}{\partial x} + E\frac{\partial u}{\partial y} + Fu$$

$$= G(x, y) = 0$$

Where A, B, C, D, E, F all are fns of x, y.

Comparing is get

$A = x + 2,\ B = (x + 4),\ C = 2(y - 1),$

The condition for parabolic PDE, of $B^2 - 4AC = 0$

i.e. $4(x + 4)^2 - 4(x + 2) \times 2(y - 1) = 0$

$\Rightarrow (x + y)^2 = 2(x + 2)(y - 1)$

$\Rightarrow x^2 + 2xy + y^2 = 8xy + 4y - 2x - 4$

$\Rightarrow x^2 + y^2 + 2x - 4y + 4 = 0$

$\Rightarrow x^2 + 2x + 1 + y^2 - 4y + 4 = 1$

$\Rightarrow (x + 1)^2 + (y - 2)^2 = 1$

$\therefore$ a = 1, b = 2, c = 1

14. Green's theorem,

$$\oint_P < Mdx + N\,dy = \iint\left(\frac{\partial N}{\partial x} - \frac{\partial M}{\partial y}\right)dx\,dy$$

Comparing with the given equation,

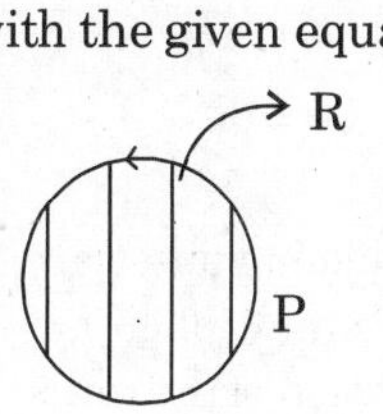

$M = 3y + e^{x^{\sin x}}$

$$\frac{\partial M}{\partial y} = 3$$

$N = 7x + \sqrt{e^v + 2}$

$$\Rightarrow \frac{\partial N}{\partial x} = 7$$

I (given integral)

$= \iint_R (7-3)\,dx\,dy,$

where $R : x^2 + y^2 = 9$

$= 4\iint_R dx\,dy = 4 \times \pi \times 3^2 = 36\pi$

$\therefore \boxed{\propto = 3b}$

15. Homogeneous linear differential equation → Cauchy Euler form

Let $z = \log x \Rightarrow x = e^z$ & $D \equiv \frac{d}{dz}$

Using this transformation,

$x^2\frac{d^2y}{dx^2} = D(D-1), x\frac{dy}{dx} = D$

The given equation becomes

$D(D-1) - 2D + 2 = 0$

$\Rightarrow [D^2 - 3D + 2]\,y = 0$

Axillary equation

$m^2 - 3m + 2 = 0$

$\Rightarrow (m-1)(m-2) = 0 \Rightarrow m = 1, 2$

$\therefore w(y_1, y_2) = \begin{vmatrix} c_1x & c_2x^2 \\ c_1 & 2c_2x \end{vmatrix} = c_1c_2x^2$

$w(y_1, y_2)(1) = c_1c_2$

$\Rightarrow 1 = c_1\,c_2$ (using the given combination)

$\therefore w(y_1, y_2)(x) = x^2.$

$\boxed{\therefore w(y_1, y_2)(y) = 4}$

16. 1st let us find the eigen values of the given matrix by solving for characteristic equation

$|A - \lambda I| = 0$

$$\Rightarrow \begin{vmatrix} 2-\lambda & 0 & 1 & 1 \\ 1 & 2-\lambda & 5 & -5 \\ 0 & 0 & 3-\lambda & 0 \\ 0 & 0 & -1 & 3-\lambda \end{vmatrix} = 0$$

$$\Rightarrow (3-\lambda)^3 \begin{vmatrix} 2-\lambda & 0 & 1 \\ 1 & 2-\lambda & -5 \\ 0 & 0 & 3-\lambda \end{vmatrix} = 0$$

$\Rightarrow (3-\lambda)^2 \times (2-\lambda)^2 = 0$

$\Rightarrow \lambda = 2, 2, 3, 3$

$\therefore$ Two distinct eigen values are there 2, 3. Now the Geometric multiplicity of the eigen value λ is the dimension of the nulls base of $|A - \lambda x|$

$$\therefore |A - 2I| = \begin{bmatrix} 0 & 0 & 1 & 1 \\ 1 & 0 & 5 & -5 \\ 0 & 0 & 1 & 0 \\ 0 & 0 & 1 & 1 \end{bmatrix} \xrightarrow{R_1 - R_4} \begin{bmatrix} 0 & 0 & 1 & 1 \\ 1 & 0 & 5 & -5 \\ 0 & 0 & 1 & 0 \\ 0 & 0 & 0 & 0 \end{bmatrix}$$

$\downarrow R_{21} \rightarrow R_{23}$

$$\begin{bmatrix} 1 & 0 & 0 & 0 \\ 0 & 1 & 0 & 0 \\ 0 & 0 & 1 & 0 \\ 0 & 0 & 0 & 1 \end{bmatrix} \xleftarrow{R_1 + 5R_3} \begin{bmatrix} 1 & 0 & 0 & -5 \\ 0 & 0 & 1 & 0 \\ 0 & 0 & 0 & 1 \\ 0 & 0 & 0 & 0 \end{bmatrix}$$

$$\xleftarrow[R_3 - R_2]{R_1 - 5R_2} \begin{bmatrix} 1 & 0 & 5 & -5 \\ 0 & 0 & 1 & 0 \\ 0 & 0 & 1 & 1 \\ 0 & 0 & 0 & 1 \end{bmatrix}$$

$\therefore$ Rank of [A - 2I] = 3

$\therefore$ Dimension of N[A – 2I] = 4 – 3 = 1

$\therefore$ Geometric multiplicity of 2 is 1.

Similarly of 3 is 1.

$\boxed{\therefore \text{Req ans. } s = 1 + 1 = 2}$

17. The given problem can be done by "Bayes Theorem". But we will solve it by basic probability (simple approach)

Let, total total population consists of 1000 people.

$\therefore$ Number of female = 30, Number of male = 70

$\therefore$ Number of male foreigner $= \frac{70}{100} \times 30 = 21$

and Number of female foreigners $= \frac{30}{100} \times 5 = 1.5$

$\boxed{\therefore \text{The required probability} = \frac{1.5}{21 + 1.5} = 0.067}$

18. $\Rightarrow f(x) = 2 + \frac{g(x)}{x}$ ...(1)

$\Rightarrow f(x) = 2 + \frac{\int_1^x f(t)\,dt}{x}$

$\Rightarrow [f(x) - 2]x = \int_1^x f(t)\,dt$

$g(x) = \int_1^x f(t)dt$ (Given

$\therefore g(1) = 0$ ...(2)

from (1) by (2)

$f(1) = 2 + 0 = 2$ ...(3)

Differentiating both sides w.r.t x

$\Rightarrow x f'(x) + f(x) - 2 = \frac{d}{dx}\int_1^x f(t)\,dt$

By Leibniz rule

$\Rightarrow f(x) \times 1 - f(1) \times 0 + \int_1^x \frac{\partial}{\partial x} f(t)dt$

$\Rightarrow xf'(x) + f(x) - 2 = f(x)$

$\Rightarrow f'(x) = \frac{2}{x}$

$\Rightarrow f(x) = 2\log x + c$ ...(4)

$\Rightarrow c = 2$

$\therefore f(x) = 2 \log x + 2$

$\Rightarrow \boxed{f(2) = 2 + \log 4}$

19. Statement P is true.

Because it is a property of a symmetric matrix that the rank of a symmetric matrix is equal to the number of non-zero eigen value.

Statement Q is also true.

It is a straight forward result from the inequality, rank (AB) ≥ rank (A) + rank (B) – n. where $A_{m \times n} . B_{n \times p}$. Here AB = 0 so rank (AB) = 0

i.e. rank (A) + rank (B) – n ≤ 0

i.e. $\boxed{\text{rank}(A) + \text{rank}(B) \le n}$

Hence, both P and Q are correct.

20. $f(x, y) = 4xy - 2x^2 - y^4 + 1$

$f_x = 4y - 4x, f_{xx} = 4, f_{xy} = 4$

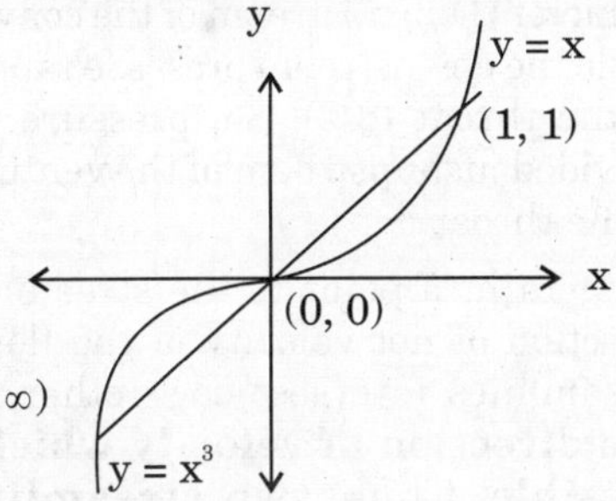

$f_y = 4x - 4y^3, f_{yy} = -12y^2$

For critical points,

$f_x = 0 \Rightarrow x = y$

$f_y = 0 \Rightarrow x = y^3 \Rightarrow x - x^3 = 0$

$\Rightarrow x(x^2 - 1) = 0 \Rightarrow x = 0, \pm 1$

$\therefore$ Critical points are (0, 0), (–1, –1), (1, 1)

Now we need to calcualte the Hesstan determinant

$$H - \begin{vmatrix} f_{xx} & f_{xy} \\ f_{yx} & f_{yy} \end{vmatrix} = f_{xx} \times f_{yy} - f_{xy}^2 \left(f_{xy} = f_{yx}\right)$$

for local minimum H > 0 (+ e),

$f_{xx} < 0$ (–ve)

$\therefore$ The Reqd. points are (1, 1), (–1, –1).

Hence 2 points

21. Given : $\int_{-1}^{1} f(x)dx \approx f(\infty) + \gamma f(\beta)$

We'll take f(x) = 1. f(x) = x′f(x) = x^2 consecutively

$\Rightarrow f(x) = 1$, gives $2 = 1 + \gamma \times 1$

i.e. $\gamma \approx 1$

$\Rightarrow f(x) = x$. as it is a add function so $\int_1^1 x\,dx = 0$

$\therefore \alpha + \gamma\beta = 0$ i.e. $\alpha = -\beta (\text{as } \gamma = 1)$

$\Rightarrow f(x) = x^2$ given, $\int_{-1}^{1} x^2\,dx = \alpha^2 + \beta^2$

$$\frac{2}{3} = 2\alpha^2 (\text{as } \alpha = -\beta)$$

$$\Rightarrow \alpha = \pm\frac{1}{\sqrt{3}}$$

$\therefore$ Let us take $\alpha = \frac{1}{\sqrt{3}}, \beta = -\frac{1}{\sqrt{3}}$

$\therefore \gamma + 3(\alpha^2 + \beta^2) - (\alpha^2 + \beta^2)$

$$= 1 + 3 \times \frac{2}{3} + \frac{1}{3\sqrt{3}} - \frac{1}{3\sqrt{3}} = 1 + 2$$

$= 3$

B : FLUID MECHANICS

22. As we know that

Vertical component of force exerted by liquid per unit length

$$\boxed{\begin{aligned} f_v &= mg \quad \text{or} \\ f_v &= \rho \times v \times g = v \times \rho g \end{aligned}}$$

As given that;

$\therefore$ weight of water supported by half of cylinder of unit length.

$$\text{length} = \frac{1}{2}$$

So, f_v = Area of cylinder × length × ρg

$$f_v = \frac{\frac{\pi}{4}D^2}{2} \times \rho g \times \phi$$

$$\boxed{f_v = \frac{\pi}{8}D^2 \times \rho g}$$

23. As we know that

$$\tau = \mu\frac{du}{dy}$$

According to the given figure :

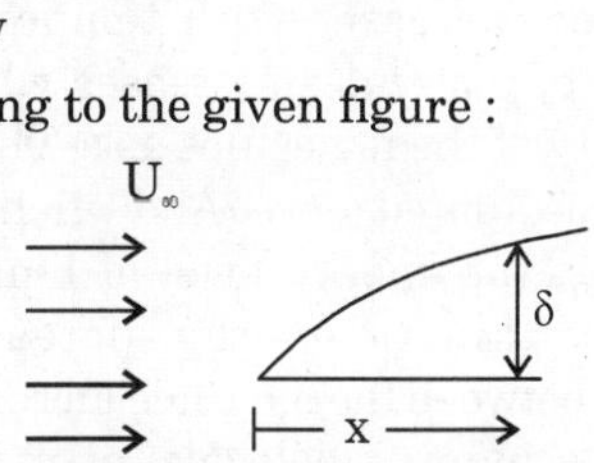

$$\tau_w = K\mu\frac{V_\infty}{\delta}$$

Where, K = 1 for linear

= 2 for parabolic

We also know that;

For Laminar flow

$$\delta = \frac{5x}{\sqrt{R_e}} : \tau_w \propto \frac{\sqrt{R_e}}{5x} \propto \frac{\sqrt{\rho v x}}{5x}$$

For Turbulent flow $\boxed{\tau_w \propto \frac{1}{\sqrt{x}}}$

So, $\boxed{\tau_w \propto x^{-1/2}}$

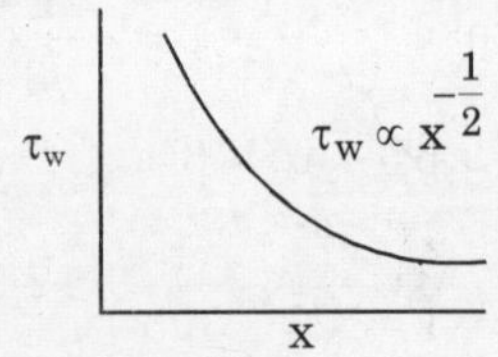

So, for the giving figure.

$\tau_A > \tau_B$ and $\tau_C = \tau_D$

24. (i) The Reynolds number **represents the ratio of inertial forces to viscous forces** (Fox and McDonald). Thus, a low Reynolds number means viscous drag dominates, and a high Reynolds number means pressure drag dominates.

(ii) Variation of wave drag with Froude number for a ship traveling through deep water.

(iii) Compressible flows have a Mach number **greater than 0.8**. The pressure strongly affects the density, and shocks are possible. Compressible flows can be either transonic $(0.8 < M < 1.2)$ or supersonic $(1.2 < M < 3.0)$. In supersonic flows, pressure effects are only transported downstream.

(iv) The **Weber number (We)** is a dimensionless number in fluid mechanics that is often useful in analyzing fluid flows where there is an interface between two different fluids, especially for multiphase flows with strongly curved surfaces.

25. Viscosity of gases increases with increase in temperature because the velocity of gas molecules increases with temperature hence collision of molecules increases which deceases the ability of the molecules as a whole to engage in the coordinated move. And when temperature increases, the energy level of liquid molecules increases (kinetic energy increases) and the distance between the molecule increases. It causes a decrease in inter-molecular attraction between them, which reduces viscosity.

26. A venturimeter is the device used for measuring the rate of a flow of a fluid flowing through a pipe. It consists of three parts

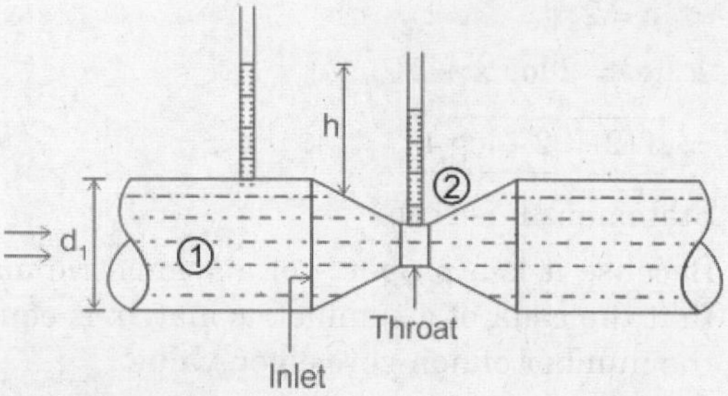

(a) A short converging part

(b) Throat, and

(c) Diverging part

It is based on the principle of Bernoulli's equation So, in the direction of flow, it consists of a converging section, a throat, and a diverging section.

There are two tappings on the venture-meter for pressure measurement; the upstream pressure tapping is located at a distance of one-half of pipe diameter (D/2) upstream of the convergent entry, while the downstream pressure tapping is located in the throat (d/2). So, pressure tappings are provided just upstream of the venture-meter and at the throat.

27. Tangent at a point to the streamline gives the direction of net velocity of the flow. If the two streamlines intersect each other, **it signifies two direction of velocity which cannot be possible**. Thus, **two streamlines cannot intersect each other**. Stream function has a **constant value along any streamline.** For a continuous flow (no sources or sinks), the volume flow rate across any closed path is equal to zero.

28. Velocity potential

$$\frac{\partial \varphi}{\partial x} = -u, \frac{\partial \varphi}{\partial y} = -v, \frac{\partial \varphi}{\partial z} = -w$$

$u = -3Ax^2,$

$v = 0, w = 0$

for incompressible

$$\frac{\partial u}{\partial x} + \frac{\partial v}{\partial y} + \frac{\partial w}{\partial z} = 0$$

$$\frac{\partial u}{\partial x} = -6Ax$$

$$\frac{\partial v}{\partial y} = 0; \frac{\partial w}{\partial z} = 0$$

for irrational flew

$$\frac{\partial v}{\partial y} = \frac{\partial v}{\partial x}$$

$0 = 0$

So, the local acceleration and convective acceleration

$$a_x = \underbrace{\frac{\partial u}{\partial t}}_{\text{local Accelereation}} + \underbrace{u\frac{\partial u}{\partial x} + v\frac{\partial u}{\partial y} + w\frac{\partial u}{\partial z}}_{\text{Convertive Acceleration}}$$

$a_x = 0 + (-3Ax^2)(-6Ax)$

$a_x = 18\,A^2x^3$

Hence, the flow has local and convective acceleration.

29. According to the question

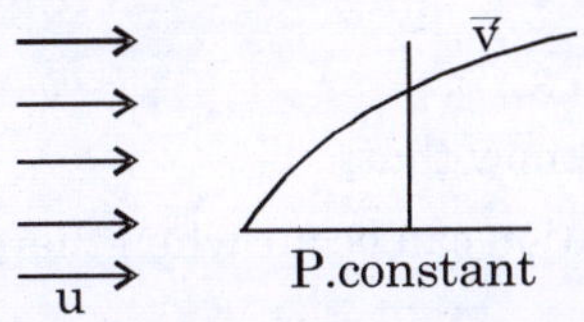

Along vertical line pressure constant.

$U, \frac{\partial u}{\partial x}$ & v are variable.

Hence, options (a, b & c) are the correct answer.

30. According to the question

Force by Jet = weight

So, mass flow rate × v = mg

or, ρAv × v = mg

By putting all values;

$$1000 \times \frac{\pi}{4} \times \left(\frac{5}{100}\right)^2 \times v^2 = 10 \times 10$$

$$v^2 = \frac{100 \times 4 \times 100^2}{1000 \times 25 \times \pi}$$

$\boxed{v = 7.13\ m/sec}$

31. According to the question,

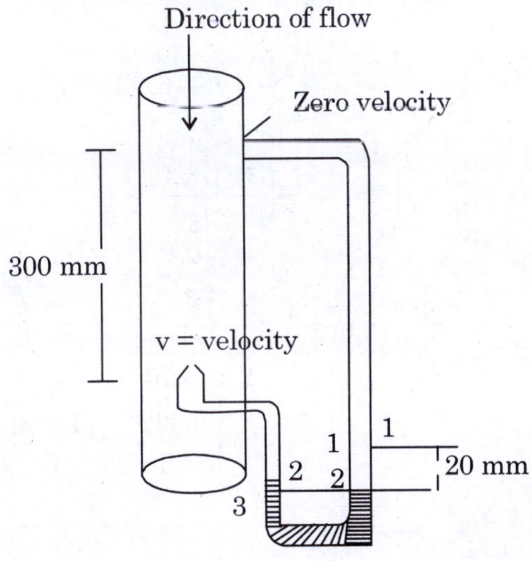

As we know that,

$$v = \sqrt{2g \times \Delta h}$$

$$\therefore\ \Delta h = x\left(\frac{\rho_m}{\rho_h} - 1\right)$$

$$= 0.02\left(\frac{11}{1} - 1\right)$$

$\boxed{\Delta h = 0.2}$

$$v = \sqrt{2 \times 10 \times 0.2}$$

$\boxed{v = 2\ m/s}$

32. As we know that,

$C = f(\rho, \lambda, \sigma)$

ρ & λ remain constant

$C' = f(\sigma)$

And, $C = \rho^a\,\lambda^b\,\sigma^c$

fundamental dimension;

$LT^{-1} = [ML^{-3}]^a\,[L]^b\,[HT^{-2}]^c$

Comparing $-1 = -2C$

$$C = \frac{1}{2}$$

So, $C = \sigma^{\frac{1}{2}}$

When new speed of propagation;

$$C' = f\left(\frac{\sigma}{2}\right)$$

$$\frac{C}{C'} = \left(\frac{\sigma}{\frac{\sigma}{2}}\right)^{\frac{1}{2}}$$

$$\frac{C}{C'} = 2^{\frac{1}{2}}$$

$$C' = \frac{C}{\sqrt{2}}$$

33. According to the question;

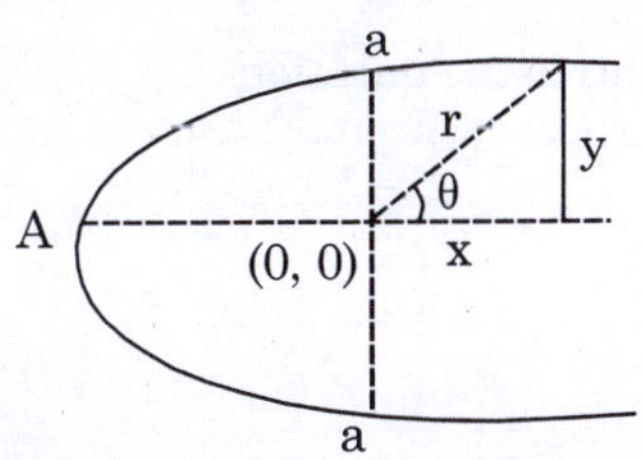

$x = r\cos\theta$

$y = r\sin\theta$

Now velocity profile

$$\varphi = u_x + \frac{m}{2\pi}\ln\sqrt{x^2+y^2} \quad (\because u_x = v_r\cos\theta)$$

$$\varphi = v_r\cos\theta + \frac{m}{2\pi}\ln\sqrt{r^2\cos^2\theta + r^2\sin^2\theta}$$

$$= v_r\cos\theta + \frac{m}{2\pi}\ln r \quad \text{...(i)}$$

Velocity in $U_r = \frac{\partial\phi}{\partial r} = U\cos\theta + \frac{m}{2\pi r}$.

for stagnation point $U_r = 0$ & $V_\theta = 0$

$$U\cos\theta + \frac{m}{2\pi r} = 0$$

$$U\cos\theta = -\frac{m}{2\pi r} \quad \text{...(ii)}$$

also $\frac{1}{r} = \frac{\partial\varphi}{\partial\theta} = V_\theta = V\sin\theta + 0$

$V_\theta = 0 : \theta = 180°$ ...(iii)

$u \neq 0$

Now, put θ value in eq (ii)

$$V\cos 180 = -\frac{m}{2\pi r}$$

$V = -\frac{m}{2\pi r}$ at point A

Point A is an stagnation steam live so stream function on a Aa' is constant and is given by

$$\varphi = U_y + \frac{m}{2\pi}\theta \quad \text{...(iv)}$$

At stagnation point A $\theta = \pi$ & $y = 0$

$$4 = \frac{m}{2\pi}\times\pi = \frac{m}{2}$$

As stream function is constant on aAa' from (iv)

$$\frac{m}{2} = U_y + \frac{m}{2\pi}\times\theta$$

At point a.$\theta = \frac{\pi}{2}$

$$\frac{m}{2} = U_y + \frac{m}{2\pi}\times\frac{\pi}{2}$$

$$y = \frac{m}{4U}$$

So, the distance between $aa' = \frac{m}{4U}\times 2 = \frac{m}{2U}$

34. According to the question

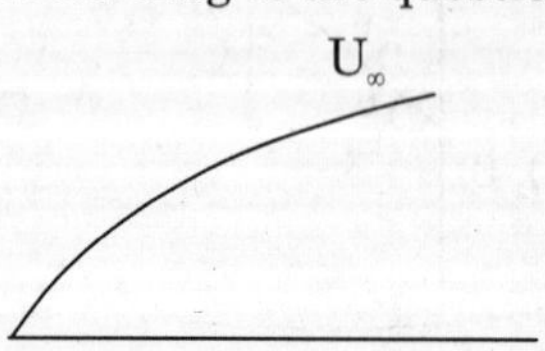

∴ Linear velocity profile;

$$\frac{u}{U} = \frac{y}{\delta}$$

∴ Momentum thickness;

$$\theta = \int_0^\delta \frac{u}{U}\left[1 - \frac{\mu}{U}\right) dy$$

$$\theta = \int_0^\delta \frac{y}{\delta}\left[1 - \frac{y}{\delta}\right] dy$$

So, $\theta = \left[\frac{y^2}{2\delta} - \frac{y^3}{3\delta^2}\right]_0^\delta$

$$\theta = \frac{\delta}{6}$$

$$\frac{\theta}{\delta} = \frac{1}{6}$$

35. As we know that,

Separation can occur when there is

$$\left(\frac{\partial v}{\partial y}\right)_{y=0} = 0$$

& this is possible when there is increase in area in the direction of flow.

36. $\frac{dp}{dx} = ?$

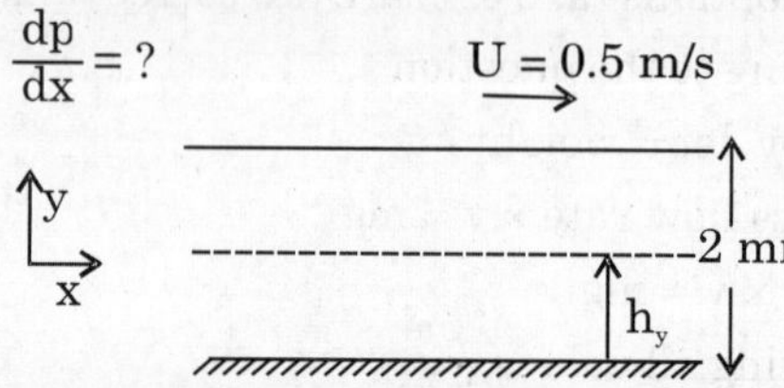

$Q_{net} = 0$

As we know that

Coutle flow;

$$U = \frac{U_y}{h} + \frac{1}{2\mu}\left(\frac{\partial P}{\partial x}\right)\left(y^2 - h_y\right)$$

We also know that

$$\therefore \quad Q = \int_0^{0.002} U\,d_A$$

So, $Q = \int \frac{U_y}{h} + \frac{1}{2\mu}\left(\frac{\partial P}{\partial x}\right)\left(y^2 - h_y\right) 1\times dy$

$$Q = \left|\frac{U}{h}\cdot\frac{y^2}{2}\right|_0^{0.002} + \frac{1}{2\mu}\left(\frac{\partial P}{\partial x}\right)\left[\frac{y^3}{3} - \frac{h_y{}^2}{2}\right]_0^{0.002}$$

$(\because Q_{net} = 0)$

$$0 = \frac{0.5}{0.002}\times\frac{(0.002)^2}{2} + \frac{1}{2\times 5\times 10^{-4}}\times\frac{\partial P}{\partial x}\times\left[\frac{0.002^3}{3} - \frac{(0.002)}{2}\right]$$

$$\boxed{\frac{\partial p}{\partial x} = 375.093\ \text{Pa/m}}$$

37. As given that

$$C_f = \frac{0.06}{(Re_x)^{0.2}}; R_{ex} = \frac{\rho v L}{\mu}$$

$$= \frac{1.2 \times 10 \times 1}{1.83 \times 10^{-5}} = 655737$$

$$C_f = \frac{0.06}{(655737)^{0.2}}$$

So, $C_f = 0.004119$

We know that,

$$\text{Drag force} = \frac{1}{2} \times \rho \times A \times C_f \times V^2$$

$$= \frac{1}{2} \times 1.2 \times 1 \times 1 \times 0.004119 \times 10^2$$

$$= 0.247 \text{ N}$$

38. As given that,

$U\ (-2x + y)\ ti + (2x + y)t\ j$

Where, $u = (-2x + y)t$

$v = (2x + y)t$

Now, by Navier stock equation;

$$\rho\left[\frac{\partial v}{\partial t} + u\frac{\partial u}{\partial x} + v\frac{\partial u}{\partial y} + w\frac{\partial u}{\partial z}\right] = -\frac{\partial p}{\partial x} \quad \text{x direction}$$

$$\rho[(-2x+y) + (-2x+y)t\ x - 2 + (2x+y) \times t] = -\frac{\partial p}{\partial x}$$

$[\therefore\ x = 1, y = 1, t = 1]$

So, $1 \times [(-2 + 1) + (-2 + 1) \times 1 \times -2 + (2 + 1) \times 1]$

$$= -\frac{\partial p}{\partial x}$$

$$\boxed{\frac{\partial p}{\partial x} = 4}$$

Similarly in y direction;

$$\rho\left[\frac{\partial v}{\partial t} + u\frac{\partial v}{\partial x} + v\frac{\partial v}{\partial y} + w\frac{\partial v}{\partial z}\right] = -\frac{\partial p}{\partial y}$$

$$\rho \times [(2x + u) - (-2x + y)\ t \times 2 + (2x + y)t \times 1]$$

$$= -\frac{\partial p}{\partial y}$$

$$1 \times \{3 - 2 + 3\} = -\frac{\partial p}{\partial y}$$

$$\boxed{\frac{\partial p}{\partial y} = -4}$$

So, pressure gradient

$$= \sqrt{4^2 + 4^2} = \sqrt{32} = 5.656$$

39. According to the question;

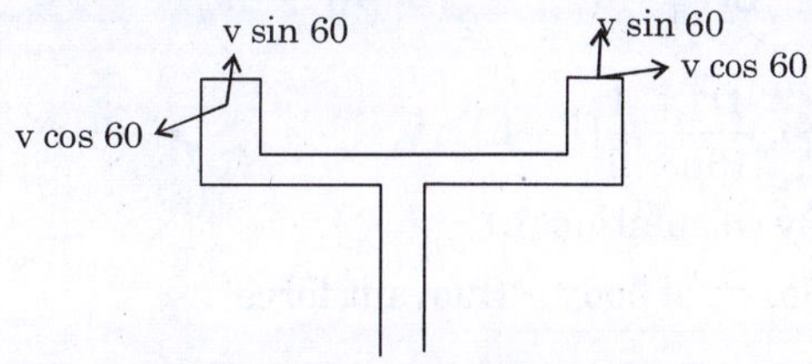

The component of V in horizontal direction will given the rotation to sprinkler

As we know that

The mass flow rate= ρAV

$$\dot{m} = 100 \times \frac{25}{(100)^2} \times 1$$

So, the force on one arm;

$$\boxed{f = \dot{m} \times v \cos 60}$$

$$= 1000 \times \frac{25}{(100)^2} \times 1 \times 1 \times \frac{1}{2}$$

$$\boxed{f = 1.25\text{N}}$$

And, Torque which is causing rotation = f × 2a

$$\Rightarrow\ 1.25 \times 2 \times \frac{10}{100} = 0.25\text{Nm}$$

40. According to the given question

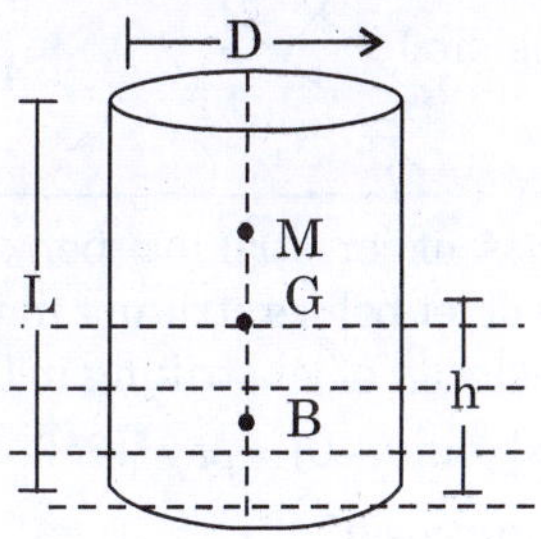

Floating in water

As given that,

$\rho_{wooden} = 600$ kg/m^3

$\rho_{water} = 1000$ kg.m^3

BM – BG = GM for stability of wooden box

$$\boxed{GM \geq 0}$$

So, BM – BG ≥ 0 ...(i)

$$\therefore\ BM = \frac{1}{V_{min}}$$

$$\frac{I}{V_{min}} - BG \geq 0$$

$$\frac{\frac{\pi}{64}D^4}{\frac{\pi}{4}D^2 \times h} - BG \geq 0 \qquad (\therefore \text{From (i)})$$

$\therefore\ BG = \frac{L}{2} - \frac{h}{2} = \frac{1}{2}(L - h)$

So, $\frac{D^2}{16h} - \frac{1}{2}(L - h) \geq 0$ ...(ii)

By calculating (h) :

So, wt of body = Buoyant force

$mg|_{wooden} = mg|_{submerged\ Body}$

$\rho \times v = \rho \times v$

$600 \times \frac{\pi}{4} D^2 \times L = 1000 \times \frac{\pi}{4} D^2 \times h$

$h = 0.6L$ put in (1)

$\frac{D^2}{16 \times 0.62} - \frac{1}{2}(L - 0.61) \geq 0$

$\frac{D^2}{9.6L} = 0.2L; \quad \frac{D^2}{L^2} = 1.92$

$\boxed{\frac{D}{L} = 1.385}$

41. According to the question

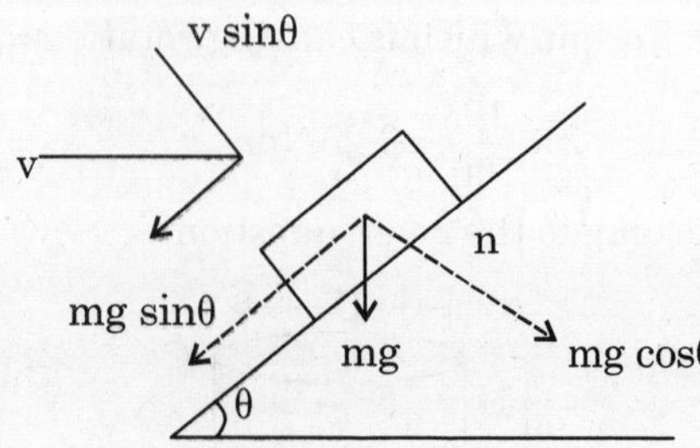

f_n = Mass of jet striking per second × [initial velocity of jet before striking in the direction of n – final velocity of jet striking in the direction of n]

$f_n = \rho av\,[v\sin\theta - 0] = \rho av^2 \sin\theta.$

$\therefore\ f_n = mg\cos\theta$

So, $mg\cos\theta = \rho av^2 \sin\theta$

$1000 \times 0.1 \times v^2 \times \sin 60 = 10 \times 10 \times \cos 60$

$v^2 = \frac{1}{\sqrt{3}}$

$\boxed{v = 0.7549}$

42. Applying Bernouli equation between A & C & assuming uniform velocity through siphon

$r = 1000\ kg/m^3.$

$\boxed{V_A = V_C}$

$z_A = 0$

$\frac{p_A}{\rho_g} + \frac{V_A^2}{2_g} + z_A = \frac{p_c}{\rho_g} + \frac{V_c^2}{2_g} + z_c$

$\frac{101 \times 1000}{1000 \times 10} = \frac{29.5 \times 1000}{1000 \times 10} + (z + 2)$

$10.1 = 2.95 + z + 2$

$\boxed{z = 5.15\ m}$

43. As given that,

p = Power = 100w;

As we know that

Velocity (v) = 2.5 m/s

As we know that

$p = f \times v$

$f = \frac{100}{2.5} = 40N$ (Shear force)

And, shear force (P) $= \mu \frac{du}{dy}$

$P = \mu \times \frac{2.5}{3} \times 100$

$\frac{F}{A} = \mu \times \frac{2.5}{3} \times 100$

$\frac{40 \times 100}{2 \times 60} = \mu \times \frac{2.5}{3} \times 100$

$\boxed{\mu = 0.4}$

C : MATERIALS SCIENCE

44. According to given question

Total number of atoms in BCC= 2

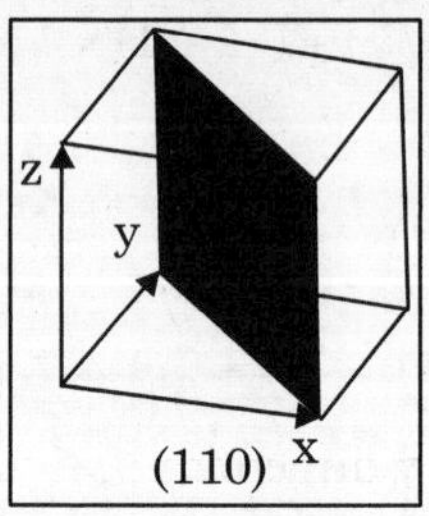

So, Area in [110] plane $= \sqrt{2}a^2$

And, number of atoms per unit area

$= 2\sqrt{2}a^2 = \frac{\sqrt{2}}{a^2}$

45. (i) **Metallic bonds have a** completely different structure than ionic and covalent bonds. Metallic bonds are formed only between metal atoms. And Cu0.5Al0.5 alloys material has covalent bonds.

(ii) The structure of Zinc-blende is simple and consists of the zinc metal-sulfur atom attached to each other via a **polar covalent bond**. The molecular or chemical formula of Zinc Sulfide is (ZnS).

(iii) **Na_2O is an ionic compound**. It consists of $2Na^+$ and O^{2-} ions which are held together by ionic bonding. The bond formed between the metal, sodium (Na), and nonmetal, oxygen (O) is ionic, due to the large gap of electronegativity difference that exists between them.

(iv) Li_4SiO_4 melt has shown that materials comprised of **biphasic mixtures** of Li_4SiO_4 and Li_2TiO_3 exhibit improved.

Hence option (c) is the correct answer.

46. In an ideal rubber, entropy change:

Heat is produced in the rubber band as it is stretched. If we stretch slowly in air at room temperature, we have a reversible, isothermal process with $q < 0$ (heat is lost to the surroundings). Then,

$$\Delta S_{stretch} = \frac{q_{rev}}{T} < 0$$

in the stretching process.

Since entropy is a state function :

$$\Delta S_{contract} = \frac{q_{rev}}{T} > 0$$

For the reverse process of slowly allowing the rubber band to contract, the rubber band has lower entropy, (is more ordered) when stretched. Under tension, the molecules in a rubber band line up and the arrangement becomes much more ordered, lowering the entropy. The criterion for spontaneity for an isolated system is $\Delta S_{sys} > 0$.

Suppose the system is the rubber band and the air in a large room. Therefore, if a constraint keeping the rubber band stretched is removed, it will spontaneously contract, with ΔS_{sys}, contract > 0.

So, we can say the primary factor responsible for elasticity up to small strains is entropy changed, but enthalpy is not change, its change only with heat source.

47. The electrical conductivity of semiconductors increases with increasing temperature because, **with increase in temperature, number of electrons from the valence bond can jump to the conduction band in semiconductors**.

48. Differential scanning calorimetry (DSC) is one of the thermal analysis techniques. In case of DSC, differences in heat flow between a specimen and reference are measured against the temperature or time while the specimen is continuously heated under controlled conditions. Function of calorimeter is to measure the heat content which comes in and out of the sample. Throughout the experiment both reference and sample are maintained at nearly constant temperature. Usually for DSC analysis temperature program for sample holder is designed in such a way that its temperature increases linearly with time. For DSC analysis only few mg materials is required. It is the most used thermal technique due to its speed, availability and simplicity. DSC technique is used for quantitative analysis. These measurements give both quantitative as well as qualitative information regarding the physical and chemical changes which involves endothermic and exothermic processes or variations in heat capacity of the sample.

In this diffrential scanning calorinetry (DBC) the heat- Flux principle and the power-compensaled in the DSC. In case of when heating pure ice temperature very from 263 K to 284 K; Heat flow into or out of a system as a function of temperature, in this experiment it performed at 1 atmospheric pressure. The DSC thermogram exhibit a spike either upward or downward.

In this experiment uses a sample and reference that are facing the same conditions and their signal is directly from each other.

49. Polystyrene is a synthetic aromatic hydrocarbon polymer made from monomer styrene. Styrene is produced from ethylene and benzene. When ethylene passes into benzene in presence of an aluminum tri-chloride catalyst, ethyl-benzene is produced, which on passing over a catalyst such as iron oxide or magnesium oxide at high temperature gets converted into styrene and hydrogen. Staudinger obtained a series of polymers which varied from polystyrenes soluble in benzene through polymers of high swelling capacity to completely insoluble products. So, benzene solvent environments will likely result in swelling of solid polystyrene.

50. As we know that,

Hardness α Brittle α Yield Strength α 1/ ductility α Scratch resistance. So, Higher the hardness value means material is brittle. Brittle material has higher yield strength. But ductility is less. Scratch resistance inversely proportional to ductility.

51. As we know that

$$N_v = N \times e^{\left(\frac{-Q_n}{RT}\right)}$$

$$N_v = 6.023 \times 10^{23} \exp\left(\frac{-4}{8.62 \times 10^{-5} \times 1000}\right)$$

$$\boxed{N_v = 4235.81}$$

52. By using the nerst equation-

$$E_{cell} = E^0 - \frac{0.0592}{n}\log(Q)$$

For $\frac{Li^+}{Li}$; n = 1 form lithium

$$4.50 = -3.04 - \frac{0.0592}{n}\log(Q)$$

$$0.0592 \log(Q) = -7.54$$

For $\frac{Zn^+}{Zn}$; n = 2 form Zinc

$$E_{cell} = -0.77 + \frac{7.54}{2}$$

$$\boxed{E_{cell} = 3V}$$

53. In a binary isobaric temperature-composition phase diagram there are two possible types of invariant reactions: "eutectic type" invariant reactions ($\beta \rightarrow \alpha + \gamma$), and "peritectic type" invariant reactions ($\alpha + \gamma \rightarrow \beta$). In a ternary system, there are "eutectic type" ($\alpha \rightarrow \beta + \gamma + \Delta$), "peritectic type" ($\alpha + \beta + \gamma \rightarrow \Delta$) and "quasi-peritectic type" ($\alpha + \beta + \gamma + \Delta$) invariants. In a system of C components, the number of types of invariant reactions is equal to C. A reaction with one reactant, such as $\alpha \rightarrow \beta + \gamma + \Delta + \varepsilon$ is clearly a "eutectic type" invariant reaction but in general, there is no standard terminology.

54. The temperature at which the liquid and crystalline states would have the same entropy has been taken to be an *ideal glass transition temperature* (below for a glass marginally). Glass has higher entropy than the corresponding crystalline phase and lower entropy than the corresponding liquid phase. This is somewhat lower than experimentally observed values of T_g. The observed transition occurs on cooling, when the decreasing atomic mobility in the liquid (manifested as increasing viscosity) prevents structural change at a rate fast enough to keep the liquid structure in internal equilibrium. Certain atomic configurations are thus frozen in. The faster the cooling, the higher the temperature at which the freezing-in occurs, and the less relaxed the glassy state produced.

55. By resolved shear stress theory;

$$\tau_R = \sigma (\cos \varphi)(\cos \lambda)$$

where, φ is the angle between the normal to the slip plane and loading axis.

A is the angle between slip direction and loading axis.

The minimum stress necessary to introduce yielding occurs when a single crystal is oriented ($\varphi = \lambda = 45°$):

$$\sigma_y = 2\tau_{crss}$$

Aluminium is under the FCC Plane. Slip plane {111} and slip direction $\langle 1\bar{1}0 \rangle$. As per the option 4 is the suitable answer.

56. Sample 1 is loaded upto E point i.e. before the yield point (elastic deformation). When we unload it, region its original shape, there is no change in size of hardness.

Sample 2 is loaded upto Point P i.e. the yield point (Plastic Deformation). When we unload it, not regain its original shape and size. Due to unloading of the sample it will show strain hardening. So hardness value will be increased.

57. As per the below curve hardness of the alloy first increased then decreased.

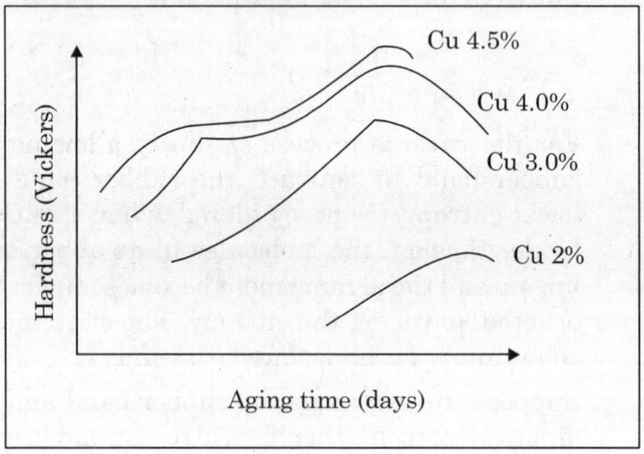

Figure 1 Ageing Curve at 130 C forAl-cu System

58. Choose a metallic adhesion layer with very low interfacial energy with the deposited thin film, this process uses an ion source to sputter the target material-either metallic or dielectric-onto the substrate, forming the thin film.

When depositing a thin metallic film on SiO? substrate, with an adhesion layer between the metal film. adhesive layer constructions are known as transfer tapes because they truly allow the user to transfer a thin layer of adhesive to a substrate. To perform an extensive optimization the student has to choose a metallic adhesion layer with very low interfacial energy with the deposited thin film.

Hence, Option (a) is the correct answer.

59. If rate kinetic of phase transformation than the structure will be finer and this is indicated by the hall-petch eqn states that decrease in grain size and with finenes in the structure the strength in increase.

$[\delta_0 = \delta + k_a^{(-1/2)}]$

Hall-Petch eqn.

Where $\delta_0 \rightarrow$ Friction stress

$\delta \rightarrow$ in stress

$a \rightarrow$ grain size

$K \rightarrow$ locking parameter

And, the thermodynamic driving force of a reaction is usually taken as the chemical potential difference between products and reactants. The forward and backward reaction rates are then related to this force. So, that thermodynamic driving force increases monotonically.

Hence option (a & c) are the correct answer.

60. As given that;

0.5 + 0.5

α x β

(α+β)

50% wt of A ?

50% wt of B

Conversion of mole (atom)% to weight%

$$C_1 = \frac{C_1' A_1}{C_1' A_1 + C_2' A_2} \times 100$$

Mole fraction of A in β = 2 × Mole fraction of B in β.

So, $m_A = 2\,m_B$

$A_a = 2A_b$

W_b in B

$$= \frac{m_B \times A_b}{m_B \times A_b + m_A \times A_a} = \frac{m_B \times A_b}{5(m_B \times A_b)} = 0.2$$

$W_A + W_B = 1$

$W_A = 0.8$ & $W_B = = 0.2$

α	x	β
50% A	(α + β)	80% A
50% B	65% of A	20% B
	35% of B	

Apply liver rule;

$$\text{So, } W_A \text{ in } x = \frac{x - 0.5}{0.3 - 0.5}$$

x = 0.65 or 65.1 of A

Conversion of weight % to mole (t_m)%

$$C_1' = \frac{C_1 A_2}{C_1 A_2 + C_2 + A_1} \times 100$$

$$m_A \text{ in } x = \frac{0.65\lambda A_b}{(0.65 \times Ab)(0.35 \times 2A_b)} = 0.48 = 0.5$$

61. As we know that

$$D_A = D_A^0 \exp\left(-\frac{E_a}{k_B T}\right)$$

Diffusion D_A at 300K = 9.16×10^{-6}

Diffusion D_A at 600K = 3026.2×10^{-6}

$$\therefore \exp(-E_a/k_\beta T) = 3600/T$$

$$\therefore \text{Time (t)} = \frac{1}{D_A}$$

$$\text{So, } \frac{3026.2 \times 10^{-6}}{9.16 \times 10^{-6}} = \frac{3600}{t}$$

$$\boxed{t = 10.90\,\text{sec}}$$

62. As given figure

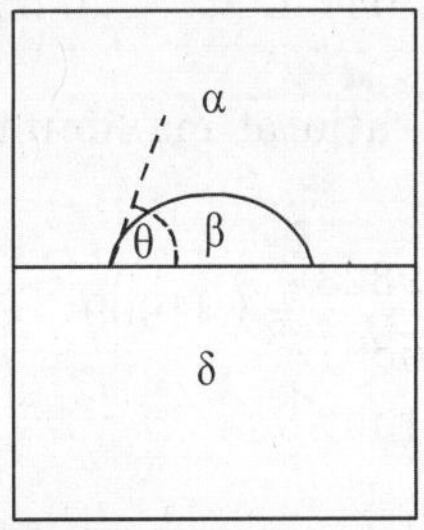

As we know that

$$\Delta G_{heterogeneous} = \Delta G_{homogeneous} * f(\theta)$$

$$f\theta = \frac{\Delta G_{heterogeneous}}{\Delta G_{homogeneous}}$$

$$f(\theta) = \frac{2 - 3\cos\theta + \cos^3\theta}{4}$$

$$\gamma_{\beta\delta} + \gamma_{\alpha\beta}\cos\theta = \gamma_{\alpha\delta}$$

$0.02 + 0.4\cos\theta = 0.3$

$\boxed{\cos\theta = 0.7}$

$$\text{So, } f(\theta) = \frac{2 - 3 * 0.7 + 0.7^3}{4} = 0.06075$$

63. As we know that;

$$\text{The resistivity } (\rho_i) = \frac{1}{\sigma_i} = \frac{1}{n_i e[\mu_x + \mu_p]}$$

$$3000 = \frac{1}{n_e \times 1.6 \times 10^{-19} \times (0.14 + .060)}$$

$n_e = 1.042 \times 10^{16}$

$$n_e = N_A \exp\left(-\frac{E_g}{k_B T}\right)$$

$$1.042 \times 10^{16} = 6.032 \times 10^{23} \exp\left(-\frac{E_g}{8.62 \times 10^{-5} \times 298}\right)$$

$\boxed{E_g = 0.46\text{ev}}$

64. As we know that,

Fraction of light beam that is not reflected or absorbed is transmitted through the material.

$I_t = I_0 (1 - R)^2 \exp(-ax)$

And, R-reflection $(R) = \left(\frac{n-1}{n+1}\right)^2$

$$R = \left(\frac{1.5-1}{1.5+1}\right)^2 = 0.04$$

Case 1, when absorption coefficient = 0.3 cm^{-1}

So, $\frac{I_t}{I_0} = (1-.04)^2 e^{-0.3\times0.5} = 0.79323$

Case 2, when absorption coefficient = 1 cm^{-1}

So, $\frac{I_t}{I_0} = (1-.04)^2 e^{-1\times0.5} = 0.55897$

Hence, the ratio of maximum and minimum transmission

$$= \frac{0.79323}{0.55897} = 1.41909$$

65. As we know that

$2d \sin\theta = n\lambda.$

$2d \sin\frac{45}{2} = 1\times1.54$

$\boxed{d = 0.2}$

So, Inter planar distance $= \left(\frac{d_{hkl}}{cubic}\right) = \frac{a}{\sqrt{h^2+k^2+l^2}}$

In the question asked for the third peak of F_{cc} = miller indices of third peak (2,2,0).

So, $2 = \frac{a}{\sqrt{2^2+2^2+0^2}}$

$\boxed{a = 5.66\text{Å}}$

D : SOLID MECHANICS

66. As in the given fig. both ends are hinged

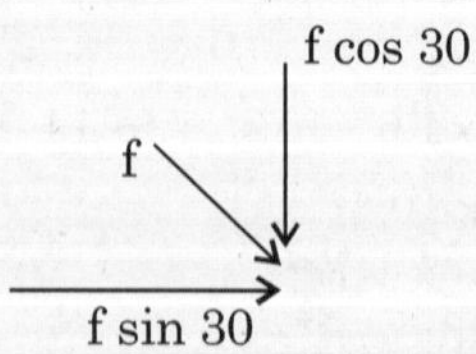

So, the vertical force required to cause buckling;

$f_v = \frac{\pi^2 EI}{l^2}$

$\boxed{f_v = f\cos60^\circ}$

$$f = \frac{f_v}{\cos60^\circ} = \frac{\pi^2\times E\times I}{L^2\times\cos30^\circ}$$

$$\boxed{f = \frac{2}{\sqrt{3}}\times\frac{\pi^2 EI}{L^2}}$$

67. According to the question

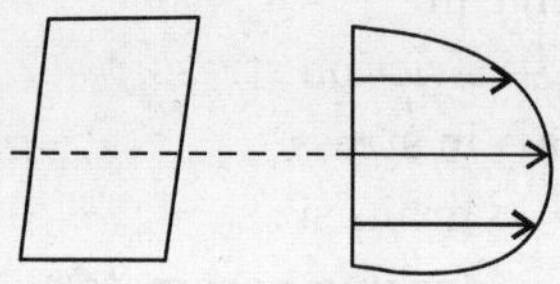

Shear stress due to transverse shear force in a linear elastic Isotropic beam of rectangular cross-section is maximum at the neutral axis.

68. According to the question;

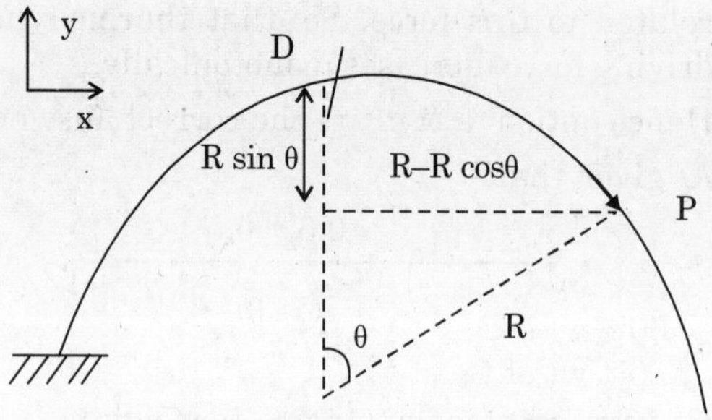

Moment $(M_y) = P\times R\sin\theta$ = Bending Moment

$M_z = PR(1-\cos\theta)$ = Twesting Moment

Force is in negative (z-direction)

shear force = $-P\hat{k}$,

Bending Moment = $PR\sin\theta\,\hat{j}$

And, Twisting Moment = $PR(1-\cos\theta)\,\hat{i}$

69. According to the question;

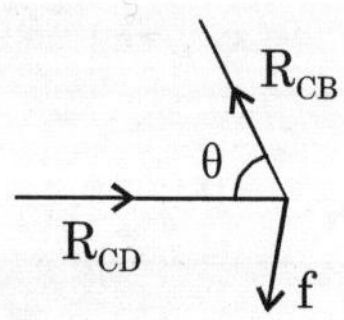

Where, R = Reaction

F = Force

$F = R_{CB}\sin\theta$

Here, Produce Tensile Stress in BC.

So, Considering joint B

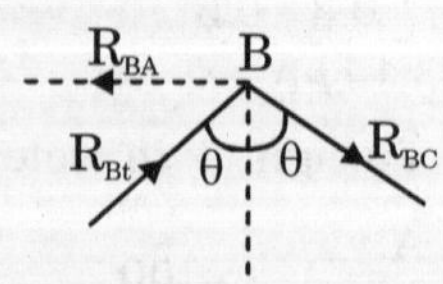

$R_{BE}\cos\theta = F_B\cos\theta$

$R_{BD} = 0$

B_D is a no force member

$R_{BA} = F_{BC} + R_{BE} \sin\theta$
will cause tensile stress in AB

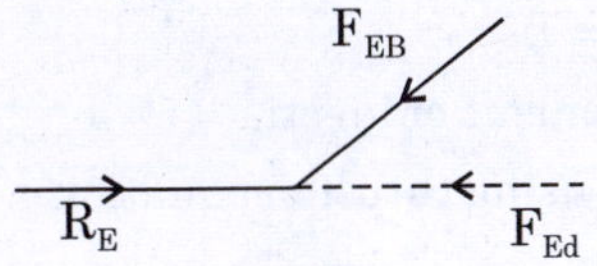

R_E & F_{ED} will cause
Compression in ED

70. Accoding to the question :

$$V_A = V_x\hat{i} + V_y\hat{j}$$

Due to translation V_x is in positive direction

$$V_x = 1\hat{i}\,m/s$$

Due to rotation velocity;

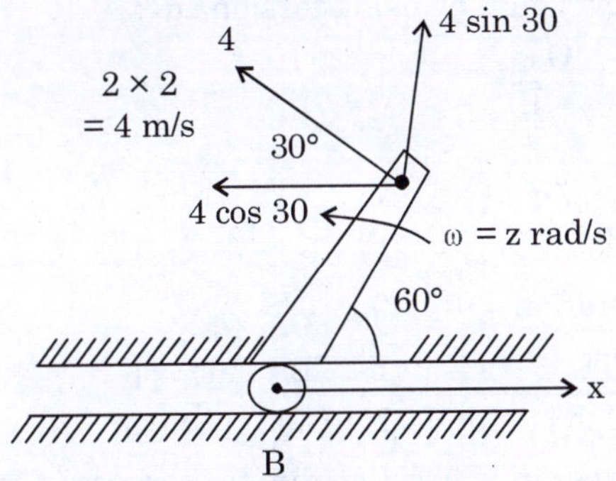

$$V_r = -4\cos 30 = -4 \times \frac{\sqrt{3}}{2} = -2\sqrt{3}\ i$$

$$V_y = 4 \sin 30 = \frac{4}{2} = 2$$

So, $\boxed{V_A = (1-2\sqrt{3})\hat{i} + 2\hat{j}}$

71. As we know that

Torque T = 2FL

T = I α

2F.C = 4 m.C² × α.

$$\alpha = \frac{F}{2mC}$$

72. As given that,
Velocity of particle along AB direction = v cos ϕ
And, velocity of particle along z direction
= – v cos ϕ sin θ
= – 2 × cos30 sin 40
= –1.1133 m/sec

73. As given that,

$$\delta_A = \Delta l_{AC}$$
$$\delta_B = \Delta l_{BC}$$

As we know that,

$$\Delta l_{AC} = \frac{P \times 2L}{AE} = \frac{2PL}{AE}$$

$$\Delta l_{BC} = \frac{P \times L}{AE} = \frac{PL}{AE}$$

$$\text{So, } \frac{\delta_A}{\delta_B} = \frac{2\dfrac{PL}{AE}}{\dfrac{PL}{AE}} = 2$$

$$\Rightarrow \boxed{\frac{\delta_A}{\delta_B} = 2}$$

74. As we know that
Maximum shear stress theory for yield FOS = 1
As we know that

$$\tau_{max} = \frac{\delta_{yt}}{2\,FOS} = \frac{\delta_{yt}}{2}$$

We also know that,

$$\therefore\ \tau_{max} = \frac{pd}{8t}$$

$$\text{So, } \frac{Pd}{8t} = \frac{\delta_{yt}}{2}$$

$$\frac{P \times 1000}{8 \times 10} = \frac{500}{2}$$

$$\boxed{P = 20\text{ MPa}}$$

75. According to the question
Point B deflection at $y_B = 0$
$\uparrow y_B$ due to $R_B + \downarrow y_B$ due to $M_0 = 0$

$$\Rightarrow \frac{R_B \times L^3}{3EI} - \left[\frac{M_0\left(\frac{L}{2}\right)^2}{2EI} + \frac{Mo\left(\frac{L}{2}\right)}{EI} \times \frac{L}{2}\right]$$

$$\Rightarrow y_B \text{ due to } M_0 = y_C + Q_C \times B_C$$

$$\Rightarrow \frac{M_0\left(\frac{L}{2}\right)^2}{2EI} + \frac{M_0\left(\frac{L}{2}\right)}{EI} \times \frac{L}{2} = \frac{3M_0L^2}{8EI}$$

$$\frac{R_B \times L^3}{3EI} - \frac{3M_0L^2}{8EI} = 0$$

$$\boxed{R_B = \frac{9M_0}{8L}}$$

76. According to the question

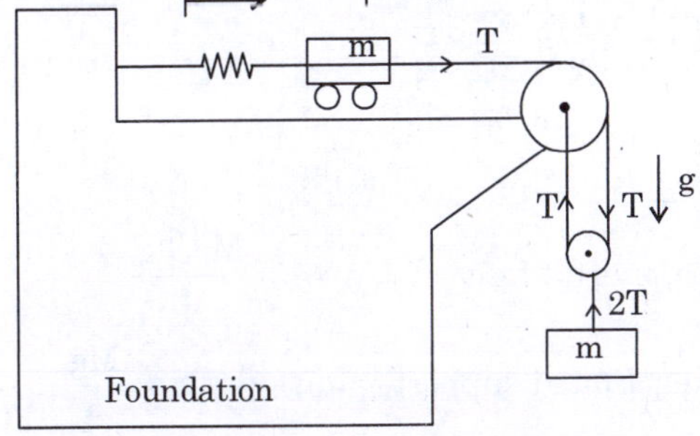

As we know that,

Potential energy spring $(P.E) = \frac{1}{2}Kx^2$

$$K.E = \frac{1}{2}m\,\dot{x}^2 + \frac{1}{2}m\left(\frac{\dot{x}}{2}\right)^2 = \frac{5}{8}m\,\dot{x}^2$$

$$\frac{d}{dt}(P.E + K.E) = 0$$

$$\frac{d}{dt}\left[\frac{1}{2}\left(Kx^2 + \frac{5}{8}m\,\dot{x}^2\right)\right] = 0$$

$$K.x.\dot{x} + \frac{5}{8}.2m.\dot{x}\,\bar{x} = 0$$

$$\dot{x}\left(Kx + \frac{5}{4}m\,\bar{x}\right) = 0$$

$\downarrow$ K_e $\quad$ $\downarrow$ M_2

$$\boxed{W_n = \sqrt{\frac{K_e}{M_e}} = \sqrt{\frac{4K}{5M}}}$$

77. Let R be the reactor at hinged support at the instant of threat force removal then

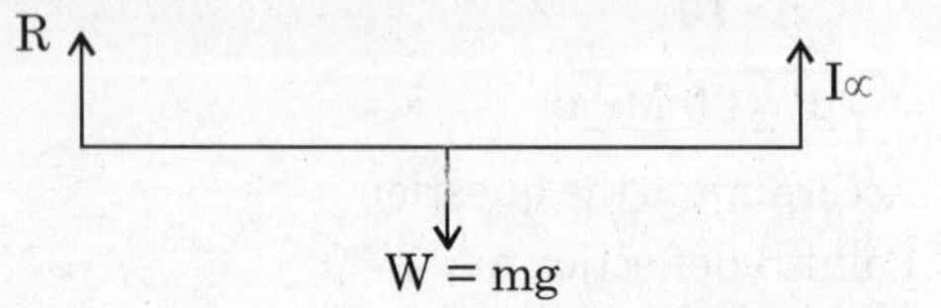

where, $I \propto$ = Inertia torque which acts in a direction opposite to rotation of rigid rod

$I = \frac{mL^2}{3}$ $\quad (\because m = \text{mass of Rod})$

About hinge moment is 0

$$I \propto = W \times \frac{L}{2}$$

$$\frac{ML^2}{3} \times \propto = W \times \frac{L}{2}$$

$$\frac{ML^2}{3} \times \propto = Mg \times \frac{L}{2}$$

So, angular acceleration of rod $\propto = \frac{3g}{2L}$

And, Linear acceleration at centre

$$(a) = \propto\frac{L}{2} = \frac{3g}{2L} \times \frac{L}{2}$$

$$\boxed{a = \frac{3g}{4}}$$

So, inertia force at centre $= \frac{M \times 3g}{4}$

Reaction at support $= Mg - \frac{3}{4}Mg = \frac{Mg}{4}$

78. As given that

For solid portion;

Torque = T

Polar moment of inertia = 50 J

Now, by using throsion equation :

$$\frac{T}{50J} = \frac{G\theta_S}{L}$$

$$\theta_H = \frac{T \times L}{50\,J \times G}$$

or Hollow portion up to section (P–Q)

Polar moment of inertia = J

$$\text{Length} = \frac{5L}{4} - L = \frac{L}{4}$$

Now again by using torsion eq. :

$$\frac{T}{J} = \frac{G\,\theta_H}{\frac{L}{4}}$$

$$\theta_H = \frac{T \times L}{4 \times GJ}$$

Rotation $= (\theta_S + \theta_H)$

$$\frac{TL}{50\,JG} + \frac{TL}{4JG} = \frac{54\,TL}{200\,GJ} = \frac{27\,TL}{100\,GJ}$$

79. Stress in x direction due to increase in temp

$= \propto\Delta TE$ (compressive)

Strain in x direction $= \frac{\propto \Delta TE}{E} = \propto \Delta T(-ve)$

So, the lateral strain or strain in j direction;

$= \nu\propto\Delta T\ (+ve)$

Strain in y direction due to temp, increase $= \propto\Delta T$

So, the total strain in y direction $= \propto\Delta T + \nu\propto\Delta T$

Change in (a) $= (\propto\Delta T + \nu\propto\Delta T)a = a(1 + \nu)\propto\Delta T$

80. According to the question :

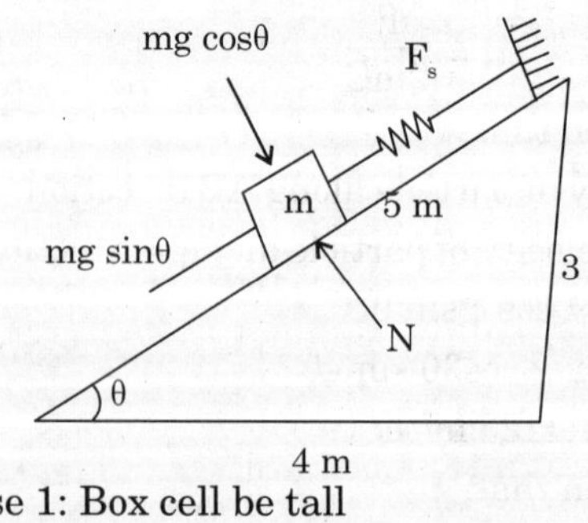

Case 1: Box cell be tall

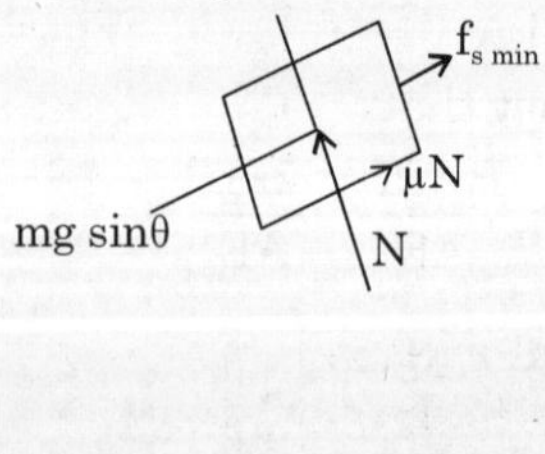

$N = mg\cos\theta$

$N = 10 \times 10 \times \frac{4}{5} = 80$

$F_{smin} = Mg\sin\theta - \mu N$

$F_{smin} = 10 \times 10 \times \frac{3}{5} - 0.25 \times 80$

$\boxed{F_{smin} = 40N}$

Now, spring force required to hold mass is 40N

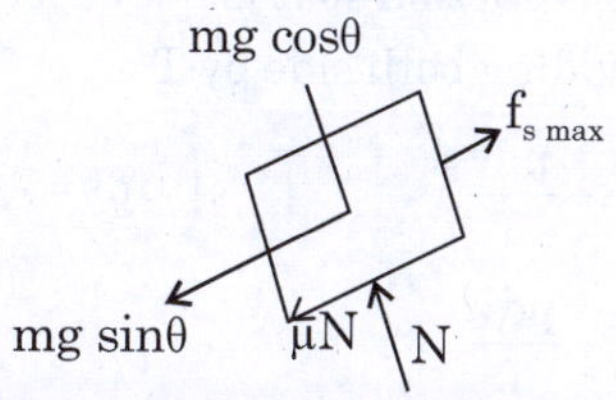

Case 2 : Mass will travel upside

N = 80

$F_{smax} = Mg\sin\theta + \mu N$

$10 \times 10 \times \frac{3}{5} + 0.25 \times 80$

$\boxed{F_{smax} = 80N}$

So equilibrium of the mass is impossible if the spring force is less than 40N or more than 80 N.

81. By taking moment about B

$R_A \times 0.6 - 1000 \times 0.6 \times 0.3 = 0$

$R_A = 300$ N

$R_A + R_B = 1000 \times 0.6$

$R_B = 300$ N

Now, Shear force diagram;

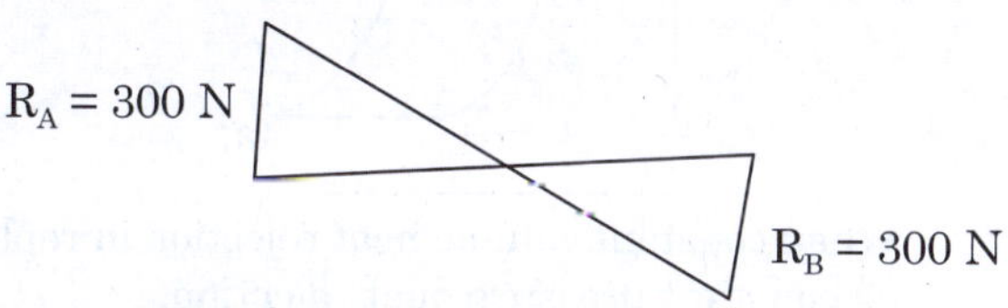

Hence, the maximum shear force = 300 N

82. As given that,

Spring deflection initial $x_1 = 3m$

After release deflection of spring $x_2 = 1m$

It means spring energy is reducing

conversion of energy; so that,

$$\frac{1}{2}m\left(v^2 - u^2\right) = \frac{1}{2}k\left(x_1^2 - x_2^2\right)$$

Energy gain By Mass = energy loss By spring

$10 \times (v^2 - 0^2) = 2000\,(3^2 - 1^2)$

$\boxed{v = 40\ m/s}$

83. According to the question;

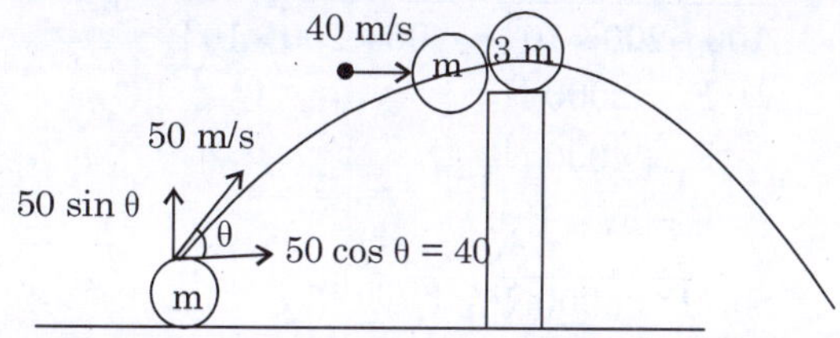

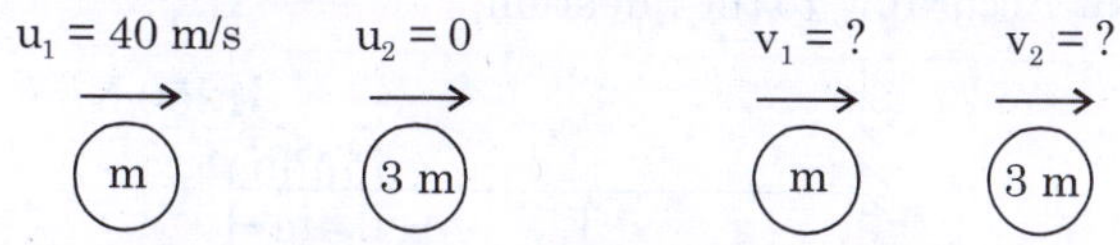

Now, from Conservation of momentum;

$40 \times m + 0 = mv_1 + 3mv_2$

$40\ m = mv_1 + 3mv_2$

$v_1 + 3v_2 = 40$ ...(i)

$\therefore$ Coeff. Restitution $(e) = \frac{v_2 - v_1}{u_1 - u_2} = \frac{v_2 - v_1}{40 - 0} = 0.3$

$v_2 - v_1 = 12$ ...(ii)

Now, solve the equation (i) & (ii), we get;

$v_1 = 1$ m/s

$v_2 = 13$ m/s

So, the speed of A immediately after the collision

$\boxed{(v_1) = 1\ m/s}$

84. According to the question :

When brakes are applied on the box, tends to move in forward direction due to inertia and friction force avoids that supping or movement

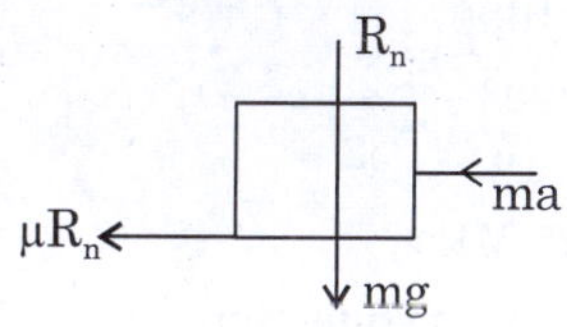

so $\mu R_n = -ma$

$\mu mg = -ma$

$a = -\mu g = -0.25 \times 10 = 2.5$

using $v^2 - u^2 = 2as$

$0^2 - 20^2 = -2 \times 2.5 \times 5$

$\boxed{s = 80m}$

85. According to the question :

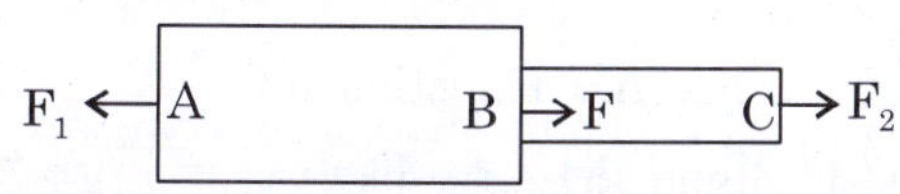

F_1 = force or reaction at A

F_2 = Reaction at C

Then compression in BC = Expansion in AB = 0.1

$$\frac{F_2 \times 1000}{100 \times 200 \times 10^3} = \frac{F_1 \times 1000}{200 \times 200 \times 10^3} = 0.1$$

$\therefore\ F_2 = 2000\ N$

$\therefore\ F_1 = 4000\ N$

$F = F_1 + F_2$

So, $\boxed{F = 6000N}$

86. According to the question;

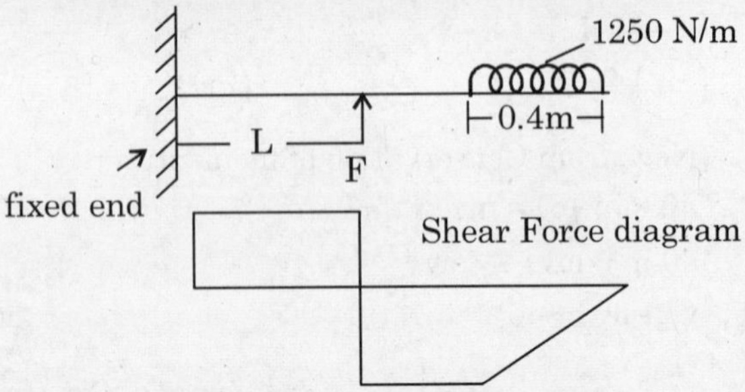

Now, taking moment about fixed end

$= 1250 \times 0.4 \times 0.8 - F \times L$

$M_{fixed} = 0 = 1250 \times 0.4 \times 0.8 - F \times L$

$$L = \frac{1250 \times 0.4 \times 0.8}{1000} = 0.4m$$

As direction of shear force is changing at F so

$M_{max} = 1250 \times 0.4 \times (0.6 - 0.2)$

$= 200\ Nm$

Now, by using bending equation;

$$\boxed{\frac{M}{I} = \frac{\sigma}{y}}$$

$$\Rightarrow \frac{200 \times 1000}{24 \times \frac{20^3}{12}} = \frac{\sigma}{\frac{20}{2}}$$

So, $\sigma = 125$ MPa

87. According to the question;

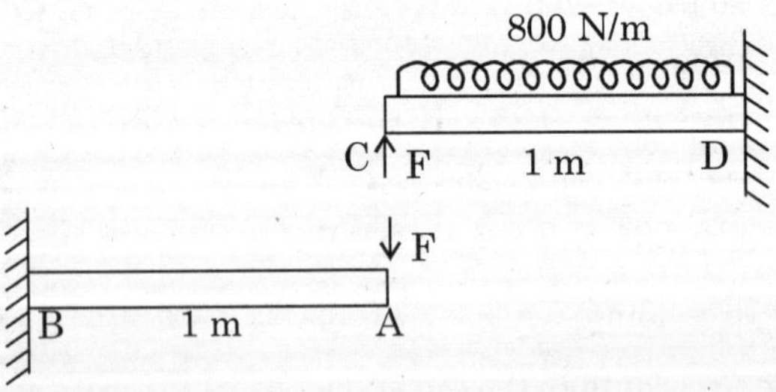

$y_A = y_C$

deflection at A = deflection at C

$$\frac{F \times 1^3}{3EI} = \frac{800 \times 1^4}{8EI} - \frac{F \times 1^3}{3EI}$$

$$\frac{2F}{3} = 100$$

$\boxed{F = 150\ N}$

Bending moment at B. $M_B = 150 \times 1$

$M_B = 150\ N - m$

E : THERMODYNAMICS

88. As we know that

Energy equation

$\delta q = du + pdv$

δq is path function so it is not exect differential

Now, dividing both side by T

$$\frac{\delta q}{T} = \frac{du}{T} + \frac{pdv}{T} \qquad \left(\therefore ds = \frac{\delta q}{T}\right)$$

$$ds = \frac{du}{dT} + \frac{pdv}{T}$$

in this equation all term i.e. q, u, t, p, s, v are point function so these are exect differential of $\frac{1}{T}$.

89. According to the question;

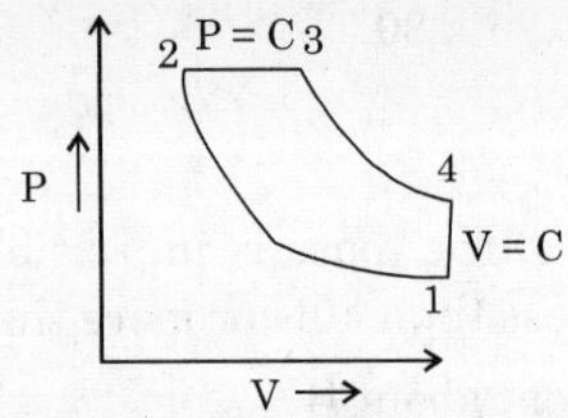

Diesel cycle;

T_1 = Same for all cycle

T_3 = also same

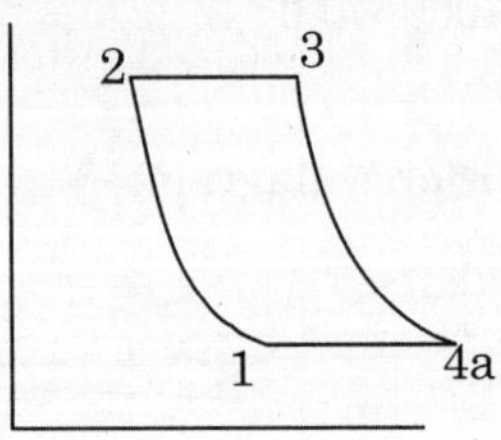

When constant volume heat rejection in replaced by constant pressure heat rejection.

In expansion process the pressure drop during 3-4a is more than pressure drop in 3–4. As both process are isentropic so

$$\frac{T_3}{T_4} = \left(\frac{P_3}{P_4}\right)^{\frac{r-1}{r}} \ \& \ \frac{T_3}{T_{4a}} = \left(\frac{P_3}{P_4}\right)^{\frac{r-1}{r}}$$

$$\frac{P_3}{P_{4a}} > \frac{P_3}{P_4}$$

$$\frac{T_3}{T_{4a}} > \frac{T_3}{T_4}$$

or

$\boxed{T_4 > T_{4a}}$

90. As we know that

$$W(\text{work done}) = \int_{v_1}^{v_2} p\,dv$$

By pdytropic process;

$$\boxed{pv^n = c}$$

So, $p = \dfrac{c}{v^n}$

$$\therefore \quad W = \int_{v_1}^{v_2}\left(\frac{c}{v^n}\right) dv$$

$$W = c\left(\frac{v_2^{1-n} - v_1^{1-n}}{1-n}\right)$$

$$C = p_2\,v_2{}^n = p_1\,v_1{}^n$$

$$W = \frac{p_2 v_2{}^n v_2^{1-n} - p_1 v_1^n v_1^{1-n}}{1-n}$$

$$\boxed{W = \frac{p_2 v_2 - p_1 v_1}{1-n}}$$

91. Carnot cycle i.e. Reversible cycle working between same temperature limits has the highest efficiency.

$$\boxed{\eta < \eta_c}$$

92. As given that,

specific volume of liquid – vapour – mixture

$$(v) = \frac{1.2}{8}\ m^3/kg$$

As we know that;

The specific enthalpy of mixture

$(h) = h_f + xh_{fg} = h_f + x(h_g - h_f)$

We also know that;

$v = v_f + x\,v_{fg}$

$0.15 = 0.001053 + x \times (1.1594 - 0.001053)$

$x = 0.128.$

then, $h = 467.13 + 0.128\,(2693.1 - 467.13)$

$= 752.05$ kJ/kg

93. As we know that

$$\text{Internal energy } (U) = \int cvdT$$

$$U = \int_{300}^{400}\left(0.7 + 0.27\times10^{-3}\right)dT$$

$$U = \left[0.7T + 0.27\times10^{-3}\frac{T^2}{2}\right]_{300}^{400}$$

$$U = 0.7\times(400-300) + 0.27 \times10^{-3}\times\left(\frac{400^2}{2} - \frac{300^2}{2}\right)$$

$$\boxed{U = 79.45\ \text{kJ/kg}}$$

94. According to the question;

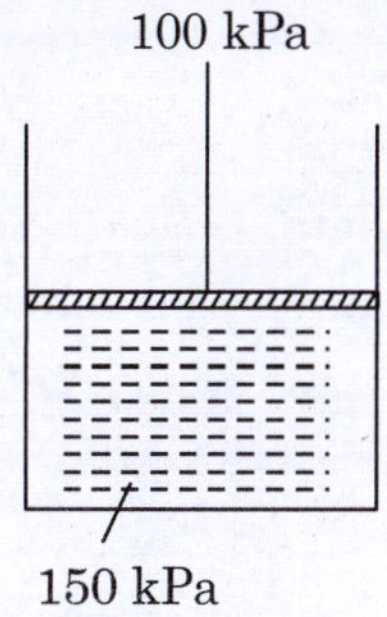

$\therefore$ 150 = atm + piston pressure

$\therefore$ 150 = 100 + piston pressure

So, piston pressure = 50

$$\frac{F}{A} = 50$$

$$\frac{mg}{A} = 50 \qquad (\therefore A = 0.5\ (\text{given}))$$

$$m = \frac{50\times0.05}{9.81}$$

$m = 0.254 \times 1000$

$m = 254.841$ Kg

95. As given that :

Reduction in network output = 20% = 0.2

initial network output = 100% = 1

final network output = 100 – 20 = 80% = 0.8

As we know that,

$$\text{Network output} = \frac{1}{\text{specific stream consumption}}$$

$$0.8 = \frac{1}{\text{specific stream consumption}}$$

So, specific steam consumption = 1.25 = 125%

Hence, specific stream consumption increased by 25%.

96. According to the question :

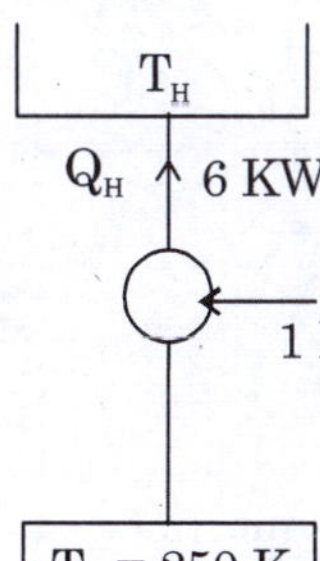

$$COP_{HP} = \left[\frac{\text{Desired}(Q_H)}{\text{Power input}(W)}\right]$$

$$\frac{Q_H}{W} = \frac{T_H}{T_H - T_L}$$

$$\frac{6}{1}=\frac{T_H}{T_H-250}$$

$$\boxed{T_H = 300k}$$

97. As we know that;

Specific heat at constant volume (C_v) :

$$C_v=\left(\frac{\partial Q}{\partial T}\right)_v$$

$$d_Q = TdS$$

$$C_v = T\left(\frac{\partial S}{\partial T}\right)_v$$

differentiating both sides w.r.t volume

$$\left(\frac{\partial C_v}{\partial v}\right)_T = \frac{\partial}{\partial v}\left(T\left(\frac{\partial S}{\partial T}\right)_v\right)_T$$

$$\left(\frac{\partial C_v}{\partial v}\right)_T = \left(T\frac{\partial^2 S}{\partial v \partial T}\right)_T \quad ...(i)$$

$$\text{Now } \left(\frac{\partial S}{\partial v}\right)_T = \left(\frac{\partial P}{\partial T}\right)_v$$

Differentiating w.r.t.T.

$$\frac{\partial^2 S}{\partial v \partial T} = \frac{\partial}{\partial T}\left[\left(\frac{\partial P}{\partial T}\right)_v\right]_v$$

$$\frac{\partial^2 S}{\partial v \partial T} = \left(\frac{\partial^2 P}{\partial T^2}\right)_v \quad ...(ii)$$

Now, putting the value from (ii) in equation (i)

$$\boxed{\left(\frac{\partial C_v}{\partial v}\right)_T = T\left(\frac{\partial^2 P}{\partial T^2}\right)_v}$$

98. h, P, T will be replaced x, y, z respectively in given eq.

$$\text{So, } \left(\frac{\partial h}{\partial P}\right)_T \times \left(\frac{\partial P}{\partial T}\right)_h \times \left(\frac{\partial T}{\partial h}\right)_P = -1$$

$$\left(\frac{\partial h}{\partial P}\right)_T \times \frac{1}{\mu} \times \frac{1}{C_P} = -1 \qquad \left\{\begin{aligned} \therefore\ \mu &= \left(\frac{\partial T}{\partial P}\right)_n \\ \&\ CP &= \left(\frac{\partial h}{\partial T}\right)_P \end{aligned}\right\}$$

$$\left(\frac{\partial h}{\partial P}\right)_T = -\mu C_P.$$

99. According to the question;

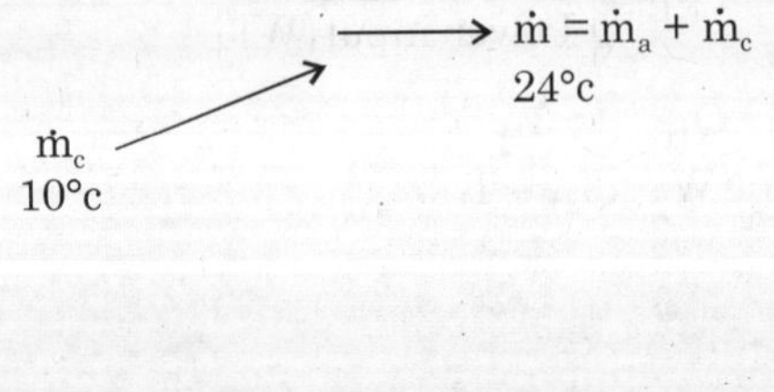

$$\dot m_a c_p(303) + \dot m_{cp}(283) = (\dot m_a + \dot m_c)\,cp \times 237$$

$$6\dot m_a = 14\dot m_c$$

$$\frac{\dot m_c}{m_a} = \frac{6}{14} = \frac{3}{7}$$

$$\boxed{\frac{\dot m_c}{m_a} = \frac{3}{7}}$$

100. As we know that,

$$P_{N_2} = P_t \times X_{N_2}$$

Where, P_t = total pressure

X = mole fraction

$$\therefore\ x_{N_2} = \frac{\text{moles of } N_2}{\text{moles of } N_2 + \text{moles of } O_2}$$

$$= \frac{\frac{0.8x}{28}}{\frac{0.8x}{28}+\frac{0.2x}{32}}$$

$$\therefore\ x_{N_2} = 0.8205$$

$$\text{So, } P_{N_2} = 300 \times 0.8205$$

$$= 246.153\ \text{KPa}$$

101. As we know that,

$$\left(P+\frac{a}{v^2}\right)(v-b) = RT$$

$$T = \frac{1}{R}\left[p+\frac{a}{v^2}\right](v-b)$$

$$T = \frac{1}{0.3}\left[1000+\frac{0.18}{0.005^2}\right]+(0.005-0.0014)$$

$$T = 206.4K$$

102. According to the question,

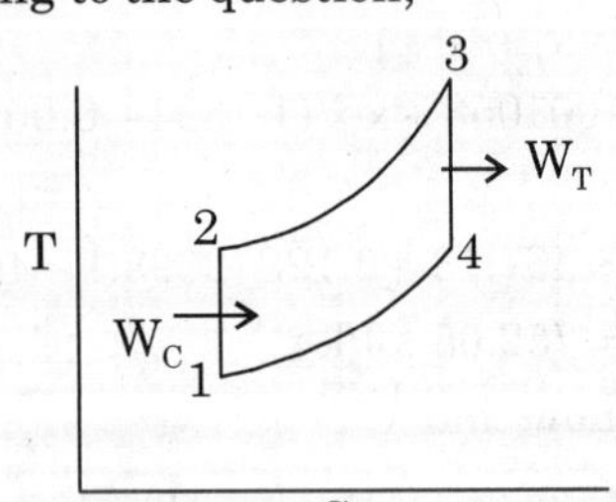

$$W_{net} = W_T - W_C$$

$$W_{net} = C_P(T_3 - T_4) - C_P(T_2 - T_1)$$

$$\therefore\ \text{ideal gas } \frac{T_2}{T_1} = \frac{T_3}{T_4}$$

T_1 & T_3 constant

$$T_4 = \frac{T_1 T_3}{T_2}$$

$$W_{net} = C_P\left[T_3 - \frac{T_1T_3}{T_2} - T_2 + T_1\right]$$

Differentiate w.r.t T_2 for maximum W_{net}.

$$\frac{dW_{net}}{dT_2} = C_P\left[0 + \frac{T_1T_3}{T_2} - 1 + 0\right] = 0$$

$$(W_{net})_{max} \text{ at } T_2 = \sqrt{T_1T_3}$$

$$\text{Pressure ratio } (P_r) = \frac{P_2}{P_1} = \left(\frac{T_2}{T_1}\right)^{\frac{r}{r-1}}$$

$$P_r = \left(\frac{\sqrt{T_1T_3}}{T_1}\right)^{\frac{r}{r-1}}$$

$$\boxed{r_P = \left(\frac{T_3}{T_1}\right)^{\frac{r}{2(r-1)}}}$$

103. As given that :

Initial condition in tank

Present P_1 = 1 MPa

Temperature T_1 = 600k

Volume $V_1 = 10\ m^3$

Now by using

PV = mRT

$1 \times 10^6 \times 10 = m_1 \times 287 \times 600$

m_1 = 58.07 kg

For final condition in tank

Pressure = (P_2)

Volume $(V_2) = 10m^3$

Temp. (T_2) = 400k

PV = mRT

$P_2 \times 10 = m_2 \times 287 \times 400$

$$m_2 = \frac{10P_2}{287 \times 400} \quad ...(i)$$

Condition of air in pipeline

mass of $m_P = (m_1 \quad m_2)$

So, the temp of $T_P = \frac{600+400}{2} = 500k$

Now, by using unsteady flow equations;

$m_1 u_1 - m_2 u_2 = m_p h_p$

$C_v (58.07 \times 600 - m_2 \times 400) = (m_1 - m_2)\ C_P \times 500$

$1 \times (58.07 \times 600 - m_2 \times 400) = (58.07 - m_2) \times 1.4 \times 500$

$$m_2 = \frac{5807}{300} = 19.35 \text{ kg}$$

Now put the value of m_2 in eq. (i)

$$\frac{P_2 \times 10}{287 \times 400} = 19.35$$

P_2 = 222214 Pa

P_2 = 0.22 MPa

104. According to the question;

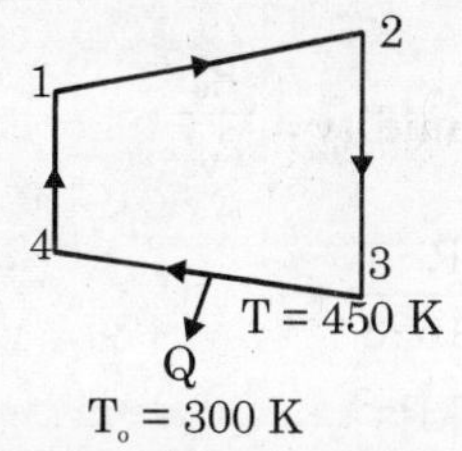

Consider steam turbine as system and atmosphere as surrounding.

ΔS_{System} = 7.5831 – 7.2605 = 0.3326 kJ/kg k

Heat rejected by steam to surrounding = 50 KJ

As Heat rejected at the surface of 450K

$$\Delta S_{surrounding} = \frac{50}{450} = 0.1111 \text{ kJ/kg k}$$

Entropy generation $\Delta S_{gen} = \Delta S_{system} + \Delta S_{surrounding}$

= 0.3326 + 0.1111 = 0.4437 kJ/kg k

So, the irreversibility = $T_o\ (\Delta s_{gen})$

= 300 × 0.4437

= 133.323 kg/kJ

105. According to the question;

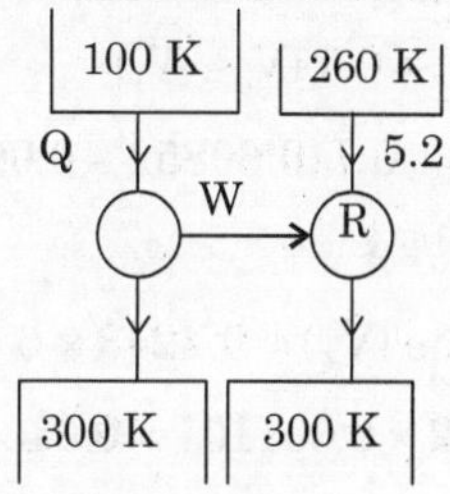

As we know that,

$$\eta_c = 1 - \frac{T_L}{T_H} = 1 - \frac{300}{1000} = 0.7 = 70\%$$

$$\eta_{actual} = \frac{0.7}{2} = 0.35$$

$$\text{Heat Engine efficiency } (\eta) = 0.35 = \frac{W}{Q}.$$

$$Q = \frac{W}{0.35} \quad ...(i)$$

We also know that,

$$(COP)_{Refrigerator} = \frac{\text{Desired}}{\text{Input}} = \frac{260}{300-260} = 6.5$$

$$(COP)_{actual} = \frac{6.5}{2} = 3.25$$

$$(COP)_{actual} = 3.25 = \frac{Q}{W} = \frac{5.2}{W}$$

So, W = 1.6 kW

Now put the value of W in eq. (i)

$$Q = \frac{1.6}{0.35} = 4.571 \text{ kW}$$

106. As we know that,

$$\text{Relative Humidity} = \frac{P_v}{P_{vs}}$$

$$\text{So, } 0.8 = \frac{P_v}{3.1698}$$

$P_v = 2.5354$ kPa

We also know that;

$$\text{Specific humidity (W)} = 0.622\frac{P_v}{P_t - P_v}$$

$$= 0.622 \times \frac{2.5354}{100 - 2.5354} \qquad (\because P_t = 100)$$

$$= 0.0161 \left[\frac{\text{Kg of water vapour}}{\text{Kg of dry air}}\right]$$

107. According to the question;

A	B
L + V	L
0.3MPa	0.6MPa

Part A, $V_A = V_f + 0.7\,(V_g - V_f)$

$= 0.001073 + 0.7\,(0.60852 - 0.001073)$

$V_A = 0.4243$ m³/kg

So, total volume $(V_A) = 0.4243 \times 6 = 2.5458$ m³

Part B, $V_B = 12 \times 0.001101 = 0.013212$ m³

Total specific volume

$$= \frac{V_A + V_B}{18} = \frac{2.5458 + 0.013212}{18}$$

$= 0.142$ m³/kg

108. According to the question :

$\longrightarrow$ [heating] $\longrightarrow$ 30°C

15°C
40% R_H $\qquad$ $R_H = ?$

$$R_H = \frac{P_v}{P_{vs}}$$

$$0.4 = \frac{P_v}{1.7057}$$

$P_v = 0.68228$ KPa

$$(R_H) = \frac{0.68228}{4.2469} \times 100 = 16.06\%$$

109. As given that,

Power generated by turbine under steady state = 16.2 Mw

At the exit turbine is saturated then

As we know that

given power = m(h – hg)

= 16(3625.34 – 2584.63)

= 16651.36 KW

= 16.651 MW

Which is not equal to power generation by turbine Under steady state condition

Assume steady state condition doesnot prevail.

inlt $\longrightarrow$ [ ] $\longrightarrow$ exit

$m_1 = 16\text{kg/sec} \qquad m_2 = ?$

$m_1 h - m_2 hg = 16.2 \times 1000$

$16 \times 3625.34 - m_2 \times 2584.63 = 16.2 \times 1000$

$m_2 = 16.17$ kg/sec.

So, the entropy generation = $[m_2 Sg - m_1 S]$

$= 16.17 \times 8.1501 - 16 \times 6.9028$

$= 21.342\ 317\ 1$ J/kg

F : POLYMER SCIENCE AND ENGINEERING

110. Interfacial polymerisation (then termed "interfacial poly-condensation") was first discovered by Emerson L. Wittbecker and Paul W. Morgan in 1959 as an alternative to the typically high-temperature and low-pressure melt polymerization technique. As opposed to melt polymerization, interfacial polymerization reactions can be accomplished using standard laboratory equipment and under atmospheric conditions.

Nylon-6, 6 is obtained by condensation Interfacial polymerisation of hexamethylene diamine with adipic acid under high pressure and at high temperature.

$$\underset{\text{Adipic acid}}{nHOOC - (CH_2)_4 - COOH} + \underset{\text{Hexamethylene diamine}}{nH_2N - (CH_2)_6 - NH_2} \longrightarrow$$

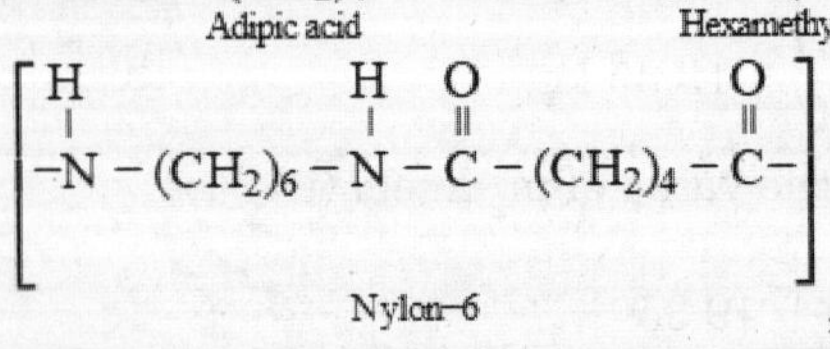

Nylon–6

111. The most common instrument used in determining the viscosity of rubber polymers and compounds is the Mooney viscometer. The

Mooney viscometer can be described as a rotation viscometer, in which the rubber is sheared between an inner rotor and an outer die. The die has a diameter of 50 mm and a height of 10 mm. The normal rotor has a diameter of 38 mm. For the measurement of high viscous materials, there is a smaller rotor with a diameter of 30 mm. The temperature of the die can be adjusted and normally 100 °C is used. The rotor is rotated with a speed of 2 rpm. This corresponds to an average shear rate of 1 s^{-1}. The torque required for driving the rotor, which is a measurement of the viscosity, is measured and presented in Mooney units. 8,3 ± 0,02 Nm corresponds to 100 Mooney units. Placing two circular pieces of the testing material in the die, one under and one above the rotor does the testing. The die is closed using a force of 11500 N and the material fills the die. Preheating is normally carried out for 1 minute after which the rotor is started. The viscosity decreases at first because of the heating and degradation of thixotropic properties. Readings are then normally taken after 4 minutes. The result is normally given as follows: 60 ML (1+4) 100 °C.

Where 60 refers to viscosity in Money units.

M : indicates that a large rotor.

L : large rotor

I : refers to a one minute pre-heating time, after the cavity is closed but before the rotor is switched on, during which the rubber warms up so the cavity temperature.

4 : refers to the time in minutes, after starting the rotor, at which the reading time is taken eat. 100°C is the temperature of the test.

Values under 40 can be designated as low, values between 40 and 80 as normal and values over 80 as high.

112. The initiator system which can be used for free radical polymerization at 5°C is $FeSO_4$ + t-butyl hydroperoxide. Polymerization at 5°C is giving the indication of cold SBR polymerization process. In the cold SBR polymerization process, use Cumene hydroperoxide and $FeSO_4$ as initiator. Whereas per sulfacte use in HOT SBR polymerization.

A free Radical Polymerization is a polymerizing approach by which successive addition of free radicals takes place to form a polymer unit. It is a type of chain-growth polymerization, with a cationic, anionic and coordination polymerization.

113. Ethylene Propylene Diene Monomer (EPDM), rubber improved weather resistance of high impact polystyrene by blending polystyrene. EPDM was dispersed in polystyrene in the place of poly butadiene which is normally used as impact additive. It was also necessary to add compatibilizer because both polymers are not miscible. Styrene-butadiene-styrene block copolymer, SBS, and styrene-ethylene-co-butylene-styrene block copolymer, SEBS, were used as compatibilizers and both contributed to increased weather-stability of composition.

114. **Thermoforming** is a manufacturing process where a plastic sheet is heated to a pliable forming temperature, formed to a specific shape in a mold, and trimmed to create a usable product. The sheet, or "film" when referring to thinner gauges and certain material types, is heated in an oven to a high-enough temperature that permits it to be stretched into or onto a mold and cooled to a finished shape. Its simplified version is vacuum forming. The discontinuous process consists of metering and mixing the various foam components and pouring the resulting reaction mixture into a mold. Generally, the mold has hinged sides for easier removal of the foam block and may have a fixed or rigid top pan to square the top of the block and provide some foam packing. So, we can say the thermoforming is a discontinuous polymer processing operation.

115. The polypropylene and polyethylene are immiscible and partially compatible and this blends have been attract a lot of attention due to their potential industrial applications for impact improvement and environmental stress cracking proprieties. **Isotactic polypropylene (PP)** has been blended with various types of polyethylene, high density (HDPE), low density (LDPE), linear low density (LLDPE), very low density (VLDPE) and ultra-low density (ULDPE).

116. The toughness of a polymer is to examine the area underneath the stress - strain curve for the particular polymer.

Both Charpy and Izod impact testing are popular methods of determining impact strength, or toughness, of a polymer material. In other words, these tests measure the total amount of energy that a material is able to absorb. This energy absorption is directly related to the brittleness of the material. Brittle materials, such as ceramics or glass, tend to have lower absorption rates than ductile materials like copper or aluminum.

Hence, options (a,c & d) are the correct answer.

117. • Polyester is a category of polymers that contain the ester functional group in every repeat unit of their main chain.

• Poly-lactic acid is a hydrophobic polymer that belongs to the class of biomaterials commonly referred as poly-α-hydroxy acids, poly-α-esters or aliphatic polyesters.

• The poly-hydroxy-butyrate polymer is a member of the polyester category, which is derived from the activity of Lactobacillus acidophilus or by using another biotechnological process at industrial scale. It can also be derived from the activity of other microbes in glucose controlled media.

• PCL is one of the polymers of the polyester series composed of repeating units of hexanoate units. PCL is synthesized by poly-condensation of 6-hydroxyhexanoic acid. It can also be synthesized by ring-opening polymerization of ε-capro-lactone using $Sn(Oct)_2$ as a catalyst.

Hence, options (b, c & d) are the correct answer.

118. In the condensation of bifunction reactants, the functionally of the reaction product is always, regardless of the extent of function.

119. As we know that,

$$\tan\delta = \frac{\text{loss modulus (E'')}}{\text{Storage modulus (E')}}$$

$$\tan 30 = \frac{E''}{E'}$$

$$\frac{1}{\sqrt{3}} = \frac{E''}{E'}$$

$$E' = \sqrt{3}E''$$

120. (P) : The physical unit of surface resistivity is ohm/Ω.

(Q) : The volume resistivity is measured in ohms-cm.

(R) : The coefficient of thermal expension measures.

(S) : The fractional change in size per degree change in temperature at a constant pressure expressed.

121. (P) : Injection blow molding is the process whereby the plastic preform is injection molded and the preform travels on the core rod to the blow mold station, where blow air enters through the core rod and lifts the hot preform material off the core rod and forms it by air pressure to the design of the female blow mold. Upon cooling to the desired shape, the blow molds open and the core rod containing the blown plastic bottle (Polymer product) indexes to the stripper station where the blown bottle is stripped from the core rod.

(Q) : Blister packaging is a form of plastic packaging that is used for small consumer goods and pharmaceuticals. It is mostly made from thermoformed plastics.

(R) : Pultrusion is a continuous process used to manufacture fiber-reinforced polymer (FRP) composite profiles with a constant cross section.

(S) : Sewage pipes are produced by horizontal extruding molten polymer through an annular opening in to several sizing, cooling devices that stabilize the final dimensions. The elements of the production line are extruder, Die, Sizing device, Cooling bath, Puller and Cutter.

Hence option (b) is the correct answer.

122. (P) : PPD(phenylenediamine) is an organic chemical that is widely used as an anti-ozonant and antioxidant in rubber tires. It is one of several p-phenylenediamine (PPD) additives used to protect various rubber materials.

(Q) : Trixylyl phosphate has good flame retardancy and can be used as a plasticizer for vinyl resins, cellulose resins, natural and synthetic rubbers.

(R) : Polybutadiene also has a major application as an impact modifier for polystyrene and acrylonitrile-butadiene-styrene resin (ABS) with about 25% of the total volume going into these applications. Typically about 7% polybutadiene is added to the polymerisation process to make these rubber-toughened resins.

(S) : Talc as a nucleating agent and reinforcing filler in poly (lactic acid) composites.

Hence option (d) is the correct answer.

123.

BOND	TYPE OF COMPOUND	FREQUENCY (CM-1)
–C– (with vertical bonds)	Alkane	2800-3000
=C–H	Alkenes, aromatics	3000-3100
≡C–H	Alkynes	3300
—O—H	Alcohols, phenols	3600-3650 (free) 3200-3500 (H-bonded) (broad)
—OH	Carboxylic acids	2500-3300
–N–H	Amines	3300-3500 (doublet for NH2)
O=C–H (–C(=O)–H)	Aldehyde	2720 and 2820
–C=C–	Alkenes	1600-1680
–C–C–	Aromatic	1500 and 1600
–C≡C–	Alkynes	2100-2270
–C(=O)–C	Aldehydes, ketones	1680-1740
–C≡N	Nitriles	2220-2260
C—N	Amines	1180-1360

A siemens per centimeter (s/cm) is a decimal multiple of the SI unit of electrical conductivity siemens per meter.

Hence, option (c) is the correct answer.

124. (P) : Expanded Polystyrene (EPS) works as an adequate thermal barrier, but if used solely, it is not an appropriate sound barrier. Even though it possesses sound absorbing properties, it doesn't have adequate mass so it won't be appropriate as an adequate acoustic panel for you to hang on your wall.

(Q) : The present invention also provides aforementioned polyether ether ketone (PEEK) matrix material in the application of making on the wear part, and wear part comprises motor bearing, gear and sealing member.

(R) : Polycarbonate safety glasses are lightweight, impact resistant, & naturally protects against harmful UV rays. It's the best option for safety and comfort.

(S) : The polymer gives good durability under thermal stress and/or harsh chemical environments, particularly in automotive under-the-hood applications. They are most commonly used in body exterior, safety applications including airbag covers, containers brake and fuel line clips, cable liners, power distribution boxes and TV cabinet.

125. (P) : LLDPE is produced by the copolymerization of ethylene with a few percent of a higher α-olefin comonomer (e.g., 1-butene, 1-hexene, and 1-octene) using Ziegler-Natta catalysts or the other coordination polymerization catalysts as described in HDPE.

(Q) : Nylon 6 is produced by the hydrolytic or catalytic ring opening polymerization of caprolactam to obtain epsilon-aminocaproic acid, which readily condenses to nylon 6 at high temperatures and a vacuum (to remove the water condensate) to high-molecular-weight polymer.

(R) : Emulsion polymerized styrene-butadiene rubber (E-SBR) is one of the most widely used polymers in the world. Styrene-butadiene or styrene-butadiene rubber (SBR) describe families of synthetic rubbers derived from styrene and butadiene (the version developed by Goodyear is called Neolite). These materials have good abrasion resistance and good aging stability when protected by additives.

(S) : The method of synthesize of polyamides can be employed to synthesize fully aromatic polyamides, known as aramids. However, due to their high melting point, they are typically prepared in solution at relative low temperature or, alternatively, by direct condensation of an aromatic di-acid with a di-amine at high temperature.

126. We know that

The weight of the sample is

$$W = \sum w_i = \sum N_i M_i$$

Mass in terms of percentage given, we use mass as w_i and divide M_i by the molecular weight and get the number of chains.

Weight average molecular weight:

$$M_w = \frac{\sum N_i M_i^2}{\sum N_i M_i} = \frac{\sum w_i M_i}{\sum w_i}$$

$$M_w = \frac{5 \times 1000 + 15 \times 20000}{5 + 15} = 17500$$

Number average molecular weight:

$$M_n = \frac{\sum N_i M_i}{\sum N_i}$$

$$M_n = \frac{\sum w_i}{\sum \frac{w_i}{M_i}} = \frac{5+15}{\frac{5}{10000} + \frac{15}{20000}} = 16000$$

So, the Polydispersity $= \frac{M_w}{M_n} = \frac{17500}{16000} = 1.09$

127. As given that :

η (Viscosity) = 6×10^{11} Poise; E(Modulus) $= 3 \times 10^6$ dyne cm^{-2}; (1 Poise = 1 dyne/cm^2.S)

As we know that,

Relaxation time $= \frac{\eta}{E}$

$$= \frac{6 \times 10^{11}}{3 \times 10^6} = 2 \times 10^5 \text{ sec}$$

$$\frac{3 \times 10^5}{24 \times 3600} = 2.31 \text{ days}$$

128. As we know that,

Total Volume = $V = V_f + V_m$;

$V_f = 0.6$ & $V_m = 0.4$

Conductivity of the composite in the longitudinal direction

$K_{composite} = K_m V_m + K_f V_f.$

$K_{composite} = 0.25 \times 0.4 + 1.05 \times 0.6 = 0.73$

129. Modulus of the composite;

$$E_{cl} = E_m V_m + E_f V_f \qquad \text{...(i)}$$

Where, E_m & E_f are the elastic modulus for matrix and fiber.

$V_m + V_f = 1$

V_m and V_f are volume fraction for matrix and fiber.

$\therefore V_m = V_f = 1$

V_m and V_f are volume fraction for matrix and fiber.

As given that

$V_m = 0.4$

$V_f = 0.6$

$E_m = 3$GPa

$E_r = 80$ GPa

By putting all value in eq. (i);

Elastic modulus (E_{cl}) = 0.4 * 3 + 0.6 * 80 = 49.2 GPa

As we know that,

Elastic Modulus = Tensile stress/Strain

Strain = Tensile stress/modulus = 110 MPa/49.2GPa

$= 0.0022 = 2.2 \times 10^{-3}$

130. Convert all the temperature in the Kelvin in from centigrade. Total weight fraction is equal to 1.

Where, $w_1 + w_2 = 1$

W_1 = nylon & W_2 = plasticizer;

T_{g1} for nylon = 50 + 273 = 323 K

T_{g2} for Plasticiser = –60 + 273 = 213K

T_g of nylon after the addition of plasticizer = 30 + 273 = 303

$$\frac{1}{T_g} = \frac{W_1}{T_{g1}} + \frac{W_2}{T_{g2}}$$

$$\frac{1}{303} = \frac{1 - W_2}{323} + \frac{W_2}{213}$$

$W_2 = 12.78\%$.

131. As we know that;

$$\text{crystallinity} = \frac{\text{enthalpy of polymer}}{\text{enthalpy of 100\% crystalline polymer}}$$

So, crystallinity $= \frac{135.6}{290} = 0.4675$ in case 1

$= 46.75\%$

After 5 year

crystallinity $= \frac{120}{290} = 0.4137 = 41.37\%$

Hence, the change in crystallinity

$46.75 - 41.37 = 5.38\%$

G : FOOD TECHNOLOGY

132. Process severity is the indication of the degree of impacts that is the development or operation of component applications in a process or system. And the use of high temperatures to preserve and ensure the **safety of food** is based on the effect of microbial destruction.

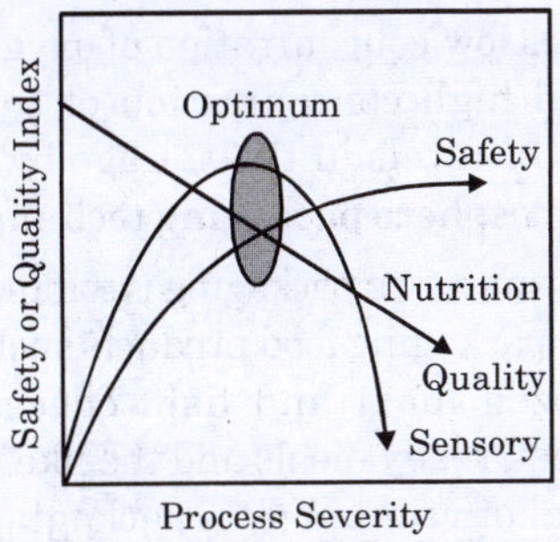

From above figure, the lines 1 and 2 in the given figure with respect to the effect of heat processing on food are safety and quality.

133. Milk is forced through a small passage at high velocity. The disintegration of the original fat globules is achieved by a combination of contributing factors such as turbulence and cavitation. The homogenization reduces fat globule size from an average of 3.5 µm in diameter to below 1 µm.

134. A characteristic feature which distinguishes Staphylococcus aureus from other pathogenic bacteria is its high tolerance to low water activity (aw) values and NaCl concentrations of up to 20%. Generally, minimal aw required for Staphylococcus aureus growth is 0.83-0.86.

135. The main (starter) cultures in yogurt are Lactobacillus bulgaricus and Streptococcus thermophilus. The function of the starter cultures is to ferment lactose (milk sugar) to produce lactic acid. The increase in lactic acid decreases pH and causes the milk to clot, or form the soft gel that is characteristic of yogurt. Yogurt was made by inoculating isolated bacteria (starter culture) usually Lactobacillus delbruckii subsp. bulgaricus and Streptococcus saliverius subsp. thermophilus into milk. After inoculation the mixture was incubated at 42°C for 6 to 8 hours and coagulation was observed and yogurt pH was measured.

136. Spray drying is one of the most convenient techniques for producing milk powders and for stabilizing milk constituents. In a dairy plant, spray drying technology is used to produce whey powder. The rate of spray drying depends on temperature of the incoming air, diameter of the whey droplet and heat transfer coefficient of hot air. Spray drying is a relatively temperature mild process. This means that damage of heat sensitive constituents is minimal. The spray-drying process helps to microencapsulate a fat, protein, and lactose matrix. It's a popular technique used for converting ice cream into solid powder form.

137. Parboiling is partial boiling of the paddy done before milling to increase its nutritional value, texture of cooked rice, and reduce the breakage in milling. Parboiling involves three steps : Soaking, steaming and drying.

Parboiling with high soaking temperature and steaming time increased the head rice yield, water uptake ratio, decreased percentage chalkiness, and enhanced the overall quality of the rice varieties.

Hence, options (b & c) are the correct answer.

138. Let's Assume that;

Total wet of paddy = 100 kg

Initial moisture = 18 kg

weight of dry matter = 100 – 18 = 82 kg

final moisture = 13

As we know that

$$\text{Content (\%)} = \frac{\text{final moistue (kg)}}{\text{final moisture} + \text{dry weight}} \times 100$$

Final moisture = 12.25 kg

Hence, amount of water removed = 18 – 12.25 = 5.75 kg

139. As we know that,

Centrifugal force (f_c) = $mr\omega^2$.

$= m \times r \times (2\pi N)^2$. $\quad \therefore \omega = 2\pi N$

$= m \times 0.1 \times (2 \times 3.14 \times 14.167)^2$

N = 850 rpm

$= \frac{850}{60} = 14.167$ rps

$= m \times 0.1 \times (2 \times 3.14 \times 14.167)^2$

$= m \times 791.54$ N

Centrifugal force (fc) in terms of gravity force.

$= \frac{m \times 791.54}{m \times g}$

$= \frac{791.54}{9.81} = 80.77$

140. As we know that;

Estimate sprolage pobalitity

$\log \frac{No}{N} = \frac{F}{D}$

$\log \frac{20}{N} = \frac{3}{1.6}$

N = 0.2667

Hence probability of spoilage would be = 2667 in 100 cans or 27 in 100 cans.

141. (P) : Degumming is the process for removal of phosphatides from crude soybean and other vegetable oils. The phosphatides are also called gums and lecithin. The latter term is also the common name for phosphatidyl choline, but common usage refers to the array of phosphatides present in all crude vegetable oils.

(Q) : In chemical refining, an alkali is used to neutralize the FFA and to remove the free fatty acidity from oil. The main purpose of Neutralization Process is to saponify the FFA by an alkaline solution and dilute the resulting soaps in a water phase. In the edible oil industry mostly caustic soda is used for the treatment with FFA's.

(R) : The bleaching process was investigated by varying the clay dosage, acid concentration and temperature. The highest removal of pigments was recorded at 7 M HCl concentration, 4 g clay dosage and 100°C temperature, and about 97.4% pigments were removed in 80 minutes.

(S) : Winterization is the process of removing compounds-such as fats, lipids, waxes, and chlorophyll-from the crude oil before the distillation process. Simply put, the winterization of oil involves taking a nonpolar substance (crude oil) and dissolving it in a polar solvent (ethanol) at sub-zero temperatures.

Hence, option(c) is the correct answer.

142. (P) : Aseptic packaging is a specialized manufacturing process in which food, pharmaceutical, or other contents are sterilized separately from packaging. The contents are then inserted into the container in a sterile environment.

(Q) : Active packaging makes non-passive contribution to product development and having active functions beyond the inert passive containment & protection of the product.

(R) : A controlled atmosphere is an agricultural storage method in which the concentrations of oxygen, carbon dioxide and nitrogen, as well as the temperature and humidity of a storage room are regulated. Both dry commodities and fresh fruit and vegetables can be stored in controlled atmospheres. Controlled atmosphere storage commonly uses low concentration of oxygen (O_2) levels and high concentration of carbon dioxide (CO_2) levels in the storage by the modified atmosphere packaging technique.

(S) : Vacuum skin packaging is commonly employed for packaging food products such as fresh and frozen meat and fish, cheese, processed meat, ready meals and the like. The ultimate goal of vacuum skin packaging is to create an attractive tight package that allows the consumer to view the product in its entirety.

143. Para-amino-benzoic acid (PABA) is a chemical that occurs naturally in the body. It is also found in several foods including grains, eggs, milk, and meat. But it is not a caramel flavor producing compound. Whereas white crystalline has solid with a characteristic, caramel-like odor and taste. In dilute solution it possesses a sweet, strawberry-like or pineapple-like flavor and odor.

144. (P) : Hammer mills consist of a series of hammers (usually four or more) hinged on a central shaft and enclosed within a rigid metal case. It produces size reduction by impact. The materials to be milled are struck by these rectangular pieces of hardened steel (ganged hammer) which rotates at high speed inside the chamber.

(Q) : Attrition mill also known as burr mill. The principle of attrition mills is that the material is rubbed between two surfaces. Both pressure and frictional forces are generated. The extent to which either of these forces predominates depends on the pressure with which both surfaces are held together and the difference in the speed of rotation of the surfaces.

(R) : Roll crushers are a type of compression crusher mostly used in heavy-duty industrial/aggregate applications. Roll crushers were once very popular in the mining industry, where they were used to reduce excess rocks, ores, and other materials.

(S) : The Rotary Knife is a common industrial application, used to cutting a web of material in to smaller lengths, without stopping the line.

145. Cryogenic liquids are liquefied gases of low boiling point such as liquid nitrogen and liquid carbon dioxide with boiling point -196°C and - 79°C respectively. Liquid nitrogen is the most commonly used cryogenic liquid in immersion freezing of food.

146. Omega-6 fatty acids (also referred to as ω-6 fatty acids or n-6 fatty acids) are a family of polyunsaturated fatty acids that have in common a final carbon-carbon double bond in the n-6 position, that is, the sixth bond, counting from the methyl end.

Table of Omega-6 fatty acids:

Common name	Lipid name	Chemical name
Linoleic acid (LA)	18 : 2 (n – 6)	all-cis-9, 12-octadecadienoic acid
Gamma-linolenic acid (GLA)	18 : 3 (n – 6)	all-cis-9, 12-octadecatrienoic acid
Calendic acid	18 : 3 (n – 6)	8E, 10 E, 12Z-octadecatrienoic acid
Eicosadienoic acid	20 : 2 (n – 6)	all-cis-11, 14-eicosadienoic acid
Dihomo-gamma-linolenic acid (DGLA)	20 : 3 (n – 6)	all-cis-8, 11, 14-eicosatrienoic acid
Arachidonic acid (AA, ARA)	20 : 4 (n – 6)	all-cis-5, 8, 11, 14-eicosatetraenoic acid
Docosadienoic acid	22 : 2 (n – 6)	all-cis-13, 16-docosatetraenoic acid
Adrenic acid	22 : 4 (n – 6)	all-cis-7, 10, 13, 16-docosatetraenoic acid
Osbond acid	22 : 5 (n – 6)	all-cis-4, 7, 10, 13, 16-docosapentaenoic acid
Tetracosatetraenoic acid	24 : 4 (n – 6)	all-cis-9, 12, 15, 18-tetracosatetraenoic acid
Tetracosapentaenoic acid	24 : 5 (n – 6)	all-cis-6, 9, 12, 15, 18-tetracosapentaenoic acid

The melting point of the fatty acids increases as the number of carbons in the chain increases.

147. The proteins that are made to subject to the process of re-naturation consist of blood serum and the enzyme called ribonuclease. There are many protein denaturation processes that are irreversible just like the egg white. A very genera outcome for getting the protein denaturation process is having the lose the activity in biology contained. When a solution of a protein is boiled, the protein frequently becomes insoluble-i.e., it is denatured-and remains insoluble even when the solution is cooled. The denaturation of the proteins of egg white by heat-as when boiling an egg-is an example of irreversible denaturation. Alpha-Helix and Beta-Pleated sheets are types of the secondary structure of the protein. They both are shaped by hydrogen bonding between the carbonyl O of one amino acid and the amino H of another.

148. (a) The Rice polisher machine is widely used in the rice mill production line to polish the white rice more bright and white.

(b) Break roll operation is the breaking open of the wheat and releasing of the endosperm from the bran gradually is a mechanical process achieved by passing the wheat between rolls that are fluted; these flutes being cut in a spiral, and the two rolls running at different speeds to produce a cutting or shearing action.

(c) Pigeonpea seed is composed of cotyledons (85%), embryo (1%), and seed coat (14%) (Faris and Singh 1990). The dietary nutrient values in the green seed, dry seed, and dhal of pigeon-pea. Rubber roll machines have been introduced recently to minimize the scouring.

(d) Maize is processed by dry or wet milling. Dry milling may or may not include Beall degermer as a preliminary step. Non-degermer dry milling is carried out on a local basis in small grist mills or in modern roller mills using sifters and purifiers.

So, in the given pair of options (a, b & d) are the correct pairs of milling equipment and the grain.

149. As we know that,

Reynolds number

$$R_e = \frac{\rho VD}{\mu}$$

$$= \frac{\text{Density} \times \text{velocity} \times \text{characteristic dimension}}{\text{viscosity}}$$

Schmidt number

$$= \frac{V(\text{kinematic viscosity of fluid})}{\text{Diffusivity}}$$

$$\text{Nusselt number} = \frac{\text{Convection heat transfer}}{\text{Conductive heat transfer}}$$

$$\text{Blot number} = \frac{\text{Internal conductive resistance}}{\text{External convective resistance}}$$

150. As we know that,

$$\text{Sauter mean diameter} = \frac{\sum n_i\, d_1^3}{\sum n_i\, d_1^2}$$

$$=\frac{5\times40^3+8\times30^3+50\times20^3+90\times17.5^3+148\times1.2^5+10\times10^3}{5\times40^2+8\times30^2+50\times20^2+90\times17.5^2+148\times12.5^2+10\times10^2}$$

$= 19.765 = 19.8\ \mu m.$

151. As we know that

Rate of heat loss

$$(Q)=\dot{m}C_P\Delta T$$

$$381.15=\dot{m}\times3.85\times1000\times(60-57)$$

$$\dot{m}=0.033\ kg/s$$

$$\text{Reynolds number (Re)}=\frac{\rho VD}{\mu}$$

$$\therefore\ \rho=\frac{M}{Volume}=\frac{M}{Area\times L}=\frac{M}{\frac{\pi}{4}D^2\times L}$$

$$\therefore\ Velocity=\left[\frac{L}{t}\right]$$

$$So,\ R_e=\frac{M\times Velocity\times D}{\frac{\pi}{4}D^2\times L\times\mu}$$

$$R_e=\frac{\frac{\mu}{t}\times4}{\pi\times D\times\mu}$$

$$R_e=\frac{\dot{m}\times4}{\pi\ D\times\mu}$$

$$=\frac{0.033\times4}{3.14\times2\times10^{-2}\times1.2\times10^{-3}}=1751.59$$

152. As given that

diameter = 3 cm

thickness = 1.5 cm

inside the radius $(r_i) = 1.5\ cm = 1.5\times10^{-2}cm$

Outside the radius $(r_o) = 1.5 + 1.5 = 3\ cm = 3\times10^{-2}m$

As we know that;

The overall heat transfer coefficient

$$\frac{1}{U_i}=\frac{r_i}{r_oh_o}+\frac{r_i\ln\frac{r_o}{r_i}}{K}+\frac{1}{h_i}$$

$$\frac{1}{25}=\frac{1.5}{3\times h_o}+\frac{1.5\times10^{-2}\times\ln\frac{3}{1.5}}{50}+\frac{1}{30}$$

$h_o = 77.42\ w/m^2k$

153. As we know that;

Relative humidity

$$(R_H)=\frac{\text{Specific Humidity}}{\text{Absolute Humidity}}\times100$$

$$50=\frac{\text{Specific Humidity of air}}{6.28}\times100$$

So, that specific humidity of air $(P_{wair}) = 6.28\times0.5$
$= 3.1$ KPa

$$\text{Humidity ratio of air}=(H)=\frac{18.02}{28.97}\times\frac{(P_{w\,air})}{P-P_{w\,air}}$$

$$=\frac{18.02}{28.97}\times\frac{3.14}{101.32-3.14}$$

$= 0.01989 = 0.020$ kg water/(Kg dry air)

H : ATMOSPHERIC AND OCEANIC SCIENCES

154. A schematic of vertical profiles of concentrations of two gases P and Q in the atmosphere near a coastal station.

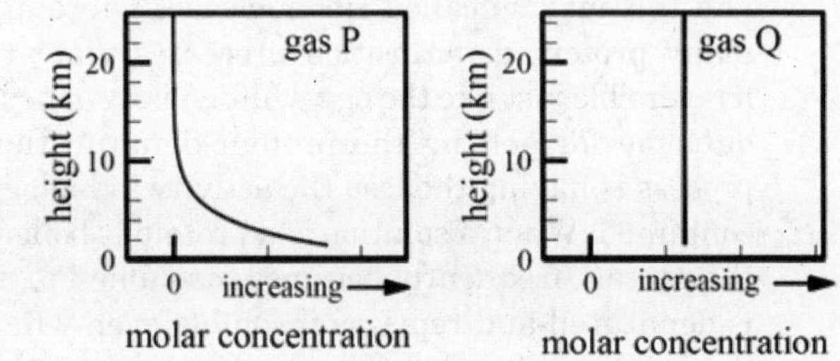

The correct pair representing P and Q, respectively, is water vapor and CO_2.

155. As per given equation;

Same dimension you can add.

$$\rho\frac{Dv}{Dt}=\frac{Kg}{m^3}.\frac{m}{s.s}=\frac{Kg}{m^2s^2}$$

from the third term

$$\mu\nabla^2v=\frac{Kg}{m^2s^2}$$

$\therefore\ Kg = M$

$\therefore\ m = L$

So, dimension of third term is

$\therefore\ s = T = [M]\ [L]^{-2}\ [T]^{-2}$

156. Tropical cyclone does not form over the equator because the coriolis force is zero at the equator. The coriolis force is simply too weak to move the air around low pressure and air prefers to flow from high to low pressure. Even though equator is a zone of wind convergence, it has nothing to do with tropical cyclone as such. Temperature is highest north and south of the equator and not exactly at the equator. Fronts are required for formation of temperate cyclone and not tropical ones.

157. The Equatorial Counter Current is an eastward flowing, wind-driven current which extends to depths of 100-150 metres (330-490 ft) in the Atlantic, Indian, and Pacific Oceans. More often called the North Equatorial Countercurrent (NECC), this current flows west-to-east at about 3-10°N in the Atlantic, Indian Ocean and Pacific basins, between the North Equatorial Current (NEC) and the South Equatorial Current (SEC).

158. In regions of rising air, the diabatic motion is balanced predominantly by condensational heating; in regions of sinking air, it is balanced mainly by the emission of thermal infrared radiation to space by water vapor. The balance between sinking diabatic motion and radiative cooling by water vapor has important implications for anvil cloud temperature and amount in the upper tropical troposphere. Once condensation occurs in adiabatic cooling associated with ascent, the latent heat of condensation is released. The rate at which saturated air cools as it expands adiabatically is smaller than the rate at which unsaturated air cools adiabatically, because part of the cooling is canceled by the latent heat released during condensation. The effects of the tertropical convergence zone on the atmospheric circulation when it has reached a steady state strong influences are seen on the horizontal wind fields at the upper and lower levels, on the vertical velocity field and on the temperature field at the middle level.

159. A Baroclinic wave is primarily responsible for the poleward transport of energy in the mid-latitude troposphere atmosphere and oceans constitutes an important branch of the global energy budget, and its role in the climate system has been the subject of many studies. In the atmosphere, the transport is affected by "eddies" and large scale meridional cells, both with zero net mass transport across latitude circles, but also partly by processes associated with a net transport of mass across latitude circles.

160. In the boreal summer half of the year there exists in the upper troposphere of the Eurasian subtropics a huge high pressure system centered over South Asia southwest of the Tibetan Plateau (TP), which was formerly known as the Tibetan High but is now referred to more frequently as the South Asian High (SAH). The bulk of the convective detrainment occurs at about 200 hPa or 350 K (12-13 km) and is associated with the divergent upper troposphere component of the Hadley circulation. At such altitudes the vertical motion is almost everywhere descending except within the clouds, due to a dominating radiative cooling. The upper tropospheric subtropical anticyclone of Asia centered over SE Tibet is known as Tibetan Anticyclone (TA).

161. As we know that

$$R_o = \frac{U}{fL}$$

Where, R_o = Ross by Number

U = Characteristic horizontal velocity scale

f = Coriolis frequency

L = Characteristic horizontal length scale

$$\text{So, } R_o = \frac{10}{10^{-4} \times 10 \times 1000}$$

$$\boxed{R_o = 0.1}$$

162. Rayleigh scattering from air molecules

$$I \propto \frac{1}{\lambda^4}$$

Ratio of scattering efficiency of Red light to Blue light

$$I_n = \frac{\lambda^4_{Blue}}{\lambda^4_{Red}} = \frac{(0.45)^4}{(0.65)^4} = 0.23$$

163. An unsaturated moist air parcel undergoes adiabatic ascent in atmosphere without mixing with surrounding air. Air is so clean that there is no possibility for heterogeneous nucleation. So, the vertical variation of water vapor pressure (shown as continuous line) and saturation water vapor pressure (shown as dotted/dashed line) of the parcel:

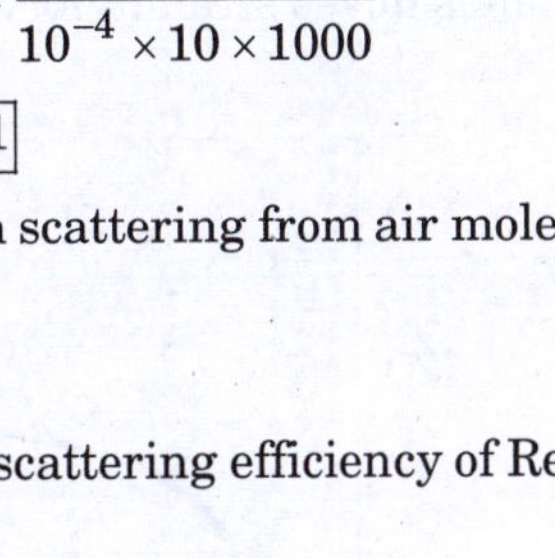

164. An infinitely long, cylindrical container of radius R rotates at the angular speed Ω. It contains water which is also in solid-body rotation with angular speed Ω = (0, 0, Ω). At time t = 0, the container suddenly stops rotating, and the contained water gradually comes to rest.

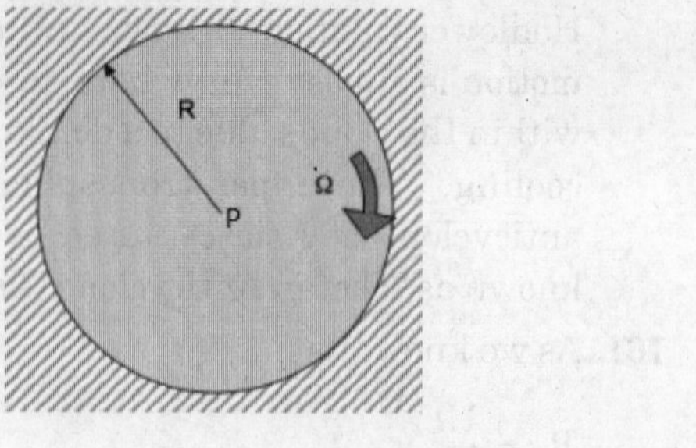

If we measure radius from the center P, the circular velocity satisfies v = Ω. This flow has vorticity ω = 2Ω everywhere. So, the circulation per unit area around a circular loop in the horizontal plane of radius r (r < R), whose center coincides with the axis of rotation is 2 Ω.

Alternative

Fluid uniform rotation with angular velocity (w) = Ω

So, the velocity field (v) = Ω × measured from the rotation axis.

Now, taking curl both side :

$\omega = \nabla \times v$

So, the circulation $\boxed{\omega = 2\Omega}$

165. Generally, temperatures decrease with increasing height because the atmosphere distributes itself according to gravity: Pressure generally decreases with height because the pressure is determined by the mass of the atmosphere above some point. At the surface, the entire mass of the atmosphere is above you bit as you go higher and higher, the amount of atmospheric mass above you decreases, so the pressure falls with height.

The relation is not so straight forward as following graph shows :

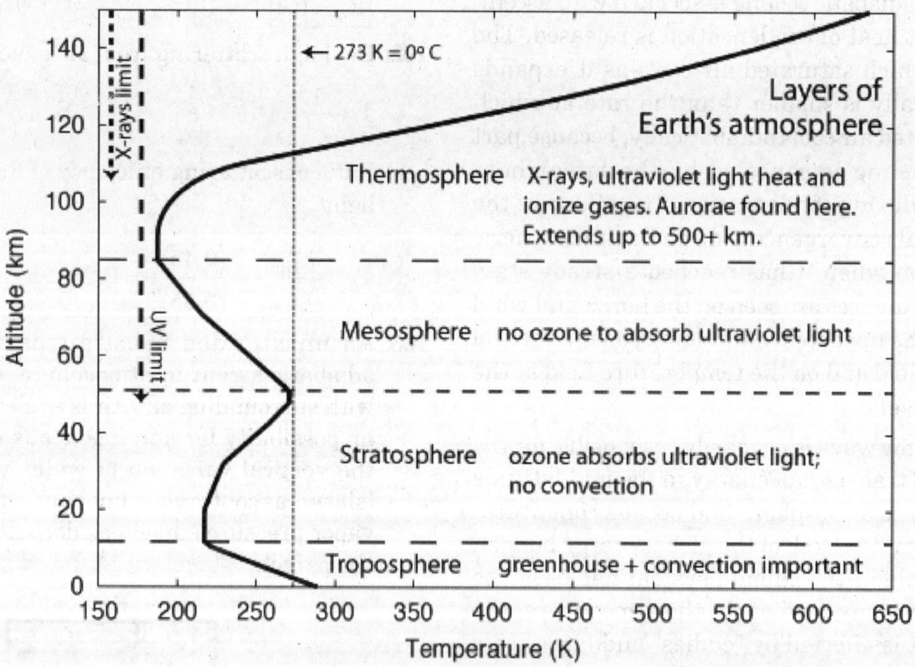

So, if the concentration of a vertically well-mixed greenhouse gas suddenly increases in this layer, then an immediate consequence is that infrared radiation leaving the top of the layer increases.

166. In the absence of biological and chemical activity, relative concentrations of dissolved gases in the surface sea water at equilibrium:

Atmospheric Gas	Mole Fraction	$k_H$mol/(L*atm)
N_2	7.81×10^{-1}	6.70×10^{-4}
O_2	2.10×10^{-1}	1.30×10^{-3}
Ar	9.34×10^{-3}	1.40×10^{-3}
CO_2	3.33×10^{-4}	3.50×10^{-2}
CH_4	2.00×10^{-6}	1.40×10^{-3}
H_2	5.00×10^{-7}	7.80×10^{-4}

The concentration of dissolved gases depends upon the solubility of gases in water. The given gases N_2, Ar and CO_2 have solubility in order Ar < N_2 < CO_2 because Ar is noble gas it doesn't react with water and is not soluble in water. Whereas CO_2 being more polar has highest solubility.

So, the ordered as $[N_2] > [CO_2] > [Ar]$.

167. Gravitational forces exerted by the Sun and the Moon are mainly responsible for ocean tides. Tidal oscillations have periods that are some integer fraction of a solar or lunar day. The solar diurnal and semidiurnal tides have 24 and 12 h periods, respectively. The lunar diurnal tidal period is about 24.8 h, while the lunar semidiurnal period is 12.4 h. And tidal amplitudes are larger during a solar eclipse compared to that during a lunar eclipse.

168. Tropical cyclones form as a result of significant convective activity, and are warm core. Meso-cyclones form as warm core cyclones over land, and can lead to tornado formation. A tropical cyclone is a rapidly rotating storm system characterized by a low-pressure center, a closed low-level flow of atmospheric circulation, strong winds, and a spiral arrangement of thunderstorms that produce heavy rain and squalls. Depending on its location and strength, a tropical cyclone is referred to by different names, including hurricane, typhoon, tropical storm, cyclonic storm, tropical depression, or simply cyclone. If the upper levels divergence is favorable for hemisphere tropical cyclone development, then there is a region of divergence aloft above the developing low pressure center. This will help pull the air that is converging at the surface upward and continue to develop the surface cyclone.

Hence, options (a, b & d) are the correct answer.

169. In the northern hemisphere, wind flows in an anticlockwise direction around areas of low pressure, while in a clockwise direction around areas of high pressure. Generally, the gradient wind is slower than the geostrophic wind. We say that it is super geostrophic or sub geostrophic, respectively. We found in class that in the Northern Hemisphere, the gradient wind "blowing" around troughs is sub geostrophic, and blowing around ridges it is super geostrophic.

170. In Biogeochemical cycle the biological carbon pump (BCP) contributes to the role of the ocean in taking up and storing carbon dioxide (CO_2) by removing the gas from the atmosphere, changing it into living matter, and distributing it to the deeper ocean layers. Without the BCP, the atmospheric concentration of CO_2 would be much higher. The levels of CO_2 in the atmosphere would be much higher if not for the biological carbon pump. So, Shutdown of the biological pump in the ocean would have resulted in higher CO_2 concentration in the atmosphere compared to present-day. Anthropogenic CO_2-driven, Ocean acidification has been reported to pose a severe threat to marine calcifying organisms because the calcification rates of marine organisms in the surface ocean layer would lead to an increase in the surface ocean CO_2.

171. As we know that

Heat Engine Efficiency

$$\eta = \frac{\text{Work output}}{\text{Energy supplied}}$$

As given that,

Energy supplied - Radiation absorbed = 2ω w/m^2

Work output - connect absorbed radiation to kinetic energy of winds - 10 w/m^2

$$\text{So, } \eta = \frac{10}{200} = 0.05$$

$$\eta(\%) = 5$$

172. As we know that,

Radius of the inertial oscillation

$$r = \frac{V}{f}$$

As given that,

v (speed of driffer) = 2 m/s

f (coriolis parameter) = 2×10^{-4} s^{-1}

$$r = \frac{2}{2 \times 10^{-4}} = 10000 \text{ m}$$

r = 10 KM

173. As given that,

$$\frac{\partial P}{\partial Z} = 5 \text{ N/m}^3$$

$$\rho = 1 \text{ kg/m}^3$$

z = 500 m

As we know that,

Rotation of the air inside the meso-cyclone is described by cyclostrophic balance

$$\frac{\partial P}{\partial 2} = \rho \frac{v^2}{z}$$

$$\text{So, } 5 = 1 \times \frac{v^2}{500}$$

$$v = 50 \text{ m/s}$$

174.

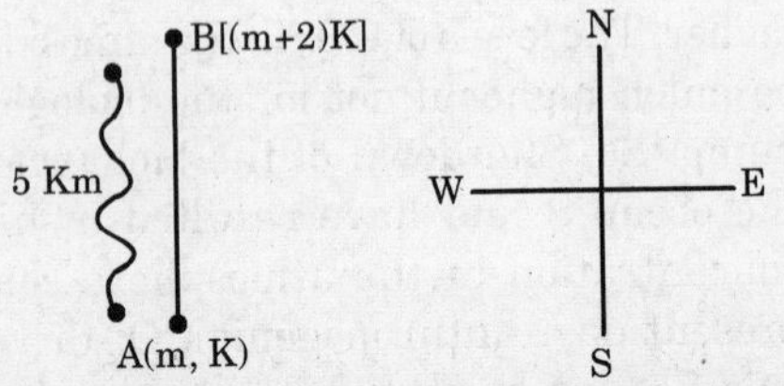

Here, given that B is 55 km north of station A. And also temperature of B is more then 2K of the station A. And also steady northerly wind blows at 1 m/s.

Let, at any time the temperature of A is mK (K = Kelven).

So, by the given condition at this time temperature of B is (m + 2)K.

So, when distance between two station is 5 K, then difference between temperature of two station is (m + 2 – m) = 2K.

Also, given wind blows at 1 m/s.

So, wind travel the distance in 2 hours is 7.2 km.

Now,

Distance	Change in temperature
5 Km	2 K
7.2 Km	?

Change in temperature

$$= \frac{2 \times 7.2}{5} K$$

$$= \frac{2 \times 72}{5 \times 10} K$$

$$= 2.88$$

175. As we know that,

$$T = \sqrt[4]{\frac{S_o(1-\alpha)}{4\sigma}}$$

$$255 = \sqrt[4]{\frac{1370 \times (1-\alpha)}{4 \times 5.67 \times 10^{-8}}}$$

$$\alpha = 0.3000$$

As given that,

New Albedo $\alpha = 0.3500$

$$\sqrt[4]{\frac{1370 \times (1-0.3500)}{4 \times 5.67 \times 10^{-8}}}$$

$$T = 250.321 \text{ K}$$

GATE - 2023
XE : ENGINEERING SCIENCES

GENERAL APTITUDE

Q. 1 to Q. 5 : Carry One Mark Each.

1. The village was nestled in a green spot, _______ the ocean and the hills.

(a) through (b) in
(c) at (d) between

2. Disagree : Protest : : Agree : _______
(By word meaning)

(a) Refuse (b) Pretext
(c) Recommend (d) Refute

3. A 'frabjous' number is defined as a 3 digit number with all digits odd, and no two adjacent digits being the same. For example, 137 is a frabjous number, while 133 is not. How many such frabjous numbers exist?

(a) 125 (b) 720
(c) 60 (d) 80

4. Which one among the following statements must be TRUE about the mean and the median of the scores of all candidates appearing for GATE 2023?

(a) The median is at least as large as the mean.
(b) The mean is at least as large as the median.
(c) At most half the candidates have a score that is larger than the median.
(d) At most half the candidates have a score that is larger than the mean.

5. In the given diagram, ovals are marked at different heights (h) of a hill. Which one of the following options P, Q, R, and S depicts the top view of the hill?

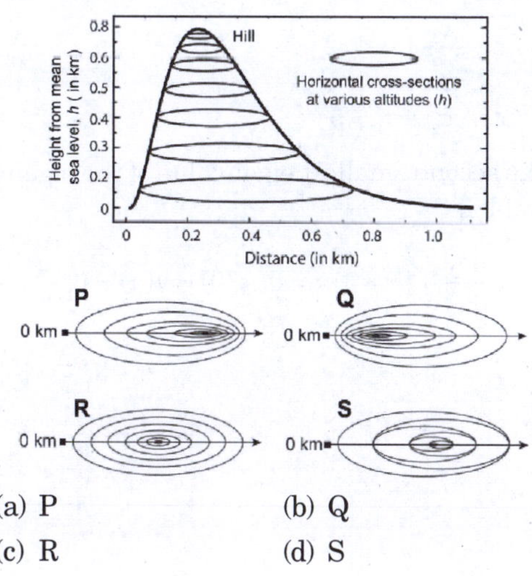

(a) P (b) Q
(c) R (d) S

Q.6 – Q.10 Carry TWO marks Each

6. Residency is a famous housing complex with many well-established individuals among its residents. A recent survey conducted among the residents of the complex revealed that all of those residents who are well established in their respective fields happen to be academicians. The survey also revealed that most of these academicians are authors of some best-selling books.

Based only on the information provided above, which one of the following statements can be logically inferred with certainty?

(a) Some residents of the complex who are well established in their fields are also authors of some best-selling books.
(b) All academicians residing in the complex are well established in their fields.
(c) Some authors of best-selling books are residents of the complex who are well established in their fields.
(d) Some academicians residing in the complex are well established in their fields.

7. Ankita has to climb 5 stairs starting at the ground, while respecting the following rules :

1. At any stage, Ankita can move either one or two stairs up.
2. At any stage, Ankita cannot move to a lower step.

Let F(N) denote the number of possible ways in which Ankita can reach the Nth stair. For example, F(1) = 1, F(2) = 2, F(3) = 3.

The value of F(5) is _______.

(a) 8 (b) 7
(c) 6 (d) 5

8. The information contained in DNA is used to synthesize proteins that are necessary for the functioning of life. DNA is composed of four nucleotides: Adenine (a), Thymine (T), Cytosine (c), and Guanine (G). The information contained in DNA can then be thought of as a sequence of these four nucleotides: A, T, C, and G. DNA has coding and non-coding regions. Coding regions—where the sequence of these nucleotides are read in groups of three to produce individual amino acids—constitute only about 2% of human DNA. For example, the triplet of nucleotides CCG codes for the amino acid glycine, while the triplet GGA codes for the amino acid proline. Multiple amino acids are then assembled to form a protein.

Based only on the information provided above, which of the following statements can be logically inferred with certainty?

(i) The majority of human DNA has no role in the synthesis of proteins.

(ii) The function of about 98% of human DNA is not understood.

(a) only (i) (b) only (ii)

(c) both (i) and (ii) (d) neither (i) nor (ii)

9. Which one of the given figures P, Q, R and S represents the graph of the following function?

$$f(x) = \big|\,|x+2| - |x-1|\,\big|$$

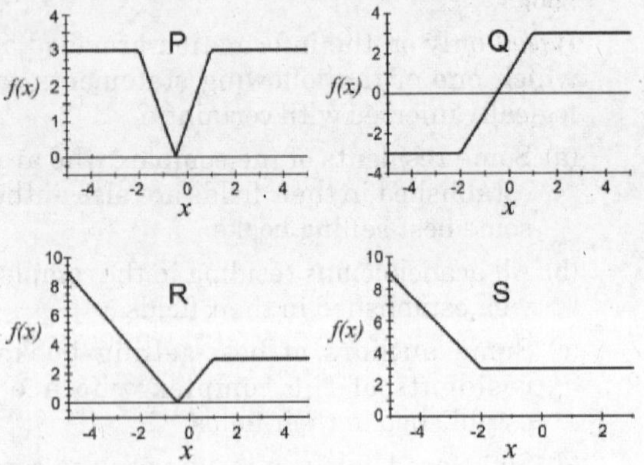

(a) P (b) Q

(c) R (d) S

10. An opaque cylinder (shown below) is suspended in the path of a parallel beam of light, such that its shadow is cast on a screen oriented perpendicular to the direction of the light beam. The cylinder can be reoriented in any direction within the light beam. Under these conditions, which one of the shadows P, Q, R, and S is NOT possible?

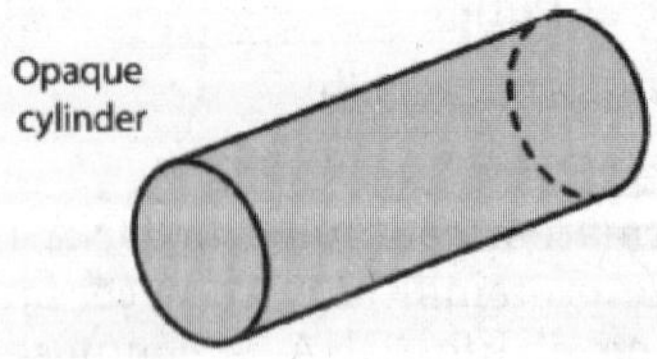

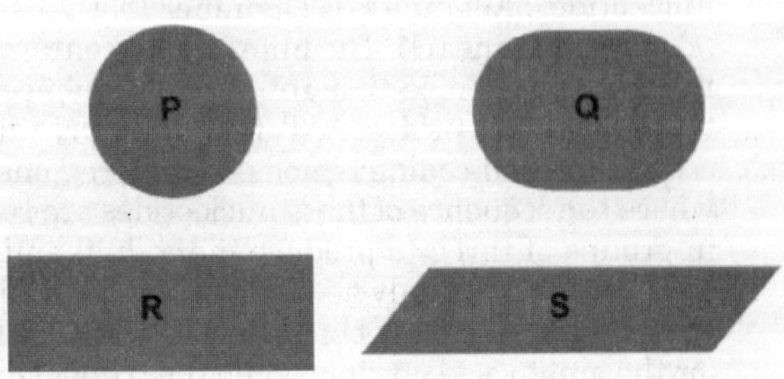

(a) P (b) Q

(c) R (d) S

ENGINEERING MATHEMATICS

Engineering Mathematics: XE-A (Compulsory)

Q.11 – Q.17 Carry ONE mark Each

11. Let A be a 3 × 3 real matrix having eigenvalues 1, 2, and 3. If $A^2 + 2A + I$, where I is the 3 × 3 identity matrix, then the eigen values of B are

(a) 4, 9, 16

(b) 1, 2, 3

(c) 1, 4, 9

(d) 4, 16, 25

12. Let $f : \mathbb{R}^2 \to \mathbb{R}$ be a function defined by

$$f(x,y) = \begin{cases} \dfrac{xy}{|x|+y}, & y \neq -|x| \\ 0, & \text{otherwise,} \end{cases}$$

Then which one of the following statement is TRUE?

(a) f is NOT continuous at (0,0).

(b) $\dfrac{\partial f}{\partial x}(0,0) = 0$, and $\dfrac{\partial f}{\partial y}(0,0) = 1$.

(c) $\dfrac{\partial f}{\partial x}(0,0) = 1$, and $\dfrac{\partial f}{\partial y}(0,0) = 0$.

(d) $\dfrac{\partial f}{\partial x}(0,0) = 1$, and $\dfrac{\partial f}{\partial y}(0,0) = 1$.

13. If the quadrature formula

$$\int_{-1}^{1} f(x)dx \approx \frac{1}{9}\left(c_1 f(-1) + c_2 f\left(\frac{1}{2}\right) + c_3 f(1)\right)$$

is exact for all polynomials of degree less than or equal to 2, then

(a) $c_1 + \dfrac{c_2}{4} + c_3 = 6$

(b) $c_1 + \dfrac{c_2}{3} + c_3 = 4$

(c) $c_1 + \dfrac{c_2}{2} + c_3 = 2$

(d) $c_1 + c_2 + c_3 = 5$

14. The second smallest eigenvalue of the eigenvalue problem

$$\frac{d^2y}{dx^2} + (\lambda - 3)y = 0,\ y(0) = y(\pi) = 0,$$

is

(a) 4

(b) 3

(c) 7

(d) 9

15. Which one of the following functions is differentiable at z = 0 but NOT differentiable at any other point in the complex plane $\mathbb{C}$?

(a) $f(z) = |z|, z \in \mathbb{C}$

(b) $f(z) = \sin(z), z \in \mathbb{C}$

(c) $f(z) = \begin{cases} e^{\frac{1}{z}}, & z \neq 0 \\ 0, & z = 0 \end{cases}$ for $z \in \mathbb{C}$

(d) $f(z) = e^{-z^2}, z \in \mathbb{C}$

16. If the polynomial

$$P(x) = a_0 + a_1x + a_2x(x-1) + a_3x(x-1)(x-2)$$

interpolates the points (0,2), (1,3), (2,2), and (3, 5), then the value of $P\left(\frac{5}{2}\right)$ is ________ (round off to 2 decimal places).

17. The value of m for which the vector field

$$\vec{F}(x,y) = \left(4x^my^2 - 2xy^m\right)\hat{i} + \left(2x^4y - 3x^2y^2\right)\hat{j}$$

is a conservative vector field, is____(in integer).

XE-A : Q.18 – Q.21 Carry TWO marks Each

18. Let

$$P = \begin{bmatrix} 4 & -2 & 2 \\ 6 & -3 & 4 \\ 3 & -2 & 3 \end{bmatrix}, \text{ and } Q = \begin{bmatrix} 3 & -2 & 2 \\ 4 & -4 & 6 \\ 2 & -3 & 5 \end{bmatrix}$$

The eigen values of both P and Q are 1,1, and 2. Which one of the following statements is TRUE?

(a) Both P and Q are diagonalizable

(b) P is diagonalizable but Q is NOT diagonalizable

(c) P is NOT diagonalizable but Q is diagonalizable

(d) Both P and Q are NOT diagonalizable

19. The surface area of the portion of the paraboloid $z = x^2 + y^2$

that lies between the planes z = 0 and $z = \frac{1}{4}$ is

(a) $\frac{\pi}{6}\left(2\sqrt{2}-1\right)$ (b) $\frac{\pi}{2}\left(2\sqrt{2}-1\right)$

(c) $\pi\left(2\sqrt{2}-1\right)$ (d) $\frac{\pi}{3}\left(2\sqrt{2}-1\right)$

20. The probability of a person telling the truth is $\frac{4}{6}$. An unbiased die is thrown by the same person twice and the person reports that the numbers appeared in both the throws are same. Then the probability that actually the numbers appeared in both the throws are same is ________ (round off to 2 decimal places).

21. Let u(x, t) be the solution of the initial boundary value problem

$$\frac{\partial u}{\partial t} - \frac{\partial^2 u}{\partial x^2} = 0,\ x \in (0,2),\ t > 0$$

$$u(x,0) = \sin(\pi x),\ x \in (0,2)$$

$$u(0,t) = u(2,t) = 0.$$

Then the value of $e^{\pi^2}\left(u\left(\frac{1}{2},1\right) - u\left(\frac{3}{2},1\right)\right)$ is ________ *(in integer)*.

FLUID MECHANICS (XE-B)

XE-B: Q.22 – Q.30 Carry ONE mark Each

22. Match the following measuring instruments with the appropriate figures.

I – Pitot probe

II – Pitot-static probe

III – Piezometer

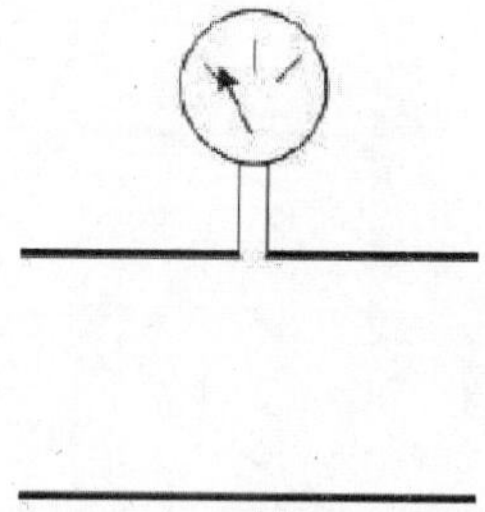

(P)

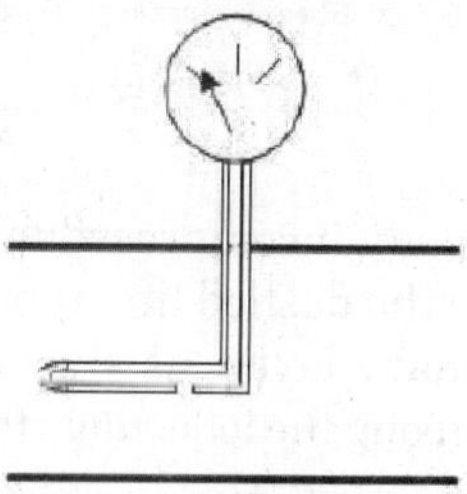

(Q)

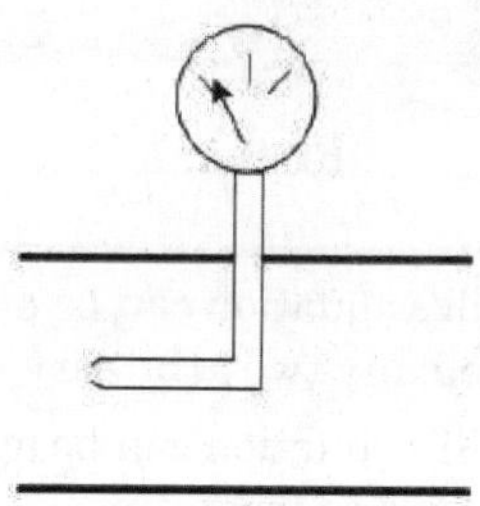

(R)

(a) I – P; II – Q; III – R

(b) I – R; II – Q; III – P

(c) I – R; II – P; III – Q

(d) I – Q; II – P; III – R

23. Among the following non-dimensional numbers, which one characterizes periodicity present in a transient flow?

(a) Froude number (b) Strouhal number

(c) Peclet number (d) Lewis number

24. For an incompressible boundary layer flow over a flat plate shown in the figure, the momentum thickness is expressed as

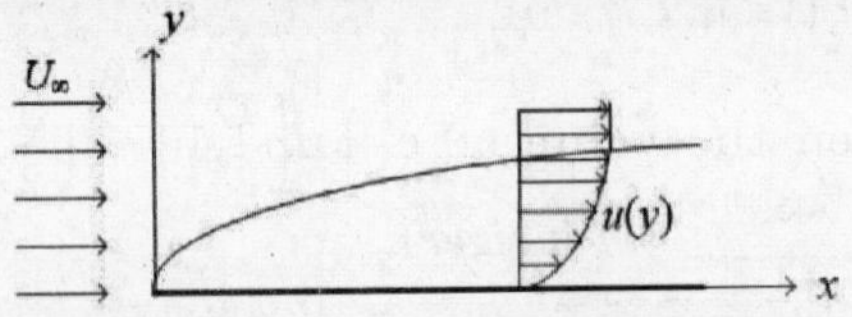

(a) $\int_0^\infty \frac{u}{U_\infty} dy$ (b) $\int_0^\infty \left(1-\frac{u}{U_\infty}\right) dy$

(c) $\int_0^\infty \frac{u}{U_\infty}\left(1-\frac{u}{U_\infty}\right) dy$ (d) $\int_0^\infty \left(1-\frac{u^2}{U_\infty^2}\right) dy$

25. Among the shear stress versus shear strain rate curves shown in the figure, which one corresponds to a shear thinning fluid?

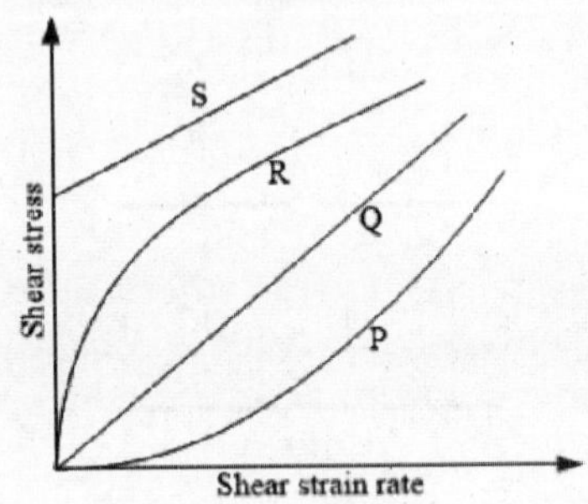

(a) P (b) Q

(c) R (d) S

26. Consider steady incompressible flow over a flat plate, where the dashed line represents the edge of the boundary layer, as shown in the figure. Which one among the following statements is true?

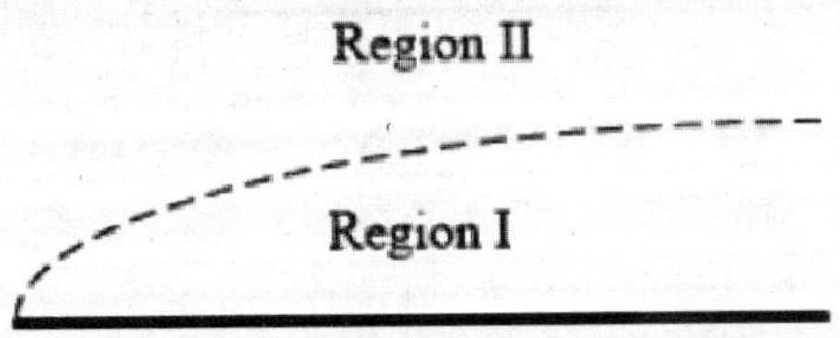

(a) Bernoulli's equation can be applied in Region I between any two arbitrary points.

(b) Bernoulli's equation can be applied in Region I only along a streamline.

(c) Bernoulli's equation cannot be applied in Region II.

(d) Bernoulli's equation cannot be applied in Region I.

27. An inviscid steady incompressible flow is formed by combining a uniform flow with velocity U_∞ and a clockwise vortex of strength K at the origin, as shown in the figure. Velocity potential (ϕ) for the combined flow in polar coordinate (r, θ) is

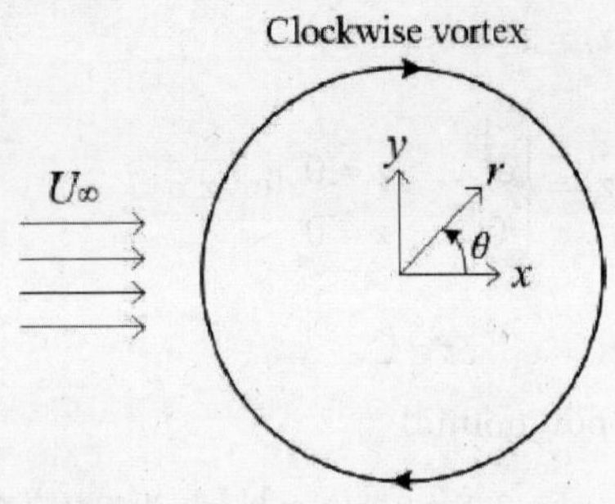

(a) $\phi = \frac{K\theta}{2\pi} - U_\infty r \cos\theta$

(b) $\phi = \frac{K\theta}{2\pi} - U_\infty r \sin\theta$

(c) $\phi = K \ln r + U_\infty r \cos\theta$

(d) $\phi = -K \ln r + U_\infty r \sin\theta$

28. Which of the following statements are true?

(i) Conservation of mass for an unsteady incompressible flow can be represented as $\nabla \cdot \bar{V} = 0$, where $\bar{V}$ denotes velocity vector.

(ii) Circulation is defined as the line integral of vorticity about a closed curve.

(iii) For some fluids, shear stress can be a nonlinear function of the shear strain rate.

(iv) Integration of the Bernoulli's equation along a streamline under steady-state leads to the Euler's equation.

(a) (i), (ii) and (iv) only

(b) (i), (ii) and (iii) only

(c) (i) and (iii) only

(d) (ii) and (iv) only

29. For a two-dimensional flow field given as $\vec{V} = -x\hat{i} + y\hat{j}$, a streamline passes through points (2, 1) and (5, p). The value of p is

(a) 5 (b) $\frac{5}{2}$

(c) $\frac{2}{5}$ (d) 2

30. A stationary object is fully submerged in a static fluid, as shown in the figure. Here, CG and CB stand for center of gravity and center of buoyancy, respectively. Which one(s) among the following statements is/are true?

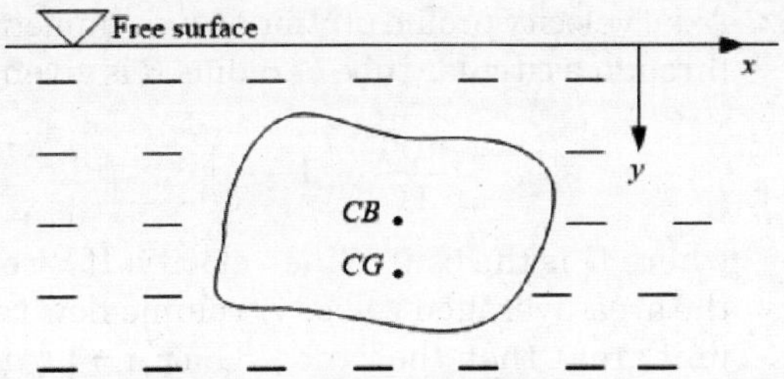

(a) The object is in stable equilibrium if $y_{CG} > y_{CB}$.

(b) The object is in stable equilibrium if $y_{CG} < y_{CB}$.

(c) The object is in neutral equilibrium if $y_{CG} = y_{CB}$.

(d) The object is in unstable equilibrium if $y_{CG} = y_{CB}$.

XE-B: Q.31 – Q.43 Carry TWO marks Each

31. Consider steady fully-developed incompressible flow of a Newtonian fluid between two infinite parallel flat plates. The plates move in the opposite directions, as shown in the figure. In the absence of body force and pressure gradient, the ratio of shear stress at the top surface (y = H) to that at the bottom surface (y = 0) is

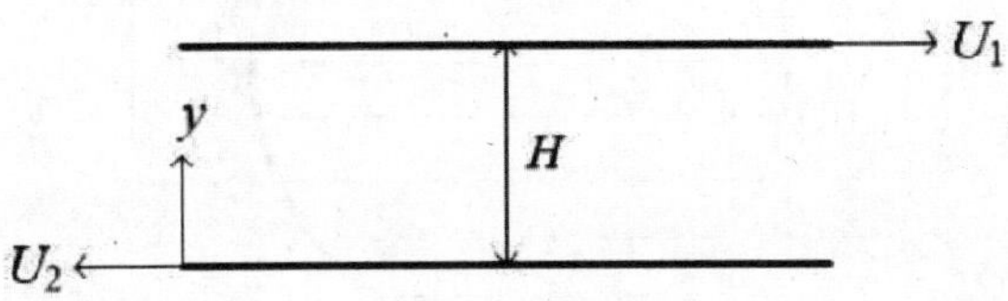

(a) 1 (b) $\frac{U_1}{U_2}$

(c) $\frac{U_1 - U_2}{U_2}$ (d) $\frac{U_1 + U_2}{U_2}$

32. A two-dimensional incompressible flow field is defined as,

$$\vec{V}(x, y) = (Axy)\hat{i} + (By^2)\hat{j}$$

where, A and B are constants. The dynamic viscosity of the Newtonian fluid is μ. In the absence of body force, which among the following expressions represents the pressure gradient at the location (5, 0) in the concerned flow field?

(a) $\mu A(5\hat{i} + \hat{j})$ (b) $\mu(-5B\hat{i} + A\hat{j})$

(c) $\mu A(-\hat{j})$ (d) $\mu A(5\hat{i})$

33. For a potential flow, the fluid velocity is given by $\vec{V}(x, y) = u\hat{i} + v\hat{j}$. The slope of the potential line at (x, y) is

(a) $\frac{u}{v}$ (b) $\frac{v}{u}$

(c) $-\frac{u}{v}$ (d) $-\frac{v}{u}$

34. Consider steady incompressible flow of a Newtonian fluid over a horizontal flat plate, as shown in the figure. The boundary layer thickness is proportional to

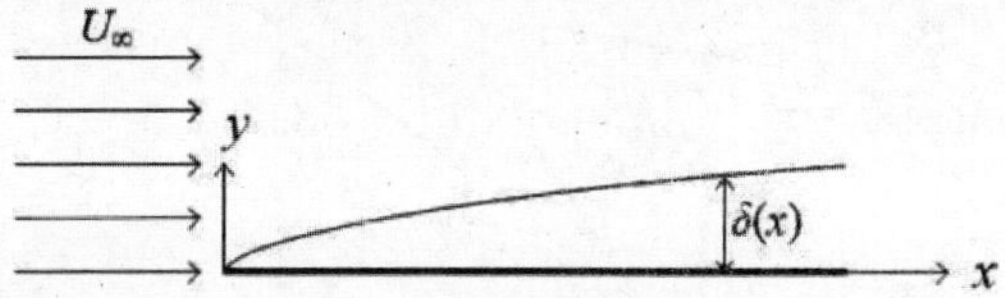

(a) $x^{1/4}$ (b) $x^{1/2}$

(c) $x^{-1/2}$ (d) x^2

35. In a steady two-dimensional compressible flow, u and v are the x- and y-components of flow velocity, respectively and ρ is the fluid density. Among the following pairs of relations, which one(s) perfectly satisfies/satisfy the definition of stream function, ψ, for this flow?

(a) $u = \frac{\partial \psi}{\partial y}$ and $v = -\frac{\partial \psi}{\partial x}$

(b) $u = -\frac{\partial \psi}{\partial x}$ and $v = -\frac{\partial \psi}{\partial y}$

(c) $\rho u = \frac{\partial \psi}{\partial y}$ and $\rho v = -\frac{\partial \psi}{\partial x}$

(d) $\rho u = -\frac{\partial \psi}{\partial y}$ and $\rho v = \frac{\partial \psi}{\partial x}$

36. A water jet (density = 1000 kg/m^3) is approaching a vertical plate, having an orifice at the center, as shown in the figure. While a part of the jet passes through the orifice, remainder flows along the plate. Neglect friction and assume both the inlet and exit jets to have circular cross-sections. If V = 5 m/s, D = 100 mm and d = 25 mm, magnitude of the horizontal force (in N, *rounded off to one decimal place*) required to hold the plate in its position is ________.

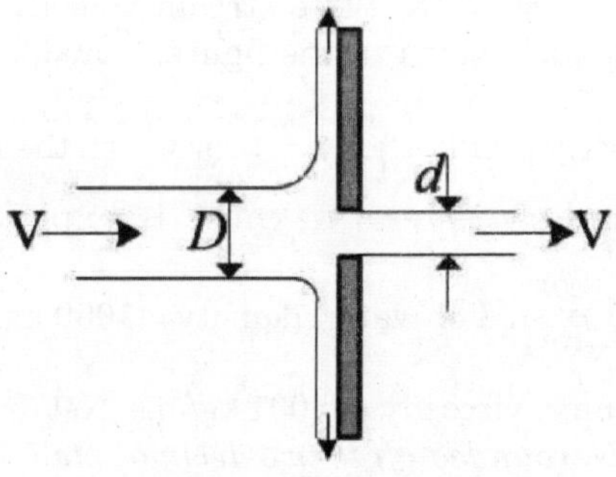

37. Water (density = 1000 kg/m^3) and alcohol (specific gravity = 0.7) enter a Y-shaped channel at flow rates of 0.2 m^3/s and 0.3 m^3/s, respectively. Their mixture leaves through the other end of the channel, as shown in the figure. The average density (in kg/m^3) of the mixture is ________.

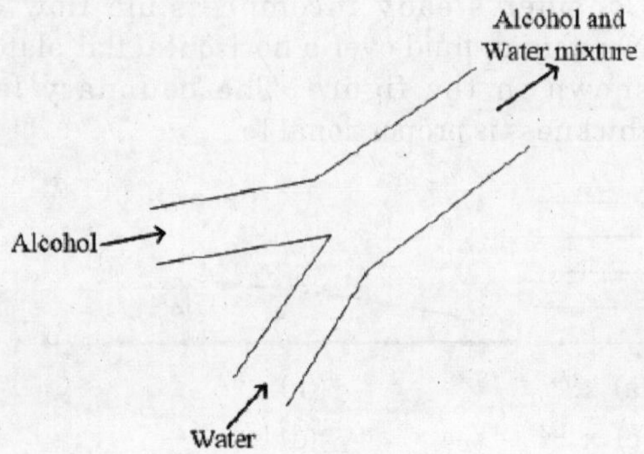

38. The velocity and acceleration of a fluid particle are given as $\vec{V} = (-\hat{i} + 2\hat{j})$ m/s and $\vec{a} = (-2\hat{i} - 4\hat{j})\,\text{m/s}^2$, respectively. The magnitude of the component of acceleration (*in m/s², rounded off to two decimal places*) of the fluid particle along the streamline is ________.

39. A hydraulic turbine with rotor diameter of 100 mm produces 200 W of power while rotating at 300 rpm. Another dynamically-similar turbine rotates at a speed of 1500 rpm. Consider both turbines to operate with the same fluid (identical density and viscosity), and neglect any gravitational effect. Then the power (*in W, rounded off to nearest integer*) produced by the second turbine is ______.

40. Water (density = 1000 kg/m³) flows steadily with a flow rate of 0.05 m³/s through a venturimeter having throat diameter of 100 mm. If the pipe diameter is 200 mm and losses are negligible, the pressure drop (*in kPa, rounded off to one decimal place) between an upstream location in the pipe and the throat (both at the same elevation)* is ______.

41. Water flows around a thin flat plate (0.25 m long, 2 m wide) with a free stream velocity (U_∞) of 1 m/s, as shown in the figure. Consider linear velocity profile $\left(\frac{u}{U_\infty} = \frac{y}{\delta}\right)$ for which the laminar boundary layer thickness is expressed as $\delta = \frac{3.5x}{\sqrt{Re_x}}$. For water, density = 1000 kg/m³ and dynamic viscosity = 0.001 kg/m.s. Net drag force (*in N, rounded off to two decimal places*) acting on the plate, neglecting the end effects, is ______.

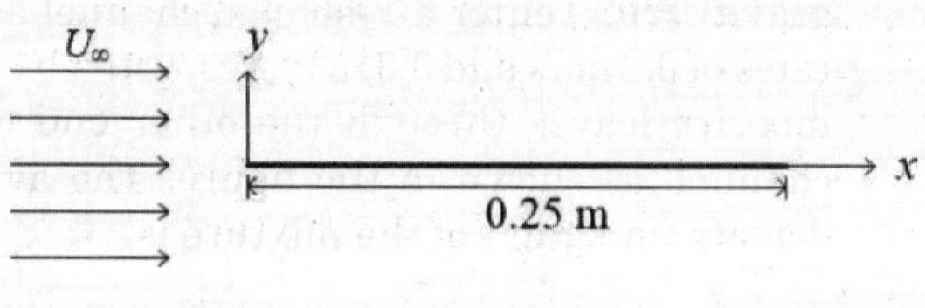

42. Axial velocity profile u(r) for an axisymmetric flow through a circular tube of radius R is given as,

$$\frac{u(r)}{U} = \left(1 - \frac{r}{R}\right)^{1/n}$$

where U is the centerline velocity. If V refers to the area-averaged velocity (volume flow rate per unit area), then the ratio V/U for n = 1 (*rounded off to two decimal places*) is ______.

43. A stationary circular pipe of radius R = 0.5 m is half filled with water (density = 1000 kg/m³), whereas the upper half is filled with air at atmospheric pressure, as shown in the figure. Acceleration due to gravity is g = 9.81 m/s². The magnitude of the force per unit length (*in kN/m, rounded off to one decimal place*) applied by water on the pipe section AB is ______.

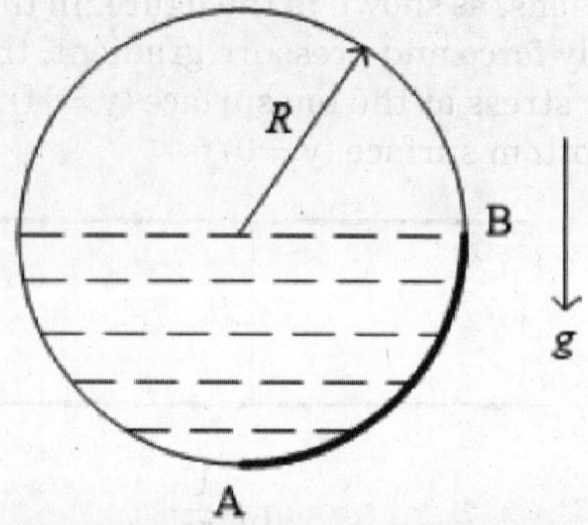

MATERIAL SCIENCE (XE-C)

XE-C: Q.44 – Q.52 Carry ONE mark Each

44. In age-hardening of an aluminium alloy, the purpose of solution treatment followed by quenching is to

(a) form martensitic structure

(b) increase the size of the precipitates

(c) form supersaturated solid solution

(d) form precipitates at the grain boundaries

45. The magnetization (M) – magnetic field (H) curves for four different materials are given below. Which one of these materials is most suitable for use as a permanent magnet?

(a)

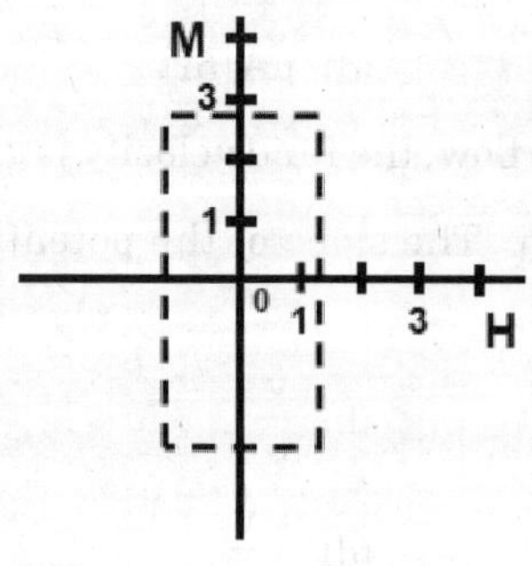

(b)

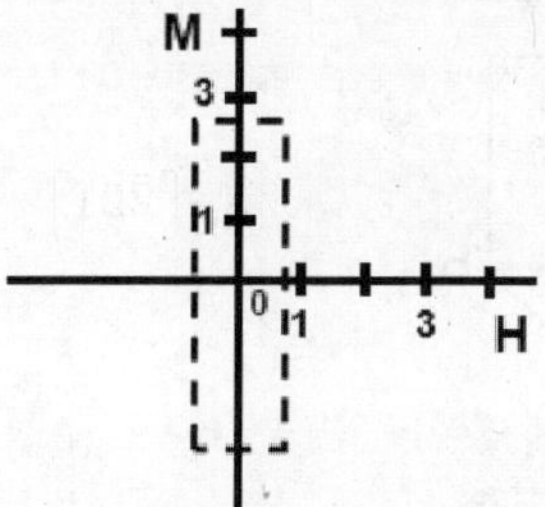

(c)

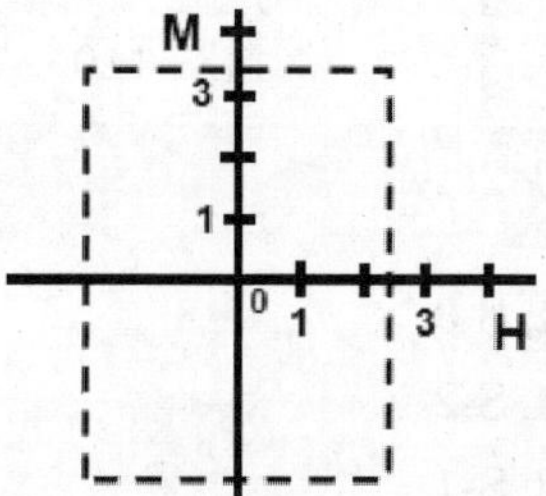

(d)

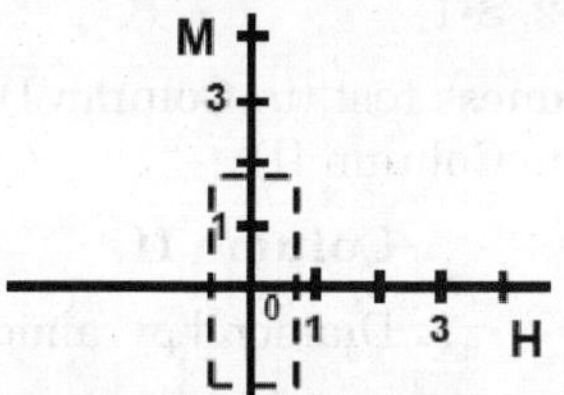

46. The band gap of a semiconducting material is ~ 2 eV. Which one of the following absorption (A) vs. energy (in eV) curves is correct?

(a)

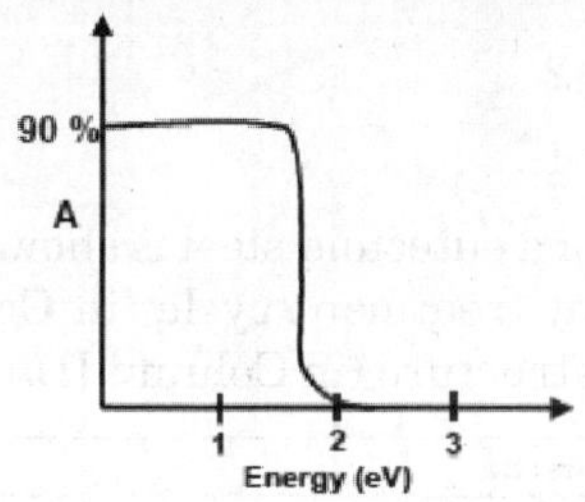

(b)

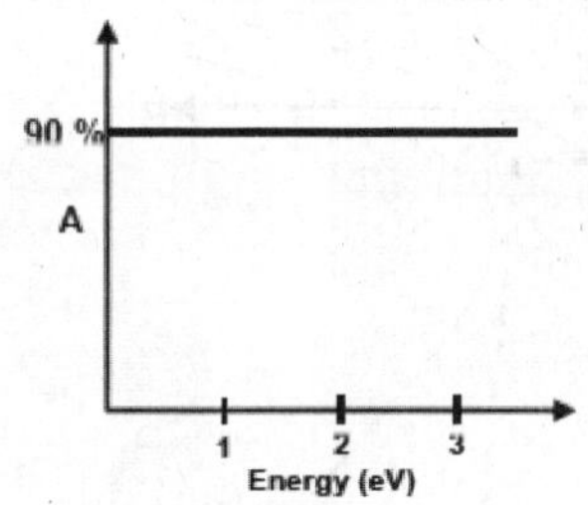

(c)

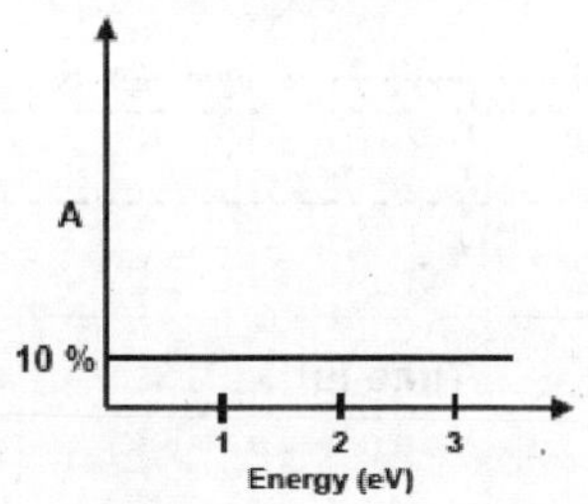

(d)

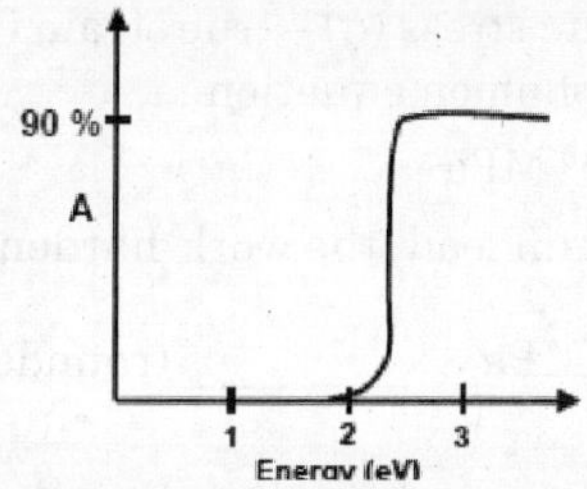

47. Figures (i) and (ii) show a binary phase diagram and the corresponding Gibbs free energy (G) vs. composition (X_B) diagram, respectively. Figure (ii) corresponds to which one of the temperatures shown in Figure (i)?

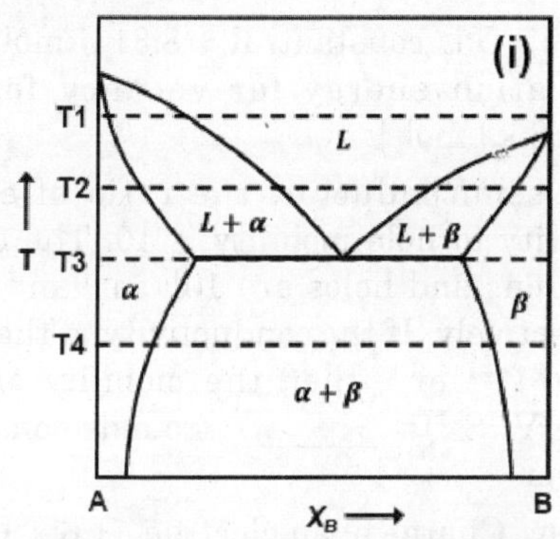

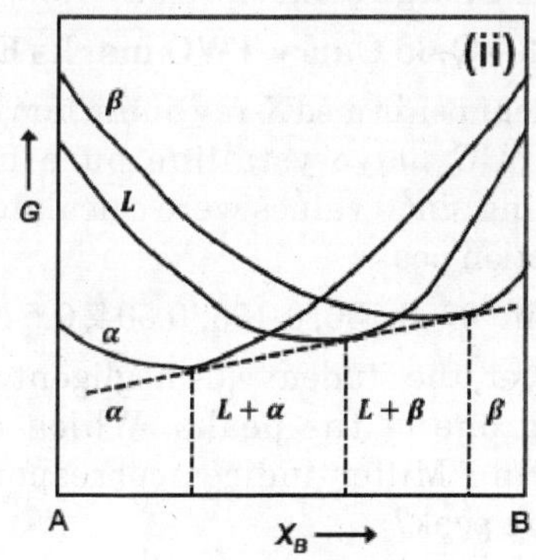

(a) T1 (b) T2

(c) T3 (d) T4

48. Aliovalent doping of $MgCl_2$ in NaCl leads to the formation of defects. Which one of the following is the correct defect reaction?

(a) $Mg^{\cdot}_{Cl} + Na_{Na} + V'_{Cl} = \varnothing$

(b) $Mg^{\cdot}_{Na} + Cl_{Cl} + V'_{Na} = \varnothing$

(c) $Mg_{Na} + Cl_{Cl} = \varnothing$

(d) $Mg'_{Na} + Cl_{Cl} + V^{\cdot}_{Na} = \varnothing$

49. A screw dislocation in a FCC crystal has Burgers vector of $\frac{a}{2}[110]$, where a is the lattice constant.

The possible slip plane(s) is/are :

(a) $(11\bar{1})$ (b) (111)

(c) $(\bar{1}11)$ (d) $(1\bar{1}1)$

50. The tensile true stress (σ) – true strain ($\in$) curve follows the Hollomon equation :

$$\sigma = 500\in^{0.15} \text{ MPa}$$

At the maximum load, the work-hardening rate $\left(\frac{d\sigma}{d\in}\right)$ is (in MPa) : ________ (rounded off to nearest integer)

51. A metal has a certain vacancy fraction at a temperature of 600 K. On increasing the temperature to 900 K, the vacancy fraction increases by a factor of ________ (rounded off to one decimal place)

Given : Gas constant, R = 8.31 J mol^{-1}K^{-1} and activation energy for vacancy formation, Q = 68 kJ mol^{-1}

52. In a semiconductor, the ratio of electronic mobility to hole mobility is 10. The density of electrons and holes are 10^{15} m^{-3} and 10^{16} m^{-3}, respectively. If the conductivity of the material is 1.6 Ω^{-1} m^{-1}, then the mobility of holes is (in m^{2}V^{-1}S^{-1}) : ________ (rounded off to nearest integer)

Given : Charge of an electron : 1.6 × 10^{-19} C

XH-C: Q.53 – Q.65 Carry TWO marks Each

53. A student performed X-ray diffraction experiment on a FCC polycrystalline pure metal. The following $\sin^2\theta$ values were calculated from the diffraction peaks.

$$\sin^2\theta = 0.136, 0.185, 0.504, 0.544$$

However, the student was negligent and missed noting one of the peaks. Which one of the following Miller indices corresponds to the missing peak?

(a) (200) (b) (220)

(c) (311) (d) (222)

54. Match the lattice planes and directions (in Column I) with the corresponding Miller indices (in Column II) :

Column I **Column II**

(P) 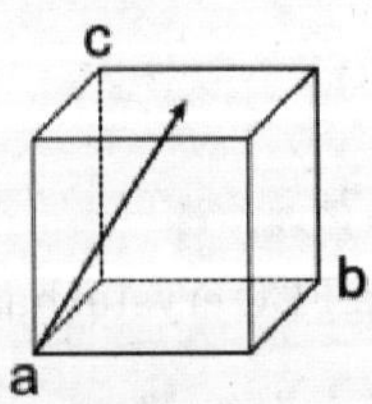 1. $(\bar{1}11)$

(Q) 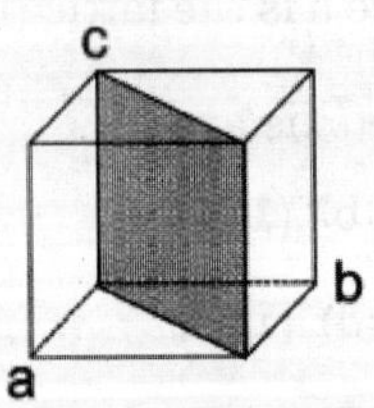 2. $[\bar{1}12]$

(R) 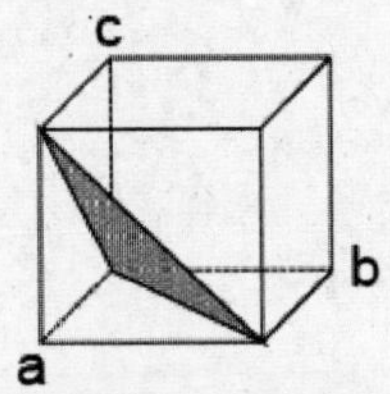 3. $[\bar{2}21]$

(S) 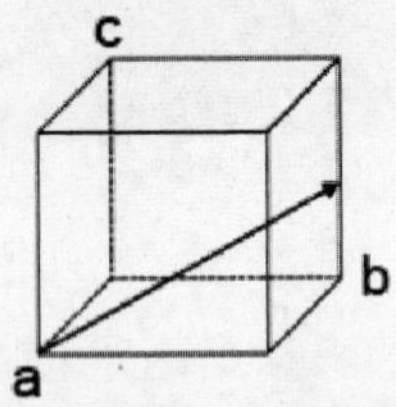 4. $[\bar{1}10]$

(a) P-2, Q-4, R-1, S-3

(b) P-3, Q-1, R-4, S-2

(c) P-2, Q-4, R-3, S-1

(d) P-3, Q-4, R-2, S-1

55. Match the hardness test (in Column I) with its indenter type (in Column II).

Column I	Column II
(P) Brinell	1. Diamond pyramidal
(Q) Rockwell	2. Diamond cone
(R) Vickers	3. Tungsten carbide sphere
	4. Steel sphere

(a) P-2, Q-4, R-1

(b) P-4, Q-2, R-3

(c) P-3, Q-4, R-2

(d) P-4, Q-2, R-1

56. TTT diagram of a eutectoid steel is shown below. Match the heat treatment cycle (in Column I) with its microstructure (in Column II).

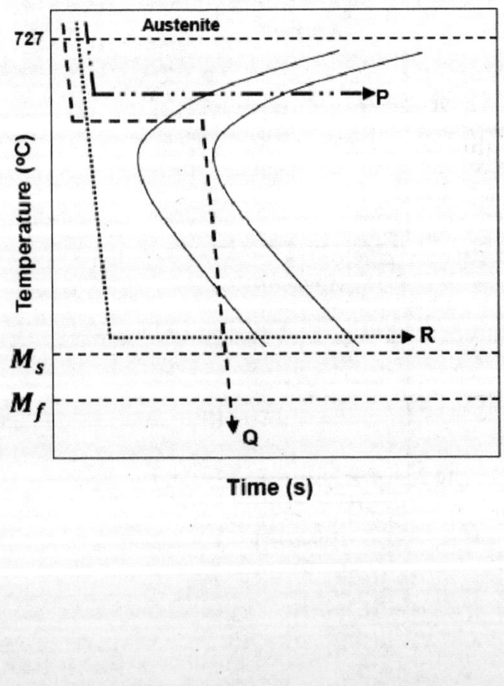

Column I	Column II
P	1. Bainite only
Q	2. Pearlite only
R	3. Pearlite + Bainite + Martensite
	4. Pearlite + Martensite

(a) P-1, Q-2, R-4
(b) P-2, Q-3, R-2
(c) P-2, Q-4, R-1
(d) P-2, Q-3, R-1

57. Which of the following statement(s) is/are true for an optical microscope?
(a) Increasing the aperture of the objective lens deteriorates the resolution
(b) Reducing the wavelength of illuminating light improves the resolution
(c) Increasing the refractive index of the medium in between the sample and the objective lens improves the resolution
(d) Reducing the wavelength of illuminating light decreases the depth of field

58. Among the 14 Bravais lattices, there is no base centred cubic unit cell. Which of the following statement(s) is/are true?
(a) The base-centred cubic unit cell is same as the simple tetragonal unit cell
(b) The base-centred cubic unit cell is same as the body centred tetragonal unit cell
(c) The base-centred cubic unit cell is same as the simple orthorhombic unit cell
(d) The base-centred cubic unit cell does not have any 3-fold rotation axis

59. Specific heat (C_v) of a material was found to depend on temperature as shown below. Which of the following statement(s) is/are true?

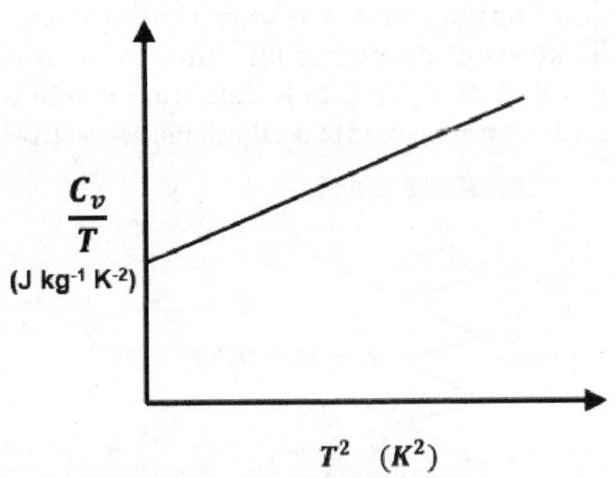

(a) The material is metallic
(b) The material is insulating
(c) The material is three dimensional
(d) The material is one dimensional

60. A pure Silicon wafer is doped with Boron by exposing it to B_2O_3 vapour at an elevated temperature. It takes 1000 seconds to reach a Boron concentration of 10^{20} *atoms* m^{-3} at a depth of 1 μm. The time taken to reach the same concentration of Boron at a depth of 2 μm is (in seconds) : ________ (rounded off to nearest integer)

Given : Boron concentration on the wafer surface remains constant.

61. The Young's modulus of a quartz piezoelectric crystal is 100 GPa. The uniaxial stress required to change its polarization by 1% is (give absolute value in GPa) ________ (rounded off to nearest integer).

62. A one-dimensional nanowire has a linear electron density of 10^8 *electrons* cm^{-1}. The Fermi energy of the system is (in eV) ______ (rounded off to two decimal places)

Given : $\frac{\hbar^2}{2m} = 0.24(eV)^2 s^2 kg^{-1}$ where 'm' is the mass of an electron.

63. Two moles of a monoatomic ideal gas at 10 atm and 300 K is expanded isothermally and reversibly to a pressure of 2 atm. The absolute value of work done by the system is (in kJ) ________ (rounded off to two decimal places)

Given : R = 8.31 J $mol^{-1}K^{-1}$, 1 atm = 101 kPa

64. An electrochemical cell consists of pure Zn electrode (anode) and a hydrogen electrode (cathode) in a dilute Zn^{+2} solution. The overall reaction is :

$$Zn(s) + 2H^+ = H_2 + Zn^{+2}$$

If the overall cell potential is + 0.690 V, then the value of $\ln\frac{[Zn^{+2}]}{[H^+]^2}$ is ________ (rounded off to two decimal places)

Given : Pressure of hydrogen gas = 1 atm;
Temperature = 298 K;
$\frac{RT}{F} = 0.0256$ V, where R is gas constant and F is Faraday constant
The standard reduction potential of :
$Zn^{+2} + 2e = Zn$ ($E° = -0.762$ V) versus Standard Hydrogen Electrode
$2H^+ + 2e = H_2$ ($E° = 0V$)

65. In a Raman spectroscopy experiment done at 300 K, a Raman line is observed at 200 cm^{-1} (~ 25 meV). The ratio of the intensity of the Stokes line to that of the Anti-Stokes line is ______ (rounded off to two decimal places)

Given : Boltzmann constant,

$$k = 8.62 \times 10^{-5} \text{ eV K}^{-1}$$

SOLID MECHANICS (XE-D)

XE-D: Q.66 – Q.74 Carry ONE mark Each

66. A plane truss is simply supported at P and R as shown. A downward force F is applied at hinge Q. The axial force developed in member PS is

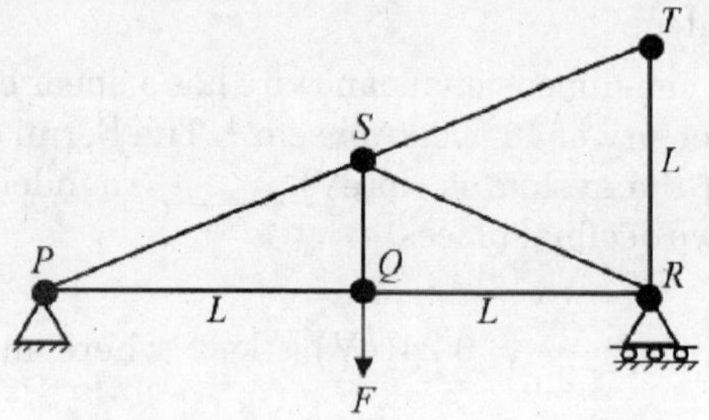

(a) $\frac{\sqrt{5}}{2}$F Tensile (b) $\frac{\sqrt{5}}{2}$F Compressive

(c) $\sqrt{5}$ F Tensile (d) $\sqrt{5}$ F Compressive

67. A massless rigid rod OP of length L is hinged frictionlessly at O. A concentrated mass m is attached to end P of the rod. Initially, the rod OP is horizontal. Then, it is released from rest. There is gravity as shown. The rod acquires an angular velocity as it swings. The clockwise angular velocity of the rod, when it first reaches the vertical position as shown, is

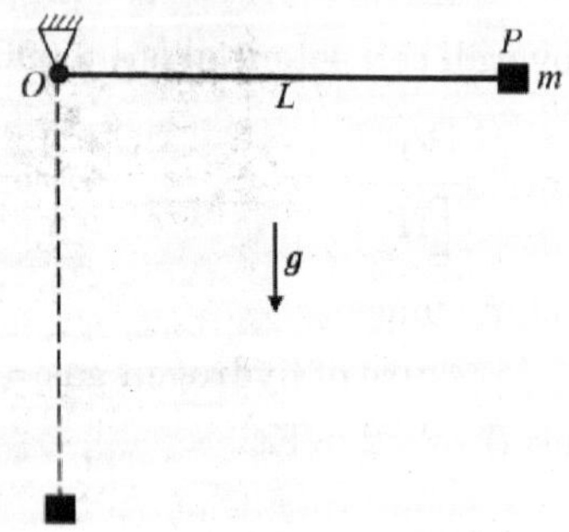

(a) $2\sqrt{\frac{g}{L}}$ (b) $\sqrt{\frac{2g}{L}}$

(c) $\sqrt{\frac{g}{L}}$ (d) $\frac{1}{2}\sqrt{\frac{g}{L}}$

68. Two equivalent descriptions of the state of stress at a point are shown in the figure. The normal stresses σ_1 and σ_2 as shown on the right must be, respectively,

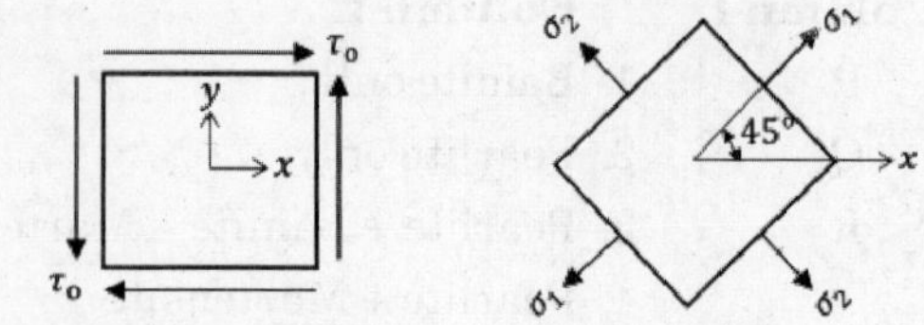

(a) τ_o and $-\tau_o$ (b) $-\tau_o$ and τ_o

(c) $\frac{\tau_o}{\sqrt{2}}$ and $-\frac{\tau_o}{\sqrt{2}}$ (d) $-\frac{\tau_o}{\sqrt{2}}$ and $-\frac{\tau_o}{\sqrt{2}}$

69. The state of strain at a point in a machine component is given as

$\varepsilon_{xx} = 2.5 \times 10^{-4}$, $\varepsilon_{yy} = 2.0 \times 10^{-4}$, $\varepsilon_{zz} = -1.5 \times 10^{-4}$,
$\varepsilon_{xy} = 2.5 \times 10^{-4}$, $\varepsilon_{yz} = -0.5 \times 10^{-4}$, $\varepsilon_{zx} = -1.0 \times 10^{-4}$.

The volumetric strain at this point is

(a) 4×10^{-4} (b) 3×10^{-4}

(c) -5×10^{-4} (d) -3×10^{-4}

70. A thin walled, closed cylindrical vessel of inside diameter d and wall thickness t contains a fluid under pressure p. The figure below shows a part of the cylindrical vessel; end caps are not shown. Consider the small element shown with sides parallel and perpendicular to the axis of the cylinder. The stresses σ_1 and σ_2 are

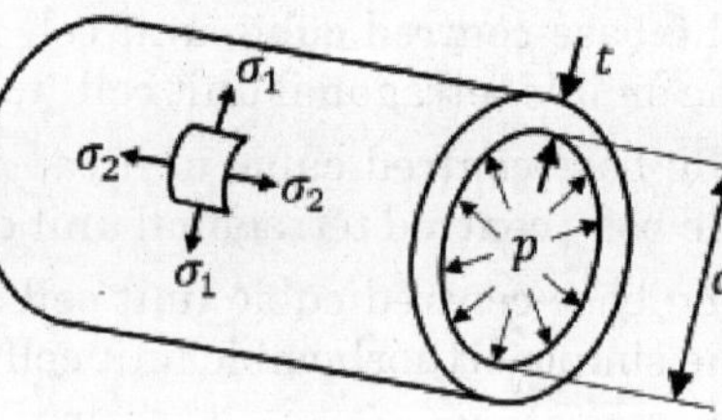

(a) $\sigma_1 = \frac{pd}{2t}$; $\sigma_2 = \frac{pd}{4t}$ (b) $\sigma_1 = \frac{pd}{t}$; $\sigma_2 = \frac{pd}{2t}$

(c) $\sigma_1 = \frac{pd}{4t}$; $\sigma_2 = \frac{pd}{2t}$ (d) $\sigma_1 = \frac{pd}{2t}$; $\sigma_2 = 0$

71. A spring mass system is shown in the figure below. Take the acceleration due to gravity as g = 9.81 m/s^2. The static deflection due to weight and the time period of oscillations, respectively, are

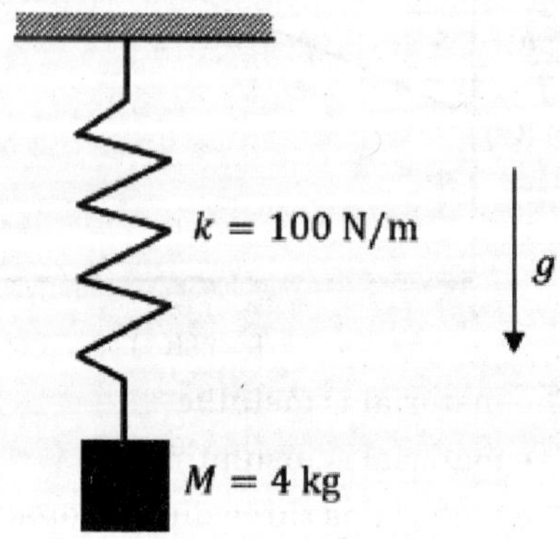

(a) 0.392 m and 1.26 s
(b) 0.392 m and 3.52 s
(c) 0.626 m and 1.26 s
(d) 0.626 m and 3.52 s

72. A rod is subjected to three forces as shown in the figure on the left. An equivalent force system with forces F_1, F_2 and moment M is shown in the figure on the right. The value of M (in N-m) is ______(rounded off to one decimal place).

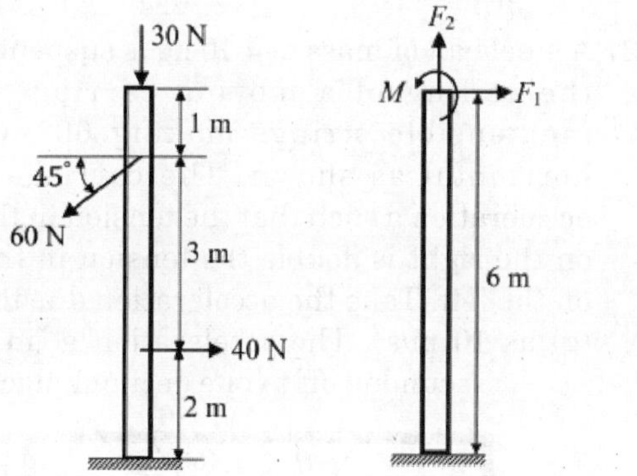

73. A simply supported beam of length 3 m is loaded as shown in the figure. The magnitude of the shear force (in kN) at the mid-point of the beam is________(rounded off to one decimal place).

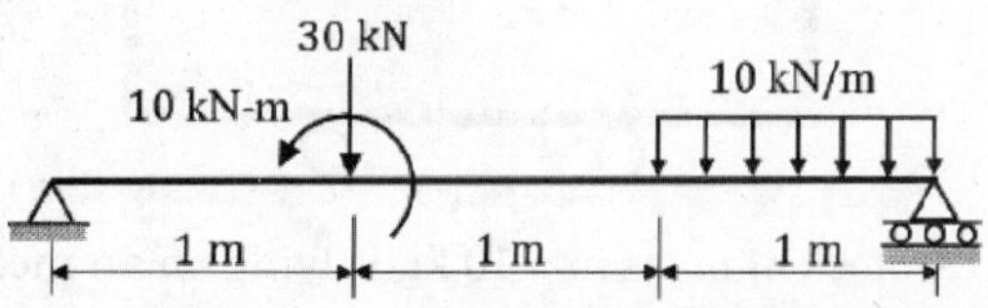

74. For a plane stress problem, the principal stresses are 100 MPa and 50 MPa. The magnitude of maximum shear stress (in MPa) in the material is________(rounded off to one decimal place).

XE-D: Q.75 – Q.87 Carry TWO marks Each

75. A solid uniform rigid disk of mass m and radius R rolls without slipping along a horizontal surface PQ. The speed of the center of the disk is v. The disk then strikes a hurdle of height $\frac{3R}{20}$ at point S. During the impact, there is no rebound or slip at S and no impulse from the surface PQ. The magnitude of the velocity of the center of the disk immediately after the impact is

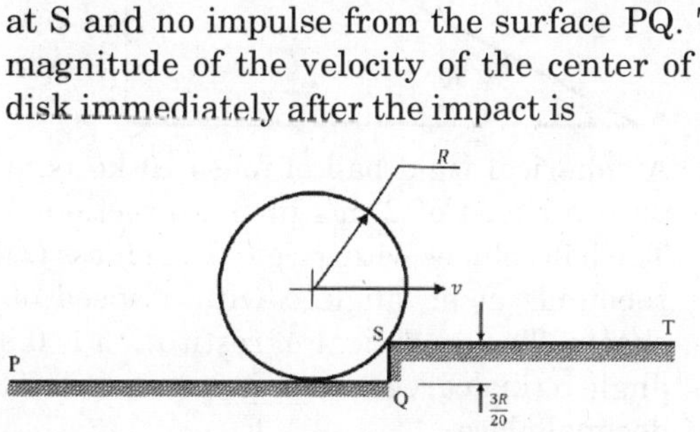

(a) 0.1 v
(b) 0.3 v
(c) 0.7 v
(d) 0.9 v

76. A cylinder made of rubber (length = L and diameter = d) is inserted in a rigid container as shown in the figure. The rubber cylinder fits snugly in the rigid container. There is no wall friction. The modulus of elasticity of the rubber is E and its Poisson's ratio is v. The cylinder is subjected to a small uniform pressure p as shown in the figure. The resulting axial strain (ε_{zz}) is

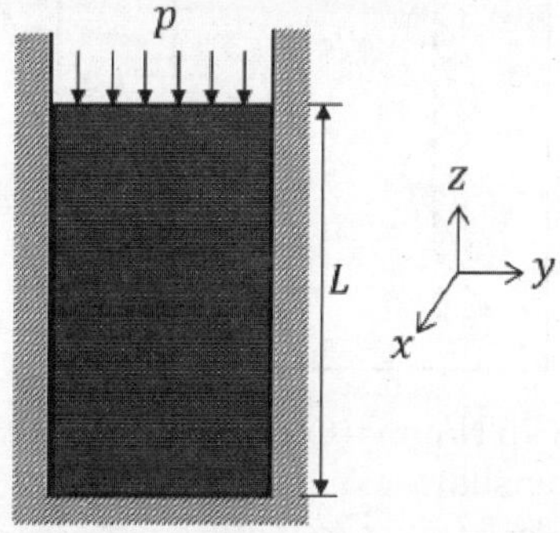

(a) $-\frac{p}{E}$

(b) $-\frac{p}{E}(1-2v)$

(c) $-\frac{p}{E}\left[\frac{(1+v)(1-2v)}{(1-v)}\right]$

(d) $-\frac{p}{E}\left[\frac{(1-v)(1-2v)}{(1+v)}\right]$

77. The state of stress at the critical location in a structure is σ_{xx} = 420 MPa, σ_{yy} = 100 MPa, $\sigma_{zz} = \sigma_{xy} = \sigma_{yz} = \sigma_{zx} = 0$. The yield stress of the material in uniaxial tension is 400 MPa. Select the correct statement among the following.

(a) The structure is safe by both Tresca (maximum shear stress) theory and von-Mises (distortion energy) theory.
(b) The structure is safe by Tresca (maximum shear stress) theory and unsafe by von-Mises (distortion energy) theory.
(c) The structure is unsafe by Tresca (maximum shear stress) theory and safe by von-Mises (distortion energy) theory.
(d) The structure is unsafe by both Tresca (maximum shear stress) theory and von-Mises (distortion energy) theory.

78. The figure shows a column of rectangular cross section 100 mm × 80 mm. It carries a load of 60 kN at a point 30 mm from the edge PQ. The values of stress component σ_{zz} on surfaces PQQ'P' and SRR'S', at points far away from both ends of the column, are respectively

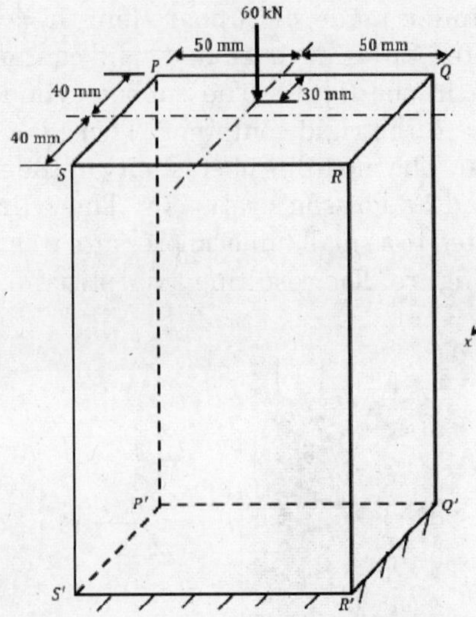

(a) 18.75 N/mm^2 (Compressive) and 3.75 N/mm^2 (Tensile)

(b) 18.75 N/mm^2 (Compressive) and 3.75 N/mm^2 (Compressive)

(c) 13.13 N/mm^2 (Compressive) and 1.88 N/mm^2 (Tensile)

(d) 13.13 N/mm^2 (Compressive) and 1.88 N/mm^2 (Compressive)

79. Consider an electric pole with dimensions as shown in the figure. Let the end R be subjected to a vertical force F. The flexural rigidity of both vertical and horizontal bars is EI. Neglect the axial deflection of the vertical bar, and all effects of self-weight. The vertical deflection at end R is

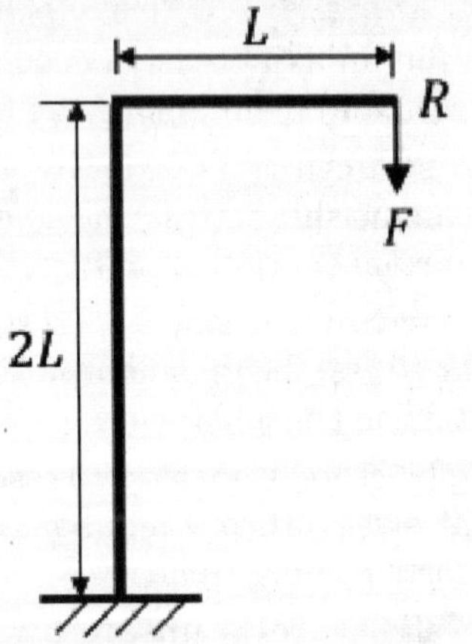

(a) $\dfrac{7FL^3}{3EI}$ (b) $\dfrac{10FL^3}{3EI}$

(c) $\dfrac{5FL^3}{3EI}$ (d) $\dfrac{8FL^3}{3EI}$

80. A uniform cantilever beam has flexural rigidity EI and length L. It is subjected to a concentrated force F and moment M = 2FL at the free end as shown. The deflection (δ) at the free end is

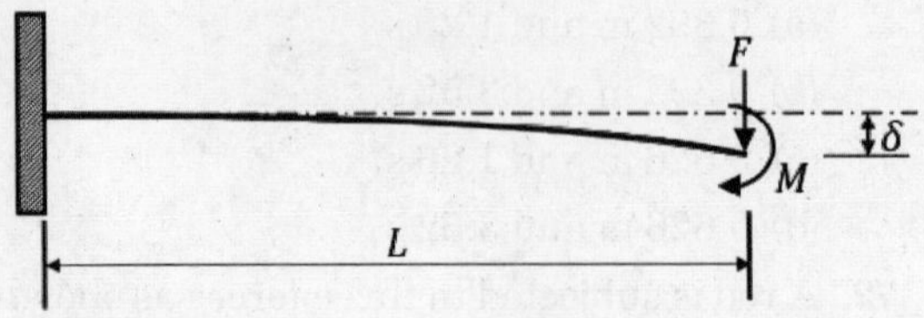

(a) $\dfrac{11FL^3}{12EI}$ (b) $\dfrac{8FL^3}{9EI}$

(c) $\dfrac{4FL^3}{3EI}$ (d) $\dfrac{7FL^3}{6EI}$

81. A steel ball of mass m = 10 kg is suspended from the ceiling of a moving carriage by two inextensible strings making 60° with the horizontal as shown. The carriage has an acceleration a such that the tension in the string on the right is double the tension in the string on the left. Take the acceleration due to gravity (g) as 10 m/s^2. The acceleration a (in m/s^2) is ______ (rounded off to one decimal place).

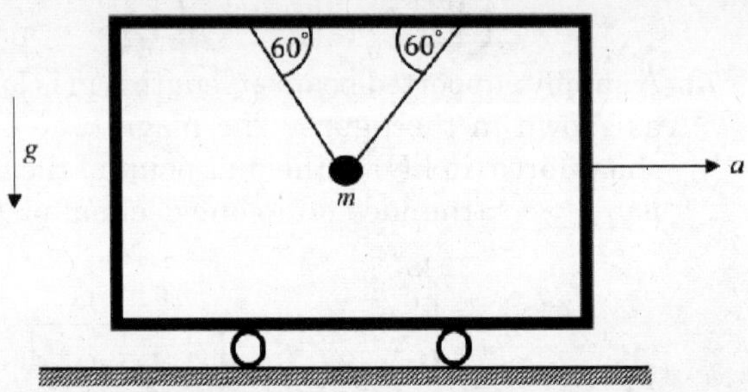

82. A block of mass m = 10 kg is lying on an inclined plane PQ. The mass is restrained from sliding down the inclined plane by a force F. The coefficient of friction between the block and the inclined plane is 0.3. Take the acceleration due to gravity as 10 m/s^2. The smallest force F (in N) required to prevent the block from sliding down is ______ (rounded off to one decimal place).

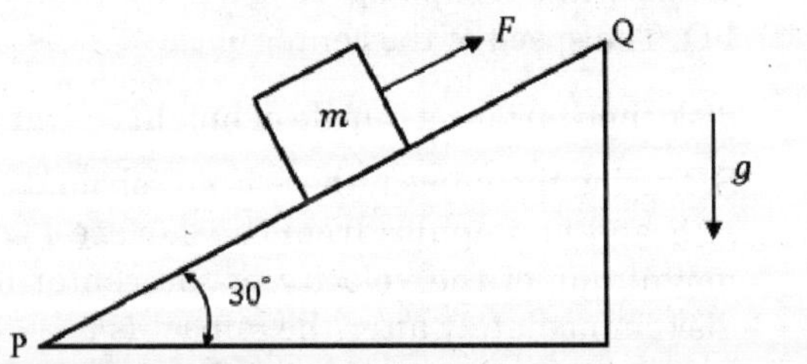

83. A spherical rigid ball of mass 10 kg is moving with a speed of 2 m/s in the direction shown. The ball collides with a rigid frictionless wall and rebounds at an angle α with a speed of v, as shown. The coefficient of restitution is 0.9. The angle α (in degrees) is ______ (rounded off to one decimal place).

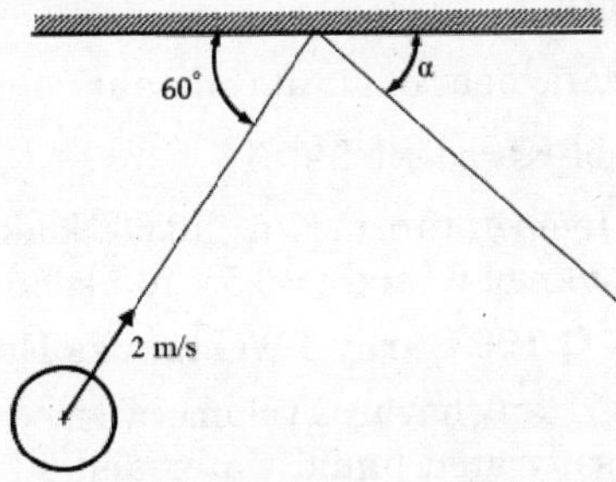

84. A thin steel plate is loaded in the x-y plane as shown in the figure. Take the Poisson's ratio of steel to be 0.3 and the modulus of elasticity of steel to be 200 GPa. The strain along the z-direction is $\varepsilon_{zz} = -3 \times 10^{-4}$. The value of σ_{yy} (in MPa) is ______(rounded off to one decimal place).

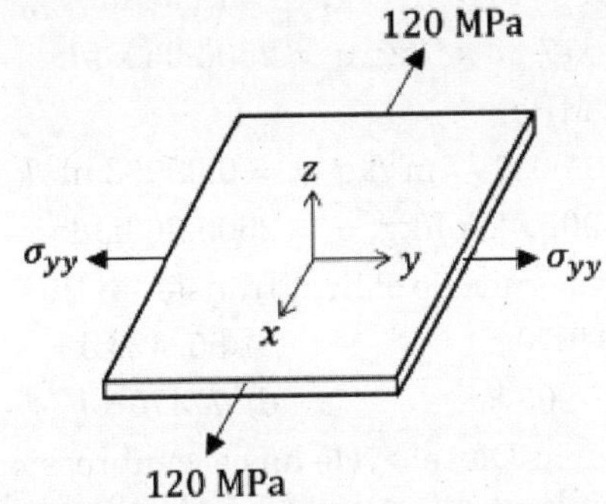

85. A composite rod made of steel and copper is fixed immovably at its ends as shown in the figure. The length of each portion of the rod is 1 m as shown. The cross-sections of both portions are the same. The moduli of elasticity of steel and copper are 200 GPa and 100 GPa, respectively. The coefficients of thermal expansion of steel and copper are 12×10^{-6} /°C and 18×10^{-6} /°C, respectively. The composite rod is initially stress free. Then, the temperature of the composite rod is increased by 100 °C. The magnitude of axial stress (in MPa) developed in the steel rod is ______ (rounded off to one decimal place).

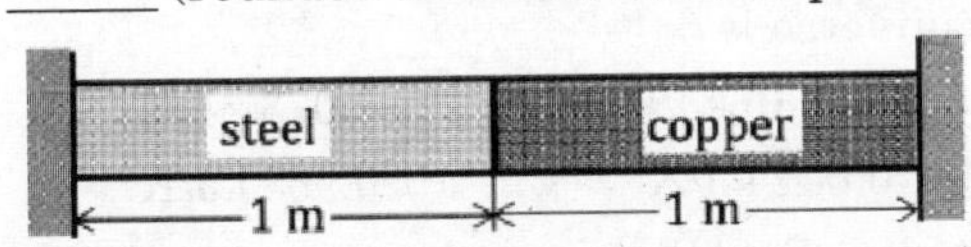

86. A slender uniform elastic rod of length 1 m and of solid circular cross-section of diameter 50 mm is originally straight. It is then loaded by equal and opposite end moments as indicated in the figure. The resulting lateral displacement of the mid-point of the rod is 10 mm (displacements are exaggerated in the figure). The maximum longitudinal strain in the rod is p × 10^{-3}, where p is______(rounded off to one decimal place).

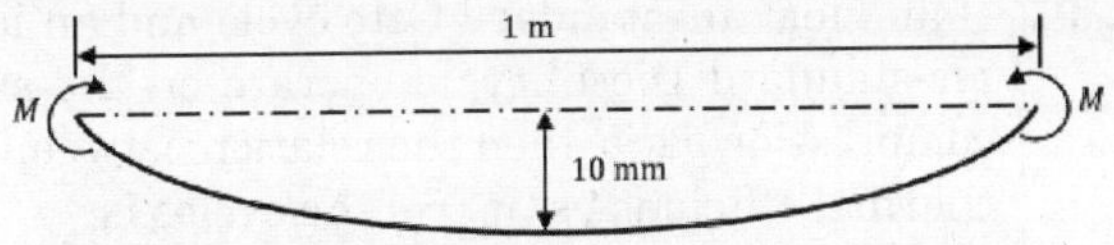

87. Consider a solid cylindrical shaft and a hollow cylindrical shaft. Both shafts are axisymmetric and elastic, and have the same cross-sectional area. The hollow shaft has an outside diameter of 150 mm and an inside diameter of 120 mm. When both the shafts are twisted by the same twisting moment, the ratio of maximum shear stress developed in the hollow shaft (τ_h) to maximum shear stress developed in the solid shaft (τ_s) will be ______(rounded off to three decimal places).

THERMODYNAMICS (XE-E)

Q.88 – Q.96 Carry ONE mark Each

88. A. The number of properties required to fix the state of a system is given by 'state postulate'.

R. The state of a simple compressible system is completely specified by two independent, intensive properties.

About the statements **A** and **R** applied to a single-phase system,

(a) **A** is correct and **R** is incorrect.

(b) **A** is incorrect and **R** is correct.

(c) Both **A** and **R** are incorrect.

(d) Both **A** and **R** are correct.

89. Which of the following is an extensive property of a system?

(a) Density (b) Pressure

(c) Temperature (d) Total mass

90. A tank of volume V contains homogeneous mixture of two ideal gases, A and B at a temperature T and a pressure P. The mixture contains n_A moles of gas A and n_B moles of gas B. If P_A and P_B are the partial pressures of gas A and gas B, respectively, then

(a) $P_A = \frac{n_A}{n_A + n_B}P,\ P_B = \frac{n_B}{n_A + n_B}P$

(b) $P_A = \frac{n_B}{n_A}P,\ P_B = \frac{n_A}{n_B}P$

(c) $P_A = \frac{n_A}{n_B}P,\ P_B = \frac{n_B}{n_A}P$

(d) $P_A = \frac{n_B}{n_A + n_B}P,\ P_B = \frac{n_A}{n_A + n_B}P$

91. If an ideal air-standard Otto cycle and an ideal air-standard Diesel cycle operate on the same compression ratio, then the relation between the thermal efficiencies (η_{th}) of the cycles is

(a) $\eta_{th,Otto} = \eta_{th,\ Diesel}$ and $\eta_{th,Otto} < 1$

(b) $\eta_{th,Otto} > \eta_{th,\ Diesel}$

(c) $\eta_{th,Otto} < \eta_{th,\ Diesel}$

(d) $\eta_{th,Otto} = \eta_{th,\ Diesel} = 1$

92. The following statements are given :

(i) The third law of thermodynamics deals with the entropy of a substance at the absolute zero temperature.

(ii) Entropy of any non-crystalline structure is zero at absolute zero temperature.

(iii) At the absolute zero temperature, the crystal structure has maximum degree of order.

(iv) The thermal energy of the substance at absolute zero temperature is maximum.

The correct option describing these statements is

(a) Only (i) is correct

(b) Only (ii) is correct

(c) Both (i) and (iii) are correct

(d) Both (i) and (iv) are correct

93. Adiabatic bulk modulus of a substance is defined as

(a) $-\frac{1}{v}\left(\frac{\partial v}{\partial P}\right)_T$ (b) $-v\left(\frac{\partial P}{\partial v}\right)_T$

(c) $-\frac{1}{v}\left(\frac{\partial v}{\partial P}\right)_s$ (d) $-v\left(\frac{\partial P}{\partial v}\right)_s$

94. An insulated rigid closed tank of 2 m^3 internal volume contains saturated liquid-vapor mixture of water at 200 °C. The quality of the mixture is 0.75. The mass of the mixture in the tank is ________ kg (*rounded off to one decimal place*).

Use the following data for water :

At 200 °C : $v_f = 0.001156$ m^3/kg, $v_{fg} = 0.12620$ m^3/kg, $v_g = 0.12736$ m^3/kg

95. A rigid closed tank contains 2 kg of an ideal gas at 500 kPa and 350 K. A valve is opened, and half of the mass of the gas is allowed to escape. Then the valve is closed. If the final pressure in the tank is 300 kPa, the final temperature in the tank is ________ K (in integer).

96. Air at 400 K and 200 kPa is heated at constant pressure to 600 K. Assuming that the internal energy is a function of temperature only, the magnitude of change in internal energy during this process is ________ kJ/kmol (*rounded off to one decimal place*).

Use the following data :

Molar specific heat of air at constant volume :

$\bar{c}_v$ (kJ/kmol-K) = a + bT + cT^2

where T is temperature in K, a = 19.686 kJ/kmol-K, b = 0.002 kJ/kmol-K^2 and c = 0.5×10^{-5} kJ/kmol-K^3.

XE-XE : Q.97 – Q.109 Carry TWO marks Each

97. A rigid closed tank having a volume of 2 m^3 contains 0.1 m^3 of saturated liquid water and 1.9 m^3 of saturated water vapor at 100 kPa. Heat is transferred to the tank until the final tank pressure reaches 2 MPa.

Following data for water is given:

At 100 kPa :

$v_f = 0.001043$ m^3/kg, $v_g = 1.694$ m^3/kg,

$u_f = 417.33$ kJ/kg, $u_g = 2506.06$ kJ/kg

At 2 MPa :

$v_f = 0.001177$ m^3/kg, $v_g = 0.09963$ m^3/kg,

$u_f = 906.42$ kJ/kg, $u_g = 2600.26$ kJ/kg

The magnitude of heat transfer in this process is

(a) 34670 kJ (b) 55842 kJ

(c) 67906 kJ (d) 77470 kJ

98. An ideal Diesel cycle has a compression ratio of 20 and cut-off ratio of 1.5. At the beginning of the compression stroke, air is at 100 kPa, 300 K. Use the cold-air-standard assumptions with property value c_p = 1.005 kJ/kg-K. Assume c_p/c_v = 1.4. For this cycle, the net work output per unit mass is

(a) 335 kJ/kg (b) 395 kJ/kg

(c) 500 kJ/kg (d) 165 kJ/kg

99. A 5 kg metal block (c_p = 0.5 kJ/kg-K) at 373 K is submerged into 10 kg of water (c_p = 4.2 kJ/kg-K) at 293 K in an insulated rigid container without spilling. Assuming thermal equilibrium is reached, the approximate entropy change of the universe is

(a) –0.565 kJ/K (b) 0.073 kJ/K

(c) 0.642 kJ/K (d) 0.963 kJ/K

100. Match the following :

Thermodynamic function/equation	**Expression**
A1 Helmholtz function (a)	B1 $\left(\frac{d}{dT}(\ln P)\right)_{sat} = \frac{h_g - h_f}{RT^2}$
A2 Gibbs function (g)	B2 u – Ts
A3 T-ds equation	B3 du = Tds – Pdv
A4 Clapeyron-Clausius equation	B4 h – Ts

(a) A1 $\rightarrow$ B2, A2 $\rightarrow$ B4, A3 $\rightarrow$ B1, A4 $\rightarrow$ B3

(b) A1 $\rightarrow$ B4, A2 $\rightarrow$ B2, A3 $\rightarrow$ B1, A4 $\rightarrow$ B3

(c) A1 $\rightarrow$ B2, A2 $\rightarrow$ B4, A3 $\rightarrow$ B3, A4 $\rightarrow$ B1

(d) A1 $\rightarrow$ B4, A2 $\rightarrow$ B2, A3 $\rightarrow$ B3, A4 $\rightarrow$ B1

101. A piston-cylinder device initially contains 1 m^3 of air at 200 kPa and 25 °C. Air expands at constant pressure while a heater of 250 W is switched on for 10 minutes. There is a heat loss of 4 kJ during this process. Assuming air as an ideal gas, the final temperature of air is ________ °C (*rounded off to one decimal place*).

Use the following data for air :

$R = 0.287$ kJ/kg-K, $c_p = 1.005$ kJ/kg-K

102. Steam at 2 MPa and 300 °C steadily enters a nozzle of inlet diameter of 20 cm. Steam leaves the nozzle with a velocity of 300 m/s. The mass flow rate of steam through the nozzle is 10 kg/s. Assume no work interaction and no change in potential energy. If the heat loss from the nozzle per kg of steam is 3 kJ, the exit enthalpy per kg of steam is ________ kJ (*rounded off to nearest integer*).

Use the following data for steam :

At 2 MPa and 300 °C :

$v = 0.12551$ m^3/kg,

$h = 3024.2$ kJ/kg

103. A rigid tank of 2 m^3 internal volume contains 5 kg of water as a saturated liquid-vapor mixture at 400 kPa. Half of the mass of the saturated liquid in the tank is drained-off while maintaining constant pressure of 400 kPa in the tank. The final quality of the mixture remaining in the tank is ________ (*rounded off to two decimal places*).

Use the following data for water :

At 400 kPa : $v_f = 0.001084$ m^3/kg,

$v_{fg} = 0.46138$ m^3/kg,

$v_g = 0.46246$ m^3/kg

104. Consider a spark ignition engine which operates on an ideal air-standard Otto cycle. It uses a fuel which produces 44 MJ/kg of heat in the engine. If the engine requires 40 mg of fuel to produce 1 kJ of work output, then the compression ratio of the Otto cycle is ________ (*rounded off to two decimal places*).

For the entire cycle, use $c_p/c_v = 1.4$

105. A refrigerator operates on an ideal vapor compression cycle between the pressure limits of 140 kPa and 800 kPa. The working fluid is the refrigerant R-134a. The refrigerant enters the compressor as saturated vapor at 140 kPa and exits at 800 kPa and 60 °C. It leaves the condenser as a saturated liquid at 800 kPa. The coefficient of performance (COP) of the refrigerator is ________ (*rounded off to two decimal places*).

Use the following property data for R-134 a :

At 140 kPa: $h_f = 27.06$ kJ/kg, $h_g = 239.19$ kJ/kg

At 800 kPa: $h_f = 95.48$ kJ/kg, $h_g = 267.34$ kJ/kg

At 800 kPa and 60 °C: $h = 296.82$ kJ/kg

106. A steam power plant operates on a simple ideal Rankine cycle. The condenser pressure is 10 kPa and the boiler pressure is 5 MPa. The steam enters the turbine at 600 °C. Mass flow rate of the steam is 50 kg/s. Neglecting the pump work, the net power output of the plant is ________ MW (rounded off to one decimal place).

Use the following property data for water :

At 10 kPa :

$h_f = 191.81$ kJ/kg, $h_{fg} = 2392.82$ kJ/kg,

$h_g = 2584.63$ kJ/kg, $s_f = 0.6492$ kJ/kg-K,

$s_{fg} = 7.5010$ kJ/kg-K, $s_g = 8.1502$ kJ/kg-K

At 5 MPa and 600 °C :

$h = 3666.47$ kJ/kg, $s = 7.2588$ kJ/kg-K

107. In an air-conditioning system, air enters at 20 °C and 30% relative humidity at a steady rate of 30 m^3/min in a humidifier and it is conditioned to 25 °C and 60% relative humidity. Assuming entire process takes place at pressure of 100 kPa, the mass flow rate of the steam added to air in the humidifier is ________ kg/min (rounded off to three decimal places).

Use the following property data :

At 20 °C and 25 °C, saturation pressures of water are 2.3392 kPa and 3.1698 kPa, respectively.

For air, $R = 0.287$ kJ/kg-K

108. An office uses a heat pump to receive 500 kJ/day heat in winter to maintain its temperature at 300 K. The ambient temperature is 280 K. If the COP of the heat pump is 60% of its theoretical maximum value, the ratio of actual work input to the minimum theoretical work input to the heat pump is ________ (rounded off to one decimal place).

109. In a liquid-vapour phase change process, $\left(\frac{dP}{dT}\right)_{sat}$ at 100 °C for saturated water is 3750 Pa/K. If the resulting change in specific volume $(v_g - v_f)$ is 1.672 m^3/kg, the enthalpy of vaporization (h_{fg}) will be ________ kJ/kg (in integer).

POLYMER SCIENCE AND ENGINEERING (XE-F)

XE-F: Q.110 – Q.118 Carry ONE mark Each

110. Which one of the monomers given is used in the synthesis of cellulose?

(a) Fructose (b) Lactic acid

(c) Galactose (d) Glucose

111. A copper wire upon loading instantaneously increases in length to l, and then continues to elongate gradually. Upon unloading, the wire retracts to length l. According to the Maxwell model, which one of the options given correctly relates the total strain E, the applied stress S, the modulus G, the material's resistance to flow η, and the elapsed time t between loading and unloading?

(a) $E = (S/G) - (S/\eta)t$

(b) $E = (S/G) \times (S/\eta)t$

(c) $E = (S/G) + (S/\eta)t$

(d) $E = (S/G) / (S/\eta)t$

112. Consider the structure of a crosslinked polymer shown in the figure. From the options given, identify the monomers that are used in the synthesis of the polymer.

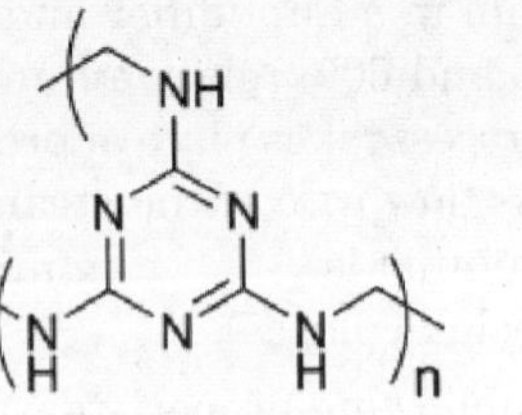

(a) Melamine and Benzaldehyde

(b) Melamine and Acetone

(c) Melamine and Formaldehyde

(d) Melamine and Ethanol

113. Among the options given, choose the most suitable compatibilizer for blending Polyvinylidene fluoride (PVDF) and Acrylonitrile butadiene styrene (ABS).

(a) Styrene-acrylonitrile (SAN)

(b) Polybutadiene (PB)

(c) Polymethyl methacrylate (PMMA)

(d) Nylon 6

114. A high molecular weight polymer passes through different zones from the hopper to the die in an extruder. Among the options given, identify the correct match between the zones and their key functions.

	Zones		Key functions
P	Metering zone	1.	High shear forces for effective mixing
Q	Compression zone	2.	Receives the charge or feed
R	Feed zone	3.	Melts the charge or feed through heat conducted by the heating element
S	Working zone	4.	The charge or feed acquires a constant flow rate imparted by the helical flight of the screw

(a) P-4; Q-3; R-2; S-1 (b) P-3; Q-4; R-1; S-2

(c) P-4; Q-1; R-2; S-3 (d) P-3; Q-1; R-2; S-4

115. Polymer wetting is improved by the addition of fillers with functional groups. In a typical case-study, natural-clay was modified with hydroxyl groups and compounded with Nylon 6 along with an antioxidant. The resulting composite exhibited poor mechanical properties. Which one among the options given explains this observation?

(a) The surface functional groups of the filler reacted with Nylon 6

(b) The antioxidant degraded during the processing

(c) The surface functional groups of the filler formed hydrogen bonds with the antioxidant

(d) The antioxidant reacted with Nylon 6

116. Among the options given, identify the correct match between the polymers and their glass transition temperatures (T_g).

	Polymer		Glass transition temperature (°C)
P	High density polyethylene	1	> 200
Q	Poly (vinyl carbazole)	2	145 to 155
R	Polymethyl methacrylate	3	–100 to –80
S	Polycarbonate	4	90 to 100

(a) P-2; Q-4; R-3; S-1 (b) P-3; Q-1; R-4; S-2

(c) P-3; Q-4; R-1; S-2 (d) P-4, Q-2; R-1; S-3

117. What is the correct order of decreasing crystallinity of the given polymers?

P. atactic-Polypropylene

Q. syndiotactic-Polystyrene

R. Nylon 6

S. Polyethylene terephthalate

(a) P > R > S > Q (b) S > Q > P > R

(c) Q > S > R > P (d) S > R > Q > P

118. Choose the correct option that best correlates the graphs with the polymerization methods.

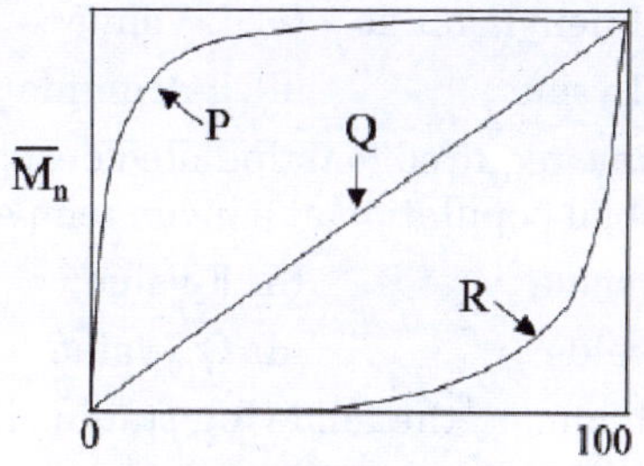

(a) P – living polymerization; Q – chain growth; R – step growth
(b) P – chain growth; Q – living polymerization; R – step growth
(c) P – step growth; Q – living polymerization; R – chain growth
(d) P – living polymerization; Q – step growth; R – chain growth

XE-F: Q.119 – Q.131 Carry TWO marks Each

119. From the options given, identify the correct match(es) between the polymer products with the most appropriate processing technique.

Polymer product	Processing technique
P. Fishing rods	1. Compression moulding
Q. Soft drink bottles	2. Thermoforming
R. Plastic sheets	3. Pultrusion
S. Plastic trays	4. Blow moulding

(a) P-3; Q-4; R-1; S-2 (b) P-3; Q-2; R-1; S-4
(c) P-1; Q-2; R-3; S-4 (d) P-3; Q-4; R-1; S-1

120. Among the options given, which agents are used to vulcanize or cure rubbers?
(a) Dicumyl peroxide (b) Zinc stearate
(c) Carbon black (d) Dinitrobenzene

121. Lipase is a natural enzyme, which cleaves carboxylic ester bonds. Among the options given, identify the polymer(s) degraded by lipase.
(a) Polypropylene (PP)
(b) Polycaprolactone (PCL)
(c) Polyvinylidene fluoride (PVDF)
(d) Polyethylene terephthalate (PET)

122. Among the options given, identify the correct pair(s) of catalyst and co-catalyst that form a Ziegler-Natta catalyst.
(a) $TiCl_3$ and $Al(CH_3CH_2)_2Cl$
(b) $ZnCl_2$ and $Al(CH_3)_3$
(c) TiO_2 and $Al(CH_3)_3$
(d) VCl_4 and $Al(CH_3CH_2)_2Cl$

123. Mechanical stress is applied on a polymer. Identify the correct match(es) between the statements (1, 2, 3, 4, 5) that describe the deformations and the regimes (P, Q, R).

P. Rubbery regime	1. Stress-relaxation takes place and the excess free energy is dissipated
Q. Region around glass transition temperature (T_g)	2. The motions of the molecules are long-range
R Sample under deformed state	3. Segmental motion of the molecules is important
	4. The maximum relaxation time is strongly dependent on the molecular weight
	5. The deformations are independent of the molecular weight and primarily depend on the local structure

(a) P-2; Q-5; R-1 (b) P-1; Q-5; R-1
(c) P-2; Q-3; R-4 (d) P-4; Q-3; R-1

124. Stress versus elongation profiles for different polymeric materials are shown in the figure. Choose the combination that best describes these profiles.

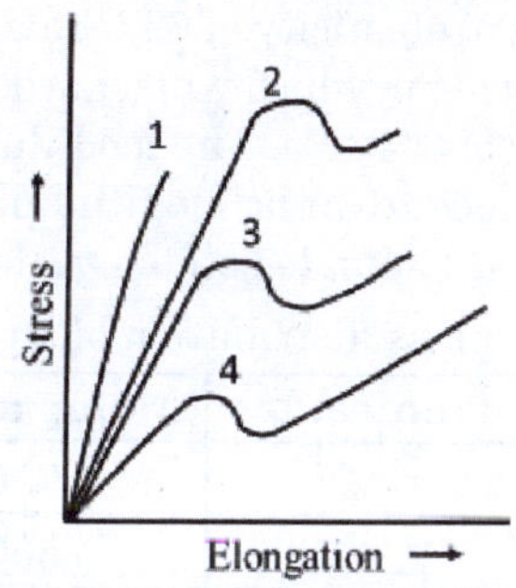

(a) 1-Nylon fibers; 2-Polyethylene; 3-Vulcanized rubber; 4-Polystyrene
(b) 1-Polyethylene; 2-Vulcanized rubber; 3-Polystyrene; 4-Nylon fibers
(c) 1-Polystyrene; 2-Nylon fibers; 3-Polyethylene; 4-Vulcanized rubber
(d) 1-Vulcanized rubber; 2-Polyethylene; 3-Nylon fibers; 4-Polystyrene

125. Among the options given, which method(s) is/are used for the synthesis of atactic polystyrene?
(a) Free radical polymerization
(b) Ring opening polymerization
(c) Polycondensation
(d) Ionic polymerization

126. A nylon sample of 0.03 m^2 cross-sectional area is subjected to a creep load of 10 kN. The load is removed after a duration of 60 s. Young's modulus and the viscosity for nylon are 1 GPa and 300 Giga Poise. The compliance of the specimen is ______ $\times 10^{-9}$ m^2/N. (Answer in integer)

127. Polyvinylidine fluoride (PVDF) was quenched from the melt in one case and in the other case, it was slowly cooled from the melt at 10 °C/min. The percentage crystallinity of the slowly cooled PVDF is 60%. The heat of fusion for the quenched PVDF is 0.5 ΔH_m (ΔH_m is the heat of fusion for the slowly cooled PVDF), and the heat of fusion for 100% crystalline PVDF is 100 J/g. The percentage crystallinity of the quenched PVDF is ______ %. (Answer in integer)

128. A polymeric material of density 0.9 g/cc and melt volume of 10 cc in an extruder has a residence time of 100 s. The output of the extruder is ______ kg/h. (Rounded off to two decimal places)

129. A polymer weighing 0.2 g is dissolved in 100 ml of benzene and has a relative viscosity of 1.5. The polymer obeys Mark-Houwink equation with constants a = 0.5 and K = 0.001. The molecular weight of the polymer is ______ $\times 10^{10}$ g/mol. (Rounded off to two decimal places).

130. A continuous and aligned glass-fiber reinforced composite consists of 40 vol% of glass-fiber having a modulus of elasticity of 69 GPa and 60 vol% of a polyester resin, which when hardened, displays a modulus of 3.4 GPa. The modulus of elasticity of this composite in the longitudinal direction is ______ GPa. (Rounded off to two decimal places)

131. The molar mass distribution of a polymer is

Number of molecules	Molar mass (g/mol)
100	7500
50	5000

The resulting weight average molecular weight of the polymer is ______ g/mol. (Answer in integer)

FOOD TECHNOLOGY (XE-G)

XE-G: Q.132 – Q.140 Carry ONE mark Each

132. Choose the correct group of fat soluble vitamins

(a) Cholecalciferol, α-Tocopherol, Menadione

(b) Thiamine, Cholecalciferol, α-Tocopherol

(c) Niacin, α-Tocopherol, Menadione

(d) Biotin, Thiamin, Niacin

133. The synthesis of thyroxine T4 in human body requires

(a) Selenium (b) Iodine

(c) Iron (d) Zinc

134. Which among the followings is NOT an essential amino acid?

(a) L-Phenylalanine (b) L-Valine

(c) L-Lysine (d) L-Arginine

135. The time required for stipulated destruction of a microbial population at a given temperature is

(a) D-value (b) F-value

(c) z-value (d) Q_{10} value

136. Which among the following statements is NOT correct?

(a) Cod fish is a major source of ω-3 fatty acids.

(b) Beetroot is a good source of β-carotene.

(c) Apple is a good source of vitamin B_{12}.

(d) Fresh sugarcane juice is a good source of polyphenol oxidase.

137. Which of the following statements is NOT correct?

(a) As the shear rate increases, the apparent viscosity decreases for a pseudoplastic fluid.

(b) As the shear rate increases, the apparent viscosity increases for a dilatant fluid.

(c) A Bingham fluid requires application of yield stress prior to any response.

(d) Rheopectic and thixotropic are two time independent fluids.

138. Calculate the efficiency in percent (rounded off to 1 decimal place) of an oil expeller which yields 37 kg oil containing 5% solid impurities from 100 kg mustard seeds. The oil content of the mustard seed is 38%.

139. Orange juice is packaged aseptically and stored under ambient conditions. The degradation of vitamin C in the juice occurs during storage and it follows first order reaction kinetics. The degradation rate constant is 5.2×10^{-3} day^{-1}. The half-life of vitamin C in days is ______ (in integer).

140. The weight of 10 kg dried cauliflower containing 5% moisture (wet basis) after rehydration is 60 kg. If the fresh cauliflower contained 87% moisture (wet basis), calculate the coefficient of rehydration (rounded off to 2 decimal places).

XE-G : Q.141 – Q.153 Carry TWO marks Each

141. Match the industrial product in Column I with fermentative organism in Column II.

Column I	Column II
P. Vinegar	1. Enterbacter aerogenes
Q. Citric acid	2. Saccharomyces cerevisiae
R. Ethanol	3. Acetobacter aceti
S. L-Lysine	4. Aspergillus niger

(a) P-3, Q-4, R-2, S-1
(b) P-1, Q-3, R-2, S-4
(c) P-3, Q-1, R-4, S-2
(d) P-2, Q-1, R-3, S-4

142. Match the enzyme in Column I with its application in food processing/reaction given in Column II.

Column I	Column II
P. Chymosin	1. Acyl glycerol restructuring
Q. Thermolysin	2. Lactose hydrolysis
R. β-Galactosidase	3. Aspartame synthesis
S. Lipase	4. Cheese manufacturing

(a) P-4, Q-3, R-2, S-1
(b) P-3, Q-1, R-2, S-4
(c) P-4, Q-2, R-3, S-1
(d) P-1, Q-4, R-2, S-3

143. Identify the Gram +ve bacteria responsible for causing food borne diseases among the followings

(a) *Campylobacter jejuni*
(b) *Clostridium botulinum*
(c) *Vibrio cholerae*
(d) *Salmonella typhi*

144. Extrusion cooking is accomplished in four different stages, which are indicated as I, II, III and IV in the figure given below. Choose the correct option representing the name of each stage.

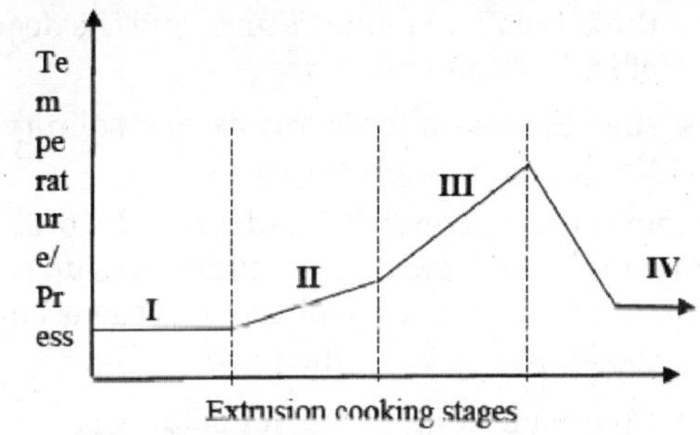

(a) I – Feeding, II – Cooking, III – Kneading, IV – Expansion
(b) I – Kneading, II – Feeding, III – Cooking, IV – Expansion
(c) I – Feeding, II – Kneading, III – Cooking, IV – Expansion
(d) I – Cooking, II – Kneading, III – Feeding, IV – Expansion

145. Match the method/ value used for measuring lipid characteristics in Column I with the corresponding properties indicated by them, in Column II.

Column I	Column II
P. Thiobarbituric acid test	1. Induction time
Q. Rancimat method	2. Degree of unsaturation
R. Peroxide value	3. Carbonyl content
S. Iodine value	4. Hydroperoxide content

(a) P-3, Q-1, R-4, S-2
(b) P-1, Q-3, R-4, S-2
(c) P-3, Q-1, R-2, S-4
(d) P-3, Q-4, R-1, S-2

146. Match the peeling technique in Column I with the vegetable, for which it is used in industry, given in Column II.

Column I	Column II
P. Knife peeling	1. Brinjal
Q. Abrasion peeling	2. Tomato
R. Flame peeling	3. Potato
S. Flash peeling	4. Cucumber

(a) P-3, Q-4, R-1, S-2
(b) P-4, Q-1, R-3, S-2
(c) P-4, Q-3, R-2, S-1
(d) P-4, Q-3, R-1, S-2

147. Match the process in Column I with the related food component in Column II.

Column I	Column II
P. Caramelization	1. Lipid
Q. Denaturation	2. Sugar
R. Oxidation	3. Pigment
S. Bleaching	4. Enzyme

(a) P-2, Q-4, R-1, S-3
(b) P-2, Q-1, R-4, S-3
(c) P-1, Q-3, R-2, S-4
(d) P-2, Q-1, R-3, S-4

148. Identify the correct statement(s) related to grain polysaccharides among the followings.

(a) Dextrin are a group of low molecular weight polysaccharides produced by dry hydrolysis of starch.
(b) Amylose is a linear polymer of D-glucose units joined by α $(1 \rightarrow 6)$ glycoside linkages.
(c) Amylopectin is a branched chain polymer of D-galactose monomer units.
(d) Retrogradation is a process of reassociation of amylose and formation of crystalline structure by gelatinized starch upon cooling.

149. Identify the correct pair(s) of governing law with respective process operation.

(a) Stoke's law – Mass transfer

(b) Kirchhoff's law – Radiation heat transfer

(c) Fourier's law – Conduction heat transfer

(d) Fick's law – Molecular diffusion

150. A hammer mill is used to grind blackgram. The size distribution of the blackgram is such that 80% passes through a 6-mesh (particle size = 3.36 mm) screen. The power requirement to produce a powder, 80% of which passes through a 45-mesh (particle size = 0.354 mm) screen is 4.5 kW. The power in kW required to produce a finer powder 80% of which passes through a 60-mesh (particle size = 0.25 mm) will be ______ (rounded off to 2 decimal places). Assume the feed rate to the mill is constant in both the cases. Use Bond's law of size reduction.

151. If D_{10} for Salmonella in egg yolk is 0.75 kGy, calculate the radiation dose in kGy (rounded off to 2 decimal places) required for reducing the Salmonella count in egg yolk by 8 log cycles.

152. The average moisture binding energy of a plant protein based snack at 8% moisture content (dry basis) is 3200 cal.mol^{-1}. If the water activity of the snack at the above moisture content is 0.30 at 30 °C, the water activity of the sample at 45 °C is _____ (rounded off to 2 decimal places). The value of Gas constant R = 1.987 cal.mol^{-1}.K^{-1}.

153. Cow milk is pasteurized at a flow rate of 1 kg.s^{-1} in a counter-current heat exchanger using hot water as the heating medium. The milk enters the heat exchanger at 15 °C and exits at 50 °C. The specific heat of cow milk is 3.5 kJ.kg^{-1}.°C^{-1} and remains constant at the inlet and exit of the heat exchanger. The inlet and exit temperatures of the hot water are 75 °C and 60 °C respectively. If the overall heat transfer coefficient is 1800 W.m^{-2}. °C^{-1}, the heat transfer surface area in m^2 is ________ (rounded off to 2 decimal places). Assume steady state conditions.

ATMOSPHERIC AND OCEAN SCIENCE (XE-H)

XE-H: Q.154 – Q.162 Carry ONE mark Each

154. The net water level elevation near the coast arising due to a tropical-cyclone-induced storm is a combination of the following factors ______.

(a) astronomical tides, and distant cyclonic storm

(b) wind-induced waves, astronomical tides, and distant cyclonic storm

(c) astronomical tides, coastal currents, and distant cyclonic storm

(d) riverine flow, astronomical tides, and distant cyclonic storm

155. The typical speeds of a tsunami wave in water depths of 10 m, 100 m, and 1000 m, respectively, are ________.

(a) approximately 100 m s^{-1}, 31 m s^{-1} and 10 m s^{-1}

(b) approximately 50 m s^{-1}, 75 m s^{-1} and 100 m s^{-1}

(c) approximately 10 m s^{-1}, 31 m s^{-1} and 100 m s^{-1}

(d) approximately 100 m s^{-1}, 10 m s^{-1} and 31 m s^{-1}

156. Which classification of tides best represents the west coast of India?

(a) Diurnal and Mixed

(b) Semidiurnal and Mixed

(c) only Diurnal

(d) only Mixed

157. The variation in geostrophic winds with height, in particular, at the atmospheric boundary layer, considering the variation in pressure gradient as a function of height, is referred to as ______.

(a) Isallobaric winds (b) Gradient winds

(c) Thermal winds (d) Cyclostrophic winds

158. Consider the following options and pick out the right choice. The solubility of a gas in sea water increases with _______.

(a) the increase of temperature, salinity and pressure

(b) the decrease of temperature, salinity and pressure

(c) the increase of temperature, and the decrease of salinity and pressure

(d) the decrease of temperature and salinity, and the increase of pressure

159. From the list given below, identify the organism type in the biological pump that takes up carbon dioxide from the atmosphere into the ocean.

(a) Zooplankton (b) Fish

(c) Phytoplankton (d) Radiolarians

160. The amount of CO_2 that can be absorbed ______ when the temperature of seawater decreases.

(a) Remains the same

(b) Increases

(c) Decreases

(d) Can either increase or decrease

161. Which among the following gases has the highest global warming potential?

(a) CO_2

(b) Water vapor

(c) Methane

(d) N_2O

162. What will happen to the speed of a balanced flow as one moves across the isobar along a particular latitude?

(a) The speed changes for a geostrophic flow and remains constant for a cyclostrophic flow

(b) The speed remains constant for a geostrophic flow and changes for a cyclostrophic flow

(c) The speed remains constant for both (geostrophic and cyclostrophic) types of flow

(d) The speed changes for both (geostrophic and cyclostrophic) types of flow

XE-H: Q.163 – Q.175 Carry TWO marks Each

163. The following illustration depicts the circulation pattern in the North Indian Ocean during the southwest (June-August) monsoon. Identify the markers numbered from 1-5 in the illustration and pick out the right choice.

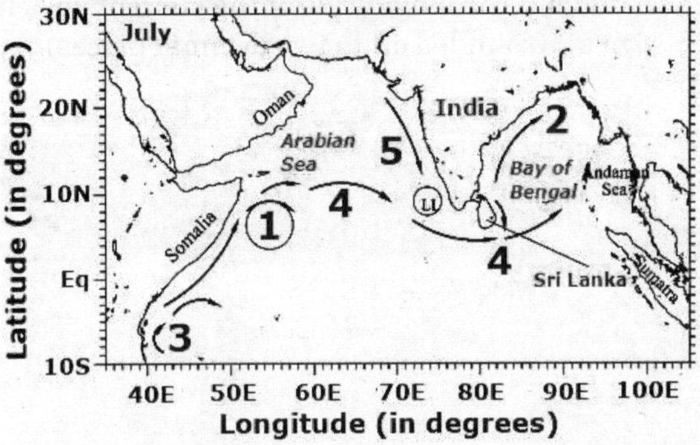

(a) 1. Great Whirl, 2. East India Coastal Current, 3. East African Coastal Current, 4. Summer (Southwest) Monsoon Current, and 5. West India Coastal Current.

(b) 1. East African Coastal Current, 2. Summer (Southwest) Monsoon Current, 3. Great Whirl, 4. South Equatorial Counter Current, and 5. East India Coastal Current.

(c) 1. Great Whirl, 2. Summer (Southwest) Monsoon Current, 3. West India Coastal Current, 4. Somali Current, and 5. East India Coastal Current.

(d) 1. East African Coastal Current, 2. West India Coastal Current, 3. Somali Current, 4. East India Coastal Current, and 5. South Equatorial Counter Current.

164. From the following list identify the region that has low chlorophyll and low nutrients.

(a) Upwelling, anticyclonic eddy

(b) Upwelling, cyclonic eddy

(c) Downwelling, anticyclonic eddy

(d) Downwelling, cyclonic eddy

165. Match the following physical phenomena from the perspective of monsoonal circulation :

Column-1	Column-2
(a) Reversal of East-West temperature gradient with seasons	(i) Over the tropics
(b) Reversal of North -South temperature gradient with seasons	(ii) Indian Ocean
(c) Surface cooling and deepening of mixed layer	(iii) Over mid-latitudes
(d) Northward movement of south equatorial current	(iv) Arabian Sea

(a) a-i, b-iii, c-ii, d-iv

(b) a-iii, b-i, c-iv, d-ii

(c) a-iii, b-ii, c-iv, d-i

(d) a-ii, b-i, c-ii, d-iv

166. Choose the correct statement(s) in context to Ekman spiral from the following :

(a) The balance of Coriolis force and pressure

(b) The balance of Coriolis force, wind shear, and frictional force

(c) Deflection of surface current to the right of the wind direction in the Northern hemisphere

(d) The increase of current velocity with depth

167. Which of the following is/are associated with winter rainfall over India?

(a) Northeast monsoon

(b) Southwest monsoon

(c) Western disturbances

(d) Agulhas Current

168. If the global albedo is increased from 0.3 to 0.4, the global radiative equilibrium temperature (in K) would decrease by ______.

(consider no greenhouse effect, solar constant = 1360 W m^{-2} and Stefan-Boltzmann constant = 5.67×10^{-8} W m^{-2} K^{-4}, rounded off to one decimal place).

169. When a parcel of dry air rises at a rate of 2 cm s^{-1} vertically, what should be the rate of heating per unit mass in J s^{-1} kg^{-1} (due to the radiation, conduction, etc.) in order to maintain the air parcel at a constant temperature?

(Consider the acceleration due to gravity as 9.8 m s^{-2}, rounded off to three decimal places).

170. Two balls each of 5 cm in diameter are placed 200 m apart on a horizontal frictionless plane at 45°N. The balls are impulsively propelled directly at each other with equal speeds. What must be the speed in m s^{-1} so that the two balls just miss each other?

(Consider the value of Ω as 7.29×10^{-5} rad s^{-1}, rounded off to two decimal places).

171. A parcel of dry air having an initial temperature of 30 °C at 1000 hPa level is lifted adiabatically. At what pressure (in hPa) its density reduces by half?

(Consider the ratio of the specific heat of dry air at a constant pressure to the specific heat of dry air at a constant volume, $C_p/C_v = 0.71$, rounded off to two decimal places).

172. A cylindrical tank containing water is rotating about the z-axis at a constant angular velocity of 10 rad s^{-1}. The schematic of the isobaric surface is shown in the following illustration, where 'A' and 'B' are two points on the isobaric surface at heights 'z_1' and 'z_2', respectively. Assuming the atmospheric pressure to be negligible and no transient flow, estimate the elevation difference in m between 'z_1' and 'z_2'.

(Consider $r_1 = 0.5$ m, $r_2 = 1.0$ m and gravitational acceleration g = 9.8 m s^{-2}, rounded off to two decimal places)

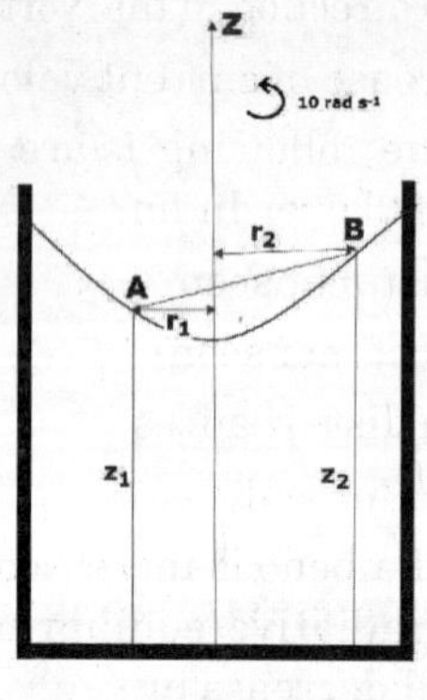

173. Consider that there are 295 million vehicles across India and they drive about 12000 km yr-1. Each vehicle consumes about 25 km lit^{-1} of petrol and the amount of carbon released per litre is 5.5 gm. What is the amount of carbon emitted into the atmosphere in mega ton per year (rounded off to two decimal places)?

174. Consider the two parallel isobars separated by a spacing of 250 km at 30° N (air density = 0.70 kg m^{-3}, $2\Omega = 14.6 \times 10^{-5}$ rad s^{-1}) with a pressure gradient of 5 hPa. Calculate the geostrophic velocity in m s^{-1} (rounded off to two decimal places).

175. Considering a gyre system in the following illustration (where 'L' and 'H' represent low- and high-pressure regions along 45° N), the average slope between the points 'A' and 'B' is 2.1 cm km-1 under no wind condition. For a steady southerly wind of 10 m s^{-1} over these regions, find out the magnitude of the current velocity in m s^{-1} (rounded off to two decimal places).

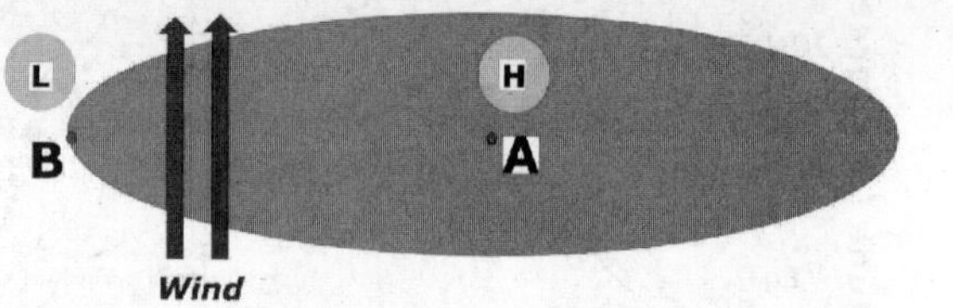

ANSWERS

General Aptitude

1. (d) **2.** (c) **3.** (d) **4.** (c) **5.** (b) **6.** (b) **7.** (a)
8. (c) **9.** (a) **10.** (d)

Engineering Mathematics

11. (a) **12.** (a) **13.** (a) **14.** (c) **15.** (a) **16.** (2.60 to 2.65)
17. (3 to 3) **18.** (b) **19.** (a) **20.** (0.26 to 0.30) **21.** (2 to 2) **22.** (b)

Fluid Mechanics

23. (b) **24.** (c) **25.** (c) **26.** (d) **27.** (a) **28.** (c) **29.** (c)
30. (a, c) **31.** (a) **32.** (c) **33.** (c) **34.** (b) **35.** (c, d) **36.** (183.0 to 185.0)
37. (820 to 820) **38.** (2.60 to 2.75) **39.** (430 to 460) **40.** (18.5 to 19.5)
41. (1.10 to 1.20) **42.** (0.32 to 0.34) **43.** (2.2 to 2.4)

Material Science

44. (c) **45.** (c) **46.** (d) **47.** (c) **48.** (b) **49.** (c, d) **50.** (350 to 400)
51. (90.0 to 105.0) **52.** (490 to 510) **53.** (b) **54.** (a) **55.** (d)
56. (c) **57.** (b, c, d) **58.** (a, d) **59.** (a, c) **60.** (3840 to 4100) **61.** (1 to 1)
62. (9.20 to 9.60) **63.** (7.90 to 8.05) **64.** (5.25 to 7.00) **65.** (2.45 to 2.80)

Solid Mechanics

66. (b) **67.** (b) **68.** (a) **69.** (b) **70.** (a) **71.** (a) **72.** (117.0 to 118.0)
73. (4.9 to 5.1) **74.** (49.5 to 50.5) **75.** (d) **76.** (c) **77.** (c)
78. (d) **79.** (a) **80.** (c) **81.** (1.8 to 2.0) **82.** (23.0 to 25.0)
83. (57.0 to 58.0) **84.** (79.5 to 80.5) **85.** (199.0 to 201.0) **86.** (1.9 to 2.1)
87. (0.350 to 0.380)

Thermodynamics

88. (d) **89.** (d) **90.** (a) **91.** (b) **92.** (c) **93.** (d) **94.** (20.5 to 21.5) **95.** (420 to 420) **96.** (4389 to 4392) **97.** (d) **98.** (a) **99.** (b) **100.** (c) **101.** (85.0 to 88.0) **102.** (2975 to 2980) **103.** (0.90 to 0.95) **104.** (8.00 to 8.30)
105. (2.45 to 2.55) **106.** (67.3 to 69.3) **107.** (0.262 to 0.290) **108.** (1.6 to 1.8)
109. (2337 to 2341)

Polymer Science and Engineering

110. (d) **111.** (c) **112.** (c) **113.** (c) **114.** (a) **115.** (c) **116.** (b) **117.** (d)
118. (b) **119.** (a, d) **120.** (a, d) **121.** (b, d) **122.** (a, d) **123.** (a, d) **124.** (c)
125. (a, d) **126.** (0.8 to 1.0 OR 3.0 to 3.0) **127.** (30 to 30) **128.** (0.31 to 0.33)
129. (6.24 to 6.26) **130.** (29.60 to 30.01) **131.** (6875 to 6875)

Food Technology

132. (a) **133.** (b) **134.** (d) **135.** (b) **136.** (c) **137.** (d) **138.** (92.0 to 93.0)
139. (132 to 134) **140.** (0.75 to 0.90) **141.** (a) **142.** (a) **143.** (b)
144. (c) **145.** (a) **146.** (d) **147.** (a) **148.** (a, d) **149.** (b, c, d) **150.** (5.20 to 6.20)
151. (6 to 7) **152.** (0.32 to 0.41) **153.** (1.80 to 2.20)

Atmospheric and Ocean Science

154. (b) **155.** (c) **156.** (b) **157.** (c) **158.** (d) **159.** (c) **160.** (b) **161.** (d)
162. (b) **163.** (a) **164.** (c) **165.** (b) **166.** (b, c) **167.** (a, c) **168.** (9.5 to 9.7)
169. (0.195 to 0.197) **170.** (20.6 to 20.9) **171.** (611.30 to 611.35) **172.** (3.82 to 3.83)
173. (0.78 to 0.78) **174.** (39.06 to 39.37) **175.** (2.10 to 2.11)

EXPLANATIONS

General Aptitude

1. Here, we are locating village near the ocean & hills. So, 'between' is appropriate proposition. So the correct sentence is "the village was nestled in a green spot, between the ocean and the hills."

2. Here, Agree is antonym of Disagree, in the same way recommend is antonym of protest. In other words, if persons are disagree then they can protest, but if they are agree then they recommend.

3. No. of odd digits are as 1, 3, 5, 7, 9

   For 3 digit number – □□□

   - First digit can be any of these five so number of selection = 5
   - Second digit will be different from first so number of selection = 4
   - Third digit will be different from second but can be similar to first so number of selection = 4

   New, total number of ways = 5 × 4 × 4 = 80

4. For mean and median of the scores, at most half of the candidates have a score that is larger than the median. There can be the case, when we have more than half of the candidates have a score that is smaller than or larger than the mean.

   So, option (c) is correct and option 'd' is not correct.

   Option 'a' and 'b' are incorrect as we can not define relation between mean and median.

5. Let's discuss each options –

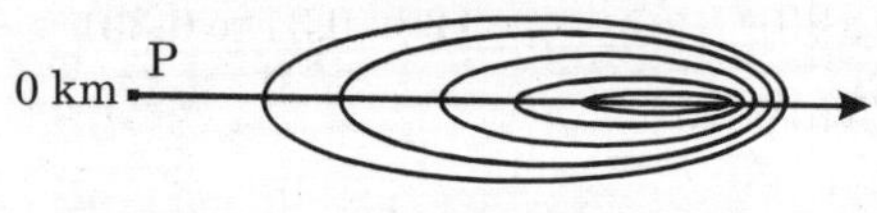

Not possible, as top of the hill is towards left side.

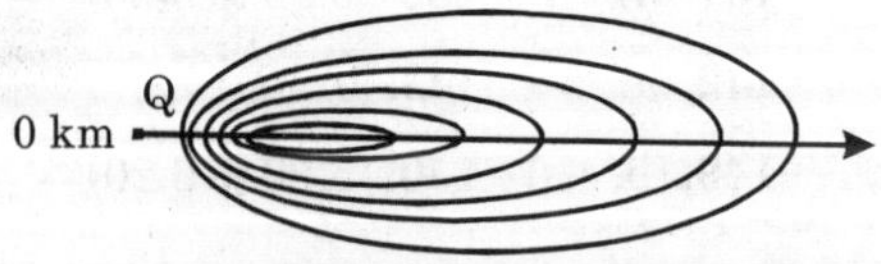

This is correct answer as positions of all the ovals are according to given hill.

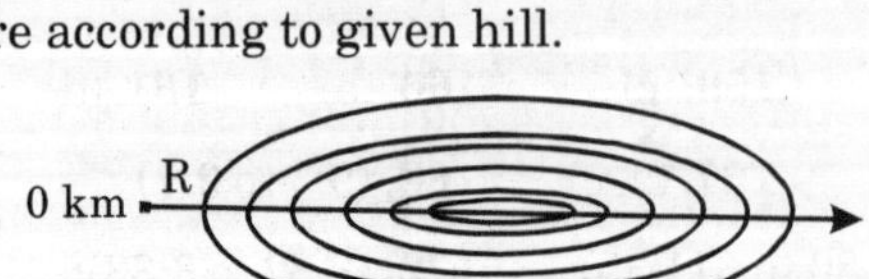

Not possible, as top of the hill is not in centre also, all ovals are not concentric.

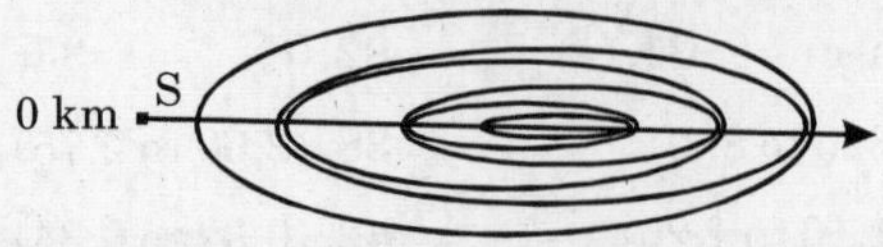

Not possible as ovals are not connected at right side.

6. According to second line of paragraph option 'b' is correctly inferred as, all academicians residing is the complex are well extablished in their fields.

7. For N = 1, only one way so F(1) = 1

   For N = 2, there are two ways as –

   – By one-one stair

   – By two stairs

   For N = 3, there are three ways as –

   – By one-one stairs

   – By first one then two stair

   – By first two then one stair

   For N = 4, 1 : One step 2 : Two steps

   (1) 1 + 1 + 1 + 1 = 4

   (2) 1 + 2 + 1 = 4

   (3) 1 + 1 + 2 = 4

   (4) 2 + 1 + 1 = 4

   (5) 2 + 2 = 4

   For N = 5, 1 : 1 step ; 2 : 2 steps

   (1) 1 + 1 + 1 + 1 + 1 = 5

   (2) 1 + 1 + 1 + 2 = 5

   (3) 1 + 1 + 2 + 1 = 5

   (4) 1 + 2 + 1 + 1 = 5

   (5) 2 + 1 + 1 + 1 = 5

   (6) 1 + 2 + 2 = 5

   (7) 2 + 1 + 2 = 5

   (8) 2 + 2 + 1 = 5

   So, 8 ways, 'a' is correct answer.

8. By using given information, we can infer that

   – Only 2% of human DNA produce amino acid and multiple amino acid are assembled to forms a protein. So majority of human DNA has no role in the synthesis of proteins.
   So, (i) is true.

   – Now remaining 98% of human DNA, has no role in the synthesis of proteins so can not be understood the role of these 98% of human DNA.
   So, both (i) & (ii) are true.

9. Given function :

$f(x) = \|x+2| - |x-1\|$

Let's find values of this function at –2, –1, 0, 1 & 2

$f(-2) = \|-2+2| - |-2-1\| = |0 - |-3|| = |-3| = 3$

$f(-1) = \|-1+2| - |-1-1\| = \|1| - |-2\|$

$= |1-2| = |-1| = 1$

$f(0) = \|0+2| - |0-1\| = |2-1| = 1$

$f(1) = \|1+2| - |1-1\| = |3-0| = 3$

$f(2) = \|2+2| - |2-1\| = |4-1| = 3$

By using values of function at different value of x, option 'a' is correct graph.

10. By taking each option

Option (a) :

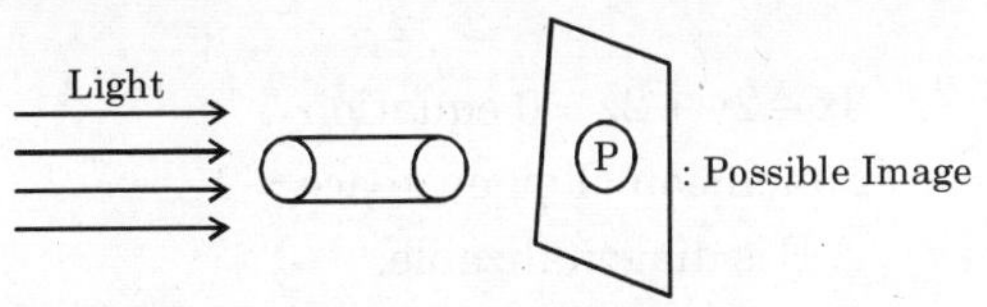

Option (b) :

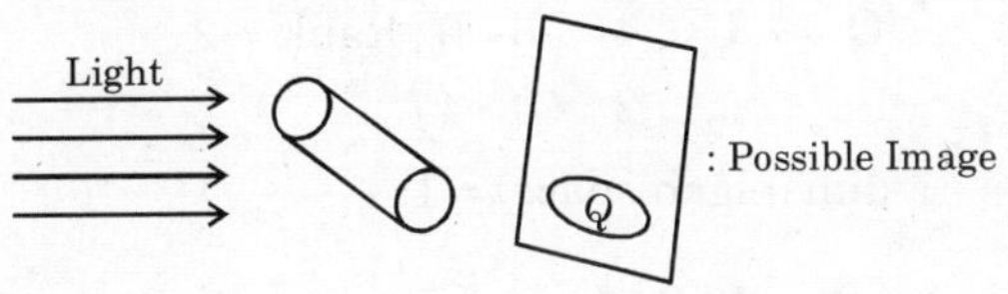

Option (c) :

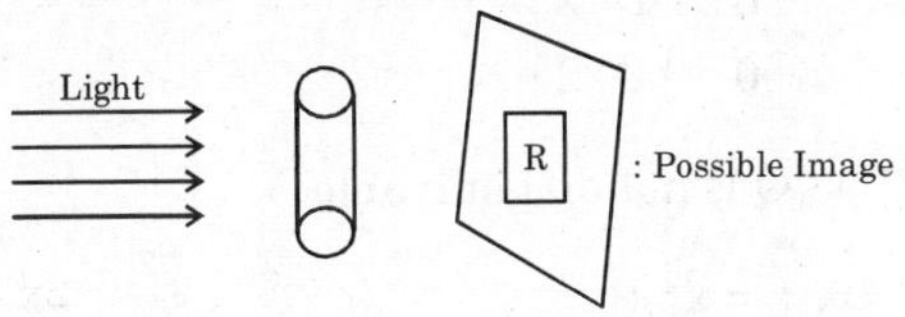

Option (d) : Not possible image in any direction.

11. I = 1 (Identity matrix)

$B = A^2 + 2A + I$

$= A(A+1) + 1(A+1) = (A+1)^2$

$B = (A+1)^2$

A is a 3 × 3 matrive having Eigen values 1, 2, and 3 so, Eigen values of B = $(1+1)^2, (2+1)^2, (3+1)^2$

= 4, 9, 16

12. $\left(\frac{\partial f}{\partial x}\right)_{(0,0)}$ & $\left(\frac{\partial f}{\partial y}\right)_{(0,0)}$

$$\left(\frac{\partial f}{\partial x}\right)_{(0,0)} = \lim_{h\to 0} \frac{\frac{h \times 0}{|h| + 0} - 0}{h}$$

$$= \lim_{h\to 0} \frac{0}{h|h|} = 0$$

$$\left(\frac{\partial f}{\partial y}\right)_{(0,0)} = \lim_{k\to 0} \frac{\frac{0 \times k}{0 + k} - 0}{k} = \frac{0}{k^2} = 0$$

$$\left(\frac{\partial f}{\partial x}\right)_{(0,0)} \;\&\; \left(\frac{\partial f}{\partial y}\right)_{(0,0)} = 0$$

Option A is the right Answer.

13. Let $f(x) = a_0 + a_1 x + a_2 x^2$

$$\int_{-1}^{1} f(x) \approx \frac{1}{9}\left(C_1 f(-1) + C_1 f\left(\frac{1}{2}\right) + C_3 f(1)\right)$$

$$9 \times \int_{-1}^{1} \left(a_0 + a_1 x + a_2 x^2\right) dx$$

$$= C_1 [a_0 - a_1 + a_2] + C_2\left[a_0 + a_1\frac{1}{2} + a_2\frac{1}{4}\right]$$

$$+ C_3 [a_0 + a_1 + a_2]$$

integrate →

$$9\left[a_0 \times 2 + \frac{2}{3}a_2\right] = a_0 [C_1 + C_2 + C_3]$$

$$+ a_1\left[-C_1 + \frac{C_2}{2} + C_3\right]$$

$$+ a_2\left[C_1 + \frac{C_2}{4} + C_3\right]$$

$$18a_0 + 6a_2 = a_0 [C_1 + C_2 + C_3] + a_1[-C_1 + \frac{C_2}{2} + C_3]$$

$$+ a_2 [C_1 + \frac{C_2}{4} + C_3]$$

Compared other side, we get

$$\Rightarrow C_1 + \frac{C_2}{4} + C_3 = 6$$

14. $\frac{d^2y}{dx^2} + (\lambda - 3)y = 0,\ y(0) = y(\pi) = 0$

Auxiliary equation

$y^2 + (\lambda - 3) - 0$

$\Rightarrow y^2 = -(\lambda - 3)$

Far Eigen values →

$(\lambda - 3) > 0$

$\boxed{\lambda > 3}$

$y = \pm i\sqrt{(\lambda - 3)}$

$\therefore y = C_1 \cos\sqrt{\lambda - 3}x + C_2 \sin\sqrt{x - 3}\, x$ ...(i)

At x = 0, y = 0

$\Rightarrow 0 = C_1(1) + C_2(0)$

$\boxed{C_1 = 0}$

Then eq (1) become

$y = C_2 \sin \sqrt{\lambda - 3}x$

At $x = \pi$, $y = 0$

$0 = C_2 \sin\left(\sqrt{\lambda - 3}\right)\pi$

$\sin\left(\sqrt{\lambda - 3}\right)\pi = 0$

$\Rightarrow \sqrt{\lambda - 3}\ \pi = n\pi$

15. Option B & option D are differentiable at all the points.

Option A

$f(z) = Z|z| = re^{i\theta} \times = r^2e^{i\theta}$

$= r^2 \cos\theta + i\, r^2 \sin\theta$

$$\boxed{\frac{\partial v}{\partial r} = \frac{1}{r}\frac{\partial v}{\partial \theta}, \frac{\partial v}{\partial r} = -\frac{1}{r}\frac{\partial v}{\partial \theta}}$$

$U_r = \frac{1}{r}V\theta$ Only when $r\cos(\theta) = 0$

$V_r = -\frac{1}{r}U\theta$ Only when $r\sin(\theta) = 0$

$U_r = 2r\cos\theta$ $\quad V_\theta = r^2\cos(\theta)$

$V_r = 2r\sin\theta$ $\quad V_\theta = -r^2\sin(\theta)$

Hence = Answer is option A.

16. $P(x) = a_0 + a_1 x + a_2 x(x-1) + a_3 x(x-1)(x-2)$

(0, 2), (1, 3, (2, 2, (3, 5)

$P\left(\frac{5}{2}\right) = -?$

$(0, 2) \rightarrow x \Rightarrow 0$

$\boxed{2 = a_0}$

$(1, 3) \rightarrow x \Rightarrow 0$

$3 = a_0 + a_1 \Rightarrow \boxed{a_1 = 1}$

$(2, 2) \rightarrow x = 2$

$2 = a_0 + 2a_1 + 2a_2$

$\boxed{a_2 = -1}$

$(3, 5) \rightarrow x = 3$

$5 = a_0 + 3a_1 + 6a_2 + 6a_3 \Rightarrow \boxed{a_3 = 1}$

Put value of a_0, a_1, a_2 and a_3 in equation P(x)

$= 2 + 1x + (-1)x(x-1) + (1)x(x-1)(x-2)$

Now,

$$P\left(\frac{5}{2}\right) = 2 + \left(\frac{5}{2}\right)\times 1 + (-1)\left(\frac{5}{2}\right)\left(\frac{3}{2}\right) + (1)\left(\frac{5}{2}\right)\left(\frac{3}{2}\right)\left(\frac{1}{2}\right)$$

$$P\left(\frac{5}{2}\right) = 2 + \frac{05}{2} + \left(-\frac{15}{4}\right) + \frac{15}{8}$$

$$P\left(\frac{5}{2}\right) = 2 + \frac{05}{2} - \frac{15}{8} \Rightarrow \boxed{P\left(\frac{5}{2}\right) = 2.63}$$

17. $\vec{F}(x,y) = \underbrace{\left(4x^m y^2 - 2xy^m\right)}_{v} + \underbrace{\left(2x^4 y - 3x^2 y^2\right)}_{v}\hat{j}$

$\therefore$ Conservative vector field $\rightarrow$ not depend on path

$\vec{v}\cdot\vec{F} = \vec{o}$

$$\frac{\delta \forall}{\delta x} - \frac{\delta v}{\delta y} = 0$$

$$\frac{\delta}{\delta x}\left(2\times 4y - 3x^2y^2\right) - \frac{\delta}{\delta y}\left(4\times my^2 - 2\times y^m\right)$$

$\Rightarrow 8x^3y - 6\times y^2 = 8x^m y - 2m \times y^{m-1} = 0$

So, m = 3

18. Eigen values are distend always diagonalizable if identical value it may or may not be diagonalizable

$$= Q - \lambda I = \begin{bmatrix} 3 & -2 & 2 \\ 6 & 4 & 4 \\ 3 & -2 & 2 \end{bmatrix} \text{ Rank} = 1$$

$3x - 2y + 2z = 0$ equation

dimension of eigen space = 2

$\therefore$ P is diagonalizable.

$$Q = \lambda I = \begin{bmatrix} 2 & -2 & 2 \\ 4 & -5 & 6 \\ 2 & -3 & 4 \end{bmatrix} \text{ Rank} = 2$$

dim (eigen space) = 1

$$\begin{bmatrix} 2 & -2 & 2 \\ 0 & -1 & 2 \\ 0 & -1 & 2 \end{bmatrix}$$

Q is not diagonlyable

19. $z = x^2 + y^2$ $\qquad \frac{\partial z}{\partial x} = 2x$

between

$z = 0$ and $z = \frac{1}{4}$ $\qquad \frac{\partial z}{\partial y} = 2y$

$$\text{Surface area (S)} = \iint_R \sqrt{1 + \left(\frac{\partial z}{\partial x}\right)^2 + \left(\frac{\partial z}{\partial y}\right)^2}\, dx\, dy$$

$$= \iint_R \sqrt{1 + 4r^2}\, r.dr\, d\theta$$

$$= \iint_R \sqrt{1 + 4r^2}.\ r.dr\, d\theta$$

$$= \int_0^{2\pi}\int_0^{\frac{1}{2}} \sqrt{1 + 4r^2}.r.dr\ d\theta$$

After solving $\Rightarrow \frac{\pi}{6}\left\{2\sqrt{2} - 1\right\}$

20. $\rightarrow$ Speaking truth P(truth) = $\frac{4}{6}$

$\rightarrow$ Getting same number P(same) = $\frac{1}{6}$

Person told that both number are same

P (saying same number) = P (same $\cap$ truth) + P(same $^c \cap$ truth^c)

Probability number appered actually $\Rightarrow$

$$\frac{P(\text{same} \cap \text{truth})}{P(\text{same} \cap \text{truth}) + \left(P\text{same}\ ^c \cap \text{truth}^c\right)}$$

on both the dies $\Rightarrow$

$$\Rightarrow \frac{\frac{1}{6} \times \frac{4}{6}}{\left(\frac{1}{6} \times \frac{4}{6}\right) + \left(\frac{5}{6} \times \frac{2}{6}\right)} \Rightarrow \frac{2}{7} \Rightarrow 0.29$$

21. $\frac{\partial U}{\partial t} - \frac{\partial^2 V}{\partial x^2} = 0$

$u = (C_1 e^{px} + c_2 e^{-px})\ c_3 e^{c^2p^2t}$

$u = (c_4 \cos px + (5 \sin px)\ c_6 e^{-c^2p^2t}$

$u = (c_7 x + c_8)c_9$

$C^2 = 1$

$U = [C_4 \cos(px) + C_5 \sin(px)]\ C_6 e^{-p^2t}$

$U(0, t) = 0 \Rightarrow C_4 = 0$

$U = C_{10} \sin(px)\ e^{-p^2t}$

$U(2, t) = 0 \Rightarrow \sin(2p) = 0\quad 2p = n\pi$

$$P = \frac{n\pi}{2}$$

$$u = \sum_{n=1}^{\infty} b_n \left(\sin \frac{n\pi x}{2}\right) e^{-\left(\frac{n\pi}{2}\right)^2 t}$$

$$U(x, 0) = (\sin \pi x) \Rightarrow \sum_{n=1}^{\infty} bn \sin\left(\frac{n\pi x}{2}\right) \times e^{-\left(\frac{nA}{2}\right)^2 \times 0}$$

$\therefore b_2 = 1\ \&\ b_n = 0 \quad \forall n \in N - \sqrt{2}\}$

$$\boxed{U(\pi, t) = \sin(\pi x) e^{-\pi^2 t}}$$

the value of

$$e^{\pi^2}\left(U\left(\frac{1}{2}, 1\right) - U\left(\frac{3}{2}, 1\right)\right)$$

$$= e^{\pi^2} \times \left[\sin\left(\frac{\pi}{2}\right) \times e^{-\pi^2} - \sin\left(\frac{3\pi}{2}\right) \times e^{-\pi^2}\right]$$

$= 2$

22. Pitot probe:

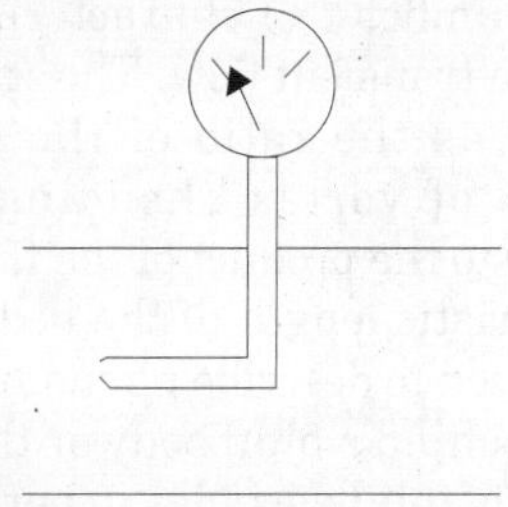

A pitot probe is a device used to measure the velocity of a fluid. It consists of a tube that is pointed directly into the flow of the fluid. The fluid flowing into the tube comes to a stop, creating a stagnation pressure. This stagnation pressure can be measured and used to calculate the velocity of the fluid.

Pitot-static probe:

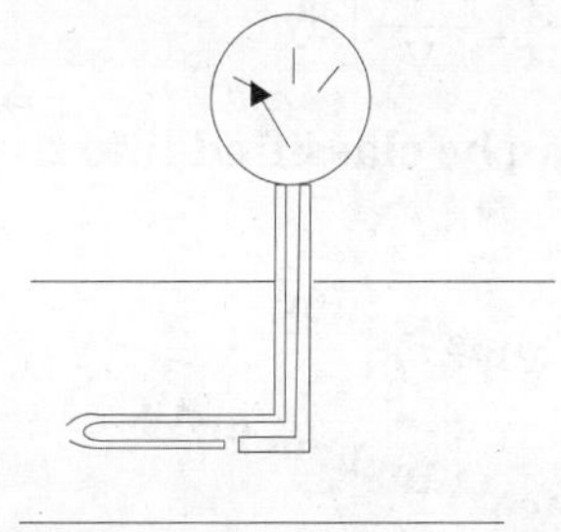

A pitot-static probe is a pitot probe that also has a set of static pressure ports. The static pressure ports are located around the pitot probe, so they measure the pressure of the fluid in the surrounding area. The difference between the stagnation pressure and the static pressure can be used to calculate the dynamic pressure of the fluid. The dynamic pressure can then be used to calculate the velocity of the fluid.

Piezometer:

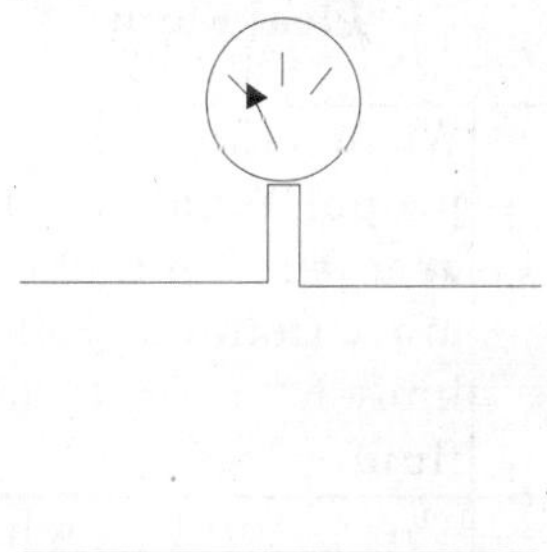

A piezometer is a device used to measure the pressure of a liquid. It consists of a small tube that is inserted into the liquid. The pressure of the liquid pushes up on the liquid in the tube, creating a column of liquid. The height of the column of liquid can be measured and used to calculate the pressure of the liquid.

Hence option (b) is the correct answer.

23. Among the given non-dimensional numbers, the strouhal number (S_t) characterizes periodicity present in a transient flow. The strouhal number is defined as the ratio of the characteristic frequency of vortex shedding (or periodic oscillation) to the product of the flow velocity and a characteristic length of the body or flow field. It's often used to describe phenomena like vortex shedding behind a bluff body or the frequency of oscillation of fluid particles in periodic flows.

24. $\theta = \int_0^{\infty} \frac{U}{U_{\infty}}\left(1-\frac{U}{U_{\infty}}\right) dy$

$(\text{Momentum Rate})_{\text{ideal few}} = (\text{Momentum Rate})_{\text{actual flow}}$

$$\int_0^{\infty} \delta U(V_{\infty} - u)dy = \int U_{\infty}^2 \theta$$

$$\theta = \int_0^{\infty} \frac{U}{V_{\infty}}\left(1-\frac{U}{V_{\infty}}\right) dy$$

25. Fluids can be classified into five type:

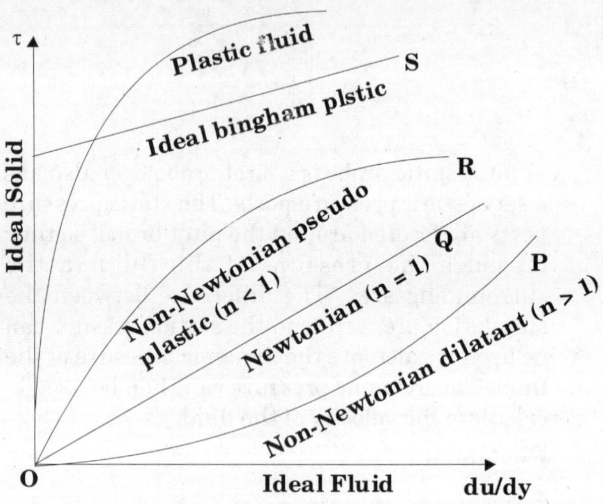

Ideal Plastic	When the shear stress is proportional to the velocity gradient and shear stress is more than the yield value, it is known as the ideal plastic fluid.
Newtonian Fluid	The fluid which obeys Newton's law of viscosity (shear stress is proportional to the rate of shear strain), is known as a Newtonian fluid.
Non-Newtonian Fluid	The fluid which doesn't obey Newton's law of viscosity, is known as a Newtonian fluid.
Ideal Fluid	A fluid is said to be ideal when it is incompressible and the non-viscous. It is an imaginary fluid that doesn't exist in reality.
Real Fluid	All the fluids are real as all the fluids possess viscosity.

26.

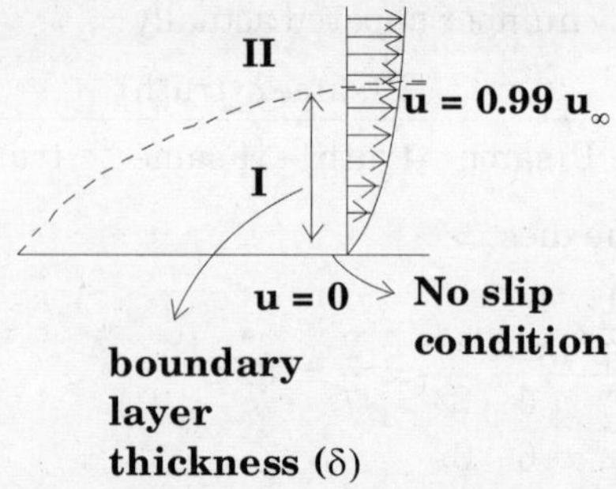

Region - I → velocity gradient will exist

$$\boxed{\frac{du}{dy} \neq 0}$$

→ flow is irrotational

Region - II → $\frac{du}{dy} = 0$

→ flow is rotational

Acc. of Bernoull's equation →

flow should be → steamline

→ vescasity ≈ 0

→ steady

→ rotational

So, D option is right.

27. Uniform flow velocity profile

$\phi = Ur \cos\theta$

Velocity profile for clockwise motion

$$\boxed{\phi = \frac{k}{2\pi}\theta}$$

Both are different direction, combination of velocity profile (ϕ)

$$= \frac{K}{2\pi}\theta - U_{\infty} \, r\cos\theta$$

28. (i) The conservation of mass for an unsteady incompressible flow can be represented as $\Delta.\vec{V} = 0$, where $\vec{V}$ denotes velocity vector. This is the continuity equation, which states that the total mass within a control volume is constant over time. Incompressible flow means that the density of the fluid does not change, so the continuity equation simplifies to $\Delta.\vec{V} = 0$.

(ii) Circulation is **not** defined as the line integral of vorticity about a closed curve. Instead, circulation is the line integral of the velocity vector along a closed curve. Mathematically, it is given by $\oint \vec{V}.dr$, where $\vec{V}$ is the velocity vector and dr is an infinitesimal vector element along the closed curve.

(iii) For some fluids, shear stress can be a nonlinear function of the shear strain rate. This is because the shear stress is caused by the friction between the fluid molecules, and the friction force can be a nonlinear function of the velocity gradient.

(iv) Integrating Bernoulli's equation along a streamline under steady-state conditions does not directly lead to Euler's equation. Bernoulli's equation and Euler's equation are distinct equations in fluid dynamics. Bernoulli's equation is a simplified form of the Euler's equation that applies along a streamline in a flow field, and it includes the effects of pressure, velocity, and gravitational potential energy. However, Bernoulli's equation assumes some simplifications such as neglecting viscous effects and considering steady, incompressible flow.

29. The velocity field given in the problem is:

V = −x*i + y*j

The streamline passing through points (2, 1) and (5, p) must have the same direction as the velocity field at both points. So, the slope of the streamline must be equal to the slope of the velocity field at both points.

The slope of the velocity field at (2, 1) is:

$$\frac{dy}{dx} = \frac{-2}{1} = -2$$

The slope of the streamline passing through points (2, 1) and (5, p) must also be equal to -2. So, we can write the following equation:

$p - 1 = -2(5 - 2)$

Solving for p, we get:

$$p = \frac{2}{5}$$

Therefore, the value of p is $\frac{2}{5}$

30.

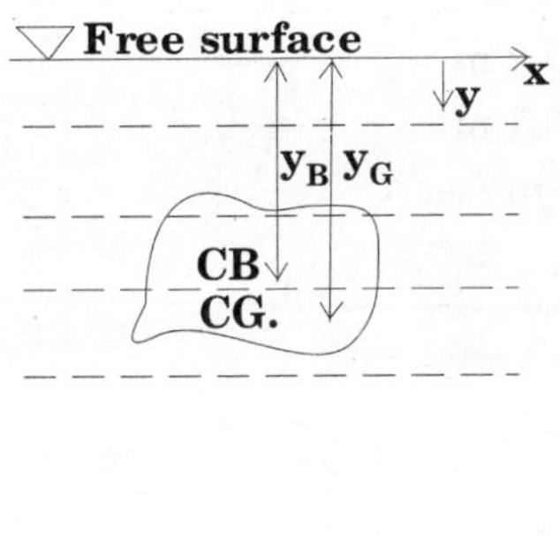

$Y_B = Y_G \rightarrow$ When it is in neutral equilibrium

$Y_B > Y_G \rightarrow$ When unstable equilibrium

$Y_B < Y_G \rightarrow$ When stable equilibrium

A and C

31. $P\underset{\substack{\downarrow \\ 0}}{x} - \underset{\substack{\downarrow \\ 0}}{\Delta P} + \mu\,\Delta\vec{V} = P.\frac{D\vec{v}}{Dt}$

body force and pressure gradien = 0

Linear velocity profile shear stress on each layer is same

The ratio of top surface and the bottom surface is 1.

32. $\vec{v}(x,y) = (Axy)\hat{i} + \left(By^2\right)\vec{v}$

$$\frac{\partial^2 v}{\partial y^2} = 2By$$

$$\underset{\downarrow}{\delta\,\vec{x}} - \Delta P + \mu.\,\Delta^2\vec{v} = \int\frac{D\vec{v}}{Dt} \quad \begin{bmatrix}\text{Navier} \\ \text{Stokes equation}\end{bmatrix}$$

Body faces
are Negligible

$$\Delta P = \frac{\partial p}{\partial x}\vec{i} + \frac{\partial p}{\partial y}\vec{j}$$

x direction →

$$-\frac{\partial p}{\partial x} + \mu\left[\frac{\partial^2 u}{dx^2} + \frac{\partial^2 u}{\partial y^2}\right] = \int\left[\frac{\partial u}{\partial t} + u.\frac{\partial u}{\partial x} + v\frac{\partial u}{\partial y}\right]$$

y direction →

$$-\frac{\partial p}{\partial y} + \mu\left[\frac{\partial^2 v}{\partial x^2} + \frac{\partial^2 v}{\partial y^2}\right] = \int\left[\frac{\partial v}{\partial t} + v.\frac{\partial v}{\partial x} + v.\frac{\partial v}{\partial y}\right]$$

$$\Rightarrow -\frac{\partial p}{\partial x}\int\left[Axy\ Ay + By^2Ax\right] \text{ at } (5, 0)$$

$$\Rightarrow -\frac{\partial p}{\partial x} + (2B)\,\mu = \int(By^2\,(2By))$$

$$\frac{\partial p}{\partial y} = +2\,B\mu$$

33. The velocity potential φ(x, y) is defined as the line integral of the velocity field along a path that starts at (x, y) and ends at any other point in the flow field. The slope of the potential line at (x, y) is then given by:

$$\frac{dy}{dx} = -\frac{\left(\frac{\partial \varphi}{\partial x}\right)}{\frac{\partial \varphi}{\partial y}} = -\frac{u}{v}$$

In this case, the velocity field is given by V(x, y) = ui + vj. So, the slope of the potential line at (x, y) is:

$$\frac{dy}{dx} = -\frac{u}{v}$$

This is the same as the negative of the ratio of the velocity components u and v. The negative sign indicates that the potential line is perpendicular to the streamlines.

34. For laminar then

$$\delta \propto \frac{x}{\sqrt{R_e}} ; \frac{\delta}{x} = \frac{(1-9)}{\sqrt{R_e}}$$

for turbulent flow

$$\delta \propto \frac{x}{R_e^{1/7}} ; \frac{\delta}{x} = \frac{0.16}{\left(R_e^{1/7}\right)}$$

for lamena flow

$$R_e = \frac{\delta VD}{\mu} \qquad \delta \propto \frac{x}{\sqrt{x}} \times \frac{\sqrt{x}}{\sqrt{x}}$$

$$R_e \propto D \qquad \delta \propto \sqrt{x}$$

$D = x$

35. Stream function is a geometric representation of the path of Fluid packet around a stationary or moving object

$$u = \frac{\partial \phi}{\partial y} ; v = -\frac{\partial \Psi}{\partial x}$$

C is the right answer.

36. In x direction →

$F_x = m_1 v_1 - m_2 v_2$

$= V(m_1 - m_2)$

$= V[A_1 V_1 - P A_2 V_2]$

$= V^2 [A_1 - A_2]$

$$F_x = 5^2 \times 1000 \left[\frac{\pi}{4} \times 0.1^2 - \frac{\pi}{4} \times 0.025^2\right]$$

$\boxed{F_x = 184N}$

37. As given that

$P_{water} = 1000 \text{ kg/m}^3$

$\int_{mix} = ?$

Specific gravity of alcohol (S) = 0.7

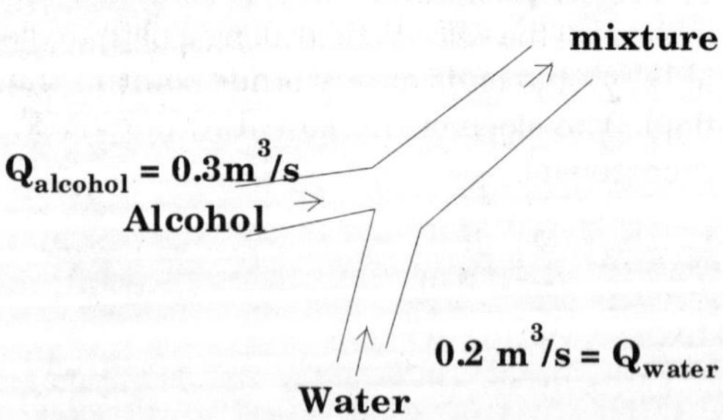

$m_{Water} + m_{Alcohol} = m_{mixture}$

$P_{water} Q_{water} + P_{Alc.} Q_{Alc} = P_{mixture} \times (Q_{water} + Q_{Alc})$

$1000 \times 0.2 + 0.7 \times 1000 \times 0.3 = P_{mixture} (0.2 + 0.3)$

$$P_{mixture} = \frac{200+210}{0.5} \Rightarrow \frac{410}{0.5}$$

$\boxed{P_{mixture} = 820 \text{ kg/m}^3}$

38. $\vec{v} = \hat{i} + 2\hat{j}$

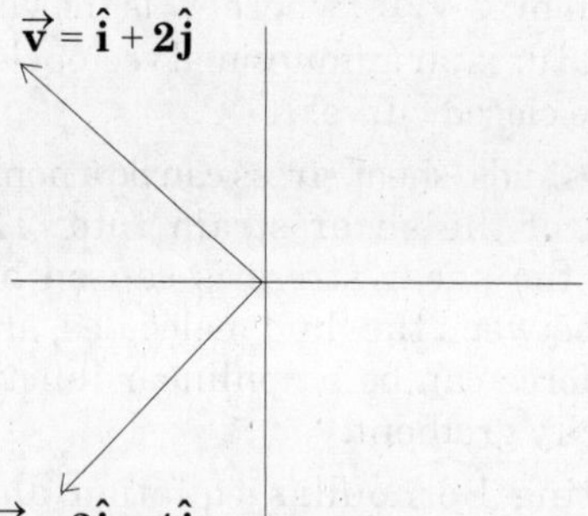

$\vec{a} = 2\hat{i} - 4\hat{j}$

$\vec{V} = \left(-\hat{i} + 2\hat{j}\right) \text{ m/s}$

$\vec{a} = \left(-2\hat{i} - 4\hat{j}\right) \text{m/s}^2$

Project n of $\vec{a}$ on $\vec{v}$

$\Rightarrow \vec{a} . \hat{v}$

$$\Rightarrow \frac{\vec{a} . \vec{v}}{|\vec{v}|}$$

$$\Rightarrow \left|\frac{2+(-8)}{\sqrt{5}}\right| = \left|\frac{6}{\sqrt{5}}\right| = 2.68 \text{ m/s}$$

39. As we know that

$P \propto N^3 \times D^5$

$$\text{So, } \frac{P_1}{P_2} = \left(\frac{D_1}{D_2}\right)^5 \times \left[\frac{N_1}{N_2}\right]^3$$

→ Same flued density equal for both the turbine

→ dynamically both are same.

$D_1^2 N_1 = D_2^2 N_2$

$$\frac{D_1}{D_2} = \left(\frac{N_2}{N_1}\right)^{\frac{1}{2}}$$

$$\frac{P_1}{P_2} = \left(\frac{N_1}{N_2}\right)^3 \times \left[\frac{N_2}{N_1}\right]^{\frac{5}{2}}$$

$$\frac{200}{P_2} = \left(\frac{300}{1500}\right)^3 \times \left[\frac{1500}{300}\right]^{\frac{5}{2}}$$

$P_2 = 447.21$

40. $d_1 = 200 \text{ mm} = 0.2 \text{ m}$

$d_2 = 100 \text{ mm} = 0.1 \text{ m}$

flow rate = $0.05 \text{ m}^3/\text{sec}$

$$\text{flow rate} = \frac{A_1 A_2 \sqrt{2gh}}{\sqrt{A_1^2 - a_2^2}} \quad \dots(1)$$

from (1) we get

$\boxed{h = 1.94\text{ m}}$ Pressure difference

$$h = \left(\frac{P_1}{P_g} + z_1\right) - \left(\frac{P_2}{P_g} + z_2\right)$$

$$h = \frac{(P_1 - P_2)}{\int_g}$$

$(P_1 - P_2) = 1.94 \times 1000 \times 9.81$

$\boxed{(P_1 - P_2) = 19.01\text{ kP}_a} = \Delta P$

41. $U_\infty = 1$ m/s $\qquad S = \dfrac{3.5x}{\sqrt{Re_x}}$

$$\frac{u}{U_\infty} = \frac{y}{\delta}$$

$$\boxed{\tau = \mu\frac{du}{dy}}\ \frac{1}{U_\infty}\frac{du}{dy} = \frac{1}{\delta}$$

$$\boxed{\frac{du}{dg} = U_\infty\frac{1}{\delta}}$$

Newton low of viscosity\

$$\boxed{\tau \Rightarrow \mu\frac{V_\infty}{S}}$$

Drag force $\boxed{F_\alpha = \dfrac{\mu U_\infty \times A \times 2}{\delta}}$

$$Re = \frac{S\,U_\infty x}{\mu} = \frac{1000 \times 1 \times 0.25}{0.001} = 2.5 \times 10^3$$

$$\delta = \frac{35 \times 0.25}{\sqrt{2.5 \times 10^3}} = 1.75 \times 10^{-3}$$

$$F_d = \frac{0.001 \times 1 \times 0.25 \times 2 \times 2}{1.75 \times 10^{-3}}$$

$\boxed{F_d = 0.57\text{ N}}$

So, forces on both side $= 2f_d = 2 \times 0.57 = 1.14$

42. $\dfrac{u(r)}{U} = \left(1 - \dfrac{r}{R}\right)^{\frac{1}{n}} \qquad n = 1$

$$Q - \int_0^R U\left(1 \quad \frac{r}{R}\right)\ (2\pi r\,dr)$$

$$Q = 2\pi U\int_0^R\left(r - \frac{r^2}{R}\right)dr = \frac{1}{3}\pi R^2 U$$

$$\left(\therefore A = \pi R^2\right)$$

$$V = \frac{Q}{A} = \frac{1.\pi UR^2}{3 \times \pi R^2}$$

$$\frac{V}{U} = \frac{1}{3} = 0.333$$

43. R = 0.5 m

S = 1000 kg/m^3

g = 9.81 m/s^2

Horizontal component of force F_H

Vertical component of force F_V

$$F = \sqrt{(F_H)^2 + (F_V)^2}$$

$F_H = S.\ g.\ A.\ y \qquad A = 0.5 \times 1$

$= 1000 \times 9.81 \times 0.5 \times \dfrac{0.5}{2} \quad y_c = \dfrac{0.5}{2}$

$\boxed{F_H} = 1.226$kN

$$F_v = g \times \frac{\pi R^2}{4} \times 1$$

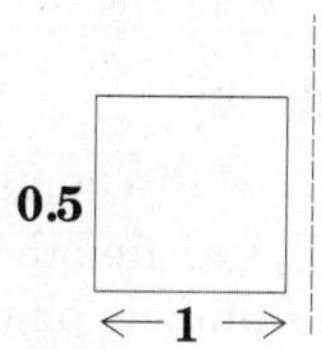

$$= 9.81 \times \frac{\pi(0.5)^2}{4}$$

$\boxed{F_v = 1.926\text{ kN}}$

$\boxed{F \approx 2.2\text{ kN}}$

44. In the age-hardening process of aluminum alloys, the first step involves solution treatment (also known as solution-zing) followed by quenching. The primary purpose of this step is to create a supersaturated solid solution of the alloying elements within the aluminum matrix. During the solution treatment, the alloy is heated to a specific temperature where the alloying elements dissolve into the aluminum lattice. This high temperature allows the alloying elements to fully dissolve into the aluminum matrix, forming a homogeneous, supersaturated solid solution. This solution is achieved because the elevated temperature provides enough energy for the atoms to overcome the barriers to diffusion and mix uniformly.

45.

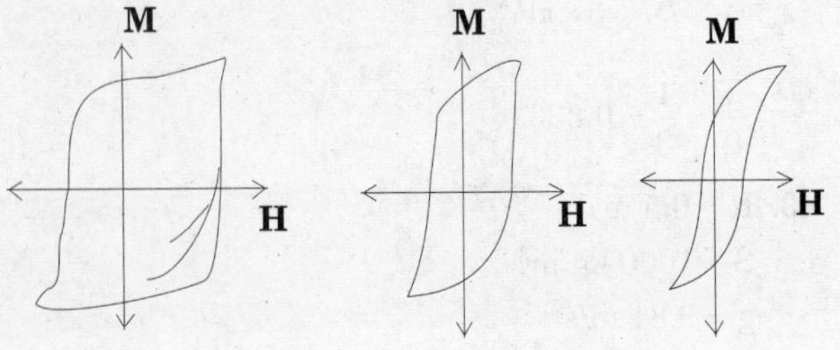

Large M & H desnable for permanent magnets

Large M & Small H is desnable for transformer & motor cores

Hence, option (c) is the correct answer.

46. T_{auc} plot

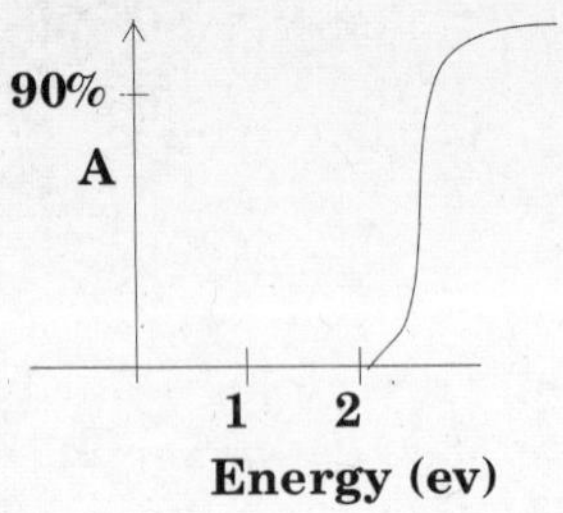

47. See the figure 2 with compares on of figure & only the temp T_3 satisfied the condition α, L + α, L + β, & β in continuous phase which only at T_3 in figure 1.

48. $Na\,Cl \rightarrow Na^+ + Cl^-$

$Mg\,Cl_2 \rightarrow Mg^{2+} + 2\,Cl^-$

doping of $Mg\,Cl_2$ in Na Cl

Mg^{2+} replace by Na^+

$Mg^\circ\,Na + Cl_{cl} + V^1 Na = \phi$

- Means + 1 electron charge

  Net charge + 2 – 1 = + 1
- Means – 1 electron charge

  Net charge + 2 – 1 = + 1
- Means – 1 electron charge

  follow the Kroger – Vink Notalion

49. As we know that

Burgers hector in fcc :

$$\bar{b} = \frac{a}{2}\,[110]$$

Screw dislocation

- lies parallel to the Burger vector
- moves (in its slep plane) in the direction perpendicular to the burger vectors)

Slep plane = Plane containing both the burgers vector & dislocation line.

Possible plane = $(\bar{1}11)$ & $(\bar{1}11)$

50. $\sigma = 500 \in^{0.15}$

$$\sigma = K \in^n$$
$$\frac{d\sigma}{d \in} = n k \in^{n-1}$$
$$n \in^{n-1} = \in^n$$
$$\in = n$$

Maximum loading

$\in = n$

$$\frac{d\sigma}{d \in} = \sigma\, n^n$$

$= 500 \times (0.15)^{0.15}$

$$\frac{d\sigma}{d \in} = 376.17 \text{ MPa}$$

51. As we know that

$$\frac{N_v}{N} = e^{-\frac{Q}{RT}}$$

$T_1 = 600\,k \qquad \frac{Nv_1}{N} = e^{-\frac{68000}{8.31 \times 600}}$ ... (1)

$T_2 = 900\,k \qquad \frac{Nv_2}{N} = e^{-\frac{68000}{8.31 \times 900}}$ ... (2)

$$\frac{Nv_1}{Nv_2} = \frac{e^{-\frac{68000}{8.31 \times 600}}}{e^{-\frac{68000}{8.31 \times 900}}} = \frac{0.000001194}{0.0001125487} = 0.0106087$$

$$\boxed{\frac{N_{v_2}}{N_{v_1}} = 94.26}$$

52. $\frac{\mu_e}{\mu_h} = 10 \qquad n_e = 10^{15}\,m^{-3}$

$n_n = 10^{16}\,m^{-3}$

$\sigma\_ = 1.6\,\pi^{-1} m^{-1}$

$\sigma = n_e e\,\mu_e + n_n\,e\,\mu_h$

$\sigma = 10^{15} \times 1.6 \times 10^{-19} \times 10\,\mu_h + 10^{16} \times 1.6 \times 10^{-19} \times \mu_h$

$\sigma = 2 \times 1.6 \times 10^{-3}\,\mu_h$

$1.6 = 2 \times 1.6 \times 10^{-2}\,\mu_h$

$$\mu_h = \frac{1.6}{2 \times 1.6 \times 10^{-3}}$$

$\mu_h = 500\ m^2\,v^{-1}\,s^{-1}$

53. $\lambda = 2d \sin\theta$

$$\lambda = \frac{2 \times 9}{\sqrt{h^2 + k^2 + l^2}} \sin\theta$$

$Sin^2\theta \propto h^2 + k^2 + l^2$

f c c Model = $h^2 + k^2 + l^2$ Peaks

3, 4, 8, 11, 12, 16

$Sin^2\theta$	sin^2% 13.6	$h^2 + k^2 + 1$	$h^2 + k^2 + l^2$
0.13 6	1	3	1
0.18 5	1.36 →	4	2.67
0.504	3.7058	8	3.66
0.544	4	11	

Student missing the value of $h^2 + k^2 + l^2 = 8$

Answer is (220).

54.

(P)

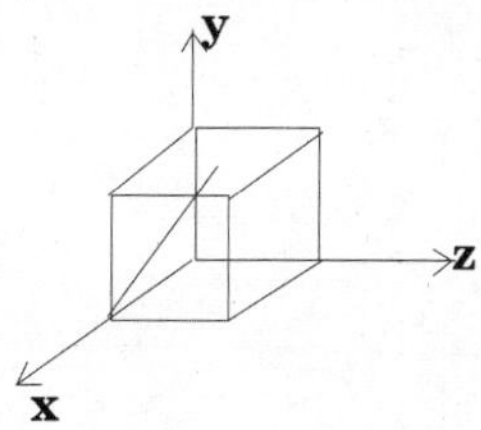

x	y	z
–a	a	$\frac{1}{2}$a
1	1	$\frac{1}{2}$

Miller Induces (1 1 2]

(Q)

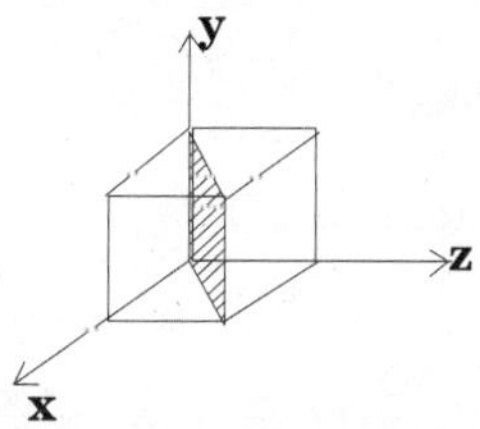

x	y	z
–1	1	∞
(1	1	0)

Mulli Induces

Similar find of the Mulli Induces.

55. • **Brinell hardness test**: This test uses a hardened steel sphere as the indenter. The indenter is forced into the surface of the material under a known load, and the diameter of the resulting indentation is measured. The Brinell hardness number (HB) is calculated from the load and the indentation diameter.

- **Rockwell hardness test**: This test uses a diamond cone or a tungsten carbide ball as the indenter. The indenter is forced into the surface of the material under a series of increasing loads. The depth of the indentation at each load is measured, and the Rockwell hardness number (HR) is calculated from the load and depth of indentation.
- **Vickers hardness test**: This test uses a diamond pyramid as the indenter. The indenter is forced into the surface of the material under a known load, and the length of the diagonal of the resulting indentation is measured. The Vickers hardness number (HV) is calculated from the load and the diagonal length of the indentation.

Hence, option (d) is the correct answer.

56.

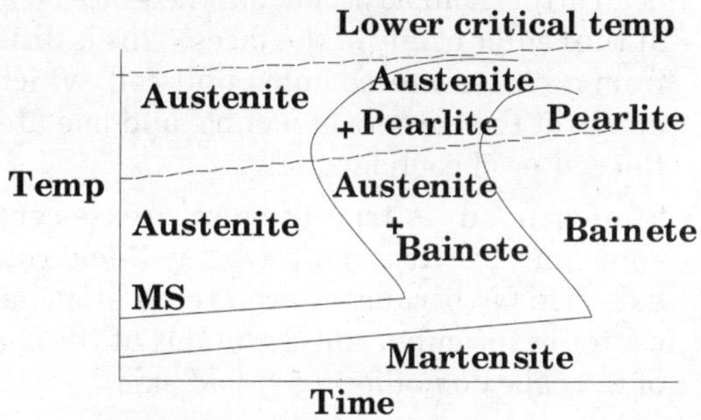

P = Pearlite Only

Q = Pearlete + Marlensuty

R = Bainite Only

57. (a) Increasing the aperture of the objective lens improves the resolution. The aperture of the objective lens is a measure of how much light can pass through the lens. A larger aperture allows more light to pass through, which improves the image brightness and contrast. However, it does not improve the resolution.

(b) Reducing the wavelength of illuminating light does indeed improve the resolution. This is based on the Rayleigh criterion, which states that the resolution of an optical system is inversely proportional to the wavelength of light used. Smaller wavelengths allow for better differentiation of fine details.

(c) Increasing the refractive index of the medium between the sample and the objective lens, often achieved using immersion oil, improves the resolution. This is because it reduces the diffraction of light as it passes from the sample into the objective lens.

(d) Reducing the wavelength of illuminating light decreases the depth of field. This is because the depth of field is inversely proportional to the wavelength of light.

Hence, options (b, c & d) are correct answer.

58 • Statement (a) is true because a base-centered cubic unit cell has atoms at the corners of a cube, and one atom at the center of each face. This is the same as a simple tetragonal unit cell, which has atoms at the corners of a square prism, and one atom at the center of the prism.

- Statement (b) is false because a body-centered tetragonal unit cell has atoms at the corners of a square prism, and one atom at the center of the prism, but the atom at the center of the prism is not shared by any other unit cells. This is different from a base-centered cubic unit cell, where the atom at the center of each face is shared by two unit cells.
- Statement (c) is false because a simple orthorhombic unit cell has atoms at the corners of an orthorhombic prism, but there are no atoms at the center of any of the faces. This is different from a base-centered cubic unit cell, which has atoms at the corners of a cube, and one atom at the center of each face.
- Statement (d) is true because a base-centered cubic unit cell does not have any 3-fold rotation axis. This is because there are no atoms at the center of the cube, and the atoms at the corners of the cube do not form a 3-fold axis.

Hence, the only two statements that are true are (a) and (d).

59. As given figure:

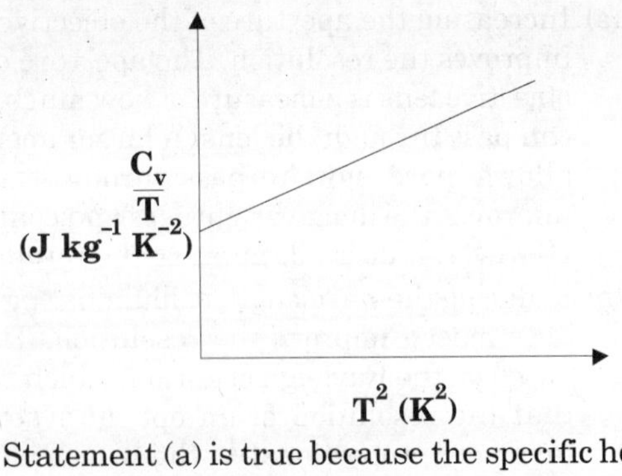

- Statement (a) is true because the specific heat of a metallic material increases with temperature, as shown in the graph. This is because the electrons in a metal are free to move, and they can absorb more energy as the temperature increases.
- Statement (b) is false because the specific heat of an insulating material decreases with temperature, as shown in the graph. This is because the electrons in an insulating material are not free to move, and they can absorb less energy as the temperature increases.
- Statement (c) is true because the specific heat of a three-dimensional material increases with temperature, as shown in the graph. This is because there are more degrees of freedom for the atoms to move in a three-dimensional material, and they can absorb more energy as the temperature increases.
- Statement (d) is false because the specific heat of a one-dimensional material is constant with temperature, as shown in the graph. This is because there are no degrees of freedom for the atoms to move in a one-dimensional material, and they can absorb the same amount of energy regardless of the temperature.

In conclusion, the only two statements that are true are (a) and (c).

60. As given that

$t_1 = 1000$ sec $\quad t_2 = ?$

$C_x = 10^{20}$ atm m^{-3} $\quad x_2 = 2$ Mm

x_1 1 μm

$$x \propto \sqrt{Dt}$$

$$\frac{x_2}{x_1} = \sqrt{\frac{t_2}{t_1}}$$

$$\frac{x_2^2}{x_1^2} = \frac{t_2}{t_1}$$

$$t_2 = 4t_1$$

$$\boxed{t_2 = 4000 \text{ sec}}$$

61. $\text{Young Modules} = \frac{\text{Stress}}{\text{Strain}}$

Stress = Young Modules × Strain

100 × 0.1

= 1 G pa.

62. As given that

$n = 10^8$ electrons C $m^{-1} = 10^{10}$ m^{-1}

$$\frac{h^2}{2m} = 0.24\ (eV)^2\ S^2\ kg^{-1}$$

Femi energy of system for 1D;

$$\boxed{E_f = \frac{h^2\pi^2 n^2}{2m \times 4}}$$

$$E_f = \frac{0.24 \times (1.6 \times 1.19)^2 \times (3.14)^2 \times \left(10^{10}\right)^2 \text{ ev}}{\left(1.6 \times 10^{-19}\right)}$$

$E_f = 9.46$ eV

63. As given that

n = 2 moles

$P_1 = 10$ atm $\quad P_2 = 2$atm

$T_1 = 300$ k $\quad T_2 = 300$ k

dQ = du + dw

dQ = du + pdv

$0 \rightarrow$ Isothermal

$\boxed{dQ = pdv}$

$$\int dw = \int_{v_1}^{v_2} pdv \qquad \boxed{p_v = nRT}$$

$$W = \int_{v_1}^{v_2} \frac{nRT}{V} dv$$

$$W = nRT \int_{v_1}^{v_2} \frac{1}{V} dv \qquad \boxed{P_1 V_1 = P_2 V_2}$$

$$\boxed{W = nRT \ln \frac{V_2}{V_1}} \qquad \boxed{\frac{P_1}{Q_2} = \frac{V_2}{V_1}}$$

$$W = n\ RT \ln \frac{P_1}{P_2}$$

$$= 2 \times 8.31 \times 300 \ln \frac{10}{2}$$

$\boxed{W = 8.03}$

64. As given that

$Z_n + 2H^+ = H_2 + Z_n^{+2}$

$Z_n \rightarrow Z_n^{+2} + 2e^-$

$2H^+ + 2e^- \rightarrow H_2$

$$\Rightarrow \boxed{\frac{\ln\left[Z^{+2}\right]}{\left[H^+\right]^2} = ?}$$

$$E_{cell} = E°_{cell} - \frac{RT}{nF} \ln \frac{\left[Zn^{2+}\right] p_{H_2}}{\left[H^+\right]^2} \qquad \dots (1)$$

$E_{cell} = 0.690$ V $\qquad E°_{Zn^{2+}} = -0.762$ V

$p_{H_2} = 1$ atm $\qquad E°_{H^+} = $ OV

$T = 298$ K $= 25°$ $\qquad n = 2$

$$\frac{RT}{F} = 0.0256V$$

$$\boxed{E^\circ_{cell} = E^\circ_{cathode} - E^\circ_{anode}}$$

$$E^\circ_{cell} = E^\circ_{H^+} - E^\circ_{Z_n^{2+}}$$

$\boxed{E°_c = 0.762V}$

After putting the values in eq (1)

we get

$$\ln \frac{\left[Z^{+2}\right]}{\left[H^+\right]^2} = 5.63$$

65. $T = 300$ k $\qquad h\nu = 25 \times 10^{-3}$ eV

$k = 8.62 \times 10^{-5}$ eV k^{-1}

$$\frac{I_s}{I_{Anti}} = e^{-\frac{hV}{kT}} = e^{-\frac{25 \times 10^{-3}}{8.62 \times 10^{-5} \times 300}}$$

$$\boxed{\frac{I_S}{I_{Anti}} = 2.61}$$

66.

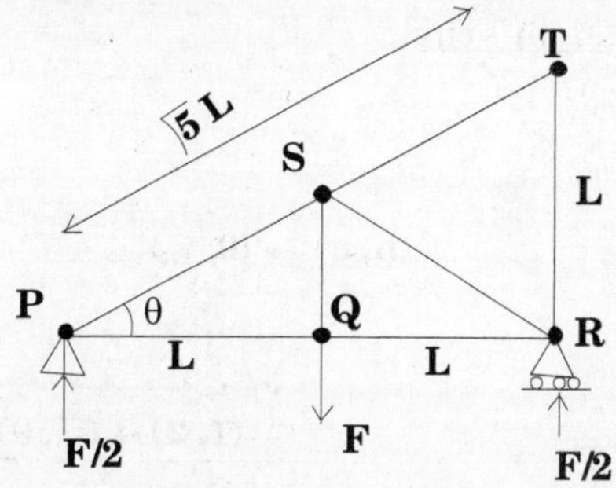

$F_{PS} = ?$

<u>Equilibrium equation</u>

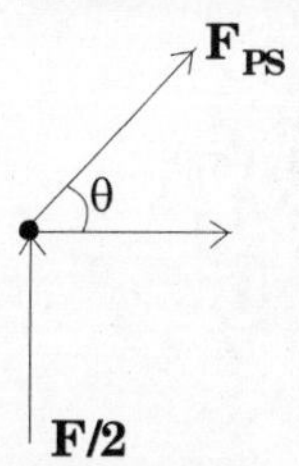

$\in F_y = 0$

$$F_{ps} \sin\theta = \frac{-F}{2}$$

$$F_{ps} \times \frac{L}{\sqrt{5}L} = \frac{-F}{2}$$

$$F_{ps} \times \frac{1}{\sqrt{5}} = \frac{-F}{2}$$

$$\boxed{F_{ps} = \frac{-F}{2}\sqrt{5}} \text{ Compressive}$$

67. As we know that

Conservation of energy

$$mg\ L = \frac{1}{2} I_0 \omega^2$$

$$mg\ L = \frac{1}{2} \times m\ L^2\ \omega^2$$

$\omega^2 = 2g/L$

So, angular velocity:

$$\boxed{\omega = \sqrt{2g / L}}$$

68. According to the Question;

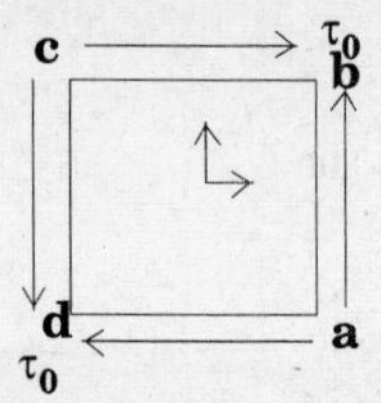

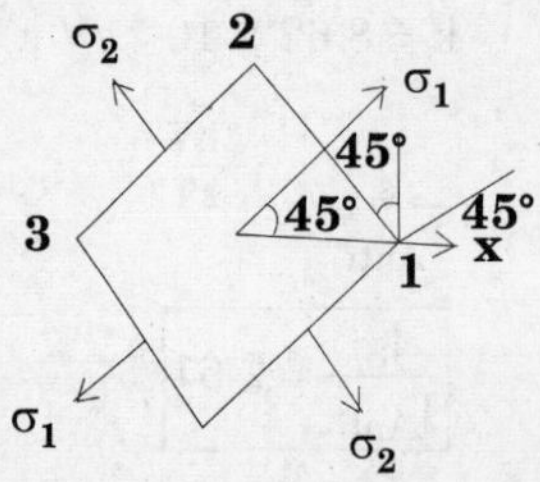

By Using Mohr's circle

ab → $(0, z_0)$

bc → $(0, z_0)$

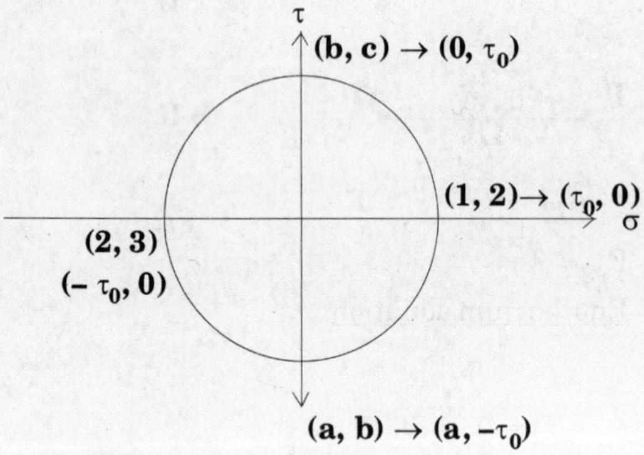

(1 2) → $(\iota_0, 0) \rightarrow (\sigma_1, 0)$

$$\boxed{\sigma_1 = \iota_0}$$

(2 3) → $(-\iota_0, 0) \rightarrow (\sigma_2, 0)$

$$\boxed{\sigma_2 = -\iota_0}$$

69. $E_{xx} = 2.5 \times 10^{-4}$

$E_{yy} = 2 \times 10^{-4}$

$E_{zz} = -1.5 \times 10^{-4}$

$E_{xy} = 2.5 \times 10^{-4}$

$E_{yz} = -0.5 \times 10^{-4}$

$E_{zx} = -1.0 \times 10^{-4}$ (No need to consider)

$E_{vs} = ?$

$E_{vs} = E_{xx} + E_{yy} + E_{zz}$

$$\boxed{E_{vs} = 3 \times 10^{-4}}$$

70. Circumferential stress

$$\boxed{\sigma_c = \sigma_1 = \frac{pd}{2t}}$$

Longitudinal stress

$$\boxed{\sigma_L = \sigma_2 = \frac{pd}{2t}}$$

71.

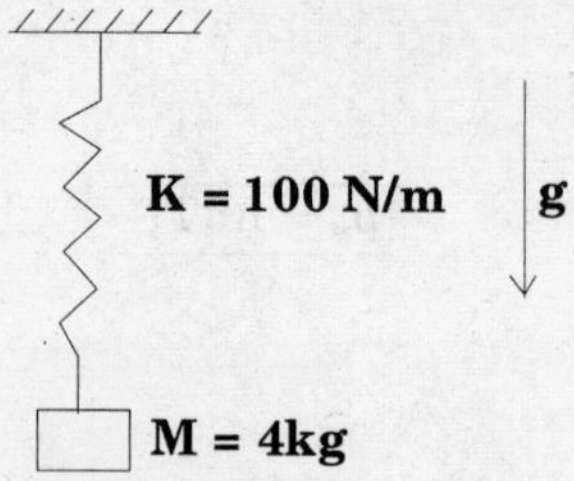

$$\boxed{g = 9.81\ m/s^2}$$

Static deflection →

$$\delta = \frac{Mg}{k} = \frac{4 \times 9.81}{100}$$

$$\boxed{\delta = 0.392}$$

Time period →

$$W_n = \sqrt{\frac{K}{M}} = \sqrt{\frac{100}{4}} = 5\ rad/sec$$

$$T_n = \frac{2\pi}{W_n} = \frac{2\pi}{5} = 1.256 \approx 1.26\ s$$

$$\boxed{T_n = 1.26\ s}\ \boxed{\delta = 0.392}$$

72.

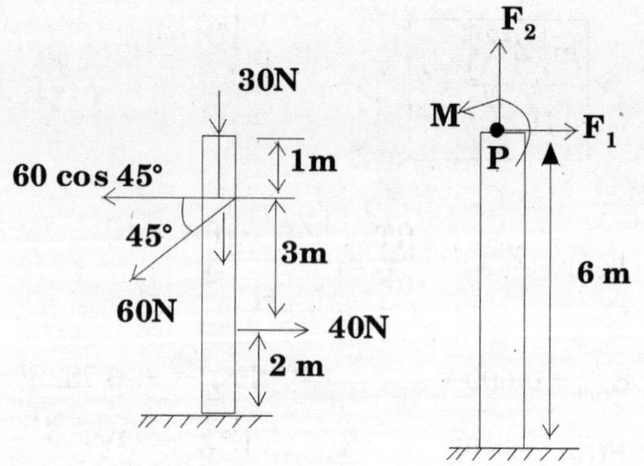

Moment $M_p = 40 \times 4 - 60 \cos 40° \times 1 = 200.01$

$$\boxed{M_p = 117.57\ N\,m}$$

73.

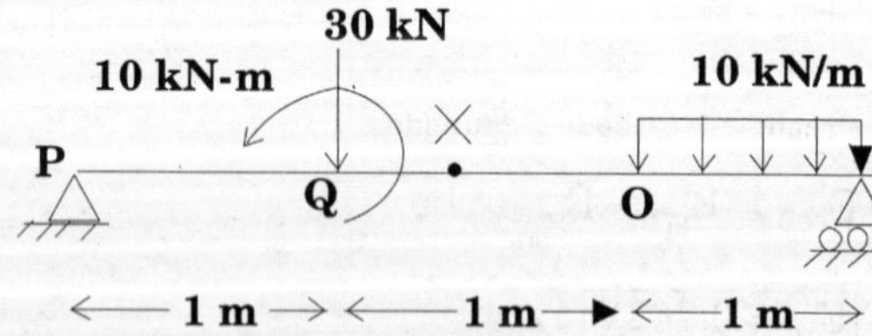

Taking moments at point S

$E\ M_s = 0$

$(-0.5) \times 10 - 30 \times 2 - 10 + R_P \times 3 = 0$

$$R_P = \frac{(5 + 60 + 10)}{3} = \frac{75}{3} = 25$$

$\boxed{R_p = 25 \text{ kN}}$

Shear force at X

$\iota_x = 25 - 30$

$\boxed{\iota_x = -5\text{k N}}$

74. $\sigma_{xx} = 100 \text{ M Pa}$

$\sigma_{yy} = 50 \text{ M Pa}$

$\sigma_{zz} = 0$ [Plane stress condition]

$\text{Max. } \tau = \left|\frac{\sigma_{xx}}{2}\right| = 50 \text{ M Pa}$

75. Applying Conservation of Momentum at point 'S'

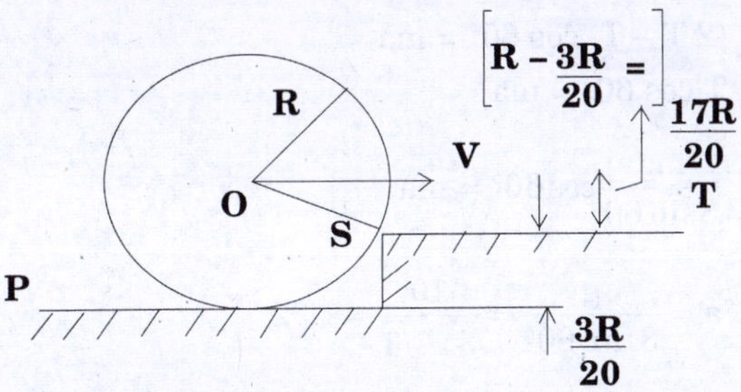

$$m\,v \times \frac{17R}{20} + I_o \cdot w = I_s \cdot w'$$

<u>After impact</u>

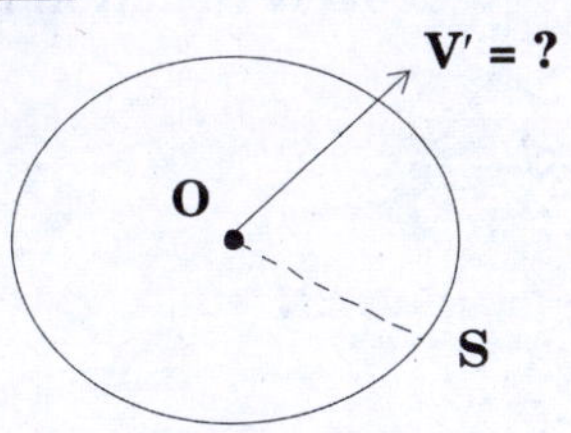

$$m\,v \times \frac{17R}{20} + \frac{mR^2}{2}\left(\frac{V}{R}\right)$$

$$= \frac{3}{2} m R^2 \frac{V'}{R}$$

$$\Rightarrow \frac{17}{20} v + \frac{V}{2} = \frac{3}{2} V'$$

$\boxed{V' = 0.9\,V}$

76.

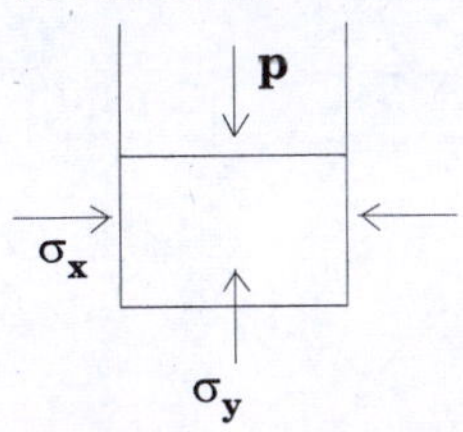

$E_{zz} = ?$

$\sigma_{zz} = -p$

$\sigma_{xx} = -px$

$\sigma_{yy} = -py$ [x and y direction are restricted]

E_{xx} & E_{yy} both are zero

$-\sigma_x = \sigma_y$

$\sigma_{xx} = \sigma_{yy} = -\sigma_x$

$E_{xx} = 0$

$$= \frac{1}{E}\left[\sigma_{xx} - \mu\,\sigma_{yy} - \mu\,\sigma_{zz}\right] = 0$$

$$= \frac{1}{E}\left[-\sigma_x - \mu\left(-\sigma_y\right) - \mu(-p)\right] = 0$$

$$\Rightarrow \boxed{\sigma x = \frac{\mu p}{1-\mu}}$$

$$\boxed{\sigma_{xx} = \sigma_{yy} = \frac{-\mu p}{1-M}}$$

$$E_{zz} = \frac{1}{E}\left[\sigma_{zz} - \mu\,\sigma_{xx} - \mu\,\sigma_{yy}\right]$$

$$= \frac{-p}{E}\left[1 - \frac{2M^2}{1-M}\right]$$

$$= \frac{-p}{E}\left[\frac{(1+\mu)(1-2\mu)}{(1-\mu)}\right]$$

77. $\sigma_{xx} = 420 \text{ M Pa} = \sigma_1$

$\sigma_{yy} = 100 \text{ M Pa} = \sigma_2$

$\sigma_{zz} = \sigma_{xy} = \sigma_{yz} = \sigma_{zx} = 0 = \sigma_3$

Yield stress (yt) = 400 M Pa

$$\iota_{max} = \left|\frac{\sigma_3 - \sigma_1}{2}\right| = \frac{420 - 0}{2} = 210 \text{ M Pa}$$

$\boxed{\iota_{max} = 210 \text{M Pa}}$

Max. shear stress theory

$$\frac{Yt}{2} = \frac{400}{2} = 200 \text{ M Pa}$$

$\boxed{\iota_{max} > \frac{4t}{2}}$ Fail / Unsafe

Van - Mises Theory

(0. 577) Yt = 2308 MPa

$\boxed{\iota_{max} < 230.8}$ safe

78. → Compressive stress

$$\sigma_{Cd} = \frac{60000}{\underbrace{100 \times 80}_{\text{Area}}} = 7.5 \text{ M Pa}$$

→ Max. bending stress

$$\sigma_{b,m} = \frac{6 \times 6000 \times 10}{100 \times (80)^2} = 5.625 \text{ MPa}$$

Nature of bending stress

at, PQ → Compressive

at, Rs → Tensile

$\sigma_{PQ} = 7.5 + 5.625 = 13.125$ MPa (Compressive)

$\sigma_{RS} = 7.5 - 5.625 = 1.875$ (Overall Nature Compressive)

79.

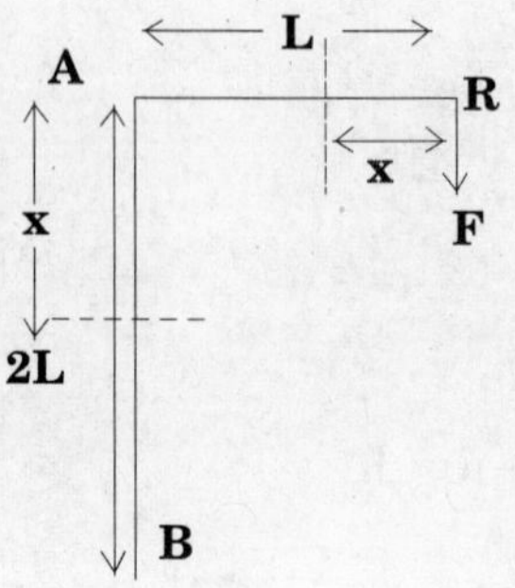

Vertical deflection = ?

at end R

$$= \int \frac{M - \frac{\Delta M}{\Delta F}}{EI} . dx$$

$\boxed{M_{RA} = F_x}$ $\boxed{M_{AB} = F.L}$

$$Y_R = \int_0^L \frac{F_x . x\, dx}{EI} + \int_0^{2L} \frac{FL.Ld}{EI}$$

$$Y_R = \frac{FL^3}{3EI} + \frac{2FL^3}{EI}$$

$$\boxed{Y_R = \frac{7FL^3}{3EI}}$$

80.

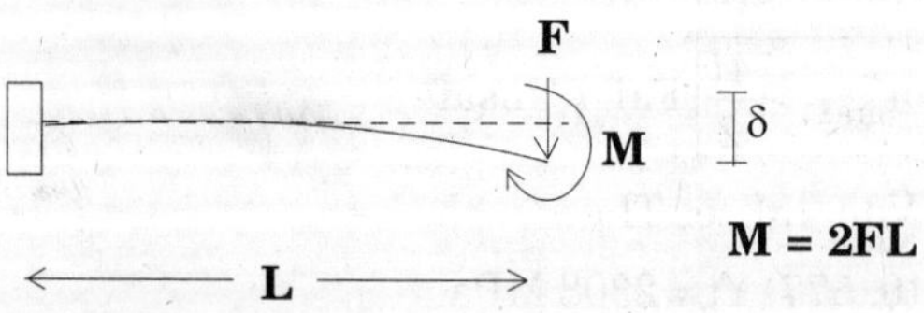

$$\delta = \frac{FL^3}{3EI} + \frac{2FL^3}{2EI}$$

$$\boxed{\delta = \frac{4}{3} \frac{FL^3}{EI}}$$

81. g = 10 m/s^2

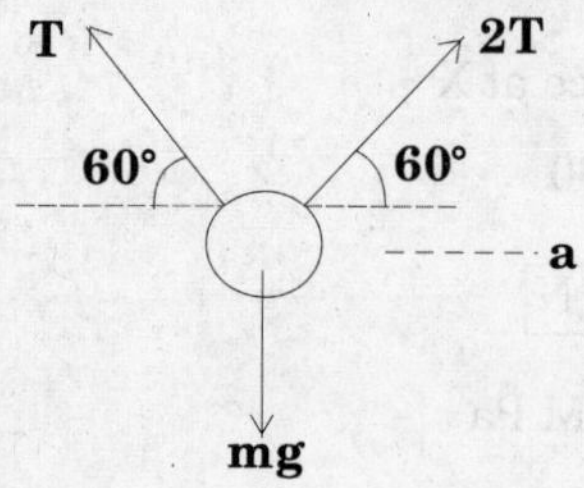

$\varepsilon F_y = 0$

$3T \sin 60° = mg$

$$T = \frac{mg}{3 \sin 60°}$$

$\varepsilon F_x = ma$

$(2T - T) \cos 60° = ma$

$T \cos 60° = ma$

$$\frac{mg}{3 \sin 60} \cos 60° = ma$$

$$a = \frac{g}{3 \tan 60°} = \frac{10}{3 \times \sqrt{3}} = 1.92$$

82. mg cos 30°

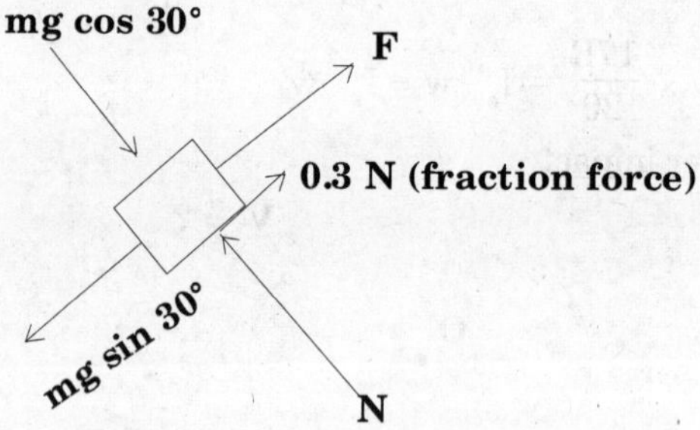

$\mu = 0.3$

g = 10 m/s^2

F = mg sin 30 – 0.3 mg cos 30

N = mg cos 30

F = 100 sin 30 – 0.3 × 100 cos 30

$\boxed{F \approx 24N}$

83. Coefficient of restitution (e) = 0.9

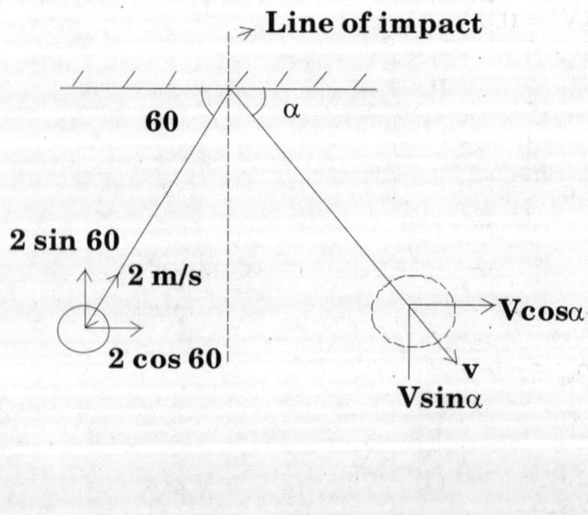

$$e = \frac{V_2 - V_1}{u_1 - u_2} \Rightarrow 0.9 = \frac{0-(-v\sin\propto)}{2\sin 60 - 0}$$

$$\boxed{V\sin\propto = 1.8\sin 60} \quad ...(1)$$

$$\Rightarrow \boxed{V\sin\propto = 2\cos 60} \quad ...(2)$$

eq (1) divide by (2)

$$\frac{V\sin\propto}{V\cos\propto} = \frac{1.8\sin 60}{2\cos 60}$$

$$\tan\propto = \frac{1.8}{2}\tan 60$$

$$\boxed{\propto = 57.32°}$$

84. $\mu = 0.3$

$E_{steel} = 200$ GPa

$\varepsilon_{zz} = -3 \times 10^{-4}$

$\sigma_{xx} = 120$ M Pa

$$\varepsilon_{zz} = \frac{1}{E}\left[\sigma_{zz} - \mu\,\sigma_{xx} - \mu\sigma_{yy}\right]$$

$\sigma_{zz} = 0$

$$\Rightarrow -3 \times 10^{-4} = \frac{1}{200000}\,[0.0.3\,(120)\,.\,0.3\,(\sigma_{yy})]$$

$$\Rightarrow -60 = -36 - 0.3\,\sigma_{yy}$$

$$\Rightarrow \sigma_{yy} = \frac{-36+60}{0.3}$$

$$\sigma_{yy} = \frac{24}{0.3} = 80\ MP_a$$

85. Coefficient of thermal expansion of steel ($\propto_{steel}$) = 12×10^{-6}

Coefficient of thermal expansion of copper ($\propto_{cu}$) = 18×10^{-6}

$\Delta T = 100°$ C (give)

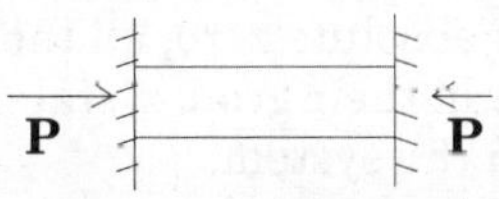

$E_{steel} = 200$ GPa   axial stress of steel rod σ_{rod} = ?

$E_{cu} = 100$ GPa

$$\Delta l_T = \Delta l_{Thermal} + \Delta l_{Mechanical} = 0$$

$$\Rightarrow \Delta l_{Thermal} = (l \propto \Delta T)_{Steel} + (l \propto \Delta T)_{Cu}$$

$$\Rightarrow \Delta l_{Mechanical} = -P\left[\left(\frac{1}{AE}\right)_{steel} + \left(\frac{1}{AE}\right)_{Cu}\right]$$

After putting the values, l will cancel out

$(12 \times 10^{-6} \times 100 + 18 \times 10^{-6} \times 100)$

$$= \frac{P}{A}\left[\frac{1}{200 \times 10^3} + \frac{1}{100 \times 10^3}\right]$$

$$\frac{P}{A} = \boxed{200\ MPa = \sigma_{rod}}$$

86. Max. longitudinal strain (ε_{max}) = $p \times 10^{-3}$

$d = 50$ mm ; $\delta = 10$ mm ; $L = 1000$ mm

$$\text{Radius of curvature } R = \frac{L^2}{8\delta} = \frac{(1000)^2}{8 \times 10}$$

$R = 12500$ mm

$$\varepsilon_{max} = \frac{y_{max}}{R} = \frac{\frac{50}{2}}{12500} = 2 \times 10^{-3}$$

$$\boxed{P = 2}$$

87.

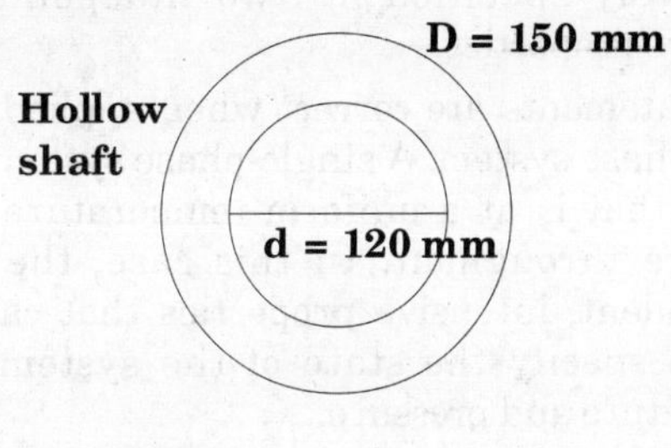

Area is same

$A_8 = A_H = A$

Torque of Hollow shaft $T_H = T$ Torque of solid shaft $(T_s) = T$

$$\boxed{T_s = T_H = T}$$

$$\frac{\tau_h}{Z_s} = ? \qquad K = \frac{d}{D} = \frac{120}{150} = 0.8$$

$$\boxed{K = 0.8}$$

$$\boxed{\tau_h = \frac{16T}{\pi(150)^3\left(1-0.8^4\right)}} \quad ...(1)$$

<u>Diameter of solid shaft</u> → $\boxed{A_s = A_H}$

$A_s = A_H$

$$\frac{\pi}{4}(D_s)^2 = \frac{\pi}{4}\left(D^2 - d^2\right) = \frac{\pi}{4}\left(D^2 - d^2\right)$$

$$\boxed{D_S = 90\text{ mm}} \qquad D_S = \frac{\pi}{4}\left(150^2 - 120^2\right)$$

$$\boxed{\tau_s = \frac{16T}{\pi(90)^3}} \quad ...(2)$$

from eq (1) and (2)

$$\frac{\iota_h}{\iota_s} = \frac{(90)^3}{(150)^3\left(1-0.8^4\right)}$$

$$\frac{\iota_h}{\iota_s} = 0.366$$

88. Statement A is the state postulate, which states that the state of a system is completely specified by a certain number of properties. The number of properties required depends on the type of system. For a simple compressible system, two independent, intensive properties are sufficient to specify the state.

Statement B is a specific case of the state postulate for a simple compressible system. It states that the state of a simple compressible system is completely specified by two independent, intensive properties.

Both statements are correct when applied to a single-phase system. A single-phase system is a system that is at a uniform temperature and pressure throughout. In this case, the two independent, intensive properties that can be used to specify the state of the system are temperature and pressure.

89. Total mass is an extensive property of a system. An extensive property is a property that depends on the amount of matter in a system. The total mass of a system is the sum of the masses of all the individual components of the system. So, if you double the amount of matter in a system, you will double the total mass of the system.

Other examples of extensive properties include volume, energy, and entropy. Intensive properties, on the other hand, do not depend on the amount of matter in a system. Examples of intensive properties include temperature, pressure, and density.

Here is a table that summarizes the difference between extensive and intensive properties:

Property	Definition	Examples
Extensive	Depends on the amount of matter in a system	Mass, volume, energy, entropy
Intensive	Does not depend on the amount of matter in a system	Temperature, pressure, density, color

90. The partial pressure of a gas in a mexline can be found:

$P_A = X_A\,P_{total}$

Mole fraction

$$Y_A = \frac{\eta_A}{\eta_A + \eta_B}$$

$$Y_B = \frac{\eta_B}{\eta_A + \eta_B}$$

$$P_A = \frac{\eta_A}{\eta_A + \eta_B} \times P$$

$$P_B = \frac{\eta_B}{\eta_B + \eta_B} \times P$$

91. If an ideal air-standard Otto cycle and an ideal air-standard Diesel cycle operate on the same compression ratio, then the relation between the thermal efficiencies (ηth) of the cycles is

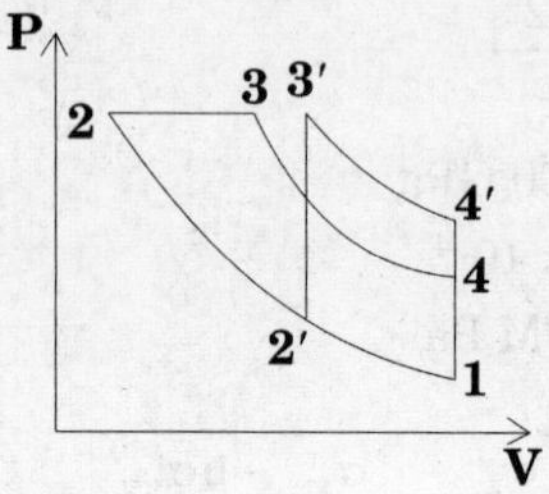

1–2–3–4 diesel cycle

1 – 2' – 3' – 4'

$$\frac{V_1}{V_2} < \frac{V_1}{V_2}$$

The efficiency of the air-standard cycle is directly proportional to compression ratio (r)

$r_{otto} < r_{diesel}$

∴ heat rejection is less in the diesel cycle than the otto cycle.

$\therefore \eta_{Diesel} > \eta_{otto}$

Hence option (c) is the correct answer.

92. Statement (i) is correct. The third law of thermodynamics states that the entropy of a pure crystalline substance approaches zero as the temperature approaches absolute zero. This is because at absolute zero, all the particles in the crystal are in their ground state, and there is no disorder in the system.

Statement (ii) is incorrect. The entropy of a non-crystalline substance is not zero at absolute zero. In fact, it is typically higher than the entropy of a crystalline substance at absolute zero. This is because non-crystalline substances have more disorder than crystalline substances.

Statement (iii) is correct. At absolute zero, the crystal structure has maximum degree of order. This is because all the particles in the crystal are in their ground state, and there is no disorder in the system.

Statement (iv) is incorrect. The thermal energy of a substance at absolute zero is zero. This is because the temperature of a substance is a measure of the average kinetic energy of the particles in the substance. At absolute zero, the particles in the substance have no kinetic energy.

Therefore, the correct option is (C). Both statements (i) and (iii) are correct.

93. As we know that,

$$\text{Bulk modulus} = \frac{\text{Stress}}{\text{Volumetric strain}}$$

$$K = \frac{-\Delta P}{\frac{\Delta V}{V}} = -V\frac{dP}{dV}$$

So, the adiabatic bulk modulus of a substance is defined as:

$$-v\left(\frac{\partial P}{\partial v}\right)_s$$

94. Specific volume of salivated liquid-vapor mexline of water at 200° C

$V = V_f + x(V_{fg})$

$= 0.001156 + 0.75 \times 0.12620$

$V = 0.095806\ m^3/kg$

Volume of tank = 2 m^3

$$\text{Mass in tank} = \frac{2}{0.095806} = 20.88\ kg$$

95. Initial | Final

$M_I = 2$ kg | $M_F = 1$ kg (given)

$P_I = 500$ k Pa | $P_F = 300$ k Pa

$T_I = 350$ k | $T_F = ?$

Initial Vol. = Final Vol.

$$\frac{M_I T_I}{P_I} = \frac{M_F T_F}{P_F}$$

$$\frac{2\times 350}{500} = \frac{1\times T_F}{300}$$

$$\boxed{T_F = 420\ k}$$

96. $T_1 = 400$ k | $P_1 = 200$ kPa

$T_2 = 600$ k | P constant

$$\boxed{C_v = a + bT + cT^2}$$

$$\Rightarrow \int_{T_1}^{T_2} d\bar{u} = \int_{T_1}^{T_2} \bar{C}_v d_T$$

a = 19.686 kJ/k mol –K

b = 0.002 kJ/k mol – K^2

c = 0.5 × 10^{-5} k J/kmol K^3

$$\Rightarrow \Delta\bar{u} = a(T_2 - T_1) + \frac{b}{2}(T_2^2 - T_1^2) + \frac{C}{3}(T_2^3 - T_1^3)$$

After putting the values we get

$$\Delta\bar{u} = 19.686 \times (600 - 400) + \frac{0.002}{2}((600)^2 - (400)^2)$$

$$+\left(\frac{0.5\times 10^{-5}}{3}\right)((600)^3 - (400)^2)$$

$\Rightarrow \Delta u = 4390$ kJ/ K mol

97. $V = 2.0\ m^3$ | $P_1 = 100$ k Pa

$V_g = 1.9\ m^3$ | $P_2 = 2$ M Pa

$V_f = 0.1\ m^3$

At 100 k Pa

$V_f = 0.00/043\ m^3/kg$ | $V_g = 1.694\ m^3/kg$

$u_f = 417.33$ kJ/kg | $u_g = 2506.06$ kJ/kg

At 2 mPa

$V_f = 0.001177\ m^3/kg$ | $V_g = 0.09963\ m^3/kg$

$u_f = 906.42$ kJ/kg | $u_g = 2600.26$ kJ/kg

First low of thermo

$Q = W + \Delta U$ | $m = m_{f1} + m_{g1}$

$Q = \Delta U$ | $u_2 = (u_f + x_2 u_{f2})_{at\ 2MPa}$

$\boxed{Q = m(u_2 - u_1)}$ | $u_1 = [u_f + n_1 u_{f2}]_{at\ 100\ kPa}$

$$x_1 = \frac{m_{g1}f}{m_{f1} + m_{g1}} \qquad m_{f1} = \frac{V_{f1}}{V_{f1}} = 95.871\ kg$$

$$m_{g1} = \frac{V_{g1}}{V_{g1}} = 1.1216\ kg$$

$$x_1 = \frac{1.1216}{96.99}$$

$$\boxed{x_1 = 0.01156}$$

$u_1 = 441.4757$ kJ/kg

$$\boxed{u_1 - 441.4757\ kJ/kg}$$

$$v_2 = \frac{\text{Total volume}}{\text{Mass}} = \frac{2}{96.99}$$

$v_2 = [v_f + x_2 v_g]$

$$\boxed{x_2 = 0.1974}$$

$u_2 = [u_f + x_2 u_{f2}]$ at 2MPa

$$\boxed{u_2 = 1241.32\ kJ/kg}$$

$Q = 96.99\ [1241.32 - 441.47]$

$\boxed{Q = 77477\ kJ} = 77470$ kJ

98. Compression ratio(r) = 20

Cut-off ratio ($\propto$) = 1.5

$P_1 = 100$ kPa ; $T_1 = 300$ K

$C_P = 1.005$ kJ/kg – K

$$r = \frac{C_P}{C_V} = 1.4$$

Net work = n × (Head supply) Q supply

<u>Heat Supply</u> →

$$Q_{supply} = C_P(T_3 - T_2)$$

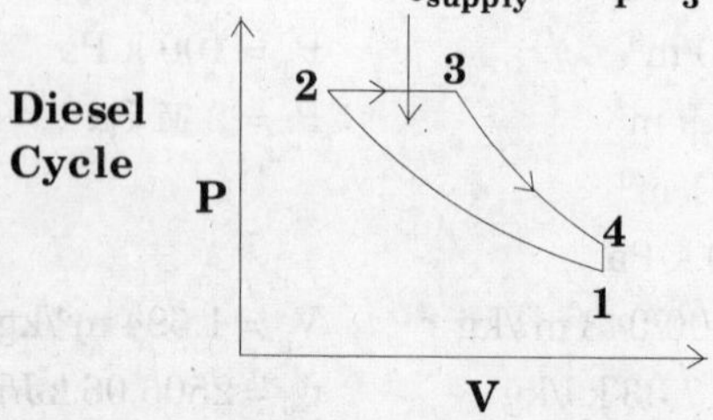

$$\frac{T_2}{T_1} = r^{r-1} = 20^{1.4-1} = 20^{0.4} = 3.31$$

$$\boxed{T_2 = 994.33k}$$

$T_3 = 1.5 \times 994.33 = 1491.5$ k

$Q_{supply} = C_p(T_3 - T_2)$

$= C_p(497.17)$

$Q_{supply} \simeq 500$ kJ/kg

$W_{out} = 500 \times \eta$

$$\text{efficiency of diesel (n)} = 1 - \frac{1}{r^{r-1}}\left(\frac{(\propto^{\propto} - 1)}{(\propto^{\propto} - 1)\propto}\right)$$

$\eta = 0.677$

$W_{out} = 500 \times .677$

$$\boxed{W_{out} = 335kJ/kg}$$

99. $m_H C_p \Delta T = m_w C_p \Delta T$

$5 \times 0.5 \times (373 - T_f) = 10 \times 4.2 \times (I_f - 293)$

$373 - T_f = (T_f - 293) \times 16.8$

$15.8\, T_f =$

$T_f = 335.15$ k

$T_f = 297.49$

The change in net entropy can be expressed as

$$\Delta S = m_M Cp \ln\left(\frac{I_f}{T_m}\right) - m_w C_p \ln\left(\frac{T_w}{I_f}\right)$$

$$= 5 \times 0.5 \times \ln\left(\frac{297.49}{373}\right) - 10 \times 4.2 \ln\left(\frac{293}{297.49}\right)$$

$= 0.073$ kJ/k

100. (A_1) : Helmholtz free energy: It is a concept in thermodynamics where the work of a closed system with constant temperature and volume is measured using thermodynamic potential. It may be described as the following equation: $F = u - T_s$.

(A_2): **Gibbs Function (G)** : Gibbs free energy is the thermodynamic potential use to calculate maximum reversible work at constant temperature and pressure. And it is an extension property.

It is expressed as:

$$\mathbf{G = H - T_S}$$

Where, H = enthalpy, T = temperature, S = entropy

The **enthalpy (H)** of a system is given by:

$H = E + PV$

Where E is the internal energy of the system, P is pressure, and V is volume.

(A_3): Fundamental T-ds equation:

From the first law : $dU = dq + dW$

From the second lawL $dS \geq \frac{dq}{T}$

Where, $dS > \frac{dq}{T}$ for irreversible system

and, $dS = \frac{dq}{T}$ for reversible system

For a closed system in which only reversible pV work is involved

$dW = -pdV$ and $dS = \frac{dq}{T}$

$\therefore dU = TdS - pdV$

(A_4):

Clapeyron- Clausius equation:

The Clausius-Clapeyron equation states that the logarithm of the vapor pressure of a substance is directly proportional to its reciprocal absolute temperature. According to it, the natural logarithm of the liquid's vapor pressure (P) is determined by the molar enthalpy of vaporization of the liquid (ΔH_{vap}), the ideal gas constant (R), and the temperature (T) of the system. Mathematically, the equation is represented by

$d(\ln P)/dT = \Delta H_{vap}/RT^2$

101. As given that

$V_1 = 1m^3$

$P_1 = 200$ k Pa $= 200 \times 10^3$ Pa $V_1 = 1m^3$ (Initial Volume)

$T_1 = 298$ K, R = 0.287 kj/kg–k)

$W_{electrical} = -250$ W (Work done on system)

Time (t) = 600 sec, $C_p = 1.005$ kJ/kg–k

Q = – 4 kJ (Heat rejected) = – 4000 J

$T_2 = ?$

As we know that

First Law of Thermodynamic

$Q = W + \Delta U$

$W = W_{electrical} + W_{displacement}$

$\Rightarrow W = W_{electrical} + \boxed{P\Delta V} \rightarrow$ Far isobaric process

$Q = W_{electrical} + P\Delta V + \Delta U$

$Q = W_{electrical} + \Delta H$

$Q = W_{electrical} + m\,C_p\,(T_2 - T_1)$

$$T_2 = T_1 + \frac{Q + (-W_{electrical})}{m\,C_p} \qquad \therefore m = \frac{P_1V_1}{RT_1}$$

$$= \frac{200 \times 10^3 \times 1}{0.287 \times 298}$$

m = 2.338 kg

$$T_2 = 298 + \frac{-4000 + 250 \times 600}{2.338 \times 1005}$$

$T_2 = 360.136\text{ k} \rightarrow T_2 = 86.98°C \approx 87°C$

102. As Given that

D = 0.2 m　　m = 10 kg/s

$C_0 = 300$ m/sec　　$W_{cv} = 0$

$\Delta p = 0$

q = – 3kJ/kg (Heat lose)

v = 0.12551 m³/kg

$h_i = 3024.2$ kJ/kg

As we know that

$q = W_{cv} + \Delta p + \Delta k\,; + \Delta h$

$q = \Delta k_i + \Delta h$

$$\boxed{q = \frac{1}{200}\left(C_o^{\,2} - C_i^{\,2}\right) + \left(h_o - h_i\right)} \quad ...(1)$$

$m_i = e_i A_i C_i$; $e_i = \dfrac{1}{V_i}$

$$C_i = \frac{m_iV_i}{A_i} \Rightarrow \frac{10 \times 0.12551}{\frac{\pi}{4}(0.2)^2}$$

$\boxed{C_i = 39.951\text{ m/sec}}$

Now by putting the all values in equation (1)

We get

$$-3 \times 10^3 = \frac{1}{200}\left((300)^2 - (39.95)^2\right) + (h_0 - 3024.2 \times 10^3)$$

$$\boxed{h_o \approx 2977 \frac{kJ}{kg}}$$

103. As given that

$V_1 = 2m^3$

$M_2 = 5$ kg

$P_1 = 400$ kPa　　$P_2 = 400$ kPa

As we know that

$$x_2 = \frac{m_{g_2}}{m_{f_2 + m_{g_2}}} \qquad \therefore m_{g_2} = m_{g_1}$$

$$m_{f_2} = \frac{1}{2} m_{f_1} \quad \text{(given)}$$

$\nu = [\nu_{f1} + x_1 V_{fg1}]$ at 600 kPa

$$x_1 = \frac{\nu - \nu_{f1}}{\nu_{fg1}}$$

$$x_1 = \frac{\frac{2}{5} - 0.001084}{0046138}$$

$\boxed{x_1 \simeq 0.864}$

Similarly;

$$x_1 = \frac{M_{g_1}}{M_{f_1} + M_{g_1}}$$

($M_{f_1} + M_{g_1}$) → **5 kg**

Where,

$M_{g1} = 0.864 \times 5 = 4.323\text{ kg} = Mg_2$

$M_{f1} = 5 - 4.323 = 0.677$ kg

$$M_{f_2} = \frac{1}{2}\,0.677 = 0.3385\text{ kg}$$

$$\text{So, } x_2 = \frac{4.323}{0.3385 + 4.323} = \frac{4.323}{4.6615}$$

$\boxed{x_2 = 0.93}$

104. As we know that

$$\eta = \frac{\text{Power output}}{\text{Heat input}}$$

m = 40 mg

$C_v = 44000$ kJ/kg

Q = Heat input $= 40 \times 10^{-3} \times 10^{-3} \times 44000 = 1.76$ kJ

$$\eta = \frac{1}{1.76} = 0.5681$$

We also know that

$$\eta = 1 - \frac{1}{C_R^{\gamma-1}} = 1 - \frac{1}{C_R^{1.4-1}} = 1 - \frac{1}{C_R^{0.4}}$$

$$0.5681 = 1 - \frac{1}{C_R^{0.4}}$$

$C_R = 8.16.$

105.

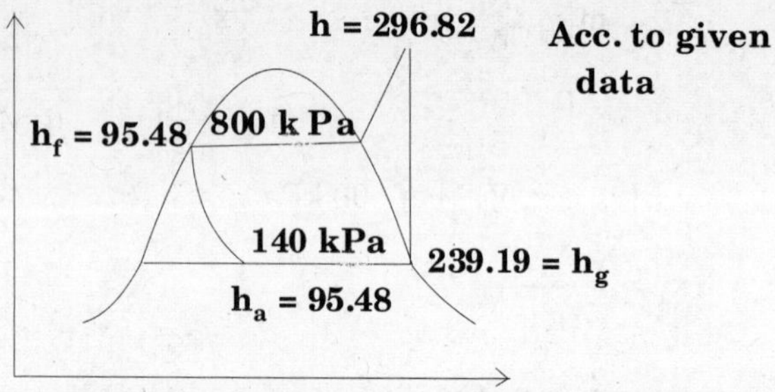

Acc. to given data

$\boxed{h_a = h_f}$ → standard vapor Compression Refrigeration Cycle

$$COP = \frac{hg - ha}{h - hg}$$

$$= \frac{2391.9 - 95.48}{296.82 - 239.19}$$

$\boxed{COP = 2.48}$

106.

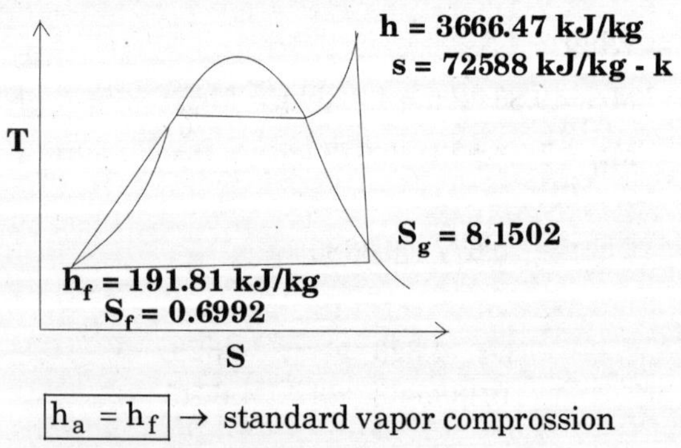

$\boxed{h_a = h_f}$ → standard vapor compression Refrigeration cycle

$h_{fg} = 2392.82$ kJ/kg

$s_{fg} = 7.5010$ kJ/kg – k

Power (P) = $W_{Net}\, m_s$ $\quad m_s = 50$ kg/s (given)

$W_{net} = W_{turtine} - W_{pump}$ ↓ o

$W_{net} = W_{turbine} = (h - h_4)$

$h_4 = h_f + x\, h_{fg}$

$S = s = 7.2588 = S_f + x\, s_{fg}$

$\Rightarrow 0.6992 + (7.5010)\, x$

$\boxed{x = 0.875}$

$h_4 = 191.81 + 2392.82\,(0.875)$

$\boxed{h_4 = 2284.32}$

$P = 50 \times [3666.47 - 2284.32]$

$P = 69107 \text{ kW} \rightarrow 69.1 \text{ M W}$

107.

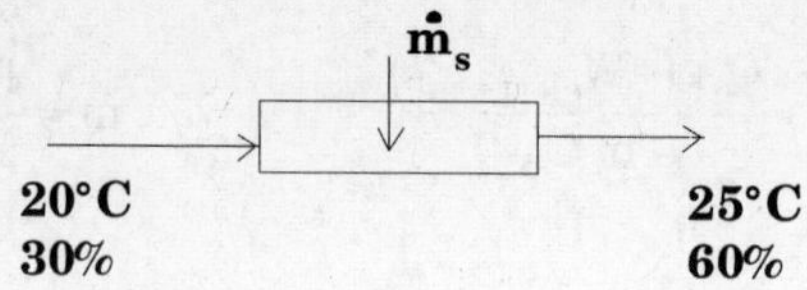

$P_t = 100$ kPa

$P_{vs_1} = 2.3392$ k Pa

$P_{vs_2} = 3.1698$ k Pa

$$w_1 = \frac{0.622\, P_{v_1}}{p_t - P_{v_1}}$$

$$= \frac{0.622 \times \left(\frac{30}{100}\right) \times 2.3392}{100 - \left(\frac{30}{100}\right) \times 2.3392}$$

$\boxed{w_1 = 0.00437 \text{ kg/kg}}$

$$w_2 = \frac{0.622 \times 0.6 \times 3.1698}{100 - 0.6 \times 3.1698}$$

$\boxed{w_2 = 0.0120 \text{ kg/kg}}$ $\quad \boxed{m_s = w_2 - w_1}$

$m_s = 0.0120 - 0.00437$

= 0.00763 kg/kg of dry air

Mass flow rate of air (m_a) =

$P_a V = m_a R_a T_a \Rightarrow (100 - 0.3 \times 2.3392) \times 30$

$= m_a \times 287 \times 293$

$\Rightarrow m_a = 35.425$ kg/mm

$m_s = 0.00763 \times m_a$

$\boxed{m_s = 0.272 \text{ kg/min}}$

108. As given that

$Q_H = 500$ kJ/day

$T_H = 300$ k

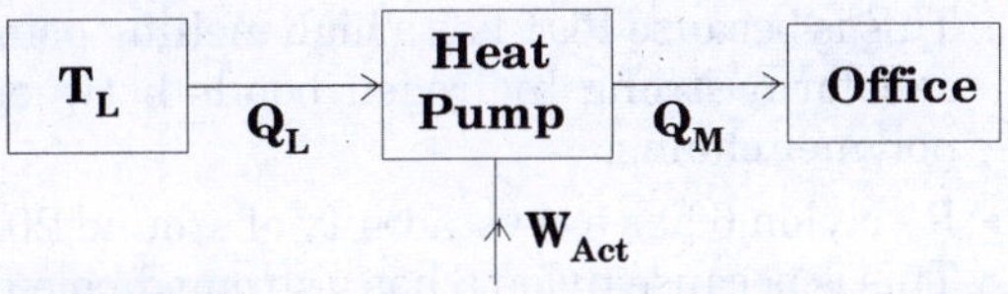

$T_L = 280$ k

$\boxed{Cop_{Act} = 0.6\ Cop_{Th}}$

$\frac{W_{Act}}{W_{Th}} = ?$

We know that →

$\boxed{COP = \frac{Q_H}{W}}$

$$\text{So, } \frac{W_{Act}}{W_{Th}} = \frac{\dfrac{Q_{H}.Act}{(Cop)\,Act}}{\dfrac{Q_{H.Th}}{(Cop)_{Th}}} \quad (\because Q_{H.Act} = Q_{H.Th})$$

$$\frac{W_{Act}}{W_{Th}} = \frac{(Cop)_{Th}}{(Cop)_{Act}} = \frac{1}{0.6} = 1.667 \simeq 1.7$$

109. As given that

$$\left(\frac{dP}{dT}\right)_{Sat} = 3750 \text{ Pa/K}$$

at 100° C → 373 k = T

$(V_g - V_f) = 1.672\ m^3/kg$

As we know that

$$\left(\frac{dP}{dT}\right)_{sat} = \frac{h_{fg}}{T\left(V_g - V_f\right)}$$

$$3750 = \frac{h_{fg}}{373(1.672)}$$

$h_{fg} = 2338710$ J/kg → 233 kJ/kg

110. Cellulose, which is a structural polysaccharide found in the cell walls of plants, is composed of glucose monomers linked together through glycosidic bonds. Therefore, glucose is the monomer used in the synthesis of cellulose.

111. According to the Maxwell model, the total strain E is equal to the instantaneous strain (S/G) plus the strain due to the material's resistance to flow (S/η)t.

The instantaneous strain is the strain that occurs immediately after the load is applied. It is caused by the elastic deformation of the material. The strain due to the material's resistance to flow is the strain that occurs over time. It is caused by the plastic deformation of the material.

The equation for the total strain is:

E = (S/G) + (S/η)t

Where:

- E is the total strain
- S is the applied stress
- G is the modulus
- η is the material's resistance to flow
- t is the elapsed time

112. The structure of a cross-linked polymer given shown below:

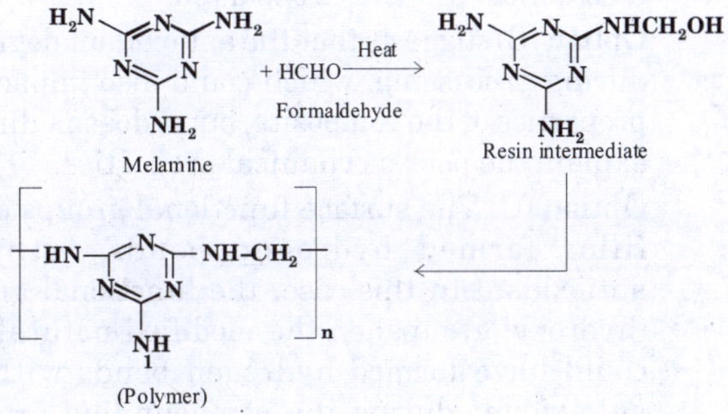

The monomers used in the synthesis of the cross-linked polymer are melamine and form aldehyde. The figure shows that the polymer is formed by the condensation reaction of melamine and formaldehyde. Melamine has three amine groups and formaldehyde has two aldehyde groups. These groups react to form methylene bridges, which cross-link the polymer chains.

113. PMMA is often used as a compatibilizer in blends of PVDF and ABS due to its ability to improve the compatibility between the two polymers. It can help enhance the mechanical properties and overall performance of the blend by promoting better adhesion and dispersion between PVDF and ABS phases.

114. The correct match between the zones and their key functions in an extruder is:

- Metering zone (P): The metering zone is responsible for controlling the flow rate of the polymer melt. This is done by the helical flights of the screw, which impart a constant shear force on the polymer melt.
- Compression zone (Q): The compression zone is responsible for melting the polymer feed and homogenizing the melt. This is done by the heat conducted by the heating element and the high shear forces generated by the screw flights.
- Feed zone (R): The feed zone is responsible for receiving the polymer feed from the hopper and conveying it into the extruder. The feed zone is typically heated to a lower temperature than the compression zone, to prevent the polymer from melting prematurely.

- Working zone (S): The working zone is the first zone in the extruder where the polymer melt is exposed to high shear forces. This is where the polymer is mixed with additives and pigments, and where the final properties of the polymer are determined.

  Hence option (a) is the correct answer.

115. Option (A) suggests that the surface functional groups of the filler reacted with Nylon 6. While this could potentially affect the properties of the composite, it doesn't directly explain the poor mechanical properties observed.

Option (B) suggests that the antioxidant degraded during processing, which could also impact the properties of the composite, but it doesn't directly explain the poor mechanical properties.

Option (C) The surface functional groups of the filler formed hydrogen bonds with the antioxidant.In this case, the functional groups (hydroxyl groups) on the modified natural clay could have formed hydrogen bonds with the antioxidant during the compounding process. Hydrogen bonding can lead to interactions between different components in a composite material, which can affect the overall mechanical properties of the composite. These interactions may disrupt the polymer matrix and filler network, leading to a decrease in mechanical strength and other related properties.

Option (D) suggests that the antioxidant reacted with Nylon 6, which again could have effects on the composite's properties, but it doesn't specifically address the poor mechanical properties.

Hence option (c) is the correct answer.

116. Glass transition and melting temperatures of common plastics

Polymer	**Tg (°C)**	**Tm**
Low-density polyethylene (LDPE)	– 100	110
High-density polyethylene (HDPE)	**– 100 to -80,**	**130**
Polypropylene (PP)	– 25	170
Acrylonitrile butadiene styrene (ABS)	100	125
Polycarbonate (PC)	**145 to 155,**	**155**
Nylon 6,6	49	250
Polyethylene Terephthalate (PET)	70	240
Poly-phenylene oxide (PPO)	**200**	**300**
Polystyrene (PS)	95	240
Poly-methyl methacrylate	**90-100**	100

117.

- S - Polyethylene terephthalate (PET) has the highest crystallinity, typically around 30-50%. This is because PET has a high melting point and can form strong hydrogen bonds between the polymer chains.
- R - Nylon 6 has a crystallinity of around 20-30%. This is because nylon 6 has a strong dipole-dipole interaction between the polymer chains.
- Q - Syndiotactic polystyrene has a crystallinity of around 10-20%. This is because syndiotactic polystyrene has a regular structure that allows the polymer chains to pack together more easily.
- P - Atactic polypropylene has the lowest crystallinity, typically around 0-10%. This is because atactic polypropylene has a random structure that makes it difficult for the polymer chains to pack together.

Here is a table that summarizes the crystallinity of the given polymers:

Polymer	Crystallinity(%g)
Polyethylene terephthalate (PET)	30-50
Nylon 6	20-30
Syndio-tactic polystyrene	10-20
Atactic polypropylene	0-10

118. The best correlation of the graphs with the polymerization methods is:

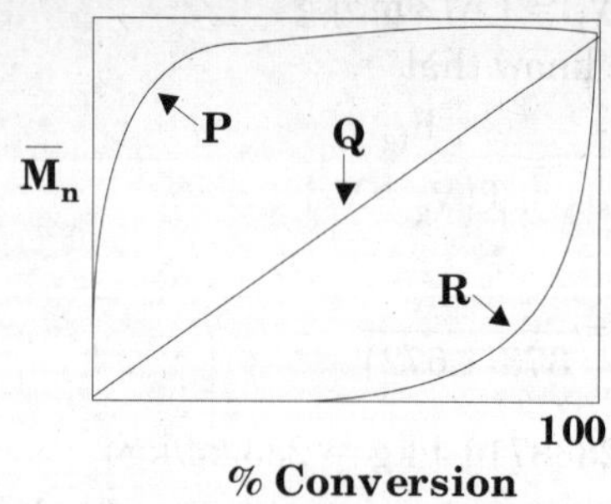

Where,

P - Chain growth:

The graph for chain growth polymerization shows a rapid increase in the molecular weight of the polymer as the reaction proceeds. This is because chain growth polymerization involves a chain reaction, where each monomer adds to the end of a growing polymer chain. The chain reaction continues until all of the monomers have been consumed.

Q - Living polymerization:

The graph for living polymerization shows a linear increase in the molecular weight of the polymer as the reaction proceeds. This is because living polymerization does not involve a chain reaction.

Instead, each monomer adds to a living polymer chain, which continues to grow until it is terminated by a chemical reaction.

R - Step growth:

The graph for step growth polymerization shows a step-wise increase in the molecular weight of the polymer as the reaction proceeds. This is because step growth polymerization involves the formation of covalent bonds between monomers. Each monomer can only form two bonds, so the molecular weight of the polymer increases in steps as each new monomer is added.

119.

1. Fishing rods are typically made using pultrusion, which is a continuous process for manufacturing composite materials with a constant cross-sectional shape, like the long, uniform shape of fishing rods.
2. Plastic sheets are often formed using thermoforming. Thermoforming is a process where a plastic sheet is heated to a pliable forming temperature, stretched onto a single-surface mold, and then cooled to create the desired shape, which is suitable for making plastic sheets of various shapes and sizes.
3. Soft drink bottles are commonly produced through blow molding. Blow molding involves inflating a hot, hollow plastic parison (a tube) inside a mold to create the shape of the bottle. This process is suitable for creating hollow containers with complex shapes.
4. Plastic trays are typically made using compression molding, where heated plastic material is placed in a mold and compressed to take the shape of the mold. This is a suitable technique for producing flat, uniform trays.

So, the correct matches are P – 3; Q – 4; R – 2; S – 1.

120. Dicumyl peroxide is a vulcanizing agent used to cure synthetic rubbers such as EPDM (ethylene propylene diene monomer) and SBR (styrene-butadiene rubber). It is a free radical initiator that breaks down under heats to form free radicals, which then attack the polymer chains and cross-link them. Dinitrobenzene is a vulcanizing agent used to cure natural rubber. It is a slow vulcanizing agent, but it produces rubber with excellent tensile strength and tear resistance. Zinc stearate is a lubricant and filler used in rubber compounding. It helps to reduce friction and improve the processing of the rubber. It is not a vulcanizing agent. Carbon black is filler used in rubber compounding to improve the strength, abrasion resistance, and aging resistance of the rubber. It is not a vulcanizing agent. Therefore, the only agents used to vulcanize or cure rubbers out of the options given are dicumyl peroxide and dinitrobenzene.

Hence options (a) and (d) are the correct answer.

121. Lipase can degrade poly-caprolactone (PCL) and polyethylene terephthalate (PET). It cannot degrade polypropylene (PP) and poly-vinylidene fluoride (PVDF).

Polypropylene is a synthetic polymer made up of repeating units of propylene. It is a non-polar polymer and does not have any carboxylic ester bonds. Poly-vinylidene fluoride is also a synthetic polymer made up of repeating units of vinylidene fluoride. It is a polar polymer and does not have any carboxylic ester bonds.

Poly-caprolactone is a synthetic polymer made up of repeating units of ε-caprolactone. It is a polyester and does have carboxylic ester bonds. Polyethylene terephthalate is also a polyester and does have carboxylic ester bonds.

Theoretically, lipase can also degrade other polyesters, such as poly-lactic acid (PLA) and poly-glycolic acid (PGA). However, the rate of degradation depends on the structure of the polyester. For example, PLA is degraded more slowly by lipase than PCL.

122. Ziegler-Natta catalysts are used to polymerize alkenes, which are hydrocarbons with a double bond. The catalyst is a mixture of a transition metal compound, typically titanium, vanadium, or zirconium, and an organo-aluminum compound. The transition metal compound provides the active site for the polymerization reaction, while the organo-aluminum compound lowers the activation energy of the reaction.

The co-catalyst is an electron donor that helps to activate the transition metal compound. The most common co-catalysts are alkyl-aluminum compounds, such as $Al(CH_3CH_2)_2Cl$.

The pairs of catalyst and co-catalyst in options (B) and (C) will not form a Ziegler-Natta catalyst. $ZnCl_2$ is not a transition metal compound, and TiO_2 is not an organoaluminum compound.

123.

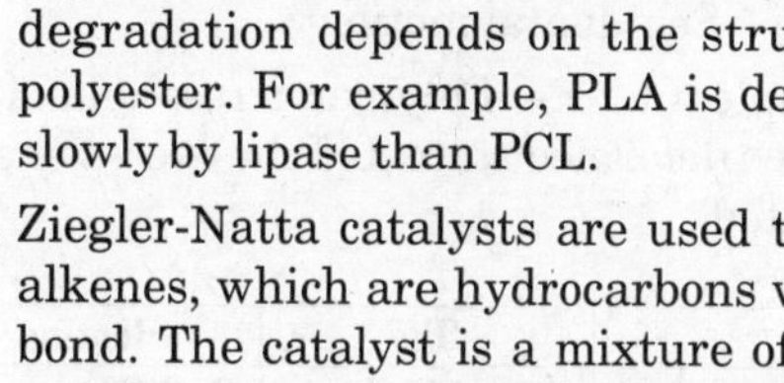

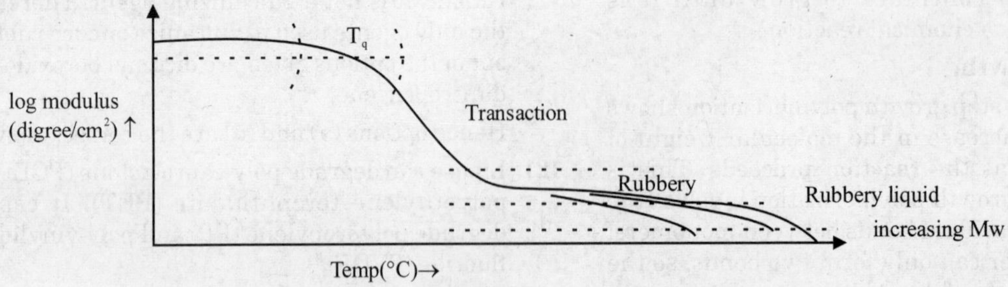

Tg > T → glassy state. restrict the motion.

When we provide heat → Segmental motion start near the glass transection temp.

→ Tm > T > Tg → Called the rubbey state – transeletion from glassy state to rubbey state

→ Motion is higher than the segmental motion (upto C_{50} - C_{60})

→ T > T_m → transection into completely lequid state but polymers are higher Molecular weight their state we called Rubbey liquid state.

P → Rubbery regeme → Motion of Molecules are long – Robber shiws relaxation behaviour.

Q → Region around Tg → depend on local structure – Segmental motion

R → Sample under → Polymer Stress relaxation takes place deformed state and the excess energy is etisspated.

124.

	Tg	Stress
Nylon fiber –	47 – 50°C	2–6 GPa
Polyethylene –	–125°C	0.1 – 1.5 GPa
Ruber –	less than 0°C	0.1 – 1.0 GPa
Polystyrene –	100°C	3 – 4 GPa

Polyoner below the glass transection temp behave as glassy state above the Tg behave as rubbey state.

– Polystyrene behaves as glassy state Means there is no elogation.

– Nylon fiber have more erystallenity they are more strong with less elastically behaviour

– Vulcanessed rubber & Polyethylene are in the same range of stress & elongation.

– So simlable answer is C.

125. Dicumyl peroxide and dinitrobenzene are both organic peroxides, which are used as vulcanizing agents for rubber. Zinc stearate is a lubricant and accelerator, while carbon black is filler. So, the correct answer is the combination of options (A) and (D).

126. Follow Maxwell creep module

η = 300 Giga poise = 30 G Pa – S

E = 1 G Pa

t = 60S

F = 10 kN

A = 0.03 m^2

As we know that

$$J = \frac{1}{E} + \frac{t}{\eta}$$

$$J = \frac{1}{1} + \frac{60}{30} = 1 + 2 = 3/GPa$$

$= 3 \times 10^{-9}$ m^2/N

127. As we know that

$$X_c(\%) = \frac{\Delta H_m}{\Delta H_m{}^\circ} \times 100\%$$

ΔH_m = heat of fusion of sample

ΔHm° = heal of fusion of 1w % Crystalline polymer.

$X_c \propto \Delta Hm$

$$\Rightarrow \frac{\text{Crystallemty slow cooled}}{\text{Crystallemty quanched}} = \frac{\Delta Hm}{0.5\ Hm}$$

$$\frac{60}{X_{c\ queche}} = \frac{1}{0.5}$$

$$X_{c\ quenched} = 30\%$$

128. As we know that

Density

$$d = \frac{Mass}{Volume}$$

Mass = d × Volume

$$\frac{Mass}{t} = d \times \frac{Volume}{t}$$

$= 0.9 \times \frac{10}{100} = 0.09 \text{ g/s}$

$\text{Output} = 0.9 \times 0.09 \times \frac{10^{-3}\text{kg}}{\frac{1}{3600}}$

$= 0.09 \times \frac{3600}{1000} = 0.324 \text{ kg/h}$

129. $[\eta] = K M^q$

$[\eta] = \frac{\eta_{sp}}{C} = \frac{\eta_{r-1}}{C}$

$C = \frac{0.2}{100} = \frac{0.2}{100} \text{g/ml} = 0.2 \text{ g/dl} = 2 \times 10^{-3} \text{ g/cm}^3$

$[\eta] = \frac{1.5-1}{2 \times 10^{-3}} = 250$

$250 = 0.001\ [M]^{0.5}$

$[M]^{0.5} = 250 \times 1000$

$[M]^{0.5 \times 2} = (250 \times 1000)^2$

$M = 62500\,000\,000$

$M = 6.25 \times 10^{10 g}/\text{mol}$

130. $V_f = 0.4 \quad E_f = 69$ GPa

$V_m = 0.6 \quad E_m = 3.4$ GPa

$E = E_f V_f + V_m E_m$

$= 0.4 \times 69 + 0.6 \times 3.4$

$E = 29.64$ GPa

131. As given that,

n1 = 100

n2 = 50

M1 = 7500g/mol

M2 = 5000g/mol

The weight average molecular weight is given as:

$M_w = (n_1M_1^2+n_2M_2^2)/(n_1M_1+n_2M_2)$

Thus by putting all the values we get,

$= (100 \times (7500)^2 + 50 \times (5000)^2)/(100 \times 7500 + 50 \times 5000) = 6875$g/mol

132. Cholecalciferol, α-Tocopherol, and Menadione are all fat-soluble vitamins:

- Cholecalciferol is vitamin D, which is essential for bone health.
- α-Tocopherol is vitamin E, which is a powerful antioxidant.
- Menadione is vitamin K, which is involved in blood clotting.

Thiamine, niacin, and biotin are water-soluble vitamins.

Here is a table of the fat-soluble vitamins:

Vitamin	Name	Function
Vitamin A	Retinol	Vision, cell growth, and reproduction
Vitamin E	Cholecalciferol	Bone health and immunity
Vitamin D	α-Tocopherol	Antioxidant
Vitamin K	Menadione	Blood clotting

133. Iodine is required for the synthesis of thyroxine T4 in the human body. Selenium, iron, and zinc are not required for the synthesis of thyroxine T4.Thyroxine T4 is a hormone produced by the thyroid gland. It is made from iodine and the amino acid tyrosine. The iodine is incorporated into the tyrosine molecules to form the thyroid hormones T3 and T4.Iodine deficiency is a common problem in many parts of the world. It can lead to hypothyroidism, which is a condition where the thyroid gland does not produce enough thyroid hormones. Hypothyroidism can cause a variety of symptoms, including fatigue, weight gain, and hair loss. Iodine deficiency can be prevented by eating foods that are rich in iodine, such as seafood, dairy products, and eggs. Iodine supplements are also available.

134. Arginine is not an essential amino acid for adults. It is synthesized in the body from ornithine and citrulline. However, arginine is an essential amino acid for infants and children, and for people with certain medical conditions, such as kidney disease.

The other three amino acids, phenylalanine, valine, and lysine, are essential amino acids for all people. They cannot be synthesized by the body and must be obtained from the diet.

135. The F value for a process is the number of minutes required to kill a known population of microorganisms in a given food under specified conditions. This F value is usually set at 12 D values to give a theoretical 12 log cycle reduction of the most heat-resistant species of mesophilic spores in a can of food.

136. Apples do not contain vitamin B12. Vitamin B12 is a water-soluble vitamin that is found in animal products, such as meat, fish, eggs, and dairy. It is also found in some fortified foods, such as breakfast cereals.

Fruits and vegetables are not good sources of vitamin B12. The only plant-based foods that contain vitamin B12 are those that have been fortified with it, such as some brands of nutritional yeast.

If you are vegan or vegetarian, it is important to make sure that you get enough vitamin B12 from fortified foods or supplements. A deficiency in vitamin B12 can lead to anemia, fatigue, and nerve damage.

Here are some good sources of vitamin B12:

- Beef liver
- Clams
- Salmon
- Tuna
- Eggs
- Dairy products
- Fortified breakfast cereals
- Nutritional yeast

137. Both thixotropic and rheopectic fluids are time-dependent fluids, meaning that their viscosity changes over time.

- Thixotropic fluids are fluids whose viscosity decreases with time under shear stress. This means that the longer the fluid is sheared, the thinner it becomes. Examples of thixotropic fluids include ketchup, paint, and blood.
- Rheopectic fluids are fluids whose viscosity increases with time under shear stress. This means that the longer the fluid is sheared, the thicker it becomes. Examples of rheopectic fluids include printer ink, gypsum paste, and some types of clay.

The time dependence of the viscosity of these fluids is due to the breakdown of their internal structure under shear stress. When the fluid is sheared, the molecules in the fluid are forced to align in a particular direction. This alignment disrupts the structure of the fluid, which makes it less viscous. Over time, the molecules in the fluid will relax and the structure will reform, which will cause the viscosity of the fluid to increase.

Therefore, the statement that rheopectic and thixotropic are two time independent fluids is incorrect. Both of these fluids are time-dependent, and their viscosity changes over time under shear stress.

138. As we know that

$$\text{Req. efficiency} = \frac{\text{Yield of oil}}{\text{Mustard seed oil content}} \times 100$$

($\because$ Yield of oil = 37 – solid imparities)

$$= 37 - 37 \times \frac{5}{100}$$

$$= 35.15$$

$$\text{Req. efficiency} = \frac{35.15}{38} \times 100$$

$$= 92.5$$

139. As given that

$K = 5.2 \times 10^{-3}\ \text{day}^{-1}$

$$\text{half life} \left(T_{\frac{1}{2}}\right) = ?$$

$n_0 = 2n$ for half life

$$\boxed{\ln\left(\frac{n_0}{n}\right) = kT}$$

$$\ln\left(\frac{2n}{n}\right) = \left(kT_{\left(\frac{1}{2}\right)}\right)$$

$$\ln(2) = kT_{1/2}$$

$$0.693 = -kT_{1/2}$$

$$\Rightarrow T_{\left(\frac{1}{2}\right)} = \frac{0.693}{k} = \frac{0.693}{5.2 \times 10^{-3}}$$

$$\boxed{T_{\left(\frac{1}{2}\right)} = 133 \text{ days}}$$

140. Wt. of dried cauliflower = 10 kg = w_d

Wt of rehydration (w_R) = 60 kg

Coeff. of rehydration

$$= \frac{w_R\,(100\% - (\text{moisture in fresh}))}{(w_d - \text{moisture in dried cauliflower}) \times 100}$$

$$= \frac{60(100-87)}{(10-0.5)\times 100} = \frac{60 \times 13}{9.5 \times 100} = \frac{780}{950} = 0.82$$

141.

- Vinegar: Vinegar is produced by the fermentation of ethanol by the bacterium Acetobacteraceti. The ethanol is produced by yeasts, such as Saccharomyces cerevisiae.
- Citric acid: Citric acid is produced by the fermentation of glucose by the fungus Aspergillusniger.
- Ethanol: Ethanol is produced by the fermentation of sugars by yeasts, such as Saccharomyces cerevisiae.
- L-Lysine: L-Lysine is produced by the fermentation of glucose by the bacterium Enterobacteraerogenes.

142.

- Chymosin (4. Cheese manufacturing): Chymosin is used to coagulate milk and make cheese. It does this by breaking down the protein casein in milk. The coagulated milk is then separated into curds and whey. The curds are then pressed and aged to form cheese.

- Thermolysin (3. Aspartame synthesis): Thermolysin is used in the production of aspartame, an artificial sweetener. Aspartame is made by combining two amino acids, aspartic acid and phenylalanine. Thermolysin is used to hydrolyze the bond between these two amino acids.
- β-Galactosidase (2. Lactose hydrolysis): β-Galactosidase is used to hydrolyze lactose, the sugar found in milk. Lactose is a disaccharide, which means that it is made up of two sugar molecules, glucose and galactose. β-Galactosidase breaks down the bond between these two sugar molecules, releasing glucose and galactose. Lactose-free milk and other dairy products are made with β-galactosidase.
- Lipase (1. Acyl glycerol restructuring): Lipase is used to hydrolyze fats and oils. This can be used to improve the texture and flavor of food, or to make it easier to digest. Lipase is also used in the production of margarine and other spreads.

143. Clostridium botulinum is a gram-positive, rod-shaped bacterium that produces a powerful neurotoxin. The toxin can cause botulism, a serious illness that can lead to paralysis. Botulism is often caused by eating food that has been contaminated with the bacteria, such as improperly canned food.Gram-positive bacteria have a thick cell wall that contains peptidoglycan, which makes them more resistant to antibiotics and other disinfectants. They are also more heat-tolerant than gram-negative bacteria.

Campylobacter jejuni, Vibrio cholerae, and Salmonella typhi are all gram-negative bacteria. They are more susceptible to antibiotics and other disinfectants than gram-positive bacteria. However, they can still cause foodborne illnesses, such as campylobacteriosis, cholera, and typhoid fever.

So, the only gram-positive bacteria on the list that is responsible for causing foodborne diseases is Clostridium botulinum.

144.

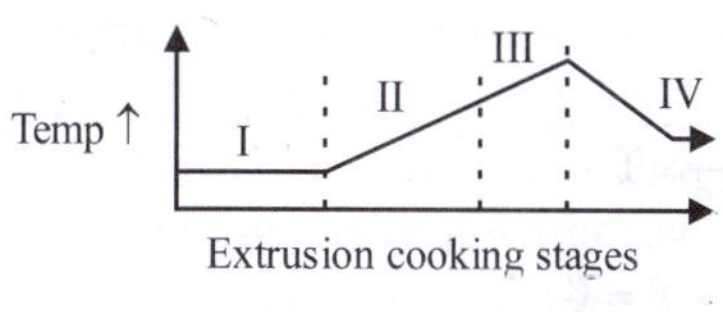

Any cooking process first stage always is feeding.

1. Feeding zone – Hydration Compression → 20 – 25° C
2. Mixing zone – Kneading Disaggregation Unfolding T ≥ 80°C
3. Melting zone – Cooking – Shearing & heating protein plastification, Bond ruptune & reformation T = 140 – 150°C
4. Cooling zone – Expansion zone – Shear alignment fiber formation & stabulzation T < 70°C

145.

- Thiobarbituric acid test (TBA) measures the amount of malondialdehyde (MDA), a carbonyl compound produced during lipid oxidation. Therefore, it is a measure of the carbonyl content of lipids.
- Rancimat method measures the time it takes for a fat to become rancid, which is a measure of its induction time.
- Peroxide value measures the amount of hydroperoxides present in a fat. Hydroperoxides are unstable compounds that can break down to form carbonyl compounds. Therefore, the peroxide value is a measure of the hydroperoxide content of lipids.
- Iodine value measures the amount of iodine that can be added to a fat. The iodine value is a measure of the degree of unsaturation of lipids.

146. Knife peeling citrus fruits: Knife peeling is used for fruits and vegetables with a thin skin that can be easily removed with a sharp knife. This method is commonly used for citrus fruits, such as oranges and lemons, as well as apples, pears, and pineapples. Cucumber has a thin skin that can be easily removed with a sharp knife.

Abrasion peeling potatoes: Potato has a thick skin that is difficult to remove with a knife. Abrasion peeling is a more effective way to remove the skin of potatoes. Abrasion peeling uses a rotating abrasive surface to remove the skin of fruits vegetables. This Method is commady used for potatoes, carrots and ginger.

Flame peeling brinjal or onions:

Brinjal has a thick, tough skin that is difficult to remove with a knife or by abrasion peeling. Flame

- This variation in geostrophic winds with height is called thermal winds.

  Thermal winds are important for a number of reasons. They can help to explain the formation of jet streams, they can affect the distribution of precipitation, and they can contribute to the development of storms.

158. The solubility of a gas in sea water increases with the decrease of temperature and salinity, and the increase of pressure.

- Temperature: The solubility of a gas in water decreases with increasing temperature. This is because the kinetic energy of the water molecules increases with temperature, making it more difficult for the gas molecules to dissolve.
- Salinity: The solubility of a gas in water decreases with increasing salinity. This is because the salt ions in seawater compete with the gas molecules for the available bonding sites on the water molecules.
- Pressure: The solubility of a gas in water increases with increasing pressure. This is because the increased pressure forces the gas molecules into the water.

159. Phytoplankton are microscopic, single-celled organisms that drift in the sunlit surface areas of the world's oceans. They are key to bringing carbon down into the ocean biological pump from the atmosphere via the process of photosynthesis. Zooplankton, fish, and radiolarians are all consumers of phytoplankton, and they do not take up carbon dioxide from the atmosphere.

160. The solubility of a gas in a liquid is the amount of the gas that can dissolve in a given volume of liquid at a given temperature and pressure. The solubility of CO_2 in water decreases with increasing temperature. This is because the kinetic energy of the CO_2 molecules increases with temperature, which makes them more likely to escape from the liquid.As a result, when the temperature of seawater decreases, the solubility of CO_2 in the water increases,this means that more CO_2 can be absorbed by the water.

This is why oceanographers are concerned about climate change. As the oceans warm, they will be able to absorb less CO_2 from the atmosphere. This will lead to an increase in atmospheric CO_2 levels, which will contribute to global warming.

161.

- Carbon dioxide (CO_2) has a global warming potential of 1 over a 100-year timescale.
- Water vapor (H_2O) has a global warming potential of 1 over a 20-year timescale, but it is the most abundant greenhouse gas in the atmosphere, so it has the greatest overall impact on climate change.
- Methane (CH_4) has a global warming potential of 28 over a 100-year timescale.
- Nitrous oxide (N_2O) has a global warming potential of 265 over a 100-year timescale.
- Sulphur hexafluoride (SF_6) has a global warming potential of 23,900 over a 100-year timescale.Sulphur hexafluoride is a synthetic gas that is used in a variety of industrial applications, such as electrical transformers and circuit breakers. It is a very potent greenhouse gas, and even a small amount of it can have a significant impact on the atmosphere.

  So, according to given options (N_2O) has a global warming potential of 265 over a 100-year timescale.

162. In a geostrophic flow, the pressure gradient force is balanced by the Coriolis force. The pressure gradient force is proportional to the change in pressure with distance, so the pressure gradient force will increase as one moves across the isobar. However, the Coriolis force is independent of distance, so the Coriolis force will not change. As a result, the speed of the geostrophic flow will remain constant.

In a cyclo-strophic flow, the pressure gradient force is balanced by the centrifugal force. The pressure gradient force is proportional to the change in pressure with distance, so the pressure gradient force will increase as one moves across the isobar. However, the centrifugal force is proportional to the square of the distance from the center of rotation, so the centrifugal force will increase as one moves across the isobar. As a result, the speed of the cyclo-strophic flow will increase.

So, the speed of a balanced flow remains constant for a geostrophic flow and changes for a cyclo-strophic flow.

163. Schematic of circulation in the Indian Ocean

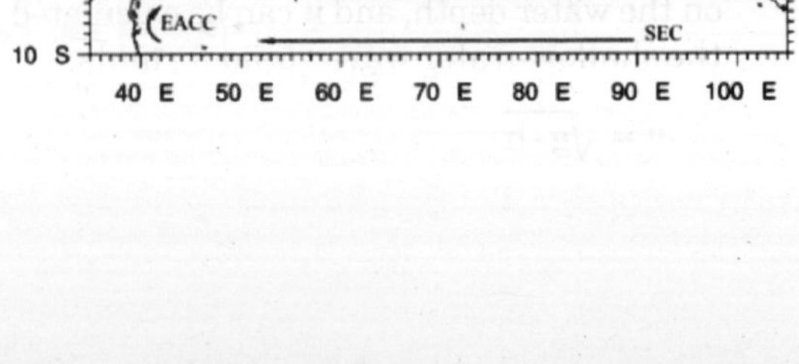

Schematic representation of the circulation, as described in the literature, in the Indian Ocean during January (winter monsoon) and July (summer monsoon). The abbreviations are as follows: SC, Somali Current; EC, Equatorial Current; SMC, Summer Monsoon Current; WMC, Winter Monsoon Current; EICC, East India Coastal Current; WICC, West India Coastal Curent; SECC, South Equatorial Counter Current, EACC, East African Coastal Current; SEC, South Equatorial Current; LH, Lakshadweep high; LL, Lakshadweep low; and GW, Great Whirl.

164. Upwelling is the process of bringing nutrient-rich water from the depths of the ocean to the surface. This process increases the amount of chlorophyll and nutrients in the surface water, which leads to increased phytoplankton growth. Cyclonic eddies are rotating currents that move in a clockwise direction in the Northern Hemisphere and counterclockwise in the Southern Hemisphere. Anti-cyclonic eddies are rotating currents that move in a counterclockwise direction in the Northern Hemisphere and clockwise in the Southern Hemisphere.Down-welling is the process of moving nutrient-poor water from the surface to the depths of the ocean. This process decreases the amount of chlorophyll and nutrients in the surface water, which leads to decreased phytoplankton growth. Therefore, the region with low chlorophyll and low nutrients is a down-welling, anti-cyclonic eddy.

165.

(a) Reversal of East-West temperature gradient with seasons: The East-West temperature gradient is the difference in temperature between the western and eastern parts of the tropics. This gradient is reversed during the monsoon season, with the western tropics becoming warmer and the eastern tropics becoming cooler. This reversal is caused by the northward movement of the Intertropical Convergence Zone (ITCZ).

(b) Reversal of North-South temperature gradient with seasons: The North-South temperature gradient is the difference in temperature between the northern and southern parts of the Indian Ocean. This gradient is reversed during the monsoon season, with the northern Indian Ocean becoming warmer and the southern Indian Ocean becoming cooler. This reversal is caused by the northward movement of the monsoon winds.

(c) Surface cooling and deepening of mixed layer: The surface cooling and deepening of the mixed layer is a phenomenon that occurs in the Arabian Sea during the monsoon season. This is caused by the northward movement of the monsoon winds, which brings cooler air from the north. The cooler air cools the surface of the Arabian Sea, which causes the mixed layer to deepen.

(d) Northward movement of south equatorial current: The northward movement of the south equatorial current is a phenomenon that occurs in the Indian Ocean during the monsoon season. This is caused by the northward movement of the monsoon winds, which pushes the south equatorial current northward.

166. The Ekman spiral is a theoretical concept that describes the pattern of water movement in response to wind stress in the ocean. It results from the balance between the Coriolis force, wind shear (the change in wind speed with height), and frictional force. In the Northern Hemisphere, the surface current is deflected to the right of the wind direction due to the Corioliseffect. This deflection gradually decreases with depth, forming a spiral-like pattern of decreasing current velocities with depth, hence the term "Ekman spiral."

167. Winter rainfall over India is primarily associated with two factors: Northeast monsoon and Western disturbances. Let me explain each of these factors:

(a). **Northeast Monsoon**: The Northeast monsoon, also known as the winter monsoon or retreating monsoon, occurs during the months of October to December. It brings rainfall to the southern and eastern parts of India, including regions like Tamil Nadu, Andhra Pradesh, and parts of Karnataka. This monsoon is caused by the reversal of wind patterns. During this period, cool, dry air from the Indian subcontinent moves towards the low-pressure areas over the Indian Ocean, resulting in moisture-laden winds blowing from the northeast. These winds pick up moisture from the Bay of Bengal and bring rainfall to the mentioned regions.

(b). **Southwest Monsoon**: The Southwest monsoon, associated with heavy rainfall, occurs during the summer months (June to September) and is responsible for the majority of the annual rainfall over India. It is characterized by moist winds from the southwest that bring rain to the western coast and northern parts of India.

(c). **Western Disturbances**: Western disturbances are extra-tropical storm systems that originate in the Mediterranean region and move eastwards. These disturbances bring winter rainfall to the northwestern parts of India, including areas like Jammu and Kashmir, Himachal Pradesh, and parts of Punjab and Haryana. As these systems move across the Indian subcontinent, they interact with the prevailing westerly flow and can lead to cloud formation and precipitation. Western disturbances are an important source of winter rainfall and snowfall in the higher altitudes.

(d). **Agulhas Current**: The Agulhas Current is a warm ocean current that flows along the eastern coast of South Africa. It is not directly associated with winter rainfall over India. Instead, it plays a role in the oceanic and climatic patterns of the southern Indian Ocean.

Hence, options(a & c) are the correct answer.

168. As we know that

$(1 - A) \propto T^4$

A = global albedoArea

$$\frac{1-0.3}{1-0.4} = \left(\frac{T_{0.3}}{T_{0.4}}\right)^4$$

$$\left(\frac{0.7}{0.6}\right)^{\frac{1}{4}} = \left(\frac{T_{0.3}}{T_{0.4}}\right)$$

$$1.0393 = \frac{T_{0.3}}{T_{0.4}}$$

$T_{0.4} = 0.9622\ T_{0.3}$

Decrease in the temp(ΔT) = $T_{0.3} - 0.9622\ T_{0.3}$

= 6.0378 k

Answer key is wrong.

169. Constant parcel temp $\left(\frac{DT}{Dt} = 0\right)$

$$W = \frac{J}{g}$$

$J = w \times g = 2 \times 10^{-2} \times 9.8$

J = 0.196 kg.

170. 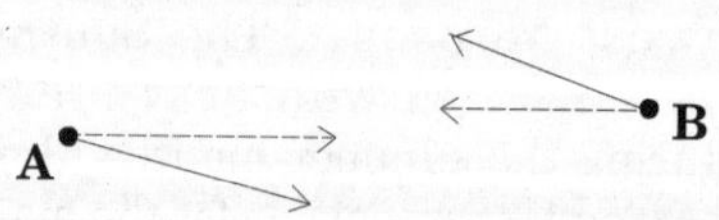

Let's assume the balls are approaching each other along east-west trajectory so,

ball A has velocity $\mu\ \hat{i}$

ball B has velocity $- \mu\ \hat{i}$

We are interested in their lateral displacement due to carioles force

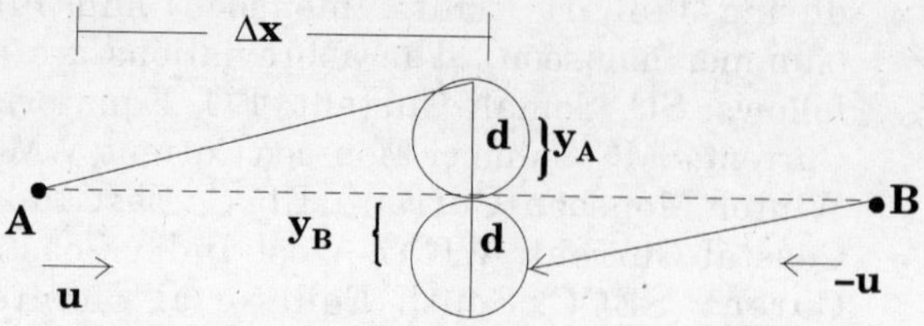

$$\frac{dv}{dt} = -2\ \pi\ i\ \sin\phi$$

displacement for b all A

$$\frac{d^2y}{dt^2} = -2\ \pi\ \mu\ \text{Sin}\ \phi / \int dt \int dt$$

$y_A = -\pi\ \bar{u}\ \sin\phi\ t^2$

$$\Delta x = \frac{2wm}{z}$$

d = 5 cm

displacement for B all B:

$y_B = \pi\ \bar{u} \sin\theta\ t^2$

$$y_A = y_B = \frac{d}{2} = \frac{5}{2} = 2.5\text{cm}$$

$$\bar{u} = \frac{\Delta x}{\Delta t} \qquad (\Delta t = t - 0]$$

$$y_B = \pi \frac{\Delta x}{\Delta t} \times \sin\phi \times t^2 \qquad \theta = 45$$

$$= \pi \frac{\Delta x}{t} \times t^2 \times \sin\phi \qquad d^x = \frac{200}{2} = 100\ \text{m}$$

$$t = \frac{y_B}{\pi \sin\phi\, \Delta x} = \frac{2.5\ \text{cm}}{7.29 \times 10^{-5} \times \frac{1}{\sqrt{2}} \times 10000}$$

t = 4.855

$$\bar{u} = \frac{\Delta x}{t} = \frac{100}{4.85} = 20.62\ \text{m/s}$$

171. As we know that

$$\left(\frac{P}{P_s}\right) = \left(\frac{P}{P_s}\right)^{\frac{Cv}{Cp}}$$

$$\left(\frac{1}{2}\right) = \left(\frac{P}{1000}\right)^{\frac{1}{0.71}}$$

$$\frac{P}{1000} = \left(\frac{1}{2}\right)^{0.71}$$

$$\frac{P}{1000} = 0.6113$$

P = 611.3 h Pa

172. As we know that

$$Z_1 = \frac{w^2 r_1^2}{2g}$$

So, $Z_2 = \frac{w^2 r_2^2}{2g}$

$$Z_2 - Z_1 = \frac{w^2 r_2^2}{2g} - \frac{w^2 r_1^2}{2g}$$

$$= \frac{10^2}{2 \times 9.8}\left[1^2 - 0.5^2\right]$$

= 3.83.

173. The amount of Carbon emitted into the atmosphere in mega ton per year

$$= \frac{295 \times 10^6 \times 5.5 \times 10^{-5} \text{kg} \times 120000}{25 \text{ KM}}$$

= 0.78

174. As we know that

Geotropic velocity $V_g = \frac{P}{2\pi \times \rho \times \sin\theta \times d}$

P = 5h pa = 500 Pa

$2\pi = 14.6 \times 10^{-5}$ rad/s

$\sin\theta = \sin 30 = 0.5$

$d = 250 \text{ KM} = 250 \times 10^3$ m

$\rho = 0.70$ kg

$$Vg = \frac{500}{14.6 \times 10^{-5} \times 0.70 \times 0.5 \times 250 \times 10^3}$$

Vg = 39.14 m/s